Distance Formula

The distance between $P_1(x_1, y_1)$ and $P_2(x_2, y_2)$ is

$$d(P_1, P_2) = \sqrt{(x_1 - x_2)^2 + (y_1 - y_2)^2}$$

Important Formulas

Slope m of a line through $P_1(x_1, y_1)$ and $P_2(x_2, y_2)$

$$m = \frac{y_2 - y_1}{x_2 - x_1}, \quad x_1 \neq x_2$$

Slope-intercept form of a line with slope m and y-intercept b

$$y = mx + b$$

Point-slope formula for a line with slope m passing through $P_1(x_1, y_1)$

$$y - y_1 = m(x - x_1)$$

Formulas for the conic sections

Parabola: $\quad x^2 = 4py$

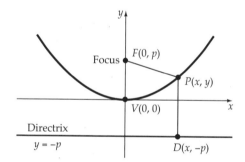

Hyperbola: $\quad \dfrac{x^2}{a^2} - \dfrac{y^2}{b^2} = 1$

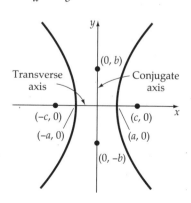

Ellipse: $\quad \dfrac{x^2}{a^2} + \dfrac{y^2}{b^2} = 1$

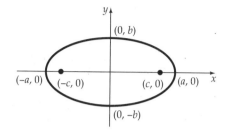

Formulas for area A, circumference C, surface area S, and volume V.

Triangle

$$A = \frac{1}{2} bh$$

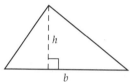

Circle

$$A = \pi r^2$$
$$C = 2\pi r$$

Parallelogram

$$A = bh$$

Sphere

$$V = \frac{4}{3} \pi r^3$$
$$S = 4\pi r^2$$

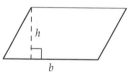

Right circular cone

$$V = \frac{1}{3} \pi r^2 h$$

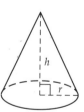

Right circular cylinder

$$V = \pi r^2 h$$
$$S = 2\pi rh + 2\pi r^2$$

College
Algebra
and
Trigonometry

Second Edition

Richard N. Aufmann/Vernon C. Barker/Richard D. Nation

Palomar College

Houghton Mifflin Company   Boston   Toronto
Dallas   Geneva, Illinois   Palo Alto   Princeton, New Jersey

Sponsoring Editor: *Maureen O'Connor*
Senior Development Editor: *Tony Palermino*
Senior Project Editor: *Toni Haluga*
Production/Design Coordinator: *Martha Drury*
Manufacturing Coordinator: *Holly Schuster*
Marketing Manager: *Michael Ginley*

Cover concept and design: *Catherine Hawkes*
Photographer: *Martin Stein*

Printed in the U.S.A.

Library of Congress Catalog Card Number: 92-72362

ISBN Numbers:
Text: 0-395-63818-6
Instructor's Annotated Edition: 0-395-63819-4
Solutions Manual: 0-395-63822-4
Student's Solutions Manual: 0-395-63821-6
Instructor's Resource Manual: 0-395-63820-8
Test Bank: 0-395-63823-2
Graphing Workbook: 0-395-65941-8

345678-VH-96 95 94 93

Contents

6 TRIGONOMETRIC FUNCTIONS   295

7 TRIGONOMETRIC IDENTITIES AND EQUATIONS   363

8 APPLICATIONS OF TRIGONOMETRY   411

9
TOPICS IN ANALYTIC GEOMETRY   453

10
SYSTEMS OF EQUATIONS   495

11
MATRICES AND DETERMINANTS   543

12   SEQUENCES, SERIES, AND PROBABILITY   593

APPENDICES   A1

Preface

The teaching of mathematics is changing in important ways. First, there is greater emphasis on doing mathematics rather than duplicating mathematics through extensive sets of drill exercises. Students are being urged to research topics and write their findings. Second, technological progress is permitting us to explore concepts in ways that would have been impractical just a few years ago.

To respond to these changes, we have written several new features for the second edition. These features include

- Essays and Projects
- Exploring Concepts with Technology
- Graphing Calculator Exercises (optional)

The *Essays and Projects* encourage students to research and write about mathematics and its applications. The essays frequently ask students to explain mathematics in their own words. Our *Exploring Concepts with Technology* extends ideas from the text using computers and calculators. In addition, we have written optional Graphing Calculator Exercises in sections where appropriate. These exercises are supplemented by our new Graphing Workbook which offers 600 additional problems that may be worked using graphing calculators or the Math Assistant graphing software.

Despite these changes, we have retained our basic philosophy which is to deliver a comprehensive and mathematically sound treatment of the topics considered essential for a college algebra and trigonometry course. To help students master these concepts, we have tried to maintain a balance among theory, application, and drill. Each definition is precisely stated and many theorems are proved. Carefully developed mathematics is complemented by abundant, creative applications that are both contemporary and representative of a wide range of disciplines. Many applications problems are accompanied by art that helps the student visualize the mathematics of the problem.

FEATURES

Interactive Presentation *College Algebra and Trigonometry* is written in a style that encourages the student to interact with the textbook. At various places throughout the text, a question in the form of (Why?) is asked of the reader. This question encourages the reader to pause and think about the current discussion and to answer the question. To make sure the student does not miss important information, the answer to the question is provided as a footnote on the same page.

Each section contains a variety of worked examples. Each example is given a name so that the student can see at a glance the type of problem being illustrated. Each example is accompanied by annotations that assist the student in moving from step to step. Following the worked example is a suggested exercise from that section's exercise set for the student to work. The exercises are color coded by number in the exercise set and the complete solution of that exercise can be found in an appendix to the text.

Extensive Exercise Sets   The exercise sets of *College Algebra and Trigonometry* were carefully developed to provide the student with a variety of exercises. The exercises range from drill and practice to interesting challenges and were chosen to illustrate the many facets of topics discussed in the text. Besides the regular exercise sets, there is a set of supplemental problems that includes material from previous chapters, presents extensions of topics, or are of the form "prove or disprove."

Applications   One way to motivate a student to an interest in mathematics is through applications. The applications in *College Algebra and Trigonometry* have been taken from agriculture, architecture, biology, business, chemistry, earth science, economics, engineering, medicine, and physics. Besides providing motivation to study mathematics, the applications provide an avenue to problem solving. The applications problems require the student to organize and implement a problem solving scheme. To help students improve their problem-solving skills, we have created many new pieces of art that effectively depict the mathematics of the problem. This art will help students to better conceptualize word problems, as well as encouraging students to become successful problem solvers.

Essays and Projects   One of our goals in writing this text has been to involve the student with the text. As mentioned earlier, we do this through various pedagogical features such as pausing at a point in the development of a concept to ask the student to answer a question. *Essays and Projects* is another feature designed to engage the student in mathematics, this time through writing. At the end of each chapter, we have provided guidelines for further investigations. Some of the guidelines ask the student to provide a historical perspective of a topic. Others ask the student to write a proof of some statement. Still others ask the student to chronicle the procedure the student used to solve a problem and to suggest extensions to that problem.

Exploring Concepts with Technology   Calculators and computers have expanded the limits of the types of problems that realistically can be solved. To take advantage of the new technologies, we have incorporated in each chapter some optional extensions of ideas presented in that chapter. These problems are not so much conceptually difficult as they are computationally messy. For each of these problems we encourage the student to use calculators or computers to investigate solutions. As the student progresses through a solution, we challenge the student to think about the pitfalls of computational solutions.

CHANGES FOR THE SECOND EDITION

We carefully reviewed each chapter and made revisions suggested by some of our colleagues who used the first edition. The changes we made were to the following chapters.

Chapter 3   The organization of this chapter was changed to allow for an earlier introduction to functions and to consolidate the graphing techniques into a single section. The graphing techniques that involve symmetry and translations are now included after the introduction of linear and quadratic functions. Symmetry and the related concepts of even and odd functions are now presented in the same section. Graphing techniques have been expanded to include the concepts of horizontal shrinking and stretching.

Chapter 6   The organization and focus of this chapter was changed completely. We included an earlier introduction to right triangle trigonometry and some of its applications. The concepts of right triangle trigonometry are then naturally extended to general trigonometric functions and then to the circular functions. The particular change in focus or emphasis occurs in the sections on graphing circular functions. The concepts of translation, reflection, stretching, and shrinking of graphs presented in Chapter 3 are used extensively to graph circular functions.

Chapter 8   Because right triangle trigonometry was moved to Chapter 6, the concepts in Chapter 8 now deal exclusively with solving oblique triangles. This allows us to concentrate on the techniques for solving these triangles and their related applications. We also rewrote the material on vectors.

Chapter 11   The algorithm for matrix multiplication has always been difficult to understand. In Chapter 11, matrix multiplication is introduced through an application. This application naturally demonstrates why matrix multiplication should be defined as it is.

In addition to refinements in our presentation of mathematical concepts, we have reviewed the exercise sets and adjusted some of these so that there is a more consistent development of skill level. Application problems were updated and contemporary applications were added. A List of Applications follows the preface.

SUPPLEMENTS FOR THE INSTRUCTOR

College Algebra and Trigonometry has an unusually complete set of teaching aids for the instructor.

Instructor's Annotated Edition   This is an exact replica of the student text with the exception that annotations for the instructor are liberally distributed throughout the text. These annotations are identified as "Instructor Notes", and they occur in red in the margin. They include teaching tips, warnings about common errors, graphing calculator suggestions and historical notes.

Solutions Manual   The Solutions Manual contains worked-out solutions for all end-of-section, supplemental, challenge and review exercises.

Instructor's Resource Manual with Chapter Tests   The Instructor's Manual contains the printed testing program, which is the first of three sources of testing material available to the user. Six printed tests (in two formats— free response and multiple choice) are provided for each chapter. In addition, there are suggestions for course sequencing, suggestions for incorporating graphing calculators and outlines of the Essays and Projects questions.

Computerized Test Generator   The Computerized Test Generator is the second source of testing material. The data base has been doubled and now contains more than 3600 test items. These questions are unique to the test generator and do not repeat items provided in the Instructor's Manual testing program. The Test Generator is designed to produce an unlimited number of tests for each chapter of the text, including cumulative tests and final exams. It is available for the IBM PC and compatible computers and the Macintosh.

Printed Test Bank   The Printed Test Bank, the third component of the testing material, is a printout of all items in the Computerized Test Generator. Instructors using the Test Generator can use the test bank to select specific items from the data base. Instructors who do not have access to a computer can use the test bank to select items to be included on a test being prepared by hand.

Texas Instruments PC-81 Graphing Calculator Software   This powerful, compact software completely emulates the look, feel and functionality of the popular TI-81 graphing software. This software is free to adopters of this texts. Users may obtain a site license free of charge so that they may install the software in computer labs. Offered in cooperation with Texas Instruments.

Houghton Mifflin Video Library   The review videos contain 32 segments that cover the essential topics in this text. These videos, professionally produced specifically for the text, offer a valuable resource for further instruction and review.

SUPPLEMENTS FOR THE STUDENT

In addition to the Student Solutions Manual, two computerized study aids, the Computer Tutor and the Math Assistant, accompany this text.

Student Solutions Manual   The Student Solutions Manual contains complete solutions to all odd-numbered problems in the text.

Computer Tutor   The Computer Tutor is an interactive instructional micro-computer program for student use. Each section in the text is supported by a lesson on the Computer Tutor. Lessons provide additional instruction and practice and can be used in several ways: (1) to cover material the student missed because of absence from class; (2) to reinforce instruction on a concept that the student has not yet mastered; (3) to review material in preparation for examinations. This tutorial is available for the IBM PC and compatible microcomputers.

Math Assistant   The Math Assistant is a collection of programs that can be used by both the instructor and the student. Some programs are instructional and allow the student to practice a skill like finding the inverse of a matrix. Other programs are computational routines that perform numerical calculations. In addition, there is a function grapher that graphs elementary functions and polar equations. The Math Assistant is available for the IBM PC, compatible microcomputers, and the Macintosh.

Graphing Workbook   The Graphing Workbook contains over 600 exercises that may be solved using a graphing calculator or the Math Assistant graphing software. These problems are designed to extend and explore such concepts as approximating roots of equations, translating graphs, and solving inequalities. Students may complete the exercises individually or in small groups.

ACKNOWLEDGMENTS

We sincerely wish to thank the following reviewers who reviewed the manuscript in various stages of development for their valuable contributions.

Gregory J. Davis, University of Wisconsin-Green Bay
Donald Buckholtz, University of Kentucky
Orville Bierman, University of Wisconsin-Eau Claire
Nancy Bowers, Pennsylvania College of Technology
Marty Hodges, Colorado Technical College
Patricia McClellan, University of North Carolina-Asheville
Karla Neal, Louisiana State University
Lou Francis Foster, Oklahoma City Community College
Jim Delany, California Polytechnic State University
Paul D. Moreland, Rio Hondo College
John A. Gosselin, University of Georgia
James Magliano, Union County College
T. F. Castillo, University of Texas at San Antonio
Paul J. Manikowski, State University of New York at Alfred
Brian Smith, Community College of Beaver County
Raymond McDaniel, Pembroke State University
Corlis P. Johnson, Mississippi State University
Michael Schramm, LeMoyne College
Eric Wakkuri, Oregon Institute of Technology

List of Applications

Hanging cable, 289
Height of the cables on a suspension bridge, 199
Maximum carrying capacity of a trough, 157
The Gateway Arch, 291
Trammel, 492

Consumer Applications

Cooking time of a pot roast, 56
Cost of insulating a ceiling, 190
Cost of insulating exterior walls, 192
Cost of painting house, 143
Diet, 572
Fairness of Keno game, 641
Lowest cost of car rentals, 104, 110, 120
Lowest cost of checking accounts, 110
Lowest cost of video rentals, 110
Monthly condominium fees, 119
Price of battery and calculator, 119
Price of book, 84
Price of computer, 83
Price of magazine subscription, 119
Price of yacht, 84
Strategies for playing Monopoly, 495
Temperature change, 240

Earth Science

Angular speed of a point on equator, 303
Asteroid crater, 69
Crop yield after successive plantings, 289
Crop yield as function of fertilizer, 142
Crop yield as function of trees planted, 141
Distance to horizon, 102
Earthquake alarms, 121
Latitude, 304
Richter scale measure of an earthquake, 272, 273, 289
Soil chemistry, 572
Tides, 350
Time of sunrise modeled by a sine function, 359
The S-wave of an earthquake, 338
The S and P waves of an earthquake, 121
Tsunami wave, 337
Variation in daylight for various latitudes, 350

Ecology and Genetics

Chromosome splits, 633
CO_2 levels, 349
Depletion of oil resources, 258
Growth of CO_2 levels, 349
Malthusian model of popular growth, 289

Population growth, 278, 283, 287
Population of squirrels in nature preserve, 289
Population of walrus colony, 289
Predator-prey interactions, 359
Rate of oil leaking, 289
The logistic model of population growth, 289

Finance

Compound interest, 281, 287, 293, 294
Compound continuously, 282
Doubling a sum of money, 282, 287
Future value of an ordinary annuity, 612, 613
Investment in bonds, 503
Present value of an ordinary annuity, 46
Simple interest and variation, 193
Simple interest investment in two accounts, 79, 83, 119
Tripling a sum of money, 287
Yield of a compound interest investment, 22

Geometry

Altitude of a triangle, 93
Area of a rectangle as a function of the length, 162
Area of a sector, 303
Area of a snowflake, 613
Area of a triangle in terms of the three sides, 143
Areas of geometric figures and factoring, 38
Areas and algebraic formulas, 38
Curve fitting-quadratic, 512, 513, 541
Curve fitting-circle, 513, 541, 542
Curve fitting-plane, 514, 541
Diagonals of a polygon, 95
Diameter of the base of a right circular cone, 101
Distance from homeplate to second base, 94
Fencing a rectangular area, 92, 94
Lengths of the sides of a triangle, 83
Length and width of a rectangle, 77, 83
Maximum area of a rectangle, 161
Open box made from a square piece of material, 141
Perimeter of a rectangle, 94, 119
Perimeter of a snowflake, 613
Radius of a circle circumscribed about a triangle, 102
Radius of a circle inscribed in a triangle, 102
Radius of a cone, 101

Ratio of the surface area of a cylinder to the surface area of a sphere, 22
Ratio of the volume of a cylinder to the volume of a sphere, 22
Right circular cylinder inscribed in a cone, 141
Spheres, 101
Surface area of a cylinder, 241
Tangent-secant theorem, 83
Volume change of a cone, 193
Volume of a cone, 119

Linear Programming

Automotive engine reconditioning, 537
Communications, 537
Diets from three food groups, 572
Maximize a farmer's profit, 536, 542
Industrial solvents, 534, 537
Manufacture of sporting goods, 537
Nutrition of farm animals, 533, 537
Sale of manufactured goods, 536

Mathematics

Approximation of pi, 314
Cantor Set, 613
Cardino's formula, 248
Combinations, 26, 28
Continued fractions, 47
Groups, 196
Mandelbrot set, 196
Metrics, 196
Monte Carlo method, 644
Napier's inequality, 291
Newton's approximation, 599
Newton's formula, 450
Pythagorean triples, 130
Random walk, 633
Stirling's formula, 289
The harmonic sum, 291
The INT function, 289
The prime number theorem, 288
The Tower of Hanoi, 620
The triangle inequality, 15
Zeller's congruence, 197

Medicine

Amount of anesthetic in a patient, 278
Bacteria growth, 258, 287
Flow of blood when blood vessel splits into two parts, 404
Optimal branching of arteries, 448
Poiseuille's Law, 241
Pulse rate of a runner, 278
Rate of healing of a skin wound, 293

The Graphing Calculator Option

Throughout the text, we have included, where appropriate, optional graphing calculator exercises. For those who wish to explore more extensively with graphing calculators and graphing software, The Graphing Workbook that accompanies this text contains approximately 600 problems that use graphing utilities to explore mathematical concepts and applications. These may be solved with graphing calculators or the Math Assistant graphing software that is available free of charge to users of the text. The chart below lists the topics covered in The Graphing Workbook and the respective sections in each text that correspond to the topic.

The Graphing Workbook also contains some supplemental investigations that extend topics in the text. Below the correlation chart, we have listed the titles of these activities and the respective chapters where the concepts are covered.

The Graphing Workbook also provides directions for how to use the most popular graphing calculators including the Texas Instruments TI-81, the TI-85, and the Casio fx-7700G models and an introduction to the Math Assistant software.

Topic in Graphing Workbook	College Algebra and Trigonometry	College Algebra	College Trigonometry	Precalculus
Absolute Value Graphs	3.1	3.1	1.2	2.1
Slope-Intercept Form	3.3	3.3	—	2.3
Quadratic Functions	3.4	3.4	—	2.4
Maximum and Minimum of Functions	3.4	3.4	—	2.4
Graphing Functions	3.5	3.5	1.4	2.5
Reflections and Translations	3.5	3.5	1.4	2.5
Stretching and Shrinking	3.5	3.5	1.4	2.5
Functions and Their Inverses	3.7	3.7	1.6	2.7
Zeros of Functions	4.3	4.3	—	3.3

continued

Topic in Graphing Workbook	College Algebra and Trigonometry	College Algebra	College Trigonometry	Precalculus
Rational Functions	4.5	4.5	—	3.5
Approximating Zeros of Functions	4.6	4.6	—	3.6
Solving Inequalities	2.5	2.5	—	1.5
Exponential Functions	5.1	5.1	7.1	4.1
Logarithmic Functions	5.3	5.3	7.3	4.3
Sine and Cosine Graphs	6.5	—	2.5	5.4
Exploring Sinusoids	6.5	—	2.5	5.4
Tangent, Cotangent, Secant, and Cosecant Graphs	6.6	—	2.6	5.5
Phase Shift of Trigonometric Functions	6.7	—	2.7	5.6
Trigonometric Identities by Graphing	7.1–7.4	—	3.1–3.4	6.1–6.4
Sums of Sines with Different Periods	7.4	—	3.4	6.4
Products of Sines and Cosines	7.4	—	3.4	6.4
Solving Trigonometric Equations by Graphing	7.6	—	3.6	6.6
Law of Cosines	8.2	—	4.2	7.2
Conics	9.1–9.3	6.1–6.3	6.1–6.3	8.1–8.3
General Polar Graphs	9.4	—	6.5	8.5
Polar Graphs of Conics and Lines	—	—	6.6	8.6
Parametric Equations	—	—	6.7	8.7
Estimating Solutions of Linear Systems of Equations	10.1	7.1	—	9.1
Breakeven Points	10.1	7.1	—	9.1
Linear Programming	10.6	7.6	—	9.6
Estimating Solutions of Quadratic Systems of Equations	10.3	7.3	—	9.3
Matrix Calculations	11.2	8.2	—	10.2
Matrix Application: Markov Chains	11.2	8.2	—	10.2
Matrix Application: Encoding and Decoding	11.2	8.2	—	10.2
Exploring Determinants	11.4	8.4	—	10.4
Arithmetic Sequences and Series	12.2	9.2	—	11.2

Topic in Graphing Workbook	College Algebra and Trigonometry	College Algebra	College Trigonometry	Precalculus
Geometric Sequences and Series	12.3	9.3	—	11.3
Geometric Series Application: Compound Interest	12.3	9.3	—	11.3
Sums of Infinite Series	12.3	9.3	—	11.3

SUPPLEMENTAL INVESTIGATIONS

These exercises can be assigned after completing the indicated chapter.

Topic in Graphing Workbook	College Algebra and Trigonometry	College Algebra	College Trigonometry	Precalculus
Correlation and Line of Best Fit	Chapter 3	Chapter 3	Chapter 1	Chapter 2
Fitting Power Functions to Data	Chapter 4	Chapter 4	—	Chapter 3
Fitting Exponential Functions to Data	Chapter 5	Chapter 5	Chapter 7	Chapter 4
Graphing Squares	Chapter 5	Chapter 5	—	Chapter 4
Polar Sun Flowers	Chapter 9	—	Chapter 6	Chapter 8
Spirals	Chapter 9	—	Chapter 6	Chapter 8
Matrix Transformations of Plane Vectors	Chapter 11	Chapter 8	—	Chapter 10

1 Fundamental Concepts

CASE IN POINT   *How Large is Infinity?*

The German mathematician Georg Cantor (1845–1918) developed the idea of the cardinality of a set. The cardinality of a finite set is the number of elements in the set. For example, the set {5, 7, 11} has a cardinality of 3. The set of natural numbers {1, 2, 3, 4, 5, 6, 7, ...} is an infinite set. Cantor denoted its cardinality by the symbol $\aleph_0$, which is read "aleph null."

The set of whole numbers consists of all the elements of the set of natural numbers and the number 0. The following display shows a one-to-one correspondence between the set of natural numbers and the set of whole numbers.

$$\{1, \quad 2, \quad 3, \quad 4, \quad 5, \quad ..., \quad n, \quad ...\}$$
$$\updownarrow \quad \updownarrow \quad \updownarrow \quad \updownarrow \quad \updownarrow \qquad\quad \updownarrow$$
$$\{0, \quad 1, \quad 2, \quad 3, \quad 4, \quad ..., \quad n-1, \quad ...\}$$

Cantor reasoned that because of this one-to-one correspondence, the set of natural numbers and the set of whole numbers both have the same cardinality, namely $\aleph_0$.

Cantor was also able to show that the set of irrational numbers has a cardinality that is different from $\aleph_0$. The idea that some infinite sets have more elements than other infinite sets was not readily accepted. Previous to Cantor's work, the philosopher Voltaire (1694–1778) expressed the following opinion:

> *We admit, in geometry, not only infinite magnitudes, that is to say magnitudes greater than any assignable magnitude, but infinite magnitudes infinitely greater, the one than the other. This astonishes our dimension of brains, which is only about six inches long, five broad, and six in depth, in the largest of heads.*

1

1.1 The Real Number System

Human beings share the desire to organize and classify. Ancient astronomers classified stars into groups called constellations. Modern astronomers continue to classify stars by such characteristics as color, mass, size, temperature, and distance from earth. In mathematics it is useful to classify numbers into groups called **sets**. The following sets of numbers are used extensively in the study of algebra:

Integers	$\{\ldots, -3, -2, -1, 0, 1, 2, 3, \ldots\}$
Rational numbers	{all terminating or repeating decimals}
Irrational numbers	{all nonterminating, nonrepeating decimals}
Real numbers	{all rational or irrational numbers}

If a decimal terminates or repeats a block of digits, then the number is a rational number. Rational numbers can also be written in the form p/q, where p and q are integers and $q \neq 0$. For example,

$$\frac{3}{4} = 0.75 \quad \text{and} \quad \frac{5}{11} = 0.\overline{45}$$

are rational numbers. The bar over the 45 means that the block repeats without end; that is, $0.\overline{45} = 0.454545\ldots$.

In its decimal form, an irrational number neither terminates nor repeats. For example, $0.272272227\ldots$ is a nonterminating, nonrepeating decimal and thus is an irrational number. One of the best-known irrational numbers is pi, denoted by the Greek symbol π. The number π is defined as the ratio of the circumference of a circle to its diameter. Often in applications, one of the rational numbers 3.14 or $\frac{22}{7}$ is used as an approximation of the irrational number π.

Every real number is either a rational number or an irrational number. If a real number is written in decimal form, it is a terminating decimal, repeating decimal, or a nonterminating and nonrepeating decimal.

Each number in a set is called an **element** of the set. Set A is a **subset** of set B if every element of set A is also an element of set B. The set of **negative integers** $\{-1, -2, -3, -4, -5, \ldots\}$ is a subset of the set of integers. The set of **positive integers** $\{1, 2, 3, 4, 5, \ldots\}$ (also known as the set of **natural numbers**) is also a subset of the integers. Figure 1.1 illustrates the subset relationships among the sets defined above.

Prime numbers and *composite numbers* play an important role in almost every branch of mathematics. A **prime number** is a positive integer greater than 1 that has no positive-integer factors other than itself and 1. The 10 smallest prime numbers are 2, 3, 5, 7, 11, 13, 17, 19, 23, and 29. Each of these numbers has only itself and 1 as factors.

A **composite number** is a positive integer greater than 1 that is not a prime number. For example, 10 is a composite number because 10 has both

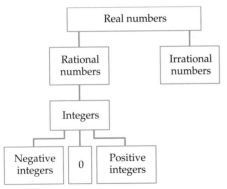

Figure 1.1

2 and 5 as factors. The 10 smallest composite numbers are 4, 6, 8, 9, 10, 12, 14, 15, 16, and 18.

EXAMPLE 1   Classify Real Numbers

Determine which of the following numbers are:

a. integers
b. rational numbers
c. irrational numbers
d. real numbers
e. prime numbers
f. composite numbers

$$-0.2, \quad 0, \quad 0.\overline{3}, \quad \pi, \quad 6, \quad 7, \quad 41, \quad 51, \quad 0.71771777177771\ldots$$

Solution

a. integers: 0, 6, 7, 41, 51
b. rational numbers: -0.2, 0, $0.\overline{3}$, 6, 7, 41, 51
c. irrational numbers: $0.71771777177771\ldots$, π
d. real numbers: -0.2, 0, $0.\overline{3}$, π, 6, 7, 41, 51, $0.71771777177771\ldots$
e. prime numbers: 7, 41
f. composite numbers: 6, 51

■ *Try Exercise* **2**, *page 8.*

Sets are often written using **set-builder notation,** which makes use of a variable and a characteristic property that the elements of the set alone possess. The set-builder notation

$$\{x^2 \,|\, x \text{ is an integer}\}$$

is read as "the set of all elements x^2 such that x is an integer." This is the set of **perfect squares:** $\{0, 1, 4, 9, 16, 25, 36, 49, \ldots\}$.

The **empty set** or **null set** is a set without any elements. The set of numbers that are both prime and also composite is an example of the null set. The null set is denoted by the symbol $\varnothing$.

Just as addition and subtraction are operations performed on real numbers, there are operations performed on sets. Two of these set operations are called *intersection* and *union*. The **intersection** of sets A and B, denoted by $A \cap B$, is the set of all elements belonging to both set A and to set B. The **union** of sets A and B, denoted by $A \cup B$, is the set of all elements belonging to set A or set B or both.

EXAMPLE 2   Find the Intersection and the Union of Two Sets

Find each of the following, given $A = \{0, 1, 4, 6, 9\}$, $B = \{1, 3, 5, 7, 9\}$ and $P = \{x \,|\, x \text{ is a prime number} < 10\}$.

a. $A \cap B$ b. $A \cap P$ c. $A \cup B$ d. $A \cup P$

Solution

a. $A \cap B = \{0, 1, 4, 6, 9\} \cap \{1, 3, 5, 7, 9\} = \{1, 9\}$ • Only 1 and 9 are common to both sets.

b. First determine that $P = \{2, 3, 5, 7\}$. Therefore:

$A \cap P = \{0, 1, 4, 6, 9\} \cap \{2, 3, 5, 7\} = \varnothing$ • There are no common elements.

c. $A \cup B = \{0, 1, 4, 6, 9\} \cup \{1, 3, 5, 7, 9\}$ • List the elements of the first set. Include elements from the second set that are not already listed.
$= \{0, 1, 3, 4, 5, 6, 7, 9\}$

d. $A \cup P = \{0, 1, 4, 6, 9\} \cup \{2, 3, 5, 7\} = \{0, 1, 2, 3, 4, 5, 6, 7, 9\}$

■ *Try Exercise* **4**, *page 8.*

Real Number Properties

Addition, multiplication, subtraction, and **division** are the operations of arithmetic. Addition of the two real numbers a and b is designated by $a + b$. If $a + b = c$, then c is the **sum** and the real numbers a and b are called the **terms.**

Multiplication of the real numbers a and b is designated by ab or $a \cdot b$. If $ab = c$, then c is the **product** and the real numbers a and b are called **factors** of c.

The number $-b$ is referred to as the **additive inverse** of b. Subtraction of the real numbers a and b is designated by $a - b$ and is defined as the sum of a and the additive inverse of b. That is,

$$a - b = a + (-b)$$

If $a - b = c$, then c is called the **difference** of a and b.

The **multiplicative inverse** or **reciprocal** of the nonzero number b is $1/b$. The division of a and b, designated by $a \div b$ with $b \neq 0$, is defined as the product of a and the reciprocal of b. That is,

$$a \div b = a\left(\frac{1}{b}\right) \quad \text{provided } b \neq 0$$

If $a \div b = c$, then c is called the **quotient** of a and b.

The notation $a \div b$ is often represented by the fractional notation a/b or $\frac{a}{b}$. The real number a is the **numerator**, and the nonzero real number b is the **denominator** of the fraction.

Properties of Real Numbers

Let a, b, and c be real numbers.

	Addition Properties	**Multiplication Properties**
Closure	$a + b$ is a unique real number.	ab is a unique real number.
Commutative	$a + b = b + a$	$ab = ba$
Associative	$(a + b) + c = a + (b + c)$	$(ab)c = a(bc)$
Identity	There exists a unique real number 0 such that $a + 0 = 0 + a = a$.	There exists a unique real number 1 such that $a \cdot 1 = 1 \cdot a = a$.
Inverse	For each real number a, there is a unique real number $-a$ such that $a + (-a) = (-a) + a = 0$.	For each *nonzero* real number a, there is a unique real number $1/a$ such that $a(1/a) = (1/a)a = 1$.
Distributive	$a(b + c) = ab + ac$	

We can identify which property of real numbers has been used to rewrite expressions by closely comparing the expressions and noting any changes.

EXAMPLE 3 **Identify Properties of Real Numbers**

Identify the property of real numbers illustrated in each of the following:

a. $(2a)b = 2(ab)$ b. $\left(\dfrac{1}{5}\right)11$ is a real number.

c. $4(x + 3) = 4x + 12$ d. $(a + 5b) + 7c = (5b + a) + 7c$

e. $\left(\dfrac{1}{2} \cdot 2\right)a = 1 \cdot a$ f. $1 \cdot a = a$

Solution

a. Associative property of multiplication
b. Closure property of multiplication of real numbers
c. Distributive property
d. Commutative property of addition
e. Inverse property of multiplication
f. Identity property of multiplication

■ *Try Exercise* **16**, *page 8.*

An **equation** is a statement of equality between two numbers or two expressions. There are four basic **properties of equality** that relate to equations.

Properties of Equality

Let a, b, and c be real numbers

Reflexive	$a = a$
Symmetric	If $a = b$, then $b = a$.
Transitive	If $a = b$ and $b = c$, then $a = c$.
Substitution	If $a = b$, then a may be replaced by b in any expression that involves a.

EXAMPLE 4   **Identify Properties of Equality**

Identify the property of equality illustrated in each of the following:

a. If $3a + b = c$, then $c = 3a + b$.

b. $5(x + y) = 5(x + y)$

c. If $4a - 1 = 7b$ and $7b = 5c + 2$, then $4a - 1 = 5c + 2$.

d. If $a = 5$ and $b(a + c) = 72$, then $b(5 + c) = 72$.

Solution

a. Symmetric   b. Reflexive   c. Transitive   d. Substitution

■ *Try Exercise* **18**, *page 8.*

The following **properties of fractions** will be used throughout this text.

Properties of Fractions

For all fractions a/b and c/d, where $b \neq 0$ and $d \neq 0$:

Equality	$\dfrac{a}{b} = \dfrac{c}{d}$ if and only if $ad = bc$
Equivalent fractions	$\dfrac{a}{b} = \dfrac{ac}{bc}, \qquad c \neq 0$
Addition	$\dfrac{a}{b} + \dfrac{c}{b} = \dfrac{a + c}{b}$
Subtraction	$\dfrac{a}{b} - \dfrac{c}{b} = \dfrac{a - c}{b}$
Multiplication	$\dfrac{a}{b} \cdot \dfrac{c}{d} = \dfrac{ac}{bd}$
Division	$\dfrac{a}{b} \div \dfrac{c}{d} = \dfrac{a}{b} \cdot \dfrac{d}{c} = \dfrac{ad}{bc}, \qquad c \neq 0$
Sign	$-\dfrac{a}{b} = \dfrac{-a}{b} = \dfrac{a}{-b}$

Remark The equality property of fractions contains the terminology "if and only if," which implies each of the following:

$$\text{If } \frac{a}{b} = \frac{c}{d}, \qquad \text{then } ad = bc.$$

$$\text{If } ad = bc, \qquad \text{then } \frac{a}{b} = \frac{c}{d}.$$

The number zero has many special properties. The following division properties of zero play an important role in this text.

Division Properties of Zero

1. For $a \neq 0$, $\dfrac{0}{a} = 0$.   (Zero divided by any nonzero number is zero.)

2. $\dfrac{a}{0}$ is undefined.   (Division by zero is undefined.)

The properties of fractions can be used to find the sum, difference, product, or quotient of fractions.

EXAMPLE 5 Compute with Fractions

Use the properties of fractions to perform the indicated operations. Assume that $a \neq 0$.

a. $\dfrac{2a}{3} - \dfrac{a}{5}$ b. $\dfrac{2a}{5} \cdot \dfrac{3a}{4}$ c. $\dfrac{5a}{6} \div \dfrac{3a}{4}$ d. $\dfrac{0}{3a}$

Solution

a. Rewrite each fraction as an equivalent fraction with a common denominator of 15 by multiplying both the numerator and the denominator of $2a/3$ by 5 and by multiplying both the numerator and the denominator of $a/5$ by 3.

$$\frac{2a}{3} - \frac{a}{5} = \frac{2a(5)}{3(5)} - \frac{a(3)}{5(3)} = \frac{10a}{15} - \frac{3a}{15} = \frac{10a - 3a}{15} = \frac{7a}{15}$$

b. $\dfrac{2a}{5} \cdot \dfrac{3a}{4} = \dfrac{(2a)(3a)}{(5)(4)} = \dfrac{6a^2}{20} = \dfrac{3a^2}{10}$

c. $\dfrac{5a}{6} \div \dfrac{3a}{4} = \dfrac{5a}{6} \cdot \dfrac{4}{3a} = \dfrac{20a}{18a} = \dfrac{10}{9}$

d. $\dfrac{0}{3a} = 0$ • Zero divided by any nonzero number is zero.

—————————————————— ■ *Try Exercise* **30**, *page 8.*

EXERCISE SET 1.1

In Exercises 1 and 2, determine which of the numbers are **a.** integers, **b.** rational numbers, **c.** irrational numbers, **d.** real numbers, **e.** prime numbers, **f.** composite numbers.

1. -3 4 $\dfrac{1}{5}$ 11   3.14   57   0.252252225...

2. $5.\overline{17}$ -4.25 $\dfrac{1}{4}$ π 21   53   0.45454545...

In Exercises 3 to 14, use $A = \{0, 1, 2, 3, 4\}$, $B = \{1, 3, 5, 11\}$, $C = \{1, 3, 6, 10\}$, and $D = \{0, 2, 4, 6, 8, 10\}$ to find the indicated intersection or union.

3. $A \cap B$ **4.** $A \cap C$ **5.** $B \cap C$

6. $B \cap D$ **7.** $A \cap D$ **8.** $C \cap D$

9. $A \cup B$ **10.** $A \cup C$ **11.** $B \cup C$

12. $B \cup D$ **13.** $A \cup D$ **14.** $C \cup D$

In Exercises 15 to 28, identify the property of real numbers or the property of equality that is illustrated.

15. $3 + (2 + 5) = (3 + 2) + 5$

16. $6 + (2 + 7) = 6 + (7 + 2)$

17. $1 \cdot a = a$

18. If $a + b = 2$, then $2 = a + b$.

19. $a(bx) = a(bx)$

20. If $x + 2y = 7$ and $7 = y$, then $x + 2(7) = 7$.

21. If $x = 2(y + z)$ and $2(y + z) = 5w$, then $x = 5w$.

22. $p(q + r) = pq + pr$

23. $m + (-m) = 0$

24. $t\left(\dfrac{1}{t}\right) = 1$

25. $7(a + b) = 7(b + a)$

26. $8(gh + 5) = 8(hg + 5)$

27. If $x + 2y = 7$ and $7 = z$, then $x + 2y = z$.

28. $5[x + (y + z)] = 5x + 5(y + z)$

In Exercises 29 to 38, use the properties of fractions to perform the indicated operations. State each answer in lowest terms. Assume a is a nonzero real number.

29. $\dfrac{2a}{7} - \dfrac{5a}{7}$ **30.** $\dfrac{2a}{5} + \dfrac{3a}{7}$

31. $\dfrac{-3a}{5} + \dfrac{a}{4}$ **32.** $\dfrac{7}{8}a - \dfrac{13}{5}a$

33. $\dfrac{-5}{7} \cdot \dfrac{2}{3}$ **34.** $\dfrac{7}{11} \cdot \dfrac{-22}{21}$

35. $\dfrac{12a}{5} \div \dfrac{-2a}{3}$ **36.** $\dfrac{2}{5} \div 3\dfrac{2}{3}$

37. $\dfrac{2a}{3} - \dfrac{4a}{5}$ **38.** $\dfrac{1}{2a} - \dfrac{3}{a}$

39. One pipe can fill a pool in 11 hours. A second pipe can fill the same pool in 15 hours. Assume the first pipe fills $\frac{1}{11}$ of the pool every hour and the second pipe fills $\frac{1}{15}$ of the pool every hour. **a.** Find the amount of the pool the two pipes together fill in 3 hours. **b.** Find the amount of the pool they fill together in x hours.

40. The relationship between the distance of an object d_0 from a curved mirror, the distance of its image d_i from the mirror, and the focal length f of the mirror is given by the **mirror equation:**

$$\frac{1}{f} = \frac{1}{d_0} + \frac{1}{d_i}$$

What is the focal length[1] f of a mirror for which $d_0 = 25$ centimeters and $d_i = -5$ centimeters?

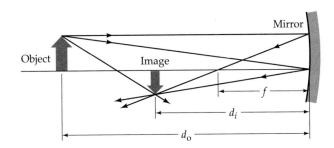

41. State the multiplicative inverse of $7\frac{3}{8}$.

42. State the multiplicative inverse of $-4\frac{2}{5}$.

43. Show by an example that the operation of subtraction of real numbers is not a commutative operation.

44. Show by an example that the operation of division of nonzero real numbers is not a commutative operation.

45. Show by an example that the operation of subtraction of real numbers is not an associative operation.

46. Show by an example that the operation of division of real numbers is not an associative operation.

In Exercises 47 to 56, classify each statement as true or false.

47. $a/0$ is the multiplicative inverse of $0/a$.

48. $(-1/\pi)$ is the multiplicative inverse of $-\pi$.

---

[1] For convex mirrors, both the focal length f and the image distance d_i are *negative* quantities.

49. If $p = q + t/2$ and $q + t/2 = \frac{1}{2}s$, then $\frac{1}{2}s = p$.

50. If $a - b = 7$, then $7 = b - a$.

51. The sum of two composite numbers is a composite number.

52. All integers are natural numbers.

53. Every real number is either a rational or an irrational number.

54. Every rational number is either even or odd.

55. 1 is the only positive integer that is not prime and not composite.

56. All repeating decimals are rational numbers.

57. Use a calculator to write each of the following rational numbers as a decimal. If the number is represented by a nonterminating decimal, then use a *bar* over the repeating portion of the decimal.

a. $\dfrac{8}{11}$ **b.** $\dfrac{33}{40}$ **c.** $\dfrac{2}{7}$ **d.** $\dfrac{5}{37}$

58. Use a calculator to determine whether 3.14 or $\frac{22}{7}$ is a closer approximation to π.

59. Use a calculator to complete the following table.

x	0.1	0.01	0.001	0.0000001
$\dfrac{\sqrt{x+9} - 3}{x}$				

Now make a guess as to the number the fraction seems to be approaching as x assumes real numbers that are closer and closer to zero.

60. Use a calculator to complete the following table:

x	0.1	0.01	0.001	0.0000001
$\dfrac{\dfrac{1}{2} - \dfrac{1}{x+2}}{x}$				

Now make a guess as to the number the fraction seems to be approaching as x assumes real numbers that are closer and closer to zero.

Supplemental Exercises

In Exercises 61 to 64, list the elements of the set.

61. $A = \{x \mid x$ is a composite number less than 11$\}$

62. $B = \{x \mid x$ is an even prime number$\}$

63. $C = \{x \mid 50 < x < 60$ and x is a prime number$\}$

64. $D = \{x \mid x$ is the smallest odd composite number$\}$

65. Which of the properties of real numbers are satisfied by the set of positive integers?

66. Which of the properties of real numbers are satisfied by the set of integers?

67. Which of the properties of real numbers are satisfied by the set of rational numbers?

68. Which of the properties of real numbers are satisfied by the set of irrational numbers?

69. To prove that the set of prime numbers has an infinite number of elements, Euclid (fl. about 300 B.C.) used a proof by contradiction. His proof was similar to the following. Assume that the set of prime numbers forms a finite set. For convenience, label the largest prime number in this finite set L. Now consider the positive integer T, which is equal to the product of all the prime numbers, plus 1. That is,

$$T = (2)(3)(5)(7)\cdots(L) + 1$$

Explain why

a. T cannot be a prime number.   (*Hint:* Which is larger, T or L?)

b. T cannot be a composite number.   (*Hint:* Which prime numbers are factors of T?)

The positive integer T must be either a prime number or a composite number. The assumption that the prime numbers form a finite set has produced the contradiction that T is not a prime number and T is not a composite number. Thus Euclid reasoned that the set of prime numbers cannot be finite.

70. In 1742 Christian Goldbach conjectured that every even number greater than 2 can be written as the sum of two prime numbers. Many mathematicians have tried to prove or disprove this conjecture without succeeding. Show that Goldbach's conjecture is true for the even numbers **a.** 12 and **b.** 30.

71. *Theorem:*   Any number of the form

$$1111\ldots1$$

can be prime only if the number of 1's is prime. For example, the numbers

11   and   1111111111111111111

are both prime numbers. The number 111 is not a prime number because $111 = 3 \cdot 37$. Explain why this is not a contradiction of the above theorem.

72. If the natural numbers n and $n + 2$ are both prime numbers, then they are said to be twin primes. For example, 11 and 13 are twin primes. It is not known whether the set of twin primes is an infinite set or a finite set. List all the twin primes less than 50.

1.2 Intervals, Absolute Value, and Distance

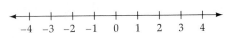

Figure 1.2

The real numbers can be represented geometrically by a **coordinate axis** called a **real number line.** Figure 1.2 shows a portion of a real number line. The number associated with a particular point on a real number line is called the **coordinate** of the point. It is customary to label those points whose coordinates are integers. The point corresponding to zero is called the **origin**, denoted 0. Numbers to the right of the origin are **positive real numbers;** numbers to the left of the origin are **negative real numbers.**

Figure 1.3

A real number line provides a picture of the real numbers. That is, each real number corresponds to one and only one point on the real number line, and each point on a real number line corresponds to one and only one real number. This type of correspondence is referred to as a **one-to-one correspondence.** The real numbers -3, $-1/2$, and 1.75 are graphed in Figure 1.3.

Certain order relationships exist between real numbers. For example, if a and b are real numbers, then:

a **equals** b (denoted by $a = b$) if $a - b = 0$.

a is **greater than** b (denoted by $a > b$) if $a - b$ is positive.

a is **less than** b (denoted by $a < b$) if $b - a$ is positive.

On a horizontal number line, the notation

$a = b$ implies that the point with coordinate a is the same point as the point with coordinate b.

$a > b$ implies that the point with coordinate a is to the right of the point with coordinate b.

$a < b$ implies that the point with coordinate a is to the left of the point with coordinate b

The **inequality** symbols $<$ and $>$ are sometimes combined with the equality symbol in the following manner:

$a \geq b$ Read "a is greater than or equal to b," which means $a > b$ or $a = b$.

$a \leq b$ Read "a is less than or equal to b," which means $a < b$ or $a = b$.

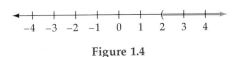

Figure 1.4

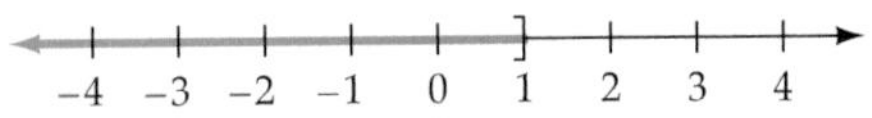

Inequalities can be used to represent subsets of real numbers. For example, the inequality $x > 2$ represents all real numbers greater than 2; Figure 1.4 shows its graph. The parenthesis at 2 means that 2 is not part of the graph.

Figure 1.5

The inequality $x \leq 1$ represents all real numbers less than or equal to 1; Figure 1.5 shows its graph. The bracket at 1 means that 1 is part of the graph.

Figure 1.6

The inequality $-1 \leq x < 3$ represents all real numbers between -1 and 3, including -1 but not including 3. Figure 1.6 shows its graph.

Subsets of real numbers can also be represented by a compact form of notation called **interval notation**. For example, $[-1, 3)$ is the interval notation for the subset of real numbers in Figure 1.6.

In general, the interval notation

(a, b) represents all real numbers between a and b, not including a and not including b. This is an **open interval**.

$[a, b]$ represents all real numbers between a and b, including a and including b. This is a **closed interval**.

$(a, b]$ represents all real numbers between a and b, not including a but including b. This is a **half-open interval**.

$[a, b)$ represents all real numbers between a and b, including a but not including b. This is a **half-open interval**.

Figure 1.7 shows the four subsets of real numbers that are associated with the four interval notations (a, b), $[a, b]$, $(a, b]$, and $[a, b)$.

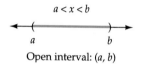
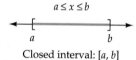
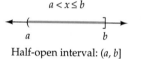
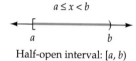

Figure 1.7
Finite intervals

Subsets of the real numbers whose graphs extend forever in one or both directions can be represented by interval notation using the **infinity symbol** ∞ or the **negative infinity symbol** $-\infty$.

As Figure 1.8 shows, the interval notation

$(-\infty, a)$ represents all real numbers less than a.

(b, ∞) represents all real numbers greater than b.

$(-\infty, a]$ represents all real numbers less than or equal to a.

$[b, \infty)$ represents all real numbers greater than or equal to b.

$(-\infty, \infty)$ represents all real numbers.

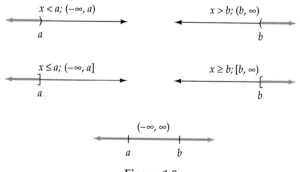

Figure 1.8
Infinite intervals

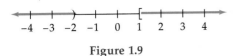

Figure 1.9

Some graphs consist of more than one interval of the real number line. Figure 1.9 is a graph of the interval $(-\infty, -2)$, along with the interval $[1, \infty)$.

The word *or* is used to denote the union of two sets. The word *and* is used to denote intersection. Thus the graph in Figure 1.9 is denoted by the inequality notation

$$x < -2 \quad \text{or} \quad x \geq 1$$

To represent this graph using interval notation, use the union symbol $\cup$ and write $(-\infty, -2) \cup [1, \infty)$.

EXAMPLE 1 **Graph Intervals and Inequalities**

Graph the following. Also write a. and b. using interval notation, and write c. and d. using inequality notation.

a. $-2 \leq x < 3$ b. $x \geq -3$ c. $[-4, -2] \cup [0, \infty)$ d. $(-\infty, 2)$

Solution

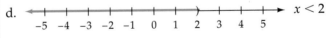

a. $[-2, 3)$

b. $[-3, \infty)$

c. $-4 \leq x \leq -2$ or $x \geq 0$

d. $x < 2$

Figure 1.10

■ *Try Exercise* **16,** *page 14.*

The absolute value of the real number a, denoted $|a|$, is the distance between a and 0 on the number line. For example, $|2| = 2$ and $|-2| = 2$. In general, if $a \geq 0$, then $|a| = a$; however, if $a < 0$, then $|a| = -a$ because $-a$ is positive when $a < 0$. This leads us to the following definition.

Definition of Absolute Value

> The **absolute value** of the real number a is defined by
>
> $$|a| = \begin{cases} a & \text{if } a \geq 0 \\ -a & \text{if } a < 0 \end{cases}$$

The following theorems can be derived by using the definition of absolute value.

Absolute Value Theorems

For all real numbers a and b,

Nonnegative	$\|a\| \geq 0$
Product	$\|ab\| = \|a\|\,\|b\|$
Quotient	$\left\|\dfrac{a}{b}\right\| = \dfrac{\|a\|}{\|b\|},\ b \neq 0$
Triangle inequality	$\|a + b\| \leq \|a\| + \|b\|$
Difference	$\|a - b\| = \|b - a\|$

The definition of absolute value and the absolute value theorems can be used to write some expressions without absolute value symbols.

EXAMPLE 2 **Evaluate Absolute Value Expressions**

Write each of the following without absolute value symbols.

a. $\|1 - \pi\|$ b. $\left\|\dfrac{2x}{\|x\| + \|x - 2\|}\right\|$, given $0 < x < 2$

Solution

a. Since $1 - \pi < 0$, $\|1 - \pi\| = -(1 - \pi) = \pi - 1$.

b. Use the quotient theorem to write the expression as a quotient of absolute values.

$$\left|\frac{2x}{|x| + |x - 2|}\right| = \frac{|2x|}{\big||x| + |x - 2|\big|}$$

Since $0 < x < 2$, $|2x| = 2x$, $|x| = x$, and $|x - 2| = -x + 2$. Substituting yields:

$$\frac{|2x|}{\big||x| + |x - 2|\big|} = \frac{2x}{|x + (-x + 2)|} = \frac{2x}{|2|} = \frac{2x}{2} = x$$

■ *Try Exercise* **56**, *page 15.*

The definition of **distance** between any two points on a real number line makes use of absolute value.

Distance Between Points on a Real Number Line

For any real numbers a and b, the distance between the graph of a and the graph of b is denoted by $d(a, b)$, where

$$d(a, b) = |a - b|$$

EXAMPLE 3   **Find the Distance Between Points**

Find the distance between the points whose coordinates are:

a.   5, −2   b.   −π, −2

Solution

a. $d(5, -2) = |5 - (-2)| = |5 + 2| = |7| = 7$

b. $d(-\pi, -2) = |-\pi - (-2)| = |-\pi + 2|$ • $-\pi + 2 < 0.$ Thus
$$= -(-\pi + 2) = \pi - 2$$ $|-\pi + 2| = -(-\pi + 2).$

■ *Try Exercise* **64**, *page 15.*

Absolute value notation and the notion of distance can also be used to describe intervals.

EXAMPLE 4   **Use Absolute Value Notation to Describe an Interval**

Use absolute value notation to describe (5, 9).

Solution   In Figure 1.11, the center of the interval is 7. The distance between every number x in the interval and 7 is less than 2. The distance between x and 7 is $|x - 7|$. Therefore the interval can be described as $|x - 7| < 2.$

■ *Try Exercise* **74**, *page 15.*

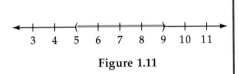

Figure 1.11

EXERCISE SET 1.2

In Exercises 1 and 2, graph each number on a real number line.

1. $-4; -2; \dfrac{7}{4}; 2.5$ **2.** $-3.5; 0; 3; \dfrac{9}{4}$

In Exercises 3 to 14, replace the □ with the appropriate symbol (<, =, or >).

3. $\dfrac{5}{2} \,\square\, 4$ **4.** $-\dfrac{3}{2} \,\square\, -3$

5. $\dfrac{2}{3} \,\square\, 0.6666$ **6.** $\dfrac{1}{5} \,\square\, 0.2$

7. $1.75 \,\square\, 2.23$ **8.** $1.25 \,\square\, 1.3$

9. $0.\overline{36} \,\square\, \dfrac{4}{11}$ **10.** $0.4 \,\square\, \dfrac{4}{9}$

11. $\dfrac{10}{5} \,\square\, 2$ **12.** $\dfrac{0}{2} \,\square\, -\dfrac{0}{5}$

13. $\pi \,\square\, 3.14159$ **14.** $\dfrac{22}{7} \,\square\, \pi$

In Exercises 15 to 26, graph each inequality and write the inequality using interval notation.

15. $3 < x < 5$ **16.** $-2 \le x < 1$

17. $x < 3$ **18.** $x \ge 4$

19. $x \ge 0$ and $x < 3$ **20.** $x > -4$ and $x \le 4$

21. $x < -3$ or $x \ge 2$ **22.** $x \le 2$ or $x > 3$

23. $x > 3$ and $x < 4$ **24.** $x > -5$ or $x < 1$

25. $x \le 3$ and $x > -1$ **26.** $x < 5$ and $x \le 2$

In Exercises 27 to 38, graph each interval and write each interval as an inequality.

27. $[-4, 1]$ **28.** $[-2, 3)$ **29.** $(1, 5)$ **30.** $(1, 4]$

31. $[2.5, \infty)$ **32.** $(-\infty, 3]$ **33.** $(-\infty, 2)$ **34.** (π, ∞)

35. $(-\infty, 2] \cup (3, \infty)$ **36.** $(-\infty, 1) \cup (4, \infty)$

37. $(-\infty, 3) \cup (3, \infty)$ **38.** $(-\infty, 1) \cup [2, \infty)$

In Exercises 39 to 46, use the given notation or graph to supply the notation or graph that is marked with a question mark.

	Inequality Notation	Interval Notation	Graph
39.	$x \leq 3$	?	?
40.	?	$(-2, \infty)$	?
41.	?	?	
42.	$-3 \leq x < -1$	?	?
43.	?	$[1, 4]$	?
44.	?	?	
45.	?	$[-2, \pi)$	?
46.	$x < 2$ or $x \geq 4$	?	?

In Exercises 47 to 60, write each expression without absolute value symbols.

47. $|4|$ **48.** $|-8|$ **49.** $|-27.4|$

50. $|3| - |-7|$ **51.** $-|-3| - |8|$ **52.** $|4| \, |-8|$

53. $|y^2 + 10|$ **54.** $|x^2 + 1|$ **55.** $|-1 - \pi|$

56. $|x + 6| + |x - 2|$, given $0 < x < 1$

57. $|x - 4| + |x + 5|$, given $2 < x < 3$

58. $|x + 1| + |x - 3|$, given $x > 5$

59. $\left| \dfrac{x + 7}{|x| + |x - 1|} \right|$, given $0 < x < 1$

60. $\left| \dfrac{x + 3}{\left| x - \dfrac{1}{2} \right| + \left| x + \dfrac{1}{2} \right|} \right|$, given $0 < x < 0.2$

In Exercises 61 to 72, find the distance between the points whose coordinates are given.

61. 8, 1   **62.** $-2, -7$ **63.** $-3, 5$

64. $-5, 8$ **65.** 16, -34 **66.** $-108, 22$

67. $-38, -5$ **68.** $\pi, 3$ **69.** $-\pi, 3$

70. $\dfrac{1}{7}, -\dfrac{1}{2}$ **71.** $\dfrac{1}{3}, \dfrac{3}{4}$ **72.** 0, -8

In Exercises 73 to 80, use absolute value notation to describe the given expression.

73. (1, 9)   **74.** (16, 24)

75. Distance between a and 2

76. Distance between b and -7

77. $d(m, n)$

78. $d(p, -q)$

79. The distance between z and 5 is greater than 4.

80. The distance between x and -2 is less than 7.

In Exercises 81 to 84, write interval notation for the given expression.

81. x is a real number and $x \neq 3$.

82. x is a real number whose square is nonnegative.

83. x is a real number whose absolute value is less than 3.

84. x is a real number whose absolute value is greater than 2.

In Exercises 85 to 87, determine whether each statement is true or false.

85. $|x|$ is a positive number.

86. $|-y| = y$

87. If $m < 0$, then $|m| = -m$.

88. For any two different real numbers x and y, the smaller of the two numbers is given by

$$\frac{1}{2}(x + y - |x - y|)$$

Verify the above statement for

a. $x = 5$ and $y = 8$

b. $x = -2$ and $y = 7$

c. $x = -4$ and $y = -7$

89. Prove that the expression in Exercise 88 yields the smaller of the numbers x and y. *Hint:* Evaluate the expression for the two cases:

$$x > y \quad \text{and} \quad x < y$$

90. The inequality $|a + b| \leq |a| + |b|$ is called the triangle inequality. For what values of a and b does

$$|a + b| = |a| + |b|?$$

Supplemental Exercises

In Exercises 91 to 94, use inequalities to describe the given statement.

91. The interest I is not greater than $120.

92. The rent R will be at least $650 a month.

93. The property has an area A that is at least 2 acres but less than 3 acres.

94. The distance D is greater than 7 miles, and it is not more than 8 miles.

In Exercises 95 to 102, use absolute value notation to describe the given statement.

95. x is closer to 2 than it is to 6.

96. x is closer to a than it is to b.

97. x is farther from 3 than it is from -7.

98. x is farther from 0 than it is from 5.

99. x is more than 2 units from 4 but less than 7 units from 4.

100. x is more than b units from a but less than c units from a.

101. x is within δ units of a.

102. x is not equal to a, but it is within δ units of a.

103. Prove the product theorem

$$|ab| = |a|\,|b|$$

104. Prove the quotient theorem

$$\left|\frac{a}{b}\right| = \frac{|a|}{|b|}$$

Hint: Make use of the product theorem.

1.3   Integer Exponents

A compact method of writing $5 \cdot 5 \cdot 5 \cdot 5$ is 5^4. The expression 5^4 is written in **exponential notation.** Similarly, we can write

$$\frac{2x}{3} \cdot \frac{2x}{3} \cdot \frac{2x}{3} \quad \text{as} \quad \left(\frac{2x}{3}\right)^3$$

and

$$(x + 2y)(x + 2y)(x + 2y)(x + 2y) \quad \text{as} \quad (x + 2y)^4$$

Exponential notation can be used to express the product of any expression that is used repeatedly as a factor.

Definition of Natural Number Exponents

If b is any real number and n is any natural number, then

$$b^n = \underbrace{b \cdot b \cdot b \cdot \dots \cdot b}_{n \text{ factors of } b}$$

In the expression b^n, b is the **base,** n is the **exponent,** and b^n is the **nth power of** b.

EXAMPLE 1   Evaluate Powers

Evaluate each of the following powers:

a. 3^2 b. $(-5)^4$ c. -5^4 d. $\left(\frac{1}{2}\right)^3$

Solution

a. $3^2 = 3 \cdot 3 = 9$ b. $(-5)^4 = (-5)(-5)(-5)(-5) = 625$

c. $-5^4 = -(5 \cdot 5 \cdot 5 \cdot 5) = -625$ d. $\left(\frac{1}{2}\right)^3 = \frac{1}{2} \cdot \frac{1}{2} \cdot \frac{1}{2} = \frac{1}{8}$

■ *Try Exercise 4, page 21.*

Remark Note the difference between $(-5)^4 = 625$ and $-5^4 = -625$. The parentheses in $(-5)^4$ indicate that the base is -5; however, the expression -5^4 means $-(5^4)$. This time the base is 5.

The exponential key $\boxed{y^x}$ (or on some calculators $\boxed{x^y}$) on a scientific calculator can be used to evaluate a positive number raised to a natural number power. The following examples illustrate the sequence of key strokes for a calculator that uses algebraic logic.

Power	Key Sequence	Calculator Display
5^2	5 $\boxed{y^x}$ 2 $\boxed{=}$	$\boxed{25.}$
2^{12}	2 $\boxed{y^x}$ 12 $\boxed{=}$	$\boxed{4096.}$

$$5^4 = 625$$
$$5^3 = 125$$
$$5^2 = 25$$
$$5^1 = 5$$
$$5^0 = ?$$
$$5^{-1} = ?$$
$$5^{-2} = ?$$

Table 1.1

Consider the sequence shown in Table 1.1. Note that each power divided by the base 5 yields the power in the row below it. For example, $625 \div 5 = 125$ and $125 \div 5 = 25$. To continue this pattern, the powers 5^0, 5^{-1}, and 5^{-2} must be defined as

$$5^0 = 5 \div 5 = 1$$

$$5^{-1} = 1 \div 5 = \frac{1}{5} = \frac{1}{5^1}$$

$$5^{-2} = \frac{1}{5} \div 5 = \frac{1}{5} \cdot \frac{1}{5} = \frac{1}{5^2}$$

These observations suggest the following definitions.

Definition of b^0

For any nonzero real number b, $b^0 = 1$.

Any nonzero real number raised to the zero power equals 1. For example,

$$7^0 = 1 \qquad \left(\frac{1}{2}\right)^0 = 1 \qquad (-3)^0 = 1 \qquad \pi^0 = 1 \qquad (a^2 + 1)^0 = 1$$

Definition of b^{-n}

If $b \neq 0$ and n is any natural number, $b^{-n} = \dfrac{1}{b^n}$ and $\dfrac{1}{b^{-n}} = b^n$.

$$3^{-2} = \frac{1}{3^2} = \frac{1}{9} \qquad \frac{1}{4^{-3}} = 4^3 = 64 \qquad \frac{5^{-2}}{7^{-1}} = \frac{7^1}{5^2} = \frac{7}{25}$$

Restriction Agreement

The expressions 0^0, 0^n where n is a negative integer, and $x/0$ are all undefined expressions. Therefore, all values of variables in this text are restricted to avoid any of these undefined expressions.

For example, in the expression

$$\frac{x^0 y^{-3}}{z-4}$$

it should be assumed that $x \neq 0$, $y \neq 0$, and $z \neq 4$.

We will make considerable use of the following properties.

Properties of Exponents

If m, n, and p are integers and a and b are real numbers, then

Product $b^m \cdot b^n = b^{m+n}$

Quotient $\dfrac{b^m}{b^n} = b^{m-n}$ if $b \neq 0$

Power $(b^m)^n = b^{mn}$ $(a^m b^n)^p = a^{mp} b^{np}$

$\left(\dfrac{a^m}{b^n}\right)^p = \dfrac{a^{mp}}{b^{np}}$ if $b \neq 0$

The properties of exponents can be used to evaluate some exponential expressions.

EXAMPLE 2   Use the Properties of Exponents

Evaluate each of the following:

a. $(5^5 \cdot 5^{-3})^{-2}$ b. $\left(\dfrac{2^2 \cdot 3^{-2}}{2^{-1} \cdot 5^2}\right)^{-1}$

Solution

a. $(5^5 \cdot 5^{-3})^{-2} = (5^{5+(-3)})^{-2} = (5^2)^{-2} = 5^{-4} = \dfrac{1}{5^4} = \dfrac{1}{625}$

b. $\left(\dfrac{2^2 \cdot 3^{-2}}{2^{-1} \cdot 5^2}\right)^{-1} = (2^3 \cdot 3^{-2} \cdot 5^{-2})^{-1} = 2^{-3} \cdot 3^2 \cdot 5^2 = \dfrac{9 \cdot 25}{8} = \dfrac{225}{8}$

■ *Try Exercise 18, page 21.*

To simplify an expression involving exponents, write the expression in a form in which *each base appears at most once* and *no powers of powers or negative exponents appear.*

EXAMPLE 3 **Simplify Exponential Expressions**

Simplify each of the following:

a. $(2x^3y^2)(3xy^5)$ b. $\left(\dfrac{2abc^2}{5a^2b}\right)^3$ c. $(3m^{-2}p^3)^4$ d. $\dfrac{x^ny^{2n}}{x^{n-1}y^n}$

Solution

a. $(2x^3y^2)(3xy^5) = (2 \cdot 3)(x^3 \cdot x)(y^2 \cdot y^5)$ • The commutative and associative properties of multiplication

$\qquad\qquad\qquad\ = 6x^4y^7$

b. $\left(\dfrac{2abc^2}{5a^2b}\right)^3 = \left(\dfrac{2c^2}{5a}\right)^3$ • The quotient property

$\qquad\qquad\ = \dfrac{8c^6}{125a^3}$ • A power property

c. $(3m^{-2}p^3)^4 = 3^4m^{-8}p^{12} = \dfrac{81p^{12}}{m^8}$ • A power property

d. $\dfrac{x^ny^{2n}}{x^{n-1}y^n} = x^{n-(n-1)}y^{2n-n} = xy^n$ • The quotient property

■ *Try Exercise* **30,** *page 21.*

Scientific Notation

The properties of exponents provide a compact method of writing very large or very small numbers and an efficient method of computing with them. A number written in **scientific notation** has the form $a \cdot 10^n$, where n is an integer and $1 \le a < 10$. The following procedure is used to change a number from its decimal form to scientific notation.

For numbers greater than 10, move the decimal point to the position to the right of the first digit. The exponent n will equal the number of places the decimal point has been moved. For numbers less than 1, move the decimal point to the right of the first nonzero digit. The exponent n will be negative, and its absolute value will equal the number of places the decimal point has been moved.

To change a number from scientific notation to its decimal form, we reverse this procedure. That is, if the exponent is positive, move the decimal point to the right the same number of places as the exponent. If the exponent is negative, move the decimal point to the left the same number of places as the absolute value of the exponent.

EXAMPLE 4 **Use Scientific Notation**

Write each decimal in scientific notation. Write each number that appears in scientific notation in its decimal form.

a. 3,770,000,000   b. 0.00000000026   c. 2.51×10^5 d. 3.221×10^{-7}

Solution

a. $3{,}770{,}000{,}000 = 3.77 \times 10^9$ b. $0.00000000026 = 2.6 \times 10^{-10}$

c. $2.51 \times 10^5 = 251{,}000$ d. $3.221 \times 10^{-7} = 0.0000003221$

■ Try Exercise 48, page 21.

Most scientific calculators display very large or very small numbers in scientific notation. The number $450{,}000^2$ is displayed as

$$\boxed{2.025 \quad 11}$$

This means $450{,}000^2 = 2.025 \times 10^{11}$.

Significant Digits

Some numbers are **exact** numbers, and some numbers are **approximate numbers.** Exact numbers are obtained by counting or from a definition. For example, 32 students, 47 cents, and 104 pages are exact numbers.

Many numbers used in scientific work are obtained by measuring; as such, they are approximations. For example, if a room is reported to have a length of 80 feet to the nearest 10 feet, this means that the actual length of the room is at least 75 feet and less than 85 feet. If the room is measured as 82 feet to the nearest foot, then the actual length is at least 81.5 feet and less than 82.5 feet.

A measurement of 80 feet is said to have 1 **significant digit.** A measurement of 82 feet is said to have 2 significant digits. A digit is a significant digit of a number if it meets any of the following conditions:

Exact: Every digit of an exact number is a significant digit.

Approximate: The significant digits of an approximate number are

1. Every nonzero digit
2. The digit 0, provided that
    a. it is between two nonzero digits or
    b. it is to the right of a nonzero digit and the number includes a decimal point

For example, consider the following approximate numbers:

57,000 has 2 significant digits because all nonzero digits are significant.

57,080 has 4 significant digits because zeros between nonzero digits are significant.

57,100.0 has 6 significant digits. The number includes a decimal point, so all zeros to the right of a nonzero digit are also significant.

0.00230 has 3 significant digits. The last zero is significant because it is to the right of a nonzero digit and the number includes a decimal point.

When a measurement is given as 700 centimeters, confusion can arise. Has this measurement been made to the nearest 1 centimeter, the nearest 10 centimeters, or the nearest 100 centimeters? To avoid confusion, write the number using scientific notation as shown on page 21.

Number	Number of Significant Digits	Measured to the Nearest
7.000×10^2	4	one-tenth
7.00×10^2	3	one
7.0×10^2	2	ten
7×10^2	1	hundred

EXERCISE SET 1.3

In Exercises 1 to 26, evaluate each expression.

1. $(-4)^3$

2. $(-2)^3$

3. -4^3

4. -2^4

5. 7^0

6. -7^0

7. $(-1)^{18}$

8. $(-1)^{19}$

9. $3^2 \cdot 3^3$

10. $2^3 \cdot 2^4$

11. $\dfrac{3^{-1}}{3^2}$

12. $\dfrac{5^6}{5^4}$

13. $2^7 \cdot 2^{-3} \cdot 2$

14. $3 \cdot 3^{-12} \cdot 3^8$

15. $\dfrac{4^{-8}}{4^{-11}}$

16. $\dfrac{5^{-1}}{5^2}$

17. $\left(\dfrac{5^{-3} \cdot 7}{3^{-2}}\right)^{-1}$

18. $\left(\dfrac{4 \cdot 5^{-1}}{2^{-3}}\right)^{-2}$

19. $\left(\dfrac{4}{9}\right)^{-2}$

20. $\left(\dfrac{4}{6}\right)^{-3}$

21. $\left(\dfrac{3^{-2}}{2^{-1} \cdot 5}\right)^2$

22. $\left(\dfrac{2^{-2} \cdot 3^2}{5}\right)^2$

23. $\dfrac{(2 \cdot 5)^2}{(2^{-1} \cdot 5)^3}$

24. $\dfrac{(2^2 \cdot 3^{-1})^3}{(3 \cdot 5)}$

25. $\left(\dfrac{2^5 \cdot 3^{-5}}{2^{-3} \cdot 5^4}\right)^0$

26. $\left(\dfrac{-3^6 \cdot 2^{-4}}{-4^{-5}}\right)^0$

In Exercises 27 to 46, simplify each exponential expression.

27. $(2x^2y^3)(3x^5y)$

28. $(3ab^4)(3ab^3c^2)$

29. $\left(\dfrac{2ab^2c^3}{5ab^2}\right)^3$

30. $\left(\dfrac{3pq^2}{-2pq^3r^2}\right)^4$

31. $\dfrac{(3xy^{-3})^2}{(2xy)^{-2}}$

32. $\dfrac{(5ab^{-2})^2}{(-3a^2b)^3}$

33. $(2x^{-3}y^0)(3^{-1}xy)^2$

34. $(-3abc^2)^2(2ab^{-1})^3$

35. $\left(\dfrac{3x}{y}\right)^{-1}$

36. $\left(\dfrac{2x}{5y}\right)^{-2}$

37. $(x^2y^{-3})^{-2}$

38. $(x^3y^{-2})^{-3}$

39. $a^{-1} + b^{-2}$

40. $a^{-1} - b^{-1}$

41. $\dfrac{4a^2(bc)^{-1}}{(-2)^2a^3b^{-2}c}$

42. $\dfrac{6abc^{-2}}{(-3)^{-1}a^{-1}bc^{-3}}$

43. $(2ab^{-3})^2(-2a^{-1}b^3)^2$

44. $(3x^{-1}y)^{-1}(3xy)$

45. $\left[\left(\dfrac{b^{-3}}{a^2}\right)^2\left(\dfrac{a^{-2}}{ab}\right)^{-1}\right]^0$

46. $\left(\dfrac{x^{-2} + x^3}{x^{-1}}\right)^0$

In Exercises 47 to 58, write each number in scientific notation.

47. 73.4

48. 25,600

49. 1,900,000

50. 21,000,000

51. 163,000,000,000

52. 521

53. 0.000032

54. 0.00000714

55. 0.007

56. 0.00095

57. 0.0000000821

58. 0.00000000072

In Exercises 59 to 70, change each number from its scientific notation to its decimal form.

59. 6.5×10^3

60. 4.2×10^4

61. 7.31×10^{-5}

62. 6.85×10^{-9}

63. 8.0×10^{10}

64. 9.008×10^{12}

65. 2.17×10^{-4}

66. 4.007×10^{-3}

67. 1.0×10^{11}

68. 1.0×10^{-5}

69. 3.75×10^0

70. 8.81×10^0

In Exercises 71 to 82, find the number of significant digits in each of the numbers.

71. 14,300

72. 17,010,000

73. 20,050.0

74. 40,900.00

75. 0.03

76. 0.0070

77. 0.00501

78. 0.000008

79. 8.3×10^3

80. 5.31×10^5

81. 1.882×10^{-7}

82. 4×10^{-5}

In Exercises 83 and 84, use a calculator to evaluate the exponential expressions.

83. 1.08^{10}

84. 1.12^8

85. The number of colors that a computer with a color monitor can display can be found using the number of **bits** the color graphics interface can display (a bit is a *binary digit*). For example, a computer with a 4-bit color interface can display $2^4 = 16$ colors. Determine how many colors each of the following can display. A computer with

  a. an 8-bit color interface

  b. a 16-bit color interface

86. It has been estimated that the human eye can detect 36,000 different colors. How many of these colors would go undetected by a human using a computer with a 24-bit interface (to the nearest 1000)? *Hint:* See Exercise 85.

87. Pluto is 5.91×10^{12} meters from the sun. The speed of light is 3.00×10^8 meters per second. Find the time it takes light from the sun to reach Pluto.

88. The earth's mean distance from the sun is 9.3×10^7 miles. This distance is called the astronomical unit (AU). Jupiter is 5.2 AU from the sun. Find the distance in miles from the sun to Jupiter.

89. A principal P invested at a yearly interest rate r compounded n times per year yields an amount A given by the formula

$$A = P\left(1 + \frac{r}{n}\right)^n$$

Find the amount after one year if \$4500 is deposited in an account with a yearly interest rate of 8 percent, compounded monthly.

90. You plan to save 1¢ the first day of a month, 2¢ the second day, and 4¢ the third day and to continue this pattern of saving twice what you saved on the previous day for every day in a month that has 30 days.   **a.** How much money will you need to save on the 30th day?   **b.** How much money will you have after 30 days? (*Hint:* Note that after 2 days you will have saved $2^2 - 1 = 3$¢. After 3 days you have $2^3 - 1 = 7$¢.)

Supplemental Exercises

In Exercises 91 to 96, write each expression as an equivalent expression in which the variables x and y occur only once. All the exponents are integers.

91. $\dfrac{x^n y^{n+2}}{x^{n-3} y}$

92. $\dfrac{x^{3n} y^{2n} y^n}{x^{-n+1} y^{-4n}}$

93. $\left(\dfrac{x^n y}{x^{1-n} y^{-1}}\right)^2$

94. $\left(\dfrac{x^n y^{2n}}{y^{3-n}}\right)^{-2}$

95. $\left(\dfrac{x^{3n} y^{2n}}{x^{-2n} y^{3n+1}}\right)^{-1}$

96. $\left(\dfrac{x^{4-n} y^{n+4}}{xy^{n-4}}\right)^2$

97. Which is larger, $3^{(3^3)}$ or $(3^3)^3$? (*Hint:* The parentheses are used to indicate which computation is to be performed first.)

98. If $3^x = y$, find 3^{x+2} in terms of y.

Archimedes (287–212 B.C.) is considered the greatest mathematician of antiquity. Archimedes discovered two remarkable relationships that exist between a cylinder and an inscribed sphere. Archimedes was so proud of his discoveries that he requested that the following figure of a sphere of radius r inscribed in a cylinder of radius r and height $2r$, be engraved on his tombstone.

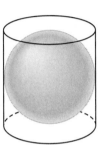

Solve the following to determine what Archimedes discovered.

99. Find the ratio of the volume of the cylinder to the volume of the sphere.

100. Find the ratio of the total surface area of the cylinder to the surface area of the sphere.

1.4   Polynomials

A **monomial** is a constant, or a variable, or a product of a constant and one or more variables, with the variables having only nonnegative integer exponents. The constant is called the **numerical coefficient** or simply the **coefficient** of the monomial. The **degree** of the monomial is the sum of the exponents of the variables. For example, $-5xy^2$ is a monomial with coefficient -5 and degree 3.

The algebraic expression $3x^{-2}$ is not a monomial because it cannot be written as a product of a constant and a variable with a *nonnegative* integer exponent.

A sum of a finite number of monomials is called a **polynomial**. Each monomial is called a **term of the polynomial**. The **degree of a polynomial** is the largest degree of the terms in the polynomial.

Terms that have exactly the same variables raised to the same powers are called **like terms.** For example, $14x^2$ and $-31x^2$ are like terms; however, $2x^3y$ and $7xy$ are not like terms because x^3y and xy are not identical.

A polynomial is said to be simplified if all its like terms have been combined. For example, the simplified form of $4x^2 + 3x + 5x$ is $4x^2 + 8x$. A simplified polynomial that has two terms is a **binomial**, and a simplified polynomial that has three terms is a **trinomial**. For example, $4x + 7$ is a binomial, and $2x^3 - 7x^2 + 11$ is a trinomial.

A nonzero constant, such as 5, is called a **constant polynomial**. It has degree zero since $5 = 5x^0$. The number 0 is defined to be a polynomial with no degree.

General Form of a Polynomial

The **general form of a polynomial** of degree n in the variable x is

$$a_nx^n + a_{n-1}x^{n-1} + \cdots + a_2x^2 + a_1x + a_0$$

where $a_n \neq 0$ and n is a nonnegative integer. The coefficient a_n is the **leading coefficient,** and a_0 is the **constant term.**

If a polynomial in the variable x is written with decreasing powers of x, then it is in **standard form.** For example, the polynomial

$$3x^2 - 4x^3 + 7x^4 - 1$$

is written in standard form as

$$7x^4 - 4x^3 + 3x^2 - 1$$

The following table shows the leading coefficient, degree, terms, and coefficients of the given polynomials.

Polynomial	Leading Coefficient	Degree	Terms	Coefficients
$9x^2 - x + 5$	9	2	$9x^2, -x, 5$	9, -1, 5
$11 - 2x$	-2	1	$-2x, 11$	-2, 11
$x^3 + 5x - 3$	1	3	$x^3, 5x, -3$	1, 5, -3

To add polynomials, we combine like terms.

EXAMPLE 1   **Add Polynomials**

Simplify: $(3x^2 + 7x - 5) + (4x^2 - 2x + 1)$.

Solution

$(3x^2 + 7x - 5) + (4x^2 - 2x + 1) = (3x^2 + 4x^2) + (7x - 2x) + [(-5) + 1]$
$$= 7x^2 + 5x - 4$$

■ *Try Exercise **24**, page 27.*

The **additive inverse of the polynomial** $3x - 7$ is

$$-(3x - 7) = -3x + 7$$

To subtract a polynomial, we add its additive inverse. For example,

$$
\begin{aligned}
(2x - 5) - (3x - 7) &= (2x - 5) + (-3x + 7) \\
&= [2x + (-3x)] + [(-5) + 7] \\
&= -x + 2
\end{aligned}
$$

The distributive property is used to find the product of polynomials. For instance, to find the product of $(3x - 4)$ and $(2x^2 + 5x + 1)$, we treat $3x - 4$ as a *single* quantity and *distribute it* over the trinomial $2x^2 + 5x + 1$, as shown in Example 2.

EXAMPLE 2   **Multiply Polynomials**

Simplify: $(3x - 4)(2x^2 + 5x + 1)$.

Solution

$$
\begin{aligned}
(3x &- 4)(2x^2 + 5x + 1) \\
&= (3x - 4)(2x^2) + (3x - 4)(5x) + (3x - 4)(1) \\
&= (3x)(2x^2) - 4(2x^2) + (3x)(5x) - 4(5x) + (3x)(1) - 4(1) \\
&= 6x^3 - 8x^2 + 15x^2 - 20x + 3x - 4 \\
&= 6x^3 + 7x^2 - 17x - 4
\end{aligned}
$$

■ *Try Exercise* **32,** *page 27.*

In the following, a vertical format has been used to find the product of $(x^2 + 6x - 7)$ and $(5x - 2)$. Note that like terms are arranged in the same vertical column.

$$
\begin{array}{r}
x^2 + 6x - 7 \\
5x - 2 \\
\hline
-2x^2 - 12x + 14 \\
5x^3 + 30x^2 - 35x \phantom{{}+ 14} \\
\hline
5x^3 + 28x^2 - 47x + 14
\end{array}
$$

If the terms of the binomials $(a + b)$ and $(c + d)$ are labeled as in Figure 1.12, then the product of the two binomials can be computed mentally by the **FOIL method.**

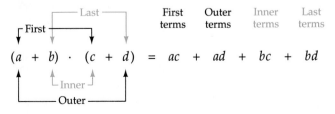

Figure 1.12

In the following illustration, we find the product of $(7x - 2)$ and $(5x + 4)$ by the FOIL method.

$$
\begin{array}{ccccccc}
& \text{First} & \text{Outer} & \text{Inner} & \text{Last} \\
(7x - 2)(5x + 4) = (7x)(5x) & + & (7x)(4) & + & (-2)(5x) & + & (-2)(4) \\
= 35x^2 & + & 28x & - & 10x & - & 8 \\
= 35x^2 + 18x - 8
\end{array}
$$

Certain products occur so frequently in algebra that they deserve special attention.

Special Product Formulas

Special Forms	**Formulas**
(Sum)(Difference)	$(x + y)(x - y) = x^2 - y^2$
(Binomial)^2	$(x + y)^2 = x^2 + 2xy + y^2$
	$(x - y)^2 = x^2 - 2xy + y^2$

The variables x and y in these special product formulas can be replaced by other algebraic expressions, as shown in Example 3.

EXAMPLE 3   **Use the Special Product Formulas**

Find each of the following special products.

a. $(7x + 10)(7x - 10)$ b. $(2y^2 + 11z)^2$

Solution

a. $(7x + 10)(7x - 10) = (7x)^2 - (10)^2 = 49x^2 - 100$

b. $(2y^2 + 11z)^2 = (2y^2)^2 + 2[(2y^2)(11z)] + (11z)^2$

$$= 4y^4 + 44y^2z + 121z^2$$

■ *Try Exercise* **60,** *page 28.*

Many applications problems require you to **evaluate polynomials.** To evaluate a polynomial, substitute the given value(s) for the variable(s) and then perform the indicated operations using the **Order of Operations Agreement.**

The Order of Operations Agreement

If grouping symbols are present, evaluate by performing the operations within the grouping symbols, innermost grouping symbol first, while observing the order given in steps 1 to 3.

1. First, evaluate each power.
2. Next, do all multiplications and divisions, working from left to right.
3. Last, do all additions and subtractions, working from left to right.

EXAMPLE 4   Evaluate a Polynomial

Evaluate the polynomial $2x^3 - 6x^2 + 7$ for $x = -4$.

Solution

$$2x^3 - 6x^2 + 7 = 2(-4)^3 - 6(-4)^2 + 7 \qquad \bullet \text{ Substitute } -4 \text{ for } x.$$

$$= 2(-64) - 6(16) + 7 \qquad \bullet \text{ Evaluate the powers.}$$

$$= -128 - 96 + 7 \qquad \bullet \text{ Perform the multiplications.}$$

$$= -217 \qquad \bullet \text{ Perform the additions and subtractions.}$$

■ *Try Exercise* **72**, *page 28.*

EXAMPLE 5   Solve an Application

The number of singles tennis matches that can be played between n tennis players is given by the polynomial $(1/2)n^2 - (1/2)n$. Find the number of singles tennis matches that can be played between 4 tennis players.

Solution

$$\frac{1}{2}n^2 - \frac{1}{2}n = \frac{1}{2}(4)^2 - \frac{1}{2}(4) \qquad \bullet \text{ Substitute 4 for } n.$$

$$= \frac{1}{2}(16) - \frac{1}{2}(4) = 8 - 2 = 6$$

Therefore 4 tennis players can play a total of 6 singles matches. (See Figure 1.13.)

■ *Try Exercise* **82**, *page 28.*

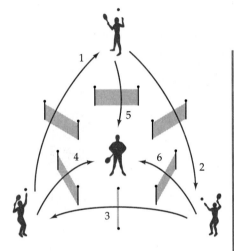

Figure 1.13
4 tennis players can play a total of 6 singles matches.

EXAMPLE 6   Solve an Application

A scientist determines that the average time in seconds that it takes a particular computer to determine whether an n-digit natural number is prime or composite is given by

$$0.002n^2 + 0.002n + 0.009, \qquad 20 \le n \le 40$$

The average time in seconds that it takes the computer to factor an n-digit number is given by

$$0.00032(1.7)^n, \qquad 20 \le n \le 40$$

Estimate the average time it takes the computer to:

a. determine whether a 30-digit number is a prime or a composite.

b. factor a 30-digit number.

Solution

a. $0.002n^2 + 0.002n + 0.009 = 0.002(30)^2 + 0.002(30) + 0.009$

$\approx 1.8 + 0.06 + 0.009 = 1.869 \approx 2$ seconds

b. $0.00032(1.7)^n = 0.00032(1.7)^{30}$

$\approx 0.00032(8{,}193{,}465.726)$

$\approx 2{,}600$ seconds

■ *Try Exercise 84, page 28.*

Remark The procedure used by the computer to determine whether a number is a prime or a composite is a **polynomial time algorithm,** because the time required can be estimated via a polynomial. The procedure used to factor a number is an **exponential time algorithm.** In the field of **computational complexity,** it is important to distinguish between polynomial time algorithms and exponential time algorithms. This example illustrates that the polynomial time algorithm can be run in about 2 seconds, whereas the exponential time algorithm requires about 44 minutes!

EXERCISE SET 1.4

In Exercises 1 to 10, match the descriptions, labeled A, B, C, ... , J, with the appropriate examples.

A. $x^3y + xy$ B. $7x^2 + 5x - 11$
C. $\frac{1}{2}x^2 + xy + y^2$ D. $4xy$
E. $8x^3 - 1$ F. $3 - 4x^2$
G. 8 H. $3x^5 - 4x^2 + 7x - 11$
I. $8x^4 - \sqrt{5}x^3 + 7$ J. 0

1. A monomial of degree 2.

2. A binomial of degree 3.

3. A polynomial of degree 5.

4. A binomial with leading coefficient of -4.

5. A zero-degree polynomial.

6. A fourth-degree polynomial that has a third-degree term.

7. A trinomial with integer coefficients.

8. A trinomial in x and y.

9. A polynomial with no degree.

10. A fourth-degree binomial.

In Exercises 11 to 16, for each polynomial determine its **a.** standard form, **b.** degree, **c.** coefficients, **d.** leading coefficient, **e.** terms.

11. $2x + x^2 - 7$ **12.** $-3x^2 - 11 - 12x^4$

13. $x^3 - 1$ **14.** $4x^2 - 2x + 7$

15. $2x^4 + 3x^3 + 5 + 4x^2$ **16.** $3x^2 - 5x^3 + 7x - 1$

In Exercises 17 to 22, determine the degree of the given polynomial.

17. $3xy^2 - 2xy + 7x$ **18.** $x^3 + 3x^2y + 3xy^2 + y^3$

19. $4x^2y^2 - 5x^3y^2 + 17xy^3$ **20.** $-9x^5y + 10xy^4 - 11x^2y^2$

21. xy **22.** $5x^2y - y^4 + 6xy$

In Exercises 23 to 34, perform the indicated operations and simplify if possible by combining like terms. Write the result in standard form.

23. $(3x^2 + 4x + 5) + (2x^2 + 7x - 2)$

24. $(5y^2 - 7y + 3) + (2y^2 + 8y + 1)$

25. $(4w^3 - 2w + 7) + (5w^3 + 8w^2 - 1)$

26. $(5x^4 - 3x^2 + 9) + (3x^3 - 2x^2 - 7x + 3)$

27. $(r^2 - 2r - 5) - (3r^2 - 5r + 7)$

28. $(7s^2 - 4s + 11) - (-2s^2 + 11s - 9)$

29. $(u^3 - 3u^2 - 4u + 8) - (u^3 - 2u + 4)$

30. $(5v^4 - 3v^2 + 9) - (6v^4 + 11v^2 - 10)$

31. $(4x - 5)(2x^2 + 7x - 8)$

32. $(5x - 7)(3x^2 - 8x - 5)$

33. $(3x^2 - 2x + 5)(2x^2 - 5x + 2)$

34. $(2y^3 - 3y + 4)(2y^2 - 5y + 7)$

In Exercises 35 to 52, use the FOIL method to find the indicated product.

35. $(2x + 4)(5x + 1)$ **36.** $(5x - 3)(2x + 7)$

37. $(y + 2)(y + 1)$ **38.** $(y + 5)(y + 3)$

39. $(4z - 3)(z - 4)$ **40.** $(5z - 6)(z - 1)$

41. $(a + 6)(a - 3)$ **42.** $(a - 10)(a + 4)$

43. $(b - 4)(b + 6)$ **44.** $(b + 5)(b - 2)$

45. $(5x - 11y)(2x - 7y)$ **46.** $(3a - 5b)(4a - 7b)$

47. $(9x + 5y)(2x + 5y)$ **48.** $(3x - 7z)(5x - 7z)$

49. $(6w - 11x)(2w - 3x)$ **50.** $(4m + 5n)(2m - 5n)$

51. $(3p + 5q)(2p - 7q)$ **52.** $(2r - 11s)(5r + 8s)$

In Exercises 53 to 58, perform the indicated operations and simplify.

53. $(4d - 1)^2 - (2d - 3)^2$ **54.** $(5c - 8)^2 - (2c - 5)^2$

55. $(r + s)(r^2 - rs + s^2)$ **56.** $(r - s)(r^2 + rs + s^2)$

57. $(3c - 2)(4c + 1)(5c - 2)$ **58.** $(4d - 5)(2d - 1)(3d - 4)$

In Exercises 59 to 68, use the special product formulas to perform the indicated operation.

59. $(3x + 5)(3x - 5)$ **60.** $(4x^2 - 3y)(4x^2 + 3y)$

61. $(3x^2 - y)^2$ **62.** $(6x + 7y)^2$

63. $(4w + z)^2$ **64.** $(3x - 5y^2)^2$

65. $[(x - 2) + y]^2$ **66.** $[(x + 3) - y]^2$

67. $[(x + 5) + y][(x + 5) - y]$

68. $[(x - 2y) + 7][(x - 2y) - 7]$

In Exercises 69 to 76, evaluate the given polynomial for the indicated value of the variable.

69. $x^2 + 7x - 1$, for $x = 3$

70. $x^2 - 8x + 2$, for $x = 4$

71. $-x^2 + 5x - 3$, for $x = -2$

72. $-x^2 - 5x + 4$, for $x = -5$

73. $3x^3 - 2x^2 - x + 3$, for $x = -1$

74. $5x^3 - x^2 + 5x - 3$, for $x = -1$

75. $1 - x^5$, for $x = -2$

76. $1 - x^3 - x^5$, for $x = 2$

In Exercises 77 to 79, evaluate the given polynomial for the indicated values of the variable.

77. $4x^2 - 5x - 4$ for   **a.** $x = 4.3$ **b.** $x = 4.4$

78. $8x^3 - 2x^2 + 6.4x - 7.1$ for   **a.** $x = 1.2$ **b.** $x = 1.3$

79. $x^3 - 2x^2 - 5x + 11$ for   **a.** $x = 0.001$ **b.** $x = 0.0001$

80. On an expressway, the recommended *safe distance* between cars in feet is given by $0.015v^2 + v + 10$ where v is the speed of the car in miles per hour. Find the safe distance when   **a.** $v = 30$ mph,   **b.** $v = 55$ mph.

81. Find the number of chess matches that can be played between a group of 150 people. Use the formula from Example 5.

82. The number of committees consisting of exactly 3 people that can be formed from a group of n people is given by the polynomial

$$\frac{1}{6}n^3 - \frac{1}{2}n^2 + \frac{1}{3}n$$

Find the number of committees consisting of exactly 3 people that can be formed from a group of 21 people.

83. If n is a positive integer, then $n!$, which is read "n factorial," is given by

$$n(n - 1)(n - 2) \cdots 2 \cdot 1$$

For example, $4! = 4 \cdot 3 \cdot 2 \cdot 1 = 24$. A statistician determines that each time a statistical package is booted up on a particular computer, the time in seconds required to compute $n!$ is given by the polynomial

$$1.9 \times 10^{-6}n^2 - 3.9 \times 10^{-3}n$$

where $1000 \le n \le 10{,}000$. Using this polynomial, estimate the time it takes the computer to calculate 4000! and 8000!. Assume the statistical package is booted up before each calculation.

84. A computer scientist determines that the time in seconds it takes a particular computer to calculate n digits of π is given by the polynomial

$$4.3 \times 10^{-6}n^2 - 2.1 \times 10^{-4}n$$

where $1000 \le n \le 10{,}000$. Estimate the time it takes the computer to calculate π to   **a.** 1000 digits   **b.** 5000 digits   **c.** 10,000 digits

Supplemental Exercises

The following special product formulas can be used to find the cube of a binomial.

$$(x + y)^3 = x^3 + 3x^2y + 3xy^2 + y^3$$
$$(x - y)^3 = x^3 - 3x^2y + 3xy^2 - y^3$$

In Exercises 85 to 90, make use of the above special product formulas to find the indicated products.

85. $(a + b)^3$ **86.** $(a - b)^3$ **87.** $(x - 1)^3$

88. $(y + 2)^3$ **89.** $(2x - 3y)^3$ **90.** $(3x + 5y)^3$

91. If P is a polynomial in x of degree n and Q is a polynomial in x of degree $n - 1$, what is the degree of
a. $P + Q$ **b.** $P - Q$ **c.** $P + P$ **d.** $P - P$

92. If R and S are each polynomials in x of degree n, what can be said about the degree of
a. $R + S$ **b.** RS

93. Many people have tried to find a method of generating prime numbers. One such attempt evaluates the polynomial $n^2 - n + 41$ for $n = 1, 2, 3, \dots$.

a. Show that this method does generate a prime number for $n = 1, 2, 3,$ and 4.

b. Find a natural number n for which the polynomial $n^2 - n + 41$ does not generate a prime number.

94. How many prime numbers are generated by the polynomial $2n$, with $n = 1, 2, 3, 4, \dots$?

95. Find each of the following products.
$(x - 1)(x + 1)$
$(x - 1)(x^2 + x + 1)$
$(x - 1)(x^3 + x^2 + x + 1)$
Then use the results to determine the following products without multiplying.

a. $(x - 1)(x^4 + x^3 + x^2 + x + 1)$
b. $(x - 1)(x^5 + x^4 + x^3 + x^2 + x + 1)$

1.5 Factoring

Writing a polynomial as a product of polynomials of lower degree is called **factoring**. Factoring is an important procedure that is often used to simplify fractional expressions and to solve equations.

In this section we consider only the factorization of polynomials that have integer coefficients. Also, we are concerned only with **factoring over the integers.** That is, we search only for polynomial factors that have integer coefficients.

The first step in any factorization of a polynomial is to use the distributive property to factor out the **greatest common factor** (GCF) of the terms of the polynomial. Given two or more exponential expressions with the same prime number base or the same variable base, the GCF is the exponential expression with the smallest exponent. For example,

$$2^3 \text{ is the GCF of } 2^3, 2^5, \text{ and } 2^8, \quad \text{and} \quad a \text{ is the GCF of } a^4 \text{ and } a.$$

The GCF of two or more monomials is the product of the GCF of each *common* base. For example, to find the GCF of $27a^3b^4$ and $18b^3c$, factor the coefficients into prime factors and then write each common base with its smallest exponent.

$$27a^3b^4 = 3^3 \cdot a^3 \cdot b^4$$

$$18b^3c = 2 \cdot 3^2 \cdot b^3 \cdot c$$

The only common bases are 3 and b. The product of these common bases with their smallest exponents is 3^2b^3. The GCF of $27a^3b^4$ and $18b^3c$ is $9b^3$.

EXAMPLE 1   **Factor Out the Greatest Common Factor**

Factor out the GCF in each of the following:

a. $10x^3 + 6x$ b. $15x^{2n} + 9x^{n+1} - 3x^n$ (where n is a positive integer)

c. $(m + 5)(x + 3) + (m + 5)(x - 10)$

Solution

a. $10x^3 + 6x = (2x)(5x^2) + (2x)(3)$ • The GCF is $2x$.

$= (2x)(5x^2 + 3)$ • Factor out the GCF.

b. $15x^{2n} + 9x^{n+1} - 3x^n = (3x^n)(5x^n) + (3x^n)(3x) - (3x^n)(1)$ • The GCF is $3x^n$.

$= 3x^n(5x^n + 3x - 1)$ • Factor out the GCF.

c. Use the distributive property to factor out $(m + 5)$.

$(m + 5)(x + 3) + (m + 5)(x - 10) = (m + 5)[(x + 3) + (x - 10)]$

$= (m + 5)(2x - 7)$ • Simplify.

■ *Try Exercise 6, page 37.*

Some polynomials can be **factored by grouping.** Pairs of terms that have a common factor are first grouped together. The process makes repeated use of the distributive property, as shown in the following factorization of $6y^3 - 21y^2 - 4y + 14$.

$6y^3 - 21y^2 - 4y + 14 = (6y^3 - 21y^2) - (4y - 14)$ • Group the first two terms and the last two terms.

$= 3y^2(2y - 7) - 2(2y - 7)$ • Factor out the GCF from each of the groups.

$= (2y - 7)(3y^2 - 2)$ • Factor out the common binomial factor.

Some trinomials of the form $x^2 + bx + c$ can be factored by a trial procedure. This method makes use of the FOIL method in reverse. For example, consider the following products:

$(x + 3)(x + 5) = x^2 + 5x + 3x + (3)(5)$ $= x^2 + 8x + 15$

$(x - 2)(x - 7) = x^2 - 7x - 2x + (-2)(-7) = x^2 - 9x + 14$

$(x + 4)(x - 9) = x^2 - 9x + 4x + (4)(-9)$ $= x^2 - 5x - 36$

The coefficient of x is the sum of the constant terms of the binomials.

The constant term of the trinomial is the product of the constant terms of the binomials.

Points to Remember to Factor $x^2 + bx + c$

1. The constant term c of the trinomial is the product of the constant terms of the binomials.
2. The coefficient b in the trinomial is the sum of the constant terms of the binomials.
3. If the constant term c of the trinomial is positive, the constant terms of the binomials have the same sign as the coefficient b of the trinomial.
4. If the constant term c of the trinomial is negative, the constant terms of the binomials have opposite signs.

EXAMPLE 2   **Factor a Trinomial of the Form $x^2 + bx + c$**

Factor $x^2 + 7x - 18$.

Solution   We must find two binomials whose first terms have a product of x^2 and whose last terms have a product of -18; also, the sum of the product of the outer terms and the product of the inner terms must be $7x$. Begin by listing the possible integer factorizations of -18.

Factors of -18	Sum of the Factors	
$1 \cdot (-18)$	$1 + (-18) = -17$	
$(-1) \cdot 18$	$(-1) + 18 = 17$	
$2 \cdot (-9)$	$2 + (-9) = -7$	
$(-2) \cdot 9$	$(-2) + 9 = 7$	• Stop. This is the desired sum.

Thus -2 and 9 are the numbers whose sum is 7 and whose product is -18. Therefore,

$$x^2 + 7x - 18 = (x - 2)(x + 9)$$

The FOIL method can be used to verify that the factorization is correct.

── ■ *Try Exercise **18**, page 37.*

The trial method can sometimes be used to factor trinomials of the form $ax^2 + bx + c$, which do not have a leading coefficient of 1. We use the factors of a and c to form trial binomial factors. Factoring trinomials of this type may require testing many factors. To reduce the number of trial factors, make use of the following points.

Points to Remember to Factor $ax^2 + bx + c$, $a > 0$

1. If the constant term of the trinomial is positive, the constant terms of the binomials have the same sign as the coefficient b in the trinomial.
2. If the constant term of the trinomial is negative, the constant terms of the binomials have opposite signs.
3. If the terms of the trinomial do not have a common factor, then neither binomial will have a common factor.

EXAMPLE 3   **Factor a Trinomial of the Form $ax^2 + bx + c$**

Factor $6x^2 - 11x + 4$.

Solution   Because the constant term of the trinomial is positive and the coefficient of the x term is negative, the constant terms of the binomials will both be negative. This time we find factors of the first term as well as factors of the constant term.

Factors of $6x^2$	Factors of 4 (both negative)
$x, 6x$	$-1, -4$
$2x, 3x$	$-2, -2$

Use these factors to write trial factors. Use the FOIL method to see whether any of the trial factors produce the correct middle term. If the terms of a trinomial do not have a common factor, then a binomial factor cannot have a common factor (point 3). Such trial factors need not be checked.

Trial Factors	Middle Term	
$(x - 1)(6x - 4)$	Common factor	• $6x$ and 4 have a common factor.
$(x - 4)(6x - 1)$	$-1x - 24x = -25x$	
$(x - 2)(6x - 2)$	Common factor	• $6x$ and 2 have a common factor.
$(2x - 1)(3x - 4)$	$-8x - 3x = -11x$	• This is the correct middle term.

Thus, $6x^2 - 11x + 4 = (2x - 1)(3x - 4)$.

_____ ■ *Try Exercise* **22,** *page 37.*

Remark   Sometimes it is impossible to factor a polynomial into the product of two polynomials having integer coefficients. Such polynomials are said to be **nonfactorable** over the integers. For example, $x^2 + 3x + 7$ is nonfactorable over the integers because there are no integers whose product is 7 and whose sum or difference is 3.

If you have difficulty factoring a trinomial, you may wish to use the following theorem. It will indicate whether the trinomial is factorable over the integers.

Factorization Theorem

> The trinomial $ax^2 + bx + c$, with integer coefficients a, b, and c, can be factored as the product of two binomials with integer coefficients if and only if $b^2 - 4ac$ is a perfect square.

EXAMPLE 4   **Apply the Factorization Theorem**

Determine whether each of the following trinomials is factorable over the integers.

a. $4x^2 + 8x - 7$ b. $6x^2 - 5x - 4$

Solution

a. The coefficients of $4x^2 + 8x - 7$ are $a = 4$, $b = 8$, and $c = -7$. Applying the factorization theorem yields

$$b^2 - 4ac = 8^2 - 4(4)(-7) = 176$$

Since 176 is not a perfect square, the trinomial is nonfactorable over the integers.

b. The coefficients of $6x^2 - 5x - 4$ are $a = 6$, $b = -5$, and $c = -4$. Thus

$$b^2 - 4ac = (-5)^2 - 4(6)(-4) = 121$$

Since 121 is a perfect square, the trinomial is factorable over the integers. Using the methods we have developed, we find

$$6x^2 - 5x - 4 = (3x - 4)(2x + 1)$$

■ *Try Exercise* **30**, *page 37.*

Some polynomials of degree greater than 2 can be factored by the trial procedure. Consider $2x^6 + 9x^3 + 9$. Because all the signs of the trinomial are positive, the coefficients of all the terms in the binomial factors must be positive.

Factors of $2x^6$	Factors of 9 (both positive)
x^3, $2x^3$	1, 9
	3, 3

The factors $(x^3 + 3)$ and $(2x^3 + 3)$ are the only trial factors whose product has the correct middle term $9x^3$. Thus $2x^6 + 9x^3 + 9 = (x^3 + 3)(2x^3 + 3)$.

Some polynomials can be factored by making use of the following factoring formulas.

Factoring Formulas

Difference of two squares	$x^2 - y^2 = (x + y)(x - y)$
Perfect square trinomials	$x^2 + 2xy + y^2 = (x + y)^2$ $x^2 - 2xy + y^2 = (x - y)^2$
Sum of cubes	$x^3 + y^3 = (x + y)(x^2 - xy + y^2)$
Difference of cubes	$x^3 - y^3 = (x - y)(x^2 + xy + y^2)$

The monomial a^2 is a square of a, and a is called a **square root** of a^2. The factoring formula

$$x^2 - y^2 = (x + y)(x - y)$$

indicates that the **difference of two squares** can be written as the product of the sum and the difference of the square roots of the squares.

EXAMPLE 5   **Factor the Difference of Squares**

Factor $49x^2 - 144$.

Solution

$49x^2 - 144 = (7x)^2 - (12)^2$ • Recognize the difference-of-squares form.

$\qquad\qquad = (7x + 12)(7x - 12)$ • The binomial factors are the sum and the difference of the square roots of the squares.

——————————————————————————— ■ *Try Exercise 38, page 37.*

Caution   The polynomial $x^2 + y^2$ is the *sum* of two squares. You may be tempted to factor it in a manner similar to the method used on the *difference* of two squares; however, $x^2 + y^2$ is nonfactorable over the integers. Also note that $x^2 + y^2 \neq (x + y)^2$.

A **perfect square trinomial** is a trinomial that is the square of a binomial. For example, $x^2 + 6x + 9$ is a perfect square trinomial because

$$(x + 3)^2 = x^2 + 6x + 9$$

Every perfect square trinomial can be factored by the trial method, but it generally is faster to factor perfect square trinomials by using the factoring formulas.

EXAMPLE 6   **Factor a Perfect Square Trinomial**

Factor $16m^2 - 40mn + 25n^2$.

Solution

$16m^2 - 40mn + 25n^2 = (4m)^2 - 2(4m)(5n) + (5n)^2$ • Recognize the perfect square trinomial form.

$= (4m - 5n)^2$

■ *Try Exercise* **48,** *page 37.*

The product of the same three factors is called a **cube.** For example, $8a^3$ is a cube because $8a^3 = (2a)^3$. The **cube root** of a cube is one of the three equal factors. To factor the sum or the difference of two cubes, you use the factoring formulas. It helps to use the following patterns, which involve the signs of the terms.

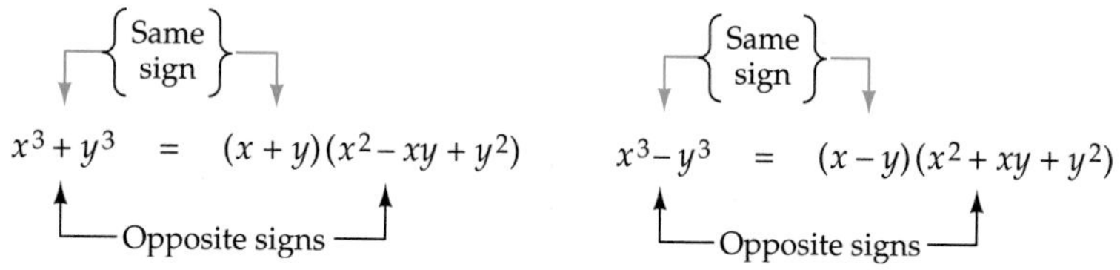

$$x^3 + y^3 = (x + y)(x^2 - xy + y^2) \qquad x^3 - y^3 = (x - y)(x^2 + xy + y^2)$$

Figure 1.14

In the factorization of the sum or difference of two cubes, the terms of the binomial factor are the cube roots of the cubes. For example,

$$8a^3 - 27b^3 = (2a)^3 - (3b)^3 = (2a - 3b)(4a^2 + 6ab + 9b^2)$$

EXAMPLE 7   **Factor the Sum or Difference of Cubes**

Factor   a. $8a^3 + b^3$ b. $a^3 - 64$

Solution

a. $8a^3 + b^3 = (2a)^3 + b^3$ • Recognize the sum-of-cubes form.

$= (2a + b)(4a^2 - 2ab + b^2)$ • Factor.

b. $a^3 - 64 = a^3 - 4^3$ • Recognize the difference-of-cubes form.

$= (a - 4)(a^2 + 4a + 16)$ • Factor.

■ *Try Exercise* **54,** *page 37.*

Here is a general factoring strategy for polynomials:

General Factoring Strategy

1. Factor out the GCF of all terms.
2. Try to factor a binomial as
   a. the difference of two squares.
   b. the sum or difference of two cubes.
3. Try to factor a trinomial:
   a. as a perfect square trinomial.
   b. using the trial method.
4. Try to factor a polynomial with more than three terms by grouping.
5. After each factorization, examine the new factors to see whether they can be factored.

EXAMPLE 8   Factor Using the General Factoring Strategy

Completely factor $x^6 + 7x^3 - 8$.

Solution   Factor $x^6 + 7x^3 - 8$ as the product of two binomials.

$$x^6 + 7x^3 - 8 = (x^3 + 8)(x^3 - 1)$$

Now factor $x^3 + 8$, which is the sum of two cubes, and factor $x^3 - 1$, which is the difference of two cubes.

$$x^6 + 7x^3 - 8 = (x + 2)(x^2 - 2x + 4)(x - 1)(x^2 + x + 1)$$

■ *Try Exercise* **64**, *page 37.*

When you are factoring by grouping, some experimentation may be necessary to find a grouping that is of the form of one of the special factoring formulas.

EXAMPLE 9   Factor by Grouping

Use the technique of grouping to factor each of the following:

a. $a^2 + 10ab + 25b^2 - c^2$ b. $p^2 + p - q - q^2$

Solution

a. $a^2 + 10ab + 25b^2 - c^2$

$= (a^2 + 10ab + 25b^2) - c^2$ • Group the terms of the perfect square trinomial.

$= (a + 5b)^2 - c^2$ • Factor the trinomial.

$= [(a + 5b) + c][(a + 5b) - c]$ • Factor the difference of squares.

$= (a + 5b + c)(a + 5b - c)$ • Simplify.

b. $p^2 + p - q - q^2 = p^2 - q^2 + p - q$ • Rearrange the terms.

$$= (p^2 - q^2) + (p - q)$$ • Regroup.

$$= (p + q)(p - q) + (p - q)$$ • Factor the difference of squares.

$$= (p - q)(p + q + 1)$$ • Factor out the common factor $(p - q)$.

■ Try Exercise 74, page 37.

EXERCISE SET 1.5

In Exercises 1 to 8, factor out the GCF from each polynomial.

1. $5x + 20$

2. $8x^2 + 12x - 40$

3. $-15x^2 - 12x$

4. $-6y^2 - 54y$

5. $10x^2y + 6xy - 14xy^2$

6. $6a^3b^2 - 12a^2b + 72ab^3$

7. $(x - 3)(a + b) + (x - 3)(a + 3b)$

8. $(x - 4)(2a - b) + (x + 4)(2a - b)$

In Exercises 9 to 14, factor by grouping in pairs.

9. $3x^3 + x^2 + 6x + 2$

10. $18w^3 + 15w^2 + 12w + 10$

11. $ax^2 - ax + bx - b$

12. $a^2y^2 - ay^3 + ac - cy$

13. $6w^3 + 4w^2 - 15w - 10$

14. $10z^3 - 15z^2 - 4z + 6$

In Exercises 15 to 28, factor each trinomial.

15. $x^2 + 7x + 12$

16. $x^2 + 9x + 20$

17. $a^2 - 10a - 24$

18. $b^2 + 12b - 28$

19. $6x^2 + 25x + 4$

20. $8a^2 - 26a + 15$

21. $51x^2 - 5x - 4$

22. $57y^2 + y - 6$

23. $6x^2 + xy - 40y^2$

24. $8x^2 + 10xy - 25y^2$

25. $x^4 + 6x^2 + 5$

26. $x^4 + 11x^2 + 18$

27. $6x^4 + 23x^2 + 15$

28. $9x^4 + 10x^2 + 1$

In Exercises 29 to 34, use the factorization theorem to determine whether the trinomials are factorable over the integers.

29. $8x^2 + 26x + 15$

30. $16x^2 + 8x - 35$

31. $4x^2 - 5x + 6$

32. $6x^2 + 8x - 3$

33. $6x^2 - 14x + 5$

34. $10x^2 - 4x - 5$

In Exercises 35 to 44, factor each difference of squares.

35. $x^2 - 9$

36. $x^2 - 64$

37. $4a^2 - 49$

38. $81b^2 - 16c^2$

39. $1 - 100x^2$

40. $1 - 121y^2$

41. $x^4 - 9$

42. $y^4 - 196$

43. $(x + 5)^2 - 4$

44. $(x - 3)^2 - 16$

In Exercises 45 to 52, factor each perfect square trinomial.

45. $x^2 + 10x + 25$

46. $y^2 + 6y + 9$

47. $a^2 - 14a + 49$

48. $b^2 - 24b + 144$

49. $4x^2 + 12x + 9$

50. $25y^2 + 40y + 16$

51. $z^4 + 4z^2w^2 + 4w^4$

52. $9x^4 - 30x^2y^2 + 25y^4$

In Exercises 53 to 60, factor each sum or difference of cubes.

53. $x^3 - 8$

54. $b^3 + 64$

55. $8x^3 - 27y^3$

56. $64u^3 - 27v^3$

57. $8 - x^6$

58. $1 + y^{12}$

59. $(x - 2)^3 - 1$

60. $(y + 3)^3 + 8$

In Exercises 61 to 80, use the general factoring strategy to completely factor each polynomial. If the polynomial does not factor, then state that it is nonfactorable over the integers.

61. $18x^2 - 2$

62. $4bx^3 + 32b$

63. $16x^4 - 1$

64. $81y^4 - 16$

65. $12ax^2 - 23axy + 10ay^2$

66. $6ax^2 - 19axy - 20ay^2$

67. $3bx^3 + 4bx^2 - 3bx - 4b$

68. $2x^6 - 2$

69. $72bx^2 + 24bxy + 2by^2$

70. $64y^3 - 16y^2z + yz^2$

71. $(w - 5)^3 + 8$

72. $5xy + 20y - 15x - 60$

73. $x^2 + 6xy + 9y^2 - 1$

74. $4y^2 - 4yz + z^2 - 9$

75. $8x^2 + 3x - 4$

76. $16x^2 + 81$

77. $5x(2x - 5)^2 - (2x - 5)^3$

78. $6x(3x + 1)^3 - (3x + 1)^4$

79. $4x^2 + 2x - y - y^2$

80. $a^2 + a + b - b^2$

Supplemental Exercises

In Exercises 81 and 82, find all positive values of k such that the trinomial is a perfect square trinomial.

81. $x^2 + kx + 16$

82. $36x^2 + kxy + 100$

In Exercises 83 and 84, find k such that the trinomial is a perfect square trinomial.

83. $x^2 + 16x + k$ **84.** $x^2 - 14xy + ky^2$

In Exercises 85 and 86, use the general strategy to completely factor each polynomial. In each exercise n represents a positive integer.

85. $x^{4n} - 1$ **86.** $x^{4n} - 2x^{2n} + 1$

In Exercises 87 to 90, write the area of the shaded portion of each geometric figure in its factored form.

87.
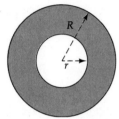

88.
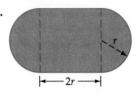

89.
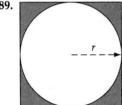

90.
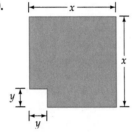

91. The ancient Greeks used geometric figures and the concept of area to illustrate many algebraic concepts. The factoring formula $x^2 - y^2 = (x + y)(x - y)$ can be illustrated by the following figure.

a. Which regions are represented by $(x + y)(x - y)$?

b. Which regions are represented by $x^2 - y^2$?

c. Explain why the area of the regions listed in **a.** must equal the area of the regions listed in **b.**

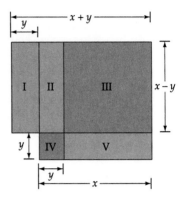

92. What algebraic formula does the following geometric figure illustrate?

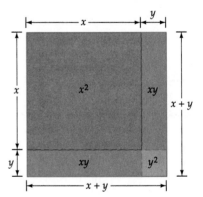

1.6 Rational Expressions

A **rational expression** is a fraction in which the numerator and denominator are polynomials. For example,

$$\frac{3}{x + 1} \quad \text{and} \quad \frac{x^2 - 4x - 21}{x^2 - 9}$$

are rational expressions.

The **domain of a rational expression** is the set of all real numbers that can be used as replacements for the variable. Any value of the variable that causes division by zero is excluded from the domain of the rational expression.

For example, the domain of

$$\frac{7x}{x^2 - 5x} \qquad x \neq 0, x \neq 5$$

is the set of all real numbers except 0 and 5. Both 0 and 5 are excluded values because the denominator $x^2 - 5x$ equals zero when $x = 0$ and also when $x = 5$. Sometimes the excluded values are specified to the right of a rational expression, as shown here. However, a rational expression is meaningful only for those real numbers that are not excluded values, regardless of whether the excluded values are specifically stated.

Rational expressions have properties similar to the properties of rational numbers.

Properties of Rational Expressions

For all rational expressions P/Q and R/S where $Q \neq 0$ and $S \neq 0$,

Equality	$\dfrac{P}{Q} = \dfrac{R}{S}$ if and only if $PS = QR$
Equivalent expressions	$\dfrac{P}{Q} = \dfrac{PR}{QR}, \qquad R \neq 0$
Sign	$-\dfrac{P}{Q} = \dfrac{-P}{Q} = \dfrac{P}{-Q}$

To **simplify a rational expression,** factor the numerator and the denominator. Then use the equivalent expressions property to eliminate factors common to both the numerator and the denominator. A rational expression is **simplified** when 1 is the only common polynomial factor of both the numerator and the denominator.

EXAMPLE 1 **Simplify a Rational Expression**

Simplify: $\dfrac{7 + 20x - 3x^2}{2x^2 - 11x - 21}$

Solution

$$\frac{7 + 20x - 3x^2}{2x^2 - 11x - 21} = \frac{(7 - x)(1 + 3x)}{(x - 7)(2x + 3)} \qquad \bullet \text{ Factor.}$$

$$= \frac{-(x - 7)(1 + 3x)}{(x - 7)(2x + 3)} \qquad \bullet \text{ Use } (7 - x) = -(x - 7).$$

$$= \frac{-\cancel{(x - 7)}(1 + 3x)}{\cancel{(x - 7)}(2x + 3)}$$

$$= \frac{-(1 + 3x)}{2x + 3} = -\frac{3x + 1}{2x + 3} \qquad x \neq 7, x \neq -\frac{3}{2}$$

■ *Try Exercise **2,** page 45.*

Caution A rational expression like $(x + 3)/3$ does not simplify to $x + 1$ because

$$\frac{x + 3}{3} = \frac{x}{3} + \frac{3}{3} = \frac{x}{3} + 1$$

Rational expressions can be simplified by dividing nonzero *factors* common to the numerator and the denominator, but not terms.

Arithmetic operations are defined on rational expressions just as they are on rational numbers.

Arithmetic Operations Defined on Rational Expressions

For all rational expressions P/Q, R/Q, and R/S where $Q \neq 0$ and $S \neq 0$,

Addition $\qquad \dfrac{P}{Q} + \dfrac{R}{Q} = \dfrac{P + R}{Q}$

Subtraction $\qquad \dfrac{P}{Q} - \dfrac{R}{Q} = \dfrac{P - R}{Q}$

Multiplication $\qquad \dfrac{P}{Q} \cdot \dfrac{R}{S} = \dfrac{PR}{QS}$

Division $\qquad \dfrac{P}{Q} \div \dfrac{R}{S} = \dfrac{P}{Q} \cdot \dfrac{S}{R} = \dfrac{PS}{QR} \quad R \neq 0$

Factoring and the equivalent expressions property of rational expressions are used in the multiplication and division of rational expressions.

EXAMPLE 2   **Divide a Rational Expression**

Simplify: $\dfrac{x^2 + 6x + 9}{x^3 + 27} \div \dfrac{x^2 + 7x + 12}{x^3 - 3x^2 + 9x}$

Solution

$\dfrac{x^2 + 6x + 9}{x^3 + 27} \div \dfrac{x^2 + 7x + 12}{x^3 - 3x^2 + 9x}$

$= \dfrac{(x + 3)^2}{(x + 3)(x^2 - 3x + 9)} \div \dfrac{(x + 4)(x + 3)}{x(x^2 - 3x + 9)}$ • Factor.

$= \dfrac{(x + 3)^2}{(x + 3)(x^2 - 3x + 9)} \cdot \dfrac{x(x^2 - 3x + 9)}{(x + 4)(x + 3)}$ • Multiply by the reciprocal.

$= \dfrac{\cancel{(x + 3)^2}\, x\, \cancel{(x^2 - 3x + 9)}}{\cancel{(x + 3)}\,\cancel{(x^2 - 3x + 9)}\,(x + 4)\cancel{(x + 3)}}$ • Simplify.

$= \dfrac{x}{x + 4}$

■ *Try Exercise* **16**, *page 45.*

Addition of rational expressions with a **common denominator** is accomplished by writing the sum of the numerators over the common denominator. For example,

$$\frac{5x}{18} + \frac{x}{18} = \frac{5x + x}{18} = \frac{6x}{18} = \frac{x}{3}$$

If the rational expressions do not have a common denominator, then they can be written as equivalent rational expressions that have a common denominator by multiplying numerator and denominator of each of the rational expressions by the required polynomials. The following procedure can be used to determine the least common denominator (LCD) of rational expressions. It is similar to the process used to find the LCD of rational numbers.

Determining the LCD of Rational Expressions

1. Factor each denominator completely and express repeated factors using exponential notation.
2. Identify the largest power of each factor in any single factorization. The LCD is the product of each factor raised to its largest power.

For example,

$$\frac{1}{x + 3} \quad \text{and} \quad \frac{5}{2x - 1}$$

have an LCD of $(x + 3)(2x - 1)$. The rational expressions

$$\frac{5x}{(x + 5)(x - 7)^3} \quad \text{and} \quad \frac{7}{x(x + 5)^2(x - 7)}$$

have an LCD of $x(x + 5)^2(x - 7)^3$.

EXAMPLE 3   **Add and Subtract Rational Expressions**

Perform the indicated operation and then simplify if possible.

a. $\dfrac{5x}{48} + \dfrac{x}{15}$ b. $\dfrac{x}{x^2 - 4} - \dfrac{2x - 1}{x^2 - 3x - 10}$

Solution

a. Determine the prime factorization of the denominators.

$$48 = 2^4 \cdot 3 \quad \text{and} \quad 15 = 3 \cdot 5$$

The desired common denominator is the product of each of the prime factors raised to its largest power. Thus the common denominator is $2^4 \cdot 3 \cdot 5 = 240$. Write each rational expression as an equivalent rational expression with a denominator of 240.

$$\frac{5x}{48} + \frac{x}{15} = \frac{5x \cdot 5}{48 \cdot 5} + \frac{x \cdot 16}{15 \cdot 16} = \frac{25x}{240} + \frac{16x}{240} = \frac{41x}{240}$$

b. Factor each denominator to determine the LCD of the rational expressions.

$$x^2 - 4 = (x + 2)(x - 2)$$

and
$$x^2 - 3x - 10 = (x + 2)(x - 5)$$

The LCD is $(x + 2)(x - 2)(x - 5)$. Forming equivalent rational expressions that have the LCD, we have

$$\frac{x}{x^2 - 4} - \frac{2x - 1}{x^2 - 3x - 10}$$

$$= \frac{x(x - 5)}{(x + 2)(x - 2)(x - 5)} - \frac{(2x - 1)(x - 2)}{(x + 2)(x - 5)(x - 2)}$$

$$= \frac{x^2 - 5x - (2x^2 - 5x + 2)}{(x + 2)(x - 2)(x - 5)} = \frac{x^2 - 5x - 2x^2 + 5x - 2}{(x + 2)(x - 2)(x - 5)}$$

$$= \frac{-x^2 - 2}{(x + 2)(x - 2)(x - 5)} = -\frac{x^2 + 2}{(x + 2)(x - 2)(x - 5)}$$

■ *Try Exercise* **30**, *page 45.*

Complex Fractions

A **complex fraction** is a fraction whose numerator or denominator contains one or more fractions. Complex fractions can be simplified by using one of the following two methods.

Methods for Simplifying Complex Fractions

Method 1: Multiply by the LCD

1. Determine the LCD of all the fractions in the complex fraction.

2. Multiply both the numerator and the denominator of the complex fraction by the LCD.

3. If possible, simplify the resulting rational expression.

Method 2: Multiply by the reciprocal of the denominator

1. Simplify the numerator to a single fraction and the denominator to a single fraction.

2. Multiply the numerator by the reciprocal of the denominator.

3. If possible, simplify the resulting rational expression.

EXAMPLE 4   **Simplify Complex Fractions**

Simplify each complex fraction.

a. $\dfrac{3 - \dfrac{2}{a}}{1 + \dfrac{4}{a}}$
   b. $\dfrac{\dfrac{2}{x - 2} + \dfrac{1}{x}}{\dfrac{3x}{x - 5} - \dfrac{2}{x - 5}}$

Solution

a. The LCD of all the fractions in the complex fraction is a. Therefore this complex fraction can be simplified by multiplying both the numerator and the denominator of the complex fraction by a.

$$\frac{3 - \dfrac{2}{a}}{1 + \dfrac{4}{a}} = \frac{\left(3 - \dfrac{2}{a}\right)a}{\left(1 + \dfrac{4}{a}\right)a} = \frac{3a - \left(\dfrac{2}{a}\right)a}{a + \left(\dfrac{4}{a}\right)a} = \frac{3a - 2}{a + 4}$$

b. First simplify the numerator to a single fraction and then simplify the denominator to a single fraction.

$$\frac{\dfrac{2}{x - 2} + \dfrac{1}{x}}{\dfrac{3x}{x - 5} - \dfrac{2}{x - 5}} = \frac{\dfrac{2 \cdot x}{(x - 2) \cdot x} + \dfrac{1 \cdot (x - 2)}{x \cdot (x - 2)}}{\dfrac{3x - 2}{x - 5}}$$ • Simplify numerator and denominator.

$$= \frac{\dfrac{2x + (x - 2)}{x(x - 2)}}{\dfrac{3x - 2}{x - 5}} = \frac{\dfrac{3x - 2}{x(x - 2)}}{\dfrac{3x - 2}{x - 5}}$$

$$= \frac{\cancel{3x - 2}}{x(x - 2)} \cdot \frac{x - 5}{\cancel{3x - 2}}$$ • Multiply by the reciprocal of the denominator.

$$= \frac{x - 5}{x(x - 2)}$$

■ *Try Exercise **42**, page 45.*

EXAMPLE 5   **Simplify a Fraction**

Simplify the fraction $\dfrac{c^{-1}}{a^{-1} + b^{-1}}$.

Solution The fraction written without negative exponents becomes

$$\frac{c^{-1}}{a^{-1} + b^{-1}} = \frac{\dfrac{1}{c}}{\dfrac{1}{a} + \dfrac{1}{b}}$$ • Using $x^{-n} = \dfrac{1}{x^n}$.

$$= \frac{\dfrac{1}{c} \cdot abc}{\left(\dfrac{1}{a} + \dfrac{1}{b}\right) abc}$$ • Multiply the numerator and the denominator by *abc*, which is the LCD.

$$= \frac{ab}{bc + ac}$$

■ *Try Exercise* **60**, *page 46.*

Caution It is a mistake to write

$$\frac{c^{-1}}{a^{-1} + b^{-1}} \quad \text{as} \quad \frac{a + b}{c}$$

because a^{-1} and b^{-1} are *terms* and cannot be treated as factors.

EXAMPLE 6 Solve an Application

The *average speed* for a round trip is given by the complex fraction

$$\frac{2}{\dfrac{1}{v_1} + \dfrac{1}{v_2}}$$

where v_1 is the average speed on the way to your destination and v_2 is the average speed on your return trip. Find the average speed for a round trip if $v_1 = 50$ mph and $v_2 = 40$ mph.

Solution Evaluate the complex fraction with $v_1 = 50$ and $v_2 = 40$.

$$\frac{2}{\dfrac{1}{v_1} + \dfrac{1}{v_2}} = \frac{2}{\dfrac{1}{50} + \dfrac{1}{40}} = \frac{2}{\dfrac{1 \cdot 4}{50 \cdot 4} + \dfrac{1 \cdot 5}{40 \cdot 5}}$$ • Substitute and simplify the denominator

$$= \frac{2}{\dfrac{4}{200} + \dfrac{5}{200}} = \frac{2}{\dfrac{9}{200}} = 2 \cdot \frac{200}{9} = \frac{400}{9} = 44\frac{4}{9}$$

The average speed of the round trip is $44\frac{4}{9}$ mph.

■ *Try Exercise* **64**, *page 46.*

Remark   The average speed of the round trip is *not* the average of v_1 and v_2. Why?[2]

EXERCISE SET 1.6

In Exercises 1 to 10, simplify each rational expression.

1. $\dfrac{x^2 - x - 20}{3x - 15}$

2. $\dfrac{2x^2 - 5x - 12}{2x^2 + 5x + 3}$

3. $\dfrac{x^3 - 9x}{x^3 + x^2 - 6x}$

4. $\dfrac{x^3 + 125}{2x^3 - 50x}$

5. $\dfrac{a^3 + 8}{a^2 - 4}$

6. $\dfrac{y^3 - 27}{-y^2 + 11y - 24}$

7. $\dfrac{x^2 + 3x - 40}{-x^2 + 3x + 10}$

8. $\dfrac{2x^3 - 6x^2 + 5x - 15}{9 - x^2}$

9. $\dfrac{4y^3 - 8y^2 + 7y - 14}{-y^2 - 5y + 14}$

10. $\dfrac{x^3 - x^2 + x}{x^3 + 1}$

In Exercises 11 to 40, simplify each expression.

11. $\left(-\dfrac{4a}{3b^2}\right)\left(\dfrac{6b}{a^4}\right)$

12. $\left(\dfrac{12x^2y}{5z^4}\right)\left(-\dfrac{25x^2z^3}{15y^2}\right)$

13. $\left(\dfrac{6p^2}{5q^2}\right)^{-1}\left(\dfrac{2p}{3q^2}\right)^2$

14. $\left(\dfrac{4r^2s}{3t^3}\right)^{-1}\left(\dfrac{6rs^3}{5t^2}\right)$

15. $\dfrac{x^2 + x}{2x + 3} \cdot \dfrac{3x^2 + 19x + 28}{x^2 + 5x + 4}$

16. $\dfrac{x^2 - 16}{x^2 + 7x + 12} \cdot \dfrac{x^2 - 4x - 21}{x^2 - 4x}$

17. $\dfrac{3x - 15}{2x^2 - 50} \cdot \dfrac{2x^2 + 16x + 30}{6x + 9}$

18. $\dfrac{y^3 - 8}{y^2 + y - 6} \cdot \dfrac{y^2 + 3y}{y^3 + 2y^2 + 4y}$

19. $\dfrac{12y^2 + 28y + 15}{6y^2 + 35y + 25} \div \dfrac{2y^2 - y - 3}{3y^2 + 11y - 20}$

20. $\dfrac{z^2 - 81}{z^2 - 16} \div \dfrac{z^2 - z - 20}{z^2 + 5z - 36}$

21. $\dfrac{a^2 + 9}{a^2 - 64} \div \dfrac{a^3 - 3a^2 + 9a - 27}{a^2 + 5a - 24}$

22. $\dfrac{6x^2 + 13xy + 6y^2}{4x^2 - 9y^2} \div \dfrac{3x^2 - xy - 2y^2}{2x^2 + xy - 3y^2}$

23. $\dfrac{p + 5}{r} + \dfrac{2p - 7}{r}$

24. $\dfrac{2s + 5t}{4t} + \dfrac{-2s + 3t}{4t}$

25. $\dfrac{x}{x - 5} + \dfrac{7x}{x + 3}$

26. $\dfrac{2x}{3x + 1} + \dfrac{5x}{x - 7}$

27. $\dfrac{5y - 7}{y + 4} - \dfrac{2y - 3}{y + 4}$

28. $\dfrac{6x - 5}{x - 3} - \dfrac{3x - 8}{x - 3}$

29. $\dfrac{4z}{2z - 3} + \dfrac{5z}{z - 5}$

30. $\dfrac{3y - 1}{3y + 1} - \dfrac{2y - 5}{y - 3}$

31. $\dfrac{x}{x^2 - 9} - \dfrac{3x - 1}{x^2 + 7x + 12}$

32. $\dfrac{m - n}{m^2 - mn - 6n^2} + \dfrac{3m - 5n}{m^2 + mn - 2n^2}$

33. $\dfrac{1}{x} + \dfrac{2}{3x - 1} \cdot \dfrac{3x^2 + 11x - 4}{x - 5}$

34. $\dfrac{2}{y} - \dfrac{3}{y + 1} \cdot \dfrac{y^2 - 1}{y + 4}$

35. $\dfrac{q + 1}{q - 3} - \dfrac{2q}{q - 3} \div \dfrac{q + 5}{q - 3}$

36. $\dfrac{p}{p + 5} + \dfrac{p}{p - 4} \div \dfrac{p + 2}{p^2 - p - 12}$

37. $\dfrac{1}{x^2 + 7x + 12} + \dfrac{1}{x^2 - 9} + \dfrac{1}{x^2 - 16}$

38. $\dfrac{2}{a^2 - 3a + 2} + \dfrac{3}{a^2 - 1} - \dfrac{5}{a^2 + 3a - 10}$

39. $\left(1 + \dfrac{2}{x}\right)\left(3 - \dfrac{1}{x}\right)$

40. $\left(4 - \dfrac{1}{z}\right)\left(4 + \dfrac{2}{z}\right)$

In Exercises 41 to 58, simplify each complex fraction.

41. $\dfrac{4 + \dfrac{1}{x}}{1 - \dfrac{1}{x}}$

42. $\dfrac{3 - \dfrac{2}{a}}{5 + \dfrac{3}{a}}$

[2] Because you were traveling slower on the return trip, the return trip took longer than the time spent going to your destination. More time was spent traveling at the slower speed. Thus the average speed is less than the average of v_1 and v_2.

43. $\dfrac{\dfrac{x}{y} - 2}{y - x}$

44. $\dfrac{3 + \dfrac{2}{x - 3}}{4 + \dfrac{1}{2 + \dfrac{1}{x}}}$

45. $\dfrac{5 - \dfrac{1}{x + 2}}{1 + \dfrac{3}{1 + \dfrac{3}{x}}}$

46. $\dfrac{\dfrac{1}{(x + h)^2} - 1}{h}$

47. $\dfrac{1 + \dfrac{1}{b - 2}}{1 - \dfrac{1}{b + 3}}$

48. $r - \dfrac{r}{r + \dfrac{1}{3}}$

49. $\dfrac{1 - \dfrac{1}{x^2}}{1 + \dfrac{1}{x}}$

50. $\dfrac{1}{\dfrac{1}{a} + \dfrac{1}{b}}$

51. $2 - \dfrac{m}{1 - \dfrac{1 - m}{-m}}$

52. $\dfrac{\dfrac{x + h + 1}{x + h} - \dfrac{x}{x + 1}}{h}$

53. $\dfrac{\dfrac{1}{x} - \dfrac{x - 4}{x + 1}}{\dfrac{x}{x + 1}}$

54. $\dfrac{\dfrac{2}{y} - \dfrac{3y - 2}{y - 1}}{\dfrac{y}{y - 1}}$

55. $\dfrac{\dfrac{1}{x + 3} - \dfrac{2}{x - 1}}{\dfrac{x}{x - 1} + \dfrac{3}{x + 3}}$

56. $\dfrac{\dfrac{x + 2}{x^2 - 1} + \dfrac{1}{x + 1}}{\dfrac{x}{2x^2 - x - 1} + \dfrac{1}{x - 1}}$

57. $\dfrac{\dfrac{x^2 + 3x - 10}{x^2 + x - 6}}{\dfrac{x^2 - x - 30}{2x^2 - 15x + 18}}$

58. $\dfrac{\dfrac{2y^2 + 11y + 15}{y^2 - 4y - 21}}{\dfrac{6y^2 + 11y - 10}{3y^2 - 23y + 14}}$

In Exercises 59 to 62, simplify each algebraic fraction. Write all answers with positive exponents.

59. $\dfrac{a^{-1} + b^{-1}}{a - b}$

60. $\dfrac{e^{-2} - f^{-1}}{ef}$

61. $\dfrac{a^{-1}b - ab^{-1}}{a^2 + b^2}$

62. $(a + b^{-2})^{-1}$

63. According to Example 6, the average speed for a round trip in which the average speed on the way to your destination was v_1 and the average speed on your return was v_2 is given by the complex fraction

$$\dfrac{2}{\dfrac{1}{v_1} + \dfrac{1}{v_2}}$$

  a. Find the average speed for a round trip by helicopter with $v_1 = 180$ mph and $v_2 = 110$ mph.

  b. Simplify the complex fraction.

64. Using Einstein's theory of relativity, the "sum" of the two speeds v_1 and v_2 is given by the complex fraction

$$\dfrac{v_1 + v_2}{1 + \dfrac{v_1 v_2}{c^2}}$$

where c is the speed of light.

  a. Evaluate this expression with $v_1 = 1.2 \times 10^8$ mph, $v_2 = 2.4 \times 10^8$ mph, and $c = 6.7 \times 10^8$ mph.

  b. Simplify the complex fraction.

65. Find the rational expression in simplest form that represents the sum of the reciprocals of the consecutive integers x and $x + 1$.

66. Find the rational expression in simplest form that represents the positive difference between the reciprocals of the consecutive even integers x and $x + 2$.

67. Find the rational expression in simplest form that represents the sum of the reciprocals of the consecutive even integers $x - 2$, x, and $x + 2$.

68. Find the rational expression in simplest form that represents the sum of the reciprocals of the squares of the consecutive even integers $x - 2$, x, and $x + 2$.

Supplemental Exercises

In Exercises 69 to 72, simplify each algebraic fraction.

69. $\dfrac{(x + 5) - x(x + 5)^{-1}}{x + 5}$

70. $\dfrac{(y + 2) + y^2(y + 2)^{-1}}{y + 2}$

71. $\dfrac{x^{-1} - 4y}{(x^{-1} - 2y)(x^{-1} + 2y)}$

72. $\dfrac{x + y}{x - y} \cdot \dfrac{x^{-1} - y^{-1}}{x^{-1} + y^{-1}}$

73. The **present value** of an ordinary annuity is given by

$$R\left[\dfrac{1 - \dfrac{1}{(1 + i)^n}}{i}\right]$$

where n is the number of payments of R dollars each invested at an interest rate of i per conversion period. Simplify the complex fraction.

74. The total resistance of the three resistances R_1, R_2, and R_3 in parallel is given by

$$\cfrac{1}{\dfrac{1}{R_1} + \dfrac{1}{R_2} + \dfrac{1}{R_3}}$$

Simplify the complex fraction.

The following complex fraction expression is called a **continued fraction.**

$$\cfrac{1}{1 + \cfrac{1}{1 + \cfrac{1}{1 + \cfrac{1}{1 + \cdots}}}}$$

In Exercises 75 to 78, the complex fractions are called **conver-gents** of the above continued fraction. Simplify each convergent, and write it as a decimal accurate to 3 decimal places.

75. $C_1 = \cfrac{1}{1 + \cfrac{1}{1}}$

76. $C_2 = \cfrac{1}{1 + \cfrac{1}{1 + \cfrac{1}{1}}}$

77. $C_3 = \cfrac{1}{1 + \cfrac{1}{1 + \cfrac{1}{1 + \cfrac{1}{1}}}}$

78. $C_4 = \cfrac{1}{1 + \cfrac{1}{1 + \cfrac{1}{1 + \cfrac{1}{1 + \cfrac{1}{1}}}}}$

79. It can be shown that if we continue to evaluate convergents as in Exercises 75 to 78, we find that the convergents get closer and closer to the irrational number

$$\frac{-1 + \sqrt{5}}{2}$$

Use a calculator to approximate this irrational number as a decimal accurate to three decimal places.

1.7 Rational Exponents and Radicals

Up to this point, the expression b^n has been defined for real numbers b and integers n, except for the restrictions listed in Section 1.3. Now we wish to extend the definition of exponents to include rational numbers, so that expressions such as $2^{1/2}$ will be meaningful. Not just any definition will do. We want a definition of rational exponents for which the properties of integer exponents are true. The following example shows the direction we can take to accomplish our goal.

If the property for multiplying exponential expressions is to hold for rational exponents, then for rational numbers p and q, $b^p b^q = b^{p+q}$. For example, $9^{1/2} \cdot 9^{1/2}$ must equal $9^{1/2+1/2} = 9^1 = 9$. Thus $9^{1/2}$ must be a square root of 9. That is, $9^{1/2} = 3$. This example suggests that $b^{1/n}$ can be defined in terms of roots according to the following definition.

Definition of $b^{1/n}$

If n is an even positive integer and $b \geq 0$, then $b^{1/n}$ is the nonnegative real number such that $(b^{1/n})^n = b$.

If n is an odd positive integer, then $b^{1/n}$ is the real number such that $(b^{1/n})^n = b$.

As examples, $25^{1/2} = 5$ because $5^2 = 25$; $(-64)^{1/3} = -4$ because $(-4)^3 = -64$.

Remark   If n is an even positive integer and $b < 0$, then $b^{1/n}$ is a complex number. We will study complex numbers in Section 1.8.

EXAMPLE 1   **Evaluate Exponential Expressions**

Evaluate each of the following:

a. $16^{1/2}$ b. $-16^{1/2}$ c. $(-16)^{1/2}$ d. $(-32)^{1/5}$

Solution

a. $16^{1/2} = 4$ because $4^2 = 16$.   b. $-16^{1/2} = -(16^{1/2}) = -4$.
c. $(-16)^{1/2}$ is not a real number.
d. $(-32)^{1/5} = -2$ because $(-2)^5 = -32$.

■ *Try Exercise 2, page 54.*

Remark   Note the difference between $-16^{1/2}$, which equals -4, and $(-16)^{1/2}$, which is not a real number.

To define expressions such as $8^{2/3}$, we will extend our definition of exponents even further. Because we want the power property $(b^p)^q = b^{pq}$ to be true for rational exponents also, we must have $(b^{1/n})^m = b^{m/n}$. With this in mind, we make the following definition.

Definition of $b^{m/n}$

> For all positive integers m and n such that m/n is in simplest form, and for all real numbers b for which $b^{1/n}$ is a real number,
>
> $$b^{m/n} = (b^{1/n})^m = (b^m)^{1/n}$$

Because $b^{m/n}$ is defined as $(b^{1/n})^m$ and also as $(b^m)^{1/n}$, we can evaluate expressions such as $8^{4/3}$ in more than one way. For example, $8^{4/3}$ can be evaluated by using either of the following:

$$8^{4/3} = (8^{1/3})^4 = 2^4 = 16$$

$$8^{4/3} = (8^4)^{1/3} = 4096^{1/3} = 16$$

Of the two methods, the $b^{m/n} = (b^{1/n})^m$ method is usually easier to apply, provided you can evaluate $b^{1/n}$.

Caution   Some calculators do not evaluate $b^{m/n}$ properly when $b < 0$ and m/n is not an integer.

EXAMPLE 2   **Evaluate Exponential Expressions**

Evaluate each of the following:

a. $8^{2/3}$ b. $32^{4/5}$ c. $(-9)^{3/2}$ d. $(-64)^{4/3}$

Solution

a. $8^{2/3} = (8^{1/3})^2 = 2^2 = 4$

b. $32^{4/5} = (32^{1/5})^4 = 2^4 = 16$

c. $(-9)^{3/2}$ is not a real number because $(-9)^{1/2}$ is not a real number.

d. $(-64)^{4/3} = [(-64)^{1/3}]^4 = [-4]^4 = 256$

■ *Try Exercise* **6,** *page 54.*

The following properties of exponents were stated in Section 1.3, but they are restated here to remind you that they have now been extended to apply to rational exponents. The bases a and b have been restricted to positive real numbers to avoid the pitfalls illustrated in Exercises 161 and 162.

Properties of Rational Exponents

If p and q represent rational numbers and a and b are positive real numbers, then

Product $b^p \cdot b^q = b^{p+q}$

Quotient $\dfrac{b^p}{b^q} = b^{p-q}$

Power $(b^p)^q = b^{pq}$ $(ab)^p = a^p b^p$ $\left(\dfrac{a}{b}\right)^p = \dfrac{a^p}{b^p}$

$b^{-p} = \dfrac{1}{b^p}$

Recall that an exponential expression is in simplest form when no powers of powers or negative exponents appear.

EXAMPLE 3 **Simplify Exponential Expressions**

Simplify each exponential expression. Assume the variables to be positive real numbers.

a. $\left(\dfrac{x^2 y^3}{x^{-3} y^5}\right)^{1/2}$ b. $(x^{1/2} - y^{1/2})^2$

Solution

a. $\left(\dfrac{x^2 y^3}{x^{-3} y^5}\right)^{1/2} = (x^5 y^{-2})^{1/2} = x^{5/2} y^{-1} = \dfrac{x^{5/2}}{y}$

b. $(x^{1/2} - y^{1/2})^2 = x - 2x^{1/2} y^{1/2} + y$

■ *Try Exercise* **38,** *page 54.*

Radicals expressed by the notation $\sqrt[n]{b}$ are also used to denote roots. The number b is the **radicand,** and the positive integer n is the **index** of the radical.

Definition of $\sqrt[n]{b}$

If n is a positive integer and b is a real number such that $b^{1/n}$ is a real number, then $\sqrt[n]{b} = b^{1/n}$.

Remark If the index n equals 2, then the radical $\sqrt[2]{b}$ is written as simply $\sqrt{b}$, and it is referred to as the **principal square root of** b or simply **the square root of** b.

The symbol $\sqrt{b}$ is reserved to represent the nonnegative square root of b. To represent the negative square root of b, write $-\sqrt{b}$. For example, $\sqrt{25} = 5$, whereas $-\sqrt{25} = -5$.

EXAMPLE 4 **Evaluate Radicals**

Evaluate each of the following radicals.

a. $\sqrt[3]{8}$ b. $\sqrt[4]{81}$ c. $\sqrt{-16}$ d. $\sqrt[5]{-32}$

Solution

a. $\sqrt[3]{8} = 8^{1/3} = 2$ b. $\sqrt[4]{81} = 81^{1/4} = 3$

c. $\sqrt{-16}$ is not a real number.   d. $\sqrt[5]{-32} = (-32)^{1/5} = -2$

■ *Try Exercise **46**, page 55.*

The expressions $(\sqrt[n]{b})^m$ and $\sqrt[n]{b^m}$ can be expressed in exponential notation according to the following definition.

Definition of $(\sqrt[n]{b})^m$

For all positive integers n, all integers m, and all real numbers b such that $\sqrt[n]{b}$ is a real number, $(\sqrt[n]{b})^m = \sqrt[n]{b^m} = b^{m/n}$.

The equations

$$b^{m/n} = \sqrt[n]{b^m} \quad \text{and} \quad b^{m/n} = (\sqrt[n]{b})^m$$

can be used to write exponential expressions such as $b^{m/n}$ in radical form. Use the denominator n as the index of the radical and the numerator m as the power of the radicand or as the power of the radical. For example,

$$(5xy)^{2/3} = (\sqrt[3]{5xy})^2 = \sqrt[3]{25x^2y^2}$$

• Use the denominator 3 as the index of the radical and the numerator 2 as the power of the radical.

$$(a^2 + 5)^{3/2} = (\sqrt{a^2 + 5})^3 = \sqrt{(a^2 + 5)^3}$$

The equations $b^{m/n} = \sqrt[n]{b^m}$ and $b^{m/n} = (\sqrt[n]{b})^m$ can also be used to write radical expressions in exponential form. For example,

$$\sqrt{(2ab)^3} = (2ab)^{3/2}$$ • Use the index 2 as the denominator of the power and the exponent 3 as the numerator of the power.

$$\sqrt[6]{m^4n^2} = (m^4n^2)^{1/6} = m^{4/6}n^{2/6} = m^{2/3}n^{1/3}$$

EXAMPLE 5   Evaluate Radical Expressions

Evaluate

a. $(\sqrt[3]{8})^4$ b. $(\sqrt[4]{9})^2$ c. $(\sqrt{7})^2$

Solution

a. $(\sqrt[3]{8})^4 = 8^{4/3} = (8^{1/3})^4 = (2)^4 = 16$ b. $(\sqrt[4]{9})^2 = 9^{2/4} = 9^{1/2} = 3$

c. $(\sqrt{7})^2 = 7^{2/2} = 7$

_____ ■ *Try Exercise* **52,** *page 55.*

Remark   You might think $\sqrt{x^2} = x$, but the following example shows that this is *not* true for all values of x.

Case 1   Let $x = 5$, then $\sqrt{5^2} = \sqrt{25} = 5 = x$

Case 2   Let $x = -5$, then $\sqrt{(-5)^2} = \sqrt{25} = 5 = -x$

In summary, if $x \geq 0$, then $\sqrt{x^2} = x$. If $x < 0$, then $\sqrt{x^2} = -x$. Thus using absolute value notation, for any real number x, $\sqrt{x^2} = |x|$.

Agreement   All variables used in radical expressions in the remainder of this text represent only nonnegative real numbers.

With this agreement in force, we will seldom need to concern ourselves with the use of absolute value symbols. That is, we shall write $\sqrt{x^2} = x$.

Properties of Radicals

> If m and n are natural numbers greater than or equal to 2, and a and b are nonnegative real numbers, then
>
> Product $\sqrt[n]{a} \cdot \sqrt[n]{b} = \sqrt[n]{ab}$
>
> Quotient $\dfrac{\sqrt[n]{a}}{\sqrt[n]{b}} = \sqrt[n]{\dfrac{a}{b}}$ $(b \neq 0)$
>
> Index $\sqrt[m]{\sqrt[n]{b}} = \sqrt[mn]{b}$ $(\sqrt[n]{b})^n = b$ $\sqrt[n]{b^n} = b$

A radical is in **simplest radical form** if it meets all the following criteria:

1. The radicand contains only powers less than the index. ($\sqrt{x^5}$ does not satisfy this requirement because $5 > 2$.)

2. The index of the radical is as small as possible. ($\sqrt[6]{x^3}$ does not satisfy this requirement because $\sqrt[6]{x^3} = x^{3/6} = x^{1/2} = \sqrt{x}$.)

3. The denominator has been rationalized. That is, no radicals appear in a denominator. ($1/\sqrt{2}$ does not satisfy this requirement.)

4. No fractions appear in the radicand. ($\sqrt{2/5}$ does not satisfy this requirement.)

EXAMPLE 6   **Simplify Radicals**

Simplify each of the following radicals.

a. $\sqrt[3]{32}$ b. $\sqrt{12y^7}$ c. $\sqrt{162x^2y^5}$ d. $\sqrt[3]{\sqrt{x^8y}}$ e. $\sqrt[4]{b^2}$

Solution

a. Factor the radicand into prime factors and simplify, using the product property of radicals and the index property $\sqrt[n]{b^n} = b$.

$$\sqrt[3]{32} = \sqrt[3]{2^5} = \sqrt[3]{2^3 \cdot 2^2} = \sqrt[3]{2^3} \cdot \sqrt[3]{2^2} = 2\sqrt[3]{4}$$

b. $\sqrt{12y^7} = \sqrt{2^2 \cdot 3 \cdot y^6 \cdot y^1} = \sqrt{(2y^3)^2(3y)} = \sqrt{(2y^3)^2}\sqrt{3y} = 2y^3\sqrt{3y}$

c. $\sqrt{162x^2y^5} = \sqrt{2 \cdot 3^4 \cdot x^2 \cdot y^4 \cdot y} = \sqrt{(3^2xy^2)^2(2y)} = 9xy^2\sqrt{2y}$

d. $\sqrt[3]{\sqrt{x^8y}} = \sqrt[6]{x^8y} = \sqrt[6]{(x^6)(x^2y)} = x\sqrt[6]{x^2y}$

e. $\sqrt[4]{b^2} = b^{2/4} = b^{1/2} = \sqrt{b}$ • Change to exponential form, reduce the exponent, and change to radical form.

■ *Try Exercise* **70**, *page 55.*

Arithmetic Operations on Radicals

Like radicals have the same radicand and the same index. For example,

$$3\sqrt[3]{x^2y} \quad \text{and} \quad -2\sqrt[3]{x^2y}$$

are like radicals. Addition and subtraction of like radicals are accomplished by using the distributive property. For example,

$$4\sqrt{3x} + 7\sqrt{3x} = (4 + 7)\sqrt{3x} = 11\sqrt{3x}$$

$$2x\sqrt[3]{y^2} - 7x\sqrt[3]{y^2} + x\sqrt[3]{y^2} = (2x - 7x + x)\sqrt[3]{y^2} = -4x\sqrt[3]{y^2}$$

Remark   The sum $2\sqrt{3} + 5\sqrt{2}$ cannot be simplified any further. The radicals are not like radicals.

It is possible to combine radicals that do not appear to be like radicals if they can be simplified to be like radicals.

EXAMPLE 7   **Combine Radicals**

Simplify: $5\sqrt{32} + 2\sqrt{128}$.

Solution

$$5\sqrt{32} + 2\sqrt{128} = 5\sqrt{2^5} + 2\sqrt{2^7} = 5 \cdot 4\sqrt{2} + 2 \cdot 8\sqrt{2}$$
$$= 20\sqrt{2} + 16\sqrt{2} = 36\sqrt{2}$$

■ Try Exercise 98, page 55.

Multiplication of radical expressions is very similar to the multiplication procedures used to multiply polynomials.

EXAMPLE 8   **Multiply Radical Expressions**

Find the product of each of the following. Simplify where possible.

a. $(\sqrt{3} + 5)(\sqrt{3} - 2)$ b. $(\sqrt{5x} - \sqrt{2y})(\sqrt{5x} + \sqrt{2y})$

Solution

a. $(\sqrt{3} + 5)(\sqrt{3} - 2) = (\sqrt{3})^2 - 2\sqrt{3} + 5\sqrt{3} - 10 = -7 + 3\sqrt{3}$

b. $(\sqrt{5x} - \sqrt{2y})(\sqrt{5x} + \sqrt{2y}) = (\sqrt{5x})^2 - (\sqrt{2y})^2 = 5x - 2y$

■ Try Exercise 102, page 55.

To **rationalize the denominator** of a fraction means to write it in an equivalent form that does not involve any radicals in its denominator.

EXAMPLE 9   **Rationalize the Denominator**

Rationalize the denominator of each of the following:

a. $\dfrac{3}{\sqrt{2}}$ b. $\dfrac{5}{\sqrt[3]{a}}$

Solution

a. $\dfrac{3}{\sqrt{2}} = \dfrac{3}{\sqrt{2}} \cdot \dfrac{\sqrt{2}}{\sqrt{2}} = \dfrac{3\sqrt{2}}{2}$ • Multiply numerator and denominator by $\sqrt{2}$.

b. $\dfrac{5}{\sqrt[3]{a}} = \dfrac{5}{\sqrt[3]{a}} \cdot \dfrac{\sqrt[3]{a^2}}{\sqrt[3]{a^2}} = \dfrac{5\sqrt[3]{a^2}}{\sqrt[3]{a^3}} = \dfrac{5\sqrt[3]{a^2}}{a}$ • Use $\sqrt[3]{a} \cdot \sqrt[3]{a^2} = \sqrt[3]{a^3} = a$.

■ Try Exercise 116, page 55.

To rationalize the denominator of a fractional expression such as

$$\frac{1}{\sqrt{m} + \sqrt{n}}$$

we make use of the conjugate of $\sqrt{m} + \sqrt{n}$, which is $\sqrt{m} - \sqrt{n}$. The product of these conjugate pairs does not involve a radical.

$$(\sqrt{m} + \sqrt{n})(\sqrt{m} - \sqrt{n}) = m - n$$

In Example 10 we use the conjugate of the denominator to rationalize the denominator.

EXAMPLE 10   Rationalize the Denominator

Rationalize the denominator of each of the following:

a. $\dfrac{2}{\sqrt{3} + \sqrt{a}}$ b. $\dfrac{a + \sqrt{5}}{a - \sqrt{5}}$

Solution

a. $\dfrac{2}{\sqrt{3} + \sqrt{a}} = \dfrac{2}{\sqrt{3} + \sqrt{a}} \cdot \dfrac{\sqrt{3} - \sqrt{a}}{\sqrt{3} - \sqrt{a}} = \dfrac{2\sqrt{3} - 2\sqrt{a}}{3 - a}$

b. $\dfrac{a + \sqrt{5}}{a - \sqrt{5}} = \dfrac{a + \sqrt{5}}{a - \sqrt{5}} \cdot \dfrac{a + \sqrt{5}}{a + \sqrt{5}} = \dfrac{a^2 + 2a\sqrt{5} + 5}{a^2 - 5}$

■ *Try Exercise* **124**, *page 55.*

Recall that a radical is in simplest radical form if

1. The radicand contains only powers less than the index.
2. The index of the radical is as small as possible.
3. The denominator has been rationalized.
4. No fractions appear in the radicand.

In the following example, we use several of the properties of real numbers and radicals to write a radical in its simplest form.

$$\sqrt{\dfrac{36x^3}{x^4}} = \sqrt{\dfrac{36}{x}} = \dfrac{\sqrt{36}}{\sqrt{x}} = \dfrac{6}{\sqrt{x}} = \dfrac{6\sqrt{x}}{\sqrt{x}\,\sqrt{x}} = \dfrac{6\sqrt{x}}{x}$$

EXERCISE SET 1.7

In Exercises 1 to 24, evaluate each expression.

1. $9^{1/2}$ **2.** $49^{1/2}$ **3.** $-9^{1/2}$ **4.** $-25^{1/2}$

5. $4^{3/2}$ **6.** $16^{3/2}$ **7.** $-64^{2/3}$ **8.** $-125^{2/3}$

9. $(-64)^{2/3}$ **10.** $(-125)^{2/3}$ **11.** $16^{-1/2}$ **12.** $9^{-1/2}$

13. $27^{-2/3}$ **14.** $4^{-3/2}$ **15.** $\left(\dfrac{9}{16}\right)^{1/2}$

16. $\left(\dfrac{4}{25}\right)^{1/2}$ **17.** $\left(\dfrac{4}{25}\right)^{3/2}$ **18.** $\left(-\dfrac{1}{27}\right)^{-2/3}$

19. $10^{3/2} \cdot 10^{1/2}$ **20.** $3^{1/2} \cdot 3^{1/2}$ **21.** $7^{-1/4} \cdot 7^{5/4}$

22. $6^{5/3} \cdot 6^{-2/3}$ **23.** $\dfrac{5^{4/3}}{5^{1/3}}$ **24.** $\dfrac{11^{5/4}}{11^{-3/4}}$

In Exercises 25 to 42, simplify each expression.

25. $(x^{1/2})(x^{3/5})$ **26.** $(y^{4/3})(y^{1/4})$

27. $(8a^3)^{2/3}$ **28.** $(27b^6)^{2/3}$

29. $(81x^4y^{12})^{1/4}$ **30.** $(625a^8b^4)^{1/4}$

31. $\dfrac{a^{3/4} \cdot b^{1/2}}{a^{1/4} \cdot b^{1/5}}$ **32.** $\dfrac{x^{1/2} \cdot y^{5/6}}{x^{3/2} \cdot y^{1/6}}$

33. $a^{1/3}(a^{5/3} + 7a^{2/3})$ **34.** $m^{3/4}(m^{1/4} - 8m^{5/4})$

35. $(p^{1/2} + q^{1/2})(p^{1/2} - q^{1/2})$ **36.** $(c + d^{1/3})(c - d^{1/3})$

37. $\left(\dfrac{m^2n^4}{m^{-2}n}\right)^{1/2}$ **38.** $\left(\dfrac{r^3s^{-2}}{rs^4}\right)^{1/2}$

39. $\dfrac{(x^{n+1/2}) \cdot x^{-n}}{x^{1/2}}$

40. $\dfrac{r^{n/2} \cdot r^{2n}}{r^{-n}}$

41. $\dfrac{r^{1/n}}{r^{1/m}}$

42. $\dfrac{s^{2/n}}{s^{-n/2}}$

In Exercises 43 to 56, make use of the properties of radicals to evaluate each radical without the aid of a calculator.

43. $\sqrt{4}$ **44.** $\sqrt{36}$ **45.** $\sqrt[3]{-216}$ **46.** $\sqrt[3]{-64}$

47. $\sqrt{\dfrac{9}{16}}$ **48.** $\sqrt{\dfrac{25}{49}}$ **49.** $\sqrt[5]{32}$ **50.** $\sqrt[6]{729}$

51. $(\sqrt[4]{4})^2$ **52.** $(\sqrt[4]{25})^2$ **53.** $(\sqrt[4]{6})^4$

54. $(\sqrt[5]{14})^5$ **55.** $(\sqrt{7})^4$ **56.** $(\sqrt{11})^4$

In Exercises 57 to 62, write each exponential expression in radical form.

57. $(3x)^{1/2}$ **58.** $(6y)^{1/3}$ **59.** $5(xy)^{1/4}$

60. $2a(bc)^{1/5}$ **61.** $(5w)^{2/3}$ **62.** $(a+b)^{3/4}$

In Exercises 63 to 68, write each radical in exponential form.

63. $\sqrt[3]{17k}$ **64.** $4\sqrt{3m}$ **65.** $\sqrt[5]{a^2}$

66. $3\sqrt[4]{5n}$ **67.** $\sqrt{\dfrac{7a}{3}}$ **68.** $\sqrt[3]{\dfrac{5b^2}{7}}$

In Exercises 69 to 84, simplify each radical.

69. $\sqrt{45}$ **70.** $\sqrt{75}$ **71.** $\sqrt[3]{24}$

72. $\sqrt[3]{135}$ **73.** $\sqrt[3]{-81}$ **74.** $\sqrt[3]{-250}$

75. $-\sqrt[3]{32}$ **76.** $-\sqrt[3]{243}$ **77.** $\sqrt{3^2 \cdot 5^3}$

78. $\sqrt{2^2 \cdot 3^5}$ **79.** $\sqrt[3]{2^3 \cdot 5^4 \cdot 7}$ **80.** $\sqrt[3]{3^4 \cdot 5^5}$

81. $\sqrt{24x^3y^2}$ **82.** $\sqrt{18x^2y^5}$ **83.** $-\sqrt[3]{16a^3y^7}$

84. $-\sqrt[3]{54c^2d^5}$

In Exercises 85 to 90, simplify each radical by writing it in exponential form.

85. $\sqrt[4]{9x^2}$ **86.** $\sqrt[4]{25a^2b^2}$ **87.** $\sqrt[6]{16m^4n^2}$

88. $\sqrt[6]{27r^3}$ **89.** $\sqrt[8]{81x^6y^2}$ **90.** $\sqrt[8]{64a^2b^4}$

In Exercises 91 to 100, simplify each expression by simplifying the radicals and combining like radicals.

91. $4\sqrt{2} + 3\sqrt{2}$ **92.** $6\sqrt{5} + 2\sqrt{5}$

93. $7\sqrt{3} - \sqrt{3}$ **94.** $8\sqrt{11} - \sqrt{11}$

95. $\sqrt{8} - 5\sqrt{2}$ **96.** $\sqrt{27} + 4\sqrt{3}$

97. $2\sqrt[3]{2} - \sqrt[3]{16}$ **98.** $5\sqrt[3]{3} + 2\sqrt[3]{81}$

99. $\sqrt{8x^3y} + x\sqrt{2xy}$ **100.** $4\sqrt{a^5b} - a^2\sqrt{ab}$

In Exercises 101 to 110, find the indicated product of the radical expressions. Express each term in simplest form.

101. $(\sqrt{5} + 8)(\sqrt{5} + 3)$ **102.** $(\sqrt{7} + 4)(\sqrt{7} - 1)$

103. $(\sqrt{2x} + 3)(\sqrt{2x} - 3)$ **104.** $(7 - \sqrt{3a})(7 + \sqrt{3a})$

105. $(5\sqrt{2y} + \sqrt{3z})^2$ **106.** $(3\sqrt{5y} - 4)^2$

107. $(\sqrt{x-3} + 5)^2$ **108.** $(\sqrt{x+7} - 3)^2$

109. $(\sqrt{2x+5} + 7)^2$ **110.** $(\sqrt{9x-2} + 11)^2$

In Exercises 111 to 126, simplify each expression.

111. $\dfrac{2}{\sqrt{2}}$ **112.** $\dfrac{3x}{\sqrt{3}}$ **113.** $\sqrt{\dfrac{5}{18}}$ **114.** $\sqrt{\dfrac{7}{40}}$

115. $\dfrac{3}{\sqrt[3]{2}}$ **116.** $\dfrac{2}{\sqrt[3]{4}}$ **117.** $\dfrac{4}{\sqrt[3]{8x^2}}$ **118.** $\dfrac{2}{\sqrt[4]{4y}}$

119. $\sqrt{\dfrac{10}{18}}$ **120.** $\sqrt{\dfrac{14}{40}}$ **121.** $\sqrt{\dfrac{2x}{27y}}$ **122.** $\sqrt{\dfrac{4c}{50d}}$

123. $\dfrac{3}{\sqrt{5} + \sqrt{x}}$ **124.** $\dfrac{5}{\sqrt{y} - \sqrt{3}}$

125. $\dfrac{\sqrt{7}}{2 - \sqrt{7}}$ **126.** $\dfrac{6\sqrt{6}}{5 + \sqrt{6}}$

In Exercises 127 to 130, rationalize the numerator of each radical expression.

127. $\dfrac{\sqrt{5}}{3}$ **128.** $\dfrac{\sqrt{7}}{4}$

129. $\dfrac{3 + \sqrt{5}}{7}$ **130.** $\dfrac{2 - \sqrt{6}}{10}$

In Exercises 131 to 138, write each radical in simplest form.

131. $\sqrt{\dfrac{20a^5}{6a}}$ **132.** $\sqrt{\dfrac{12b^4}{10b}}$ **133.** $\sqrt{\dfrac{45xy^2}{10y^3}}$

134. $\sqrt{\dfrac{14xz^5}{6xz}}$ **135.** $\sqrt{\dfrac{30xy}{4xy^2}}$ **136.** $\sqrt{\dfrac{48a}{20a^4}}$

137. $\dfrac{\sqrt[3]{24xy^2}}{\sqrt[3]{2x^2y}}$ **138.** $\dfrac{\sqrt{60}}{\sqrt{14a}}$

In Exercises 139 to 148, use a calculator to evaluate each exponential expression or radical. Express your answers accurate to three significant digits.

139. $8^{1/5}$ **140.** $10^{1/10}$ **141.** $(-12)^{3/5}$

142. $(-15)^{2/3}$ **143.** $\sqrt{437}$ **144.** $\sqrt{511}$

145. $\sqrt{7.81 \times 10^4}$ **146.** $\sqrt{6.23 \times 10^{-6}}$

147. $\sqrt[3]{-251}$ **148.** $\sqrt[3]{-344}$

Supplemental Exercises

149. The percent P of light that will pass through an opaque material is given by the equation $P = 10^{-kd}$, where d is the thickness of the material in centimeters and k is a constant that depends on the material. Find the percent

(to the nearest 1 percent) of light that will pass through opaque glass for which
a. $k = 0.15$ and $d = 0.6$ centimeter.
b. $k = 0.15$ and $d = 1.2$ centimeters.

150. The number of hours h needed to cook a pot roast that weighs p pounds can be approximated by using the formula $h = 0.9(p)^{0.6}$.
a. Find the time (to the nearest hundredth of an hour) required to cook a 12-pound pot roast.
b. If pot roast A weighs twice as much as pot roast B, then pot roast A should be cooked for a period of time that is how many times longer than the time at which pot roast B is cooked?

In Exercises 151 to 154, find the value of p for which the statement is true.

151. $a^{2/5}a^p = a^2$

152. $b^{-3/4}b^{2p} = b^3$

153. $\dfrac{x^{-3/4}}{x^{3p}} = x^4$

154. $(x^4x^{2p})^{1/2} = x$

In Exercises 155 to 158, factor each expression over the set of real numbers. *Example:*

$$x^2 - 5 = x^2 - (\sqrt{5})^2 = (x + \sqrt{5})(x - \sqrt{5})$$

155. $x^2 - 7$

156. $y^2 - 11$

157. $x^2 + 6\sqrt{2}x + 18$

158. $y^2 - 8\sqrt{5}y + 80$

159. *Prove:* $\sqrt{a^2 + b^2} \neq a + b$. (*Hint:* Find a counterexample.)

160. When does $\sqrt[3]{a^3 + b^3} = a + b$? (*Hint:* Cube each side of the equation.)

161. Which step in the following demonstration is the incorrect step?

I.	$3 = 3$	
II.	$= 3^{2/2}$	• Because $1 = 2/2$
III.	$= (3^2)^{1/2}$	• Because $b^{m/n} = (b^m)^{1/n}$
IV.	$= 9^{1/2}$	• Because $3^2 = 9$
V.	$= [(-3)^2]^{1/2}$	• Substitution of $(-3)^2$ for 9
VI.	$= (-3)^{2(1/2)}$	• The power property $(b^p)^q = b^{pq}$
VII.	$= (-3)^1$	
VIII.	$= -3$	

162. Which step in the following demonstration is the incorrect step?

I.	$4 = 16^{1/2}$
II.	$= [(-4)(-4)]^{1/2}$
III.	$= (-4)^{1/2}(-4)^{1/2}$

1.8 Complex Numbers

There is no real number whose square is a negative number. For example, there is no real number x such that $x^2 = -1$. In the seventeenth century a new number, called an **imaginary number,** was defined. The square of an imaginary number is a negative real number. The letter i was chosen to represent an imaginary number whose square is -1.

Definition of i

The number i, called the **imaginary unit,** is a number such that

$$i^2 = -1$$

Many of the solutions to equations in the remainder of this text will involve radicals such as $\sqrt{-a}$, where a is a positive real number. The expression $\sqrt{-a}$, with $a > 0$, is defined as follows:

Definition of $\sqrt{-a}$

> For any positive real number a,
> $$\sqrt{-a} = i\sqrt{a}.$$

Remark This definition with $a = 1$ implies that $\sqrt{-1} = i$. It is often used to write the square root of a negative real number as the product of the imaginary unit i and a positive real number. For example,

$$\sqrt{-4} = i\sqrt{4} = 2i \quad \text{and} \quad \sqrt{-7} = i\sqrt{7}$$

Definition of a Complex Number

> If a and b are real numbers and i is the imaginary unit, then $a + bi$ is called a **complex number**. The real number a is called the **real part** and the real number b is called the **imaginary part** of the complex number.

Remark Even though b is a real number, it is called the imaginary part of the complex number $a + bi$. For example, the complex number $3 + 8i$ has the real number 8 as its imaginary part.

The real numbers are a subset of the complex numbers. This can be observed by letting $b = 0$. Then $a + bi = a + 0i = a$, which is a real number. It can be shown that the associative, commutative, distributive, and identity properties also apply to complex numbers. Any number that can be written in the form $0 + bi = bi$, where b is a nonzero real number, is an **imaginary number** (or a pure imaginary number). For example, i, $3i$, and $-0.5i$ are all imaginary numbers.

A complex number is in **standard form** when it is written in the form $a + bi$.

EXAMPLE 1 **Write Complex Numbers in Standard Form**

Write each complex number in standard form.

a. $3 + \sqrt{-4}$ b. $\sqrt{-37} - 3$

Solution Use the definition $\sqrt{-a} = i\sqrt{a}$.

a. $3 + \sqrt{-4} = 3 + i\sqrt{4} = 3 + 2i$ • $a + bi$ form with $a = 3$ and $b = 2$.

b. $\sqrt{-37} - 3 = i\sqrt{37} - 3 = -3 + i\sqrt{37}$

■ *Try Exercise* **2,** *page 61.*

Remark   The expression $\sqrt{a}i$ is often written as $i\sqrt{a}$ so that it is not mistaken for $\sqrt{ai}$.

Definition of Addition and Subtraction of Complex Numbers

If $a + bi$ and $c + di$ are complex numbers, then

$$\text{Addition} \quad (a + bi) + (c + di) = (a + c) + (b + d)i$$

$$\text{Subtraction} \quad (a + bi) - (c + di) = (a - c) + (b - d)i$$

To add two complex numbers, add their real parts to produce the real part of the sum and add their imaginary parts to produce the imaginary part of the sum.

EXAMPLE 2   **Add or Subtract Complex Numbers**

Perform the indicated operation.

a. $(4 + 2i) + (3 + 7i)$ b. $i - (3 - 4i)$

Solution

a. $(4 + 2i) + (3 + 7i) = (4 + 3) + (2 + 7)i = 7 + 9i$

b. $i - (3 - 4i) = (0 + 1i) - (3 - 4i)$ • $0 = 0 + 1i$

$\quad\quad = (0 - 3) + [1 - (-4)]i = -3 + 5i$

————————————————————————————— ■ *Try Exercise* **16**, *page 61.*

Definition of Multiplication of Complex Numbers

If $a + bi$ and $c + di$ are complex numbers, then

$$(a + bi)(c + di) = (ac - bd) + (ad + bc)i$$

Because every complex number can be written as a sum of two terms, it is natural to perform multiplication on complex numbers in a manner consistent with the operation of multiplication defined on binomials and the definition $i^2 = -1$. Thus to multiply complex numbers, it is not necessary to memorize the definition of multiplication.

EXAMPLE 3   **Multiply Complex Numbers**

Simplify: $(3 + 5i)(2 - 4i)$

Solution

$$(3 + 5i)(2 - 4i) = 6 - 12i + 10i - 20i^2$$

$$= 6 - 12i + 10i - 20(-1) \quad \bullet \text{ Substitute } (-1) \text{ for } i^2.$$

$$= 6 - 12i + 10i + 20 \quad \bullet \text{ Simplify.}$$

$$= 26 - 2i$$

■ *Try Exercise 24, page 61.*

The complex numbers $a + bi$ and $a - bi$ are called **complex conjugates** or **conjugates** of each other. The conjugate of the complex number z is denoted by $\bar{z}$. For example,

$$\overline{3 + 2i} = 3 - 2i \quad \text{and} \quad \overline{7 - 11i} = 7 + 11i$$

The following are some theorems about complex numbers and their conjugates.

Conjugate Theorems

If z and w are complex numbers, with conjugates $\bar{z}$ and $\bar{w}$, then

1. $z + \bar{z}$ is a real number.
2. $z \cdot \bar{z}$ is a real number.
3. $\bar{z} = z$ if and only if z is a real number.
4. $\overline{z + w} = \bar{z} + \bar{w}$
5. $\overline{z \cdot w} = \bar{z} \cdot \bar{w}$
6. $\overline{\overline{z}} = z$
7. $\overline{z^n} = (\bar{z})^n$ for all natural numbers n.

The second theorem states that the product of a complex number and its conjugate is always a real number. To verify this theorem, we make use of the complex number $z = a + bi$ and its conjugate $\bar{z} = a - bi$.

Prove that $z \cdot \bar{z}$ is a real number.

Proof: Let $z = a + bi$, and $\bar{z} = a - bi$.

$$z \cdot \bar{z} = (a + bi)(a - bi) = a^2 - abi + abi - b^2 i^2$$

$$= a^2 - b^2(-1)$$

$$= a^2 + b^2, \text{ a real number}$$

The fact that the product of a complex number and its conjugate is always a real number can be used to find the quotient of two complex numbers. For example, to find the quotient $(a + bi)/(c + di)$, multiply the numerator and denominator by the conjugate of the denominator.

EXAMPLE 4   **Divide Complex Numbers**

Find the quotient $\dfrac{3 + 2i}{5 - i}$

Solution

$$\frac{3 + 2i}{5 - i} = \frac{(3 + 2i)(5 + i)}{(5 - i)(5 + i)}$$
   • Multiply numerator and denominator by $5 + i$, which is the conjugate of the denominator.

$$= \frac{15 + 3i + 10i + 2i^2}{25 + 1}$$

$$= \frac{13 + 13i}{26} = \frac{1}{2} + \frac{1}{2}i$$ • Write in standard form.

■ *Try Exercise* **34**, *page 61.*

The following powers of i illustrate a pattern:

$$i^1 = i \qquad\qquad\qquad i^5 = i^4 \cdot i = (1)i = i$$

$$i^2 = -1 \qquad\qquad\qquad i^6 = i^4 \cdot i^2 = (1)(-1) = -1$$

$$i^3 = i^2 \cdot i = (-1)i = -i \qquad i^7 = i^4 \cdot i^3 = (1)(-i) = -i$$

$$i^4 = i^2 \cdot i^2 = (-1)(-1) = 1 \qquad i^8 = (i^4)^2 = 1^2 = 1$$

Because $i^4 = 1$, $(i^4)^n = 1$ for any integer n. Thus it is possible to evaluate powers of i by factoring out powers of i^4, as shown in the following example:

$$i^{25} = (i^4)^6(i) = 1^6(i) = i$$

The following theorem can be used to evaluate powers of i. Essentially it makes use of division to eliminate powers of i^4.

Powers of i

If n is a positive integer, then $i^n = i^r$, where r is the remainder of the division of n by 4.

EXAMPLE 5   **Evaluate Powers of i**

Evaluate: i^{543}

Solution   Use the theorem on powers of i.

$$i^{543} = i^3 = -i$$ • Remainder of $543 \div 4$ is 3.

■ *Try Exercise* **54**, *page 61.*

Caution To compute $\sqrt{a}\,\sqrt{b}$ when both a and b are negative numbers, write each radical in terms of i before multiplying. For example,

Correct method $\sqrt{-1}\,\sqrt{-1} = i \cdot i = i^2 = -1$

Incorrect method $\sqrt{-1}\,\sqrt{-1} = \sqrt{(-1)(-1)} = \sqrt{1} = 1$

EXAMPLE 6 Simplify Products Involving Radicals with Negative Radicands

Simplify each of the following:

a. $\sqrt{-16}\,\sqrt{-25}$ b. $\sqrt{-9}\,\sqrt{-7}$ c. $(2 + \sqrt{-5})(2 - \sqrt{-5})$

Solution

a. $\sqrt{-16}\,\sqrt{-25} = (4i)(5i) = 20i^2 = -20$

b. $\sqrt{-9}\,\sqrt{-7} = (3i)(i\sqrt{7}) = 3i^2\sqrt{7} = -3\sqrt{7}$

c. $(2 + \sqrt{-5})(2 - \sqrt{-5}) = (2 + i\sqrt{5})(2 - i\sqrt{5}) = 4 + 5 = 9$

■ *Try Exercise* **68,** *page 61.*

EXERCISE SET 1.8

In Exercises 1 to 10, write the complex number in standard form.

1. $2 + \sqrt{-9}$ **2.** $3 + \sqrt{-25}$

3. $4 - \sqrt{-121}$ **4.** $5 - \sqrt{-144}$

5. $8 + \sqrt{-3}$ **6.** $9 - \sqrt{-75}$

7. $\sqrt{-16} + 7$ **8.** $\sqrt{-49} + 3$

9. $\sqrt{-81}$ **10.** $-\sqrt{-100}$

In Exercises 11 to 28, simplify and write the complex number in standard form.

11. $(2 + 5i) + (3 + 7i)$ **12.** $(1 - 3i) + (6 + 2i)$

13. $(-5 - i) + (9 - 2i)$ **14.** $5 + (3 - 2i)$

15. $(8 - 6i) - (10 - i)$ **16.** $(-3 + i) - (-8 + 2i)$

17. $(7 - 3i) - (-5 - i)$ **18.** $7 - (3 - 2i)$

19. $8i - (2 - 3i)$ **20.** $(4i - 5) - 2$

21. $3(2 + 7i) + 5(2 - i)$ **22.** $8(4 - i) - (4 - 3i)$

23. $(2 + 3i)(4 - 5i)$ **24.** $(5 - 3i)(-2 - 4i)$

25. $(5 + 7i)(5 - 7i)$ **26.** $(-3 - 5i)(-3 + 5i)$

27. $(8i + 11)(-7 + 5i)$ **28.** $(9 - 12i)(15i + 7)$

In Exercises 29 to 48, write each expression as a complex number in standard form.

29. $\dfrac{4 + i}{3 + 5i}$ **30.** $\dfrac{5 - i}{4 + 5i}$ **31.** $\dfrac{1}{7 - 3i}$ **32.** $\dfrac{1}{-8 + i}$

33. $\dfrac{3 + 2i}{3 - 2i}$ **34.** $\dfrac{5 - 7i}{5 + 7i}$ **35.** $\dfrac{2i}{11 + i}$ **36.** $\dfrac{3i}{5 - 2i}$

37. $\dfrac{6 + i}{i}$ **38.** $\dfrac{5 - i}{-i}$

39. $(3 - 5i)^2$ **40.** $(-5 + 7i)^2$

41. $(1 - i) - 2(4 + i)^2$ **42.** $(4 - i) - 5(2 + 3i)^2$

43. $(1 - i)^3$ **44.** $(2 + i)^3$

45. $(2i)(8i)$ **46.** $(-5)(7i)$

47. $(5i)^2(-3i)$ **48.** $(-6i)(-5i)^2$

In Exercises 49 to 64, simplify and write the complex number as i, $-i$, 1, or -1.

49. i^3 **50.** $-i^3$ **51.** i^5 **52.** $-i^5$

53. i^{10} **54.** i^{28} **55.** $-i^{40}$ **56.** i^{40}

57. i^{223} **58.** i^{553} **59.** i^{2001} **60.** i^{5000}

61. i^{5042} **62.** i^0 **63.** i^{-1} **64.** $i^{10,000}$

In Exercises 65 to 72, simplify each product.

65. $\sqrt{-1}\,\sqrt{-4}$ **66.** $\sqrt{-16}\,\sqrt{-49}$

67. $\sqrt{-64}\,\sqrt{-5}$ **68.** $\sqrt{-3}\,\sqrt{-121}$

69. $(3 + \sqrt{-2})(3 - \sqrt{-2})$ **70.** $(4 + \sqrt{-81})(4 - \sqrt{-81})$

71. $(5 + \sqrt{-16})^2$ **72.** $(3 - \sqrt{-144})^2$

In Exercises 73 to 80, evaluate

$$\frac{-b \pm \sqrt{b^2 - 4ac}}{2a}$$

for the given values of a, b, and c. Write your final answer as a complex number in standard form.

73. $a = 3, b = -3, c = 3$ **74.** $a = 1, b = -3, c = 10$

75. $a = 2, b = 4, c = 4$ **76.** $a = 4, b = -4, c = 2$

77. $a = 2, b = 6, c = 6$ **78.** $a = 6, b = -5, c = 5$

79. $a = 2, b = 1, c = 3$ **80.** $a = 3, b = 2, c = 4$

The **absolute value of the complex number** $a + bi$ is denoted by $|a + bi|$ and defined as the real number $\sqrt{a^2 + b^2}$. In Exercises 81 to 88, find the indicated absolute value of each complex number.

81. $|3 + 4i|$ **82.** $|5 + 12i|$ **83.** $|2 - 5i|$ **84.** $|4 - 4i|$

85. $|7 - 4i|$ **86.** $|11 - 2i|$ **87.** $|-3i|$ **88.** $|18i|$

Supplemental Exercises

In Exercises 89 to 92, use the complex number $z = a + bi$ and its conjugate $\bar{z} = a - bi$ to establish each result.

89. Prove that the absolute value of a complex number and the absolute value of its conjugate are equal.

90. Prove that the difference of a complex number and its conjugate is a pure imaginary number.

91. Prove that the conjugate of the sum of two complex numbers equals the sum of the conjugates of the two numbers.

92. Prove that the conjugate of the product of two complex numbers equals the product of the conjugates of the two numbers.

93. Show that if $x = 1 + i\sqrt{3}$, then $x^2 - 2x + 4 = 0$.

94. Show that if $x = 1 - i\sqrt{3}$, then $x^2 - 2x + 4 = 0$.

95. A set T is closed under the operation of addition if the sum of any two elements of T is also an element of T. Is $T = \{1, -1, i, -i\}$ closed under the operation of addition?

96. A set T is closed under the operation of multiplication if the product of any two elements of T is also an element of T. Is $T = \{1, -1, i, -i\}$ closed under the operation of multiplication?

97. Simplify

$$[(3 + \sqrt{5}) + (7 - \sqrt{3})i][(3 + \sqrt{5}) - (7 - \sqrt{3})i]$$

98. Simplify $[2 - (3 - \sqrt{5})i][2 + (3 - \sqrt{5})i]$.

99. Simplify $\left(\dfrac{-1}{2} + \dfrac{\sqrt{3}}{2}i\right)^3$.

100. Simplify $(a + bi)^3$, where a and b are real numbers.

101. Simplify $i + i^2 + i^3 + i^4 + \cdots + i^{28}$.

102. Simplify $i + i^2 + i^3 + i^4 + \cdots + i^{100}$.

The product $(a + bi)(a - bi) = a^2 + b^2$ can be used to factor the sum of two squares over the set of complex numbers. *Example:*

$$x^2 + 25 = x^2 + 5^2 = (x + 5i)(x - 5i)$$

In Exercises 103 to 106, factor each polynomial over the set of complex numbers.

103. $x^2 + 9$ **104.** $y^2 + 121$

105. $4x^2 + 81$ **106.** $144y^2 + 625$

In Exercises 107 to 110, evaluate the polynomial for the given value of x.

107. $x^2 + 36; \quad x = 6i$ **108.** $x^2 + 100; \quad x = -10i$

109. $x^2 - 6x + 10; \quad x = 3 + i$

110. $x^2 + 10x + 29; \quad x = -5 + 2i$

Exploring Concepts with Technology

The Mandelbrot Replacement Procedure

The following procedure is called the **Mandelbrot replacement procedure.**

Pick a complex number s.

1. Square s and add the result to s.
2. Square the last result and add it to s.
3. Repeat step 2.

The number s is referred to as the seed of the procedure. The number s is a seed in the sense that each seed produces a different sequence of numbers. Some seeds produce sequences that grow without bound. Some seeds produce sequences that grow toward some constant. Still other seeds yield sequences that are cyclic. Consider the following illustrations.

■ Let the seed $s = 1$.

$$1^2 + 1 = 2, \qquad 2^2 + 1 = 5, \qquad 5^2 + 1 = 26, \qquad 26^2 + 1 = 677$$

As the replacement procedure continues, we get larger and larger numbers.

■ Let the seed $s = -1$.

$$(-1)^2 + (-1) = 0, \qquad 0^2 + (-1) = -1, \qquad (-1)^2 + (-1) = 0, \ldots$$

As the replacement procedure continues, the results *cycle:* $0, -1, 0, -1, 0, \ldots$.

■ Let the seed $s = 0.25$.

$$(0.25)^2 + 0.25 = 0.3125, \qquad (0.3125)^2 + 0.25 = 0.34765625,$$

$$(0.34765625)^2 + 0.25 \approx 0.3708648682, \ldots$$

1. Use your calculator to continue the Mandelbrot replacement procedure. What number do you have after **a.** 25 applications of step 2? **b.** 50 applications of step 2? **c.** 75 applications of step 2? **d.** What constant do you think the sequence of numbers is approaching?

■ Let the seed $s = i$.

$$i^2 + i = -1 + i, \qquad (-1 + i)^2 + i = -i, \ldots$$

2. **a.** What is the next number produced by the Mandelbrot replacement procedure?
    b. What happens as the procedure is continued?

The Mandelbrot replacement procedure can be used to determine special kinds of numbers called **attractors**. The attractors produced by the Mandelbrot replacement procedure are an essential part of the Mandelbrot set, which is the basis for fractal geometry. Figure 1.15 represents a computer investigation into the Mandelbrot Set.

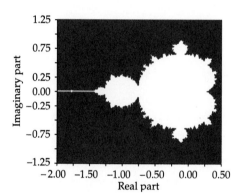

Figure 1.15

Chapter Review

1.1   The Real Number System

- The following sets of numbers are used extensively in the study of algebra:

Integers	$\{\ldots, -3, -2, -1, 0, 1, 2, 3, \ldots\}$
Rational numbers	{all terminating or repeating decimals}
Irrational numbers	{all nonterminating, nonrepeating decimals}
Real numbers	{all rational or irrational numbers}

1.2   Intervals, Absolute Value, and Distance

- The absolute value of the real number a is defined by

$$|a| = \begin{cases} a & \text{if } a \geq 0 \\ -a & \text{if } a < 0 \end{cases}$$

- For any real numbers a and b, the distance between the graph of a and the graph of b is denoted by $d(a, b)$, where $d(a, b) = |a - b|$.

1.3   Integer Exponents

- If b is any real number and n is any natural number, then

$$b^n = \underbrace{b \cdot b \cdot b \cdots \cdots b}_{n \text{ factors of } b}$$

- For any nonzero base b, $b^0 = 1$. If $b \neq 0$ and n is any natural number, then $b^{-n} = 1/b^n$ and $1/b^{-n} = b^n$.

- A number is written in scientific notation if it is of the form $a \cdot 10^n$, where n is an integer and a is a decimal such that $1 \leq a < 10$.

1.4   Polynomials

- Special product formulas are as follows:

Special Forms	Formulas
(Sum) (difference)	$(x + y)(x - y) = x^2 - y^2$
(Binomial)^2	$(x + y)^2 = x^2 + 2xy + y^2$
	$(x - y)^2 = x^2 - 2xy + y^2$

1.5   Factoring

- Factoring formulas are as follows:

Difference of two squares	$x^2 - y^2 = (x + y)(x - y)$
Perfect square trinomials	$x^2 + 2xy + y^2 = (x + y)^2$
	$x^2 - 2xy + y^2 = (x - y)^2$
Sum of cubes	$x^3 + y^3 = (x + y)$ $\cdot (x^2 - xy + y^2)$
Difference of cubes	$x^3 - y^3 = (x - y)$ $\cdot (x^2 + xy + y^2)$

- To factor a polynomial, use the general factoring strategy.

1.6   Rational Expressions

- A rational expression is a fraction in which the numerator and denominator are polynomials. The properties of rational expressions are used to simplify a rational expression

and to find the sum, difference, product, and quotient of two rational expressions.

- Complex fractions can be simplified by one of the following:
  Method 1: Multiply both the numerator and the denominator by the LCD of all the fractions in the complex fraction.

  Method 2: Simplify the numerator to a single fraction and the denominator to a single fraction. Multiply the numerator by the reciprocal of the denominator.

1.7   Rational Exponents and Radicals

- If n is an even positive integer and $b \geq 0$, then $b^{1/n}$ is the nonnegative real number such that $(b^{1/n})^n = b$. If n is an odd positive integer, then $b^{1/n}$ is the real number such that $(b^{1/n})^n = b$.

- For all integers m and n such that m/n is reduced to lowest terms, and for all real numbers b for which $b^{1/n}$ is a real number, $b^{m/n} = (b^{1/n})^m = (b^m)^{1/n}$.

- If p and q represent rational numbers, and a and b are positive real numbers, then

$$b^p \cdot b^q = b^{p+q} \qquad \frac{b^p}{b^q} = b^{p-q} \qquad (b^p)^q = b^{pq}$$

$$(ab)^p = a^p b^p \qquad \left(\frac{a}{b}\right)^p = \frac{a^p}{b^p} \qquad b^{-p} = \frac{1}{b^p}$$

- If n is a positive integer and b is a real number such that $b^{1/n}$ is a real number, then $\sqrt[n]{b} = b^{1/n}$.

- If m and n are natural numbers greater than or equal to 2, and a and b are nonnegative real numbers, then

$$\sqrt[n]{a} \cdot \sqrt[n]{b} = \sqrt[n]{ab}$$

$$\frac{\sqrt[n]{a}}{\sqrt[n]{b}} = \sqrt[n]{\frac{a}{b}} \quad (b \neq 0)$$

$$\sqrt[m]{\sqrt[n]{b}} = \sqrt[mn]{b} \qquad (\sqrt[n]{b})^n = b \qquad \sqrt[n]{b^n} = b$$

1.8   Complex Numbers

- The number i, called the imaginary unit, is the number such that $i^2 = -1$.

- If a and b are real numbers and i is the imaginary unit, then $a + bi$ is a complex number. The complex numbers $a + bi$ and $a - bi$ are complex conjugates or conjugates of each other.

Essays and Projects

1. All odd numbers can be written in the form $2k - 1$, for some integer k. Explain, by writing a paragraph and giving the supporting mathematics, why the product of two odd numbers must be an odd number.

2. Write a paragraph that explains why $ax^2 + bx + c$ is non-factorable over the integers if a, b, and c are all odd.

3. In Euclid's proof that the set of prime numbers has an infinite number of elements (See Exercise 69, page 9), Euclid made use of a number of the form

$$T = 2 \cdot 3 \cdot 5 \cdot 7 \cdot 11 \cdot \cdots \cdot L + 1$$

where each of the numbers $2, 3, 5, 7, \ldots L$, is a prime number. Some people have conjectured that every number of this type would be a prime number. For example,

$2 + 1 = 3$, a prime number

$2 \cdot 3 + 1 = 7$, a prime number

$2 \cdot 3 \cdot 5 + 1 = 31$, a prime number

$2 \cdot 3 \cdot 5 \cdot 7 + 1 = 211$, a prime number

In a few sentences, explain your position on this conjecture.

4. The **Fundamental Theorem of Arithmetic** states that each composite number can be expressed as a product of prime numbers in exactly one way (disregarding the order of the factors). By using the fundamental theorem of arithmetic and **indirect reasoning,** we can show that $\sqrt{2}$ is not a rational number. The proof is as follows.

Assume that there is a rational number a/b such that $\sqrt{2} = a/b$. Then we can show that

$$\sqrt{2} = \frac{a}{b}$$

$$2 = \frac{a^2}{b^2}$$

$$2b^2 = a^2$$

By the fundamental theorem of arithmetic, the positive integers $2b^2$ and a^2 must have the same prime factorization. Squares have prime factors that occur in pairs, so

we know that a^2 has an even number of prime factors. Using the same reasoning, we know that b^2 has an even number of prime factors. Because 2 is a prime number, $2b^2$ must have an odd number of prime factors. Now we have a contradiction, because by the fundamental theorem of arithmetic, it is impossible for a number to have an odd number of prime factors and an even number of prime factors. Thus we conclude that $\sqrt{2}$ is not a rational number.

Write a similar proof showing that for every prime p, $\sqrt{p}$ is not a rational number.

5. **Prime triples** are consecutive prime numbers that differ by 2. For example, 3, 5, and 7 are prime triples. Write a paragraph that explains why there are no other prime triples.

6. Recall that if n is a positive integer, then $n!$ means

$$n(n - 1)(n - 2) \cdots 2 \cdot 1$$

Write a sentence that explains why none of the following consecutive integers is a prime number.

$$5! + 2, \quad 5! + 3, \quad 5! + 4, \quad 5! + 5$$

How many numbers are in the following list of consecutive integers? How many numbers in the following list are prime numbers?

$$k! + 2, \quad k! + 3, \quad k! + 4, \cdots k! + (k - 1), \quad k! + k$$

Write a few sentences that explain why this exercise is so remarkable. *Hint:* Compare the results of this exercise with the information in Exercise 69 on page 9.

7. Prime numbers of the form $2^p - 1$, where p is a prime number, are called *Mersenne primes*. Write an essay on Mersenne primes. Include some of the history of Mersenne primes and the role they play in mathematics today.

8. Write an essay that explains how Cantor (see page 1) was able to prove that the set of irrational numbers has a cardinality that is larger than the set of rational numbers. One source of information is *From Zero to Infinity,* by Constance Reid, Thomas Y. Crowell Company, New York, 1964.

True/False Exercises

In Exercises 1 to 10, answer true or false. If the statement is false, give an example to show that the statement is false.

1. If a and b are real numbers, then $|a - b| = |b - a|$.

2. If a is a real number, then $a^2 \geq a$.

3. The set of rational numbers is closed under the operation of addition.

4. The set of irrational numbers is closed under the operation of addition.

5. Let $x \oplus y$ denote the average of the two real numbers x and y. That is,

$$x \oplus y = \frac{x + y}{2}$$

The operation $\oplus$ is an associative operation because

$(x \oplus y) \oplus z = x \oplus (y \oplus z)$ for all real numbers x, y, and z.

6. Using interval notation, we write the inequality $x > a$ as $[a, \infty)$.

7. If n is a real number, then $\sqrt{n^2} = n$.

8. Every real number is a complex number.

9. The sum of a complex number z and its conjugate $\bar{z}$ is a real number.

10. The product of a complex number z and its conjugate $\bar{z}$ is a real number.

Chapter Review Exercises

In Exercises 1 to 4, classify each number as one or more of the following: integer, rational number, irrational number, real number, prime number, composite number.

1. 3   2. $\sqrt{7}$ 3. $-\dfrac{1}{2}$ 4. $0.\overline{5}$

In Exercises 5 and 6, use $A = \{1, 5, 7\}$ and $B = \{2, 3, 5, 11\}$ to find the indicated intersection or union.

5. $A \cup B$ 6. $A \cap B$

In Exercises 7 to 14, identify the real number property or the property of equality that is illustrated.

7. $5(x + 3) = 5x + 15$

8. $a(3 + b) = a(b + 3)$

9. $(6c)d = 6(cd)$

10. $\sqrt{2} + 3$ is a real number.

11. $7 + 0 = 7$

12. $1x = x$

13. If $7 = x$, then $x = 7$.

14. If $3x + 4 = y$, and $y = 5z$, then $3x + 4 = 5z$.

In Exercises 15 and 16, graph each inequality and write the inequality using interval notation.

15. $-4 < x \leq 2$ 16. $x \leq -1$ or $x > 3$

In Exercises 17 and 18, graph each interval and write each interval as an inequality.

17. $[-3, 2)$ 18. $(-1, \infty)$

In Exercises 19 to 22, write each real number without absolute value symbols.

19. $|7|$ 20. $|2 - \pi|$

21. $|4 - \pi|$ 22. $|-11|$

In Exercises 23 and 24, find the distance on the real number line between the points whose coordinates are given.

23. -3, 14   24. $\sqrt{5}$, $-\sqrt{2}$

In Exercises 25 and 26, evaluate each expression.

25. $-5^2 + (-11)$ 26. $\dfrac{(2^2 \cdot 3^{-2})^2}{3^{-1} \cdot 2^3}$

In Exercises 27 and 28, simplify each expression.

27. $(3x^2y)(2x^3y)^2$ 28. $\left(\dfrac{2a^2b^3c^{-2}}{3ab^{-1}}\right)^2$

In Exercises 29 and 30, write each number in scientific notation.

29. 620,000   30. 0.0000017

In Exercises 31 and 32, change each number from scientific notation to decimal form.

31. 3.5×10^4 32. 4.31×10^{-7}

In Exercises 33 to 36, perform the indicated operation and express each result as a polynomial in standard form.

33. $(2a^2 + 3a - 7) + (-3a^2 - 5a + 6)$

34. $(5b^2 - 11) - (3b^2 - 8b - 3)$

35. $(2x^2 + 3x - 5)(3x^2 - 2x + 4)$

36. $(3y - 5)^3$

In Exercises 37 to 40, completely factor each polynomial.

37. $3x^2 + 30x + 75$

38. $25x^2 - 30xy + 9y^2$

39. $20a^2 - 4b^2$

40. $16a^3 + 250$

In Exercises 41 and 42, simplify each rational expression.

41. $\dfrac{6x^2 - 19x + 10}{2x^2 + 3x - 20}$

42. $\dfrac{4x^3 - 25x}{8x^4 + 125x}$

In Exercises 43 to 46, perform the indicated operation and simplify if possible.

43. $\dfrac{10x^2 + 13x - 3}{6x^2 - 13x - 5} \cdot \dfrac{6x^2 + 5x + 1}{10x^2 + 3x - 1}$

44. $\dfrac{15x^2 + 11x - 12}{25x^2 - 9} \div \dfrac{3x^2 + 13x + 12}{10x^2 + 11x + 3}$

45. $\dfrac{x}{x^2 - 9} + \dfrac{2x}{x^2 + x - 12}$

46. $\dfrac{3x}{x^2 + 7x + 12} - \dfrac{x}{2x^2 + 5x - 3}$

In Exercises 47 and 48, simplify each complex fraction.

47. $\dfrac{2 + \dfrac{1}{x - 5}}{3 - \dfrac{2}{x - 5}}$

48. $\dfrac{1}{2 + \dfrac{3}{1 + \dfrac{4}{x}}}$

In Exercises 49 and 50, evaluate each exponential expression.

49. $25^{1/2}$

50. $-27^{2/3}$

In Exercises 51 to 54, simplify each expression.

51. $x^{2/3} \cdot x^{3/4}$

52. $\left(\dfrac{8x^{5/4}}{x^{1/2}}\right)^{2/3}$

53. $\left(\dfrac{x^2 y}{x^{1/2} y^{-3}}\right)^{1/2}$

54. $(x^{1/2} - y^{1/2})(x^{1/2} + y^{1/2})$

In Exercises 55 to 64, simplify each radical expression. Assume the variables are positive real numbers.

55. $\sqrt{48a^2 b^7}$

56. $\sqrt{12a^3 b}$

57. $\sqrt{72x^2 y}$

58. $\sqrt{18x^3 y^5}$

59. $\sqrt{\dfrac{54xy^3}{10x}}$

60. $-\sqrt{\dfrac{24xyz^3}{15z^6}}$

61. $\dfrac{7x}{\sqrt[3]{2x^2}}$

62. $\dfrac{5y}{\sqrt[3]{9y}}$

63. $\sqrt[3]{-135x^2 y^7}$

64. $\sqrt[3]{-250xy^6}$

In Exercises 65 and 66, write the complex number in standard form and give its conjugate.

65. $3 - \sqrt{-64}$

66. $\sqrt{-4} + 6$

In Exercises 67 to 70, simplify and write the complex number in standard form.

67. $(3 + 7i) + (2 - 5i)$

68. $(6 - 8i) - (9 - 11i)$

69. $(5 + 3i)(2 - 5i)$

70. $\dfrac{4 + i}{7 - 2i}$

In Exercises 71 to 74, simplify and write each complex number as i, $-i$, 1, or -1.

71. i^{20}

72. i^{57}

73. $\dfrac{1}{i^{28}}$

74. i^{-200}

Chapter Test

1. For real numbers a, b, and c, identify the property that is illustrated by $(a + b)c = ac + bc$.

2. Given $A = \{0, 2, 4, 6, 8\}$ and $B = \{1, 3, 5, 7, 9\}$, find $A \cup B$.

3. Write $|-3| - |-6|$ without absolute value symbols.

4. Find the distance between the points -12 and -5 on the number line.

5. Simplify: $(-2x^0 y^{-2})^2 (-3x^2 y^{-1})^{-2}$

6. Simplify: $\dfrac{(2a^{-1} bc^{-2})^2}{(3^{-1} b)(2^{-1} ac^{-2})^3}$

7. Write 0.00137 in scientific notation.

8. Simplify: $(x - 2y)(x^2 - 2x + y)$

9. Evaluate the polynomial $3y^3 - 2y^2 - y + 2$ for $y = -3$.

10. Factor: $7x^2 + 34x - 5$

11. Factor: $3ax - 12bx - 2a + 8b$

12. Factor: $16x^4 - 2xy^3$

13. Factor: $x^4 + x^3 y - x - y$

14. Simplify: $\dfrac{x^4 - 2x^3 - x + 2}{x^3 - x^2 - x + 1}$

15. Simplify: $\dfrac{x}{x^2 + x - 6} - \dfrac{2}{x^2 - 5x + 6}$

16. Simplify: $\dfrac{2x^2 + 3x - 2}{x^2 - 3x} \div \dfrac{2x^2 - 7x + 3}{x^3 - 3x^2}$

17. Simplify: $\dfrac{3}{a + b} \cdot \dfrac{a^2 - b^2}{2a - b} - \dfrac{5}{a}$

18. Simplify: $x - \dfrac{x}{x + \dfrac{1}{2}}$

19. Simplify: $\dfrac{x^{1/3}y^{-3/4}}{x^{-1/2}y^{3/2}}$

20. Simplify: $3x\sqrt[3]{81xy^4} - 2y\sqrt[3]{3x^4y}$

21. Simplify: $\dfrac{x}{\sqrt[4]{2x^3}}$

22. Simplify: $\dfrac{3}{\sqrt{x} + 2}$

23. Simplify $3(2 - 3i) - 4(1 - 5i)$ and write the complex number in standard form.

24. Simplify $(2 - 3i)(4 + i)$ and write the complex number in standard form.

25. Write $\dfrac{2i}{-2 + 3i}$ as a complex number in standard form.

Equations and Inequalities

CASE IN POINT *The Mystery of the Dinosaurs*

Part of the rim of a crater has recently been discovered in the northern Yucatán peninsula of Mexico. This crater may have been created by an asteroid that struck the earth about 65 million years ago. Many scientists think the impact of this asteroid released a huge amount of iridium-laden dust into the atmosphere. They speculate the dust was so thick that it shut out most of the sunlight and resulted in a "global winter" that caused the extinction of the dinosaurs and other prehistoric life forms.

We can use the equation-solving techniques in this chapter, along with the following theorems, to determine the diameter of the crater.

Theorem A: If two chords intersect within a circle, then the product of the lengths of the segments of one chord is equal to the product of the lengths of the segments of the other chord.

Theorem B: The perpendicular bisector of a chord of a circle passes through the center of the circle.

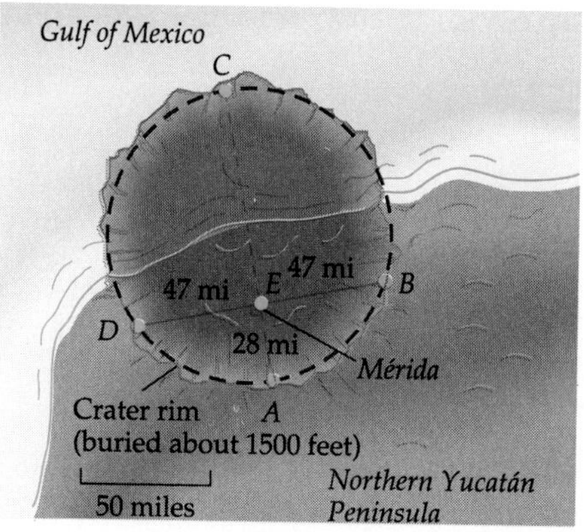

Applying theorem A to the chords shown in Figure 2.1 gives us

$$[d(B, E)] [d(E, D)] = [d(A, E)] [d(E, C)]$$

Substituting $d(B, E) = d(E, D) = 47$ miles and $d(A, E) = 28$ miles produces

$$47 \cdot 47 = 28[d(E, C)]$$

$$\frac{47 \cdot 47}{28} = d(E, C)$$

$$79 \approx d(E, C)$$

Since $\overline{AC}$ is a perpendicular bisector of chord $\overline{BD}$, we know by theorem B that $\overline{AC}$ passes through the center of the circle. Therefore, the length of the diameter of the crater is

$$d(A, C) = [d(A, E)] + [d(E, C)] \approx 28 + 79 = 107 \text{ miles}$$

Gulf of Mexico

47 mi 47 mi

28 mi

Mérida

Crater rim
(buried about 1500 feet)

Northern Yucatán Peninsula

50 miles

Figure 2.1

2.1 Linear Equations

An equation is a statement about the equality of two expressions. If either of the expressions contains a variable, the equation may be a true statement for some values of the variable and a false statement for other values of the variable. For example, $3x + 2 = 14$ is a true statement when x is 4, but it is false for any number except 4.

A number is said to **satisfy** an equation if substituting the number for the variable produces an equation that is a true statement. To **solve** an equation means to find all values of the variable that satisfy the equation. The values that make the equation true are called **solutions** or **roots** of the equation. For example, 5 is a solution or root of $2x - 10 = 0$ because $2(5) - 10 = 0$ is a true statement. The **solution set** of an equation is the set of all solutions of the equation. The solution set of $x^2 - 5x + 6 = 0$ is $\{2, 3\}$ because 2 and 3 are the only numbers that satisfy the equation.

Equivalent equations have the same solution set. The process of solving an equation is generally accomplished by producing *simpler* but equivalent equations until the solutions are easy to observe. To produce these simpler equivalent equations, we often apply the following properties.

Addition and Multiplication Properties of Equality

For real numbers a, b, and c,

1. $a = b$ and $a + c = b + c$ are equivalent equations.
2. If $c \neq 0$, then $a = b$ and $ac = bc$ are equivalent equations.

Essentially, these properties state that an equivalent equation is produced by either adding the same expression to each side of an equation or multiplying each side of an equation by the same nonzero expression. Because subtraction is defined in terms of addition, equivalent equations are produced when the same expression is subtracted from each side of an equation. Similarly, because division is defined in terms of multiplication, equivalent expressions are produced by dividing each side of an equation by the same nonzero expression.

Definition of a Linear Equation

A **linear equation** in the single variable x is an equation that can be written in the form

$$ax + b = 0$$

where a and b are real numbers, with $a \neq 0$.

The addition and multiplication properties of equality can be used to solve a linear equation.

EXAMPLE 1   **Solve a Linear Equation**

Solve $\frac{3}{4}x - 6 = 0$.

Solution
$$\frac{3}{4}x - 6 = 0$$

$$\frac{3}{4}x - 6 + 6 = 0 + 6 \qquad \bullet \text{ Add 6 to each side.}$$

$$\frac{3}{4}x = 6$$

$$\left(\frac{4}{3}\right)\left(\frac{3}{4}x\right) = \left(\frac{4}{3}\right)(6) \qquad \bullet \text{ Multiply each side} \atop \text{by } \frac{4}{3}.$$

$$x = 8$$

Check by substituting 8 for x in the original equation.

$$\frac{3}{4}x - 6 = 0$$

$$\frac{3}{4}(8) - 6 \stackrel{?}{=} 0$$

$$0 = 0 \quad \text{True}$$

Thus 8 satisfies the original equation. The solution set is $\{8\}$.

—————————————————————————— ■ *Try Exercise 2, page 74.*

If an equation involves fractions, it is helpful to multiply each side of the equation by the LCD of all the denominators to produce an equivalent equation that does not contain fractions.

EXAMPLE 2   **Solve by Clearing Fractions**

Solve $\frac{2}{3}x + 10 - \frac{x}{5} = \frac{36}{5}$.

Solution
$$\frac{2}{3}x + 10 - \frac{x}{5} = \frac{36}{5}$$

$$15\left(\frac{2}{3}x + 10 - \frac{x}{5}\right) = 15\left(\frac{36}{5}\right) \qquad \bullet \text{ Multiply each side of} \atop \text{the equation by 15.}$$

$$10x + 150 - 3x = 108 \qquad \bullet \text{ Simplify.}$$

$$7x + 150 = 108$$

$$7x + 150 - 150 = 108 - 150 \qquad \bullet \text{ Subtract 150 from} \atop \text{each side.}$$

$$7x = -42$$

$$\frac{7x}{7} = \frac{-42}{7} \qquad \bullet \text{ Divide each side} \atop \text{by 7.}$$

$$x = -6 \qquad \bullet \text{ Check as before.}$$

—————————————————————————— ■ *Try Exercise 12, page 74.*

EXAMPLE 3   **Solve an Equation by Applying Properties**

Solve $(x + 2)(5x + 1) = 5x(x + 1)$.

Solution

$$(x + 2)(5x + 1) = 5x(x + 1)$$

$$5x^2 + 11x + 2 = 5x^2 + 5x \qquad \bullet \text{ Simplify each product.}$$

$$11x + 2 = 5x \qquad \bullet \text{ Subtract } 5x^2 \text{ from each side.}$$

$$6x + 2 = 0 \qquad \bullet \text{ Subtract } 5x \text{ from each side.}$$

$$6x = -2 \qquad \bullet \text{ Subtract 2 from each side.}$$

$$x = -\frac{1}{3} \qquad \bullet \text{ Divide each side of the equation by 6.}$$

■ *Try Exercise 20, page 74.*

An equation that has no solutions is called a **contradiction.** The equation $x = x + 1$ is a contradiction. No number is equal to itself increased by 1.

An equation that is true for some values of the variable but not true for other values of the variable is called a **conditional equation.** For example, $x + 2 = 8$ is a conditional equation because it is true for $x = 6$ and false for any number not equal to 6.

An **identity** is an equation that is true for *every* real number for which all terms of the equation are defined. Examples of identities include the equations $x + x = 2x$, and $(x + 3)^2 = x^2 + 6x + 9$.

EXAMPLE 4   **Verify an Identity**

Verify the identity $\frac{3(x^3 - 8)}{x - 2} = 3x^2 + 6x + 12, x \neq 2$.

Solution   Simplify the left side of the equation.

$$\frac{3(x^3 - 8)}{x - 2} = \frac{3(x - 2)(x^2 + 2x + 4)}{x - 2} \qquad \bullet \text{ Factor the difference of cubes and simplify.}$$

$$= 3(x^2 + 2x + 4)$$

$$= 3x^2 + 6x + 12$$

Since we have shown that it is possible to write the left side of the equation exactly as the right side, we have verified the identity.

■ *Try Exercise 30, page 74.*

The multiplication property of equality states that you can multiply each side of an equation by the same *nonzero* number. If you multiply each side of an equation by an algebraic expression that involves a variable, however, you must restrict the variable so that the expression is not equal to zero.

EXAMPLE 5 **Solve Equations That Have Restrictions**

Solve the following equations:

a. $\dfrac{x}{x-3} = \dfrac{9}{x-3} - 5$ b. $1 + \dfrac{x}{x-5} = \dfrac{5}{x-5}$

Solution

a. First, note that the denominator $x - 3$ would equal zero if $x = 3$. To produce a simpler equivalent equation, multiply each side by $x - 3$, with the restriction that $x \neq 3$.

$$(x-3)\left(\dfrac{x}{x-3}\right) = (x-3)\left(\dfrac{9}{x-3} - 5\right)$$

$$x = (x-3)\left(\dfrac{9}{x-3}\right) - (x-3)5$$

$$x = 9 - 5x + 15$$

$$6x = 24$$

$$x = 4$$

Check by substituting 4 for x in the original equation. The solution set is {4}.

b. To produce a simpler equivalent equation, multiply each side of the equation by $x - 5$, with the restriction that $x \neq 5$.

$$(x-5)\left(1 + \dfrac{x}{x-5}\right) = (x-5)\left(\dfrac{5}{x-5}\right)$$

$$(x-5)1 + (x-5)\left(\dfrac{x}{x-5}\right) = 5$$

$$x - 5 + x = 5$$

$$2x = 10$$

$$x = 5$$

Although we have obtained 5 as a proposed solution, 5 is *not* a solution of the original equation since it contradicts our restriction $x \neq 5$. Substitution of 5 for x in the original equation results in denominators of 0. In this case, the solution set of the original equation is the empty set.

■ *Try Exercise* **38**, *page 74.*

A **literal equation** is an equation that involves more than one variable. To solve a literal equation for one of its variables, you treat all the other variables as constants.

EXAMPLE 6   **Solve a Literal Equation**

Solve $xy - z = yz$ for y.

Solution   To solve for y, first isolate the terms that involve the variable y on the left side of the equation.

$$xy - z = yz$$

$$xy - yz = z$$ • Subtract yz and add z to each side of the equation to isolate the terms that contain y.

$$y(x - z) = z$$ • Factor y from each term on the left side of the equation.

$$y = \frac{z}{x - z}$$ • Divide each side of the equation by $x - z$; $x - z \neq 0$.

────────────────── ■ *Try Exercise* **66**, *page 75.*

Remark   In Example 6, the restriction $x - z \neq 0$ is necessary to ensure that each side of the equation is divided by a *nonzero* expression.

EXERCISE SET 2.1 _____

In Exercises 1 to 28, solve and check each equation.

1. $2x + 10 = 40$
2. $-3y + 20 = 2$

3. $5x + 2 = 2x - 10$
4. $4x - 11 = 7x + 20$

5. $2(x - 3) - 5 = 4(x - 5)$

6. $5(x - 4) - 7 = -2(x - 3)$

7. $4(2r - 17) + 5(3r - 8) = 0$

8. $6(5s - 11) - 12(2s + 5) = 0$

9. $\frac{3}{4}x + \frac{1}{2} = \frac{2}{3}$
10. $\frac{x}{4} - 5 = \frac{1}{2}$

11. $\frac{2}{3}x - 5 = \frac{1}{2}x - 3$
12. $\frac{1}{2}x + 7 - \frac{1}{4}x = \frac{19}{2}$

13. $0.2x + 0.4 = 3.6$
14. $0.04x - 0.2 = 0.07$

15. $x + 0.08(60) = 0.20(60 + x)$

16. $6(t + 1.5) = 12t$

17. $\frac{3}{5}(n + 5) - \frac{3}{4}(n - 11) = 0$

18. $-\frac{5}{7}(p + 11) + \frac{2}{5}(2p - 5) = 0$

19. $3(x + 5)(x - 1) = (3x + 4)(x - 2)$

20. $5(x + 4)(x - 4) = (x - 3)(5x + 4)$

21. $5[x - (4x - 5)] = 3 - 2x$

22. $6[3y - 2(y - 1)] - 2 + 7y = 0$

23. $\frac{40 - 3x}{5} = \frac{6x + 7}{8}$

24. $\frac{12 + x}{-4} = \frac{5x - 7}{3} + 2$

25. $0.08x + 0.12(4000 - x) = 432$

26. $0.075y + 0.06(10,000 - y) = 727.50$

27. $0.115x + 0.0975(8000 - x) = 823.75$

28. $0.145x + 0.109(4000) = 0.12(4000 + x)$

In Exercises 29 to 36, determine whether the equation is an identity, a conditional equation, or a contradiction.

29. $-3(x - 5) = -3x + 15$
30. $2x + \frac{1}{3} = \frac{6x + 1}{3}$

31. $2y + 7 = 3(y - 1)$
32. $x^2 + 10x = x(x + 10)$

33. $\frac{4y + 7}{4} = y + 7$
34. $(x + 3)^2 = x^2 + 9$

35. $2x + 5 = x + 9 + x$

36. $(x - 3)(x + 4) = x^2 + 4x - 11$

In Exercises 37 to 56, solve and check each equation.

37. $\frac{3}{x + 2} = \frac{5}{2x - 7}$
38. $\frac{4}{y + 2} = \frac{7}{y - 4}$

39. $\frac{30}{10 + x} = \frac{20}{10 - x}$
40. $\frac{6}{8 + x} = \frac{4}{8 - x}$

41. $\dfrac{3x}{x + 4} = 2 - \dfrac{12}{x + 4}$

42. $\dfrac{8}{2m + 1} - \dfrac{1}{m - 2} = \dfrac{5}{2m + 1}$

43. $2 + \dfrac{9}{r - 3} = \dfrac{3r}{r - 3}$

44. $\dfrac{t}{t - 4} + 3 = \dfrac{4}{t - 4}$

45. $\dfrac{5}{x - 3} - \dfrac{3}{x - 2} = \dfrac{4}{x - 3}$

46. $\dfrac{4}{x - 1} + \dfrac{7}{x + 7} = \dfrac{5}{x - 1}$

47. $\dfrac{2x + 5}{3x - 1} = 1$

48. $\dfrac{4x - 1}{3x + 2} = \dfrac{5}{6}$

49. $(y + 3)^2 = (y + 4)^2 + 1$

50. $(z - 7)^2 = (z - 2)^2 + 9$

51. $\dfrac{x}{x - 3} = \dfrac{x + 4}{x + 2}$

52. $\dfrac{x}{x - 5} = \dfrac{x + 7}{x + 1}$

53. $\dfrac{x + 3}{x + 5} = \dfrac{x - 3}{x - 4}$

54. $\dfrac{x - 6}{x + 4} = \dfrac{x - 1}{x + 2}$

55. $\dfrac{4x - 3}{2x} = \dfrac{2x - 4}{x - 2}$

56. $\dfrac{x + 3}{x + 1} = \dfrac{x + 6}{x + 4}$

In Exercises 57 to 68, solve each literal equation for x. State any necessary restrictions.

57. $2x + 3y = 6$

58. $4x - 7y = -15$

59. $2cx - d = 5(x - c)$

60. $2rx + 7 = 8(r - x)$

61. $\dfrac{x}{a} + \dfrac{y}{b} = 1$

62. $y = mx + b$

63. $y - y_1 = m(x - x_1)$

64. $Ax + By + C = 0$

65. $\dfrac{3}{x + 4} = l + w$

66. $\dfrac{2}{x - 2} = \dfrac{y}{x - 3}$

67. $mx - m^2 - nx = -n^2$

68. $px - p^2 + 6pq = 3qx + 9q^2$

In Exercises 69 to 74, use a calculator to solve each equation. State your final answer as a decimal with three significant digits.

69. $2.77x - 5.47 = 9.68$

70. $3.21x + 7.14 = 7.82x$

71. $\dfrac{1.62}{3.14x} = \dfrac{2.66}{9.21x}$

72. $\dfrac{3.84}{2.45x} = \dfrac{-1.92}{4.46 - 5.78x}$

73. $\dfrac{30.45x + 12.45}{6.71 - 2.34x} = 1.86$

74. $8.53x + 7.34(125 - 2.00x) = 108$

Supplemental Exercises

In Exercises 75 to 78, determine whether the given pair of equations is equivalent.

75. $3x - 11 = -5$, $\dfrac{3x - 11}{x - 2} = \dfrac{-5}{x - 2}$

76. $3x - 9 = x - 3$, $\dfrac{3x - 9}{x - 3} = \dfrac{x - 3}{x - 3}$

77. $\dfrac{1}{t} = \dfrac{1}{a} + \dfrac{1}{b}$, $t = \dfrac{ab}{a + b}$,   where t is a variable and a and b are nonzero constants.

78. $\dfrac{2}{x} = \dfrac{1}{x - 1}$, $2(x - 1) = x$

79. Let a, b, and c be real constants. Show that an equation of the form $ax + b = c$ has $x = \dfrac{c - b}{a}$ $(a \neq 0)$ as its solution.

80. Let a, b, c, and d be real constants. Show that an equation of the form $ax + b = cx + d$ has $x = \dfrac{d - b}{a - c}$ $(a - c \neq 0)$ as its solution.

In Exercises 81 to 86, solve each equation for x.

81. $\sqrt{7}x - 3 = 7$

82. $\sqrt{8}x + 2 = 14$

83. $\sqrt{3}x - 5 = \sqrt{27}x + 2$

84. $\sqrt{20}x + 14 = \sqrt{5}x - 8$

85. $a^2x - b = b^2x + a$ (assume $a \neq \pm b$)

86. $a^3x - a^2 + ab = b^2 - b^3x$ (assume $a \neq -b$)

2.2 Formulas and Applications

A **formula** is an equation or inequality that expresses known relationships between two or more variables. Following is a table of formulas from geometry that will be used throughout this text. In each formula, the variable P represents perimeter, A represents area, C represents circumference of a circle, and V represents volume.

Formulas from Geometry

Rectangle	Square	Triangle	Circle
$P = 2l + 2w$	$P = 4s$	$P = a + b + c$	$C = \pi d = 2\pi r$
$A = lw$	$A = s^2$	$A = \frac{1}{2}bh$	$A = \pi r^2$

Rectangular Solid	Cube	Right Circular Cone	Cylinder
$V = lwh$	$V = s^3$	$V = \frac{1}{3}\pi r^2 h$	$V = \pi r^2 h$

It is often necessary to solve a formula for a specified variable. The process consists of first isolating all terms that contain the specified variable on one side of the equation and all terms that do not contain the specified variable on the other side.

EXAMPLE 1   **Solve a Formula for a Specified Variable**

Solve $2l + 2w = P$ for l.

Solution

$$2l + 2w = P$$

$$2l = P - 2w \quad \bullet \text{ Subtract } 2w \text{ from each side to isolate the } 2l \text{ term.}$$

$$l = \frac{P - 2w}{2} \quad \bullet \text{ Divide each side by 2.}$$

■ *Try Exercise* **2**, *page 82.*

Remark   The solution can also be expressed as $l = P/2 - w$.

People with good problem-solving skills generally work application problems by applying specific techniques in a series of small steps.

Guidelines for Solving Application Problems

1. Read the problem carefully. If necessary, reread the problem several times.

2. When appropriate, draw a sketch and label parts of the drawing with the specific information given in the problem.

3. Determine the unknown quantities, and label them with variables. Write down any equation that relates the variables.

4. Use the information from step 3, along with a known formula or some additional information given in the problem, to write an equation.

5. Solve the equation obtained in step 4, and check to see whether these results satisfy all the conditions of the original problem.

EXAMPLE 2   **Solve an Application**

The length of a rectangle is 2 feet longer than three times its width. If the perimeter of the rectangle is 92 feet, find the width and the length of the rectangle.

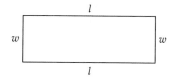

Figure 2.2

Solution

1. Read the problem carefully.
2. Draw a rectangle as shown in Figure 2.2.
3. Label the length of the rectangle l and the width of the rectangle w. The problem states that the length l is 2 feet greater than three times the width w. Thus l and w are related by the equation

$$l = 3w + 2$$

4. Since the problem involves the length, width, and perimeter of a rectangle, we use the geometric formula $2l + 2w = P$. To write an equation that involves only constants and a single variable (say, w), substitute 92 for P and $3w + 2$ for l.

$$2l + 2w = P$$
$$2(3w + 2) + 2w = 92$$

5. Solve for the unknown w.

$$6w + 4 + 2w = 92$$
$$8w + 4 = 92$$
$$8w = 88$$
$$w = 11$$

Since the length l is two more than three times the width,

$$l = 3(11) + 2 = 35$$

A check verifies that 35 is two more than three times 11. Also, twice the length (70) plus twice the width (22) gives the perimeter (92). The width of the rectangle is 11 feet, and its length is 35 feet.

■ *Try Exercise* **22,** *page 83.*

Many *uniform motion* problems can be solved by using the formula $d = rt$, where d is the distance traveled, r is the rate of speed, and t is the time.

EXAMPLE 3   **Solve a Uniform Motion Problem**

A runner runs a course at a constant speed of 6 mph. One hour after the runner begins, the cyclist starts on the same course at a constant speed of 15 mph. How long after the runner starts does the cyclist overtake the runner?

Solution   If we represent the time the runner has spent on the course by t, then the time the cyclist takes to overtake the runner is $t - 1$. The following table organizes the information and helps us determine how to write the distances each person travels.

	rate r	$\cdot$	time t	$=$	distance d
Runner	6	$\cdot$	t	$=$	$6t$
Cyclist	15	$\cdot$	$t - 1$	$=$	$15(t - 1)$

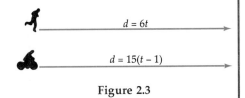

d = 6t

d = 15(t – 1)

Figure 2.3

Figure 2.3 indicates that the runner and the cyclist cover the same distance. Thus

$$6t = 15(t - 1)$$
$$6t = 15t - 15$$
$$-9t = -15$$
$$t = 1\frac{2}{3}$$

A check will verify that the cyclist does overtake the runner $1\frac{2}{3}$ hours after the runner starts.

■ *Try Exercise* **30**, *page 83.*

Some application problems require formulas from other disciplines. The next example uses a formula from geometry to solve an application of mathematics to archaeology.

EXAMPLE 4   **Solve an Application**

An archaeologist uncovers a portion of a wheel as shown in Figure 2.4. Use the given measurements to determine the diameter of the original wheel.

Solution   The tangent–secant theorem from geometry states: If a tangent segment and a secant segment are drawn to a circle, then the square of the length of the tangent segment is equal to the product of the length of the secant segment and the portion of the secant segment that is outside the circle. Applying this theorem to Figure 2.4 gives us:

$$[d(A, B)]^2 = [d(A, C)][d(A, D)]$$

Using the dimensions shown in Figure 2.4 yields

$$35^2 = [d(A, C)] \cdot 21$$
$$\frac{35^2}{21} = d(A, C)$$
$$58\frac{1}{3} = d(A, C)$$

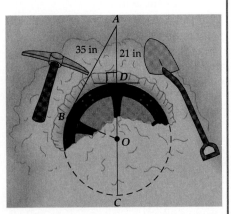

Figure 2.4

Segment $\overline{AC}$ passes through the center of the circle. Why?[1] Therefore the length of the diameter is

$$d(D, C) = d(A, C) - d(A, D) = 58\frac{1}{3} - 21 = 37\frac{1}{3} \text{ inches}$$

■ *Try Exercise* **36,** *page 83.*

Many business applications can be solved by using linear equations.

EXAMPLE 5 Solve a Business Application

It costs a tennis shoe manufacturer $26.55 to produce a pair of tennis shoes that sells for $49.95. How many pairs of tennis shoes must the manufacturer sell to make a profit of $14,274.00?

Solution The *profit* is equal to the *revenue* minus the *cost*. If x equals the number of pairs of tennis shoes to be sold, then the revenue will be $49.95x$ and the cost will be $26.55x$. Therefore,

$$\text{profit} = \text{revenue} - \text{cost}$$

$$14,274.00 = 49.95x - 26.55x$$

$$14,274.00 = 23.40x$$

$$610 = x$$

The manufacturer must sell 610 pairs of tennis shoes to make the desired profit.

■ *Try Exercise* **38,** *page 83.*

Simple interest problems can be solved by using the formula $I = Prt$, where I is the interest, P is the principal, r is the simple interest rate per period, and t is the number of periods.

EXAMPLE 6 Solve a Simple Interest Problem

An accountant invests part of a $6000 bonus in a 5 percent simple interest account and the remainder of the money at 8.5 percent simple interest. Together the investments earn $370 per year. Find the amount invested at each rate.

Solution Let x be the amount invested at 5 percent. The remainder of the money is $6000 - x$, which will be the amount invested at 8.5 percent. Using $I = Prt$, with $t = 1$ year, yields

---

[1]If a segment is tangent to a circle, then the segment is perpendicular to the radius that is drawn to the point of tangency.

Interest at 5% = $x \cdot 0.05 = 0.05x$

Interest at 8.5% = $(6000 - x) \cdot (0.085) = 510 - 0.085x$

The interest earned on the two accounts equals $370.

$$0.05x + (510 - 0.085x) = 370$$

$$-0.035x + 510 = 370$$

$$-0.035x = -140$$

$$x = 4000$$

Therefore, the accountant invested $4000 at 5 percent and the remaining $2000 at 8.5 percent. Check as before.

■ *Try Exercise* **42**, *page 83.*

Percent mixture problems involve combining solutions or alloys that have different concentrations of a common substance. Percent mixture problems can be solved by using the formula $pA = Q$, where p is the percent of concentration, A is the amount of the solution or alloy, and Q is the quantity of a substance in the solution or alloy. For example, in 4 liters of a 25% acid solution, p is the percent of acid (25%), A is the amount of solution (4 liters), and Q is the amount of acid in the solution, which equals $(0.25) \cdot (4)$ liters = 1 liter.

EXAMPLE 7 **Solve a Percent Mixture Problem**

A chemist mixes an 11% hydrochloric acid with a 6% hydrochloric acid solution. How many milliliters (ml) of each solution should the chemist use to make a 600-milliliter solution that is 8% hydrochloric acid?

Solution   Let x be the number of milliliters of the 11% solution. Since the final solution will have a total of 600 milliliters of fluid, $600 - x$ is the number of milliliters of the 6% solution. Use a table to organize the given information.

% Solution	% of Concentration p	.	Amount (ml) of Solution A	=	Quantity (ml) of Acid Q
11	0.11	.	x	=	$0.11x$
6	0.06	.	$600 - x$	=	$0.06(600 - x)$
8	0.08	.	600	=	$0.08(600)$

Because all the hydrochloric acid in the final solution comes from either the 11% solution or the 6% solution, the number of milliliters of hydrochloric acid in the 11% solution added to the number of milliliters of hydrochloric acid in the 6% solution must equal the number of milliliters of hydrochloric acid in the 8% solution.

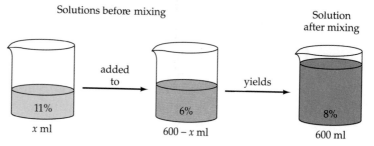

Solutions before mixing

Solution after mixing

added to

yields

11%

6%

8%

x ml

$600 - x$ ml

600 ml

Figure 2.5

$$\left(\begin{array}{c}\text{ml of acid in}\\\text{11% solution}\end{array}\right) + \left(\begin{array}{c}\text{ml of acid in}\\\text{6% solution}\end{array}\right) = \left(\begin{array}{c}\text{ml of acid in}\\\text{8% solution}\end{array}\right)$$

$$0.11x \quad + \quad 0.06(600 - x) \quad = \quad 0.08(600)$$

$$0.11x + 36 - 0.06x = 48$$

$$0.05x + 36 = 48$$

$$0.05x = 12$$

$$x = 240$$

Therefore the chemist should use 240 milliliters of the 11% solution and 360 milliliters of the 6% solution to make a 600-milliliter solution that is 8% hydrochloric acid.

■ *Try Exercise* **46**, *page 84.*

To solve a *work problem,* use the equation

Rate of work × time worked = part of task completed

For example, if a painter can paint a wall in 15 minutes, then the painter can paint $\frac{1}{15}$ of the wall in 1 minute. The painter's *rate of work* is $\frac{1}{15}$ of the wall each minute. In general, if a task can be completed in x minutes, then the rate of work is $1/x$ of the task each minute.

EXAMPLE 8 **Solve a Work Problem**

Pump A can fill a pool in 6 hours and pump B can fill the same pool in 3 hours. How long will it take to fill the pool if both pumps are used?

Solution   Since pump A fills the pool in 6 hours, $\frac{1}{6}$ represents the part of the pool filled by pump A in 1 hour. Since pump B fills the pool in 3 hours, $\frac{1}{3}$ represents the part of the pool filled by pump B in 1 hour.

Let $t = $ the number of hours to fill the pool together. Then

$$t \cdot \frac{1}{6} = \frac{t}{6} \quad \bullet \text{ Part of the pool filled by pump } A.$$

$$t \cdot \frac{1}{3} = \frac{t}{3} \quad \bullet \text{ Part of the pool filled by pump } B.$$

$$\begin{pmatrix} \text{Part filled} \\ \text{by pump } A \end{pmatrix} + \begin{pmatrix} \text{Part filled} \\ \text{by pump } B \end{pmatrix} = \begin{pmatrix} 1 \text{ filled} \\ \text{pool} \end{pmatrix}$$

$$\frac{t}{6} \quad + \quad \frac{t}{3} \quad = \quad 1$$

Multiplying each side of the equation by 6 produces

$$t + 2t = 6$$

$$3t = 6$$

$$t = 2$$

Check   Pump A fills $\frac{2}{6}$ or $\frac{1}{3}$ of the pool in 2 hours and pump B fills $\frac{2}{3}$ of the pool in 2 hours, so 2 hours is the time required to fill the pool if both pumps are used.

■ *Try Exercise* **56,** *page 84.*

EXERCISE SET 2.2

In Exercises 1 to 18, solve the formula for the specified variable.

1. $V = \dfrac{1}{3}\pi r^2 h; \quad h$ (geometry)

2. $P = S - Sdt; \quad t$ (business)

3. $I = Prt; \quad t$ (business)

4. $A = P + Prt; \quad P$ (business)

5. $F = \dfrac{Gm_1 m_2}{d^2}; \quad m_1$ (physics)

6. $A = \dfrac{1}{2}h(b_1 + b_2); \quad b_1$ (geometry)

7. $s = v_0 t - 16t^2; \quad v_0$ (physics)

8. $\dfrac{1}{f} = \dfrac{1}{d_o} + \dfrac{1}{d_i}; \quad f$ (astronomy)

9. $Q_w = m_w c_w (T_f - T_w); \quad T_w$ (physics)

10. $T\Delta t = Iw_f - Iw_i; \quad I$ (physics)

11. $a_n = a_1 + (n - 1)d; \quad d$ (mathematics)

12. $y - y_1 = m(x - x_1); \quad x$ (mathematics)

13. $S = \dfrac{a_1}{1 - r}; \quad r$ (mathematics)

14. $\dfrac{P_1 V_1}{T_1} = \dfrac{P_2 V_2}{T_2}; \quad V_2$ (chemistry)

15. $\dfrac{w_1}{w_2} = \dfrac{f_2 - f}{f - f_1}; \quad f_1$ (hydrostatics)

16. $v = \dfrac{v_1 + v_2}{1 + \dfrac{v_1 v_2}{c^2}}; \quad v_1$ (physics)

17. $f_{LC} = f_v \dfrac{v + v_{LC}}{v}; \quad v_{LC}$ (physics)

18. $F_1 d_1 + F_2 d_2 = F_3 d_3 + F_4 d_4; \quad F_3$ (physics)

In Exercises 19 to 60, solve by using the Guidelines for Solving Application Problems.

19. One-fifth of a number plus one-fourth of the number is five less than one-half the number. What is the number?

20. The numerator of a fraction is 4 less than the denominator. If the numerator is increased by 14 and the denominator is decreased by 10, the resulting number is 5. What is the original fraction?

21. The length of a rectangle is 3 feet less than twice the width of the rectangle. If the perimeter of the rectangle is 174 feet, find the width and the length.

22. The width of a rectangle is 1 meter more than half the length of the rectangle. If the perimeter of the rectangle is 110 meters, find the width and the length.

23. A triangle has a perimeter of 84 centimeters. Each of the two longer sides of the triangle is three times as long as the shortest side. Find the length of each side of the triangle.

24. A triangle has a perimeter of 161 miles. Each of the two smaller sides of the triangle is two-thirds the length of the longest side. Find the length of each side of the triangle.

25. Find two consecutive natural numbers whose sum is 1745.

26. Find three consecutive odd integers whose sum is 2001.

27. The difference of the squares of two consecutive positive even integers is 76. Find the integers.

28. The product of two consecutive integers is 90 less than the product of the next two integers. Find the four integers.

29. Running at an average rate of 6 meters per second, a sprinter ran to the end of a track and then jogged back to the starting point at an average rate of 2 meters per second. The total time for the sprint and the jog back was 2 minutes 40 seconds. Find the length of the track.

30. A motorboat left a harbor and traveled to an island at an average rate of 15 knots. The average speed on the return trip was 10 knots. If the total trip took 7.5 hours, how far is the harbor from the island?

31. A plane leaves an airport traveling at an average speed of 240 kilometers per hour. How long will it take a second plane traveling the same route at an average speed of 600 kilometers per hour to catch up with the first plane if it leaves 3 hours later?

32. A plane leaves Chicago headed for Los Angeles at 540 mph. One hour later, a second plane leaves Los Angeles headed for Chicago at 660 mph. If the air route from Chicago to Los Angeles is 1800 miles, how long will it take for the planes to pass by each other? How far from Chicago will they be at that time?

33. Use the tangent–secant theorem to find $d(A, B)$, given that $d(A, C) = 4$ and $d(C, D) = 9$.

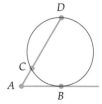

34. Use the tangent–secant theorem to find the length of the radius of the circle with center E, given that $d(A, B) = 5$ and $d(A, C)$ is 2.

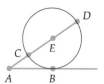

35. A student has test scores of 80, 82, 94, and 71. What score does the student need on the next test to produce an average score of 85?

36. A student has test scores of 90, 74, 82, and 90. The next examination is the final examination, which will count as two tests. What score does the student need on the final examination to produce an average score of 85?

37. It costs a manufacturer of sunglasses $8.95 to produce sunglasses that sell for $29.99. How many sunglasses must the manufacturer sell to make a profit of $17,884?

38. It costs a restaurant owner 18 cents per glass for orange juice, which is sold for 75 cents per glass. How many glasses of orange juice must the restaurant owner sell to make a profit of $2337?

39. The price of a computer fell 20 percent this year. If the computer now costs $750, how much did it cost last year?

40. The price of a magazine subscription rose 4 percent this year. If the subscription now costs $26, how much did it cost last year?

41. An investment adviser invested $14,000 in two accounts. One investment earned 8 percent annual simple interest, and the other investment earned 6.5 percent annual simple interest. The amount of interest earned for 1 year was $1024. How much was invested in each account?

42. A total of $7500 is deposited into two simple interest accounts. On one account the annual simple interest rate is 5 percent, and on the second account the annual simple interest rate is 7 percent. The amount of interest earned for 1 year was $405. How much was invested in each account?

43. An investment of $2500 is made at an annual simple interest rate of 5.5 percent. How much additional money must be invested at an annual simple interest rate of 8 percent so that the total interest earned is 7 percent of the total investment?

44. An investment of $4600 is made at an annual simple interest rate of 6.8 percent. How much additional money must be invested at an annual simple interest rate of 9 percent so that the total interest earned is 8 percent of the total investment?

45. How many grams of pure silver must a silversmith mix with a 45% silver alloy to produce 200 grams of a 50% alloy?

46. How many liters of a 40% sulfuric acid solution should be mixed with 4 liters of a 24% sulfuric acid solution to produce a 30% solution?

47. How many liters of water should be evaporated from 160 liters of a 12% saline solution so that the solution that remains is a 20% saline solution?

48. A radiator contains 6 liters of a 25% antifreeze solution. How much should be drained and replaced with pure antifreeze to produce a 33% antifreeze solution?

49. A ballet performance brought in $61,800 on the sale of 3000 tickets. If the tickets sold for $14 and $25, how many of each were sold?

50. A vending machine contains $41.25. The machine contains 255 coins, which consist of only nickels, dimes, and quarters. If the machine contains twice as many dimes as nickels, how many of each type of coin does the machine contain?

51. A coffee shop decides to blend a coffee that sells for $12 per pound with a coffee that sells for $9 per pound to produce a blend that will sell for $10 per pound. How much of each should be used to yield 20 pounds of the new blend?

52. A bag contains 42 coins, with a total weight of 246 grams. If the bag contains only gold coins that weigh 8 grams each and silver coins that weigh 5 grams each, how many gold and how many silver coins are in the bag?

53. How much pure gold should be melted with 15 grams of 14 karat gold to produce 18 karat gold? *Hint:* A karat is a measure of the purity of gold in an alloy. Pure gold measures 24 karats. An alloy that measures x karats is $x/24$ gold. For example, 18 karat gold is $18/24 = 3/4$ gold.

54. How much 14 karat gold should be melted with 4 ounces of pure gold to produce 18 karat gold? (*Hint:* See Exercise 53.)

55. An electrician can install the electric wires in a house in 14 hours. A second electrician requires 18 hours. How long would it take both electricians working together to install the wires?

56. Printer A can print a report in 3 hours. Printer B can print the same report in 4 hours. How long would it take both printers working together to print the report?

57. A worker can build a fence in 8 hours. With the help of an assistant, the fence can be built in 5 hours. How long would it take the assistant to build the fence alone?

58. A roofer and an assistant can repair a roof together in 6 hours. The assistant can complete the repair alone in 14 hours. If both the roofer and the assistant work to-gether for 2 hours and then the assistant is left alone to finish the job, how much longer will the assistant need to finish the repairs?

59. A book and a bookmark together sell for $10.10. If the price of the book is $10.00 more than the price of the bookmark, find the price of the book and the price of the bookmark.

60. Three people decide to share the cost of a yacht. By bringing in an additional partner, they can reduce the cost for each by $4000. What is the total cost of the yacht?

Supplemental Exercises

The *Archimedean law of the lever* states that for a lever to be in a state of balance with respect to a point called the fulcrum, the sum of the downward forces times their respective distances from the fulcrum on one side of the fulcrum must equal the sum of the downward forces times their respective distances from the fulcrum on the other side of the fulcrum. The accompanying figure shows this relationship.

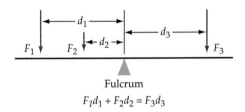

Fulcrum

$$F_1 d_1 + F_2 d_2 = F_3 d_3$$

61. A 100-pound person 8 feet from the fulcrum and a 40-pound person 5 feet from the fulcrum balance with a 160-pound person on a teeter-totter. How far from the fulcrum is the 160-pound person?

62. A lever 21 feet long has a force of 117 pounds applied to one end of the lever and a force of 156 pounds to the other end. Where should the fulcrum be located to produce a state of balance?

63. How much force applied 5 feet from the fulcrum is needed to lift 400 pounds that are 0.5 feet on the other side of the fulcrum?

64. Two workers need to lift a 1440-pound rock. They use a 6-foot steel bar with the fulcrum 1 foot from the rock, as the accompanying figure shows. One worker applies

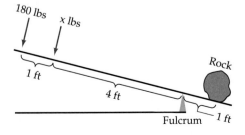

180 pounds to the other end of the lever. How much force will the second worker need to apply 1 foot from that end to lift the rock?

65. One pump can fill a pool in 10 hours. A second pump can fill the same pool in 8 hours. The pool has a drain that can drain the pool in 6 hours. Both pumps work together for 2 hours with the drain closed. The drain is then opened. How much longer will it take the two pumps to fill the pool?

66. If a pump can fill a pool in A hours and another pump can fill the pool in B hours, then the formula

$$T = \frac{AB}{A + B}$$

can be used to determine the total time T in hours that will be required to fill the pool if the pumps work together. Now consider the case where the pool is filled by three pumps. One can fill it in A hours, a second in B hours, and a third in C hours. Derive a formula for the total time T needed to fill the pool in terms of A, B, and C.

67. Two seconds after firing a rifle at a target, the shooter hears the impact of the bullet. Sound travels at 1100 feet per second and the bullet at 1865 feet per second. Determine the distance to the target.

68. Sound travels through sea water 4.62 times faster than through air. The sound of an exploding mine on the surface of the water and partially submerged reaches a ship through the water 4 seconds before it reaches the ship through the air. How far is the ship from the explosion? Use 1100 feet per second as the speed of sound through the air.

69. If a parade 2 miles long is proceeding at 3 mph, how long will it take a runner, jogging at 6 mph, to travel from the front of the parade to the end of the parade?

70. How long would the runner in Exercise 69 take to jog from the end of the parade to the start of the parade?

71. The work of the ancient Greek mathematician Diophantus had great influence on later European number theorists. Nothing is known about his personal life, except for the information given in the following epigram. "Diophantus passed $\frac{1}{6}$ of his life in childhood, $\frac{1}{12}$ in youth, and $\frac{1}{7}$ more as a bachelor. Five years after his marriage was born a son who died four years before his father, at $\frac{1}{2}$ his father's (final) age." How old was Diophantus when he died?

72. The relationship between the Fahrenheit temperature (F) and the Celsius temperature (C) is given by the formula

$$F = \frac{9}{5}C + 32$$

At what temperature will a Fahrenheit thermometer and a Celsius thermometer read the same?

2.3   Quadratic Equations

A **quadratic equation** in x is an equation that can be written in the **standard quadratic form** $ax^2 + bx + c = 0, \quad a \neq 0$.

Several methods can be used to solve quadratic equations. If the quadratic polynomial $ax^2 + bx + c$ can be factored over the integers, then the equation can be solved by factoring and using the **zero product property.**

Zero Product Property

If A and B are algebraic expressions, then

$$AB = 0 \quad \text{if and only if} \quad A = 0 \text{ or } B = 0.$$

This property states that when the product of two factors equals zero, then at least one of the factors is zero.

EXAMPLE 1   **Solve by Using the Zero Product Property**

Solve each of the following quadratic equations:

a. $3x^2 + 10x - 8 = 0$ b. $x^2 + 10x + 25 = 0$

Solution

a. $3x^2 + 10x - 8 = 0$

$(3x - 2)(x + 4) = 0$ • Factor.

$3x - 2 = 0$ or $x + 4 = 0$ • Apply the zero product property.

$3x = 2$ or $x = -4$

$x = \dfrac{2}{3}$ or $x = -4$ • Check as before.

The solution set of $3x^2 + 10x - 8 = 0$ is $\left\{-4, \dfrac{2}{3}\right\}$.

b. $x^2 + 10x + 25 = 0$

$(x + 5)^2 = 0$ • Factor.

$x + 5 = 0$ or $x + 5 = 0$ • Apply the zero product property.

$x = -5$ or $x = -5$ • Check as before.

The solution set of $x^2 + 10x + 25 = 0$ is $\{-5\}$.

_____ ■ *Try Exercise* **4**, *page 93.*

Remark   In Example 1b, the solution or root -5 is called a **double solution** or a **double root** of the equation because the application of the zero product property produced the two identical equations $x + 5 = 0$, both of which have a root of -5.

The quadratic equation $x^2 = c$ can be solved by factoring and applying the zero product property to yield the roots $\sqrt{c}$ and $-\sqrt{c}$.

$$x^2 = c$$

$$x^2 - c = 0$$

$$(x + \sqrt{c})(x - \sqrt{c}) = 0$$

$$x + \sqrt{c} = 0 \qquad \text{or} \qquad x - \sqrt{c} = 0$$

$$x = -\sqrt{c} \qquad \text{or} \qquad x = \sqrt{c}$$

This result is known as the square root theorem, which we will use to solve quadratic equations that can be written in the form $A^2 = B$.

The Square Root Theorem

If A and B are algebraic expressions such that

$$A^2 = B, \qquad \text{then} \quad A = \pm\sqrt{B}.$$

EXAMPLE 2 **Solve by Using the Square Root Theorem**

Use the square root theorem to solve each quadratic equation.

a. $(x + 1)^2 = 49$ b. $(x - 3)^2 = -28$

Solution

a. $(x + 1)^2 = 49$

$x + 1 = \pm\sqrt{49}$ • Apply the square root theorem.

$x + 1 = \pm 7$

$x = -1 \pm 7$

Thus $x = -1 - 7 = -8$ or $x = -1 + 7 = 6$.
The solution set of $(x + 1)^2 = 49$ is $\{-8, 6\}$.

b. $(x - 3)^2 = -28$

$x - 3 = \pm\sqrt{-28}$ • Apply the square root theorem.

$x - 3 = \pm 2i\sqrt{7}$

$x = 3 \pm 2i\sqrt{7}$

The solution set of $(x - 3)^2 = -28$ is $\{3 - 2i\sqrt{7}, 3 + 2i\sqrt{7}\}$.

■ *Try Exercise* **18**, *page 93.*

Consider the following binomial squares and their perfect square trinomial products.

Square of a Binomial		Perfect Square Trinomial
$(x + 6)^2$	=	$x^2 + 12x + 36$
$(x - 3)^2$	=	$x^2 - 6x + 9$

In each perfect square trinomial, the coefficient of x^2 is 1, and the constant term of the perfect square trinomial is the square of half the coefficient of its x term.

$$x^2 + 12x + 36, \quad \left(\frac{1}{2} \cdot 12\right)^2 = 36$$

$$x^2 - 6x + 9, \quad \left(\frac{1}{2}(-6)\right)^2 = 9$$

Adding, to a binomial of the form $x^2 + bx$, the constant that makes that binomial a perfect square trinomial is called **completing the square.** For example, to complete the square of $x^2 + 14x$, add

$$\left(\frac{1}{2} \cdot 14\right)^2 = 49$$

to produce the perfect square trinomial $x^2 + 14x + 49$.

Completing the square is a powerful method because it can be used to solve any quadratic equation.

EXAMPLE 3   Solve by Completing the Square

Solve $x^2 = 2x - 6$.

Solution $x^2 - 2x = -6$ • Isolate the constant term.

$x^2 - 2x + 1 = -6 + 1$ • Complete the square.

$(x - 1)^2 = -5$ • Factor and simplify.

$x - 1 = \pm\sqrt{-5}$ • Apply the square root theorem.

$x = 1 \pm i\sqrt{5}$

The solution set is $\{1 - i\sqrt{5}, 1 + i\sqrt{5}\}$.

■ *Try Exercise* **34,** *page 93.*

Completing the square by adding the square of half the coefficient of the x term requires that the coefficient of the x^2 term be 1. If the coefficient of the x^2 term is not 1, multiply each term on each side of the equation by the reciprocal of the coefficient of x^2.

EXAMPLE 4   Solve by Completing the Square

Solve $2x^2 + 8x - 15 = 0$.

Solution $2x^2 + 8x - 15 = 0$

$2x^2 + 8x = 15$ • Isolate the constant term.

$\dfrac{1}{2}(2x^2 + 8x) = \dfrac{1}{2}(15)$ • Multiply both sides of the equation by the reciprocal of the leading coefficient.

$x^2 + 4x = \dfrac{15}{2}$

$x^2 + 4x + 4 = \dfrac{15}{2} + 4$ • Add the square of half the x coefficient to both sides.

$(x + 2)^2 = \dfrac{23}{2}$ • Factor and simplify.

$x + 2 = \pm\sqrt{\dfrac{23}{2}}$ • Apply the square root theorem.

$x = -2 \pm \dfrac{\sqrt{46}}{2}$ • Add -2 to each side of the equation, and rationalize the denominator.

$x = \dfrac{-4 \pm \sqrt{46}}{2}$

The solution set is $\left\{ \dfrac{-4 - \sqrt{46}}{2}, \dfrac{-4 + \sqrt{46}}{2} \right\}$.

■ *Try Exercise* **38,** *page 93.*

Completing the square on $ax^2 + bx + c = 0$, $(a \neq 0)$, produces a formula for x in terms of the coefficients a, b, and c. The formula is known as the **quadratic formula,** and applying it is another way to solve quadratic equations.

The Quadratic Formula

If $ax^2 + bx + c = 0$, $a \neq 0$, then

$$x = \frac{-b \pm \sqrt{b^2 - 4ac}}{2a}$$

Proof: We assume a is a positive real number. If a were a negative real number, then we could multiply each side of the equation by -1 to make it positive.

$ax^2 + bx + c = 0 \quad (a \neq 0)$ • Given

$ax^2 + bx = -c$ • Isolate the constant term.

$x^2 + \dfrac{b}{a}x = -\dfrac{c}{a}$ • Multiply each term on each side of the equation by $\dfrac{1}{a}$.

$x^2 + \dfrac{b}{a}x + \left(\dfrac{b}{2a}\right)^2 = \left(\dfrac{b}{2a}\right)^2 - \dfrac{c}{a}$ • Complete the square.

$\left(x + \dfrac{b}{2a}\right)^2 = \left(\dfrac{b}{2a}\right)^2 - \dfrac{c}{a}$ • Factor the left side.

$\left(x + \dfrac{b}{2a}\right)^2 = \dfrac{b^2}{4a^2} - \dfrac{4a}{4a} \cdot \dfrac{c}{a}$ • Simplify the right side.

$x + \dfrac{b}{2a} = \pm\sqrt{\dfrac{b^2 - 4ac}{4a^2}}$ • Apply the square root theorem.

$x + \dfrac{b}{2a} = \pm\dfrac{\sqrt{b^2 - 4ac}}{2a}$ • Since $a > 0$, $\sqrt{4a^2} = 2a$.

$x = -\dfrac{b}{2a} \pm \dfrac{\sqrt{b^2 - 4ac}}{2a}$ • Add $-\dfrac{b}{2a}$ to each side.

$x = \dfrac{-b \pm \sqrt{b^2 - 4ac}}{2a}$

As a general rule, you should first try to solve quadratic equations by factoring. If the factoring process proves difficult, then solve by using the quadratic formula.

EXAMPLE 5   **Solve by Using the Quadratic Formula**

Solve $4x^2 - 8x + 1 = 0$.

Solution   The coefficients are $a = 4$, $b = -8$, and $c = 1$.

$$x = \frac{-(-8) \pm \sqrt{(-8)^2 - 4(4)(1)}}{2(4)} = \frac{8 \pm \sqrt{48}}{8} = \frac{8 \pm 4\sqrt{3}}{8} = \frac{2 \pm \sqrt{3}}{2}$$

The solution set is $\left\{ \dfrac{2 - \sqrt{3}}{2}, \dfrac{2 + \sqrt{3}}{2} \right\}$.

■ *Try Exercise 48, page 93.*

In the quadratic formula

$$x = \frac{-b \pm \sqrt{b^2 - 4ac}}{2a}$$

the expression $b^2 - 4ac$ is called the **discriminant** of the quadratic formula. If $b^2 - 4ac \geq 0$, then $\sqrt{b^2 - 4ac}$ is a real number; if $b^2 - 4ac < 0$, then $\sqrt{b^2 - 4ac}$ is not a real number. Thus the sign of the discriminant determines whether the roots of a quadratic equation are real numbers or nonreal numbers.

The Discriminant and Roots of a Quadratic Equation

The quadratic equation $ax^2 + bx + c = 0$, with real coefficients and $a \neq 0$, has discriminant $b^2 - 4ac$.

If $b^2 - 4ac > 0$, then the quadratic equation has *two distinct real roots*.

If $b^2 - 4ac = 0$, then the quadratic equation has *a real root* that is a double root.

If $b^2 - 4ac < 0$, then the quadratic equation has *two distinct nonreal roots*.

By examining the discriminant, it is possible to determine whether the roots of a quadratic equation are real numbers without actually finding the roots.

EXAMPLE 6   **Use the Discriminant to Classify Roots**

Classify the roots of each quadratic equation as real numbers or nonreal numbers.

a. $2x^2 - 5x + 1 = 0$ b. $3x^2 + 6x + 7 = 0$ c. $x^2 + 6x + 9 = 0$

Solution

a. $2x^2 - 5x + 1 = 0$ has coefficients $a = 2$, $b = -5$, and $c = 1$.

$$b^2 - 4ac = (-5)^2 - 4(2)(1) = 25 - 8 = 17$$

Because the discriminant 17 is *positive*, $2x^2 - 5x + 1 = 0$ has *two distinct real roots*.

b. $3x^2 + 6x + 7 = 0$ has coefficients $a = 3$, $b = 6$, and $c = 7$.

$$b^2 - 4ac = 6^2 - 4(3)(7) = 36 - 84 = -48$$

Because the discriminant -48 is *negative*, $3x^2 + 6x + 7 = 0$ has *two distinct nonreal roots*.

c. $x^2 + 6x + 9 = 0$ has coefficients $a = 1$, $b = 6$, and $c = 9$.

$$b^2 - 4ac = 6^2 - 4(1)(9) = 36 - 36 = 0$$

Because the discriminant is 0, $x^2 + 6x + 9 = 0$ has *a real root*. The root is a double root.

■ *Try Exercise* **58**, *page 93.*

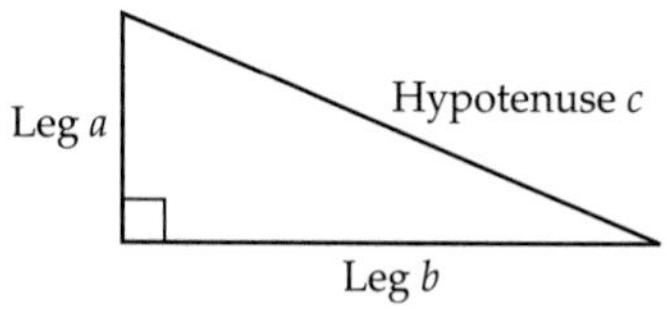

Leg a
Hypotenuse c
Leg b

Figure 2.6

A **right triangle** contains one 90° angle. The side opposite the 90° angle is called the **hypotenuse.** The two other sides are called **legs.** See Figure 2.6.

The Pythagorean Theorem

If a and b denote the lengths of the legs of a right triangle and c the length of the hypotenuse, then

$$c^2 = a^2 + b^2$$

The Pythagorean Theorem states that the square of the length of the hypotenuse of a right triangle is equal to the sum of the squares of the lengths of the two legs. This theorem is often used to solve applications that involve right triangles.

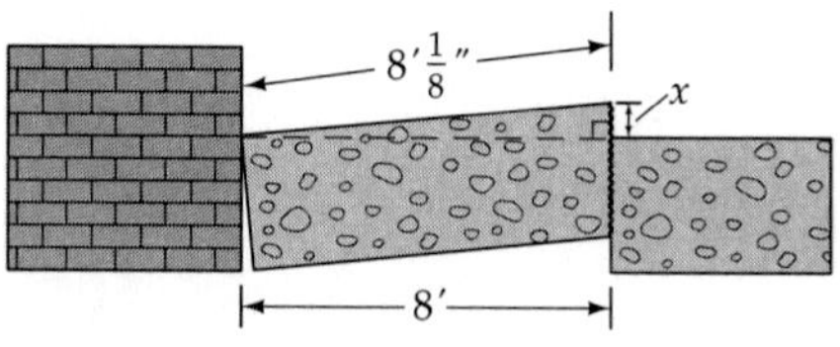

$8'\frac{1}{8}''$
x
$8'$

Figure 2.7

EXAMPLE 7 **Solve a Construction Application**

Concrete slabs often crack and buckle if proper expansion joints are not installed. Suppose a concrete slab expands as a result of an increase in temperature, as shown in Figure 2.7. Determine the height x, to the nearest inch, that the concrete will rise as a consequence of this expansion.

Solution   Use the Pythagorean Theorem.

$$\left(8 \text{ feet} + \frac{1}{8} \text{ inch}\right)^2 = x^2 + (8 \text{ feet})^2$$

$$(96.125)^2 = x^2 + (96)^2 \quad \bullet \text{ Change units to inches.}$$

$$(96.125)^2 - (96)^2 = x^2$$

$$\sqrt{(96.125)^2 - (96)^2} = x \qquad \bullet \text{ Only the positive root}$$
$$\text{is taken since } x > 0.$$

$$4.9 \approx x$$

Thus, to the nearest inch, the concrete will rise 5 inches.

——————————————————————— ■ *Try Exercise* **68**, *page 93.*

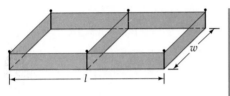

Figure 2.8

EXAMPLE 8   **Solve a Geometric Application**

A veterinarian wishes to use 132 feet of chain-link fencing to enclose a rectangular region and subdivide the region into two smaller rectangles, as shown in Figure 2.8. If the total enclosed area is 576 square feet, find the dimensions of the enclosed region.

Solution   Let w be the width of the enclosed region. Then $3w$ represents the amount of fencing used to construct the three widths. The amount of fencing left for the two lengths is $132 - 3w$. Thus each length must be half of the remaining fencing, or $\frac{132 - 3w}{2}$.

Now we have variable expressions in w for both the width and the length. Substituting these into the area formula $lw = A$ produces

$$\left(\frac{132 - 3w}{2}\right)(w) = 576 \quad \bullet \text{ Substitute.}$$

$$132w - 3w^2 = 1152 \quad \bullet \text{ Simplify.}$$

$$-3w^2 + 132w - 1152 = 0$$

$$w^2 - 44w + 384 = 0 \qquad \bullet \text{ Divide each term by } -3.$$

Although this quadratic formula can be solved by factoring, the following solution makes use of the quadratic formula. What reason can you give for using the quadratic formula rather than factoring?[2]

$$w = \frac{-(-44) \pm \sqrt{(-44)^2 - 4(1)(384)}}{2(1)} \quad \bullet \text{ Apply the quadratic formula.}$$

$$w = \frac{44 \pm \sqrt{400}}{2} = \frac{44 \pm 20}{2} = 12 \text{ or } 32$$

[2] Factoring $w^2 - 44w + 384$ may be time-consuming because 384 has several integer factors.

Thus there are two valid solutions to the problem:

1. If the width $w = 12$ feet, then the length is $\dfrac{132 - 3(12)}{2} = 48$ feet.

2. If the width $w = 32$ feet, then the length is $\dfrac{132 - 3(32)}{2} = 18$ feet.

■ *Try Exercise **74**, page 94.*

EXERCISE SET 2.3

In Exercises 1 to 16, solve each quadratic equation by factoring and applying the zero product property.

1. $x^2 - 2x - 15 = 0$　　**2.** $y^2 + 3y - 10 = 0$

3. $8y^2 + 189y - 72 = 0$　**4.** $12w^2 - 41w + 24 = 0$

5. $3x^2 - 7x = 0$　　　　**6.** $5x^2 = -8x$

7. $8 + 14t - 15t^2 = 0$　**8.** $12 - 26w + 10w^2 = 0$

9. $12 - 21s - 6s^2 = 0$　**10.** $-144 + 320y + 9y^2 = 0$

11. $(x - 5)^2 - 9 = 0$　　**12.** $(3x + 4)^2 - 16 = 0$

13. $(2x - 5)^2 - (4x - 11)^2 = 0$

14. $(5x + 3)^2 - (x + 7)^2 = 0$

15. $14x = x^2 + 49$

16. $41x = 12x^2 + 35$

In Exercises 17 to 28, use the square root theorem to solve each quadratic equation.

17. $x^2 = 81$　　　　　　**18.** $y^2 = 225$

19. $2x^2 = 48$　　　　　**20.** $3x^2 = 144$

21. $3x^2 + 12 = 0$　　　**22.** $4y^2 + 20 = 0$

23. $(x - 5)^2 = 36$　　　**24.** $(x + 4)^2 = 121$

25. $(x - 8)^2 = (x + 1)^2$　**26.** $(x + 5)^2 = (2x + 1)^2$

27. $x^2 = (x + 1)^2$　　　**28.** $4x^2 = (2x + 3)^2$

In Exercises 29 to 42, solve by completing the square.

29. $x^2 + 6x + 1 = 0$　　**30.** $x^2 + 8x - 10 = 0$

31. $x^2 - 2x - 15 = 0$　　**32.** $x^2 + 2x - 8 = 0$

33. $x^2 + 10x = 0$　　　**34.** $x^2 - 6x = 0$

35. $x^2 + 3x - 1 = 0$　　**36.** $x^2 + 7x - 2 = 0$

37. $2x^2 + 4x - 1 = 0$　　**38.** $2x^2 + 10x - 3 = 0$

39. $3x^2 - 8x + 1 = 0$　　**40.** $4x^2 - 4x + 15 = 0$

41. $5 - 6x - 3x^2 = 0$　　**42.** $2 + 10x - 5x^2 = 0$

In Exercises 43 to 56, solve by using the quadratic formula.

43. $x^2 - 2x - 15 = 0$　　**44.** $x^2 - 5x - 24 = 0$

45. $x^2 + x - 1 = 0$　　　**46.** $x^2 + x + 1 = 0$

47. $2x^2 + 4x + 1 = 0$　　**48.** $2x^2 + 4x - 1 = 0$

49. $3x^2 - 5x + 3 = 0$　　**50.** $3x^2 - 5x + 4 = 0$

51. $\dfrac{1}{2}x^2 + \dfrac{3}{4}x - 1 = 0$　**52.** $\dfrac{2}{3}x^2 - 5x + \dfrac{1}{2} = 0$

53. $\sqrt{2}x^2 + 3x + \sqrt{2} = 0$　**54.** $2x^2 + \sqrt{5}x - 3 = 0$

55. $x^2 = 3x - 5$　　　　**56.** $-x^2 = 7x - 1$

In Exercises 57 to 62, determine the discriminant of the quadratic equation, and then classify the roots of the equation as **a.** two distinct real numbers, **b.** one real number (which is a double root), or **c.** two distinct nonreal numbers. Do not solve the equations.

57. $2x^2 - 5x - 7 = 0$　　**58.** $x^2 + 3x - 11 = 0$

59. $3x^2 - 2x + 10 = 0$　　**60.** $x^2 + 3x + 3 = 0$

61. $x^2 - 20x + 100 = 0$　**62.** $4x^2 + 12x + 9 = 0$

In Exercises 63 to 66, find all values of k such that each quadratic equation has exactly one real root. *Hint:* The quadratic equation $ax^2 + bx + c = 0$ has exactly one real root if and only if $b^2 - 4ac = 0$.

63. $16x^2 + kx + 9 = 0$　　**64.** $x^2 + kx + 81 = 0$

65. $y^2 - 7y + k = 0$　　　**66.** $x^2 + 15x + k = 0$

67. The length of each side of a square is 54 inches. Find the length of the diagonal of the square. Round to the nearest tenth of an inch.

68. A concrete slab cracks and expands as a result of an increase in temperature, as shown in the following figure. Determine the height x, to the nearest inch, that the concrete will rise as a consequence of this expansion.

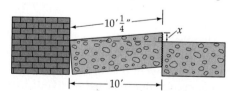

69. The length of each side of an equilateral triangle is 31 centimeters. Find the altitude of the triangle. Round to the nearest tenth of a centimeter.

70. How far, to the nearest foot, is it from homeplate to second base on a baseball diamond? *Hint:* The distance between home plate and first base is 90 feet.

71. The perimeter of a rectangle is 27 centimeters and its area is 35 square centimeters. Find the length and the width of the rectangle.

72. The perimeter of a rectangle is 34 feet and its area is 60 square feet. Find the length and the width of the rectangle.

73. A gardener wishes to use 600 feet of fencing to enclose a rectangular region and subdivide the region into two smaller rectangles. The total enclosed area is 15,000 square feet. Find the dimensions of the enclosed region.

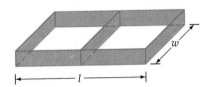

74. A farmer wishes to use 400 yards of fencing to enclose a rectangular region and subdivide the region into three smaller rectangles. The total enclosed area is 4800 square yards. Find the dimensions of the enclosed region.

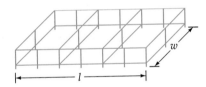

75. The sum of the squares of two consecutive positive even integers is 244. Find the numbers.

76. The sum of the squares of three consecutive integers is 302. Find the numbers.

77. Find a positive real number that is 5 larger than its reciprocal.

78. Find a positive real number that is 2 smaller than its reciprocal.

79. A salesperson drove the first 105 miles of a trip in 1 hour more than it took to drive the last 90 miles. The average rate during the last 90 miles was 10 mph faster than the average rate during the first 105 miles. Find the average rate for each portion of the trip.

80. A car and a bus both completed a 240-mile trip. The car averaged 10 mph faster than the bus and completed the trip in 48 minutes less time than the bus. Find the average rate, in miles per hour, of the bus.

81. A mason can build a wall in 6 hours less than an apprentice. Together they can build the wall in 4 hours. How long would it take the apprentice working alone to build the wall?

82. Pump A can fill a pool in 2 hours less time than pump B. Together the pumps can fill the pool in 2 hours 24 minutes. Find how long it takes pump A to fill the pool.

83. The height s in feet of a diver t seconds after diving off a 65-foot cliff is given by

$$s = -16t^2 + 65.$$

How long will it take the diver to hit the water?

84. A football player kicks a football downfield. The height s in feet of the football t seconds after it leaves the kicker's foot is given by

$$s = -16t^2 + 88t + 2.$$

Find the "hang time."

Supplemental Exercises

The following theorem is known as the *sum and product of the roots theorem.*

If $a \neq 0$ and r_1 and r_2 are roots of

$$ax^2 + bx + c = 0$$

then the sum of the roots $r_1 + r_2 = -\dfrac{b}{a}$, and the

product of the roots $r_1 r_2 = \dfrac{c}{a}$.

In Exercises 85 to 90, use the sum and product of the roots theorem to determine whether the given numbers are roots of the quadratic equation.

85. $x^2 - 5x - 24 = 0$, $-3, 8$

86. $x^2 + 4x - 21 = 0$, $-7, 3$

87. $2x^2 - 7x - 30 = 0$, $-5/2, 6$

88. $9x^2 - 12x - 1 = 0$, $(2 + \sqrt{5})/3, (2 - \sqrt{5})/3$

89. $x^2 - 2x + 2 = 0$, $1 + i, 1 - i$

90. $x^2 - 4x + 12 = 0$, $2 + 3i, 2 - 3i$

In Exercises 91 to 98, use the quadratic formula to solve each equation for the indicated variable in terms of the other variables. Assume that none of the denominators is zero.

91. $s = -\frac{1}{2}gt^2 + v_0 t + s_0$,   for t

92. $S = 2\pi rh + 2\pi r^2$,   for r

93. $-xy^2 + 4y + 3 = 0$,   for y

94. $D = \dfrac{n}{2}(n - 3)$,   for n

95. $3x^2 + xy + 4y^2 = 0$,   for x

96. $3x^2 + xy + 4y^2 = 0$,   for y

97. $x = y^2 + y - 8$,   for y

98. $P = \dfrac{E^2 R}{(r + R)^2}$,   for R

99. Prove that the equation $ax^2 + bx + c = 0$ with real coefficients such that $ac < 0$ has two distinct real roots.

100. Prove that the equation $ax^2 + c = 0$ with real coefficients such that $ac > 0$ has two distinct nonreal roots.

101. A rectangle is a "golden rectangle" provided its length l and its width w satisfy the equation

$$\frac{l}{w} = \frac{w}{l - w}$$

   a. Solve the above formula for w.

b. If the length l of a golden rectangle measures 101 feet, what is the width of the rectangle?

102. Use the quadratic formula to prove the sum and product of the roots theorem stated prior to Exercise 85.

103. The sum S of the first n natural numbers $1, 2, 3, \ldots, n$ is given by the formula

$$S = \frac{n}{2}(n + 1)$$

How many consecutive natural numbers starting with 1 produce a sum of 253?

104. The number of diagonals D of a polygon with n sides is given by the formula

$$D = \frac{n}{2}(n - 3)$$

Determine the number of sides of a polygon with 464 diagonals.

2.4   Other Types of Equations

Some equations that are neither linear nor quadratic can be solved by the various techniques presented in this section. For instance, the **third-degree equation,** or **cubic equation,** in Example 1 can be solved by factoring the polynomial on the left side of the equation and using the zero product property.

EXAMPLE 1   **Solve an Equation by Factoring**

Solve $x^3 - 16x = 0$.

   Solution $x^3 - 16x = 0$

   $x(x^2 - 16) = 0$ • Factor out the GCF, x.

   $x(x + 4)(x - 4) = 0$ • Factor the difference of squares.

Set each factor equal to zero.

   $x = 0$ or $x + 4 = 0$ or $x - 4 = 0$

   $x = 0$ or $x = -4$ or $x = 4$

A check will show that -4, 0, and 4 are roots of the original equation. The solution set is $\{-4, 0, 4\}$.

■ *Try Exercise 6, page 101.*

Caution   If you had attempted to solve Example 1 by dividing each side by x, you would have produced the equation $x^2 - 16 = 0$, which has roots of only -4 and 4. In this case the division of each side of the equation by the variable x has not produced an equivalent equation. Why?[3] To avoid this common mistake, factor out any variable factors that are common to each term instead of dividing each side of the equation by the factor.

Some equations that involve radical expressions can be solved by using the following result.

The Power Principle

> If P and Q are algebraic expressions and n is a positive integer, then every solution of $P = Q$ is a solution of $P^n = Q^n$.

EXAMPLE 2   **Solve a Radical Equation**

Use the power principle to solve $\sqrt{x + 4} = 3$.

Solution

$$\sqrt{x + 4} = 3$$
$$(\sqrt{x + 4})^2 = 3^2 \quad \bullet \text{ Apply the power principle with } n = 2.$$
$$x + 4 = 9$$
$$x = 5$$

Check: $$\sqrt{x + 4} = 3$$
$$\sqrt{5 + 4} \stackrel{?}{=} 3 \quad \bullet \text{ Substitute 5 for } x.$$
$$\sqrt{9} \stackrel{?}{=} 3$$
$$3 = 3 \quad \bullet \text{ 5 checks.}$$

The solution set is $\{5\}$.

■ *Try Exercise 14, page 101.*

Caution   Some care must be taken when using the power principle, because the equation $P^n = Q^n$ may have more solutions than the original equation $P = Q$. As an example, consider $x = 3$. The only solution is the real number 3. Square each side of the equation to produce $x^2 = 9$, which has both 3 and -3 as solutions. The -3 is called an **extraneous solution** because it is not a solution of the original equation $x = 3$. In general, any solution of $P^n = Q^n$ that is not a solution of $P = Q$ is called an extraneous solution. Extraneous solutions *may* be introduced whenever we raise each side of an equation to an *even* power.

---

[3] In order for us to divide each side of an equation by a variable, the variable must be restricted so that it is not equal to 0. However, $x = 0$ is a solution of the original equation.

EXAMPLE 3   **Solve a Radical Equation**

Solve $x = 2 + \sqrt{2 - x}$. Check all proposed solutions.

Solution

$$x = 2 + \sqrt{2 - x}$$

$$x - 2 = \sqrt{2 - x} \qquad \bullet \text{ Isolate the radical.}$$

$$(x - 2)^2 = (\sqrt{2 - x})^2 \qquad \bullet \text{ Square each side of the equation.}$$

$$x^2 - 4x + 4 = 2 - x$$

$$x^2 - 3x + 2 = 0 \qquad \bullet \text{ Collect and combine like terms.}$$

$$(x - 2)(x - 1) = 0 \qquad \bullet \text{ Factor.}$$

$$x - 2 = 0 \quad \text{or} \quad x - 1 = 0$$

$$x = 2 \quad \text{or} \quad x = 1 \quad \bullet \text{ Proposed solutions}$$

Check for $x = 2$: $x = 2 + \sqrt{2 - x}$

$$2 \stackrel{?}{=} 2 + \sqrt{2 - (2)} \qquad \bullet \text{ Substitute 2 for } x.$$

$$2 \stackrel{?}{=} 2 + \sqrt{0}$$

$$2 = 2 \qquad \bullet \text{ 2 is a solution.}$$

Check for $x = 1$: $x = 2 + \sqrt{2 - x}$

$$1 \stackrel{?}{=} 2 + \sqrt{2 - (1)} \qquad \bullet \text{ Substitute 1 for } x.$$

$$1 \stackrel{?}{=} 2 + \sqrt{1}$$

$$1 \neq 3 \qquad \bullet \text{ 1 is not a solution.}$$

The check shows that 1 is not a solution. It is an extraneous solution that we created by squaring each side of the equation. The solution set is {2}.

──────────────────────── ■ *Try Exercise* **16,** *page 101.*

In Example 4 it will be necessary to square $(1 + \sqrt{2x - 5})$. Recall the special product formula $(x + y)^2 = x^2 + 2xy + y^2$. Using this special product formula to square $(1 + \sqrt{2x - 5})$ produces

$$(1 + \sqrt{2x - 5})^2 = 1 + 2\sqrt{2x - 5} + (2x - 5)$$

EXAMPLE 4   **Solve a Radical Equation**

Solve $\sqrt{x + 1} - \sqrt{2x - 5} = 1$. Check all proposed solutions.

Solution   First write an equivalent equation in which one radical is isolated on one side of the equation.

$$\sqrt{x + 1} - \sqrt{2x - 5} = 1$$

$$\sqrt{x + 1} = 1 + \sqrt{2x - 5}$$

The next step is to square each side. Using the result from the discussion preceding this example, we have

$$x + 1 = 1 + 2\sqrt{2x - 5} + (2x - 5)$$

$$-x + 5 = 2\sqrt{2x - 5} \qquad \text{• Isolate the remaining radical.}$$

The right side still contains a radical, so square each side again.

$$x^2 - 10x + 25 = 4(2x - 5)$$

$$x^2 - 10x + 25 = 8x - 20$$

$$x^2 - 18x + 45 = 0$$

$$(x - 3)(x - 15) = 0$$

$$x = 3 \quad \text{or} \quad x = 15 \quad \text{• Proposed solutions}$$

Check for $x = 3$: $\sqrt{x + 1} - \sqrt{2x - 5} = 1$

$$\sqrt{3 + 1} - \sqrt{2(3) - 5} \stackrel{?}{=} 1$$

$$\sqrt{4} - \sqrt{1} \stackrel{?}{=} 1$$

$$2 - 1 \stackrel{?}{=} 1$$

$$1 = 1 \qquad \text{• 3 is a solution.}$$

Check for $x = 15$: $\sqrt{x + 1} - \sqrt{2x - 5} = 1$

$$\sqrt{15 + 1} - \sqrt{2(15) - 5} \stackrel{?}{=} 1$$

$$\sqrt{16} - \sqrt{25} \stackrel{?}{=} 1$$

$$4 - 5 \stackrel{?}{=} 1$$

$$-1 \neq 1 \qquad \text{• 15 is not a solution.}$$

Therefore, the solution set is $\{3\}$.

■ *Try Exercise* **20,** *page 101.*

Some equations that involve fractional exponents can be solved by raising each side to a reciprocal power. For example, to solve $x^{1/3} = 4$, raise each side to the third power to find that $x = 64$. Be sure to check all proposed solutions to determine whether they are actual solutions or extraneous solutions.

EXAMPLE 5   **Solve Equations That Involve Fractional Exponents**

Solve $(x^2 + 4x + 52)^{3/2} = 512$.

Solution   Because the equation involves a three-halves power, start by raising each side of the equation to the two-thirds power.

$$[(x^2 + 4x + 52)^{3/2}]^{2/3} = 512^{2/3} \qquad \text{• The reciprocal of } \frac{3}{2} \text{ is } \frac{2}{3}.$$

$$x^2 + 4x + 52 = 64 \qquad \bullet \text{ Think: } 512^{2/3} = (\sqrt[3]{512})^2 = 8^2 = 64.$$

$$x^2 + 4x - 12 = 0 \qquad \bullet \text{ Subtract 64 from each side.}$$

$$(x - 2)(x + 6) = 0 \qquad \bullet \text{ Factor.}$$

$$x - 2 = 0 \quad \text{or} \quad x + 6 = 0$$

$$x = 2 \quad \text{or} \quad x = -6$$

A check will verify that 2 and -6 are both solutions of the original equation. The solution set is $\{-6, 2\}$.

_____ ■ *Try Exercise* **32,** *page 101.*

The equation $4x^4 - 25x^2 + 36 = 0$ is said to be **quadratic in form,** which means it can be written in the form

$$au^2 + bu + c = 0 \qquad a \neq 0$$

where u is an algebraic expression involving x. For example, if we make the substitution $u = x^2$ (which implies $u^2 = x^4$), then our original equation can be written as

$$4u^2 - 25u + 36 = 0.$$

This quadratic equation can be solved for u, and then, using the relationship $u = x^2$, we can find the solutions of the original equation.

EXAMPLE 6   **Solve an Equation That Is Quadratic in Form**

Solve $4x^4 - 25x^2 + 36 = 0$.

Solution   Make the substitutions $u = x^2$ and $u^2 = x^4$ to produce the quadratic equation $4u^2 - 25u + 36 = 0$. Factor the quadratic polynomial on the left side of the equation.

$$(4u - 9)(u - 4) = 0$$

$$4u - 9 = 0 \qquad \text{or} \qquad u - 4 = 0$$

$$u = \frac{9}{4} \qquad \text{or} \qquad u = 4$$

Substitute x^2 for u to produce

$$x^2 = \frac{9}{4} \qquad \text{or} \qquad x^2 = 4$$

$$x = \pm\sqrt{\frac{9}{4}} \qquad \text{or} \qquad x = \pm\sqrt{4}$$

$$x = \pm\frac{3}{2} \qquad \text{or} \qquad x = \pm 2 \qquad \bullet \text{ Check as before.}$$

The solution set is $\{-2, -\frac{3}{2}, \frac{3}{2}, 2\}$.

_____ ■ *Try Exercise* **42,** *page 101.*

Following is a table of equations that are quadratic in form, along with an appropriate substitution that will allow them to be written in the form $au^2 + bu + c = 0$.

Equations That Are Quadratic in Form

Original Equation	Substitution	$au^2 + bu + c = 0$ Form
$x^4 - 8x^2 + 15 = 0$	$u = x^2$	$u^2 - 8u + 15 = 0$
$x^6 + x^3 - 12 = 0$	$u = x^3$	$u^2 + u - 12 = 0$
$x^{1/2} - 9x^{1/4} + 20 = 0$	$u = x^{1/4}$	$u^2 - 9u + 20 = 0$
$2x^{2/3} + 7x^{1/3} - 4 = 0$	$u = x^{1/3}$	$2u^2 + 7u - 4 = 0$
$15x^{-2} + 7x^{-1} - 2 = 0$	$u = x^{-1}$	$15u^2 + 7u - 2 = 0$

EXAMPLE 7   **Solve an Equation That Is Quadratic in Form**

Solve $3x^{2/3} - 5x^{1/3} - 2 = 0$.

Solution   Substituting u for $x^{1/3}$ gives us

$$3u^2 - 5u - 2 = 0$$

$$(3u + 1)(u - 2) = 0$$

$$3u + 1 = 0 \qquad \text{or} \qquad u - 2 = 0$$

$$u = -\frac{1}{3} \qquad \text{or} \qquad u = 2$$

$$x^{1/3} = -\frac{1}{3} \qquad \text{or} \qquad x^{1/3} = 2 \quad \bullet \text{ Replace } u \text{ with } x^{1/3}.$$

$$x = -\frac{1}{27} \qquad \text{or} \qquad x = 8 \quad \bullet \text{ Cube each side.}$$

A check will verify that both proposed solutions are actual solutions. The solution set is $\{-\frac{1}{27}, 8\}$.

■ *Try Exercise* **52**, *page 101.*

It is possible to solve equations that are quadratic in form without making a formal substitution. For example, to solve $x^4 + 5x^2 - 36 = 0$, factor the equation and apply the zero product property.

$$x^4 + 5x^2 - 36 = 0$$

$$(x^2 + 9)(x^2 - 4) = 0$$

$$x^2 + 9 = 0 \qquad \text{or} \qquad x^2 - 4 = 0$$

$$x^2 = -9 \qquad \text{or} \qquad x^2 = 4$$

$$x = \pm 3i \qquad \text{or} \qquad x = \pm 2$$

EXERCISE SET 2.4

In Exercises 1 to 12, factor to solve each equation.

1. $x^3 - 25x = 0$

2. $x^3 - x = 0$

3. $x^3 - 2x^2 - x + 2 = 0$

4. $x^3 - 4x^2 - 2x + 8 = 0$

5. $2x^5 - 18x^3 = 0$

6. $x^4 - 36x^2 = 0$

7. $x^4 - 3x^3 - 40x^2 = 0$

8. $x^4 + 3x^3 - 8x - 24 = 0$

9. $x^4 - 16x^2 = 0$

10. $x^4 - 16 = 0$

11. $x^3 - 8 = 0$

12. $x^3 + 8 = 0$

In Exercises 13 to 30, use the power principle to solve each radical equation. Check all proposed solutions.

13. $\sqrt{x - 4} - 6 = 0$

14. $\sqrt{10 - x} = 4$

15. $x = 3 + \sqrt{3 - x}$

16. $x = \sqrt{5 - x} + 5$

17. $\sqrt{3x - 5} - \sqrt{x + 2} = 1$

18. $\sqrt{6 - x} + \sqrt{5x + 6} = 6$

19. $\sqrt{2x + 11} - \sqrt{2x - 5} = 2$

20. $\sqrt{x + 7} - 2 = \sqrt{x - 9}$

21. $\sqrt{x + 7} + \sqrt{x - 5} = 6$

22. $x = \sqrt{12x - 35}$

23. $2x = \sqrt{4x + 15}$

24. $\sqrt[3]{7x - 3} = \sqrt[3]{2x + 7}$

25. $\sqrt[3]{2x^2 + 5x - 3} = \sqrt[3]{x^2 + 3}$

26. $\sqrt[4]{x^2 + 20} = \sqrt[4]{9x}$

27. $\sqrt{3\sqrt{5x + 16}} = \sqrt{5x - 2}$

28. $\sqrt{4\sqrt{2x - 5}} = \sqrt{x + 5}$

29. $\sqrt{3x + 1} + \sqrt{2x - 1} = \sqrt{10x - 1}$

30. $\sqrt{x - 3} + \sqrt{x + 3} = \sqrt{9 - x}$

In Exercises 31 to 40, solve each equation that involves fractional exponents. Check all proposed solutions.

31. $(3x + 5)^{1/3} = (-2x + 15)^{1/3}$

32. $(4z + 7)^{1/3} = 2$

33. $(x + 4)^{2/3} = 9$

34. $(x - 5)^{3/2} = 125$

35. $(4x)^{2/3} = (30x + 4)^{1/3}$

36. $z^{2/3} = (3z - 2)^{1/3}$

37. $4x^{3/4} = x^{1/2}$

38. $x^{3/5} = 2x^{1/5}$

39. $(3x - 5)^{2/3} + 6(3x - 5)^{1/3} = -8$

40. $2(x + 1)^{1/2} - 11(x + 1)^{1/4} + 12 = 0$

In Exercises 41 to 60, find all the real solutions of each equation by first rewriting each equation as a quadratic equation.

41. $x^4 - 9x^2 + 14 = 0$

42. $x^4 - 10x^2 + 9 = 0$

43. $2x^4 - 11x^2 + 12 = 0$

44. $6x^4 - 7x^2 + 2 = 0$

45. $x^6 + x^3 - 6 = 0$

46. $6x^6 + x^3 - 15 = 0$

47. $21x^6 + 22x^3 = 8$

48. $-3x^6 + 377x^3 - 250 = 0$

49. $x^{1/2} - 3x^{1/4} + 2 = 0$

50. $2x^{1/2} - 5x^{1/4} - 3 = 0$

51. $3x^{2/3} - 11x^{1/3} - 4 = 0$

52. $6x^{2/3} - 7x^{1/3} - 20 = 0$

53. $9x^4 = 30x^2 - 25$

54. $4x^4 - 28x^2 = -49$

55. $x^{2/5} - 1 = 0$

56. $2x^{2/5} - x^{1/5} = 6$

57. $\dfrac{1}{x^2} + \dfrac{3}{x} - 10 = 0$

58. $10\left(\dfrac{x - 2}{x}\right)^2 + 9\left(\dfrac{x - 2}{x}\right) - 9 = 0$

59. $9x - 52\sqrt{x} + 64 = 0$

60. $8x - 38\sqrt{x} + 9 = 0$

In Exercises 61 to 64, solve each equation. Round each solution to the nearest hundredth.

61. $x^4 - 3x^2 + 1 = 0$

62. $x - 4\sqrt{x} + 1 = 0$

63. $x^2 - \sqrt{9x^2 - 1} = 0$

64. $2x^2 = \sqrt{10x^2 - 3}$

Supplemental Exercises

In Exercises 65 to 70, solve for x in terms of the other variables.

65. $x^2 + y^2 = 9$

66. $\dfrac{x^2}{a^2} + \dfrac{y^2}{b^2} = 1$

67. $\sqrt{x} - \sqrt{y} = \sqrt{z}$

68. $x - y = \sqrt{x^2 + y^2 + 5}$

69. $x + y = \sqrt{x^2 - y^2 + 7}$

70. $x + \sqrt{x} = -y$

71. Solve $(\sqrt{x} - 2)^2 - 5\sqrt{x} + 14 = 0$ for x. (*Hint:* Use the substitution $u = \sqrt{x} - 2$, and then rewrite so that the equation is quadratic in terms of the variable u.)

72. Solve $(\sqrt[3]{x} + 3)^2 - 8\sqrt[3]{x} = 12$ for x. (*Hint:* Use the substitution $u = \sqrt[3]{x} + 3$, and then rewrite so that the equation is quadratic in terms of the variable u.)

73. A conical funnel has a height h of 4 inches and a lateral surface area L of 15π square inches. Find the radius r of the cone. (*Hint:* Use the formula $L = \pi r \sqrt{r^2 + h^2}$.)

74. As flour is poured onto a table, it forms a right circular cone whose height is one-third the diameter of the base. What is the diameter of the base when the cone has a volume of 192 cubic inches?

75. A silver sphere has a diameter of 8 millimeters, and a second silver sphere has a diameter of 12 millimeters. The spheres are melted down and recast to form a single cube. What is the length s of each edge of the cube? Round your answer to the nearest tenth of a millimeter.

76. The period of a pendulum T is the time it takes a pendulum to complete one swing from left to right and back. For a pendulum near the surface of the earth,

$$T = 2\pi\sqrt{\dfrac{L}{32}}$$

where T is measured in seconds and L is the length of the pendulum in feet. Find the length of a pendulum that has a period of 4 seconds. Round to the nearest tenth of a foot.

77. On a ship, the distance d that you can see to the horizon is given by $d = 1.5\sqrt{h}$, where h is the height of your eye measured in feet above sea level and d is measured in miles. How high is the eye level of a navigator who can see 14 miles to the horizon? Round to the nearest foot.

78. The radius r of a circle inscribed in a triangle with sides of length a, b, and c is given by

$$r = \sqrt{\frac{(s - a)(s - b)(s - c)}{s}}$$

where $s = \frac{1}{2}(a + b + c)$. **a.** Find the length of the radius of a circle inscribed in a triangle with sides of 5 inches, 6 inches, and 7 inches. **b.** The radius of a circle inscribed in an equilateral triangle measures 2 inches. What is the length of each side of the equilateral triangle?

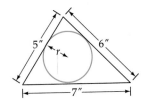

79. The radius r of a circle that is circumscribed about a triangle with sides a, b, and c is given by

$$r = \frac{abc}{4\sqrt{s(s - a)(s - b)(s - c)}}$$

where $s = \frac{1}{2}(a + b + c)$. **a.** Find the radius of a circle that is circumscribed about a triangle with sides of length 7 inches, 10 inches, and 15 inches. **b.** A circle with radius 5 inches is circumscribed about an equilat-

eral triangle. What is the length of each side of the equilateral triangle?

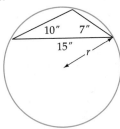

For Exercises 80 and 81, the depth s from the opening of a well to the water can be determined by measuring the total time between the instant you drop a stone and the time you hear it hit the water. The time (in seconds) it takes the stone to hit the water is given by $\sqrt{s}/4$, where s is measured in feet. The time (also in seconds) required for the sound of the impact to travel up to your ears is given by $s/1100$. Thus the total time T (in seconds) between the instant you drop a stone and the moment you hear its impact is

$$T = \frac{\sqrt{s}}{4} + \frac{s}{1100}$$

80. One of the world's deepest water wells is 7320 feet deep. Find the time between the instant a stone is dropped and the time you hear it hit the water if the surface of the water is 7100 feet below the opening of the well. Round your answer to the nearest tenth of a second.

81. Solve $T = \dfrac{\sqrt{s}}{4} + \dfrac{s}{1100}$ for s.

82. Use the result of Exercise 81 to determine the depth from the opening of a well to the water level if the time between the instant you drop a stone and the moment you hear its impact is 3 seconds. Round your answer to the nearest foot.

<h2>2.5   Inequalities</h2>

In Section 1.2 we used the concept of an inequality to describe the order of real numbers on the real number line, and we also used inequalities to represent subsets of real numbers. In this section we consider inequalities that involve a variable. In particular, we consider how to determine which real values of the variable make the inequality a true statement.

The set of all solutions of an inequality is called the **solution set of the inequality.** For example, the solution set of $x + 1 > 4$ is the set of all real numbers greater than 3. **Equivalent inequalities** have the same solution set. We can solve an inequality by producing *simpler* but equivalent inequalities until the solutions are found. To produce these simpler but equivalent inequalities, we apply the following properties.

Properties of Inequalities

For real numbers a, b, and c,

1. $a < b$ and $a + c < b + c$ are equivalent inequalities. (*Adding the same number to each side of an inequality preserves the order of the inequality.*)
2. If $c > 0$, then $a < b$ and $ac < bc$ are equivalent inequalities. (*Multiplying each side of an inequality by the same positive number preserves the order of the inequality.*)
3. If $c < 0$, then $a < b$ and $ac > bc$ are equivalent inequalities. (*Multiplying each side of an inequality by the same negative number reverses the order of the inequality.*)

Note the difference between Properties 2 and 3. Property 2 states that an equivalent inequality is produced by multiplying each side of an inequality by the same *positive* number, provided that the inequality symbol is not changed. However, Property 3 states that an equivalent inequality is produced by multiplying each side of an inequality by the same *negative* number, provided that the inequality symbol is reversed.

Because subtraction is defined in terms of addition, subtracting the same number from each side of an inequality preserves the order of the inequality. Because division is defined in terms of multiplication, dividing each side of an inequality by the same *positive* number preserves the order of the inequality, and dividing each side of an inequality by the same *negative* number reverses the order of the inequality.

EXAMPLE 1   **Solve an Inequality**

Solve $2(x + 3) < 4x + 10$.

Solution

$$2(x + 3) < 4x + 10$$

$$2x + 6 < 4x + 10 \quad \bullet \text{ Use the distributive property.}$$

$$-2x < 4 \quad \bullet \text{ Subtract } 4x \text{ and } 6 \text{ from each side of the inequality.}$$

$$x > -2 \quad \bullet \text{ Divide each side by } -2 \text{ and reverse the inequality symbol.}$$

Thus the original inequality is true for all real numbers greater than -2. The solution set is $\{x \mid x > -2\}$. Using interval notation, the solution set is written as $(-2, \infty)$.

■ *Try Exercise **8**, page 110.*

EXAMPLE 2   **Solve an Application That Involves an Inequality**

You can rent a car from Company A for $26 per day plus $0.09 a mile. Company B charges $12 per day plus $0.14 a mile. Find the number of miles for which it is cheaper to rent from Company A if you rent a car for 1 day.

Solution   Let m equal the number of miles the car is to be driven. Then the cost of renting the car will be

$$\$26 + \$0.09m \quad \text{from Company A}$$

$$\$12 + \$0.14m \quad \text{from Company B}$$

If renting from Company A is to be cheaper than renting from Company B, then we must have

$$26 + 0.09m < 12 + 0.14m$$

Solving for m produces

$$14 < 0.05m$$

$$\frac{14}{0.05} < m$$

$$280 < m$$

Renting from Company A is cheaper if you drive over 280 miles per day.

■ *Try Exercise* **12,** *page 110.*

EXAMPLE 3   **Solve an Application That Involves Inequalities**

A photographic developer needs to be kept at a temperature between 15°C and 25°C. What is that temperature range in degrees Fahrenheit (°F)?

Solution   The formula that relates the Celsius temperature (C) to the Fahrenheit temperature (F) is

$$C = \frac{5}{9}(F - 32)$$

We are given that

$$15 < C < 25$$

Substituting $\frac{5}{9}(F - 32)$ for C yields

$$15 < \frac{5}{9}(F - 32) < 25$$

$$27 < \quad F - 32 \quad < 45 \quad \bullet \text{ Multiply each of the three parts of the inequality by 9/5.}$$

$$59 < \quad\quad F \quad\quad < 77 \quad \bullet \text{ Add 32 to each of the three parts of the inequality.}$$

Thus the developer needs to be kept between 59°F and 77°F.

■ *Try Exercise* **22,** *page 110.*

Critical Value Method for Solving Inequalities

Any value of x that causes a polynomial in x to equal zero is called a **zero** of the polynomial. For example, -4 and 1 are both zeros of the polynomial $x^2 + 3x - 4$, because $(-4)^2 + 3(-4) - 4 = 0$ and $1^2 + 3 \cdot 1 - 4 = 0$.

A Sign Property of Polynomials

> Nonzero polynomials in x have the property that for any value of x between two consecutive real zeros, either all values of the polynomial are positive or all values of the polynomial are negative.

In our work with inequalities that involve polynomials, the real zeros of the polynomial are also referred to as **critical values of the inequality,** because on a number line they separate the real numbers that make an inequality involving a polynomial true from those that make it false. In Example 4 we use critical values and the Sign Property of Polynomials to solve an inequality.

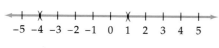

Figure 2.9

EXAMPLE 4   **Solve a Polynomial Inequality**

Solve $x^2 + 3x - 4 < 0$.

Solution   Factoring the polynomial $x^2 + 3x - 4$ produces the equivalent inequality

$$(x + 4)(x - 1) < 0$$

Thus the zeros of the polynomial $x^2 + 3x - 4$ are -4 and 1. They are the critical values of the inequality $x^2 + 3x - 4 < 0$. They separate the real number line into the three intervals shown in Figure 2.9.

To determine the intervals on which $x^2 + 3x - 4 < 0$, pick a number called a **test value** from each of the three intervals and then determine whether $x^2 + 3x - 4 < 0$ for each of these test values. For example, in the interval $(-\infty, -4)$, pick a test value of, say, -5. Then

$$x^2 + 3x - 4 = (-5)^2 + 3(-5) - 4 = 6$$

Since 6 is not less than 0, by the Sign Property of Polynomials, no number in the interval $(-\infty, -4)$ makes $x^2 + 3x - 4 < 0$.

Now pick a test value from the interval $(-4, 1)$, say, 0. When $x = 0$,

$$x^2 + 3x - 4 = 0^2 + 3(0) - 4 = -4$$

Since -4 is less than 0, by the Sign Property of Polynomials, all numbers in the interval $(-4, 1)$ make $x^2 + 3x - 4 < 0$.

If we pick a test value of 2 from the interval $(1, \infty)$, then

$$x^2 + 3x - 4 = (2)^2 + 3(2) - 4 = 6$$

Since 6 is not less than 0, by the Sign Property of Polynomials, no number in the interval $(1, \infty)$ makes $x^2 + 3x - 4 < 0$.

| The following table is a summary of our work:

Interval	$(-\infty, -4)$	$(-4, 1)$	$(1, \infty)$
Test value x	-5	0	2
$x^2 + 3x - 4 \overset{?}{<} 0$	$(-5)^2 + 3(-5) - 4 < 0$ $6 < 0$ False	$(0)^2 + 3(0) - 4 < 0$ $-4 < 0$ True	$(2)^2 + 3(2) - 4 < 0$ $6 < 0$ False

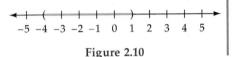

Figure 2.10

The solution set of $x^2 + 3x - 4 < 0$ is graphed in Figure 2.10. Note that in this case the critical values -4 and 1 are not included in the solution set because they do not make $x^2 + 3x - 4$ less than 0.

■ *Try Exercise* **30**, *page* 110.

To avoid the arithmetic in Example 4, we often use a *sign diagram*. For example, note that the factor $(x + 4)$ is negative for all $x < -4$ and positive for all $x > -4$. The factor $(x - 1)$ is negative for all $x < 1$ and positive for all $x > 1$. These results are shown in Figure 2.11.

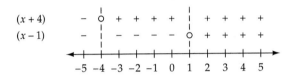

Figure 2.11
Sign diagram for $(x + 4)(x - 1)$.

To determine on which intervals the product $(x + 4)(x - 1)$ is negative, we examine the sign diagram to see where the factors have opposite signs. Since this occurs only on the interval $(-4, 1)$, where $(x + 4)$ is positive and $(x - 1)$ is negative, the original inequality is true only on the interval $(-4, 1)$.

Following is a summary of the steps used to solve polynomial inequalities by the critical value method.

Solving a Polynomial Inequality by the Critical Value Method

1. Write the inequality so that one side of the inequality is a nonzero polynomial and the other side is 0.

2. Find the real zeros of the polynomial.[4] They are the critical values of the original inequality.

3. Use a sign diagram or test values to determine which of the intervals formed by the critical values are to be included in the solution set.

[4] In Chapter 4, additional methods are developed to find the zeros of a polynomial. For the present, however, we will find the zeros by factoring or by using the quadratic formula.

EXAMPLE 5   **Use the Critical Value Method to Solve an Application**

A manufacturer of tennis racquets finds that the yearly revenue R from a particular type of racquet is given by $R = 160x - x^2$, where x is the price in dollars of each racquet. Find the interval in terms of x for which the yearly revenue is greater than \$6000. That is, solve

$$160x - x^2 > 6000$$

Solution   Write the inequality in such a way that 0 appears on the right side of the inequality.

$160x - x^2 - 6000 > 0$

$x^2 - 160x + 6000 < 0$ • Arrange the terms in descending powers. Multiply each side of the inequality by -1.

$(x - 60)(x - 100) < 0$ • Factor the left side.

Use the zero product property to find the zeros.

$(x - 60)(x - 100) = 0$ • Replace the inequality with an equals sign.

$x = 60$ or $x = 100$ • Set each factor equal to 0 and solve for x.

The zeros are 60 and 100. They separate the real number line into the intervals $(-\infty, 60)$, $(60, 100)$, and $(100, \infty)$. The sign diagram in Figure 2.12 shows that the inequality $(x - 60)(x - 100) < 0$ is true on the interval $(60, 100)$ and that it is false on the other intervals. Thus the revenue is greater than \$6000 per year when the price of each racquet is between \$60 and \$100.

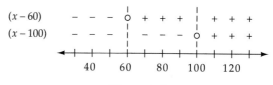

Figure 2.12

———————————————————————— ■ *Try Exercise* **38**, *page 110.*

Remark   In Example 5 it is easy to see that x cannot equal zero because no revenue will be produced if the racquets are given away. Also, if the price of the racquets increases above \$100, the revenue becomes less than \$6000. Why does this seem reasonable?[5]

—————————

[5] As the price increases above \$100, the number of people who decide to purchase the racquet decreases. The manufacturer makes more money on each sale, but since there are fewer sales, the revenue decreases.

Rational Inequalities

A rational expression is the quotient of two polynomials. **Rational inequalities** involve rational expressions, and they can be solved by an extension of the critical value method. The **critical values of a rational expression** are the numbers that cause the numerator of the rational expression to equal zero or the denominator of the rational expression to equal zero.

Rational expressions also have the property that they remain either positive for all values of the variable between consecutive critical values or negative for all values of the variable between consecutive critical values.

EXAMPLE 6   **Solve a Rational Inequality**

Solve $\dfrac{(x - 2)(x + 3)}{x - 4} \geq 0$.

Solution   The critical values include the zeros of the numerator, which are 2 and −3, and the zero of the denominator, which is 4. These three critical numbers separate the real number line into four intervals.

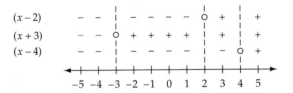

Figure 2.13

The sign diagram in Figure 2.13 shows the sign of each of the factors $(x - 2)$, $(x + 3)$, and $(x - 4)$ on each of the four intervals. The sign diagram shows that the rational expression is positive on the two intervals $(-3, 2)$ and $(4, \infty)$. The critical values −3 and 2 are solutions because they satisfy the original inequality. However, the critical value 4 is not a solution because the denominator $x - 4$ is zero when $x = 4$. Therefore the inequality's solution set is $[-3, 2] \cup (4, \infty)$. The graph of the solution set is shown in Figure 2.14.

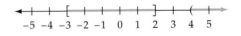

Figure 2.14

■ *Try Exercise* **40,** *page 110.*

EXAMPLE 7   **Solve a Rational Inequality**

Solve $\dfrac{3x + 4}{x + 1} \le 2$.

Solution   Write the inequality so that 0 appears on the right side of the inequality.

$$\frac{3x + 4}{x + 1} \le 2$$

$$\frac{3x + 4}{x + 1} - 2 \le 0$$

Write the left side as a rational expression.

$$\frac{3x + 4}{x + 1} - \frac{2(x + 1)}{x + 1} \le 0 \quad \bullet \text{ The LCD is } x + 1.$$

$$\frac{3x + 4 - 2x - 2}{x + 1} \le 0 \quad \bullet \text{ Simplify.}$$

$$\frac{x + 2}{x + 1} \le 0$$

The critical values of the above inequality are -2 and -1 because the numerator $x + 2$ is equal to zero when $x = -2$, and the denominator $x + 1$ is equal to zero when $x = -1$. The critical values -2 and -1 separate the real number line into the three intervals $(-\infty, -2)$, $(-2, -1)$, and $(-1, \infty)$.

All values of x on the interval $(-2, -1)$, make $(x + 2)/(x + 1)$ negative, as desired. On the other intervals the quotient $(x + 2)/(x + 1)$ is positive. See the sign diagram in Figure 2.15.

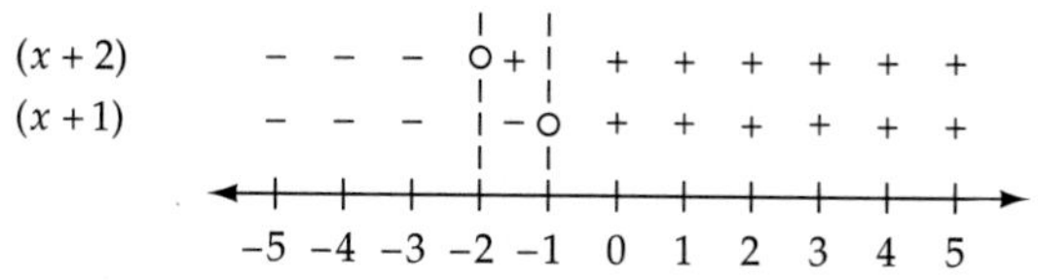

$(x + 2)$ $-$ $-$ $-$ $\circ$ $+$ $+$ $+$ $+$ $+$ $+$
$(x + 1)$ $-$ $-$ $-$ $-\circ$ $+$ $+$ $+$ $+$ $+$ $+$

$$\begin{array}{ccccccccccc} & & & & & & & & & & \\ -5 & -4 & -3 & -2 & -1 & 0 & 1 & 2 & 3 & 4 & 5 \end{array}$$

Figure 2.15

The solution set is $[-2, -1)$. The graph of the solution set is shown in Figure 2.16. Note that -2 is included in the solution set because $(x + 2)/(x + 1) = 0$ when $x = -2$. However, -1 is not included in the solution set because the denominator $(x + 1)$ is zero when $x = -1$.

$$\begin{array}{ccccccccccc} & & & [&) & & & & & & \\ -5 & -4 & -3 & -2 & -1 & 0 & 1 & 2 & 3 & 4 & 5 \end{array}$$

Figure 2.16

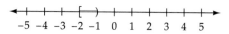 ■ *Try Exercise* **48,** *page 110.*

EXAMPLE 1   **Solve an Absolute Value Equation**

Solve the equation $|2x - 5| = 21$.

Solution $|2x - 5| = 21$ implies $2x - 5 = 21$ or $2x - 5 = -21$. Solving each of these equations produces

$$2x - 5 = 21 \qquad \text{or} \qquad 2x - 5 = -21$$
$$2x = 26 \qquad\qquad\qquad 2x = -16$$
$$x = 13 \qquad\qquad\qquad x = -8$$

Therefore the solution set of $|2x - 5| = 21$ is $\{-8, 13\}$.

■ *Try Exercise* **10**, *page 115.*

Remark   Some absolute value equations have an empty solution set. For example, $|x + 2| = -5$ is false for all values of x. Note that the left side of the equation is an absolute value. Because the absolute value of any real number is nonnegative, the equation is never true.

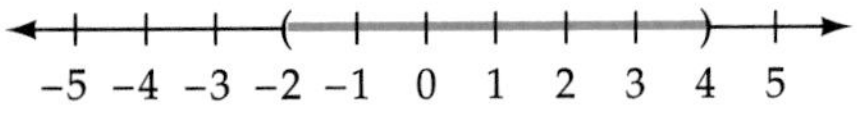

Absolute Value Inequalities

The solution set of the absolute value inequality $|x - 1| < 3$ is the set of all real numbers whose distance from 1 is *less* than 3. Therefore the solution set consists of all numbers greater than -2 and less than 4. See Figure 2.18. In interval notation, the solution set is $(-2, 4)$.

The solution set of the absolute value inequality $|x - 1| > 3$ is the set of all real numbers whose distance from 1 is *greater* than 3. Therefore the solution set consists of all real numbers less than -2 *or* greater than 4. See Figure 2.19. In interval notation, the solution set is $(-\infty, -2) \cup (4, \infty)$.

The following properties are used to solve absolute value inequalities.

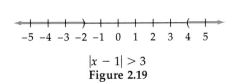

$|x - 1| < 3$

Figure 2.18

$|x - 1| > 3$

Figure 2.19

Properties of Absolute Value Inequalities

For any variable expression E and any nonnegative real number k,

$$|E| \leq k \qquad \text{if and only if} \qquad -k \leq E \leq k$$
$$|E| \geq k \qquad \text{if and only if} \qquad E \leq -k \ \text{ or } \ E \geq k$$

EXAMPLE 2   **Solve an Absolute Value Inequality**

Solve $|2 - 3x| < 7$.

Solution $|2 - 3x| < 7$ implies $-7 < 2 - 3x < 7$. Solving this inequality yields

In Exercises 55 to 64, determine the set of all real numbers x such that y will be a real number. (*Hint:* $\sqrt{a}$ is a real number if and only if $a \geq 0$.)

55. $y = \sqrt{x + 9}$

56. $y = \sqrt{x - 3}$

57. $y = \sqrt{9 - x^2}$

58. $y = \sqrt{25 - x^2}$

59. $y = \sqrt{x^2 - 16}$

60. $y = \sqrt{x^2 - 81}$

61. $y = \sqrt{x^2 - 2x - 15}$

62. $y = \sqrt{x^2 + 4x - 12}$

63. $y = \sqrt{x^2 + 1}$

64. $y = \sqrt{x^2 - 1}$

Supplemental Exercises

In Exercises 65 to 68, use the critical value method to solve each inequality. Use interval notation to write each solution set.

65. $\dfrac{(x - 3)^2}{(x - 6)^2} > 0$

66. $\dfrac{(x - 1)^2}{(x - 4)^4} \geq 0$

67. $\dfrac{(x - 4)^2}{(x + 3)^3} \geq 0$

68. $\dfrac{(2x - 7)}{(x - 1)^2 (x + 2)^2} \geq 0$

In Exercises 69 to 74, determine the set of all real numbers x such that y will be a real number.

69. $y = \sqrt[4]{x^3 - 3x}$

70. $y = \sqrt[6]{x^4 - 4x^3 + 4x^2}$

71. $y = \sqrt[6]{5 + x^2}$

72. $y = \sqrt[6]{(x + 3)^6}$

73. $y = \sqrt{x(x + 2)(x - 5)}$

74. $y = \sqrt{\dfrac{x - 3}{(x + 2)(x - 4)}}$

In Exercises 75 to 78, find the values of k such that the given equation will have at least one real solution.

75. $x^2 + kx + 6 = 0$

76. $x^2 + kx + 11 = 0$

77. $2x^2 + kx + 7 = 0$

78. $-3x^2 + kx - 4 = 0$

79. The equation $s = -16t^2 + v_0 t + s_0$ gives the height s in feet above ground level, at the time t seconds, of an object thrown directly upward from a height s_0 feet above the ground and with an initial velocity of v_0 feet per second. A ball is thrown directly upward from ground level with an initial velocity of 64 feet per second. Find the time interval for which the ball has a height of more than 48 feet.

80. A ball is thrown directly upward from a height of 32 feet above the ground with an initial velocity of 80 feet per second. Find the time interval for which the ball will be more than 96 feet above the ground. (*Hint:* See Exercise 79.)

81. In any triangle, the sum of the lengths of the two shorter sides must be greater than the length of the longest side. Find all possible values of x if a triangle has sides of length

  a. $x, x + 5$, and $x + 9$ **b.** $x, x^2 + x$, and $2x^2 + x$

  c. $\dfrac{1}{x + 2}, \dfrac{1}{x + 1}$, and $\dfrac{1}{x}$

82. Find the solution set of $a^2 > a$, where $a > 0$.

83. If $a > b > 0$, show that $\dfrac{1}{a} < \dfrac{1}{b}$.

2.6   Absolute Value Equations and Inequalities

Recall that the absolute value of a real number x is the distance between the number x and 0 on the real number line. For example, the solution set of the absolute value equation $|x| = 3$ is the set of all real numbers that are 3 units from 0. Figure 2.17 illustrates that there are only two numbers that are 3 units from 0: 3 and -3. Therefore the solution set of $|x| = 3$ is $\{-3, 3\}$.

Recall from Section 1.2 that $|a - b|$ is the distance on a real number line between the graph of a and the graph of b. Thus $|x - 2|$ is the distance between the real number x and 2. The equation $|x - 2| = 5$ is satisfied by real numbers x that are 5 units from 2. Therefore $|x - 2| = 5$ has $\{-3, 7\}$ as its solution set.

The following property is used to solve absolute value equations.

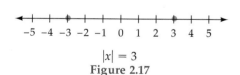

$|x| = 3$

Figure 2.17

A Property of Absolute Value Equations

For any variable expression E and any nonnegative real number k,

$$|E| = k \qquad \text{if and only if} \qquad E = k \quad \text{or} \quad E = -k$$

EXAMPLE 1   Solve an Absolute Value Equation

Solve the equation $|2x - 5| = 21$.

Solution $|2x - 5| = 21$ implies $2x - 5 = 21$ or $2x - 5 = -21$. Solving each of these equations produces

$$2x - 5 = 21 \quad \text{or} \quad 2x - 5 = -21$$
$$2x = 26 \qquad\qquad\qquad 2x = -16$$
$$x = 13 \qquad\qquad\qquad x = -8$$

Therefore the solution set of $|2x - 5| = 21$ is $\{-8, 13\}$.

■ Try Exercise 10, page 115.

Remark   Some absolute value equations have an empty solution set. For example, $|x + 2| = -5$ is false for all values of x. Note that the left side of the equation is an absolute value. Because the absolute value of any real number is nonnegative, the equation is never true.

Absolute Value Inequalities

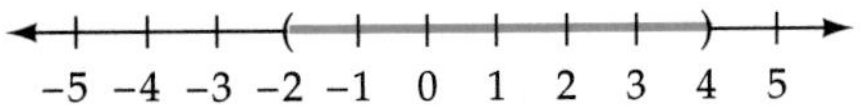

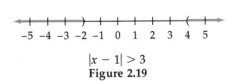

$|x - 1| < 3$
Figure 2.18

The solution set of the absolute value inequality $|x - 1| < 3$ is the set of all real numbers whose distance from 1 is *less* than 3. Therefore the solution set consists of all numbers greater than -2 and less than 4. See Figure 2.18. In interval notation, the solution set is $(-2, 4)$.

$|x - 1| > 3$
Figure 2.19

The solution set of the absolute value inequality $|x - 1| > 3$ is the set of all real numbers whose distance from 1 is *greater* than 3. Therefore the solution set consists of all real numbers less than -2 *or* greater than 4. See Figure 2.19. In interval notation, the solution set is $(-\infty, -2) \cup (4, \infty)$.

The following properties are used to solve absolute value inequalities.

Properties of Absolute Value Inequalities

For any variable expression E and any nonnegative real number k,

$$|E| \leq k \quad \text{if and only if} \quad -k \leq E \leq k$$
$$|E| \geq k \quad \text{if and only if} \quad E \leq -k \quad \text{or} \quad E \geq k$$

EXAMPLE 2   Solve an Absolute Value Inequality

Solve $|2 - 3x| < 7$.

Solution $|2 - 3x| < 7$ implies $-7 < 2 - 3x < 7$. Solving this inequality yields

EXAMPLE 7   **Solve a Rational Inequality**

Solve $\dfrac{3x + 4}{x + 1} \leq 2$.

Solution   Write the inequality so that 0 appears on the right side of the inequality.

$$\frac{3x + 4}{x + 1} \leq 2$$

$$\frac{3x + 4}{x + 1} - 2 \leq 0$$

Write the left side as a rational expression.

$$\frac{3x + 4}{x + 1} - \frac{2(x + 1)}{x + 1} \leq 0 \quad \bullet \text{ The LCD is } x + 1.$$

$$\frac{3x + 4 - 2x - 2}{x + 1} \leq 0 \quad \bullet \text{ Simplify.}$$

$$\frac{x + 2}{x + 1} \leq 0$$

The critical values of the above inequality are -2 and -1 because the numerator $x + 2$ is equal to zero when $x = -2$, and the denominator $x + 1$ is equal to zero when $x = -1$. The critical values -2 and -1 separate the real number line into the three intervals $(-\infty, -2)$, $(-2, -1)$, and $(-1, \infty)$.

All values of x on the interval $(-2, -1)$, make $(x + 2)/(x + 1)$ negative, as desired. On the other intervals the quotient $(x + 2)/(x + 1)$ is positive. See the sign diagram in Figure 2.15.

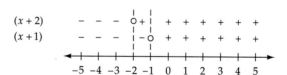

Figure 2.15

The solution set is $[-2, -1)$. The graph of the solution set is shown in Figure 2.16. Note that -2 is included in the solution set because $(x + 2)/(x + 1) = 0$ when $x = -2$. However, -1 is not included in the solution set because the denominator $(x + 1)$ is zero when $x = -1$.

Figure 2.16

■ *Try Exercise 48, page 110.*

EXERCISE SET 2.5

In Exercises 1 to 10, use the properties of inequalities to solve each inequality.

1. $2x + 3 < 11$

2. $3x - 5 > 16$

3. $x + 4 > 3x + 16$

4. $5x + 6 < 2x + 1$

5. $-6x + 1 \geq 19$

6. $-5x + 2 \leq 37$

7. $-3(x + 2) \leq 5x + 7$

8. $-4(x - 5) \geq 2x + 15$

9. $-4(3x - 5) > 2(x - 4)$

10. $3(x + 7) \leq 5(2x - 8)$

11. A bank offers two checking account plans. Under plan A, you pay $5.00 a month plus $0.01 a check. Under plan B, you pay $1.00 a month plus $0.08 a check. Under what conditions is it less expensive to use plan A?

12. You can rent a car for the day from Company A for $19.00 plus $0.12 a mile. Company B charges $12.00 plus $0.21 a mile. Find the number of miles m (to the nearest mile) per day for which it is cheaper to rent from Company A.

13. A sales clerk has a choice between two payment plans. Plan A pays $100.00 a week plus $8.00 a sale. Plan B pays $250.00 a week plus $3.50 a sale. How many sales per week must be made for plan A to yield the greater paycheck?

14. A video store offers two rental plans. Plan A requires a $15.00 yearly membership fee and charges $1.49 per video per day. Plan B does not have a membership fee but charges $1.99 per video per day. How many videos can be rented per year if plan B is to be the least expensive of the plans?

In Exercises 15 to 20, use the properties of inequalities to solve each inequality.

15. $-2 < 4x + 1 \leq 17$

16. $-16 < 2x + 5 < 9$

17. $10 \geq 3x - 1 \geq 0$

18. $0 \leq 2x + 6 \leq 54$

19. $20 > 8x - 2 \geq -5$

20. $4 \leq 10x + 1 \leq 51$

21. The average daily minimum-maximum temperature range for the city of Palm Springs during the month of September is 68 to 104 degrees Fahrenheit. What is the corresponding temperature range measured on the Celsius temperature scale? (*Hint:* Let F be the average daily temperature. Then $68 \leq F \leq 104$. Now substitute $\frac{9}{5}C + 32$ for F and solve the resulting inequality for C.)

22. The average daily minimum-maximum temperature range for the city of Palm Springs during the month of January is 41 to 68 degrees Fahrenheit. What is the corresponding temperature range measured on the Celsius temperature scale?

23. The sum of three consecutive even integers is between 36 and 54. Find all possible sets of integers that satisfy these conditions.

24. The sum of three consecutive odd integers is between 63 and 81. Find all possible sets of integers that satisfy these conditions.

In Exercises 25 to 36, use the critical value method to solve each inequality. Use interval notation to write each solution set.

25. $x^2 + 7x > 0$

26. $x^2 - 5x \leq 0$

27. $x^2 - 16 \leq 0$

28. $x^2 - 49 > 0$

29. $x^2 + 7x + 10 < 0$

30. $x^2 + 5x + 6 < 0$

31. $x^2 - 3x \geq 28$

32. $x^2 < -x + 30$

33. $6x^2 - 4 \leq 5x$

34. $12x^2 + 8x \geq 15$

35. $8x^2 \geq 2x + 15$

36. $12x^2 - 16x < -5$

37. The monthly revenue R for a product is given by $R = 420x - 2x^2$, where x is the price in dollars of each unit produced. Find the interval in terms of x for which the monthly revenue is greater than zero.

38. A shoe manufacturer finds that the monthly revenue R from a particular style of running shoe is given by $R = 312x - 3x^2$, where x is the price in dollars of each pair of shoes sold. Find the interval in terms of x for which the monthly revenue is greater than or equal to $5925.

In Exercises 39 to 54, use the critical value method to solve each inequality. Write each solution set in interval notation.

39. $\dfrac{x + 4}{x - 1} < 0$

40. $\dfrac{x - 2}{x + 3} > 0$

41. $\dfrac{x - 5}{x + 8} \geq 0$

42. $\dfrac{x - 4}{x + 6} \leq 0$

43. $\dfrac{x}{2x + 7} \geq 0$

44. $\dfrac{x}{3x - 5} \leq 0$

45. $\dfrac{(x + 1)(x - 4)}{x - 2} < 0$

46. $\dfrac{x(x - 4)}{x + 5} > 0$

47. $\dfrac{x + 2}{x - 5} \leq 2$

48. $\dfrac{3x + 1}{x - 2} \geq 4$

49. $\dfrac{6x^2 - 11x - 10}{x} > 0$

50. $\dfrac{3x^2 - 2x - 8}{x - 1} \geq 0$

51. $\dfrac{x^2 - 6x + 9}{x - 5} \leq 0$

52. $\dfrac{x^2 + 10x + 25}{x + 1} \geq 0$

53. $\dfrac{x^2 - 3x - 4}{x + 1} \geq 0$

54. $\dfrac{x^2 + 6x + 9}{x + 3} \leq 0$

$$-7 < 2 - 3x < 7$$

$$-9 < \quad -3x \quad < 5 \qquad \bullet \text{ Add } -2 \text{ to each of the three parts of the inequality.}$$

$$3 > \quad x \quad > -\frac{5}{3} \qquad \bullet \text{ Multiply each part of the inequality by } -\frac{1}{3} \text{ and reverse the inequality symbols.}$$

In interval notation, the solution set is given by $(-5/3, 3)$.

■ *Try Exercise* **30,** *page 115.*

EXAMPLE 3 **Solve an Absolute Value Inequality**

Solve $|4x - 3| \geq 5$.

Solution

$|4x - 3| \geq 5$ implies $4x - 3 \leq -5$ or $4x - 3 \geq 5$. Solving each of these inequalities produces

$$4x - 3 \leq -5 \qquad \text{or} \qquad 4x - 3 \geq 5$$

$$4x \leq -2 \qquad\qquad\qquad 4x \geq 8$$

$$x \leq -\frac{1}{2} \qquad\qquad\qquad x \geq 2$$

Therefore the solution set is $(-\infty, -1/2] \cup [2, \infty)$.

■ *Try Exercise* **34,** *page 115.*

Remark Some absolute value inequalities have a solution set that consists of all real numbers. For example, $|x + 9| \geq 0$ is true for all values of x. The left side of the equation is an absolute value. Because the absolute value of any real number is nonnegative, the equation is always true. The interval notation for the set of all real numbers is $(-\infty, \infty)$.

The graph of the solution set of $|x - a| < \delta$ is called the **delta neighborhood** of a. The delta symbol, δ, is used to represent a positive real number, and a represents a constant. Example 4 shows why the graph of the solution set of $|x - a| < \delta$ is called a delta neighborhood of a. The graph consists of all points on a number line that are within a distance δ of a.

EXAMPLE 4 **Solve an Absolute Value Inequality**

Solve $|x - a| < \delta$ for x. Assume $\delta > 0$.

Solution $|x - a| < \delta$ means $-\delta < x - a < \delta$. Adding a to each of the three parts of the inequality produces $a - \delta < x < a + \delta$. Therefore the solution set of $|x - a| < \delta$ is the open interval $(a - \delta, a + \delta)$. The graph of $|x - a| < \delta$ is shown in Figure 2.20.

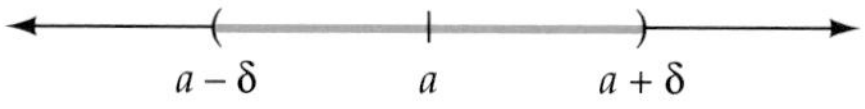

■ *Try Exercise* **46,** *page 115.*

$$|x - a| < \delta$$
Figure 2.20

We can also solve absolute value inequalities by using the critical value method. To solve an absolute value inequality of the form $|P| > k$, where P is a polynomial such as $x^2 - 5$, we first find all values of x such that the left side of the inequality is *equal* to the right side.

EXAMPLE 5   **Solve an Absolute Value Inequality**

Solve $|x^2 - 5| > 4$.

Solution $|x^2 - 5| = 4$ implies $x^2 - 5 = 4$ or $x^2 - 5 = -4$. Solving each of these equations produces

$$x^2 - 5 = 4 \qquad \text{or} \qquad x^2 - 5 = -4$$
$$x^2 = 9 \qquad\qquad\qquad x^2 = 1$$
$$x = \pm 3 \qquad\qquad\qquad x = \pm 1$$

The four values -3, -1, 1, and 3 separate a real number line into the 5 intervals as shown in Figure 2.21. They are the critical values of the inequality because they separate the real numbers that make the inequality true from those that make it false.

Figure 2.21

We can use a test value from each of the intervals to determine on which intervals the original inequality is true. For example, if we choose -4 from the interval $(-\infty, -3)$, then the inequality $|x^2 - 5| > 4$ is true because

$$|(-4)^2 - 5| = |16 - 5| = |11| = 11$$

and 11 is greater than 4. Therefore the interval $(-\infty, -3)$ is part of the solution set. Continuing in a similar manner produces the results shown in the following table:

Interval	$(-\infty, -3)$	$(-3, -1)$	$(-1, 1)$	$(1, 3)$	$(3, \infty)$
Test value x	-4	-2	0	2	4
$\mid x^2 - 5 \mid \overset{?}{>} 4$	$\mid(-4)^2 - 5\mid > 4$ $11 > 4$ True	$\mid(-2)^2 - 5\mid > 4$ $1 > 4$ False	$\mid(0)^2 - 5\mid > 4$ $5 > 4$ True	$\mid(2)^2 - 5\mid > 4$ $1 > 4$ False	$\mid(4)^2 - 5\mid > 4$ $11 > 4$ True

Thus the solution set is $(-\infty, -3) \cup (-1, 1) \cup (3, \infty)$.

■ *Try Exercise* **48**, *page 115.*

EXERCISE SET 2.6

In Exercises 1 to 22, solve each absolute value equation for x.

1. $|x| = 4$
2. $|x| = 7$
3. $|x - 5| = 2$
4. $|x - 8| = 3$
5. $|x + 6| = 1$
6. $|x + 9| = 5$
7. $|x + 14| = 20$
8. $|x - 3| = 14$
9. $|2x - 5| = 11$
10. $|2x - 3| = 21$
11. $|2x + 6| = 10$
12. $|2x + 14| = 60$
13. $\left|\dfrac{x - 4}{2}\right| = 8$
14. $\left|\dfrac{x + 3}{4}\right| = 6$
15. $|2x + 5| = -8$
16. $|4x - 1| = -17$
17. $2|x + 3| + 4 = 34$
18. $3|x - 5| - 16 = 2$
19. $|2x - a| = b \quad (b > 0)$
20. $3|x - d| = c \quad (c > 0)$
21. $|x - a| = \delta \quad (\delta > 0)$
22. $|x + m| = m \quad (m > 0)$

In Exercises 23 to 46, use interval notation to express the solution set of each inequality.

23. $|x| < 4$
24. $|x| > 2$
25. $|x - 1| < 9$
26. $|x - 3| < 10$
27. $|x + 3| > 30$
28. $|x + 4| < 2$
29. $|2x - 1| > 4$
30. $|2x - 9| < 7$
31. $|x + 3| \geq 5$
32. $|x - 10| \geq 2$
33. $|3x - 10| \leq 14$
34. $|2x - 5| \geq 1$
35. $|4 - 5x| \geq 24$
36. $|3 - 2x| \leq 5$
37. $|x - 5| \geq 0$
38. $|x - 7| \geq 0$
39. $|x - 4| \leq 0$
40. $|2x + 7| \leq 0$
41. $|5x - 1| < -4$
42. $|2x - 1| < -9$
43. $|2x + 7| \geq -5$
44. $|3x + 11| \geq -20$
45. $|x - 3| < b \quad (b > 0)$
46. $|x - c| < d \quad (d > 0)$

In Exercises 47 to 54, use the critical value method to solve each inequality. Use interval notation to write the solution sets.

47. $|x^2 - 1| < 1$
48. $|x^2 - 2| > 1$
49. $|x^2 - 10| < 6$
50. $|x^2 + 4| \geq 10$
51. $|x^2 + 7x + 11| \geq 1$
52. $|x^2 - 5x + 6| \leq 1$
53. $|x^2 - 21.5| \geq 4.5$
54. $|x^2 - 6.5| \leq 2.5$

In Exercises 55 to 60, determine whether the statement is true or false. If it is false, explain why.

55. $|x + 2| = |x| + |2|$
56. $|x - 5| = |x| - |5|$
57. $|x - 7| \geq 0$
58. $|x| |5| = |5x|$
59. If $t < 0$, then $|t| = -t$.
60. The absolute value of any real number is a positive number.

Supplemental Exercises

In Exercises 61 to 68, find the values of x that make the equation true.

61. $|x + 4| = x + 4$
62. $|x - 1| = x - 1$
63. $|x + 7| = -(x + 7)$
64. $|x - 3| = -(x - 3)$
65. $|2x + 7| = 2x + 7$
66. $|3x - 11| = -3x + 11$
67. $|x - 2| + |x + 4| = 8$
68. $|x + 1| - |x + 3| = 4$

In Exercises 69 to 80, use interval notation to express the solution set of each inequality.

69. $1 < |x| < 5$
70. $2 < |x| < 3$
71. $3 \leq |x| < 7$
72. $0 < |x| \leq 3$
73. $0 < |x - a| < \delta \quad (\delta > 0)$
74. $0 < |x - 5| < 2$
75. $2 < |x - 6| < 4$
76. $1 \leq |x - 3| < 5$
77. $\left|1 - \dfrac{3x}{4}\right| \geq 6$
78. $\left|2 + \dfrac{3x}{5}\right| < 10$
79. $|x| > |x - 1|$
80. $|x - 2| \leq |x + 4|$

81. Write an absolute value inequality to represent all real numbers within **a.** 8 units of 3; **b.** k units of j (assume $k > 0$).

82. Write an absolute value inequality to represent all real numbers that are more than **a.** 5 units away from 1; **b.** k units away from j (assume $k > 0$).

83. The length of the sides of a square have been measured accurately to within 0.01 foot. This measured length is 4.25 feet.
   a. Write an absolute value inequality that describes the relationship between the actual length of each side of the square s and its measured length.
   b. Solve for s the absolute value inequality you found in part **a.**

Exploring Concepts with Technology

Which Mean Do You Mean?

If $a \leq c \leq b$, then c is said to be a **mean** of a and b. The following chart shows expressions that can be used to calculate the **arithmetic mean,** the **geometric mean,** the **harmonic mean,** and the **root mean square** of a and b.

The arithmetic mean is often used to find the average of two test scores. It can be determined by dividing the sum of the scores by 2. For instance, scores of 70, and 80, have an arithmetic mean of $(70 + 80)/2 = 75$.

Example 6, page 44, illustrated that the arithmetic mean did not determine the average speed for a round trip where v_1 was the average speed on the way to a destination and v_2 was the average speed on the return trip. In this situation the average speed for the round trip is given by

$$\frac{2v_1 v_2}{v_1 + v_2}$$

This expression is called the harmonic mean of v_1 and v_2.

The geometric mean will be used when we develop the concept of a geometric sequence. The root mean square is used in circuit analysis.

1. Use a calculator or a computer to complete the chart.

a	b	Arithmetic Mean $\dfrac{a + b}{2}$	Geometric Mean $\sqrt{ab}$	Harmonic Mean $\dfrac{2ab}{a + b}$	Root Mean Square $\sqrt{\dfrac{a^2 + b^2}{2}}$
8	10				
9	10				
40	50				
49	50				
100	200				
199	200				

2. In each case, which mean was the smallest? Which mean was the largest? Rank the means from smallest to largest.

3. How do the means compare when $a = b$?

The following chart shows expressions for calculating the arithmetic mean, the geometric mean, the harmonic mean, and the root mean square of a, b, and c.

4. Use a calculator or a computer to complete the chart.

a	b	c	Arithmetic Mean $\dfrac{a + b + c}{3}$	Geometric Mean $\sqrt[3]{abc}$	Harmonic Mean $\dfrac{3abc}{ab + ac + bc}$	Root Mean Square $\sqrt{\dfrac{a^2 + b^2 + c^2}{3}}$
8	9	10				
40	50	60				
100	100	100				

5. Rank the means from smallest to largest.

6. Propose formulas that could be used to calculate each of the four means for the numbers a, b, c, and d.

Chapter Review

2.1 Linear Equations

- A number is said to satisfy an equation if a substitution of the number for the variable results in an equation that is a true statement. To solve an equation means to find all values of the variable that satisfy the equation. These values that make the equation true are called solutions or roots of the equation. The set of all solutions of an equation is called the solution set of the equation. Equivalent equations have the same solution set.

- A linear equation in the variable x is an equation that can be written in the form $ax + b = 0$, where a and b are real numbers, with $a \neq 0$. A literal equation is an equation that involves more than one variable.

2.2 Formulas and Applications

- A formula is an equation or inequality that expresses known relationships between two or more variables. Application problems are best solved by using the guidelines developed in this section.

2.3 Quadratic Equations

- A quadratic equation in the variable x is an equation that can be written in the form $ax^2 + bx + c = 0$, where $a \neq 0$. If the quadratic polynomial in a quadratic equation is factorable over the set of integers, then the equation can be solved by factoring and using the zero product property (see page 85). Every quadratic equation can be solved by completing the square or the quadratic formula.

- **The Quadratic Formula**

  If $ax^2 + bx + c = 0$, $a \neq 0$, then $x = \dfrac{-b \pm \sqrt{b^2 - 4ac}}{2a}$.

2.4 Other Types of Equations

- **The Power Principle**

  If P and Q are algebraic expressions and n is a positive integer, then every solution of $P = Q$ is a solution of $P^n = Q^n$.

- An equation is said to be quadratic in form if it can be written in the form $au^2 + bu + c = 0$, $a \neq 0$, where u is an expression involving x.

2.5 Inequalities

- The set of all solutions of an inequality is the solution set of the inequality. Equivalent inequalities have the same solution set. To solve an inequality, use the Properties of Inequalities or the critical value method.

2.6 Absolute Value Equations and Inequalities

- Absolute value equations and inequalities can be solved by applying the following properties:

  For any variable expression E and any nonnegative real number k,

  $$|E| = k \quad \text{if and only if} \quad E = k \ \text{ or } \ E = -k$$
  $$|E| \leq k \quad \text{if and only if} \quad -k \leq E \leq k$$
  $$|E| \geq k \quad \text{if and only if} \quad E \leq -k \ \text{ or } \ E \geq k$$

Essays and Projects

1. One of the most famous of the unsolved mathematical problems is known as *Fermat's Last Theorem*. Write an essay on Fermat's Last Theorem. How is Fermat's Last Theorem related to the Pythagorean theorem?

2. Consider the quadratic equation

$$y = 2x(1 - x)$$

Let x be any number between 0 and 1. Evaluate y for that value of x, substitute that value of y back in for x, and evaluate for the next y value. Continuing this process over and over will produce a sequence of numbers that are *attracted* to 0.5. Verify the above statements for

   a. $x = 0.2$, and

   b. $x = 0.713$.

   c. Write an essay on *attractors*. Explain how the topic of attractors is related to *chaos*. An excellent source of information on attractors is *The Mathematical Tourist* by Ivars Peterson (New York: Freeman, 1988).

3. The right side of the following equation is a continued fraction.

   a. Write an essay on continued fractions and their applications.

   b. Evaluate the following continued fraction by using a substitution that allows the equation to be written as a quadratic equation.

$$x = \cfrac{1}{1 + \cfrac{1}{1 + \cfrac{1}{1 + \cfrac{1}{1 + \cdots}}}}$$

4. Write an essay that explains at least three different proofs of the Pythagorean theorem. Include the proof attributed to President Garfield. Also explain what the converse of the Pythagorean theorem is, and give some of its applications.

5. Consult a geometry text and then write an essay that explains what is meant by the phrase *power of a point P for the circle O*. Include a drawing that illustrates this concept. Use the power of a point terminology to state the Tangent–Secant Theorem on page 78 in Example 4.

True/False Exercises

In Exercises 1 to 10, answer true or false. If the statement is false, give an example to show that the statement is false.

1. If $x^2 = 9$, then $x = 3$.

2. The equations

$$x = \sqrt{12 - x} \quad \text{and} \quad x^2 = 12 - x$$

   are equivalent equations.

3. Adding the same constant to each side of a given equation produces an equation that is equivalent to the given equation.

4. If $a > b$, then $-a < -b$.

5. If $a \neq 0$, $b \neq 0$, and $a > b$, then $\dfrac{1}{a} > \dfrac{1}{b}$.

6. The discriminant of $ax^2 + bx + c = 0$ is $\sqrt{b^2 - 4ac}$.

7. If $\sqrt{a} + \sqrt{b} = c$, then $a + b = c^2$.

8. The solution set of $|x - a| < b$ with $b > 0$ is given by the interval $(a - b, a + b)$.

9. The only quadratic equation that has roots of 4 and -4 is $x^2 - 16 = 0$.

10. Every quadratic equation $ax^2 + bx + c = 0$ with real coefficients such that $ac < 0$ has two distinct real roots.

Chapter Review Exercises

In Exercises 1 to 30, solve each equation.

1. $x - 2(5x - 3) = -3(-x + 4)$

2. $3x - 5(2x - 7) = -4(5 - 2x)$

3. $\dfrac{4x}{3} - \dfrac{4x - 1}{6} = \dfrac{1}{2}$

4. $\dfrac{3x}{4} - \dfrac{2x - 1}{8} = \dfrac{3}{2}$

5. $\dfrac{x}{x + 2} + \dfrac{1}{4} = 5$

6. $\dfrac{y - 1}{y + 1} - 1 = \dfrac{2}{y}$

7. $x^2 - 5x + 6 = 0$

8. $6x^2 + x - 12 = 0$

9. $3x^2 - x - 1 = 0$

10. $x^2 - x + 1 = 0$

11. $3x^3 - 5x^2 = 0$

12. $2x^3 - 8x = 0$

13. $6x^4 - 23x^2 + 20 = 0$ **14.** $3x + 16\sqrt{x} - 12 = 0$

15. $\sqrt{x^2 - 15} = \sqrt{-2x}$ **16.** $\sqrt{x^2 - 24} = \sqrt{2x}$

17. $\sqrt{3x + 4} + \sqrt{x - 3} = 5$

18. $\sqrt{2x + 2} - \sqrt{x + 2} = \sqrt{x - 6}$

19. $\sqrt{4 - 3x} - \sqrt{5 - x} = \sqrt{5 + x}$

20. $\sqrt{3x + 9} - \sqrt{2x + 4} = \sqrt{x + 1}$

21. $\dfrac{1}{(y + 3)^2} = 1$ **22.** $\dfrac{1}{(2s - 5)^2} = 4$

23. $|x - 3| = 2$ **24.** $|x + 5| = 4$

25. $|2x + 1| = 5$ **26.** $|3x - 7| = 8$

27. $(x + 2)^{1/2} + x(x + 2)^{3/2} = 0$

28. $x^2(3x - 4)^{1/4} + (3x - 4)^{5/4} = 0$

29. $(2x - 1)^{2/3} + (2x - 1)^{1/3} = 12$

30. $6(x + 1)^{1/2} - 7(x + 1)^{1/4} - 3 = 0$

In Exercises 31 to 48, solve each inequality. Express your solutions sets by using interval notation.

31. $-3x + 4 \geq -2$ **32.** $-2x + 7 \leq 5x + 1$

33. $x^2 + 3x - 10 \leq 0$ **34.** $x^2 - 2x - 3 > 0$

35. $61 \leq \dfrac{9}{5}C + 32 \leq 95$

36. $30 < \dfrac{5}{9}(F - 32) < 65$

37. $x^3 - 7x^2 + 12x \leq 0$

38. $x^3 + 4x^2 - 21x > 0$

39. $\dfrac{x + 3}{x - 4} > 0$ **40.** $\dfrac{x(x - 5)}{x + 7} \leq 0$

41. $\dfrac{2x}{3 - x} \leq 10$ **42.** $\dfrac{x}{5 - x} \geq 1$

43. $|3x - 4| < 2$ **44.** $|2x - 3| \geq 1$

45. $0 < |x| < 2$ **46.** $0 < |x| \leq 1$

47. $0 < |x - 2| < 1$

48. $0 < |x - a| < b \quad (b > 0)$

In Exercises 49 to 54, solve each equation for the indicated unknown.

49. $V = \pi r^2 h$, for h **50.** $P = \dfrac{A}{1 + rt}$, for t

51. $A = \dfrac{h}{2}(b_1 + b_2)$, for b_1 **52.** $P = 2(l + w)$, for w

53. $e = mc^2$, for m **54.** $F = G\dfrac{m_1 m_2}{s^2}$, for m_1

55. One-half of a number minus one-fourth of the number is four more than one-fifth of the number. What is the number?

56. The length of a rectangle is 9 feet less than twice the width of the rectangle. The perimeter of the rectangle is 54 feet. Find the width and the length.

57. A motorboat left a harbor and traveled to an island at an average rate of 8 knots. The average speed on the return trip was 6 knots. If the total trip took 7 hours, how far is it from the harbor to the island?

58. The price of a magazine subscription rose 5% this year. If the subscription now costs $21, how much did the subscription cost last year?

59. A total of $5500 is deposited into two simple interest accounts. On one account the annual simple interest rate is 4%, and on the second account the annual simple interest rate is 6%. The amount of interest earned for 1 year was $295. How much is invested in each account?

60. A calculator and a battery together sell for $21. The price of the calculator is $20 more than the price of the battery. Find the price of the calculator and the price of the battery.

61. Eighteen owners share the maintenance cost of a condominium complex. If six more units are sold, the maintenance cost will be reduced by $12 per month for each of the present owners. What is the total monthly maintenance cost for the condominium complex?

62. The perimeter of a rectangle is 40 inches and its area is 96 square inches. Find the length and the width of the rectangle.

63. A mason can build a wall in 9 hours less than an apprentice. Together they can build the wall in 6 hours. How long would it take the apprentice working alone to build the wall?

64. An art show brought in $33,196 on the sale of 4526 tickets. The adult tickets sold for $8 and the student tickets sold for $2. How many of each type of ticket were sold?

65. As sand is poured from a chute, it forms a right circular cone whose height is one-fourth the diameter of the base. What is the diameter of the base when the cone has a volume of 144 cubic feet?

66. A manufacturer of calculators finds that the monthly revenue R from a particular style of calculator is given by $R = 72x - 2x^2$, where x is the price in dollars of each calculator. Find the interval, in terms of x, for which the monthly revenue is greater than $576.

Chapter Test

1. Solve $3 - \dfrac{x}{4} = \dfrac{3}{5}$.

2. Solve $\dfrac{3}{x + 2} - \dfrac{3}{4} = \dfrac{5}{x + 2}$.

3. Solve $ax - c = c(x - d)$ for x.

4. Solve $x^2 + 4x - 1 = 0$ by completing the square.

5. Solve $3x^2 + 2x - 9 = 0$.

6. Solve $x^4 + 4x^3 - x - 4 = 0$.

7. Solve $\sqrt{x - 2} + 5 = \sqrt{3 - x}$.

8. Solve $3x^{2/3} + 10x^{1/3} - 8 = 0$.

9. Solve $(x - 3)^{2/3} = 16$.

10. Solve $-3(x + 2) \le 4 - 7x$.

11. Solve $\dfrac{x^2 + x - 12}{x + 1} \ge 0$.

12. Solve $-3 \le 3x - 9 \le 4$.

13. Solve $|2x + 7| = 5$.

14. Solve $|x + 4| < 3$.

15. Solve $|3x - 2| \ge 7$.

16. A boat has a speed of 5 mph in still water. The boat can travel 21 miles with the current in the same time in which it can travel 9 miles against the current. Find the rate of the current.

17. A total of $9000 was deposited into two simple interest accounts. On one account the annual simple interest rate is 8.2%, and on the second account the annual simple interest rate is 6.5%. The amount of interest earned for 1 year was $695.50. How much was invested in each account?

18. A radiator contains 6 liters of a 20% antifreeze solution. How much should be drained and replaced with pure antifreeze to produce a 50% antifreeze solution?

19. A worker can cover a parking lot with asphalt in 10 hours. With the help of an assistant, the work can be done in 6 hours. How long would it take the assistant working alone to cover the parking lot with asphalt?

20. You can rent a car for the day from Company A for $18 plus $0.10 a mile. Company B charges $10 plus $0.18 a mile. At what point, in terms of miles driven per day, is it cheaper to rent from Company A?

3 Functions and Graphs

Earthquake Alarm

Earthquake alarms are now available at your local home improvement store. The principle on which the alarm works exploits the fact that an earthquake generates more than one type of wave. One of the waves is referred to as a *primary wave*, or *P wave*, which travels at the rate of 5 miles per second. Another wave is the *secondary wave*, or *S wave*. The *S* wave is responsible for most of the structural damage caused by an earthquake. It travels at the rate of 3 miles per second. The alarm is triggered by the *P* wave, which alerts residents to the fact that the *S* wave is about to hit.

The amount of time between when the alarm sounds and when the *S* wave hits is a *function* of the distance from the location of the alarm to the epicenter of the earthquake. We can use the formula $t = d/r$ to determine a formula for the function. Let d be the distance to the epicenter. The time required for the *P* wave to reach a specific location is $d/5$. The time required for the *S* wave to reach the same location is $d/3$. The amount of time t between the two waves is

$$t = \frac{d}{3} - \frac{d}{5} = \frac{2}{15}d$$

If you are located 15 miles from the epicenter, you will have only

$$\frac{2}{15} \cdot 15 = 2 \text{ seconds}$$

of warning. However if you are located 90 miles from the epicenter, you will have a warning time of

$$\frac{2}{15} \cdot 90 = 12 \text{ seconds}$$

Mathematicians often express the time t as $t(d)$ to indicate that t is a function of the distance d. In functional notation, the warning time is

$$t(d) = \frac{2}{15}d.$$

3.1   A Two-Dimensional Coordinate System and Graphs

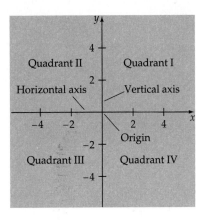

Figure 3.1

Each point on a coordinate axis is associated with a number called its coordinate. Each point on a flat **two-dimensional** surface, called a **coordinate plane,** or xy-plane is associated with an **ordered pair** of numbers called **coordinates** of the point. Ordered pairs are denoted by (a, b), where the real number a is the **x-coordinate** or **abscissa** and the real number b is the **y-coordinate** or **ordinate**.

The coordinates of a point are determined by the point's position relative to a horizontal coordinate axis called the **x-axis** and a vertical coordinate axis called the **y-axis**. The axes intersect at the point $(0, 0)$, called the **origin**. In Figure 3.1, the axes are labeled such that positive numbers appear to the right of the origin on the x-axis and above the origin on the y-axis. The four regions formed by the axes are called **quadrants** and are numbered counterclockwise. This two-dimensional coordinate system is referred to as a **Cartesian, coordinate system** in honor of René Descartes (1596–1650).

To **plot a point** $P(a, b)$ means to draw a dot at its location in the coordinate plane. In Figure 3.2, we have plotted the points $(4, 3)$, $(-3, 1)$, $(-2, -3)$, $(3, -2)$, and $(0, 1)$.

Caution   In Section 1.2, the notation (a, b) was used to denote an interval on a one-dimensional number line. In this section, (a, b) denotes an ordered pair in a two-dimensional plane. This should not cause confusion in future sections, because as each mathematical topic is introduced, it will be clear whether a one-dimensional or a two-dimensional coordinate system is involved.

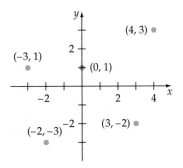

Figure 3.2

Equality of Ordered Pairs

> The ordered pairs (a, b) and (c, d) are equal if and only if
> $$a = c \quad \text{and} \quad b = d$$

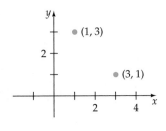

Figure 3.3

The order in which the coordinates of an ordered pair are listed is important. Figure 3.3 shows that $(1, 3)$ and $(3, 1)$ do not denote the same point.

The distance between two points on a horizontal line is the absolute value of the difference between the x-coordinates of the two points. The distance between two points on a vertical line is the absolute value of the difference between the y-coordinates of the two points. For example, as shown in Figure 3.4, the distance d between the points with coordinates $(1, 2)$ and $(1, -3)$ is

$$d = |2 - (-3)| = 5.$$

If two points are not on a horizontal or vertical line, then a *distance formula* for the distance between the two points can be developed as follows.

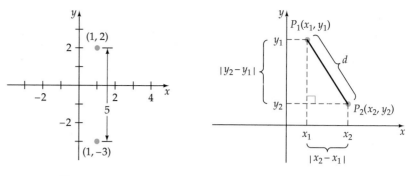

Figure 3.4   Figure 3.5

The distance between the points $P_1(x_1, y_1)$ and $P_2(x_2, y_2)$ in Figure 3.5 is the length of the hypotenuse of a right triangle whose sides are horizontal and vertical line segments that measure $|x_2 - x_1|$ and $|y_2 - y_1|$, respectively. Applying the Pythagorean Theorem to this triangle produces

$$d^2 = |x_2 - x_1|^2 + |y_2 - y_1|^2$$

$$d = \sqrt{|x_2 - x_1|^2 + |y_2 - y_1|^2} \qquad \bullet \text{ The square root theorem. Since } d \text{ is nonnegative, the negative root is not listed.}$$

$$= \sqrt{(x_2 - x_1)^2 + (y_2 - y_1)^2} \qquad \bullet \text{ Since } |x_2 - x_1|^2 = (x_2 - x_1)^2 \text{ and } |y_2 - y_1|^2 = (y_2 - y_1)^2$$

Thus we have established the following theorem.

The Distance Formula

The distance d between the points $P_1(x_1, y_1)$ and $P_2(x_2, y_2)$ is

$$d = \sqrt{(x_2 - x_1)^2 + (y_2 - y_1)^2}$$

The distance d between the points whose coordinates are $P_1(x_1, y_1)$ and $P_2(x_2, y_2)$ is denoted by $d(P_1, P_2)$. To find the distance $d(P_1, P_2)$ between the points $P_1(-3, 4)$ and $P_2(7, 2)$, we apply the distance formula to the points $P_1(-3, 4)$ and $P_2(7, 2)$. Thus $x_1 = -3$, $y_1 = 4$, $x_2 = 7$, and $y_2 = 2$.

$$d(P_1, P_2) = \sqrt{(x_2 - x_1)^2 + (y_2 - y_1)^2}$$

$$= \sqrt{[7 - (-3)]^2 + (2 - 4)^2}$$

$$= \sqrt{104} = 2\sqrt{26} \approx 10.2$$

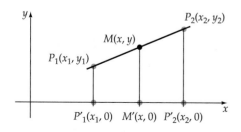

Figure 3.6

The **midpoint** M of a line segment is the point on the line segment that is equidistant from the endpoints $P_1(x_1, y_1)$ and $P_2(x_2, y_2)$ of the segment. See Figure 3.6. To determine a formula for the midpoint $M(x, y)$ of the line segment from $P_1(x_1, y_1)$ to $P_2(x_2, y_2)$, draw vertical lines through P_1, M, and P_2. Note that the vertical lines intersect the x-axis at $P_1'(x_1, 0)$, $M'(x, 0)$, and $P_2'(x_2, 0)$. A theorem from geometry (if three parallel lines intercept equal parts on one transversal, they intercept equal parts on every transversal) gives us $d(P_1', M') = d(M', P_2')$. Now, $d(P_1', P_2') = x_2 - x_1$ (assuming that $x_1 < x_2$). Since M' is the midpoint of line segment from P_1' to P_2', the x-coordinate of M' is

$$x = x_1 + \frac{1}{2}(x_2 - x_1) = x_1 + \frac{1}{2}x_2 - \frac{1}{2}x_1$$

$$= \frac{1}{2}x_1 + \frac{1}{2}x_2 = \frac{x_1 + x_2}{2}$$

Thus the x-coordinate of M is $(x_1 + x_2)/2$. A similar argument can be used to show that the y-coordinate of M is $(y_1 + y_2)/2$. In the above proof, we assumed that $x_1 < x_2$. If $x_1 > x_2$, we obtain the same result. Thus we have established the following theorem.

The Midpoint Formula

The midpoint M of the line segment from $P_1(x_1, y_1)$ to $P_2(x_2, y_2)$ is given by

$$\left(\frac{x_1 + x_2}{2}, \frac{y_1 + y_2}{2} \right)$$

The midpoint formula states that the x-coordinate of the midpoint of a line segment is the *average* of the x-coordinates of the endpoints of the line segment and that the y-coordinate of the midpoint of a line segment is the average of the y-coordinates of the endpoints of the line segment.

The midpoint M of the line segment connecting $P_1(-2, 6)$ and $P_2(3, 4)$ is

$$M = \left(\frac{x_1 + x_2}{2}, \frac{y_1 + y_2}{2} \right) = \left(\frac{(-2) + 3}{2}, \frac{6 + 4}{2} \right) = \left(\frac{1}{2}, 5 \right)$$

The Cartesian coordinate system makes it possible to combine the concepts and methods of algebra and geometry in a way that is useful to both branches of mathematics. One mathematical tool that makes this possible is called a *graph*.

The Graph of an Equation

The **graph of an equation** in the two variables x and y is the set of all points whose coordinates satisfy the equation.

To graph an equation, plot points whose coordinates satisfy the equation.

EXAMPLE 1   **Graph by Plotting Points**

Graph $-2x + y = 1$.

Solution   First solve the equation for y.

$$-2x + y = 1$$

$$y = 2x + 1$$

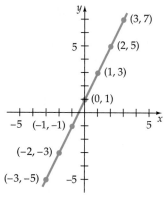

$-2x + y = 1$
Figure 3.7

Now choose a value for x and use the equation to determine the corresponding y value. For example, if $x = -3$, then the corresponding y value is

$$y = 2(-3) + 1 = -6 + 1 = -5.$$

Continuing in this manner produces the following table:

When x is:	-3	-2	-1	0	1	2	3
y is:	-5	-3	-1	1	3	5	7

The table represents the ordered pairs

$(-3, -5)$ $(-2, -3)$ $(-1, -1)$ $(0, 1)$ $(1, 3)$ $(2, 5)$ and $(3, 7)$

Now plot the ordered pairs as points on a Cartesian coordinate system. The points lie on a line, as shown in Figure 3.7.

■ *Try Exercise **26**, page 129.*

EXAMPLE 2   **Graph by Plotting Points**

Graph $y = |x - 2|$.

Solution   This equation is already solved for y, so start by choosing an x value and using the equation to determine the corresponding y value. For example, if $x = -3$, then $y = |(-3) - 2| = |-5| = 5$. Continuing in this manner produces the following table:

When x is:	-3	-2	-1	0	1	2	3	4	5
y is:	5	4	3	2	1	0	1	2	3

Now plot the points listed in the table. The points form a V shape, as shown in Figure 3.8.

■ *Try Exercise **30**, page 129.*

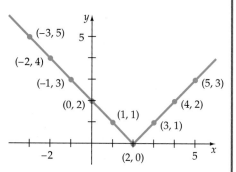

$y = |x - 2|$
Figure 3.8

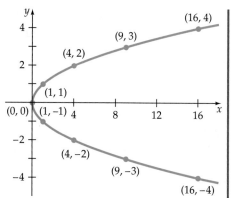

$y^2 = x$
Figure 3.9

EXAMPLE 3   **Graph by Plotting Points**

Graph $y^2 = x$.

Solution   Solving this equation for y yields:

$$y = \pm\sqrt{x}$$

Choose several x-values and use the equation to determine the corresponding y-values.

When x is:	0	1	4	9	16
y is:	0	± 1	± 2	± 3	± 4

Plot the points as shown in Figure 3.9. The graph is a *parabola*.

■ *Try Exercise **32**, page 129.*

Intercepts

Any point that has an x- or a y-coordinate of zero is called an **intercept** of the graph of an equation because it is at these points that the graph intersects the x- or the y-axis.

Definition of x-Intercepts and y-Intercepts

If $(x_1, 0)$ satisfies an equation, then the point $(x_1, 0)$ is called an **x-intercept** of the graph of the equation.
If $(0, y_1)$ satisfies an equation, then the point $(0, y_1)$ is called a **y-intercept** of the graph of the equation.

To find the x-intercepts of the graph of an equation, let $y = 0$ and solve the equation for x. To find the y-intercepts of the graph of an equation, let $x = 0$ and solve the equation for y.

EXAMPLE 4   **Find Intercepts and Graph an Equation**

Graph $y = x^2 - 2x - 3$.

Solution   To find any y-intercepts, let $x = 0$ and solve for y.

$$y = 0^2 - 2(0) - 3 = -3$$

The y intercept is $(0, -3)$. To find the x-intercepts, let $y = 0$ and solve for x.

$$0 = x^2 - 2x - 3$$
$$0 = (x - 3)(x + 1)$$
$$(x - 3) = 0 \quad \text{or} \quad (x + 1) = 0$$
$$x = 3 \quad \text{or} \quad x = -1$$

Thus the x-intercepts are $(3, 0)$ and $(-1, 0)$. To find a few additional points on the graph, we use the equation and determine that

When x is:	1	2	4
y is:	-4	-3	5

Drawing a smooth curve through the intercepts and the points $(1, -4)$, $(2, -3)$, $(4, 5)$ produces the graph of the parabola in Figure 3.10.

■ *Try Exercise **40**, page 129.*

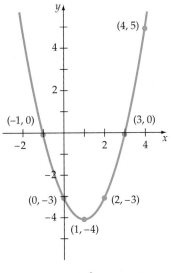

$y = x^2 - 2x - 3$
Figure 3.10

Remark   Since every x-intercept $(x_1, 0)$ has 0 for its y-coordinate, it is convenient to refer to the x-intercept as the real number x_1. Also, since every y-intercept $(0, y_1)$ has 0 for its x-coordinate, it is convenient to refer to the y-intercept as the real number y_1. For example, to say that $3x + 5y = 15$ has an x-intercept of 5 and a y-intercept of 3 means that the graph of $3x + 5y = 15$ intercepts the x-axis at $(5, 0)$ and the y-axis at $(0, 3)$.

Circles and Their Graphs

Frequently you will sketch graphs by plotting points, however, some graphs can be sketched by merely recognizing the form of the equation. A *circle* is an example of a curve whose graph can be sketched after you have inspected its equation.

Definition of a Circle

> A **circle** is the set of points in a plane that are a fixed distance from a specified point. The distance is the **radius** of the circle, and the specified point is the **center** of the circle.

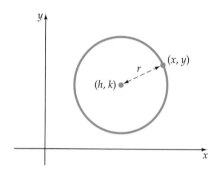

Figure 3.11

The standard form of the equation of a circle is derived by using this definition. To derive the standard form, we use the distance fomula. Figure 3.11 is a circle with center (h, k) and radius r. The point (x, y) is on the circle if and only if it is a distance of r units from the center (h, k). Thus (x, y) is on the circle if and only if

$$\sqrt{(x - h)^2 + (y - k)^2} = r$$

$$(x - h)^2 + (y - k)^2 = r^2 \quad \bullet \text{ Square each side}$$

Standard Form of the Equation of a Circle

> The **standard form of the equation of a circle** with center at (h, k) and radius r is
>
> $$(x - h)^2 + (y - k)^2 = r^2$$

For example, the equation $(x - 3)^2 + (y + 1)^2 = 4$ is the equation of a circle. The standard form of the equation is

$$(x - 3)^2 + (y - (-1))^2 = 2^2$$

from which it can be determined that $h = 3$, $k = -1$, and $r = 2$. Thus the graph is a circle centered at $(3, -1)$ with a radius of 2.

If a circle is centered at the origin $(0, 0)$ (that is, if $h = 0$ and $k = 0$), then the standard form of the equation of the circle simplifies to

$$x^2 + y^2 = r^2$$

For example, the graph of $x^2 + y^2 = 9$ is a circle with center at the origin and radius $r = 3$.

EXAMPLE 5   **Find the Standard Form**

Find the standard form of the equation of a circle that has center $C(-4, -2)$ and contains the point $P(-1, 2)$.

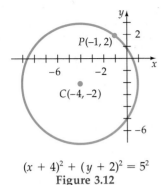

$(x + 4)^2 + (y + 2)^2 = 5^2$

Figure 3.12

Solution   See the sketch of the circle in Figure 3.12. Since the point P is on the circle, the radius r of the circle must equal the distance from C to P. Thus

$$r = \sqrt{(-1 - (-4))^2 + (2 - (-2))^2}$$
$$= \sqrt{9 + 16} = \sqrt{25} = 5$$

Using the standard form with $h = -4$, $k = -2$, and $r = 5$, we obtain

$$(x + 4)^2 + (y + 2)^2 = 5^2$$

■ *Try Exercise* **60,** *page 129.*

If we rewrite $(x + 4)^2 + (y + 2)^2 = 5^2$ by squaring and combining like terms, we produce

$$x^2 + 8x + 16 + y^2 + 4y + 4 = 25$$
$$x^2 + y^2 + 8x + 4y - 5 = 0$$

The above form of the equation is known as the **general form of the equation of a circle.** By completing the square, it is always possible to write the equation $x^2 + y^2 + Ax + By + C = 0$ in the form

$$(x - h)^2 + (y - k)^2 = s$$

for some number s. If $s > 0$, the graph is a circle with radius $r = \sqrt{s}$. If $s = 0$, the graph is the point (h, k), and if $s < 0$, the equation has no real solutions and there is no graph.

EXAMPLE 6   **Find the Center and Radius of a Circle**

Find the center and the radius of the circle that is determined by the equation

$$x^2 + y^2 - 6x + 4y - 3 = 0$$

Solution   First rearrange and group the terms as shown.

$$(x^2 - 6x) + (y^2 + 4y) = 3$$

Now complete the square of $(x^2 - 6x)$ and $(y^2 + 4y)$.

$$(x^2 - 6x + 9) + (y^2 + 4y + 4) = 3 + 9 + 4 \quad \bullet \text{ Add 9 and 4 to each side of the equation.}$$

$$(x - 3)^2 + (y + 2)^2 = 16$$
$$(x - 3)^2 + (y - (-2))^2 = 4^2$$

The above standard form implies that the graph of the equation is a circle with center $(3, -2)$ and radius 4.

■ *Try Exercise* **68,** *page 129.*

EXERCISE SET 3.1

In Exercises 1 to 4, plot the points whose coordinates are given on a Cartesian coordinate system.

1. $(2, 4)$, $(0, -3)$, $(-2, 1)$, $(-5, -3)$

2. $(-3, -5)$, $(-4, 3)$, $(0, 2)$, $(-2, 0)$

3. $(20, 5)$, $(-10, 5)$, $(14, 8)$, $(15, -15)$

4. $(100, 200)$, $(-50, 300)$, $(-100, -250)$

In Exercises 5 to 16, find the distance between the points whose coordinates are given.

5. $(6, 4)$, $(-8, 11)$

6. $(-5, 8)$, $(-10, 14)$

7. $(-4, -20)$, $(-10, 15)$

8. $(40, 32)$, $(36, 20)$

9. $(5, -8)$, $(0, 0)$

10. $(0, 0)$, $(5, 13)$

11. $(\sqrt{3}, \sqrt{8})$, $(\sqrt{12}, \sqrt{27})$

12. $(\sqrt{125}, \sqrt{20})$, $(6, 2\sqrt{5})$

13. (a, b), $(-a, -b)$

14. $(a - b, b)$, $(a, a + b)$

15. $(x, 4x)$, $(-2x, 3x)$ given that $x < 0$

16. $(x, 4x)$, $(-2x, 3x)$ given that $x > 0$

17. Find all points on the x-axis that are 10 units from $(4, 6)$. (*Hint:* First write the distance formula with $(4, 6)$ as one of the points and $(x, 0)$ as the other point.)

18. Find all points on the y-axis that are 12 units from $(5, -3)$.

In Exercises 19 to 24, find the midpoint of the line segment with the following endpoints.

19. $(1, -1)$, $(5, 5)$

20. $(-5, -2)$, $(6, 10)$

21. $(6, -3)$, $(6, 11)$

22. $(4, 7)$, $(-10, 7)$

23. $(1.75, 2.25)$, $(-3.5, 5.57)$

24. $(-8.2, 10.1)$, $(-2.4, -5.7)$

In Exercises 25 to 38, graph each equation by plotting points that satisfy the equation.

25. $x - y = 4$

26. $2x + y = -1$

27. $y = 0.25x^2$

28. $3x^2 + 2y = -4$

29. $y = -2|x - 3|$

30. $y = |x + 3| - 2$

31. $y = x^2 - 3$

32. $y = x^2 + 1$

33. $y = \dfrac{1}{2}(x - 1)^2$

34. $y = 2(x + 2)^2$

35. $y = x^2 + 2x - 8$

36. $y = x^2 - 2x - 8$

37. $y = -x^2 + 2$

38. $y = -x^2 - 1$

In Exercises 39 to 48, find the x- and the y-intercepts, of the graph of each equation. Use the intercepts to draw the graph of the equation.

39. $2x + 5y = 12$

40. $3x - 4y = 15$

41. $x = -y^2 + 5$

42. $x = y^2 - 6$

43. $x = |y| - 4$

44. $x = y^3 - 2$

45. $x^2 + y^2 = 4$

46. $x^2 = y^2$

47. $|x| + |y| = 4$

48. $|x - 4y| = 8$

In Exercises 49 to 58, determine the center and radius of the circle with the given equation.

49. $x^2 + y^2 = 36$

50. $x^2 + y^2 = 49$

51. $x^2 + y^2 = 10^2$

52. $x^2 + y^2 = 4^2$

53. $(x - 1)^2 + (y - 3)^2 = 7^2$

54. $(x - 2)^2 + (y - 4)^2 = 5^2$

55. $(x + 2)^2 + (y + 5)^2 = 25$

56. $(x + 3)^2 + (y + 5)^2 = 121$

57. $(x - 8)^2 + y^2 = \dfrac{1}{4}$

58. $x^2 + (y - 12)^2 = 1$

In Exercises 59 to 66, find an equation of a circle that satisfies the given conditions. Write your answer in standard form.

59. Center $(4, 1)$, radius $r = 2$

60. Center $(5, -3)$, radius $r = 4$

61. Center $\left(\dfrac{1}{2}, \dfrac{1}{4}\right)$, radius $r = \sqrt{5}$

62. Center $\left(0, \dfrac{2}{3}\right)$, radius $r = \sqrt{11}$

63. Center $(0, 0)$, passing through $(-3, 4)$

64. Center $(0, 0)$, passing through $(5, 12)$

65. Center $(1, 3)$, passing through $(4, -1)$

66. Center $(-2, 5)$, passing through $(1, 7)$

In Exercises 67 to 76, find the center and the radius of each of the following circles whose equations are written in the general form.

67. $x^2 + y^2 - 6x + 5 = 0$

68. $x^2 + y^2 - 6x - 4y + 12 = 0$

69. $x^2 + y^2 - 4x - 10y + 20 = 0$

70. $x^2 + y^2 + 4x - 2y - 11 = 0$

71. $x^2 + y^2 - 14x + 8y + 56 = 0$

72. $x^2 + y^2 - 10x + 2y + 25 = 0$

73. $4x^2 + 4y^2 + 4x - 63 = 0$

74. $9x^2 + 9y^2 - 6y - 17 = 0$

75. $x^2 + y^2 - x + \dfrac{2}{3}y + \dfrac{1}{3} = 0$

76. $x^2 + y^2 - 2x + 2y + \dfrac{7}{4} = 0$

Supplemental Exercises

In Exercises 77 to 88, graph the set of all points whose x- and y-coordinates satisfy the given conditions.

77. $x = 3$

78. $y = 2$

79. $x = 1, y \geq 1$

80. $y = -3, x \geq -2$

81. $y \leq 3$

82. $x \geq 2$

83. $xy \geq 0$

84. $|y| \geq 1, \dfrac{x}{y} \leq 0$

85. $|x| = 2, |y| = 3$

86. $|x| = 4, |y| = 1$

87. $|x| \leq 2, y \geq 2$

88. $x \geq 1, |y| \leq 3$

In Exercises 89 to 92, find the other endpoint of the line segment that has the given endpoint and midpoint.

89. Endpoint (5, 1), midpoint (9, 3)

90. Endpoint (4, −6), midpoint (−2, 11)

91. Endpoint (−3, −8), midpoint (2, −7)

92. Endpoint (5, −4), midpoint (0, 0)

93. Use the distance formula to determine whether the points given by (1, −4), (3, 2), (−3, 4), and (−5, −2) are the vertices of a square.

94. Use the distance formula to determine whether the points given by (2, −1), (5, 0), (6, 3), and (3, 2) are the vertices of a parallelogram, a rhombus, or a square.

95. Find a formula for the set of all points (x, y) for which the distance from (x, y) to (3, 4) is 5.

96. Find a formula for the set of all points (x, y) for which the distance from (x, y) to (−5, 12) is 13.

97. Find a formula for the set of all points (x, y) for which the sum of the distances from (x, y) to (4, 0) and from (x, y) to (−4, 0) is 10.

98. Find a formula for the set of all points for which the absolute value of the differences of the distance from (x, y) to (0, 4) and from (x, y) to (0, −4) is 6.

99. Three positive integers x, y, and z are **Pythagorean triples** if they satisfy the equation $x^2 + y^2 = z^2$. Use a calcula-

tor to determine which of the following are Pythagorean triples.

   a. 24, 7, 25   **b.** 80, 39, 89

   c. 420, 29, 421   **d.** 52, 165, 173

100. Pythagorean triples x, y, and z (see Exercise 99) are produced by the equations

$$x = m^2 - n^2, \qquad y = 2mn, \qquad z = m^2 + n^2$$

where m and n are positive integers, with $m > n$. Use the foregoing equations and the following values of m and n to produce the three positive integers x, y, and z. Then use a calculator to verify that they are Pythagorean triples.

   a. $m = 2, n = 1$ **b.** $m = 4, n = 3$

   c. $m = 8, n = 3$ **d.** $m = 15, n = 8$

101. Find an equation of a circle that has a diameter with endpoints (2, 3) and (−4, 11). Write your answer in standard form.

102. Find an equation of a circle that has a diameter with endpoints (7, −2) and (−3, 5). Write your answer in standard form.

103. Find an equation of a circle that has its center at (7, 11) and is tangent to the x-axis. Write your answer in standard form.

104. Find an equation of a circle that has its center at (−2, 3) and is tangent to the y-axis. Write your answer in standard form.

105. Find an equation of a circle that is tangent to both axes, has its center in the second quadrant, and has a radius of 3.

106. Find an equation of a circle that is tangent to both axes, has its center in the third quadrant, and has a diameter of $\sqrt{5}$.

3.2 Functions

Table 3.1

Score	Grade
90–100	A
80–89	B
70–79	C
60–69	D
0–59	F

In many situations in science, business, and mathematics, a correspondence exists between two sets of objects. The correspondence is often described by a table, an equation, or a graph. For example, Table 3.1 describes a grading scale that defines a correspondence between the set of percent scores and the set of letter grades. For any percent score, the table assigns *only one* letter grade. For example, a score of 84% receives a letter grade of B. It is convenient to record this example as the ordered pair (84, B).

The equation $d = 16t^2$ indicates the distance (d) a rock falls (neglecting air resistance) and the time (t) it has been falling. According to this equation, in 3 seconds a rock will fall 144 feet, which we denote by the ordered pair (3, 144). For each nonnegative number t, the equation assigns *only one* nonnegative value for the distance. Therefore the equation defines a correspondence between the set of nonnegative real numbers and the set of nonnegative real numbers. Several of the other ordered pairs determined by $d = 16t^2$ are (0, 0), (1, 16), (2, 64), and $(\frac{5}{2}, 100)$.

Correspondences can be represented by graphs such as Figure 3.13 which describes a correspondence between a nonnegative number x and its positive and negative square root. The graph assigns to each positive number x two real numbers y_1 and y_2. For example, the points (9, 3) and (9, −3) are both on the graph.

Table 3.1, the equation $d = 16t^2$, and the graph in Figure 3.13 all express a correspondence between two sets. The first set is the *domain D* of the correspondence; the second set is the *range R* of the correspondence. For example, the domain of the correspondence defined by Table 3.1 is the set of percent scores {0, 1, 2, ..., 100}, and its range is the set of grades {A, B, C, D, F}. In this chapter we are primarily interested in correspondences in which each element in the domain of the correspondence is assigned to *only one* element in the range. Such correspondences are called functions.

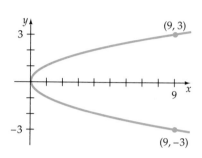

Figure 3.13

Definition of a Function

A **function** f from a set D to a set R is a correspondence, or rule, that pairs each element of D with exactly one element of R. The set D is called the **domain** of f, and the set R is called the **range** of f.

The correspondence described by Table 3.1 and that described by the equation $d = 16t^2$ are both functions. However, the correspondence given by the graph in Figure 3.13 is *not* a function, because some elements of its domain are paired with more than one element of its range.

Functions are designated by letters or a combination of letters, such as f, g, A, log, or tan. If x is an element of the domain of a function f, then $f(x)$, which is read "f of x," is the element in the range of f that is associated with x. For example, if $f(x) = x + 3$ and 2 is in the domain of f then $f(2) = 5$. The **value** of the function f at $x = 2$ is 5, so 5 is an element of the range of f. The process of determining the value of $f(x)$ is referred to as *evaluating the function f at x.*

EXAMPLE 1   **Evaluate Functions**

For the function f defined by $f(x) = x^2 - 11$, evaluate each of the following:

a. $f(7)$ b. $f(-5)$ c. $f(3h)$ d. $f(w + 3)$

Solution

a. $f(7) = 7^2 - 11 = 49 - 11 = 38$

b. $f(-5) = (-5)^2 - 11 = 25 - 11 = 14$

c. $f(3h) = (3h)^2 - 11 = 9h^2 - 11$

d. $f(w + 3) = (w + 3)^2 - 11 = w^2 + 6w + 9 - 11 = w^2 + 6w - 2$

■ *Try Exercise 2, page 138.*

If we solve the equation $3x + y = 5$ for y, we obtain

$$y = -3x + 5$$

Because the value of the variable y depends on the value of the variable x, we call y the **dependent variable** and x the **independent variable.** Using $f(x)$ as a symbol for the dependent variable produces the equation

$$f(x) = -3x + 5$$

Not all equations, however, define a function. Consider the equation

$$y^2 = 25 - x^2$$

This equation does not define a function because if x is 3, y can be 4 or -4. Thus there is an x that is not paired with *exactly* one element of R. The value 3 is paired with 4 and -4.

To determine whether an equation defines a function,

1. Solve (if possible) for the dependent variable.

2. Determine whether each single value of the independent variable produces exactly one value of the dependent variable.

The specific letters used for the independent and dependent variables are not important. For example, the equation $s = t^2$ represents the same function as the equation $y = x^2$. Traditionally, the letter x is used for the independent variable and y for the dependent variable.

Remark   Anytime we use the phrase "y is a function of x" or a similar phrase with different letters, the variable that follows "function of" is the independent variable. The definition of function requires that for each value of the independent variable there be exactly one value of the dependent variable.

EXAMPLE 2   **Identify a Function Given by an Equation**

Identify which of the following equations define y as a function of x.

a. $3x + y = 1$ b. $-4x^2 + y^2 = 9$

Solution

a. $3x + y = 1$ • Solve for y.

$$y = -3x + 1$$

Because $-3x + 1$ is a unique real number for each value of x, this equation defines y as a function of x.

b. $-4x^2 + y^2 = 9$

$$y^2 = 4x^2 + 9$$

$$y = \pm\sqrt{4x^2 + 9}$$

The right side of this equation, $\pm\sqrt{4x^2 + 9}$, produces two values of y for each value of x. For example, when $x = 0$, $y = 3$ or $y = -3$. Therefore this equation does not define y as a function of x.

——————————————————————————— ■ *Try Exercise* **14,** *page 138.*

If the domain of a function is not given, then the domain of the function can be determined from its equation.

Domain of a Function

Unless otherwise stated, the domain of a function defined by an equation is all real numbers except:

a. those numbers for which the denominator of the equation is zero; and

b. those numbers for which the value of the function is not a real number.

EXAMPLE 3   **Find the Domain of a Function**

Find the domain of the functions represented by the following equations.

a. $f(x) = \sqrt{x + 1}$ b. $S(t) = \dfrac{1}{t - 4}$

Solution

a. To avoid taking the square root of a negative number, we must require that the radicand $x + 1$ be greater than or equal to zero. However, $x + 1 \geq 0$ implies that $x \geq -1$. Thus the domain of the function defined by $f(x) = \sqrt{x + 1}$ is $\{x \mid x \geq -1\}$.

b. The real number 4 must be excluded from the domain since it causes division by zero. Therefore the domain of S is all real numbers except 4. In set notation, the domain is $\{t \mid t \neq 4\}$.

——————————————————————————— ■ *Try Exercise* **22,** *page 139.*

The following alternative definition of a function lends itself to a geometric interpretation of a function.

Alternative Definition of a Function

A **function** is a set of ordered pairs in which no two ordered pairs that have the same first component have different second components. The set of first components of the ordered pairs is the **domain** of the function. The set of second components is the **range** of the function.

Remark When using the notation $y = f(x)$, the ordered pairs of the function f can be written as (x, y) or $(x, f(x))$.

You can determine whether a finite set of ordered pairs is a function by examining the x and y values. For example, the set of ordered pairs

$$\{(1, 2), (2, 4), (3, 6), (4, 8)\}$$

is a function because each first component is paired with exactly one second component. The domain of this function is $\{1, 2, 3, 4\}$ and the range is $\{2, 4, 6, 8\}$.

However, the set of ordered pairs

$$\{(0, 0), (1, 1), (4, 2), (1, -1), (4, -2)\}$$

is *not* a function because there are ordered pairs that have the same first component but *different* second components. For example, $(1, 1)$ and $(1, -1)$ are two such ordered pairs.

Graph of a Function

The **graph of a function** is the graph of all the ordered pairs of the function.

To graph a function given by an equation, plot ordered pair solutions of the equation. For example, to graph $f(x) = 2x - 1$, think of this as the equation $y = 2x - 1$. Some of the ordered pair solutions of this equation are $(-1, -3)$, $(0, -1)$, and $(2, 3)$. The graph is shown in Figure 3.14.

To graph $f(x) = x^2 - 1$, think of the equation $y = x^2 - 1$. Some of the ordered pair solutions are $(-2, 3)$, $(-1, 0)$, $(0, -1)$, $(1, 0)$, and $(2, 3)$. The graph is shown in Figure 3.15.

The domain of a function can be determined by observing the x-coordinates of the points on its graph. The range of a function can be determined by observing the y-coordinates of the points on its graph.

The definition that a function is a set of ordered pairs in which no two ordered pairs that have the same first component have different second components implies that any vertical line intersects the graph of a function at no more than one point. This is known as the **vertical line test.**

The Vertical Line Test for Functions

A graph is the graph of a function if and only if no vertical line intersects the graph at more than one point.

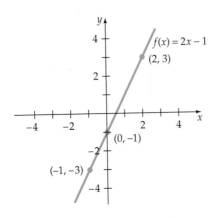

Figure 3.14

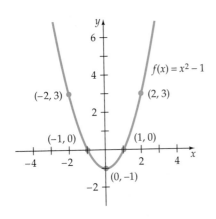

Figure 3.15

EXAMPLE 4 **Use the Vertical Line Test**

Which of the following graphs are graphs of functions?

a.

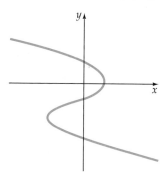

Figure 3.16

b.

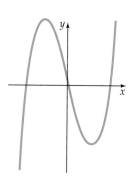

Figure 3.17

Solution

a. This graph *is not* the graph of a function since some vertical lines intersect the graph in more than one point.
b. This graph is the graph of a function since every vertical line intersects the graph in at most one point.

— ■ *Try Exercise* **40**, *page 139.*

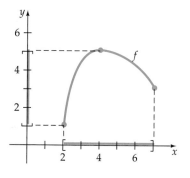

Figure 3.18

EXAMPLE 5 **Find Domain and Range from a Graph**

Find the domain and the range of the function f defined by the graph in Figure 3.18.

Solution   Because the graph extends from $x = 2$ on the left to $x = 7$ on the right, the domain of f is the interval [2, 7]. Using set notation, the domain is written as $\{x \mid 2 \le x \le 7\}$. Because the graph extends from a height of $y = 1$ to a height of $y = 5$, the range is the interval [1, 5]. Using set notation, the range is written as $\{y \mid 1 \le y \le 5\}$.

— ■ *Try Exercise* **52**, *page 139.*

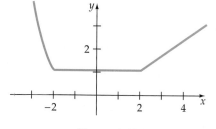

Figure 3.19

Consider the graph in Figure 3.19. As a point on the graph moves from left to right, this graph falls for values of $x \le -2$, it remains at the same height from $x = -2$ to $x = 2$, and it rises for $x \ge 2$. The function represented by the graph is said to be **decreasing** on the interval $(-\infty, -2]$, **constant** on the interval $[-2, 2]$, and **increasing** on the interval $[2, \infty)$.

The concepts of a function increasing, decreasing, or remaining constant are made more precise by the following definition.

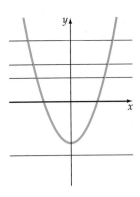

Figure 3.20
Some horizontal lines intersect this graph at more than one point. It is *not* the graph of a one-to-one function.

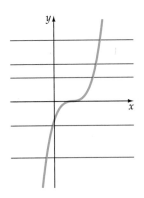

Figure 3.21
Every horizontal line intersects this graph at most once. This is the graph of a one-to-one function.

Definition of Increasing, Decreasing, and Constant Functions

If a and b are elements of an interval I that is a subset of the domain of a function f, then

$$f \text{ is } \textbf{increasing} \text{ on } I \text{ if } f(a) < f(b) \text{ whenever } a < b.$$

$$f \text{ is } \textbf{decreasing} \text{ on } I \text{ if } f(a) > f(b) \text{ whenever } a < b.$$

$$f \text{ is } \textbf{constant} \text{ on } I \text{ if } f(a) = f(b) \text{ for all } a \text{ and } b.$$

Recall that a function is a set of ordered pairs in which no two ordered pairs that have the same first component have different second components. This means that given any x, there is only one y that can be paired with that x. A **one-to-one function** satisfies the additional condition that given any y, there is only one x that can be paired with that given y. In a manner similar to the vertical line test, we can state a horizontal line test for one-to-one functions.

Horizontal Line Test for a One-to-one Function

If any horizontal line intersects the graph of a function at most once, then the graph is the graph of a one-to-one function.

For example, some horizontal lines intersect the graph in Figure 3.20 at more than one point. It is *not* the graph of a one-to-one function. Every horizontal line intersects the graph in Figure 3.21 at most once. This is the graph of a one-to-one function.

Piecewise-defined functions are functions represented by more than one equation. To graph a piecewise-defined function, graph each equation over the indicated domain.

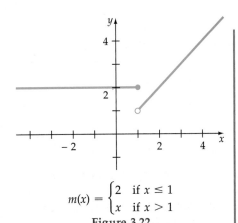

$$m(x) = \begin{cases} 2 & \text{if } x \leq 1 \\ x & \text{if } x > 1 \end{cases}$$

Figure 3.22

EXAMPLE 6   **Graph Piecewise-Defined Functions**

Graph the function $m(x) = \begin{cases} 2 & \text{if } x \leq 1 \\ x & \text{if } x > 1 \end{cases}$

Solution   First draw the graph of $m(x) = 2$ for all domain values $x \leq 1$. This gives you the horizontal ray shown in the graph. The solid dot indicates that the point (1, 2) *is* part of the graph. Then draw the graph of $m(x) = x$ for the domain values $x > 1$. The open dot at (1, 1) indicates that it *is not* part of the graph. See Figure 3.22.

■ *Try Exercise* **80,** *page 140.*

The **greatest integer function** is the piecewise function defined by

$$f(x) = \begin{cases} \vdots & \vdots \\ -1 & \text{if } -1 \le x < 0 \\ 0 & \text{if } 0 \le x < 1 \\ 1 & \text{if } 1 \le x < 2 \\ \vdots & \vdots \end{cases}$$

It can be written more compactly as $f(x) = [\![x]\!]$ by using the following definition.

Definition of the Greatest Integer Function $f(x) = [\![x]\!]$

For any real number x, the greatest integer function $f(x) = [\![x]\!]$ is equal to x if x is an integer. It is equal to the greatest integer less than x if x is not an integer.

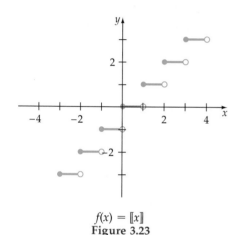

$f(x) = [\![x]\!]$
Figure 3.23

For example,

$$[\![3]\!] = 3 \qquad [\![2.72]\!] = 2 \qquad [\![0.954]\!] = 0 \qquad [\![-3.7]\!] = -4$$

The greatest integer function is a **step function.** Figure 3.23 shows that the graph of $f(x) = [\![x]\!]$ resembles a set of steps. The domain of the greatest integer function is the set of all real numbers, and its range is the set of integers.

Functions are an important aspect of applied mathematics.

EXAMPLE 7 **Solve an Application**

A car was purchased for $16,500. Assuming the car depreciates at a constant rate of $2200 per year (*straight-line depreciation*) for the first 7 years, write the value v of the car as a function of time, and calculate the value of the car 3 years after purchase.

Solution   Let t represent the number of years that have passed since the car was purchased. Then $2200t$ is the amount that the car has depreciated after t years. The value of the car at time t is given by

$$v(t) = 16,500 - 2200t \quad 0 \le t \le 7$$

When $t = 3$, the value of the car is

$$v(3) = 16,500 - 2200(3) = 16,500 - 6600 = \$9900$$

■ *Try Exercise* **92,** *page 141.*

Often in applied mathematics, formulas are used to determine the functional relationship that exists between two variables.

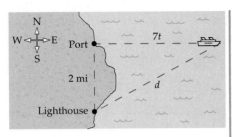

Figure 3.24

EXAMPLE 8   Solve an Application

A lighthouse is 2 miles south of a port. A ship leaves port and sails east at a rate of 7 mph. Express the distance d between the ship and the lighthouse as a function of time, given that the ship has been sailing for t hours.

Solution   Draw a diagram and label it as shown in Figure 3.24. Note that because distance = (rate)(time) and the rate is 7, in t hours the ship has sailed a distance of $7t$.

$$[d(t)]^2 = (7t)^2 + 2^2 \quad \bullet \text{ The Pythagorean Theorem}$$
$$[d(t)]^2 = 49t^2 + 4$$
$$d(t) = \sqrt{49t^2 + 4} \quad \bullet \text{ The } \pm \text{ sign is not used since}$$
$$\phantom{d(t) = \sqrt{49t^2 + 4} \quad \bullet \ } d \text{ must be nonnegative.}$$

■ *Try Exercise* **100**, *page 142.*

EXERCISE SET 3.2

In Exercises 1 to 10, find the indicated functional values.

1. Given $f(x) = 3x - 1$, find

  a. $f(2)$ **b.** $f(-1)$ **c.** $f(0)$

  d. $f\left(\dfrac{2}{3}\right)$ **e.** $f(k)$ **f.** $f(k + 2)$

2. Given $g(x) = 2x^2 + 3$, find

  a. $g(3)$ **b.** $g(-1)$ **c.** $g(0)$

  d. $g\left(\dfrac{1}{2}\right)$ **e.** $g(c)$ **f.** $g(c + 5)$

3. Given $A(w) = \sqrt{w^2 + 5}$, find

  a. $A(0)$ **b.** $A(2)$ **c.** $A(-2)$

  d. $A(4)$ **e.** $A(r + 1)$ **f.** $A(-c)$

4. Given $J(t) = 3t^2 - t$, find

  a. $J(-4)$ **b.** $J(0)$ **c.** $J\left(\dfrac{1}{3}\right)$

  d. $J(-c)$ **e.** $J(x + 1)$ **f.** $J(x + h)$

5. Given $f(x) = \dfrac{1}{|x|}$, find

  a. $f(2)$ **b.** $f(-2)$ **c.** $f\left(\dfrac{-3}{5}\right)$

  d. $f(-\pi)$ **e.** $f(c^2 + 4)$ **f.** $f(2 + h)$

6. Given $T(x) = 5$, find

  a. $T(-3)$ **b.** $T(0)$ **c.** $T\left(\dfrac{2}{7}\right)$

  d. $T(\pi)$ **e.** $T(x + h)$ **f.** $T(3k + 5)$

7. Given $L(t) = -\sqrt{16 - t^2}$, find

  a. $L(0)$ **b.** $L(2)$ **c.** $L(\sqrt{2})$

  d. $L\left(\dfrac{5}{2}\right)$ **e.** $L(a)$ **f.** $L(-a)$

8. Given $a(t) = -16t^2 + 20t - 2$, find

  a. $a(0)$ **b.** $a(2)$ **c.** $a(-2)$

  d. $a(3)$ **e.** $a(k - 1)$ **f.** $a(x + h)$

9. Given $s(x) = \dfrac{x}{|x|}$, find

  a. $s(4)$ **b.** $s(5)$ **c.** $s(-2)$

  d. $s(-3)$ **e.** $s(t), \ t > 0$ **f.** $s(t), \ t < 0$

10. Given $r(x) = \dfrac{x}{x + 4}$, find

  a. $r(0)$ **b.** $r(-1)$ **c.** $r(-3)$

  d. $r\left(\dfrac{1}{2}\right)$ **e.** $r(0.1)$ **f.** $r(10,000)$

In Exercises 11 to 20, identify the equations that define y as a function of x.

11. $2x + 3y = 7$ **12.** $5x + y = 8$

13. $-x + y^2 = 2$ **14.** $x^2 - 2y = 2$

15. $y = 4 \pm \sqrt{x}$ **16.** $x^2 + y^2 = 9$

17. $y = \sqrt[3]{x}$ **18.** $y = |x| + 5$

19. $y^2 = x^2$ **20.** $y^3 = x^3$

In Exercises 21 to 32, determine the domain of the function represented by the given equation.

21. $f(x) = 3x - 4$

22. $f(x) = -2x + 1$

23. $f(x) = x^2 + 2$

24. $f(x) = 3x^2 + 1$

25. $f(x) = \dfrac{4}{x + 2}$

26. $f(x) = \dfrac{6}{x - 5}$

27. $f(x) = \sqrt{7 + x}$

28. $f(x) = \sqrt{4 - x}$

29. $f(x) = \sqrt{4 - x^2}$

30. $f(x) = \sqrt{12 - x^2}$

31. $f(x) = \dfrac{1}{\sqrt{x + 4}}$

32. $f(x) = \dfrac{1}{\sqrt{5 - x}}$

In Exercises 33 to 38, identify the sets of the ordered pairs (x, y) that define y as a function of x.

33. $\{(2, 3), (5, 1), (-4, 3), (7, 11)\}$

34. $\{(5, 10), (3, -2), (4, 7), (5, 8)\}$

35. $\{(4, 4), (6, 1), (5, -3)\}$

36. $\{(2, 2), (3, 3), (7, 7)\}$

37. $\{(1, 0), (2, 0), (3, 0)\}$

38. $\left\{\left(-\dfrac{1}{3}, \dfrac{1}{4}\right), \left(-\dfrac{1}{4}, \dfrac{1}{3}\right), \left(-\dfrac{1}{4}, \dfrac{2}{3}\right)\right\}$

In Exercises 39 to 48, use the vertical line test for functions to determine which of the graphs are graphs of functions.

39.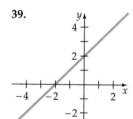

40.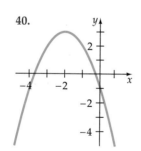

41.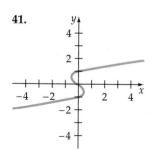

42.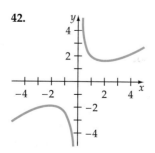

43.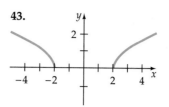

44.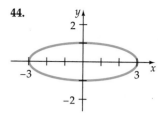

45.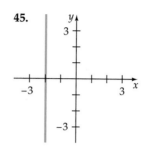

46.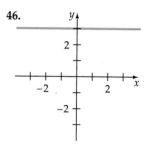

47.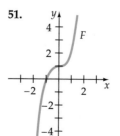

48.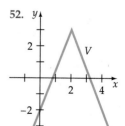

In Exercises 49 to 58 find the domain and range of the function defined by the graph.

49.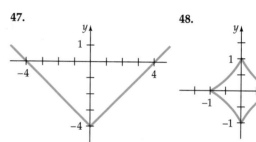

50.

51.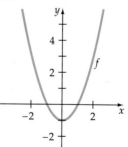

52.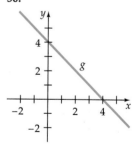

53.

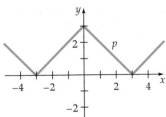

54.

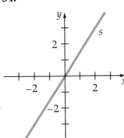

55.

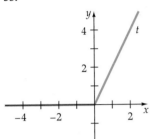

56.

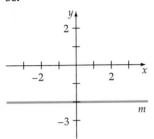

57.

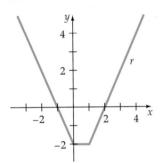

58.

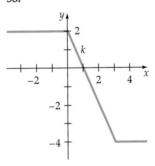

In Exercises 59 to 68, use the indicated graph to identify the intervals over which the function is increasing, constant, or decreasing.

59. f as shown in Exercise 49.

60. g as shown in Exercise 50.

61. F as shown in Exercise 51.

62. V as shown in Exercise 52.

63. p as shown in Exercise 53.

64. s as shown in Exercise 54.

65. t as shown in Exercise 55.

66. m as shown in Exercise 56.

67. r as shown in Exercise 57.

68. k as shown in Exercise 58.

In Exercises 69 to 78, use the indicated graph to determine whether the function is a one-to-one function.

69. f as shown in Exercise 49.

70. g as shown in Exercise 50.

71. F as shown in Exercise 51.

72. V as shown in Exercise 52.

73. p as shown in Exercise 53.

74. s as shown in Exercise 54.

75. t as shown in Exercise 55.

76. m as shown in Exercise 56.

77. r as shown in Exercise 57.

78. k as shown in Exercise 58.

In Exercises 79 to 88, sketch the graph of each function.

79. $f(x) = \begin{cases} |x| & \text{if } x \leq 1 \\ 2 & \text{if } x > 1 \end{cases}$

80. $g(x) = \begin{cases} -4 & \text{if } x \leq 0 \\ x^2 - 4 & \text{if } 0 < x \leq 1 \\ -x & \text{if } x > 1 \end{cases}$

81. $J(x) = \begin{cases} 4 & \text{if } x \leq -2 \\ x^2 & \text{if } -2 < x < 2 \\ -x + 6 & \text{if } x \geq 2 \end{cases}$

82. $K(x) = \begin{cases} 1 & \text{if } x \leq -2 \\ x^2 - 3 & \text{if } -2 < x < 2 \\ \dfrac{1}{2}x & \text{if } x \geq 2 \end{cases}$

83. $L(x) = \left[\!\left[\dfrac{1}{3}x \right]\!\right]$ for $-6 \leq x \leq 6$

84. $M(x) = [\![x]\!] + 2$ for $0 \leq x \leq 4$

85. $N(x) = [\![-x]\!]$ for $-3 \leq x \leq 3$

86. $P(x) = [\![x]\!] + x$ for $0 \leq x \leq 4$

87. $Q(x) = \begin{cases} 2 & \text{if } x < 1 \\ [\![x]\!] & \text{if } x \geq 1 \end{cases}$

88. $R(x) = \begin{cases} [\![-x]\!] & \text{if } x < 0 \\ [\![x]\!] & \text{if } x \geq 0 \end{cases}$

89. A rectangle has length of l feet and a perimeter of 50 feet.
   a. Write the width w of the rectangle as a function of its length.
   b. Write the area A of the rectangle as a function of its length.

90. The sum of two numbers is 20. Let x represent one of the numbers.

a. Write the second number y as a function of x.

b. Write the product P of the two numbers as a function of x.

91. A bus was purchased for $80,000. Assuming the bus depreciates at a rate of $6500 per year (*straight-line depreciation*) for the first 10 years, write the value v of the bus as a function of the time t (measured in years) for $0 \leq t \leq 10$.

92. A boat was purchased for $44,000. Assuming the boat depreciates at a rate of $4200 per year (*straight-line depreciation*) for the first 8 years, write the value v of the boat as a function of the time t (measured in years) for $0 \leq t \leq 8$.

93. A watch manufacturer charges $19.95 per watch if fewer than 50 watches are ordered. If more than 50 but fewer than 200 watches are ordered, the manufacturer reduces the charge per watch by $0.05 for each watch over 50. Thus, for example, if 60 watches are ordered, the charge per watch is $19.45. Find a function that describes the correspondence between the number of watches x ordered and the cost C of the order if $50 < x < 200$.

94. A particular type of avocado tree yields 240 avocados per tree if only 30 trees are planted per acre. For each additional tree planted per acre, the yield of a tree decreases by 10 avocados. Find the function that describes the correspondence between the yield per acre and the number of trees planted.

95. A manufacturer produces a product at a cost of $22.80 per unit. The manufacturer has a fixed cost of $400.00 per day. Each unit retails for $37.00. Let x represent the number of units produced in a 5-day period.

a. Write the total cost C as a function of x.

b. Write the revenue R as a function of x.

c. Write the profit P as a function of x. (*Hint:* The profit function is given by $P(x) = R(x) - C(x)$.)

96. An open box is to be made from a square piece of cardboard having dimensions 30 inches by 30 inches by cutting out squares of area x^2 from each corner, as shown in the figure. Express the volume V of the box as a function of x.

97. A cone has an altitude of 15 centimeters and a radius of 3 centimeters. A right circular cylinder of radius r and height h is inscribed in the cone as shown in the figure. Use similar triangles to write h as a function of r.

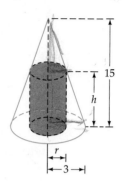

98. Water is running out of a conical funnel that has an altitude of 20 inches and a radius of 10 inches, as shown in the figure.

a. Write the radius r of the water as a function of its depth h.

b. Write the volume V of the water as a function of its depth h.

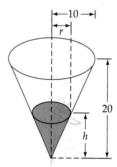

99. For the first minute of flight, a hot air balloon rises vertically at a rate of 3 meters per second. If t is the time in seconds that the balloon has been airborne, write the distance d between the balloon and a point on the ground 50 meters from the point of lift-off as a function of t.

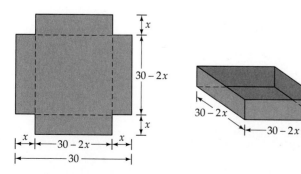

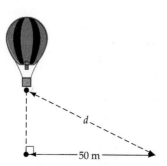

100. An athlete swims from point A to point B at the rate of 2 mph and runs from point B to point C at a rate of 8 mph. Use the dimensions in the figure to write the time t required to reach point C as a function of x.

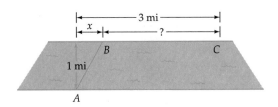

101. An airplane consumes fuel at the rate of

$$f(s) = \frac{1}{425}\left(\frac{8000}{s} + s\right), \quad 200 \le s \le 400$$

gallons per mile when flying at a speed of s miles per hour. Complete the following table by evaluating f (to the nearest hundredth of a gallon) for the indicated speeds.

s	200	250	300	350	400
$f(s)$					

102. A business finds that the number of feet f of pipe it can sell per week is a function of the price p in cents per foot as given by

$$f(p) = \frac{320,000}{p + 25}, \quad 40 \le p \le 90$$

Complete the following table by evaluating f (to the nearest 100 feet) for the indicated values of p.

p	40	50	60	75	90
$f(p)$					

103. The yield Y of apples per tree is related to the amount x of a particular type of fertilizer applied (in pounds per year) by the function

$$Y(x) = 400[1 - 5(x - 1)^{-2}] \quad 5 \le x \le 20$$

Complete the following table by evaluating Y (to the nearest apple) for the indicated applications.

x	5	10	12.5	15	20
$Y(x)$					

104. A manufacturer finds that the cost C in dollars of producing x items of a product is given by

$$C(x) = (225 + 1.4\sqrt{x})^2, \quad 100 \le x \le 1000$$

Complete the following table by evaluating C (to the nearest dollar) for the indicated number of items.

x	100	200	500	750	1000
$C(x)$					

105. If $f(x) = x^2 - x - 5$ and $f(c) = 1$, find c.

106. If $g(x) = -2x^2 + 4x - 1$ and $g(c) = -4$, find c.

107. Determine whether 1 is in the range of

$$f(x) = \frac{x - 1}{x + 1}$$

108. Determine whether 0 is in the range of

$$g(x) = \frac{1}{x - 3}$$

Graphing Calculator Exercises*

109. Sketch the graph of $f(x) = \dfrac{[\![x]\!]}{|x|}$ for $-4 \le x \le 4$, and $x \ne 0$.

110. Sketch the graph of $f(x) = \dfrac{[\![2x]\!]}{|x|}$ for $-4 \le x \le 4$, and $x \ne 0$.

111. Sketch the graph of $f(x) = \sqrt{x^2}$.

112. Sketch the graph of $f(x) = \dfrac{1}{2}x^3 - x - 40$.

113. Sketch the graph of $f(x) = x^2 - 2|x| - 3$.

114. Sketch the graph of $f(x) = x^2 - |2x - 3|$.

115. Sketch the graph of $f(x) = |x^2 - 1| - |x - 2|$.

116. Sketch the graph of $f(x) = |x^2 - 2x| - 3$.

Supplemental Exercises

The notation $f(x)\big|_a^b$ is used to denote the difference $f(b) - f(a)$. That is,

$$f(x)\big|_a^b = f(b) - f(a)$$

In Exercises 117 to 122, evaluate $f(x)\big|_a^b$ for the given function f and the indicated values of a and b.

117. $f(x) = x^2 - x; f(x)\big|_2^3$

118. $f(x) = -3x + 2; f(x)\big|_4^7$

119. $f(x) = 2x^3 - 3x^2 - x; f(x)\big|_0^2$

120. $f(x) = \sqrt{8 - x}; f(x)\big|_0^8$

121. $f(x) = x^2 - 4; f(x)\big|_{-3}^3$

122. $f(x) = 2|6 - x| + 3; f(x)\big|_1^{10}$

*Additional graphing calculator exercises appear in the Graphing Workbook as described in the front of this textbook.

In Exercises 123 to 126, each function has two or more independent variables.

123. Given $f(x, y) = 3x + 5y - 2$, find

    a. $f(1, 7)$ **b.** $f(0, 3)$ **c.** $f(-2, 4)$

    d. $f(4, 4)$ **e.** $f(k, 2k)$ **f.** $f(k + 2, k - 3)$

124. Given $g(x, y) = 2x^2 - |y| + 3$, find

    a. $g(3, -4)$ **b.** $g(-1, 2)$

    c. $g(0, -5)$ **d.** $g\left(\dfrac{1}{2}, -\dfrac{1}{4}\right)$

    e. $g(c, 3c), c > 0$ **f.** $g(c + 5, c - 2), c < 0$

125. The area of a triangle with sides a, b, and c is given by the function

$$A(a, b, c) = \sqrt{s(s - a)(s - b)(s - c)}$$

where s is the semiperimeter

$$s = \frac{a + b + c}{2}$$

Find $A(5, 8, 11)$.

126. The cost in dollars to hire a house painter is given by the function

$$C(h, g) = 15h + 14g$$

where h is the number of hours it takes to paint the house and g is the number of gallons of paint required to paint the house. Find $C(18, 11)$.

127. Let $g(x)$ be the xth digit in the decimal representation of π. For example, $g(1) = 3$. Find $g(6)$.

128. Let $d(x)$ be the xth digit in the decimal representation of $\sqrt{2}$. Find $d(7)$.

129. Let $P(n)$ be the nth prime number. For example, $P(1) = 2$. Find $P(7)$.

130. Let $T(n)$ be the total number of digits it takes to number from page 1 to page n of a book. Find $T(250)$.

A **fixed point** of a function is a number a such that $f(a) = a$. In Exercises 131 and 132, find all fixed points for the given function.

131. $f(x) = x^2 + 3x - 3$

132. $g(x) = \dfrac{x}{x + 5}$

In Exercises 133 and 134, sketch the graph of the piecewise function.

133. $s(x) = \begin{cases} 1 & \text{if } x \text{ is an integer} \\ 2 & \text{if } x \text{ is not an integer} \end{cases}$

134. $v(x) = \begin{cases} 2x - 2 & \text{if } x \neq 3 \\ 1 & \text{if } x = 3 \end{cases}$

3.3   Linear Functions

Much of the remainder of this text is concerned with the study of various functions and their graphs. The following function is important because it is both simple and useful.

Definition of a Linear Function

> A linear function is a function that can be represented by an equation of the form
>
> $$f(x) = mx + b$$
>
> where m and b are real constants.

The graph of a linear function is a nonvertical straight line. If $m = 0$, then $f(x) = mx + b$ simplifies to $f(x) = b$, which is called a **constant function.** The graph of a constant function is a horizontal line, as shown in Figure 3.25.

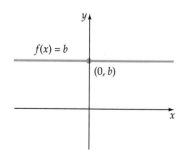

Figure 3.25

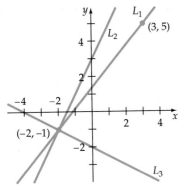

Figure 3.26

The graphs shown in Figure 3.26 are the graphs of linear functions for various values of m. The graphs intersect at the point $(-2, -1)$ but the graphs differ in *steepness*. The steepness of a line is called the **slope** of the line and is denoted by the symbol m. The slope of a line is the ratio of the change in y values of any two points on the line to the change in the x values of the same two points. For example, the graph of the line L_1 in Figure 3.26 passes through the points $(-2, -1)$ and $(3, 5)$. The change in the y values of these two points is determined by subtracting the two y-coordinates.

$$\text{Change in } y = 5 - (-1) = 6$$

The change in the x values is determined by subtracting the two x-coordinates.

$$\text{Change in } x = 3 - (-2) = 5$$

The slope m of L_1 is the ratio of the change in the y values of the two points to the change in the x values of the two points. That is,

$$m = \frac{\text{change in } y}{\text{change in } x} = \frac{6}{5}$$

Since the slope of a nonvertical line can be calculated by using any two arbitrary points on the line, we have the following formula.

Slope of a Nonvertical Line

The slope m of the line passing through the points $P_1(x_1, y_1)$ and $P_2(x_2, y_2)$ with $x_1 \neq x_2$ is given by

$$m = \frac{y_2 - y_1}{x_2 - x_1}$$

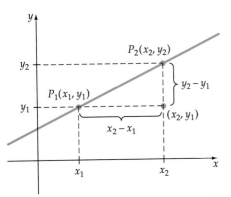

Figure 3.27

Since the numerator $y_2 - y_1$ is the vertical **rise** and the denominator $x_2 - x_1$ is the horizontal **run** from P_1 to P_2, slope is often referred to as the *rise over the run* or the *change in y divided by the change in x*. See Figure 3.27. Lines that have a positive slope slant upward from left to right. Lines that have a negative slope slant downward from left to right.

EXAMPLE 1   **Find the Slope of a Line**

Find the slope of the line passing through the points whose coordinates are given.   a.   (1, 2) and (3, 6)   b.   (−3, 4) and (1, −2)

Solution

a.   The slope of the line passing through (1, 2) and (3, 6) is

$$m = \frac{y_2 - y_1}{x_2 - x_1} = \frac{6 - 2}{3 - 1} = \frac{4}{2} = 2$$

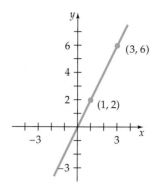

Figure 3.28

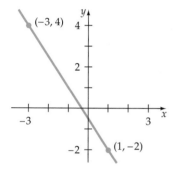

Figure 3.29

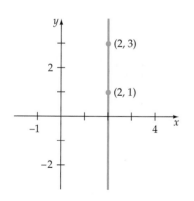

Figure 3.30

Since $m > 0$, the line slants upward from left to right. See the graph of the line in Figure 3.28.

b. The slope of the line passing through $(-3, 4)$ and $(1, -2)$ is

$$m = \frac{y_2 - y_1}{x_2 - x_1} = \frac{-2 - 4}{1 - (-3)} = \frac{-6}{4} = -\frac{3}{2}$$

Since $m < 0$, the line slants downward from left to right. See the graph of the line in Figure 3.29.

■ *Try Exercise* **2,** *page 152.*

The definition of slope does not apply to vertical lines. Consider, for example, the points $(2, 1)$ and $(2, 3)$ on the vertical line in Figure 3.30. Applying the definition of slope to this line produces

$$m = \frac{3 - 1}{2 - 2}$$

which is undefined because it requires division by zero. Since division by zero is undefined, we say that the slope of any vertical line is undefined.

When computing the slope of a line, it does not matter which point we label P_1 and which P_2 because

$$\frac{y_2 - y_1}{x_2 - x_1} = \frac{y_1 - y_2}{x_1 - x_2}$$

For example, the slope of a line that passes through $(1, 2)$ and $(5, 10)$ is given by

$$\frac{10 - 2}{5 - 1} = \frac{8}{4} = 2, \quad \text{or} \quad \frac{2 - 10}{1 - 5} = \frac{-8}{-4} = 2$$

In functional notation, the points P_1 and P_2 can be represented by

$$(x_1, f(x_1)) \quad \text{and} \quad (x_2, f(x_2)).$$

And in this notation, the slope formula

$$m = \frac{y_2 - y_1}{x_2 - x_1} \quad \text{is expressed as} \quad m = \frac{f(x_2) - f(x_1)}{x_2 - x_1}$$

The linear function $f(x) = mx + b$ is often written as $y = mx + b$. The equation $y = mx + b$ is called the **slope-intercept form** because of the following theorem.

Slope-Intercept Form

The graph of $f(x) = mx + b$ has slope m and y intercept $(0, b)$.

Proof: The slope of the graph of $f(x) = mx + b$ is given by

$$\frac{f(x_2) - f(x_1)}{x_2 - x_1} = \frac{(mx_2 + b) - (mx_1 + b)}{x_2 - x_1} = \frac{m(x_2 - x_1)}{x_2 - x_1} = m, \quad x_1 \neq x_2$$

The y-intercept of the graph of $f(x) = mx + b$ is found by letting $x = 0$ and solving for y:

$$y = m(0) + b = b$$

Thus $(0, b)$ is the y-intercept, and m is the slope of the graph of $y = mx + b$.

If an equation is written in the form $y = mx + b$, then its graph can be drawn by first plotting the y-intercept $(0, b)$ and then using its slope m to determine another point on the line.

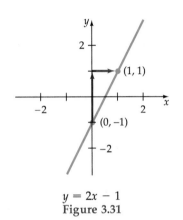

$y = 2x - 1$
Figure 3.31

EXAMPLE 2   Graph a Linear Equation

Graph $y = 2x - 1$.

Solution The equation $y = 2x - 1$ is in slope-intercept form, with $b = -1$ and $m = 2$ or $\frac{2}{1}$. To graph the equation, first plot the y-intercept $(0, -1)$ and then use the slope to plot a second point, which is two units up and one unit over from the y-intercept. (See Figure 3.31.)

■ *Try Exercise* **16**, *page 152.*

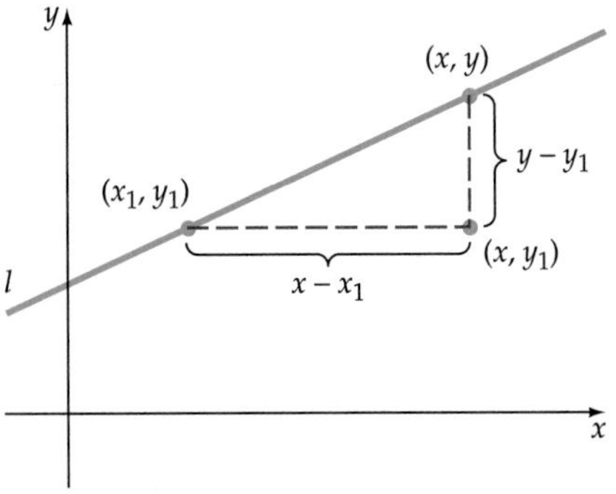

The previous example was concerned with sketching the graph of a linear equation. Let us now consider the problem of finding an equation of a line, provided we know its slope and at least one point on the line. Figure 3.32 suggests that if (x_1, y_1) is a point on a line l of slope m, and (x, y) is *any other* point on the line, then

$$\frac{y - y_1}{x - x_1} = m$$

Multiplying each side of this equation by $x - x_1$ produces

$$y - y_1 = m(x - x_1)$$

This equation is called the **point-slope form** of the line l.

Figure 3.32

The slope of line l is $m = \dfrac{y - y_1}{x - x_1}$.

Point-Slope Form

The graph of the equation

$$y - y_1 = m(x - x_1)$$

is a line that has slope m and passes through (x_1, y_1).

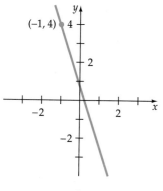

$y = -3x + 1$
Figure 3.33

EXAMPLE 3   **Use the Point-Slope Form**

Find an equation of a line with slope -3 that passes through $(-1, 4)$. Write your answer in slope-intercept form.

Solution   Use the point-slope form with $m = -3$, $x_1 = -1$, and $y_1 = 4$.

$$y - y_1 = m(x - x_1)$$
$$y - 4 = -3(x - (-1))$$
$$y - 4 = -3x - 3$$
$$y = -3x + 1$$

Figure 3.33 shows the graph of $y = -3x + 1$, which has slope -3 and passes through $(-1, 4)$.

■ *Try Exercise **28**, page 152.*

To determine an equation of a nonvertical line that passes through two points, first determine the slope of the line and then use the coordinates of either one of the points in the point-slope form.

EXAMPLE 4   **Find an Equation of a Line Through Two Points**

Find an equation of a line that passes through $(-3, 2)$ and $(2, -4)$. Write the equation in the form $y = mx + b$.

Solution   The slope is

$$m = \frac{y_2 - y_1}{x_2 - x_1} = \frac{-4 - 2}{2 - (-3)} = -\frac{6}{5}$$

Use the point-slope form with $x_1 = -3$, and $y_1 = 2$.

$$y - y_1 = m(x - x_1)$$
$$y - 2 = -\frac{6}{5}(x - (-3))$$
$$y - 2 = -\frac{6}{5}x - \frac{18}{5}$$
$$y = -\frac{6}{5}x - \frac{8}{5}$$

■ *Try Exercise **40**, page 152.*

Remark   The same result is obtained by using $x_1 = 2$, and $y_1 = -4$.

An equation of the form $Ax + By + C = 0$, where A, B, and C are real numbers and both A and B are not zero, is called the **general form** of the equation of a line. For example, the equation $y = -\frac{6}{5}x - \frac{8}{5}$ in Example 4 is written in general form as $6x + 5y + 8 = 0$.

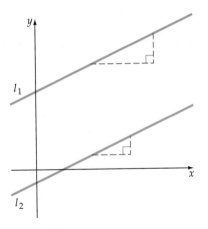

Figure 3.34

Parallel Lines and Perpendicular Lines

Two lines in a plane that have no points in common are said to be **parallel**. In Figure 3.34, two right triangles have been constructed by using the lines l_1 and l_2 and drawing line segments parallel to the coordinate axes. The lines l_1 and l_2 are parallel if and only if the right triangles are similar. Also, the triangles are similar if and only if corresponding sides are proportional. This suggests the following theorem.

Parallel Lines and Slope

Two nonvertical lines are parallel if and only if their slopes are equal.

Recall that the phrase "if and only if" in this theorem means that both of the following are true:

1. If the two nonvertical lines l_1 and l_2 are parallel, then their slopes are equal.
2. If the two nonvertical lines l_1 and l_2 have equal slopes, then they are parallel.

EXAMPLE 5   **Find an Equation of a Parallel Line**

Find the general form of the equation of the line l that passes through the point $(-2, 3)$ and is parallel to the graph of $3x + 4y - 12 = 0$.

Solution   First express the equation in its slope-intercept form.

$$3x + 4y - 12 = 0$$

$$y = -\frac{3}{4}x + 3$$

Since parallel lines have equal slopes, the slope of the line l is $m = -\frac{3}{4}$. Use the point-slope form with $m = -\frac{3}{4}$, $x_1 = -2$, and $y_1 = 3$.

$$y - y_1 = m(x - x_1)$$

$$y - 3 = -\frac{3}{4}(x + 2)$$

$$4y - 12 = -3x - 6$$

$$3x + 4y - 6 = 0$$

■ *Try Exercise* **48**, *page 152.*

Two lines that intersect at a right angle (90°) are **perpendicular** to each other. For example, every vertical line is perpendicular to every horizontal line. If the lines l_1 and l_2 are neither vertical nor horizontal, then you can determine whether they are perpendicular by applying the following theorem.

Perpendicular Lines

Two lines with slopes m_1 and m_2 are perpendicular if and only if

$$m_1 m_2 = -1$$

Remark $m_1 m_2 = -1$ means that m_1 and m_2 are *negative reciprocals* of each other; that is, $m_1 = -1/m_2$.

EXAMPLE 6   Find an Equation of a Perpendicular Line

Find the equation of the line l through (2, 1) that is perpendicular to the graph of $3x + y - 6 = 0$. Write the answer in the general form of the equation of a line.

Solution Find the slope m_1 of the given line. The slope m_2 of the line l is the negative reciprocal of m_1.

$$3x + y - 6 = 0$$

$$y = -3x + 6 \quad \text{Thus } m_1 = -3.$$

$$m_2 = -\frac{1}{m_1} = -\frac{1}{-3} = \frac{1}{3}$$

Now use the point-slope form to find an equation of the line l, and write the equation in its general form.

$$y - y_1 = m(x - x_1)$$

$$y - 1 = \frac{1}{3}(x - 2)$$

$$3y - 3 = x - 2$$

$$-x + 3y - 1 = 0$$

■ *Try Exercise* **52,** *page 152.*

Remark The general form $-x + 3y - 1 = 0$ is not unique. For instance, it could also be expressed as $x - 3y + 1 = 0$.

The equation

$$\frac{x}{a} + \frac{y}{b} = 1$$

is referred to as the intercept form of a line because its graph is a line through the x-intercept $(a, 0)$ and the y-intercept $(0, b)$. In algebra courses, all the following forms of the equation of a line are introduced and studied. The derivation of the **two-point form** and the **intercept form** are left for Exercise Set 3.3. They are listed here to provide a complete summary.

Forms of Linear Equations

General form:	$Ax + By + C = 0$
Vertical line:	$x = a$
Horizontal line:	$y = b$
Slope-intercept form:	$y = mx + b$
Point-slope form:	$y - y_1 = m(x - x_1)$
Two-point form:	$y - y_1 = \left(\dfrac{y_2 - y_1}{x_2 - x_1} \right)(x - x_1)$
Intercept form:	$\dfrac{x}{a} + \dfrac{y}{b} = 1$

EXAMPLE 7   Solve a Business Application

A business purchases a duplicating machine for $4200. It is estimated that after 8 years the duplicating machine will have a value v of $400. If straight-line depreciation is used, find

a. A linear function that expresses the value of the duplicating machine v as a function of the machine's age x, where $0 \le x \le 8$.

b. The value of the machine after $2\frac{1}{2}$ years.

Solution

a. Since the value of the machine is determined by straight-line depreciation, we need to find a linear function $v(x)$ such that $v(0) = 4200$ and $v(8) = 400$. The slope m of the line given by the linear function v is found by computing the ratio of the change in v to the change in x (dividing the change in v by the change in x). That is,

$$m = \frac{v(0) - v(8)}{0 - 8} = \frac{4200 - 400}{0 - 8} = \frac{3800}{-8} = -475$$

Now, using the point-slope formula $v - v_1 = m(x - x_1)$ with $m = -475$, $v_1 = 4200$, and $x_1 = 0$, we get

$$v - 4200 = -475(x - 0)$$

$$v - 4200 = -475x$$

$$v = -475x + 4200$$

Using functional notation, we have $v(x) = -475x + 4200$.

b. To find the value of the duplicating machine after $2\frac{1}{2}$ years, evaluate $v(x)$ with $x = 2.5$.

$$v(2.5) = -475(2.5) + 4200 = -1187.50 + 4200 = 3012.50$$

The value of the duplicating machine after $2\frac{1}{2}$ years will be \$3012.50.

*■ Try Exercise **60**, page 153.*

If a manufacturer produces x units of a product that sells for p dollars per unit, then the **cost function** $C(x)$, the **revenue function** $R(x)$, and the **profit function** $P(x)$, are defined as follows:

$C(x) =$ cost of producing and selling x units

$R(x) = xp =$ revenue from the sale of x units at p dollars each

$P(x) =$ profit from selling x units

Since profit equals the revenue less the cost, we have

$$P(x) = R(x) - C(x)$$

The value of x for which $R(x) = C(x)$ is called the **break-even point.** At the break-even point, $P(x) = 0$.

EXAMPLE 8   Solve a Business Application

A manufacturer finds that the costs incurred in the manufacture and sale of a particular type of calculator are \$180,000 plus \$27 per calculator.

a. Determine the profit function $P(x)$, given that x calculators are manufactured and sold at \$59 each.

b. Determine the break-even point.

Solution

a. The cost function is $C(x) = 27x + 180,000$. The revenue function is $R(x) = 59x$. Thus the profit function is

$$P(x) = R(x) - C(x)$$
$$= 59x - (27x + 180,000)$$
$$= 32x - 180,000$$

b. At the break-even point, $R(x) = C(x)$.

$$59x = 27x + 180,000$$
$$32x = 180,000$$
$$x = 5625$$

The manufacturer will break even when 5625 calculators are sold.

*■ Try Exercise **62**, page 153.*

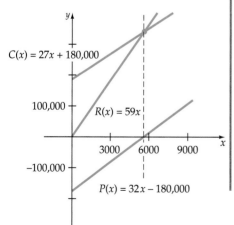

Figure 3.35

Remark The graphs of $C(x)$, $R(x)$, and $P(x)$ are shown in Figure 3.35. Observe that the graphs of $C(x)$ and $R(x)$ intersect at the break-even point, where $x = 5625$ and $P(5625) = 0$.

EXERCISE SET 3.3

In Exercises 1 to 10, find the slope of the line that passes through the given points.

1. (3, 4) and (1, 7)

2. (−2, 4) and (5, 1)

3. (4, 0) and (0, 2)

4. (−3, 4) and (2, 4)

5. (0, 0) and (0, 4)

6. (0, 0) and (3, 0)

7. (−3, 4) and (−4, −2)

8. (−5, −1) and (−3, 4)

9. $\left(-4, \frac{1}{2}\right)$ and $\left(\frac{7}{3}, \frac{7}{2}\right)$

10. $\left(\frac{1}{2}, 4\right)$ and $\left(\frac{7}{4}, 2\right)$

In Exercises 11 to 14, find the slope of a line that passes through the given points.

11. $(3, f(3))$ and $(3 + h, f(3 + h))$

12. $(-2, f(-2 + h))$ and $(-2 + h, f(-2 + h))$

13. $(0, f(0))$ and $(h, f(h))$

14. $(a, f(a))$ and $(a + h, f(a + h))$

In Exercises 15 to 26, graph the lines whose equations are given by finding the slope and y-intercept of each line.

15. $y = 2x - 4$

16. $y = -x + 1$

17. $y = -\frac{1}{3}x + 4$

18. $y = \frac{2}{3}x - 2$

19. $y = 3$

20. $y = x$

21. $y = 2x$

22. $y = -3x$

23. $2x + y = 5$

24. $x - y = 4$

25. $4x + 3y - 12 = 0$

26. $2x + 3y + 6 = 0$

In Exercises 27 to 54, find the equation of the indicated line. Write the equation in the form $y = mx + b$.

27. y-intercept (0, 3), slope 1

28. y-intercept (0, 5), slope −2

29. y-intercept (0, −1), slope 3

30. y-intercept (0, −2), slope −4

31. y-intercept $\left(0, \frac{1}{2}\right)$, slope $\frac{3}{4}$

32. y-intercept $\left(0, \frac{3}{4}\right)$, slope $-\frac{2}{3}$

33. y-intercept (0, 4), slope 0

34. y-intercept (0, −1), slope $\frac{1}{2}$

35. Through (1, 3), slope 2

36. Through (2, −1), slope 3

37. Through (−3, 2), slope −4

38. Through (−5, −1), slope −3

39. Through (3, 1) and (−1, 4)

40. Through (5, −6) and (2, −8)

41. Through (7, 11) and (2, −1)

42. Through (−5, 6) and (−3, −4)

43. Through $\left(\frac{1}{2}, 0\right)$ and $\left(\frac{3}{4}, \frac{1}{5}\right)$

44. Through $\left(-\frac{3}{4}, \frac{1}{5}\right)$ and $\left(\frac{2}{3}, \frac{1}{4}\right)$

45. Through (1.5, 2.4) and (3.6, −5.1)

46. Through (−4, −3.5) and (2.5, 6.5)

47. Through (1, 3) parallel to $3x + 4y = -24$

48. Through (2, −1) parallel to $x + y = 10$

49. Through (−3, 4) parallel to $2x - y = 7$

50. Through (−1, −5) parallel to $x + 3y = 9$

51. Through (1, 2) perpendicular to $x + y = 4$

52. Through (−3, 4) perpendicular to $2x - y = 7$

53. Through (−2, 1) perpendicular to $3x - 5y = 11$

54. Through (−3, 5) perpendicular to $4x - 7y = 8$

55. Use the graph to find the slope of the line that passes through P_1 and P_2.

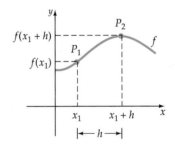

56. Use the graph to find the slope of the line that passes through P_3 and P_4.

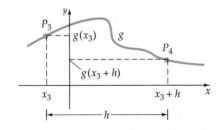

57. Write a linear function that expresses the relationship between the Celsius temperature, C, and the Fahrenheit temperature, F. Use the fact that water freezes at 32°F, which is 0°C, and boils at 212°F, which is 100°C.

58. Use the result of Exercise 57 to find

a. The Celsius temperature that corresponds to 81°F.

b. The Fahrenheit temperature that corresponds to 60°C.

59. A business purchases a computer for $8280. After 12 years the computer will be obsolete and will have no value.

a. Find a linear function that expresses the value V of the computer in terms of the number of years n that it has been used, where $0 \le n \le 12$.

b. Evaluate $V(3)$ to determine the value of the computer after 3 years.

60. A car has a purchase price of $40,090. It is estimated that after 15 years the car will have a value V of $8500.

a. Find a linear function that expresses the value V of the car in terms of the car's age n, where $0 \le n \le 15$.

b. Evaluate $V(4)$ to determine the value of the car after 4 years.

In Exercises 61 to 64, determine the profit function for the given revenue function and cost function. Also determine the break-even point.

61. $R(x) = 92.50x;$ $C(x) = 52x + 1782$

62. $R(x) = 124x;$ $C(x) = 78.5x + 5005$

63. $R(x) = 259x;$ $C(x) = 180x + 10{,}270$

64. $R(x) = 14{,}220x;$ $C(x) = 8010x + 1{,}602{,}180$

65. In business, *marginal cost* is a phrase used to represent the rate of change or slope of a cost function that relates the cost C to the number of units x produced. If a cost function is given by $C(x) = 8x + 275$, find

a. $C(0)$ **b.** $C(1)$ **c.** $C(10)$ **d.** marginal cost

66. In business, *marginal revenue* is a phrase used to represent the rate of change or slope of a revenue function that relates the revenue R to the number of units x sold. If a revenue function is given by the function $R(x) = 210x$, find

a. $R(0)$ **b.** $R(1)$ **c.** $R(10)$ **d.** marginal revenue

67. A rental company purchases a truck for $19,500. The truck requires an average of $6.75 per day in maintenance.

a. Find the linear function that expresses the total cost C of owning the truck after t days.

b. The truck rents for $55.00 a day. Find the linear function that expresses the revenue R when the truck has been rented for t days.

c. The profit after t days, $P(t)$, is given by the function $P(t) = R(t) - C(t)$. Find the linear function $P(t)$.

d. Use the function $P(t)$ that you obtained in part **c** to determine how many days it will take the company to break even on the purchase of the truck.

68. A magazine company had a profit of $98,000 per year when it had 32,000 subscribers. When it obtained 35,000 subscribers, it had a profit of $117,500. Assume that the profit P is a linear function of the number of subscribers s.

a. Find the function P.

b. What will the profit be if the company obtains 50,000 subscribers?

c. What is the number of subscribers needed to break even?

Supplemental Exercises

69. Use the point-slope form to derive the following equation, which is called the two-point form.

$$y - y_1 = \left(\frac{y_2 - y_1}{x_2 - x_1} \right)(x - x_1)$$

70. Use the two-point form from Exercise 69 to show that the line with intercepts $(a, 0)$ and $(0, b)$, $a \ne 0$ and $b \ne 0$, has the equation

$$\frac{x}{a} + \frac{y}{b} = 1$$

In Exercises 71 to 74, use the two-point form to find an equation of the line that passes through the indicated points. Write your answers in slope-intercept form.

71. $(5, 1)$, $(4, 3)$ **72.** $(2, 7)$, $(-1, 6)$

73. $(-11, 8)$, $(7, -5)$ **74.** $(-3, 4)$, $(-7, -11)$

In Exercises 75 to 80, use the equation from Exercise 70 (called the intercept form) to write an equation of a line with the indicated intercepts.

75. x-intercept $(3, 0)$, y-intercept $(0, 5)$

76. x-intercept $(-2, 0)$, y-intercept $(0, 7)$

77. x-intercept $\left(\frac{1}{2}, 0 \right)$, y-intercept $\left(0, \frac{1}{4} \right)$

78. x-intercept $\left(\frac{2}{3}, 0 \right)$, y-intercept $\left(0, \frac{7}{8} \right)$

79. x-intercept $(a, 0)$, y-intercept $(0, 3a)$, point on the line $(5, 2)$, $a \ne 0$

80. x-intercept $(-b, 0)$, y-intercept $(0, 2b)$, point on the line $(-3, 10)$, $b \ne 0$

3.4 Quadratic Functions

In the previous section, we studied linear functions and their applications. Many applications cannot be accurately modeled by a linear function. For example, to determine the distance that a rock will fall in t seconds requires a *quadratic function*.

Definition of a Quadratic Function

A **quadratic function** is a function that can be represented by an equation of the form

$$f(x) = ax^2 + bx + c$$

where a, b, and c are real numbers and $a \neq 0$.

The graph of $f(x) = ax^2 + bx + c$ is a **parabola**. If b and c are both zero, then $f(x) = ax^2 + bx + c$ simplifies to $f(x) = ax^2$. The graph of $f(x) = ax^2$ is a parabola that

1. Opens up if $a > 0$.
2. Opens down if $a < 0$.

The **vertex of a parabola** is the lowest point of a parabola that opens up or the highest point of a parabola that opens down. The graph of $f(x) = ax^2$ has its vertex at the origin $(0, 0)$.

Quadratic functions can be graphed by plotting points and drawing a smooth curve through these points. The graphs of $f(x) = x^2$, $g(x) = 2x^2$, and $h(x) = -\frac{1}{2}x^2$ are shown in Figure 3.36.

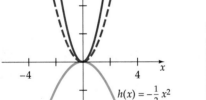

Figure 3.36

Standard Form of Quadratic Functions

Every quadratic function f given by $f(x) = ax^2 + bx + c$ can be written in the **standard form**

$$f(x) = a(x - h)^2 + k, \quad a \neq 0$$

The graph of f is a parabola with vertex (h, k). The parabola opens up if $a > 0$, and it opens down if $a < 0$.

The standard form is useful because it readily gives information about the vertex of the graph of the function. For example, the graph of $f(x) = 2(x - 4)^2 - 3$ is a parabola with vertex $(4, -3)$. Since a is the positive number 2, the parabola opens upward.

If a quadratic function is not written in standard form, you can find its standard form by completing the square.

EXAMPLE 1 Find the Standard Form of a Parabola

Use the technique of completing the square to find the standard form of the quadratic function $g(x) = 2x^2 - 12x + 19$. Sketch the graph.

Solution

$$g(x) = 2x^2 - 12x + 19$$

$$= 2(x^2 - 6x) + 19 \qquad \bullet \text{ Factor 2 from the variable terms.}$$

$$= 2(x^2 - 6x + 9 - 9) + 19 \qquad \bullet \text{ Complete the square.}$$

$$= 2(x^2 - 6x + 9) - 2(9) + 19 \qquad \bullet \text{ Regroup.}$$

$$= 2(x - 3)^2 - 18 + 19 \qquad \bullet \text{ Factor and simplify.}$$

$$= 2(x - 3)^2 + 1 \qquad \bullet \text{ Standard form. See Figure 3.37.}$$

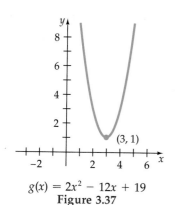

$g(x) = 2x^2 - 12x + 19$

Figure 3.37

■ *Try Exercise 10, page 160.*

The formula for the vertex of the graph of $f(x) = ax^2 + bx + c$ can be determined by the method of completing the square.

$$f(x) = ax^2 + bx + c$$

$$= (ax^2 + bx) + c \qquad \bullet \text{ Regroup.}$$

$$= a\left(x^2 + \frac{b}{a}x\right) + c \qquad \bullet \text{ Since } bx = a\frac{b}{a}x, \text{ factor an } a \text{ out.}$$

$$= a\left(x^2 + \frac{b}{a}x + \frac{b^2}{4a^2} - \frac{b^2}{4a^2}\right) + c \qquad \bullet \text{ Complete the square.}$$

$$= a\left(x^2 + \frac{b}{a}x + \frac{b^2}{4a^2}\right) - \frac{b^2}{4a} + c \qquad \bullet \text{ Multiply } a \text{ times } -\frac{b^2}{4a^2} \text{ to produce the } -\frac{b^2}{4a} \text{ term.}$$

$$= a\left(x + \frac{b}{2a}\right)^2 + c - \frac{b^2}{4a} \qquad \bullet \text{ Factor.}$$

$$= a\left(x - \left(-\frac{b}{2a}\right)\right)^2 + \frac{4ac - b^2}{4a}$$

From the standard form, the vertex of the graph of $f(x) = ax^2 + bx + c$ is

$$\left(-\frac{b}{2a}, \frac{4ac - b^2}{4a}\right)$$

Although the *y*-coordinate of the vertex is $\frac{4ac - b^2}{4a}$, it is often easier to determine the *y* value of the vertex point by evaluating *f* at $x = -\frac{b}{2a}$. Thus we have the following formula.

Vertex Formula

The vertex of the graph of $f(x) = ax^2 + bx + c$ is $\left(-\dfrac{b}{2a}, f\left(-\dfrac{b}{2a}\right)\right)$.

Remark This formula can be used to write the standard form of a quadratic function. Let $h = -\dfrac{b}{2a}$ and $k = f\left(-\dfrac{b}{2a}\right)$.

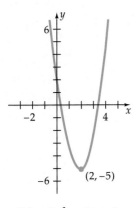

$f(x) = 2x^2 - 8x + 3.$
Figure 3.38

EXAMPLE 2 **Find the Vertex and Standard Form of a Quadratic Function**

Use the vertex formula to find the vertex and standard form of $f(x) = 2x^2 - 8x + 3$.

Solution

$$h = -\frac{b}{2a} = -\frac{-8}{2(2)} = 2 \qquad \bullet \; x\text{-coordinate of the vertex}$$

$$k = f\left(-\frac{b}{2a}\right) = 2(2)^2 - 8(2) + 3 = -5 \quad \bullet \; y\text{-coordinate of the vertex}$$

The vertex is $(2, -5)$. The standard form is $f(x) = 2(x - 2)^2 - 5$.

■ *Try Exercise* **20**, *page* 160.

In many applications it is important to be able to determine the maximum or minimum value of a function.

Maximum or Minimum Value of a Quadratic Function

If $a > 0$, then the vertex (h, k) is the lowest point on the graph of $f(x) = a(x - h)^2 + k$, and the y-coordinate k of the vertex is the **minimum value** of the function f. See Figure 3.39.

If $a < 0$, then the vertex (h, k) is the highest point on the graph of $f(x) = a(x - h)^2 + k$, and the y-coordinate k is the **maximum value** of the function f. See Figure 3.39.

In either case, the maximum or minimum is achieved when $x = h$.

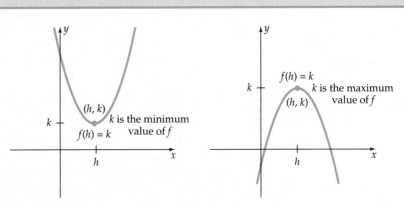

Figure 3.39

EXAMPLE 3   **Find the Maximum or Minimum of a Quadratic Function**

Find the maximum or minimum value of each quadratic function. State whether the value is a maximum or a minimum.

a. $F(x) = -2x^2 + 8x - 1$ b. $G(x) = x^2 - 3x + 1$

Solution   The maximum or minimum value of a quadratic function is the y-coordinate of the vertex of the graph of the function.

a. $h = -\dfrac{b}{2a} = -\dfrac{8}{2(-2)} = 2$ • *x*-coordinate of the vertex

$k = F\left(-\dfrac{b}{2a}\right) = -2(2)^2 + 8(2) - 1 = 7$ • *y*-coordinate of the vertex

Because $a < 0$, the function has a maximum value. The maximum value is 7.

b. $h = -\dfrac{b}{2a} = -\dfrac{-3}{2(1)} = \dfrac{3}{2}$ • *x*-coordinate of the vertex

$k = G\left(-\dfrac{b}{2a}\right) = \left(\dfrac{3}{2}\right)^2 - 3\left(\dfrac{3}{2}\right) + 1$

$= -\dfrac{5}{4}$ • *y*-coordinate of the vertex

Because $a > 0$, the function has a minimum value. The minimum value is $-\dfrac{5}{4}$.

■ *Try Exercise* **30,** *page 160.*

EXAMPLE 4   **Determine Maximum or Minimum Values**

A long sheet of tin 20 inches wide is to be made into a trough by bending up two sides until they are perpendicular to the bottom. How many inches should be turned up so that the trough will achieve its maximum carrying capacity?

Solution   The trough is shown in Figure 3.40. If *x* is the number of inches to be turned up on each side, then the width of the base is $20 - 2x$ inches. The maximum carrying capacity of the trough will occur when the cross-sectional area is a maximum. The cross-sectional area $A(x)$ is given by

$A(x) = x(20 - 2x)$ • Area = (length) (width)

$= -2x^2 + 20x$

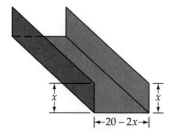

Figure 3.40

To find when A obtains its maximum value, find the x-coordinate of the vertex of the graph of A. Using the vertex formula with $a = -2$ and $b = 20$, we get

$$x = -\frac{b}{2a} = -\frac{20}{2(-2)} = 5$$

Therefore, the maximum carrying capacity will be achieved when $x = 5$ inches are turned up.

■ *Try Exercise* **40**, *page 161.*

EXAMPLE 5 **Solve a Business Application**

The owners of a travel agency have determined that they can sell all 160 tickets for a tour if they charge $8 (their cost) for each ticket. For each $0.25 increase in the price of a ticket, they estimate they will sell 1 ticket less. A business manager determines that their cost function is $C(x) = 8x$ and that the customer's price per ticket is

$$p(x) = 8 + 0.25(160 - x)$$
$$= 48 - 0.25x$$

where x represents the number of tickets sold. Determine the maximum profit and the cost per ticket that yields the maximum profit.

Solution   The profit from selling x tickets is $P(x) = R(x) - C(x)$, where $P(x)$, $R(x)$, and $C(x)$ are the profit function, the revenue function, and the cost function as defined in Section 3.3. Thus

$$P(x) = R(x) - C(x)$$
$$= x[p(x)] - C(x)$$
$$= x(48 - 0.25x) - 8x$$
$$= 40x - 0.25x^2$$

The profit function is a quadratic function. It graphs as a parabola that opens down. Thus the maximum profit occurs when

$$x = -\frac{b}{2a} = -\frac{40}{2(-0.25)} = 80$$

The maximum profit is determined by evaluating $P(x)$ with $x = 80$.

$$P(80) = 40(80) - 0.25(80)^2 = 1600$$

To find the price per ticket that yields the maximum profit, we evaluate $p(x)$ with $x = 80$.

$$p(80) = 48 - 0.25(80) = 28$$

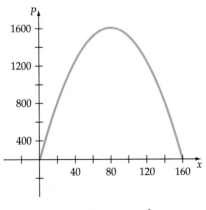

$P(x) = 40x - 0.25x^2$
Figure 3.41

Thus the travel agency can expect a maximum profit of $1600 when 80 people take the tour at a ticket price of $28 per person. The graph of the profit function is shown in Figure 3.41. We have shown only that portion of the graph that lies in quadrant I. Why?[1]

■ *Try Exercise 56, page 161.*

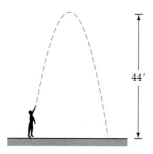

Figure 3.42

EXAMPLE 6   Solve an Application

In Figure 3.42, a ball is thrown upward with an initial velocity of 48 feet per second. If the ball started its flight at a height of 8 feet, then its height s at time t can be determined by the function $s(t) = -16t^2 + 48t + 8$, where $s(t)$ is measured in feet above ground level and t is the number of seconds of flight.

a. Determine the time it takes the ball to attain its maximum height.

b. Determine the maximum height the ball attains.

c. Determine the time it takes the ball to hit the ground.

Solution

a. The graph of the function $s(t) = -16t^2 + 48t + 8$ is a parabola that opens downward. Therefore s will attain its maximum value at the vertex of its graph. Using the vertex formula with $a = -16$ and $b = 48$, we get

$$t = -\frac{b}{2a} = -\frac{48}{2(-16)} = \frac{3}{2}$$

Therefore the ball attains its maximum height one and one-half seconds into its flight.

b. When $t = \frac{3}{2}$, the height of the ball is

$$s\left(\frac{3}{2}\right) = -16\left(\frac{3}{2}\right)^2 + 48\left(\frac{3}{2}\right) + 8 = 44 \text{ feet}$$

c. The ball will hit the ground when its height $s(t) = 0$. Therefore, solve $-16t^2 + 48t + 8 = 0$ for t.

$$-16t^2 + 48t + 8 = 0$$

$$-2t^2 + 6t + 1 = 0 \qquad \text{• Divide each side by 8.}$$

$$t = \frac{-(6) \pm \sqrt{6^2 - 4(-2)(1)}}{2(-2)} \qquad \text{• The quadratic formula}$$

$$= \frac{-6 \pm \sqrt{44}}{-4} = \frac{-3 \pm \sqrt{11}}{-2}$$

Using a calculator to approximate the positive root, we find that the ball will hit the ground in $t \approx 3.16$ seconds.

■ *Try Exercise 58, page 162.*

---

[1] Since x represents the number of tickets sold, x must be greater than or equal to zero but less than or equal to 160. $P(x)$ is nonnegative for $0 \le x \le 160$.

EXERCISE SET 3.4

In Exercises 1 to 8, match the graphs in **a** through **h** with the quadratic function.

1. $f(x) = x^2 - 3$ **2.** $f(x) = x^2 + 2$

3. $f(x) = (x - 4)^2$ **4.** $f(x) = (x + 3)^2$

5. $f(x) = -2x^2 + 2$ **6.** $f(x) = -\frac{1}{2}x^2 + 3$

7. $f(x) = (x + 1)^2 + 3$ **8.** $f(x) = -2(x - 2)^2 + 2$

g.

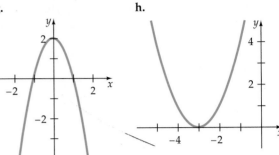

h.

a.

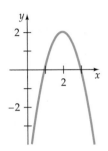

b.

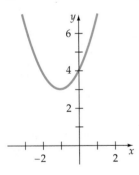

c.

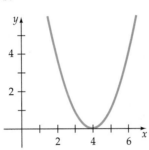

d.

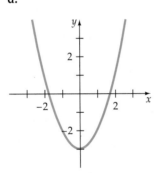

e.

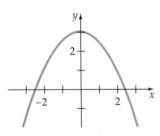

f.

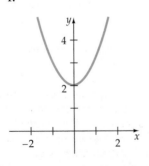

In Exercises 9 to 18, use the method of completing the square to find the standard form of the quadratic function, and then sketch its graph.

9. $f(x) = x^2 + 4x + 1$ **10.** $f(x) = x^2 + 6x - 1$

11. $f(x) = x^2 - 8x + 5$ **12.** $f(x) = x^2 - 10x + 3$

13. $f(x) = x^2 + 3x + 1$ **14.** $f(x) = x^2 + 7x + 2$

15. $f(x) = -x^2 + 4x + 2$ **16.** $f(x) = -x^2 - 2x + 5$

17. $f(x) = -3x^2 + 3x + 7$ **18.** $f(x) = -2x^2 - 4x + 5$

In Exercises 19 to 28, use the vertex formula to determine the vertex of the graph of the quadratic function and write the quadratic function in standard form.

19. $f(x) = x^2 - 10x$ **20.** $f(x) = x^2 - 6x$

21. $f(x) = x^2 - 10$ **22.** $f(x) = x^2 - 4$

23. $f(x) = -x^2 + 6x + 1$ **24.** $f(x) = -x^2 + 4x + 1$

25. $f(x) = 2x^2 - 3x + 7$ **26.** $f(x) = 3x^2 - 10x + 2$

27. $f(x) = -4x^2 + x + 1$ **28.** $f(x) = -5x^2 - 6x + 3$

In Exercises 29 to 38, find the maximum or minimum value of the quadratic function. State whether this value is a maximum or a minimum.

29. $f(x) = x^2 + 8x$ **30.** $f(x) = -x^2 - 6x$

31. $f(x) = -x^2 + 6x + 2$ **32.** $f(x) = -x^2 + 10x - 3$

33. $f(x) = 2x^2 + 3x + 1$ **34.** $f(x) = 3x^2 + x - 1$

35. $f(x) = 5x^2 - 11$ **36.** $f(x) = 3x^2 - 41$

37. $f(x) = -\frac{1}{2}x^2 + 6x + 17$ **38.** $f(x) = -\frac{3}{4}x^2 - \frac{2}{5}x + 7$

39. The height of an arch is given by the equation

$$h(x) = -\frac{3}{64}x^2 + 27 \quad -24 \le x \le 24$$

where $|x|$ is the horizontal distance in feet from the center of the arch.

a. What is the maximum height of the arch?

b. What is the height of the arch 10 feet to the right of center?

c. How far from the center is the arch 8 feet tall?

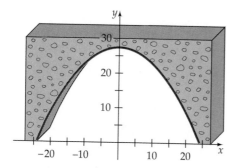

40. A company sells tennis rackets at a price of $60 per racket if 40 or less are ordered. If a buyer orders between 41 and 400 rackets, then the price of every racket is reduced $0.10 for each racket over 40 that is ordered. Find the size of the order that will give the company the most money.

41. The profit function P for a company selling x items is given by $P(x) = -3x^2 + 96x - 368$.

a. What value of x will maximize the profit?

b. What is the maximum profit?

42. The sum of the length l and the width w of a rectangular area is 240 meters.

a. Write w as a function of l.

b. Write the area A as a function of l.

c. Find the dimensions that produce the greatest area.

43. A veterinarian uses 600 feet of chain-link fencing to enclose a rectangular region and also to subdivide the region into two smaller rectangular regions by placing a fence parallel to one of the sides, as shown in the figure.

a. Write the width w as a function of the length l.

b. Write the total area A as a function of l.

c. Find the dimensions that produce the greatest enclosed area.

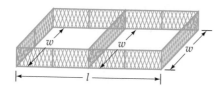

44. A farmer uses 1200 feet of fence to enclose a rectangular region and also to subdivide the region into three smaller rectangular regions by placing two fences parallel to one of the sides. Find the dimensions that produce the greatest enclosed area.

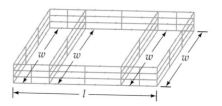

In Exercises 45 to 48, determine the y- and x-intercepts (if any) of the quadratic function.

45. $f(x) = x^2 + 6x$

46. $f(x) = -x^2 + 4x$

47. $f(x) = -3x^2 + 5x - 6$

48. $f(x) = 2x^2 + 3x + 4$

In Exercises 49 and 50, determine the number of units x that produce a maximum revenue for the given revenue function. Also determine the maximum revenue.

49. $R(x) = 296x - 0.2x^2$

50. $R(x) = 810x - 0.6x^2$

In Exercises 51 and 52, determine the number of units x that produce a maximum profit for the given profit function. Also determine the maximum profit.

51. $P(x) = -0.01x^2 + 1.7x - 48$

52. $P(x) = -\dfrac{x^2}{14,000} + 1.68x - 4000$

In Exercises 53 and 54, determine the profit function for the given revenue function and cost function. Also determine the break-even point.

53. $R(x) = x(102.50 - 0.1x)$; $C(x) = 52.50x + 1840$

54. $R(x) = x(210 - 0.25x)$; $C(x) = 78x + 6399$

55. A charter bus company has determined that its cost of providing x people a tour is

$$C(x) = 180 + 2.50x$$

A full tour consists of 60 people. The ticket price per person is $15 plus $0.25 for each unsold ticket. Determine **a.** the revenue function, **b.** the profit function, **c.** the company's maximum profit, and **d.** the number of ticket sales that yields the maximum profit.

56. An air freight company has determined that its cost of delivering x parcels per flight is

$$C(x) = 2025 + 7x$$

The price it charges to send x parcels is

$$p(x) = 22 - 0.01x$$

Determine **a.** the revenue function, **b.** the profit function, **c.** the company's maximum profit, **d.** the price per parcel that yields the maximum profit, and **e.** the minimum number of parcels the air freight company must ship to break even.

57. If the initial velocity of a projectile is 128 feet per second, then its height h is a function of time given by the equation $h(t) = -16t^2 + 128t$.

   a. Find the time t when the projectile achieves its maximum height.

   b. Find the maximum height of the projectile.

   c. Find the time t when the projectile hits the ground.

58. The height of a projectile with an initial velocity of 64 feet per second and an initial height of 80 feet is a function of time given by $h(t) = -16t^2 + 64t + 80$.

   a. Find the maximum height of the projectile.

   b. Find the time t when the projectile achieves its maximum height.

   c. Find the time t when the projectile has a height of 0 feet.

Supplemental Exercises

59. Find the quadratic function whose graph has a minimum at $(2, 1)$ and passes through $(0, 4)$.

60. Find the quadratic function whose graph has a maximum at $(-3, 2)$ and passes through $(0, -5)$.

61. A wire 32 inches long is bent so that it has the shape of a rectangle. The length of the rectangle is x and the width is w.

   a. Write w as a function of x.

   b. Write the area A of the rectangle as a function of x.

62. Use the function A from Exercise 61b to prove that the area A is greatest if the rectangle is a square.

63. Show that the function $f(x) = x^2 + bx - 1$ has a real zero for any value b.

64. Show that the function $g(x) = -x^2 + bx + 1$ has a real zero for any value b.

65. What effect does increasing the constant c have on the graph of $f(x) = ax^2 + bx + c$?

66. If $a > 0$, what effect does decreasing the coefficient a have on the graph of $f(x) = ax^2 + bx + c$?

67. Find two numbers whose sum is 8 and whose product is a maximum.

68. Find two numbers whose difference is 12 and whose product is a minimum.

Figure 3.43

3.5   Additional Graphing Techniques

Some equations can be graphed by using the concept of symmetry.

Definition of Symmetry with Respect to a Line

A graph is **symmetric with respect to a line** L if for each point P on the graph there is a point P' on the graph such that the line L is the perpendicular bisector of the line segment PP'.

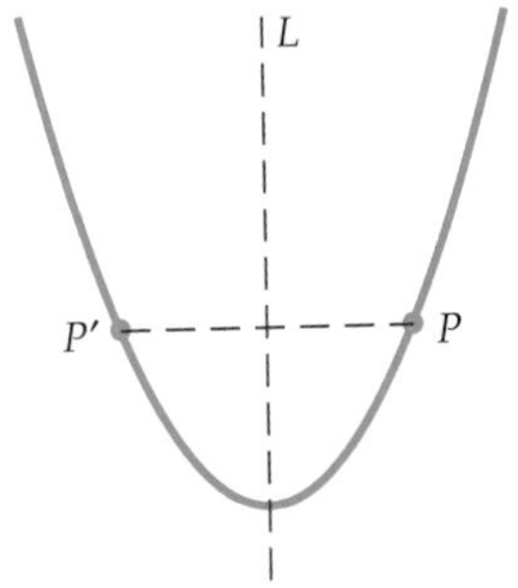

In Figure 3.43, the graph in color is symmetric with respect to the line L. The line L is called a **line of symmetry.** The points P and P' are reflections or images of each other with respect to the line L.

Every quadratic function of the form $f(x) = ax^2 + bx + c$ when graphed results in a parabola with vertex (h, k). The parabola is symmetric with respect to the vertical line $x = h$, which is called the **axis** of the parabola.

The graph in Figure 3.44 is symmetric with respect to the line l. Note that the graph has the property that if the paper is folded along the dotted

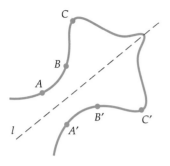

Figure 3.44

line l, the point A' will coincide with the point A, the point B' will coincide with the point B, and the point C' will coincide with the point C. One part of the graph is a *mirror image* of the rest of the graph across the line l.

A graph is **symmetric with respect to the y-axis** if, whenever the point given by (x, y) is on the graph, then $(-x, y)$ is also on the graph. The graph in Figure 3.45 is symmetric with respect to the y-axis. A graph is **symmetric with respect to the x-axis** if, whenever the point given by (x, y) is on the graph, then $(x, -y)$ is also on the graph. The graph in Figure 3.46 is symmetric with respect to the x-axis.

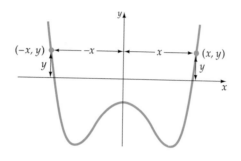

Figure 3.45
Symmetry with respect to the y-axis

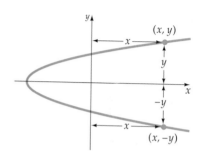

Figure 3.46
Symmetry with respect to the x-axis

The following *tests* enable us to determine, by examining its equation, whether a graph will be symmetric to a coordinate axis.

Tests for Symmetry with Respect to a Coordinate Axis

> The graph of an equation is symmetric with respect to the
> 1. y-axis if the replacement of x with $-x$ leaves the equation unaltered.
> 2. x-axis if the replacement of y with $-y$ leaves the equation unaltered.

EXAMPLE 1 **Determine Symmetries of a Graph**

Determine whether the graph of the given equations has symmetry with respect to either the x- or the y-axis.   a. $y = x^2 + 2$ b. $x = |y| - 2$

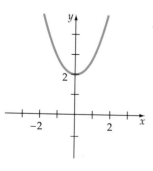

$y = x^2 + 2$
Figure 3.47

Solution

a.   The equation $y = x^2 + 2$ is unaltered by the replacement of x with $-x$. That is, the simplification of $y = (-x)^2 + 2$ yields the original equation $y = x^2 + 2$. Thus the graph of $y = x^2 + 2$ is symmetric with respect to the y-axis. However, the equation $y = x^2 + 2$ is *altered* by the replacement of y with $-y$. That is, the simplification of $-y = x^2 + 2$ *does not* yield the original equation $y = x^2 + 2$. The graph of $y = x^2 + 2$ is not symmetric with respect to the x-axis. See Figure 3.47.

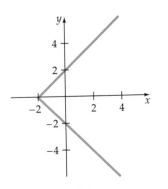

$x = |y| - 2$
Figure 3.48

b. The equation $x = |y| - 2$ *is altered* by the replacement of x with $-x$. That is, the simplification of $-x = |y| - 2$ *does not* yield the original equation $x = |y| - 2$. This implies that the graph of $x = |y| - 2$ is not symmetric with respect to the y-axis. However, the equation $x = |y| - 2$ is unaltered by the replacement of y with $-y$. That is, the simplification of $x = |-y| - 2$ yields the original equation $x = |y| - 2$. The graph of $x = |y| - 2$ is symmetric with respect to the x-axis. See Figure 3.48.

■ *Try Exercise* **14**, *page 172.*

Symmetry with Respect to a Point

> A graph is **symmetric with respect to a point** Q if for each point P on the graph there is a point P' on the graph such that Q is the midpoint of the line segment PP'.

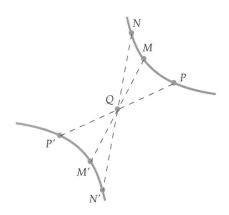

Figure 3.49
Symmetry with respect to the point Q

The graph in Figure 3.49 is symmetric with respect to the point Q. For any point P on the graph, there exists a point P' on the graph such that Q is the midpoint of $P'P$.

Frequently when discussing symmetry with respect to a point, the origin is used. A graph is symmetric with respect to the origin if, whenever the point given by (x, y) is on the graph, then $(-x, -y)$ is also on the graph. The graph in Figure 3.50 is symmetric with respect to the origin.

To determine whether the graph of an equation is symmetric with respect to the origin, we use the following test.

Test for Symmetry with Respect to the Origin

> The graph of an equation is symmetric with respect to the origin if the replacement of x with $-x$ and of y with $-y$ leaves the equation unaltered.

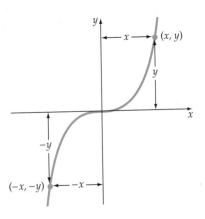

Figure 3.50

EXAMPLE 2   Determine Symmetry with Respect to the Origin

Determine whether the graph of each of the following equations has symmetry with respect to the origin: a. $xy = 4$ b. $y = x^3 + 1$

Solution

a. The equation $xy = 4$ is unaltered by the replacement of x with $-x$ and y with $-y$. That is, the simplification of $(-x)(-y) = 4$ yields the original equation $xy = 4$. Thus the graph of $xy = 4$ is symmetric with respect to the origin. See Figure 3.51.

b. The equation $y = x^3 + 1$ *is altered* by the replacement of x with $-x$ and y with $-y$. That is, the simplification of $-y = (-x)^3 + 1$ *does not* yield the original equation $y = x^3 + 1$. Thus the graph of $y = x^3 + 1$ is not symmetric with respect to the origin. See Figure 3.52.

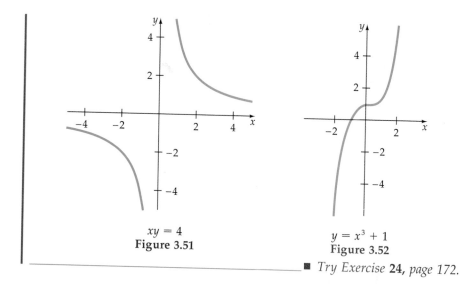

$xy = 4$
Figure 3.51

$y = x^3 + 1$
Figure 3.52

■ *Try Exercise* **24,** *page 172.*

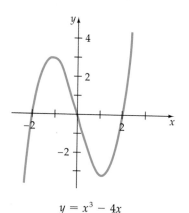

$y = x^3 - 4x$
Figure 3.53

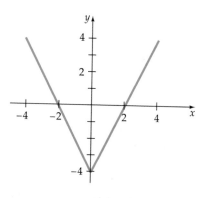

$y = 2|x| - 4$
Figure 3.54

EXAMPLE 3 **Use Symmetry to Graph**

Graph each of the following: a. $y = x^3 - 4x$ b. $y = 2|x| - 4$

Solution

a.   Since the equation $y = x^3 - 4x$ is unaltered by replacing x with $-x$ and y with $-y$, the graph is symmetric with respect to the origin. Further testing shows that the graph is not symmetric with respect to the x-axis, or the y-axis. The following table lists some ordered pairs that satisfy the equation. We have used only ordered pairs whose x-coordinate is nonnegative because we will use symmetry with respect to the origin to graph the portion of the graph to the left of the y-axis.

x	0	$\frac{1}{2}$	1	$\frac{3}{2}$	2	$\frac{5}{2}$	3
$y = x^3 - 4x$	0	$-\frac{15}{8}$	-3	$-\frac{21}{8}$	0	$\frac{45}{8}$	15

Since the graph is symmetric with respect to the origin, we know that if (x, y) is on the graph, then its image through the origin, $(-x, -y)$ is also on the graph. For example, since the point $(1, -3)$ is on the graph, its image through the origin $(-1, 3)$ is also on the graph. Continuing in this manner produces the graph in Figure 3.53. The intercepts are $(-2, 0)$, $(0, 0)$, and $(2, 0)$.

b.   Since the equation is unaltered by replacing x with $-x$, the graph is symmetric with respect to the y-axis. Thus we need to choose only values of x for which $x \geq 0$.

x	0	1	2	3	4
$y = 2|x| - 4$	-4	-2	0	2	4

Plotting these points and their images with respect to the y-axis produces the graph in Figure 3.54. The graph has a y-intercept of $(0, -4)$ and x-intercepts of $(-2, 0)$, and $(2, 0)$.

■ *Try Exercise* **32,** *page 172.*

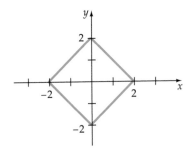

$|x| + |y| = 2$
Figure 3.55

Some graphs have more than one symmetry. For example, the graph of $|x| + |y| = 2$ has symmetry with respect to the x-axis, the y-axis, and the origin. Figure 3.55 is the graph of $|x| + |y| = 2$.

Even and Odd Functions

Some functions are classified as either *even* or *odd*. Knowing a function is even or odd is an aid to drawing its graph. The following definition provides a procedure that can be used to determine whether a function is an even or odd function.

Definition of Even and Odd Functions

The function f is an **even function** if
$$f(-x) = f(x) \quad \text{for all } x \text{ in the domain of } f$$

The function f is an **odd function** if
$$f(-x) = -f(x) \quad \text{for all } x \text{ in the domain of } f$$

EXAMPLE 4   **Identify Even or Odd Functions**

Determine whether the following functions are even, odd, or neither.

a. $f(x) = x^3$ b. $F(x) = |x|$ c. $h(x) = x^4 + 2x$

Solution   Replace x with $-x$ and simplify.

a. $f(-x) = (-x)^3 = -x^3 = -(x^3) = -f(x)$

This function is an odd function because $f(-x) = -f(x)$.

b. $F(-x) = |-x| = |x| = F(x)$

This function is an even function because $F(-x) = F(x)$.

c. $h(-x) = (-x)^4 + 2(-x) = x^4 - 2x$

This function is neither an even nor an odd function because
$$h(-x) = x^4 - 2x,$$
which is not equal to either $h(x)$ or $-h(x)$.

■ *Try Exercise **44**, page 172.*

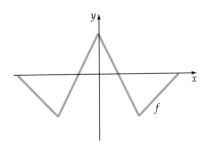

Figure 3.56

The following properties are a result of the tests for symmetry.

■   The graph of an even function is symmetric with respect to the y-axis.

■   The graph of an odd function is symmetric with respect to the origin.

The graph of f in Figure 3.56 is symmetric with respect to the y-axis. It is the graph of an even function. The graph of g in Figure 3.57 is symmetric

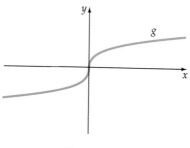

Figure 3.57

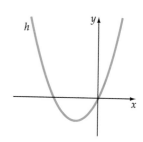

Figure 3.58

with respect to the origin. It is the graph of an odd function. The graph of h in Figure 3.58 is not symmetric to the y-axis and is not symmetric to the origin. It is neither an even nor an odd function.

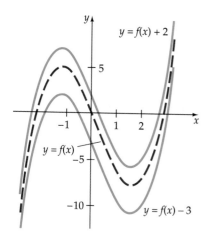

Figure 3.59

Translations of Graphs

The shape of a graph may be exactly the same as the shape of another graph; only their position in the xy-plane may differ. For example, the graph of $y = f(x) + 2$ is the graph of $y = f(x)$ with each point moved up vertically 2 units. The graph of $y = f(x) - 3$ is the graph of $y = f(x)$ with each point moved down vertically 3 units. See Figure 3.59.

The graphs of $y = f(x) + 2$ and $y = f(x) - 3$ in Figure 3.59 are called **vertical translations** of the graph of $y = f(x)$.

Vertical Translations

If f is a function and c is a positive constant, then

$y = f(x) + c$ is the graph of $y = f(x)$ shifted up *vertically* c units.

$y = f(x) - c$ is the graph of $y = f(x)$ shifted down *vertically* c units.

In Figure 3.60, the graph of $y = h(x + 3)$ is the graph of $y = h(x)$ with each point shifted to the left horizontally 3 units. Similarly, the graph of $y = h(x - 3)$ is the graph of $y = h(x)$ with each point shifted to the right horizontally 3 units.

The graphs of $y = h(x + 3)$ and $y = h(x - 3)$ in Figure 3.60 are called **horizontal translations** of the graph of $y = h(x)$.

Horizontal Translations

If f is a function and c is a positive constant, then

$y = f(x + c)$ is the graph of $y = f(x)$ shifted left *horizontally* c units.

$y = f(x - c)$ is the graph of $y = f(x)$ shifted right *horizontally* c units.

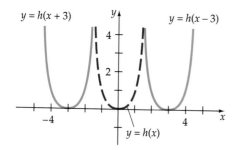

Figure 3.60

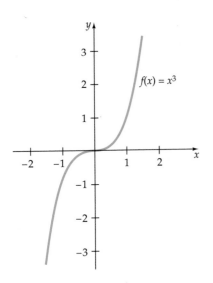

$f(x) = x^3$

Figure 3.61

EXAMPLE 5 Graph by Using Translations

Use vertical and horizontal translations of the graph of $f(x) = x^3$ shown in Figure 3.61, to sketch the graph of each of the following.

a. $g(x) = x^3 - 2$ b. $h(x) = (x + 1)^3$

Solution

a. The graph of $g(x) = x^3 - 2$ is the graph of $f(x) = x^3$ shifted down vertically 2 units. See Figure 3.62.

b. The graph of $h(x) = (x + 1)^3$ is the graph of $f(x) = x^3$ shifted to the left horizontally 1 unit. See Figure 3.63.

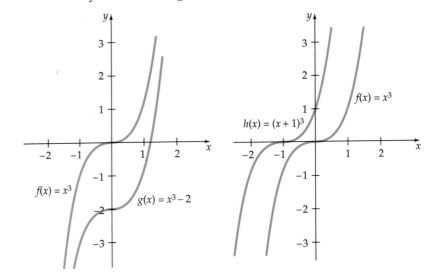

Figure 3.62   **Figure 3.63**

■ *Try Exercise* **58**, *page 172*.

Reflections of Graphs

The graph of $y = -f(x)$ cannot be obtained from the graph of $y = f(x)$ by a combination of vertical and/or horizontal shifts. Figure 3.64 illustrates that the graph of $y = -f(x)$ is the reflection of the graph of $y = f(x)$ across the x-axis.

Figure 3.65 illustrates that the graph of $y = f(-x)$ is the reflection of the graph of $y = f(x)$ across the y-axis.

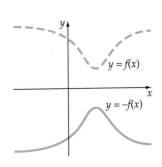

Figure 3.64

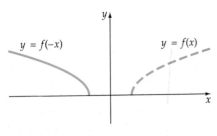

Figure 3.65

Reflections

The graph of

$y = -f(x)$ is the graph of $y = f(x)$ reflected across the x-axis.

$y = f(-x)$ is the graph of $y = f(x)$ reflected across the y-axis.

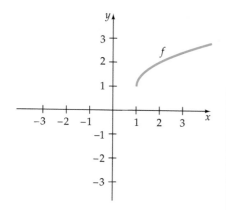

Figure 3.66

EXAMPLE 6   **Graph by Using Reflections**

Use reflections of the graph of $f(x) = \sqrt{x - 1} + 1$ shown in Figure 3.66, to sketch the graph of each of the following.

a. $g(x) = -(\sqrt{x - 1} + 1)$ b. $h(x) = \sqrt{-x - 1} + 1$

Solution

a. The graph of $g(x) = -(\sqrt{x - 1} + 1)$ is the graph of $f(x) = \sqrt{x - 1} + 1$ reflected across the x-axis. See Figure 3.67.

b. The graph of $h(x) = \sqrt{-x - 1} + 1$ is the graph of $f(x) = \sqrt{x - 1} + 1$ reflected across the y-axis. See Figure 3.68.

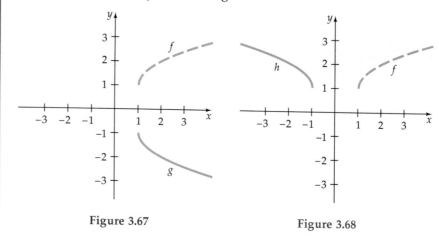

Figure 3.67

Figure 3.68

■ *Try Exercise* **60,** *page 173.*

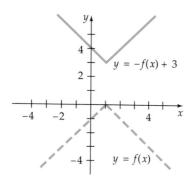

Figure 3.69

Some graphs of functions can be sketched by using a combination of translations and reflections. For instance, the graph of $y = -f(x) + 3$ in Figure 3.69 was obtained by reflecting the graph of $y = f(x)$ in Figure 3.69 with respect to the x-axis and then shifting that graph up 3 units.

Shrinking and Stretching of Graphs

The graph of the equation $y = c \cdot f(x)$ for $c \neq 1$ vertically shrinks or stretches the graph of $y = f(x)$. To determine the points on the graph of $y = c \cdot f(x)$, multiply each y-coordinate of the points on the graph of $y = f(x)$ by c. For example, Figure 3.70 shows that the graph of $y = \frac{1}{2}|x|$ can be obtained by plotting points that have a y-coordinate that is one half of the y-coordinate of those found on the graph of $y = |x|$.

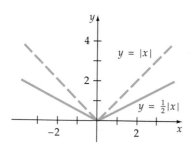

Figure 3.70

If $0 < c < 1$, then the graph of $y = c \cdot f(x)$ is obtained by **shrinking** the graph of $y = f(x)$. Figure 3.70 illustrates the vertical shrinking of the graph of $y = |x|$ toward the x-axis to form the graph of $y = \frac{1}{2}|x|$.

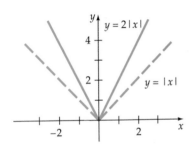

Figure 3.71

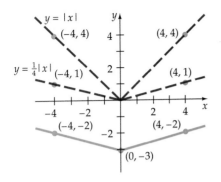

$H(x) = \frac{1}{4}|x| - 3$
Figure 3.72

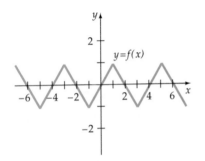

Figure 3.73

If $c > 1$, then the graph of $y = c \cdot f(x)$ is obtained by **stretching** the graph of $y = f(x)$. For example, if $f(x) = |x|$, then we obtain the graph of

$$y = 2f(x) = 2|x|$$

by stretching the graph of f away from the x-axis. See Figure 3.71.

EXAMPLE 7   Graph By Using Vertical Shrinking and Shifting

Graph the function $H(x) = \frac{1}{4}|x| - 3$.

Solution   The graph of $y = |x|$ has a V shape that has its lowest point at $(0, 0)$ and passes through $(4, 4)$ and $(-4, 4)$. The graph of $y = \frac{1}{4}|x|$ is a shrinking of the graph of $y = |x|$. The y-coordinates $(0, 0)$, $(4, 1)$, and $(-4, 1)$ are obtained by multiplying the y-coordinates of the ordered pairs $(0, 0)$, $(4, 4)$, and $(4, -4)$ by $1/4$. To find the points on the graph of H, we still need to subtract 3 from each y-coordinate. Thus the graph of H is a V shape that has its lowest point at $(0, -3)$ and passes through $(4, -2)$ and $(-4, -2)$. See Figure 3.72.

■ *Try Exercise* **62**, *page 173.*

Some functions can be graphed by using a horizontal shrinking or stretching of a given graph. The procedure makes use of the following concept.

Horizontal Shrinking and Stretching

If $a > 0$ and the graph of $y = f(x)$ contains the point (x, y), then the graph of $y = f(ax)$ contains the point $\left(\frac{1}{a}x, y\right)$.

If $a > 1$, then the graph of $y = f(ax)$ is a **horizontal shrinking** of the graph of $y = f(x)$. If $0 < a < 1$, then the graph of $y = f(ax)$ is a **horizontal stretching** of the graph of $y = f(x)$.

EXAMPLE 8   Graph by Using Horizontal Shrinking and Stretching

Use the graph of $y = f(x)$ shown in Figure 3.73 to sketch the graph of each of the following:

a. $y = f(2x)$ b. $y = f\left(\frac{1}{3}x\right)$

Solution

a. Since $2 > 1$, the graph of $y = f(2x)$ is a horizontal contraction (shrinking) of the graph of $y = f(x)$. We sketch the graph of $y = f(2x)$ by contracting each point on the graph of $y = f(x)$ toward the y-axis by a factor of $1/2$. For example, the point $(2, 0)$ on the graph of $y = f(x)$ becomes the point $(1, 0)$ on the graph of $y = f(2x)$. See Figure 3.74.

b. Since $0 < \dfrac{1}{3} < 1$, the graph of $y = f\left(\dfrac{1}{3}x\right)$ is a horizontal dilation (stretching) of the graph of $y = f(x)$. We sketch the graph of $y = f\left(\dfrac{1}{3}x\right)$ by moving each point on the graph of $y = f(x)$ away from the y-axis by a factor of 3. For example, the point $(1, 1)$ on the graph of $y = f(x)$ becomes the point $(3, 1)$ on the graph of $y = f\left(\dfrac{1}{3}x\right)$. See Figure 3.75.

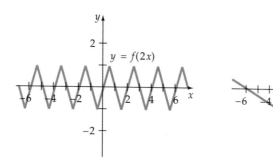

Figure 3.74

Figure 3.75

■ *Try Exercise* **64**, *page 173.*

EXERCISE SET 3.5

In Exercises 1 to 6, plot the image of the given point with respect to the

a. y-axis. Label this point A.

b. x-axis. Label this point B.

c. origin. Label this point C.

1. $P(5, -3)$ **2.** $Q(-4, 1)$ **3.** $R(-2, 3)$

4. $S(-5, 3)$ **5.** $T(-4, -5)$ **6.** $U(5, 1)$

In Exercises 7 and 8, sketch a graph that is the mirror image of the given graph with respect to the x-axis.

7.

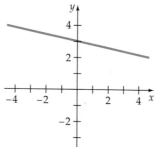

8.

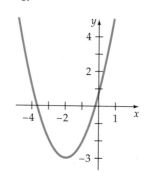

In Exercises 9 and 10, sketch a graph that is symmetric to the given graph with respect to the y-axis.

9.   **10.**

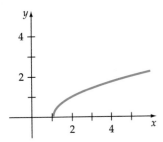   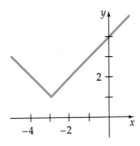

In Exercises 11 and 12, sketch a graph that is symmetric to the given graph with respect to the origin.

11.   **12.**

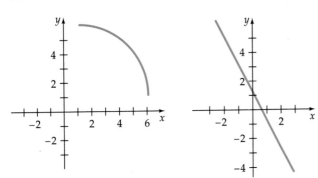

In Exercises 13 to 21, determine whether the graph of each equation is symmetric with respect to the   **a.** x-axis, **b.** y-axis.

13. $y = 2x^2 - 5$ **14.** $x = 3y^2 - 7$ **15.** $y = x^3 + 2$

16. $y = x^5 - 3x$ **17.** $x^2 + y^2 = 9$ **18.** $x^2 - y^2 = 10$

19. $x^2 = y^4$ **20.** $xy = 8$ **21.** $|x| - |y| = 6$

In Exercises 22 to 30, determine whether the graph of each equation is symmetric with respect to the origin.

22. $y = x + 1$ **23.** $y = 3x - 2$ **24.** $y = x^3 - x$

25. $y = -x^3$ **26.** $y = \dfrac{9}{x}$ **27.** $x^2 + y^2 = 10$

28. $x^2 - y^2 = 4$ **29.** $y = \dfrac{x}{|x|}$ **30.** $|y| = |x|$

In Exercises 31 to 42, use symmetry to graph the given equations. Label each intercept.

31. $y = x^2 - 1$ **32.** $x = y^2 - 1$

33. $y = x^3 - x$ **34.** $y = -x^3$

35. $xy = 4$ **36.** $xy = -8$

37. $y = 2|x - 4|$ **38.** $y = |x - 2| - 1$

39. $y = (x - 2)^2 - 4$ **40.** $y = (x - 1)^2 - 4$

41. $y = x - |x|$ **42.** $|y| = |x|$

In Exercises 43 to 56, identify whether the given function is an even function, an odd function, or neither.

43. $g(x) = x^2 - 7$ **44.** $h(x) = x^2 + 1$

45. $F(x) = x^5 + x^3$ **46.** $G(x) = 2x^5 - 10$

47. $H(x) = 3|x|$ **48.** $T(x) = |x| + 2$

49. $f(x) = 1$ **50.** $k(x) = 2 + x + x^2$

51. $r(x) = \sqrt{x^2 + 4}$ **52.** $u(x) = \sqrt{3 - x^2}$

53. $s(x) = 16x^2$ **54.** $v(x) = 16x^2 + x$

55. $w(x) = 4 + \sqrt[3]{x}$ **56.** $z(x) = \dfrac{x^3}{x^2 + 1}$

57. Use the graph of $f(x) = \sqrt{4 - x^2}$ to sketch the graph of each of the following.
   a. $y = f(x) + 3$ **b.** $y = f(x - 3)$

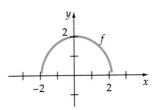

58. Use the graph of $g(x) = |x|$ to sketch the graph of each of the following.
   a. $y = g(x) - 2$ **b.** $y = g(x - 3)$

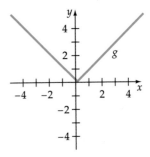

59. Use the graph of $F(x) = (x - 1)^{2/3}$ to sketch the graph of each of the following.

a. $y = -F(x)$ **b.** $y = F(-x)$

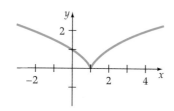

60. Use the graph of $E(x) = |x - 1| + 1$ to sketch the graph of each of the following.
  a. $y = -E(x)$ **b.** $y = E(-x)$

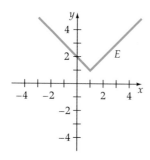

61. Use the graph of $m(x) = x^2 - 2x - 3$ to sketch the graph of $y = -\frac{1}{2}m(x) + 3$.

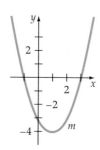

62. Use the graph of $n(x) = -x^2 - 2x + 8$ to sketch the graph of $y = \frac{1}{2}n(x) + 1$.

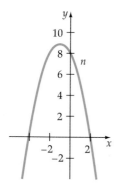

63. Use the graph of $y = f(x)$ to sketch the graph of each of the following.
  a. $y = f(2x)$ **b.** $f\left(\frac{1}{3}x\right)$

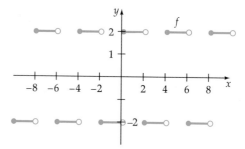

64. Use the graph of $y = g(x)$ to sketch the graph of each of the following.
  a. $y = g(2x)$ **b.** $y = g\left(\frac{1}{2}x\right)$

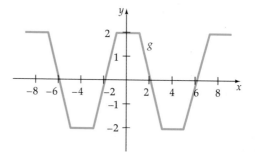

65. Use the graph of $y = h(x)$ to sketch the graph of each of the following.
  a. $y = h(2x)$ **b.** $y = h\left(\frac{1}{2}x\right)$

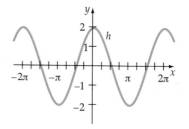

66. Use the graph of $y = j(x)$ to sketch the graph of each of the following.

a. $y = j(2x)$ **b.** $y = j\left(\dfrac{1}{3}x\right)$

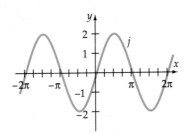

75. Sketch the graph of $V(x) = [\![cx]\!]$ for **a.** $c = 1$, **b.** $c = \frac{1}{2}$, **c.** $c = 2$, and $0 \le x \le 6$.

76. Sketch the graph of $W(x) = [\![cx]\!] - cx$ for **a.** $c = 1$, **b.** $c = \frac{1}{3}$, **c.** $c = 3$, and $0 \le x \le 6$.

Supplementary Exercises

77. Use the graph of $f(x) = 2/(x^2 + 1)$ to determine an equation for the graphs shown in a and b.

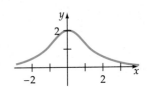

a. 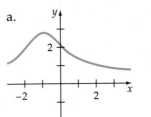   **b.**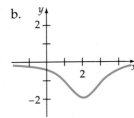

78. Use the graph of $f(x) = x\sqrt{2 + x}$ to determine an equation for the graphs shown in a and b.

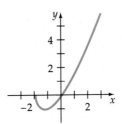

a. 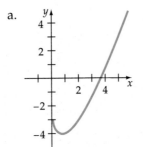   **b.**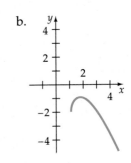

In Exercises 79 to 82, express the height of the given rectangle as a function of x. The midpoint of the upper base of the rectangle is on the graph of f, and the midpoint of the lower base is on the graph of g.

Graphing Calculator Exercises*

67. On the same coordinate axes sketch the graph of
$$G(x) = \sqrt[3]{x} + c$$
for $c = 0, -1$, and 3.

68. On the same coordinate axes sketch the graph of
$$H(x) = \sqrt[3]{x + c}$$
for $c = 0, -1$, and 3.

69. On the same coordinate axes sketch the graph of
$$J(x) = |2(x + c) - 3| - |x + c|$$
for $c = 0, -1$, and 2.

70. On the same coordinate axes sketch the graph of
$$K(x) = |x - 1| - |x| + c$$
for $c = 0, -1$, and 2.

71. On the same coordinate axes sketch the graph of
$$L(x) = cx^2$$
for $c = 1, \frac{1}{2}$, and 2.

72. On the same coordinate axes sketch the graph of
$$M(x) = c\sqrt{x^2 - 4}$$
for $c = 1, \frac{1}{3}$, and 3.

73. On the same coordinate axes sketch the graph of
$$S(x) = c(|x - 1| - |x|)$$
for $c = 1, \frac{1}{4}$, and 4.

74. On the same coordinate axes sketch the graph of
$$T(x) = c\left(\frac{x}{|x|}\right)$$
for $c = 1, \frac{2}{3}$, and $\frac{3}{2}$.

*Additional graphing calculator exercises appear in the Graphing Workbook as described in the front of this textbook.

79.

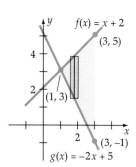

80.

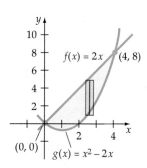

81.

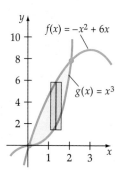

82.

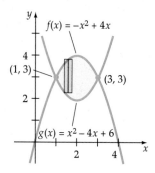

3.6 The Algebra of Functions

Functions can be defined in terms of other functions. For example, the function defined by $h(x) = x^2 + 8x$ is the sum of

$$f(x) = x^2 \quad \text{and} \quad g(x) = 8x$$

Thus if we are given any two functions, f and g, we can define the four new functions $f + g$, $f - g$, fg, and f/g as follows.

Operations on Functions

For all values of x for which both $f(x)$ and $g(x)$ are defined, we define the following.

Sum: $(f + g)(x) = f(x) + g(x)$

Difference: $(f - g)(x) = f(x) - g(x)$

Product: $(fg)(x) = f(x) \cdot g(x)$

Quotient: $\left(\dfrac{f}{g}\right)(x) = \dfrac{f(x)}{g(x)} \quad g(x) \neq 0$

Domain of $f + g$, $f - g$, fg, f/g

For the given functions f and g, the domains of $f + g$, $f - g$, and $f \cdot g$ consist of all real numbers formed by the intersection of the domains of f and g. The domain of f/g is the set of all real numbers formed by the intersection of the domains of f and g, except for those real numbers x such that $g(x) = 0$.

EXAMPLE 1   **Determine the Domain of a Function**

If $f(x) = \sqrt{x - 1}$ and $g(x) = x^2 - 4$, find the domain of each of the following: $f + g, f - g, fg$, and $\dfrac{f}{g}$.

Solution   Note that f has the domain $\{x \mid x \geq 1\}$ and g has the domain of all real numbers. Therefore the domain of $f + g$, $f - g$, and fg is $\{x \mid x \geq 1\}$. Since $g(x) = 0$ when $x = -2$ or $x = 2$, neither -2 nor 2 is in the domain of f/g. The domain of f/g is $\{x \mid x \geq 1 \text{ and } x \neq 2\}$.

■ *Try Exercise* **10**, *page 179.*

EXAMPLE 2   **Evaluate Functions**

Let $f(x) = x^2 - 9$ and $g(x) = 2x + 6$. Find each of the following.

a. $(f + g)(5)$ b. $(fg)(-1)$ c. $\left(\dfrac{f}{g}\right)(4)$

Solution

a. $(f + g)(x) = f(x) + g(x)$
$$= (x^2 - 9) + (2x + 6)$$
$$= x^2 + 2x - 3$$
Therefore, $(f + g)(5) = (5)^2 + 2(5) - 3 = 25 + 10 - 3 = 32$.

b. $(fg)(x) = f(x) \cdot g(x)$
$$= (x^2 - 9)(2x + 6)$$
$$= 2x^3 + 6x^2 - 18x - 54$$
Therefore, $(fg)(-1) = 2(-1)^3 + 6(-1)^2 - 18(-1) - 54$
$$= -2 + 6 + 18 - 54 = -32$$

c. $\left(\dfrac{f}{g}\right)(x) = \dfrac{f(x)}{g(x)} = \dfrac{x^2 - 9}{2x + 6}$
$$= \dfrac{\cancel{(x + 3)}(x - 3)}{2\cancel{(x + 3)}}$$
$$= \dfrac{x - 3}{2}, \qquad x \neq -3$$
Therefore, $\left(\dfrac{f}{g}\right)(4) = \dfrac{4 - 3}{2} = \dfrac{1}{2}$.

■ *Try Exercise* **14**, *page 179.*

The Difference Quotient

The expression

$$\frac{f(x + h) - f(x)}{h}, \quad h \neq 0$$

is called the **difference quotient** of f. It enables us to study the manner in which a function changes in value as the independent variable changes.

EXAMPLE 3 **Determine a Difference Quotient**

Determine the difference quotient of $f(x) = x^2 + 7$.

Solution

$$\frac{f(x + h) - f(x)}{h} = \frac{[(x + h)^2 + 7] - [x^2 + 7]}{h}$$

• Apply the difference quotient.

$$= \frac{[x^2 + 2xh + h^2 + 7] - [x^2 + 7]}{h}$$

$$= \frac{x^2 + 2xh + h^2 + 7 - x^2 - 7}{h}$$

$$= \frac{2xh + h^2}{h} = \frac{\cancel{h}(2x + h)}{\cancel{h}} = 2x + h$$

--- ■ *Try Exercise* **30**, *page 180.*

Composition of Functions

Composition of functions is yet another method of constructing a function from two given functions. The process consists of using the range element of one function as the domain element of another function.

Composite functions occur in many situations. For example, suppose the manufacturing cost (in dollars) per compact disc player is given by

$$m(x) = \frac{180x + 2600}{x}$$

where x is the number of compact disc players to be manufactured. An electronics outlet agrees to sell the compact discs by marking up the manufacturing cost per player $m(x)$ by 30%. Note that the selling price s will be a function of $m(x)$. More specifically,

$$s[m(x)] = 1.30(m(x))$$

Simplifying $s[m(x)]$ produces

$$s[m(x)] = 1.30\left(\frac{180x + 2600}{x}\right)$$

$$= 1.30(180) + 1.30\frac{2600}{x} = 234 + \frac{3380}{x}$$

The function produced in this manner is referred to as the composition of m by s. The notation $s \circ m$ is used to denote this composition function. That is,

$$(s \circ m)(x) = 234 + \frac{3380}{x}$$

Composition of Functions

For the functions f and g, the **composite function** or **composition** of f by g, is given by

$$(g \circ f)(x) = g[f(x)]$$

for all x in the domain of f such that $f(x)$ is in the domain of g.

If f and g are specified by equations, you can use substitution to find equations that specify $(g \circ f)$ and $(f \circ g)$.

EXAMPLE 4   Form Composite Functions

If $f(x) = x^2 - 3x$ and $g(x) = 2x + 1$, find   a. $(g \circ f)$ b. $(f \circ g)$.

Solution

a. $(g \circ f) = g[f(x)] = 2(f(x)) + 1$ • Substitute $f(x)$ for x in g.

$\qquad = 2(x^2 - 3x) + 1$ • $f(x) = x^2 - 3x$

$\qquad = 2x^2 - 6x + 1$

b. $(f \circ g) = f[g(x)] = (g(x))^2 - 3(g(x))$ • Substitute $g(x)$ for x in f.

$\qquad = (2x + 1)^2 - 3(2x + 1)$ • $g(x) = 2x + 1$

$\qquad = 4x^2 - 2x - 2$

■ *Try Exercise* **38**, *page 180.*

Note that in this example $(f \circ g) \neq (g \circ f)$. In general, the composition of function is not a commutative operation.

Caution   Some care must be used when forming the composition of functions. For instance, if $f(x) = x + 1$ and $g(x) = \sqrt{x - 4}$, then

$$(g \circ f)(2) = g[f(2)] = g(3) = \sqrt{3 - 4} = \sqrt{-1}$$

which is not a real number. We can avoid this problem by imposing suitable restrictions on the domain of f so that the range of f is part of the domain of g. If the domain of f is restricted to $[3, \infty)$, then the range of f is $[4, \infty)$. But this is precisely the domain of g. Note that $2 \notin [3, \infty)$, and thus we avoid the problem of $(g \circ f)(2)$ not being a real number.

To evaluate $(f \circ g)(c)$ for some constant c, you can use either of the following methods.

Method 1   First evaluate $g(c)$. Then substitute this result for x in the function f and evaluate.

Method 2   First determine $f[g(x)]$ and then substitute c for x.

EXAMPLE 5   **Evaluate Composite Functions**

Evaluate $(f \circ g)(3)$, where $f(x) = 2x - 5$ and $g(x) = 4x^2 + 1$.

Solution

Method 1 $(f \circ g)(3) = f[g(3)]$ • Evaluate $g(3)$.

$\qquad\qquad\qquad = f[4(3)^2 + 1]$

$\qquad\qquad\qquad = f(37)$ • Substitute in f.

$\qquad\qquad\qquad = 2(37) - 5 = 69$

Method 2 $(f \circ g)(x) = 2[g(x)] - 5$ • Form $f[(g(x)]$.

$\qquad\qquad\qquad = 2[4x^2 + 1] - 5$

$\qquad\qquad\qquad = 8x^2 + 2 - 5 = 8x^2 - 3$

$\qquad (f \circ g)(3) = 8(3)^2 - 3 = 69$ • Substitute 3 for x.

■ *Try Exercise* **54**, *page 180.*

Remark   Both methods 1 and 2 produced the same result. Although method 2 required more work, it is the better method to use if you have to evaluate $(f \circ g)(x)$ for several values of x.

EXERCISE SET 3.6

In Exercises 1 to 12, use the given functions f and g to find $f + g, f - g, fg$, and f/g. State the domain of each.

1. $f(x) = x^2 - 2x - 15$, $g(x) = x + 3$

2. $f(x) = x^2 - 25$, $g(x) = x - 5$

3. $f(x) = 2x + 8$, $g(x) = x + 4$

4. $f(x) = 5x - 15$, $g(x) = x - 3$

5. $f(x) = x^3 + 2x^2 + 7x$, $g(x) = x$

6. $f(x) = x^2 - 5x - 8$, $g(x) = -x$

7. $f(x) = 2x^2 + 4x - 7$, $g(x) = 2x^2 + 3x - 5$

8. $f(x) = 6x^2 + 10$, $g(x) = 3x^2 + x - 10$

9. $f(x) = \sqrt{x - 3}$, $g(x) = x$

10. $f(x) = \sqrt{x - 4}$, $g(x) = -x$

11. $f(x) = \sqrt{4 - x^2}$, $g(x) = 2 + x$

12. $f(x) = \sqrt{x^2 - 9}$, $g(x) = x - 3$

In Exercises 13 to 28, evaluate the indicated function, where $f(x) = x^2 - 3x + 2$ and $g(x) = 2x - 4$.

13. $(f + g)(5)$

14. $(f + g)(-7)$

15. $(f + g)\left(\dfrac{1}{2}\right)$

16. $(f + g)\left(\dfrac{2}{3}\right)$

17. $(f - g)(-3)$

18. $(f - g)(24)$

19. $(f - g)(-1)$

20. $(f - g)(0)$

21. $(fg)(7)$

22. $(fg)(-3)$

23. $(fg)\left(\dfrac{2}{5}\right)$

24. $(fg)(-100)$

25. $\left(\dfrac{f}{g}\right)(-4)$

26. $\left(\dfrac{f}{g}\right)(11)$

27. $\left(\dfrac{f}{g}\right)\left(\dfrac{1}{2}\right)$

28. $\left(\dfrac{f}{g}\right)\left(\dfrac{1}{4}\right)$

In Exercises 29 to 36, find the difference quotient of the given function.

29. $f(x) = 2x + 4$

30. $f(x) = 4x - 5$

31. $f(x) = x^2 - 6$

32. $f(x) = x^2 + 11$

33. $f(x) = 2x^2 + 4x - 3$

34. $f(x) = 2x^2 - 5x + 7$

35. $f(x) = -4x^2 + 6$

36. $f(x) = -5x^2 - 4x$

In Exercises 37 to 48, find $g \circ f$ and $f \circ g$ for the given functions f and g.

37. $f(x) = 3x + 5, \quad g(x) = 2x - 7$

38. $f(x) = 2x - 7, \quad g(x) = 3x + 2$

39. $f(x) = x^2 + 4x - 1, \quad g(x) = x + 2$

40. $f(x) = x^2 - 11x, \quad g(x) = 2x + 3$

41. $f(x) = x^3 + 2x, \quad g(x) = -5x$

42. $f(x) = -x^3 - 7, \quad g(x) = x + 1$

43. $f(x) = \dfrac{2}{x + 1}, \quad g(x) = 3x - 5$

44. $f(x) = \sqrt{x + 4}, \quad g(x) = \dfrac{1}{x}$

45. $f(x) = \dfrac{1}{x^2}, \quad g(x) = \sqrt{x - 1}$

46. $f(x) = \dfrac{6}{x - 2}, \quad g(x) = \dfrac{3}{5x}$

47. $f(x) = \dfrac{3}{|5 - x|}, \quad g(x) = -\dfrac{2}{x}$

48. $f(x) = |2x + 1|, \quad g(x) = 3x^2 - 1$

In Exercises 49 to 64, evaluate each composite function, where $f(x) = 2x + 3$, $g(x) = x^2 - 5x$, and $h(x) = 4 - 3x^2$.

49. $(g \circ f)(4)$

50. $(f \circ g)(4)$

51. $(f \circ g)(-3)$

52. $(g \circ f)(-1)$

53. $(g \circ h)(0)$

54. $(h \circ g)(0)$

55. $(f \circ f)(8)$

56. $(f \circ f)(-8)$

57. $(h \circ g)\left(\dfrac{2}{5}\right)$

58. $(g \circ h)\left(-\dfrac{1}{3}\right)$

59. $(g \circ f)(\sqrt{3})$

60. $(f \circ g)(\sqrt{2})$

61. $(g \circ f)(2c)$

62. $(f \circ g)(3k)$

63. $(g \circ h)(k + 1)$

64. $(h \circ g)(k - 1)$

In Exercises 65 to 68, evaluate each function. Give your answers accurate to four significant digits. Use $f(x) = x^2$, $g(x) = \sqrt{x}$, and $h(x) = 1/x$.

65. $g(644.5)$

66. $h(0.2354)$

67. $(f \circ h)(427.4)$

68. $(h \circ f)(9101)$

Supplemental Exercises

In Exercises 69 to 74, show that $(g \circ f)(x) = x$ and $(f \circ g)(x) = x$.

69. $f(x) = 2x + 3, \quad g(x) = \dfrac{x - 3}{2}$

70. $f(x) = 4x - 5, \quad g(x) = \dfrac{x + 5}{4}$

71. $f(x) = \dfrac{4}{x + 1}, \quad g(x) = \dfrac{4 - x}{x}$

72. $f(x) = \dfrac{2}{1 - x}, \quad g(x) = \dfrac{x - 2}{x}$

73. $f(x) = x^3 - 1, \quad g(x) = \sqrt[3]{x + 1}$

74. $f(x) = -x^3 + 2, \quad g(x) = \sqrt[3]{2 - x}$

75. Let x be the number of computer monitors to be manufactured. The manufacturing cost (in dollars) per computer monitor is given by the function

$$m(x) = \frac{60x + 34{,}000}{x}$$

A computer store will sell the monitors by marking up the manufacturing cost per monitor $m(x)$ by 45%. Thus the selling price s is a function of $m(x)$ given by the equation

$$s[m(x)] = 1.45(m(x)).$$

a. Express the selling price as a function of the number of monitors to be manufactured. That is, find $s \circ m$.

b. Find $(s \circ m)(24{,}650)$.

76. The number of bookcases b that a factory can produce per day is a function of the number of hours t it operates.

$$b(t) = 40t \quad \text{for } 0 \le t \le 12$$

The daily cost c to manufacture b bookcases is given by the function

$$c(b) = 0.1b^2 + 90b + 800$$

Evaluate each of the following and interpret your answers.

a. $b(5)$

b. $c(5)$

c. $(c \circ b)(t)$

d. $(c \circ b)(10)$

77. Let $f(x) = \dfrac{x + 1}{x}$. Evaluate each of the following:

a. $f(1)$

b. $(f \circ f)(1)$

c. $(f \circ f \circ f)(1)$

d. $(f \circ f \circ f \circ f)(1)$

78. Let $f(x) = \sqrt{x}$. Evaluate each of the following:

a. $f(65536)$

b. $(f \circ f)(65536)$

c. $(f \circ f \circ f)(65536)$

d. $(f \circ f \circ f \circ f)(65536)$

3.7 Inverse Functions

The operations of addition and subtraction are said to be **inverse operations** because one *undoes* the other. For example, if you start with a number (say, 10) and then add 4 and subtract 4, you will have the number 10 that you started with. Some functions are inverses of each other in the sense that one undoes the other. The function $g(x) = \frac{1}{2}x - 4$ undoes the function $f(x) = 2x + 8$. To illustrate, let $x = 10$. Then

$$f(x) = f(10) = 2(10) + 8 = 28$$

Now evaluate $g[f(10)] = g(28)$.

$$g[f(10)] = g(28) = \frac{1}{2}(28) - 4 = 14 - 4 = 10$$

Thus we started with $x = 10$, we evaluated $f(10)$, we evaluated $g[f(10)]$, and the end result was the number that we started with, 10. This was not a coincidence.

Definition of an Inverse Function

> If f is a one-to-one function with domain X and range Y, and g is a function with domain Y and range X, then g is the **inverse function** of f if and only if
>
> $$(f \circ g)(x) = x \quad \text{for all } x \text{ in the domain of } g$$
>
> and
>
> $$(g \circ f)(x) = x \quad \text{for all } x \text{ in the domain of } f$$

EXAMPLE 1   **Verify That Functions Are Inverse Functions**

Verify that $g(x) = \dfrac{1}{2}x - 4$ is the inverse of $f(x) = 2x + 8$.

Solution   We need to show that $(f \circ g)(x) = x$ and that $(g \circ f)(x) = x$.

$$(f \circ g)(x) = f[g(x)] = f\left[\frac{1}{2}x - 4\right] = 2\left[\frac{1}{2}x - 4\right] + 8 = x - 8 + 8 = x$$

Also,

$$(g \circ f)(x) = g[f(x)] = g[2x + 8] = \frac{1}{2}[2x + 8] - 4 = x + 4 - 4 = x$$

Therefore the function g is the inverse function of f. This work also shows that f is the inverse function of g.

■ *Try Exercise* **2,** *page 186.*

The definition of an inverse function requires f to be a one-to-one function. The reason for this restriction is now explained.

If a one-to-one function is given as a set of ordered pairs, then its inverse is the set of ordered pairs with their components interchanged. For example, the inverse of

$$\{(4, 7), (5, 2), (6, 11)\} \quad \text{is} \quad \{(7, 4), (2, 5), (11, 6)\}$$

Now consider the function j defined by $j(x) = x^2 - 1$. Some of the ordered pairs of j are

$$(-2, 3), \quad (-1, 0), \quad (0, -1), \quad (1, 0), \quad \text{and} \quad (2, 3)$$

The inverse of j contains the ordered pairs

$$(3, -2), \quad (0, -1), \quad (-1, 0), \quad (0, 1), \quad \text{and} \quad (3, 2)$$

This set of ordered pairs does *not* satisfy the definition of a function, because there are ordered pairs with the same first component and *different* second components. For example, the ordered pairs $(3, -2)$ and $(3, 2)$ both have 3 as their first component, but they have different second components. This example illustrates that not all functions have inverses that are functions.

Figure 3.76 is the graph of the function j. The horizontal line test indicates that j is *not* the graph of a *one-to-one* function. The horizontal line test can be used to show that the function $h(x) = \frac{1}{2}x^3$ is a one-to-one function. See Figure 3.77. Some of the ordered pairs of h are

$$(-2, -4), \quad \left(-1, -\frac{1}{2}\right), \quad (0, 0), \quad \left(1, \frac{1}{2}\right), \quad \text{and} \quad (2, 4)$$

Because h is a one-to-one function, given any y in the range of h, there corresponds exactly one x in the domain of h. Thus interchanging the coordinates of each ordered pair defined by h yields a set of ordered pairs that is a function. This function with the coordinates interchanged is the inverse function of h. The one-to-one property is exactly what is required for a function to have an inverse function.

Condition for a Function to Have an Inverse Function

A function f has an inverse function if and only if it is a one-to-one function.

The inverse of the function f is often denoted by f^{-1}. In Example 1, we verified that g was the inverse of f, so in this case the function g could be written as f^{-1}.

Caution The notation f^{-1} for an inverse function does not mean $1/f$. The function denoted by $1/f$ is called the **reciprocal function** and is an entirely different function from f^{-1}. For example, in Example 1 we showed that $f^{-1}(x) = g(x) = \frac{1}{2}x - 4$, whereas $1/f(x) = 1/(2x + 8)$.

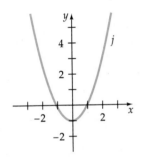

$j(x) = x^2 - 1$
Figure 3.76

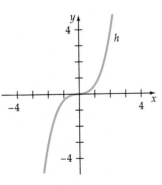

$h(x) = \frac{1}{2}x^3$
Figure 3.77

If a one-to-one function f is defined by an equation, then we use the following method to find the equation of the inverse f^{-1}.

Find the Equation for f^{-1}

To find the inverse f^{-1} of the one-to-one function f:
1. Substitute y for $f(x)$.
2. Interchange x and y.
3. Solve, if possible, for y in terms of x.
4. Substitute $f^{-1}(x)$ for y.
5. Verify that the domain of f is the range of f^{-1} and that the range of f is the domain of f^{-1}.

EXAMPLE 2   **Find the Inverse of a One-to-One Function**

Find the inverse of the one-to-one function $f(x) = 2x - 6$.

Solution   Begin by substituting y for $f(x)$:

$$y = 2x - 6$$

Interchanging x and y yields $x = 2y - 6$.

$$x + 6 = 2y \quad \bullet \text{ Solve for } y.$$

$$\frac{x + 6}{2} = y$$

This equation can be written as

$$y = \frac{1}{2}x + 3.$$

In inverse notation,

$$f^{-1}(x) = \frac{1}{2}x + 3.$$

In this example, the function f has a domain of all real numbers and a range of all real numbers, so the inverse f^{-1} also has a domain of all real numbers and a range of all real numbers.

———————————————————————— ■ *Try Exercise 10, page 186.*

Sometimes it is necessary to collect on the same side of the equation the terms that contain a y factor and then factor those terms to solve for y.

EXAMPLE 3   **Find the Inverse of a One-to-One Function**

Find the inverse of the function defined by $g(x) = \dfrac{2x}{x + 3}$.

Solution

$$y = \frac{2x}{x + 3}$$ • Replace $g(x)$ with y.

$$x = \frac{2y}{y + 3}$$ • Interchange x and y.

$$x(y + 3) = 2y$$ • Multiply by $(y + 3)$.

$$xy + 3x = 2y$$

$$xy - 2y = -3x$$ • Collect on one side the terms that contain a factor of y.

$$y(x - 2) = -3x$$ • Factor out the y.

$$y = \frac{-3x}{x - 2}$$ • Solve for y.

$$g^{-1}(x) = \frac{-3x}{x - 2} \quad \text{or} \quad \frac{3x}{2 - x}$$

■ *Try Exercise* **18,** *page 186.*

The function defined by $f(x) = x^2 - 4x$ graphs as a parabola that opens upward. It is not a one-to-one function and therefore does not have an inverse function. However, the function $G(x) = x^2 - 4x$ with domain restricted to $\{x \mid x \geq 2\}$ is a one-to-one function. It has an inverse function denoted by G^{-1}.

EXAMPLE 4   Find the Inverse Function and State Its Domain and Range

Find the inverse G^{-1} of the function $G(x) = x^2 - 4x$, for $x \geq 2$. State the domain and range of both G and G^{-1}.

Solution   First note that the domain of G is given as $\{x \mid x \geq 2\}$. The graph of G in Figure 3.78 shows that G has the range $\{y \mid y \geq -4\}$. Because the domain of G^{-1} is the range of G and the range of G^{-1} is the domain of G^{-1}, G^{-1} has the domain $\{x \mid x \geq -4\}$ and the range $\{y \mid y \geq 2\}$.

Now we proceed to find G^{-1}. The method shown uses the technique of completing the square.

$$G(x) = x^2 - 4x \quad \text{for } x \geq 2$$

$$y = x^2 - 4x$$

$$x = y^2 - 4y$$ • Interchange x and y.

$$x + 4 = y^2 - 4y + 4$$ • To complete the square of $y^2 - 4y$, we need to add 4 to each side.

$$x + 4 = (y - 2)^2$$ • Factor.

$$\pm\sqrt{x + 4} = y - 2$$ • Apply the Square Root Theorem.

$$2 \pm \sqrt{x + 4} = y$$

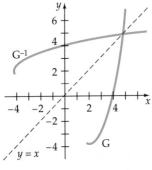

$$G(x) = x^2 - 4x, \quad x \geq 2$$
$$G^{-1}(x) = 2 + \sqrt{x + 4}$$
Figure 3.78

The range of G^{-1} is $\{y \mid y \geq 2\}$. Recall that the radical $\sqrt{x + 4}$ is a nonnegative number. Therefore, to make $G^{-1}(x) = 2 \pm \sqrt{x + 4}$ a real number greater than or equal to 2 requires that we consider only the nonnegative square root. Thus G^{-1} is given by

$$G^{-1}(x) = 2 + \sqrt{x + 4}$$

■ *Try Exercise* **32**, *page 186.*

The graphs of G and G^{-1} are shown in Figure 3.78. The graphs are symmetric with respect to the line $y = x$. This is always the case for the graph of a function and its inverse.

Symmetry Property of f and f^{-1}

The graph of a function f and the graph of the inverse function f^{-1} are symmetric with respect to the line given by $y = x$.

The symmetry property of f and f^{-1} can be used to graph the inverse of a one-to-one function.

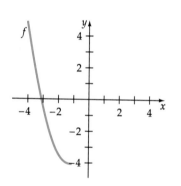

Figure 3.79

EXAMPLE 5   **Graph the Inverse of a Function**

Graph f^{-1} if f is the function defined by the graph in Figure 3.79.

Solution   Sketch the graph of f^{-1} by drawing the reflection of f with respect to the line given by $y = x$. See Figure 3.80.

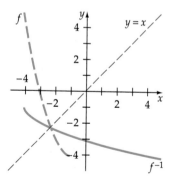

Figure 3.80

■ *Try Exercise* **36**, *page 186.*

Remark   In Figure 3.80 the diagonal line given by $y = x$ is not a part of the graph of f or its inverse f^{-1}. It is included to illustrate that the graph of f and f^{-1} are symmetric with respect to this diagonal line.

The concept of an inverse function will play an important role as new functions are introduced in the following chapters.

EXERCISE SET 3.7

In Exercises 1 to 8, verify that f and g are inverse functions by showing that $(f \circ g)(x) = x$ and $(g \circ f)(x) = x$.

1. $f(x) = 2x + 1, g(x) = \dfrac{x - 1}{2}$

2. $f(x) = \dfrac{1}{2}x - 3, g(x) = 2x + 6$

3. $f(x) = 3x - 5, g(x) = \dfrac{x + 5}{3}$

4. $f(x) = -2x + 1, g(x) = -\dfrac{1}{2}x + \dfrac{1}{2}$

5. $f(x) = \dfrac{1}{x + 1}, g(x) = \dfrac{1 - x}{x}$

6. $f(x) = \dfrac{1}{x} + 1, g(x) = \dfrac{1}{x - 1}$

7. $f(x) = \sqrt[3]{x - 1}, g(x) = x^3 + 1$
8. $f(x) = x^3 - 2, g(x) = \sqrt[3]{x + 2}$

In Exercises 9 to 24, find the inverse of the given function.

9. $f(x) = 4x + 1$

10. $g(x) = \dfrac{2}{3}x + 4$

11. $F(x) = -6x + 1$

12. $h(x) = -3x - 2$

13. $j(t) = 2t + 1$

14. $m(s) = -3s + 8$

15. $f(v) = 1 - v^3$

16. $u(t) = 2t^3 + 5$

17. $f(x) = \dfrac{-3x}{x + 4}$

18. $G(x) = \dfrac{3x}{x - 5}$

19. $M(t) = \dfrac{t - 5}{t}$

20. $P(v) = \dfrac{2v}{v + 1}$

21. $r(t) = \dfrac{1}{t^2}, t < 0$

22. $F(x) = \dfrac{1}{x}, x > 0$

23. $J(x) = x^2 + 4, x \ge 0$

24. $N(x) = 2x^2 + 1, x \le 0$

In Exercises 25 to 34, find the inverse of f. State the domain and range of both f and f^{-1}.

25. $f(x) = x^2 + 3, x \ge 0$

26. $f(x) = x^2 - 4, x \ge 0$

27. $f(x) = \sqrt{x}, x \ge 0$

28. $f(x) = \sqrt{16 - x}, x \le 16$

29. $f(x) = \sqrt{9 - x^2}, 0 \le x \le 3$

30. $f(x) = \sqrt{16 - x^2}, -4 \le x \le 0$

31. $f(x) = x^2 - 4x + 1, x \ge 2$

32. $f(x) = x^2 + 6x - 6, x \ge -3$

33. $f(x) = x^2 + 8x - 9, x \le -4$
34. $f(x) = x^2 - 2x - 2, x \le 1$

In Exercises 35 to 40, graph f^{-1} if f is the function defined by the graph.

35.

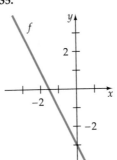

36.

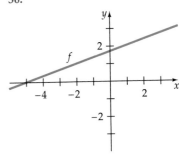

37.

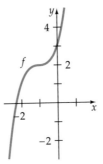

38.

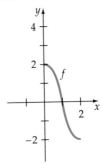

39.

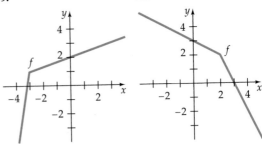

40.

In Exercises 41 to 48, graph each function f and its inverse f^{-1} on the same coordinate plane. Note that the graphs are symmetric with respect to the line $y = x$.

41. $f(x) = 3x + 3, f^{-1}(x) = \dfrac{1}{3}x - 1$

42. $f(x) = x - 4, f^{-1}(x) = x + 4$

43. $f(x) = \dfrac{1}{2}x, f^{-1}(x) = 2x$

44. $f(x) = 2x - 4, f^{-1}(x) = \dfrac{1}{2}x + 2$

45. $f(x) = x^2 + 2, x \ge 0, f^{-1}(x) = \sqrt{x - 2}, x \ge 2$

46. $f(x) = x^2 - 3, x \ge 0, f^{-1}(x) = \sqrt{x + 3}, x \ge -3$

47. $f(x) = (x - 2)^2, x \le 2, f^{-1}(x) = -\sqrt{x} + 2, x \ge 0$

48. $f(x) = (x + 3)^2, x \ge -3, f^{-1}(x) = \sqrt{x} - 3, x \ge 0$

Supplemental Exercises

In Exercises 49 to 52, find the inverse of the given function.

49. $f(x) = ax + b, a \ne 0$

50. $f(x) = ax^2 + bx + c; a \ne 0, x > -\dfrac{b}{2a}$

51. $f(x) = \dfrac{x - 1}{x + 1}, x \ne -1$

52. $f(x) = \dfrac{2 - x}{x + 2}, x \ne -2$

Only one-to-one functions have inverses that are functions. In Exercises 53 to 60, determine whether or not the given function is a one-to-one function.

53. $f(x) = x^2 + 8$
54. $v(s) = s^2 - 4$

55. $p(t) = \sqrt{9 - t}$
56. $v(t) = \sqrt{16 + t}$

57. $G(x) = -\sqrt{x}$
58. $K(x) = 1 - \sqrt{x - 5}$

59. $F(x) = |x| + x$
60. $T(x) = |x| - x$

In Exercises 61 to 64, assume that the given function has an inverse function.

61. If $f(5) = 2$, find $f^{-1}(2)$.
62. If $v(3) = 11$, find $v^{-1}(11)$.

63. If $s(4) = 60$, find $s^{-1}(60)$.
64. If $F(-8) = 5$, find $F^{-1}(5)$.

65. If the ordered pair (a, b) belongs to the graph of the function f, then (b, a) belongs to the graph of f^{-1}. Prove that the points $P(a, b)$ and $Q(b, a)$ are symmetric with respect to the graph of the line $y = x$ by using the definition of symmetry with respect to a line.

66. Graph $f(x) = -x + 3$. Use the graph to explain why f is its own inverse.

67. Reflect the point $(5, 2)$ about the line given by $y = 2x$. What are the coordinates of the image? If the point (a, b) is reflected about the line given by $y = mx$, what are the coordinates of the image?

3.8 Variation and Applications

Many real-life situations involve variables that are related by a type of function called a **variation**. For example, a fish jumping or a stone thrown into a pond generates circular ripples whose circumference and diameter are increasing. The equation $C = \pi d$ expresses the relationship between the circumference C of a circle and its diameter d. If d increases, then C increases. In fact, if d doubles in size, then C also doubles in size. The circumference C is said to **vary directly** as the diameter d.

Definition of Direct Variation

The variable y **varies directly** as the variable x, or y **is directly proportional** to x, if and only if

$$y = kx,$$

where k is a constant called the **constant of proportionality** or the **variation constant**.

Direct variations occur in many daily applications. For example, the cost of a newspaper is 25 cents. The cost C to purchase n newspapers is directly proportional to the number n. That is, $C = 25n$. In this example the variation constant is 25.

To solve a problem that involves a variation, we typically write a general equation that relates the variables and then use given information to solve for the variation constant.

EXAMPLE 1   **Solve a Direct Variation**

The distance sound travels varies directly as the time it travels. If sound travels 1340 meters in 4 seconds, find the distance sound will travel in 5 seconds.

Solution   Write an equation that relates the distance d to the time t. Since d varies directly as t, our equation is $d = kt$. Because $d = 1340$ when $t = 4$, we obtain

$$1340 = k \cdot 4 \quad \text{which implies} \quad k = \frac{1340}{4} = 335$$

Therefore, the specific equation that relates the distance d sound travels in t seconds is $d = 335t$. To find the distance sound travels in 5 seconds, replace t with 5 to produce

$$d = 335(5) = 1675$$

Under the same conditions, sound will travel 1675 meters in 5 seconds.

■ *Try Exercise 22, page 192.*

Direct Variation as the nth Power

> If y **varies directly as the nth power** of x, then
>
> $$y = kx^n$$
>
> where k is a constant.

EXAMPLE 2   **Solve a Variation of the Form $y = kx^2$**

The distance s that an object falls from rest (neglecting air resistance) varies directly as the square of the time t that it has been falling. If an object falls 64 feet in 2 seconds, how far will it fall in 10 seconds?

Solution   Since s varies directly as the square of t, $s = kt^2$. The variable s is 64 when t equals 2, so

$$64 = k \cdot 2^2 \quad \text{which implies} \quad k = \frac{64}{4} = 16$$

The specific equation that relates the distance s an object falls in t seconds is $s = 16t^2$. Letting $t = 10$ yields

$$s = 16(10^2) = 16(100) = 1600$$

Under the same conditions, an object will fall 1600 feet in 10 seconds.

■ *Try Exercise **24**, page 192.*

Definition of Inverse Variation

The variable y **varies inversely** as the variable x, or y **is inversely proportional to** x, if and only if

$$y = \frac{k}{x}$$

where k is the variation constant.

EXAMPLE 3   **Solve an Inverse Variation**

Boyle's Law states that the pressure P of a sample of gas at a constant temperature varies inversely as the volume V. The pressure of a gas in a balloon with volume 8 cubic inches is found to be 12 pounds per square inch. If the pressure is reduced, the volume of the balloon increases to 20 cubic inches. Find the new pressure of the gas.

   Solution   Since P varies inversely as V, $P = k/V$. The pressure P equals 12 when the volume V is 8, so

$$12 = \frac{k}{8} \quad \text{which implies} \quad k = 96$$

Consequently, the specific formula for P is $P = 96/V$. When the volume is 20 cubic inches, we have

$$P = \frac{96}{20} = 4.8 \text{ pounds per square inch}$$

■ *Try Exercise **28**, page 192.*

Inverse Variation as the nth Power

If y **varies inversely as the nth power** of x, then

$$y = \frac{k}{x^n}$$

where k is a constant.

Some variations involve more than two variables.

Definition of Joint Variation

The variable z **varies jointly as the variables** x **and** y if and only if

$$z = kxy$$

where k is a constant.

EXAMPLE 4   Solve a Joint Variation

The cost of insulating the ceiling of a house varies jointly with the thickness of the insulation and the area of the ceiling. It costs \$175 to insulate a 2100-square-foot ceiling with insulation 4 inches thick. Find the cost of insulating a 2400-square-foot ceiling with insulation that is 6 inches thick.

Solution   Since the cost C varies jointly as the area A of the ceiling and the thickness T of the insulation, we know $C = kAT$. Using the fact that $C = 175$ when $A = 2100$ and $T = 4$ gives us

$$175 = k(2100)\,(4) \quad \text{which implies} \quad k = \frac{175}{(2100)\,(4)} = \frac{1}{48}$$

Consequently, the specific formula for C is $C = \dfrac{1}{48}\,AT$. Now when $A = 2400$ and $T = 6$, we have

$$C = \frac{1}{48}\,(2400)\,(6) = 300$$

Thus the cost of insulating the 2400-square-foot ceiling with 6-inch insulation is \$300.

━━━━━━━━━━━━━━ ■ *Try Exercise* **30,** *page 193.*

EXAMPLE 5   Solve a Joint Variation

The load L that can be safely supported by a horizontal beam of given length varies jointly as the width w and the square of the depth d. See Figure 3.81. If a beam with width 2 inches and depth 4 inches can safely support a load of 400 pounds, determine the load that a beam (of the same length and material) of width 4 inches and depth 6 inches can safely support.

Solution   The general formula for the safe load is $L = kwd^2$. Substitution of known values produces

$$400 = k(2)\,(4^2) = 32k$$

Solving for k yields

$$k = \frac{400}{32} = 12.5$$

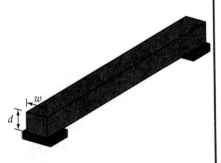

Figure 3.81

Therefore $L = 12.5(wd^2)$. Substituting 4 for w and 6 for d produces

$$L = 12.5(4)(6^2) = 1800 \text{ pounds}$$

■ *Try Exercise* **32,** *page 193.*

Combined variations involve more than one type of variation.

EXAMPLE 6   **Solve a Combined Variation**

The weight that a horizontal beam with rectangular cross section can safely support varies jointly as the width and square of the depth of the cross section and inversely as the length of the beam. See Figure 3.82. If a 4-inch by 4-inch beam 10 feet long safely supports a load of 256 pounds, what load L can be safely supported by a beam made of the same material and with a width w of 4 inches, a depth d of 6 inches, and a length l of 16 feet?

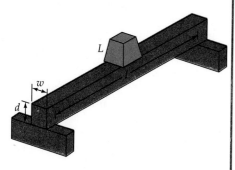

Figure 3.82

Solution   The general variation equation is $L = k\dfrac{wd^2}{l}$. Using the given data yields

$$256 = k\frac{4(4^2)}{10}$$

Solving for k produces $k = 40$, so the specific formula for L is

$$L = 40\frac{wd^2}{l}.$$

Substituting 4 for w, 6 for d, and 16 for l gives

$$L = 40\frac{4(6^2)}{16} = 360 \text{ pounds}$$

■ *Try Exercise* **34,** *page 193.*

EXERCISE SET 3.8

In Exercises 1 to 12, write an equation that represents the relationship between the given variables. Use k as the variation constant.

1. d varies directly as t.

2. r varies directly as the square of s.

3. y varies inversely as x.

4. p is inversely proportional to q.

5. m varies jointly as n and p.

6. t varies jointly as r and the cube of s.

7. V varies jointly as l, w, and h.

8. u varies directly as v and inversely as the square of w.

9. A is directly proportional to the square of s.

10. A varies jointly as h and the square of r.

11. F varies jointly as m_1 and m_2 and inversely as the square of d.

12. T varies jointly as t and r and the square of a.

In Exercises 13 to 20, write the equation that expresses the relationship between the variables, and then use the given data to solve for the variation constant.

13. y varies directly as x and $y = 64$ when $x = 48$.

14. m is directly proportional to n and $m = 92$ when $n = 23$.

15. r is directly proportional to the square of t and $r = 144$ when $t = 108$.

16. C varies directly as r and $C = 94.2$ when $r = 15$.

17. T varies jointly as r and the square of s and $T = 210$ when $r = 30$ and $s = 5$.

18. u varies directly as v and inversely as the square root of w and $u = 0.04$ when $v = 8$ and $w = 0.04$.

19. V varies jointly as l, w, and h and $V = 240$ when $l = 8$ and $w = 6$ and $h = 5$.

20. t varies directly as the cube of r and inversely as the square root of s and $t = 10$ when $r = 5$ and $s = 0.09$.

21. **Charles's Law** states that the volume V occupied by a gas (at a constant pressure) is directly proportional to its absolute temperature T. An experiment with a balloon shows that the volume of the balloon is 0.85 liters at 270 K (absolute temperature).[2] What will the volume of the balloon be when its temperature is 324 K?

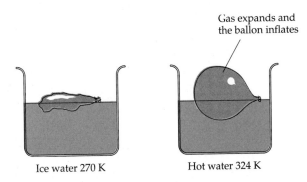

Gas expands and the ballon inflates

Ice water 270 K   Hot water 324 K

22. **Hooke's Law** states that the distance a spring stretches varies directly as the weight on the spring. A weight of 80 pounds stretches a spring 6 inches. How far will a weight of 100 pounds stretch the spring?

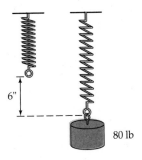

6"

80 lb

[2] Absolute temperature is measured on the Kelvin scale. A degree on the Kelvin scale is the same measure as a degree on the Celsius scale; however, 0 on the Kelvin scale corresponds to −273 on the Celsius scale.

23. The pressure a liquid exerts at a given point on a submarine is directly proportional to the depth of the point below the surface of the liquid. If the pressure at a depth of 3 feet is 187.5 pounds per square foot, find the pressure at a depth of 7 feet.

24. The range of a projectile is directly proportional to the square of its velocity. If a motorcyclist can make a jump of 140 feet by coming off a ramp at 60 mph, find the distance the motorcyclist could expect to jump if the speed coming off the ramp were increased to 65 mph.

25. The period T (the time it takes a pendulum to make one complete oscillation) varies directly as the square root of its length L. A pendulum 3 feet long has a period of 1.8 seconds.

   a. Find the period of a pendulum 10 feet long.

   b. What is the length of a pendulum that *beats seconds* (that is, it has a 2-second period)?

26. The area of a projected picture on a movie screen varies directly as the square of the distance from the projector to the screen. If a distance of 20 feet produces a picture with an area of 64 square feet, what distance produces an area of 100 square feet?

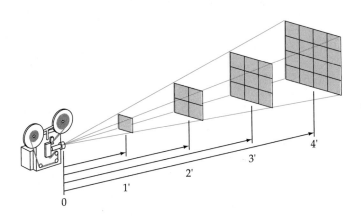

27. The loudness, measured in decibels, of a stereo speaker is inversely proportional to the square of the distance of the listener from the speaker. The loudness is 28 decibels at a distance of 8 feet. What is the loudness when the listener is 4 feet from the speaker?

28. The illumination a source of light provides is inversely proportional to the square of the distance from the source. If the illumination at a distance of 10 feet from the source is 50 footcandles, what is the illumination at a distance of 15 feet from the source?

29. The cost of insulating the exterior walls of a house varies jointly with the thickness of the insulation and the area of the walls. It cost $180 to insulate 1600 square feet of walls with insulation 3 inches thick. Find the cost of in-

sulating 2200 square feet of exterior walls with insulation 4 inches thick.

30. The simple interest earned in a given time period varies jointly with the principal and the interest rate. Investing a principal of $2600 at an interest rate of 4% yields $78. During the same time period, determine the simple interest earned on $4400 at an interest rate of 6%.

31. The volume V of a right circular cone varies jointly as the square of the radius r and the height h. Tell what happens to V when

   a. r is tripled.

   b. h is tripled.

   c. Both r and h are tripled.

32. The load L that a horizontal beam can safely support varies jointly as the width w and the square of the depth d. If a beam with width 2 inches and depth 6 inches safely supports up to 200 pounds, how many pounds can a beam of the same length that has width 4 inches and depth 4 inches be expected to support?

33. The **Ideal Gas Law** states that the volume V of a gas varies jointly as the number of moles of gas n and the absolute temperature T and inversely as the pressure P. What happens to V when n is tripled and P is reduced by a factor of one-half?

34. The maximum load a cylindrical column of circular cross section can support varies directly as the fourth power of the diameter and inversely as the square of the height. If a column 2 feet in diameter and 10 feet high supports up to 6 tons, how much of a load does a column 3 feet in diameter and 14 feet high support?

35. A meteorite approaching the earth has a velocity that varies inversely as the square root of the distance from the center of the earth. The meteorite has a velocity of 3 miles per second at 4900 miles from the center of the earth. Find the velocity of the meteorite when it is 4225 miles from the center of the earth.

36. A commuter airline has found that the average number of passengers per month between any two cities on its service routes is directly proportional to each city's population and inversely proportional to the square of the distance between them. Alameda has a population of 50,000 and is 80 miles from Baltic, which has a population of 64,000. Airline records indicate that an average of 3840 passengers per month travel between Alameda and Baltic. Estimate the average number of passengers per day that the airline could expect between Alameda and Crystal Lake, which has a population of 144,000 and is 160 miles from Alameda.

37. The frequency f of vibration of a piano string varies directly as the square root of the tension T on the string and inversely as the length L of the string. The middle a

string has a frequency of 440 vibrations per second. Find the frequency of a string that has 1.25 times as much tension and is six-fifths as long.

38. The load L a horizontal beam can safely support varies jointly as the width b and the square of the depth d and inversely as the length l. If a 12-foot beam with width 4 inches and depth 8 inches safely supports 800 pounds, how many pounds can a 16-foot beam that has width 3.5 inches and depth 6 inches be expected to support?

39. The force needed to keep a car from skidding on a curve varies jointly as the weight of the car and the square of the speed and inversely as the radius of the curve. It takes 2800 pounds of force to keep a 1800-pound car from skidding on a curve with radius 425 feet at 45 mph. What force is needed to keep the same car from skidding when it takes a similar curve with radius 450 feet at 55 mph?

Supplemental Exercises

40. **Kepler's Third Law** states that the time T needed for a planet to make one complete revolution about the sun is directly proportional to the $\frac{3}{2}$ power of the average distance d between the planet and the sun. The earth, which averages 93 million miles from the sun, completes one revolution in 365 days. Find the average distance from the sun to Mars if Mars completes one revolution about the sun in 686 days.

41. The weight W of an object varies inversely as the square of the distance d from the object to the center of the earth. At what altitude will a rocket weigh half of what it weighs at sea level? Assume sea level to be 4000 miles from the center of the earth.

42. If $f(x)$ varies directly as x, prove that $f(x_2) = f(x_1)\dfrac{x_2}{x_1}$.

43. Use the formula in Exercise 42 to solve the following direct variation *without* solving for the variation constant. The distance a spring stretches varies directly as the force applied. An experiment shows that a force of 17 kilograms stretches the spring 8.5 centimeters. How far will a 22-kilogram force stretch the spring?

44. If $f(x)$ varies inversely as x, prove that $f(x_2) = f(x_1)\dfrac{x_1}{x_2}$.

45. Use the formula in Exercise 44 to solve the following inverse variation *without* solving for the variation constant. The volume of a gas varies inversely as pressure (assuming the temperature remains constant). An experiment shows that a particular gas has a volume of 2.4 liters under a pressure of 280 grams per square centimeter. What volume will the gas have when a pressure of 330 grams per square centimeter is applied?

Exploring Concepts with Technology

Graph Functions and Interpret the Results

Some functions are difficult or impossible to graph even with a computer or a graphing calculator. For example, consider the function given by the equation

$$y = \sqrt{(x - 1)^2(x - 2)} + 1$$

Verify that the point $(1, 1)$ is a solution of the equation. Now use a computer or a graphing calculator to graph the function.

1. Does your graph include the isolated point at $(1, 1)$, as shown in the following figure? If your graphing utility failed to include the point $(1, 1)$, explain at least one reason for the omission of this isolated point.

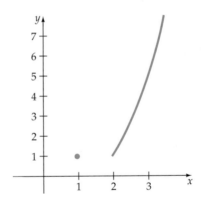

2. The equation

$$y = \frac{(x - 2)(x + 1)}{(x - 2)}$$

   graphs as a line with a y-intercept of 1, a slope of 1, and a hole at $(2, 3)$. Does your graphing utility show the hole at $(2, 3)$?

3. The *Dirichlet function* is a two part function given by

$$f(x) = \begin{cases} 0 & \text{if } x \text{ is rational} \\ 1 & \text{if } x \text{ is irrational} \end{cases}$$

   Write a sentence that describes the graph of this function.

4. Use a computer or graphing calculator to graph $f(x) = 3x^{5/3} - 6x^{4/3} + 2$ for $-2 \le x \le 10$. Compare your graph with the graph on page 195. Does your graph include the part to the left of the y-axis? If not, how might

you enter the function in such a way that your graphing utility would include this part?

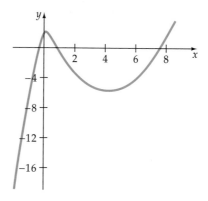

Chapter Review

3.1 A Two-Dimensional Coordinate System and Graphs

- **The Distance Formula** The distance d between the points represented by (x_1, y_1) and (x_2, y_2) is $d = \sqrt{(x_2 - x_1)^2 + (y_2 - y_1)^2}$.

- The standard form of the equation of a circle with center at (h, k) and radius r is $(x - h)^2 + (y - k)^2 = r^2$.

3.2 Functions

- **Definition of a Function**
  A function f from a set D to a set R is a correspondence, or rule, that pairs each element of D with exactly one element of R. The set D is called the domain of f, and the set R is called the range of f.

- **Alternative Definition of a Function**
  A function is a set of ordered pairs in which no two ordered pairs that have the same first component have different second components.

- A graph is the graph of a function if and only if no vertical line intersects the graph at more than one point. If any horizontal line intersects the graph of a function at most once, then the graph is the graph of a one-to-one function.

3.3 Linear Functions

- A function is a linear function if it can be written in the form $f(x) = mx + b$, where m and b are real constants.

- The slope m of the line passing through the points $P_1(x_1, y_1)$ and $P_2(x_2, y_2)$ with $x_1 \neq x_2$ is given by

$$m = \frac{y_2 - y_1}{x_2 - x_1}$$

- The graph of the equation $y = mx + b$ has slope m and y intercept $(0, b)$.

- Two nonvertical lines are parallel if and only if their slopes are equal. Two lines with slopes m_1 and m_2 are perpendicular if and only if $m_1 m_2 = -1$.

3.4 Quadratic Functions

- A quadratic function is a function that can be represented by an equation of the form $f(x) = ax^2 + bx + c$, where a, b, and c are real numbers and $a \neq 0$.

- The vertex of the graph of $f(x) = ax^2 + bx + c$ is

$$\left(-\frac{b}{2a}, f\left(-\frac{b}{2a} \right) \right)$$

- Every quadratic function $f(x) = ax^2 + bx + c$ can be written in the standard form $f(x) = a(x - h)^2 + k$, $a \neq 0$. The graph of f is a parabola with vertex (h, k). The parabola is symmetric with respect to the vertical line $x = h$, which is called the axis of the parabola. The parabola opens up if $a > 0$; it opens down if $a < 0$.

3.5 Additional Graphing Techniques

- The graph of an equation is symmetric with respect to the —y-axis if the replacement of x with $-x$ leaves the equation unaltered.

— x-axis if the replacement of y with $-y$ leaves the equation unaltered.

— origin if the replacement of x with $-x$ and y with $-y$ leaves the equation unaltered.

— line $y = x$ if the replacement of x with y and y with x leaves the equation unaltered.

- If f is a function and c is a positive constant, then
  — $y = f(x) + c$ is the graph of $y = f(x)$ shifted *vertically* c units up.
  — $y = f(x) - c$ is the graph of $y = f(x)$ shifted *vertically* c units down.
  — $y = f(x + c)$ is the graph of $y = f(x)$ shifted *horizontally* c units to the left.
  — $y = f(x - c)$ is the graph of $y = f(x)$ shifted *horizontally* c units to the right.

- The graph of
  — $y = -f(x)$ is the graph of $y = f(x)$ reflected across the x-axis
  — $y = f(-x)$ is the graph of $y = f(x)$ reflected across the y-axis.

3.6   The Algebra of Functions

- For all values of x for which both $f(x)$ and $g(x)$ are defined, we define the following functions:

  Sum: $(f + g)(x) = f(x) + g(x)$

  Difference: $(f - g)(x) = f(x) - g(x)$

Product: $(fg)(x) = f(x) \cdot g(x)$

Quotient: $\left(\dfrac{f}{g}\right)(x) = \dfrac{f(x)}{g(x)}, \quad g(x) \neq 0$

For the functions f and g, the composite function, or composition, of f by g is given by $(g \circ f)(x) = g[f(x)]$ for all x in the domain of f such that $f(x)$ is in the domain of g.

3.7   Inverse Functions

- If f is a one-to-one function with domain X and range Y, and g is a function with domain Y and range X, then g is the inverse function of f if and only if $(f \circ g)(x) = x$ for all x in the domain of g, and $(g \circ f)(x) = x$ for all x in the domain of f.

- A function f has an inverse function if and only if it is a one-to-one function. The graph of a function f and the graph of the inverse function f^{-1} are symmetric with respect to the line given by $y = x$.

3.8   Variation and Applications

- The variable y varies directly as the variable x if and only if $y = kx$, where k is a constant called the variation constant.

- The variable y varies inversely as the variable x if and only if $y = k/x$, where k is the variation constant.

- The variable z varies jointly as the variables x and y if and only if $z = kxy$, where k is the variation constant.

Essays and Projects

1. The mathematician Benoit Mandelbrot created a new type of geometry called *fractal geometry*. Write an essay on Benoit Mandelbrot and fractal geometry. Include information on the *Mandelbrot set*, and explain how the Mandelbrot set is related to the Mandelbrot replacement procedure explained on page 62.

2. Write an essay on the development of the cubic formula. An excellent source of information is the chapter "Cardano and the Solution of the Cubic" in *Journey Through Genius*, by William Dunham, (New York: Wiley, 1990).

3. The distance formula

$$d = \sqrt{(x_2 - x_1)^2 + (y_2 - y_1)^2}$$

is sometimes referred to as the *Euclidean metric*. Write an essay that explains the mathematical meaning of the term *metric*. A topology text may be helpful. Be sure to include

the four properties of a metric. Give some examples of metrics. Make a guess as to why $d = |x_2 - x_1| + |y_2 - y_1|$ is referred to as the *Manhattan metric*.

4. Write an essay that explains the difference between *static symmetry* and *dynamic symmetry*. Illustrate the use of dynamic symmetry in art and architecture. One source of information on dynamic symmetry is given in *The Golden Section and Related Curiosa*, by Garth E. Runion (Glenview, Ill.: Scott, Foresman, 1972).

5. Let $f(x) = x$, $g(x) = \dfrac{1}{x}$, $h(x) = 1 - x$,

$$j(x) = \frac{1}{1 - x}, \qquad k(x) = \frac{x}{x - 1}, \qquad l(x) = \frac{x - 1}{x}.$$

The composition of any two of these functions is one of the six functions. For example,

$$h \circ k = h(k) = h\left(\frac{x}{x-1}\right) = 1 - \left(\frac{x}{x-1}\right)$$

$$= \frac{x - 1 - x}{x - 1} = \frac{-1}{x - 1} = \frac{1}{1 - x} = j$$

Since $h \circ k = j$, we have placed a j in the row labeled h and the column labeled k of the following composition chart.

$\circ$	f	g	h	j	k	l
f						
g						
h				j		
j						
k						
l						

Fill in all the missing entries in the composition chart. The operation of composition defined on the above functions forms a mathematical structure called a *group*. Read about groups in an abstract algebra text, and then write an essay that defines the concept of a group. Also answer the following questions about the composition group. **a.** What is the identity element of the group? **b.** Is the group an *abelian* group?

6. A formula known as Zeller's Congruence makes use of the greatest integer function $[\![x]\!]$ to determine the day of the week on which a given date fell or will fall. To use Zeller's Congruence, we first compute the integer z given by

$$z = \left[\!\!\left[\frac{13m - 1}{5}\right]\!\!\right] + \left[\!\!\left[\frac{y}{4}\right]\!\!\right] + \left[\!\!\left[\frac{c}{4}\right]\!\!\right] + d + y - 2c$$

The variables c, y, d, and m are defined as follows:

c = the century

y = the year

d = the day of the month

m = the month, using 1 for March, 2 for April, ..., 10 for December. January and February are assigned the values 11 and 12 of the previous year.

For example, for the date September 12, 1991, we use $c = 19$, $y = 91$, $d = 12$, and $m = 7$. The remainder of z divided by 7 gives the day of the week. A remainder of 0 represents a Sunday, a remainder of 1 a Monday, ..., a remainder of 6 a Saturday.

a. Verify that December 7, 1941 was a Sunday.

b. Verify that January 1, 2000 will fall on a Saturday.

c. Determine on what day of the week Independence Day (July 4, 1776) fell?

True/False Exercises

In Exercises 1 to 10, answer true or false. If the statement is false, give an example to show that the statement is false.

1. Let f be any function. Then $f(a) = f(b)$ implies that $a = b$.

2. Every function has an inverse function.

3. If $(f \circ g)(a) = a$ and $(g \circ f)(a) = a$ for some constant a, then f and g are inverse functions.

4. Let f be a function such that $f(x) = f(x + 4)$ for all real numbers x. If $f(2) = 3$, then $f(18) = 3$.

5. For all functions f, $[f(x)]^2 = f[f(x)]$.

6. Let f be any function. Then for all a and b such that $f(b) \neq 0$ and $b \neq 0$,

$$\frac{f(a)}{f(b)} = \frac{a}{b}.$$

7. The **identity function** $f(x) = x$ is its own inverse.

8. If f is the function given by $f(x) = |x|$, then $f(a + b) = f(a) + f(b)$ for all real numbers a and b.

9. If f is the function given by $f(x) = |x|$, then $f(ab) = f(a)f(b)$ for all real numbers a and b.

10. If f is a one-to-one function and a and b are real numbers in the domain of f with $a < b$, then $f(a) \neq f(b)$.

Chapter Review Exercises

In Exercises 1 and 2, find the distance between the points whose coordinates are given.

1. $(-3, 2)$, $(7, 11)$ **2.** $(5, -4)$, $(-3, -8)$

In Exercises 3 and 4, find the midpoint of the line segment with the given endpoints.

3. $(2, 8)$, $(-3, 12)$ **4.** $(-4, 7)$, $(8, -11)$

In Exercises 5 and 6, determine the center and radius of the circle with the given equation.

5. $(x - 3)^2 + (y + 4)^2 = 81$

6. $x^2 + y^2 + 10x + 4y + 20 = 0$

In Exercises 7 and 8, find the equation in standard form of a circle that satisfies the given conditions.

7. Center $C = (2, -3)$,   radius $r = 5$

8. Center $C = (-5, 1)$,   passing through $(3, 1)$

9. If $f(x) = 3x^2 + 4x - 5$, find

  a. $f(1)$ **b.** $f(-3)$ **c.** $f(t)$

  d. $f(x + h)$ **e.** $3f(t)$ **f.** $f(3t)$

10. If $g(x) = \sqrt{64 - x^2}$, find

  a. $g(3)$ **b.** $g(-5)$ **c.** $g(8)$

  d. $g(-x)$ **e.** $2g(t)$ **f.** $g(2t)$

11. If $f(x) = x^2 + 4x$ and $g(x) = x - 8$, find

  a. $(f \circ g)(3)$ **b.** $(g \circ f)(-3)$

  c. $(f \circ g)(x)$ **d.** $(g \circ f)(x)$

12. If $f(x) = 2x^2 + 7$ and $g(x) = |x - 1|$, find

  a. $(f \circ g)(-5)$ **b.** $(g \circ f)(-5)$

  c. $(f \circ g)(x)$ **d.** $(g \circ f)(x)$

13. If $f(x) = 4x^2 - 3x - 1$, find the difference quotient

$$\frac{f(x + h) - f(x)}{h}$$

14. If $g(x) = x^3 - x$, find the difference quotient

$$\frac{g(x + h) - g(x)}{h}$$

In Exercises 15 to 20, sketch the graph of f. Find the interval(s) in which f is **a.** increasing, **b.** constant, **c.** decreasing.

15. $f(x) = |x - 3| - 2$ **16.** $f(x) = x^2 - 5$

17. $f(x) = |x + 2| - |x - 2|$ **18.** $f(x) = [\![x + 3]\!]$

19. $f(x) = \dfrac{1}{2}x - 3$ **20.** $f(x) = \sqrt[3]{x}$

In Exercises 21 to 24, determine the domain of the function represented by the given equation.

21. $f(x) = -2x^2 + 3$ **22.** $f(x) = \sqrt{6 - x}$

23. $f(x) = \sqrt{25 - x^2}$ **24.** $f(x) = \dfrac{3}{x^2 - 2x - 15}$

In Exercises 25 and 26, find the slope-intercept form of the equation of the line through the two points.

25. $(-1, 3)$ $(4, -7)$ **26.** $(0, 0)$ $(7, 11)$

27. Find the slope-intercept form of the equation of the line that is parallel to the graph of $3x - 4y = 8$ and passes through $(2, 11)$.

28. Find the slope-intercept form of the equation of the line that is perpendicular to the graph of $2x = -5y + 10$ and passes through $(-3, -7)$.

In Exercises 29 to 34, use the method of completing the square to write each quadratic equation in its standard form.

29. $f(x) = x^2 + 6x + 10$ **30.** $f(x) = 2x^2 + 4x + 5$

31. $f(x) = -x^2 - 8x + 3$ **32.** $f(x) = 4x^2 - 6x + 1$

33. $f(x) = -3x^2 + 4x - 5$ **34.** $f(x) = x^2 - 6x + 9$

In Exercises 35 to 40, find the vertex of the graph of the quadratic function.

35. $f(x) = 3x^2 - 6x + 11$ **36.** $h(x) = 4x^2 - 10$

37. $k(x) = -6x^2 + 60x + 11$ **38.** $m(x) = 14 - 8x - x^2$

39. $s(t) = -16t^2 + 1050$ **40.** $d(t) = 2t^2 - 10t$

In Exercises 41 and 42, sketch a graph that is symmetric to the given graph with respect to the **a.** x-axis, **b.** y-axis, **c.** origin.

41.   **42.**

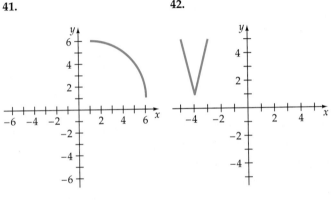

In Exercises 43 to 50, determine whether the graph of each equation is symmetric with respect to the **a.** x-axis, **b.** y-axis, **c.** origin.

43. $y = x^2 - 7$

44. $x = y^2 + 3$

45. $y = x^3 - 4x$

46. $y^2 = x^2 + 4$

47. $\dfrac{x^2}{3^2} + \dfrac{y^2}{4^2} = 1$

48. $xy = 8$

49. $|y| = |x|$

50. $|x + y| = 4$

In Exercises 51 to 56, sketch the graph of g. **a.** Find the domain and the range of g. **b.** State whether g is even, odd, or neither even nor odd.

51. $g(x) = -x^2 + 4$

52. $g(x) = -2x - 4$

53. $g(x) = |x - 2| + |x + 2|$

54. $g(x) = \sqrt{16 - x^2}$

55. $g(x) = x^3 - x$

56. $g(x) = 2[\![x]\!]$

In Exercises 57 to 62, first write the function in standard form, and then make use of translations to graph the function.

57. $F(x) = x^2 + 4x - 7$

58. $A(x) = x^2 - 6x - 5$

59. $P(x) = 3x^2 - 4$

60. $G(x) = 2x^2 - 8x + 3$

61. $W(x) = -4x^2 - 6x + 6$

62. $T(x) = -2x^2 - 10x$

63. On the same coordinate axes sketch the graph of $p(x) = \sqrt{x} + c$ for $c = 0, -1,$ and 2.

64. On the same coordinate axes sketch the graph of $q(x) = \sqrt{x + c}$ for $c = 0, -1$ and 2.

65. On the same coordinate axes sketch the graph of $r(x) = c\sqrt{9 - x^2}$ for $c = 1, \frac{1}{2},$ and -2.

66. On the same coordinate axes sketch the graph of $s(t) = [\![cx]\!]$ for $c = 1, \frac{1}{4},$ and 4.

In Exercises 67 and 68, sketch the graph of the function.

67. $f(x) = \begin{cases} x & \text{if } x \le 0 \\ \dfrac{1}{2}x & \text{if } x > 0 \end{cases}$

68. $g(x) = \begin{cases} -2 & \text{if } x < -3 \\ \dfrac{2}{3}x & \text{if } -3 \le x \le 3 \\ 2 & \text{if } x > 3 \end{cases}$

In Exercises 69 and 70, use the given functions f and g to find $f + g, f - g, fg,$ and f/g. State the domain of each.

69. $f(x) = x^2 - 9, \quad g(x) = x + 3$

70. $f(x) = x^3 + 8, \quad g(x) = x^2 - 2x + 4$

In Exercises 71 to 74, determine whether the given functions are inverses.

71. $F(x) = 2x - 5, \quad G(x) = \dfrac{x + 5}{2}$

72. $h(x) = \sqrt{x}, \quad k(x) = x^2, \quad x \ge 0$

73. $l(x) = \dfrac{x + 3}{x}, \quad m(x) = \dfrac{3}{x - 1}$

74. $p(x) = \dfrac{x - 5}{2x}, \quad q(x) = \dfrac{2x}{x - 5}$

In Exercises 75 to 78, find the inverse of the function. Sketch the graph of the function and its inverse on the same set of coordinates axes.

75. $f(x) = 3x - 4$

76. $g(x) = -2x + 3$

77. $h(x) = -\frac{1}{2}x - 2$

78. $k(x) = \dfrac{1}{x}$

79. Find two numbers whose sum is 50 and whose product is a maximum.

80. Find two numbers whose difference is 10 and whose sum of their squares is a minimum.

81. The roadway of the Golden Gate Bridge is 220 feet above the water. The height h of a rock dropped from the bridge is a function of the time t it has fallen. If h is measured in feet and t is measured in seconds, then the function is given by $h(t) = -16t^2 + 220$. Use the function to find the time it will take the rock to hit the water.

82. The suspension cables of the Golden Gate Bridge approximate the shape of a parabola. If h is the height of the cables above the roadway in feet, and $|x|$ is the distance in feet from the center of the bridge, then the parabolic shape of the cables is represented by

$$h(x) = \frac{1}{8820}x^2 + 25, \quad -2100 \le x \le 2100$$

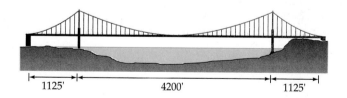

1125'   4200'   1125'

a. Find the height of the cables 1050 feet from the center of the bridge.

b. The towers that support the cables are 2100 feet from the center of the bridge. Find the height (above the roadway) of the towers that support the cables.

Chapter Test

1. Find the midpoint of the line segment with endpoints $(-2, 3)$ and $(4, -1)$.

2. Determine the x- and y-intercepts, and then graph the equation $x = 2y^2 - 4$.

3. Graph the equation $y = |x + 2| + 1$.

4. Find the center and radius of the circle that has the general form $x^2 - 4x + y^2 + 2y - 4 = 0$.

5. Determine the domain of the function $f(x) = -\sqrt{x^2 - 16}$.

6. A boat was purchased for $60,000. Assuming that the boat depreciates at a rate of $8000 per year (straight-line depreciation) for the first 6 years, write the value v of the boat as a function of the time t (measured in years) for $0 \le t \le 6$.

7. Graph $f(x) = -2|x - 2| + 1$. Identify the intervals over which the function is **a.** increasing, **b.** constant, or **c.** decreasing.

8. Graph the function $f(x) = x^2 + 2$. From the graph, find the domain and range of the function.

9. Sketch the graph of the function $f(x) = -(x - 2)^2 + 1$.

10. Use the graph of $f(x) = |x|$ to graph $y = -f(x + 2) - 1$.

11. Which of the following are odd functions?

   a. $f(x) = x^4 - x^2$ **b.** $f(x) = x^3 - x$

   c. $f(x) = x - 1$

12. Find the slope-intercept form of the equation of the line that passes through $(4, -2)$ and is perpendicular to the graph of $3x - 2y = 4$.

13. Find the maximum or minimum value of the function $f(x) = x^2 - 4x - 8$. State whether this value is a maximum or a minimum value.

14. Let $f(x) = x^2 - 1$ and $g(x) = x - 2$. Find $(f + g)$ and $\left(\dfrac{f}{g}\right)$.

15. Find the difference quotient of the function $f(x) = x^2 + 1$.

16. Evaluate $(f \circ g)$, where $f(x) = x^2 - 2x + 1$ and $g(x) = \sqrt{x - 2}$.

17. Find the inverse of $f(x) = x^2 - 9$, $x \ge 0$. State the domain and range of both f and f^{-1}.

18. Find the inverse of the function given by the equation $f(x) = 2x - 3$. Graph f and f^{-1} on the same coordinate axes.

19. Hooke's Law states that the distance a spring stretches varies directly as the weight on the spring. A weight of 40 pounds stretches a spring 1.5 inches. How far will a weight of 225 pounds stretch the spring?

20. The illumination that a source of light provides is inversely proportional to the square of the distance from the source. If the illumination at a distance of 8 feet is 20 lumens, what is the illumination at a distance of 15 feet from the source?

Polynomial and Rational Functions

CASE IN POINT **1, 2, 4, 8, 16, ?**

For each circle, count the number of points on the circle and the maximum number of regions formed by the chords. Your results should agree with the following.

Number of points	1	2	3	4	5	6
Number of regions	1	2	4	8	16	?

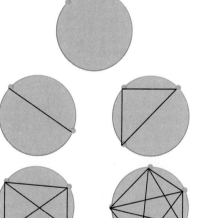

There appears to be a pattern. Guess the maximum number of regions you expect for a circle with six points. Check your guess by counting the maximum number of regions formed by the chords that connect six points on a large circle. Your drawing should show that for six points on the circle, there are only 31 regions, not 32 as you may have guessed. This example illustrates that conclusions based on inductive reasoning may prove to be incorrect.

The concepts developed in later chapters of this text can be used to show that the maximum number of regions R formed by connecting n points on a circle is given by

$$R(n) = (n^4 - 6n^3 + 23n^2 - 18n + 24)/24$$

Verify that $R(n)$ yields the correct number of regions for $n = 1, 2, 3, 4, 5,$ and 6. Use the polynomial $R(n)$ to determine the maximum number of regions formed for $n = 7$, and for $n = 8$.

4.1 Polynomial Division and Synthetic Division

If $P(x)$ is a polynomial, then the values of x for which $P(x)$ is equal to 0 are called the **zeros** of $P(x)$ or the **roots** of $P(x) = 0$. Much of the work in this chapter concerns finding the zeros of a polynomial. Sometimes the zeros of a polynomial can be determined by dividing the polynomial by another polynomial. Dividing a polynomial by another polynomial is similar to the long division process used for dividing positive integers. For example, to divide $(x^2 + 9x - 16)$ by $(x - 3)$, we use the following procedure.

$$
\begin{array}{r}
x + 12 \\
x - 3 \overline{)\, x^2 + 9x - 16} \\
\underline{x^2 - 3x \phantom{{}-16}} \\
12x - 16 \\
\underline{12x - 36} \\
20
\end{array}
$$

Thus $(x^2 + 9x - 16) \div (x - 3) = x + 12$, with a remainder of 20.

In this example, $x^2 + 9x - 16$ is called the **dividend**, $x - 3$ is the **divisor**, $x + 12$ is the **quotient**, and 20 is the **remainder**. The dividend is equal to the product of the divisor and the quotient, plus the remainder. That is,

$$
\underbrace{x^2 + 9x - 16}_{\text{Dividend}} = \underbrace{(x - 3)}_{\text{Divisor}} \cdot \underbrace{(x + 12)}_{\text{Quotient}} + \underbrace{20}_{\text{Remainder}}
$$

The above result is a special case of a theorem known as the *Division Algorithm for Polynomials.*

The Division Algorithm for Polynomials

> If $P(x)$ and $D(x)$ are polynomials such that $D(x) \neq 0$, then there exist unique polynomials $Q(x)$ and $R(x)$ such that $P(x) = D(x)Q(x) + R(x)$, where either $R(x) = 0$ or the degree of $R(x)$ is less than the degree of $D(x)$.

The polynomial $P(x)$ is the dividend, $D(x)$ is the divisor, $Q(x)$ is the quotient and the polynomial $R(x)$ is the remainder.

$$
\underbrace{P(x)}_{\text{Dividend}} = \underbrace{D(x)}_{\text{Divisor}} \cdot \underbrace{Q(x)}_{\text{Quotient}} + \underbrace{R(x)}_{\text{Remainder}}
$$

Multiplying both sides of $P(x) = D(x)Q(x) + R(x)$ by $1/D(x)$ produces the fractional form

$$\frac{P(x)}{D(x)} = Q(x) + \frac{R(x)}{D(x)}$$

Remark If $R(x) = 0$, then $D(x)$ is a factor of $P(x)$. If the degree of $D(x)$ is greater than the degree of $P(x)$, then $Q(x) = 0$, and $R(x) = P(x)$.

EXAMPLE 1 **Divide Polynomials**

Perform the indicated division.

$$\frac{x^4 + 3x^2 - 6x - 10}{x^2 + 3x - 5}$$

Solution

$$
\begin{array}{r}
x^2 - 3x \;\; + 17 \\
x^2 + 3x - 5 \overline{\smash{\big)}\, x^4 + 0x^3 + \;\; 3x^2 - \;\; 6x - 10} \\
\underline{x^4 + 3x^3 - \;\; 5x^2} \\
-3x^3 + \;\; 8x^2 - \;\; 6x \\
\underline{-3x^3 - \;\; 9x^2 + 15x} \\
17x^2 - 21x - 10 \\
\underline{17x^2 + 51x - 85} \\
-72x + 75
\end{array}
$$

• Writing $0x^3$ for the missing term helps us align like terms in the same column.

Thus $\dfrac{x^4 + 3x^2 - 6x - 10}{x^2 + 3x - 5} = x^2 - 3x + 17 + \dfrac{-72x + 75}{x^2 + 3x - 5}.$

■ *Try Exercise **6**, page 208.*

Synthetic Division

The procedure for dividing a polynomial by a binomial of the form $x - c$ can be condensed by a method called **synthetic division**. To understand the synthetic division method, consider the following division.

$$
\begin{array}{r}
3x^2 - 2x \;\; + 3 \\
x - 2 \overline{\smash{\big)}\, 3x^3 - 8x^2 + 7x + 2} \\
\underline{3x^3 - 6x^2} \\
-2x^2 + 7x \\
\underline{-2x^2 + 4x} \\
3x + 2 \\
\underline{3x - 6} \\
8
\end{array}
$$

No essential data are lost by omitting the variables since the position of a term indicates the power of the term.

$$
\begin{array}{r}
3 \quad -2 \quad 3 \\
-2\overline{)3 \quad -8 \quad 7 \quad 2} \\
\underline{3 \quad -6} \\
-2 \quad 7 \\
\underline{-2 \quad 4} \\
3 \quad 2 \\
\underline{3 \quad -6} \\
8
\end{array}
$$

The coefficients shown in color are duplicates of those directly above them. Omitting these repeated coefficients (in color) enables us to condense the vertical spacing.

$$
\begin{array}{r}
3 \quad -2 \quad 3 \\
-2\overline{)3 \quad -8 \quad 7 \quad 2} \\
\underline{-6 \quad 4 \quad -6} \\
-2 \quad 3 \quad 8
\end{array}
$$

The coefficients in color in the top row can be omitted since they are duplicates of those in the bottom row. The leading coefficient of the quotient (top row) can be written in the bottom row with the coefficients of the other terms in order to condense the vertical spacing even more.

$$
\begin{array}{r|rrrr}
-2 & 3 & -8 & 7 & 2 \\
 & & -6 & 4 & -6 \\
\hline
 & 3 & -2 & 3 & 8
\end{array}
$$

So that we may add the numbers in each column instead of subtracting them, we change the sign of the divisor. This changes the sign of each number in the second row.

$$
\begin{array}{r|rrrr}
2 & 3 & -8 & 7 & 2 \\
 & & 6 & -4 & 6 \\
\hline
 & 3 & -2 & 3 & 8
\end{array}
$$

Coefficients of the quotient   Remainder

The following example illustrates step by step the synthetic division procedure.

$$\frac{2x^3 - 9x^2 + 5}{x - 3} = ?$$

Coefficients
of the dividend

$$3 \,\big|\; 2 \quad -9 \quad 0 \quad 5$$ • Synthetic division form with 0 inserted for the missing x term.

$$3 \,\big|\; 2 \quad -9 \quad 0 \quad 5$$ • Bring down the leading coefficient 2.
$$\qquad 2$$

$$3 \,\big|\; 2 \quad -9 \quad 0 \quad 5$$ • Multiply $3 \cdot 2$ and place the product (6) in the middle row and in the next column to the right.
$$\qquad\quad 6$$
$$\qquad 2$$

$$3 \,\big|\; 2 \quad -9 \quad 0 \quad 5$$ • Add -9 and 6 and place the sum in the bottom row.
$$\qquad\quad 6$$
$$\qquad 2 \quad -3$$

$$3 \,\big|\; 2 \quad -9 \quad 0 \quad 5$$ • Repeat the previous steps for columns 3 and 4.
$$\qquad\quad 6 \quad -9 \quad -27$$
$$\qquad 2 \quad -3 \quad -9 \quad -22$$

Remainder

Coefficients
of the quotient

$$\frac{2x^3 - 9x^2 + 5}{x - 3} = 2x^2 - 3x - 9 + \frac{-22}{x - 3}$$

Remark   The synthetic division method shown in the previous example is used only to divide by a polynomial of the form $x - c$, where the coefficient of x is 1. To divide a polynomial by a polynomial that is not a binomial, use the long division method.

EXAMPLE 2   **Use Synthetic Division to Divide Polynomials**

Use synthetic division to perform the indicated division.

$$\frac{x^4 - 4x^2 + 7x + 15}{x + 4}$$

Solution   Since the divisor is $x + 4$, we perform the synthetic division with $c = -4$.

$$
\begin{array}{r|rrrrr}
-4 & 1 & 0 & -4 & 7 & 15 \\
   &   & -4 & 16 & -48 & 164 \\
\hline
   & 1 & -4 & 12 & -41 & 179
\end{array}
$$

The quotient is $x^3 - 4x^2 + 12x - 41$ and the remainder is 179.

$$\frac{x^4 - 4x^2 + 7x + 15}{x + 4} = x^3 - 4x^2 + 12x - 41 + \frac{179}{x + 4}$$

■ *Try Exercise* **12**, *page 208.*

The following theorem shows that synthetic division can be used to find the value $P(c)$ for any polynomial function P and constant c.

The Remainder Theorem

If a polynomial $P(x)$ is divided by $x - c$, then the remainder is $P(c)$.

Proof: The Division Algorithm states that

$$P(x) = (x - c)Q(x) + R(x)$$

where $R(x)$ is zero or the degree of $R(x)$ is less than the degree of $x - c$. Since the degree of $x - c$ is 1, the remainder $R(x)$ must be some constant—say, r. Therefore,

$$P(x) = (x - c)Q(x) + r$$

The above equality evaluated at $x = c$ produces

$$P(c) = (c - c)Q(c) + r = (0)Q(c) + r = r$$

EXAMPLE 3   **Use the Remainder Theorem to Evaluate a Polynomial**

Use the Remainder Theorem to evaluate $P(x) = 2x^3 - 3x^2 + 4x - 1$ for $x = -1$ and $x = 3$.

Solution   Perform synthetic divisions and examine the remainders.

$$
\begin{array}{r|rrrr}
-1 & 2 & -3 & 4 & -1 \\
   &   & -2 & 5 & -9 \\
\hline
   & 2 & -5 & 9 & -10
\end{array}
$$

The remainder is -10. By the Remainder Theorem, $P(-1) = -10$.

$$
\begin{array}{r|rrrr}
3 & 2 & -3 & 4 & -1 \\
  &   & 6 & 9 & 39 \\
\hline
  & 2 & 3 & 13 & 38
\end{array}
$$

The remainder is 38. By the Remainder Theorem, $P(3) = 38$.

■ *Try Exercise* **32,** *page 208.*

The following theorem is a result of the Remainder Theorem.

The Factor Theorem

A polynomial $P(x)$ has a factor $(x - c)$ if and only if $P(c) = 0$.

Proof: Part 1: Given $P(x)$ has a factor of $(x - c)$, show that $P(c) = 0$. If $(x - c)$ is a factor of $P(x)$, then $P(x) = (x - c) \cdot Q(x)$ for some $Q(x)$. Thus the division of $P(x)$ by $(x - c)$ has a remainder of zero, and the Remainder Theorem implies that $P(c) = 0$.

Part 2: Given $P(c) = 0$, show that $(x - c)$ is a factor of $P(x)$. The division algorithm applied to the polynomial $P(x)$ with divisor $(x - c)$ produces

$$P(x) = (x - c)Q(x) + R(x)$$

Since $P(c) = 0$, the Remainder Theorem implies that $R(x) = 0$. Thus

$$P(x) = (x - c)Q(x)$$

which shows $(x - c)$ is a factor of $P(x)$.

EXAMPLE 4   **Find a Factor of a Polynomial**

Determine whether $(x + 5)$ is a factor of

$$P(x) = x^4 + x^3 - 21x^2 - x + 20$$

Solution

$$
\begin{array}{r|rrrrr}
-5 & 1 & 1 & -21 & -1 & 20 \\
   &   & -5 & 20 & 5 & -20 \\
\hline
   & 1 & -4 & -1 & 4 & 0
\end{array}
$$

The remainder 0 implies that $(x + 5)$ is a factor of $P(x)$.

■ *Try Exercise* **42,** *page 208.*

Remark   Because $(x + 5)$ is a factor of $P(x)$, it is also true that -5 is a zero of $P(x)$. Why?[1] The quotient $Q(x)$ is also a factor of $P(x)$. From the last line of

---

[1] Because the Factor Theorem states that a polynomial $P(x)$ has a factor $(x - c)$ if and only if $P(c) = 0$.

the synthetic division, $Q(x) = x^3 - 4x^2 - x + 4$. Thus,

$$P(x) = (x + 5)(x^3 - 4x^2 - x + 4)$$

The polynomial $Q(x) = x^3 - 4x^2 - x + 4$ is called a **reduced polynomial** because it is 1 degree less than the degree of $P(x)$. Reduced polynomials play an important role in Section 4.3.

EXERCISE SET 4.1

In Exercises 1 to 10, use long division to divide the first polynomial by the second.

1. $5x^3 + 6x^2 - 17x + 20, \quad x + 3$

2. $6x^3 + 15x^2 - 8x + 2, \quad x + 4$

3. $2x^4 + 15x^3 + 7x^2 - 135x - 225, \quad 2x + 5$

4. $6x^4 + 3x^3 - 11x^2 - 3x + 9, \quad 2x - 3$

5. $3x^4 + x^3 - 99x^2 - 30, \quad 3x^2 + x + 1$

6. $2x^4 - x^3 - 23x^2 + 9x + 45, \quad 2x^2 - x - 5$

7. $20x^4 - 3x^2 + 9, \quad 5x^2 - 2$

8. $24x^5 + 20x^3 - 16x^2 - 15, \quad 6x^2 + 5$

9. $x^3 + 5x^2 + 6x - 19, \quad x^2 + x - 4$

10. $2x^4 + 3x^3 - 7x - 10, \quad x^2 - 2x - 5$

In Exercises 11 to 30, use synthetic division to divide the first polynomial by the second.

11. $4x^3 - 5x^2 + 6x - 7, \quad x - 2$

12. $5x^3 + 6x^2 - 8x + 1, \quad x - 5$

13. $4x^3 - 2x + 3, \quad x + 1$

14. $6x^3 - 4x^2 + 17, \quad x + 3$

15. $x^5 - 10x^3 + 5x - 1, \quad x - 4$

16. $6x^4 - 2x^3 - 3x^2 - x, \quad x - 5$

17. $x^5 - 1, \quad x - 1$

18. $x^4 + 1, \quad x + 1$

19. $8x^3 - 4x^2 + 6x - 3, \quad x - \dfrac{1}{2}$

20. $12x^3 + 5x^2 + 5x + 6, \quad x + \dfrac{3}{4}$

21. $x^8 + x^6 + x^4 + x^2 + 4, \quad x - 2$

22. $-x^7 - x^5 - x^3 - x - 5, \quad x + 1$

23. $x^6 + x - 10, \quad x + 3$

24. $2x^5 - 3x^4 - 5x^2 - 10, \quad x - 4$

25. $3x^2 - 4x + 5, x - 0.3$ 26. $2x^2 - 12x + 1, x + 0.4$

27. $2x^3 - 11x^2 - 17x + 3, x$ 28. $5x^4 - 2x^2 + 6x - 1, x$

29. $x + 8, x + 2$ 30. $3x - 17, x - 3$

In Exercises 31 to 40, use the Remainder Theorem to find $P(c)$.

31. $P(x) = 3x^3 + x^2 + x - 5, c = 2$

32. $P(x) = 2x^3 - x^2 + 3x - 1, c = 3$

33. $P(x) = 4x^4 - 6x^2 + 5, c = -2$

34. $P(x) = 6x^3 - x^2 + 4x, c = -3$

35. $P(x) = -2x^3 - 2x^2 - x - 20, c = 10$

36. $P(x) = -x^3 + 3x^2 + 5x + 30, c = 8$

37. $P(x) = -x^4 + 1, c = 3$

38. $P(x) = x^5 - 1, c = 1$

39. $P(x) = x^4 - 10x^3 + 2, c = 3$

40. $P(x) = x^5 + 20x^2 - 1, c = -5$

In Exercises 41 to 52, use synthetic division and the Factor Theorem to determine whether the given binomial is a factor of $P(x)$.

41. $P(x) = x^3 + 2x^2 - 5x - 6, x - 2$

42. $P(x) = x^3 + 4x^2 - 27x - 90, x + 6$

43. $P(x) = 2x^3 + x^2 - 2x - 1, x + 1$

44. $P(x) = 3x^3 + 4x^2 - 27x - 36, x - 4$

45. $P(x) = x^4 - 25x^2 + 144, x + 3$

46. $P(x) = x^4 - 25x^2 + 144, x - 3$

47. $P(x) = x^5 + 2x^4 - 22x^3 - 50x^2 - 75x, x - 5$

48. $P(x) = 9x^4 - 6x^3 - 23x^2 - 4x + 4, x + 1$

49. $P(x) = 16x^4 - 8x^3 + 9x^2 + 14x - 4, x - \dfrac{1}{4}$

50. $P(x) = 10x^4 + 9x^3 - 4x^2 + 9x + 6, x + \dfrac{1}{2}$

51. $P(x) = x^2 - 4x - 1, x - (2 + \sqrt{5})$

52. $P(x) = x^2 - 4x - 1, x - (2 - \sqrt{5})$

In Exercises 53 to 62, use synthetic division to show that c is a zero of $P(x)$.

53. $P(x) = 3x^3 - 8x^2 - 10x + 28, c = 2$

54. $P(x) = 4x^3 - 10x^2 - 8x + 6, c = 3$

55. $P(x) = x^4 - 1, c = 1$

56. $P(x) = x^3 + 8, c = -2$

57. $P(x) = 3x^4 + 8x^3 + 10x^2 + 2x - 20, c = -2$

58. $P(x) = x^4 - 2x^2 - 100x - 75, c = 5$

59. $P(x) = 2x^3 - 18x^2 - 50x + 66, c = 11$

60. $P(x) = 2x^4 - 34x^3 + 70x^2 - 153x + 45, c = 15$

61. $P(x) = 3x^2 - 8x + 4, c = \dfrac{2}{3}$

62. $P(x) = 5x^2 + 12x + 4, c = -\dfrac{2}{5}$

Graphing Calculator Exercises*

You can use a graph to factor some polynomials. For example, the graph of $y = x^2 - x - 12$ intersects the x-axis at $x = -3$, and $x = 4$. Thus $x^2 - x - 12$ has -3 and 4 as zeros. Hence $x^2 - x - 12$ has factors of $(x + 3)$ and $(x - 4)$. Use a graphing calculator to factor each of the following.

63. $x^3 - 7x + 6$

64. $x^3 + 6x^2 + 3x - 10$

65. $x^4 + 2x^3 - 13x^2 - 38x - 24$

66. $x^4 + 2x^3 - 7x^2 - 8x + 12$

Supplemental Exercises

67. Use the Factor Theorem to prove that for any positive odd integer n, $x^n + 1$ has $x + 1$ as a factor.

68. Use the Factor Theorem to prove that for any positive integer n, $x^n - 1$ has $x - 1$ as a factor.

69. Find the remainder of $5x^{48} + 6x^{10} - 5x + 7$ divided by $x - 1$.

70. Find the remainder of $18x^{80} - 6x^{50} + 4x^{20} - 2$ divided by $x + 1$.

71. Prove that $P(x) = 4x^4 + 7x^2 + 12$ has no factor of the form $x - c$, where c is a real number.

72. Prove that $P(x) = -5x^6 - 4x^2 - 10$ has no factor of the form $x - c$, where c is a real number.

73. Use synthetic division to show that $(x - i)$ is a factor of $x^3 - 3x^2 + x - 3$.

74. Use synthetic division to show that $(x + 2i)$ is a factor of $x^4 - 2x^3 + x^2 - 8x - 12$.

4.2 Graphs of Polynomial Functions

Table 4.1 summarizes information developed in Chapter 3 about graphs of polynomial functions of degree 0, 1, or 2. Polynomial functions of degree 3 or higher can be graphed by the technique of plotting points. However, some additional knowledge about polynomial functions will make graphing easier.

TABLE 4.1

Polynomial Function $P(x)$	Graph
$P(x) = a$ (degree 0)	Horizontal line through $(0, a)$
$P(x) = ax + b$ (degree 1)	Line with y-intercept $(0, b)$ and slope a
$P(x) = ax^2 + bx + c$ (degree 2)	Parabola with vertex $\left(-\dfrac{b}{2a}, P\left(-\dfrac{b}{2a}\right)\right)$

All polynomial functions have graphs that are **smooth continuous curves.** The terms *smooth* and *continuous* are defined rigorously in calculus, but for the present a smooth curve is a curve that does not have sharp corners, as shown in Figure 4.1(a). A continuous curve does not have a break or hole, as shown in Figure 4.1(b). See Figure 4.1 on page 210.

*Additional graphing calculator exercises appear in the Graphing Workbook as described in the front of this textbook.

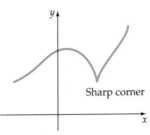
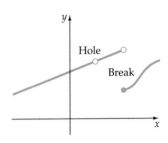

a. Continuous, but not smooth   b. Not continuous

Figure 4.1

The Leading Term Test

The graph of a polynomial function may have several up and down fluctuations; however, the graph of every polynomial function will eventually increase or decrease without bound as the graph moves far to the left or far to the right. The **leading term** $a_n x^n$ is said to **dominate** the polynomial function $P(x) = a_n x^n + a_{n-1} x^{n-1} + \cdots + a_1 x + a_0$ as $|x|$ becomes large, because the absolute value of $a_n x^n$ will be much larger than the absolute value of any of the other terms. Because of this condition, you can determine the far-left and far-right behavior of the polynomial by examining the leading coefficient a_n and the degree n of the polynomial.

Table 4.2 indicates the far-left and the far-right behavior of a polynomial function P with leading term $a_n x^n$.

TABLE 4.2   The graph of polynomial $P(x)$ with leading term $a_n x^n$ has the following far-right and far-left behavior.

	n is even	n is odd
$a_n > 0$	Up to left and up to right	Down to left and up to right
$a_n < 0$	Down to left and down to right	Up to left and down to right

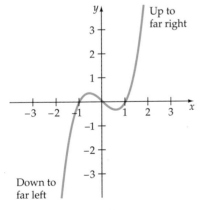

$P(x) = x^3 - x$
Figure 4.2

EXAMPLE 1   Determine the Left and Right Behavior of a Polynomial Function

Examine the leading term to determine the far-left and the far-right behavior of the graphs of each of the following polynomial functions.

a. $P(x) = x^3 - x$ b. $S(x) = \dfrac{1}{2}x^4 - \dfrac{5}{2}x^2 + 2$

c. $T(x) = -2x^3 + x^2 + 7x - 6$ d. $U(x) = -x^4 + 8x^2 + 9$

Solution

a. Since $a_n = 1$ is *positive* and $n = 3$ is *odd*, the graph of P goes down to its far left and up to its far right. See Figure 4.2.

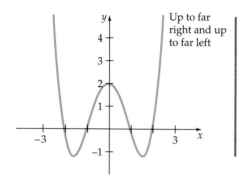

$$S(x) = \frac{1}{2}x^4 - \frac{5}{2}x^2 + 2$$

Figure 4.3

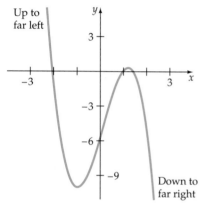

$$T(x) = -2x^3 + x^2 + 7x - 6$$

Figure 4.4

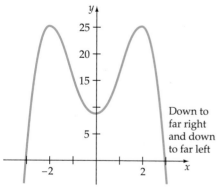

$$U(x) = -x^4 + 8x^2 + 9$$

Figure 4.5

b.   Since $a_n = \frac{1}{2}$ is *positive* and $n = 4$ is *even*, the graph of S goes up to its far left and up to its far right. See Figure 4.3.

c.   Since $a_n = -2$ is *negative* and $n = 3$ is *odd*, the graph of T goes up to its far left and down to its far right. See Figure 4.4.

d.   Since $a_n = -1$ is *negative* and $n = 4$ is *even*, the graph of U goes down to its far left and down to its far right. See Figure 4.5.

■ *Try Exercise **2**, page 214.*

Recall that the Remainder Theorem states that if the polynomial $P(x)$ is divided by $x - c$, then the remainder is $P(c)$. Because the arithmetic in synthetic division is often easier than the arithmetic involved in evaluating a polynomial at $x = c$ by substituting c for x, we will often use synthetic division to evaluate a polynomial.

EXAMPLE 2   Use Synthetic Division to Evaluate a Polynomial Function

Use synthetic division to find points on the graph of the polynomial function $P(x) = 2x^3 + 5x^2 - x - 5$, and then sketch the graph of $P(x)$.

Solution   Choose convenient values of x, say, $-3, -2, -1, 1,$ and 2.

For $x = -3$,

$$\begin{array}{r|rrrr} -3 & 2 & 5 & -1 & -5 \\ & & -6 & 3 & -6 \\ \hline & 2 & -1 & 2 & -11 \end{array}$$

The remainder is -11. Thus, by the Remainder Theorem, $P(-3) = -11$.

For $x = -2$,

$$\begin{array}{r|rrrr} -2 & 2 & 5 & -1 & -5 \\ & & -4 & -2 & 6 \\ \hline & 2 & 1 & -3 & 1 \end{array} \quad P(-2) = 1.$$

For $x = -1$,

$$\begin{array}{r|rrrr} -1 & 2 & 5 & -1 & -5 \\ & & -2 & -3 & 4 \\ \hline & 2 & 3 & -4 & -1 \end{array} \quad P(-1) = -1.$$

For $x = 1$

$$\begin{array}{r|rrrr} 1 & 2 & 5 & -1 & -5 \\ & & 2 & 7 & 6 \\ \hline & 2 & 7 & 6 & 1 \end{array} \quad P(1) = 1.$$

For $x = 2$,

$$\begin{array}{r|rrrr} 2 & 2 & 5 & -1 & -5 \\ & & 4 & 18 & 34 \\ \hline & 2 & 9 & 17 & 29 \end{array} \quad P(2) = 29.$$

To find the y-intercept of the graph of $P(x) = 2x^3 + 5x^2 - x - 5$, substitute 0 for x to produce $P(0) = -5$. Thus the y-intercept is $(0, -5)$. Drawing a smooth continuous curve through the points $(-3, -11)$, $(-2, 1)$, $(-1, -1)$, $(0, -5)$, $(1, 1)$ and $(2, 29)$ produces the graph in Figure 4.6.

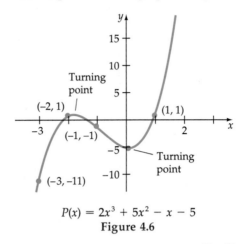

$$P(x) = 2x^3 + 5x^2 - x - 5$$
Figure 4.6

■ *Try Exercise **12**, page 214.*

Figure 4.6 illustrates the graph of a polynomial function of degree 3 with two **turning points,** points where the function changes from an increasing function to a decreasing function or vice versa. In general, the graph of a polynomial function of degree n has at most $n - 1$ turning points. Determining the exact location of turning points generally requires concepts and techniques from calculus.

Zeros and *x*-Intercepts

If a polynomial function can be factored into linear factors, then it can be graphed by plotting its x-intercepts and a few intermediate points. Before this procedure is illustrated, it is important to review some equivalent terminology.

If P is a polynomial function and c is a real number, then the following statements are equivalent in the sense that if any one statement is true, then they are all true, and if any one statement is false, then they are all false.

■ $(x - c)$ is a *factor* of P.

■ $x = c$ is a *solution* or *root* of the equation $P(x) = 0$.

■ $x = c$ is a *zero* of P.

■ $(c, 0)$ is an *x-intercept* of the graph of $y = P(x)$.

EXAMPLE 3   **Find Intercepts and Graph a Polynomial**

Sketch the graph of $P(x) = (x + 1)(x - 1)(2x - 3)$.

Solution   Because $(x + 1)$, $(x - 1)$ and $(2x - 3)$ are factors of P,

$$(-1, 0), \ (1, 0), \quad \text{and} \quad \left(\frac{3}{2}, 0\right)$$

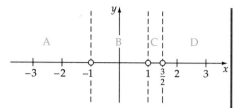

Figure 4.7

are x-intercepts of the graph of P. The x-intercepts separate the x-axis into the four intervals shown in Figure 4.7.

Synthetic division can now be used to determine additional points. We will use convenient x-values from each of the intervals. First, we rewrite $P(x)$ in its standard form.

$$P(x) = (x - 1)(x + 1)(2x - 3)$$
$$= 2x^3 - 3x^2 - 2x + 3$$

$x = -2$: $\quad -2 \begin{array}{|rrrr} 2 & -3 & -2 & 3 \\ & -4 & 14 & -24 \\ \hline 2 & -7 & 12 & -21 \end{array}$ • $(-2, -21)$ is a point on the graph of P.

$x = 1.2$: $\quad 1.2 \begin{array}{|rrrr} 2 & -3 & -2 & 3 \\ & 2.4 & -0.72 & -3.264 \\ \hline 2 & -0.6 & -2.72 & -0.264 \end{array}$ • $(1.2, -0.264)$ is a point on the graph of P.

$x = 2$: $\quad 2 \begin{array}{|rrrr} 2 & -3 & -2 & 3 \\ & 4 & 2 & 0 \\ \hline 2 & 1 & 0 & 3 \end{array}$ • $(2, 3)$ is a point on the graph of P.

$x = 3$: $\quad 3 \begin{array}{|rrrr} 2 & -3 & -2 & 3 \\ & 6 & 9 & 21 \\ \hline 2 & 3 & 7 & 24 \end{array}$ • $(3, 24)$ is a point on the graph of P.

Because $P(0) = 3$, the y-intercept is $(0, 3)$.

Sketch the graph by drawing a smooth continuous curve through each of the points determined above and the x-intercepts, as shown in Figure 4.8.

■ *Try Exercise* **32**, *page 215.*

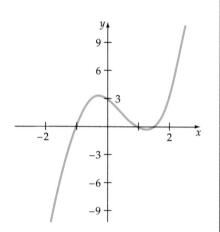

$$P(x) = (x - 1)(x + 1)(2x - 3)$$
$$= 2x^3 - 3x^2 - 2x + 3$$

Figure 4.8

The graph of every polynomial function P is a smooth continuous curve, and if the graph of P changes sign on an interval, then the function P must equal zero at least once in the interval. This result is known as the Zero Location Theorem. Although we will not prove the Zero Location Theorem, we will often use it in our search for the zeros of polynomial functions.

The Zero Location Theorem

Let $P(x)$ be a polynomial. If $a < b$, and if $P(a)$ and $P(b)$ have opposite signs, then there is at least one value c between a and b such that $P(c) = 0$.

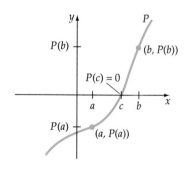

$P(a) < 0, P(b) > 0$

Figure 4.9

If, for instance, the value of a polynomial function is negative at $x = a$ and positive at $x = b$, then there must be at least one value c between a and b where the polynomial is zero. See Figure 4.9.

EXAMPLE 4   **Apply the Zero Location Theorem**

Use the Zero Location Theorem to verify that $P(x) = x^3 - x - 25$ has a real zero between $a = 3$ and $b = 4$.

Solution

$$
\begin{array}{r|rrrr}
3 & 1 & 0 & -1 & -25 \\
  &   & 3 & 9 & 24 \\
\hline
  & 1 & 3 & 8 & -1 \\
\end{array}
$$
• $P(3)$ is negative.

$$
\begin{array}{r|rrrr}
4 & 1 & 0 & -1 & -25 \\
  &   & 4 & 16 & 60 \\
\hline
  & 1 & 4 & 15 & 35 \\
\end{array}
$$
• $P(4)$ is positive.

Because $P(3)$ and $P(4)$ have opposite signs, P must have a real zero between 3 and 4.

■ *Try Exercise* **48**, *page 215.*

EXERCISE SET 4.2

In Exercises 1 to 10, examine the leading term and determine the far-left and the far-right behavior of the graph of the polynomial function.

1. $P(x) = 3x^4 - 2x^2 - 7x + 1$

2. $P(x) = -2x^3 - 6x^2 + 5x - 1$

3. $P(x) = 5x^5 - 4x^3 - 17x^2 + 2$

4. $P(x) = -6x^4 - 3x^3 + 5x^2 - 2x + 5$

5. $P(x) = 2 - 3x - 4x^2$

6. $P(x) = -16 + x^4$

7. $P(x) = \dfrac{1}{2}(x^3 + 5x^2 - 2)$

8. $P(x) = -\dfrac{1}{4}(x^4 + 3x^3 - 2x + 6)$

9. $P(x) = -\dfrac{2}{3}(x + 1)^3$

10. $P(x) = \dfrac{1}{5}(x - 1)^4$

In Exercises 11 to 20, use synthetic division to find the indicated functional values for the given polynomial function.

11. $f(x) = 3x^2 - 10x + 1$

   a. $f(2)$ **b.** $f(-2)$ **c.** $f(3)$

   d. $f(-3)$ **e.** $f(4)$ **f.** $f(-6)$

12. $g(x) = 2x^3 - 5x^2 + 4x + 10$

   a. $g(1)$ **b.** $g(-1)$ **c.** $g(2)$

   d. $g(3)$ **e.** $g(0)$ **f.** $g(-5)$

13. $h(x) = -3x^3 - 2x + 10$

   a. $h(3)$ **b.** $h(-3)$ **c.** $h(-1)$

   d. $h(1)$ **e.** $h(2)$ **f.** $h(-2)$

14. $j(x) = x^4 - 2x^3 - 3x^2 + x - 8$

   a. $j(2)$ **b.** $j(-2)$ **c.** $j(4)$

   d. $j(3)$ **e.** $j(0)$ **f.** $j(-1)$

15. $k(x) = -x^4 - 8x^3 + 2x^2 - 5x + 8$

   a. $k(-1)$ **b.** $k(-2)$ **c.** $k(3)$

   d. $k(-3)$ **e.** $k(4)$ **f.** $k(5)$

16. $f(t) = t^3 - t - 21$

   a. $f(2)$ **b.** $f(-2)$ **c.** $f(3)$

   d. $f(-3)$ **e.** $f(-4)$ **f.** $f(6)$

17. $s(t) = -16t^2 + 40t + 110$

   a. $s(2)$ **b.** $s(3)$ **c.** $s(4)$

   d. $s\left(\dfrac{1}{2}\right)$ **e.** $s\left(\dfrac{3}{2}\right)$ **f.** $s\left(\dfrac{5}{2}\right)$

18. $F(x) = 2x^5 - 3x^3 - 2x^2 + x + 7$

   a. $F(1)$ **b.** $F(-2)$ **c.** $F(3)$

   d. $F(-3)$ **e.** $F(0)$ **f.** $F(4)$

19. $P(x) = -3x^5 + x^3 + 80$

   a. $P(2)$ **b.** $P(-1)$ **c.** $P(-2)$

   d. $P(-3)$ **e.** $P\left(\dfrac{1}{3}\right)$ **f.** $P(-6)$

20. $v(t) = -x^3 + 2x^2 - 8x - 6$

   a. $v(-2)$ **b.** $v(-1)$ **c.** $v\left(\dfrac{1}{2}\right)$

   d. $v(1)$ **e.** $v(2)$ **f.** $v(7)$

In Exercises 21 to 30, find the real zeros of the polynomial function.

21. $P(x) = (x + 2)(x - 3)(2x + 7)$
22. $P(x) = (x - 5)(x - 1)(4x + 1)$
23. $P(x) = x(x - 1)(5x - 2)$
24. $P(x) = x(x - 4)(x - 1)(x + 7)$
25. $P(x) = (3x + 7)(2x - 11)(x + 5)^2$
26. $P(x) = (x - 3)^2(2x - 1)$
27. $P(x) = x^3 + x^2 - 6x$ **28.** $P(x) = 2x^3 - 7x^2 - 15x$
29. $P(x) = x^3 - 1$ **30.** $P(x) = x^3 + 8$

In Exercises 31 to 46, use the sketching techniques of this section to sketch the graph of each polynomial function.

31. $P(x) = (x - 1)(x + 1)(x - 3)$
32. $P(x) = (x - 2)(x + 3)(x + 1)$
33. $P(x) = -(x + 4)(x - 1)(x + 2)$
34. $P(x) = -x(x + 3)(x - 3)$
35. $P(x) = (x - 3)(x - 1)(2x + 7)$
36. $P(x) = (3x - 1)(x + 3)(x + 1)$
37. $P(x) = x^2(x^2 - 4)$ **38.** $P(x) = -x^2(x^2 - 1)$
39. $P(x) = (x - 2)^2(x + 1)$ **40.** $P(x) = (x - 3)(x + 1)^2$
41. $P(x) = x^3 + 2x^2 - 3x$ **42.** $P(x) = x^3 - 6x^2 + 9x$
43. $P(x) = x^3 - x^2 + x - 1$
44. $P(x) = 2x^3 - x^2 - 20x + 28$
45. $P(x) = x^4 + 3x^3 + 4x^2$

46. $P(x) = 2x^4 + 6x^3 - 25x^2 - 43x + 30$

In Exercises 47 to 52, use the Zero Location Theorem to verify that P has a zero between a and b.

47. $P(x) = 2x^3 + 3x^2 - 23x - 42;\quad a = 3, b = 4$
48. $P(x) = 4x^3 - x^2 - 6x + 1;\quad a = 0, b = 1$
49. $P(x) = 3x^3 + 7x^2 + 3x + 7;\quad a = -3, b = -2$
50. $P(x) = 2x^3 - 21x^2 - 2x + 21;\quad a = 10, b = 11$
51. $P(x) = 4x^4 + 7x^3 - 11x^2 + 7x - 15;\quad a = 1, b = 1\dfrac{1}{2}$
52. $P(x) = 5x^3 - 16x^2 - 20x + 64;\quad a = 3, b = 3\dfrac{1}{2}$

Graphing Calculator Exercises*

53. On the same coordinate axes, sketch the graph of the function $f(x) = x^n$ over the interval $-1 \le x \le 1$, for each value of n.

   a. $n = 2$ **b.** $n = 4$ **c.** $n = 6$

54. Use the result of Exercise 53 to make a conjecture about the graph of $y = x^n$, over the interval $-1 \le x \le 1$, where n is a large, positive, even integer.

55. Graph $P(x) = x^5 - x^4 - 2x^3$.
56. Graph $P(x) = -x^5 + 4x^3$.

Supplemental Exercises

57. Let $f(x) = x^3 + c$. On the same coordinate axes, sketch the graph of f for each value of c.

   a. $c = 0$ **b.** $c = 2$ **c.** $c = -3$

58. Let $f(x) = ax^3$. On the same coordinate axes, sketch the graph of f for each value of a.

   a. $a = 2$ **b.** $a = \frac{1}{2}$ **c.** $a = -1$

59. Let $f(x) = (x - h)^3$. On the same coordinate axes, sketch the graph of f for each value of h.

   a. $h = 2$ **b.** $h = -1$ **c.** $h = -5$

60. Explain how the graph of $f(x) = a(x - h)^3 + c$ compares with the graph of $g(x) = x^3$.

4.3   Zeros of Polynomial Functions

Recall that if $P(x)$ is a polynomial function, then the values of x for which $P(x)$ is to equal 0 are called the *zeros* of $P(x)$ or the *roots* of the equation $P(x) = 0$. A zero of a polynomial may be a **multiple zero.** For example, the

*Additional graphing calculator exercises appear in the Graphing Workbook as described in the front of this textbook.

polynomial $x^2 + 6x + 9$ can be expressed in factored form as $(x + 3)(x + 3)$. Setting each factor equal to zero yields $x = -3$ in both cases. Thus $x^2 + 6x + 9$ has a zero of -3 that occurs twice. The following definition will be most useful when we are discussing multiple zeros.

Definition of Multiple Zeros of a Polynomial

If a polynomial $P(x)$ has $(x - r)$ as a factor exactly k times, then r is a **zero of multiplicity k** of the polynomial $P(x)$.

The polynomial

$$P(x) = (x - 5)(x - 5)(x + 2)(x + 2)(x + 2)(x + 4)$$

has

- 5 as a zero of multiplicity 2
- -2 as a zero of multiplicity 3
- -4 as a zero of multiplicity 1

A zero of multiplicity 1 is generally referred to as a **simple zero.**

When searching for the zeros of a polynomial function, it is important that we know how many zeros to expect. This question is answered completely in Section 4.4. For the work in this section, the following result is valuable.

Number of Zeros of a Polynomial Function

A polynomial function P of degree n has at most n zeros, where each zero of multiplicity k is counted k times.

The rational zeros of polynomials with integer coefficients can be found with the aid of the following theorem.

The Rational Zero Theorem

If $P(x) = a_n x^n + a_{n-1} x^{n-1} + \cdots + a_1 x + a_0$ has integer coefficients, and p/q (where p and q have no common prime factors) is a rational zero of $P(x)$, then p is a factor of a_0 and q is a factor of a_n.

Proof: Since p/q is a zero of $P(x)$,

$$a_n \left(\frac{p}{q}\right)^n + a_{n-1} \left(\frac{p}{q}\right)^{n-1} + \cdots + a_1 \left(\frac{p}{q}\right) + a_0 = 0$$

Multiplying both sides by q^n produces

$$a_n p^n + a_{n-1} p^{n-1} q + \cdots + a_1 p q^{n-1} + a_0 q^n = 0$$

which can be written as

$$p(a_n p^{n-1} + a_{n-1} p^{n-2} q + \cdots + a_1 q^{n-1}) = -a_0 q^n$$

This implies that p is a factor of the integer a_0q^n. Also, since p and q have no common prime factors, p is a factor of a_0. A similar procedure may be used to establish that q is a factor of a_n.

The Rational Zero Theorem often is used to make a list of all possible rational zeros of a polynomial. The list consists of all rational numbers of the form p/q, where p is an integer factor of the constant term a_0, and q is an integer factor of the leading coefficient a_n.

EXAMPLE 1 **Apply the Rational Zero Theorem**

Use the Rational Zero Theorem to list all possible rational zeros of

$$4x^4 + 5x^3 + 7x^2 - 34x + 8$$

Solution List all integers p that are factors of 8 and all integers q that are factors of 4.

p: • Factors of 8 $\pm1, \pm2, \pm4, \pm8$

q: • Factors of 4 $\pm1, \pm2, \pm4$

Form all possible rational numbers using ±1, ±2, ±4, or ±8 as the numerator and ±1, ±2, or ±4 for the denominator. By the Rational Zero Theorem, the possible rational zeros are

$$\pm1, \pm2, \pm4, \pm8, \pm\frac{1}{2}, \pm\frac{1}{4}$$

■ *Try Exercise* **12,** *page 222.*

Remark It is not necessary to list a factor that is already listed in reduced form. For example, $\pm\frac{4}{2}$ is not listed because it is equal to ±2.

Caution The Rational Zero Theorem gives the *possible* rational zeros. That is, if $P(x)$ has a rational zero p/q, where p and q have no common prime factors, then p is a factor of a_0 and q is a factor of a_n. However, it is possible that $P(x)$ has no rational zeros.

Upper and Lower Bounds for Real Zeros

A real number b is called an **upper bound** of the zeros of the polynomial function P if no zero is greater than b. A real number a is called a **lower bound** of the zeros of P if no zero is less than a. The following theorem is often used to find positive upper bounds and negative lower bounds for the real zeros of a polynomial function.

Upper- and Lower-Bound Theorem

Upper bound If $b > 0$ and all the numbers in the bottom row of the synthetic division of P by $x - b$ are either positive or zero, then b is an upper bound for the real zeros of P.

Lower bound If $a < 0$ and the numbers in the bottom row of the synthetic division of P by $x - a$ alternate in sign (the number zero can be considered positive or negative), then a is a lower bound for the real zeros of P.

Upper and lower bounds are not unique. For example, if b is an upper bound for the real zeros of P, then any number greater than b is also an upper bound. Also, if a is a lower bound for the real zeros of P, then any number less than a is also a lower bound.

EXAMPLE 2   **Find Upper and Lower Bounds**

According to the upper- and lower-bound theorem, what is the smallest positive integer that is an upper bound and the largest negative integer that is a lower bound of $P(x) = 2x^3 + 7x^2 - 4x - 14$?

Solution   To find the smallest positive integer upper bound, use synthetic division with 1, 2, ..., as test values.

$$
\begin{array}{r|rrrr}
1 & 2 & 7 & -4 & -14 \\
  &   & 2 & 9 & 5 \\
\hline
  & 2 & 9 & 5 & -9
\end{array}
\qquad
\begin{array}{r|rrrr}
2 & 2 & 7 & -4 & -14 \\
  &   & 4 & 22 & 36 \\
\hline
  & 2 & 11 & 18 & 22
\end{array}
$$
• All positive signs

Thus 2 is the smallest positive integer upper bound.
    Now find the largest negative integer lower bound.

$$
\begin{array}{r|rrrr}
-1 & 2 & 7 & -4 & -14 \\
   &   & -2 & -5 & 9 \\
\hline
   & 2 & 5 & -9 & -5
\end{array}
\qquad
\begin{array}{r|rrrr}
-2 & 2 & 7 & -4 & -14 \\
   &   & -4 & -6 & 20 \\
\hline
   & 2 & 3 & -10 & 6
\end{array}
$$

$$
\begin{array}{r|rrrr}
-3 & 2 & 7 & -4 & -14 \\
   &   & -6 & -3 & 21 \\
\hline
   & 2 & 1 & -7 & 7
\end{array}
\qquad
\begin{array}{r|rrrr}
-4 & 2 & 7 & -4 & -14 \\
   &   & -8 & 4 & 0 \\
\hline
   & 2 & -1 & 0 & -14
\end{array}
$$
• Alternating signs

Thus −4 is the largest negative integer lower bound.

■ *Try Exercise 24, page 222.*

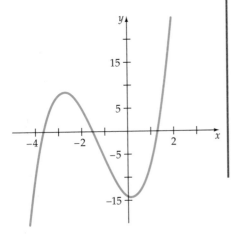

$P(x) = 2x^3 + 7x^2 - 4x - 14$
Figure 4.10

Remark   Since −4 is a lower bound and 2 is an upper bound, the real zeros of $2x^3 + 7x^2 - 4x - 14$ must be in the interval $(-4, 2)$. The graph of P is shown in Figure 4.10. Notice that its x-intercepts are between −4 and 2.

Descartes' Rule of Signs

Descartes' Rule of Signs is another theorem often used to obtain information about the zeros of a polynomial. In Descartes' Rule of Signs, the number of *variations in sign* of the coefficients of a polynomial $P(x)$ or $P(-x)$ refers to sign changes of the coefficients from positive to negative or negative to positive as we examine successive terms of the polynomial. The terms of the polynomial are assumed to be in descending powers of x. For example, the polynomial

$$P(x) = +3x^4 - 5x^3 - 7x^2 + x - 7$$

$$\qquad\qquad 1 \qquad\qquad 2 \quad 3$$

has three variations of sign. The polynomial

$$P(-x) = +3(-x)^4 - 5(-x)^3 - 7(-x)^2 + (-x) - 7$$

$$= +\ 3x^4\ +\ 5x^3\ -\ 7x^2\ -\ x\ -\ 7$$

$$\qquad\qquad 1$$

has one variation in sign.

Terms that have a coefficient of 0 are not counted as a variation of sign and may be ignored. For example,

$$P(x) = -x^5 + 4x^2 + 1$$

$$\qquad 1$$

has one variation in sign.

Descartes' Rule of Signs

Let $P(x)$ be a polynomial with real coefficients and with the terms arranged in decreasing powers of x.

1. The number of positive real zeros of $P(x)$ is equal to the number of variations in sign of $P(x)$ or is equal to that number decreased by an even integer.
2. The number of negative real zeros of $P(x)$ is equal to the number of variations in sign of $P(-x)$ or is equal to that number decreased by an even integer.

The proof of Descartes' Rule of Signs is beyond the scope of this course and not given in this text.

EXAMPLE 3 **Apply Descartes' Rule of Signs**

Determine both the number of possible positive and the number of possible negative real zeros of each of the following polynomials.

a. $x^4 - 5x^3 + 5x^2 + 5x - 6$ b. $2x^5 + 3x^3 + 5x^2 + 8x + 7$

Solution

a.
$$P(x) = x^4 - 5x^3 + 5x^2 + 5x - 6$$

$$\underbrace{\qquad}_{1} \quad \underbrace{\qquad}_{2} \qquad \underbrace{\qquad}_{3}$$

There are three variations of sign. By Descartes' Rule of Signs, there are either three or one positive real zeros. Now examine the variations of sign of $P(-x)$.

$$P(-x) = x^4 + 5x^3 + 5x^2 - 5x - 6$$

$$\underbrace{\qquad}_{1}$$

There is one variation of sign of $P(-x)$. By Descartes' Rule of Signs, there is one negative real zero.

b. $P(x) = 2x^5 + 3x^3 + 5x^2 + 8x + 7$ has no variation of sign, so there are no positive real zeros.

$$P(-x) = -2x^5 - 3x^3 + 5x^2 - 8x + 7$$

$$\underbrace{\qquad}_{1} \quad \underbrace{\qquad}_{2} \quad \underbrace{\qquad}_{3}$$

$P(-x)$ has three variations of sign, so there are either three or one negative real zeros.

■ *Try Exercise* **36**, *page 222.*

In applying Descartes' Rule of Signs, we count each zero of multiplicity k as k zeros. For instance, the polynomial

$$P(x) = x^2 - 10x + 25$$

has two variations in sign. Thus by Descartes' Rule of Signs it must have either two or zero positive real zeros. Factoring the polynomial produces $(x - 5)^2$, from which it can be observed that 5 is a positive zero of multiplicity two.

Example 4 uses the theorems of this section to determine the zeros of polynomial functions.

EXAMPLE 4   **Find the Zeros of a Polynomial**

Find the zeros of each of the following polynomials.

a. $P(x) = 3x^4 + 23x^3 + 56x^2 + 52x + 16$ b. $P(x) = 3x^3 + 13x^2 - 16$

Solution

a. By Descartes' Rule of Signs there are no positive real zeros, and there are either four, two, or zero negative real zeros. By the Rational Zero Theorem, the possible negative rational zeros are

$$\frac{p}{q}: \quad -1, -2, -4, -8, -16, -\frac{1}{3}, -\frac{2}{3}, -\frac{4}{3}, -\frac{8}{3}, -\frac{16}{3}.$$

We use synthetic division to test possible zeros. We show only the work for values which are zeros.

$x = -4$:

$$
\begin{array}{r|rrrrr}
-4 & 3 & 23 & 56 & 52 & 16 \\
   &   & -12 & -44 & -48 & -16 \\
\hline
   & 3 & 11 & 12 & 4 & 0
\end{array}
$$
• -4 is a zero.

Since -4 is a zero, the factors of $P(x)$ are $(x + 4)$ and the reduced polynomial $(3x^3 + 11x^2 + 12x + 4)$. Thus

$$P(x) = (x + 4)(3x^3 + 11x^2 + 12x + 4)$$

All remaining zeros must be zeros of $3x^3 + 11x^2 + 12x + 4$. The Rational Zero Theorem indicates that the only possible negative rational zeros are

$$\frac{p}{q}: \quad -1, -2, -4, -\frac{1}{3}, -\frac{2}{3}, -\frac{4}{3}$$

Synthetic division is again used to test possible zeros.

$$
\begin{array}{r|rrrr}
-2 & 3 & 11 & 12 & 4 \\
   &   & -6 & -10 & -4 \\
\hline
   & 3 & 5 & 2 & 0
\end{array}
$$
• -2 is a zero.

Since -2 is a zero, $(x + 2)$ is also a factor of P. Thus we have

$$P(x) = (x + 4)(x + 2)(3x^2 + 5x + 2)$$

All the remaining zeros must be zeros of the reduced polynomial $3x^2 + 5x + 2$.

$$3x^2 + 5x + 2 = 0$$

$$(3x + 2)(x + 1) = 0$$

$$x = -\frac{2}{3} \quad \text{or} \quad -1$$

The zeros of $3x^4 + 23x^3 + 56x^2 + 52x + 16$ are $-4, -2, -1, -\frac{2}{3}$.

b. By Descartes' Rule of Signs $P(x) = 3x^3 + 13x^2 - 16$ has one positive real zero and either two or zero negative real zeros. By the Rational Zero Theorem, the possible rational zeros are

$$\frac{p}{q}: \quad \pm 1, \pm 2, \pm 4, \pm 8, \pm 16, \pm \frac{1}{3}, \pm \frac{2}{3}, \pm \frac{4}{3}, \pm \frac{8}{3}, \pm \frac{16}{3}$$

Synthetic division is used to test these possible zeros.

$x = 1$:

$$
\begin{array}{r|rrrr}
1 & 3 & 13 & 0 & -16 \\
  &   & 3 & 16 & 16 \\
\hline
  & 3 & 16 & 16 & 0
\end{array}
$$
• 1 is the positive real zero

Find the zeros of the reduced polynomial $3x^2 + 16x + 16$ to determine the remaining zeros.

$$3x^2 + 16x + 16 = 0$$

$$(3x + 4)(x + 4) = 0$$

$$3x + 4 = 0 \qquad \text{or} \qquad x + 4 = 0$$

$$x = -\frac{4}{3} \qquad \text{or} \qquad x = -4$$

The zeros of $3x^3 + 13x^2 - 16$ are $1, -\frac{4}{3}, -4$.

■ *Try Exercise* **48**, *page 222*.

EXERCISE SET 4.3

In Exercises 1 to 10, find the zeros of the polynomial and state the multiplicity of each zero.

1. $P(x) = (x - 3)^2(x + 5)$

2. $P(x) = (x + 4)^3(x - 1)^2$

3. $P(x) = x^2(3x + 5)^2$

4. $P(x) = x^3(2x + 1)(3x - 12)^2$

5. $P(x) = (x^2 - 4)(x + 3)^2$

6. $P(x) = (x + 4)^3(x^2 - 9)^2$

7. $P(x) = (x^2 - 3x - 10)^2$

8. $P(x) = (x^3 - 4x)(2x - 7)^2$

9. $P(x) = x^4 - 10x^2 + 9$

10. $P(x) = x^4 - 12x^2 + 32$

In Exercises 11 to 22, use the Rational Zero Theorem to list possible rational zeros for each polynomial.

11. $x^3 + 3x^2 - 6x - 8$ **12.** $x^3 - 19x - 30$

13. $2x^3 + x^2 - 25x + 12$ **14.** $3x^3 + 11x^2 - 6x - 8$

15. $6x^4 + 23x^3 + 19x^2 - 8x - 4$

16. $6x^4 + 23x^3 + 15x^2 - 23x - 21$

17. $2x^3 + 9x^2 - 2x - 9$

18. $2x^4 + 11x^3 + 21x^2 + 17x + 5$

19. $4x^4 - 12x^3 - 3x^2 + 12x - 7$

20. $x^5 - x^4 - 7x^3 + 7x^2 - 12x - 12$

21. $x^5 - 32$ **22.** $x^4 - 1$

In Exercises 23 to 34, find the smallest positive integer and the largest negative integer that by the upper- and lower-bound theorem are upper and lower bounds for the real zeros of the following polynomials.

23. $x^3 + 3x^2 - 6x - 6$ **24.** $x^3 - 19x - 28$

25. $2x^3 + x^2 - 25x + 10$ **26.** $3x^3 + 11x^2 - 6x - 9$

27. $6x^4 + 23x^3 + 19x^2 - 8x - 4$

28. $6x^4 + 23x^3 + 15x^2 - 23x - 21$

29. $2x^3 + 9x^2 - 2x - 9$

30. $2x^4 + 11x^3 + 21x^2 + 17x + 5$

31. $4x^4 - 12x^3 - 3x^2 + 12x - 7$

32. $x^5 - x^4 - 7x^3 + 7x^2 - 12x - 12$

33. $x^5 - 32$ **34.** $x^4 - 1$

In Exercises 35 to 46, use Descartes' Rule of Signs to state the number of possible positive and negative real zeros of each polynomial.

35. $x^3 + 3x^2 - 6x - 8$ **36.** $x^3 - 19x - 30$

37. $2x^3 + x^2 - 25x + 12$ **38.** $3x^3 + 11x^2 - 6x - 8$

39. $6x^4 + 23x^3 + 19x^2 - 8x - 4$

40. $6x^4 + 23x^3 + 15x^2 - 23x - 21$

41. $2x^3 + 9x^2 - 2x - 9$

42. $2x^4 + 11x^3 + 21x^2 + 17x + 5$

43. $4x^4 - 12x^3 - 3x^2 + 12x - 7$

44. $x^5 - x^4 - 7x^3 + 7x^2 - 12x - 12$

45. $x^5 - 32$ **46.** $x^4 - 1$

In Exercises 47 to 64, find the zeros of each polynomial.

47. $x^3 + 3x^2 - 6x - 8$ **48.** $x^3 - 19x - 30$

49. $2x^3 + x^2 - 25x + 12$ **50.** $3x^3 + 11x^2 - 6x - 8$

51. $6x^4 + 23x^3 + 19x^2 - 8x - 4$

52. $6x^4 + 23x^3 + 15x^2 - 23x - 21$

53. $2x^3 + 9x^2 - 2x - 9$

54. $2x^4 + 11x^3 + 21x^2 + 17x + 5$

55. $2x^4 - 9x^3 - 2x^2 + 27x - 12$

56. $3x^3 - x^2 - 6x + 2$

57. $x^3 - 3x - 2$

58. $3x^4 - 4x^3 - 11x^2 + 16x - 4$

59. $x^4 - 5x^2 - 2x$

60. $x^3 - 2x + 1$

61. $x^4 + x^3 - 3x^2 - 5x - 2$

62. $6x^4 - 17x^3 - 11x^2 + 42x$

63. $2x^4 - 17x^3 + 4x^2 + 35x - 24$

64. $x^5 + 5x^4 + 10x^3 + 10x^2 + 5x + 1$

Supplemental Exercises

In Exercises 65 to 70, verify that each polynomial has no rational zeros.

65. $x^4 - 2x^3 + 11x^2 - 2x + 10$

66. $x^4 - 2x^3 + 21x^2 - 2x + 20$

67. $2x^4 + x^2 + 5$

68. $4x^4 + 14x^2 + 5$

69. $x^4 - 4x^3 + 14x^2 - 4x + 13$

70. $x^6 + 3x^4 + 3x^2 + 1$

In Exercises 71 to 74, determine whether the given polynomial satisfies the following theorem.

Theorem Let $P(x) = a_n x^n + a_{n-1} x^{n-1} + \cdots + a_1 x + a_0$ be a polynomial with integer coefficients and $n \geq 2$. If a_n, a_0, and $P(1)$ are all odd, then $P(x)$ has no rational zeros.

71. $x^5 + 2x^4 + x^3 - x^2 + x + 945$

72. $5x^3 - 2x^2 - x + 1815$

73. $3x^4 - 5x^3 + 6x^2 - 2x + 9009$

74. $15x^7 - 4x^3 + x^2 - 6075$

75. Prove that $\sqrt{2}$ is an irrational number. (*Hint:* First show that $P(x) = x^2 - 2$ has a positive zero. Then show that P has no rational zeros.)

4.4 The Fundamental Theorem of Algebra

The German mathematician Carl Friedrich Gauss (1777–1855) was the first to prove that every polynomial has at least one complex zero. This concept is so basic to the study of algebra that it is called the **Fundamental Theorem of Algebra.** The proof of the Fundamental Theorem is beyond the scope of this text; however, it is important to understand the theorem and its consequences. In each of the following theorems, keep in mind that the terms *complex coefficients* and *complex zeros* include real coefficients and real zeros because the set of real numbers is a subset of the set of complex numbers.

The Fundamental Theorem of Algebra

If $P(x)$ is a polynomial of degree $n \geq 1$ with complex coefficients, then $P(x)$ has at least one complex zero.

Let $P(x)$ be a polynomial of degree $n \geq 1$, with complex coefficients. The Fundamental Theorem implies that $P(x)$ has a complex zero—say, c_1. The Factor Theorem implies that

$$P(x) = (x - c_1)Q(x)$$

where $Q(x)$ is a polynomial of degree 1 less than the degree of $P(x)$. Recall that the polynomial $Q(x)$ is called a reduced polynomial. Assuming the degree of $Q(x)$ is 1 or more, the Fundamental Theorem implies that it must also have a zero. A continuation of this reasoning process leads to the following theorem, which is a corollary of the Fundamental Theorem.

The Number of Zeros of a Polynomial

If $P(x)$ is a polynomial of degree $n \geq 1$ with complex coefficients, then $P(x)$ has exactly n complex zeros, provided each zero is counted according to its multiplicity.

Even though every polynomial of nth degree has exactly n zeros, the zeros may not be distinct. For example, the third-degree polynomial

$$x^3 - 5x^2 + 3x + 9$$

factors into

$$(x + 1)(x - 3)(x - 3)$$

which has zeros -1, 3, and 3. The zero 3 is a zero of multiplicity 2.

Although the Fundamental Theorem and its corollary give information about the existence and the number of zeros of a polynomial, they do not provide a method of actually finding the zeros. If a polynomial has real coefficients, then the following theorem can help us determine the zeros of the polynomial.

The Conjugate Pair Theorem

If $a + bi$ ($b \neq 0$) is a complex zero of the polynomial $P(z)$, *with real coefficients*, then the conjugate $a - bi$ is also a complex zero of the polynomial.

Proof: If the complex number z is a zero of the polynomial $P(z)$, then

$$P(z) = a_n z^n + a_{n-1} z^{n-1} + \cdots + a_1 z + a_0 = 0$$

where each coefficient a_i is a real number. Since the complex number on the left side of the equation is equal to the complex number on the right side, their conjugates are also equal. That is,

$$\overline{a_n z^n + a_{n-1} z^{n-1} + \cdots + a_1 z + a_0} = \overline{0} = 0.$$

Because the conjugate of a sum of complex numbers is the sum of the conjugates (see Exercise 91, Exercise Set 1.8),

$$\overline{a_n z^n} + \overline{a_{n-1} z^{n-1}} + \cdots + \overline{a_1 z} + \overline{a_0} = 0$$

Because the conjugate of a product of complex numbers is the product of the conjugates (see Exercise 92, Exercise Set 1.8),

$$\overline{a_n}\,\overline{z^n} + \overline{a_{n-1}}\,\overline{z^{n-1}} + \cdots + \overline{a_1}\,\overline{z} + \overline{a_0} = 0$$

It can also be shown that if z is a complex number, $\overline{z^n} = \overline{z}^n$ for every positive integer n. This, combined with the fact that for each real number a, $\overline{a} = a$, produces the following:

$$a_n \overline{z}^n + a_{n-1} \overline{z}^{n-1} + \cdots + a_1 \overline{z} + a_0 = 0.$$

This equation can be written as $P(\overline{z}) = 0$, and the theorem is established.

EXAMPLE 1   **Use the Conjugate Pair Theorem to Find Zeros**

Find all the zeros of $x^4 - 4x^3 + 14x^2 - 36x + 45$ given that $2 + i$ is a zero.

Solution   Because the coefficients are real numbers and $2 + i$ is a zero, the Conjugate Pair Theorem implies that $2 - i$ must also be a zero. Using synthetic division with $2 + i$ and then $2 - i$, we have,

$$
\begin{array}{r|rrrrr}
2+i & 1 & -4 & 14 & -36 & 45 \\
 & & 2+i & -5 & 18+9i & -45 \\
\hline
 & 1 & -2+i & 9 & -18+9i & 0 \\
\end{array}
$$

$$
\begin{array}{r|rrrr}
2-i & 1 & -2+i & 9 & -18+9i \\
 & & 2-i & 0 & 18-9i \\
\hline
 & 1 & 0 & 9 & 0 \\
\end{array}
$$

• The coefficients of the reduced polynomial

• The coefficients of the next reduced polynomial

The resulting reduced polynomial is $x^2 + 9$, which has $3i$ and $-3i$ as zeros. Therefore the four zeros of $x^4 - 4x^3 + 14x^2 - 36x + 45$ are $2 + i, 2 - i, 3i,$ and $-3i$.

■ *Try Exercise 2, page 227.*

Factors of a Polynomial

The following theorem is a result of the Conjugate Pair Theorem.

Linear and Quadratic Factors of a Polynomial

Every polynomial with real coefficients and positive degree n can be written as the product of linear and quadratic factors with real coefficients, where the quadratic factors have no real zeros.

Proof: If P is a polynomial with real coefficients of degree n, then it has precisely n complex zeros $c_1, c_2, \ldots, c_n$. It may be written in the factored form

$$P(x) = a(x - c_1)(x - c_2) \cdots (x - c_n)$$

where a is the leading coefficient of P. If any zero c_k is a real number, then $(x - c_k)$ is a linear factor as referred to in the statement of the theorem. If any zero c_k is a complex number — say,

$$c_k = a + bi, \quad b \neq 0$$

then by the Conjugate Pair Theorem,

$$c_j = a - bi$$

is also a zero of P. The product of $(x - c_k)$ and $(x - c_j)$ is a quadratic factor with real coefficients of 1, $-2a$, and $a^2 + b^2$ as demonstrated below.

$$(x - c_k)(x - c_j) = [x - (a + bi)][x - (a - bi)]$$
$$= x^2 - 2ax + (a^2 + b^2)$$

Thus $x^2 - 2ax + (a^2 + b^2)$ is a quadratic factor with real coefficients.

A quadratic factor with no real zeros is said to be **irreducible over the reals.**

EXAMPLE 2   **Factor a Polynomial into Linear and Quadratic Factors**

Write each polynomial as a product of linear factors and quadratic factors that are irreducible over the reals.

a. $P(x) = x^3 - 3x^2 + x - 3$ b. $P(x) = x^3 - 6x^2 + 13x - 10$

Solution

a. Factoring by grouping produces

$$P(x) = x^3 - 3x^2 + x - 3 = (x^3 - 3x^2) + (x - 3)$$
$$= x^2(x - 3) + 1(x - 3) = (x - 3)(x^2 + 1)$$

Since each binomial factor is irreducible over the reals, the factorization is complete.

b. Since $x^3 - 6x^2 + 13x - 10$ cannot be factored by grouping, synthetic division is used to determine zeros that also determine factors. By the Rational Zero Theorem, we know that ± 1, ± 2, ± 5, and ± 10 are possible rational zeros. Testing each of these, we find

$$
\begin{array}{r|rrrr}
2 & 1 & -6 & 13 & -10 \\
  &   & 2 & -8 & 10 \\
\hline
  & 1 & -4 & 5 & 0
\end{array}
\quad \bullet \ \ 2 \text{ is a zero}
$$

Using the quadratic formula, we find that the reduced polynomial $x^2 - 4x + 5$ has zeros of $2 \pm i$, so it cannot be factored using real numbers. Thus $x^3 - 6x^2 + 13x - 10$ factors into

$$(x - 2)(x^2 - 4x + 5)$$

which is a product of a linear and a quadratic factor that is irreducible over the reals.

■ *Try Exercise* **20**, *page 227.*

Many of the problems in this section and in Section 4.3 dealt with the process of finding the zeros of a given polynomial. Example 3 considers the reverse process, finding a polynomial when the zeros are given.

EXAMPLE 3 **Determine a Polynomial Given Its Zeros**

Find each of the following:

a. A polynomial of degree 3 that has 1, 2, and -3 as zeros

b. A polynomial of degree 4 that has real coefficients and zeros $2i$ and $3 - 7i$

Solution

a. Since 1, 2, and -3 are zeros, $(x - 1)$, $(x - 2)$, and $(x + 3)$ are factors. Multiplying these factors produces a polynomial that has the indicated zeros.

$$(x - 1)(x - 2)(x + 3) = (x^2 - 3x + 2)(x + 3) = x^3 - 7x + 6$$

b. By the Conjugate Pair Theorem, the polynomial also must have $-2i$ and $3 + 7i$ as zeros. The product of the factors $x - 2i$, $x - (-2i)$, $x - (3 - 7i)$, and $x - (3 + 7i)$ produces the desired polynomial.

$$(x - 2i)(x + 2i)[x - (3 - 7i)][x - (3 + 7i)]$$
$$= (x^2 + 4)(x^2 - 6x + 58)$$
$$= x^4 - 6x^3 + 62x^2 - 24x + 232$$

■ *Try Exercise 40, page 228.*

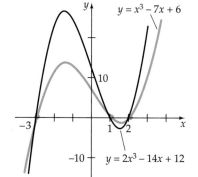

$y = x^3 - 7x + 6$

$y = 2x^3 - 14x + 12$

Figure 4.11

A polynomial that has a given set of zeros is not unique. For example, $x^3 - 7x + 6$ has zeros 1, 2, and -3, but so does any nonzero multiple of that polynomial, such as $2x^3 - 14x + 12$. This concept is illustrated in Figure 4.11. The graphs of the two polynomials are different; however, they have the same x-intercepts.

EXERCISE SET 4.4

In Exercises 1 to 12, use the given zero to find the remaining zeros of each polynomial.

1. $2x^3 - 5x^2 + 6x - 2$; $1 + i$
2. $3x^3 - 29x^2 + 92x + 34$; $5 + 3i$
3. $x^3 + 3x^2 + x + 3$; $-i$
4. $x^4 - 6x^3 + 71x^2 - 146x + 530$; $2 + 7i$
5. $x^5 - x^4 - 3x^3 + 3x^2 - 10x + 10$; $i\sqrt{2}$
6. $x^4 - 4x^3 + 14x^2 - 4x + 13$; $2 - 3i$
7. $12x^3 - 28x^2 + 23x - 5$; $\frac{1}{3}$
8. $8x^4 - 2x^3 + 199x^2 - 50x - 25$; $-5i$
9. $x^4 - 4x^3 + 19x^2 - 30x + 50$; $1 + 3i$
10. $12x^4 - 52x^3 + 19x^2 - 13x + 4$; $\frac{1}{2}i$
11. $x^5 - x^4 - 4x^3 - 4x^2 - 5x - 3$; i
12. $x^5 - 3x^4 + 7x^3 - 13x^2 + 12x - 4$; $-2i$

In Exercises 13 to 18, find all the zeros of the polynomial. (*Hint:* First determine the rational zeros.)

13. $x^4 + x^3 - 2x^2 + 4x - 24$
14. $x^4 - 3x^3 + 5x^2 - 27x - 36$
15. $2x^4 + x^3 + 39x^2 + 136x - 78$
16. $x^3 - 13x^2 + 65x - 125$
17. $x^5 - 9x^4 + 34x^3 - 58x^2 + 45x - 13$
18. $x^4 - 4x^3 + 53x^2 - 196x + 196$

In Exercises 19 to 28, factor each polynomial into linear factors and/or quadratic factors that are irreducible over the reals.

19. $x^3 - x^2 - 2x$ 20. $6x^3 - 23x^2 - 4x$
21. $x^3 + 9x$ 22. $x^3 + 10x$
23. $x^4 + 2x^2 - 24$ 24. $x^4 - 8x^2 - 20$

25. $x^4 + 3x^2 + 2$ **26.** $x^5 + 11x^3 + 18x$

27. $x^4 - 2x^3 + x^2 - 8x - 12$

28. $x^4 + 2x^3 + 6x^2 + 32x + 40$

In Exercises 29 to 38, find a polynomial of lowest degree that has the given zeros.

29. $4, -3, 2$ **30.** $-1, 1, -5$

31. $3, 2i, -2i$ **32.** $0, i, -i$

33. $3 + i, 3 - i, 2 + 5i, 2 - 5i$

34. $2 + 3i, 2 - 3i, -5, 2$

35. $6 + 5i, 6 - 5i, 2, 3, 5$

36. $\dfrac{1}{2}, 4 - i, 4 + i$

37. $\dfrac{3}{4}, 2 + 7i, 2 - 7i$

38. $\dfrac{1}{4}, -\dfrac{1}{5}, i, -i$

In Exercises 39 to 46, find a polynomial $P(x)$ with real coefficients that has the indicated zeros and satisfies the given conditions.

39. Zeros: $2 - 5i, -4$, degree 3

40. Zeros: $3 + 2i, 7$, degree 3

41. Zeros: $4 + 3i, 5 - i$, degree 4

42. Zeros: $i, 3 - 5i$, degree 4

43. Zeros: $-1, 2, 3$, degree 3, $P(1) = 12$

44. Zeros: $3i, 2$, degree 3, $P(3) = 27$

45. Zeros: $3, -5, 2 + i$, degree 4, $P(1) = 48$

46. Zeros: $\frac{1}{2}, 1 - i$, degree 3, $P(4) = 140$

Graphing Calculator Exercises*

In Exercises 47 to 50, use the graph of the polynomial to determine the number of real zeros of the polynomial.

47. $P(x) = x^3 + 3x^2 + x + 3$

48. $P(x) = x^4 - 2x^3 + 3x^2 - 2x + 2$

49. $P(x) = x^3 - 3x - 2$

50. $P(x) = x^4 + 4x^3 - 2x^2 - 12x + 9$

Supplemental Exercises

51. Verify that $x^3 - x^2 - ix^2 - 9x + 9 + 9i$ has $1 + i$ as a zero and that its conjugate $1 - i$ is not a zero. Explain why this does not contradict the Conjugate Pair Theorem.

52. Verify that $x^3 - x^2 - ix^2 - 20x + ix + 20i$ has a zero of i but not of its conjugate $-i$. Explain why this does not contradict the Conjugate Pair Theorem.

53. Show that 2 is a zero of multiplicity 3 of the polynomial

$$P(x) = x^5 - 6x^4 + 21x^3 - 62x^2 + 108x - 72$$

and express $P(x)$ as a product of linear factors and/or quadratic factors that are irreducible over the reals.

54. Show that -1 is a zero of multiplicity 4 of the polynomial

$$P(x) = x^6 + 5x^5 + 11x^4 + 14x^3 + 11x^2 + 5x + 1$$

and express $P(x)$ as a product of linear factors and/or quadratic factors that are irreducible over the reals.

55. Find a polynomial $P(x)$ of degree 5 such that 1 is a zero of multiplicity 2, 2 is a zero of multiplicity 3, and $P(-1) = -54$.

56. Find a polynomial $P(x)$ of degree 5 such that -4 is a zero of multiplicity 4, $\frac{1}{2}$ is a zero of multiplicity 1, and $P(1) = 125$.

4.5   Rational Functions and Their Graphs

If $P(x)$ and $Q(x)$ are polynomials, then the function F given by

$$F(x) = \frac{P(x)}{Q(x)}$$

is called a **rational function.** The domain of F is the set of all real numbers except for those for which $Q(x) = 0$. For example, the domain of

$$F(x) = \frac{x^2 - x - 5}{x(2x - 5)(x + 3)}$$

*Additional graphing calculator exercises appear in the Graphing Workbook as described in the front of this textbook.

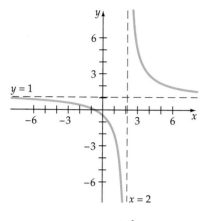

$$G(x) = \frac{x + 1}{x - 2}$$

Figure 4.12

is the set of all real numbers except 0, 5/2, and -3.

The graph of $G(x) = \frac{x + 1}{x - 2}$ is given in Figure 4.12. The graph shows that G has the following behavior:

1. The graph does not exist at $x = 2$. That is, 2 is not in the domain of G.
2. The graph has an x-intercept at $(-1, 0)$ and a y-intercept at $(0, -\frac{1}{2})$.
3. The functional values $G(x)$ approach 1 as x increases or decreases without bound.
4. The functional values $G(x)$ increase without bound as x approaches 2 from the right.
5. The functional values $G(x)$ decrease without bound as x approaches 2 from the left.

When discussing graphs that increase or decrease without bound, it is convenient to use mathematical notation. The notation

$$f(x) \to \infty \text{ as } x \to a^+$$

means that the functional values $f(x)$ increase without bound as x approaches a from the right. Recall that the symbol ∞ does not represent a real number but is used merely to describe the concept of a variable taking on larger and larger values without bound. See Figure 4.13 (a).

The notation

$$f(x) \to \infty \text{ as } x \to a^-$$

means that the functional values $f(x)$ increase without bound as x approaches a from the left. See Figure 4.13 (b).

The notation

$$f(x) \to -\infty \text{ as } x \to a^+$$

means that the functional values $f(x)$ decrease without bound as x approaches a from the right. See Figure 4.13 (c).

The notation

$$f(x) \to -\infty \text{ as } x \to a^-$$

means that the functional values $f(x)$ decrease without bound as x approaches a from the left. See Figure 4.13 (d).

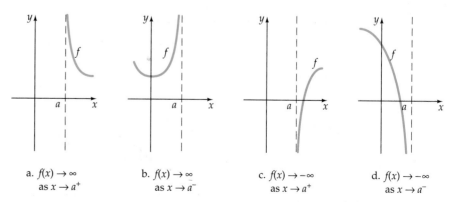

a. $f(x) \to \infty$
   as $x \to a^+$

b. $f(x) \to \infty$
   as $x \to a^-$

c. $f(x) \to -\infty$
   as $x \to a^+$

d. $f(x) \to -\infty$
   as $x \to a^-$

Figure 4.13

Asymptotes

Each graph in Figure 4.13 approaches a vertical line through $(a, 0)$ as $x \to a^+$ or a^-. The line is said to be a *vertical asymptote* to the graph.

Definition of a Vertical Asymptote

The line $x = a$ is a **vertical asymptote** of the graph of a function F provided that

$$F(x) \to \infty \qquad \text{or} \qquad F(x) \to -\infty$$

as x approaches a from either left or right.

In Figure 4.12, the line $x = 2$ is a vertical asymptote of the graph of G. Note that the graph of G in Figure 4.12 also approaches the horizontal line $y = 1$ as $x \to \infty$ and as $x \to -\infty$. The line $y = 1$ is a *horizontal asymptote* of the graph of G.

Definition of a Horizontal Asymptote

The line $y = b$ is a **horizontal asymptote** of the graph of a function F provided that

$$F(x) \to b \text{ as } x \to \infty \text{ or } x \to -\infty$$

Figure 4.14 illustrates some of the ways in which the graph of a rational function may approach its horizontal asymptote. It is common practice to display the asymptotes of the graph of a rational function by using dashed lines. Although a rational function may have several vertical asymptotes, it can have at most one horizontal asymptote. The graph of a rational function will never intersect any of its vertical asymptotes. Why?[2] However, the graph may intersect its horizontal asymptote.

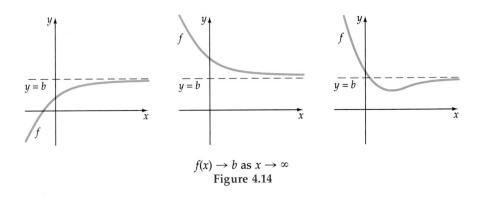

$$f(x) \to b \text{ as } x \to \infty$$
Figure 4.14

[2] If $x = a$ is a vertical asymptote of a rational function R, then $R(a)$ is undefined.

Geometrically, a line is an asymptote to a curve if the distance between the line and a point $P(x, y)$ on the curve approaches zero as the distance between the origin and the point P increases without bound.

Vertical asymptotes of the graph of a rational function can be found by using the following theorem.

Theorem on Vertical Asymptotes

If the real number a is a zero of the denominator $Q(x)$, then the graph of $F(x) = P(x)/Q(x)$, where $P(x)$ and $Q(x)$ have no common factors, has the vertical asymptote $x = a$.

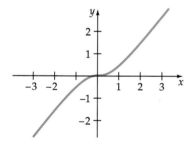

$$f(x) = \frac{x^3}{x^2 + 1}$$

Figure 4.15

EXAMPLE 1 **Find the Vertical Asymptotes of a Rational Function**

Find the vertical asymptotes of each rational function.

a. $f(x) = \dfrac{x^3}{x^2 + 1}$ b. $g(x) = \dfrac{x}{x^2 - x - 6}$

Solution

a. To find the vertical asymptotes, set the denominator equal to zero. The denominator $x^2 + 1$ has no real zeros, so the graph of f has no vertical asymptotes. See Figure 4.15.

b. The denominator $x^2 - x - 6 = (x - 3)(x + 2)$ has zeros of 3 and -2. The numerator has no common factors with the denominator, so $x = 3$ and $x = -2$ are both vertical asymptotes of the graph of g, as shown in Figure 4.16.

■ *Try Exercise **2**, page 239.*

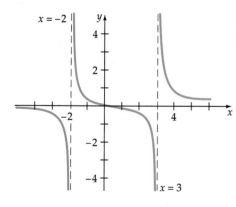

$$g(x) = \frac{x}{x^2 - x - 6}$$

Figure 4.16

The following theorem implies that a horizontal asymptote can be determined by examining the leading terms of the numerator and the denominator of a rational function.

Theorem on Horizontal Asymptotes

Let
$$F(x) = \frac{a_n x^n + a_{n-1} x^{n-1} + \cdots + a_1 x + a_0}{b_m x^m + b_{m-1} x^{m-1} + \cdots + b_1 x + b_0}$$

be a rational function with numerator of degree n and denominator of degree m.

1. If $n < m$, then the x-axis is the horizontal asymptote of the graph of F.

2. If $n = m$, then the line $y = a_n/b_m$ is the horizontal asymptote of the graph of F.

3. If $n > m$, the graph of F has no horizontal asymptote.

EXAMPLE 2   Find the Horizontal Asymptote of a Rational Function

Find the horizontal asymptote of each rational function.

a. $f(x) = \dfrac{2x + 3}{x^2 + 1}$ b. $g(x) = \dfrac{4x^2 + 1}{3x^2}$ c. $h(x) = \dfrac{x^3 + 1}{x - 2}$

Solution

a. The degree of the numerator $2x + 3$ is less than the degree of the denominator $x^2 + 1$. By the Theorem on Horizontal Asymptotes, the x-axis is the horizontal asymptote of f. See the graph of f in Figure 4.17.

b. The numerator $4x^2 + 1$ and denominator $3x^2$ of g are both of degree 2. By the Theorem on Horizontal Asymptotes, the line $y = \frac{4}{3}$ is the horizontal asymptote of g. See the graph of g in Figure 4.18.

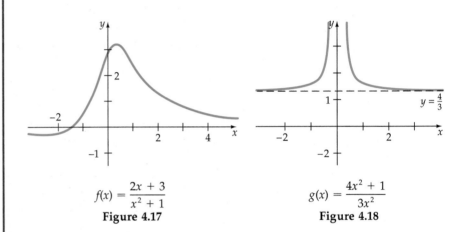

$$f(x) = \dfrac{2x + 3}{x^2 + 1}$$

Figure 4.17

$$g(x) = \dfrac{4x^2 + 1}{3x^2}$$

Figure 4.18

c. The degree of the numerator $x^3 + 1$ is larger than the degree of the denominator $x - 2$, so by the Theorem on Horizontal Asymptotes, the graph of h has no horizontal asymptote.

■ *Try Exercise 6, page 239.*

The proof of the Theorem on Horizontal Asymptotes makes use of the technique employed in the following verification. To verify that

$$k(x) = \dfrac{5x^2 + 4}{3x^2 + 8x + 7}$$

has a horizontal asymptote of $y = \frac{5}{3}$, divide the numerator and the denominator by the largest power of the variable x (x^2 in this case).

$$k(x) = \dfrac{\dfrac{5x^2 + 4}{x^2}}{\dfrac{3x^2 + 8x + 7}{x^2}} = \dfrac{5 + \dfrac{4}{x^2}}{3 + \dfrac{8}{x} + \dfrac{7}{x^2}}, \quad x \neq 0$$

As x increases without bound or decreases without bound, the fractions $\frac{4}{x^2}$, $\frac{8}{x}$, and $\frac{7}{x^2}$ approach zero. Thus

$$k(x) \rightarrow \frac{5 + 0}{3 + 0 + 0} \text{ as } x \rightarrow \pm\infty$$

and hence the line $y = \frac{5}{3}$ is a horizontal asymptote of the graph of the rational function k.

The zeros and vertical asymptotes of a rational function F divide the x-axis into intervals. In each interval,

- $F(x)$ is positive for all x in the interval, or
- $F(x)$ is negative for all x in the interval.

For example, consider the rational function

$$g(x) = \frac{x + 1}{x^2 + 2x - 3}$$

which has vertical asymptotes of $x = -3$ and $x = 1$ and a zero of -1. These three numbers divide the x-axis into the four intervals $(-\infty, -3)$, $(-3, -1)$, $(-1, 1)$, and $(1, \infty)$. Note in Figure 4.19 that the graph of g is

- Negative for all x such that $x < -3$.
- Positive for all x such that $-3 < x < -1$.
- Negative for all x such that $-1 < x < 1$.
- Positive for all x such that $x > 1$.

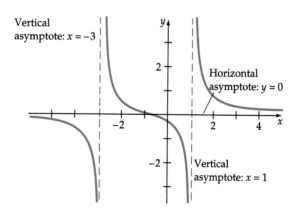

$$g(x) = \frac{x + 1}{x^2 + 2x - 3}$$

Figure 4.19

General Graphing Procedure

If $F(x) = P(x)/Q(x)$, where $P(x)$ and $Q(x)$ are polynomials that have no common factor, then the following general procedure offers useful guidelines for graphing F.

General Procedure for Graphing Rational Functions That Have No Common Factors

1. *Asymptotes:* Find the real zeros of the denominator $Q(x)$. For each zero a, draw the dashed line $x = a$. Each line is a vertical asymptote of the graph of F. Graph any horizontal asymptotes. These can be found by using the Theorem on Horizontal Asymptotes. If the degree of the numerator $P(x)$ is larger than the degree of the denominator $Q(x)$, then the graph of F does not have a horizontal asymptote.

2. *Intercepts:* Find the real zeros of the numerator $P(x)$. For each zero a, plot the point $(a, 0)$. Each such point is an x-intercept of the graph of F. Evaluate $F(0)$. Plot $(0, F(0))$, the y-intercept of the graph of F.

3. *Additional points:* Plot at least two points that lie in the intervals between and beyond the vertical asymptotes and the x-intercepts.

4. *Behavior near asymptotes:* If $x = a$ is a vertical asymptote, determine whether $F(x) \to \infty$ or $F(x) \to -\infty$ as $x \to a^-$ and also as $x \to a^+$.

5. *Complete the sketch:* Use all the information obtained above to sketch the graph of F. Plot additional points if necessary to gain additional knowledge about the function.

EXAMPLE 3   **Graph a Rational Function**

Sketch the graph of $f(x) = \dfrac{4x^2}{x^2 + 3}$.

Solution   *Asymptotes:* The denominator $x^2 + 3$ has no real zeros, so the graph of f has no vertical asymptotes.

The numerator and denominator both have degree 2. The leading coefficients of the numerator and denominator are 4 and 1, respectively. By the Theorem on Horizontal Asymptotes, the graph of f has the horizontal asymptote $y = \frac{4}{1} = 4$.

Intercepts: The numerator $4x^2$ has 0 as its only zero. Therefore the graph of f has an x-intercept at the origin. Since $f(0) = 0$, f has the y-intercept $(0, 0)$.

Additional points: The intervals determined by the x-intercept are $x < 0$ and $x > 0$. Generally, it is necessary to determine points in all intervals. However, since f is an even function, its graph is symmetrical with respect to the y-axis. The following table lists a few points for $x > 0$. Symmetry can be used to locate corresponding points for $x < 0$.

x	1	2	6
$f(x)$	1	$\frac{16}{7} \approx 2.29$	$\frac{48}{13} \approx 3.69$

Behavior near asymptotes: As x increases or decreases without bound, $f(x)$ approaches the horizontal asymptote $y = 4$.

To determine whether the graph of f intersects the horizontal asymptote at any point, solve the equation $f(x) = 4$.
There are no solutions of $f(x) = 4$ because

$$\frac{4x^2}{x^2 + 3} = 4 \quad \text{implies} \quad 4x^2 = 4x^2 + 12$$

This is not possible. Thus the graph of f does not intersect the horizontal asymptote but approaches it from below as x increases or decreases without bound.

Vertical Asymptote	Horizontal Asymptote	x-Intercept	y-Intercept	Additional Points
None	$y = 4$	$(0, 0)$	$(0, 0)$	$(1, 1), (2, 2.29), (6, 3.69)$

Complete the sketch: The previous information can now be used to finish the sketch. The completed graph is shown in Figure 4.20.

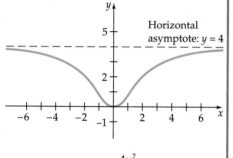

Horizontal asymptote: $y = 4$

$$f(x) = \frac{4x^2}{x^2 + 3}$$

Figure 4.20

■ *Try Exercise* **10,** *page 239.*

EXAMPLE 4   **Graph a Rational Function**

Sketch the graph of $h(x) = \dfrac{x^2 + 1}{x^2 + x - 2}$.

Solution   *Asymptotes:* The denominator $x^2 + x - 2 = (x + 2)(x - 1)$ has zeros -2 and 1; since there are no common factors of the numerator and the denominator, the lines $x = -2$ and $x = 1$ are vertical asymptotes.
The numerator and denominator both have degree 2. The leading coefficients of the numerator and denominator are both 1. Thus h has the horizontal asymptote $y = \frac{1}{1} = 1$.

Intercept(s):   The numerator $x^2 + 1$ has no real zeros, so the graph of h has no x-intercepts. Since $h(0) = -0.5$, h has the y-intercept $(0, -0.5)$.

Additional points: The intervals determined by the vertical asymptotes are $(-\infty, -2)$, $(-2, 1)$, and $(1, \infty)$. Plot a few points from each interval:

x	-5	-3	-1	0.5	2	3	4
$h(x)$	$\frac{13}{9}$	2.5	-1	-1	1.25	1	$\frac{17}{18}$

The graph of h will intersect the horizontal asymptote $y = 1$ exactly once. This can be determined by solving the equation $h(x) = 1$.

$$\frac{x^2 + 1}{x^2 + x - 2} = 1$$

$$x^2 + 1 = x^2 + x - 2 \quad \bullet \text{ Multiply both sides by } x^2 + x - 2.$$

$$1 = x - 2$$

$$3 = x$$

The only solution is $x = 3$. Therefore the graph of h intersects the horizontal asymptote at $(3, 1)$.

Behavior near asymptotes: As x approaches -2 from the left, the denominator $(x + 2)(x - 1)$ approaches 0 but remains positive. The numerator $x^2 + 1$ approaches 5, which is positive, so the quotient $h(x)$ increases without bound. Stated in mathematical notation,

$$h(x) \rightarrow \infty \text{ as } x \rightarrow -2^-$$

Similarly, it can be determined that

$$h(x) \rightarrow -\infty \text{ as } x \rightarrow -2^+$$

$$h(x) \rightarrow -\infty \text{ as } x \rightarrow 1^-$$

$$h(x) \rightarrow \infty \quad \text{ as } x \rightarrow 1^+$$

Vertical Asymptotes	Horizontal Asymptote	x-Intercept	y-Intercept	Additional Points
$x = -2$, $x = 1$	$y = 1$	None	$(0, -0.5)$	$(-5, 1.\overline{4})$, $(-3, 2.5)$, $(-1, -1)$, $(0.5, -1)$ $(2, 1.25)$, $(3, 1)$, $(4, 0.9\overline{4})$

Complete the sketch: Use the information to obtain the graph sketched in Figure 4.21.

■ *Try Exercise* **26,** *page 240.*

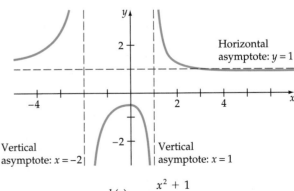

$$h(x) = \frac{x^2 + 1}{x^2 + x - 2}$$

Figure 4.21

Slant Asymptotes

Some rational functions have an asymptote that is neither vertical nor horizontal but slanted.

Theorem on Slant Asymptotes

The rational function given by $F(x) = P(x)/Q(x)$, where $P(x)$ and $Q(x)$ have no common factors, has a **slant asymptote** if the degree of the polynomial $P(x)$ in the numerator is one greater than the degree of the polynomial $Q(x)$ in the denominator.

To find the slant asymptote, use division to express $F(x)$ in the form

$$F(x) = \frac{P(x)}{Q(x)} = (mx + b) + \frac{r(x)}{Q(x)}$$

where the degree of $r(x)$ is less than the degree of $Q(x)$. Since

$$\frac{r(x)}{Q(x)} \to 0 \quad \text{as} \quad x \to \pm\infty$$

we know that $F(x) \to mx + b$ as $x \to \pm\infty$.

The line represented by $y = mx + b$ is called the slant asymptote of the graph of F.

EXAMPLE 5   Find the Slant Asymptote of a Rational Function

Find the slant asymptote of $f(x) = \dfrac{2x^3 + 5x^2 + 1}{x^2 + x + 3}$.

Solution   Because the degree of the numerator $2x^3 + 5x^2 + 1$ is exactly one larger than the degree of the denominator $x^2 + x + 3$ and f is in simplest form, f has a slant asymptote. To find the asymptote, divide $2x^3 + 5x^2 + 1$ by $x^2 + x + 3$.

$$
\begin{array}{r}
2x + 3 \\
x^2 + x + 3 \overline{)\,2x^3 + 5x^2 + 0x + 1} \\
\underline{2x^3 + 2x^2 + 6x} \\
3x^2 - 6x + 1 \\
\underline{3x^2 + 3x + 9} \\
-9x - 8
\end{array}
$$

Therefore

$$f(x) = \frac{2x^3 + 5x^2 + 1}{x^2 + x + 3} = (2x + 3) + \frac{-9x - 8}{x^2 + x + 3}$$

and the line $y = 2x + 3$ is the slant asymptote for the graph of f. Figure 4.22 shows the graph of f and its slant asymptote.

■ *Try Exercise 34, page 240.*

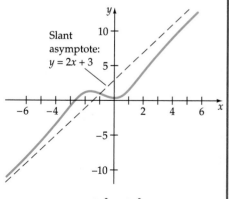

Slant asymptote: $y = 2x + 3$

$$f(x) = \frac{2x^3 + 5x^2 + 1}{x^2 + x + 3}$$

Figure 4.22

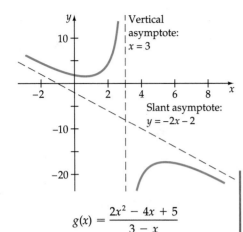

$$g(x) = \frac{2x^2 - 4x + 5}{3 - x}$$

Figure 4.23

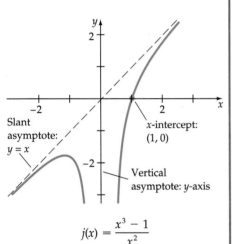

$$j(x) = \frac{x^3 - 1}{x^2}$$

Figure 4.24

Remark The function f in Example 5 does not have a vertical asymptote because the denominator $x^2 + x + 3$ does not have any real zeros. However, the function

$$g(x) = \frac{2x^2 - 4x + 5}{3 - x}$$

has both a slant asymptote and a vertical asymptote. Why?[3] Figure 4.23 shows the graph of g and its asymptotes.

EXAMPLE 6 **Graph a Rational Function That Has a Slant Asymptote**

Sketch the graph of $j(x) = \frac{x^3 - 1}{x^2}$.

Solution Asymptotes: The denominator x^2 has 0 as its only zero. Since there are no common factors of the numerator and the denominator the y-axis is the vertical asymptote of the graph of j.
The degree of the numerator $x^3 - 1$ is exactly one more than the degree of the denominator x^2, so j has a slant asymptote. Dividing $x^3 - 1$ by x^2 shows that j can be expressed as

$$j(x) = \frac{x^3}{x^2} - \frac{1}{x^2} = x - \frac{1}{x^2}$$

From this we see that $j(x) \to x$ as $x \to \pm\infty$. Therefore j has a slant asymptote of $y = x$.

Intercepts: The numerator $x^3 - 1$ has a real zero of 1. Therefore $(1, 0)$ is the only x-intercept of the graph of j. Since $j(0)$ is undefined, the graph of j does not have a y-intercept.

Additional points: The intervals determined by the vertical asymptote and the x-intercept are $x < 0$, $0 < x < 1$, and $x > 1$. The following table lists a few points from each interval.

x	-5	-2	-1	-0.5	0.5	0.8	2	5
$j(x)$	-5.04	-2.25	-2	-4.5	-3.5	-0.7625	1.75	4.96

Vertical Asymptote	Slant Asymptote	x-Intercept	y-Intercept	Additional Points
y-axis	$y = x$	$(1, 0)$	None	$(-5, -5.04)$, $(-2, -2.25)$, $(-1, -2)$ $(-0.5, -4.5)$, $(0.5, -3.5)$, $(0.8, -0.7625)$ $(2, 1.75)$, $(5, 4.96)$

Complete the sketch: Use all the previous information to complete the sketch of j as shown in Figure 4.24.

■ *Try Exercise* **40,** *page 240.*

[3] It has a slant asymptote because the degree of the numerator is one greater than the degree of the denominator and they have no common factors. It has a vertical asymptote because the denominator equals zero when $x = 3$.

If a rational function has a numerator and denominator that have a common factor, then you should reduce the rational function to lowest terms before you apply the general procedure for sketching the graph of a rational function.

EXAMPLE 7 Graph a Rational Function That Has a Common Factor

Sketch the graph of $f(x) = \dfrac{x^2 - 3x - 4}{x^2 - 6x + 8}$.

Solution Factor the numerator and denominator to obtain

$$f(x) = \frac{x^2 - 3x - 4}{x^2 - 6x + 8}$$

$$= \frac{(x + 1)(x - 4)}{(x - 2)(x - 4)}, \quad x \neq 2, x \neq 4$$

Thus for all x values other than $x = 4$, the graph of f is the same as the graph of

$$G(x) = \frac{x + 1}{x - 2}$$

Figure 4.12 shows a graph of G. The graph of f will be the same as this graph, except that it will have an open circle at $(4, 2.5)$ to indicate that it is undefined for $x = 4$. See the graph of f in Figure 4.25.

■ *Try Exercise* **50**, *page 240.*

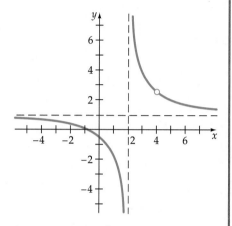

$f(x) = \dfrac{x^2 - 3x - 4}{x^2 - 6x + 8}$

Figure 4.25

EXERCISE SET 4.5

In Exercises 1 to 4, find all vertical asymptotes of each rational function.

1. $F(x) = \dfrac{2x - 1}{x^2 + 3x}$

2. $F(x) = \dfrac{3x^2 + 5}{x^2 - 4}$

3. $F(x) = \dfrac{x^2 + 11}{6x^2 - 5x - 4}$

4. $F(x) = \dfrac{3x - 5}{x^3 - 8}$

In Exercises 5 to 8, find the horizontal asymptote of each rational function.

5. $F(x) = \dfrac{4x^2 + 1}{x^2 + x + 1}$

6. $F(x) = \dfrac{3x^3 - 27x^2 + 5x - 11}{x^5 - 2x^3 + 7}$

7. $F(x) = \dfrac{15{,}000x^3 + 500x - 2000}{700 + 500x^3}$

8. $F(x) = 6000\left(1 - \dfrac{25}{(x + 5)^2}\right)$

In Exercises 9 to 32, determine the vertical and horizontal asymptotes and sketch the graph of the rational function F. Label all intercepts and asymptotes.

9. $F(x) = \dfrac{1}{x + 4}$

10. $F(x) = \dfrac{1}{x - 2}$

11. $F(x) = \dfrac{-4}{x - 3}$

12. $F(x) = \dfrac{-3}{x + 2}$

13. $F(x) = \dfrac{4}{x}$

14. $F(x) = \dfrac{-4}{x}$

15. $F(x) = \dfrac{x}{x + 4}$

16. $F(x) = \dfrac{x}{x - 2}$

17. $F(x) = \dfrac{x + 4}{2 - x}$

18. $F(x) = \dfrac{x + 3}{1 - x}$

19. $F(x) = \dfrac{1}{x^2 - 9}$

20. $F(x) = \dfrac{-2}{x^2 - 4}$

21. $F(x) = \dfrac{1}{x^2 + 2x - 3}$

22. $F(x) = \dfrac{1}{x^2 - 2x - 8}$

23. $F(x) = \dfrac{x}{9 - x^2}$

24. $F(x) = \dfrac{x}{x^2 - 16}$

25. $F(x) = \dfrac{x^2}{x^2 + 4x + 4}$

26. $F(x) = \dfrac{x^2}{x^2 - 6x + 9}$

27. $F(x) = \dfrac{10}{x^2 + 2}$

28. $F(x) = \dfrac{-20}{x^2 + 4}$

29. $F(x) = \dfrac{2x^2 - 2}{x^2 - 9}$

30. $F(x) = \dfrac{6x^2 - 5}{2x^2 + 6}$

31. $F(x) = \dfrac{x^2 + x + 4}{x^2 + 2x - 1}$

32. $F(x) = \dfrac{2x^2 - 14}{x^2 - 6x + 5}$

In Exercises 33 to 36, find the slant asymptote of each rational function.

33. $F(x) = \dfrac{3x^2 + 5x - 1}{x + 4}$

34. $F(x) = \dfrac{x^3 - 2x^2 + 3x + 4}{x^2 - 3x + 5}$

35. $F(x) = \dfrac{x^3 - 1}{x^2}$

36. $F(x) = \dfrac{4000 + 20x + 0.0001x^2}{x}$

In Exercises 37 to 46, determine the vertical and slant asymptotes and sketch the graph of the rational function F.

37. $F(x) = \dfrac{x^2 - 4}{x}$

38. $F(x) = \dfrac{x^2 + 10}{2x}$

39. $F(x) = \dfrac{x^2 - 3x - 4}{x + 3}$

40. $F(x) = \dfrac{x^2 - 4x - 5}{2x + 5}$

41. $F(x) = \dfrac{2x^2 + 5x + 3}{x - 4}$

42. $F(x) = \dfrac{4x^2 - 9}{x + 3}$

43. $F(x) = \dfrac{x^2 - x}{x + 2}$

44. $F(x) = \dfrac{x^2 + x}{x - 1}$

45. $F(x) = \dfrac{x^3 + 1}{x^2 - 4}$

46. $F(x) = \dfrac{x^3 - 1}{3x^2}$

In Exercises 47 to 56, sketch the graph of the rational function F. (*Hint:* First examine the numerator and denominator to determine whether there are any common factors.)

47. $F(x) = \dfrac{x^2 + x}{x + 1}$

48. $F(x) = \dfrac{x^2 - 3x}{x - 3}$

49. $F(x) = \dfrac{2x^3 + 4x^2}{2x + 4}$

50. $F(x) = \dfrac{x^2 - x - 12}{x^2 - 2x - 8}$

51. $F(x) = \dfrac{-2x^3 + 6x}{2x^2 - 6x}$

52. $F(x) = \dfrac{x^3 + 3x^2}{x(x + 3)(x - 1)}$

53. $F(x) = \dfrac{x^2 - 3x - 10}{x^2 + 4x + 4}$

54. $F(x) = \dfrac{2x^2 + x - 3}{x^2 - 2x + 1}$

55. $F(x) = \dfrac{x^3 + x^2 - 14x - 24}{x + 2}$

56. $F(x) = \dfrac{2x^3 + 5x^2 - 4x - 3}{x - 1}$

Graphing Calculator Exercises*

57. The cost C in dollars to remove p percent of the salt in a tank of sea water is given by

$$C(p) = \dfrac{2000p}{100 - p}, \quad 0 \le p < 100$$

 a. Find the cost of removing 40 percent of the salt.

 b. Find the cost of removing 80 percent of the salt.

 c. Sketch the graph of C.

58. The temperature F (measured in degrees Fahrenheit) of a dessert placed in a freezer for t hours is given by the rational function

$$F(t) = \dfrac{60}{t^2 + 2t + 1}, \quad t \ge 0$$

 a. Find the temperature of the dessert after it has been in the freezer for 1 hour.

 b. Find the temperature of the dessert after 4 hours.

 c. Sketch the graph of F.

59. A large electronics firm finds that the number of computers it can produce per week after t weeks of production is approximated by

$$C(t) = \dfrac{2000t^2 + 20{,}000t}{t^2 + 10t + 25}, \quad 0 \le t \le 50$$

 a. Find the number of computers it produced during the first week.

 b. Find the number of computers it produced during the tenth week.

 c. What is the equation of the horizontal asymptote of the graph of C?

 d. Sketch the graph of C and then use the graph to estimate how many weeks pass until the firm can produce 1900 computers in a single week.

*Additional graphing calculator exercises appear in the Graphing Workbook as described in the front of this textbook.

60. The cost of publishing x books is given by

$$C(x) = 40{,}000 + 20x + 0.0001x^2$$

The average cost per book is given by

$$A(x) = \frac{C(x)}{x} = \frac{40{,}000 + 20x + 0.0001x^2}{x}$$

where $1000 \le x \le 100{,}000$.

a. What is the average cost per book if 5000 books are published?

b. What is the average cost per book if 10,000 books are published?

c. What is the equation of the slant asymptote of the graph of the average cost function?

d. Graph A and estimate the number of books that should be published to minimize the average cost per book.

61. One of Poiseuille's Laws states that the resistance R encountered by blood flowing through a blood vessel is given by the rational function

$$R(r) = C\frac{L}{r^4}$$

where C is a positive constant determined by the viscosity of the blood, L is the length of the blood vessel, and r is the radius.

a. Explain the meaning of $R(r) \to \infty$ as $r \to 0$.

b. Explain the meaning of $R(r) \to 0$ as $r \to \infty$.

c. Graph R for $0 < r \le 4$ millimeters, given that $C = 1$ and $L = 100$ millimeters.

62. A cylindrical soft drink can is to be made so that it will have a volume of 354 milliliters. If r is the radius of the can in centimeters, then the total surface area A of the can is given by the rational function

$$A(r) = \frac{2\pi r^3 + 708}{r}$$

a. Use the graph of A to estimate the value of r that produces the minimum value of A.

b. Does the graph of A have a slant asymptote?

c. Explain the meaning of the following statement as it applies to the graph of A:

$$\text{As } r \to \infty, A \to 2\pi r^2.$$

Supplemental Exercises

63. Determine the point where the graph of

$$F(x) = \frac{2x^2 + 3x + 4}{x^2 + 4x + 7}$$

intersects its horizontal asymptote.

64. Determine the point where the graph of

$$F(x) = \frac{3x^3 + 2x^2 - 8x - 12}{x^2 + 4}$$

intersects its slant asymptote.

65. Determine the two points where the graph of

$$F(x) = \frac{x^3 + x^2 + 4x + 1}{x^3 + 1}$$

intersects its horizontal asymptote.

66. Give an example of a rational function that intersects its slant asymptote at two points.

4.6 Approximation of Zeros

Finding the zeros of polynomials can be difficult. For example, finding the zeros of the polynomial $x^3 - 9x - 12$ is challenging because it does not have rational zeros. Thus the Rational Zeros Theorem is not helpful. The Zero Location Theorem from Section 4.2 can be used to establish that there is an irrational zero of $x^3 - 9x - 12$ in the interval between $x = 3$ and $x = 4$. For convenience, the Zero Location Theorem is now restated.

The Zero Location Theorem

> Let P be a polynomial with real coefficients. If $a < b$ and if $P(a)$ and $P(b)$ have opposite signs, then there is at least one value c between a and b such that $P(c) = 0$. See Figure 4.26.

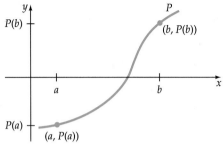

Figure 4.26

EXAMPLE 1   Apply the Zero Location Theorem

Show that $P(x) = x^3 - 9x - 12$ has a zero between 3 and 4.

Solution   Since there is one variation of sign of $P(x)$, Descartes' Rule of Signs implies that there is one positive zero. Compare the sign of $P(3)$ with the sign of $P(4)$.

$$P(3) = -12 \quad \text{and} \quad P(4) = 16$$

Because $P(3)$ and $P(4)$ have opposite signs, we know by the Zero Location Theorem that P has a zero between 3 and 4.

■ *Try Exercise **2**, page 245.*

The Bisection Method

The **bisection method** is often used to approximate the real zeros of a polynomial. The strategy of the bisection method is to establish that exactly one real zero of a polynomial P is in an interval (a, b) and then reduce the size of the interval.

Consider for example a polynomial P such that $P(a) < 0$ and $P(b) > 0$. In Figure 4.26, the graph of $y = P(x)$ is below the x-axis at $x = a$ and above the x-axis at $x = b$. Because the graph of $y = P(x)$ is continuous, there must be a zero of P between a and b. Now bisect the interval (a, b) and denote the midpoint of the interval by x_1, where

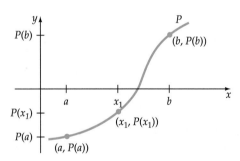

Figure 4.27

$$x_1 = \frac{a + b}{2}$$

If $P(x_1) = 0$, then x_1 is a zero of P. However, if $P(x_1) < 0$, as in Figure 4.27, then by the Zero Location Theorem, P has a zero between x_1 and b. The zero is now located in the interval (x_1, b), whose length is half the length of the original interval.

Evaluate $P(x_2)$, where x_2 is the midpoint of this new interval. Suppose that this time $P(x_2) > 0$, as in Figure 4.28. By the Zero Location Theorem, P has a zero between x_1 and x_2. Once again we have located the zero in an interval half the length of the previous interval. Each application of the bisection method that locates the zero in a smaller interval is referred to as an **iteration**.

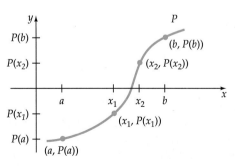

Figure 4.28

By repeating this bisection procedure, you can approximate a zero as closely as desired, because each iteration halves the length of the interval in which the zero lies. Three iterations of the bisection method reduce the length of the initial interval that contains the zero by a factor of $2^3 = 8$. Ten iterations reduce the length of the initial interval by a factor of 2^{10}, which is about 1000.

EXAMPLE 2   Use the Bisection Method to Approximate a Zero

Use three iterations of the bisection method to approximate the real zero of $P(x) = x^3 - 9x - 12$, which is located between 3 and 4. Determine the maximum error of the approximation.

Solution   Use the bisection method to decrease the size of the interval in which a zero of the function is located. We will use the midpoint of this shorter interval to approximate the zero. Then the maximum error of the approximation can be only half of the length of the interval.

First iteration: The zero is in the interval $I_1 = (3, 4)$. The midpoint of I_1 is 3.5. Because $P(3.5) = -0.625$ is negative and $P(4)$ is positive, we know by the Zero Location Theorem that the zero is located between 3.5 and 4.

Second iteration: The zero is in the interval $I_2 = (3.5, 4)$. The midpoint of I_2 is 3.75. Because $P(3.5) = -0.625$ is negative and $P(3.75) = 6.984375$ is positive, we know by the Zero Location Theorem that the zero is located between 3.5 and 3.75.

Third iteration: The zero is in the interval $I_3 = (3.5, 3.75)$. The midpoint of I_3 is 3.625. Because $P(3.5) = -0.625$ is negative and $P(3.625) = 3.009765625$ is positive, we know by the Zero Location Theorem that the zero is located between 3.5 and 3.625.

The midpoint of $(3.5, 3.625)$, which is 3.5625, is used as the approximation of the zero. The maximum error of the approximation is one-half of the length of the interval $(3.5, 3.625)$.

$$\frac{3.625 - 3.5}{2} = 0.0625$$

The approximate real zero of $P(x) = x^3 - 9x - 12$ found by three iterations of the bisection method is 3.5625 and has a maximum error of 0.0625.

■ *Try Exercise 8, page 245.*

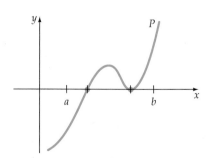

Figure 4.29

Caution   Sometimes the bisection method is not applicable. For example, the polynomial P in Figure 4.29 has two zeros in the interval (a, b). The bisection method is not applicable for finding the root closest to b. To apply the Zero Location Theorem, we must isolate a single zero between a and b such that $P(a)$ and $P(b)$ have opposite signs.

Variation of the Bisection Method

The following iteration procedure is a variation of the bisection method. It can be used to approximate a real zero of a polynomial to within 10^{-k} units by applying exactly k iterations.

Subdivide into Ten Equal Subintervals

The following method will approximate a real zero of a polynomial to within 10^{-k} units. Use the Zero Location Theorem to determine consecutive integers a and b such that the polynomial P has a single zero between a and b.

1. Subdivide the interval (a, b) into ten equal subintervals. Use the Zero Location Theorem to determine which of these subintervals contains the zero.

2. Replace a and b with the endpoints of the subinterval from step 1. Repeat step 1 k times, and stop. The midpoint of the resulting interval will be within 10^{-k} units of the zero.

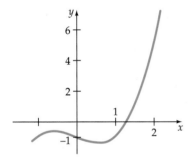

$P(x) = x^3 - x - 1$

Figure 4.30

EXAMPLE 3   **Use the Variation of the Bisection Method to Approximate a Zero**

Approximate the real zero of $P(x) = x^3 - x - 1$ to within 10^{-2}.

Solution   Descartes' Rule of Signs implies that there is one positive zero. Because $P(1) = -1$ and $P(2) = 5$, we know by the Zero Location Theorem that P has a zero in the interval $(1, 2)$. See Figure 4.30.

Step 1:   Subdivide the interval $(1, 2)$ into ten equal subintervals, as shown in Figure 4.31. Evaluate $P(x)$ for $x = 1.1, 1.2, \ldots$ until there is a change of sign of P. Because $P(1.3) < 0$ and $P(1.4) > 0$, there is a zero in the interval $(1.3, 1.4)$.

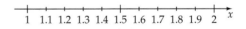

Figure 4.31

Step 2:   Let $a = 1.3$ and $b = 1.4$; repeat the iteration process.
Step 1':   Subdivide the interval $(1.3, 1.4)$ into ten equal subintervals. Evaluate $P(x)$ for $x = 1.31, 1.32, \ldots$ until there is a change of sign of P. Because $P(1.32) < 0$ and $P(1.33) > 0$, there is a zero in the interval $(1.32, 1.33)$. See Figure 4.32.
Step 2':   Stop. After two iterations, we know that P has a zero in the interval $(1.32, 1.33)$. Every real number in the interval $(1.32, 1.33)$ is within

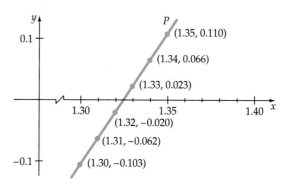

Figure 4.32

10^{-2} of the zero. The midpoint of this interval is 1.325, which is the desired approximation.

■ *Try Exercise 14, page 245.*

EXERCISE SET 4.6

In Exercises 1 to 6, use the Zero Location Theorem to verify the given statement.

1. $P(x) = x^3 - 2x - 5$ has a zero between 2 and 3.

2. $P(x) = x^3 + 18x - 30$ has a zero between 1 and 2.

3. $P(x) = x^3 - 36x - 96$ has a zero between 7 and 8.

4. $P(x) = x^3 - 27x - 90$ has a zero between 6 and 7.

5. $P(x) = 3x^4 - 6x^2 + 8x - 3$ has a zero between -2 and -1 and a zero between 0 and 1.

6. $P(x) = x^4 - 8x^3 + 25x^2 - 36x + 8$ has a zero between 0 and 1 and a zero between 3 and 4.

In Exercises 7 to 12, the given polynomials have one zero that satisfies the given condition. Locate the zero between two consecutive integers and then use three iterations of the bisection method to approximate the zero. Use the midpoint of the third iteration to approximate the zero.

7. $P(x) = x^3 - 2x - 5, x > 0$

8. $P(x) = x^3 + 18x - 10, x > 0$

9. $P(x) = 2x^3 + 3x^2 - 8, x > 0$

10. $P(x) = x^3 - 4x^2 + 10, x < 0$

11. $P(x) = x^4 + x - 5, x < 0$

12. $P(x) = x^4 + x - 20, x > 0$

In Exercises 13 to 18, use the variation of the bisection method to approximate the indicated zero of the given polynomial to the nearest hundredth of a unit. Use the midpoint of the interval of the second iteration as the approximation of the indicated zero.

13. $P(x) = x^3 + 10x^2 - 8, 0 < x < 1$

14. $P(x) = x^3 - 2x - 20, 2 < x < 3$

15. $P(x) = x^4 - x^2 - 1, 1 < x < 2$

16. $P(x) = x^4 + x - 10, 1 < x < 2$

17. $P(x) = x^5 - 4x^3 - 5, x > 0$

18. $P(x) = 3x^4 + x^2 - 7, x < 0$

Supplemental Exercises

19. Use the bisection method to approximate $\sqrt{2}$ to the nearest tenth. (*Hint:* $\sqrt{2}$ is the positive zero of the polynomial $x^2 - 2$.)

20. Use the bisection method to approximate $\sqrt[3]{5}$ to the nearest tenth. (*Hint:* $\sqrt[3]{5}$ is the real zero of the polynomial $x^3 - 5$.)

In Exercises 21 and 22, use the bisection method to approximate to the nearest tenth the x-coordinate of the point(s) of intersection of the graphs of the given polynomials. *Hint:* The graphs intersect where $P(x) = Q(x)$; therefore, find the roots of $P(x) - Q(x) = 0$ or the zeros of $P(x) - Q(x)$.

21. $P(x) = x^3 - 5$ and $Q(x) = -2x^3 + x^2 - 1$

22. $P(x) = x^4 + x^2 + 2$ and $Q(x) = x^4 + x^3 - 5x^2 + 1$

Exploring Concepts with Technology

Wiggle Functions

Consider the rational function

$$f(x) = \frac{2(x + 3)^2(x + 1)^3}{5(x + 4)^3(x + 2)^2(x - 2)}$$

and its graph as shown in the following figure.

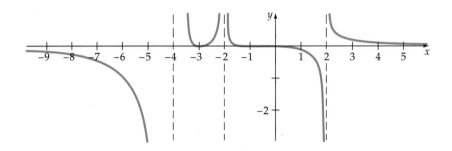

X-Intercepts   The numerator is zero when $x + 3 = 0$, or $x + 1 = 0$. Thus the x-intercepts are $(-3, 0)$ and $(-1, 0)$. Observe the behavior of the graph near the intercepts.

Even Exponent	Odd Exponent
$(x + 3)^2$	$(x + 1)^3$
At $x = -3$ the graph does not cross the x-axis.	At $x = -1$ the graph does cross the x-axis.

Vertical Asymptotes   The denominator is zero when $x + 4 = 0$, when $x + 2 = 0$, and when $x - 2 = 0$. Thus the vertical asymptotes are $x = -4$, $x = -2$, and $x = 2$. Observe the behavior of the graph near the asymptotes.

Odd Exponent	Even Exponent	Odd Exponent
$(x + 4)^3$	$(x + 2)^2$	$(x - 2)^1$
The graph tends to $-\infty$ and ∞.	The graph tends to ∞ from each side.	The graph tends to $-\infty$ and ∞.

Horizontal Asymptote   The numerator is of degree 5. The denominator is of degree 6. Thus the theorem on horizontal asymptotes indicates that the x-axis is a horizontal asymptote.

Sketch the graph of each of the rational functions on page 247. Use a computer or a graphing calculator to check your results. It may be helpful to use the zoom feature on a graphing calculator to see the behavior of the graph near its intercepts.

1. $j(x) = \dfrac{3(x - 1)^2(x - 3)(x + 5)^2}{(x + 2)^3(x - 2)^2}$

2. $k(x) = \dfrac{6x^3(x - 4)^2}{(x + 1)^2(2x - 5)^3}$

Chapter Review

4.1   Polynomial Division and Synthetic Division

- *The Remainder Theorem*   If a polynomial $P(x)$ is divided by $(x - c)$, then the remainder is $P(c)$.

- *The Factor Theorem*   A polynomial $P(x)$ has a factor $(x - c)$ if and only if $P(c) = 0$.

4.2   Graphs of Polynomial Functions

- Characteristics and properties used in graphing polynomial functions:

   1. Continuity — Polynomial functions are smooth continuous curves.
   2. Leading-term test — Determines the behavior of the graph of a polynomial function at the far right or the far left.
   3. Zeros of the function determine the x-intercepts.

4.3   Zeros of Polynomial Functions

- Values of x that satisfy $P(x) = 0$ are called zeros of P.

- Definition of Multiple Zeros of a Polynomial   If a polynomial function has $(x - r)$ as a factor exactly k times, then r is said to be a zero of multiplicity k of the polynomial P.

- Number of Zeros of a Polynomial Function   A polynomial of degree n has at most n zeros, where each zero of multiplicity k is counted k times.

- *The Rational Zero Theorem*   If $P(x) = a_nx^n + a_{n-1}x^{n-1} + \cdots + a_1x + a_0$ has integer coefficients, and p/q (where p and q have no common prime factors) is a rational zero of $P(x)$, then p is a factor of a_0 and q is a factor of a_n.

4.4   The Fundamental Theorem of Algebra

- *The Fundamental Theorem of Algebra*   If $P(x)$ is a polynomial of degree $n \geq 1$ with complex coefficients, then $P(x)$ has at least one complex zero.

- *The Number of Zeros of a Polynomial*   If $P(x)$ is a polynomial of degree $n \geq 1$ with complex coefficients, then $P(x)$ has exactly n complex zeros, provided that each zero is counted according to its multiplicity.

- *The Conjugate Pair Theorem*   If $a + bi$ ($b \neq 0$) is a complex zero of the polynomial $P(x)$, with real coefficients, then the conjugate $a - bi$ is also a complex zero of the polynomial.

4.5   Rational Functions and Their Graphs

- If $P(x)$ and $Q(x)$ are polynomials, then the function F given by

$$F(x) = \frac{P(x)}{Q(x)}$$

   is called a rational function.

- **General Procedure for Graphing Rational Functions That Have No Common Factors:**

   1. Find the zeros in the denominator. The vertical asymptotes will occur at these points. Graph any horizontal asymptote.
   2. Find the zeros in the numerator. These points will give the x-intercepts.
   3. Find additional points that lie in the intervals between the x-intercepts and the vertical asymptotes.
   4. Determine the behavior near the asymptotes.

4.6   Approximation of Zeros

- *The Zero Location Theorem*   Let P be a polynomial with real coefficients. If $a < b$ and if $P(a)$ and $P(b)$ have opposite signs, then there is at least one value c between a and b such that $P(c) = 0$.

- The bisection method makes repeated use of the Zero Location Theorem to approximate the zeros of a polynomial.

Essays and Projects

1. Write an equation and draw the graph of a function that has a parabolic asymptote. Write a theorem that can be used to determine when a rational function will have a parabolic asymptote.

2. Consider the polynomial

$$x^n + C_1 x^{n-1} + C_2 x^{n-2} + \cdots + C_n$$

which has the zeros $r_1, r_2, \ldots, r_n$. The following theorems illustrate important relationships that exist between the zeros of the polynomial and the roots of the polynomial.

$$r_1 + r_2 + r_3 + \cdots + r_{n-1} + r_n = -C_1$$

$$r_1 r_2 + r_1 r_3 + \cdots + r_{n-2} r_{n-1} + r_{n-1} r_n = C_2$$

$$r_1 r_2 r_3 + r_1 r_2 r_4 + \cdots + r_{n-2} r_{n-1} r_n = -C_3$$

$$\vdots$$

$$r_1 r_2 r_3 r_4 \cdots r_{n-1} r_n = (-1)^n C_n$$

State the above theorems in words. Illustrate the theorems for a polynomial of degree 4.

3. Write an essay about Carl Friedrich Gauss (1777–1855). Include information about his work as it applies to the zeros of a polynomial.

4. Write an essay about René Descartes (1596–1650). The proof of Descartes' Rule of Signs is too advanced for consideration in this text. However, prove the special case of Descartes' Rule of Signs for polynomials with no negative coefficients.

5. Write an essay about Evariste Galois (1811–1832). Include information about his work with polynomials of degree n where $n \geq 5$.

6. The following formula

$$x = \sqrt[3]{\sqrt{\left(\frac{n}{2}\right)^2 + \left(\frac{m}{3}\right)^3} + \frac{n}{2}} - \sqrt[3]{\sqrt{\left(\frac{n}{2}\right)^2 + \left(\frac{m}{3}\right)^3} - \frac{n}{2}}$$

is known as Cardano's Formula. It can be used to find one solution of the cubic

$$x^3 + mx = n$$

Use Cardano's Formula to find a solution of $x^3 + 6x = 4$. Write your answer in radical form, and then find an approximate result accurate to two decimal places. Use a calculator to check your result.

True/False Exercises

In Exercises 1 to 14, answer true or false. If the statement is false, give an example.

1. The complex zeros of a polynomial with complex coefficients always occur in conjugate pairs.

2. Descartes' Rule of Signs indicates that $x^3 - x^2 + x - 1$ must have three positive zeros.

3. The polynomial $2x^5 + x^4 - 7x^3 - 5x^2 + 4x + 10$ has two variations in sign.

4. If 4 is an upper bound of the zeros of the polynomial P, then 5 is also an upper bound of the zeros of P.

5. The graph of every rational function has a vertical asymptote.

6. The graph of the rational function

$$F(x) = \frac{x^2 - 4x + 4}{x^2 - 5x + 6}$$

has a vertical asymptote of $x = 2$.

7. If 7 is a zero of the polynomial P, then $x - 7$ is a factor of P.

8. According to the Zero Location Theorem, the polynomial function $P(x) = x^3 + 6x - 2$ has a real zero between 0 and 1.

9. Synthetic division can be used to show that $3i$ is a zero of $x^3 - 2x^2 + 9x - 18$.

10. Every fourth-degree polynomial with complex coefficients has exactly four complex zeros, provided that each zero is counted according to its multiplicity.

11. The graph of a rational function never intersects any of its vertical asymptotes.

12. The graph of a rational function can have at most one horizontal asymptote.

13. Descartes' Rule of Signs indicates that the polynomial function $P(x) = x^3 + 2x^2 + 4x - 7$ does have a positive zero.

14. Every polynomial has at least one real zero.

Chapter Review Exercises

In Exercises 1 to 6, use long division to divide the first polynomial by the second.

1. $x^3 + 5x^2 + 2x - 17, x^2 + x + 3$

2. $2x^3 - 5x + 1, x^2 + 4$

3. $-x^4 + 2x^2 - 12x - 3, x^3 + x$

4. $x^3 - 5x^2 - 6x - 11, x^2 - 6x - 1$

5. $6x^4 + 8x^3 - 47x^2 + 19x + 5, 2x^2 + 6x - 5$

6. $x^4 + 3x^3 - 6x^2 - 13x + 15, x^2 + 2x - 3$

In Exercises 7 to 12, use synthetic division to divide the first polynomial by the second.

7. $4x^3 - 11x^2 + 5x - 2, x - 3$

8. $5x^3 - 18x + 2, x - 1$

9. $3x^3 - 5x + 1, x + 2$

10. $2x^3 + 7x^2 + 16x - 10, x - \dfrac{1}{2}$

11. $3x^3 - 10x^2 - 36x + 55, x - 5$

12. $x^4 + 9x^3 + 6x^2 - 65x - 63, x + 7$

In Exercises 13 to 16, use the Remainder Theorem to find $P(c)$.

13. $P(x) = x^3 + 2x^2 - 5x + 1, c = 4$

14. $P(x) = -4x^3 - 10x + 8, c = -1$

15. $P(x) = 6x^4 - 12x^2 + 8x + 1, c = -2$

16. $P(x) = 5x^5 - 8x^4 + 2x^3 - 6x^2 - 9, c = 3$

In Exercises 17 to 20, use synthetic division to show that c is a zero of the given polynomial.

17. $x^3 + 2x^2 - 26x + 33, c = 3$

18. $2x^4 + 8x^3 - 8x^2 - 31x + 4, c = -4$

19. $x^5 - x^4 - 2x^2 + x + 1, c = 1$

20. $2x^3 + 3x^2 - 8x + 3, c = \dfrac{1}{2}$

In Exercises 21 to 26, graph the polynomial function.

21. $P(x) = x^3 - x$

22. $P(x) = -x^3 - x^2 + 8x + 12$

23. $P(x) = x^4 - 6$ **24.** $P(x) = x^5 - x$

25. $P(x) = x^4 - 10x^2 + 9$ **26.** $P(x) = x^5 - 5x^3$

In Exercises 27 to 32, use the Rational Zero Theorem to list all possible rational zeros for each polynomial.

27. $x^3 - 7x - 6$ **28.** $2x^3 + 3x^2 - 29x - 30$

29. $15x^3 - 91x^2 + 4x + 12$

30. $x^4 - 12x^3 + 52x^2 - 96x + 64$

31. $x^3 + x^2 - x - 1$ **32.** $6x^5 + 3x - 2$

In Exercises 33 to 36, use Descartes' Rule of Signs to state the number of possible positive and negative real zeros of each polynomial.

33. $x^3 + 3x^2 + x + 3$

34. $x^4 - 6x^3 - 5x^2 + 74x - 120$

35. $x^4 - x - 1$

36. $x^5 - 4x^4 + 2x^3 - x^2 + x - 8$

In Exercises 37 to 42, find the zeros of the polynomial.

37. $x^3 + 6x^2 + 3x - 10$ **38.** $x^3 - 10x^2 + 31x - 30$

39. $6x^4 + 35x^3 + 72x^2 + 60x + 16$

40. $2x^4 + 7x^3 + 5x^2 + 7x + 3$

41. $x^4 - 4x^3 + 6x^2 - 4x + 1$ **42.** $2x^3 - 7x^2 + 22x + 13$

43. Find a third-degree polynomial with zeros of 4, -3, and $1/2$.

44. Find a fourth-degree polynomial with zeros of 2, -3, i, and $-i$.

45. Find a fourth-degree polynomial with real coefficients that has zeros of 1, 2, and $5i$.

46. Find a fourth-degree polynomial with real coefficients that has -2 as a zero of multiplicity 2 and also has $1 + 3i$ as a zero.

In Exercises 47 to 50, find the vertical, horizontal, and slant asymptotes for each rational function.

47. $f(x) = \dfrac{3x + 5}{x + 2}$ **48.** $f(x) = \dfrac{2x^2 + 12x + 2}{x^2 + 2x - 3}$

49. $f(x) = \dfrac{2x^2 + 5x + 11}{x + 1}$ **50.** $f(x) = \dfrac{6x^2 - 1}{2x^2 + x + 7}$

In Exercises 51 to 58, graph each rational function.

51. $f(x) = \dfrac{3x - 2}{x}$ **52.** $f(x) = \dfrac{x + 4}{x - 2}$

53. $f(x) = \dfrac{6}{x^2 + 2}$ **54.** $f(x) = \dfrac{4x^2}{x^2 + 1}$

55. $f(x) = \dfrac{2x^3 - 4x + 6}{x^2 - 4}$ **56.** $f(x) = \dfrac{x}{x^3 - 1}$

57. $f(x) = \dfrac{3x^2 - 6}{x^2 - 9}$ **58.** $f(x) = \dfrac{-x^3 + 6}{x^2}$

In Exercises 59 to 62, the given polynomials have one zero that satisfies the given conditions. Use the Zero Location Theorem and the bisection method to approximate the zero. Use the midpoint of the second iteration to approximate the zero.

59. $x^3 - x - 1 = 0, \quad x > 0$ **60.** $x^3 - 3x - 6 = 0, \quad x > 0$

61. $x^4 + x^2 - 1 = 0, \quad x > 0$ **62.** $x^4 - 2x^2 - 2 = 0, \quad x > 0$

Chapter Test

1. Use synthetic division to divide:

$$(3x^3 + 5x^2 + 4x - 1) \div (x + 2)$$

2. Use the Remainder Theorem to find $P(-2)$ if

$$P(x) = -3x^3 + 7x^2 + 2x - 5$$

3. Show that $x - 1$ is a factor of

$$x^4 - 4x^3 + 7x^2 - 6x + 2$$

4. Examine the leading term of the function given by the equation $P(x) = -3x^3 + 2x^2 - 5x + 2$ and determine the far-left and far-right behavior of the graph of the polynomial function.

5. Find the real zeros of the polynomial function given by the equation $3x^3 + 7x^2 - 6x = 0$.

6. Use the Zero Location Theorem to verify that

$$P(x) = 2x^3 - 3x^2 - x + 1$$

   has a zero between 1 and 2.

7. Find the zeros of the function given by the equation

$$P(x) = (x^2 - 4)^2(2x - 3)(x + 1)^3$$

   and state the multiplicity of each.

8. Use the Rational Zero Theorem to list the possible rational zeros for the function given by the equation

$$P(x) = 6x^3 - 3x^2 + 2x - 3$$

9. Find the smallest positive integer and the largest negative integer that are upper and lower bounds for the polynomial given by the equation

$$P(x) = 2x^4 + 5x^3 - 23x^2 - 38x + 24$$

10. Use Descartes' Rule of Signs to state the number of possible positive and negative real zeros of

$$P(x) = x^4 - 3x^3 + 2x^2 - 5x + 1$$

11. Find the zeros of the polynomial given by the equation $P(x) = 2x^3 - 3x^2 - 11x + 6$.

12. Given that $-i$ is a zero of $P(x) = 2x^4 - 3x^3 - 3x - 2$, find the remaining zeros.

13. Find all the zeros of the polynomial

$$P(x) = x^5 - 6x^4 + 14x^3 - 14x^2 + 5x$$

14. Find a polynomial of lowest degree that has real coefficients and zeros $1 + i$, 3, and 0.

15. Find all vertical asymptotes of the rational function given by the equation

$$f(x) = \frac{3x^2 - 2x + 1}{x^2 - 5x + 6}$$

16. Find all horizontal asymptotes of the rational function given by the equation

$$f(x) = \frac{3x^2 - 2x + 1}{2x^2 - 1}$$

17. Graph $f(x) = \dfrac{x^2 - 1}{x^2 - 2x - 3}$.

18. Graph $f(x) = \dfrac{2x^2 + 2x + 1}{x + 1}$.

19. Graph $f(x) = \dfrac{x}{x^2 + 1}$.

20. Approximate the zero in the interval $1 < x < 2$ of the polynomial $P(x) = x^3 - 5x + 3$ to within one-tenth of a unit.

5 Exponential and Logarithmic Functions

CASE IN POINT *Perfect Numbers*

A natural number is a **perfect number** if it is the sum of its proper factors. The proper factors of 6, are 1, 2, and 3. Thus 6 is a perfect number because $1 + 2 + 3 = 6$. Euclid knew of the first four perfect numbers: 6, 28, 496, and 8128. He also discovered that the exponential expression $2^{n-1}(2^n - 1)$ generates a perfect number whenever $2^n - 1$ is a prime. For instance,

$$\text{If } n = 2, \quad 2^{2-1}(2^2 - 1) = 2(3) = 6$$

$$\text{If } n = 3, \quad 2^{3-1}(2^3 - 1) = 4(7) = 28$$

$$\text{If } n = 5, \quad 2^{5-1}(2^5 - 1) = 16(31) = 496$$

$$\text{If } n = 7, \quad 2^{7-1}(2^7 - 1) = 64(127) = 8128$$

The fifth, sixth, and seventh perfect numbers were not discovered until 1603. In exponential form, they are $2^{12}(2^{13} - 1)$, $2^{16}(2^{17} - 1)$, and $2^{18}(2^{19} - 1)$.

Ancient Greek philosophers observed that 6 has one digit, 28 has two digits, 496 has three digits, and 8128 has four digits. They also observed that the last digit of the first four perfect numbers alternate 6, 8, 6, 8. They conjectured that each successive perfect number would have one digit more than its predecessor and that the last digit would continue to alternate between 6 and 8. Evaluate $2^{12}(2^{13} - 1)$, $2^{16}(2^{17} - 1)$, and $2^{18}(2^{19} - 1)$ to see if their conjectures were correct up to the seventh perfect number.

The search for perfect numbers still continues. To date, only thirty-two perfect numbers have been discovered. The largest one is $2^{756,838}(2^{756,839} - 1)$, which when expanded contains approximately 456,000 digits.

5.1 Exponential Functions and Their Graphs

The real number b^x is defined for every positive base b and every *rational* number x. For example,

$$2^3 = 8, \qquad 2^{-4} = \frac{1}{2^4} = \frac{1}{16}, \qquad \text{and} \qquad 2^{2/3} = \sqrt[3]{2^2} = \sqrt[3]{4}$$

To define powers of the form b^x, where b is a positive *real* number and x is a real number, we will require a definition that includes powers with irrational exponents, such as $2^{\sqrt{2}}$, 3^π, and $10^{-\sqrt{5}}$.

For our purposes it is convenient to define $2^{\sqrt{2}}$ as the unique real number that we can approximate as closely as desired using an exponent that takes on closer and closer rational approximations of $\sqrt{2}$.

Using a scientific calculator, $2^{\sqrt{2}} = 2.6651441$ (to the nearest ten millionth).

Definition of Exponential Functions

> The **exponential function f with base b** is defined by
>
> $$f(x) = b^x$$
>
> where b is a positive constant other than 1 and x is any real number.

The following graphs are representative of the graphs of exponential functions.

EXAMPLE 1   **Graph Exponential Functions**

Sketch the graph of the following.

a. $f(x) = 2^x$ and b. $f(x) = \left(\frac{1}{2}\right)^x$

Solution

a.

x	$f(x) = 2^x$
-2	$2^{-2} = \dfrac{1}{4}$
-1	$2^{-1} = \dfrac{1}{2}$
0	$2^0 = 1$
1	$2^1 = 2$
2	$2^2 = 4$
3	$2^3 = 8$

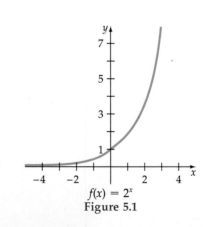

$f(x) = 2^x$
Figure 5.1

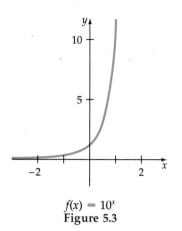

$f(x) = 10^x$
Figure 5.3

b. Note: $\left(\dfrac{1}{2}\right)^x = (2^{-1})^x = 2^{-x}$

x	$f(x) = \left(\dfrac{1}{2}\right)^x = 2^{-x}$
-3	$2^3 = 8$
-2	$2^2 = 4$
-1	$2^1 = 2$
0	$2^0 = 1$
1	$2^{-1} = \dfrac{1}{2}$
2	$2^{-2} = \dfrac{1}{4}$

$f(x) = \left(\tfrac{1}{2}\right)^x$
Figure 5.2

■ *Try Exercise* **26**, *page 257.*

To graph an exponential function over a certain portion of its domain, you may need to use different scales on the x- and y-axes. For example, the graph of $f(x) = 10^x$, for $-2 \le x \le 1$, is shown in Figure 5.3. Observe that each unit on the y-axis represents a distance of five units and that each unit on the x-axis represents one unit.

Properties of $f(x) = b^x$

For positive real numbers b, $b \ne 1$, the exponential function defined by $f(x) = b^x$ has the following properties:

1. f has the set of real numbers as its domain.
2. f has the set of positive real numbers as its range.
3. f has a graph with a y-intercept of $(0,1)$.
4. f has a graph asymptotic to the x-axis.
5. f is a one-to-one function.
6. f is an increasing function if $b > 1$. See Figure 5.4(a) on page 254.
7. f is a decreasing function if $0 < b < 1$. See Figure 5.4(b) on page 254.

Many applications involve functions defined as

$$f(x) = ab^{p(x)}$$

where a is a constant and p is a function of x. Functions of this type can be graphed by plotting points. Be sure to make use of symmetry when possible.

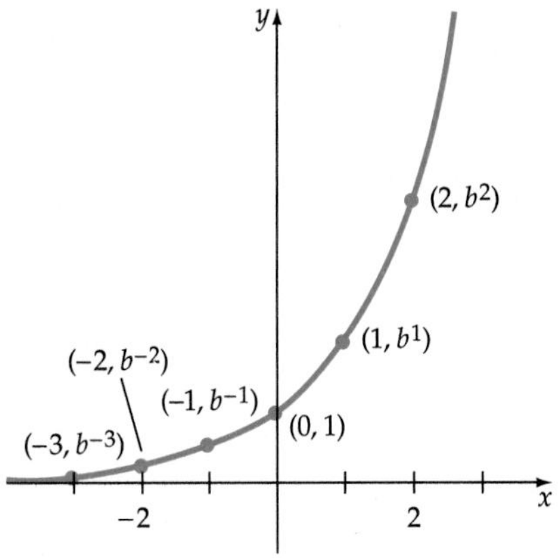

a. $f(x) = b^x, b > 0$

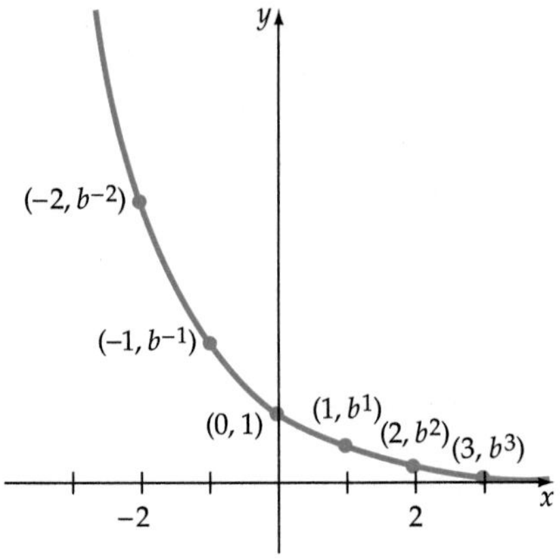

b. $f(x) = b^x, 0 < b < 1$

Figure 5.4

EXAMPLE 2   **Graph a Function of the Form $f(x) = ab^{p(x)}$**

Sketch the graph of $f(x) = 3 \cdot 2^{-x^2}$.

Solution   Since $f(x) = f(-x)$, f is an even function. We can sketch its graph by plotting points to the right of the y-axis and then use symmetry to determine points to the left of the y-axis. See Figure 5.5.

x	$f(x) = 3 \cdot 2^{-x^2}$
0	$3 \cdot 2^{-0^2} = 3$
1	$3 \cdot 2^{-1^2} = \dfrac{3}{2}$
2	$3 \cdot 2^{-2^2} = \dfrac{3}{16}$

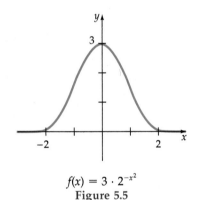

$f(x) = 3 \cdot 2^{-x^2}$
Figure 5.5

■ *Try Exercise* **36**, *page 257.*

Remark   A calculator could be used to compute additional points on the graph of f. However, writing f in the fractional form

$$f(x) = 3 \cdot \frac{1}{2^{x^2}}$$

enables us to determine that as $|x|$ increases without bound, the denominator increases without bound while the numerator remains constant. Therefore, $f(x)$ approaches 0 as $|x|$ increases without bound. This implies that the graph of f has the x-axis as a horizontal asymptote. Also, the maximum value of f is reached when the denominator 2^{x^2} is its smallest. This occurs when $x = 0$. Thus the maximum value of f is 3. The graph of f is a *bell-shaped* curve similar to the graph of the normal distribution curve, which plays a major role in statistics.

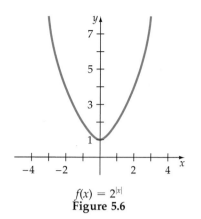

$f(x) = 2^{|x|}$
Figure 5.6

TABLE 5.1

Value of n	Value of $\left(1 + \dfrac{1}{n}\right)^n$
1	2
10	2.59374246
100	2.704813829
1000	2.716923932
10,000	2.718145927
100,000	2.718268237
1,000,000	2.718280469
10,000,000	2.718281693

EXAMPLE 3 **Graph a Function of the Form $f(x) = b^{p(x)}$**

Sketch the graph of $f(x) = 2^{|x|}$.

Solution

x	$f(x) = 2^{|x|}$
0	$2^{|0|} = 1$
1	$2^{|1|} = 2$
2	$2^{|2|} = 4$
3	$2^{|3|} = 8$

Because the function defined by $f(x) = 2^{|x|}$ is an even function, the points to the left of the y-axis can be determined by using symmetry. See Figure 5.6.

■ *Try Exercise* **38**, *page 257.*

The irrational number π is often used in applications that involve circles. Using techniques developed in a calculus course, we can verify that as n increases without bound,

$$\left(1 + \frac{1}{n}\right)^n$$

approaches an irrational number that is denoted by e. The number e often occurs in applications involving growth or decay. It is denoted by e in honor of the mathematician Leonhard Euler (1707–1783). Although e has a nonterminating and nonrepeating decimal representation, Euler was able to compute e to several decimal places by using large values of n to evaluate $(1 + 1/n)^n$. The entries in Table 5.1 illustrate the process. The value of e accurate to eight decimal places is 2.71828183.

The Natural Exponential Function

For all real numbers x, the function defined by

$$f(x) = e^x$$

is called the **natural exponential function.**

To evaluate e^x for specific values of x, you use a calculator with an $\boxed{e^x}$ key. For example,

$$e^2 \approx 7.389056099$$

$$e^{4.21} \approx 67.35653981$$

$$e^{-1.8} \approx 0.165298888$$

To graph the natural exponential function, use a calculator to approximate e^x for the desired domain values. Then plot the resulting points and connect them with a smooth curve.

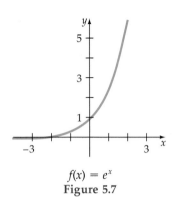

$f(x) = e^x$

Figure 5.7

EXAMPLE 4   **Graph the Natural Exponential Function**

Graph $f(x) = e^x$.

Solution   The values in the table have been rounded to the nearest hundredth. Plot the points and then connect the points with a smooth curve. Since $e > 1$, we know by the properties of exponential functions, that the graph of $f(x) = e^x$ is an increasing function. To the far left the graph is asymptotic to the x-axis. The y-intercept is $(0, 1)$. See Figure 5.7.

x	$f(x) = e^x$
-3	0.05
-2	0.14
-1	0.37
1	2.72
2	7.39

■ *Try Exercise* **44,** *page 257.*

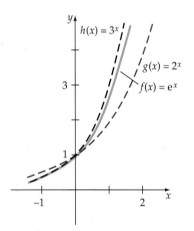

Figure 5.8

Note in Figure 5.8 how the graph of $f(x) = e^x$ compares with the graphs of $g(x) = 2^x$ and $h(x) = 3^x$. You may have anticipated that the graph of f would be between the graph of g and h because e is between 2 and 3.

Functions of the type

$$f(x) = \frac{b^x + b^{-x}}{2}$$

occur in calculus. These functions are difficult to graph by plotting points. Example 5 uses the technique of *averaging the y values* of known graphs to sketch the desired graph.

EXAMPLE 5   **Graph by Averaging y Values**

Graph $f(x) = \dfrac{2^x + 2^{-x}}{2}$.

Solution   Note that f is the average of 2^x and 2^{-x}. Therefore, sketch the graph of f by drawing a curve *halfway between* the graph of $y = 2^x$ and that of $y = 2^{-x}$. See Figure 5.9.

■ *Try Exercise* **46,** *page 257.*

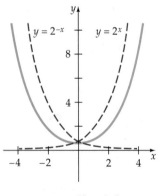

$$f(x) = \frac{2^x + 2^{-x}}{2}$$

Figure 5.9

From the sketch it appears that the graph of f is symmetric with respect to the y-axis, which would imply that f is an even function. This is indeed the case because

$$f(-x) = \frac{2^{-x} + 2^{-(-x)}}{2} = \frac{2^{-x} + 2^x}{2} = f(x)$$

EXERCISE SET 5.1

In Exercises 1 to 12, evaluate each power accurate to six significant digits.

1. $3^{\sqrt{2}}$ **2.** $5^{\sqrt{3}}$ **3.** $10^{\sqrt{7}}$ **4.** $10^{\sqrt{11}}$

5. $\sqrt{3}^{\sqrt{2}}$ **6.** $\sqrt{5}^{\sqrt{7}}$ **7.** $e^{5.1}$ **8.** $e^{-3.2}$

9. $e^{\sqrt{3}}$ **10.** $e^{\sqrt{5}}$ **11.** $e^{-0.031}$ **12.** $e^{-0.42}$

In Exercises 13 to 24, evaluate each functional value, accurate to six significant digits, given that $f(x) = 3^x$ and $g(x) = e^x$.

13. $f(\sqrt{15})$ **14.** $f(\pi)$ **15.** $f(e)$ **16.** $f(-\sqrt{15})$

17. $g(\sqrt{7})$ **18.** $g(\pi)$ **19.** $g(e)$ **20.** $g(-3.4)$

21. $f[g(2)]$ **22.** $f[g(-1)]$ **23.** $g[f(2)]$ **24.** $g[f(-1)]$

In Exercises 25 to 34, sketch the graph of each exponential function.

25. $f(x) = 3^x$ **26.** $f(x) = 4^x$

27. $f(x) = \left(\dfrac{3}{2}\right)^x$ **28.** $f(x) = \left(\dfrac{4}{3}\right)^x$

29. $f(x) = \left(\dfrac{1}{3}\right)^x$ **30.** $f(x) = \left(\dfrac{2}{3}\right)^x$

31. $f(x) = \left(\dfrac{1}{2}\right)^{-x}$ **32.** $f(x) = \left(\dfrac{1}{3}\right)^{-x}$

33. $f(x) = \dfrac{5^x}{2}$ **34.** $f(x) = \dfrac{10^x}{10}$

In Exercises 35 to 48, sketch the graph of each function.

35. $f(x) = \left(\dfrac{1}{3}\right)^{|x|}$ **36.** $f(x) = 3^{-x^2}$

37. $f(x) = 3^{x^2}$ **38.** $f(x) = 2^{-|x|}$

39. $f(x) = 2^{x-3}$ **40.** $f(x) = 2^{x+3}$

41. $f(x) = 3^x - 1$ **42.** $f(x) = 3^x + 1$

43. $f(x) = -e^x$ **44.** $f(x) = \dfrac{1}{2}e^x$

45. $f(x) = \dfrac{3^x + 3^{-x}}{2}$ **46.** $f(x) = \dfrac{e^x + e^{-x}}{2}$

47. $f(x) = -(3^x)$ **48.** $f(x) = -(0.5^x)$

In Exercises 49 to 52, graph each pair of functions on the same set of coordinate axes.

49. $f(x) = 2^x$, $g(x) = 2^{-x}$

50. $f(x) = \left(\dfrac{2}{3}\right)^x$, $g(x) = \left(\dfrac{2}{3}\right)^{-x}$

51. $f(x) = 2^{x+1}$, $g(x) = 2^x + 1$

52. $f(x) = 3^x$, $g(x) = \left(\dfrac{1}{3}\right)^{-x}$

53. Complete the following table:

Value of n	Value of $\left(1 + \dfrac{1}{n}\right)^n$
5	
50	
500	
5000	
50,000	
500,000	
5,000,000	

54. Complete the following table.

  a. How do the entries in this table compare with those in Exercise 53?

  b. As n approaches 0, what value does $(1 + n)^{1/n}$ appear to approach?

Value of n	Value of $\left(1 + n\right)^{1/n}$
0.2	
0.02	
0.002	
0.0002	
0.00002	
0.000002	
0.0000002	

Graphing Calculator Exercises*

55. Graph $f = x^x$ on $(0, 3]$. Estimate

  a. the minimum value of f on this interval and

  b. the behavior of f as x approaches 0 from the right

56. Graph the **normal distribution curve** defined by

$$f(x) = \frac{1}{\sqrt{2\pi}}e^{-x^2/2}$$

Estimate the maximum value of f (nearest tenth).

* Additional graphing calculator exercises appear in the Graphing Workbook as described in the front of this textbook.

In Exercises 57 to 64, graph f. State the domain and range of f using interval notation. When necessary, estimate values to the nearest tenth. Also state whether f is an even function, an odd function, or neither an even nor an odd function.

57. $f(x) = \dfrac{e^x - e^{-x}}{e^x + e^{-x}}$

58. $f(x) = x^2 e^x$

59. $f(x) = \dfrac{4x^2}{e^{|x|}}$

60. $f(x) = \sqrt{\dfrac{|x|}{1 + e^x}}$

61. $f(x) = \dfrac{e^{|x|}}{1 + e^x}$

62. $f(x) = 1 + e^{(x^3 - x^2 - 2x)}$

63. $f(x) = \sqrt{1 - e^x}$

64. $f(x) = \sqrt{e^x - e^{-x}}$

In Exercises 65 and 66, use $f(x) = e^x$ and $g(x) = e^{-x}$ to graph the given equations. State the domain and range of y using interval notation.

65. a. $y = (f + g)(x)$ **b.** $y = (f - g)(x)$

66. a. $y = (f \cdot g)(x)$ **b.** $y = (f/g)(x)$

In Exercises 67 and 68, graph the equation given $f(x) = e^x$ and f^{-1} is the inverse of f. State the domain and range of y using interval notation.

67. $y = f^{-1}(x)$

68. $y = f(f^{-1}(x))$

Supplemental Exercises

69. Evaluate $h(x) = (-2)^x$ for **a.** $x = 1$, **b.** $x = 2$, **c.** $x = 1.5$. Explain why h is not an exponential function.

70. Graph $j(x) = 1^x$. Explain why j is not an exponential function.

71. Graph $f(x) = e^x$, and then sketch the graph of f reflected about the graph of the line given by $y = x$.

72. Graph $g(x) = 10^x$, and then sketch the graph of g reflected about the graph of the line given by $y = x$.

73. Prove that the hyperbolic sine function

$$F(x) = \frac{e^x - e^{-x}}{2}$$

is an odd function.

74. Prove that $G(x) = e^x$ is neither an odd function nor an even function.

75. The number of bacteria present in a culture is given by $N(t) = 10{,}000(2^t)$, where $N(t)$ is the number of bacteria present after t hours. Find the number of bacteria present when **a.** $t = 1$ hour, **b.** $t = 2$ hours, **c.** $t = 5$ hours.

76. The production function for an oil well is given by the function $B(t) = 100{,}000(e^{-0.2t})$, where $B(t)$ is the number of barrels of oil the well can produce per month after t years. Find the number of barrels of oil the well can produce per month when **a.** $t = 1$ year, **b.** $t = 2$ years, **c.** $t = 5$ years.

77. Which of the powers e^{π} and π^e is larger?

78. Let $f(x) = x^{(x^x)}$ and $g(x) = (x^x)^x$. Which is larger, $f(3)$ or $g(3)$?

79. Graph $f(x) = \dfrac{2^x - 2^{-x}}{2}$. (*Hint:* $\dfrac{2^x - 2^{-x}}{2}$ is the average of 2^x and -2^{-x}).

80. Graph $f(x) = \dfrac{e^x - e^{-x}}{2}$.

5.2   Logarithms and Logarithmic Properties

Every exponential function is a one-to-one function and therefore has an inverse function. Sometimes we can determine the inverse of a function represented by an equation by interchanging the variables and then solving for the dependent variable. If we attempt to use this procedure for $f(x) = b^x$, we get

$$f(x) = b^x$$

$$y = b^x$$

$$x = b^y \quad \bullet \text{ Interchange the variables.}$$

None of our previous methods can be used to solve the equation $x = b^y$ for the exponent y. Thus we must develop a new procedure. One method would be merely to write

$$y = \text{the power of } b \text{ that produces } x$$

This procedure would work, but it is not concise. We need compact notation to represent y as the exponent of b that produces x. For historical reasons, we use the notation in the following definition.

Definition of a Logarithm

If $x > 0$ and b is a positive constant ($b \neq 1$), then

$$y = \log_b x \quad \text{if and only if} \quad b^y = x$$

In the equation $y = \log_b x$, y is referred to as the **logarithm**, b is the **base**, and x is the **argument.**

The notation $\log_b x$ is read "the logarithm (or log) base b of x." The definition of a logarithm indicates that *a logarithm is an exponent.*

The equations

$$y = \log_b x \qquad \text{and} \qquad b^y = x$$

are different ways of expressing the same thing.

$$y = \log_b x \text{ is the logarithmic form of } b^y = x$$
$$b^y = x \text{ is the exponential form of } y = \log_b x.$$

EXAMPLE 1   **Change from Logarithmic to Exponential Form**

Write each of the following equations in its exponential form:

a. $2 = \log_7 x$ b. $3 = \log_{10}(x + 8)$ c. $\log_5 125 = x$

Solution   Use the definition $y = \log_b x$ if and only if $b^y = x$.

a.

Logarithms are exponents

$$2 = \log_7 x \quad \text{if and only if} \quad 7^2 = x$$

Base

b. $3 = \log_{10}(x + 8)$ if and only if $10^3 = (x + 8)$.
c. $\log_5 125 = x$ if and only if $5^x = 125$.

■ *Try Exercise 2, page 266.*

EXAMPLE 2   **Change from Exponential to Logarithmic Form**

Write each of the following equations in its logarithmic form.

a. $x = 25^{1/2}$ b. $\dfrac{1}{16} = x^{-4}$ c. $27^x = 3$

Solution   Use $x = b^y$ if and only if $y = \log_b x$.

a.

$$\overbrace{\phantom{x = 25^{1/2} \text{ if and only if } \tfrac{1}{2}}}^{\text{Exponent}}$$

$$x = 25^{1/2} \quad \text{if and only if} \quad \tfrac{1}{2} = \log_{25} x$$

$$\underbrace{\phantom{x = 25^{1/2} \text{ if and only if } \tfrac{1}{2} = \log_{25} x}}_{\text{Base}}$$

b. $\dfrac{1}{16} = x^{-4}$ if and only if $-4 = \log_x \dfrac{1}{16}$.

c. $27^x = 3$ if and only if $\log_{27} 3 = x$.

_____ ■ *Try Exercise* **12,** *page 266.*

Some logarithms can be evaluated by using the definition of a logarithm and the following theorem.

Equality of Exponents Theorem

> If b is a positive real number ($b \neq 1$) such that $b^x = b^y$, then $x = y$.

EXAMPLE 3   **Evaluate Logarithms**

Evaluate each logarithm.

a. $\log_2 32 = x$ b. $\log_5 125 = x$ c. $\log_{10} \dfrac{1}{100} = x$

Solution

a. $\log_2 32 = x$ if and only if $2^x = 32$ • Change to exponential form.

$$2^x = 2^5 \quad \text{• Factor.}$$

$$x = 5 \quad \text{• Equality of Exponents Theorem}$$

b. $\log_5 125 = x$ if and only if $5^x = 125$

$$5^x = 5^3$$

$$x = 3$$

c. $\log_{10} \dfrac{1}{100} = x$ if and only if $10^x = \dfrac{1}{100}$

$$10^x = 10^{-2}$$

$$x = -2$$

_____ ■ *Try Exercise* **22,** *page 266.*

Since logarithms are exponents, they have many properties that can be established by using the properties of exponents.

Properties of Logarithms

In the following properties, b, M, and N are positive real numbers ($b \neq 1$), and p is any real number.

$$\log_b b = 1$$

$$\log_b 1 = 0$$

$$\log_b(b^p) = p \qquad \bullet \text{ An inverse property}$$

$$\log_b(MN) = \log_b M + \log_b N \qquad \bullet \text{ Product property}$$

$$\log_b\left(\frac{M}{N}\right) = \log_b M - \log_b N \qquad \bullet \text{ Quotient property}$$

$$\log_b(M^p) = p \log_b M \qquad \bullet \text{ Power property}$$

$$\log_b M = \log_b N \quad \text{implies } M = N \qquad \bullet \text{ One-to-one property}$$

$$M = N \quad \text{implies } \log_b M = \log_b N \qquad \bullet \text{ Logarithm of each side property}$$

$$b^{\log_b p} = p \quad (\text{for } p > 0) \qquad \bullet \text{ An inverse property}$$

The first three properties of logarithms can be proved by using the definition of a logarithm. That is,

$$\log_b b = 1 \qquad \text{because} \qquad b^1 = b$$

$$\log_b 1 = 0 \qquad \text{because} \qquad b^0 = 1$$

$$\log_b(b^p) = p \qquad \text{because} \qquad b^p = b^p$$

To prove the product property $\log_b(MN) = \log_b M + \log_b N$, let

$$\log_b M = x \qquad \text{and} \qquad \log_b N = y$$

Writing each equation in its equivalent exponential form produces

$$M = b^x \qquad \text{and} \qquad N = b^y$$

Forming the product of the respective right and left sides produces

$$MN = b^x b^y \qquad \text{or} \qquad MN = b^{x+y}$$

Applying the definition of a logarithm yields the equivalent form

$$\log_b(MN) = x + y$$

Since $x = \log_b M$ and $y = \log_b N$, this becomes

$$\log_b(MN) = \log_b M + \log_b N$$

The proofs of the quotient property and the power property are similar to the proof of the product property. They are left as exercises.

The logarithm of each side property will be used several times in this chapter. It states that if two positive real numbers are equal, their logarithms base b are also equal.

Proof: Let $M > 0$, $N > 0$ and $M = N$.

$$\log_b M = \log_b M \quad \bullet \text{ The reflexive property}$$

$$\log_b M = \log_b N \quad \bullet \text{ Substitute } N \text{ for } M.$$

The proofs of the one-to-one property and the inverse property are left as exercises.

The properties of logarithms are often used to rewrite logarithms and expressions that involve logarithms.

EXAMPLE 4   **Rewrite Logarithmic Expressions**

Use the properties of logarithms to express the following logarithms in terms of logarithms of x, y, and z.

a. $\log_b xy^2$ b. $\log_b \dfrac{x^2 \sqrt{y}}{z^5}$

Solution

a. $\log_b xy^2 = \log_b x + \log_b y^2 \quad \bullet \text{ Product property}$

$\qquad\qquad = \log_b x + 2 \log_b y \quad \bullet \text{ Power property}$

b. $\log_b \dfrac{x^2 \sqrt{y}}{z^5} = \log_b x^2 \sqrt{y} - \log_b z^5 \qquad\qquad \bullet \text{ Quotient property}$

$\qquad\qquad = \log_b x^2 + \log_b \sqrt{y} - \log_b z^5 \quad \bullet \text{ Product property}$

$\qquad\qquad = 2 \log_b x + \dfrac{1}{2} \log_b y - 5 \log_b z \quad \bullet \text{ Power property}$

―――――――――――――――――――――――――― ■ *Try Exercise **32**, page 266.*

Sometimes it is possible to use known logarithmic values and the properties of logarithms to evaluate logarithms.

EXAMPLE 5   **Evaluate Logarithms**

Given $\log_8 2 \approx 0.3333$, $\log_8 3 \approx 0.5283$, and $\log_8 5 \approx 0.7740$, evaluate the following:

a. $\log_8 15$ b. $\log_8 \dfrac{5}{2}$ c. $\log_8 \sqrt[3]{9}$

Solution

a. $\log_8 15 = \log_8(3 \cdot 5) = \log_8 3 + \log_8 5 \quad \bullet \text{ Product property}$

$\qquad\qquad \approx 0.5283 + 0.7740 = 1.3023$

b. $\log_8 \dfrac{5}{2} = \log_8 5 - \log_8 2$ • Quotient property

   $\approx 0.7740 - 0.3333 = 0.4407$

c. $\log_8 \sqrt[3]{9} = \log_8 3^{2/3} = \dfrac{2}{3} \log_8 3$ • Power property

   $\approx \dfrac{2}{3}(0.5283) = 0.3522$

■ *Try Exercise* **42,** *page 266.*

The properties of logarithms are also used to rewrite expressions that involve logarithms as a single logarithm.

EXAMPLE 6   **Rewrite Logarithmic Expressions**

Use the properties of logarithms to rewrite the following expressions as a single logarithm.

a. $2 \log_b x + \dfrac{1}{2} \log_b(x + 4)$ b. $4 \log_b(x + 2) - 3 \log_b(x - 5)$

Solution

a. $2 \log_b x + \dfrac{1}{2} \log_b(x + 4) = \log_b x^2 + \log_b(x + 4)^{1/2}$ • Power property

   $= \log_b x^2(x + 4)^{1/2}$ • Product property

b. $4 \log_b(x + 2) - 3 \log_b(x - 5) = \log_b(x + 2)^4 - \log_b(x - 5)^3$ • Power property

   $= \log_b \dfrac{(x + 2)^4}{(x - 5)^3}$ • Quotient property

■ *Try Exercise* **52,** *page 266.*

Definition of Common Logarithm

Logarithms with a base of 10 are called **common logarithms.** It is customary to write $\log_{10} x$ as $\log x$.

Definition of Natural Logarithm

Logarithms with a base of e are called **natural logarithms.** They are often used in calculus. It is customary to write $\log_e x$ as $\ln x$.

Most scientific calculators have a key marked $\boxed{\log}$ for evaluating common logarithms and a key marked $\boxed{\ln}$ for evaluating natural logarithms. For example,

$$\log 24 \approx 1.380211242$$
$$\ln 81 \approx 4.394449155$$
$$\log 0.58 \approx -0.236572006$$

If you use a scientific calculator to try to evaluate the logarithm of a negative number, it may give you an error indication. Recall that the definition of $y = \log_b x$ required x to be a positive real number.

Appendix II contains a table of common logarithms and a table of natural logarithms. Use the tables to evaluate common and natural logarithms if a scientific calculator is not available. Appendix I explains the use of the common logarithmic table. All the properties of logarithms apply to both common and natural logarithms.

Logarithms that are not common logarithms or natural logarithms can be evaluated by using the following theorem.

Change-of-Base Formula

If x, a, and b are positive real numbers with $a \neq 1$ and $b \neq 1$, then

$$\log_b x = \frac{\log_a x}{\log_a b}$$

Proof: Let $\log_b x = y$

Then $b^y = x$ • By definition of $\log_b x$

Now take the logarithm with base a of each side.

$$\log_a(b^y) = \log_a x$$ • Logarithm of each side property

$$y \log_a b = \log_a x$$ • Power property

$$y = \frac{\log_a x}{\log_a b}$$ • Solve for y by dividing by $\log_a b$.

$$\log_b x = \frac{\log_a x}{\log_a b}$$ • Substitute $\log_b x$ for y.

EXAMPLE 7   **Use the Change-of-Base Formula**

Evaluate each of the following logarithms.

a. $\log_3 18$ b. $\log_{12} 400$

Solution   In each case we use the change-of-base formula with $a = 10$.

That is, we will evaluate these logarithms by using the $\boxed{\texttt{log}}$ key on a scientific calculator.

a. $\log_3 18 = \dfrac{\log 18}{\log 3} \approx 2.63093$ b. $\log_{12} 400 = \dfrac{\log 400}{\log 12} \approx 2.41114$

_____ ■ *Try Exercise* **62,** *page 266.*

Remark We could also have evaluated the logarithms by using the $\boxed{\texttt{ln}}$ key. For example, for part a,

$$\log_3 18 = \frac{\ln 18}{\ln 3} \approx 2.63093$$

The change-of-base formula can be employed to evaluate common logarithms by using natural logarithms and to evaluate natural logarithms by using common logarithms. For example, if we substitute e for a and 10 for b in the change-of-base formula, we get

$$\log x = \frac{\ln x}{\ln 10} \approx 0.4343 \ln x$$

Substituting e for b and 10 for a in the change-of-base formula yields

$$\ln x = \frac{\log x}{\log e} \approx \frac{\log x}{0.4343}$$

Antilogarithms

Given $M = \log N$, it is often necessary to determine the value of N. In this case the number N is called the **antilogarithm of** M.

Definition of Antilogarithms

> If M and N are real numbers with $N > 0$, such that
> $$\log_b N = M$$
> then N is the **antilogarithm of** M for the base b.

Rewriting $\log_b N = M$ as $N = b^M$, we have a formula to evaluate N, the antilogarithm of M.

$$N = b^M$$

For instance,

$$\text{if } \log_4 N = 1.2251 \quad \text{then} \quad N = 4^{1.2251} \approx 5.4649$$

$$\text{if } \log_7 N = -1.3041 \quad \text{then} \quad N = 7^{-1.3041} \approx 0.0791$$

For the special cases of common logarithms and natural logarithms, we have

$$\text{if } \log N = 2.3571 \quad \text{then} \quad N = 10^{2.3571} \approx 227.5621$$

$$\text{if } \ln N = 1.0892 \quad \text{then} \quad N = e^{1.0892} \approx 2.9719$$

Antilogarithms can also be found by using the tables in Appendix II. Appendix I explains the process of using these tables to find logarithms and antilogarithms.

EXERCISE SET 5.2

In Exercises 1 to 10, change each equation to its exponential form.

1. $\log_{10} 100 = 2$

2. $\log_{10} 1000 = 3$

3. $\log_5 125 = 3$

4. $\log_5 \dfrac{1}{25} = -2$

5. $\log_3 81 = 4$

6. $\log_3 1 = 0$

7. $\log_b r = t$

8. $\log_b(s + t) = r$

9. $-3 = \log_3 \dfrac{1}{27}$

10. $-1 = \log_7 \dfrac{1}{7}$

In Exercises 11 to 20, change each equation to its logarithmic form.

11. $2^4 = 16$

12. $3^5 = 243$

13. $7^3 = 343$

14. $7^{-4} = \dfrac{1}{2401}$

15. $10{,}000 = 10^4$

16. $\dfrac{1}{1000} = 10^{-3}$

17. $b^k = j$

18. $p = m^n$

19. $b^1 = b$

20. $b^0 = 1$

In Exercises 21 to 30, evaluate each logarithm. Do not use a calculator.

21. $\log_{10} 1{,}000{,}000$

22. $\log_{10} \dfrac{1}{1000}$

23. $\log_2 32$

24. $\log_3 243$

25. $\log_{3/2} \dfrac{27}{8}$

26. $\log_{0.5} 16$

27. $\log_5 \dfrac{1}{25}$

28. $\log_{0.3} \dfrac{100}{9}$

29. $\log_b 1$

30. $\log_b b$

In Exercises 31 to 40, write the given logarithm in terms of logarithms of x, y, and z.

31. $\log_b xyz$

32. $\log_b x^2 y^3$

33. $\log_3 \dfrac{x}{z^4}$

34. $\log_5 \dfrac{x^2}{yz^3}$

35. $\log_b \dfrac{\sqrt{x}}{y^3}$

36. $\log_b \dfrac{\sqrt{x}}{\sqrt[3]{z}}$

37. $\log_b x \sqrt[3]{\dfrac{y^2}{z}}$

38. $\log_b \sqrt[3]{x^2 z \sqrt{y}}$

39. $\log_7 \dfrac{\sqrt{x + z^2}}{x^2 - y}$

40. $\log_5 \left(\dfrac{x^2 - y}{z^2} \right)$

In Exercises 41 to 50, evaluate the logarithm using the values $\log_7 2 \approx 0.3562$, $\log_7 3 \approx 0.5646$, and $\log_7 5 \approx 0.8271$ and the properties of logarithms. Do not use a calculator.

41. $\log_7 6$

42. $\log_7 20$

43. $\log_7 9$

44. $\log_7 4$

45. $\log_7 \dfrac{2}{5}$

46. $\log_7 \dfrac{3}{2}$

47. $\log_7 30$

48. $\log_7 45$

49. $\log_7 14$

50. $\log_7 \dfrac{7^2}{3}$

In Exercises 51 to 60, write each logarithmic expression as a single logarithm.

51. $\log_{10}(x + 5) + 2 \log_{10} x$

52. $5 \log_3 x - 4 \log_3 y + 2 \log_3 z$

53. $\dfrac{1}{2}[3 \log_b(x - y) + \log_b(x + y) - \log_b z]$

54. $\log_b(y^3 z^2) - 3 \log_b(x\sqrt{y}) + 2 \log_b \left(\dfrac{x}{z} \right)$

55. $\log_8(x^2 - y^2) - \log_8(x - y)$

56. $\log_4(x^3 - y^3) - \log_4(x - y)$

57. $4 \ln(x - 3) + 2 \ln x$

58. $3 \ln z - 2 \ln(z + 1)$

59. $\ln x - \ln y + \ln z$

60. $\dfrac{1}{2} \log x + 2 \log y$

In Exercises 61 to 70, use the change-of-base formula to approximate the logarithm accurate to five significant digits.

61. $\log_7 20$

62. $\log_5 37$

63. $\log_{11} 8$

64. $\log_{50} 22$

65. $\log_6 0.045$

66. $\log_4 \sqrt{7}$

67. $\log_{0.5} 5$

68. $\log_{0.2} 17$

69. $\log_\pi e$

70. $\log_\pi \sqrt{15}$

In Exercises 71 to 82, approximate the antilogarithm N accurate to three significant digits.

71. $\log N = 0.4857$

72. $\log N = 0.9557$

73. $\log N = 3.5038$

74. $\log N = 7.8476$

75. $\log N = -2.4760$

76. $\log N = -4.3536$

77. $\ln N = 2.001$

78. $\ln N = 2.262$

79. $\ln N = 0.693$

80. $\ln N = 0.531$

81. $\ln N = -1.204$

82. $\ln N = -0.511$

Supplemental Exercises

In Exercises 83 to 88, find all the real numbers that are solutions of the given inequality. Use interval notation to write your answers.

83. $0 \le \log x \le 1000$

84. $-3 \le \log x \le -2$

85. $e \le \ln x \le e^3$

86. $-2 \le \ln x \le 3$

87. $-\log x > 0$

88. $100 - 10 \log(x + 1) > 0$

89. Verify the quotient property of logarithms. (*Hint:* Use a method similar to the proof of the product property.)

90. Verify the power property of logarithms.

91. Give the reason for each step in the proof of the inverse property of logarithms.

$$\log_b x = \log_b x \qquad \underline{\quad ? \quad}$$

$$b^{\log_b x} = x \qquad \underline{\quad ? \quad}$$

92. Give the reason for each step in the proof of the one-to-one property of logarithms.

$$\log_b M = \log_b N \qquad \text{Given}$$

$$b^{\log_b N} = M \qquad \underline{\quad ? \quad}$$

$$N = M \qquad \underline{\quad ? \quad}$$

5.3 Logarithmic Functions and Their Graphs

Section 5.2 developed the concept of a logarithm and the properties of logarithms. With this background, we can now introduce the concept of a *logarithmic function*.

Definition of a Logarithmic Function

The **logarithmic function** f with base b is defined by

$$f(x) = \log_b x$$

where b is a positive constant $b \ne 1$, and x is any *positive* real number.

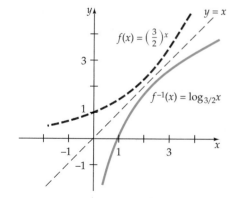

Figure 5.10

Recall that logarithms were defined so that we could write the inverse of $g(x) = b^x$ in a convenient manner. Thus the logarithmic function given by $f(x) = \log_b x$ is the inverse of the exponential function $g(x) = b^x$. The graph of $y = \log_b x$ can be obtained by reflecting the graph of $y = b^x$ across the graph of the line $y = x$. This is illustrated in Figure 5.10 for the exponential function $f(x) = \left(\frac{3}{2}\right)^x$ and its inverse $f^{-1}(x) = \log_{3/2} x$.

If you rewrite a logarithmic function in its equivalent exponential form, then the logarithmic function can be graphed by plotting points.

EXAMPLE 1 **Graph a Logarithmic Function**

Graph $f(x) = \log_2 x$.

Solution Changing $y = \log_2 x$ to its exponential form $2^y = x$ enables us to evaluate x for convenient integer values of y.

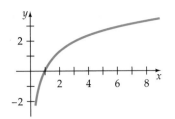

$f(x) = \log_2 x$
Figure 5.11

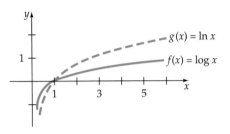

Figure 5.12

y	-2	-1	0	1	2	3
$x = 2^y$	$2^{-2} = \frac{1}{4}$	$2^{-1} = \frac{1}{2}$	$2^0 = 1$	$2^1 = 2$	$2^2 = 4$	$2^3 = 8$

The above table was constructed by first picking a y value and then computing the corresponding x value. Drawing a smooth curve through the resulting points produces the graph of $f(x) = \log_2 x$ shown in Figure 5.11.

■ *Try Exercise* **2,** *page 272.*

The logarithmic functions defined by $f(x) = \log x$ and $g(x) = \ln x$ are often used in applications and advanced mathematics. Their graphs can be drawn by using the techniques developed in Example 1. However, most scientific calculators have the ⌑log⌑ and the ⌑ln⌑ keys that can be used to determine points on the graphs. For example, the following table was made by entering values of x and then evaluating $\log x$ and $\ln x$ by pressing the ⌑log⌑ and the ⌑ln⌑ keys. Figure 5.12 shows the graph of $f(x) = \log x$ and the graph of $g(x) = \ln x$.

x	0.5	1	5	10
$f(x) = \log x$	-0.3010	0	0.6990	1
$g(x) = \ln x$	-0.6931	0	1.6094	2.3026

Properties of $f(x) = \log_b x$

> For all positive real numbers $b \neq 1$, the function defined by $f(x) = \log_b x$ has the following properties:
>
> 1. f has the set of positive real numbers as its domain.
> 2. f has the set of real numbers as its range.
> 3. f has a graph with an x-intercept of $(1, 0)$.
> 4. f has a graph asymptotic to the y-axis.
> 5. f is a one-to-one function.
> 6. f is an increasing function if $b > 1$. See Figure 5.13a.
> 7. f is a decreasing function if $0 < b < 1$. See Figure 5.13b.

a.
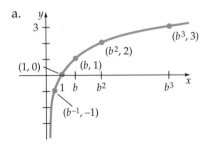

b.
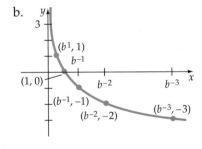

Figure 5.13

Many applied problems involve functions that are defined in terms of

$$f(x) = \log_b c(x)$$

where c is a function of x. Example 2 illustrates that the graph of such a function may differ considerably from the graph of $f(x) = \log_b x$.

EXAMPLE 2   Graph a Function That Is Logarithmic in Form

Graph $f(x) = \log_2|x|$.

Solution   This graph can be sketched by using the method of plotting points, but the procedure developed in the following discussion enables us to sketch the graph with little or no computation.

The domain of $f(x) = \log_2 x$ is the set of positive real numbers. Since $|x| > 0$ for all $x \neq 0$, the domain of $f(x) = \log_2|x|$ is the set of all nonzero real numbers. If $x > 0$, then $|x| = x$, and the graph of $f(x) = \log_2|x|$ will be the same as the graph of $f(x) = \log_2 x$ shown in Figure 5.11.

For $x < 0$, we use the fact that $f(x) = \log_2|x|$ is an even function. Its graph is symmetrical to the y-axis; thus we reflect the right-hand part of the graph across the y-axis to produce the graph in Figure 5.14.

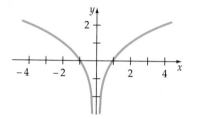

$f(x) = \log_2|x|$
Figure 5.14

■ *Try Exercise* **10**, *page 272.*

EXAMPLE 3   Graph a Function That Is Logarithmic in Form

Graph $f(x) = \log_4(-2x)$.

Solution   Writing $y = \log_4(-2x)$ in its exponential form produces $4^y = -2x$ or $(-\frac{1}{2})4^y = x$. Choosing convenient values for y yields the following table:

y	-2	-1	0	1	2
$x = \left(-\dfrac{1}{2}\right)4^y$	$\left(-\dfrac{1}{2}\right)4^{-2} = -\dfrac{1}{32}$	$\left(-\dfrac{1}{2}\right)4^{-1} = -\dfrac{1}{8}$	$\left(-\dfrac{1}{2}\right)4^0 = -\dfrac{1}{2}$	$\left(-\dfrac{1}{2}\right)4^1 = -2$	$\left(-\dfrac{1}{2}\right)4^2 = -8$

The domain of $f(x) = \log_4(-2x)$ is the set of all negative real numbers because $-2x > 0$ if and only if $x < 0$. See Figure 5.15.

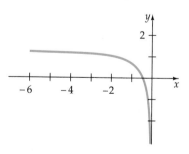

$f(x) = \log_4(-2x)$
Figure 5.15

■ *Try Exercise* **12**, *page 272.*

EXAMPLE 4   **Graph a Function That Is Logarithmic in Form**

Graph $f(x) = \log \sqrt{x}$.

Solution   Since $f(x) = \log \sqrt{x}$ is equivalent to $f(x) = \frac{1}{2} \log x$, the graph of $f(x) = \log \sqrt{x}$ can be obtained by *shrinking* the graph of $y = \log x$. That is, we determine points (a, b) on the graph of $y = \log x$. Then we plot the points $(a, \frac{1}{2}b)$ to sketch the graph of $f(x) = \log \sqrt{x}$.

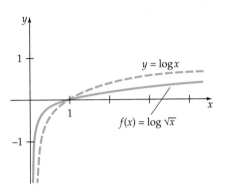

Figure 5.16

Figure 5.16 shows the graph of $f(x) = \log \sqrt{x}$. Note that the point $(a, \frac{1}{2}b)$ is on the graph of $f(x) = \log \sqrt{x}$ if and only if the point (a, b) is on the graph of $f(x) = \log x$.

■ *Try Exercise* **14,** *page 272.*

Horizontal and/or vertical translations of the graph of the logarithmic function $f(x) = \log_b x$ sometimes can be used to obtain the graph of functions that involve logarithms.

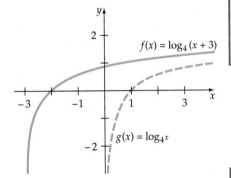

Figure 5.17

EXAMPLE 5   **Use Translations to Graph**

Graph   a. $f(x) = \log_4(x + 3)$ b. $f(x) = \log_4 x + 3$.

Solution

a. The graph of $f(x) = \log_4(x + 3)$ can be obtained by shifting the graph of $g(x) = \log_4 x$ three units to the left. Figure 5.17 shows the graph of $g(x) = \log_4 x$ and the graph of $f(x) = \log_4(x + 3)$.

b. The graph of $f(x) = \log_4 x + 3$ can be obtained by shifting the graph of $g(x) = \log_4 x$ three units upward. Figure 5.18 shows the graph of $g(x) = \log_4 x$ and the graph of $f(x) = \log_4 x + 3$.

■ *Try Exercise* **20,** *page 272.*

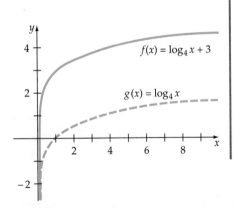

Figure 5.18

Logarithm functions and their graphs are useful in classifying data by associating very large differences or very small differences with small positive numbers. An example of mapping very large numbers to small positive numbers is the *Richter scale* for classifying the magnitude of an earthquake.

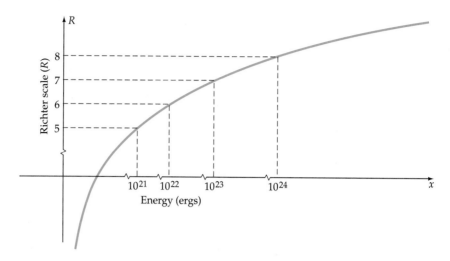

Figure 5.19

The amount of energy released in a moderate earthquake is on the order of 10^{20} ergs. The intervals on the *x*-axis in Figure 5.19 have been reduced, and the distance between consecutive numbers on the vertical axis have been expanded to illustrate the mapping of the measure of the energy of an earthquake to the Richter number that represents the magnitude of the earthquake.

Remark Observe that the scale on which the *x*-axis is calibrated is not linear. Each value on the *x*-axis is ten times the previous value. These amounts of energy are then mapped to consecutive integers on the vertical axis. Thus an increase of one unit on the Richter scale corresponds to a tenfold increase in the energy in an earthquake.

On the Richter scale, seismologists measure the magnitude R of an earthquake by the formula

$$R = \log \frac{I}{I_0}$$

where I_0 is the measure of a *zero-level earthquake,* and I is the intensity of the earthquake being measured.

EXAMPLE 6 **Find the Richter Scale Measure of an Earthquake**

Find the Richter scale measure of

a. An earthquake that had an intensity of $I = 10{,}000 I_0$

b. The San Francisco earthquake of 1906, which had an intensity of $I = 199{,}526{,}000 I_0$

Solution

a. $R = \log \dfrac{10{,}000 I_0}{I_0} = \log 10{,}000 = \log 10^4 = 4$ (on the Richter scale)

b. $R = \log \dfrac{199{,}526{,}000 I_0}{I_0} = \log 199{,}526{,}000 \approx 8.3$ (on the Richter scale)

■ *Try Exercise **42**, page 273.*

EXAMPLE 7   **Compare the Richter Scale Measures of Earthquakes**

If an earthquake has an intensity 100 times the intensity of a second earthquake, then how much larger is the Richter scale measure of the larger earthquake than that of the smaller?

Solution   Let I represent the intensity of the smaller earthquake and let $100I$ represent the intensity of the larger earthquake. The Richter scale measures of the small earthquake R_1 and the larger earthquake R_2 are given by

$$R_1 = \log \frac{I}{I_0} \quad \text{and} \quad R_2 = \log \frac{100I}{I_0}.$$

Using the properties of logarithms, we can write R_2 as

$$R_2 = \log 100\frac{I}{I_0} = \log 100 + \log \frac{I}{I_0} = \log 10^2 + R_1 = 2 + R_1$$

Thus an earthquake that is 100 times as intense as a smaller earthquake will have a Richter scale measure that is 2 more than that of the smaller earthquake.

■ *Try Exercise **44**, page 273.*

EXERCISE SET 5.3

In Exercises 1 to 28, graph the following equations. If possible, make use of translations.

1. $f(x) = \log_3 x$
2. $f(x) = \log_5 x$
3. $f(x) = \log_{1/2} x$
4. $f(x) = \log_{1/4} x$
5. $f(x) = -2 \ln x$
6. $f(x) = -\log x$
7. $f(x) = \log_4 x^2$
8. $f(x) = 2 \log_5 x$
9. $f(x) = |\ln x|$
10. $f(x) = \ln|x|$
11. $f(x) = -|\ln x|$
12. $f(x) = -|\log_2 x^3|$
13. $f(x) = \log \sqrt[3]{x}$
14. $f(x) = \ln \sqrt{x}$
15. $f(x) = 3 + \log_2 x$
16. $f(x) = -2 + \log_4 x$
17. $f(x) = \log(x + 10)$
18. $f(x) = \ln(x + 3)$
19. $f(x) = \ln(x - 5)$
20. $f(x) = \log(x - 2)$
21. $f(x) = \log_5(x + 2)^2$
22. $f(x) = \log_5(x - 1)^2$
23. $f(x) = -\ln(x - 4)$
24. $f(x) = \ln(x + 3)^{-1}$

25. $f(x) = 4 - 2 \ln x$
26. $f(x) = x + \ln x$
27. $f(x) = \log(4 - x)$
28. $f(x) = -\ln(e - x)$

In Exercises 29 to 34, determine the domain of each of the logarithmic functions defined by the following equations.

29. $J(x) = \dfrac{1}{\ln x}$
30. $K(x) = \log(x^2 + 4)$
31. $L(x) = \log(x^2 - 9)$
32. $f(x) = \dfrac{1}{\ln(4x + 10)}$
33. $h(x) = \log(x^2) + 4$
34. $Q(x) = \log(|x| + 1)$

In Exercises 35 to 40, find the range of each of the logarithmic functions defined by the following equations.

35. $J(x) = \dfrac{1}{\ln x}$
36. $K(x) = \log(x^2 + 4)$

37. $L(x) = \log(x^2 - 9)$ **38.** $V(x) = |\ln x|$

39. $R(x) = \ln|x|$ **40.** $B(x) = 4 + \log(x^2)$

41. A mathematics class takes a final exam. The average grade on the final exam is 75. At monthly intervals, after the final exam, the class takes equivalent forms of the final exam. The average grade is given by the model $S = 75 - 9.5 \ln(t + 1)$, where t is the number of months after the final exam.

   a. Find the original average grade.

   b. Find the average grade on an equivalent test six months later.

   c. How many months will elapse before the average grade falls below 50?

42. What will an earthquake measure on the Richter scale if it has an intensity of $I = 100,000 I_0$?

43. The Colombia earthquake of 1906 had an intensity of $I = 398,107,000 I_0$. What did it measure on the Richter scale?

44. If an earthquake has an intensity 1000 times the intensity of a second earthquake, then how much larger is the Richter scale measure of the larger earthquake than that of the smaller?

$\boxed{\bigvee}$ **Graphing Calculator Exercises***

45. Graph $f(x) = \dfrac{e^x - e^{-x}}{2}$ and $g(x) = \ln(x + \sqrt{x^2 + 1})$ on the same coordinate axes. Use the same scale on both the x and the y-axis. What appears to be the relationship between f and g?

46. On the same coordinate axes, graph $f(x) = \dfrac{e^x + e^{-x}}{2}$ for $x \geq 0$ and $g(x) = \ln(x + \sqrt{x^2 - 1})$ for $x \geq 1$. Use the

same scale on both the x and the y-axis. What appears to be the relationship between f and g?

47. Graph $f(x) = e^{-x}(\ln x)$ for $1 \leq x \leq e^2$.

48. Graph $g(x) = \log[\![x]\!]$ for $1 \leq x \leq 10$. Recall that $[\![x]\!]$ represents the greatest integer function.

In Exercises 49 to 54, graph each function to determine its domain and its range.

49. $f(x) = \sqrt{\log x}$ **50.** $f(x) = \sqrt{\ln x^3}$

51. $f(x) = 100 - \ln \sqrt{1 - x^2}$ **52.** $f(x) = 10 + |\ln(x - e)|$

53. $f(x) = \log(\log x)$ **54.** $f(x) = |\ln(-\ln x)|$

Supplemental Exercises

55. Given $f(x) = \ln x$, evaluate

   a. $f(e^3)$,   **b.** $f(e^{\ln 4})$,   **c.** $f(e^{3 \ln 3})$.

56. Given $f(x) = \log_5 x$, evaluate

   a. $f(5^2)$,   **b.** $f(5^{\log_5 4})$,   **c.** $f(5^{3 \log_5 3})$.

57. Explain why the graph of $F(x) = \log_b x^2$ and the graph of $G(x) = 2 \log_b x$ are not identical.

58. Explain why the graph of $F(x) = |\log_b x|$ and the graph of $G(x) = \log_b |x|$ are not identical.

59. The Coalinga, California, earthquake of May 2, 1983, had a Richter scale measure of 6.5. Find the Richter scale measure of an earthquake that has an intensity 200 times the intensity of the Coalinga quake.

60. The earthquake that occurred just south of Concepción, Chile, on May 22, 1960, had a Richter scale measure of 9.5. Find the Richter scale measure of an earthquake that has an intensity one-half the intensity of this quake.

5.4   Exponential and Logarithmic Equations

If a variable appears as an exponent in a term of an equation, then the equation is called an **exponential equation**. Example 1 uses the Equality of Exponents Theorem to solve exponential equations.

> *EXAMPLE 1*   **Solve Exponential Equations**
>
> Solve $49^{2x} = \dfrac{1}{7}$.

---

*Additional graphing calculator exercises appear in the Graphing Workbook as described in the front of this textbook.

Solution Write each side of the equation as a power of the same base, and then equate the exponents.

$$49^{2x} = \frac{1}{7}$$

$(7^2)^{2x} = 7^{-1}$ • Write each side as a power of 7.

$7^{4x} = 7^{-1}$

$4x = -1$ • Equate the exponents.

$$x = -\frac{1}{4}$$

■ *Try Exercise **2**, page 278.*

In Example 1, we were able to write each side of the equation as a power of the same base. If this is difficult to do, then consider taking the logarithm of both sides of the equation.

EXAMPLE 2 **Solve Exponential Equations**

Solve the following exponential equations: a. $5^x = 40$ b. $3^{2x-1} = 5^{x+2}$

Solution Start by taking the logarithm of each side of the equation.

a. $5^x = 40$

$\log(5^x) = \log 40$

$x \log 5 = \log 40$

$$x = \frac{\log 40}{\log 5}$$ • Exact solution

$x \approx 2.29203$ • Decimal approximation

b. $3^{2x-1} = 5^{x+2}$

$\ln 3^{2x-1} = \ln 5^{x+2}$

$(2x - 1)\ln 3 = (x + 2)\ln 5$ • Power property

$2x \ln 3 - \ln 3 = x \ln 5 + 2 \ln 5$ • Distributive property

Collecting terms that involve the variable x on the left side yields

$2x \ln 3 - x \ln 5 = 2 \ln 5 + \ln 3$ • Solve for x

$x(2 \ln 3 - \ln 5) = 2 \ln 5 + \ln 3$

$$x = \frac{2 \ln 5 + \ln 3}{2 \ln 3 - \ln 5}$$ • Exact solution

$x \approx 7.34533$ • Decimal approximation

■ *Try Exercise **12**, page 278.*

Logarithmic Equations

Equations that involve logarithms are called **logarithmic equations.** The properties of logarithms, along with the definition of a logarithm, are valuable aids to solving a logarithmic equation.

EXAMPLE 3   **Solve a Logarithmic Equation**

Solve $\log 2x - \log(x - 3) = 1$.

Solution

$$\log 2x - \log(x - 3) = 1$$

$$\log \frac{2x}{x - 3} = 1 \qquad \bullet \text{ Quotient property}$$

$$\frac{2x}{x - 3} = 10^1 \qquad \bullet \text{ Definition of logarithm}$$

$$2x = 10x - 30$$

$$-8x = -30$$

$$x = \frac{15}{4}$$

Check the solution by substituting 15/4 into the original equation.

─────────────────────────────────────── ■ *Try Exercise **24**, page 278.*

Caution   Be sure to check solutions. Solving an equation by using the properties of logarithms may introduce an extraneous solution, as shown in Example 4.

EXAMPLE 4   **Solve a Logarithmic Equation**

Solve $\ln(3x + 8) = \ln(2x + 2) + \ln(x - 2)$.

Solution

$$\ln(3x + 8) = \ln(2x + 2) + \ln(x - 2)$$

$$\ln(3x + 8) = \ln(2x + 2)(x - 2) \qquad \bullet \text{ Product property}$$

$$\ln(3x + 8) = \ln(2x^2 - 2x - 4)$$

$$3x + 8 = 2x^2 - 2x - 4 \qquad \bullet \begin{array}{l}\text{One-to-one property} \\ \text{of logarithms}\end{array}$$

$$0 = 2x^2 - 5x - 12$$

$$0 = (2x + 3)(x - 4)$$

Thus $-3/2$ and 4 are possible solutions. The number $-3/2$ does not check

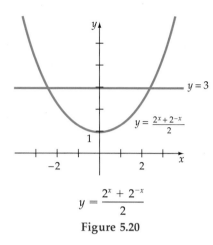

$$y = \frac{2^x + 2^{-x}}{2}$$

Figure 5.20

in the original equation. Why?[1] It can be shown that 4 checks, so the only solution is 4.

■ *Try Exercise 32, page 278.*

The solutions of the equation $(2^x + 2^{-x})/2 = 3$ are represented by the *x*-coordinates of the points of intersection of the graph of $y = 3$ and the graph of $y = (2^x + 2^{-x})/2$, as shown in Figure 5.20. Example 5 uses an algebraic method to solve $(2^x + 2^{-x})/2 = 3$.

EXAMPLE 5   **Solve an Equation Involving $b^x + b^{-x}$**

Solve $\dfrac{2^x + 2^{-x}}{2} = 3$.

Solution   Multiplying each side by 2 produces

$$2^x + 2^{-x} = 6$$

$$2^{2x} + 2^0 = 6 \cdot (2^x) \quad \text{• Multiply by } 2^x \text{ to clear negative exponents.}$$

$$(2^x)^2 - 6(2^x) + 1 = 0 \qquad \text{• Write in quadratic form.}$$

Substituting *u* for 2^x produces the quadratic equation

$$(u)^2 - 6(u) + 1 = 0$$

By the quadratic formula,

$$u = \frac{6 \pm \sqrt{36 - 4}}{2} = \frac{6 \pm 4\sqrt{2}}{2} = 3 \pm 2\sqrt{2}$$

Replacing *u* with 2^x produces

$$2^x = 3 \pm 2\sqrt{2}$$

Now take the common logarithm of each side.

$$\log 2^x = \log(3 \pm 2\sqrt{2})$$

$$x \log 2 = \log(3 \pm 2\sqrt{2}) \qquad \text{• Power property of logarithms}$$

$$x = \frac{\log(3 \pm 2\sqrt{2})}{\log 2} \approx \pm 2.54$$

■ *Try Exercise 42, page 278.*

Remark   If natural logarithms had been used in Example 5, then the exact solutions would have been

$$x = \frac{\ln(3 \pm 2\sqrt{2})}{\ln 2}$$

[1] If $x = -3/2$, the original equation becomes $\ln(7/2) = \ln(-1) + \ln(-7/2)$. This cannot be true because the function $f(x) = \ln x$ is not defined for negative values of *x*.

The pH of a Solution

Whether an aqueous solution is acidic or basic depends on its hydronium-ion concentration. Thus acidity is a function of hydronium-ion concentration. Since these hydronium-ion concentrations may be very small, it is convenient to measure acidity in terms of **pH**, which is defined as the negative of the common logarithm of the molar hydronium-ion concentration, M. As a mathematical formula, this is stated as

$$pH = -\log[H_3O^+]$$

EXAMPLE 6   **Find the pH of a Solution**

Find the pH of the following:

a. Orange juice with $[H_3O^+] = 2.80 \times 10^{-4}$ M
b. milk with $[H_3O^+] = 3.97 \times 10^{-7}$ M

Solution

a. $pH = -\log[H_3O^+] = -\log(2.80 \times 10^{-4}) \approx -(-3.55) = 3.55$

The orange juice has a pH of 3.55 (to the nearest hundredth).

b. $pH = -\log[H_3O^+] = -\log(3.97 \times 10^{-7}) \approx -(-6.40) = 6.40$

The milk has a pH of 6.40 (to the nearest hundredth).

■ *Try Exercise 50, page 278.*

The difference in hydronium-ion concentration between the two numbers shown in Figure 5.21 is

$$0.00028 - 0.000000397 = 0.000279603$$

Figure 5.21 shows how the pH function *maps* small positive numbers that are relatively close together on the hydronium-ion concentration axis into numbers (3.55 and 6.40) that are farther apart on the pH axis. The pH of pure water is 7.0.

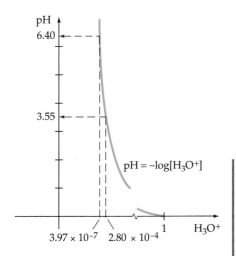

Figure 5.21

EXAMPLE 7   **Find the Hydronium-Ion Concentration**

Determine the hydronium-ion concentration of a sample of blood with pH = 7.41.

Solution   Substitute 7.41 for the pH and solve for H_3O^+.

$$pH = -\log[H_3O^+]$$
$$7.41 = -\log[H_3O^+] \quad \bullet \text{ Substitute 7.41 for pH.}$$
$$-7.41 = \log[H_3O^+] \quad \bullet \text{ Multiply both sides by } -1.$$
$$10^{-7.41} = [H_3O^+] \quad \bullet \text{ Definition of } y = \log_b x.$$
$$3.9 \times 10^{-8} \approx [H_3O^+]$$

The hydronium-ion concentration of the blood sample is 3.9×10^{-8} M.

■ *Try Exercise 52, page 278.*

In Exercises 1 to 40, solve for x.

1. $2^x = 64$

2. $3^x = 243$

3. $8^x = 512$

4. $25^x = 3125$

5. $49^x = \dfrac{1}{343}$

6. $9^x = \dfrac{1}{243}$

7. $2^{5x+3} = \dfrac{1}{8}$

8. $3^{4x-7} = \dfrac{1}{9}$

9. $\left(\dfrac{2}{5}\right)^x = \dfrac{8}{125}$

10. $\left(\dfrac{2}{5}\right)^x = \dfrac{25}{4}$

11. $5^x = 70$

12. $6^x = 50$

13. $3^{-x} = 120$

14. $7^{-x} = 63$

15. $\left(\dfrac{3}{5}\right)^x = 0.92$

16. $\left(\dfrac{7}{3}\right)^x = 22$

17. $10^{2x+3} = 315$

18. $10^{6-x} = 550$

19. $e^x = 10$

20. $e^{x+1} = 20$

21. $\left(1 + \dfrac{0.08}{12}\right)^{12x} = 1.5$

22. $\left(1 + \dfrac{0.05}{365}\right)^{365x} = 2$

23. $\log_2 x + \log_2(x - 4) = 2$

24. $\log_3 x + \log_3(x + 6) = 3$

25. $\log(5x - 1) = 2 + \log(x - 2)$

26. $1 + \log(3x - 1) = \log(2x + 1)$

27. $\log \sqrt{x^3 - 17} = \dfrac{1}{2}$

28. $\log(x^3) = (\log x)^2$

29. $\log(\log x) = 1$

30. $2 \ln \dfrac{e}{\sqrt{3}} = 3 - \ln x$

31. $\dfrac{1}{3} \ln 125 + \dfrac{1}{2} \ln x = \ln x$

32. $\ln x = \dfrac{1}{2} \ln\left(2x + \dfrac{5}{2}\right) + \dfrac{1}{2} \ln 2$

33. $\ln(e^{3x}) = 6$

34. $\log_b(b^{5x+2}) = 4$

35. $\ln x^2 = \ln 9$

36. $\log_x 9 = 3$

37. $\log_x 8 = 2$

38. $4 \log_x 2 - \dfrac{1}{2} \log_x 4 = 2 - \dfrac{1}{3} \log_x 8$

39. $e^{\ln(x-1)} = 4$

40. $10^{\log(2x+7)} = 8$

In Exercises 41 to 44, use common logarithms to solve for x.

41. $\dfrac{10^x - 10^{-x}}{2} = 20$

42. $\dfrac{10^x + 10^{-x}}{2} = 8$

43. $\dfrac{10^x + 10^{-x}}{10^x - 10^{-x}} = 5$

44. $\dfrac{10^x - 10^{-x}}{10^x + 10^{-x}} = \dfrac{1}{2}$

In Exercises 45 to 48, use natural logarithms to solve for x.

45. $\dfrac{e^x + e^{-x}}{2} = 15$

46. $\dfrac{e^x - e^{-x}}{2} = 15$

47. $\dfrac{1}{e^x - e^{-x}} = 4$

48. $\dfrac{e^x + e^{-x}}{e^x - e^{-x}} = 3$

49. Find the pH of a sample of lemon juice that has a hydronium-ion concentration of 6.3×10^{-3}.

50. An *acidic solution* has a pH of less than 7, whereas a *basic solution* has a pH of greater than 7. Household ammonia has an hydronium-ion concentration of 1.26×10^{-12}. Determine the pH of the ammonia, and state whether it is an acid or a base.

51. Find the hydronium-ion concentration of beer, which has a pH of 4.5.

52. Normal rain has a pH of 5.6. A recent acid rain had a pH of 3.1. Find the hydronium-ion concentration of this rain.

53. The population P of a city grows exponentially according to the function

$$P(t) = 8500(1.1)^t, \qquad 0 \le t \le 8$$

where t is measured in years.

a. Find the population at time $t = 0$ and also at time $t = 2$.

b. When, to the nearest year, will the population reach 15,000?

54. After a race, a runner's pulse rate R in beats per minute decreases according to the function $R(t) = 145e^{-0.092t}$ $(0 \le t \le 15)$ where t is measured in minutes.

a. Find the runner's pulse rate at the end of the race and also 1 minute after the end of the race.

b. How long, to the nearest minute, after the end of the race will the runner's pulse rate be 80 beats per minute?

55. A can of soda at 79°F is placed in a refrigerator that maintains a constant temperature of 36°F. The temperature T of the soda t minutes after it is placed in the refrigerator is

$$T(t) = 36 + 43e^{-0.058t}$$

a. Find the temperature of the soda 10 minutes after it is placed in the refrigerator.

b. When, to the nearest minute, will the temperature of the soda be 45°F?

56. During surgery, a patient's circulatory system requires at least 50 milligrams of an anesthetic. The amount of anesthetic present t hours after 80 milligrams of anesthetic are administered is

$$A(t) = 80(0.727)^t$$

a. How much of the anesthetic is present in the patient's circulatory system 30 minutes after the anesthetic is administered?

b. How long, to the nearest minute, can the operation last if the patient does not receive additional anesthetic?

Graphing Calculator Exercises*

In Exercises 57 to 64, determine the *number* of solutions of the given equation. (*Hint:* Graph the function on the left and right sides of the equation on the same coordinate axes. Then determine the *number* of intersections of their graphs to find the *number* of solutions to the equation.)

57. $2^x = \log x$

58. $10^{-x} = \log x$

59. $e^x - 4 = \ln x$

60. $x = -\log x$

61. $\ln x = x^2 - 5$

62. $\log x = x^3$

63. $\dfrac{2^x + 2^{-x}}{2} - 2 = \log|x|$

64. $\dfrac{2^x + 2^{-x}}{2} = \dfrac{3^x + 3^{-x}}{2}$

Supplemental Exercises

65. The following argument shows that $0.125 > 0.25$. Find the incorrect step.

$$3 > 2$$
$$3(\log 0.5) > 2(\log 0.5)$$
$$\log 0.5^3 > \log 0.5^2$$
$$0.5^3 > 0.5^2$$
$$0.125 > 0.25$$

66. The following argument shows that $4 = 6$. Find the incorrect step.

$$4 = \log_2 16$$
$$= \log_2(8 + 8)$$
$$= \log_2 8 + \log_2 8$$
$$= 3 + 3$$
$$= 6$$

67. A common mistake that students make is to write $\log(x + y)$ as $\log x + \log y$. For what values of x and y does $\log(x + y) = \log x + \log y$? (*Hint:* Solve for x in terms of y.)

68. Which is larger, 500^{501} or 506^{500}? (*Hint:* Let $x = 500^{501}$ and $y = 506^{500}$ and then compare $\ln x$ with $\ln y$.)

69. Explain why the functions $F(x) = (1.4)^x$ and $G(x) = e^{0.336x}$ essentially represent the same functions.

70. Find the constant k that will make $f(t) = (2.2)^t$ and $g(t) = e^{-kt}$ essentially represent the same function.

71. Solve $e^{1/x} > 2$. Write your answer in interval notation.

72. Solve $\log(x^2) > (\log x)^2$. Write your answer in interval notation.

5.5 Applications of Exponential and Logarithmic Functions

In many applications, a quantity N grows or decays according to the function $N(t) = N_0 e^{kt}$. In this function, N is a function of time t, and N_0 is the value of N at time $t = 0$. If k is a *positive* constant, then $N(t) = N_0 e^{kt}$ is called an exponential **growth function.** If k is a *negative* constant, then $N(t) = N_0 e^{kt}$ is called an exponential **decay function.** The following examples will give you an understanding of how growth and decay functions arise naturally in the investigation of certain phenomena.

   Interest is money paid for the use of money. The interest I is called **simple interest** if it is a fixed percent r per time period t of the amount of money invested. The amount of money invested is called the **principal** P.

*Additional graphing calculator exercises appear in the Graphing Workbook as described in the front of this textbook.

Simple interest is computed using the formula $I = Prt$. For example, if $1000 is invested at 12% for 3 years, the simple interest is

$$I = Prt = \$1000(0.12)\,(3) = \$360$$

The balance after t years is $B = P + I = P + Prt$. In the previous example, the $1000 invested for 3 years produced $360 interest. Thus the balance after 3 years is $1360.

Compound Interest

In many financial transactions, interest is added to the principal at regular intervals so that interest is paid on interest as well as on the principal. Interest earned in this manner is called **compound interest.** For example, if $1000 is invested at 12% annual interest compounded annually for 3 years, then the total interest after 3 years is

First-year interest	$1000(0.12) = \$120.00$
Second-year interest	$1120(0.12) = \$134.40$
Third-year interest	$1254.40(0.12) \approx \$150.53$

$$\$404.93 \quad \bullet \text{ Total interest}$$

This method of computing the balance can be tedious and time-consuming. A *compound interest formula* that can be used to determine the balance due after t years of compounding can be developed as follows.

Note that if P dollars is invested at an interest rate of r per year, then the balance after one year is $B_1 = P + Pr = P(1 + r)$, where Pr represents the interest earned for the year. Observe that B_1 is the product of the original principal P and $(1 + r)$. If the amount B_1 is reinvested for another year, then the balance after the second year is

$$B_2 = (B_1)\,(1 + r) = P(1 + r)\,(1 + r) = P(1 + r)^2$$

Successive reinvestments lead to the following results.

Number of years	Balance
3	$B_3 = P(1 + r)^3$
4	$B_4 = P(1 + r)^4$
.	.
.	.
n	$B_n = P(1 + r)^n$

The equation $B_t = P(1 + r)^t$ is valid if r is the interest rate paid during each of the t years.

If r is an annual interest rate and n is the number of compounding periods per year, then the interest rate each period is r/n and the number of compounding periods after t years is nt. Thus the compound interest formula is expressed as follows:

The Compound Interest Formula

A principal P invested at an annual interest rate r, expressed as a decimal and compounded n times per year for t years, produces the balance

$$B = P\left(1 + \frac{r}{n}\right)^{nt}$$

EXAMPLE 1   **Solve a Compound Interest Application**

Find the balance if $1000 is invested at an annual interest rate of 10%, for 2 years compounded   a.   annually   b.   daily   c.   hourly

Solution

a.   Use the compound interest formula, with $P = 1000$, $r = 0.1$, $t = 2$, and $n = 1$.

$$B = \$1000\left(1 + \frac{0.1}{1}\right)^{1 \cdot 2} = \$1000(1.1)^2 = \$1210.00$$

b.   Since there are 365 days in a year, use $n = 365$.

$$B = \$1000\left(1 + \frac{0.1}{365}\right)^{365 \cdot 2} \approx \$1000(1.000273973)^{730} \approx \$1221.37$$

c.   Since there are 8760 hours in a year, use $n = 8760$.

$$B = \$1000\left(1 + \frac{0.1}{8760}\right)^{8760 \cdot 2} \approx \$1000(1.000011416)^{17520} \approx \$1221.40$$

■ *Try Exercise 4, page 287.*

Remark   As the number of compounding periods increases, the balance seems to approach some upper limit. Even if the interest is compounded each *second*, the balance to the nearest cent remains $1221.40.

To **compound continuously** means to increase the number of compounding periods without bound.

To derive a continuous compounding interest formula, substitute $1/m$ for r/n in the compound interest formula

$$B = P\left(1 + \frac{r}{n}\right)^{nt} \tag{1}$$

to produce

$$B = P\left(1 + \frac{1}{m}\right)^{nt} \tag{2}$$

This substitution is motivated by the desire to express $(1 + r/n)^n$ as $[(1 + 1/m)^m]^r$, which approaches e^r as m gets large without bound.

Solving the equation $1/m = r/n$ for n yields $n = mr$, so the exponent nt can be written as mrt. Therefore Equation (2) can be expressed as

$$B = P\left(1 + \frac{1}{m}\right)^{mrt} = P\left[\left(1 + \frac{1}{m}\right)^{m}\right]^{rt} \tag{3}$$

By the definition of e, we know that as m gets larger without bound,

$$\left(1 + \frac{1}{m}\right)^{m} \text{ approaches } e$$

Thus, using continuous compounding, Equation (3) simplifies to $B = Pe^{rt}$.

Continuous Compounding Interest Formula

> If an account with principal P and annual interest rate r is compounded continuously for t years, then the balance is $B = Pe^{rt}$.

EXAMPLE 2   **Compound Continuously**

Find the balance after 4 years if \$800 is invested at an annual rate of 6% compounded continuously.

Solution   Use the continuous compounding formula.

$$B = Pe^{rt} = 800e^{0.06(4)} = 800e^{0.24}$$

$$\approx \$800(1.27124915) = \$1017.00 \quad \bullet \text{ To the nearest cent}$$

■ *Try Exercise 6, page 287.*

EXAMPLE 3   **Doubling a Sum of Money**

Find the time it takes for money invested at an annual rate of r to double.

Solution   Use $B = Pe^{rt}$ with $B = 2P$, twice the principle P.

$$2P = Pe^{rt}$$

$$2 = e^{rt}$$

$$\ln 2 = rt \quad \bullet \text{ Take the natural logarithm of each side.}$$

$$\frac{\ln 2}{r} = t \quad \bullet \text{ Solve for } t.$$

The time it takes for money to double when interest is compounded continuously at an annual rate of r is $t = \dfrac{\ln 2}{r}$.

■ *Try Exercise 10, page 287.*

Exponential Growth

Given any two points on the graph of $N(t) = N_0 e^{kt}$, you can use the given data to solve for the constants N_0 and k.

EXAMPLE 4   Find the Exponential Growth Equation That Models Given Data

Find the exponential growth function for a town whose population was 16,400 in 1970 and 20,200 in 1980.

Solution   We need to determine N_0 and k in $N(t) = N_0 e^{kt}$. If we represent the year 1970 by $t = 0$, then our given data are $N(0) = 16,400$ and $N(10) = 20,200$. Because N_0 is defined to be $N(0)$, we know $N_0 = 16,400$. To determine k, substitute $t = 10$ and $N_0 = 16,400$ into $N(t) = N_0 e^{kt}$ to produce

$$N(10) = 16,400e^{k \cdot 10}$$

$$20,200 = 16,400e^{10k} \quad \bullet \text{ Substitute 20,200 for } N(10).$$

$$\frac{20,200}{16,400} = e^{10k}$$

To solve this equation for k, take the natural logarithm of each side.

$$\ln\left(\frac{20,200}{16,400}\right) = \ln e^{10k}$$

$$\ln\left(\frac{20,200}{16,400}\right) = 10k \qquad \bullet \text{ Use } \log_b(b^p) = p.$$

$$\tfrac{1}{10}\ln\left(\frac{20,200}{16,400}\right) = k$$

$$0.0208 \approx k$$

The exponential growth equation is $N(t) = 16,400e^{0.0208t}$.

_____ ■ *Try Exercise 16, page 287.*

EXAMPLE 5   Solve a Population Application

Use the exponential growth equation from Example 4 to

a.   Estimate the population of the town in the year 1995.

b.   Estimate when the population will be double its 1970 population.

Solution

a. Since 1970 is the year represented by $t = 0$, we let $t = 25$ represent the year 1995.

$$N(t) = 16,400e^{0.0208t}$$

$$N(25) = 16,400e^{0.0208 \cdot 25}$$

$$\approx 27,600 \qquad \bullet \text{ To the nearest 100}$$

b. We need to solve for t when $N(t) = 2 \cdot 16,400 = 32,800$.

$$32,800 = 16,400e^{0.0208t}$$

$$2 = e^{0.0208t} \qquad \bullet \text{ Divide each side by 16,400.}$$

$$\ln 2 = 0.0208t \qquad \bullet \text{ Take the natural logarithm of each side.}$$

$$\frac{\ln 2}{0.0208} = t$$

$$33 \approx t \qquad \bullet \text{ To the nearest year.}$$

The population will double by the year 2003.

■ *Try Exercise* **18**, *page 287.*

Exponential Decay

Many radioactive materials *decrease* exponentially. This decrease, called radioactive decay, is measured in terms of **half-life,** which is defined as the time required for the disintegration of half the atoms in a sample of a radioactive substance. Following are the half-lives of selected radioactive isotopes.

Isotope	Half-Life
Carbon (^{14}C)	5730 years
Radium (^{226}Ra)	1660 years
Polonium (^{210}Po)	138 days
Phosphorus (^{32}P)	14 days
Polonium (^{214}Po)	1/10,000th of a second

EXAMPLE 6 **Find the Exponential Decay Equation That Models Given Data**

Find the exponential decay function for the amount of phosphorus (^{32}P) that remains in a sample after t days.

Solution When $t = 0$, $N(0) = N_0 e^{k(0)} = N_0$. Thus $N(0) = N_0$. Also, because the phosphorus has a half-life of 14 days, $N(14) = 0.5N_0$. To find k, substitute $t = 14$ into $N(t) = N_0 e^{kt}$ and solve for k.

$$N(14) = N_0 \cdot e^{k \cdot 14}$$

$$0.5N_0 = N_0 e^{14k} \qquad \bullet \text{ Substitute } 0.5N_0 \text{ for } N(14).$$

$$0.5 = e^{14k} \qquad \bullet \text{ Divide each side by } N_0.$$

$$\ln 0.5 = 14k \qquad \bullet \text{ Take the natural logarithm of each side.}$$

$$\frac{1}{14} \ln 0.5 = k \qquad \bullet \text{ Solve for } k.$$

$$-0.0495 \approx k$$

The exponential decay function is $N(t) = N_0 e^{-0.0495t}$.

■ *Try Exercise* **20,** *page 287.*

Remark Since $e^{-0.0495} \approx (0.5)^{1/14}$, the decay function $N(t) = N_0 e^{-0.0495t}$ can also be written as $N(t) = N_0 (0.5)^{t/14}$. In this form it is easy to see that if t is increased by 14, N will decrease by a factor of 0.5.

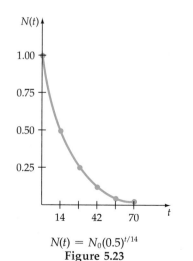

Figure 5.22

EXAMPLE 7   Solve a Radioactive Decay Application

Use $N(t) = N_0(0.5)^{t/14}$ to estimate the amount of phosphorus (^{32}P) that remains in a sample after 50 days. See Figure 5.22.

Solution

$$N(t) = N_0(0.5)^{t/14}$$

$$N(50) = N_0(0.5)^{50/14} \approx 0.0841N_0$$

After 50 days, approximately 8% of the original phosphorus (^{32}P) remains.

■ *Try Exercise* **22,** *page 287.*

The following tables and Figure 5.23 show the percent of phosphorus (^{32}P) that remains after 0, 14, 28, 42, 56, and 70 days.

Days t	0	14	28	42	56	70
Percent (^{32}P) remaining $N(t)$	100	50	25	12.5	6.25	3.125

Carbon Dating

The bone tissue in all living animals contains both carbon-12, which is non-radioactive, and carbon-14, which is radioactive with a half-life of approximately 5730 years. As long as the animal is alive, the ratio of carbon-14 to carbon-12 remains constant. When the animal dies ($t = 0$), the carbon-14 begins to decay. Thus a bone that has a smaller ratio of carbon-14 to carbon-12 is older than a bone that has a larger ratio. The amount of carbon-14 present at time t is

$$N(t) = N_0(0.5)^{t/5730}$$

where N_0 is the amount of carbon-14 present in the bone at time $t = 0$.

$N(t)$

1.00

0.75

0.50

0.25

14   42   70 t

$N(t) = N_0(0.5)^{t/14}$
Figure 5.23

EXAMPLE 8   **Solve a Carbon-Dating Application**

Determine the age of a bone if it now contains 85 percent of the carbon-14 it had when $t = 0$.

Solution   Let t be the time at which $N(t) = 0.85N_0$.

$$0.85N_0 = N_0(0.5)^{t/5730}$$

$$0.85 = (0.5)^{t/5730} \qquad \bullet \text{ Divide each side by } N_0.$$

$$\ln 0.85 = \ln (0.5)^{t/5730} \qquad \bullet \text{ Take the natural logarithm of each side}$$

$$\ln 0.85 = \frac{t}{5730} \ln 0.5 \qquad \bullet \text{ Power property}$$

$$5730 \frac{\ln 0.85}{\ln 0.5} = t \qquad \bullet \text{ Solve for } t.$$

$$1340 \approx t \qquad \bullet \text{ To the nearest 10 years}$$

The bone is about 1340 years old.

■ *Try Exercise* **24**, *page 288.*

The Decibel Scale

The range of sound intensities that the human ear can detect is so large that a special *decibel scale* (named after the inventor of the telephone, Alexander Graham Bell) is used to measure and compare different sound intensities. Specifically, the *intensity level N* of sound measured in decibels is directly proportional to the *power I* of the sound measured in watts per square centimeter. That is,

$$N(I) = 10 \log\left(\frac{I}{I_0}\right)$$

where I_0 is the power of sound that is barely audible to the human ear. By international agreement, I_0 is the constant 10^{-16} watts per square centimeter.

EXAMPLE 9   **Solve a Decibel Scale Application**

The power of normal conversation is 10^{-10} watts per square centimeter. What is the intensity level N, in decibels, of normal conversation?

Solution   Evaluate $N(10^{-10})$.

$$N(I) = 10 \log\left(\frac{I}{10^{-16}}\right)$$

$$N(10^{-10}) = 10 \log\left(\frac{10^{-10}}{10^{-16}}\right) \qquad \bullet \text{ Substitute } 10^{-10} \text{ for } I.$$

$$= 10 \log(10^6) \qquad \bullet \text{ Since } \frac{10^{-10}}{10^{-16}} = 10^{-10-(-16)} = 10^6$$

$$= 10(6) = 60$$

The intensity level of normal conversation is 60 decibels.

■ *Try Exercise* **28**, *page 288.*

EXERCISE SET 5.5

1. If $8000 is invested at an annual interest rate of 5 percent and compounded annually, find the balance after **a.** 4 years,   **b.** 7 years.

2. If $22,000 is invested at an annual interest rate of 4.5 percent and compounded annually, find the balance after **a.** 2 years,   **b.** 10 years.

3. If $38,000 is invested at an annual interest rate of 6.5 percent for 4 years, find the balance if the interest is compounded   **a.** annually,   **b.** daily,   **c.** hourly.

4. If $12,500 is invested at an annual interest rate of 8 percent for 10 years, find the balance if the interest is compounded   **a.** annually,   **b.** daily,   **c.** hourly.

5. Find the balance if $15,000 is invested at an annual rate of 10 percent for 5 years, compounded continuously.

6. Find the balance if $32,000 is invested at an annual rate of 8 percent for 3 years, compounded continuously.

7. How long will it take $4000 to double if it is invested in a certificate of deposit that pays 7.84% annual interest compounded continuously? Round to the nearest tenth of a year.

8. How long will it take $25,000 to double if it is invested in a savings account that pays 5.88% annual interest compounded continuously? Round to the nearest tenth of a year.

9. Use the Continuous Compounding Interest Formula to derive an expression for the time it will take money to triple when invested at an annual interest rate of r compounded continuously.

10. How long will it take $1000 to triple if it is invested at an annual interest rate of 5.5% compounded continuously? Round to the nearest year.

11. How long will it take $6000 to triple if it is invested in a savings account that pays 7.6% annual interest compounded continuously? Round to the nearest year.

12. How long will it take $10,000 to triple if it is invested in a savings account that pays 5.5% annual interest compounded continuously? Round to the nearest year.

13. The number of bacteria $N(t)$ present in a culture at time t hours is given by $N(t) = 2200(2)^t$. Find the number of bacteria present when **a.** $t = 0$ hours,   **b.** $t = 3$ hours.

14. The population of a town grows exponentially according to the function $f(t) = 12,400(1.14)^t$, for $0 \le t \le 5$ years. Find the population of the town when t is **a.** 3 years,   **b.** 4.25 years.

15. Find the growth function for a town whose population was 22,600 in 1980 and 24,200 in 1985. Use $t = 0$ to represent the year 1980.

16. Find the growth function for a town whose population was 53,700 in 1982 and 58,100 in 1988. Use $t = 0$ to represent the year 1982.

17. The function $P(t) = 9700(e^{0.08t})$ yields an estimate of the population of a city at time t years after 1985.

   a. Estimate the population of the city in 1993.

   b. Estimate the year the population will be double its 1985 population.

18. The function $P(t) = 15,600(e^{0.09t})$ yields an estimate of the population of a city at time t years after 1984.

   a. Estimate the population of the city in 1994.

   b. Estimate the year the population will be double its 1984 population.

19. Radium (^{226}Ra) has a half-life of 1660 years. Find the decay function for the amount of radium (^{226}Ra) that remains in a sample after t years.

20. Polonium (^{210}Po) has a half-life of 138 days. Find the decay function for the amount of polonium (^{210}Po) that remains in a sample after t days.

21. Use $N(t) = N_0(0.5)^{t/1660}$ to estimate the percentage of radium (^{226}Ra) that remains in a sample after 2250 years.

22. Use $N(t) = N_0(0.5)^{t/138}$, where t is measured in days, to estimate the percentage of polonium (^{210}Po) that remains in a sample after 2 years.

23. Determine the age of a bone if it now contains 77 percent of its original amount of carbon-14.

24. Determine the age of a bone if it now contains 65 percent of its original amount of carbon-14.

25. Newton's Law of Cooling states that if an object at temperature T_0 is placed into an environment at constant temperature A, then the temperature of the object will be $T(t)$ after t minutes according to the function given by $T(t) = A + (T_0 - A)e^{-kt}$, where k is a constant that depends on the object.

  a. Determine the constant k (to the nearest thousandth) for a canned soda drink that takes 5 minutes to cool from 75°F to 65°F after being placed in a refrigerator that maintains a constant temperature of 34°F.

  b. What will be the temperature (to the nearest degree) of the soda drink after 30 minutes?

  c. When (to the nearest minute) will the temperature of the soda drink be 36°F?

  d. When will the temperature of the soda drink be exactly 34°F?

26. Solve the sound intensity equation $N = 10 \log\left(\dfrac{I}{I_0}\right)$ for I.

27. How much more powerful is a sound that measures 120 decibels than a sound (at the same frequency) that measures 110 decibels?

28. The power of a band is 3.4×10^{-5} watts per square centimeter. What is the band's intensity level N in decibels?

29. If the power of a sound is doubled, what is the increase in the intensity level? *Hint:* Find $N(2I) - N(I)$.

30. According to a software company, the users of its typing tutorial can expect to type $N(t)$ words per minute after t hours of practice with the product, according to the function $N(t) = 100(1.04 - 0.99^t)$.

  a. How many words per minute can a student expect to type after 2 hours of practice?

  b. How many words per minute can a student expect to type after 40 hours of practice?

  c. According to the function N, how many hours (to the nearest 1 hour) of practice will be required before a student can expect to type 60 words per minute?

31. In the city of Whispering Palms, the number of people $P(t)$ exposed to a rumor in t hours is given by the function $P(t) = 80,000(1 - e^{-0.0005t})$.

  a. Find the number of hours until 10 percent of the population have heard the rumor.

  b. Find the number of hours until 50 percent of the population have heard the rumor.

32. A lawyer has determined that the number of people $P(t)$ who have been exposed to a news item after t days is given by the function $P(t) = 1,200,000(1 - e^{-0.03t})$.

  a. How many days after a major crime has been reported have 40 percent of the population heard of the crime.

  b. A defense lawyer knows it will be very difficult to pick an unbiased jury after 80 percent of the population have heard of the crime. After how many days will 80 percent of the population have heard of the crime?

33. An automobile depreciates according to the function $V(t) = V_0(1 - r)^t$, where $V(t)$ is the value in dollars after t years, V_0 is the original value, and r is the yearly depreciation rate. A car has a yearly depreciation rate of 20 percent. Determine in how many years the car will depreciate to half its original value.

34. The current $I(t)$ (measured in amperes) of a circuit is given by the function $I(t) = 6(1 - e^{-2.5t})$, where t is the number of seconds after the switch is closed.

  a. Find the current when $t = 0$.

  b. Find the current when $t = 0.5$.

  c. Solve the equation for t.

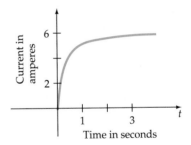

Time in seconds

Supplemental Exercises

35. The Prime Number Theorem states that the number of prime numbers $P(n)$ less than a number n can be approximated by the function

$$P(n) = \frac{n}{\ln n}$$

  a. The actual number of prime numbers less than 100 is 25. Compute $P(100)$ and $P(100)/25$.

  b. The actual number of prime numbers less than 10,000 is 1229. Compute $P(10,000)$ and $P(10,000)/1229$.

  c. The actual number of prime numbers less than 1,000,000 is 78,498. Compute $P(1,000,000)$, and then compute the ratio $P(1,000,000)/78,498$.

36. The number $n!$ (which is read "n factorial") is defined as

$$n! = n(n - 1)(n - 2) \cdots 1$$

for all positive integers n. Thus, $4! = 4 \cdot 3 \cdot 2 \cdot 1 = 24$.

Stirling's Formula (after James Stirling, 1692–1770)

$$n! \approx \left(\frac{n}{e}\right)^n \sqrt{2\pi n}$$

is often used to approximate very large factorials. Use Stirling's Formula to approximate 10!, and then compute the ratio of Stirling's approximation of 10! divided by the actual value of 10!, which is 3,628,800.

37. A farmer knows that planting the same crop in the same field year after year reduces the yield. If the yield on each succeeding year's crop is 90 percent of the preceding year's yield, then the yield $Y(t)$ at any time t is given by the function $Y(t) = Y_0(0.90)^t$, where Y_0 is the yield when $t = 0$. In how many years (to the nearest year) will the yield be 60 percent of Y_0?

38. Crude oil leaks from a tank at a rate that depends on the amount of oil that remains in the tank. Since 1/8 of the oil in the tank leaks out every 2 hours, the volume of oil $V(t)$ in the tank at t hours is given by the function $V(t) = V_0(0.875)^{t/2}$, where $V_0 = 350,000$ gallons is the number of gallons in the tank at the time the tank started to leak ($t = 0$).

   a. How many gallons does the tank hold after 3 hours?

   b. How many gallons does the tank hold after 5 hours?

   c. How long will it take until 90 percent of the oil has leaked from the tank?

39. How many times stronger is an earthquake that measures 6 on the Richter scale than one that measures 3 on the Richter scale?

40. How many times stronger was the Chile earthquake of 1960, which measured 9.5 on the Richter scale, than the San Francisco earthquake of 1906, which measured 8.3 on the Richter scale?

41. Logarithms and a function called the integer function (denoted by *INT*) can be used to determine the number of digits in a number written in exponential notation. The *INT* function is illustrated in the following examples:

$$INT(8.75) = 8 \qquad INT(102.003) = 102$$

$$INT(55) = 55 \qquad INT(e) = 2$$

Note that the *INT* function removes the decimal portion of a real number and returns the integer part of the real number as its output. The number of digits $N(x)$ in the number b^x, with $0 < b < 10$ and x a positive integer, is given by the function $N(x) = INT(x \log b) + 1$.

   a. Find the number of digits in 3^{200}.

   b. Find the number of digits in 7^{4005}.

   c. The largest known prime number in 1980 was the number $2^{44,497} - 1$. Find the number of digits in this prime number.

   d. The largest known prime number as of 1983 was $2^{132,049} - 1$. Find the number of digits in this prime number.

42. How many years will it take the price of goods to double if the annual rate of inflation is 5 percent per year? Use continuous compounding.

43. The current rate of inflation will cause the price of goods to double in the next 10 years. Determine the current rate of inflation. Use continuous compounding.

44. The height h in feet of any point P on the cable shown is a function of the horizontal distance in feet from point P to the origin given by the function

$$h(x) = \frac{20}{2}(e^{x/20} + e^{-x/20}) \quad -40 \le x \le 40$$

   a. What is the height of the cable at point P if P is directly above the origin?

   b. What is the height of the cable at point P if P is 25 feet to the right of the origin?

   c. How far to the right or left of the origin is the cable 30 feet in height? (*Hint:* Use the method developed in Example 5 of Section 5.4.)

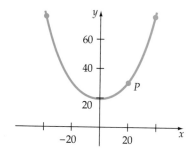

Exploring Concepts with Technology

The Logistic Model for Population Growth

A population that grows or decays according to the function $P(t) = P_0 e^{kt}$ is called a **Malthusian model.** This formula models the growth or decay of many populations with unlimited resources. However, if such factors as limited food supply and other limited resources affect the growth of the population, then it may be necessary to model the population growth by using the following function, which is known as the **logistic law**:

$$P(t) = \frac{mP_0}{P_0 + (m - P_0)e^{-kt}}$$

where m is the maximum possible population and k is a positive constant.
   Assume that the world's population growth satisfies the logistics law with $m = 80$ billion. If P was 4 billion in 1976 (think of this as the year $t = 0$) and 5 billion in 1986 ($t = 10$), find the constant k.

1. a. Use the value of k to graph the logistics model of population growth. Describe the graph. Note that the graph of $y = m$ is a horizontal asymptote of the graph.
   b. Use the value of k and $P_0 = 4$ billion to graph the Malthusian model of population growth.
   c. Compare the logistic model with the Malthusian model.

2. Find the world's predicted population (according to the logistic law) for the year 2000 and for the year 3000.

3. The world's population reached 3 billion in 1961. How does this compare with the result you got by using the logistic law and the value of k obtained above.

4. The population of squirrels in a nature reserve satisfies the logistic law, with $P_0 = 1500$, $k = 0.29$, and $P(2) = 2500$.
   a. What is the maximum number of squirrels (to the nearest thousand) that the reserve can support? (*Hint:* Compute m.)
   b. Find the number of squirrels when $t = 10$.

5. The population of walruses in a colony satisfies the logistic law, with $P_0 = 800$, $P(1) = 900$, and $k = 0.14$.
   a. What is the maximum number of walruses (to the nearest hundred) that the colony can support? (*Hint:* Compute m.)
   b. Find the number of walruses when $t = 5$ years.

Chapter Review

5.1   *Exponential Functions and Their Graphs*

- For all positive real numbers b ($b \neq 1$), the exponential function defined by $f(x) = b^x$ has the following properties:

   —f has the set of real numbers as its domain.
   —f has the set of positive real numbers as its range.
   —f has a graph with a y-intercept of $(0, 1)$.
   —f has a graph asymptotic to the x-axis.
   —f is a one-to-one function.
   —f is an increasing function if $b > 1$.
   —f is a decreasing function if $0 < b < 1$.

- As n increases without bound, $(1 + 1/n)^n$ approaches an irrational number denoted by e. The value of e accurate to eight decimal places is 2.71828183.

- The function defined by $f(x) = e^x$ is called the natural exponential function.

5.2   *Logarithms and Logarithmic Properties*

- *Definition of a Logarithm*   If $x > 0$ and b is a positive constant ($b \neq 1$), then

$$y = \log_b x \quad \text{if and only if} \quad b^y = x$$

In the equation $y = \log_b x$, y is referred to as the logarithm, b is the base, and x is the argument.

- *Change-of-Base Formula*   If $x > 0$, $a > 0$, $b > 0$, and neither a nor b equals 1, then

$$\log_b x = \frac{\log_a x}{\log_a b}$$

- If M, N, and b are positive real numbers with $b \neq 1$, then $\log_b M = \log_b N$ if and only if $M = N$.

5.3   *Logarithmic Functions and Their Graphs*

- For all positive real numbers b, $b \neq 1$, the function defined by $f(x) = \log_b x$ has the following properties:
  - f has the set of positive real numbers as its domain.
  - f has the set of real numbers as its range.
  - f has a graph with an x-intercept of $(1, 0)$.
  - f has a graph asymptotic to the y-axis.
  - f is a one-to-one function.
  - f is an increasing function if $b > 1$.
  - f is a decreasing function if $0 < b < 1$.

5.4   *Exponential and Logarithmic Equations*

- *Equality of Exponents Theorem*   If b is a positive real number ($b \neq 1$) such that $b^x = b^y$, then $x = y$.

5.5   *Applications of Exponential and Logarithmic Functions*

- The function defined by $N(t) = N_0 e^{kt}$ is called an exponential growth function if k is positive, and it is called an exponential decay function if k is negative.

- *The Compound Interest Formula*   A principal P invested at an annual interest rate r compounded n times per year for t years produces the balance

$$B = P\left(1 + \frac{r}{n}\right)^{nt}$$

- *Continuous Compounding Interest Formula*   If an account with principal P and annual interest rate r is compounded continuously for t years, then the balance is $B = Pe^{rt}$.

Essays and Projects

1. Write an essay about John Napier (1550–1617). Include information about the work of Napier in inventing logarithms.

2. Write an essay about Henry Briggs (1561–1630). Include information about his work related to the development of logarithms.

3. The sum

$$S_n = 1 + \frac{1}{2} + \frac{1}{3} + \frac{1}{4} + \cdots + \frac{1}{n}$$

is called the *harmonic sum*. As n increases, S_n increases, but it increases very slowly. To illustrate this concept, consider the following inequality.

$$\ln n < S_n < 1 + \ln n$$

Use this inequality to determine the smallest value of n such that   **a.** $S_n > 1000$,   **b.** $S_n > 10{,}000$.

4. For $a > b > 0$, find an inequality that shows the relationship between the arithmetic mean of a and b, the geometric mean of a and b, and the quantity

$$\frac{a - b}{\ln a - \ln b}$$

5. The Gateway Arch in St. Louis is about 625 feet tall and 600 feet across at its base. If the arch is placed on a coordinate grid so that its vertex is $(0, 625)$ and it intersects the x-axis at $(-300, 0)$ and $(300, 0)$, then the equation of the arch is given by

$$y = 694 - 69\left(\frac{e^{x/100} + e^{-x/100}}{2}\right), \quad -300 \leq x \leq 300$$

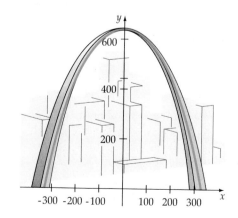

Use a computer or graphing calculator to graph the equation.

The equation of a parabola that passes through the points $(-300, 0)$, $(0, 625)$, and $(300, 0)$ is $y = -\dfrac{1}{144}x^2 + 625$. Use your graphing utility to graph the parabola on the same grid as the arch. Estimate the maximum vertical difference between the height of the arch and the height of the parabola.

6. Sometimes it is convenient to graph a function using a logarithmic scale. Graph paper that makes use of a logarithmic scale is called log-log paper or semi-log paper. Write an essay that explains how to construct log-log paper and also how to construct semi-log paper. Explain what the graph of $y = 10^x$ looks like when drawn on semi-log paper.

True/False Exercises

In Exercises 1 to 14, answer true or false. If the statement is false, give an example to show that the statement is false.

1. If $7^x = 40$, then $\log_7 40 = x$.

2. If $\log_4 x = 3.1$, then $4^{3.1} = x$.

3. If $f(x) = \log x$ and $g(x) = 10^x$, then $f[g(x)] = x$ for all real numbers x.

4. If $f(x) = \log x$ and $g(x) = 10^x$, then $g[f(x)] = x$ for all real numbers x.

5. The exponential function $h(x) = b^x$ is an increasing function.

6. The logarithmic function $j(x) = \log_b x$ is an increasing function.

7. The exponential function $h(x) = b^x$ is a one-to-one function.

8. The logarithmic function $j(x) = \log_b x$ is a one-to-one function.

9. The graph of
$$f(x) = \frac{2^x + 2^{-x}}{2}$$
is symmetric with respect to the y-axis.

10. The graph of
$$f(x) = \frac{2^x - 2^{-x}}{2}$$
is symmetric with respect to the origin.

11. If $x > 0$ and $y > 0$, then $\log(x + y) = \log x + \log y$.

12. If $x > 0$, then $\log x^2 = 2 \log x$.

13. If M and N are positive real numbers, then
$$\ln\left(\frac{M}{N}\right) = \ln M - \ln N$$

14. For all $p > 0$, $e^{\ln p} = p$.

Chapter Review Exercises

In Exercises 1 to 12, solve each equation. Do not use a calculator.

1. $\log_5 25 = x$

2. $\log_3 81 = x$

3. $\ln e^3 = x$

4. $\ln e^\pi = x$

5. $3^{2x+7} = 27$

6. $5^{x-4} = 625$

7. $2^x = \dfrac{1}{8}$

8. $27(3^x) = 3^{-1}$

9. $\log x^2 = 6$

10. $\dfrac{1}{2}\log|x| = 5$

11. $10^{\log 2x} = 14$

12. $e^{\ln x^2} = 64$

In Exercises 13 to 18, use a calculator to evaluate each power. Give your answers accurate to six significant digits.

13. $7^{\sqrt{2}}$

14. $3^{\sqrt{5}}$

15. $e^{1.7}$

16. $e^{-2.2}$

17. $10^{1.135}$

18. $10^{-\sqrt{10}}$

In Exercises 19 to 32, sketch the graph of each function.

19. $f(x) = (2.5)^x$

20. $f(x) = \left(\dfrac{1}{4}\right)^x$

21. $f(x) = 3^{|x|}$

22. $f(x) = 4^{-|x|}$

23. $f(x) = 2^x - 3$

24. $f(x) = 2^{(x-3)}$

25. $f(x) = \dfrac{4^x + 4^{-x}}{2}$

26. $f(x) = \dfrac{3^x - 3^{-x}}{2}$

27. $f(x) = \dfrac{1}{3} \log x$

28. $f(x) = 3 \log x^{1/3}$

29. $f(x) = -x + \log x$

30. $f(x) = 2^{-x} \log x$

31. $f(x) = -\dfrac{1}{2} \ln x$

32. $f(x) = -\ln|x|$

In Exercises 33 to 36, change each logarithmic equation to its exponential form.

33. $\log_4 64 = 3$

34. $\log_{1/2} 8 = -3$

35. $\log_{\sqrt{2}} 4 = 4$

36. $\ln 1 = 0$

In Exercises 37 to 40, change each exponential equation to its logarithmic form.

37. $5^3 = 125$

38. $2^{10} = 1024$

39. $10^0 = 1$

40. $8^{1/2} = 2\sqrt{2}$

In Exercises 41 to 44, write the given logarithm in terms of logarithms of x, y, and z.

41. $\log_b \dfrac{x^2 y^3}{z}$

42. $\log_b \dfrac{\sqrt{x}}{y^2 z}$

43. $\ln xy^3$

44. $\ln \dfrac{\sqrt{xy}}{z^4}$

In Exercises 45 to 48, write each logarithmic expression as a single logarithm.

45. $2 \log x + \dfrac{1}{3} \log(x + 1)$

46. $5 \log x - 2 \log(x + 5)$

47. $\dfrac{1}{2} \ln 2xy - 3 \ln z$

48. $\ln x - (\ln y - \ln z)$

In Exercises 49 to 52, use the change-of-base formula and a calculator to approximate each logarithm accurate to six significant digits.

49. $\log_5 101$

50. $\log_3 40$

51. $\log_4 0.85$

52. $\log_8 0.3$

In Exercises 53 to 56, use a calculator to approximate N to three significant digits.

53. $\log N = 2.47$

54. $\log N = -0.48$

55. $\ln N = 51$

56. $\ln N = -0.09$

In Exercises 57 to 72, solve each equation for x. Give exact answers. Do not use a calculator.

57. $4^x = 30$

58. $5^{x+1} = 41$

59. $\ln 3x - \ln(x - 1) = \ln 4$

60. $\ln 3x + \ln 2 = 1$

61. $e^{\ln(x+2)} = 6$

62. $10^{\log(2x+1)} = 31$

63. $\dfrac{4^x + 4^{-x}}{4^x - 4^{-x}} = 2$

64. $\dfrac{5^x + 5^{-x}}{2} = 8$

65. $\log(\log x) = 3$

66. $\ln(\ln x) = 2$

67. $\log \sqrt{x - 5} = 3$

68. $\log x + \log(x - 15) = 1$

69. $\log_4(\log_3 x) = 1$

70. $\log_7(\log_5 x^2) = 0$

71. $\log_5 x^3 = \log_5 16x$

72. $25 = 16^{\log_4 x}$

73. Find the pH of tomatoes that have a hydronium-ion concentration of 6.28×10^{-5}.

74. Find the hydronium-ion concentration of rainwater that has a pH of 5.4.

75. Find the balance when \$16,000 is invested at an annual rate of 8 percent for 3 years if the interest is compounded

   a. monthly,   **b.** continuously.

76. Find the balance when \$19,000 is invested at an annual rate of 6 percent for 5 years if the interest is compounded

   a. daily,   **b.** continuously.

77. The scrap value S of a product with an expected life span of n years is given by $S(n) = P(1 - r)^n$, where P is the original purchase price of the product and r is the annual rate of depreciation. A taxicab is purchased for \$12,400 and is expected to last 3 years. What is its scrap value if it depreciates at a rate of 29 percent per year?

78. A skin wound heals according to the function given by $N(t) = N_0 e^{-0.12t}$, where N is the number of square centimeters of unhealed skin t days after the injury, and N_0 is the number of square centimeters covered by the original wound.

   a. What percentage of the wound will be healed after 10 days?

   b. How many days will it take for 50 percent of the wound to heal?

   c. How long will it take for 90 percent of the wound to heal?

In Exercises 79 to 82, find the exponential growth/decay function $N(t) = N_0 e^{kt}$ that satisfies the given conditions.

79. $N(0) = 1$, $N(2) = 5$

80. $N(0) = 2$, $N(3) = 11$

81. $N(1) = 4$, $N(5) = 5$

82. $N(-1) = 2$, $N(0) = 1$

Chapter Test

1. Given that $f(x) = 2.5^x$, use a calculator to evaluate $f(\sqrt{3})$ accurate to six significant digits.

2. Given that $f(x) = e^{x/2}$, use a calculator to evaluate $f(2.7)$ to six significant digits.

3. Graph $f(x) = 3^{-x/2}$.

4. Graph $f(x) = \left(\dfrac{3}{2}\right)^x$.

5. Graph $f(x) = e^{x/2}$.

6. Write $\log_b (5x - 3) = c$ in exponential form.

7. Write $3^{x/2} = y$ in logarithmic form.

8. Write $\log_b \dfrac{x^2 y^4}{z^3}$ in terms of logarithms of x, y, and z.

9. Write $\log_b \dfrac{z^2}{y^3 \sqrt{x}}$ in terms of logarithms of x, y, and z.

10. Write $\log_{10}(2x + 3) - 3\log_{10}(x - 2)$ as a single logarithm.

11. Given $\log_b 2 = 0.2774$ and $\log_b 7 = 0.7445$, evaluate $\log_b 14$.

12. Use the change-of-base formula and a calculator to approximate $\log_4 12$ to five significant digits.

13. Graph $f(x) = \log(x - 1)$.

14. Graph $f(x) = \log x + 2$.

15. Graph $f(x) = -\ln(x + 1)$.

16. Solve $3^{2x-5} = \dfrac{1}{27}$.

17. Solve $5^x = 22$.

18. Solve $\log(x + 99) - \log(3x - 2) = 2$.

19. Find the balance if $20,000 is invested at an annual rate of 7.8 percent for 5 years, compounded continuously.

20. The scrap value S of a product with an expected life of n years is given by $S(n) = P(1 - r)^n$, where P is the original purchase price of the product and r is the annual rate of depreciation. A computer system is purchased for $8400 and is expected to last 4 years. What is its scrap value if the computer system depreciates at a rate of 22 percent per year?

6

Trigonometric Functions

CASE IN POINT *Tsunamis and Earthquakes*

Imagine an undersea earthquake. The energy from the earthquake would be translated to the water as water waves. These water waves are called *tsunamis* or *tidal waves*. Although the phrase *tidal waves* is still used to describe these waves, *tsunami* is the preferred description because the waves have nothing to do with the tides.

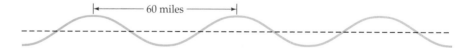
|← 60 miles →|

In the open ocean, the distance between crests of a tsunami may be as great as 60 miles and the height of the wave no more than 2 feet as shown in the figure above. As the depth of the ocean decreases however, the water wave slows down. As it slows, the height of the wave increases. When a tsunami reaches the shore, the wave height can be quite high with crests 100 feet above the normal tide level.

The earthquake that generated the tsunamis also creates waves within the earth. Two of the wave types that are created are the *primary* or *P wave* and the *secondary* or *S wave*. These two waves are quite different. The P wave is very much like a sound wave. It alternately compresses and dilates the substances within the earth. These waves can travel through solid rock and water.

S waves are slower than P waves and are more like water waves. As an S wave travels through the earth, it shears the rock sideways at right angles to the direction of travel. The S wave causes much of the structural damage associated with earthquakes. Wave phenomena exhibited by tsunami and earthquake waves can be described by trigonometric functions, the subject of this chapter.

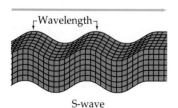
Compressions

Dilatations
P-wave

Wavelength
S-wave

295

6.1 Measuring Angles

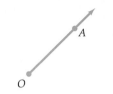

Figure 6.1

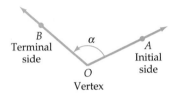

Figure 6.2

Early Babylonians noticed that the seasons repeated about every 360 days. Thinking that the earth was at the center of the universe, they assumed the universe made one complete revolution in 360 days. Our concept of measuring angles in degrees is an outgrowth of those early beliefs.

A **ray** originates at a point and extends infinitely, as Figure 6.1 shows. An **angle** is formed by rotating a ray about its endpoint. The initial position of the ray is called the **initial side** of the angle. The position of the ray after it has been rotated is called the **terminal side** of the angle. The point at which the two rays meet is called the **vertex** of the angle. See Figure 6.2.

There are different ways to name an angle. One way uses Greek letters. In Figure 6.2, the name of the angle is α. This is written $\angle\alpha$. The angle can also be named by giving the letter corresponding to the vertex—in this case, $\angle O$. The angle can also be named by including the vertex and the points on the rays that form the angle. This is written $\angle AOB$, where the vertex is always given as the middle letter.

Angles formed by a counterclockwise rotation are considered **positive** angles; angles formed by a clockwise rotation are considered **negative** angles. See Figure 6.3.

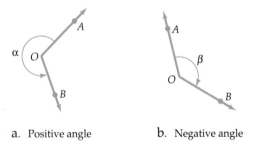

a. Positive angle   b. Negative angle

Figure 6.3

The measure of an angle is the amount of rotation of the ray. Consider a circle whose circumference is divided into 360 equal parts. Using the center of the circle as the beginning point, draw rays through two consecutive points on the circle. The measure of the angle between these two rays is one *degree*.

Definition of Degree

An angle formed by rotating a ray $\frac{1}{360}$ of a complete revolution has a measure of one **degree**. The symbol for degree is °.

There are a number of different types of angles, each classified by the measure of the angle. See Figure 6.4.

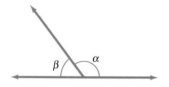

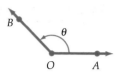

a. Straight angle (α = 180°)   b. Right angle (β = 90°)   c. Acute angle $0° < \theta < 90°$ d. Obtuse angle $90° < \theta < 180°$

Figure 6.4

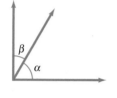

a. Complementary angles
    $\alpha + \beta = 90°$

Two nonnegative angles are **complementary angles** (Figure 6.5a) if the sum of the measures of the angles is 90°. Each angle is the *complement* of the other angle. Two nonnegative angles are **supplementary angles** (Figure 6.5b) if the sum of the measures of the angles is 180°. Each angle is the *supplement* of the other angle.

Angles greater than 360° and less than 0° can also be measured. See Figure 6.6.

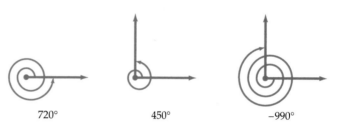

720°   450°   −990°

b. Supplementary angles
    $\alpha + \beta = 180°$

Figure 6.5

Figure 6.6

In the DMS (**D**egree, **M**inute, **S**econd) system, the degree is subdivided into 60 equal smaller units called **minutes.** Each minute is further divided into 60 smaller units called **seconds.**

$$\text{One minute } (1') = \left(\frac{1}{60}\right)^{\circ} \quad \bullet \text{ One sixtieth of a degree}$$

$$\text{One second } (1'') = \left(\frac{1}{60}\right)' \quad \bullet \text{ One sixtieth of a minute}$$

From these two equations we have

$$60' = 1° \quad \text{and} \quad 60'' = 1'$$

Therefore, 1° = 3600″. Why?[1]

Degrees may also be divided into smaller units by using decimal degrees. In the decimal degree system,

$$29.76° \text{ means } 29° \text{ plus 76 hundredths of } 1°$$

[1] $1° = 60' = 60' \cdot \frac{60''}{1'} = 3600''$

EXAMPLE 1 Change from DMS to Decimal Degrees

Change 126°16′ to decimal degrees. Round to the nearest thousandths of a degree.

Solution Use the conversion formula $1° = 60′$ to change minutes to decimal degrees.

$$126°16′ = 126° + 16′\left(\frac{1°}{60′}\right) \qquad \bullet\ 1° = 60′, \text{ therefore } 1 = \frac{1°}{60′}$$

$$\approx 126° + 0.267° = 126.267°$$

■ *Try Exercise* **16,** *page 303.*

EXAMPLE 2 Change from Decimal Degrees to DMS

Change −36.78° to the DMS system of measurement.

Solution

$$-36.78° = -(36° + 0.78°)$$

$$= -\left[36° + 0.78°\left(\frac{60′}{1°}\right)\right] \qquad \bullet\ 1° = 60′, \text{ so } \frac{60′}{1°} = 1$$

$$= -[36° + 46.8′]$$

$$= -[36° + 46′ + 0.8′]$$

$$= -\left[36° + 46′ + 0.8′\left(\frac{60″}{1′}\right)\right] \qquad \bullet\ 1′ = 60″, \text{ so } \frac{60″}{1′} = 1$$

$$= -[36° + 46′ + 48″] = -36°46′48″$$

■ *Try Exercise* **22,** *page 303.*

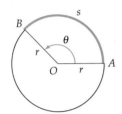

Figure 6.7

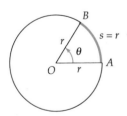

Figure 6.8

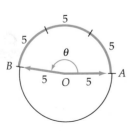

Figure 6.9

Another commonly used angle measurement is the *radian*. To define a radian, first consider a circle of radius r and two radii OA and OB. The angle θ formed by the two radii is a **central angle.** The portion of the circle between A and B is an **arc** of the circle and is written $\overset{\frown}{AB}$. We say that $\overset{\frown}{AB}$ *subtends* the angle θ. The length of $\overset{\frown}{AB}$ is s (see Figure 6.7). A radian is the measure of a central angle subtended by a certain arc.

Definition of Radian

One **radian** is the measure of the central angle subtended by an arc of length r on a circle of radius r. See Figure 6.8.

For example, an arc of length 15 centimeters on a circle with a radius of 5 centimeters will subtend an angle of 3 radians, as shown in Figure 6.9. The same result can be found by dividing 15 centimeters by 5 centimeters. To find the measure in radians of any central angle θ, divide the length s of the arc that subtends θ by the radius of the circle.

Radian Measure

Given an arc of length s on a circle of radius r, the radian measure of the central angle subtended by the arc is given by $\theta = \dfrac{s}{r}$.

Using the formula for radian measure, we find that an arc of length 12 cm on a circle of radius 8 cm subtends a central angle θ whose measure is given by

$$\theta = \frac{s}{r} = \frac{12 \text{ cm}}{8 \text{ cm}} = \frac{3}{2} \text{ radians}$$

Note that units of measurement (in this case, centimeters) are *not* part of the result. The radian measure of a central angle formed by an arc of length 12 miles on a circle of radius 8 miles would be the same, 3/2 radians. We say that radian is a *dimensionless* quantity because there are no units of measurement associated with a radian.

Recall that the circumference of a circle is given by the equation $C = 2\pi r$. The radian measure of the central angle θ subtended by the circumference is $\theta = \dfrac{2\pi r}{r} = 2\pi$. In degree measure, the central angle θ is 360°.

Thus we have the equation $360° = 2\pi$ radians. Dividing each side of the equation by 2 gives $180° = \pi$ radians. From this last equation, we have the following conversion factors for changing between degree measure and radian measure.

Radian-Degree Conversion Factors

$$1 \text{ radian} = \left(\frac{180}{\pi}\right)^{\circ} \qquad 1° = \left(\frac{\pi}{180}\right) \text{ radians}$$

Remark   Using a calculator, we have

$$1 \text{ radian} \approx 57.29577951° \qquad \text{and} \qquad 1° \approx 0.017453292 \text{ radians}$$

EXAMPLE 3   **Convert Degree Measure to Radian Measure**

Convert 300° to radians.

Solution

$$300° = 300\left(\frac{\pi}{180}\right) \text{ radians}$$

$$= \frac{5}{3}\pi \text{ radians} \qquad \bullet \text{ Exact answer}$$

$$\approx 5.23598776 \text{ radians} \qquad \bullet \text{ Approximate answer}$$

■ *Try Exercise* **30,** *page 303.*

Remark   Often we will express the radian measure of an angle in terms of π.

EXAMPLE 4   Convert Radian Measure to Degree Measure

Convert $-\frac{3}{4}\pi$ radians to degrees.

Solution

$$-\frac{3}{4}\pi \text{ radians} = -\frac{3}{4}\pi\left(\frac{180}{\pi}\right)^{\circ} = -135°$$

■ *Try Exercise 34, page 303.*

The table to the left lists the degree and radian measure of selected angles. Figure 6.10 illustrates each angle as measured from the positive *x*-axis.

Degrees	Radians
0	0
30	$\pi/6$
45	$\pi/4$
60	$\pi/3$
90	$\pi/2$
120	$2\pi/3$
135	$3\pi/4$
150	$5\pi/6$
180	π
210	$7\pi/6$
225	$5\pi/4$
240	$4\pi/3$
270	$3\pi/2$
300	$5\pi/3$
315	$7\pi/4$
330	$11\pi/6$
360	2π

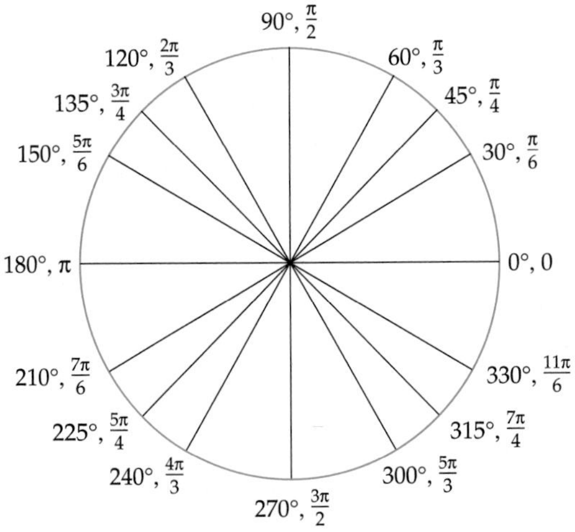

Figure 6.10
Degree and radian measures of selected angles.

Consider a circle of radius *r*. By solving the formula $\theta = s/r$ for *s*, we have an equation for arc length.

Arc Length

Let *r* be the length of the radius of a circle *C* and θ the radian measure of a central angle of *C*. Then the length of the arc *s* that subtends the central angle is $s = r\theta$. See Figure 6.11.

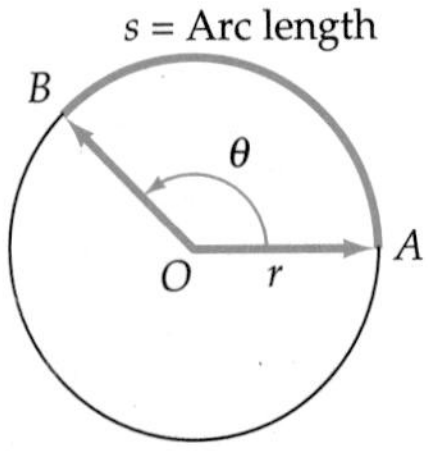

Figure 6.11

EXAMPLE 5   Find the Length of an Arc

Find the length of an arc that subtends a central angle of 120° in a circle of radius 10 cm.

Solution   The formula $s = r\theta$ requires that θ be expressed in radians. We first convert $120°$ to radian measure and then use the formula $s = r\theta$.

$$\theta = 120° = 120\left(\frac{\pi}{180}\right) \text{ radians} = \frac{2\pi}{3} \text{ radians}$$

$$s = r\theta = 10\left(\frac{2\pi}{3}\right) = \frac{20\pi}{3} \text{ cm}$$

■ *Try Exercise* **50,** *page 303.*

There is a wide range of application problems that are solved by using radian measure.

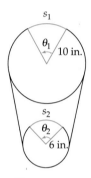

Figure 6.12

EXAMPLE 6   **Solve an Application Involving Radians**

A pulley with a radius of 10 inches uses a belt to drive a pulley with a radius of 6 inches. Find the angle through which the smaller pulley turns as the 10-inch pulley makes one revolution.

Solution   Use the formula $s = r\theta$. As the 10-inch pulley turns through an angle θ_1, a point on that pulley moves s_1 inches, where $s_1 = 10\theta_1$. See Figure 6.12. At the same time, the 6-inch pulley turns through an angle of θ_2 and a point on that pulley moves s_2 inches, where $s_2 = 6\theta_2$. Assuming the belt does not slip on the pulleys, we have $s_1 = s_2$. Thus,

$$10\theta_1 = 6\theta_2$$

$$10(2\pi) = 6\theta_2 \quad \bullet \; \theta_1 = 2\pi \text{ radians}$$

$$\frac{10}{3}\pi = \theta_2 \quad \bullet \text{ Solve for } \theta_2.$$

The 6-inch pulley turns through an angle of $\dfrac{10}{3}\pi$ radians.

■ *Try Exercise* **54,** *page 303.*

Linear and Angular Speed

A car traveling at a speed of 55 miles per hour covers a distance of 55 miles in 1 hour. **Linear speed** is *distance* traveled per unit time. In equation form, $v = s/t$ where v is the speed, s is the distance traveled and t is the time.

The floppy disk in a computer disk drive revolving at 300 revolutions per minute (rpm) makes 300 complete revolutions in 1 minute. **Angular speed** is the *angle* through which a point on a circle moves per unit time. In equation form, $\omega = \theta/t$ where ω is the angular speed, θ is the measure of the angle through which a point has moved, and t is the time. Some common units of angular speed are revolutions per second, revolutions per minute (rpm), radians per second, and radians per minute.

EXAMPLE 7   **Solve an Angular Speed Application Problem**

A hard disk in a computer rotates at 3600 rpm. Find the angular speed in radians per second.

Solution   As a point on the disk makes 1 revolution, the angle through which the point moves is 2π radians. Thus 3600 rpm is $3600(2\pi)$ radians per minute. To find the radians per second, convert minutes to seconds.

$$\omega = \frac{3600(2\pi)\ \text{radians}}{1\ \text{min}} = \frac{3600(2\pi)\ \text{radians}}{1\ \text{min}} \cdot \frac{1\ \text{min}}{60\ \text{sec}}$$

$$= 120\pi\ \text{radians/sec} \approx 377\ \text{radians/sec}$$

■ *Try Exercise 56, page 303.*

The tire on a car traveling along a road has both linear speed and angular speed. The relationship between linear and angular speed can be expressed by an equation.

Assume that the wheel in Figure 6.13 is rolling without slipping. As the wheel moves a distance s, point A moves through an angle θ. The arc length subtending angle θ is also s, the distance traveled by the wheel. From the equations for linear and angular speed, we have

$$v = \frac{s}{t} = \frac{r\theta}{t} = r\frac{\theta}{t} \quad \bullet\ s = r\theta$$

$$= r\omega \quad \bullet\ \omega = \theta/t$$

The equation $v = r\omega$ gives the linear speed of a point on a rotating body in terms of a distance r from the axis of rotation and the angular speed ω.

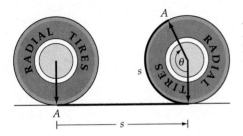

Figure 6.13

EXAMPLE 8   **Solve a Linear Speed Application Problem**

An R 14 label on an automobile tire indicates that the radius of the tire is 14 inches. Find, to the nearest mile per hour, the speed of an automobile with an R 14 tire that is rotating 600 rpm.

Solution   We need to change the units to miles per hour. First change 600 revolutions per minute to 36,000 revolutions per hour by multiplying by 60. Next multiply by 2π to change ω to radians per hour: $\omega = 72,000\pi$ radians per hour. We will use the equation $v = r\omega$ to find the speed. Because the answer is to be in miles per hour, the radius 14 inches must be changed to miles.

$$14\ \text{inches} = 14\ \text{inches} \cdot \frac{1\ \text{mile}}{63,360\ \text{inches}} = \frac{14}{63,360}\ \text{miles} \quad \bullet\ 63,360\ \text{inches} = 1\ \text{mile}$$

$$v = r\omega = \frac{14}{63,360} \cdot 72,000\pi \approx 50\ \text{mph}$$

■ *Try Exercise 60, page 303.*

EXERCISE SET 6.1

In Exercises 1 to 12, find the complement and supplement of each angle.

1. 15°   **2.** 87°   **3.** 70°15′

4. 22°43′   **5.** 56°33′15″   **6.** 19°42′05″

7. 1   **8.** 0.5   **9.** $\pi/4$

10. $\pi/3$ **11.** $\pi/2$ **12.** $\pi/6$

In Exercises 13 to 18, convert the DMS measure of each angle to decimal degree measure to the nearest thousandth of a degree.

13. 78°8′   **14.** 5°39′   **15.** 16°44″

16. 35°42″   **17.** 47°20′18″   **18.** 20°4′45″

In Exercises 19 to 24, convert the decimal degree measure of each angle to the DMS system of measurement.

19. 110.4°   **20.** 36.6°   **21.** −66.72°

22. 55.44°   **23.** −7.05°   **24.** 342.17°

In Exercises 25 to 30, convert the measure of each angle to exact radian measure.

25. 15°   **26.** 165°   **27.** 315°

28. −210°   **29.** −225°   **30.** −330°

In Exercises 31 to 36, convert the radian measure of each angle to exact degree measure.

31. $\pi/6$ **32.** $\pi/9$ **33.** $3\pi/8$

34. $11\pi/18$ **35.** $11\pi/3$ **36.** $6\pi/5$

In Exercises 37 to 42, convert radians to degrees or degrees to radians. Round answers to the nearest hundredth.

37. 1.5   **38.** −2.3   **39.** 133°

40. 327°   **41.** 5.25   **42.** −90°

In Exercises 43 to 46, find the measure in radians and degrees of the central angle of a circle subtended by the given arc.

43. $r = 2$ in, $s = 8$ in   **44.** $r = 7$ ft, $s = 4$ ft

45. $r = 5.2$ cm, $s = 12.4$ cm   **46.** $r = 35.8$ m, $s = 84.3$ m

In Exercises 47 to 50, find the length of the arc of a circle with the given radius and central angle.

47. $r = 8$ in, $\theta = \pi/4$ **48.** $r = 3$ ft, $\theta = 7\pi/2$

49. $r = 25$ cm, $\theta = 42°$ **50.** $r = 5$ m, $\theta = 144°$

51. Find the number of radians in $1\frac{1}{2}$ revolutions.

52. Find the number of radians in $\frac{3}{8}$ revolution.

53. A pulley with a radius of 14 inches uses a belt to drive a pulley with a radius of 28 inches. The 14-inch pulley turns through an angle of 150°. Find the angle through which the 28-inch pulley turns.

54. A pulley with a diameter of 1.2 meters uses a belt to drive a pulley with a diameter of 0.8 meter. The 1.2-meter pulley turns through an angle of 240°. Find the angle through which the 0.8-meter pulley turns.

55. Find the angular speed of the second hand on a clock in radians per second.

56. Find the angular speed, in radians per second, of a point on the equator of the earth.

57. A wheel is rotating at 50 rpm. Find the angular speed in radians per second.

58. A wheel is rotating at 200 rpm. Find the angular speed in radians per second.

59. The turntable of a record player turns at 45 rpm. Find the angular speed in radians per second.

60. A car with a wheel of radius 14 inches is moving with a speed of 55 mph. Find the angular speed of the wheel in radians per second.

61. A car with a tire of radius 15 inches is rotating at 450 rpm. Find the speed of the automobile to the nearest mile per hour.

62. A truck with a tire of radius 18 inches is rotating at 500 rpm. Find the speed of the truck to the nearest mile per hour.

Supplemental Exercises

A **sector** of a circle is the figure bounded by radii OA and OB and the intercepted arc AB. The area of the sector is given by $A = \frac{1}{2}r^2\theta$, where r is the radius of the circle and θ is the measure of the central angle in radians. In Exercises 63 to 68, find the area, to the nearest square unit, of the sector of a circle with the given radius and central angle.

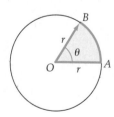

63. $r = 5$ in, $\theta = \pi/3$ radians

64. $r = 2.8$ ft, $\theta = 5\pi/2$ radians

65. $r = 120$ cm, $\theta = 0.65$ radians

66. $r = 30$ ft, $\theta = 62°$

67. $r = 20$ m, $\theta = 125°$

68. $r = 25$ cm, $\theta = 220°$

69. The minute hand on the clock atop city hall measures 6 ft 3 inches from the tip to its axle.

　a. Through what angle (in radians) does the minute hand pass between 9:12 A.M. and 9:48 A.M.?

　b. What distance to the nearest tenth of a foot, does the tip of the minute hand travel during this period?

70. At a time when the earth was 93,000,000 miles from the sun, using a transit, you observed through a properly smoked glass that the diameter of the sun occupied an arc of 31′. Calculate the approximate diameter of the sun to the nearest ten thousand.

71. A merry-go-round horse is 11.6 meters from the center. The merry-go-round makes $14\frac{1}{4}$ revolutions per ride in 5 minutes. (a) How many meters, to the nearest meter, does the horse travel? (b) How fast is it moving in meters per second?

72. a. A car with 13-inch radius tires makes an 8-mile trip. Find the number of revolutions the tire makes on the 8-mile trip.

　b. A car with 15-inch radius tires makes an 8-mile trip. Find the number of revolutions the tire makes on the 8-mile trip.

73. A water wheel has a 10-foot radius. When the wheel makes 18 revolutions per minute, what is the speed of the river in feet per second?

74. A pulley with a 50-centimeter diameter drives a pulley with a 20-centimeter diameter. The larger pulley makes 30 revolutions per minute. What is the linear speed of a point on the circumference of the smaller pulley?

75. Find the area of the shaded portion of the graph shown. The radius of the circle is 9 inches.

76. Latitude describes the position of a point on the earth's surface in relation to the equator. A point on the equator has a latitude of 0°. The north pole has a latitude of 90°. The radius of the earth is approximately 3960 miles. Assuming that the earth is a perfect sphere, find the distance along the earth's surface that subtends a central angle of latitude (a) 1°, (b) 1′, and (c) 1″. Express your answer to 3 significant digits.

6.2   Trigonometric Functions of Acute Angles

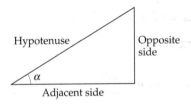

a. Adjacent and opposite sides of ∠α

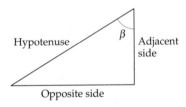

b. Adjacent and opposite sides of ∠β

Figure 6.14

The study of trigonometry, which means "triangle measurement," began more than 2000 years ago, partially as a means of solving surveying problems. Early trigonometry used the length of a chord of a circle as the value of a *trigonometric function*. In the sixteenth century, right triangles were used to define a trigonometric function. We will use a modification of this approach.

　When working with right triangles, it is convenient to refer to the side *opposite* an angle or the side *adjacent* to (next to) an angle. Figure 6.14a shows the sides opposite and adjacent to the angle α. For angle β, the opposite and adjacent sides are shown as in Figure 6.14b. In both cases, the hypotenuse remains the same.

　Consider an angle θ in the right triangle shown in Figure 6.15. Let x and y represent the lengths, respectively, of the adjacent and opposite sides of the triangle, and let r be the length of the hypotenuse. Six possible ratios can be formed:

$$\frac{y}{r}, \frac{x}{r}, \frac{y}{x}, \frac{r}{y}, \frac{r}{x}, \frac{x}{y}$$

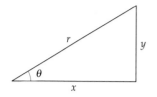

Figure 6.15

Each ratio defines a value of a trigonometric function of the acute angle θ. The functions are **sine** (sin), **cosine** (cos), **tangent** (tan), **cosecant** (csc), **secant** (sec), and **cotangent** (cot).

Trigonometric Functions of An Acute Angle

Let θ be an acute angle of a right triangle. The values of the six trigonometric functions of θ are

$$\sin \theta = \frac{\text{length of opposite side}}{\text{length of hypotenuse}} = \frac{y}{r} \qquad \cos \theta = \frac{\text{length of adjacent side}}{\text{length of hypotenuse}} = \frac{x}{r}$$

$$\tan \theta = \frac{\text{length of opposite side}}{\text{length of adjacent side}} = \frac{y}{x} \qquad \cot \theta = \frac{\text{length of adjacent side}}{\text{length of opposite side}} = \frac{x}{y}$$

$$\sec \theta = \frac{\text{length of hypotenuse}}{\text{length of adjacent side}} = \frac{r}{x} \qquad \csc \theta = \frac{\text{length of hypotenuse}}{\text{length of opposite side}} = \frac{r}{y}$$

We will write opp, adj, and hyp as abbreviations for *the length of the* opposite side, adjacent side, and hypotenuse, respectively.

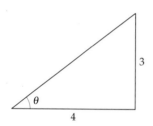

Figure 6.16

EXAMPLE 1   **Evaluate Trigonometric Functions**

Find the values of the six trigonometric functions of θ for the triangle given in Figure 6.16.

Solution   Use the Pythagorean Theorem to find the length of the hypotenuse.

$$r = \sqrt{3^2 + 4^2} = \sqrt{25} = 5$$

From the definitions of the trigonometric functions,

$$\sin \theta = \frac{\text{opp}}{\text{hyp}} = \frac{3}{5} \qquad\qquad \cos \theta = \frac{\text{adj}}{\text{hyp}} = \frac{4}{5}$$

$$\tan \theta = \frac{\text{opp}}{\text{adj}} = \frac{3}{4} \qquad\qquad \cot \theta = \frac{\text{adj}}{\text{opp}} = \frac{4}{3}$$

$$\sec \theta = \frac{\text{hyp}}{\text{adj}} = \frac{5}{4} \qquad\qquad \csc \theta = \frac{\text{hyp}}{\text{opp}} = \frac{5}{3}$$

■ *Try Exercise **6**, page 312.*

Given the value of one trigonometric function, it is possible to find the value of any of the remaining trigonometric functions.

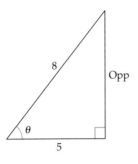

Figure 6.17

EXAMPLE 2   Find the Value of a Trigonometric Function

Given $\cos \theta = \dfrac{5}{8}$, find $\tan \theta$.

Solution $\cos \theta = \dfrac{5}{8} = \dfrac{\text{adj}}{\text{hyp}}$

Draw a triangle with adjacent side of length 5 units and hypotenuse of length 8 units. (Figure 6.17) Use the Pythagorean Theorem to find the length of the opposite side.

$$8^2 = (\text{opp})^2 + 5^2$$
$$(\text{opp})^2 = 39$$
$$\text{opp} = \sqrt{39}$$
$$\tan \theta = \frac{\text{opp}}{\text{adj}} = \frac{\sqrt{39}}{5}$$

■ *Try Exercise* **16**, *page 312.*

In Example 1, the lengths of the legs of the triangle were given and you were asked to find the values of the six trigonometric functions of the angle θ. Often we will want to find the value of a trigonometric function when we are given *the measure of an angle* rather than the measure of the sides of a triangle. For most angles, advanced mathematical methods are required to evaluate a trigonometric function. For some *special angles* however, the value of a trigonometric function can be found by geometric methods. These special acute angles are 30°, 45°, and 60°.

First we will find the values of the six trigonometric functions of 45°. (This discussion is based on angles measured in degrees. Radian measure could have been used without changing the results.) Figure 6.18 shows a right triangle with angles 45°, 45°, and 90°. Because $\angle A = \angle B$, the lengths of the sides opposite these angles are equal. Let the length of each equal side be denoted by a. From the Pythagorean theorem,

$$r^2 = a^2 + a^2 = 2a^2$$
$$r = \sqrt{2a^2} = a\sqrt{2}$$

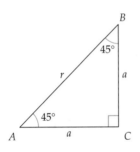

Figure 6.18

The values of the six trigonometric functions of 45° are

$$\sin 45° = \frac{a}{a\sqrt{2}} = \frac{1}{\sqrt{2}} = \frac{\sqrt{2}}{2} \qquad \cos 45° = \frac{a}{a\sqrt{2}} = \frac{1}{\sqrt{2}} = \frac{\sqrt{2}}{2}$$

$$\tan 45° = \frac{a}{a} = 1 \qquad\qquad \cot 45° = \frac{a}{a} = 1$$

$$\sec 45° = \frac{a\sqrt{2}}{a} = \sqrt{2} \qquad\qquad \csc 45° = \frac{a\sqrt{2}}{a} = \sqrt{2}$$

The values of the trigonometric functions of the special angles 30° and 60° can be found by drawing an equilateral triangle and bisecting one of the angles, as Figure 6.19 shows. The angle bisector also bisects one of the sides. Thus the length of the side opposite the 30° angle is one-half the length of the hypotenuse of triangle OAB.

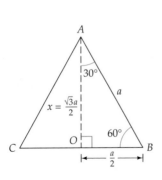

Figure 6.19

Let a denote the length of the hypotenuse. Then the length of the side opposite the 30° angle is $a/2$. The length of the adjacent side to the 30° angle, x, is found by using the Pythagorean theorem.

$$a^2 = \left(\frac{a}{2}\right)^2 + x^2$$

$$a^2 = \frac{a^2}{4} + x^2$$

$$\frac{3a^2}{4} = x^2 \qquad \bullet \text{ Subtracting } \frac{a^2}{4} \text{ from each side.}$$

$$x = \frac{\sqrt{3}\,a}{2} \qquad \bullet \text{ Solve for } x.$$

The values of the six trigonometric functions of 30° are

$$\sin 30° = \frac{a/2}{a} = \frac{1}{2} \qquad\qquad \cos 30° = \frac{\sqrt{3}\,a/2}{a} = \frac{\sqrt{3}}{2}$$

$$\tan 30° = \frac{a/2}{\sqrt{3}\,a/2} = \frac{1}{\sqrt{3}} = \frac{\sqrt{3}}{3} \qquad \cot 30° = \frac{\sqrt{3}\,a/2}{a/2} = \sqrt{3}$$

$$\sec 30° = \frac{a}{\sqrt{3}\,a/2} = \frac{2}{\sqrt{3}} = \frac{2\sqrt{3}}{3} \qquad \csc 30° = \frac{a}{a/2} = 2$$

The values of the trigonometric functions of 60° can be found by again using Figure 6.19. The length of the side opposite the 60° angle is $\sqrt{3}\,a/2$ and the length of the side adjacent to the 60° angle is $a/2$. The values of the trigonometric functions of 60° are

$$\sin 60° = \frac{\sqrt{3}\,a/2}{a} = \frac{\sqrt{3}}{2} \qquad\qquad \cos 60° = \frac{a/2}{a} = \frac{1}{2}$$

$$\tan 60° = \frac{\sqrt{3}\,a/2}{a/2} = \sqrt{3} \qquad\qquad \cot 60° = \frac{a/2}{\sqrt{3}\,a/2} = \frac{1}{\sqrt{3}} = \frac{\sqrt{3}}{3}$$

$$\sec 60° = \frac{a}{a/2} = 2 \qquad\qquad \csc 60° = \frac{a}{\sqrt{3}\,a/2} = \frac{2}{\sqrt{3}} = \frac{2\sqrt{3}}{3}$$

Table 6.1 summarizes the values of the trigonometric functions for the special angles 30°($\pi/6$), 45°($\pi/4$), and 60°($\pi/3$).

TABLE 6.1   Trigonometric Functions of Special Angles

θ	$\sin \theta$	$\cos \theta$	$\tan \theta$	$\csc \theta$	$\sec \theta$	$\cot \theta$
30°; $\dfrac{\pi}{6}$	$\dfrac{1}{2}$	$\dfrac{\sqrt{3}}{2}$	$\dfrac{\sqrt{3}}{3}$	2	$\dfrac{2\sqrt{3}}{3}$	$\sqrt{3}$
45°; $\dfrac{\pi}{4}$	$\dfrac{\sqrt{2}}{2}$	$\dfrac{\sqrt{2}}{2}$	1	$\sqrt{2}$	$\sqrt{2}$	1
60°; $\dfrac{\pi}{3}$	$\dfrac{\sqrt{3}}{2}$	$\dfrac{1}{2}$	$\sqrt{3}$	$\dfrac{2\sqrt{3}}{3}$	2	$\dfrac{\sqrt{3}}{3}$

EXAMPLE 3   **Evaluate a Trigonometric Expression**

Find the exact value of $\sin^2 45° + \cos^2 60°$.

Solution   Substitute the values of $\sin 45°$ and $\cos 60°$ into the expression and simplify. *Note:* $\sin^2\theta = (\sin\ \theta)(\sin\ \theta) = (\sin\ \theta)^2$ and $\cos^2\theta = (\cos\ \theta)(\cos\ \theta) = (\cos\ \theta)^2$.

$$\sin^2 45° + \cos^2 60° = \left(\frac{\sqrt{2}}{2}\right)^2 + \left(\frac{1}{2}\right)^2 = \frac{2}{4} + \frac{1}{4} = \frac{3}{4}$$

■ *Try Exercise* **34,** *page 312.*

Appendix III gives the table of values of the trigonometric functions for some angles with degree measure between 0° and 90° (radian measure between 0 and $\pi/2$). For more accuracy, a calculator or interpolation can be used. (Interpolation is discussed in Appendix I.)

From the definition of the sine and cosecant functions,

$$(\sin\theta)(\csc\theta) = \frac{y}{r} \cdot \frac{r}{y} = 1 \quad \text{or} \quad (\sin\theta)(\csc\theta) = 1$$

By rewriting the last equation, we can express the sine and cosecant functions in the following forms:

$$\sin\theta = \frac{1}{\csc\theta} \quad \text{and} \quad \csc\theta = \frac{1}{\sin\theta}$$

The sine and cosecant functions are called **reciprocal** functions. The cosine and secant are also reciprocal functions, as are the tangent and cotangent functions. Why?[2]

To find the values of the trigonometric functions using a scientific calculator, first note that the calculator has only the function keys $\boxed{\sin}$, $\boxed{\cos}$, and $\boxed{\tan}$; the values of the cosecant, secant, and cotangent are found by using the *reciprocal* key, $1/x$.

Table 6.2 shows each trigonometric function and its reciprocal. These relationships are true for all values of the variable θ for which the functions are defined.

TABLE 6.2   Trigonometric Functions and Their Reciprocals

$$\sin\theta = \frac{1}{\csc\theta} \qquad \cos\theta = \frac{1}{\sec\theta} \qquad \tan\theta = \frac{1}{\cot\theta}$$

$$\csc\theta = \frac{1}{\sin\theta} \qquad \sec\theta = \frac{1}{\cos\theta} \qquad \cot\theta = \frac{1}{\tan\theta}$$

---

[2] $\sec\theta = r/x$ and $\cos\theta = x/r$; therefore, $(\sec\theta)(\cos\theta) = 1$ and $\sec\theta = \frac{1}{\cos\theta}$.

$\tan\theta = y/x$ and $\cot\theta = x/y$; therefore, $(\cot\theta)(\tan\theta) = 1$ and $\cot\theta = \frac{1}{\tan\theta}$.

Here are some examples of evaluating a trigonometric function using a calculator. When an angle is written with the degree symbol, make sure your calculator is in degree mode. When an angle is *not* written with the degree symbol, the angle is assumed to be in radians. In this case, your calculator should be in radian mode. *Many needless errors are made because the correct mode of the calculator was not selected. Be careful!*

The keystrokes given below are typical of those required for an algebraic calculator. Consult your users manual for the keystrokes for your particular calculator.

Trigonometric Function	Key Sequence	Calculator Display
cos 33°	33 `cos`	0.83867057
sin 1	1 `sin`	0.84147098
sec 58°	58 `cos` `1/x`	1.88707992

The calculation of sec 58° used the reciprocal key, `1/x`, to find the reciprocal of cos 58°. By pressing this key when the value of cos 58° is shown in the display, we are performing the mathematical equivalent of

$$\frac{1}{\cos 58°} = \sec 58°$$

As a final reminder, degree mode was used for cos 33° and sec 58°; radian mode was used for sin 1. Many values of trigonometric functions are irrational numbers. Often the number in the display of your calculator is an approximation to the actual value.

Applications of Right Triangles

One of the major reasons for the development of trigonometry was to solve applications problems. In this section we will consider some applications involving right triangles. A more extensive examination of application problems appears later in the text.

In some application problems, a horizontal line of sight is used as a reference line. An angle measured above the line of sight is called an **angle of elevation,** and an angle measured below the line of sight is called an **angle of depression.** See Figure 6.20.

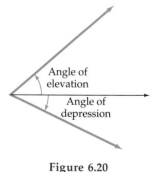

Angle of elevation

Angle of depression

Figure 6.20

EXAMPLE 4   **Solve an Angle of Elevation Problem**

Redwood trees are among the tallest of all trees. From a point 115 feet from the base of a redwood tree, the angle of elevation to the top of the tree is 64.3°. Find the height of the tree to the nearest foot.

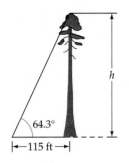

Figure 6.21

Solution   From Figure 6.21, the length of the adjacent side of the angle is known (115 feet). Because we need to determine the height of the tree (length of the opposite side), we use the tangent function. Let h represent the length of the opposite side.

$$\tan 64.3° = \frac{\text{opp}}{\text{adj}} = \frac{h}{115}$$

$$h = 115 \tan 64.3° \approx 238.952$$

The height of the redwood tree is approximately 239 feet.

■ *Try Exercise **62**, page 313.*

Remark   Because the cotangent function involves the sides adjacent to and opposite an angle, we could have solved Example 4 by using the cotangent function. The solution would have been

$$\cot 64.3° = \frac{\text{adj}}{\text{opp}} = \frac{115}{h}$$

$$h = \frac{115}{\cot 64.3°} \approx 238.952$$

The accuracy of a calculator is sometimes beyond the limits of measurement. In the last example, the distance from the base of the tree was given as 115 ft (three significant digits) whereas the height of the tree was shown as 238.947 ft (6 significant digits). When using approximate numbers, we will use the following conventions when calculating with trigonometric functions.

Significant Digits for Trigonometric Calculations

Angle Measure to the Nearest	Significant Digits of the Lengths
Degree	Two
Tenths of a degree	Three
Hundredths of a degree	Four

EXAMPLE 5   Solve an Angle of Depression Problem

DME (Distance Measuring Equipment) is standard avionic equipment on a commercial airplane. This equipment measures the distance from a plane to a radar station. If the distance from a plane to a radar station is 160 miles and the angle of depression is 33°, find the number of ground miles from a point directly below the plane to the radar station.

Solution   From Figure 6.22, the length of the hypotenuse is known (160 miles). The length of the side opposite the angle of 57° is unknown.

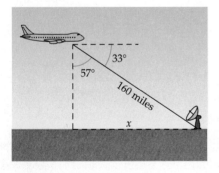

Figure 6.22

The sine function involves the hypotenuse and the opposite side, x, of the angle.

$$\sin 57° = \frac{\text{opp}}{\text{hyp}} = \frac{x}{160}$$

$$x = 160 \sin 57°$$

The plane is approximately 130 ground miles from the radar station.

■ Try Exercise 64, page 313.

EXAMPLE 6   Solve an Angle of Elevation Problem

An observer notes that the angle of elevation from point A to the top of a space shuttle is 27.2°. From a point 17.5 meters further from the space shuttle, the angle of elevation is 23.9°. Find the height of the space shuttle.

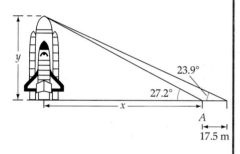

Figure 6.23

Solution   From Figure 6.23, let x denote the distance from point A to the base of the space shuttle, and let y denote the height of the space shuttle. Then

$$(1) \ \tan 27.2° = \frac{y}{x} \quad \text{and} \quad (2) \ \tan 23.9° = \frac{y}{x + 17.5}$$

Solving Equation (1) for x, $\left(x = \frac{y}{\tan 27.2°} = y \cot 27.2° \right)$, and substituting into Equation (2), we have

$$\tan 23.9° = \frac{y}{y \cot 27.2° + 17.5}$$

$$y = \tan 23.9°(y \cot 27.2° + 17.5) \quad \bullet \text{ Solve for } y.$$

$$y - y \tan 23.9° \cot 27.2° = \tan 23.9°(17.5)$$

$$y = \frac{\tan 23.9°(17.5)}{1 - \tan 23.9° \cot 27.2°}$$

$$\approx 56.2993$$

The approximate height of the space shuttle is 56.3 meters.

■ Try Exercise 68, page 313.

Remark   The intermediate calculations for the last two examples were not rounded off. This ensures better accuracy for the final result. Using the conventions stated earlier, we round off only the last result.

EXERCISE SET 6.2

For Exercises 1 to 10, find the values of the six trigonometric functions of θ for a right triangle with the given sides. Let y denote the opposite side, x the adjacent side, and r the hypotenuse.

1. $y = 12, x = 5$ **2.** $y = 7, x = 3$

3. $y = 4, r = 7$ **4.** $y = 3, r = 9$

5. $y = 5, x = 2$ **6.** $y = 5, r = 8$

7. $y = \sqrt{3}, x = 2$ **8.** $y = \sqrt{5}, r = \sqrt{10}$

9. $x = \sqrt{7}, r = \sqrt{15}$ **10.** $x = \sqrt{11}, r = 2\sqrt{3}$

For Exercises 11 to 13, let θ be an acute angle of a right triangle and $\sin \theta = 3/5$. Find the following:

11. $\tan \theta$ **12.** $\sec \theta$ **13.** $\cos \theta$

For Exercises 14 to 16, let θ be an acute angle of a right triangle and $\tan \theta = 4/3$. Find the following:

14. $\sin \theta$ **15.** $\cot \theta$ **16.** $\sec \theta$

For Exercises 17 to 19, let β be an acute angle of a right triangle and $\sec \beta = 13/12$. Find the following:

17. $\cos \beta$ **18.** $\cot \beta$ **19.** $\csc \beta$

For Exercises 20 to 22, let θ be an acute angle of a right triangle and $\cos \theta = 2/3$. Find the following:

20. $\sin \theta$ **21.** $\sec \theta$ **22.** $\tan \theta$

For Exercises 23 to 38, find the *exact* value of each expression.

23. $\sin 45° + \cos 45°$ **24.** $\csc 45° - \sec 45°$

25. $\sin 30° \cos 60° - \tan 45°$ **26.** $\csc 60° \sec 30° + \cot 45°$

27. $\sin 30° \cos 60° + \tan 45°$

28. $\sec 30° \cos 30° - \tan 60° \cot 60°$

29. $2 \sin 60° - \sec 45° \tan 60°$

30. $\sec 45° \cot 30° + 3 \tan 60°$

31. $\sin \dfrac{\pi}{3} + \cos \dfrac{\pi}{6}$ **32.** $\csc \dfrac{\pi}{6} - \sec \dfrac{\pi}{3}$

33. $\sin \dfrac{\pi}{4} + \tan \dfrac{\pi}{6}$ **34.** $\sin \dfrac{\pi}{3} \cos \dfrac{\pi}{4} - \tan \dfrac{\pi}{4}$

35. $\sec \dfrac{\pi}{3} \cos \dfrac{\pi}{3} - \tan \dfrac{\pi}{6}$ **36.** $\cos \dfrac{\pi}{4} \tan \dfrac{\pi}{6} + 2 \tan \dfrac{\pi}{3}$

37. $2 \csc \dfrac{\pi}{4} - \sec \dfrac{\pi}{3} \cos \dfrac{\pi}{6}$ **38.** $3 \tan \dfrac{\pi}{4} + \sec \dfrac{\pi}{6} \sin \dfrac{\pi}{3}$

In Exercises 29 to 54, find the value of the trigonometric function to 4 decimal places.

39. $\tan 32°$ **40.** $\sec 88°$ **41.** $\cos 63°20'$

42. $\cot 55°50'$ **43.** $\cos 34.7°$ **44.** $\tan 81.3°$

45. $\sec 5.9°$ **46.** $\sin \dfrac{\pi}{5}$ **47.** $\tan \dfrac{\pi}{7}$

48. $\sec \dfrac{3\pi}{8}$ **49.** $\csc 1.2$ **50.** $\sin 0.45$

51. $\cos 1.25$ **52.** $\tan \dfrac{3}{4}$ **53.** $\sec \dfrac{5}{8}$

54. $\cot \dfrac{3}{5}$

55. A 12-foot ladder is resting against a wall and makes an angle of 52° with the ground. Find the height to which the ladder will reach on the wall.

56. Find the distance AB across the marsh shown in the accompanying figure.

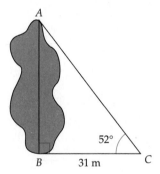

57. Show that the slope of a line that makes an angle θ with the positive x-axis equals $\tan \theta$.

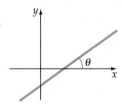

58. Television screens are measured by the length of the diagonal of the screen. Find the width of a 19-inch television screen if the diagonal makes an angle of 38° with the base of the screen.

59. At 1:00 P.M., a boat is 40 km due east of a lighthouse and traveling 10 km/h in a direction that is 30° south of an east-west line. At what time will the boat be closest to the lighthouse?

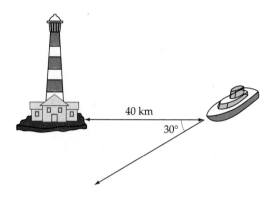

60. At 3:00 P.M., a boat is 12.5 miles due west of a radar station and traveling at 11 mph in a direction that is 57.3° south of an east-west line. At what time will the boat be closest to the radar station?

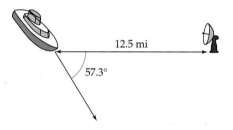

61. For best illumination of a piece of art, a lighting specialist for an art gallery recommends that a ceiling-mounted light be 6 ft from a piece of art and that the angle of depression of the light be 38°. How far from a wall should the light be placed so that the recommendations of the specialist are met?

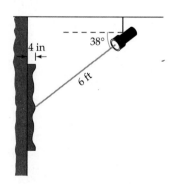

62. The angle of elevation from a point 116 meters from the base of the Eiffel Tower to the top of the tower is 68.9°. Find the approximate height of the tower.

63. An airplane traveling at 240 mph is descending at an angle of depression of 6°. How many miles will the plane descend in 4 minutes?

64. A submarine traveling 9.0 mph is descending at a angle of depression of 5°. How many minutes does it take the submarine to reach a depth of 80 feet?

65. From a point 300 feet from the base of a Roman aqueduct in southern France, the angle of elevation to the top of the aqueduct is 78°. Find the height of the aqueduct.

66. The angle of depression of one side of a lake, measured from a balloon 2500 feet above the lake as shown in the figure, is 43°. The angle of depression to the opposite side of the lake is 27°. Find the width of the lake.

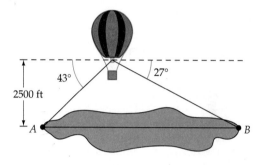

67. The angle of elevation to the top of the Egyptian pyramid Cheops is 36.4°, measured from a point 350 feet from the base of the pyramid. The angle of elevation of a face of the pyramid is 51.9°. Find the height of Cheops.

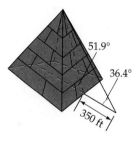

68. Two buildings are 240 feet apart. The angle of elevation from the top of the shorter building to the top of the other building is 22°. If the shorter building is 80 feet high, how high is the taller building?

69. From a point A on a line from the base of the Washington Monument, the angle of elevation to the top of the monument is 42.0°. From a point 100 feet away and on the same line, the angle to the top is 37.8°. Find the approximate height of the Washington Monument.

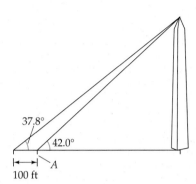

70. The angle of elevation to the top of a radio antenna on the top of a building is 53.4°. After moving 200 feet closer to the building, the angle of elevation is 64.3°. Find the height of the building if the height of the antenna is 180 feet.

Supplemental Exercises

71. A circle is inscribed in a regular hexagon with each side 6.0 meters long. Find the radius of the circle.

72. Show that the area A of the triangle given in the figure is $A = \frac{1}{2}ab \sin \theta$.

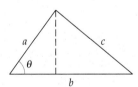

73. If an angle of 27° has been measured to the nearest degree, then the actual measure of the angle θ is such that $26.5° \le \theta < 27.5$. From a distance of exactly 100 meters from the base of a tree, the angle of elevation is measured as 27° to the nearest degree. Find the range in which the height of the tree must fall.

74. Let B denote the base of a clock tower. The angle of elevation from a point A to the top of a clock tower is 56.3°. On a line perpendicular to AB and 25 feet from A, the angle of elevation is 53.3°. Find the height of the clock tower.

75. Find the length of the longest piece of wood that can be slid around the corner of the hallway in the figure. (*Hint:*

The angle θ must be 45°.)

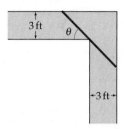

76. In Exercise 75, suppose that the hall is 8 feet high. Find the length of the longest piece of wood that can be taken around the corner.

77. At a point A at a distance from the base B of a flagpole, the angle of elevation to the top of the pole is 28.1°. On a line perpendicular to AB and 85 feet from A, the angle of elevation to the top of the pole is 23.5°. Find the height of the pole.

78. Show that the perimeter P of a regular n-sided polygon (n-gon) inscribed in a circle of radius 1 is $P = 2n \sin \frac{180°}{n}$.

79. Show that the area A of a regular n-gon inscribed in a circle of radius 1 is $A = \frac{n}{2} \sin \frac{360}{n}$.

80. Using the triangle given in the figure at the right, show that

   a. $\sin^2\theta + \cos^2\theta = 1$

   b. $\tan \theta = \frac{\sin \theta}{\cos \theta}$

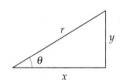

81. Show that the value of $\cos \theta$ does not depend on the lengths of the sides of a triangle.

82. Let P_n denote the perimeter of a regular n-gon inscribed in a circle of radius 1. Use the result from Exercise 78 to complete the following table.

n	10	50	100	1000	10,000
P_n					

Note that as n increases, P_n approximates 2π. Explain.

83. Let A_n denote the area of a regular n-gon inscribed in a circle of radius 1. Use the result from Exercise 79 to complete the following table.

n	10	50	100	1000	10,000
A_n					

Note that as n increases, A_n approximates π. Explain.

Trigonometric Functions of Any Angle

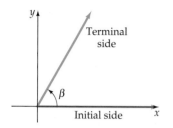

Figure 6.24

The application of trigonometry would be quite limited if all angles had to be acute angles. Fortunately, this is not the case. In this section we extend the definition of a trigonometric function to include any angle.

We begin by placing an angle in a coordinate system. An angle is in **standard position** when the initial side of the angle coincides with the positive x-axis and the vertex is at the origin. The angle β in Figure 6.24 is in standard position.

Consider angle θ in Figure 6.25 in standard position and a point $P(x, y)$ on the terminal side of the angle. We define the trigonometric functions of any angle according to the following definitions.

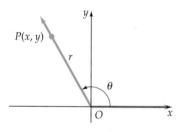

Figure 6.25

Definition of the Trigonometric Functions of Any Angle

Let $P(x, y)$ be any point, except the origin, on the terminal side of an angle θ in standard position. Let $r = d(O, P)$, the distance from the origin to P. The six trigonometric functions of θ are

$$\sin \theta = \frac{y}{r} \qquad \cos \theta = \frac{x}{r} \qquad \tan \theta = \frac{y}{x}, \quad x \neq 0$$

$$\csc \theta = \frac{r}{y}, y \neq 0 \qquad \sec \theta = \frac{r}{x}, x \neq 0 \qquad \cot \theta = \frac{x}{y}, \quad y \neq 0$$

where $r = \sqrt{x^2 + y^2}$.

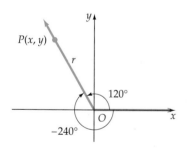

Figure 6.26

Remark   Angle θ can be positive or negative. Note from Figure 6.26 that $\sin 120° = \sin(-240°)$ because $P(x, y)$ is on the terminal side of each angle. In a similar way, the value of any trigonometric function at $120°$ is equal to the value of that function at $-240°$.

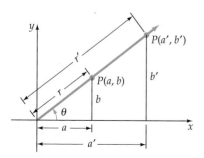

Figure 6.27

The value of a trigonometric function is independent of the point chosen on the terminal side of the angle. Consider any two points on the terminal side of an angle θ in standard position, as shown in Figure 6.27. The right triangles formed are similar triangles, so the ratios of the corresponding sides are equal. Thus, for example, $\frac{b}{a} = \frac{b'}{a'}$. Because $\tan \theta = \frac{b}{a} = \frac{b'}{a'}$ we have $\tan \theta = \frac{b'}{a'}$. Therefore, the value of the tangent function is independent of the point chosen on the terminal side of the angle. By a similar argument, we can show that the value of any trigonometric function is independent of the point chosen on the terminal side of the angle.

Any point in a rectangular coordinate system (except the origin) can determine an angle in standard position. For example, $P(-4, 3)$ in

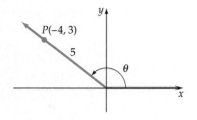

Figure 6.28

Figure 6.28 is a point in the second quadrant and determines an angle θ in standard position with $r = \sqrt{(-4)^2 + 3^2} = 5$. The values of the trigonometric functions of θ are

$$\sin \theta = \frac{3}{5} \qquad \cos \theta = \frac{-4}{5} = -\frac{4}{5} \qquad \tan \theta = \frac{3}{-4} = -\frac{3}{4}$$

$$\csc \theta = \frac{5}{3} \qquad \sec \theta = \frac{5}{-4} = -\frac{5}{4} \qquad \cot \theta = \frac{-4}{3} = -\frac{4}{3}$$

As this example shows, the sign of a trigonometric function depends on the quadrant in which the terminal side of the angle lies. For example, if θ is an angle whose terminal side lies in quadrant III and $P(x, y)$ is on the terminal side of θ, then both x and y are negative and therefore $\frac{y}{x}$ and $\frac{x}{y}$ are positive. (See Figure 6.29.) Since $\tan \theta = \frac{y}{x}$ and $\cot \theta = \frac{x}{y}$, only the values of the tangent and cotangent function are positive in quadrant III.

Table 6.3 lists the sign of the six trigonometric functions in each quadrant. Figure 6.30 is a graphical display of the contents of Table 6.3.

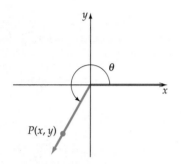

Figure 6.29

Table 6.3   Signs of the Trigonometric Functions

	Terminal Side of θ in Quadrant			
Sign of	*I*	*II*	*III*	*IV*
$\sin \theta$ and $\csc \theta$	positive	positive	negative	negative
$\cos \theta$ and $\sec \theta$	positive	negative	negative	positive
$\tan \theta$ and $\cot \theta$	positive	negative	positive	negative

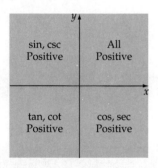

Figure 6.30

EXAMPLE 1   **Evaluate Trigonometric Functions**

Find the value of each of the six trigonometric functions of an angle θ whose terminal side contains the point $P(-3, -2)$.

Solution   The angle is sketched in Figure 6.31. Find r by using the equation $r = \sqrt{x^2 + y^2}$, where $x = -3$ and $y = -2$.

$$r = \sqrt{(-3)^2 + (-2)^2} = \sqrt{9 + 4} = \sqrt{13}$$

Now use the definitions for each trigonometric function.

$$\sin \theta = \frac{-2}{\sqrt{13}} = -\frac{2\sqrt{13}}{13} \qquad \cos \theta = \frac{-3}{\sqrt{13}} = -\frac{3\sqrt{13}}{13} \qquad \tan \theta = \frac{-2}{-3} = \frac{2}{3}$$

$$\csc \theta = \frac{\sqrt{13}}{-2} = -\frac{\sqrt{13}}{2} \qquad \sec \theta = \frac{\sqrt{13}}{-3} = -\frac{\sqrt{13}}{3} \qquad \cot \theta = \frac{-3}{-2} = \frac{3}{2}$$

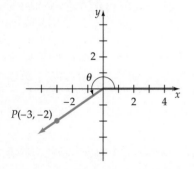

Figure 6.31

■ *Try Exercise **6**, page 320.*

It is also possible to find the values of any trigonometric function if the value of one of the functions is known.

EXAMPLE 2 Evaluate Trigonometric Functions

Given an angle θ whose terminal side is in quadrant III, and given $\cos \theta = -\dfrac{\sqrt{3}}{2}$, find the value of each of the remaining five trigonometric functions of θ.

Solution   Using the definition of $\cos \theta$, we have

$$\cos \theta = -\frac{\sqrt{3}}{2} = \frac{x}{r}$$

Let $x = -\sqrt{3}$ and $r = 2$. Now solve the equation $r = \sqrt{x^2 + y^2}$ for y.

$$2 = \sqrt{(-\sqrt{3})^2 + y^2}$$
$$4 = 3 + y^2$$
$$1 = y^2$$
$$y = \pm 1$$

Because the terminal side of θ is in quadrant III, $y = -1$. The values of the remaining five trigonometric functions are

$$\sin \theta = \frac{-1}{2} = -\frac{1}{2} \qquad \tan \theta = \frac{-1}{-\sqrt{3}} = \frac{\sqrt{3}}{3} \qquad \csc \theta = \frac{2}{-1} = -2$$

$$\sec \theta = \frac{2}{-\sqrt{3}} = -\frac{2\sqrt{3}}{3} \qquad \cot \theta = \frac{-\sqrt{3}}{-1} = \sqrt{3}$$

■ *Try Exercise* **18,** *page 320.*

Given a *point* on the terminal side of an angle θ, we can determine the value of any trigonometric function from its definition. In many cases, however, we must evaluate a trigonometric function given an *angle* θ. For most angles, it is necessary to have a calculator or tables such as appear in Appendix III to find the value of the trigonometric function given some arbitrary angle.

The value of a trigonometric function of an angle θ, where $0° \le \theta \le 90°$, can be found from the tables in Appendix III. To find the value of a trigonometric function for some other angle, a *reference angle* is used.

Reference Angle

The **reference angle** α for an angle θ in standard position is the positive acute angle formed by the terminal side of θ and the *x*-axis.

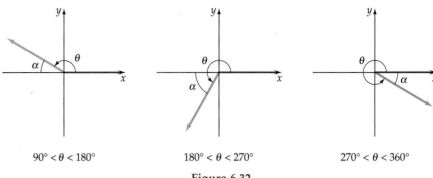

$$90° < \theta < 180° \qquad 180° < \theta < 270° \qquad 270° < \theta < 360°$$

Figure 6.32

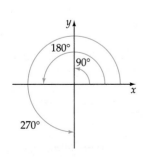

Figure 6.33

Figure 6.32 shows the reference angle α for three different angles θ whose terminal sides lie in the second, third, and fourth quadrants. The reference angle for an angle in the first quadrant is the given angle. From Figure 6.33, the measure of the reference angle for 223° is $223° - 180° = 43°$. The measure of the reference angle for 2 radians is $\pi - 2 \approx 1.1416$.

Reference angles are used to find the value of a trigonometric function when tables such as those in Appendix III are used. For example,

- $\cos 223° = -\cos 43° \approx -0.7314$ ($\cos \theta$ is < 0 in quadrant III).
- $\tan 2 \approx -\tan(1.1416) \approx -2.185$ ($\tan \theta$ in < 0 in quadrant II).

Using Table 6.1 and reference angles, it is possible to find the exact value of some trigonometric functions. For example, the measure of the reference angle for 330° is $360° - 330° = 30°$. Thus $\sin 330° = -\sin 30° = -\dfrac{1}{2}$; ($\sin \theta$ is negative in quadrant IV). In a similar manner, the measure of the reference angle for $-120°$ is 60°. Thus $\tan(-120°) = \tan 60° = \sqrt{3}$; ($\tan \theta$ is positive in quadrant III).

EXAMPLE 3   **Find the Exact Value of a Trigonometric Expression**

Find the exact value of $\sin 225° - \cos 300° \tan 150°$.

Solution

$$\sin 225° = -\sin 45° = -\frac{\sqrt{2}}{2}; \quad \cos 300° = \cos 60° = \frac{1}{2};$$

$$\tan 150° = -\tan 30° = -\frac{\sqrt{3}}{3}$$

$$\sin 225° - \cos 300° \tan 150° = -\frac{\sqrt{2}}{2} - \left(\frac{1}{2}\right)\left(-\frac{\sqrt{3}}{3}\right) = -\frac{3\sqrt{2}}{6} + \frac{\sqrt{3}}{6}$$

■ *Try Exercise* **54,** *page 320.*

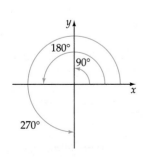

Figure 6.34

A **quadrantal angle** is an angle whose terminal side coincides with the x- or y-axis as shown in Figure 6.34. The value of a trigonometric function of a quadrantal angle can be found by choosing any point on the terminal

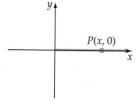

Figure 6.35

side of the angle and then applying the definition of that trigonometric function.

The terminal side of $0°$ coincides with the positive x-axis. Let $P(x, 0)$, $x \neq 0$, be any point on the x-axis as shown in Figure 6.35. Then $y = 0$ and $r = x$. The values of the six trigonometric functions of $0°$ are

$$\sin 0° = \frac{0}{r} = 0 \quad \cos 0° = \frac{x}{r} = \frac{x}{x} = 1 \quad \tan 0° = \frac{0}{x} = 0$$

$$\csc 0° \text{ is undefined} \quad \sec 0° = \frac{r}{x} = \frac{x}{x} = 1 \quad \cot 0° \text{ is undefined}$$

Why are $\csc 0°$ and $\cot 0°$ undefined?[3]

In like manner, the values of the trigonometric functions of the other quadrantal angles can be found. The results are shown in Table 6.4.

TABLE 6.4   Values of Trigonometric Functions for Quadrantal Angles

θ	$\sin \theta$	$\cos \theta$	$\tan \theta$	$\csc \theta$	$\sec \theta$	$\cot \theta$
$0°$	0	1	0	undefined	1	undefined
$90°$	1	0	undefined	1	undefined	0
$180°$	0	-1	0	undefined	-1	undefined
$270°$	-1	0	undefined	-1	undefined	0

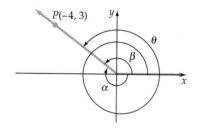

Figure 6.36

Angles that have the same terminal sides are called **coterminal angles.** Figure 6.36 shows the point $P(-4, 3)$ on the terminal side of an angle β in standard position. The same point is also on the terminal side of the negative angle α and the angle $\theta = \beta + 360°$. Because the value of a trigonometric function is defined in terms of a point on the terminal side of an angle, $\sin \beta = \sin \alpha = \sin \theta$. The values of the other trigonometric functions also are equal for the three angles.

We determine the reference angle for an angle β greater than $360°$ by first finding a positive coterminal angle θ whose measure is less than $360°$ and then finding the reference angle α for that angle. For example, to find $\tan 830°$ to the nearest ten-thousandth, first find a coterminal angle θ where $0° \leq \theta < 360°$. Because $830°$ is more than two complete revolutions and less than three revolutions, subtract $720°$ ($2 \cdot 360°$) from $830°$. $\theta = 830° - 720° = 110°$. The reference angle α for θ is $\alpha = 180° - 110° = 70°$. Now determine $\tan 830°$.

$$\tan 830° = -\tan 70° \quad \bullet \; \tan \theta < 0 \text{ in quadrant II}$$

$$\approx -2.7475 \quad \bullet \; \text{From the table in Appendix III}$$

The value of a trigonometric function can be found by using a calculator *without* first finding a coterminal angle or a reference angle. It is only when using the table in Appendix III that we must find a reference angle. Using a calculator, we would determine $\tan 830°$ by pressing 830 $\boxed{\tan}$. The result in the display would be $\boxed{-2.747477419}$.

---

[3] $P(x, 0)$ is a point on the terminal side of $0°$. Thus $\csc 0° = \frac{r}{0}$, which is undefined. Similarly, $\cot 0° = \frac{x}{0}$, which is undefined.

EXERCISE SET 6.3

In Exercises 1 to 8, find the value of each of the six trigonometric functions for the angle whose terminal side passes through the given point.

1. $P(2, 3)$ **2.** $P(3, 7)$ **3.** $P(-2, 3)$

4. $P(-3, 5)$ **5.** $P(-8, -5)$ **6.** $P(-6, -9)$

7. $P(-5, 0)$ **8.** $P(0, 2)$

In Exercises 9 to 14, let θ be an angle in standard position. State the quadrant in which the terminal side of θ lies.

9. $\sin \theta > 0$, $\cos \theta > 0$ **10.** $\tan \theta < 0$, $\sin \theta < 0$

11. $\cos \theta > 0$, $\tan \theta < 0$ **12.** $\sin \theta < 0$, $\cos \theta > 0$

13. $\sin \theta < 0$, $\cos \theta < 0$ **14.** $\tan \theta < 0$, $\cos \theta < 0$

In Exercises 15 to 24, find the value of each expression.

15. $\sin \theta = -1/2$, $180° < \theta < 270°$; find $\tan \theta$.

16. $\cot \theta = -1$, $90° < \theta < 180°$; find $\cos \theta$.

17. $\csc \theta = \sqrt{2}$, $\dfrac{\pi}{2} < \theta < \pi$; find $\cot \theta$.

18. $\sec \theta = 2\sqrt{3}/3$, $\dfrac{3\pi}{2} < \theta < 2\pi$; find $\sin \theta$.

19. $\sin \theta = -1/2$ and $\cos \theta > 0$; find $\tan \theta$.

20. $\tan \theta = 1$ and $\sin \theta < 0$; find $\cos \theta$.

21. $\cos \theta = 1/2$ and $\tan \theta = \sqrt{3}$; find $\csc \theta$.

22. $\tan \theta = 1$ and $\sin \theta = -\sqrt{2}/2$; find $\sec \theta$.

23. $\cos \theta = -1/2$ and $\sin \theta = \sqrt{3}/2$; find $\cot \theta$.

24. $\sec \theta = 2\sqrt{3}/3$ and $\sin \theta = -1/2$; find $\cot \theta$.

In Exercises 25 to 36, use a calculator to evaluate the following expressions to 4 decimal places.

25. $\sin 127°$ **26.** $\sin(-257°)$ **27.** $\cos(-116°)$

28. $\cot 398°$ **29.** $\sec 578°$ **30.** $\sec 740°$

31. $\sin(-\pi/5)$ **32.** $\cos (3\pi/7)$ **33.** $\csc (9\pi/5)$

34. $\tan(-4.12)$ **35.** $\sec(-4.45)$ **36.** $\csc 0.34$

In Exercises 37 to 48, use the table in Appendix III to find the following expressions to 4 decimal places.

37. $\tan 99°$ **38.** $\cos 192°$ **39.** $\sec 468°$

40. $\csc 550°$ **41.** $\cot(-173°)$ **42.** $\sin(-289°)$

43. $\cos 148°20'$ **44.** $\tan(-305°10')$ **45.** $\sec 5.3727$

46. $\csc 4.2499$ **47.** $\sin(-624°40')$ **48.** $\cos(953°)$

In Exercises 49 to 56, find the exact value of each expression.

49. $\sin 210° - \cos 330° \tan 330°$

50. $\tan 225° + \sin 240° \cos 60°$

51. $\sin^2 30° + \cos^2 30°$

52. $\cos \pi \sin(7\pi/4) - \tan(11\pi/6)$

53. $\sin(3\pi/2) \tan(\pi/4) - \cos(\pi/3)$

54. $\cos(7\pi/4) \tan(4\pi/3) + \cos(7\pi/6)$

55. $\sin^2(5\pi/4) + \cos^2(5\pi/4)$

56. $\tan^2(7\pi/4) - \sec^2(7\pi/4)$

Supplemental Exercises

In Exercises 57 to 62, find two values of θ, $0° \le \theta < 360°$, that satisfy the given trigonometric equation.

57. $\sin \theta = 1/2$ **58.** $\tan \theta = -\sqrt{3}$

59. $\cos \theta = -\sqrt{3}/2$ **60.** $\tan \theta = 1$

61. $\csc \theta = -\sqrt{2}$ **62.** $\cot \theta = -1$

In Exercises 63 to 68, find two values of θ, $0 \le \theta < 2\pi$, that satisfy the given trigonometric equation.

63. $\tan \theta = -1$ **64.** $\cos \theta = 1/2$

65. $\tan \theta = -\sqrt{3}/3$ **66.** $\sec \theta = -2\sqrt{3}/3$

67. $\sin \theta = \sqrt{3}/2$ **68.** $\cos \theta = -1/2$

If $P(x, y)$ is a point on the terminal side of an acute angle θ in standard position and $r = \sqrt{x^2 + y^2}$, then $\sin \theta = \dfrac{y}{r}$ and $\cos \theta = \dfrac{x}{r}$. Using these definitions, we find that

$$\cos^2\theta + \sin^2\theta = \left(\frac{x}{r}\right)^2 + \left(\frac{y}{r}\right)^2 = \frac{x^2}{r^2} + \frac{y^2}{r^2} = \frac{x^2 + y^2}{r^2} = \frac{r^2}{r^2} = 1$$

Hence $\cos^2\theta + \sin^2\theta = 1$ for all acute angles θ. This important identity is actually true for all angles θ. We will show this later. In the meantime, use the definitions of the trigonometric functions to prove the identities in Exercise 69 to 78 for the acute angle θ.

69. $1 + \tan^2\theta = \sec^2\theta$ **70.** $\cot^2\theta + 1 = \csc^2\theta$

71. $\tan \theta = \dfrac{\sin \theta}{\cos \theta}$ **72.** $\cot \theta = \dfrac{\cos \theta}{\sin \theta}$

73. $\cos(90° - \theta) = \sin \theta$ **74.** $\sin(90° - \theta) = \cos \theta$

75. $\tan(90° - \theta) = \cot \theta$ **76.** $\cot(90° - \theta) = \tan \theta$

77. $\sin(\theta + \pi) = -\sin \theta$ **78.** $\cos(\theta + \pi) = -\cos \theta$

For each of the angles given in Exercises 79 to 84, find the coordinates of a point on the terminal side of angle θ in standard position to the nearest ten-thousandth.

79. $\theta = 78°$ **80.** $\theta = 165°$ **81.** $\theta = 3$

82. $\theta = 2$ **83.** $\theta = -68°$ **84.** $\theta = -1$

85. Complete the table below.

θ	0°	15°	30°	45°	60°	75°	90°	105°	120°	135°	150°	165°	180°
$\sin \theta$													

Use this table to complete the following sentences where $0° \leq \theta \leq 180°$.

a. The maximum value of $\sin \theta$ is _____.

b. For $0° \leq \theta \leq 90°$, the value of $\sin \theta$ _____.

c. For $90° \leq \theta \leq 180°$, the value of $\sin \theta$ _____.

86. Complete the following table.

θ	0°	15°	30°	45°	60°	75°	90°	105°	120°	135°	150°	165°	180°
$\cos \theta$													

Use this table to complete the following sentences where $0° \leq \theta \leq 180°$.

a. The maximum value of $\cos \theta$ is _____.

b. For $0° \leq \theta \leq 180°$, the value of $\cos \theta$ _____.

c. The minimum value of $\cos \theta$ is _____.

d. See Exercise 85. For $0 \leq \theta \leq 90°$, $\sin \theta$ _____ as $\cos \theta$ _____.

e. See Exercise 85. For $90° \leq \theta \leq 180°$, $\sin \theta$ _____ as $\cos \theta$ _____.

6.4 Trigonometric Functions of Real Numbers

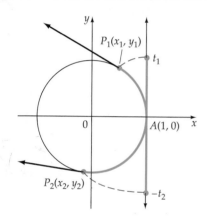

Figure 6.37

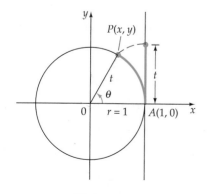

Figure 6.38

In the seventeenth century, applications of trigonometry were extended to problems in physics and engineering. These kinds of problems required trigonometric functions whose domains were sets of real numbers rather than sets of angles. During this time, the definitions of trigonometric functions were extended to real numbers by using a correspondence between an angle and a number.

Consider a circle given by the equation $x^2 + y^2 = 1$, called a **unit circle,** and a coordinate line tangent to the unit circle at (1, 0). We define a function W that pairs a real number t on the coordinate line with a point $P(x, y)$ on the unit circle. This function is called the *wrapping function* because it is analogous to wrapping a line around a circle.

As shown in Figure 6.37, the positive part of the coordinate line is wrapped around the unit circle in a counterclockwise direction. The negative part of the coordinate line is wrapped around the circle in a clockwise direction. The wrapping function is defined by the equation $W(t) = P(x, y)$, where t is a real number and $P(x, y)$ is the point on the unit circle that corresponds to t.

Through the wrapping function, each real number t defines an arc $\widehat{AP}$ that subtends a central angle with a measure of θ radians. The length of the arc $\widehat{AP}$ is t (see Figure 6.38). From the equation $s = r\theta$ for the arc length of a circle, we have (with $t = s$) $t = r\theta$. For a unit circle, $r = 1$ and the equation becomes $t = \theta$. Thus on a unit circle, *the measure of a central angle and the length of an arc can be represented by the same real number t.*

EXAMPLE 1 **Evaluate the Wrapping Function**

Find the values of x and y such that $W\left(\dfrac{\pi}{3}\right) = P(x, y)$.

Solution The point $\pi/3$ on the coordinate line tangent to the unit circle at $A(1, 0)$ is shown in Figure 6.39. From the wrapping function,

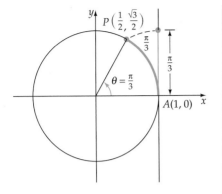

Figure 6.39

$W(\pi/3)$ is the point P on the unit circle for which arc $\overset{\frown}{AP}$ subtends an angle θ, the measure of which is $\pi/3$ radians. The coordinates of P can be determined from the definition of $\cos\theta$ and $\sin\theta$ given in Section 6.3 and Table 6.1.

$$\cos\theta = \frac{x}{r} \qquad\qquad \sin\theta = \frac{y}{r}$$

$$\cos\frac{\pi}{3} = \frac{x}{1} = x \qquad \sin\frac{\pi}{3} = \frac{y}{r} = y \qquad \bullet\ \theta = \frac{\pi}{3}; r = 1$$

$$\frac{1}{2} = x \qquad\qquad \frac{\sqrt{3}}{2} = y \qquad\qquad \bullet\ \cos\frac{\pi}{3} = \frac{1}{2}; \sin\frac{\pi}{3} = \frac{\sqrt{3}}{2}$$

From these equations, $x = \frac{1}{2}$ and $y = \frac{\sqrt{3}}{2}$. Therefore, $W\!\left(\frac{\pi}{3}\right) = P\!\left(\frac{1}{2}, \frac{\sqrt{3}}{2}\right)$.

■ *Try Exercise **10**, page 328.*

To determine $W\!\left(\frac{\pi}{2}\right)$, recall that the circumference of a unit circle is 2π. One-fourth the circumference is $\frac{1}{4}(2\pi) = \frac{\pi}{2}$ (see Figure 6.40). Thus $W\!\left(\frac{\pi}{2}\right) = P(0, 1)$.

Note from the last two examples that for the given real number t, $\cos t = x$ and $\sin t = y$. That is, for a real number t and $W(t) = P(x, y)$, the value of the cosine of t is the x-coordinate of P and the value of the sine of t is the y-coordinate of P. This suggests the following definition.

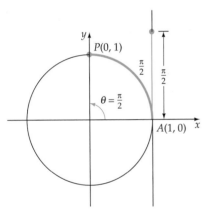

Figure 6.40

Definition of the Trigonometric Functions of Real Numbers

Let W be the wrapping function, t be a real number, and $W(t) = P(x, y)$. Then

$$\sin t = y \qquad\qquad \cos t = x \qquad\qquad \tan t = \frac{y}{x}, x \neq 0$$

$$\csc t = \frac{1}{y}, y \neq 0 \qquad \sec t = \frac{1}{x}, x \neq 0 \qquad \cot t = \frac{x}{y}, y \neq 0$$

Remark Trigonometric functions of real numbers are frequently called *circular functions* to distinguish them from trigonometric functions of angles.

The *trigonometric functions of real numbers* (or circular functions) look remarkably like the trigonometric functions defined in the last section. The difference between the two is that of domain: In one case, the domains are sets of *real numbers*; in the other case, the domains are sets of *angles*. However, there are similarities between the two functions.

Consider an angle θ (in radians) in standard position as shown in Figure 6.41. Let $P(x, y)$ and $P'(x', y')$ be two points on the terminal side of θ,

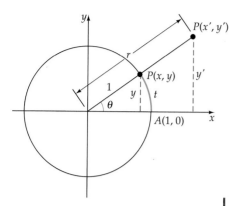

Figure 6.41

where $x^2 + y^2 = 1$ and $(x')^2 + (y')^2 = r^2$. Let t be the length of the arc from $A(1, 0)$ to $P(x, y)$. Then

$$\sin \theta = \frac{y'}{r} = \frac{y}{1} = \sin t$$

Thus the value of the sine function of θ, measured in radians, is equal to the value of the sine of the real number t. Similar arguments can be given to show corresponding results for the other five trigonometric functions. With this in mind, we can assert that *the value of a trigonometric function at the real number t is its value at an angle of t radians.*

EXAMPLE 2   **Find the Exact Value of a Trigonometric Function of a Real Number**

Find the exact value of each of the following.

a. $\cos \dfrac{\pi}{4}$ b. $\sin\left(-\dfrac{7\pi}{6}\right)$ c. $\tan\left(-\dfrac{5\pi}{4}\right)$ d. $\sec\left(\dfrac{5\pi}{3}\right)$

Solution   The value of a trigonometric function at the real number t is its value at an angle of t radians. Using Table 6.1, we have

a. $\cos \dfrac{\pi}{4} = \dfrac{\sqrt{2}}{2}$

b. $\sin\left(-\dfrac{7\pi}{6}\right) = \sin\left(\dfrac{\pi}{6}\right) = \dfrac{1}{2}$ • Reference angle for $-\dfrac{7\pi}{6}$ is $\dfrac{\pi}{6}$ and $\sin t > 0$ in quadrant II.

c. $\tan\left(-\dfrac{5\pi}{4}\right) = -\tan\left(\dfrac{\pi}{4}\right) = -1$ • Reference angle for $-\dfrac{5\pi}{4}$ is $\dfrac{\pi}{4}$ and $\tan t < 0$ in quadrant II.

d. $\sec\left(\dfrac{5\pi}{3}\right) = \sec\left(\dfrac{\pi}{3}\right) = 2$ • Reference angle for $\dfrac{5\pi}{3}$ is $\dfrac{\pi}{3}$ and $\sec t > 0$ in quadrant IV.

■ *Try Exercise* **16**, *page 328.*

Properties of Trigonometric Functions of Real Numbers

The domain and range of the trigonometric functions can be found from the definition of these functions. If t is any real number and $P(x, y)$ is the point corresponding to $W(t)$, then by definition $\cos t = x$ and $\sin t = y$. Thus the domain of the sine and cosine functions is the set of real numbers. Because the radius of the unit circle is 1, we have

$$-1 \le x \le 1 \quad \text{and} \quad -1 \le y \le 1$$

Therefore, with $x = \cos t$ and $y = \sin t$, we have

$$-1 \le \cos t \le 1 \quad \text{and} \quad -1 \le \sin t \le 1$$

The range of the cosine and sine functions is $[-1, 1]$.

Using the definition of tangent and secant,

$$\tan t = \frac{y}{x} \quad \text{and} \quad \sec t = \frac{1}{x}$$

The domain of the tangent and secant functions is all real numbers t except those for which the x-coordinate of $W(t)$ is zero. The x-coordinate is zero when $t = \pm\frac{\pi}{2}$, $t = \pm\frac{3\pi}{2}$, $t = \pm\frac{5\pi}{2}$, and in general when $t = (2n + 1)\pi/2$, where n is an integer. Thus the domain of the tangent and secant functions is the set of all real numbers t except $t = (2n + 1)\pi/2$, where n is an integer.

The range of the tangent function is all real numbers. The range of the secant function is all real numbers y such that $|y| \geq 1$.

Using the definition of cotangent and cosecant,

$$\cot t = \frac{x}{y} \quad \text{and} \quad \csc t = \frac{1}{y}$$

The domain of the cotangent and cosecant functions is all real numbers t except those for which the y-coordinate of $W(t)$ is zero. The y-coordinate is zero when $t = \pm\pi$, $t = \pm 2\pi$, $t = \pm 3\pi$, and in general when $t = n\pi$, where n is an integer. Thus the domain of the cotangent and cosecant functions is the set of all real numbers t except $t = n\pi$, where n is an integer.

The range of the cotangent function is all real numbers. The range of the cosecant function is all real numbers y such that $|y| \geq 1$.

Table 6.5 lists the domain and range for all six trigonometric functions.

TABLE 6.5   Domain and Range of the Trigonometric Functions (n is an integer)

Function	Domain	Range
$y = \sin t$	$\{t\,|\,-\infty < t < \infty\}$	$\{y\,|\,-1 \leq y \leq 1\}$
$y = \cos t$	$\{t\,|\,-\infty < t < \infty\}$	$\{y\,|\,-1 \leq y \leq 1\}$
$y = \tan t$	$\{t\,|\,-\infty < t < \infty,\, t \neq (2n + 1)\pi/2\}$	$\{y\,|\,-\infty < y < \infty\}$
$y = \csc t$	$\{t\,|\,-\infty < t < \infty,\, t \neq n\pi\}$	$\{y\,|\,y \geq 1,\, y \leq -1\}$
$y = \sec t$	$\{t\,|\,-\infty < t < \infty,\, t \neq (2n + 1)\pi/2\}$	$\{y\,|\,y \geq 1,\, y \leq -1\}$
$y = \cot t$	$\{t\,|\,-\infty < t < \infty,\, t \neq n\pi\}$	$\{y\,|\,-\infty < y < \infty\}$

Consider the points t and $-t$ on the coordinate line tangent to the unit circle at the point $(1, 0)$. The points $W(t)$ and $W(-t)$ are symmetric with respect to the x-axis. Therefore, if $P_1(x, y)$ are the coordinates of $W(t)$, then $P_2(x, -y)$ are the coordinates of $W(-t)$. See Figure 6.42.

From the definitions of the trigonometric functions, we have

$$\sin t = y \quad \text{and} \quad \sin(-t) = -y \quad \text{and} \quad \cos t = x \quad \text{and} \quad \cos(-t) = x$$

Substituting $\sin t$ for y and $\cos t$ for x yields

$$\sin(-t) = -\sin t \quad \text{and} \quad \cos(-t) = \cos t$$

Thus the sine is an odd function and the cosine is an even function. Since $\csc t = 1/\sin t$ and $\sec t = 1/\cos t$, it follows that

$$\csc(-t) = -\csc t \quad \text{and} \quad \sec(-t) = \sec t$$

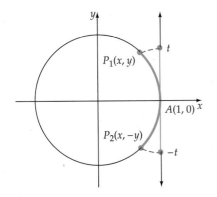

Figure 6.42

These equations state that the cosecant is an odd function and the secant an even function.

From the definition of the tangent function, we have $\tan t = y/x$ and $\tan(-t) = -y/x$. Substituting $\tan t$ for y/x yields $\tan(-t) = -\tan t$. Since $\cot t = 1/\tan t$, it follows that $\cot(-t) = -\cot t$. Thus the tangent and cotangent are odd functions.

Even and Odd Trigonometric Functions

The odd trigonometric functions are $y = \sin t$, $y = \csc t$, $y = \tan t$, and $y = \cot t$. The even trigonometric functions are $y = \cos t$ and $y = \sec t$.

EXAMPLE 3 Determine Whether a Function is Even, Odd, or Neither

Is the function defined by $f(x) = x - \tan x$ even, odd, or neither?

Solution Find $f(-x)$ and compare it to $f(x)$.

$$f(-x) = (-x) - \tan(-x) = -x + \tan x \quad \bullet \ \tan(-x) = -\tan x$$

$$= -(x - \tan x)$$

$$= -f(x)$$

The function defined by $f(x) = x - \tan x$ is an odd function.

■ *Try Exercise 36, page 329.*

Let W be the wrapping function, t be a point of the coordinate line tangent to the unit circle at $(1, 0)$, and $W(t) = P(x, y)$. Because the circumference of the unit circle is 2π (Why?)[4], $W(t + 2\pi) = W(t) = P(x, y)$. Thus the value of the wrapping function repeats itself in 2π units. A function that repeats itself is said to be *periodic*.

Periodic Function

Let p be a constant. If $f(t) = f(t + p)$ for all t in the domain of f, then f is a **periodic function**. The **period** of f is the smallest positive value of p for which $f(t + p) = f(t)$.

The wrapping function is periodic and the period is 2π.

Recalling the definitions of $\cos t$ and $\sin t$:

$$\cos t = x \quad \text{and} \quad \sin t = y$$

where $W(t) = P(x, y)$. Because $W(t + 2\pi) = W(t) = P(x, y)$ for all t,

$$\cos(t + 2\pi) = x \quad \text{and} \quad \sin(t + 2\pi) = y$$

---

[4] For a unit circle, $r = 1$. Thus $c = 2\pi \cdot 1 = 2\pi$.

Thus $\cos t$ and $\sin t$ have period 2π. Since

$$\sec t = \frac{1}{\cos t} = \frac{1}{\cos(t + 2\pi)} = \sec(t + 2\pi) \quad \text{and}$$

$$\csc t = \frac{1}{\sin t} = \frac{1}{\sin(t + 2\pi)} = \csc(t + 2\pi),$$

$\sec t$ and $\csc t$ have a period of 2π.

Period of $\cos t$, $\sin t$, $\sec t$, and $\csc t$

The period of $\cos t$, $\sin t$, $\sec t$, and $\csc t$ is 2π.

Although it is true that $\tan t = \tan(t + 2\pi)$, the period of $\tan t$ is not 2π. Recall that the period of a function is the *smallest* value of p for which $f(t) = f(t + p)$.

If W is the wrapping function (see Figure 6.43) and $W(t) = P(x, y)$, then $W(t + \pi) = P(-x, -y)$. Because

$$\tan t = \frac{y}{x} \quad \text{and} \quad \tan(t + \pi) = \frac{-y}{-x} = \frac{y}{x} = \tan t$$

we have $\tan(t + \pi) = \tan t$ for all t. A similar argument applies to $\cot t$.

Period of $\tan t$ and $\cot t$

The period of $\tan t$ and $\cot t$ is π.

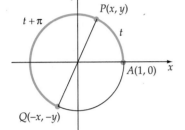

Figure 6.43

Trigonometric Function Identities

Recall that any equation that is true for every number in the domain of the equation is an identity. The statement

$$\csc t = \frac{1}{\sin t}, \quad \sin t \neq 0$$

is an identity because the two expressions produce the same result for all values of t for which the functions are defined.

The **ratio identities** are obtained by writing the tangent and cotangent functions in terms of the sine and cosine functions.

$$\tan t = \frac{y}{x} = \frac{\sin t}{\cos t} \quad \text{and} \quad \cot t = \frac{x}{y} = \frac{\cos t}{\sin t} \quad \bullet \ x = \cos t \text{ and } y = \sin t$$

The **Pythagorean identities** are based on the equation of a unit circle, $x^2 + y^2 = 1$, and on the definitions of the sine and cosine functions.

$$x^2 + y^2 = 1$$

$$\cos^2 t + \sin^2 t = 1 \quad \bullet \ \text{Replace } x \text{ by } \cos t \text{ and } y \text{ by } \sin t$$

Dividing each term of $\cos^2 t + \sin^2 t = 1$ by $\cos^2 t$, we have

$$\frac{\cos^2 t}{\cos^2 t} + \frac{\sin^2 t}{\cos^2 t} = \frac{1}{\cos^2 t} \qquad \bullet \ \cos t \neq 0$$

$$1 + \tan^2 t = \sec^2 t \qquad \bullet \ \frac{\sin t}{\cos t} = \tan t$$

Dividing each term of $\cos^2 t + \sin^2 t = 1$ by $\sin^2 t$, we have

$$\frac{\cos^2 t}{\sin^2 t} + \frac{\sin^2 t}{\sin^2 t} = \frac{1}{\sin^2 t} \qquad \bullet \ \sin t \neq 0$$

$$\cot^2 t + 1 = \csc^2 t \qquad \bullet \ \frac{\cos t}{\sin t} = \cot t$$

Here is a summary of the Fundamental Trigonometric Identities:

Fundamental Trigonometric Identities

The reciprocal identities are:

$$\sin t = \frac{1}{\csc t} \qquad \cos t = \frac{1}{\sec t} \qquad \tan t = \frac{1}{\cot t}$$

The ratio identities are

$$\tan t = \frac{\sin t}{\cos t} \qquad \cot t = \frac{\cos t}{\sin t}$$

The Pythagorean identities are

$$\cos^2 t + \sin^2 t = 1 \qquad 1 + \tan^2 t = \sec^2 t \qquad 1 + \cot^2 t = \csc^2 t$$

EXAMPLE 4   **Use the Unit Circle to Verify an Identity**

By using the unit circle and the definitions of the trigonometric functions, show that $\sin(t + \pi) = -\sin t$.

Solution   Sketch a unit circle, and let P be the point on the unit circle such that $W(t) = P(x, y)$, as shown in Figure 6.44. Draw a diameter from P and label the endpoint Q. For any line through the origin, if $P(x, y)$ is a point on the line, then $Q(-x, -y)$ is also a point on the line. Because PQ is a diameter, the length of arc PQ is π. Thus the length of the arc AQ is $t + \pi$. Therefore, $W(t + \pi) = Q(-x, -y)$. From the definition of $\sin t$, we have

$$\sin t = y \qquad \text{and} \qquad \sin(t + \pi) = -y$$

Thus $\sin(t + \pi) = -\sin t$.

■ *Try Exercise **42**, page 329.*

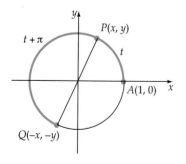

Figure 6.44

Using identities and basic algebra concepts, we can rewrite trigonometric expressions in different forms.

EXAMPLE 5   Simplify a Trigonometric Expression

Write the expression $\dfrac{1}{\sin^2 t} + \dfrac{1}{\cos^2 t}$ as a single term.

Solution   Express each fraction in terms of a common denominator. The common denominator is $\sin^2 t \cos^2 t$.

$$\frac{1}{\sin^2 t} + \frac{1}{\cos^2 t} = \frac{1}{\sin^2 t}\frac{\cos^2 t}{\cos^2 t} + \frac{1}{\cos^2 t}\frac{\sin^2 t}{\sin^2 t}$$

$$= \frac{\cos^2 t + \sin^2 t}{\sin^2 t \cos^2 t} = \frac{1}{\sin^2 t \cos^2 t} \quad \bullet \ \cos^2 t + \sin^2 t = 1$$

■ *Try Exercise* **60,** *page 329.*

Remark   Because $\dfrac{1}{\sin^2 t \cos^2 t} = \dfrac{1}{\sin^2 t} \cdot \dfrac{1}{\cos^2 t} = \csc^2 t \sec^2 t$, we could have written the answer to Example 5 in terms of the cosecant and secant functions.

EXAMPLE 6   Write a Trigonometric Expression in Terms of a Given Function

For $\pi/2 < t < \pi$, write $\tan t$ in terms of $\sin t$.

Solution   Write $\tan t = \dfrac{\sin t}{\cos t}$. Now solve $\cos^2 t + \sin^2 t = 1$ for $\cos t$.

$$\cos^2 t + \sin^2 t = 1$$

$$\cos^2 t = 1 - \sin^2 t$$

$$\cos t = \pm\sqrt{1 - \sin^2 t}$$

Because $\pi/2 < t < \pi$, $\cos t$ is negative. Therefore, $\cos t = -\sqrt{1 - \sin^2 t}$. Thus

$$\tan t = -\frac{\sin t}{\sqrt{1 - \sin^2 t}} \quad \bullet \ \frac{\pi}{2} < t < \pi$$

■ *Try Exercise* **86,** *page 329.*

EXERCISE SET 6.4

In Exercises 1 to 12, find $W(t)$ for each given t.

1. $t = \pi/6$ **2.** $t = \pi/4$ **3.** $t = 7\pi/6$

4. $t = 4\pi/3$ **5.** $t = 5\pi/3$ **6.** $t = -\pi/6$

7. $t = 11\pi/6$ **8.** $t = 0$ **9.** $t = \pi$

10. $t = -7\pi/4$ **11.** $t = -2\pi/3$ **12.** $t = -\pi$

In Exercises 13 to 22, find the exact value of each of the following.

13. $\tan(11\pi/6)$ **14.** $\cot(2\pi/3)$

15. $\cos(-2\pi/3)$ **16.** $\sec(-5\pi/6)$

17. $\csc(-\pi/3)$ **18.** $\tan(12\pi)$

19. $\sin(3\pi/2)$

20. $\cos(7\pi/3)$

21. $\sec(-7\pi/6)$

22. $\sin(-5\pi/3)$

In Exercises 23 to 32, find an approximate value of each of the following. Round answers to the nearest ten-thousandth.

23. $\sin 1.22$

24. $\cos 4.22$

25. $\csc(-1.05)$

26. $\sin(-0.55)$

27. $\tan 11\pi/12$

28. $\cos 2\pi/5$

29. $\cos(-\pi/5)$

30. $\csc 8.2$

31. $\sec 1.55$

32. $\cot 2.11$

In Exercises 33 to 40, determine whether the function defined by each equation is even, odd, or neither.

33. $f(x) = -4 \sin x$

34. $f(x) = -2 \cos x$

35. $G(x) = \sin x + \cos x$

36. $F(x) = \tan x + \sin x$

37. $S(x) = \dfrac{\sin x}{x}, x \neq 0$

38. $C(x) = \dfrac{\cos x}{x}, x \neq 0$

39. $v(x) = 2 \sin x \cos x$

40. $w(x) = x \tan x$

In Exercises 41 to 48, use the unit circle to verify each identity.

41. $\cos(-t) = \cos t$

42. $\tan(t - \pi) = \tan t$

43. $\cos(t + \pi) = -\cos t$

44. $\sin(-t) = -\sin t$

45. $\sin(t - \pi) = -\sin t$

46. $\sec(-t) = \sec t$

47. $\csc(-t) = -\csc t$

48. $\tan(-t) = -\tan t$

In Exercises 49 to 64, use the trigonometric identities to write each expression in terms of single trigonometric function or a constant. Your answers may vary.

49. $\tan t \cos t$

50. $\cot t \sin t$

51. $\dfrac{\csc t}{\cot t}$

52. $\dfrac{\sec t}{\tan t}$

53. $1 - \sec^2 t$

54. $1 - \csc^2 t$

55. $\tan t - \dfrac{\sec^2 t}{\tan t}$

56. $\dfrac{\csc^2 t}{\cot t} - \cot t$

57. $\dfrac{1 - \cos^2 t}{\tan^2 t}$

58. $\dfrac{1 - \sin^2 t}{\cot^2 t}$

59. $\dfrac{1}{1 - \cos t} + \dfrac{1}{1 + \cos t}$

60. $\dfrac{1}{1 - \sin t} + \dfrac{1}{1 + \sin t}$

61. $\dfrac{\tan t + \cot t}{\tan t}$

62. $\dfrac{\csc t - \sin t}{\csc t}$

63. $\sin^2 t(1 + \cot^2 t)$

64. $\cos^2 t(1 + \tan^2 t)$

In Exercises 65 to 76, perform the indicated operation and simplify.

65. $\cos t - \dfrac{1}{\cos t}$

66. $\tan t + \dfrac{1}{\tan t}$

67. $\cot t + \dfrac{1}{\cot t}$

68. $\sin t - \dfrac{1}{\sin t}$

69. $(1 - \sin t)^2$

70. $(1 - \cos t)^2$

71. $(\sin t - \cos t)^2$

72. $(\sin t + \cos t)^2$

73. $(1 - \sin t)(1 + \sin t)$

74. $(1 - \cos t)(1 + \cos t)$

75. $\dfrac{\sin t}{1 + \cos t} + \dfrac{1 + \cos t}{\sin t}$

76. $\dfrac{1 - \sin t}{\cos t} - \dfrac{1}{\tan t + \sec t}$

In Exercises 77 to 84, factor the expression.

77. $\cos^2 t - \sin^2 t$

78. $\sec^2 t - \csc^2 t$

79. $\tan^2 t - \tan t - 6$

80. $\cos^2 t + 3 \cos t - 4$

81. $2 \sin^2 t - \sin t - 1$

82. $4 \cos^2 t + 4 \cos t + 1$

83. $\cos^4 t - \sin^4 t$

84. $\sec^4 t - \csc^4 t$

85. Write $\sin t$ in terms of $\cos t$, $0 < t < \pi/2$.

86. Write $\tan t$ in terms of $\sec t$, $3\pi/2 < t < 2\pi$.

87. Write $\csc t$ in terms of $\cot t$, $\pi/2 < t < \pi$.

88. Write $\sec t$ in terms of $\tan t$, $\pi < t < 3\pi/2$.

Supplemental Exercises

In Exercises 89 to 92, use the trigonometric identities to find the value of the function.

89. Given $\csc t = \sqrt{2}$, $0 < t < \pi/2$; find $\cos t$.

90. Given $\cos t = 1/2$, $3\pi/2 < t < 2\pi$; find $\sin t$.

91. Given $\sin t = 1/2$, $\pi/2 < t < \pi$; find $\tan t$.

92. Given $\cot t = \sqrt{3}/3$, $\pi < t < 3\pi/2$; find $\cos t$.

93. Use the unit circle and the triangles shown in the accompanying figure to write the value of each function in terms of the length of a line segment.

   a. $\sin \phi$ b. $\cos \phi$ c. $\tan \phi$

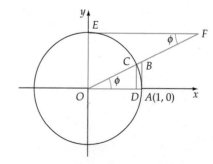

94. Use the graph above to write the value of each function in terms of the length of a line segment.

   a. $\csc \phi$ b. $\sec \phi$ c. $\cot \phi$

In Exercises 95 to 98, simplify the first expression to the second expression.

95. $\dfrac{\sin^2 t + \cos^2 t}{\sin^2 t}$; $\csc^2 t$ **96.** $\dfrac{\sin^2 t + \cos^2 t}{\cos^2 t}$; $\sec^2 t$

97. $(\cos t - 1)(\cos t + 1)$; $-\sin^2 t$

98. $(\sec t - 1)(\sec t + 1)$; $\tan^2 t$

99. Given that f is a periodic function with period 3 and that $f(2) = -1$, find the value of $f(14)$.

100. Given that f is a periodic function with period 2 and that $f(1) = 4$ and $f(2) = -2$, find $f(49) + f(50)$.

101. If f is a periodic function with period 2 and g is a periodic function with period 3, find the period of $f + g$.

102. If f and g are periodic functions with periods p_1 and p_2, respectively, is the product of f and g a periodic function? Explain.

6.5   Graphs of the Sine and Cosine Functions

The Graph of the Sine Function

The trigonometric functions can be graphed on a rectangular coordinate system by plotting the points whose coordinates belong to the function. We begin with the graph of the sine function.

Table 6.6 lists some ordered pairs (x, y), where $y = \sin x$, $0 \leq x \leq 2\pi$. In Figure 6.45 the points are plotted and a smooth curve is drawn through the points.

TABLE 6.6

x	0	$\pi/4$	$\pi/2$	$3\pi/4$	π	$5\pi/4$	$3\pi/2$	$7\pi/4$	2π
$y = \sin x$	0	0.7	1	0.7	0	-0.7	-1	-0.7	0

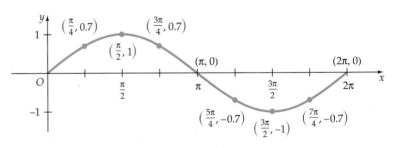

$y = \sin x$, $0 \leq x \leq 2\pi$
Figure 6.45

Because the domain of the sine function is the real numbers and the period is 2π, the graph of $y = \sin x$ is drawn by repeating the portion shown in Figure 6.45. The part of the graph that corresponds to one period (2π) is one cycle of the graph of $y = \sin x$ (see Figure 6.46).

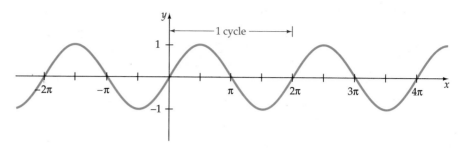

$y = \sin x$
Figure 6.46

The maximum value M reached by $\sin x$ is 1 and the minimum value m is -1. The amplitude of the graph of $y = \sin x$ is given by

$$\text{amplitude} = \frac{1}{2}(M - m)$$

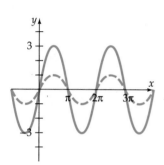

$y = 3 \sin x$
Figure 6.47

For $y = \sin x$, the amplitude is 1. (Why?)[5]
  Recall that the graph of $y = a \cdot f(x)$ is obtained by *stretching* ($|a| > 1$) or *shrinking* ($0 < |a| < 1$) the graph of $y = f(x)$. Figure 6.47 shows the graph of $y = 3 \sin x$ that was drawn by stretching the graph of $y = \sin x$. The amplitude of $y = 3 \sin x$ is 3 because

$$\text{amplitude} = \frac{1}{2}(M - m) = \frac{1}{2}[3 - (-3)] = 3$$

Note that for $y = \sin x$ and $y = 3 \sin x$, the amplitude of the graph was the coefficient of $\sin x$. This suggests the following theorem.

Amplitude of $y = a \sin x$

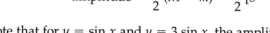

The amplitude of $y = a \sin x$ is $|a|$.

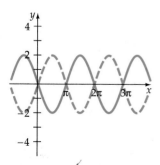

$y = -2 \sin x$
Figure 6.48

EXAMPLE 1   **Graph $y = a \sin x$**

Graph $y = -2 \sin x$.

  Solution   The amplitude of $y = -2 \sin x$ is 2. The graph of $y = -f(x)$ is a *reflection* across the x-axis of $y = f(x)$. Thus the graph of $y = -2 \sin x$ is a reflection across the x-axis of $y = 2 \sin x$. See Figure 6.48.

■ *Try Exercise* **20**, *page 336.*

[5] From Figure 6.46, Amplitude $= \frac{1}{2}[1 - (-1)] = \frac{1}{2}(2) = 1.$

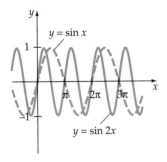

Figure 6.49

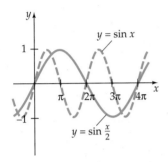

Figure 6.50

The graphs of $y = \sin x$ and $y = \sin 2x$ are shown in Figure 6.49. Because one cycle of the graph of $y = \sin 2x$ is completed in an interval of length π, the period of $y = \sin 2x$ is π.

The graphs of $y = \sin x$ and $y = \sin(x/2)$ are shown in Figure 6.50. Because one cycle of the graph of $y = \sin(x/2)$ is completed in an interval of length 4π, the period of $y = \sin(x/2)$ is 4π.

Generalizing the last two examples, one cycle of $y = \sin bx$, $b > 0$, is completed as bx varies from 0 to 2π. Algebraically, one cycle of $y = \sin bx$ is completed as bx varies from 0 to 2π. Therefore,

$$0 \le bx \le 2\pi$$

$$0 \le x \le \frac{2\pi}{b}$$

The length of the interval, $2\pi/b$, is the period of $y = \sin bx$. Now we consider the case when the coefficient of x is negative. If $b > 0$, then using the fact that the sine is an odd function, we have $y = \sin(-bx) = -\sin bx$ and thus the period is still $2\pi/b$. This gives the following theorem.

Period of $y = \sin bx$

> The period of $y = \sin bx$ is $2\pi/|b|$.

Table 6.7 gives the amplitude and period of several sine functions.

TABLE 6.7

Function	Amplitude	Period
$y = a \sin bx$	$\lvert a \rvert$	$\dfrac{2\pi}{\lvert b \rvert}$
$y = 3 \sin(-2x)$	$\lvert 3 \rvert = 3$	$\dfrac{2\pi}{2} = \pi$
$y = -\sin \dfrac{x}{3}$	$\lvert -1 \rvert = 1$	$\dfrac{2\pi}{1/3} = 6\pi$
$y = -2 \sin \dfrac{3x}{4}$	$\lvert -2 \rvert = 2$	$\dfrac{2\pi}{3/4} = \dfrac{8\pi}{3}$

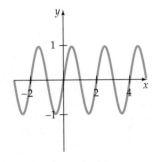

$y = \sin \pi x$
Figure 6.51

EXAMPLE 2   **Graph $y = \sin bx$**

Graph $y = \sin \pi x$.

Solution

$$\text{Amplitude} = 1 \qquad \text{Period} = \frac{2\pi}{b} = \frac{2\pi}{\pi} = 2 \quad \bullet \; b = \pi$$

The graph is sketched in Figure 6.51.

■ *Try Exercise **30**, page 336.*

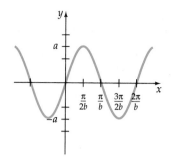

$y = a \sin bx$
Figure 6.52

Figure 6.52 shows the graph of $y = a \sin bx$ for both a and b positive. Note from the graph that

- The amplitude is a.
- The period is $2\pi/b$.
- The zeros are 0, π/b, and $2\pi/b$.
- The maximum value is a when $x = \dfrac{\pi}{2b}$, and the minimum value is $-a$ when $x = \dfrac{3\pi}{2b}$.
- If $a < 0$, the graph is reflected across the x-axis.

EXAMPLE 3   **Graph $y = a \sin bx$**

Graph $y = -\dfrac{1}{2} \sin \dfrac{x}{3}$.

Solution

$$\text{Amplitude} = \left| -\frac{1}{2} \right| = \frac{1}{2} \qquad \text{Period} = \frac{2\pi}{1/3} = 6\pi \quad \bullet \; b = \frac{1}{3}$$

The zeros in the interval $0 \le x \le 6\pi$ are 0, $\dfrac{\pi}{1/3} = 3\pi$, and $\dfrac{2\pi}{1/3} = 6\pi$. Because $-\dfrac{1}{2} < 0$, the graph is the graph of $y = \dfrac{1}{2} \sin \dfrac{x}{3}$ reflected across the x-axis as sketched in Figure 6.53.

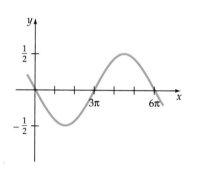

$y = -\frac{1}{2} \sin \frac{x}{3}$
Figure 6.53

■ *Try Exercise* **38**, *page 336.*

The Graph of the Cosine Function

Table 6.8 lists some of the ordered pairs (x, y), where $y = \cos x, 0 \le x \le 2\pi$. In Figure 6.54 the points are plotted and a smooth curve is drawn through the points.

TABLE 6.8

x	0	$\pi/4$	$\pi/2$	$3\pi/4$	π	$5\pi/4$	$3\pi/2$	$7\pi/4$	2π
$y = \cos x$	1	0.7	0	-0.7	-1	-0.7	0	0.7	1

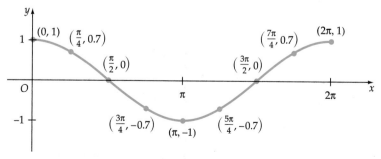

$y = \cos x, 0 \le x \le 2\pi$
Figure 6.54

Because the domain of $y = \cos x$ is the real numbers and the period is 2π, the graph of $y = \cos x$ is drawn by repeating the portion shown in Figure 6.54. The part of the graph corresponding to one period (2π) is one cycle of $y = \cos x$ (see Figure 6.55).

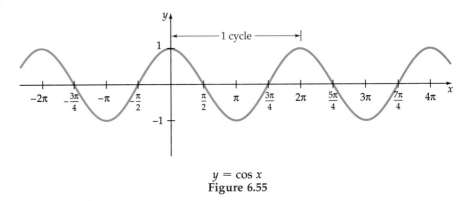

$$y = \cos x$$
Figure 6.55

The following two theorems concerning cosine functions can be developed using methods that are analogous to those we used to determine the amplitude and period of a sine function.

Amplitude of $y = a \cos x$

The amplitude of $y = a \cos x$ is $|a|$.

Period of $y = \cos bx$

The period of $y = \cos bx$ is $2\pi/|b|$.

Table 6.9 gives the amplitude and period of several cosine functions.

TABLE 6.9

Function	Amplitude	Period
$y = a \cos bx$	$\|a\|$	$\dfrac{2\pi}{\|b\|}$
$y = 2 \cos 3x$	$\|2\| = 2$	$\dfrac{2\pi}{3}$
$y = -3 \cos \dfrac{2x}{3}$	$\|-3\| = 3$	$\dfrac{2\pi}{2/3} = 3\pi$

EXAMPLE 4   **Graph $y = \cos bx$**

Graph $y = \cos \dfrac{2\pi}{3}x$.

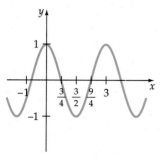

$$y = \cos \frac{2\pi}{3}x$$

Figure 6.56

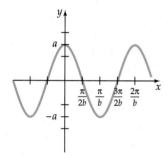

$$y = a \cos bx$$
Figure 6.57

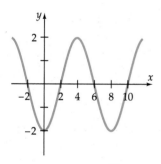

$$y = -2 \cos \frac{\pi x}{4}$$

Figure 6.58

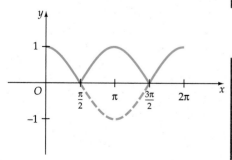

$$y = |\cos x|, \ 0 \le x \le 2\pi$$
Figure 6.59

Solution

$$\text{Amplitude} = 1 \qquad \text{Period} = \frac{2\pi}{b} = \frac{2\pi}{2\pi/3} = 3 \quad \bullet \ b = \frac{2\pi}{3}$$

The graph is sketched in Figure 6.56.

■ *Try Exercise* **32,** *page 336.*

Figure 6.57 shows the graph of $y = a \cos bx$ for both a and b positive. Note from the graph that

- The amplitude is a.

- The period is $\frac{2\pi}{b}$.

- The zeros are $\frac{\pi}{2b}$ and $\frac{3\pi}{2b}$.

- The maximum value is a when $x = 0$, and the minimum value is $-a$ when $x = \pi/b$.

- If $a < 0$, then the graph is reflected across the x-axis.

EXAMPLE 5 **Graph a Cosine Function**

Graph $y = -2 \cos \dfrac{\pi x}{4}$.

Solution

$$\text{Amplitude} = |-2| = 2 \qquad \text{Period} = \frac{2\pi}{\pi/4} = 8 \quad \bullet \ b = \pi/4$$

The zeros in the interval $0 \le x \le 8$ are $\dfrac{\pi}{2\pi/4} = 2$ and $\dfrac{3\pi}{2\pi/4} = 6$. Because $-2 < 0$, the graph is the graph of $y = 2 \cos \dfrac{\pi x}{4}$ reflected across the x-axis as sketched in Figure 6.58.

■ *Try Exercise* **46,** *page 336.*

EXAMPLE 6 **Graph the Absolute Value of the Cosine Function**

Graph $y = |\cos x|$, where $0 \le x \le 2\pi$.

Solution Because $|\cos x| \ge 0$, the graph of $y = |\cos x|$ is drawn by reflecting the negative portion of the graph of $y = \cos x$ across the x-axis. The graph is the solid graph shown in Figure 6.59.

■ *Try Exercise* **52,** *page 336.*

EXERCISE SET 6.5

In Exercises 1 to 16, state the amplitude and period of the function defined by each equation.

1. $y = 2 \sin x$

2. $y = -\dfrac{1}{2} \sin x$

3. $y = \sin 2x$

4. $y = \sin \dfrac{2x}{3}$

5. $y = \dfrac{1}{2} \sin 2\pi x$

6. $y = 2 \sin \dfrac{\pi x}{3}$

7. $y = -2 \sin \dfrac{x}{2}$

8. $y = -\dfrac{1}{2} \sin \dfrac{x}{2}$

9. $y = \dfrac{1}{2} \cos x$

10. $y = -3 \cos x$

11. $y = \cos \dfrac{x}{4}$

12. $y = \cos 3x$

13. $y = 2 \cos \dfrac{\pi x}{3}$

14. $y = \dfrac{1}{2} \cos 2\pi x$

15. $y = -3 \cos \dfrac{2x}{3}$

16. $y = \dfrac{3}{4} \cos 4x$

In Exercises 17 to 54, graph the function defined by each equation.

17. $y = \dfrac{1}{2} \sin x$

18. $y = \dfrac{3}{2} \cos x$

19. $y = 3 \cos x$

20. $y = -\dfrac{3}{2} \sin x$

21. $y = -\dfrac{7}{2} \cos x$

22. $y = 3 \sin x$

23. $y = -4 \sin x$

24. $y = -5 \cos x$

25. $y = \cos 3x$

26. $y = \sin 4x$

27. $y = \sin \dfrac{3x}{2}$

28. $y = \cos \pi x$

29. $y = \cos \dfrac{\pi}{2} x$

30. $y = \sin \dfrac{3\pi}{4} x$

31. $y = \sin 2\pi x$

32. $y = \cos 3\pi x$

33. $y = 4 \cos \dfrac{x}{2}$

34. $y = 2 \cos \dfrac{3x}{4}$

35. $y = -2 \cos \dfrac{x}{3}$

36. $y = -\dfrac{4}{3} \cos 3x$

37. $y = 2 \sin \pi x$

38. $y = \dfrac{1}{2} \sin \dfrac{\pi x}{3}$

39. $y = \dfrac{3}{2} \cos \dfrac{\pi x}{2}$

40. $y = \cos \dfrac{\pi x}{3}$

41. $y = 4 \sin \dfrac{2\pi x}{3}$

42. $y = 3 \cos \dfrac{3\pi x}{2}$

43. $y = 2 \cos 2x$

44. $y = \dfrac{1}{2} \sin 2.5x$

45. $y = -2 \sin 1.5x$

46. $y = -\dfrac{3}{4} \cos 5x$

47. $y = \left| 2 \sin \dfrac{x}{2} \right|$

48. $y = \left| \dfrac{1}{2} \sin 3x \right|$

49. $y = |-2 \cos 3x|$

50. $y = \left| -\dfrac{1}{2} \cos \dfrac{x}{2} \right|$

51. $y = -\left| 2 \sin \dfrac{x}{3} \right|$

52. $y = -\left| 3 \sin \dfrac{2x}{3} \right|$

53. $y = -|3 \cos \pi x|$

54. $y = -\left| 2 \cos \dfrac{\pi x}{2} \right|$

In Exercises 55 to 60, find an equation of each graph.

55.

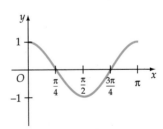

56.

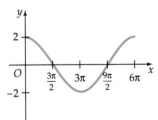

57.

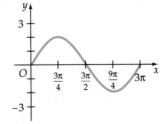

58.

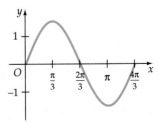

59.

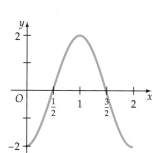

60.

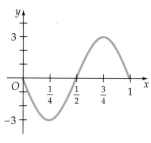

61. Sketch the graph of $y = 2 \sin \dfrac{2x}{3}$, $-3\pi \le x \le 6\pi$.

62. Sketch the graph of $y = -3 \cos \dfrac{3x}{4}$, $-2\pi \le x \le 4\pi$.

63. Sketch the graphs of

$$y_1 = 2 \cos \dfrac{x}{2} \quad \text{and} \quad y_2 = 2 \cos x$$

on the same set of axes for $-2\pi \le x \le 4\pi$.

64. Sketch the graphs of

$$y_1 = \sin 3\pi x \quad \text{and} \quad y_2 = \sin \dfrac{\pi x}{3}$$

on the same set of axes for $-2 \le x \le 4$.

Graphing Calculator Exercises*

65. Graph $y = \cos^2 x$.

66. Graph $y = 3^{\cos^2 x} \cdot 3^{\sin^2 x}$.

67. Graph $y = \cos|x|$.

68. Graph $y = \sin|x|$.

69. Graph $y = \dfrac{1}{2}x \sin x$.

70. Graph $y = \dfrac{1}{2}x + \sin x$.

71. Graph $y = -x \cos x$.

72. Graph $y = -x + \cos x$.

73. Complete the table for $f(x) = \dfrac{\sin x}{x}$.

x	-0.1	-0.05	-0.01	-0.001	0.001	0.01	0.05	0.1
$\dfrac{\sin x}{x}$								

*Additional graphing calculator exercises appear in the Graphing Workbook as described in the front of this textbook.

What conclusion might you draw about the value of $f(x)$ as $x \to 0$? Does the graph of f have a vertical asymptote at 0?

74. Complete the table for $f(x) = \dfrac{\cos x}{x}$.

x	-0.1	-0.05	-0.01	-0.001	0.001	0.01	0.05	0.1
$\dfrac{\cos x}{x}$								

What conclusion might you draw about the value of $f(x)$ as $x \to 0$? Does the graph of f have a vertical asymptote at 0?

75. Graph $y = e^{\sin x}$. What is the maximum value of $e^{\sin x}$? What is the minimum value of $e^{\sin x}$? Is the function defined by $y = e^{\sin x}$ a periodic function? If so, what is the period?

76. Graph $y = e^{\cos x}$. What is the maximum value of $e^{\cos x}$? What is the minimum value of $e^{\cos x}$? Is the function defined by $y = e^{\cos x}$ a periodic function? If so, what is the period?

Supplemental Exercises

In Exercises 77 to 80, write an equation for a sine function with the given information.

77. Amplitude = 2; period = 3π

78. Amplitude = 5; period = $2\pi/3$

79. Amplitude = 4; period = 2

80. Amplitude = 2.5; period = 3.2

In Exercises 81 to 84, write an equation for a cosine function with the given information.

81. Amplitude = 3; period = $\pi/2$

82. Amplitude = 0.8; period = 4π

83. Amplitude = 3; period = 2.5

84. Amplitude = 4.2; period = 1

85. A tidal wave that is caused by an earthquake under the ocean is called a **tsunami wave**. The formula $f(t) = A \cos Bt$ can be used to model the equation of a tsunami. Find the equation of a tsunami that has a amplitude of 60 feet, a period of 20 seconds, and travels 120 feet per second.

86. The electricity supplied to your home is called *alternating current* and can be expressed by the equation $I = A \sin \omega t$, where I is the number of amperes of current at time t seconds. Write the equation of household current whose graph is given in the figure below. Calculate I when $t = 0.5$ seconds.

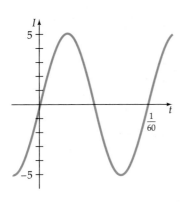

87. The secondary wave or S-wave of an earthquake travels about 3 miles per second and can be approximated by a sine curve. Assuming that an S-wave has an amplitude of 4 feet and a period of $\frac{2\pi}{3}$, write the equation for the S-wave. Find the displacement of the S wave when $t = 0.75$ seconds.

6.6 Graphs of the Other Trigonometric Functions

The Tangent Function

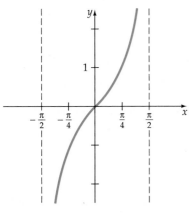

$y = \tan x,\ -\pi/2 < x < \pi/2$
Figure 6.60

Figure 6.60 shows the graph of $y = \tan x$ for $-\pi/2 < x < \pi/2$. The lines $x = \dfrac{\pi}{2}$ and $x = -\dfrac{\pi}{2}$ are vertical asymptotes for the graph of $y = \tan x$. From Section 6.4, the period of $y = \tan x$ is π. Therefore, the portion of the graph shown in Figure 6.60 is repeated along the x-axis as shown in Figure 6.61.

Because the tangent function is unbounded, there is no amplitude for the tangent function. The graph of $y = a \tan x$ is drawn by stretching ($|a| > 1$) or shrinking ($|a| < 1$) the graph of $y = \tan x$. If $a < 0$, then the graph is reflected across the x-axis. Figure 6.62 shows the graph of three tangent functions. Because $\tan \pi/4 = 1$, the point $(\pi/4, a)$ is convenient to plot as a guide for the graph of the equation $y = a \tan x$.

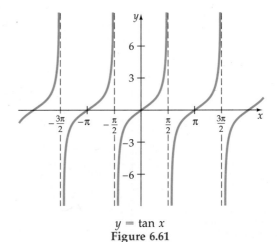

$y = \tan x$
Figure 6.61

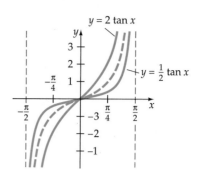

Figure 6.62

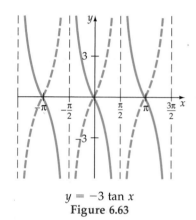

$y = -3 \tan x$
Figure 6.63

EXAMPLE 1   **Graph $y = a \tan x$**

Graph $y = -3 \tan x$.

   Solution   The graph of $y = -3 \tan x$ is the reflection across the x-axis of the graph of $y = 3 \tan x$ as shown by the solid graph in Figure 6.63

────────────────────────────────── ■ *Try Exercise* **22**, *page 344.*

   Because the period of $y = \tan x$ is π and the graph of one segment is completed as $-\pi/2 < x < \pi/2$, the period of $y = \tan bx$ ($b > 0$) is π/b and one segment of the graph will be completed as $-\dfrac{\pi}{2b} < x < \dfrac{\pi}{2b}$. If $b > 0$, then $\tan(-bx) = -\tan bx$ and one cycle of the graph is still completed as $-\dfrac{\pi}{2b} < x < \dfrac{\pi}{2b}$.

Period of $y = \tan bx$

> The period of $y = \tan bx$ is $\dfrac{\pi}{|b|}$.

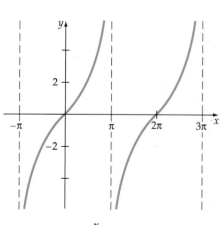

$y = 2 \tan \dfrac{x}{2},\ -\pi < x < 3\pi$

Figure 6.64

EXAMPLE 2   **Graph $y = a \tan bx$**

Graph $y = 2 \tan \dfrac{x}{2}$.

   Solution   Period $= \dfrac{\pi}{b} = \dfrac{\pi}{1/2} = 2\pi$. Graph one period for values of x such that $-\pi < x < \pi$. This curve is repeated along the x-axis as shown in Figure 6.64.

────────────────────────────────── ■ *Try Exercise* **30**, *page 344.*

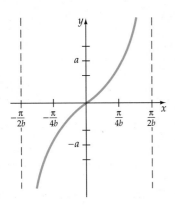

$$y = a \tan bx, \quad -\frac{\pi}{2b} < x < \frac{\pi}{2b}$$

Figure 6.65

Figure 6.65 shows one cycle of the graph of $y = a \tan bx$ for both a and b positive. Note from the graph that

- The period is π/b.

- $x = 0$ is a zero.

- The graph passes through $\left(-\frac{\pi}{4b}, -a\right)$ and $\left(\frac{\pi}{4b}, a\right)$.

- If $a < 0$, the graph is reflected across the x-axis.

The Cotangent Function

Figure 6.66 shows the graph of $y = \cot x$ for $0 < x < \pi$. The lines $x = 0$ and $x = \pi$ are vertical asymptotes for the graph of $y = \cot x$. From Section 6.4, the period of $y = \cot x$ is π. Therefore, the graph cycle shown in Figure 6.66 is repeated along the x-axis as shown in Figure 6.67. As with the graph of $y = \tan x$, the graph of $y = \cot x$ is unbounded and there is no amplitude. The graph of $y = a \cot x$ is drawn by stretching $(|a| > 1)$ or shrinking $(|a| < 1)$ the graph of $y = \cot x$. The graph is reflected across the x-axis when $a < 0$. Figure 6.68 shows the graphs of two cotangent functions.

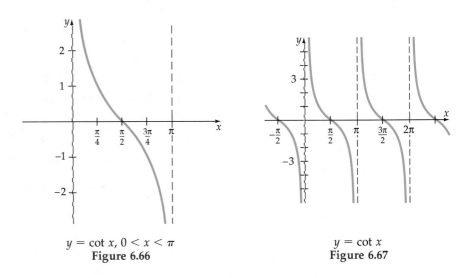

$y = \cot x, 0 < x < \pi$
Figure 6.66

$y = \cot x$
Figure 6.67

Because the period of $y = \cot x$ is π, the period of $y = \cot bx$ $(b > 0)$ is π/b. For $b > 0$, $\cot(-bx) = -\cot bx$ and a period is still π/b. Therefore one cycle of the graph $y = \cot bx$ is completed as $0 < x < \pi/b$. Normally we will show the graph of a cotangent function in this interval.

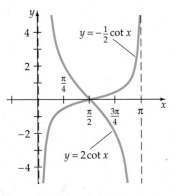

Figure 6.68

Period of $y = \cot bx$

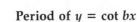

The period of $y = \cot bx$ is $\dfrac{\pi}{|b|}$.

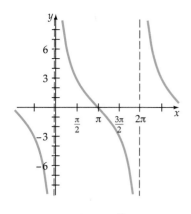

$$y = 3 \cot \frac{x}{2}$$

Figure 6.69

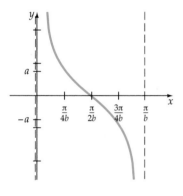

$$y = a \cot bx, \quad 0 < x < \frac{\pi}{b}$$

Figure 6.70

EXAMPLE 3 **Graph $y = a \cot bx$**

Graph $y = 3 \cot \dfrac{x}{2}$.

Solution Period $= \dfrac{\pi}{b} = \dfrac{\pi}{1/2} = 2\pi$. Sketch the graph for values of x where $0 < x < 2\pi$. This curve is repeated along the x-axis as shown in Figure 6.69.

■ *Try Exercise 32, page 344.*

Figure 6.70 shows one cycle of the graph of $y = a \cot bx$ for both a and b positive. Note from the graph that

- The period is $\dfrac{\pi}{b}$.

- $x = \dfrac{\pi}{2b}$ is a zero.

- The graph passes through $\left(\dfrac{\pi}{4b}, a\right)$ and $\left(\dfrac{3\pi}{4b}, -a\right)$.

- If $a < 0$, the graph is reflected across the x-axis.

The Cosecant Function

Because $\csc x = \dfrac{1}{\sin x}$, the value of $\csc x$ is the reciprocal of the value of $\sin x$. Therefore, $\csc x$ is undefined when $\sin x = 0$ or when $x = n\pi$, where n is an integer. The graph of $y = \csc x$ has vertical asymptotes at $n\pi$. Because $y = \csc x$ has period 2π, the graph will be repeated along the x-axis every 2π units. A graph of $y = \csc x$ is shown in Figure 6.71.

The graph of $y = \sin x$ has been sketched as a dashed curve in Figure 6.71. Note the relationships among the zeros of $y = \sin x$ and the asymptotes of $y = \csc x$. Also note that because $|\sin x| \le 1$, $\dfrac{1}{|\sin x|} \ge 1$. Thus

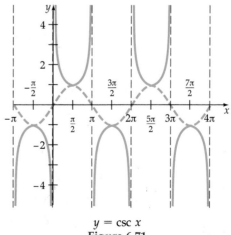

$$y = \csc x$$
Figure 6.71

the range of $y = \csc x$ is $|y| \geq 1$. The general procedure for graphing $y = a \csc bx$ is first to graph $y = a \sin bx$. Then sketch the graph of the cosecant function using reciprocal values determined by the graph of $y = a \sin bx$.

EXAMPLE 4   **Graph $y = a \csc bx$**

Graph $y = 2 \csc \dfrac{\pi x}{2}$.

Solution   First sketch the graph of $y = 2 \sin \dfrac{\pi x}{2}$ and draw vertical asymptotes through the zeros. Now sketch the graph of $y = 2 \csc \dfrac{\pi x}{2}$, using the asymptotes as guides for the graph as shown in Figure 6.72.

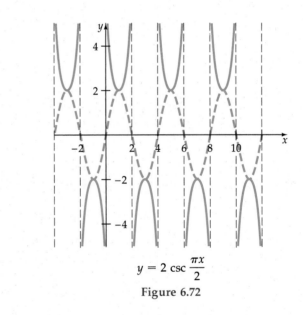

$$y = 2 \csc \frac{\pi x}{2}$$

Figure 6.72

■ *Try Exercise* **38**, *page 344.*

Figure 6.73 shows one cycle of the graph of $y = a \csc bx$ for both a and b positive. Note from the graph that

- The period is $\dfrac{2\pi}{b}$.

- The vertical asymptotes of $y = a \csc bx$ are the zeros of $y = a \sin bx$.

- The graph passes through $\left(\dfrac{\pi}{2b}, a\right)$, and $\left(\dfrac{3\pi}{2b}, -a\right)$.

- If $a < 0$, then the graph is reflected across the x-axis.

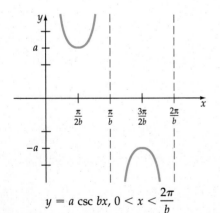

$y = a \csc bx, 0 < x < \dfrac{2\pi}{b}$

Figure 6.73

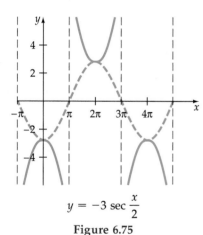

$y = \sec x$

Figure 6.74

The Secant Function

Because $\sec x = \dfrac{1}{\cos x}$, the value of $\sec x$ is the reciprocal of the value of $\cos x$. Therefore, $\sec x$ is undefined when $\cos x = 0$ or when $x = \dfrac{\pi}{2} + n\pi$, n an integer. The graph of $y = \sec x$ has vertical asymptotes at $\dfrac{\pi}{2} + n\pi$.

Because $y = \sec x$ has period 2π, the graph will be replicated along the x-axis every 2π units. A graph of $y = \sec x$ is shown in Figure 6.74.

The graph of $y = \cos x$ has been sketched as a dashed curve in Figure 6.74. Note the relationships among the zeros of $y = \cos x$ and the asymptotes of $y = \sec x$. Also note that because $|\cos x| \leq 1$, $\dfrac{1}{|\cos x|} \geq 1$. Thus the range of $y = \sec x$ is $|y| \geq 1$. The general procedure for graphing $y = a \sec bx$ is first to graph $y = a \cos bx$. Then sketch the graph of the secant function using reciprocal values determined by the graph of $y = a \cos bx$.

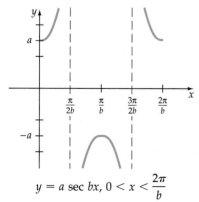

$y = -3 \sec \dfrac{x}{2}$

Figure 6.75

EXAMPLE 5 **Graph $y = a \sec bx$**

Graph $y = -3 \sec \dfrac{x}{2}$.

Solution First sketch the graph of $y = -3 \cos \dfrac{x}{2}$ and draw vertical asymptotes through the zeros. Now sketch the graph of $y = -3 \sec \dfrac{x}{2}$, using the asymptotes as guides for the graph as shown in Figure 6.75.

───────────────────────── ■ *Try Exercise 42, page 344.*

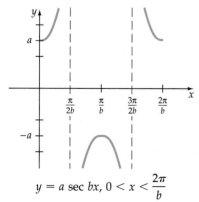

$y = a \sec bx, \; 0 < x < \dfrac{2\pi}{b}$

Figure 6.76

Figure 6.76 shows one cycle of the graph of $y = a \sec bx$ for both a and b positive. Note from the graph that

■ The period is $\dfrac{2\pi}{b}$.

■ The vertical asymptotes of $y = a \sec bx$ are the zeros of $y = a \cos bx$.

■ The graph passes through $(0, a)$, $(\pi/b, -a)$, and $(2\pi/b, a)$.

■ If $a < 0$, then the graph is reflected across the x-axis.

EXERCISE SET 6.6

1. For what values of x is $y = \tan x$ undefined?

2. For what values of x is $y = \cot x$ undefined?

3. For what values of x is $y = \sec x$ undefined?

4. For what values of x is $y = \csc x$ undefined?

In Exercises 5 to 20, state the period of each equation.

5. $y = \sec x$ 6. $y = \cot x$ 7. $y = \tan x$

8. $y = \csc x$ 9. $y = 2 \tan \dfrac{x}{2}$ 10. $y = \dfrac{1}{2} \cot 2x$

11. $y = \csc 3x$ 12. $y = \csc \dfrac{x}{2}$

13. $y = -\tan 3x$ 14. $y = -3 \cot \dfrac{2x}{3}$

15. $y = -3 \sec \dfrac{x}{4}$ 16. $y = -\dfrac{1}{2} \csc 2x$

17. $y = \cot \pi x$ 18. $y = \cot \dfrac{\pi x}{3}$

19. $y = 2 \csc \dfrac{\pi x}{2}$ 20. $y = -3 \cot \pi x$

In Exercises 21 to 40, sketch the graph of each equation.

21. $y = 3 \tan x$ 22. $y = \dfrac{1}{3} \tan x$

23. $y = \dfrac{3}{2} \cot x$ 24. $y = 4 \cot x$

25. $y = 2 \sec x$ 26. $y = \dfrac{3}{4} \sec x$

27. $y = \dfrac{1}{2} \csc x$ 28. $y = 2 \csc x$

29. $y = 2 \tan \dfrac{x}{2}$ 30. $y = -3 \tan 3x$

31. $y = -3 \cot \dfrac{x}{2}$ 32. $y = \dfrac{1}{2} \cot 2x$

33. $y = -2 \csc \dfrac{x}{3}$ 34. $y = \dfrac{3}{2} \csc 3x$

35. $y = \dfrac{1}{2} \sec 2x$ 36. $y = -3 \sec \dfrac{2x}{3}$

37. $y = -2 \sec \pi x$ 38. $y = 3 \csc \dfrac{\pi x}{2}$

39. $y = 3 \tan 2\pi x$ 40. $y = -\dfrac{1}{2} \cot \dfrac{\pi x}{2}$

41. Graph $y = 2 \csc 3x$ from -2π to 2π.

42. Graph $y = \sec \dfrac{x}{2}$ from -4π to 4π.

43. Graph $y = 3 \sec \pi x$ from -2 to 4.

44. Graph $y = \csc \dfrac{\pi x}{2}$ from -4 to 4.

45. Graph $y = 2 \cot 2x$ from $-\pi$ to π.

46. Graph $y = \dfrac{1}{2} \tan \dfrac{x}{2}$ from -4π to 4π.

47. Graph $y = 3 \tan \pi x$ from -2 to 2.

48. Graph $y = \cot \dfrac{\pi x}{2}$ from -4 to 4.

In Exercises 49 to 54, find an equation of each solid graph.

49.
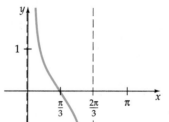

50.
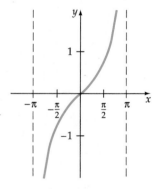

51.
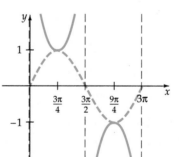

52.
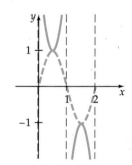

53.
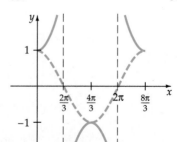

54.
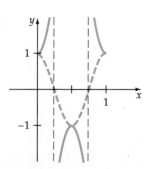

Graphing Calculator Exercises*

In Exercises 55 to 60, graph each equation.

55. $y = \tan |x|$ **56.** $y = \sec |x|$

57. $y = |\csc x|$ **58.** $y = |\cot x|$

59. $y = \tan x \cos x$ **60.** $y = \cot x \sin x$

61. Graph $y = \tan x$ and $x = \tan y$ on the same coordinate axes.

62. Graph $y = \sin x$ and $x = \sin y$ on the same coordinate axes.

Supplemental Exercises

In Exercises 63 to 70, write an equation of the form $y = \tan bx$, $y = \cot bx$, $y = \sec bx$, or $y = \csc bx$ that satisfies the given conditions.

63. Tangent, period: $\pi/3$ **64.** Cotangent, period: $\pi/2$

65. Secant, period: $3\pi/4$ **66.** Cosecant, period: $5\pi/2$

67. Cotangent, period: 2   **68.** Tangent, period: 0.5

69. Cosecant, period: 1.5   **70.** Secant, period: 3

6.7 Translation and Addition of Ordinates

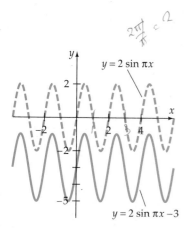

Figure 6.77

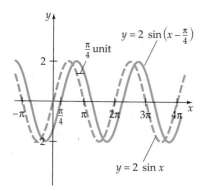

Figure 6.78

Translation of Trigonometric Functions

Recall that the graph of $y = f(x) \pm c$ is a *vertical translation* of the graph of $y = f(x)$. For $c > 0$, the graph of $y = f(x) - c$ is shifted c units down; the graph of $y = f(x) + c$ is shifted c units up. The graph in Figure 6.77 is a graph of the equation $y = 2 \sin \pi x - 3$, which is a vertical translation of $y = 2 \sin \pi x$ three units down. Note that subtracting three from $2 \sin \pi x$ changes neither its amplitude nor its period.

Also, the graph of $y = f(x \pm c)$ is a *horizontal translation* of the graph of $y = f(x)$. For $c > 0$, the graph of $y = f(x - c)$ is shifted c units to the right; the graph of $y = f(x + c)$ is shifted c units to the left. The graph in Figure 6.78 is a graph of the equation $y = 2 \sin\left(x - \dfrac{\pi}{4}\right)$, which is the graph of $y = 2 \sin x$ translated $\dfrac{\pi}{4}$ units to the right. Note that neither the period nor the amplitude was affected. The horizontal shift of the graph of a trigonometric function is called its **phase shift.**

Because one cycle of $y = a \sin x$ is completed for $0 \le x \le 2\pi$, one cycle of the graph of $y = a \sin(bx + c)$, where $b > 0$, is completed for $0 \le bx + c \le 2\pi$. Solving this inequality for x, we have

$$0 \le bx + c \le 2\pi$$

$$-c \le bx \le -c + 2\pi$$

$$-\frac{c}{b} \le x \le -\frac{c}{b} + \frac{2\pi}{b}$$

*Additional graphing calculator exercises appear in the Graphing Workbook as described in the front of this textbook.

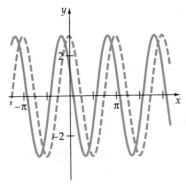

$$y = 3\cos\left(2x + \frac{\pi}{3}\right)$$

Figure 6.79

The number $-\frac{c}{b}$ is the phase shift for $y = a\sin(bx + c)$. The graph of the equation $y = a\sin(bx + c)$ is the graph of $y = a\sin bx$ shifted $-\frac{c}{b}$ units horizontally. Similar arguments apply to the remaining trigonometric functions.

EXAMPLE 1   Graph $y = a\cos(bx + c)$

Graph $y = 3\cos\left(2x + \frac{\pi}{3}\right)$.

Solution   The phase shift is $-\frac{c}{b} = -\frac{\pi/3}{2} = -\frac{\pi}{6}$. The graph of the equation $y = 3\cos\left(2x + \frac{\pi}{3}\right)$ is the graph of $y = 3\cos 2x$ shifted $\frac{\pi}{6}$ units to the left as shown in Figure 6.79.

■ *Try Exercise* **20**, *page 348.*

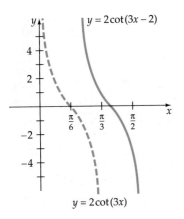

Figure 6.80

EXAMPLE 2   Graph $y = a\cot(bx + c)$

Graph $y = 2\cot(3x - 2)$.

Solution   The phase shift is $-\frac{c}{b} = -\frac{-2}{3} = \frac{2}{3}$.   • $3x - 2 = 3x + (-2)$

The graph of $y = 2\cot(3x - 2)$ is the graph of $y = 2\cot(3x)$ shifted $\frac{2}{3}$ units to the right as shown in Figure 6.80.

■ *Try Exercise* **22**, *page 348.*

The graph of trigonometric function may be the combination of a vertical translation and a phase shift.

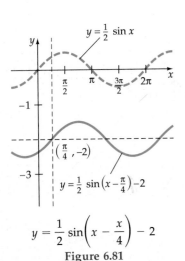

$$y = \frac{1}{2}\sin\left(x - \frac{x}{4}\right) - 2$$

Figure 6.81

EXAMPLE 3   Graph $y = a\sin(bx + c) + d$

Graph $y = \frac{1}{2}\sin\left(x - \frac{\pi}{4}\right) - 2$.

Solution   The phase shift is $-\frac{c}{b} = -\frac{-\pi/4}{1} = \frac{\pi}{4}$. The vertical shift is 2 units down. The graph of $y = \frac{1}{2}\sin\left(x - \frac{\pi}{4}\right) - 2$ is the graph of $y = \frac{1}{2}\sin x$ shifted $\frac{\pi}{4}$ units to the right and 2 units down as shown in Figure 6.81.

■ *Try Exercise* **40**, *page 348.*

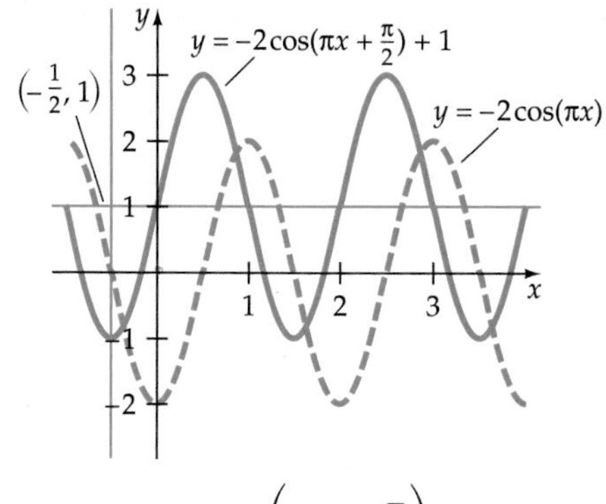

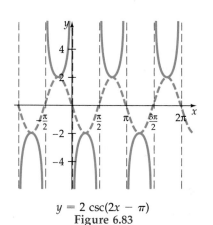

$$y = -2 \cos\left(\pi x + \frac{\pi}{2}\right) + 1$$

Figure 6.82

EXAMPLE 4   Graph $y = a\cos(bx + c) + d$

Graph $y = -2\cos\left(\pi x + \dfrac{\pi}{2}\right) + 1$

Solution   The phase shift is $-\dfrac{c}{b} = -\dfrac{\pi/2}{\pi} = -\dfrac{1}{2}$. The vertical shift is 1 unit up. The graph of $y = -2\cos\left(\pi x + \dfrac{\pi}{2}\right) + 1$ is the graph of $y = -2\cos \pi x$ shifted $\dfrac{1}{2}$ units to the left, and then shifted 1 unit up as shown in Figure 6.82.

■ *Try Exercise* **42**, *page 348.*

Similar techniques are used to graph secant and cosecant functions.

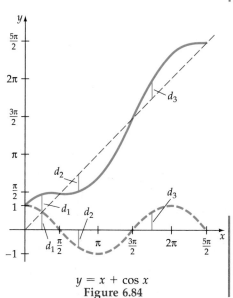

$$y = 2 \csc(2x - \pi)$$

Figure 6.83

EXAMPLE 5   Graph a Cosecant Function

Graph $y = 2\csc(2x - \pi)$.

Solution   The phase shift is $-\dfrac{c}{b} = -\dfrac{-\pi}{2} = \dfrac{\pi}{2}$. The graph of $y = 2\csc(2x - \pi)$ is the graph of $y = 2\csc 2x$ shifted $\dfrac{\pi}{2}$ units to the right.

Sketch the graph of the equation $y = 2\sin 2x$ shifted $\dfrac{\pi}{2}$ units to the right as a dashed curve. Use this graph to draw the graph of $y = 2\csc(2x - \pi)$ as shown in Figure 6.83.

■ *Try Exercise* **48**, *page 349.*

Addition of Ordinates

Given two functions g and h, the sum of the functions is the function f defined by $f(x) = g(x) + h(x)$. The graph of the sum f can be obtained by graphing g and h separately and then geometrically adding the y coordinates of each function for a given value of x. It is convenient, when we are drawing the graph of the sum of two functions, to pick zeros of the function. This technique is illustrated in Examples 6 and 7.

EXAMPLE 6   Graph the Sum of Two Functions

Graph $y = x + \cos x$.

Solution   Graph $g(x) = x$ and $h(x) = \cos x$ on the same coordinate grid. Then add the y-coordinates geometrically point by point. Figure 6.84 shows

$$y = x + \cos x$$

Figure 6.84

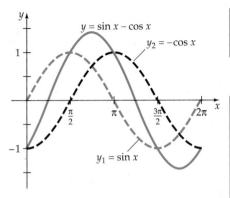

$y = \sin x - \cos x$
Figure 6.85

the results of adding, by using a ruler, the y-coordinates of the two functions for selected values of x.

■ *Try Exercise 52, page 349.*

EXAMPLE 7   Graph the Difference of Two Functions

Graph $y = \sin x - \cos x$ for $0 \le x \le 2\pi$.

Solution   Graph $g(x) = \sin x$ and $h(x) = -\cos x$ on the same coordinate grid. For selected values of x, add $g(x)$ and $h(x)$ geometrically. Now draw a smooth curve through the points. See Figure 6.85.

■ *Try Exercise 56, page 349.*

EXERCISE SET 6.7

In Exercises 1 to 8, find the amplitude, phase shift, and period for the graph of each equation.

1. $y = 2 \sin\left(x - \dfrac{\pi}{2}\right)$ **2.** $y = -3 \sin(x + \pi)$

3. $y = \cos\left(2x - \dfrac{\pi}{4}\right)$ **4.** $y = \dfrac{3}{4} \cos\left(\dfrac{x}{2} + \dfrac{\pi}{3}\right)$

5. $y = -4 \sin\left(\dfrac{2x}{3} + \dfrac{\pi}{6}\right)$ **6.** $y = \dfrac{3}{2} \sin\left(\dfrac{x}{4} - \dfrac{3\pi}{4}\right)$

7. $y = \dfrac{5}{4} \cos(3x - 2\pi)$ **8.** $y = 6 \cos\left(\dfrac{x}{3} - \dfrac{\pi}{6}\right)$

In Exercises 9 to 16, find the phase shift and the period for the graph of each equation.

9. $y = 2 \tan\left(2x - \dfrac{\pi}{4}\right)$ **10.** $y = \dfrac{1}{2} \tan\left(\dfrac{x}{2} - \pi\right)$

11. $y = -3 \csc\left(\dfrac{x}{3} + \pi\right)$ **12.** $y = -4 \csc\left(3x - \dfrac{\pi}{6}\right)$

13. $y = 2 \sec\left(2x - \dfrac{\pi}{8}\right)$ **14.** $y = 3 \sec\left(\dfrac{x}{4} - \dfrac{\pi}{2}\right)$

15. $y = -3 \cot\left(\dfrac{x}{4} + 3\pi\right)$ **16.** $y = \dfrac{3}{2} \cot\left(2x - \dfrac{\pi}{4}\right)$

In Exercises 17 to 32, graph one cycle of each equation.

17. $y = \sin\left(x - \dfrac{\pi}{2}\right)$ **18.** $y = \sin\left(x + \dfrac{\pi}{6}\right)$

19. $y = \cos\left(\dfrac{x}{2} + \dfrac{\pi}{3}\right)$ **20.** $y = \cos\left(2x - \dfrac{\pi}{3}\right)$

21. $y = \tan\left(x + \dfrac{\pi}{4}\right)$ **22.** $y = \tan(x - \pi)$

23. $y = 2 \cot\left(\dfrac{x}{2} - \dfrac{\pi}{8}\right)$ **24.** $y = \dfrac{3}{2} \cot\left(3x + \dfrac{\pi}{4}\right)$

25. $y = \sec\left(x + \dfrac{\pi}{4}\right)$ **26.** $y = \csc(2x + \pi)$

27. $y = \csc\left(\dfrac{x}{3} - \dfrac{\pi}{2}\right)$ **28.** $y = \sec\left(2x + \dfrac{\pi}{6}\right)$

29. $y = -2 \sin\left(\dfrac{x}{3} - \dfrac{2\pi}{3}\right)$ **30.** $y = -\dfrac{3}{2} \sin\left(2x + \dfrac{\pi}{4}\right)$

31. $y = -3 \cos\left(3x + \dfrac{\pi}{4}\right)$ **32.** $y = -4 \cos\left(\dfrac{3x}{2} + 2\pi\right)$

In Exercises 33 to 50, graph each equation using translations.

33. $y = \sin x + 1$ **34.** $y = -\sin x + 1$

35. $y = -\cos x - 2$ **36.** $y = 2 \sin x + 3$

37. $y = \sin 2x - 2$ **38.** $y = -\cos \dfrac{x}{2} + 2$

39. $y = 4 \cos(\pi x - 2) + 1$

40. $y = 2 \sin\left(\dfrac{\pi x}{2} + 1\right) - 2$

41. $y = -\sin(\pi x + 1) - 2$

42. $y = -3 \cos(2\pi x - 3) + 1$

43. $y = \sin\left(x - \dfrac{\pi}{2}\right) - \dfrac{1}{2}$

44. $y = -2 \cos\left(x + \dfrac{\pi}{3}\right) + 3$

45. $y = \tan \dfrac{x}{2} - 4$

46. $y = \cot 2x + 3$

47. $y = \sec 2x - 2$

48. $y = \csc \dfrac{x}{3} + 4$

49. $y = \csc \dfrac{x}{2} - 1$

50. $y = \sec\left(x - \dfrac{\pi}{2}\right) + 1$

In Exercises 51 to 56, graph the given functions by using the addition of ordinates.

51. $y = x - \sin x$

52. $y = \dfrac{x}{2} + \cos x$

53. $y = x + \sin 2x$

54. $y = \dfrac{2x}{3} - \sin x$

55. $y = \sin x + \cos x$

56. $y = -\sin x + \cos x$

In Exercises 57 to 62, find an equation of the trigonometric function from the accompanying solid graph.

57.

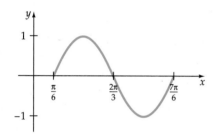

58.

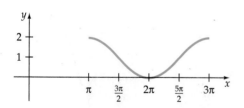

59.

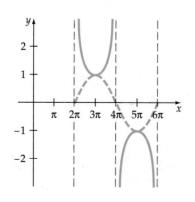

60.

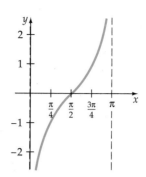

61.

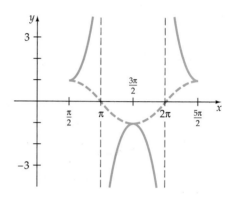

62.

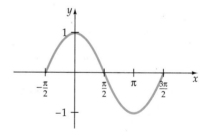

63. Because of seasonal changes in vegetation, carbon dioxide (CO_2) levels, as a product of photosynthesis, rise and fall during the year. Besides the naturally occurring CO_2 from plants, additional CO_2 is given off as pollutants. A reasonable model of CO_2 levels in a city for the years 1972–1992 is given by $y = 2.3 \sin 2\pi t + 1.25t + 315$, where t is the number of years since 1972 and y is the concentration of CO_2 in parts per million (ppm). Find the difference in CO_2 levels between the beginning of 1972 and 1992.

64. Some environmentalists contend that the rate of growth of atmospheric CO_2 is given by $y = 2.54e^{0.112t} + \sin 2\pi t + 315$. Use this model to find the difference between CO_2 levels from the beginning of 1972 to 1992.

65. The paddle wheel on a river boat is shown in the accompanying figure. Write an equation for the position of a paddle relative to the water at time t. The radius of the paddle wheel is 7 feet and the distance from the center of the paddle wheel to the water is 5 feet. Assume that the paddle wheel rotates at 5 rpm and that the paddle is at its highest point at $t = 0$. Graph the equation for $0 \le t \le 0.20$ minutes.

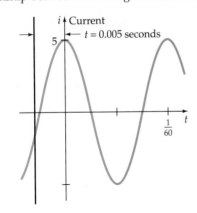

66. The graphs of the voltage and the amperage of an alternating household circuit are shown in the accompanying figure. Note that there is a phase shift between the graph of the voltage and the graph of the current. The current is said to *lag* the voltage. Write an equation that shows the relationship between the voltage and the current.

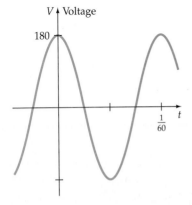

67. The beacon of a lighthouse 400 meters from a straight sea wall rotates at 6 rpm. Using the accompanying figure, write an equation expressing the distance s measured in

meters in terms of time t. Assume that when $t = 0$, the beam is perpendicular to the sea wall. Sketch a graph of the equation for $0 \le t \le 10$ seconds.

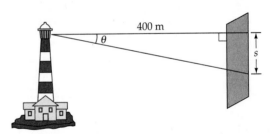

68. The duration of daylight for a region is dependent not only on the time of year but also on the latitude of the region. The graph gives the daylight hours for a one-year period for various latitudes. Assuming that a sine function can model these curves, write an equation for each curve.

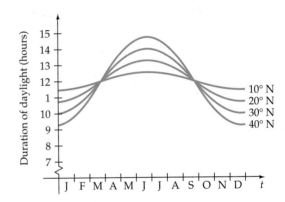

69. During a 24-hour day, the tides raise and lower the depth of water at the pier as shown in the figure. Write an equation in the form $f(t) = A \cos Bt + k$, and find the depth of the water at 6 P.M.

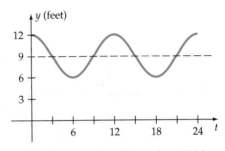

t is the number of hours from 6AM

70. During a summer day, the ground temperature at a desert location was recorded and graphed as a function of time as shown in the figure. The graph can be approxi-

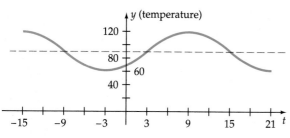

t is the number of hours from 6AM

mated by $f(x) = A\cos(bx + c) + k$. Find the equation, and approximate the temperature at 1:00 P.M.

Graphing Calculator Exercises*

71. $y = \sin x - \cos \dfrac{x}{2}$

72. $y = 2\sin 2x - \cos x$

73. $y = 2\cos x + \sin \dfrac{x}{2}$

74. $y = -\dfrac{1}{2}\cos 2x + \sin \dfrac{x}{2}$

75. $y = \dfrac{x}{2}\sin x$

76. $y = x\cos x$

77. $y = x\sin \dfrac{x}{2}$

78. $y = \dfrac{x}{2}\cos \dfrac{x}{2}$

79. $y = x\sin\left(x + \dfrac{\pi}{2}\right)$

80. $y = x\cos\left(x - \dfrac{\pi}{2}\right)$

When two sound waves have approximately the same frequency, the sound waves interfere with one another and produce phenomena called *beats*, which are heard as variations in the loudness of the sound. A piano tuner can use these phenomena to tune a piano. By striking a tuning fork and then tapping the corresponding key on a piano, the piano tuner listens for beats and adjusts the tension in the string until the beats disappear. Graph the equations in 81 to 84 that are based on beats.

81. $y = \sin(5\pi x) \cdot \sin\left(-\dfrac{\pi}{2}x\right)$

82. $y = \sin(9\pi x) \cdot \sin\left(-\dfrac{\pi}{2}x\right)$

83. $y = \sin(13\pi x) \cdot \sin\left(-\dfrac{\pi}{2}x\right)$

84. $y = \sin(17\pi x) \cdot \sin\left(-\dfrac{\pi}{2}x\right)$

Supplemental Exercises

85. Find an equation of the sine function with amplitude 2, period π, and phase shift $\pi/3$.

86. Find an equation of the cosine function with amplitude 3, period 3π, and phase shift $-\pi/4$.

87. Find an equation of the tangent function with period 2π and phase shift $\pi/2$.

88. Find an equation of the cotangent function with period $\pi/2$ and phase shift $-\pi/4$.

89. Find an equation of the secant function with period 4π and phase shift $3\pi/4$.

90. Find an equation of the cosecant function with period $3\pi/2$ and phase shift $\pi/4$.

91. If $g(x) = \sin^2 x$ and $h(x) = \cos^2 x$, find $g(x) + h(x)$.

92. If $g(x) = 2\sin x - 3$ and $h(x) = 4\cos x + 2$, find the sum $g(x) + h(x)$.

93. If $g(x) = x^2 + 2$ and $h(x) = \cos x$, find $g[h(x)]$.

94. If $g(x) = \sin x$ and $h(x) = x^2 + 2x + 1$, find $h[g(x)]$.

In Exercises 91 to 94, sketch the graph of each equation.

95. $y = \dfrac{\sin x}{x}$

96. $y = 2 + \sec \dfrac{x}{2}$

97. $y = |x|\sin x$

98. $y = |x|\cos x$

6.8 Simple Harmonic Motion—An Application of the Sine and Cosine Functions

Many phenomena occur in nature that can be modeled by periodic functions, including vibrations in buildings, sound waves, electromagnetic waves, and vibrations of a swing or in a spring. These phenomena can be

*Additional graphing calculator exercises appear in the Graphing Workbook as described in the front of this textbook.

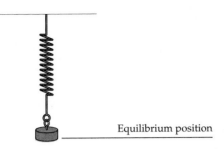

Equilibrium position

Figure 6.86

described by the *sinusoidal* functions, which are the sine and cosine functions or the sum of these two functions.

We will consider a mass on a spring to illustrate vibratory motion. Assume that we have placed a mass on a spring and allowed the spring to come to rest, as shown in Figure 6.86. The system is said to be in equilibrium when the mass is at rest. The point of rest is called the origin of the system. We consider the distance above the equilibrium point as positive and the distance below the equilibrium point as negative.

If the mass is now lifted a distance a and released, the mass will oscillate up and down in periodic motion. If there is no fraction, the motion repeats itself in a certain period of time. The distance a is called the displacement from the origin. The number of times the mass oscillates in 1 second is called the frequency of the motion, and the time one oscillation takes is the period of the motion. For small oscillations, this period is a constant and the motion is referred to as simple harmonic motion. Figure 6.87 shows the position y of the mass for one oscillation for $t = 0$, $p/4$, $p/2$, $3p/4$, and p when the period is p.

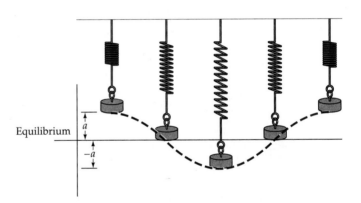

Figure 6.87

Remark   Note that if we were to graph the displacement y as a function of t and draw a smooth line through the points, we would have a cosine curve.

There is a relationship between the frequency and the period. Assume that we have a mass that will make two oscillations (an oscillation is a back-and-forth motion) in 1 second. The time for one oscillation is $\frac{1}{2}$ second. Thus the period is $\frac{1}{2}$ second. The frequency and the period are related by the formula $f = 1/\text{period}$.

The maximum displacement from the equilibrium position is called the amplitude of the motion. Vibratory motion can be quite complicated. However, the simple harmonic motion with the mass on the spring can be described by the following equation.

Definition of Simple Harmonic Motion

Simple harmonic motion is motion that can be modeled by one of the following equations:

$$y = a \cos 2\pi ft \quad \text{or} \quad y = a \sin 2\pi ft$$

where $|a|$ is the amplitude (maximum displacement), f is the frequency, $1/f$ is the period, y is the displacement, and t is the time.

Remark   We have been given two equations of simple harmonic motion. The cosine function is used if the displacement from the origin is at a maximum at time $t = 0$. The sine function is used if the displacement at time $t = 0$ is zero.

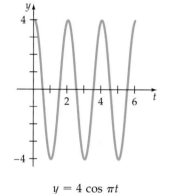

$y = 4 \cos \pi t$
Figure 6.88

EXAMPLE 1   **Find the Equation of Motion of a Mass on a Spring**

A mass on a spring has been displaced 4 centimeters above the equilibrium point and released. The mass is vibrating with a frequency of $\frac{1}{2}$ cycle per second. Write the equation of motion and graph three cycles of the displacement as a function of time.

Solution   Since the maximum displacement is 4 centimeters when $t = 0$, use $y = a \cos 2\pi ft$. See Figure 6.88.

$$y = a \cos 2\pi ft \qquad \bullet \text{ Equation for simple harmonic motion}$$

$$= 4 \cos 2\pi\left(\frac{1}{2}\right)t \qquad \bullet a = 4, f = \tfrac{1}{2}.$$

$$= 4 \cos \pi t$$

■ *Try Exercise 20, page 355.*

From physical laws determined by experiment, the frequency of oscillation of a mass on a spring is given by

$$f = \frac{1}{2\pi} \sqrt{\frac{k}{m}}$$

where k is a spring constant determined by experiment and m is the mass. The motion of the mass on the spring can then be described by

$$y = a \cos 2\pi ft = a \cos 2\pi\left(\frac{1}{2\pi} \sqrt{\frac{k}{m}}\right)t$$

$$= a \cos \sqrt{\frac{k}{m}}\,t$$

The equation of motion for zero displacement at $t = 0$ is

$$y = a \sin \sqrt{\frac{k}{m}}\, t$$

EXAMPLE 2   Find the Equation of Motion of a Mass on a Spring

A mass of 2 units is in equilibrium suspended from a spring. The mass is pulled down 0.5 units and released. Find the period, frequency, and amplitude of the resulting motion. Write the equation of the motion if $k = 18$, and graph two cycles of the displacement as a function of time.

Solution   At the start of the motion, the displacement is at a maximum but in the negative direction. The resulting motion is described by the cosine function, using $a = -0.5$, $k = 18$, $m = 2$. See Figure 6.89.

$$y = a \cos \sqrt{\frac{k}{m}}\, t = -0.5 \cos \sqrt{\frac{18}{2}}\, t \qquad \bullet \text{ Substitute for } a, k, \text{ and } m.$$

$$= -0.5 \cos 3t \qquad\qquad \bullet \text{ Equation of motion}$$

Period: $\dfrac{2\pi}{|b|} = \dfrac{2\pi}{3}$

Frequency: $\dfrac{1}{\text{period}} = \dfrac{3}{2\pi},$

Amplitude: 0.5

■ *Try Exercise* **28,** *page 356.*

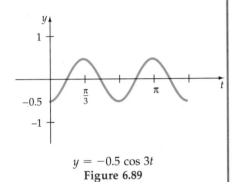

$y = -0.5 \cos 3t$
Figure 6.89

A simple pendulum is a system that exhibits approximate simple harmonic motion. A simple pendulum consists of a mass suspended from a string that is attached to a fixed point. If the mass is displaced through an angle θ and released, the pendulum will oscillate back and forth in a plane. It can be shown from physical laws that if the angle θ is small, the equation of motion is approximated by the equation

$$y = a \cos 2\pi ft \qquad \text{or} \qquad y = a \sin 2\pi ft$$

where y is the displacement at time t, a is the maximum displacement, and f is the frequency of motion. The period of the motion is $1/f$.

From measurements taken from experiments on a simple pendulum,

$$f = \frac{1}{2\pi} \sqrt{\frac{g}{l}}$$

where g is the gravitational constant 32 feet per second squared, and l is the length of the pendulum in feet. See Figure 6.90

Thus the equation for the motion of a pendulum is given by

$$y = a \cos \sqrt{\frac{g}{l}}\, t \qquad \text{or} \qquad y = a \sin \sqrt{\frac{g}{l}}\, t$$

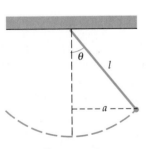

Figure 6.90

EXAMPLE 3 **Find the Period and Frequency of the Motion of the Pendulum**

Find the period and frequency of a pendulum with a length of 8 feet. Graph two cycles of the motion if the maximum displacement is 1.5 feet and the displacement is zero at $t = 0$.

Solution Since the displacement is zero at $t = 0$, we use $y = a \sin \sqrt{\frac{g}{l}} t$ to model the displacement of the pendulum. Using $a = 1.5$, $g = 32$, and $l = 8$, we have

$$y = 1.5 \sin \sqrt{\frac{32}{8}} t$$

$$= 1.5 \sin 2t$$

The period of the motion is $\frac{2\pi}{2} = \pi$. The frequency is the reciprocal of the period or $\frac{1}{\pi}$ and the amplitude is 1.5. See the graph in Figure 6.91.

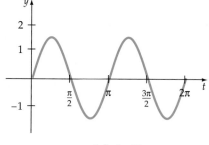

$y = 1.5 \sin 2t$
Figure 6.91

■ *Try Exercise* **30,** *page 356.*

EXERCISE SET 6.8

In Exercises 1 to 8, find the amplitude, period, and frequency of the harmonic motion.

1. $y = 2 \sin 2t$

2. $y = \frac{2}{3} \cos \frac{t}{3}$

3. $y = 3 \cos \frac{2t}{3}$

4. $y = 4 \sin 3t$

5. $y = 4 \cos \pi t$

6. $y = 2 \sin \frac{\pi t}{3}$

7. $y = \frac{3}{4} \sin \frac{\pi t}{2}$

8. $y = 5 \cos 2\pi t$

In Exercises 9 to 12, write an equation of motion and graph the amplitude as a function of time for the following given values. Assume that the motion is harmonic motion and that maximum displacement occurs at $t = 0$.

9. Frequency = 1.5 cycles per second, $a = 4$ inches

10. Frequency = 0.8 cycle per second, $a = 4$ centimeters

11. Period = 1.5 seconds, $a = \frac{3}{2}$ feet

12. Period = 0.6 second, $a = 1$ meter

In Exercises 13 to 18, find an equation of simple harmonic motion with the given conditions. Assume zero displacement at $t = 0$.

13. Amplitude 2 centimeters, period π seconds

14. Amplitude 4 inches, period $\pi/2$ seconds

15. Amplitude 1 inch, period 2 seconds

16. Amplitude 3 centimeters, period 1 second

17. Amplitude 2 centimeters, frequency 1 second

18. Amplitude 4 inches, frequency 4 seconds

In Exercises 19 to 26, write an equation for simple harmonic motion. Assume that the maximum displacement occurs when $t = 0$.

19. Amplitude $\frac{1}{2}$ centimeters, frequency $2/\pi$ cycles per second

20. Amplitude 3 inches, frequency $1/\pi$ cycles per second

21. Amplitude 2.5 inches, frequency 0.5 cycles per second

22. Amplitude 5 inches, frequency $\frac{1}{8}$ cycles per second

23. Amplitude $\frac{1}{2}$ inch, period 3 seconds

24. Amplitude 5 centimeters, period 5 seconds

25. Amplitude 4 inches, period $\pi/2$ seconds

26. Amplitude 2 centimeters, period π seconds

27. A mass of 32 units is in equilibrium suspended from a spring. The mass is pulled down 2 feet and released. Find the period, frequency, and amplitude of the resulting motion. Write an equation of motion. Let $k = 8$.

28. A mass of 27 units is in equilibrium suspended from a spring. The mass is pulled down 1.5 feet and released. Find the period, frequency, and amplitude of the resulting motion. Write an equation of motion. Let $k = 3$.

29. A pendulum 6 feet long is displaced a distance of 1 foot and released. Write an equation for the displacement as a function of time. Find the period and frequency of the motion.

30. A pendulum 20 feet long is displaced a distance of 4 feet and released. Write an equation for the displacement as a function of time. Find the period and frequency of the motion.

31. A mass of 5 units is suspended from a spring. The spring is compressed 6 inches and released. Find the period, frequency, and amplitude of the resulting motion. The constant $k = 2$. Write an equation of motion.

32. A pendulum of 3 feet long is displaced a distance of 6 inches and released. Find the period and frequency of the motion and write an equation of motion.

Graphing Calculator Exercises*

In all mechanical systems, friction is present. Consequently, a mass on the end of a spring will not continue to oscillate indefinitely. The result is *damped harmonic motion*. An equation of damped harmonic motion is $y = e^{-kt} \cos \omega t$, where $k > 0$. Graph the following damped harmonic equations.

33. $y = e^{-0.5t} \cos 2\pi t$

34. $y = e^{-0.2t} \cos 3\pi t$

35. $y = e^{-0.75t} \cos 2\pi t$

36. $y = e^{-t} \cos 2\pi t$

Supplemental Exercises

37. A pendulum with a length of 4 feet is released with an initial displacement of 6 inches. Write an equation for the displacement and find the frequency of the motion.

38. A mass of 0.5 unit is suspended from a spring with a constant of 32. The mass is displaced a distance of 10 inches and released. Write an equation for the motion and find the frequency of the motion.

39. A weight on a spring is displaced 6 inches from its equilibrium position and then released. The weight oscillated with a frequency of 1.5 cycles per second. Find the period and an equation of the motion.

40. A weight on a spring is displaced 9 inches from its equilibrium position and then released. The weight oscillated with a frequency of 2 cycles per second. Find the period and an equation of the motion.

41. What effect does doubling the length of a pendulum have on the period of the pendulum?

42. The length of a pendulum is changed until the period is cut in half. Express the new length of the pendulum in terms of its original length.

Exploring Concepts with Technology

Vertical Asymptotes and Removable Singularities

Recall that a function f has a vertical asymptote at a if $|f(x)| \to \infty$ as $x \to a^+$ or $x \to a^-$. Consider the function defined by

$$f(x) = \frac{\sin x}{x}, \quad x \neq 0 \tag{1}$$

Sketch a graph of this equation for $-2\pi \leq x \leq 2\pi$. As $x \to 0^+$ or $x \to 0^-$, does $|f(x)| \to \infty$? Does f have a vertical asymptote at 0?

*Additional graphing calculator exercises appear in the Graphing Workbook as described in the front of this textbook.

The function defined by $g(x) = \dfrac{1}{x}$ has a vertical asymptote at 0 because $\left|\dfrac{1}{x}\right| \to \infty$ as $x \to 0^+$ and $x \to 0^-$. However, consider the function defined by

$$f(x) = \sin\left(\frac{1}{x}\right), \quad x \neq 0 \tag{2}$$

Sketch a graph of this equation for $-2\pi \leq x \leq 2\pi$. As $x \to 0^+$ or $x \to 0^-$, does $|f(x)| \to \infty$? Does f have a vertical asymptote at 0? Try graphing this equation for $-1 \leq x \leq 1$ and then for $-0.1 \leq x \leq 0.1$. What conclusions can you reach from the graphs you have sketched?

Discuss the behavior of the graphs of Equations (1) and (2) as $x \to \infty$ and $x \to -\infty$.

Examine similar situations using the cosine function. Consider the function defined by

$$f(x) = \frac{\cos x}{x}, \quad x \neq 0 \tag{3}$$

Sketch a graph of this equation for $-2\pi \leq x \leq 2\pi$. As $x \to 0^+$ or $x \to 0^-$, does $|f(x)| \to \infty$? Does f have a vertical asymptote at 0?

Now consider the function defined by

$$f(x) = \cos\left(\frac{1}{x}\right), \quad x \neq 0 \tag{4}$$

Sketch a graph of this equation for $-2\pi \leq x \leq 2\pi$. As $x \to 0^+$ or $x \to 0^-$, does $|f(x)| \to \infty$? Does f have a vertical asymptote at 0? Try graphing this equation for $-1 \leq x \leq 1$ and then for $-0.1 \leq x \leq 0.1$. What conclusions can you reach from the graphs you have sketched? Discuss the behavior of the graphs of Equations (3) and (4) as $x \to \infty$ and $x \to -\infty$.

A function f is said to have a *removable singularity* at a if $f(a)$ is not defined but $f(x) \to b$ as $x \to a$, where b is a real number. Do any of the functions defined in Equations 1–4 above have removable singularities?

Examine the following functions for asymptotes, behavior of the graph as $x \to \infty$ and $x \to -\infty$, and removable singularities.

1. $f(x) = x\sin\left(\dfrac{1}{x}\right)$

2. $f(x) = \dfrac{\tan x}{x}$

3. $f(x) = \dfrac{\sin\left(\dfrac{1}{x}\right)}{x}$

Chapter Review

6.1 Measuring Angles

- An angle is formed by rotating a ray about a point.
- An angle is in standard position when its initial side is along the positive x-axis and its vertex is at the origin of the coordinate axes.
- The length of the arc s subtended by the central angle θ (in radians) on a circle of radius r is given by $s = r\theta$.
- Angular speed is given by $\omega = \dfrac{\theta}{t}$.
- α and β are complementary angles when $\alpha + \beta = 90°$; they are supplementary angles when $\alpha + \beta = 180°$.
- Angle α is an acute angle when $0° < \alpha < 90°$; it is an obtuse angle when $90° < \alpha < 180°$.

6.2 Trigonometric Functions of Acute Angles

- Let θ be an acute angle of a right triangle. The six trigonometric functions of θ are given by

$$\sin \theta = \frac{\text{opp}}{\text{hyp}} \qquad \csc \theta = \frac{\text{hyp}}{\text{opp}}$$

$$\cos \theta = \frac{\text{adj}}{\text{hyp}} \qquad \sec \theta = \frac{\text{hyp}}{\text{adj}}$$

$$\tan \theta = \frac{\text{opp}}{\text{adj}} \qquad \cot \theta = \frac{\text{adj}}{\text{opp}}$$

6.3 Trigonometric Functions of Any Angle

- Let $P(x, y)$ be a point on the terminal side of an angle θ in standard position. The six trigonometric functions of θ are

$$\sin \theta = \frac{y}{r} \qquad \csc \theta = \frac{r}{y}, \quad y \neq 0$$

$$\cos \theta = \frac{x}{r} \qquad \sec \theta = \frac{r}{x}, \quad x \neq 0$$

$$\tan \theta = \frac{y}{x}, \quad x \neq 0 \qquad \cot \theta = \frac{x}{y}, \quad y \neq 0$$

6.4 Trigonometric Functions of Real Numbers

- The wrapping function pairs a real number with a point on the unit circle.
- Let W be the wrapping function, t be a real number, and

$W(t) = P(x, y)$. Then the trigonometric functions of the real number t are defined as follows:

$$\cos t = x \qquad\qquad \sec t = \frac{1}{x}, x \neq 0$$

$$\sin t = y \qquad\qquad \csc t = \frac{1}{y}, y \neq 0$$

$$\tan t = \frac{y}{x}, x \neq 0 \qquad \cot t = \frac{x}{y}, y \neq 0$$

- $\sin t$, $\cos t$, $\sec t$, and $\csc t$ have period 2π.
- $\tan t$ and $\cot t$ have period π.
- $\sin t$, $\csc t$, $\tan t$, and $\cot t$ are odd functions.
- $\cos t$ and $\sec t$ are even functions.

Domain and Range of Each Trigonometric Function
(n is an integer)

Function	Domain	Range
$\sin t$	$\{t \mid -\infty < t < \infty\}$	$\{y \mid -1 \le y \le 1\}$
$\cos t$	$\{t \mid -\infty < t < \infty\}$	$\{y \mid -1 \le y \le 1\}$
$\tan t$	$\{t \mid -\infty < t < \infty, t \neq (2n + 1)\pi/2\}$	$\{y \mid -\infty < y < \infty\}$
$\csc t$	$\{t \mid -\infty < t < \infty, t \neq n\pi\}$	$\{y \mid y \ge 1, y \le -1\}$
$\sec t$	$\{t \mid -\infty < t < \infty, t \neq (2n + 1)\pi/2\}$	$\{y \mid y \ge 1, y \le -1\}$
$\cot t$	$\{t \mid -\infty < t < \infty, t \neq n\pi\}$	$\{y \mid -\infty < y < \infty\}$

6.5 Graphs of the Sine and Cosine Functions

- The graph of $y = a \sin bx$ and that of $y = a \cos bx$ both have an amplitude of $|a|$ and a period of $\dfrac{2\pi}{|b|}$. The graph of each for $a > 0$ and $b > 0$ is given below.

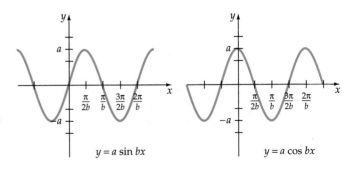

$$y = a \sin bx \qquad\qquad y = a \cos bx$$

6.6   Graphs of the Other Trigonometric Functions

- The period of $y = a \sec bx$ and of $y = a \csc bx$ is $\dfrac{2\pi}{|b|}$.

- The period of $y = a \tan bx$ and of $y = a \cot bx$ is $\dfrac{\pi}{|b|}$.

6.7   Translation and Addition of Ordinates

- Phase shift is a horizontal translation of the graph of a trigonometric function. If $y = f(bx + c)$, where f is a trigonometric function, then the phase shift is $-c/b$.

- Addition of ordinates is a method of graphing the sum of two functions by graphically adding the value of their y-coordinates.

6.8   Simple Harmonic Motion—An Application of the Sine and Cosine Functions

- The equations of simple harmonic motion are $y = a \cos 2\pi ft$ and $y = a \sin 2\pi ft$, where a is the amplitude and f is the frequency.

Essays and Projects

1. Microwave ovens heat food by using waves from the portion of the electromagnetic spectrum called microwaves. Discuss how microwaves are used to heat and cook food. Include in your discussion the range of frequencies of microwaves. Some microwave ovens are rated 600 watts and others 750 watts. What role does watt rating play for microwave ovens?

2. A computer is sometimes rated in *megahertz*. One **hertz** is one cycle (for example, of a sine wave) in one second. One megahertz is 1,000,000 hertz. Explain why a computer that is rated 25 megahertz operates more slowly than a computer that is rated 40 megahertz.

3. The graph of one arch of $y = \sin x$ is shown in the accompanying figure. Write a few paragraphs, supplying figures if necessary, that describe a method to approximate the area under the curve and above the x-axis. The rectangles sketched under the curve may give you a hint. There are quite a number of other methods, however.

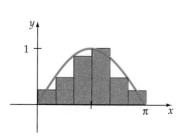

4. Write a short history of the sine function. Include references to its earliest known use, to how it was measured, and to how the word *sine* evolved.

5. Assume that the time of sunrise can be modeled by a sine function. Let time t be the hour (as a decimal) at which the sun rises, and let d be the day of the year with $d = 1$ as January 1. Using an almanac, find the time of sunrise for the longest and shortest days of the year for your location. This information can be used to calculate the amplitude of the sine function. Determine the period of the function, using 365 days in 1 year. Now write the equation of the sine function, ignoring daylight saving time. Compare the sunrise times given by the equation with the actual values given in an almanac for at least one day in each season.

6. A *square wave* is a wave form that is similar to a sine wave except that it is rectangular in shape rather than smooth. Research how square waves are generated using the sum of certain sine functions. What are the applications of square waves?

7. Predator-prey interactions can produce cyclic population growth for both the predator population and the prey population. Consider an animal reserve where the rabbit population r is given by

$$r(t) = 850 + 210 \sin\left(\frac{\pi}{6}t\right)$$

and the wolf population w is given by

$$w(t) = 120 + 30 \sin\left(\frac{\pi}{6}t - 2\right)$$

where t is the number of months after March 1, 1990. Graph $r(t)$ and $w(t)$ on the same coordinate system for $0 \le t \le 24$. Write an essay that explains a possible relationship between the two populations.

Write an equation that could be used to model the rabbit population shown by the following graph where t is measured in months.

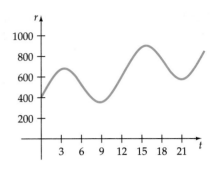

True/False Exercises

In Exercises 1 to 14, answer true or false. If the statement is false, give an example to show that the statement is false.

1. An angle is in standard position when the vertex is at the origin of a coordinate system.

2. The angle θ is in radians in standard position with the terminal side in the second quadrant. The reference angle of θ is $\pi - \theta$.

3. In the formula $s = r\theta$, the angle θ must be measured in radians.

4. If $\tan \theta < 0$ and $\cos \theta > 0$, then the terminal side θ is in the third quadrant.

5. $\sec^2\theta + \tan^2\theta = 1$ is an identity.

6. The amplitude of the graph of $y = 2 \tan x$ is 2.

7. The period of $y = \cos x$ is π.

8. The graph of $y = \sin x$ is symmetric to the origin.

9. For any acute angle θ, $\sin \theta + \cos(90° - \theta) = 1$.

10. $\sin(x + y) = \sin x + \sin y$.

11. $\sin^2 x = \sin x^2$.

12. The phase shift of $f(x) = 2 \sin\left(2x - \dfrac{\pi}{3}\right)$ is $\dfrac{\pi}{3}$.

13. The measure of one radian is more than 50 times the measure of one degree.

14. The measure of one radian differs depending on the radius of the circle used.

Chapter Review Exercises

1. Convert $37°34'$ to decimal degrees.

2. Convert $-39.38°$ to DMS measure.

3. Convert 2 radians to the nearest hundredth of a degree.

4. Convert $315°$ to radian measure.

5. Find the length of the arc on a circle of radius 3 meters that subtends an angle of $75°$.

6. Find the radian measure of the angle subtended by an arc of length 12 centimeters on a circle whose radius is 40 centimeters.

7. A car with a 16-inch radius wheel is moving with a speed of 50 miles per hour. Find the angular speed of the wheel in radians per second.

In Exercises 8 to 11, let θ be an acute angle of a right triangle and $\csc \theta = \dfrac{3}{2}$. Evaluate each of the following.

8. $\cos \theta$ 9. $\cot \theta$

10. $\sin \theta$ 11. $\sec \theta$

12. Find the values of the six trigonometric functions of an angle in standard position with the point $P(1, -3)$ on the terminal side of the angle.

13. Find the exact value of

    a. $\sec 150°$ c. $\cot(-225°)$

    b. $\tan(-3\pi/4)$ d. $\cos 2\pi/3$

14. Find each of the following to the nearest ten-thousandth.

  a. cos 123°   **b.** cot 4.22   **c.** sec 612°   **d.** tan 2π/5

15. Given $\cos \phi = -\sqrt{3}/2$, $180° < \phi < 270°$, find the exact value of **a.** $\sin \phi$ **b.** $\tan \phi$.

16. Given $\tan \phi = -\sqrt{3}/3$, $90° < \phi < 180°$, find the exact value of **a.** $\sec \phi$ **b.** $\csc \phi$

17. Given $\sin \phi = -\sqrt{2}/2$, $270° < \phi < 360°$, find the exact value of **a.** $\cos \phi$ **b.** $\cot \phi$

18. Let W be the wrapping function. Evaluate

  a. $W(\pi)$ **b.** $W\left(-\dfrac{\pi}{3}\right)$ **c.** $W\left(\dfrac{5\pi}{4}\right)$ **d.** $W(28\pi)$

19. Is the function defined by $f(x) = \sin(x)\tan(x)$ even, odd, or neither?

In Exercises 20 to 21, use the unit circle to show that each equation is an identity.

20. $\cos(\pi + t) = -\cos t$ **21.** $\tan(-t) = -\tan t$

In Exercises 22 to 27, use trigonometric identities to write each expression in terms of a single trigonometric function.

22. $1 + \dfrac{\sin^2\phi}{\cos^2\phi}$

23. $\dfrac{\tan\phi + 1}{\cot\phi + 1}$

24. $\dfrac{\cos^2\phi + \sin^2\phi}{\csc \phi}$

25. $\sin^2\phi(\tan^2\phi + 1)$

26. $1 + \dfrac{1}{\tan^2\phi}$

27. $\dfrac{\cos^2\phi}{1 - \sin^2\phi} - 1$

In Exercises 28 to 33, state the amplitude (if there is one), period, and phase shift of the graph of each equation.

28. $y = 3\cos(2x - \pi)$ **29.** $y = 2\tan 3x$

30. $y = -2\sin\left(3x + \dfrac{\pi}{3}\right)$ **31.** $y = \cos\left(2x - \dfrac{2\pi}{3}\right) + 2$

32. $y = -4\sec\left(4x - \dfrac{3\pi}{2}\right)$ **33.** $y = 2\csc\left(x - \dfrac{\pi}{4}\right) - 3$

In Exercises 34 to 51, graph each equation.

34. $y = 2\cos \pi x$ **35.** $y = -\sin\dfrac{2x}{3}$

36. $y = 2\sin\dfrac{3x}{2}$ **37.** $y = \cos\left(x - \dfrac{\pi}{2}\right)$

38. $y = \dfrac{1}{2}\sin\left(2x + \dfrac{\pi}{4}\right)$ **39.** $y = 3\cos 3(x - \pi)$

40. $y = -\tan\dfrac{x}{2}$ **41.** $y = 2\cot 2x$

42. $y = \tan\left(x - \dfrac{\pi}{2}\right)$ **43.** $y = -\cot\left(2x + \dfrac{\pi}{4}\right)$

44. $y = -2\csc\left(2x - \dfrac{\pi}{3}\right)$ **45.** $y = 3\sec\left(x + \dfrac{\pi}{4}\right)$

46. $y = 3\sin 2x - 3$ **47.** $y = 2\cos 3x + 3$

48. $y = -\cos\left(3x + \dfrac{\pi}{2}\right) + 2$ **49.** $y = 3\sin\left(4x - \dfrac{2\pi}{3}\right) - 3$

50. $y = 2 - \sin 2x$ **51.** $y = \sin x - \sqrt{3}\cos x$

52. If a car climbs a hill that has a constant angle of 4.5° for a distance of 0.5 miles, how many vertical feet has it climbed?

53. A tree casts a shadow of 8.5 feet when the angle of elevation of the sun is 55.3°. Find the height of the tree.

54. Find the angle α formed by the intersection of a diagonal of a face of a cube and the diagonal of the cube from the same vertex.

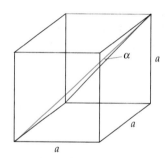

55. Find the height of a building if the angle of elevation to the top of the building changes from 18° to 37° as the observer moves a distance of 80 feet towards the building.

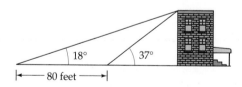

56. Find the amplitude, period, and frequency of the harmonic motion given by the equation $y = 2.5\sin 50t$.

57. A pendulum 5 feet long is displaced a distance of 9 inches and released. Write an equation for the displacement as a function of time. Find the period and frequency of the motion.

58. A mass of 5 kilograms is in equilibrium suspended from a spring. The mass is pulled down 0.5 feet and released. Find the period, frequency, and amplitude of the resulting motion. Write an equation of motion. Assume $k = 20$.

Chapter Test

1. Convert $150°$ to exact radian measure.

2. Find the supplement of the angle whose radian measure is $\frac{11}{12}\pi$. Express your answer in terms of π.

3. Find the length of an arc that subtends a central angle of $75°$ in a circle of radius 10 centimeters.

4. A wheel is rotating at 6 revolutions per second. Find the angular speed in radians per second.

5. A wheel with a diameter of 16 centimeters is rotating at 10 radians per second. Find the linear speed of a point on the edge of the wheel.

6. If θ is an acute angle of a right triangle and $\tan \theta = \frac{3}{7}$, find $\sec \theta$.

7. Find the value of $\csc 67°$ to the nearest ten thousandth.

8. Find the exact value of $\tan \frac{\pi}{6} \cos \frac{\pi}{3} - \sin \frac{\pi}{2}$.

9. Find $W\left(\frac{11\pi}{6}\right)$.

10. Express $\dfrac{\sec^2 t - 1}{\sec^2 t}$ in terms of a single trigonometric function.

11. State the period of $y = -4 \tan 3x$.

12. State the amplitude, period, and phase shift for the equation $y = -3 \cos\left(2x + \frac{\pi}{2}\right)$.

13. Graph $y = 3 \cos \frac{1}{2}x$.

14. Graph $y = -2 \sec \frac{1}{2}x$.

15. Graph $y = 2 \cot\left(x + \frac{3\pi}{2}\right)$.

16. Graph $y = 2 \sin\left(2x - \frac{\pi}{2}\right) - 1$.

17. Graph $y = 2 - \sin \frac{x}{2}$.

18. Graph $y = \sin x - \cos 2x$.

19. The angle of elevation from point A to the top of a tree is $42.2°$ degree. At point B, 5.2 meters from A and on a line through the base of the tree and A, the angle of elevation is $37.4°$. Find the height of the tree.

20. Write the equation for simple harmonic motion, given that the amplitude is 13 feet, the period is 5 seconds, and the displacement is zero when $t = 0$.

7 Trigonometric Identities and Equations

CASE IN POINT Resonance Phenomena in a Physical System

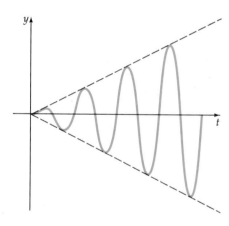

Many vibrating systems can be described in terms of simple harmonic motion. Simple harmonic motion can be described by equations of the form $y = a \cos 2\pi ft$. In pure simple harmonic motion, the amplitude of the vibrations remain constant. In an actual physical system, however, energy is lost, and the amplitude of the vibrations gradually decreases. When energy is put into a vibrating system, the amplitude of the vibrations may increase. For example, the amplitude of a swing increases if impulses of energy are applied to the system at the proper times. *Resonance* is the building up of larger amplitudes of a vibrating system when small impulses are applied at the proper time.

An equation for a motion with resonance is $y = at \cos 2\pi ft$. The graph at the left shows how the amplitude of the vibrations in a resonance system increases with time t.

Resonance phenomena can occur in almost any physical system. It is possible for soldiers marching across a bridge to cause large vibrations if the frequency of march is the same as the natural frequency of the bridge. Another example of resonance is the Tacoma Narrows Bridge collapse, which occurred on July 2, 1940. A wind coming down a canyon created vibrations in the bridge that increased in amplitude as the day passed. Vibrations in the roadway of the concrete and steel structure reached an amplitude of 8 feet, and the bridge collapsed.

The Tacoma Narrows Bridge at Puget Sound, Washington. Wide World.

Solution   Multiply the numerator and denominator of the left side of the identity by the conjugate of $1 + \cos x$.

$$\frac{\sin x}{1 + \cos x} = \frac{\sin x}{1 + \cos x} \cdot \frac{1 - \cos x}{1 - \cos x} = \frac{\sin x(1 - \cos x)}{1 - \cos^2 x}$$

$$= \frac{\sin x(1 - \cos x)}{\sin^2 x} = \frac{1 - \cos x}{\sin x}$$

■ *Try Exercise* **46,** *page 367.*

EXAMPLE 5   **Verify a Trigonometric Identity**

Verify the identity $\dfrac{\sin x + \tan x}{1 + \cos x} = \tan x$.

Solution   Rewrite the left side of the identity in terms of sines and cosines.

$$\frac{\sin x + \tan x}{1 + \cos x} = \frac{\sin x + \dfrac{\sin x}{\cos x}}{1 + \cos x} \qquad \bullet \ \tan x = \frac{\sin x}{\cos x}$$

$$= \frac{\dfrac{\sin x \cos x + \sin x}{\cos x}}{1 + \cos x} \qquad \bullet \ \text{Write the numerator with}$$
$$\text{a common denominator.}$$

$$= \frac{\sin x \cos x + \sin x}{\cos x(1 + \cos x)} \qquad \bullet \ \text{Simplify.}$$

$$= \frac{\sin x \cancel{(1 + \cos x)}}{\cos x \cancel{(1 + \cos x)}}$$

$$= \tan x$$

■ *Try Exercise* **56,** *page 367.*

EXERCISE SET 7.1

In Exercises 1 to 12, state whether each statement is an identity. If the statement is an identity, verify. If not, explain why.

1. $(\sin x + \cos x)^2 = \sin^2 x + \cos^2 x$

2. $\tan 2x = 2 \tan x$

3. $\cos(x + 30°) = \cos x + \cos 30°$

4. $\sqrt{1 - \sin^2 x} = \cos x$

5. $\tan^4 x - \sec^4 x = \tan^2 x + \sec^2 x$

6. $\sqrt{1 + \tan^2 x} = |\sec x|$

7. $\tan^4 x - 1 = \sec^2 x$

8. $\sin^3 x + \cos^3 x = (\sin x + \cos x)(1 + \sin x \cos x)$

9. $2 \sin 30° = \sin 60°$

10. $\cot x \csc x \sec x = 1$

11. $\sin^4 x + \cos^2 x = 1 + \sin^2 x$

12. $\sec^2 x + \csc^2 x = 2 + \cot^2 x + \tan^2 x$

In Exercises 13 to 66, verify the identities.

13. $\tan x \csc x \cos x = 1$

EXAMPLE 1 **Verify a Trigonometric Identity**

Verify the identity $\sec x - \cos x = \sin x \tan x$.

Solution Rewrite the left side of the equation. Use the Pythagorean identity $\sin^2 x + \cos^2 x = 1$, which implies $1 - \cos^2 x = \sin^2 x$.

$$\sec x - \cos x = \frac{1}{\cos x} - \cos x = \frac{1 - \cos^2 x}{\cos x} \quad \bullet \text{ Write as a single fraction with a common denominator}$$

$$= \frac{\sin^2 x}{\cos x} = \sin x \cdot \frac{\sin x}{\cos x} \quad \bullet\ 1 - \cos^2 x = \sin^2 x$$

$$= \sin x \tan x \quad \bullet\ \frac{\sin x}{\cos x} = \tan x$$

Since the left side of the identity has been rewritten to be the right, we have verified the identity.

■ *Try Exercise **14**, page 367.*

EXAMPLE 2 **Verify a Trigonometric Identity**

Verify the identity $1 - 2\sin^2 x = 2\cos^2 x - 1$.

Solution Rewrite the right side of the equation.

$$2\cos^2 x - 1 = 2(1 - \sin^2 x) - 1 \quad \bullet\ \cos^2 x = 1 - \sin^2 x$$

$$= 2 - 2\sin^2 x - 1$$

$$= 1 - 2\sin^2 x$$

■ *Try Exercise **24**, page 367.*

EXAMPLE 3 **Verify a Trigonometric Identity**

Verify the identity $\csc^2 x - \cos^2 x \csc^2 x = 1$.

Solution Simplify the left side of the equation.

$$\csc^2 x - \cos^2 x \csc^2 x = \csc^2 x(1 - \cos^2 x) \quad \bullet \text{ Factor out } \csc^2 x.$$

$$= \csc^2 x \sin^2 x$$

$$= \frac{1}{\sin^2 x} \cdot \sin^2 x = 1 \quad \bullet\ \csc^2 x = \frac{1}{\sin^2 x}$$

■ *Try Exercise **36**, page 367.*

EXAMPLE 4 **Verify a Trigonometric Identity**

Verify the identity $\dfrac{\sin x}{1 + \cos x} = \dfrac{1 - \cos x}{\sin x}$.

Solution Multiply the numerator and denominator of the left side of the identity by the conjugate of $1 + \cos x$.

$$\frac{\sin x}{1 + \cos x} = \frac{\sin x}{1 + \cos x} \cdot \frac{1 - \cos x}{1 - \cos x} = \frac{\sin x(1 - \cos x)}{1 - \cos^2 x}$$

$$= \frac{\sin x(1 - \cos x)}{\sin^2 x} = \frac{1 - \cos x}{\sin x}$$

■ *Try Exercise* **46,** *page 367.*

EXAMPLE 5   **Verify a Trigonometric Identity**

Verify the identity $\dfrac{\sin x + \tan x}{1 + \cos x} = \tan x.$

Solution Rewrite the left side of the identity in terms of sines and cosines.

$$\frac{\sin x + \tan x}{1 + \cos x} = \frac{\sin x + \dfrac{\sin x}{\cos x}}{1 + \cos x} \qquad \bullet \ \tan x = \frac{\sin x}{\cos x}$$

$$= \frac{\dfrac{\sin x \cos x + \sin x}{\cos x}}{1 + \cos x} \qquad \bullet \ \text{Write the numerator with a common denominator.}$$

$$= \frac{\sin x \cos x + \sin x}{\cos x(1 + \cos x)} \qquad \bullet \ \text{Simplify.}$$

$$= \frac{\sin x \cancel{(1 + \cos x)}}{\cos x \cancel{(1 + \cos x)}}$$

$$= \tan x$$

■ *Try Exercise* **56,** *page 367.*

EXERCISE SET 7.1

In Exercises 1 to 12, state whether each statement is an identity. If the statement is an identity, verify. If not, explain why.

1. $(\sin x + \cos x)^2 = \sin^2 x + \cos^2 x$

2. $\tan 2x = 2 \tan x$

3. $\cos(x + 30°) = \cos x + \cos 30°$

4. $\sqrt{1 - \sin^2 x} = \cos x$

5. $\tan^4 x - \sec^4 x = \tan^2 x + \sec^2 x$

6. $\sqrt{1 + \tan^2 x} = |\sec x|$

7. $\tan^4 x - 1 = \sec^2 x$

8. $\sin^3 x + \cos^3 x = (\sin x + \cos x)(1 + \sin x \cos x)$

9. $2 \sin 30° = \sin 60°$

10. $\cot x \csc x \sec x = 1$

11. $\sin^4 x + \cos^2 x = 1 + \sin^2 x$

12. $\sec^2 x + \csc^2 x = 2 + \cot^2 x + \tan^2 x$

In Exercises 13 to 66, verify the identities.

13. $\tan x \csc x \cos x = 1$

7 Trigonometric Identities and Equations

CASE IN POINT   *Resonance Phenomena in a Physical System*

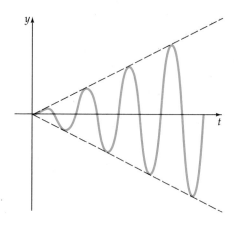

Many vibrating systems can be described in terms of simple harmonic motion. Simple harmonic motion can be described by equations of the form $y = a \cos 2\pi ft$. In pure simple harmonic motion, the amplitude of the vibrations remain constant. In an actual physical system, however, energy is lost, and the amplitude of the vibrations gradually decreases. When energy is put into a vibrating system, the amplitude of the vibrations may increase. For example, the amplitude of a swing increases if impulses of energy are applied to the system at the proper times. *Resonance* is the building up of larger amplitudes of a vibrating system when small impulses are applied at the proper time.

An equation for a motion with resonance is $y = at \cos 2\pi ft$. The graph at the left shows how the amplitude of the vibrations in a resonance system increases with time t.

Resonance phenomena can occur in almost any physical system. It is possible for soldiers marching across a bridge to cause large vibrations if the frequency of march is the same as the natural frequency of the bridge. Another example of resonance is the Tacoma Narrows Bridge collapse, which occurred on July 2, 1940. A wind coming down a canyon created vibrations in the bridge that increased in amplitude as the day passed. Vibrations in the roadway of the concrete and steel structure reached an amplitude of 8 feet, and the bridge collapsed.

The Tacoma Narrows Bridge at Puget Sound, Washington. Wide World.

7.1 The Fundamental Trigonometric Identities

An equation that is true for all replacements of the variables for which all terms of the equation are defined is called an **identity**. Trigonometric identities are used to simplify a trigonometric expression or to write a trigonometric expression in an equivalent form.

As for algebraic equations, the domain of a trigonometric identity is all values of the variable for which the expression is meaningful. For example, the identity

$$\frac{\sin x \cos x}{\sin x} = \cos x$$

is true for all real numbers x such that $\sin x \neq 0$. Because $\sin x = 0$ when $x = n\pi$ and n is an integer, the domain of the equation must exclude all integral multiples of π. To verify an identity means to show that each side of the equation represents the same expression. For instance, simplifying the left side of the equation, we now have $\cos x = \cos x$. The left and right sides of the equation are the same expression.

There is no one method that can be used to verify every identity. Generally we work on only one side of the equation. The following list provides some suggestions for verifying trigonometric identities.

1. If one side of the identity is more complex than the other, simplify the more complex side.
2. Perform algebraic operations in the identity such as
   a. Squaring
   b. Factoring
   c. Adding fractions
   d. Multiplying the numerator and denominator by a nonzero factor.
3. Rewrite the identity in terms of sine and cosine functions.
4. Rewrite one side of the identity in terms of a single function.
5. Make use of previously established identities.

Table 7.1 lists the eight fundamental identities established in the previous chapter. These identities will be valuable when we verify other trigonometric identities.

TABLE 7.1   Fundamental Trigonometric Identities

Reciprocal identities	$\sin x = \dfrac{1}{\csc x}$	$\cos x = \dfrac{1}{\sec x}$	$\tan x = \dfrac{1}{\cot x}$
Ratio identities	$\tan x = \dfrac{\sin x}{\cos x}$	$\cot x = \dfrac{\cos x}{\sin x}$	
Pythagorean identities	$\sin^2 x + \cos^2 x = 1$	$\tan^2 x + 1 = \sec^2 x$	$1 + \cot^2 x = \csc^2 x$

14. $\sin x \cot x \sec x = 1$

15. $\dfrac{4 \sin^2 x - 1}{2 \sin x + 1} = 2 \sin x - 1$

16. $\dfrac{\sin^2 x - 2 \sin x + 1}{\sin x - 1} = \sin x - 1$

17. $(\sin x - \cos x)(\sin x + \cos x) = 1 - 2 \cos^2 x$

18. $(\tan x)(1 - \cot x) = \tan x - 1$

19. $\dfrac{1}{\sin x} - \dfrac{1}{\cos x} = \dfrac{\cos x - \sin x}{\sin x \cos x}$

20. $\dfrac{1}{\sin x} + \dfrac{3}{\cos x} = \dfrac{\cos x + 3 \sin x}{\sin x \cos x}$

21. $\dfrac{\cos x}{1 - \sin x} = \sec x + \tan x$

22. $\dfrac{\sin x}{1 - \cos x} = \csc x + \cot x$

23. $\dfrac{1 - \tan^4 x}{\sec^2 x} = 1 - \tan^2 x$

24. $\sin^4 x - \cos^4 x = \sin^2 x - \cos^2 x$

25. $\dfrac{1 + \tan^3 x}{1 + \tan x} = 1 - \tan x + \tan^2 x$

26. $\dfrac{\cos x \tan x - \sin x}{\cot x} = 0$

27. $\dfrac{\sin x - 2 + \dfrac{1}{\sin x}}{\sin x - \dfrac{1}{\sin x}} = \dfrac{\sin x - 1}{\sin x + 1}$

28. $\dfrac{\sin x}{1 - \cos x} - \dfrac{\sin x}{1 + \cos x} = 2 \cot x$

29. $(\sin x + \cos x)^2 = 1 + 2 \sin x \cos x$

30. $(\tan x + 1)^2 = \sec^2 x + 2 \tan x$

31. $\dfrac{\cos x}{1 + \sin x} = \sec x - \tan x$

32. $\dfrac{\sin x}{1 + \cos x} = \csc x - \cot x$

33. $\csc x = \dfrac{\cot x + \tan x}{\sec x}$

34. $\sec x = \dfrac{\cot x + \tan x}{\csc x}$

35. $\dfrac{\cos x \tan x + 2 \cos x - \tan x - 2}{\tan x + 2} = \cos x - 1$

36. $\dfrac{2 \sin x \cot x + \sin x - 4 \cot x - 2}{2 \cot x + 1} = \sin x - 2$

37. $\sec x - \tan x = \dfrac{1 - \sin x}{\cos x}$

38. $\cot x - \csc x = \dfrac{\cos x - 1}{\sin x}$

39. $\sin^2 x - \cos^2 x = 2 \sin^2 x - 1$

40. $\sin^2 x - \cos^2 x = 1 - 2 \cos^2 x$

41. $\dfrac{1}{\sin^2 x} + \dfrac{1}{\cos^2 x} = \csc^2 x \sec^2 x$

42. $\dfrac{1}{\tan^2 x} - \dfrac{1}{\cot^2 x} = \csc^2 x - \sec^2 x$

43. $\sec x - \cos x = \sin x \tan x$

44. $\tan x + \cot x = \sec x \csc x$

45. $\dfrac{\dfrac{1}{\sin x} + 1}{\dfrac{1}{\sin x} - 1} = \tan^2 x + 2 \tan x \sec x + \sec^2 x$

46. $\dfrac{\dfrac{1}{\sin x} + \dfrac{1}{\cos x}}{\dfrac{1}{\sin x} - \dfrac{1}{\cos x}} = \dfrac{\cos^2 x - \sin^2 x}{1 - 2 \cos x \sin x}$

47. $\sin^4 x - \cos^4 x = 2 \sin^2 x - 1$

48. $\sin^6 x + \cos^6 x = \sin^4 x - \sin^2 x \cos^2 x + \cos^4 x$

49. $\dfrac{1}{1 - \cos x} = \dfrac{1 + \cos x}{\sin^2 x}$

50. $1 + \sin x = \dfrac{\cos^2 x}{1 - \sin x}$

51. $\dfrac{\sin x}{1 - \sin x} - \dfrac{\cos x}{1 - \sin x} = \dfrac{1 - \cot x}{\csc x - 1}$

52. $\dfrac{\tan x}{1 + \tan x} - \dfrac{\cot x}{1 + \tan x} = 1 - \cot x$

53. $\dfrac{1}{1 + \cos x} - \dfrac{1}{1 - \cos x} = -2 \cot x \csc x$

54. $\dfrac{1}{1 - \sin x} - \dfrac{1}{1 + \sin x} = 2 \tan x \sec x$

55. $\dfrac{\dfrac{1}{\sin x} + \csc x}{\dfrac{1}{\sin x} - \sin x} = \dfrac{2}{\cos^2 x}$

56. $\dfrac{\dfrac{1}{\tan x} + \cot x}{\dfrac{1}{\tan x} + \tan x} = \dfrac{2}{\sec^2 x}$

57. $\sqrt{\dfrac{1 + \sin x}{1 - \sin x}} = \dfrac{1 + \sin x}{\cos x}, \quad \cos x > 0$

58. $\dfrac{\cos x + \cot x \sin x}{\cot x} = 2 \sin x$

59. $\dfrac{\sin^3 x + \cos^3 x}{\sin x + \cos x} = 1 - \sin x \cos x$

60. $\dfrac{1 - \sin x}{1 + \sin x} - \dfrac{1 + \sin x}{1 - \sin x} = -4 \sec x \tan x$

61. $\dfrac{\sec x - 1}{\sec x + 1} - \dfrac{\sec x + 1}{\sec x - 1} = -4 \csc x \cot x$

62. $\dfrac{1}{1 - \cos x} - \dfrac{\cos x}{1 + \cos x} = 2 \csc^2 x - 1$

63. $\dfrac{1 + \sin x}{\cos x} - \dfrac{\cos x}{1 - \sin x} = 0$

64. $(\sin x + \cos x + 1)^2 = 2(\sin x + 1)(\cos x + 1)$

65. $\dfrac{\sec x + \tan x}{\sec x - \tan x} = \dfrac{(\sin x + 1)^2}{\cos^2 x}$

66. $\dfrac{\sin^3 x - \cos^3 x}{\sin x + \cos x} = \dfrac{\csc^2 x - \cot x - 2 \cos^2 x}{1 - \cot^2 x}$

$\bigvee$ **Graphing Calculator Exercises***

In Exercises 67 to 70, establish the identity by graphing the function on each side of the equation and showing that the graphs are identical.

67. $\sin 2x = 2 \sin x \cos x$ **68.** $\sin^2 x + \cos^2 x = 1$

69. $\sin x + \cos x = \sqrt{2} \sin(x + \pi/4)$

70. $\cos 2x = 2 \cos^2 x - 1$

Supplemental Exercises

71. Express $\cos x$ in terms of $\sin x$.

72. Express $\tan x$ in terms of $\cos x$.

73. Express $\sec x$ in terms of $\sin x$.

74. Express $\csc x$ in terms of $\sec x$.

In Exercises 75 to 80, verify the identity.

75. $\dfrac{1 - \sin x + \cos x}{1 + \sin x + \cos x} = \dfrac{\cos x}{\sin x + 1}$

76. $\dfrac{1 - \tan x + \sec x}{1 + \tan x - \sec x} = \dfrac{1 + \sec x}{\tan x}$

77. $\dfrac{2 \sin^4 x + 2 \sin^2 x \cos^2 x - 3 \sin^2 x - 3 \cos^2 x}{2 \sin^2 x}$
$$= 1 - \dfrac{3}{2} \csc^2 x$$

78. $\dfrac{4 \tan x \sec^2 x - 4 \tan x - \sec^2 x + 1}{4 \tan^3 x - \tan^2 x} = 1$

79. $\dfrac{\sin x(\tan x + 1) - 2 \tan x \cos x}{\sin x - \cos x} = \tan x$

80. $\dfrac{\sin^2 x \cos x + \cos^3 x - \sin^3 x \cos x - \sin x \cos^3 x}{1 - \sin^2 x}$
$$= \dfrac{\cos x}{1 + \sin x}$$

81. Verify the identity $\sin^4 x + \cos^4 x = 1 - 2 \sin^2 x \cos^2 x$ by completing the square of the left side of the identity.

82. Verify the identity $\tan^4 x + \sec^4 x = 1 + 2 \tan^2 x \sec^2 x$ by completing the square of the left side of the identity.

7.2   Sum and Difference Identities

There are several useful identities relating the sum and difference of two angles $(\alpha \pm \beta)$. We begin by finding an identity for $\cos(\alpha - \beta)$.

In Figure 7.1, angles α and β are drawn in standard position, with OA and OB as the terminal sides of α and β, respectively. The coordinates of A are $(\cos \alpha, \sin \alpha)$, and the coordinates of B are $(\cos \beta, \sin \beta)$. The angle $(\alpha - \beta)$ is formed by the terminal sides of the angles α and β (angle AOB).

An angle equal in measure to angle $(\alpha - \beta)$ is placed in standard position in the same figure (angle COD). From geometry, if two central angles of a circle have the same measure, then their chords are also equal in mea-

*Additional graphing calculator exercises appear in the Graphing Workbook as described in the front of this textbook.

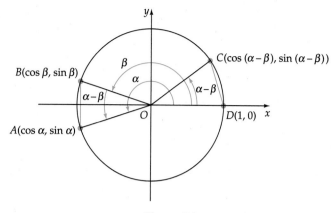

Figure 7.1

sure. Thus the chords AB and CD are equal in length. Using the distance formula, we can calculate the lengths of the chords AB and CD.

$$d(A, B) = \sqrt{(\cos \alpha - \cos \beta)^2 + (\sin \alpha - \sin \beta)^2}$$

$$d(C, D) = \sqrt{[\cos(\alpha - \beta) - 1]^2 + [\sin(\alpha - \beta) - 0]^2}$$

Since $d(A, B) = d(C, D)$, we have

$$\sqrt{(\cos \alpha - \cos \beta)^2 + (\sin \alpha - \sin \beta)^2}$$
$$= \sqrt{[\cos(\alpha - \beta) - 1]^2 + [\sin(\alpha - \beta)]^2}$$

Squaring each side of the equation and simplifying, we obtain

$$(\cos \alpha - \cos \beta)^2 + (\sin \alpha - \sin \beta)^2$$
$$= [\cos(\alpha - \beta) - 1]^2 + [\sin(\alpha - \beta)]^2$$

$$\cos^2\alpha - 2 \cos \alpha \cos \beta + \cos^2\beta + \sin^2\alpha - 2 \sin \alpha \sin \beta + \sin^2\beta$$
$$= \cos^2(\alpha - \beta) - 2 \cos(\alpha - \beta) + 1 + \sin^2(\alpha - \beta)$$

$$\cos^2\alpha + \sin^2\alpha + \cos^2\beta + \sin^2\beta - 2 \cos \alpha \cos \beta - 2 \sin \alpha \sin \beta$$
$$= \cos^2(\alpha - \beta) + \sin^2(\alpha - \beta) + 1 - 2 \cos(\alpha - \beta)$$

Simplifying by using $\sin^2\theta + \cos^2\theta = 1$, we have

$$2 - 2 \sin \alpha \sin \beta - 2 \cos \alpha \cos \beta = 2 - 2 \cos(\alpha - \beta)$$

Solving for $\cos(\alpha - \beta)$ gives us

$$\cos(\alpha - \beta) = \cos \alpha \cos \beta + \sin \alpha \sin \beta.$$

To derive an identity for $\cos(\alpha + \beta)$, write $\cos(\alpha + \beta)$ as $\cos(\alpha - (-\beta))$.

$$\cos(\alpha + \beta) = \cos(\alpha - (-\beta)) = \cos \alpha \cos(-\beta) + \sin \alpha \sin(-\beta)$$

Recall that $\cos(-\beta) = \cos \beta$ and $\sin(-\beta) = -\sin \beta$. Substituting into the previous equation, we obtain the identity

$$\cos(\alpha + \beta) = \cos \alpha \cos \beta - \sin \alpha \sin \beta$$

If we apply the identity $\cos(\alpha - \beta) = \cos \alpha \cos \beta + \sin \alpha \sin \beta$ to the expression $\cos(90° - \beta)$, we have

$$\cos(90° - \beta) = \cos 90° \cos \beta + \sin 90° \sin \beta$$

$$= 0 \cdot \cos \beta + 1 \cdot \sin \beta$$

which gives

$$\cos(90° - \beta) = \sin \beta$$

Thus the sine of an angle is equal to the cosine of its complement. Using $\cos(90° - \beta) = \sin \beta$ with $\beta = 90° - \alpha$, we have

$$\cos \alpha = \cos[90° - (90° - \alpha)] = \sin(90° - \alpha)$$

Therefore,

$$\cos \alpha = \sin(90° - \alpha)$$

Any pair of trigonometric functions f_1 and f_2 for which $f_1(x) = f_2(90° - x)$ is said to be **cofunctions**. We can use the ratio identities to show that the tangent and cotangent functions are cofunctions.

$$\tan(90° - \theta) = \frac{\sin(90° - \theta)}{\cos(90° - \theta)} \qquad\qquad \cot(90° - \theta) = \frac{\cos(90° - \theta)}{\sin(90° - \theta)}$$

$$= \frac{\cos \theta}{\sin \theta} = \cot \theta \qquad\qquad\qquad\qquad = \frac{\sin \theta}{\cos \theta} = \tan \theta$$

The secant and cosecant functions are also cofunctions.

To derive an identity for $\sin(\alpha + \beta)$, substitute $\alpha + \beta$ for θ in the cofunction identity $\sin \theta = \cos(90° - \theta)$.

$$\sin \theta = \cos(90° - \theta)$$

$$\sin(\alpha + \beta) = \cos[90° - (\alpha + \beta)]$$

$$= \cos[(90° - \alpha) - \beta] \quad \bullet \text{ Rewrite as the difference of two angles.}$$

$$= \cos(90° - \alpha) \cos \beta + \sin(90° - \alpha) \sin \beta$$

$$= \sin \alpha \cos \beta + \cos \alpha \sin \beta$$

Therefore,

$$\sin(\alpha + \beta) = \sin \alpha \cos \beta + \cos \alpha \sin \beta$$

We can also derive an identity for $\sin(\alpha - \beta)$ by rewriting $(\alpha - \beta)$ as $[\alpha + (-\beta)]$.

$$\sin(\alpha - \beta) = \sin[\alpha + (-\beta)]$$

$$= \sin \alpha \cos(-\beta) + \cos \alpha \sin(-\beta)$$

$$= \sin \alpha \cos \beta - \cos \alpha \sin \beta \qquad \bullet \begin{aligned} &\cos(-\beta) = \cos \beta \\ &\sin(-\beta) = -\sin \beta \end{aligned}$$

Thus,

$$\sin(\alpha - \beta) = \sin \alpha \cos \beta - \cos \alpha \sin \beta .$$

The identity for $\tan(\alpha + \beta)$ is a result of the identity $\tan \theta = \dfrac{\sin \theta}{\cos \theta}$ and the identities for $\sin(\alpha + \beta)$ and $\cos(\alpha + \beta)$.

$$\tan(\alpha + \beta) = \frac{\sin(\alpha + \beta)}{\cos(\alpha + \beta)} = \frac{\sin \alpha \cos \beta + \cos \alpha \sin \beta}{\cos \alpha \cos \beta - \sin \alpha \sin \beta}$$

$$= \frac{\dfrac{\sin \alpha \cos \beta}{\cos \alpha \cos \beta} + \dfrac{\cos \alpha \sin \beta}{\cos \alpha \cos \beta}}{\dfrac{\cos \alpha \cos \beta}{\cos \alpha \cos \beta} - \dfrac{\sin \alpha \sin \beta}{\cos \alpha \cos \beta}}$$

• Multiply both the numerator and the denominator by $\dfrac{1}{\cos \alpha \cos \beta}$ and simplify.

Therefore,

$$\tan(\alpha + \beta) = \frac{\tan \alpha + \tan \beta}{1 - \tan \alpha \tan \beta}$$

The tangent function is an odd function, so $\tan(-\theta) = -\tan \theta$. Rewriting $(\alpha - \beta)$ as $[\alpha + (-\beta)]$ allows us to derive an identity for $\tan(\alpha - \beta)$.

$$\tan(\alpha - \beta) = \tan[\alpha + (-\beta)] = \frac{\tan \alpha + \tan(-\beta)}{1 - \tan \alpha \tan(-\beta)}$$

Therefore,

$$\tan(\alpha - \beta) = \frac{\tan \alpha - \tan \beta}{1 + \tan \alpha \tan \beta}$$

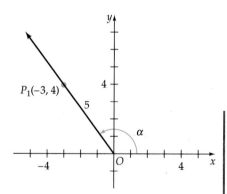

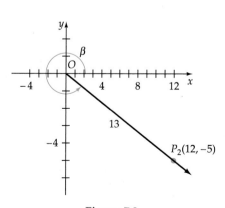

Figure 7.2

EXAMPLE 1   **Evaluate a Trigonometric Function**

Given $\tan \alpha = -\dfrac{4}{3}$ for α in quadrant II and $\tan \beta = -\dfrac{5}{12}$ for β in quadrant IV, find a. $\tan(\alpha - \beta)$ and b. $\sin(\alpha + \beta)$ (See Figure 7.2)

Solution   Because $\tan \alpha = \dfrac{y}{x} = -\dfrac{4}{3}$ and the terminal side of α is in Quadrant II, $P_1(-3, 4)$ is a point on the terminal side of α. Similarly, $P_2(12, -5)$ is a point on the terminal side of β. Using the Pythagorean Theorem, the length of the line segment OP_1 is 5 and the length of OP_2 is 13.

a. $\tan(\alpha - \beta) = \dfrac{\tan \alpha - \tan \beta}{1 + \tan \alpha \tan \beta}$

$$= \frac{-\dfrac{4}{3} - \left(-\dfrac{5}{12}\right)}{1 + \left(-\dfrac{4}{3}\right)\left(-\dfrac{5}{12}\right)} = \frac{-\dfrac{11}{12}}{\dfrac{14}{9}} = -\frac{33}{56}$$

b. $\sin(\alpha + \beta) = \sin \alpha \cos \beta + \cos \alpha \sin \beta$

$$= \frac{4}{5} \cdot \frac{12}{13} + \frac{-3}{5} \cdot \frac{-5}{13} = \frac{48}{65} + \frac{15}{65} = \frac{63}{65}$$

■ *Try Exercise* **32**, *page 374.*

EXAMPLE 2   **Verify an Identity**

Verify the identity $\cos(\pi - \theta) = -\cos \theta$.

   Solution   Use the identity for $\cos(\alpha - \beta)$.

$\cos(\pi - \theta) = \cos \pi \cos \theta + \sin \pi \sin \theta = -1 \cdot \cos \theta + 0 \cdot \sin \theta = -\cos \theta$

■ *Try Exercise* **44**, *page 374.*

EXAMPLE 3   **Verify an Identity**

Verify the identity $\dfrac{\cos 4\theta}{\sin \theta} - \dfrac{\sin 4\theta}{\cos \theta} = \dfrac{\cos 5\theta}{\sin \theta \cos \theta}$.

   Solution   Subtract the fractions on the left side of the equation.

$$\frac{\cos 4\theta}{\sin \theta} - \frac{\sin 4\theta}{\cos \theta} = \frac{\cos 4\theta \cos \theta - \sin 4\theta \sin \theta}{\sin \theta \cos \theta}$$

$$= \frac{\cos(4\theta + \theta)}{\sin \theta \cos \theta} = \frac{\cos 5\theta}{\sin \theta \cos \theta} \quad \bullet \text{ Use the identity for } \cos(\alpha + \beta).$$

■ *Try Exercise* **56**, *page 374.*

EXAMPLE 4   **Verify an Identity**

Verify the identity $\sin(\alpha + \beta) \cdot \sin(\alpha - \beta) = \sin^2\alpha - \sin^2\beta$.

   Solution   Work on the left side of the identity.

$\sin(\alpha + \beta) \cdot \sin(\alpha - \beta)$

$$= (\sin \alpha \cos \beta + \cos \alpha \sin \beta)(\sin \alpha \cos \beta - \cos \alpha \sin \beta)$$

$$= \sin^2\alpha \cos^2\beta - \cos^2\alpha \sin^2\beta$$

$$= \sin^2\alpha(1 - \sin^2\beta) - (1 - \sin^2\alpha) \sin^2\beta$$

$$= \sin^2\alpha - \sin^2\alpha \sin^2\beta - \sin^2\beta + \sin^2\alpha \sin^2\beta$$

$$= \sin^2\alpha - \sin^2\beta$$

■ *Try Exercise* **60**, *page 374.*

The identities we developed in this section are used frequently in problems involving trigonometric functions. Following is a list of the identities for convenient reference.

Sum or Difference of Two Angle Identities

$$\cos(\alpha - \beta) = \cos \alpha \cos \beta + \sin \alpha \sin \beta$$

$$\cos(\alpha + \beta) = \cos \alpha \cos \beta - \sin \alpha \sin \beta$$

$$\sin(\alpha - \beta) = \sin \alpha \cos \beta - \cos \alpha \sin \beta$$

$$\sin(\alpha + \beta) = \sin \alpha \cos \beta + \cos \alpha \sin \beta$$

$$\tan(\alpha + \beta) = \frac{\tan \alpha + \tan \beta}{1 - \tan \alpha \tan \beta}$$

$$\tan(\alpha - \beta) = \frac{\tan \alpha - \tan \beta}{1 + \tan \alpha \tan \beta}$$

Cofunction Identities

$$\sin(90° - \theta) = \cos \theta \quad \cos(90° - \theta) = \sin \theta \quad \tan(90° - \theta) = \cot \theta$$

$$\csc(90° - \theta) = \sec \theta \quad \sec(90° - \theta) = \csc \theta \quad \cot(90° - \theta) = \tan \theta$$

The cofunction identities are also true when 90° is replaced by $\pi/2$ radians.

EXERCISE SET 7.2

In Exercises 1 to 18, find the exact value of the expression.

1. $\sin(45° + 30°)$

2. $\sin(330° + 45°)$

3. $\cos(45° - 30°)$

4. $\cos(120° - 45°)$

5. $\tan(45° + 135°)$

6. $\tan(240° - 45°)$

7. $\sin\left(\dfrac{5\pi}{4} - \dfrac{\pi}{6}\right)$

8. $\sin\left(\dfrac{4\pi}{3} + \dfrac{\pi}{4}\right)$

9. $\cos\left(\dfrac{3\pi}{4} + \dfrac{\pi}{6}\right)$

10. $\cos\left(\dfrac{\pi}{4} - \dfrac{\pi}{3}\right)$

11. $\tan\left(\dfrac{\pi}{6} + \dfrac{\pi}{4}\right)$

12. $\tan\left(\dfrac{11\pi}{6} - \dfrac{\pi}{4}\right)$

13. $\cos 212° \cos 122° + \sin 212° \sin 122°$

14. $\sin 167° \cos 107° - \cos 167° \sin 107°$

15. $\sin \dfrac{5\pi}{12} \cos \dfrac{\pi}{4} - \cos \dfrac{5\pi}{12} \sin \dfrac{\pi}{4}$

16. $\cos \dfrac{\pi}{12} \cos \dfrac{\pi}{4} - \sin \dfrac{\pi}{12} \sin \dfrac{\pi}{4}$

17. $\dfrac{\tan 7\pi/12 - \tan \pi/4}{1 + \tan 7\pi/12 \tan \pi/4}$

18. $\dfrac{\tan \pi/6 + \tan \pi/3}{1 - \tan \pi/6 \tan \pi/3}$

In Exercises 19 to 30, write each expression in terms of a single trigonometric function.

19. $\sin 7x \cos 2x - \cos 7x \sin 2x$

20. $\sin x \cos 3x + \cos x \sin 3x$

21. $\cos x \cos 2x + \sin x \sin 2x$

22. $\cos 4x \cos 2x - \sin 4x \sin 2x$

23. $\sin 7x \cos 3x - \cos 7x \sin 3x$

24. $\cos x \cos 5x - \sin x \sin 5x$

25. $\cos 4x \cos(-2x) - \sin 4x \sin(-2x)$

26. $\sin(-x) \cos 3x - \cos(-x) \sin 3x$

27. $\sin \dfrac{x}{3} \cos \dfrac{2x}{3} + \cos \dfrac{x}{3} \sin \dfrac{2x}{3}$

28. $\cos \dfrac{3x}{4} \cos \dfrac{x}{4} + \sin \dfrac{3x}{4} \sin \dfrac{x}{4}$

29. $\dfrac{\tan 3x + \tan 4x}{1 - \tan 3x \tan 4x}$

30. $\dfrac{\tan 2x - \tan 3x}{1 + \tan 2x \tan 3x}$

In Exercises 31 to 42, find the exact value of the given functions.

31. Given $\tan \alpha = -\frac{4}{3}$, α in quadrant II and $\tan \beta = \frac{15}{8}$, β in quadrant III, find   **a.** $\sin(\alpha - \beta)$, **b.** $\cos(\alpha + \beta)$, and **c.** $\tan(\alpha - \beta)$.

32. Given $\tan \alpha = \frac{24}{7}$, α in quadrant I and $\sin \beta = -\frac{8}{17}$, β in quadrant III, find   **a.** $\sin(\alpha + \beta)$, **b.** $\cos(\alpha + \beta)$, and **c.** $= \tan(\alpha - \beta)$.

33. Given $\sin \alpha = \frac{3}{5}$, α in quadrant I and $\cos \beta = -\frac{5}{13}$, β in quadrant II, find   **a.** $\sin(\alpha - \beta)$, **b.** $\cos(\alpha + \beta)$, and **c.** $\tan(\alpha - \beta)$.

34. Given $\sin \alpha = \frac{24}{25}$, α in quadrant II and $\cos \beta = -\frac{4}{5}$, β in quadrant III, find   **a.** $\cos(\beta - \alpha)$, **b.** $\sin(\alpha + \beta)$, and **c.** $\tan(\alpha + \beta)$.

35. Given $\sin \alpha = -\frac{4}{5}$, α in quadrant III and $\cos \beta = -\frac{12}{13}$, β in quadrant II, find   **a.** $\sin(\alpha - \beta)$, **b.** $\cos(\alpha + \beta)$, and **c.** $\tan(\alpha + \beta)$.

36. Given $\sin \alpha = -\frac{7}{25}$, α in quadrant IV and $\cos \beta = \frac{8}{17}$, β in quadrant IV, find   **a.** $\sin(\alpha + \beta)$, **b.** $\cos(\alpha - \beta)$, and **c.** $\tan(\alpha + \beta)$.

37. Given $\cos \alpha = \frac{15}{17}$, α in quadrant I and $\sin \beta = -\frac{3}{5}$, β in quadrant III, find   **a.** $\sin(\alpha + \beta)$, **b.** $\cos(\alpha - \beta)$, and **c.** $\tan(\alpha - \beta)$.

38. Given $\cos \alpha = -\frac{7}{25}$, α in quadrant II and $\sin \beta = -\frac{12}{13}$, β in quadrant IV, find   **a.** $\sin(\alpha + \beta)$, **b.** $\cos(\alpha + \beta)$, and **c.** $\tan(\alpha - \beta)$.

39. Given $\cos \alpha = -\frac{3}{5}$, α in quadrant III and $\sin \beta = \frac{5}{13}$, β in quadrant I, find   **a.** $\sin(\alpha - \beta)$, **b.** $\cos(\alpha + \beta)$, and **c.** $\tan(\alpha + \beta)$.

40. Given $\cos \alpha = \frac{8}{17}$, α in quadrant IV and $\sin \beta = -\frac{24}{25}$, β in quadrant III, find   **a.** $\sin(\alpha - \beta)$, **b.** $\cos(\alpha + \beta)$, and **c.** $\tan(\alpha + \beta)$.

41. Given $\sin \alpha = \frac{3}{5}$, α in quadrant I and $\tan \beta = \frac{5}{12}$, β in quadrant III, find   **a.** $\sin(\alpha + \beta)$, **b.** $\cos(\alpha - \beta)$, and **c.** $\tan(\alpha - \beta)$.

42. Given $\tan \alpha = \frac{15}{8}$, α in quadrant I and $\tan \beta = -\frac{7}{24}$, β in quadrant IV, find   **a.** $\sin(\alpha - \beta)$, **b.** $\cos(\alpha - \beta)$, and **c.** $\tan(\alpha + \beta)$.

In Exercises 43 to 66, verify the identities.

43. $\cos\left(\dfrac{\pi}{2} - \theta\right) = \sin \theta$ **44.** $\cos(\theta + \pi) = -\cos \theta$

45. $\sin\left(\theta + \dfrac{\pi}{2}\right) = \cos \theta$ **46.** $\sin(\theta + \pi) = -\sin \theta$

47. $\tan\left(\theta + \dfrac{\pi}{4}\right) = \dfrac{\tan \theta + 1}{1 - \tan \theta}$

48. $\tan 2\theta = \dfrac{2 \tan \theta}{1 - \tan^2\theta}$ **49.** $\cos\left(\dfrac{3\pi}{2} - \theta\right) = -\sin \theta$

50. $\sin\left(\dfrac{3\pi}{2} + \theta\right) = -\cos \theta$ **51.** $\cot\left(\dfrac{\pi}{2} - \theta\right) = \tan \theta$

52. $\cot(\pi + \theta) = \cot \theta$ **53.** $\csc(\pi - \theta) = \csc \theta$

54. $\sec\left(\dfrac{\pi}{2} - \theta\right) = \csc \theta$

55. $\sin 6x \cos 2x - \cos 6x \sin 2x = 2 \sin 2x \cos 2x$

56. $\cos 5x \cos 3x + \sin 5x \sin 3x = \cos^2x - \sin^2x$

57. $\cos(\alpha + \beta) + \cos(\alpha - \beta) = 2 \cos \alpha \cos \beta$

58. $\cos(\alpha - \beta) - \cos(\alpha + \beta) = 2 \sin \alpha \sin \beta$

59. $\sin(\alpha + \beta) + \sin(\alpha - \beta) = 2 \sin \alpha \cos \beta$

60. $\sin(\alpha - \beta) - \sin(\alpha + \beta) = -2 \cos \alpha \sin \beta$

61. $\dfrac{\cos(\alpha - \beta)}{\sin(\alpha + \beta)} = \dfrac{\cot \alpha + \tan \beta}{1 + \cot \alpha \tan \beta}$

62. $\dfrac{\sin(\alpha + \beta)}{\sin(\alpha - \beta)} = \dfrac{1 + \cot \alpha \tan \beta}{1 - \cot \alpha \tan \beta}$

63. $\sin(\pi/2 + \alpha - \beta) = \cos \alpha \cos \beta + \sin \alpha \sin \beta$

64. $\cos(\pi/2 + \alpha + \beta) = -(\sin \alpha \cos \beta + \cos \alpha \sin \beta)$

65. $\sin 3x = 3 \sin x - 4 \sin^3x$

66. $\cos 3x = 4 \cos^3x - 3 \cos x$

Graphing Calculator Exercises*

In Exercises 67 to 70, establish the identity by graphing the function on each side of the equation and showing that the graphs are identical.

67. $\sin\left(\dfrac{\pi}{2} - x\right) = \cos x$ **68.** $\cos(x + \pi) = -\cos x$

69. $\sin 7x \cos 2x - \cos 7x \sin 2x = \sin 5x$

70. $\sin 3x = 3 \sin x - 4 \sin^3x$

Supplemental Exercises

In Exercises 71 to 77, verify the identities.

71. $\sin(x - y) \cdot \sin(x + y) = \sin^2x \cos^2y - \cos^2x \sin^2y$

72. $\sin(x + y + z) = \sin x \cos y \cos z + \cos x \sin y \cos z$
$+ \cos x \cos y \sin z - \sin x \sin y \sin z$

*Additional graphing calculator exercises appear in the Graphing Workbook as described in the front of this textbook.

73. $\cos(x + y + z) = \cos x \cos y \cos z - \sin x \sin y \cos z$
$$- \sin x \cos y \sin z - \cos x \sin y \sin z$$

74. $\dfrac{\sin(x + y)}{\sin x \sin y} = \cot x + \cot y$

75. $\dfrac{\cos(x - y)}{\cos x \sin y} = \cot y + \tan x$

76. $\dfrac{\sin(x + h) - \sin x}{h} = \cos x \dfrac{\sin h}{h} + \sin x \dfrac{\cos h - 1}{h}$

77. $\dfrac{\cos(x + h) - \cos x}{h} = \cos x \dfrac{\cos h - 1}{h} - \sin x \dfrac{\sin h}{h}$

78. The drag (resistance) on a fish when it is swimming is 2 to 3 times the drag when it is gliding. To compensate for this, some fish swim in a saw-tooth pattern as shown in the accompanying figure. The ratio of the amount of energy the fish expends when swimming upward at angle β and then gliding down at angle α to the energy it expends swimming horizontally is given by

$$E_R = \frac{k \sin \alpha + \sin \beta}{k \sin(\alpha + \beta)}$$

where k is a value such that $2 \le k \le 3$ and k depends on the assumptions we make about the amount of drag experienced by the fish. Find E_R for $k = 2$, $\alpha = 10°$, and $\beta = 20°$.

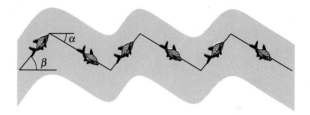

7.3 Double- and Half-Angle Identities

Double-Angle Identities

By using the sum identities, we can derive identities for $f(2\alpha)$ where f is a trigonometric function. These are called the *double-angle* identities. To find the sine of a double angle, substitute α for β in the identity for $\sin(\alpha + \beta)$.

$$\sin(\alpha + \beta) = \sin \alpha \cos \beta + \cos \alpha \sin \beta$$

$$\sin(\alpha + \alpha) = \sin \alpha \cos \alpha + \cos \alpha \sin \alpha \quad \bullet \text{ Let } \beta = \alpha.$$

$$\sin 2\alpha = 2 \sin \alpha \cos \alpha.$$

A double-angle identity for cosine is derived in a similar manner.

$$\cos(\alpha + \beta) = \cos \alpha \cos \beta - \sin \alpha \sin \beta$$

$$\cos(\alpha + \alpha) = \cos \alpha \cos \alpha - \sin \alpha \sin \alpha \quad \bullet \text{ Let } \beta = \alpha.$$

$$\cos 2\alpha = \cos^2\alpha - \sin^2\alpha$$

There are two alternative forms of the double-angle identity for $\cos 2\alpha$. Using $\cos^2\alpha = 1 - \sin^2 x$, we can rewrite the identity for $\cos 2\alpha$ as follows:

$$\cos 2\alpha = \cos^2\alpha - \sin^2\alpha$$

$$\cos 2\alpha = (1 - \sin^2\alpha) - \sin^2\alpha \quad \bullet \cos^2\alpha = 1 - \sin^2\alpha$$

$$\cos 2\alpha = 1 - 2 \sin^2\alpha$$

We can also rewrite $\cos 2\alpha$ as

$$\cos 2\alpha = \cos^2\alpha - \sin^2\alpha$$

$$\cos 2\alpha = \cos^2\alpha - (1 - \cos^2\alpha) \quad \bullet \ \sin^2\alpha = 1 - \cos^2\alpha$$

$$\cos 2\alpha = 2\cos^2\alpha - 1$$

The double-angle identity for the tangent function is derived from the identity for $\tan(\alpha + \beta)$ with $\beta = \alpha$.

$$\tan(\alpha + \beta) = \frac{\tan \alpha + \tan \beta}{1 - \tan \alpha \tan \beta}$$

$$\tan(\alpha + \alpha) = \frac{\tan \alpha + \tan \alpha}{1 - \tan \alpha \tan \alpha} \quad \bullet \ \text{Let } \beta = \alpha.$$

$$\tan 2\alpha = \frac{2 \tan \alpha}{1 - \tan^2\alpha}$$

EXAMPLE 1   **Evaluate a Trigonometric Function**

For an angle α in quadrant I, $\sin \alpha = \dfrac{4}{5}$. Find $\sin 2\alpha$.

Solution   Use the identity $\sin 2\alpha = 2 \sin \alpha \cos \alpha$. Find $\cos \alpha$ by substituting for $\sin \alpha$ in $\sin^2\alpha + \cos^2\alpha = 1$ and solving for $\cos \alpha$.

$$\cos \alpha = \sqrt{1 - \sin^2\alpha} = \sqrt{1 - \left(\frac{4}{5}\right)^2} = \frac{3}{5} \quad \bullet \ \cos \alpha > 0 \text{ if } \alpha \text{ is in quadrant I}$$

Substitute the values of $\sin \alpha$ and $\cos \alpha$ in the double-angle formula for $\sin 2\alpha$.

$$\sin 2\alpha = 2 \sin \alpha \cos \alpha = 2\left(\frac{4}{5}\right)\left(\frac{3}{5}\right) = \frac{24}{25}$$

■ *Try Exercise* **26,** *page 380.*

EXAMPLE 2   **Verify a Double-Angle Identity**

Verify the identity $\csc 2\alpha = \dfrac{1}{2}(\tan \alpha + \cot \alpha)$.

Solution   Work on the right-hand side of the equation.

$$\frac{1}{2}(\tan \alpha + \cot \alpha) = \frac{1}{2}\left(\frac{\sin \alpha}{\cos \alpha} + \frac{\cos \alpha}{\sin \alpha}\right) = \frac{1}{2}\left(\frac{\sin^2\alpha + \cos^2\alpha}{\cos \alpha \sin \alpha}\right)$$

$$= \frac{1}{2\cos \alpha \sin \alpha} = \frac{1}{\sin 2\alpha} = \csc 2\alpha$$

■ *Try Exercise* **54,** *page 380.*

EXAMPLE 3 **Verify a Double-Angle Identity**

Verify the identity $\sin^2 x = \dfrac{1}{2}(1 - \cos 2x)$.

Solution Work on the right side of the equation.

$$\frac{1}{2}(1 - \cos 2x) = \frac{1}{2}[1 - (1 - 2\sin^2 x)] = \frac{1}{2}(1 - 1 + 2\sin^2 x) = \sin^2 x$$

■ *Try Exercise* **58,** *page 381.*

Half-Angle Identities

An identity for one-half an angle, $\alpha/2$, is called a half-angle identity. To derive a half-angle identity for $\sin \alpha/2$, we solve for $\sin^2 \theta$ in the following double-angle identity for $\cos 2\theta$.

$$\cos 2\theta = 1 - 2\sin^2 \theta$$

$$\sin^2 \theta = \frac{1 - \cos 2\theta}{2}$$

Substitute $\alpha/2$ for θ and take square root of both sides of the equation.

$$\sin^2 \frac{\alpha}{2} = \frac{1 - \cos 2\left(\dfrac{\alpha}{2}\right)}{2}$$

$$\sin \frac{\alpha}{2} = \pm\sqrt{\frac{1 - \cos \alpha}{2}}$$

The sign of the radical is determined by the quadrant in which the terminal side of angle $\alpha/2$ lies.

In a similar manner, we derive an identity for $\cos \alpha/2$.

$$\cos 2\theta = 2\cos^2 \theta - 1$$

$$\cos^2 \theta = \frac{1 + \cos 2\theta}{2}$$

Substitute $\alpha/2$ for θ and take the square root of both sides of the equation.

$$\cos^2 \frac{\alpha}{2} = \frac{1 + \cos 2\left(\dfrac{\alpha}{2}\right)}{2}$$

$$\cos \frac{\alpha}{2} = \pm\sqrt{\frac{1 + \cos \alpha}{2}}$$

Two different identities for $\tan \alpha/2$ are possible.

$$\tan \frac{\alpha}{2} = \frac{\sin \dfrac{\alpha}{2}}{\cos \dfrac{\alpha}{2}} = \frac{\sin \dfrac{\alpha}{2}}{\cos \dfrac{\alpha}{2}} \cdot \frac{2 \cos \dfrac{\alpha}{2}}{2 \cos \dfrac{\alpha}{2}}$$

$$= \frac{2 \sin \dfrac{\alpha}{2} \cos \dfrac{\alpha}{2}}{2 \cos^2 \dfrac{\alpha}{2}}$$

$$= \frac{\sin 2\left(\dfrac{\alpha}{2}\right)}{2\left(\pm \sqrt{\dfrac{1 + \cos \alpha}{2}}\right)^2} \qquad \bullet\ \cos \frac{\alpha}{2} = \pm \sqrt{\frac{1 + \cos \alpha}{2}}$$

$$\tan \frac{\alpha}{2} = \frac{\sin \alpha}{1 + \cos \alpha}$$

To obtain an equivalent identity for $\tan \alpha/2$, multiply by the conjugate of the denominator.

$$\tan \frac{\alpha}{2} = \frac{\sin \alpha}{1 + \cos \alpha} \cdot \frac{1 - \cos \alpha}{1 - \cos \alpha}$$

$$= \frac{\sin \alpha (1 - \cos \alpha)}{1 - \cos^2 \alpha}$$

$$= \frac{\sin \alpha (1 - \cos \alpha)}{\sin^2 \alpha}$$

$$\tan \frac{\alpha}{2} = \frac{1 - \cos \alpha}{\sin \alpha}$$

EXAMPLE 4   **Verify a Half-Angle Identity**

Verify the identity $2 \csc x \cos^2 \dfrac{x}{2} = \dfrac{\sin x}{1 - \cos x}$.

Solution   Work on the left side of the identity.

$$2 \csc x \cos^2 \frac{x}{2} = 2 \csc x \left(\frac{1 + \cos x}{2}\right) \qquad \bullet\ \cos^2 \frac{x}{2} = \frac{1 + \cos x}{2}$$

$$= \frac{1 + \cos x}{\sin x} \qquad \bullet\ \csc x = \frac{1}{\sin x}$$

$$= \frac{1 + \cos x}{\sin x} \cdot \frac{1 - \cos x}{1 - \cos x} \qquad \bullet\ \text{Multiply the numerator and denominator by the conjugate of the numerator.}$$

$$= \frac{1 - \cos^2 x}{\sin x(1 - \cos x)}$$

$$= \frac{\sin^2 x}{\sin x(1 - \cos x)} \qquad \bullet \; 1 - \cos^2 x = \sin^2 x$$

$$= \frac{\sin x}{1 - \cos x}$$

■ *Try Exercise **72**, page 381.*

EXAMPLE 5   **Verify a Half-Angle Identity**

Verify the identity $\tan \dfrac{\alpha}{2} = \sin \alpha + \cos \alpha \cot \alpha - \cot \alpha$.

Solution   Work on the left side of the identity.

$$\tan \frac{\alpha}{2} = \frac{1 - \cos \alpha}{\sin \alpha} = \frac{\sin^2 \alpha + \cos^2 \alpha - \cos \alpha}{\sin \alpha} \qquad \bullet \; 1 = \sin^2 \alpha + \cos^2 \alpha$$

$$= \frac{\sin^2 \alpha}{\sin \alpha} + \frac{\cos^2 \alpha}{\sin \alpha} - \frac{\cos \alpha}{\sin \alpha} \qquad \bullet \; \text{Write each numerator over the common denominator.}$$

$$= \sin \alpha + \cos \alpha \cot \alpha - \cot \alpha$$

■ *Try Exercise **78**, page 381.*

Here is a summary of the double-angle and half-angle identities:

Double-Angle Identities

$$\sin 2\alpha = 2 \sin \alpha \cos \alpha$$

$$\cos 2\alpha = \cos^2 \alpha - \sin^2 \alpha = 1 - 2 \sin^2 \alpha = 2 \cos^2 \alpha - 1$$

$$\tan 2\alpha = \frac{2 \tan \alpha}{1 - \tan^2 \alpha}$$

Half-Angle Identities

$$\sin \frac{\alpha}{2} = \pm \sqrt{\frac{1 - \cos \alpha}{2}}$$

$$\cos \frac{\alpha}{2} = \pm \sqrt{\frac{1 + \cos \alpha}{2}}$$

$$\tan \frac{\alpha}{2} = \frac{\sin \alpha}{1 + \cos \alpha} = \frac{1 - \cos \alpha}{\sin \alpha}$$

EXERCISE SET 7.3

In Exercises 1 to 8, write the trigonometric expressions in terms of a single trigonometric function.

1. $2 \sin 2\alpha \cos 2\alpha$

2. $2 \sin 3\theta \cos 3\theta$

3. $1 - 2 \sin^2 5\beta$

4. $2 \cos^2 2\beta - 1$

5. $\cos^2 3\alpha - \sin^2 3\alpha$

6. $\cos^2 6\alpha - \sin^2 6\alpha$

7. $\dfrac{2 \tan 3\alpha}{1 - \tan^2 3\alpha}$

8. $\dfrac{2 \tan 4\theta}{1 - \tan^2 4\theta}$

In Exercises 9 to 24, use the half-angle identities to evaluate the trigonometric expressions.

9. $\sin 75°$

10. $\cos 105°$

11. $\tan 67.5°$

12. $\tan 165°$

13. $\cos 157.5°$

14. $\sin 112.5°$

15. $\sin 22.5°$

16. $\cos 67.5°$

17. $\sin \dfrac{7\pi}{8}$

18. $\cos \dfrac{5\pi}{8}$

19. $\cos \dfrac{5\pi}{12}$

20. $\sin \dfrac{3\pi}{8}$

21. $\tan \dfrac{7\pi}{12}$

22. $\tan \dfrac{3\pi}{8}$

23. $\cos \dfrac{\pi}{12}$

24. $\sin \dfrac{\pi}{8}$

In Exercises 25 to 36, find the exact value of $\sin 2\theta$, $\cos 2\theta$, and $\tan 2\theta$ given the following information.

25. $\cos \theta = -\dfrac{4}{5}$, θ is in quadrant II.

26. $\cos \theta = \dfrac{24}{25}$, θ is in quadrant IV.

27. $\sin \theta = \dfrac{8}{17}$, θ is in quadrant II.

28. $\sin \theta = -\dfrac{9}{41}$, θ is in quadrant III.

29. $\tan \theta = -\dfrac{24}{7}$, θ is in quadrant IV.

30. $\tan \theta = \dfrac{4}{3}$, θ is in quadrant I.

31. $\sin \theta = \dfrac{15}{17}$, θ is in quadrant I.

32. $\sin \theta = -\dfrac{3}{5}$, θ is in quadrant III.

33. $\cos \theta = \dfrac{40}{41}$, θ is in quadrant IV.

34. $\cos \theta = \dfrac{4}{5}$, θ is in quadrant IV.

35. $\tan \theta = \dfrac{15}{8}$, θ is in quadrant III.

36. $\tan \theta = -\dfrac{40}{9}$, θ is in quadrant II.

In Exercises 37 to 48, find the value of the sine, cosine, and tangent of $\alpha/2$ given the following information.

37. $\sin \alpha = \dfrac{5}{13}$, α is in quadrant II.

38. $\sin \alpha = -\dfrac{7}{25}$, α is in quadrant III.

39. $\cos \alpha = -\dfrac{8}{17}$, α is in quadrant III.

40. $\cos \alpha = \dfrac{12}{13}$, α is in quadrant I.

41. $\tan \alpha = \dfrac{4}{3}$, α is in quadrant I.

42. $\tan \alpha = -\dfrac{8}{15}$, α is in quadrant II.

43. $\cos \alpha = \dfrac{24}{25}$, α is in quadrant IV.

44. $\sin \alpha = -\dfrac{9}{41}$, α is in quadrant IV.

45. $\sec \alpha = \dfrac{17}{15}$, α is in quadrant I.

46. $\csc \alpha = -\dfrac{5}{3}$, α is in quadrant IV.

47. $\cot \alpha = \dfrac{8}{15}$, α is in quadrant III.

48. $\sec \alpha = -\dfrac{13}{5}$, α is in quadrant II.

In Exercises 49 to 94, verify the given identity.

49. $\sin 3x \cos 3x = \dfrac{1}{2} \sin 6x$

50. $\cos 8x = \cos^2 4x - \sin^2 4x$

51. $\sin^2 x + \cos 2x = \cos^2 x$

52. $\dfrac{\cos 2x}{\sin^2 x} = \cot^2 x - 1$

53. $\dfrac{1 + \cos 2x}{\sin 2x} = \cot x$

54. $\dfrac{1}{1 - \cos 2x} = \dfrac{1}{2} \csc^2 x$

55. $\dfrac{\sin 2x}{1 - \sin^2 x} = 2 \tan x$

56. $\dfrac{\cos^2 x - \sin^2 x}{2 \sin x \cos x} = \cot 2x$

57. $1 - \tan^2 x = \dfrac{\cos 2x}{\cos^2 x}$ **58.** $\tan 2x = \dfrac{2 \sin x \cos x}{\cos^2 x - \sin^2 x}$

59. $\sin 2x - \tan x = \tan x \cos 2x$

60. $\sin 2x - \cot x = -\cot x \cos 2x$

61. $\cos^4 x - \sin^4 x = \cos 2x$

62. $\sin 4x = 4 \sin x \cos^3 x - 4 \cos x \sin^3 x$

63. $\cos^2 x - 2 \sin^2 x \cos^2 x - \sin^2 x + 2 \sin^4 x = \cos^2 2x$

64. $2 \cos^4 x - \cos^2 x - 2 \sin^2 x \cos^2 x + \sin^2 x = \cos^2 2x$

65. $\cos 4x = 1 - 8 \cos^2 x + 8 \cos^4 x$

66. $\sin 4x = 4 \sin x \cos x - 8 \cos x \sin^3 x$

67. $\cos 3x - \cos x = 4 \cos^3 x - 4 \cos x$

68. $\sin 3x + \sin x = 4 \sin x - 4 \sin^3 x$

69. $\sin^3 x + \cos^3 x = (\sin x + \cos x)\left(1 - \dfrac{1}{2} \sin 2x\right)$

70. $\cos^3 x - \sin^3 x = (\cos x - \sin x)\left(1 + \dfrac{1}{2} \sin 2x\right)$

71. $\sin^2 \dfrac{x}{2} = \dfrac{\sec x - 1}{2 \sec x}$ **72.** $\cos^2 \dfrac{x}{2} = \dfrac{\sec x + 1}{2 \sec x}$

73. $\tan \dfrac{x}{2} = \csc x - \cot x$ **74.** $\tan \dfrac{x}{2} = \dfrac{\tan x}{\sec x + 1}$

75. $2 \sin \dfrac{x}{2} \cos \dfrac{x}{2} = \sin x$

76. $\cos^2 \dfrac{x}{2} - \sin^2 \dfrac{x}{2} = \cos x$

77. $\left(\cos \dfrac{x}{2} + \sin \dfrac{x}{2}\right)^2 = 1 + \sin x$

78. $\tan^2 \dfrac{x}{2} = \dfrac{\sec x - 1}{\sec x + 1}$

79. $\sin^2 \dfrac{x}{2} \sec x = \dfrac{1}{2}(\sec x - 1)$

80. $\cos^2 \dfrac{x}{2} \sec x = \dfrac{1}{2}(\sec x + 1)$

81. $\cos^2 \dfrac{x}{2} - \cos x = \sin^2 \dfrac{x}{2}$

82. $\sin^2 \dfrac{x}{2} + \cos x = \cos^2 \dfrac{x}{2}$

83. $\sin^2 \dfrac{x}{2} - \cos^2 \dfrac{x}{2} = -\cos x$

84. $\cos^2 \dfrac{x}{2} - \sin^2 \dfrac{x}{2} = \dfrac{1}{2} \csc x \sin 2x$

85. $\sin 2x - \cos x = (\cos x)(2 \sin x - 1)$

86. $\dfrac{\cos 2x}{\sin^2 x} = \csc^2 x - 2$

87. $\tan 2x = \dfrac{2}{\cot x - \tan x}$

88. $\dfrac{2 \cos 2x}{\sin 2x} = \cot x - \tan x$

89. $2 \tan \dfrac{x}{2} = \dfrac{\sin^2 x + 1 - \cos^2 x}{\sin x(1 + \cos x)}$

90. $\dfrac{1}{2} \csc^2 \dfrac{x}{2} = \csc^2 x + \cot x \csc x$

91. $\csc 2x = \dfrac{1}{2} \csc x \sec x$ **92.** $\sec 2x = \dfrac{\sec^2 x}{2 - \sec^2 x}$

93. $\cos \dfrac{x}{5} = 1 - 2 \sin^2 \dfrac{x}{10}$ **94.** $\sec^2 \dfrac{x}{2} = \dfrac{2}{1 + \cos x}$

Graphing Calculator Exercises*

In Exercises 95 to 98, establish the identity by graphing the function on each side of the equation and showing that the graphs are identical.

95. $\sin^2 x + \cos 2x = \cos^2 x$ **96.** $\dfrac{\sin 2x}{1 - \sin^2 x} = 2 \tan x$

97. $2 \sin \dfrac{x}{2} \cos \dfrac{x}{2} = \sin x$

98. $\left(\cos \dfrac{x}{2} + \sin \dfrac{x}{2}\right)^2 = 1 + \sin x$

Supplemental Exercises

In Exercises 99 to 102, verify the identities.

99. $\dfrac{\sin^3 x + \cos^3 x}{\sin x + \cos x} = 1 - \dfrac{1}{2} \sin 2x$

100. $\cos^4 x = \dfrac{1}{8} \cos 4x + \dfrac{1}{2} \cos 2x + \dfrac{3}{8}$

101. $\sin \dfrac{x}{2} - \cos \dfrac{x}{2} = \sqrt{1 - \sin x},\ 0° \le x \le 90°$

102. $\dfrac{\sin x - \sin 2x}{\cos x + \cos 2x} = -\tan \dfrac{x}{2}$

103. If $x + y = 90°$; verify $\sin(x - y) = -\cos 2x$.

104. If $x + y = 90°$; verify $\sin(x - y) = \cos 2y$.

105. If $x + y = 180°$; verify $\sin(x - y) = -\sin 2x$.

106. If $x + y = 180°$; verify $\cos(x - y) = -\cos 2x$.

---

* Additional graphing calculator exercises appear in the Graphing Workbook as described in the front of this textbook.

7.4 Identities Involving the Sum of Trigonometric Functions

Some applications require that a product of trigonometric functions be written as a sum or difference of these functions. Other applications require that the sum or difference of trigonometric functions be represented as a product of these functions. The product-to-sum identities are particularly useful in these types of problems.

The Product-to-Sum Identities

The product-to-sum identities can be derived by using the sum and difference identities. Adding the identities for $\sin(\alpha + \beta)$ and $\sin(\alpha - \beta)$, we have

$$\sin(\alpha + \beta) = \sin \alpha \cos \beta + \cos \alpha \sin \beta$$

$$\underline{\sin(\alpha - \beta) = \sin \alpha \cos \beta - \cos \alpha \sin \beta}$$

$$\sin(\alpha + \beta) + \sin(\alpha - \beta) = 2 \sin \alpha \cos \beta$$

Solving for $\sin \alpha \cos \beta$, we obtain the first product-to-sum identity:

$$\sin \alpha \cos \beta = \frac{1}{2}[\sin(\alpha + \beta) + \sin(\alpha - \beta)].$$

The identity for $\cos \alpha \sin \beta$ is obtained when $\sin(\alpha - \beta)$ is subtracted from $\sin(\alpha + \beta)$. The result is

$$\cos \alpha \sin \beta = \frac{1}{2}[\sin(\alpha + \beta) - \sin(\alpha - \beta)].$$

In like manner, the identities for $\cos(\alpha + \beta)$ and $\cos(\alpha - \beta)$ are used to derive the identities for $\cos \alpha \cos \beta$ and $\sin \alpha \sin \beta$.

$$\cos \alpha \cos \beta = \frac{1}{2}[\cos(\alpha + \beta) + \cos(\alpha - \beta)]$$

$$\sin \alpha \sin \beta = \frac{1}{2}[\cos(\alpha - \beta) - \cos(\alpha + \beta)]$$

EXAMPLE 1 **Verify an Identity**

Verify the identity $\cos 2x \sin 5x = \frac{1}{2}(\sin 7x + \sin 3x)$.

Solution

$$\cos 2x \sin 5x = \frac{1}{2}[\sin(2x + 5x) - \sin(2x - 5x)] \quad \bullet \text{ Use the product-to-sum identity: } \cos \alpha \sin \beta.$$

$$= \frac{1}{2}[\sin 7x - \sin(-3x)]$$

$$= \frac{1}{2}(\sin 7x + \sin 3x) \quad \bullet \sin(-3x) = -\sin 3x$$

■ *Try Exercise 36, page 387.*

The Sum-to-Product Identities

The sum-to-product identities can be derived from the product-to-sum identities. To derive the sum-to-product identity for $\sin x + \sin y$, we first let $x = \alpha + \beta$ and $y = \alpha - \beta$. Then

$$x + y = \alpha + \beta + \alpha - \beta \qquad \text{and} \qquad x - y = \alpha + \beta - (\alpha - \beta)$$

$$x + y = 2\alpha \qquad\qquad\qquad\qquad x - y = 2\beta$$

$$\alpha = \frac{x + y}{2} \qquad\qquad\qquad\qquad \beta = \frac{x - y}{2}$$

Substituting the expression for α and β in the product-to-sum identity,

$$\frac{1}{2}[\sin(\alpha + \beta) + \sin(\alpha - \beta)] = \sin \alpha \cos \beta$$

we have

$$\sin\left(\frac{x + y}{2} + \frac{x - y}{2}\right) + \sin\left(\frac{x + y}{2} - \frac{x - y}{2}\right) = 2 \sin \frac{x + y}{2} \cos \frac{x - y}{2}$$

Simplifying the left side, we have a sum-to-product identity.

$$\sin x + \sin y = 2 \sin \frac{x + y}{2} \cos \frac{x - y}{2}$$

In like manner, three other sum-to-product identities can be derived from the other product-to-sum identities. The proofs of these identities are left as exercises.

$$\sin x - \sin y = 2 \cos \frac{x + y}{2} \sin \frac{x - y}{2}$$

$$\cos x + \cos y = 2 \cos \frac{x + y}{2} \cos \frac{x - y}{2}$$

$$\cos x - \cos y = -2 \sin \frac{x + y}{2} \sin \frac{x - y}{2}$$

EXAMPLE 2   **Write the Difference of Trigonometric Expressions as a Product**

Write $\sin 4\theta - \sin \theta$ as the product of two functions.

Solution

$$\sin 4\theta - \sin \theta = 2 \cos \frac{4\theta + \theta}{2} \sin \frac{4\theta - \theta}{2}$$

$$= 2 \cos \frac{5\theta}{2} \sin \frac{3\theta}{2}$$

■ *Try Exercise **22**, page 387.*

EXAMPLE 3   Verify a Sum-to-Product Identity

Verify the identity $\dfrac{\sin 6x + \sin 2x}{\sin 6x - \sin 2x} = \tan 4x \cot 2x$.

Solution

$$\frac{\sin 6x + \sin 2x}{\sin 6x - \sin 2x} = \frac{2 \sin \dfrac{6x + 2x}{2} \cos \dfrac{6x - 2x}{2}}{2 \cos \dfrac{6x + 2x}{2} \sin \dfrac{6x - 2x}{2}} = \frac{\sin 4x \cos 2x}{\cos 4x \sin 2x}$$

$$= \tan 4x \cot 2x$$

■ *Try Exercise* **44**, *page 387.*

Functions of the Form $f(x) = a \sin x + b \cos x$

The function given by $f(x) = a \sin x + b \cos x$ can be written in the form $f(x) = k \sin(x + \alpha)$. This form of the function is useful in graphing and engineering applications because the amplitude, period, and phase shift can be readily calculated.

Let $P(a, b)$ be a point on a coordinate plane, and let α represent an angle in standard position. See Figure 7.3. To rewrite $y = a \sin x + b \cos x$, multiply and divide the expression $a \sin x + b \cos x$ by $\sqrt{a^2 + b^2}$.

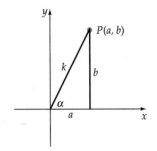

Figure 7.3

$$a \sin x + b \cos x = \frac{\sqrt{a^2 + b^2}}{\sqrt{a^2 + b^2}}(a \sin x + b \cos x)$$

$$= \sqrt{a^2 + b^2}\left(\frac{a}{\sqrt{a^2 + b^2}} \sin x + \frac{b}{\sqrt{a^2 + b^2}} \cos x\right) \quad (1)$$

From the definition of the sine and cosine of an angle in standard position, let

$$k = \sqrt{a^2 + b^2}, \quad \cos \alpha = \frac{a}{\sqrt{a^2 + b^2}}, \quad \text{and} \quad \sin \alpha = \frac{b}{\sqrt{a^2 + b^2}}.$$

Substitute these expressions into Equation (1). We have

$$a \sin x + b \cos x = k(\cos \alpha \sin x + \sin \alpha \cos x).$$

Now, using the identity for the sum of two angles, we have

$$a \sin x + b \cos x = k \sin(x + \alpha).$$

Thus $a \sin x + b \cos x = k \sin(x + \alpha)$, where $k = \sqrt{a^2 + b^2}$ and α is the angle for which $\sin \alpha = b/\sqrt{a^2 + b^2}$ and $\cos \alpha = a/\sqrt{a^2 + b^2}$.

EXAMPLE 4 **Rewrite $a \sin x + b \cos x$**

Rewrite $\sin x + \cos x$ in the form $k \sin(x + \alpha)$.

Solution Since $a = 1$, $b = 1$, we have $k = \sqrt{1^2 + 1^2} = \sqrt{2}$, $\sin \alpha = \dfrac{1}{\sqrt{2}}$, and $\cos \alpha = \dfrac{1}{\sqrt{2}}$. Thus $\alpha = 45°$.

$$\sin x + \cos x = k \sin(x + \alpha) = \sqrt{2} \sin(x + 45°)$$

■ *Try Exercise 62, page 387.*

The equation $y = a \sin x + b \cos x$ can be graphed by the addition of ordinates. However, it is easier to graph the equation as a sine function with a phase shift.

EXAMPLE 5 **Use an Identity to Graph a Trigonometric Function**

Graph $y = -\sin x + \sqrt{3} \cos x$.

Solution First we write $f(x)$ as $k \sin(x + \alpha)$. Let $a = -1$ and $b = \sqrt{3}$; then $k = \sqrt{(-1)^2 + (\sqrt{3})^2} = 2$. The point $P(-1, \sqrt{3})$ is in the second quadrant. Let α be an angle with P on its terminal side. Let β be the reference angle for α. Then

$$\sin \beta = \frac{\sqrt{3}}{2}$$

$$\beta = \frac{\pi}{3}$$

$$\alpha = \pi - \beta = \pi - \frac{\pi}{3} = \frac{2\pi}{3}$$

Substituting 2 for k and $\dfrac{2\pi}{3}$ for α in $y = k \sin(x + \alpha)$, we have

$$y = 2 \sin\left(x + \frac{2\pi}{3}\right).$$

The phase shift is $-\dfrac{c}{b} = -\dfrac{\frac{2\pi}{3}}{1} = -\dfrac{2\pi}{3}$. The graph of $y = 2 \sin\left(x + \dfrac{2\pi}{3}\right)$ is the graph of $y = 2 \sin x$ shifted $\dfrac{2\pi}{3}$ units to the left. See Figure 7.4.

■ *Try Exercise 70, page 387.*

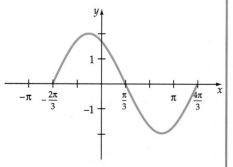

$f(x) = -\sin x + \sqrt{3} \cos x$
Figure 7.4

We now list the identities that have been discussed.

Product-to-Sum Identities

$$\sin \alpha \cos \beta = \frac{1}{2}[\sin(\alpha + \beta) + \sin(\alpha - \beta)]$$

$$\cos \alpha \sin \beta = \frac{1}{2}[\sin(\alpha + \beta) - \sin(\alpha - \beta)]$$

$$\cos \alpha \cos \beta = \frac{1}{2}[\cos(\alpha + \beta) + \cos(\alpha - \beta)]$$

$$\sin \alpha \sin \beta = \frac{1}{2}[\cos(\alpha - \beta) - \cos(\alpha + \beta)]$$

Sum-to-Product Identities

$$\sin x + \sin y = 2 \sin \frac{x + y}{2} \cos \frac{x - y}{2}$$

$$\cos x - \cos y = -2 \sin \frac{x + y}{2} \sin \frac{x - y}{2}$$

$$\sin x - \sin y = 2 \cos \frac{x + y}{2} \sin \frac{x - y}{2}$$

$$\cos x + \cos y = 2 \cos \frac{x + y}{2} \cos \frac{x - y}{2}$$

Sums of the Form $a \sin x + b \cos x$

$$a \sin x + b \cos x = k \sin(x + \alpha),$$

$$\text{where} \quad k = \sqrt{a^2 + b^2}, \quad \sin \alpha = \frac{b}{\sqrt{a^2 + b^2}}, \text{ and } \cos \alpha = \frac{a}{\sqrt{a^2 + b^2}}.$$

EXERCISE SET 7.4 _____

In Exercises 1 to 8, write each expression as the sum or difference of two functions.

1. $2 \sin x \cos 2x$

2. $2 \sin 4x \sin 2x$

3. $\cos 6x \sin 2x$

4. $\cos 3x \cos 5x$

5. $2 \sin 5x \cos 3x$

6. $2 \sin 2x \cos 6x$

7. $\sin x \sin 5x$

8. $\cos 3x \sin x$

In Exercises 9 to 16, evaluate each of the expressions. Do not use a calculator or tables.

9. $\cos 75° \cos 15°$

10. $\sin 105° \cos 15°$

11. $\cos 157.5° \sin 22.5°$

12. $\sin 195° \cos 15°$

13. $\sin \frac{13\pi}{12} \cos \frac{\pi}{12}$

14. $\sin \frac{11\pi}{12} \sin \frac{7\pi}{12}$

15. $\sin \dfrac{\pi}{12} \cos \dfrac{7\pi}{12}$

16. $\cos \dfrac{17\pi}{12} \sin \dfrac{7\pi}{12}$

In Exercises 17 to 32, write each expression as the product of two functions.

17. $\sin 4\theta + \sin 2\theta$

18. $\cos 5\theta - \cos 3\theta$

19. $\cos 3\theta + \cos \theta$

20. $\sin 7\theta - \sin 3\theta$

21. $\cos 6\theta - \cos 2\theta$

22. $\cos 3\theta + \cos 5\theta$

23. $\cos \theta + \cos 7\theta$

24. $\sin 3\theta + \sin 7\theta$

25. $\sin 5\theta + \sin 9\theta$

26. $\cos 5\theta - \cos \theta$

27. $\cos 2\theta - \cos \theta$

28. $\sin 2\theta + \sin 6\theta$

29. $\cos \dfrac{\theta}{2} - \cos \theta$

30. $\sin \dfrac{3\theta}{4} + \sin \dfrac{\theta}{2}$

31. $\sin \dfrac{\theta}{2} - \sin \dfrac{\theta}{3}$

32. $\cos \theta + \cos \dfrac{\theta}{2}$

In Exercises 33 to 48, verify the identities.

33. $2 \cos \alpha \cos \beta = \cos(\alpha + \beta) + \cos(\alpha - \beta)$

34. $2 \sin \alpha \sin \beta = \cos(\alpha - \beta) - \cos(\alpha + \beta)$

35. $2 \cos 3x \sin x = 2 \sin x \cos x - 8 \cos x \sin^3 x$

36. $\sin 5x \cos 3x = \sin 4x \cos 4x + \sin x \cos x$

37. $2 \cos 5x \cos 7x = \cos^2 6x - \sin^2 6x + 2 \cos^2 x - 1$

38. $\sin 3x \cos x = \sin x \cos x(3 - 4 \sin^2 x)$

39. $\sin 3x - \sin x = 2 \sin x - 4 \sin^3 x$

40. $\cos 5x - \cos 3x = -8 \sin^2 x(2 \cos^3 x - \cos x)$

41. $\sin 2x + \sin 4x = 2 \sin x \cos x(4 \cos^2 x - 1)$

42. $\cos 3x + \cos x = 4 \cos^3 x - 2 \cos x$

43. $\dfrac{\sin 3x - \sin x}{\cos 3x - \cos x} = -\cot 2x$

44. $\dfrac{\cos 5x - \cos 3x}{\sin 5x + \sin 3x} = -\tan x$

45. $\dfrac{\sin 5x + \sin 3x}{4 \sin x \cos^3 x - 4 \sin^3 x \cos x} = 2 \cos x$

46. $\dfrac{\cos 4x - \cos 2x}{\sin 2x - \sin 4x} = \tan 3x$

47. $\sin(x + y) \cos(x - y) = \sin x \cos x + \sin y \cos y$

48. $\sin(x + y) \sin(x - y) = \sin^2 x - \sin^2 y$

In Exercises 49 to 58, write the given equation in the form $y = k \sin(x + \alpha)$, where the measure of α is in degrees.

49. $y = -\sin x - \cos x$

50. $y = \sqrt{3} \sin x - \cos x$

51. $y = \dfrac{1}{2} \sin x - \dfrac{\sqrt{3}}{2} \cos x$

52. $y = \dfrac{\sqrt{3}}{2} \sin x - \dfrac{1}{2} \cos x$

53. $y = \dfrac{1}{2} \sin x - \dfrac{1}{2} \cos x$

54. $y = -\dfrac{\sqrt{3}}{2} \sin x - \dfrac{1}{2} \cos x$

55. $y = 8 \sin x + 15 \cos x$

56. $y = -7 \sin x + 24 \cos x$

57. $y = 8 \sin x - 3 \cos x$

58. $y = -4 \sin x + 7 \cos x$

In Exercises 59 to 66, write the given equations in the form $k \sin(x + \alpha)$ where the measure of α is in radians.

59. $y = -\sin x + \cos x$

60. $y = -\sqrt{3} \sin x - \cos x$

61. $y = \dfrac{\sqrt{3}}{2} \sin x + \dfrac{1}{2} \cos x$

62. $y = \sin x + \sqrt{3} \cos x$

63. $y = -4 \sin x + 9 \cos x$

64. $y = -3 \sin x + 5 \cos x$

65. $y = -5 \sin x + 5 \cos x$

66. $y = 3 \sin x - 3 \cos x$

In Exercises 67 to 76, graph one cycle of the equations.

67. $y = -\sin x - \sqrt{3} \cos x$

68. $y = -\sqrt{3} \sin x + \cos x$

69. $y = 2 \sin x + 2 \cos x$

70. $y = \sin x + \sqrt{3} \cos x$

71. $y = -\sqrt{3} \sin x - \cos x$

72. $y = -\sin x + \cos x$

73. $y = 2 \sin x - 5 \cos x$

74. $y = -6 \sin x - 10 \cos x$

75. $y = 3 \sin x - 4 \cos x$

76. $y = -5 \sin x + 9 \cos x$

Graphing Calculator Exercises*

In Exercises 77 to 82, establish the identity by graphing the function on each side of the equation and showing that the graphs are identical.

77. $\sin 3x - \sin x = 2 \sin x - 4 \sin^3 x$

78. $\dfrac{\sin 3x - \sin x}{\cos 3x - \cos x} = -\dfrac{1}{\tan 2x}$

79. $2 \sin x + 2 \cos x = 2\sqrt{2} \sin\left(x + \dfrac{\pi}{4}\right)$

80. $-\sqrt{3} \sin x - \cos x = 2 \sin\left(x - \dfrac{5\pi}{6}\right)$

81. $\dfrac{\sqrt{3}}{2} \sin x - \dfrac{1}{2} \cos x = \sin\left(x - \dfrac{\pi}{6}\right)$

82. $\dfrac{\sqrt{3}}{2} \sin x + \dfrac{1}{2} \cos x = \sin\left(x + \dfrac{\pi}{6}\right)$

*Additional graphing calculator exercises appear in the Graphing Workbook as described in the front of this textbook.

Supplemental Exercises

83. Derive the sum-to-product identity:

$$\cos x + \cos y = 2 \cos \frac{x+y}{2} \cos \frac{x-y}{2}$$

84. Derive the product-to-sum identity:

$$\sin x \sin y = \frac{1}{2}[\cos(x-y) - \cos(x+y)]$$

85. If $x + y = 180°$, show that $\sin x + \sin y = 2 \sin x$.

86. If $x + y = 360°$, show that $\cos x + \cos y = 2 \cos x$.

In Exercises 87 to 92, verify the identities.

87. $\sin 2x + \sin 4x + \sin 6x = 4 \sin 3x \cos 2x \cos x$

88. $\sin 4x - \sin 2x + \sin 6x = 4 \cos 3x \sin 2x \cos x$

89. $\dfrac{\cos 10x + \cos 8x}{\sin 10x - \sin 8x} = \cot x$

90. $\dfrac{\sin 10x + \sin 2x}{\cos 10x + \cos 2x} = \dfrac{2 \tan 3x}{1 - \tan^2 3x}$

91. $\dfrac{\sin 2x + \sin 4x + \sin 6x}{\cos 2x + \cos 4x + \cos 6x} = \tan 4x$

92. $\dfrac{\sin 2x + \sin 6x}{\cos 6x - \cos 2x} = -\cot 2x$

93. Verify that $\cos^2 x - \sin^2 x = \cos 2x$ by using a product-to-sum identity.

94. Verify that $2 \sin x \cos x = \sin 2x$ by using a product-to-sum identity.

95. Verify that $a \sin x + b \cos x = k \cos(x - \alpha)$, where $k = \sqrt{a^2 + b^2}$ and $\tan \alpha = a/b$.

96. Verify that $a \sin cx + b \cos cx = k \sin(cx + \alpha)$, where $k = \sqrt{a^2 + b^2}$ and $\tan \alpha = b/a$.

In Exercises 97 to 102, find the amplitude, phase shift, and period and then graph the equation.

97. $y = \sin \dfrac{x}{2} - \cos \dfrac{x}{2}$

98. $y = -\sqrt{3} \sin \dfrac{x}{2} + \cos \dfrac{x}{2}$

99. $y = \sqrt{3} \sin 2x - \cos 2x$

100. $y = -\sin 2x + \cos 2x$

101. $y = \sin \pi x + \sqrt{3} \cos \pi x$

102. $y = 3 \sin 2\pi x - 4 \cos 2\pi x$

103. Two nonvertical lines intersect in a plane. The slope of l_1 is m_1 and the slope of l_2 is m_2. Show that the tangent of the angle θ formed by the smallest positive angle from line l_1 to line l_2 is given by

$$\tan \theta = \frac{m_2 - m_1}{1 + m_1 m_2}.$$

104. Use the equation from Exercise 103 to find the angle from the line $y = x + 5$ to the line $y = 3x - 4$.

105. Use the equation from Exercise 103 to find the angle from the line $y = -3x/2 - 4$ to the line $y = 2x/3 + 3$.

7.5   Inverse Trigonometric Functions

Because the graph of $y = \sin x$ fails the horizontal line test, it is not the graph of a one-to-one function. Therefore, it does not have an inverse function. Figure 7.5 shows the graph of $y = \sin x$ and the graph of the inverse relation $x = \sin y$. Note that the graph of $x = \sin y$ does not satisfy the vertical line test and therefore is not the graph of a function.

If the domain of $y = \sin x$ is restricted to $-\dfrac{\pi}{2} \leq x \leq \dfrac{\pi}{2}$, the graph of $y = \sin x$ satisfies the horizontal line test and therefore is the graph of a one-to-one function. The graph of $y = \sin x$ and that of $x = \sin y$ are shown in Figure 7.6. To find the inverse of the function defined by $y = \sin x$, with $-\dfrac{\pi}{2} \leq x \leq \dfrac{\pi}{2}$, interchange x and y. Then solve for y.

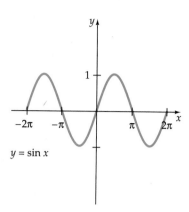

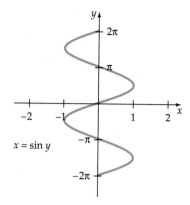

Figure 7.5

$$y = \sin x \quad \bullet \ -\frac{\pi}{2} \leq x \leq \frac{\pi}{2}$$

$$x = \sin y \quad \bullet \ \text{Interchange } x \text{ and } y.$$

$$y = ? \qquad \bullet \ \text{Solve for } y.$$

Unfortunately, there is no algebraic solution for y. Thus we establish new notation and write

$$y = \sin^{-1} x$$

which is read "y is the inverse sine of x." Some textbooks use the notation arcsin x instead of $\sin^{-1} x$.

Caution The -1 in $\sin^{-1} x$ is not an exponent. The -1 is used to denote the inverse function. To use -1 as an exponent for a sine function, enclose the function in parentheses.

$$(\sin x)^{-1} = \frac{1}{\sin x} = \csc x \qquad \sin^{-1} x \neq \frac{1}{\sin x}$$

Definition of $\sin^{-1} x$

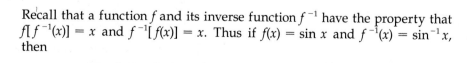

$$y = \sin^{-1} x \text{ if and only if } x = \sin y$$

$$\text{where } -1 \leq x \leq 1 \text{ and } -\frac{\pi}{2} \leq y \leq \frac{\pi}{2}$$

Recall that a function f and its inverse function f^{-1} have the property that $f[f^{-1}(x)] = x$ and $f^{-1}[f(x)] = x$. Thus if $f(x) = \sin x$ and $f^{-1}(x) = \sin^{-1} x$, then

$$f[f^{-1}(x)] = \sin(\sin^{-1} x) = x, \quad \text{where } -1 \leq x \leq 1$$

$$f^{-1}[f(x)] = \sin^{-1}(\sin x) = x, \quad \text{where } -\frac{\pi}{2} \leq x \leq \frac{\pi}{2}$$

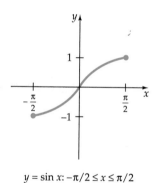

$y = \sin x \colon -\pi/2 \leq x \leq \pi/2$

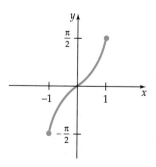

$x = \sin y \colon -\pi/2 \leq y \leq \pi/2$

Figure 7.6

It is convenient to think of the value of an inverse trigonometric function as an angle. For example, $y = \sin^{-1}(\frac{1}{2})$ is the angle in the interval $[-\frac{\pi}{2}, \frac{\pi}{2}]$ whose sine is $\frac{1}{2}$.

EXAMPLE 1   **Evaluate an Inverse Sine Function**

Find the exact value of $y = \sin^{-1}\left(-\dfrac{\sqrt{3}}{2}\right)$.

Solution

$$y = \sin^{-1}\left(-\frac{\sqrt{3}}{2}\right) \qquad \bullet \ -\frac{\pi}{2} \le y \le \frac{\pi}{2}$$

$$-\frac{\sqrt{3}}{2} = \sin y \qquad \bullet \ y = \sin^{-1}x \text{ if and only if } x = \sin y$$

$$-\frac{\pi}{3} = y \qquad \bullet \ \sin\left(-\frac{\pi}{3}\right) = -\frac{\sqrt{3}}{2}$$

■ *Try Exercise* **2,** *page 397.*

Because the graph of $y = \cos x$ fails the horizontal line test, it is not the graph of a one-to-one function. Therefore, it does not have an inverse function. Figure 7.7 shows the graph of $y = \cos x$ and the graph of the inverse relation $x = \cos y$. Note that the graph of $x = \cos y$ does not satisfy the vertical line test and therefore is not the graph of a function.

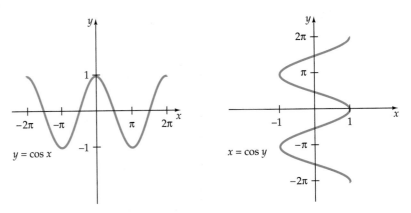

Figure 7.7

If the domain of $y = \cos x$ is restricted to $0 \le x \le \pi$, the graph of $y = \cos x$ satisfies the horizontal line test and therefore is the graph of a one-to-one function. The graph of $y = \cos x$ and that of $x = \cos y$ are shown in Figure 7.8. To find the inverse of the function defined by $y = \cos x$, with $0 \le x \le \pi$, interchange x and y. Then solve for y.

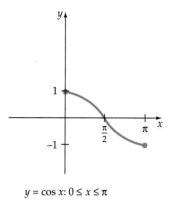

$y = \cos x : 0 \le x \le \pi$

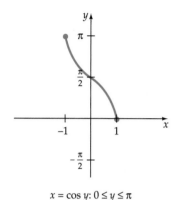

$x = \cos y : 0 \le y \le \pi$

Figure 7.8

$y = \cos x \quad \bullet \ 0 \le x \le \pi$

$x = \cos y \quad \bullet$ Interchange x and y.

$y = ? \quad \bullet$ Solve for y.

As in the case for the inverse sine function, there is no algebraic solution for y. Thus the notation for the inverse cosine function becomes $y = \cos^{-1}x$. We can write the following definition of the inverse cosine function.

Definition of $\cos^{-1}x$

$$y = \cos^{-1}x \text{ if and only if } x = \cos y$$

$$\text{where } -1 \le x \le 1 \text{ and } 0 \le y \le \pi$$

A function f and its inverse f^{-1} have the property that $f[f^{-1}(x)] = x$ and $f^{-1}[f(x)] = x$. Thus, if $f(x) = \cos x$ and $f^{-1}(x) = \cos^{-1}x$, then

$$f[f^{-1}(x)] = \cos(\cos^{-1}x) = x, \quad \text{where } -1 \le x \le 1$$
$$f^{-1}[f(x)] = \cos^{-1}(\cos x) = x, \quad \text{where } 0 \le x \le \pi$$

EXAMPLE 2   **Evaluate an Inverse Cosine Function**

Evaluate $y = \cos^{-1}\left(\cos \dfrac{5\pi}{4}\right)$.

Solution   Recall that $\cos^{-1}(\cos x) = x$ when $0 \le x \le \pi$. Because $\dfrac{5\pi}{4}$ is not in this interval, we must first evaluate $\cos \dfrac{5\pi}{4}$.

$$y = \cos^{-1}\left(\cos \frac{5\pi}{4}\right) = \cos^{-1}\left(-\frac{\sqrt{2}}{2}\right) \quad \bullet \ \cos \frac{5\pi}{4} = -\frac{\sqrt{2}}{2}$$

$$-\frac{\sqrt{2}}{2} = \cos y, \quad 0 \le y \le \pi \qquad \bullet \ y = \cos^{-1}x \text{ if and only if}$$
$$\qquad\qquad\qquad\qquad\qquad\qquad\qquad x = \cos y.$$

$$\frac{3\pi}{4} = y \qquad\qquad\qquad\qquad \bullet \ y \text{ is the angle whose cosine is } -\frac{\sqrt{2}}{2}.$$

■ *Try Exercise* **44**, *page 397*.

Because the graphs of $y = \tan x$, $y = \csc x$, $y = \sec x$, and $y = \cot x$ fail the horizontal line test, these functions are not one-to-one functions. Therefore, these functions do not have inverse functions. If the domains of each of these functions are restricted in a certain way, however, the graphs will satisfy the horizontal line test. Thus each of these functions will have an inverse function over a restricted domain. Table 7.2 shows the restricted function and the inverse function for $\tan x$, $\csc x$, $\sec x$, and $\cot x$.

TABLE 7.2

	$y = \tan x$	$y = \tan^{-1}x$	$y = \csc x$	$y = \csc^{-1}x$
Domain	$-\dfrac{\pi}{2} < x < \dfrac{\pi}{2}$	$-\infty < x < \infty$	$-\dfrac{\pi}{2} \le x \le \dfrac{\pi}{2},$ $x \ne 0$	$x \le -1$ or $x \ge 1$
Range	$-\infty < y < \infty$	$-\dfrac{\pi}{2} < y < \dfrac{\pi}{2}$	$y \le -1,$ or $y \ge 1$	$-\dfrac{\pi}{2} \le y \le \dfrac{\pi}{2},$ $y \ne 0$
Asymptotes	$x = -\dfrac{\pi}{2}, x = \dfrac{\pi}{2}$	$y = -\dfrac{\pi}{2}, y = \dfrac{\pi}{2}$	$x = 0$	$y = 0$

	$y = \sec x$	$y = \sec^{-1}x$	$y = \cot x$	$y = \cot^{-1}x$
Domain	$0 \le x \le \pi,$ $x \ne \dfrac{\pi}{2}$	$x \le -1$ or $x \ge 1$	$0 < x < \pi$	$-\infty < x < \infty$
Range	$y \le -1$ or $y \ge 1$	$0 \le y \le \pi,$ $y \ne \dfrac{\pi}{2}$	$-\infty < y < \infty$	$0 < y < \pi$
Asymptotes	$x = \dfrac{\pi}{2}$	$y = \dfrac{\pi}{2}$	$x = 0, x = \pi$	$y = 0, y = \pi$

The choice of the ranges for $y = \sec^{-1} x$ and $y = \csc^{-1} x$ is not universally accepted. For example, some calculus texts use $[0, \frac{\pi}{2}) \cup [\pi, \frac{3\pi}{2})$ as the range of $y = \sec^{-1} x$. This definition has some advantages and some disadvantages that are explained in more advanced mathematics courses.

EXAMPLE 3 **Evaluate an Inverse Tangent Function**

Find the exact value of $y = \tan^{-1}\left(-\dfrac{\sqrt{3}}{3}\right)$.

Solution

$$y = \tan^{-1}\left(-\frac{\sqrt{3}}{3}\right) \quad \bullet \; -\frac{\pi}{2} < y < \frac{\pi}{2}$$

$$-\frac{\sqrt{3}}{3} = \tan y \qquad \bullet \; y = \tan^{-1}x \text{ if and only if } x = \tan y$$

$$-\frac{\pi}{6} = y \qquad \bullet \; \tan\left(-\frac{\pi}{6}\right) = -\frac{\sqrt{3}}{3}$$

■ *Try Exercise* **62**, *page 397.*

A calculator may not have keys for the inverse secant, cosecant, and cotangent functions. The following procedure shows an identity for the inverse cosecant function in terms of the inverse sine function. If we need to determine y, which is the angle (or number) whose cosecant is x, we can rewrite $y = \csc^{-1}x$ as follows.

$$y = \csc^{-1}x \quad \bullet \text{ Domain: } x \le -1 \text{ or } x \ge 1$$
$$\bullet \text{ Range: } -\pi/2 \le y \le \pi/2, \, y \ne 0$$

$$\csc y = x \qquad \bullet \text{ Definition of inverse function}$$

$$\frac{1}{\sin y} = x \qquad \bullet \text{ Substitute } 1/\sin x \text{ for } \csc y.$$

$$\sin y = \frac{1}{x} \qquad \bullet \text{ Solve for } \sin y.$$

$$y = \sin^{-1}\frac{1}{x}$$

$$\csc^{-1}x = \sin^{-1}\frac{1}{x}$$

Thus $\csc^{-1}x$ is the same as $\sin^{-1}\dfrac{1}{x}$. There are similar expressions for $\sec^{-1}x$ and $\cot^{-1}x$.

$$\sec^{-1}x = \cos^{-1}\frac{1}{x} \qquad \cot^{-1}x = \begin{cases} \tan^{-1}\dfrac{1}{x}, \, x > 0 \\[2ex] \tan^{-1}\dfrac{1}{x} + \pi, \, x < 0 \end{cases}$$

EXAMPLE 4   **Evaluate an Inverse Trigonometric Expression**

Evaluate $\sin\left[\cos^{-1}\left(-\dfrac{3}{5}\right)\right]$.

Solution   Let $y = \cos^{-1}\left(-\dfrac{3}{5}\right)$; then

$$\cos y = -\frac{3}{5} \quad \text{and} \quad \sin\left[\cos^{-1}\left(-\frac{3}{5}\right)\right] = \sin y$$

The goal is to find $\sin y$ knowing $\cos y = -\dfrac{3}{5}$. Thus we need an identity that relates $\sin y$ and $\cos y$.

$$\sin^2 y + \cos^2 y = 1$$

$$\sin^2 y = 1 - \cos^2 y$$

$$= 1 - \left(-\frac{3}{5}\right)^2 = \frac{16}{25} \quad \bullet \text{ Solve for } y$$

$$\sin y = \pm\sqrt{\frac{16}{25}} = \pm\frac{4}{5}$$

Because $\cos y = -\dfrac{3}{5}$, we know $\dfrac{\pi}{2} < y < \pi$ and thus $\sin y > 0$. Therefore,

$$\sin y = \frac{4}{5}$$

■ *Try Exercise* **70**, *page 397.*

EXAMPLE 5   **Evaluate an Inverse Trigonometric Expression**

Evaluate $\sin\left(\sin^{-1}\dfrac{3}{5} + \cos^{-1}\dfrac{5}{13}\right)$.

Solution   Let $x = \sin^{-1}\dfrac{3}{5}$ and let $y = \cos^{-1}\dfrac{5}{13}$; then $\sin x = \dfrac{3}{5}$ and $\cos y = \dfrac{5}{13}$ with $0 < x < \dfrac{\pi}{2}$ and $0 < y < \dfrac{\pi}{2}$. Therefore,

$$\cos x = \sqrt{1 - \sin^2 x} = \sqrt{1 - \left(\frac{3}{5}\right)^2} = \frac{4}{5}$$

$$\sin y = \sqrt{1 - \cos^2 y} = \sqrt{1 - \left(\frac{5}{13}\right)^2} = \frac{12}{13}$$

$$\sin\left(\sin^{-1}\frac{3}{5} + \cos^{-1}\frac{5}{13}\right) = \sin(x + y) = \sin x \cos y + \cos x \sin y$$

$$= \frac{3}{5}\cdot\frac{5}{13} + \frac{4}{5}\cdot\frac{12}{13}$$

$$= \frac{15}{65} + \frac{48}{65} = \frac{63}{65}$$

■ *Try Exercise 84, page 398.*

EXAMPLE 6   **Solve an Inverse Trigonometric Equation**

Solve the inverse trigonometric equation $\sin^{-1}\frac{3}{5} + \cos^{-1}x = \pi$.

  Solution   Solve for $\cos^{-1}x$, and then take the cosine of both sides of the equation.

$$\sin^{-1}\frac{3}{5} + \cos^{-1}x = \pi$$

$$\cos^{-1}x = \pi - \sin^{-1}\frac{3}{5}$$

$$\cos(\cos^{-1}x) = \cos\left(\pi - \sin^{-1}\frac{3}{5}\right)$$

$$x = \cos(\pi - \alpha) \qquad\qquad \bullet \text{ Let } \alpha = \sin^{-1}\frac{3}{5}.$$

$$= \cos\pi\cos\alpha + \sin\pi\sin\alpha$$

$$= (-1)\cos\alpha + (0)\sin\alpha$$

$$= -\cos\alpha$$

$$= -\frac{4}{5} \qquad\qquad \bullet \cos\alpha = \frac{4}{5} \text{ (see Figure 7.9)}$$

■ *Try Exercise 94, page 398.*

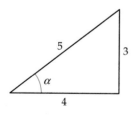

Figure 7.9

EXAMPLE 7   **Verify an Inverse Trigonometric Identity**

Verify the identity $\sin^{-1}x + \cos^{-1}x = \frac{\pi}{2}$.

  Solution   Let $\alpha = \sin^{-1}x$ and $\beta = \cos^{-1}x$, which implies that $\sin\alpha = x$ and $\cos\beta = x$. Then

$$\cos\alpha = \sqrt{1 - \sin^2\alpha} = \sqrt{1 - x^2} \quad \text{and} \quad \sin\beta = \sqrt{1 - \cos^2\beta} = \sqrt{1 - x^2}$$

Working with the left side of the identity produces

$$\sin^{-1}x + \cos^{-1}x = \alpha + \beta$$

$$= \sin^{-1}[\sin(\alpha + \beta)]$$

$$= \sin^{-1}[\sin \alpha \cos \beta + \cos \alpha \sin \beta]$$

$$= \sin^{-1}[x \cdot x + \sqrt{1 - x^2}\sqrt{1 - x^2}]$$

$$= \sin^{-1}[x^2 + 1 - x^2]$$

$$= \sin^{-1}1$$

$$= \frac{\pi}{2}$$

■ *Try Exercise* **102**, *page 398.*

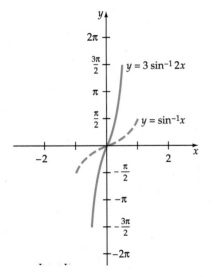

Figure 7.10

The inverse trigonometric functions can be graphed by using the procedures of stretching, shrinking, and translations that have been developed in previous chapters. Say we want to graph $y = 3 \sin^{-1}2x$. Because the domain of $y = \sin^{-1}x$ is all values of x such that $-1 \le x \le 1$, the domain of $y = 3 \sin^{-1}2x$ is

$$-1 \le 2x \le 1 \quad \text{or} \quad -\frac{1}{2} \le x \le \frac{1}{2}$$

Because $-\frac{\pi}{2} \le \sin^{-1}2x \le \frac{\pi}{2}$, the range of $3 \sin^{-1}2x$ is

$$-\frac{3\pi}{2} \le 3 \sin^{-1}2x \le \frac{3\pi}{2}$$

In Figure 7.10, we show the graph of $y = \sin^{-1}x$ as a dotted curve. The graph of $y = 3 \sin^{-1}2x$ is shown with the domain and range adjusted according to the above calculations.

EXAMPLE 8   **Graph an Inverse Cosine Function**

Graph $y = 0.5 \cos^{-1}(x + 2)$.

Solution   The graph of $y = 0.5 \cos^{-1}x$ is the graph of $y = \cos^{-1}x$ with each y-coordinate $\frac{1}{2}$ of its previous value. The graph of $y = 0.5 \cos^{-1}(x + 2)$ is the graph of $y = 0.5 \cos^{-1}x$ shifted left horizontally 2 units. See Figure 7.11.

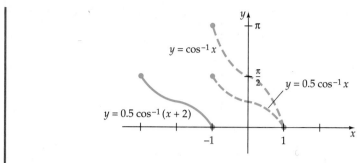

Figure 7.11

■ *Try Exercise* **106,** *page 398.*

EXERCISE SET 7.5

In Exercises 1 to 12, find the exact radian value.

1. $\sin^{-1}1$

2. $\sin^{-1}\dfrac{\sqrt{2}}{2}$

3. $\cos^{-1}\left(-\dfrac{\sqrt{3}}{2}\right)$

4. $\cos^{-1}\left(-\dfrac{1}{2}\right)$

5. $\tan^{-1}(-1)$

6. $\tan^{-1}\sqrt{3}$

7. $\cot^{-1}\dfrac{\sqrt{3}}{3}$

8. $\cot^{-1}1$

9. $\sec^{-1}2$

10. $\sec^{-1}\dfrac{2\sqrt{3}}{3}$

11. $\csc^{-1}(-\sqrt{2})$

12. $\csc^{-1}(-2)$

In Exercises 13 to 20, find the exact degree value.

13. $\sin^{-1}\left(-\dfrac{\sqrt{3}}{2}\right)$

14. $\sin^{-1}\dfrac{1}{2}$

15. $\cos^{-1}\left(-\dfrac{1}{2}\right)$

16. $\cos^{-1}\dfrac{\sqrt{3}}{2}$

17. $\tan^{-1}\dfrac{\sqrt{3}}{3}$

18. $\tan^{-1}1$

19. $\cot^{-1}\sqrt{3}$

20. $\cot^{-1}(-1)$

In Exercises 21 to 28, find the approximate radian value to 4 significant digits.

21. $\sin^{-1}0.4555$

22. $\sin^{-1}0.8700$

23. $\cos^{-1}(-0.2357)$

24. $\cos^{-1}(-0.1298)$

25. $\tan^{-1}(-1.4344)$

26. $\tan^{-1}(-5.2691)$

27. $\cot^{-1}0.9823$

28. $\cot^{-1}4.2317$

In Exercises 29 to 40, find the approximate degree value to the nearest tenth of a degree.

29. $\sin^{-1}(-0.2781)$

30. $\sin^{-1}(-0.9650)$

31. $\cos^{-1}0.5555$

32. $\cos^{-1}0.1598$

33. $\tan^{-1}(-2.0440)$

34. $\tan^{-1}10.0050$

35. $\cot^{-1}(-0.9752)$

36. $\cot^{-1}1.0578$

37. $\sec^{-1}(-3.4785)$

38. $\sec^{-1}9.4455$

39. $\csc^{-1}1.0056$

40. $\csc^{-1}(-10.9856)$

In Exercises 41 to 60, evaluate the given expression.

41. $\cos\left(\cos^{-1}\dfrac{1}{2}\right)$

42. $\cos(\cos^{-1}2)$

43. $\tan(\tan^{-1}2)$

44. $\tan\left(\tan^{-1}\dfrac{1}{2}\right)$

45. $\sin\left(\tan^{-1}\dfrac{3}{4}\right)$

46. $\cos\left(\sin^{-1}\dfrac{5}{13}\right)$

47. $\tan\left(\sin^{-1}\dfrac{\sqrt{2}}{2}\right)$

48. $\sin\left[\cos^{-1}\left(-\dfrac{\sqrt{3}}{2}\right)\right]$

49. $\cos[\sec^{-1}2]$

50. $\sin^{-1}(\sin 2)$

51. $\sin^{-1}\left(\sin\dfrac{\pi}{6}\right)$

52. $\sin^{-1}\left(\sin\dfrac{5\pi}{6}\right)$

53. $\cos^{-1}\left(\sin\dfrac{\pi}{4}\right)$

54. $\cos^{-1}\left(\cos\dfrac{5\pi}{4}\right)$

55. $\sin^{-1}\left(\tan\dfrac{\pi}{3}\right)$

56. $\cos^{-1}\left(\tan\dfrac{2\pi}{3}\right)$

57. $\tan^{-1}\left(\sin\dfrac{\pi}{6}\right)$

58. $\cot^{-1}\left(\cos\dfrac{2\pi}{3}\right)$

59. $\sin^{-1}\left[\cos\left(-\dfrac{2\pi}{3}\right)\right]$

60. $\cos^{-1}\left[\tan\left(-\dfrac{\pi}{3}\right)\right]$

In Exercises 61 to 96, solve each equation.

61. $y = \sin^{-1}\left(-\dfrac{\sqrt{3}}{2}\right)$

62. $y = \tan^{-1}\dfrac{\sqrt{3}}{3}$

63. $y = \cos^{-1}(-0.5669)$

64. $y = \csc^{-1}(2.3033)$

65. $y = \cot^{-1}(-1.0886)$

66. $y = \sec^{-1}(-2.9071)$

67. $y = \cos^{-1}\dfrac{\pi}{4}$

68. $y = \tan^{-1}\left(-\dfrac{\pi}{3}\right)$

69. $y = \cos\left(\sin^{-1}\dfrac{7}{25}\right)$

70. $y = \tan\left(\cos^{-1}\dfrac{3}{5}\right)$

71. $y = \sec\left(\tan^{-1}\dfrac{12}{5}\right)$ **72.** $y = \csc\left(\sin^{-1}\dfrac{12}{13}\right)$

73. $y = \sin^{-1}\left(\sin\dfrac{2\pi}{3}\right)$ **74.** $y = \tan^{-1}\left(\tan\dfrac{5\pi}{4}\right)$

75. $y = \cos^{-1}\left[\cos\left(-\dfrac{\pi}{6}\right)\right]$ **76.** $y = \sin^{-1}\left(\sin\dfrac{5\pi}{3}\right)$

77. $y = \tan^{-1}\left(\tan\dfrac{3\pi}{4}\right)$ **78.** $y = \cos^{-1}\left(\cos\dfrac{5\pi}{6}\right)$

79. $y = \cos\left(2\sin^{-1}\dfrac{\sqrt{2}}{2}\right)$ **80.** $y = \tan\left(2\sin^{-1}\dfrac{\sqrt{3}}{2}\right)$

81. $y = \sin\left(2\sin^{-1}\dfrac{4}{5}\right)$ **82.** $y = \cos(2\tan^{-1}1)$

83. $\sin^{-1}x = \cos^{-1}\dfrac{5}{13}$ **84.** $\tan^{-1}x = \sin^{-1}\dfrac{24}{25}$

85. $\sin^{-1}(y-1) = \dfrac{\pi}{2}$ **86.** $\cos^{-1}\left(y - \dfrac{1}{2}\right) = \dfrac{\pi}{3}$

87. $\tan^{-1}\left(y + \dfrac{\sqrt{2}}{2}\right) = \dfrac{\pi}{4}$ **88.** $\sin^{-1}(y-2) = -\dfrac{\pi}{6}$

89. $y = \sin\left(\sin^{-1}\dfrac{2}{3} + \cos^{-1}\dfrac{1}{2}\right)$

90. $y = \cos\left(\sin^{-1}\dfrac{3}{4} + \cos^{-1}\dfrac{5}{13}\right)$

91. $y = \tan\left(\cos^{-1}\dfrac{1}{2} - \sin^{-1}\dfrac{3}{4}\right)$

92. $y = \sec\left(\cos^{-1}\dfrac{2}{3} + \sin^{-1}\dfrac{2}{3}\right)$

93. $\sin^{-1}\dfrac{3}{5} + \cos^{-1}x = \dfrac{\pi}{4}$ **94.** $\sin^{-1}x + \cos^{-1}\dfrac{4}{5} = \dfrac{\pi}{6}$

95. $\sin^{-1}\dfrac{\sqrt{2}}{2} - \cos^{-1}x = \dfrac{2\pi}{3}$

96. $\cos^{-1}x + \sin^{-1}\dfrac{\sqrt{3}}{2} = \dfrac{\pi}{2}$

In Exercises 97 to 100, evaluate each expression.

97. $y = \cos(\sin^{-1}x)$ **98.** $y = \tan(\cos^{-1}x)$

99. $y = \sin(\sec^{-1}x)$ **100.** $y = \sec(\sin^{-1}x)$

In Exercises 101 to 104, verify the identity.

101. $\sin^{-1}x + \sin^{-1}(-x) = 0$

102. $\cos^{-1}x + \cos^{-1}(-x) = \pi$

103. $\tan^{-1}x + \tan^{-1}\dfrac{1}{x} = \dfrac{\pi}{2}, x > 0$

104. $\sec^{-1}\dfrac{1}{x} + \csc^{-1}\dfrac{1}{x} = \dfrac{\pi}{2}$

In Exercises 105 to 112, graph each equation.

105. $y = 2\sin^{-1}x$ **106.** $y = \dfrac{1}{2}\sin^{-1}x$

107. $y = 2\cos^{-1}\dfrac{x}{2}$ **108.** $y = \dfrac{1}{2}\cos^{-1}2x$

109. $y = 2\sin^{-1}(x-2)$ **110.** $y = 3\sin^{-1}(x+1)$

111. $y = 2\cos^{-1}(x+3)$ **112.** $y = \dfrac{1}{3}\cos^{-1}(x-2)$

Graphing Calculator Exercises*

113. Graph $f(x) = \cos^{-1}x$ and $g(x) = \sin^{-1}\sqrt{1-x^2}$ on the same coordinate axes. Does $f(x) = g(x)$ on the interval $[-1, 1]$?

114. Graph $y = \cos(\cos^{-1}x)$ on $[-1, 1]$. Graph $y = \cos^{-1}(\cos x)$ on $[0, \pi]$.

In Exercises 115 to 122, graph the equations.

115. $y = \csc^{-1}2x$ **116.** $y = 0.5\sec^{-1}\dfrac{x}{2}$

117. $y = \sec^{-1}(x-1)$ **118.** $y = \sec^{-1}(x+\pi)$

119. $y = 2\tan^{-1}2x$ **120.** $y = \tan^{-1}(x-1)$

121. $y = \cot^{-1}\dfrac{x}{3}$ **122.** $y = 2\cot^{-1}(x-1)$

Supplemental Exercises

In Exercises 123 to 126, verify the identities.

123. $\cos(\sin^{-1}x) = \sqrt{1-x^2}$ **124.** $\sec(\sin^{-1}x) = \dfrac{\sqrt{1-x^2}}{1-x^2}$

125. $\tan(\csc^{-1}x) = \dfrac{\sqrt{x^2-1}}{x^2-1}, x > 1$

126. $\sin(\cot^{-1}x) = \dfrac{\sqrt{x^2+1}}{x^2+1}$

In Exercises 127 to 130, solve for y in terms of x.

127. $5x = \tan^{-1}3y$ **128.** $2x = \dfrac{1}{2}\sin^{-1}2y$

129. $x - \dfrac{\pi}{3} = \cos^{-1}(y-3)$ **130.** $x + \dfrac{\pi}{2} = \tan^{-1}(2y-1)$

In Exercises 131 and 132, use the following formula. In dot-matrix printing, the *blank-area factor* is the ratio of the blank

*Additional graphing calculator exercises appear in the Graphing Workbook as described in the front of this textbook.

area (unprinted area) to the total area of the line. If circular dots are used to print, then the blank-area factor is given by

$$\frac{A}{SD} = 1 - \frac{1}{2}\left[1 - \left(\frac{S}{D}\right)^2 + \frac{D}{S}\sin^{-1}\left(\frac{S}{D}\right)\right]$$

where $A = A_1 + A_2$, with A_1 and A_2 the areas of the regions shown in the figure, S equals the distance between centers of overlapping dots, and D is the diameter of a dot.

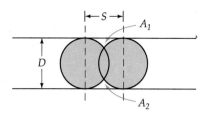

131. Calculate the blank-area factor where $D = 0.2$ millimeters and $S = 0.1$ millimeters.

132. Calculate the blank-area factor where $D = 0.16$ millimeters and $S = 0.1$ millimeters.

7.6 Trigonometric Equations

Consider the equation $\sin x = 1/2$. The graph of $y = \sin x$ along with the line $y = 1/2$ is shown in Figure 7.12. The intersections of the two graphs are the solutions of $\sin x = 1/2$. The solutions in the interval $0 \le x < 2\pi$ are $x = \pi/6$ and $5\pi/6$.

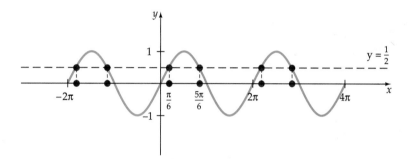

Figure 7.12

If we remove the restriction $0 \le x < 2\pi$, there are many more possible solutions. Because the sine function is periodic with a period of 2π, other solutions are obtained by adding $2k\pi$, k an integer, to either of the above solutions. Thus, the solutions of $\sin x = 1/2$ are

$$x = \frac{\pi}{6} + 2k\pi, \quad k \text{ an integer}$$

$$x = \frac{5\pi}{6} + 2k\pi, \quad k \text{ an integer}$$

The identity for $\sin(x + y)$ can be used to check these solutions.

$$\sin\left(\frac{\pi}{6} + 2k\pi\right) = \sin\frac{\pi}{6}\cos 2k\pi + \cos\frac{\pi}{6}\sin 2k\pi$$

$$= \sin\frac{\pi}{6}(1) + \cos\frac{\pi}{6}(0) = \sin\frac{\pi}{6} = \frac{1}{2}$$

A similar check will show that $x = 5\pi/6 + 2k\pi$ are also solutions.

Algebraic methods and trigonometric identities are used frequently to find the solutions of trigonometric equations. Algebraic methods that are often employed are solving by factoring, solving by using the quadratic formula, and squaring each side of the equation.

EXAMPLE 1   **Solve a Trigonometric Equation by Factoring**

Solve $2 \sin^2 x \cos x - \cos x = 0$ where $0 \le x < 2\pi$.

Solution

$2 \sin^2 x \cos x - \cos x = 0$ • Factor $\cos x$ from each term.

$\cos x(2 \sin^2 x - 1) = 0$ • Use the Principle of Zero Products

$\cos x = 0 \quad \text{or} \quad 2 \sin^2 x - 1 = 0$

$$x = \frac{\pi}{2}, \frac{3\pi}{2} \qquad \sin^2 x = \frac{1}{2}$$

$$\sin x = \pm\frac{\sqrt{2}}{2}$$

$$x = \frac{\pi}{4}, \frac{3\pi}{4}, \frac{5\pi}{4}, \frac{7\pi}{4}$$

The solutions in the interval $0 \le x < 2\pi$ are $\dfrac{\pi}{4}, \dfrac{\pi}{2}, \dfrac{3\pi}{4}, \dfrac{5\pi}{4}, \dfrac{3\pi}{2}$, and $\dfrac{7\pi}{4}$.

■ *Try Exercise* **14**, *page 403.*

Squaring both sides of an equation may not produce an equivalent equation. Thus, when this method is used, the proposed solutions must be checked to eliminate any extraneous solutions.

EXAMPLE 2   **Solve a Trigonometric Equation by Squaring**

Solve $\sin x + \cos x = 1$ where $0 \le x < 2\pi$.

Solution

$\sin x + \cos x = 1$ • Solve for $\sin x$.

$\sin x = 1 - \cos x$

$\sin^2 x = (1 - \cos x)^2$ • Square each side.

$\sin^2 x = 1 - 2 \cos x + \cos^2 x$

$1 - \cos^2 x = 1 - 2 \cos x + \cos^2 x$ • $\sin^2 x = 1 - \cos^2 x$

$2 \cos^2 x - 2 \cos x = 0$

Factor the left side and set each factor equal to zero.

$$2 \cos x(\cos x - 1) = 0$$

$$2 \cos x = 0 \quad \text{or} \quad \cos x = 1$$

$$x = \frac{\pi}{2}, \frac{3\pi}{2} \qquad x = 0$$

Squaring each side of an equation may introduce extraneous solutions. Therefore, we must check the solutions. A check will show that 0 and $\frac{\pi}{2}$ are solutions but $\frac{3\pi}{2}$ is not a solution.

_____ ■ *Try Exercise **52**, page 403.*

EXAMPLE 3 **Solve a Trigonometric Equation by Using the Quadratic Formula**

Solve the equation $3 \cos^2 x - 5 \cos x - 4 = 0, 0° \le x < 360°$.

Solution This equation is quadratic in form and cannot be factored easily. However, we can use the quadratic formula to solve for $\cos x$.

$$3 \cos^2 x - 5 \cos x - 4 = 0 \qquad \bullet \ a = 3, b = -5, c = -4.$$

$$\cos x = \frac{-(-5) \pm \sqrt{(-5)^2 - 4(3)(-4)}}{(2)(3)}$$

$$= \frac{5 \pm \sqrt{73}}{6}$$

The equation $\cos x = \dfrac{5 + \sqrt{73}}{6}$ has no solution. Why?[1] Thus

$$\cos x = \frac{5 - \sqrt{73}}{6}$$

$$x \approx 126.2° \text{ or } 233.8°$$

_____ ■ *Try Exercise **56**, page 403.*

Caution The condition $0° \le x < 360°$ requires that you list both 126.2° and 233.8° as solutions of the equation.

When solving equations that contain multiple angles, we must be sure all the solutions of the equation are found for the given interval. For example, find all solutions of $\sin 2x = \frac{1}{2}$ where $0 \le x < 2\pi$. We first solve for $2x$.

$$\sin 2x = \frac{1}{2}$$

$$2x = \frac{\pi}{6} + 2k\pi \quad \text{or} \quad 2x = \frac{5\pi}{6} + 2k\pi \qquad \bullet \ k \text{ is an integer.}$$

[1] The range of the cosine function is $-1 \le \cos x \le 1$. The value $(5 + \sqrt{73})/6$ is not in the range of the function.

Solving for x, we have $x = \pi/12 + k\pi$ or $x = 5\pi/12 + k\pi$. Substituting integers for k, we obtain

$$k = 0: \qquad x = \frac{\pi}{12} \quad \text{or} \quad x = \frac{5\pi}{12}$$

$$k = 1: \qquad x = \frac{13\pi}{12} \quad \text{or} \quad x = \frac{17\pi}{12}$$

$$k = 2: \qquad x = \frac{25\pi}{12} \quad \text{or} \quad x = \frac{29\pi}{12}$$

Note that for $k \geq 2$, $x \geq 2\pi$ and the solutions to $\sin 2x = \frac{1}{2}$ are not in the interval $0 \leq x < 2\pi$. Thus for $0 \leq x < 2\pi$ the solutions are $\pi/12$, $5\pi/12$, $13\pi/12$, and $17\pi/12$.

EXAMPLE 4   Solve a Trigonometric Equation

Solve $\sin 3x = 1$.

Solution   The equation $\sin 3x = 1$ implies

$$3x = \frac{\pi}{2} + 2k\pi, \quad k \text{ an integer}$$

$$x = \frac{\pi}{6} + \frac{2k\pi}{3}, \quad k \text{ an integer}$$

Because there are no restrictions on x, the solutions are $\frac{\pi}{6} + \frac{2k\pi}{3}$, k an integer.

——————————————————————— ■ *Try Exercise* **66**, *page 404.*

EXAMPLE 5   Solve a Trigonometric Equation

Solve $\sin^2 2x - \dfrac{\sqrt{3}}{2} \sin 2x + \sin 2x - \dfrac{\sqrt{3}}{2} = 0$, $0° \leq x < 360°$.

Solution   Factor the left side of the equation by grouping, and then set each factor equal to zero.

$$\sin^2 2x - \frac{\sqrt{3}}{2} \sin 2x + \sin 2x - \frac{\sqrt{3}}{2} = 0$$

$$\sin 2x \left(\sin 2x - \frac{\sqrt{3}}{2} \right) + \left(\sin 2x - \frac{\sqrt{3}}{2} \right) = 0$$

$$(\sin 2x + 1) \left(\sin 2x - \frac{\sqrt{3}}{2} \right) = 0$$

$$\sin 2x + 1 = 0 \quad \text{or} \quad \sin 2x - \frac{\sqrt{3}}{2} = 0$$

$$\sin 2x = -1 \qquad \qquad \sin 2x = \frac{\sqrt{3}}{2}$$

The equation $\sin 2x = -1$ implies that $2x = 270° + 360° \cdot k$, k an integer. Thus $x = 135° + 180° \cdot k$. The solutions of this equation with $0° \le x < 360°$ are 135° and 315°. Similarly, the equation $\sin 2x = \dfrac{\sqrt{3}}{2}$ implies

$$2x = 60° + 360° \cdot k \quad \text{or} \quad 2x = 120° + 360° \cdot k$$

$$x = 30° + 180° \cdot k \qquad\qquad x = 60° + 180° \cdot k$$

The solutions with $0° \le x < 360°$ are 30°, 60°, 210°, 240°. Combining the solutions from each equation, we have 30°, 60°, 135°, 210°, 240°, and 315° as our solutions.

■ *Try Exercise* **84**, *page 404.*

EXERCISE SET 7.6

In Exercises 1 to 22, solve the equation for all values in the interval $0 \le x < 2\pi$.

1. $\sec x - \sqrt{2} = 0$

2. $2 \sin x = \sqrt{3}$

3. $\tan x - \sqrt{3} = 0$

4. $\cos x - 1 = 0$

5. $2 \sin x \cos x = \sqrt{2} \cos x$

6. $2 \sin x \cos x = \sqrt{3} \sin x$

7. $\sin^2 x - 1 = 0$

8. $\cos^2 x - 1 = 0$

9. $4 \sin x \cos x - 2\sqrt{3} \sin x - 2\sqrt{2} \cos x + \sqrt{6} = 0$

10. $\sec^2 x + \sqrt{3} \sec x - \sqrt{2} \sec x - \sqrt{6} = 0$

11. $\csc x - \sqrt{2} = 0$

12. $3 \cot x + \sqrt{3} = 0$

13. $2 \sin^2 x + 1 = 3 \sin x$

14. $2 \cos^2 x + 1 = -3 \cos x$

15. $4 \cos^2 x - 3 = 0$

16. $2 \sin^2 x - 1 = 0$

17. $2 \sin^3 x = \sin x$

18. $4 \cos^3 x = 3 \cos x$

19. $4 \sin^2 x + 2\sqrt{3} \sin x - \sqrt{3} = 2 \sin x$

20. $\tan^2 x + \tan x - \sqrt{3} = \sqrt{3} \tan x$

21. $\sin^4 x = \sin^2 x$

22. $\cos^4 x = \cos^2 x$

In Exercises 23 to 60, solve the following equations, where $0° \le x < 360°$. Round to the nearest tenth of a degree.

23. $\cos x - 0.75 = 0$

24. $\sin x + 0.432 = 0$

25. $3 \sin x - 5 = 0$

26. $4 \cos x - 1 = 0$

27. $3 \sec x - 8 = 0$

28. $4 \csc x + 9 = 0$

29. $\cos x + 3 = 0$

30. $\sin x - 4 = 0$

31. $3 - 5 \sin x = 4 \sin x + 1$

32. $4 \cos x - 5 = \cos x - 3$

33. $\dfrac{1}{2} \sin x + \dfrac{2}{3} = \dfrac{3}{4} \sin x + \dfrac{3}{5}$

34. $\dfrac{2}{5} \cos x - \dfrac{1}{2} = \dfrac{1}{3} - \dfrac{1}{2} \cos x$

35. $3 \tan^2 x - 2 \tan x = 0$

36. $4 \cot^2 x + 3 \cot x = 0$

37. $3 \cos x + \sec x = 0$

38. $5 \sin x - \csc x = 0$

39. $\tan^2 x = 3 \sec^2 x - 2$

40. $\csc^2 x - 1 = 3 \cot^2 x + 2$

41. $2 \sin^2 x = 1 - \cos x$

42. $\cos^2 x + 4 = 2 \sin x - 3$

43. $3 \cos^2 x + 5 \cos x - 2 = 0$

44. $2 \sin^2 x + 5 \sin x + 3 = 0$

45. $2 \tan^2 x - \tan x - 10 = 0$

46. $2 \cot^2 x - 7 \cot x + 3 = 0$

47. $3 \sin x \cos x - \cos x = 0$

48. $\tan x \sin x - \sin x = 0$

49. $2 \sin x \cos x - \sin x - 2 \cos x + 1 = 0$

50. $6 \cos x \sin x - 3 \cos x - 4 \sin x + 2 = 0$

51. $2 \sin x - \cos x = 1$

52. $\sin x + 2 \cos x = 1$

53. $2 \sin x - 3 \cos x = 1$

54. $\sqrt{3} \sin x + \cos x = 1$

55. $3 \sin^2 x - \sin x - 1 = 0$

56. $2 \cos^2 x - 5 \cos x - 5 = 0$

57. $2 \cos x - 1 + 3 \sec x = 0$

58. $3 \sin x - 5 + \csc x = 0$

59. $\cos^2 x - 3 \sin x + 2 \sin^2 x = 0$

60. $\sin^2 x = 2 \cos x + 3 \cos^2 x$

In Exercises 61 to 70, solve the trigonometric equations.

61. $\tan 2x - 1 = 0$ **62.** $\sec 3x - \dfrac{2\sqrt{3}}{3} = 0$

63. $\sin 5x = 1$ **64.** $\cos 4x = -\dfrac{\sqrt{2}}{2}$

65. $\sin 2x - \sin x = 0$ **66.** $\cos 2x = -\dfrac{\sqrt{3}}{2}$

67. $\sin\left(2x + \dfrac{\pi}{6}\right) = -\dfrac{1}{2}$ **68.** $\cos\left(2x - \dfrac{\pi}{4}\right) = -\dfrac{\sqrt{2}}{2}$

69. $\sin^2 \dfrac{x}{2} + \cos x = 1$ **70.** $\cos^2 \dfrac{x}{2} - \cos x = 1$

In Exercises 71 to 84, solve each equation where $0 \le x < 2\pi$.

71. $\cos 2x = 1 - 3 \sin x$ **72.** $\cos 2x = 2 \cos x - 1$

73. $\sin 4x - \sin 2x = 0$ **74.** $\sin 4x - \cos 2x = 0$

75. $\tan \dfrac{x}{2} = \sin x$ **76.** $\tan \dfrac{x}{2} = 1 - \cos x$

77. $\sin 2x \cos x + \cos 2x \sin x = 0$

78. $\cos 2x \cos x - \sin 2x \sin x = 0$

79. $\sin x \cos 2x - \cos x \sin 2x = \dfrac{\sqrt{3}}{2}$

80. $\cos 2x \cos x + \sin 2x \sin x = -1$

81. $\sin 3x - \sin x = 0$

82. $\cos 3x + \cos x = 0$

83. $2 \sin x \cos x + 2 \sin x - \cos x - 1 = 0$

84. $2 \sin x \cos x - 2\sqrt{2} \sin x - \sqrt{3} \cos x + \sqrt{6} = 0$

Graphing Calculator Exercises*

In Exercises 85 to 88, use a graphing calculator to solve the equations.

85. $\cos x = x$ when $0 \le x < 2\pi$

86. $2 \sin x = x$ when $0 \le x < 2\pi$

87. $\sin 2x = \dfrac{1}{x}$ when $-4 \le x \le 4$

88. $\cos x = \dfrac{1}{x}$ when $0 \le x \le 5$

89. Solve $\cos x = x^3 - x$ by graphing each side and finding all points of intersection. Explore why this method may yield an incorrect result for this equation.

* Additional graphing calculator exercises appear in the Graphing Workbook as described in the front of this textbook.

90. Approximate the largest value of k for which the equation $\sin x \cos x = k$ has a solution.

Supplemental Exercises

In Exercises 91 to 100, solve the trigonometric equations for $0 \le x < 2\pi$.

91. $\sqrt{3} \sin x + \cos x = \sqrt{3}$

92. $\sin x - \cos x = 1$

93. $-\sin x + \sqrt{3} \cos x = \sqrt{3}$

94. $-\sqrt{3} \sin x - \cos x = 1$

95. $\cos 5x - \cos 3x = 0$

96. $\cos 5x - \cos x - \sin 3x = 0$

97. $\sin 3x + \sin x = 0$

98. $\sin 3x + \sin x - \sin 2x = 0$

99. $\cos 4x + \cos 2x = 0$

100. $\cos 4x + \cos 2x - \cos 3x = 0$

101. Find the area of the sector OAB in the accompanying figure in terms of r and θ.

102. Find the area of triangle OAB in the figure in terms of r and θ.

103. Find the area of the shaded part of the figure. Find that area in terms of r and θ.

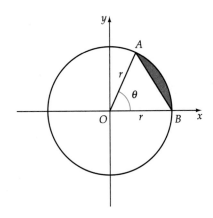

104. It is theorized that the system of blood vessels in primates has evolved so that it has an optimal structure. The case of a blood vessel splitting into two vessels, as shown in the accompanying figure, assumes that each new branch carries an equal amount of blood. A model of the angle θ is given by the equation $\cos \theta = 2^{(x-4)/(x+4)}$. The value of x is such that $1 \le x \le 2$ and depends on assumptions about the thickness of the blood vessels. Assuming this is an accurate model, find the values of the angle θ.

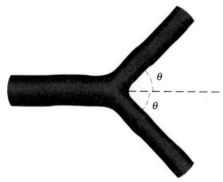

105. As bus A_1 makes a left turn, the back B of the bus moves to the right. If bus A_2 were waiting at a stop light while A_1 turned left, as shown in the figure, there is a chance the two buses would scrape against one another. For a

bus 28 feet long and 8 feet wide, the movement of the back of the bus to the right can be approximated by

$$x = \sqrt{(4 + 18 \cot \theta)^2 + 100} - (4 + 18 \cot \theta)$$

where θ is the angle the bus driver has turned the front of the bus. Find the value of x for $\theta = 20°$ and $\theta = 30°$.

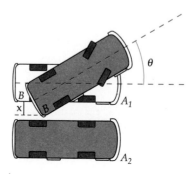

Exploring Concepts with Technology

Composition of Inverse Trigonometric Functions

Composition of functions is a method of constructing a function from any two given functions. The process consists of using the range element of one function as the domain element of another function.

The composition of functions which are inverses of each other is the identity function for appropriate values. The definition of inverse functions is given below.

Definition of Inverse Function

If f is a one-to-one function with domain X and range Y, and g is a function with domain Y and range X, then g is the inverse function of f if and only if

$$(f \circ g) = x \quad \text{for all } x \text{ in the domain of } g$$

and

$$(g \circ f) = x \quad \text{for all } x \text{ in the domain of } f$$

A graphing calculator was used to graph both $y = \sin^{-1}(\sin x)$ and $y = \sin(\sin^{-1} x)$. The results are shown in the following graphs.

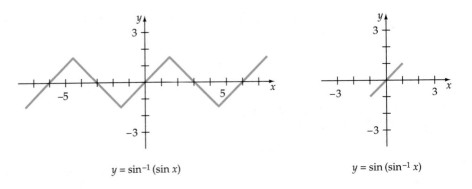

$$y = \sin^{-1}(\sin x) \qquad\qquad y = \sin(\sin^{-1} x)$$

The composition of inverse functions should give the identity function $y = x$. Why, then, are these graphs different?
Graph the following functions.

1. $y = \cos^{-1}(\cos x)$ and $y = \cos(\cos^{-1}x)$
2. $y = \tan^{-1}(\tan x)$ and $y = \tan(\tan^{-1}x)$
3. $y = \csc^{-1}(\csc x)$ and $y = \csc(\csc^{-1}x)$

Does the composition of the functions result in the identity function? If not, explain.

Chapter Review

7.1   The Fundamental Trigonometric Identities

- Trigonometric identities are verified by using algebraic methods and previously proved identities.

$$\sin x = \frac{1}{\csc x} \qquad \cos x = \frac{1}{\sec x} \qquad \tan x = \frac{1}{\cot x}$$

$$\tan x = \frac{\sin x}{\cos x} \qquad \cot x = \frac{\cos x}{\sin x}$$

$$\sin^2 x + \cos^2 x = 1 \qquad \tan^2 x + 1 = \sec^2 x \qquad 1 + \cot^2 x = \csc^2 x$$

7.2   Sum and Difference Identities

- Sum and difference identities for the cosine function are

$$\cos(\alpha - \beta) = \cos \alpha \cos \beta + \sin \alpha \sin \beta$$

$$\cos(\alpha + \beta) = \cos \alpha \cos \beta - \sin \alpha \sin \beta$$

- Sum and difference identities for the sine function are

$$\sin(\alpha - \beta) = \sin \alpha \cos \beta - \cos \alpha \sin \beta$$

$$\sin(\alpha + \beta) = \sin \alpha \cos \beta + \cos \alpha \sin \beta$$

- Sum and difference identities for the tangent function are

$$\tan(\alpha + \beta) = \frac{\tan \alpha + \tan \beta}{1 - \tan \alpha \tan \beta}$$

$$\tan(\alpha - \beta) = \frac{\tan \alpha - \tan \beta}{1 + \tan \alpha \tan \beta}$$

7.3   Double- and Half-Angle Identities

- The double-angle identities are

$$\sin 2\alpha = 2 \sin \alpha \cos \alpha$$

$$\cos 2\alpha = \cos^2 \alpha - \sin^2 \alpha$$

$$= 1 - 2 \sin^2 \alpha$$

$$= 2 \cos^2 \alpha - 1$$

$$\tan 2\alpha = \frac{2 \tan \alpha}{1 - \tan^2 \alpha}$$

- The half-angle identities are

$$\sin \frac{\alpha}{2} = \pm \sqrt{\frac{1 - \cos \alpha}{2}}$$

$$\cos \frac{\alpha}{2} = \pm \sqrt{\frac{1 + \cos \alpha}{2}}$$

$$\tan \frac{\alpha}{2} = \frac{\sin \alpha}{1 + \cos \alpha} = \frac{1 - \cos \alpha}{\sin \alpha}$$

7.4 Identities Involving the Sum of Trigonometric Functions

- The product-to-sum identities are

$$\sin \alpha \cos \beta = \frac{1}{2}[\sin(\alpha + \beta) + \sin(\alpha - \beta)]$$

$$\cos \alpha \sin \beta = \frac{1}{2}[\sin(\alpha + \beta) - \sin(\alpha - \beta)]$$

$$\cos \alpha \cos \beta = \frac{1}{2}[\cos(\alpha + \beta) + \cos(\alpha - \beta)]$$

$$\sin \alpha \sin \beta = \frac{1}{2}[\cos(\alpha - \beta) - \cos(\alpha + \beta)]$$

- The sum-to-product identities are

$$\sin x + \sin y = 2 \sin \frac{x + y}{2} \cos \frac{x - y}{2}$$

$$\cos x - \cos y = -2 \sin \frac{x + y}{2} \sin \frac{x - y}{2}$$

$$\sin x - \sin y = 2 \cos \frac{x + y}{2} \sin \frac{x - y}{2}$$

$$\cos x + \cos y = 2 \cos \frac{x + y}{2} \cos \frac{x - y}{2}$$

- For sums of the form $a \sin x + b \cos x$,

$$a \sin x + b \cos x = k \sin(x + \alpha)$$

where $k = \sqrt{a^2 + b^2}$, $\sin \alpha = \dfrac{b}{\sqrt{a^2 + b^2}}$, and

$\cos \alpha = \dfrac{a}{\sqrt{a^2 + b^2}}$.

7.5 Inverse Trigonometric Functions

- The inverse of $y = \sin x$ is $y = \sin^{-1} x$, with $-1 \le x \le 1$ and $-\pi/2 \le y \le \pi/2$.
- The inverse of $y = \cos x$ is $y = \cos^{-1} x$, with $-1 \le x \le 1$ and $0 \le y \le \pi$.
- The inverse of $y = \tan x$ is $y = \tan^{-1} x$, with $-\infty < x < \infty$ and $-\pi/2 < y < \pi/2$.
- The inverse of $y = \cot x$ is $y = \cot^{-1} x$, with $-\infty < x < \infty$ and $0 < y < \pi$.
- The inverse of $y = \csc x$ is $y = \csc^{-1} x$, with $x \le -1$ or $x \ge 1$ and $-\pi/2 \le y \le \pi/2$, $y \ne 0$.
- The inverse of $y = \sec x$ is $y = \sec^{-1} x$, with $x \le -1$ or $x \ge 1$ and $0 \le y \le \pi$, $y \ne \pi/2$.

7.6 Trigonometric Equations

- Algebraic methods and identities are used to solve trigonometric equations. Since the trigonometric functions are periodic, there may be an infinite number of solutions.

Essays and Projects

1. The terms *fundamental* and *overtones* are used in music. Write an essay in which you draw on your understanding of trigonometry to explain these terms.

2. Honeybees build their combs by using hexagonal cells that are constructed in such a way as to minimize the amount of wax required. Write an essay on the mathematics of the honeycomb. Include formulas for the volume of a cell and the surface area of a cell.

3. Write an essay that explains the meaning of the versin function, the vercos function, and the exsecant function. Include information about the history of these functions.

4. Explain why the following is not an identity.

$$\frac{\sin \theta \cos \theta}{\cos \theta} = \sin \theta$$

5. A camera is horizontally x feet from a diver, as shown in the figure.

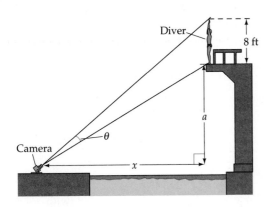

If the extended length of the diver is 8 feet, show that when the diver is a feet above the camera lens, the angle θ subtended at the lens by the diver is

$$\theta = \cot^{-1}\frac{x}{8+a} - \cot^{-1}\frac{x}{a}$$

Use a computer or a graphing calculator to determine the value of x that maximizes θ when the diver is 20 feet above the camera lens.

6. Explain what is meant by the term *cylindrical coordinates.* How would the point $(1, 2, \sqrt{3})$ be expressed in cylindrical coordinates? Include a drawing. Explain what is meant by the term *spherical coordinates.* How would the point $(2, 2, 2)$ be expressed in spherical coordinates? Include a drawing.

7. Write an essay about Menelaus of Alexandria and his work in the area of trigonometry. Include a statement of the theorem known as *Menelaus's Theorem.*

8. Write an essay about the notation and the names that we use for the trigonometric functions. Also include information about the first use of the law of sines and of the law of cosines.

9. In spherical trigonometry, each side of a triangle is a portion of a *great circle.* Explain what a great circle is and

make a drawing of a spherical triangle. Associated with every spherical triangle is a *trihedral angle.* Explain what a trihedral angle is and show its location with respect to the spherical triangle in your drawing. In a plane triangle, the sum of the measures of the angles is 180°. What is the sum of the measures of the angles of a spherical triangle? Right spherical triangles are solved by using *Napier's rules of circular parts.* What are Napier's rules of circular parts?

10. Consider a square as shown. Start at the point $(1, 0)$ and travel around the square for a distance t ($t \geq 0$). Let $P(x, y)$ be the

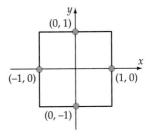

point on the square determined by traveling counter-clockwise a distance of t units from $(1, 0)$. We define the square sine of t, denoted by ssin t, to be the y value of point P. The square cosine of t, denoted by scos t, is defined to be the x value of point P. For example,

$$\text{ssin } 0.4 = 0.4, \quad \text{scos } 0.4 = 1,$$

$$\text{scos } 1.2 = 0.8, \quad \text{scos } 5.3 = -0.7$$

Draw a graph of $y = \text{ssin } t$, $0 \leq t \leq 8$. Which of the properties of the sine function given on page 333 hold for the function $f(t) = \text{ssin } t$? Let

$$\text{stan } t = \frac{\text{ssin } t}{\text{scos } t}, \quad \text{scos } t \neq 0$$

Draw a graph of stan t for $0 \leq t \leq 8$.

True/False Exercises

In Exercises 1 to 12, answer true or false. If the statement is false, give an example to show that the statement is false.

1. $\dfrac{\tan \alpha}{\tan \beta} = \dfrac{\alpha}{\beta}$

2. $\dfrac{\sin x}{\cos y} = \tan \dfrac{x}{y}$

3. $\sin^{-1} x = \csc x$

4. $\sin 2\alpha = 2 \sin \alpha$ for all α

5. $\sin(\alpha + \beta) = \sin \alpha + \sin \beta$

6. An equation that has an infinite number of solutions is an identity.

7. If $\tan \alpha = \tan \beta$, then $\alpha = \beta$.

8. $\cos^{-1}(\cos x) = x$

9. $\cos(\cos^{-1}x) = x$

10. $\csc^{-1}\dfrac{1}{\alpha} = \dfrac{1}{\csc \alpha}$

11. If $0° \leq \theta \leq 90°$, then $\cos \theta = \sin(180° - \theta)$.

12. $\sin^2 \theta = \sin \theta^2$

Chapter Review Exercises

In Exercises 1 to 10, find the exact value.

1. $\cos(45° + 30°)$

2. $\tan(210° - 45°)$

3. $\sin\left(\dfrac{2\pi}{3} + \dfrac{\pi}{4}\right)$

4. $\sec\left(\dfrac{4\pi}{3} - \dfrac{\pi}{4}\right)$

5. $\sin(60° - 135°)$

6. $\cos\left(\dfrac{5\pi}{3} - \dfrac{7\pi}{4}\right)$

7. $\sin\left(22\dfrac{1}{2}\right)°$

8. $\cos 105°$

9. $\tan\left(67\dfrac{1}{2}\right)°$

10. $\sin 112.5°$

In Exercises 11 to 14, find the exact value of the given functions.

11. Given $\sin \alpha = \frac{1}{2}$, α in quadrant I, and $\cos \beta = \frac{1}{2}$, β in quadrant IV, find a. $\cos(\alpha - \beta)$, b. $\tan 2\alpha$, c. $\sin \beta/2$.

12. Given $\sin \alpha = \frac{\sqrt{3}}{2}$, α in quadrant II, and $\cos \beta = -\frac{1}{2}$, β in quadrant III, find a. $\sin(\alpha + \beta)$, b. $\sec 2\beta$ c. $\cos \frac{\alpha}{2}$.

13. Given $\sin \alpha = -\frac{1}{2}$, α in quadrant IV, and $\cos \beta = -\frac{\sqrt{3}}{2}$, β in quadrant III, find a. $\sin(\alpha - \beta)$, b. $\tan 2\alpha$, c. $\cos \beta/2$.

14. Given $\sin \alpha = \frac{\sqrt{2}}{2}$, α in quadrant I, and $\cos \beta = \frac{\sqrt{3}}{2}$, β in quadrant IV, find a. $\cos(\alpha - \beta)$, b. $\tan 2\beta$, c. $\sin 2\alpha$.

In Exercises 15 to 20, write the given expression as a single trigonometric function.

15. $2 \sin 3x \cos 3x$

16. $\dfrac{\tan 2x + \tan x}{1 - \tan 2x \tan x}$

17. $\sin 4x \cos x - \cos 4x \sin x$

18. $\cos^2 2\theta - \sin^2 2\theta$

19. $1 - 2 \sin^2 \dfrac{\beta}{2}$

20. $\pm\sqrt{\dfrac{1 - \cos 4\theta}{2}}$

In Exercises 21 to 24, evaluate each expression.

21. $\sin 47° \sin 22°$

22. $\cos 14° \cos 92°$

23. $2 \sin \dfrac{\pi}{3} \cos \dfrac{2\pi}{3}$

24. $2 \cos \dfrac{\pi}{4} \cos \dfrac{3\pi}{2}$

In Exercises 25 to 28, write each expression as the product of two functions.

25. $\cos 2\theta - \cos 4\theta$

26. $\sin 3\theta - \sin 5\theta$

27. $\sin 6\theta + \sin 2\theta$

28. $\sin 5\theta - \sin \theta$

In Exercises 29 to 46, verify the identities.

29. $\dfrac{1}{\sin x - 1} + \dfrac{1}{\sin x + 1} = -2 \tan x \sec x$

30. $\dfrac{\sin x}{1 - \cos x} = \csc x + \cot x, \quad 0 < x < \dfrac{\pi}{2}$

31. $\dfrac{1 + \sin x}{\cos^2 x} = \tan^2 x + 1 + \tan x \sec x$

32. $\cos^2 x - \sin^2 x - \sin 2x = \dfrac{\cos^2 2x - \sin^2 2x}{\cos 2x + \sin 2x}$

33. $\dfrac{1}{\cos x} - \cos x = \tan x \sin x$

34. $\sin(270° - \theta) - \cos(270° - \theta) = \sin \theta - \cos \theta$

35. $\sin\left(\dfrac{\pi}{4} - \alpha\right) = \dfrac{\sqrt{2}}{2}(\cos \alpha - \sin \alpha)$

36. $\sin(180° - \alpha + \beta) = \sin \alpha \cos \beta - \cos \alpha \sin \beta$

37. $\dfrac{\sin 4x - \sin 2x}{\cos 4x - \cos 2x} = -\cot 3x$

38. $2 \sin x \sin 3x = (1 - \cos 2x)(1 + 2 \cos 2x)$

39. $\sin x - \cos 2x = (2 \sin x - 1)(\sin x + 1)$

40. $\cos 4x = 1 - 8 \sin^2 x + 8 \sin^4 x$

41. $\tan 4x = \dfrac{4 \tan x - 4 \tan^3 x}{1 - 6 \tan^2 x + \tan^4 x}$

42. $\dfrac{\sin 2x - \sin x}{\cos 2x + \cos x} = \dfrac{1 - \cos x}{\sin x}$

43. $2 \cos 4x \sin 2x = 2 \sin 3x \cos 3x - 2 \sin x \cos x$

44. $2 \sin x \sin 2x = 4 \cos x \sin^2 x$

45. $\cos(x + y) \cos(x - y) = \cos^2 x + \cos^2 y - 1$

46. $\cos(x + y) \sin(x - y) = \sin x \cos x - \sin y \cos y$

In Exercises 47 to 52, solve the equation.

47. $y = \sec\left(\sin^{-1} \dfrac{12}{13}\right)$

48. $y = \cos\left(\sin^{-1} \dfrac{3}{5}\right)$

49. $2 \sin^{-1}(x - 1) = \dfrac{\pi}{3}$

50. $y = \cos\left[\sin^{-1}\left(-\dfrac{3}{5}\right) + \cos^{-1} \dfrac{5}{13}\right]$

51. $\sin^{-1} x + \cos^{-1} \dfrac{4}{5} = \dfrac{\pi}{2}$

52. $y = \cos\left[2 \sin^{-1}\left(\dfrac{3}{5}\right)\right]$

In Exercises 53 to 54, solve each equation for $0° \le x < 360°$.

53. $4 \sin^2 x + 2\sqrt{3} \sin x - 2 \sin x - \sqrt{3} = 0$

54. $2 \sin x \cos x - \sqrt{2} \cos x - 2 \sin x + \sqrt{2} = 0$

In Exercises 55 and 56, solve the trigonometric equation.

55. $3 \cos^2 x + \sin x = 1$

56. $\tan^2 x - 2 \tan x - 3 = 0$

In Exercises 57 and 58, solve each equation for $0 \leq x < 2\pi$.

57. $\sin 3x \cos x - \cos 3x \sin x = \dfrac{1}{2}$

58. $\cos\left(2x - \dfrac{\pi}{3}\right) = -\dfrac{\sqrt{3}}{2}$

In Exercises 59 to 62, find the amplitude and phase shift of each function. Graph each function.

59. $f(x) = \sqrt{3} \sin x + \cos x$ **60.** $f(x) = -2 \sin x - 2 \cos x$

61. $f(x) = -\sin x - \sqrt{3} \cos x$ **62.** $f(x) = \dfrac{\sqrt{3}}{2} \sin x - \dfrac{1}{2} \cos x$

In Exercises 63 to 66, graph each equation.

63. $f(x) = 2 \cos^{-1} x$

64. $f(x) = \sin^{-1}(x - 1)$

65. $f(x) = \sin^{-1} \dfrac{x}{2}$

66. $f(x) = \sec^{-1} 2x$

Chapter Test

1. Verify the identity $1 + \sin^2 x \sec^2 x = \sec^2 x$.

2. Verify the identity

$$\dfrac{1}{\sec x - \tan x} - \dfrac{1}{\sec x + \tan x} = 2 \tan x$$

3. Verify the identity $\cos^3 x + \cos x \sin^2 x = \cos x$.

4. Verify the identity $\csc x - \cot x = \dfrac{1 - \cos x}{\sin x}$.

5. Find the exact value of $\sin 195°$.

6. Given $\sin \alpha = -\dfrac{3}{5}$, α in quadrant III and $\cos \beta = -\dfrac{\sqrt{2}}{2}$, β in quadrant II, find $\sin(\alpha + \beta)$.

7. Verify the identity $\sin\left(\theta - \dfrac{3\pi}{2}\right) = \cos \theta$.

8. Write $\cos 6x \sin 3x + \sin 6x \cos 3x$ in terms of a single trigonometric function.

9. Find the exact value of $\cos 2\theta$ given that $\sin \theta = \dfrac{4}{5}$ and θ is in quadrant II.

10. Verify the identity $\tan \dfrac{\theta}{2} + \dfrac{\cos \theta}{\sin \theta} = \csc \theta$.

11. Verify the identity $\sin^2 2x + 4 \cos^4 x = 4 \cos^2 x$.

12. Find the exact value of $\sin 15° \cos 75°$.

13. Write $y = -\dfrac{\sqrt{3}}{2} \sin x + \dfrac{1}{2} \cos x$ in the form $y = k \sin(x + \alpha)$, where α is measured in radians.

14. Find the exact degree value of $\cos^{-1}\left(-\dfrac{\sqrt{3}}{2}\right)$.

15. Find the approximate radian measure of $\cos^{-1} 0.7644$.

16. Find the exact value of $\sin\left(\cos^{-1} \dfrac{12}{13}\right)$.

17. Graph $y = \sin^{-1}(x + 2)$.

18. Solve $3 \sin x - 2 = 0$, where $0° \leq x < 360°$.

19. Solve $\sin x \cos x - \dfrac{\sqrt{3}}{2} \sin x = 0$, where $0° \leq x < 360°$.

20. Solve $\sin 2x + \sin x - 2 \cos x - 1 = 0$, where $0° \leq x < 360°$.

8

Applications of Trigonometry

CASE IN POINT   *The Electromagnetic Spectrum*

Electromagnetic waves are generated by oscillations of an electrically charged particle. All electromagnetic waves travel through space at approximately 186,000 miles per second, or 300,000 kilometers per second. This speed is frequently referred to as the speed of light.

Visible light, however, makes up only a small portion of what is called the *electromagnetic spectrum*. The electromagnetic spectrum consists of bands of electromagnetic waves of different wavelengths. The major types of waves are gamma rays, x-rays, ultraviolet light, visible light, infrared rays, microwaves, and radio waves.

The electromagnetic spectrum is shown below. The lengths of the waves are given in meters, and the frequencies of the waves are given in hertz. One **hertz** is 1 wave cycle per second. No sharp line can be drawn between various portions of the spectrum. The only difference between the waves is in their wavelength or frequency.

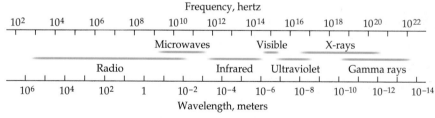

The Electromagnetic Spectrum

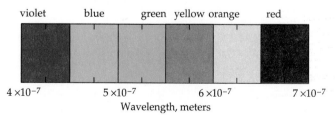

The Visible Light Spectrum

8.1 The Law of Sines

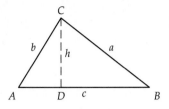

Figure 8.1

Solving a triangle involves finding the lengths of all sides and the measure of all angles in the triangle. In this section and the next, we develop formulas for solving an *oblique triangle,* which is a triangle that does not contain a right angle. The *Law of Sines* can be used to solve oblique triangles in which either two angles and a side or two sides and an angle opposite one of the sides are known. In Figure 8.1, altitude CD is drawn from C. The length of the altitude is h. Triangles ACD and BCD are right triangles.

Using the definition of the sine of an angle of a right triangle, we have from Figure 8.1

$$\sin B = \frac{h}{a} \qquad\qquad \sin A = \frac{h}{b}$$

$$h = a \sin B \quad (1) \qquad h = b \sin A \quad (2)$$

Equating the values of h in Equations (1) and (2), we obtain

$$a \sin B = b \sin A$$

Dividing each side of the equation by $\sin A \sin B$, we obtain

$$\frac{a}{\sin A} = \frac{b}{\sin B}$$

Similarly, when an altitude is drawn to a different side, the following formulas result:

$$\frac{c}{\sin C} = \frac{b}{\sin B} \quad \text{and} \quad \frac{c}{\sin C} = \frac{a}{\sin A}$$

The Law of Sines

If A, B, and C are the measures of the angles of a triangle, and a, b, and c are the lengths of the sides opposite these angles, then

$$\frac{a}{\sin A} = \frac{b}{\sin B} = \frac{c}{\sin C}$$

EXAMPLE 1 **Solve a Triangle Using the Law of Sines**

Solve triangle ABC if $A = 42°$, $B = 63°$, and $C = 18$ centimeters.

Solution   Find C by using the fact that the sum of the interior angles of a triangle is 180°.

$$A + B + C = 180°$$

$$42° + 63° + C = 180°$$

$$C = 75°$$

Use the Law of Sines to find a.

$$\frac{a}{\sin A} = \frac{c}{\sin C}$$

$$\frac{a}{\sin 42°} = \frac{18}{\sin 75°}$$

$$a = \frac{18 \sin 42°}{\sin 75°} \approx 12 \text{ centimeters}$$

Use the Law of Sines again, this time to find b.

$$\frac{b}{\sin B} = \frac{c}{\sin C}$$

$$\frac{b}{\sin 63°} = \frac{18}{\sin 75°}$$

$$b = \frac{18 \sin 63°}{\sin 75°} \approx 17 \text{ centimeters}$$

The solution is $C = 85°$, $a \approx 12$ centimeters, and $b \approx 17$ centimeters.

■ *Try Exercise* **4,** *page 417.*

When you are given two sides of a triangle and an angle opposite one of them, you sometimes find that the triangle is not unique. Some information may result in two triangles, and some may result in no triangle at all. The case of knowing two sides and an angle opposite one of them is called the *ambiguous case* of the Law of Sines.

Suppose sides a and c and the nonincluded angle A of a triangle are known and we are then asked to solve triangle ABC. First determine h, the height of the triangle, which is the perpendicular distance from point B to side b. Solving $\sin A = \frac{h}{c}$ for h, we have $h = c \sin A$. The relationships among h, a (the side opposite $\angle A$), and c determine whether there are 0, 1, or 2 triangles.

First consider the case in which $\angle A$ is an acute angle (Figure 8.2). There are four possible situations.

1. $a < h$; there is no possible triangle.
2. $a = h$; there is one triangle, a right triangle.
3. $h < a < c$; there are two possible triangles. One has all acute angles, the second has one obtuse angle.

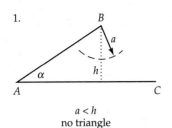

1.

$a < h$
no triangle

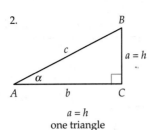

2.

$a = h$
one triangle

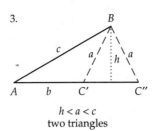

3.

$h < a < c$
two triangles

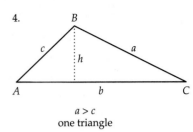

4.

$a > c$
one triangle

Figure 8.2
Case 1: A an Acute Angle

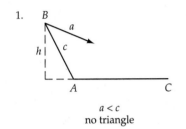

1.

$a < c$
no triangle

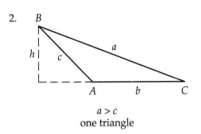

2.

$a > c$
one triangle

Figure 8.3
Case 2: A an Obtuse Angle

4. $a > c$; there is one triangle, which is not a right triangle.

Now consider the case in which $\angle A$ is an obtuse angle (Figure 8.3). Here, there are two possible situations.

1. $a \leq c$; there is no triangle.

2. $a > c$; there is one triangle.

To find the number of possible solutions of the triangle in Figure 8.4, where the measure of $\angle A = 57°$ and sides are $a = 15$ feet and $c = 20$ feet, we must first find h.

$$h = 20 \sin 57° \approx 17$$

Since $a < h$, no triangle is formed, and thus there is no solution.
Note the result when the Law of Sines is used to find C.

$$\frac{a}{\sin A} = \frac{c}{\sin C}$$

$$\frac{15}{\sin 57°} = \frac{20}{\sin C}$$

$$\sin C = \frac{20 \sin 57°}{15}$$

$$\approx 1.1182$$

Because 1.1182 is not in the range of the sine function, there is no solution of the equation. Thus, there is no triangle for these values of A, a, and c.

Figure 8.4

EXAMPLE 2   **Solve a Triangle Using the Law of Sines**

Given the triangle ABC with $B = 32°$, $a = 42$, and $b = 30$, find A.

Solution   Find h to determine whether this is the ambiguous case.

$$h = 42 \sin 32° \approx 22$$

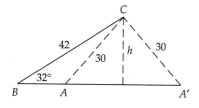

Figure 8.5

We have $h < b < a$; thus there are two solutions. Use the Law of Sines to find the value of A and A'. See Figure 8.5.

$$\frac{b}{\sin B} = \frac{a}{\sin A}$$

$$\frac{30}{\sin 32°} = \frac{42}{\sin A}$$

$$\sin A = \frac{42 \sin 32°}{30}$$

$$\approx 0.7419$$

$$A' \approx 48°$$

$$A \approx 132°$$

• The two angles with measure between 0° and 180° that have a sine of 0.7419 are approximately 48° and 132°.

Angle $A' \approx 48°$ or $A \approx 132°$.

■ *Try Exercise* **18,** *page 417.*

EXAMPLE 3   **Solve an Application Using the Law of Sines**

A radio antenna 85 feet high is located on top of an office building. At a distance AD from the base of the building, the angle of elevation to the top of the antenna is 26°, and the angle of elevation to the bottom of the antenna is 16°. Find the height of the building.

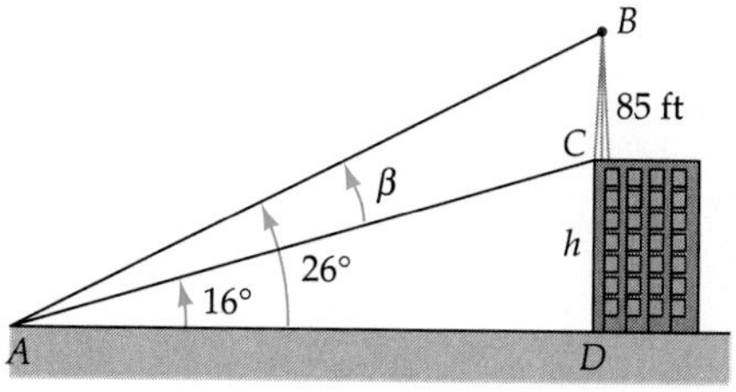

Solution   Sketch the diagram. See Figure 8.6. Find B and β.

$$B = 90° - 26° = 64°$$

$$\beta = 26° - 16° = 10°$$

Since we know the length BC and the measure of β, we can use the Law of Sines to find length AC.

$$\frac{BC}{\sin \beta} = \frac{AC}{\sin B}$$

$$\frac{85}{\sin 10°} = \frac{AC}{\sin 64°}$$

$$AC = \frac{85 \sin 64°}{\sin 10°}$$

Having found AC, we can now find the height of the building.

$$\sin 16° = \frac{h}{AC}$$

$$h = AC \sin 16°$$

$$= \frac{85 \sin 64°}{\sin 10°} \sin 16° \approx 121 \text{ ft}$$ • Substitute for AC.

Figure 8.6

Using the rounding convention, the height of the building to two significant digits is 120 feet.

■ *Try Exercise* **26,** *page 417.*

In navigation and surveying problems, there are two commonly used methods for specifying direction. The angular direction in which a craft is pointed is called the **heading.** Heading is expressed in terms of an angle measured clockwise from north. Figure 8.7 shows a heading of 65° and a heading of 285°.

The angular direction used to locate one object in relation to another object is called the **bearing.** Bearing is expressed in terms of the acute angle formed by a north–south line and the line of direction. Figure 8.8 shows a bearing of N38°W and a bearing of S15°E.

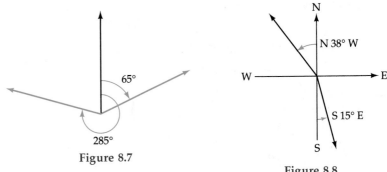

Figure 8.7

Figure 8.8

EXAMPLE 4   Solve an Application

A ship with a heading of 330° first sighted a lighthouse (point *B*) at a bearing of N65°E. After traveling 8.5 miles, the ship observed the lighthouse at a bearing of S50°E. Find the distance from the ship to the lighthouse when the first sighting is made.

Solution From Figure 8.9 we see that the measure of $\angle CAB = 65° + 30° = 95°$, the measure of $\angle BCA = 50° - 30° = 20°$, and $B = 180° - 95° - 20° = 65°$. Use the Law of Sines to find *c*.

$$\frac{b}{\sin B} = \frac{c}{\sin C}$$

$$\frac{8.5}{\sin 65°} = \frac{c}{\sin 20°}$$

$$c = \frac{8.5 \sin 20°}{\sin 65°} \approx 3.2$$

The lighthouse was 3.2 miles (two significant digits) from the ship when the first sighting was made.

■ *Try Exercise* **32,** *page 418.*

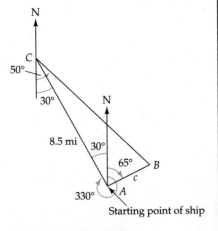

Figure 8.9

EXERCISE SET 8.1

In Exercises 1 to 12, solve the triangles.

1. $A = 42°, B = 61°, a = 12$
2. $B = 25°, C = 125°, b = 5.0$
3. $A = 110°, C = 32°, b = 12$
4. $B = 28°, C = 78°, c = 44$
5. $A = 132°, a = 22, b = 16$
6. $B = 82°, b = 6, c = 3$
7. $A = 82.0°, B = 65.4°, b = 36.5$
8. $B = 54.8°, C = 72.6°, a = 14.4$
9. $A = 33.8°, C = 98.5°, c = 102$
10. $B = 36.9°, C = 69.2°, a = 166$
11. $C = 114.2°, c = 87.2, b = 12.1$
12. $A = 54.32°, a = 24.42, c = 16.92$

In Exercises 13 to 24, solve the triangles that exist.

13. $A = 37°, c = 40, a = 28$
14. $B = 32°, c = 14, b = 9$
15. $C = 65°, b = 10, c = 8.0$
16. $A = 42°, a = 12, c = 18$
17. $A = 30°, a = 1.0, b = 2.4$
18. $B = 22.6°, b = 5.55, a = 13.8$
19. $A = 14.8°, c = 6.35, a = 4.80$
20. $C = 37.9°, b = 3.50, c = 2.84$
21. $C = 47.2°, a = 8.25, c = 5.80$
22. $B = 52.7°, b = 12.3, c = 16.3$
23. $B = 117.32°, b = 67.25, a = 15.05$
24. $A = 49.22°, a = 16.92, c = 24.62$

25. The angle of elevation of a balloon from one observer is 67°, and the angle of elevation from another observer, 220 ft away, is 31°. If the balloon is in the same vertical plane as the two observers and in between them, find the distance of the balloon from the first observer.

26. The longer side of a parallelogram is 6.0 meters. The measure of $\angle BAD$ is 56° and α is 35°. Find the length of the longer diagonal.

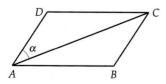

27. To find the distance across a canyon, a surveying team locates points A and B on one side of the canyon and point C on the other side of the canyon. The distance between A and B is 85 yards. The measure of $\angle CAB$ is 68°, and the measure of $\angle CBA$ is 75°. Find the distance across the canyon.

28. Two observers, in the same vertical plane as a kite and a distance of 30 feet apart, observe the kite at an angle of 62° and 78°, as shown in the diagram. Find the height of the kite.

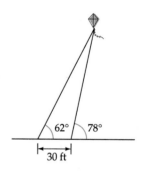

29. A telephone pole 35 feet high is situated on an 11° slope from the horizontal. The angle CAB is 32°. Find the length of the guy wire AC.

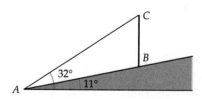

30. Three roads intersect in such a way as to form a triangular piece of land. See the accompanying figure. Find the lengths of the other two sides of the land.

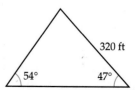

31. A surveying team determines the height of a hill by placing a 12-foot pole at the top of the hill and measuring the angles of elevation to the bottom and the top of the pole.

They find the angles of elevation shown in the accompanying figure. Find the height of the hill.

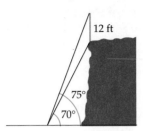

12 ft

75°

70°

32. Two fire lookouts are located on mountains 20 miles apart. Lookout B is at a bearing of S65°E from A. A fire was sighted at a bearing of N50°E from A and at a bearing of N8°E from B. Find the distance of the fire from lookout A.

33. A navigator on a ship sights a lighthouse at a bearing of N36°E. After traveling 8.0 miles at a heading of 332°, the ship sights the lighthouse at a bearing of S82°E. How far is the ship from the lighthouse at the second sighting?

34. The navigator on a ship traveling due east at 8 mph sights a lighthouse at a bearing of S55°E. One hour later the lighthouse is sighted at a bearing of S25°W. Find the closest the ship came to the lighthouse.

35. An airplane flew 450 miles at a bearing of N65°E from airport A to airport B. The pilot then flew at a bearing of S38°E to airport C. Find the distance from A to C if the bearing from airport A to airport C is S60°E.

36. A 12-foot solar panel is to be installed on a roof with a 15° pitch. Find the length of the brace d if the panel must be installed to make a 40° angle with the horizontal.

12 ft

d

15°

Supplemental Exercises

37. A house B is located at a bearing of N67°E, from house A. A house C is 300 meters at a bearing of S68°E, from house A. House B is located at a bearing of 349° from house C. Find the distance from house A to house B.

38. For any triangle ABC, show that

$$\frac{a - b}{b} = \frac{\sin A - \sin B}{\sin B}$$

39. For any triangle ABC, show that

$$\frac{a + b}{b} = \frac{\sin A + \sin B}{\sin B}$$

40. Show that for any triangle ABC,

$$\frac{a - b}{a + b} = \frac{\sin A - \sin B}{\sin A + \sin B}$$

8.2   The Law of Cosines and Area

The *Law of Cosines* can be used to solve triangles in which two sides and the included angle (the angle between the two sides) are known or in which three sides are known. Consider the triangle in Figure 8.10. The height BD is drawn from B perpendicular to the x-axis. The triangle BDA is a right triangle, and the coordinates of B are $(a \cos C, a \sin C)$. The coordinates of A are $(b, 0)$. Using the distance formula, we can find the distance c.

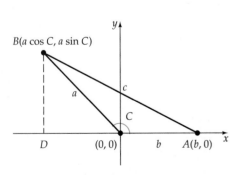

Figure 8.10

$$c = \sqrt{(a \cos C - b)^2 + (a \sin C - 0)^2}$$
$$c^2 = a^2 \cos^2 C - 2ab \cos C + b^2 + a^2 \sin^2 C$$
$$c^2 = a^2(\cos^2 C + \sin^2 C) + b^2 - 2ab \cos C$$
$$c^2 = a^2 + b^2 - 2ab \cos C$$

The Law of Cosines

If A, B, and C are the measures of the angles of a triangle, and a, b, and c are the lengths of sides opposite these angles, then

$$c^2 = a^2 + b^2 - 2ab \cos C$$

$$a^2 = b^2 + c^2 - 2bc \cos A$$

$$b^2 = a^2 + c^2 - 2ac \cos B$$

EXAMPLE 1   **Solve a Triangle Using the Law of Cosines**

In triangle ABC, $B = 110.0°$, $a = 10.0$ centimeters, and $c = 15.0$ centimeters. See Figure 8.11. Find b.

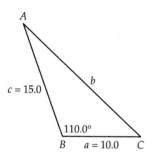

$c = 15.0$

b

$110.0°$

B $a = 10.0$ C

A

Figure 8.11

Solution   The Law of Cosines can be used because two sides and the included angle are known.

$$b^2 = a^2 + c^2 - 2ac \cos B$$

$$= 10.0^2 + 15.0^2 - 2(10.0)(15.0) \cos 110.0°$$

$$b = \sqrt{10.0^2 + 15.0^2 - 2(10.0)(15.0) \cos 110.0°}$$

$$b \approx 20.7 \text{ cm}$$

———————————————— ■ *Try Exercise* **12,** *page 423.*

EXAMPLE 2   **Solve a Triangle Using the Law of Cosines**

In triangle ABC, $a = 32$ feet, $b = 20$ feet, and $c = 40$ feet. Find B.

Solution $b^2 = a^2 + c^2 - 2ac \cos B$

$$\cos B = \frac{a^2 + c^2 - b^2}{2ac} \qquad \bullet \text{ Solve for } \cos B.$$

$$= \frac{32^2 + 40^2 - 20^2}{2(32)(40)} \qquad \bullet \text{ Substitute for } a, b, \text{ and } c \\ \text{and solve for angle } B.$$

$$B = \cos^{-1}\left(\frac{32^2 + 40^2 - 20^2}{2(32)(40)}\right)$$

$$B \approx 30° \qquad\qquad \bullet \text{ (nearest degree)}$$

———————————————— ■ *Try Exercise* **18,** *page 424.*

EXAMPLE 3   **Solve an Application Using the Law of Cosines**

A car traveled 3.0 miles at a heading of 78°. The road turned and the car traveled another 4.3 miles at a heading of 138°. Find the distance and the bearing of the car from the starting point.

Solution   Sketch a diagram. First find B.

$$B = 78° + (180° - 138°) = 120°$$

Use the Law of Cosines to find b.

$$b^2 = a^2 + c^2 - 2ac \cos B$$

$$= 4.3^2 + 3.0^2 - 2(4.3)(3.0) \cos 120°$$

$$b = \sqrt{3.0^2 + 4.3^2 - 2(3.0)(4.3) \cos 120°}$$

$$b \approx 6.4 \text{ mi}$$

Find A.

$$\cos A = \frac{b^2 + c^2 - a^2}{2bc}$$

$$A = \cos^{-1}\left(\frac{b^2 + c^2 - a^2}{2bc}\right) \approx \cos^{-1}\left(\frac{6.4^2 + 3.0^2 - 4.3^2}{(2)(6.4)(3.0)}\right)$$

$$A \approx 35°$$

The bearing of the present position of the car from the starting point A can be determined by calculating the measure of angle α in Figure 8.12.

$$\alpha = 180° - (78° + 35°) = 67°$$

The distance is approximately 6.4 miles, and the bearing to the nearest degree is S67°E.

■ *Try Exercise* **48**, *page 424.*

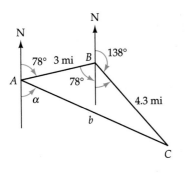

Figure 8.12

Remark   The measure of A in Example 3 can also be determined by using the Law of Sines.

Area

The formula $A = \frac{1}{2}bh$ is for the area of a triangle when the base and height are given. In this section, we will find the area of triangles when the height is not given. We will use K for the area of a triangle since A is often used to represent the measure of an angle.

Consider the areas of the acute and obtuse triangles in Figure 8.13.

Height of each triangle: $h = c \sin A$

Area of each triangle: $K = \frac{1}{2}bh$

$$K = \frac{1}{2}bc \sin A$$ • Substitute for h.

Thus we have established the following theorem.

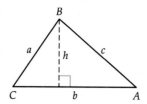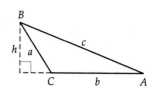

Figure 8.13
a. Acute triangle.   b. Obtuse triangle.

Area of a Triangle

The area K of triangle ABC is one-half the product of the lengths of any two sides and the sine of the included angle. Thus

$$K = \frac{1}{2}bc \sin A$$

$$K = \frac{1}{2}ab \sin C$$

$$K = \frac{1}{2}ac \sin B$$

EXAMPLE 4   **Find the Area of a Triangle**

Given angle $A = 62°$, $b = 12$ meters, and $c = 5.0$ meters, find the area of triangle ABC.

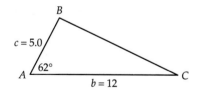

Figure 8.14

Solution   In Figure 8.14, two sides and the included angle of the triangle are given. Using the formula for area, we have

$$K = \frac{1}{2}bc \sin A = \frac{1}{2}(12)(5.0)(\sin 62°) \approx 26 \text{ m}^2.$$

━━━━━━━━━━━━━━━━━━━━━━━━━━━━━ ■ *Try Exercise* **26**, *page 424.*

When two angles and an included side are given, the Law of Sines is used to derive a formula for the area of a triangle. First, solve for c in the Law of Sines.

$$\frac{c}{\sin C} = \frac{b}{\sin B}$$

$$c = \frac{b \sin C}{\sin B}$$

Substitute for c in the first formula for area:

$$K = \frac{1}{2}bc \sin A$$

$$= \frac{1}{2}b\left(\frac{b \sin C}{\sin B}\right) \sin A$$

$$K = \frac{b^2 \sin C \sin A}{2 \sin B}$$

In like manner, the following two alternative formulas can be derived for the area of a triangle:

$$K = \frac{a^2 \sin B \sin C}{2 \sin A} \quad \text{and} \quad K = \frac{c^2 \sin A \sin B}{2 \sin C}$$

EXAMPLE 5   Find the Area of a Triangle

Given $A = 32°$, $C = 77°$, and $a = 14$ inches. Find the area of triangle ABC.

Solution To use the area formula, we need to know two angles and the included side. Therefore, we need to determine the measure of angle B.

$$B = 180° - 32° - 77° = 71°$$

Thus

$$K = \frac{a^2 \sin B \sin C}{2 \sin A} = \frac{14^2 \sin 71° \sin 77°}{2 \sin 32°} \approx 170 \text{ in}^2$$

■ *Try Exercise 28, page 424.*

The Law of Cosines can be used to derive *Heron's formula* for the area of a triangle in which three sides of the triangle are given.

Heron's Formula for Finding the Area of a Triangle

If a, b, and c are the lengths of the sides of a triangle, then the area K of the triangle is

$$K = \sqrt{s(s - a)(s - b)(s - c)}, \quad \text{where } s = \frac{1}{2}(a + b + c)$$

Remark Since s is one half of the perimeter of the triangle, it is called the **semi-perimeter.**

EXAMPLE 6   Find an Area by Heron's Formula

Find the area of the triangle with $a = 7.0$ meters, $b = 15$ meters, and $c = 12$ meters.

Solution   Calculate the semi-perimeter *s*.

$$s = \frac{a + b + c}{2} = \frac{7.0 + 15 + 12}{2} = 17$$

Use Heron's formula

$$K = \sqrt{s(s - a)(s - b)(s - c)}$$
$$= \sqrt{17(17 - 7.0)(17 - 15)(17 - 12)}$$
$$= \sqrt{1700} \approx 41 \text{ m}^2$$

■ *Try Exercise* **36**, *page 424.*

EXAMPLE 7   **Solve an Application Using Heron's Formula**

A commercial piece of real estate is priced at $6.50 per square foot. Find the cost of a triangular piece of commercial property measuring 260 feet by 320 feet by 440 feet.

Solution $$s = \frac{a + b + c}{2} = \frac{260 + 320 + 440}{2} = 510$$

$$K = \sqrt{s(s - a)(s - b)(s - c)}$$
$$= \sqrt{510(510 - 260)(510 - 320)(510 - 440)}$$
$$= \sqrt{1,695,750,000} \quad \bullet \text{ Area of property}$$

Find the cost of the property by multiplying the cost per square foot by the area.

$$\text{Cost} = 6.50(\sqrt{1,695,750,000}) \approx 267,667$$

The cost of the commercial property is approximately $267,667.

■ *Try Exercise* **56**, *page 425.*

EXERCISE SET 8.2

In Exercises 1 to 14, find the third side of the triangle.

1. $a = 12, b = 18, C = 44°$

2. $b = 30, c = 24, A = 120°$

3. $a = 120, c = 180, B = 56°$

4. $a = 400, b = 620, C = 116°$

5. $b = 60, c = 84, A = 13°$

6. $a = 122, c = 144, B = 48°$

7. $a = 9.0, b = 7.0, C = 72°$

8. $b = 12, c = 22, A = 55°$

9. $a = 4.6, b = 7.2, C = 124°$

10. $b = 12.3, c = 14.5, A = 6.5°$

11. $a = 25.9, c = 33.4, B = 84°$

12. $a = 14.2, b = 9.30, C = 9.20°$

13. $a = 122, c = 55.9, B = 44.2°$

14. $b = 444.8, c = 389.6, A = 78.44°$

In Exercises 15 to 24, given three sides of a triangle, find the specified angle.

15. $a = 25, b = 32, c = 40$; find A.

16. $a = 60, b = 88, c = 120$; find B.

17. $a = 8.0, b = 9.0, c = 12$; find C.

18. $a = 108$, $b = 132$, $c = 160$; find A.

19. $a = 80.0$, $b = 92.0$, $c = 124$; find B.

20. $a = 166$, $b = 124$, $c = 139$; find B.

21. $a = 1025$, $b = 625.0$, $c = 1420$; find C.

22. $a = 4.7$, $b = 3.2$, $c = 5.9$; find A.

23. $a = 32.5$, $b = 40.1$, $c = 29.6$; find B.

24. $a = 112.4$, $b = 96.80$, $c = 129.2$; find C.

In Exercises 25 to 36, find the area of the given triangle.

25. $A = 105°$, $b = 12$, $c = 24$

26. $B = 127°$, $a = 32$, $c = 25$

27. $A = 42°$, $B = 76°$, $c = 12$

28. $B = 102°$, $C = 27°$, $a = 8.5$

29. $a = 10$, $b = 12$, $c = 14$

30. $a = 32$, $b = 24$, $c = 36$

31. $B = 54.3°$, $a = 22.4$, $b = 26.9$

32. $C = 18.2°$, $b = 13.4$, $a = 9.84$

33. $A = 116°$, $B = 34°$, $c = 8.5$

34. $B = 42.8°$, $C = 76.3°$, $c = 17.9$

35. $a = 3.6$, $b = 4.2$, $c = 4.8$

36. $a = 13.3$, $b = 15.4$, $c = 10.2$

37. A plane leaves airport A and travels 560 miles to airport B at a bearing of N32°E. The plane leaves airport B and travels to airport C 320 miles away at a bearing of S72°E. Find the distance from airport A to airport C.

38. A developer has a triangular lot at the intersection of two streets. The streets meet at an angle of 72°, and the lot has 300 feet of frontage along one street and 416 feet of frontage along the other street. Find the length of the third side of the lot.

39. Two ships left a port at the same time. One ship traveled at a speed of 18 mph at a heading of 318°. The other ship traveled at a speed of 22 mph at a heading of 198°. Find the distance between the two ships after 10 hours of travel.

40. Find the distance across a lake, using the measurements shown in the figure.

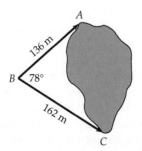

41. A regular hexagon is inscribed in a circle with a radius of 40 centimeters. Find the length of one side of the hexagon.

42. A regular pentagon is inscribed in a circle with a radius of 25 inches. Find the length of one side of the pentagon.

43. The lengths of the diagonals of a parallelogram are 20 inches and 32 inches. The diagonals intersect at an angle of 35°. Find the lengths of the sides of the parallelogram. (*Hint:* The diagonals of a parallelogram bisect one another.)

44. The sides of a parallelogram are 10 feet and 14 feet. The longer diagonal of the parallelogram is 18 feet. Find the length of the shorter diagonal of the parallelogram. (*Hint:* The diagonals of a parallelogram bisect one another.)

45. The sides of a parallelogram are 30 centimeters and 40 centimeters. The shorter diagonal of the parallelogram is 44 centimeters. Find the length of the longer diagonal of the parallelogram. (*Hint:* The diagonals of a parallelogram bisect one another.)

46. A triangular city lot has sides of 224 feet, 182 feet, and 165 feet. Find the angle between the longer two sides of the lot.

47. A plane traveling at 180 mph passes 400 feet directly over an observer. The plane is traveling along a path with an angle of elevation of 14°. Find the distance of the plane from the observer 10 seconds after the plane has passed directly overhead.

48. A ship leaves a port at a speed of 16 mph at a heading of 32°. One hour later another ship leaves the port at a speed of 22 mph at a heading of 254°. Find the distance between the ships 4 hours after the first ship leaves the port.

49. Find the area of a triangular piece of land that is bounded by sides of 236 meters, 620 meters, and 814 meters.

50. Find the area of a parallelogram whose diagonals are 24 inches and 32 inches and intersect at an angle of 40°.

51. Find the area of a parallelogram with sides of 12 meters and 18 meters and with one angle of 70°.

52. Find the area of a parallelogram with sides of 8 feet and 12 feet. The shorter diagonal is 10 feet.

53. Find the area of a square inscribed in a circle with a radius of 9 inches.

54. Find the area of a regular hexagon inscribed in a circle with a radius of 24 centimeters.

55. A commercial piece of real estate is priced at $2.20 per square foot. Find the cost of a triangular lot measuring 212 feet by 185 feet by 240 feet.

56. An industrial piece of real estate is priced at $4.15 per square foot. Find the cost of a triangular lot measuring 324 feet by 516 feet by 412 feet.

57. Find the number of acres in a pasture whose shape is a triangle measuring 800 feet by 1020 feet by 680 feet. (An acre is 43,560 square feet.)

58. Find the number of acres in a housing tract whose shape is a triangle measuring 420 yards by 540 yards by 500 yards. (An acre is 4840 square yards.)

Supplemental Exercises

59. Find the measure of the angle formed by the sides $P_1 P_2$ and $P_1 P_3$ of a triangle with the vertices at $P_1(-2, 4)$, $P_2(2, 1)$, and $P_3(4, -3)$.

60. The sides of a parallelogram are x and y, and the diagonals are w and z. Show that $w^2 + z^2 = 2x^2 + 2y^2$.

61. A rectangular box has dimensions of length 10 feet, width 4 feet, and height 3 feet. Find the angle between the diagonal of the bottom of the box and the diagonal of the end of the box.

62. A regular pentagon is inscribed in a circle with a radius of 4 inches. Find the perimeter of the pentagon.

63. An equilateral triangle is inscribed in a circle with a radius of 10 centimeters. Find the perimeter of the triangle.

64. Given a triangle ABC, prove that

$$a^2 = b^2 + c^2 - 2bc \cos A$$

65. Use the Law of Cosines to show that

$$\cos A = \frac{(b + c - a)(b + c + a)}{2bc} - 1$$

66. Prove that $K = xy \sin A$ for a parallelogram, where x and y are the lengths of adjacent sides and A is the measure of the angle between side x and side y.

67. Show that the area of the parallelogram in the figure is $K = 2ab \sin C$.

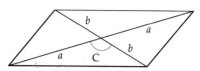

68. Given a regular hexagon inscribed in a circle with a radius of 10 inches, find the area of a sector (see the figure).

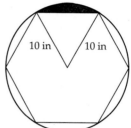

69. Find the volume of the triangular prisim piece of aluminum shown in the figure.

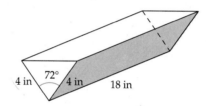

70. Show that the area of the circumscribed triangle in the figure is $K = rs$, where $s = \frac{a + b + c}{2}$.

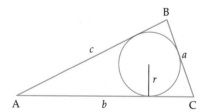

8.3 Vectors

In scientific applications, some measurements such as area, mass, distance, speed, and time are completely described by a real number and a unit. Examples include, 30 square feet (area), 25 meters/second (speed), and 5 hours (time). These measurements are **scalar quantities,** and the number used to indicate the magnitude of the measurement is called a **scalar.** Two other examples of scalar quantities are volume and temperature.

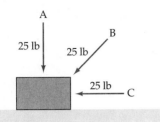

Figure 8.15

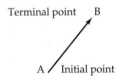

Figure 8.16

Figure 8.17

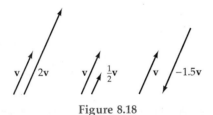

Figure 8.18

For other quantities, besides the numerical and unit description, it is also necessary to include a *direction* to describe the quantity completely. For example, applying a force of 25 pounds at various angles to a small metal box will influence how the box moves. In Figure 8.15, applying the 25-pound force straight down (A) will not move the box to the left. However, applying a 25-pound force (C) parallel to the floor will move the box along the floor.

Vector quantities have a *magnitude* (numerical and unit description) and a *direction*. Force is a vector quantity. Velocity is another. Velocity includes the speed (magnitude) and a direction. A velocity of 40 mph east is different from a velocity of 40 mph north. Displacement is another vector quantity; it consists of distance (a scalar) moved in a certain direction; for example, a displacement of 13 cm at an angle of 15° from the positive *x*-axis.

Definition of a Vector

> A **vector** is a directed line segment. The length of the line segment is the magnitude of the vector, and the direction of the vector is measured by an angle.

The point *A* for the vector in Figure 8.16 is called the initial point (or tail) of the vector, and the point *B* is the terminal point (or head) of the vector. An arrow over the letters $\overleftrightarrow{AB}$, an arrow over a single letter $\overleftrightarrow{V}$, or boldface type (**AB** or **V**) is used to denote a vector. The magnitude of the vector is the length of the line segment and is denoted by $\|\overleftrightarrow{AB}\|$, $\|\overleftrightarrow{V}\|$, $\|\mathbf{AB}\|$, or $\|\mathbf{V}\|$.

Equivalent vectors have the same magnitude and the same direction. From this definition, the location of a vector is *not* important; only its magnitude and its direction are of importance. The vectors in Figure 8.17 are equivalent. They have the same magnitude and direction.

Multiplying a vector by a positive real number (other than 1) changes the magnitude of the vector but not its direction. If **v** is any vector, then 2**v** is the vector that has the same direction as **v** but is twice the magnitude of **v**. The multiplication of 2 and **v** is called the **scalar multiplication** of vector **v** and the scalar 2. Multiplying a vector by a negative number *a* reverses the direction of the vector and multiplies the magnitude of the vector by $|a|$. See Figure 8.18.

The sum of two vectors, called the **resultant** vector, is the single equivalent vector that will have the same effect as the application of the two vectors. For example, a displacement 40 m along the positive *x*-axis and then 30 m in the positive *y* direction is equivalent to a vector of magnitude 50 m at an angle of approximately 53° to the positive *x*-axis. See Figure 8.19.

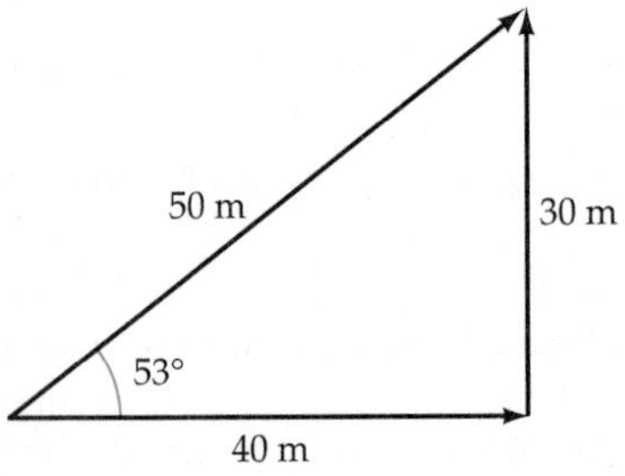

Vectors can be added graphically by using the *triangle method* or the *parallelogram method*. In the triangle method shown in Figure 8.20, the tail of **V** is placed at the head of **U**. The vector connecting the tail of **U** with the head of **V** is the sum **U** + **V**.

The parallelogram method of adding two vectors graphically places the tails of the two vectors **U** and **V** together, as in Figure 8.21. Complete the parallelogram so that **U** and **V** are sides of the parallelogram. The diagonal beginning at the tails of the two vectors is **U** + **V**.

Figure 8.19

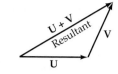

Figure 8.20

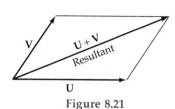

Figure 8.21

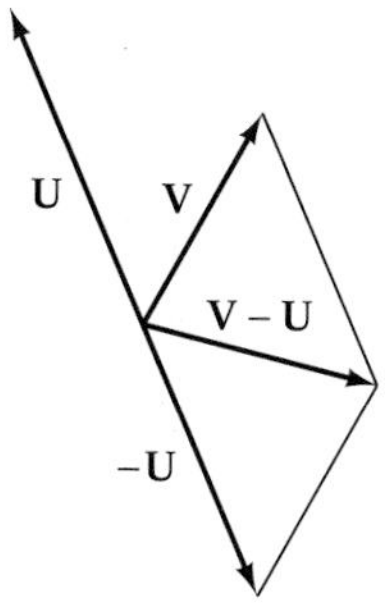

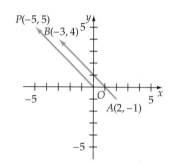

Figure 8.22

To find the difference between two vectors, first rewrite the expression as $\mathbf{V} - \mathbf{U} = \mathbf{V} + (-\mathbf{U})$. The difference is shown geometrically in Figure 8.22.

Vectors in a Coordinate Plane

By introducing a coordinate plane, it is possible to develop an analytic approach to vectors. Recall from our discussion about equivalent vectors that a vector can be moved in the plane as long as *the magnitude and direction* are not changed.

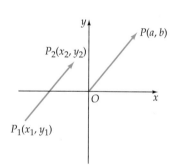

Figure 8.23

With this in mind, consider **AB**, whose initial point is $A(2, -1)$ and whose terminal point is $B(-3, 4)$. If this vector is moved so that the initial point is at the origin O, the terminal point becomes $P(-5, 5)$ as shown in Figure 8.23. The vector **OP** is equivalent to vector **AB**.

In Figure 8.24, let $P_1(x_1, y_1)$ be the initial point of a vector and $P_2(x_2, y_2)$ its terminal point. Then an equivalent vector **OP** with the initial point at the origin and terminal point $P(a, b)$, where $a = x_2 - x_1$ and $b = y_2 - y_1$. The vector **OP** can be denoted by $\mathbf{v} = \langle a, b \rangle$; a and b are called the **components** of the vector.

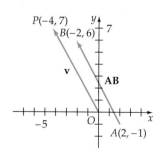

Figure 8.24

EXAMPLE 1   Find the Components of a Vector

Find the components of a vector whose tail is the point $A(2, -1)$ and whose head is the point $B(-2, 6)$. Write an equivalent vector **v** with initial point at the origin in terms of its components.

Solution   The vector is sketched as **AB** in Figure 8.25. The components of the vector are given by

$$a = x_2 - x_1 = -2 - 2 = -4 \qquad \text{and} \qquad b = y_2 - y_1 = 6 - (-1) = 7$$

A vector equivalent to **AB** with the initial point at the origin is $\mathbf{v} = \langle -4, 7 \rangle$.

■ *Try Exercise* **6**, *page 437.*

Figure 8.25

The magnitude and direction of a vector can be found from its components. The head of vector **v** sketched in Figure 8.25 is the ordered pair

$(-4, 7)$. From the Pythagorean Theorem,

$$\|\mathbf{v}\| = \sqrt{(-4)^2 + 7^2}$$
$$= \sqrt{16 + 49} = \sqrt{65}$$

Let θ be the angle made by the positive x-axis and $\mathbf{v}$. Let α be the reference angle for θ. Then

$$\tan \alpha = \left| \frac{b}{a} \right| = \left| \frac{7}{-4} \right| = \frac{7}{4}$$

$$\alpha = \tan^{-1} \frac{7}{4} \approx 60° \qquad \bullet\ \alpha \text{ is the reference angle}$$

$$\theta = 180° - 60° = 120° \qquad \bullet\ \theta \text{ is the angle made by the vector and the positive } x\text{-axis.}$$

The magnitude of $\mathbf{v}$ is $\sqrt{65}$, and its direction is $120°$ as measured from the positive x-axis. The angle between a vector and the positive x-axis is called the **direction angle** of the vector. Because $\mathbf{AB}$ in Figure 8.25 is equivalent to $\mathbf{v}$, $\|\mathbf{AB}\| = \sqrt{65}$ and the direction angle of $\mathbf{AB}$ is also $120°$.

Expressing vectors in terms of components provides a convenient method for performing operations on vectors.

Fundamental Vector Operations

If $\mathbf{v} = \langle a, b \rangle$ and $\mathbf{w} = \langle c, d \rangle$ are two vectors and k is a real number, then

1. $\|\mathbf{v}\| = \sqrt{a^2 + b^2}$
2. $\mathbf{v} + \mathbf{w} = \langle a, b \rangle + \langle c, d \rangle = \langle a + c, b + d \rangle$
3. $k\mathbf{v} = k\langle a, b \rangle = \langle ka, kb \rangle$

In terms of components, the zero vector $\mathbf{0} = \langle 0, 0 \rangle$. The additive inverse of a vector $\mathbf{v} = \langle a, b \rangle$ is given by $-\mathbf{v} = \langle -a, -b \rangle$.

EXAMPLE 2   **Perform Operations on Vectors**

Given $\mathbf{v} = \langle -2, 3 \rangle$ and $\mathbf{w} = \langle 4, -1 \rangle$, find each of the following:

a. $\|\mathbf{w}\|$ b. $\mathbf{v} + \mathbf{w}$ c. $-3\mathbf{v}$ d. $2\mathbf{v} - 3\mathbf{w}$

Solution

a. $\|\mathbf{w}\| = \sqrt{4^2 + (-1)^2} = \sqrt{17}$ c. $-3\mathbf{v} = -3\langle -2, 3 \rangle = \langle 6, -9 \rangle$

b. $\mathbf{v} + \mathbf{w} = \langle -2, 3 \rangle + \langle 4, -1 \rangle$ d. $2\mathbf{v} - 3\mathbf{w} = 2\langle -2, 3 \rangle - 3\langle 4, -1 \rangle$
$\qquad = \langle -2 + 4, 3 + (-1) \rangle$ $\qquad\qquad = \langle -4, 6 \rangle - \langle 12, -3 \rangle$
$\qquad = \langle 2, 2 \rangle$ $\qquad\qquad = \langle -16, 9 \rangle$

■ *Try Exercise 20, page 438.*

A **unit vector** is a vector whose magnitude is 1. For example, the vector $\mathbf{v} = \left\langle \frac{3}{5}, -\frac{4}{5} \right\rangle$ is a unit vector because

$$\|\mathbf{v}\| = \sqrt{\left(\frac{3}{5}\right)^2 + \left(-\frac{4}{5}\right)^2} = \sqrt{\frac{9}{25} + \frac{16}{25}} = \sqrt{\frac{25}{25}} = 1$$

Given any nonzero vector $\mathbf{v}$, we can obtain a unit vector in the direction of $\mathbf{v}$ by dividing each component of $\mathbf{v}$ by the magnitude of $\mathbf{v}$, $\|\mathbf{v}\|$.

EXAMPLE 3 **Find a Unit Vector**

Find a unit vector $\mathbf{u}$ in the direction of $\mathbf{v} = \langle -4, 2 \rangle$.

Solution   Find the magnitude of $\mathbf{v}$.

$$\|\mathbf{v}\| = \sqrt{(-4)^2 + 2^2} = \sqrt{16 + 4} = \sqrt{20} = 2\sqrt{5}$$

Divide each component of $\mathbf{v}$ by $\|\mathbf{v}\|$.

$$\mathbf{u} = \left\langle \frac{-4}{2\sqrt{5}}, \frac{2}{2\sqrt{5}} \right\rangle = \left\langle \frac{-2}{\sqrt{5}}, \frac{1}{\sqrt{5}} \right\rangle = \left\langle -\frac{2\sqrt{5}}{5}, \frac{\sqrt{5}}{5} \right\rangle.$$

A unit vector in the direction of $\mathbf{v}$ is $\mathbf{u}$.

■ *Try Exercise* **8,** *page 437.*

Two unit vectors, one parallel to the x-axis and one parallel to the y-axis, are of special importance. See Figure 8.26.

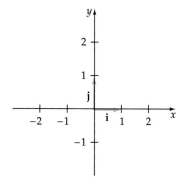

Figure 8.26

Definition of Unit Vectors i and j

$$\mathbf{i} = \langle 1, 0 \rangle \qquad \mathbf{j} = \langle 0, 1 \rangle$$

The vector $\mathbf{v} = \langle 3, 4 \rangle$ can be written in terms of the unit vectors $\mathbf{i}$ and $\mathbf{j}$ as shown in Figure 8.27.

$$\langle 3, 4 \rangle = \langle 3, 0 \rangle + \langle 0, 4 \rangle \quad \bullet \text{ Vector Addition Property}$$
$$= 3\langle 1, 0 \rangle + 4\langle 0, 1 \rangle \quad \bullet \text{ Scalar multiplication of a vector}$$
$$= 3\mathbf{i} + 4\mathbf{j} \quad \bullet \text{ Definition of i and j}$$

By means of scalar multiplication and addition of vectors, any vector can be expressed in terms of the unit vectors $\mathbf{i}$ and $\mathbf{j}$. Let $\mathbf{v} = \langle a_1, a_2 \rangle$. Then

$$\mathbf{v} = \langle a_1, a_2 \rangle$$
$$= a_1\langle 1, 0 \rangle + a_2\langle 0, 1 \rangle$$
$$= a_1\mathbf{i} + a_2\mathbf{j}$$

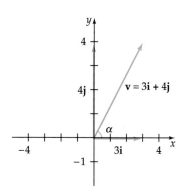

Figure 8.27

This gives the following result.

Representation of a Vector in Terms of i and j

If $\mathbf{v}$ is a vector and $\mathbf{v} = \langle a_1, a_2 \rangle$, then $\mathbf{v} = a_1\mathbf{i} + a_2\mathbf{j}$.

The rules for addition and scalar multiplication of vectors can be restated in terms of **i** and **j**. If $\mathbf{v} = a_1\mathbf{i} + a_2\mathbf{j}$ and $\mathbf{w} = b_1\mathbf{i} + b_2\mathbf{j}$, then

$$\begin{aligned} \mathbf{v} + \mathbf{w} &= (a_1\mathbf{i} + a_2\mathbf{j}) + (b_1\mathbf{i} + b_2\mathbf{j}) \\ &= (a_1 + b_1)\mathbf{i} + (a_2 + b_2)\mathbf{j} \\ k\mathbf{v} &= k(a_1\mathbf{i} + a_2\mathbf{j}) \\ &= ka_1\mathbf{i} + ka_2\mathbf{j} \end{aligned}$$

EXAMPLE 4 **Operate on Vectors Written in Terms of i and j**

Given $\mathbf{v} = 3\mathbf{i} - 4\mathbf{j}$ and $\mathbf{w} = 5\mathbf{i} + 3\mathbf{j}$, find $3\mathbf{v} - 2\mathbf{w}$

Solution

$$\begin{aligned} 3\mathbf{v} - 2\mathbf{w} &= 3(3\mathbf{i} - 4\mathbf{j}) - 2(5\mathbf{i} + 3\mathbf{j}) \\ &= (9\mathbf{i} - 12\mathbf{j}) - (10\mathbf{i} + 6\mathbf{j}) \\ &= (9 - 10)\mathbf{i} + (-12 - 6)\mathbf{j} \\ &= -\mathbf{i} - 18\mathbf{j} \end{aligned}$$

■ *Try Exercise* **26,** *page 438.*

The components a_1 and a_2 of the vector $\mathbf{v} = \langle a_1, a_2 \rangle$ can be expressed in terms of the magnitude of **v** and the direction angle of **v** (the angle that **v** makes with the positive x-axis). Consider the vector **v** in Figure 8.28. Then

$$\|\mathbf{v}\| = \sqrt{(a_1)^2 + (a_2)^2}$$

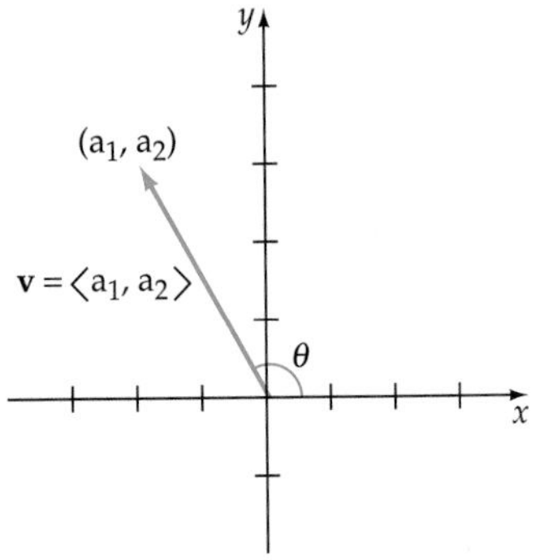

Figure 8.28

From the definitions of sine and cosine, we have

$$\cos\theta = \frac{a_1}{\|\mathbf{v}\|} \quad \text{and} \quad \sin\theta = \frac{a_2}{\|\mathbf{v}\|}$$

Rewriting the last two equations, we find that the components of **v** are

$$a_1 = \|\mathbf{v}\|\cos\theta \quad \text{and} \quad a_2 = \|\mathbf{v}\|\sin\theta$$

Horizontal and Vertical Components of a Vector

Let $\mathbf{v} = \langle a_1, a_2 \rangle$, where $\mathbf{v} \neq 0$, the zero vector. Then

$$a_1 = \|\mathbf{v}\|\cos\theta \quad \text{and} \quad a_2 = \|\mathbf{v}\|\sin\theta$$

where θ is the angle between the positive x-axis and **v**.

The **horizontal component** of **v** is $\|\mathbf{v}\|\cos\theta$. The **vertical component** of **v** is $\|\mathbf{v}\|\sin\theta$.

Any nonzero vector can be written in terms of its horizontal and vertical components. Let $\mathbf{v} = a_1\mathbf{i} + a_2\mathbf{j}$. Then

$$\mathbf{v} = a_1\mathbf{i} + a_2\mathbf{j}$$
$$= (\|\mathbf{v}\| \cos \theta)\mathbf{i} + (\|\mathbf{v}\| \sin \theta)\mathbf{j}$$
$$= \|\mathbf{v}\|(\cos \theta\mathbf{i} + \sin \theta\mathbf{j})$$

$\|\mathbf{v}\|$ is the magnitude of $\mathbf{v}$, and the vector $\cos \theta\mathbf{i} + \sin \theta\mathbf{j}$ is a unit vector (Why?)[1] The last equation shows that any vector $\mathbf{v}$ can be written as the product of its magnitude and a unit vector in the direction of $\mathbf{v}$.

EXAMPLE 5   **Find the Horizontal and Vertical Components of a Vector**

Find the approximate horizontal and vertical components of a vector $\mathbf{v}$ of magnitude 10 meters with direction angle 228°. Write the vector in the form $\mathbf{v} = a_1\mathbf{i} + a_2\mathbf{j}$.

Solution

$$a_1 = 10 \cos 228° \approx -6.7$$
$$a_2 = 10 \sin 228° \approx -7.4$$

The approximate horizontal and vertical components are -6.7 and -7.4, respectively.

$$\mathbf{v} \approx -6.7\mathbf{i} - 7.4\mathbf{j}$$

■ *Try Exercise 36, page 438.*

Application Problems Using Vectors

Consider an object on which two vectors are acting simultaneously. This occurs when a boat is moving in a current or an airplane is flying in a wind. The **airspeed** of a plane is the speed of the plane as if there were no wind. The actual velocity of a plane is the velocity relative to the ground. The magnitude of the actual velocity is the **ground speed.**

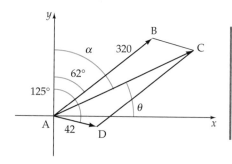

Figure 8.29

EXAMPLE 6   **Solve an Application Involving Airspeed**

An airplane is traveling with an airspeed of 320 mph and a heading of 62°. A wind of 42 mph is blowing at a heading of 125°. Find the ground speed and the course of the airplane.

Solution   Sketch a diagram similar to Figure 8.29 showing the relevant vector. **AB** represents the heading and the airspeed, **AD** represents

[1] $\|\cos \theta\mathbf{i} + \sin \theta\mathbf{j}\| = \sqrt{\cos^2\theta + \sin^2\theta} = \sqrt{1} = 1$

the wind velocity, and **AC** represents the course and the ground speed. By vector addition, **AC** = **AB** + **AD**. From the figure,

$$\mathbf{AB} = 320(\cos 28°\mathbf{i} + \sin 28°\mathbf{j})$$

$$\mathbf{AD} = 42[\cos(-35°)\mathbf{i} + \sin(-35°)\mathbf{j}]$$

$$\mathbf{AC} = 320(\cos 28°\mathbf{i} + \sin 28°\mathbf{j}) + 42[\cos(-35°)\mathbf{i} + \sin(-35°)\mathbf{j}]$$

$$\approx (282.5\mathbf{i} + 150.2\mathbf{j}) + (34.4\mathbf{i} - 24.1\mathbf{j})$$

$$= 316.9\mathbf{i} + 126.1\mathbf{j}$$

AC is the course of the plane. The ground speed is ‖**AC**‖. The heading is $\alpha = 90° - \theta$.

$$\|\mathbf{AC}\| = \sqrt{(316.9)^2 + (126.1)^2} \approx 340$$

$$\alpha = 90° - \theta = 90° - \tan^{-1}\left(\frac{126.1}{316.9}\right) \approx 68°$$

The ground speed is approximately 340 mph at a heading of 68°.

■ *Try Exercise* **40**, *page 438.*

There are numerous problems involving force (a vector) that can be solved by using vectors. One type involves objects that are resting on a ramp. For these problems, we frequently try to find the components of a force vector relative to the ramp rather than to the x-axis.

EXAMPLE 7   **Solve an Application Involving Force**

A 110-pound box is on a 24° ramp. Find the magnitude of the component of the force that is parallel to the ramp.

Solution   The force-of-gravity vector (110 pounds) is the sum of two components, one parallel to the ramp and the other perpendicular (called the *normal component*) to the ramp. (See Figure 8.30) **AB** is the vector that represents the force tending to move the box down the plane. Because triangle *DAB* is a right triangle and ∠*ADB* is 24°,

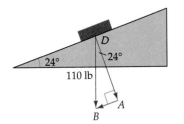

Figure 8.30

$$\sin 24° = \frac{\|\mathbf{AB}\|}{110}$$

$$\|\mathbf{AB}\| = 110 \sin 24° \approx 45$$

The component of force parallel to the ramp is approximately 45 pounds.

■ *Try Exercise* **42**, *page 438.*

Dot Product

We have considered the product of a real number (scalar) and a vector. We now turn our attention to the product of two vectors. Finding the *dot product* of two vectors is one way to multiply a vector by a vector. The dot product of two vectors is a real number and *not* a vector.[2] The dot product is also called the *inner product* or the *scalar product*. This product is useful in engineering and physics.

Definition of Dot Product

Given $\mathbf{v} = \langle a, b \rangle$ and $\mathbf{w} = \langle c, d \rangle$, the dot product of $\mathbf{v}$ and $\mathbf{w}$ is given by

$$\mathbf{v} \cdot \mathbf{w} = ac + bd$$

EXAMPLE 8 **Find the Dot Product of Two Vectors**

Find the dot product of $\mathbf{v} = \langle 6, -2 \rangle$ and $\mathbf{w} = \langle -2, 4 \rangle$.

Solution

$$\mathbf{v} \cdot \mathbf{w} = 6(-2) + (-2)4 = -12 - 8 = -20$$

■ *Try Exercise 50, page 438.*

If the vectors in Example 8 were given in terms of the vectors $\mathbf{i}$ and $\mathbf{j}$, then $\mathbf{v} = 6\mathbf{i} - 2\mathbf{j}$ and $\mathbf{w} = -2\mathbf{i} + 4\mathbf{j}$. In this case,

$$\mathbf{v} \cdot \mathbf{w} = (6\mathbf{i} - 2\mathbf{j}) \cdot (-2\mathbf{i} + 4\mathbf{j})$$
$$= 6(-2) + (-2)4 = -20$$

Caution From Example 8, *the dot product of two vectors is a real number, not a vector.*

Properties of the Dot Product

In the following properties, $\mathbf{u}$, $\mathbf{v}$, and $\mathbf{w}$ are vectors and a is a scalar.

1. $\mathbf{v} \cdot \mathbf{w} = \mathbf{w} \cdot \mathbf{v}$
2. $\mathbf{u} \cdot (\mathbf{v} + \mathbf{w}) = \mathbf{u} \cdot \mathbf{v} + \mathbf{u} \cdot \mathbf{w}$
3. $a(\mathbf{u} \cdot \mathbf{v}) = (a\mathbf{u}) \cdot \mathbf{v} = \mathbf{u} \cdot (a\mathbf{v})$
4. $\mathbf{v} \cdot \mathbf{v} = \|\mathbf{v}\|^2$
5. $\mathbf{0} \cdot \mathbf{v} = 0$

The proofs of these properties follow from the definition of dot product. Here is the proof of the fourth property. Let $\mathbf{v} = a\mathbf{i} + b\mathbf{j}$.

$$\mathbf{v} \cdot \mathbf{v} = (a\mathbf{i} + b\mathbf{j}) \cdot (a\mathbf{i} + b\mathbf{j})$$
$$= a^2 + b^2 = \|\mathbf{v}\|^2$$

---

[2] Another product of two vectors is the *cross product*. The cross product of two vectors is another vector. It is not used in this text.

Rewriting the fourth property of the dot product yields an alternative way of expressing the magnitude of a vector.

Magnitude of a Vector in Terms of the Dot Product

$$\text{If } \mathbf{v} = \langle a, b \rangle, \text{ then } \|\mathbf{v}\| = \sqrt{\mathbf{v} \cdot \mathbf{v}}.$$

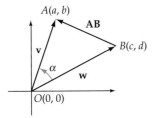

Figure 8.31

The Law of Cosines can be used to derive an alternative formula for the dot product. Consider the vectors $\mathbf{v} = \langle a, b \rangle$ and $\mathbf{w} = \langle c, d \rangle$ as shown in Figure 8.31.

Using the Law of Cosines for triangle OAB, we have

$$\|\mathbf{AB}\|^2 = \|\mathbf{v}\|^2 + \|\mathbf{w}\|^2 - 2\|\mathbf{v}\|\|\mathbf{w}\| \cos \alpha$$

By the distance formula, $\|\mathbf{AB}\|^2 = (a - c)^2 + (b - d)^2$, $\|\mathbf{v}\|^2 = a^2 + b^2$, and $\|\mathbf{w}\|^2 = c^2 + d^2$. Thus

$$(a - c)^2 + (b - d)^2 = (a^2 + b^2) + (c^2 + d^2) - 2\|\mathbf{v}\|\|\mathbf{w}\| \cos \alpha$$

$$a^2 - 2ac + c^2 + b^2 - 2bd + d^2 = a^2 + b^2 + c^2 + d^2 - 2\|\mathbf{v}\|\|\mathbf{w}\| \cos \alpha$$

$$-2ac - 2bd = -2\|\mathbf{v}\|\|\mathbf{w}\| \cos \alpha$$

$$ac + bd = \|\mathbf{v}\|\|\mathbf{w}\| \cos \alpha$$

$$\mathbf{v} \cdot \mathbf{w} = \|\mathbf{v}\|\|\mathbf{w}\| \cos \alpha \qquad \bullet \ \mathbf{v} \cdot \mathbf{w} = ac + bd$$

Alternative Formula for the Dot Product

If $\mathbf{v}$ and $\mathbf{w}$ are two nonzero vectors and α is the smallest positive angle between $\mathbf{v}$ and $\mathbf{w}$, then $\mathbf{v} \cdot \mathbf{w} = \|\mathbf{v}\|\|\mathbf{w}\| \cos \alpha$.

Solving the alternative formula for the dot product for $\cos \alpha$, we have

$$\cos \alpha = \frac{\mathbf{v} \cdot \mathbf{w}}{\|\mathbf{v}\|\|\mathbf{w}\|}$$

This gives the following result.

Angle Between Two Vectors

If $\mathbf{v}$ and $\mathbf{w}$ are two nonzero vectors and α is the smallest positive angle between $\mathbf{v}$ and $\mathbf{w}$, then $\cos \alpha = \dfrac{\mathbf{v} \cdot \mathbf{w}}{\|\mathbf{v}\|\|\mathbf{w}\|}$.

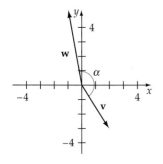

Figure 8.32

EXAMPLE 9 Find the Angle Between Two Vectors

Find the measure of smallest positive angle between the vectors $\mathbf{v} = 2\mathbf{i} - 3\mathbf{j}$ and $\mathbf{w} = -\mathbf{i} + 5\mathbf{j}$ as shown in Figure 8.32.

Solution Use the equation for the angle between two vectors.

$$\cos \alpha = \frac{\mathbf{v} \cdot \mathbf{w}}{\|\mathbf{v}\|\,\|\mathbf{w}\|}$$

$$= \frac{(2\mathbf{i} - 3\mathbf{j}) \cdot (-\mathbf{i} + 5\mathbf{j})}{\left(\sqrt{2^2 + (-3)^2}\right)\left(\sqrt{(-1)^2 + 5^2}\right)}$$

$$= \frac{-2 - 15}{\sqrt{13}\,\sqrt{26}} = \frac{-17}{\sqrt{338}}$$

$$\alpha = \cos^{-1}\left(\frac{-17}{\sqrt{338}}\right) \approx 157.6°$$

The angle between the two vectors is approximately 157.6°.

■ Try Exercise 60, page 438.

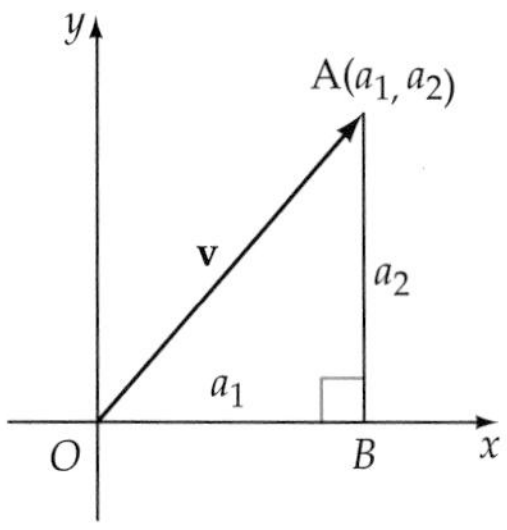

Consider $\mathbf{v} = \langle a_1, a_2 \rangle$ as shown in Figure 8.33 and the right triangle formed by drawing a line from the tip of $\mathbf{v}$ to the x-axis. The first component a_1 is the magnitude (length) of the line segment OB on the x-axis. In some application problems, it is necessary to find the magnitude of the line segment along a line other than the x-axis. The ramp problem in Example 7 is such an instance.

Let $\mathbf{v} = \langle a_1, a_2 \rangle$ and $\mathbf{w} = \langle b_1, b_2 \rangle$ be two nonzero vectors. Two possible configurations, one for which α is an acute angle and one for which α is an obtuse angle, are shown in Figure 8.34. In each case, a right triangle is formed by drawing a line segment from the tip of $\mathbf{v}$ to a line through $\mathbf{w}$.

Figure 8.33

Definition of the Scalar Projection of v on w

If $\mathbf{v}$ and $\mathbf{w}$ are two nonzero vectors and α is the smallest positive angle between $\mathbf{v}$ and $\mathbf{w}$, then the scalar projection of $\mathbf{v}$ on $\mathbf{w}$, $\text{proj}_{\mathbf{w}}\mathbf{v}$, is given by

$$\text{proj}_{\mathbf{w}}\mathbf{v} = \|\mathbf{v}\| \cos \alpha.$$

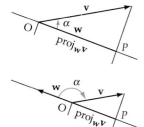

Figure 8.34

From the dot product, $\mathbf{v} \cdot \mathbf{w} = \|\mathbf{v}\|\|\mathbf{w}\| \cos \alpha$. Solving for $\|\mathbf{v}\| \cos \alpha$, which is $\text{proj}_{\mathbf{w}}\mathbf{v}$, we have

$$\text{proj}_{\mathbf{w}}\mathbf{v} = \frac{\mathbf{v} \cdot \mathbf{w}}{\|\mathbf{w}\|}$$

Remark When the angle α between the two vectors is an acute angle, $\text{proj}_{\mathbf{w}}\mathbf{v}$ is a positive number; $\text{proj}_{\mathbf{w}}\mathbf{v}$ is a negative number when α is an obtuse angle.

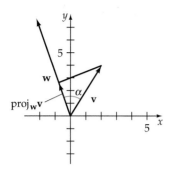

Figure 8.35

EXAMPLE 10   Find the Projection of v on w

Given $\mathbf{v} = 2\mathbf{i} + 4\mathbf{j}$ and $\mathbf{w} = -2\mathbf{i} + 8\mathbf{j}$ as shown in Figure 8.35, find $\text{proj}_{\mathbf{w}}\mathbf{v}$.

Solution   Use the equation $\text{proj}_{\mathbf{w}}\mathbf{v} = \dfrac{\mathbf{v} \cdot \mathbf{w}}{\|\mathbf{w}\|}$.

$$\text{proj}_{\mathbf{w}}\mathbf{v} = \frac{(2\mathbf{i} + 4\mathbf{j}) \cdot (-2\mathbf{i} + 8\mathbf{j})}{\sqrt{(-2)^2 + 8^2}} = \frac{28}{\sqrt{68}} = \frac{14\sqrt{17}}{17}$$

■ *Try Exercise 62, page 438.*

Parallel and Perpendicular Vectors

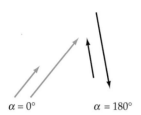

Figure 8.36

Two vectors are *parallel* when the angle α between the vectors is 0° or 180°, as shown in Figure 8.36. When the angle α is 0°, the vectors point in the same direction; the vectors point in opposite directions when α is 180°.

Let $\mathbf{v} = a_1\mathbf{i} + b_1\mathbf{j}$, c be a real number, and $\mathbf{w} = c\mathbf{v}$. Because $\mathbf{w}$ is a constant multiple of $\mathbf{v}$, $\mathbf{w}$ and $\mathbf{v}$ are parallel vectors. When $c > 0$, the vectors point in the same direction. When $c < 0$, the vectors point in opposite directions. In Exercise 88 on page 439, you are asked to show that for $\mathbf{v}$ and $\mathbf{w}$ as given, $\cos \alpha = \pm 1$ depending on the sign of c. Thus $\alpha = 0°$ or $\alpha = 180°$.

Two vectors are *perpendicular* when the angle between the vectors is 90°. See Figure 8.37. Perpendicular vectors are referred to as orthogonal. If $\mathbf{v}$ and $\mathbf{w}$ are two nonzero orthogonal vectors, then from the formula for the angle between two vectors and the fact that $\cos \alpha = 0$ (Why?)[3], we have

Figure 8.37

$$0 = \frac{\mathbf{v} \cdot \mathbf{w}}{\|\mathbf{v}\| \|\mathbf{w}\|}$$

If a fraction equals zero, the numerator must be zero. Thus, for orthogonal vectors $\mathbf{v}$ and $\mathbf{w}$, $\mathbf{v} \cdot \mathbf{w} = 0$. This gives the following result.

Condition for Perpendicular Vectors

Two nonzero vectors $\mathbf{v}$ and $\mathbf{w}$ are orthogonal if and only if $\mathbf{v} \cdot \mathbf{w} = 0$.

Work: An Application of the Dot Product

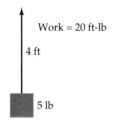

Figure 8.38

When a 5-pound force is used to lift a box from the ground a distance of 4 feet, *work* is done. The amount of work is the product of the force on the box and the distance the box is moved. In this case, the work is 20 foot-pounds. When the box is lifted, the force and the displacement vector (the direction and the distance in which the box was moved) were in the same direction. (See Figure 8.38.)

Now consider a sled being pulled by a child along the ground by a rope attached to the sled, as shown in Figure 8.39. The force vector (along the rope)

Figure 8.39

---

[3] If $\mathbf{v}$ and $\mathbf{w}$ are orthogonal, then $\alpha = 90°$ and $\cos 90° = 0$.

is *not* in the same direction as the displacement vector (parallel to the ground). In this case, the dot product is used to determine the work done by the force.

Definition of Work

> The work W done by a force $\mathbf{F}$ applied along a displacement $\mathbf{s}$ is
> $$W = \mathbf{F} \cdot \mathbf{s}$$

In the case of the child pulling the sled 7 feet, the work done is

$$W = \mathbf{F} \cdot \mathbf{s}$$
$$= \|\mathbf{F}\| \|\mathbf{s}\| \cos \alpha \quad \bullet \; \alpha \text{ is the angle between } \mathbf{F} \text{ and } \mathbf{s}.$$
$$= (25)(7) \cos 37° \approx 140 \text{ foot-pounds}$$

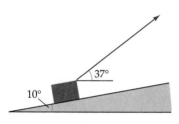

Figure 8.40

EXAMPLE 11 **Solve a Work Problem**

A force of 50 pounds on a rope is used to drag a box up a ramp that is inclined 10°. If the rope makes an angle of 37° with the ground, find the amount of work done in moving the box 15 feet along the ramp. See Figure 8.40.

Solution   We will provide two solutions to this example.

1. From the last section, $\mathbf{u} = \cos 10°\mathbf{i} + \sin 10°\mathbf{j}$ is a unit vector parallel to the ramp. Multiplying $\mathbf{u}$ by 15 (the magnitude of the displacement vector) gives $\mathbf{s} = 15(\cos 10°\mathbf{i} + \sin 10°\mathbf{j})$. Similarly, the force vector is $\mathbf{F} = 50(\cos 37°\mathbf{i} + \sin 37°\mathbf{j})$. The work done is given by the dot product.

$$W = \mathbf{F} \cdot \mathbf{s} = 50(\cos 37°\mathbf{i} + \sin 37°\mathbf{j}) \cdot 15(\cos 10°\mathbf{i} + \sin 10°\mathbf{j})$$
$$= [50 \cdot 15](\cos 37° \cos 10° + \sin 37° \sin 10°) \approx 668.3 \text{ ft-lbs}$$

2. When we write the work equation as $W = \|\mathbf{F}\| \|\mathbf{s}\| \cos \alpha$, α is the angle between the force and the displacement. Thus $\alpha = 37° - 10° = 27°$. The work done is

$$W = \|\mathbf{F}\| \|\mathbf{s}\| \cos \alpha = 50 \cdot 15 \cdot \cos 27° \approx 668.3 \text{ foot-pounds}.$$

■ *Try Exercise 70, page 438.*

EXERCISE SET 8.3

In Exercises 1 to 6, find the components of a vector with the given initial and terminal points. Write an equivalent vector in terms of its components.

1. $P_1(-3, 0)$; $P_2(4, -1)$ **2.** $P_1(5, -1)$; $P_2(3, 1)$

3. $P_1(4, 2)$; $P_2(-3, -3)$ **4.** $P_1(0, -3)$; $P_2(0, 4)$

5. $P_1(2, -5)$; $P_2(2, 3)$ **6.** $P_1(3, -2)$; $P_2(3, 0)$

In Exercises 7 to 14, find the magnitude and direction of each vector. Find the unit vector in the direction of the given vector.

7. $\mathbf{v} = \langle -3, 4 \rangle$ **8.** $\mathbf{v} = \langle 6, 10 \rangle$

9. $\mathbf{v} = \langle 20, -40 \rangle$ **10.** $\mathbf{v} = \langle -50, 30 \rangle$

11. $\mathbf{v} = 2\mathbf{i} - 4\mathbf{j}$ **12.** $\mathbf{v} = -5\mathbf{i} + 6\mathbf{j}$

13. $\mathbf{v} = 42\mathbf{i} - 18\mathbf{j}$ **14.** $\mathbf{v} = -22\mathbf{i} - 32\mathbf{j}$

In Exercises 15 to 23, perform the indicated operations where $\mathbf{u} = \langle -2, 4 \rangle$ and $\mathbf{v} = \langle -3, -2 \rangle$.

15. $3\mathbf{u}$ **16.** $-4\mathbf{v}$ **17.** $2\mathbf{u} - \mathbf{v}$

18. $4\mathbf{v} - 2\mathbf{u}$ **19.** $\frac{2}{3}\mathbf{u} + \frac{1}{6}\mathbf{v}$ **20.** $\frac{3}{4}\mathbf{u} - 2\mathbf{v}$

21. $\|\mathbf{u}\|$ **22.** $\|\mathbf{v} + 2\mathbf{u}\|$ **23.** $\|3\mathbf{u} - 4\mathbf{v}\|$

In Exercises 24 to 32, perform the indicated operations where $\mathbf{u} = 3\mathbf{i} - 2\mathbf{j}$ and $\mathbf{v} = -2\mathbf{i} + 3\mathbf{j}$.

24. $-2\mathbf{u}$ **25.** $4\mathbf{v}$ **26.** $3\mathbf{u} + 2\mathbf{v}$

27. $6\mathbf{u} + 2\mathbf{v}$ **28.** $\frac{1}{2}\mathbf{u} - \frac{3}{4}\mathbf{v}$ **29.** $\frac{2}{3}\mathbf{v} + \frac{3}{4}\mathbf{u}$

30. $\|\mathbf{v}\|$ **31.** $\|\mathbf{u} - 2\mathbf{v}\|$ **32.** $\|2\mathbf{v} + 3\mathbf{u}\|$

In Exercises 33 to 36, find the horizontal and vertical components of each vector. Write an equivalent vector in the form $\mathbf{v} = a_1\mathbf{i} + a_2\mathbf{j}$.

33. Magnitude = 5, direction angle = 27°

34. Magnitude = 4, direction angle = 127°

35. Magnitude = 4, direction angle = $\pi/4$

36. Magnitude = 2, direction angle = $8\pi/7$

37. A plane is flying at an airspeed of 340 mph at a heading of 124°. A wind of 45 mph is blowing from the west. Find the ground speed of the plane.

38. A person who can row 2.6 mph in still water wants to row due east across a river. The river is flowing from the north at a rate of 0.8 mph. Determine the heading of the boat that will be required for it to travel due east across the river.

39. A pilot is flying at a heading of 96° at 225 mph. A 50 mph wind is blowing from the southwest at a heading of 37°. Find the ground speed and course of the plane.

40. The captain of a boat is steering at a heading of 327° at 18 mph. The current is flowing at 4 mph at a heading of 60°. Find the course of the boat.

41. Find the magnitude of force necessary to keep a 3000-pound car from sliding down a ramp inclined at an angle of 5.6°.

42. A 120-pound force keeps an 800-pound object from sliding down an inclined ramp. Find the angle of the ramp.

43. A 25-pound box is resting on a ramp that is inclined 9.0°. Find the magnitude of the normal component of force.

44. Find the magnitude of the normal component of force for a 50-pound crate that is resting on a ramp that is inclined 12°.

In Exercises 45 to 52, find the dot product of the vectors.

45. $\mathbf{v} = \langle 3, -2 \rangle$; $\mathbf{w} = \langle 1, 3 \rangle$ **46.** $\mathbf{v} = \langle 2, 4 \rangle$; $\mathbf{w} = \langle 0, 2 \rangle$

47. $\mathbf{v} = \langle 4, 1 \rangle$; $\mathbf{w} = \langle -1, 4 \rangle$ **48.** $\mathbf{v} = \langle 2, -3 \rangle$; $\mathbf{w} = \langle 3, 2 \rangle$

49. $\mathbf{v} = \mathbf{i} + 2\mathbf{j}$; $\mathbf{w} = -\mathbf{i} + \mathbf{j}$ **50.** $\mathbf{v} = 5\mathbf{i} + 3\mathbf{j}$; $\mathbf{w} = 4\mathbf{i} - 2\mathbf{j}$

51. $\mathbf{v} = 6\mathbf{i} - 4\mathbf{j}$; $\mathbf{w} = -2\mathbf{i} - 3\mathbf{j}$

52. $\mathbf{v} = -4\mathbf{i} + 2\mathbf{j}$; $\mathbf{w} = -2\mathbf{i} - 4\mathbf{j}$

In Exercises 53 to 60, find the angle between the two vectors. State which pair of vectors is orthogonal.

53. $\mathbf{v} = \langle 2, -1 \rangle$; $\mathbf{w} = \langle 3, 4 \rangle$ **54.** $\mathbf{v} = \langle 1, -5 \rangle$; $\mathbf{w} = \langle -2, 3 \rangle$

55. $\mathbf{v} = \langle 0, 3 \rangle$; $\mathbf{w} = \langle 2, 2 \rangle$ **56.** $\mathbf{v} = \langle -1, 7 \rangle$; $\mathbf{w} = \langle 3, -2 \rangle$

57. $\mathbf{v} = 5\mathbf{i} - 2\mathbf{j}$; $\mathbf{w} = 2\mathbf{i} + 5\mathbf{j}$

58. $\mathbf{v} = 8\mathbf{i} + \mathbf{j}$; $\mathbf{w} = -\mathbf{i} + 8\mathbf{j}$

59. $\mathbf{v} = 5\mathbf{i} + 2\mathbf{j}$; $\mathbf{w} = -5\mathbf{i} - 2\mathbf{j}$

60. $\mathbf{v} = 3\mathbf{i} - 4\mathbf{j}$; $\mathbf{w} = 6\mathbf{i} - 12\mathbf{j}$

In Exercises 61 to 68, find $\text{proj}_\mathbf{w}\mathbf{v}$.

61. $\mathbf{v} = \langle 6, 7 \rangle$; $\mathbf{w} = \langle 3, 4 \rangle$ **62.** $\mathbf{v} = \langle -7, 5 \rangle$; $\mathbf{w} = \langle -4, 1 \rangle$

63. $\mathbf{v} = \langle -3, 4 \rangle$; $\mathbf{w} = \langle 2, 5 \rangle$ **64.** $\mathbf{v} = \langle 2, 4 \rangle$; $\mathbf{w} = \langle -1, 5 \rangle$

65. $\mathbf{v} = 2\mathbf{i} + \mathbf{j}$; $\mathbf{w} = 6\mathbf{i} + 3\mathbf{j}$

66. $\mathbf{v} = 5\mathbf{i} + 2\mathbf{j}$; $\mathbf{w} = -5\mathbf{i} - 2\mathbf{j}$

67. $\mathbf{v} = 3\mathbf{i} - 4\mathbf{j}$; $\mathbf{w} = -6\mathbf{i} + 12\mathbf{j}$

68. $\mathbf{v} = 2\mathbf{i} + 2\mathbf{j}$; $\mathbf{w} = -4\mathbf{i} - 2\mathbf{j}$

69. A 150-pound box is dragged 15 feet along a level floor. Find the work done if a force of 75 pounds at an angle of 32° is used.

70. A 100-pound force is pulling a sled loaded with bricks that weighs 400 pounds. The force is at an angle of 42° with the displacement. Find the work done in moving the sled 25 feet.

71. A rope is being used to pull a box up a ramp that is inclined at 15°. The rope exerts a force of 75 pounds on the box, and it makes an angle of 30° with the plane of the ramp. Find the work done in moving the box 12 feet.

72. A dock worker exerts a force on a box sliding down the ramp of a truck. The ramp makes an angle of 48° with the road, and the worker exerts a 50-pound force parallel to the road. Find the work done in sliding the box 6 feet.

Supplemental Exercises

73. For $\mathbf{u} = \langle -1, 1 \rangle$, $\mathbf{v} = \langle 2, 3 \rangle$, and $\mathbf{w} = \langle 5, 5 \rangle$, find the sum of the three vectors geometrically by using the triangle method of adding vectors.

74. For $\mathbf{u} = \langle 1, 2 \rangle$, $\mathbf{v} = \langle 3, -2 \rangle$, and $\mathbf{w} = \langle -1, 4 \rangle$, find $\mathbf{u} + \mathbf{v} - \mathbf{w}$ geometrically by using the triangle method of adding vectors.

75. Find a vector that has the initial point $(3, -1)$ and is equivalent to $\mathbf{v} = 2\mathbf{i} - 3\mathbf{j}$.

76. Find a vector that has the initial point $(-2, 4)$ and is equivalent to $\mathbf{v} = \langle -1, 3 \rangle$.

77. If $\mathbf{v} = 2\mathbf{i} - 5\mathbf{j}$ and $\mathbf{w} = 5\mathbf{i} + 2\mathbf{j}$ have the same initial point, is $\mathbf{v}$ perpendicular to $\mathbf{w}$? Why or why not?

78. If $\mathbf{v} = \langle 5, 6 \rangle$ and $\mathbf{w} = \langle 6, 5 \rangle$ have the same initial point, is $\mathbf{v}$ perpendicular to $\mathbf{w}$? Why or why not?

79. Let $\mathbf{v} = \langle -2, 7 \rangle$. Find a vector perpendicular to $\mathbf{v}$.

80. Let $\mathbf{w} = 4\mathbf{i} + \mathbf{j}$. Find a vector perpendicular to $\mathbf{w}$.

In Example 7, if the box were to be kept from sliding down the ramp, it would be necessary to provide a force of 45 pounds parallel to the ramp but pointed *up* the ramp. Some of this force would be provided by a frictional force between the box and the ramp. The force of friction is $\mathbf{F}_\mu = \mu\mathbf{N}$, where $\mathbf{N}$ is the normal component of the force of gravity. In Exercises 81 and 82, find the frictional force.

81. A 50-pound box is resting on a ramp inclined at 12°. Find the force of friction if the coefficient of friction, μ, is 0.13.

82. A car weighing 2500 pounds is resting on a ramp that is inclined 15°. Find the frictional force if the coefficient of friction, μ, is 0.21.

83. Is the dot product an associative operation? That is, given nonzero vectors $\mathbf{u}$, $\mathbf{v}$, and $\mathbf{w}$, does

$$(\mathbf{u} \cdot \mathbf{v}) \cdot \mathbf{w} = \mathbf{u} \cdot (\mathbf{v} \cdot \mathbf{w})?$$

84. Prove that $\mathbf{v} \cdot \mathbf{w} = \mathbf{w} \cdot \mathbf{v}$.

85. Prove that $c(\mathbf{v} \cdot \mathbf{w}) = (c\mathbf{v}) \cdot \mathbf{w}$.

86. Show that the dot product of two nonzero vectors is positive if the angle between the vectors is an acute angle and that the dot product is negative if the angle between the two vectors is an obtuse angle.

87. Consider the following two situations. (1) A rope is being used to pull a box up a ramp inclined at an angle α. The rope exerts a force $\mathbf{F}$ on the box, and the rope makes an angle θ with the ramp. The box is pulled s feet. (2) A rope is being used to pull a box along a level floor. The rope exerts the same force $\mathbf{F}$ on the box. The box is pulled the same s feet. In which case is more work done?

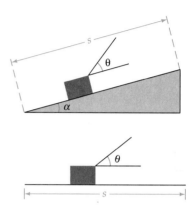

88. Let $\mathbf{v} = c\mathbf{v}$, where c is a nonzero real number and $\mathbf{w}$ is a nonzero vector. Show that $\dfrac{\mathbf{v} \cdot \mathbf{w}}{\|\mathbf{v}\|\|\mathbf{w}\|} = \pm 1$ and that the result is 1 when $c > 0$ and -1 when $c < 0$.

89. Prove that $\|\mathbf{v} - \mathbf{w}\|^2 = \|\mathbf{v}\|^2 + \|\mathbf{w}\|^2 - 2\mathbf{v} \cdot \mathbf{w}$.

90. What is the relationship between $\mathbf{v}$ and $\mathbf{w}$ if **a.** $\text{proj}_\mathbf{w}\mathbf{v} = 0$ and **b.** $\text{proj}_\mathbf{w}\mathbf{v} = \|\mathbf{v}\|$?

8.4 Complex Numbers and DeMoivre's Theorem

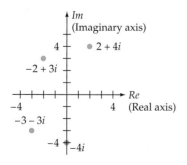

Figure 8.41

Real numbers are graphed as points on a number line. Complex numbers can be graphed in a coordinate plane called an **Argand diagram.** The horizontal axis of the coordinate plane is called the **real axis;** the vertical axis is called the **imaginary axis.** This coordinate system is called the **complex plane.**

A complex number written in the form $z = a + bi$ is written in **standard** or **rectangular form.** The graph of $a + bi$ is associated with the point $P(a, b)$ in the complex plane. Figure 8.41 shows the graphs of several complex numbers.

The length of the line segment from the origin to the point $(-3, 4)$ in the complex plane is the *absolute value* of $z = -3 + 4i$. See Figure 8.42.

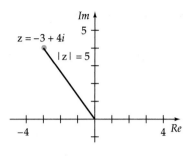

Figure 8.42

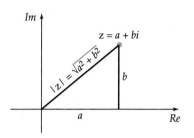

Figure 8.43

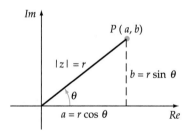

Figure 8.44

From the Pythagorean Theorem, the absolute value of $z = -3 + 4i$ is

$$\sqrt{(-3)^2 + 4^2} = \sqrt{25} = 5$$

Definition of the Absolute Value of a Complex Number

> The absolute value of the complex number $z = a + bi$, denoted by $|z|$, is
>
> $$|z| = |a + bi| = \sqrt{a^2 + b^2}$$

Thus $|z|$ is the distance from the origin to z (see Figure 8.43).

A complex number $z = a + bi$ can be written in terms of trigonometric functions. Consider the complex number graphed in Figure 8.44. We can write a and b in terms of the sine and the cosine.

$$\cos \theta = \frac{a}{r} \qquad \sin \theta = \frac{b}{r}$$

$$a = r \cos \theta \qquad b = r \sin \theta$$

where $r = |z| = \sqrt{a^2 + b^2}$. Substituting for a and b in $z = a + bi$, we obtain

$$z = r \cos \theta + ir \sin \theta = r(\cos \theta + i \sin \theta)$$

The expression $z = r(\cos \theta + i \sin \theta)$ is known as the trigonometric form of a complex number. The notation $\cos \theta + i \sin \theta$ is often abbreviated as cis θ using the c from $\cos \theta$, the imaginary unit i, and the s from $\sin \theta$.

Trigonometric Form of a Complex Number

> The complex number $z = a + bi$ can be written in trigonometric form as
>
> $$z = r(\cos \theta + i \sin \theta) = r \text{ cis } \theta$$
>
> where $a = r \cos \theta$, $b = r \sin \theta$, $r = \sqrt{a^2 + b^2}$, and $\tan \theta = \frac{b}{a}$.

In this text, we will usually write the trigonometric form of a complex number in its abbreviated form $z = r$ cis θ. The value of r is called the **modulus** of the complex number z, and the angle θ is called the **argument** of the complex number z. The modulus r and the argument θ of a complex number $z = a + bi$ are given by

$$r = \sqrt{a^2 + b^2} \quad \text{and} \quad \cos \theta = \frac{a}{r}, \qquad \sin \theta = \frac{b}{r}$$

We can also write $\alpha = \tan^{-1} \left| \frac{b}{a} \right|$, where α is the reference angle for θ. Because of the periodic nature of the sine and cosine functions, the trigonometric form of a complex number is not unique. Since $\cos \theta = \cos(\theta + 2k\pi)$

and $\sin\theta = \sin(\theta + 2k\pi)$, where k is an integer, the following complex numbers are equal.

$$r \text{ cis } \theta = r \text{ cis}(\theta + 2k\pi) \quad \text{for } k \text{ an integer}$$

For example, $2 \text{ cis } \dfrac{\pi}{6} = 2 \text{ cis}\left(\dfrac{\pi}{6} + 2\pi\right)$.

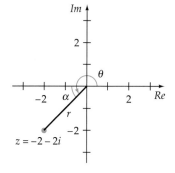

Figure 8.45

EXAMPLE 1   Write a Complex Number in Trigonometric Form

Write $z = -2 - 2i$ in trigonometric form.

Solution   Find the modulus and the argument of z. Then substitute these values in the trigonometric form of z.

$$r = \sqrt{(-2)^2 + (-2)^2} = \sqrt{8} = 2\sqrt{2}$$

To determine θ, we first determine α.

$$\alpha = \tan^{-1}\left|\dfrac{b}{a}\right|$$

• α is the reference angle of angle θ as shown in Figure 8.45.

$$\alpha = \tan^{-1}\left|\dfrac{-2}{-2}\right| = \tan^{-1}1 = 45°$$

$$\theta = 180° + 45° = 225°$$

• Since z is in the third quadrant, $180° < \theta < 270°$.

The trigonometric form is

$$z = r \text{ cis } \theta$$

$$= 2\sqrt{2} \text{ cis } 225°$$

• $r = 2\sqrt{2}, \theta = 225°$

■ *Try Exercise **12**, page 446.*

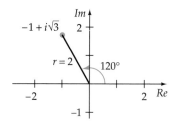

Figure 8.46

EXAMPLE 2   Write a Complex Number in Standard Form

Write $z = 2 \text{ cis } 120°$ in standard form.

Solution   Write z in the form $r(\cos\theta + i\sin\theta)$ and then evaluate $\cos\theta$ and $\sin\theta$. See Figure 8.46.

$$z = 2 \text{ cis } 120° = 2(\cos 120° + i\sin 120°) = 2\left(-\dfrac{1}{2} + \dfrac{\sqrt{3}}{2}i\right) = -1 + i\sqrt{3}$$

■ *Try Exercise **26**, page 447.*

Let z_1 and z_2 be two complex numbers written in trigonometric form. The product of z_1 and z_2 can be found by using several trigonometric identities. If $z_1 = r_1(\cos\theta_1 + i\sin\theta_1)$, and $z_2 = r_2(\cos\theta_2 + i\sin\theta_2)$, then

$$z_1z_2 = r_1(\cos\theta_1 + i\sin\theta_1) \cdot r_2(\cos\theta_2 + i\sin\theta_2)$$

$$= r_1r_2(\cos\theta_1\cos\theta_2 + i\cos\theta_1\sin\theta_2 + i\sin\theta_1\cos\theta_2 + i^2\sin\theta_1\sin\theta_2)$$

$$= r_2r_2[(\cos\theta_1\cos\theta_2 - \sin\theta_1\sin\theta_2) + i(\sin\theta_1\cos\theta_2 + \cos\theta_1\sin\theta_2)]$$

$$= r_1r_2[\cos(\theta_1 + \theta_2) + i\sin(\theta_1 + \theta_2)] \quad \bullet \text{ Identities for } \cos(\theta_1 + \theta_2)$$
$$\text{and } \sin(\theta_1 + \theta_2)$$

$$z_1z_2 = r_1r_2\,\text{cis}(\theta_1 + \theta_2)$$

The modulus for the product of two complex numbers in trigonometric form is the product of the moduli of the two complex numbers, and the argument of the product is the sum of the arguments of the two numbers.

Similarly, the quotient of z_1 and z_2 can be found by using several trigonometric identities. If $z_1 = r_1(\cos\theta_1 + i\sin\theta_1)$, and $z_2 = r_2(\cos\theta_2 + i\sin\theta_2)$, then

$$\frac{z_1}{z_2} = \frac{r_1(\cos\theta_1 + i\sin\theta_1)}{r_2(\cos\theta_2 + i\sin\theta_2)}$$

$$= \frac{r_1(\cos\theta_1 + i\sin\theta_1)(\cos\theta_2 - i\sin\theta_2)}{r_2(\cos\theta_2 + i\sin\theta_2)(\cos\theta_2 - i\sin\theta_2)}$$

$$= \frac{r_1(\cos\theta_1\cos\theta_2 - i\cos\theta_1\sin\theta_2 + i\sin\theta_1\cos\theta_2 - i^2\sin\theta_1\sin\theta_2)}{r_2(\cos^2\theta_2 - i^2\sin^2\theta_2)}$$

$$= \frac{r_1[(\cos\theta_1\cos\theta_2 + \sin\theta_1\sin\theta_2) + i(\sin\theta_1\cos\theta_2 - \cos\theta_1\sin\theta_2)]}{r_2(\cos^2\theta_2 + \sin^2\theta_2)}$$

$$= \frac{r_1}{r_2}[\cos(\theta_1 - \theta_2) + i\sin(\theta_1 - \theta_2)] \quad \bullet \text{ Identities for } \cos(\theta_1 - \theta_2),$$
$$\sin(\theta_1 - \theta_2), \text{ and } \cos^2\theta_2 + \sin^2\theta_2$$

$$\frac{z_1}{z_2} = \frac{r_1}{r_2}\,\text{cis}(\theta_1 - \theta_2)$$

The modulus for the quotient of two complex numbers in trigonometric form is the quotient of the moduli of the two complex numbers, and the argument of the quotient is the difference of the arguments of the two numbers. All these relationships can be summarized as follows:

Product and Quotient of Complex Numbers Written in Trigonometric Form

Let $z_1 = r_1(\cos\theta_1 + i\sin\theta_1)$ and $z_2 = r_2(\cos\theta_2 + i\sin\theta_2)$. Then

$$z_1z_2 = r_1r_2[\cos(\theta_1 + \theta_2) + i\sin(\theta_1 + \theta_2)]$$

$$z_1z_2 = r_1r_2\,\text{cis}(\theta_1 + \theta_2) \quad \bullet \text{ Using cis notation}$$

and

$$\frac{z_1}{z_2} = \frac{r_1}{r_2}[\cos(\theta_1 - \theta_2) + i\sin(\theta_1 - \theta_2)]$$

$$\frac{z_1}{z_2} = \frac{r_1}{r_2}\,\text{cis}(\theta_1 - \theta_2) \quad \bullet \text{ Using cis notation}$$

EXAMPLE 3 **Find the Product or Quotient of Complex Numbers**

a. Find the product of $z_1 = -1 + i\sqrt{3}$ and $z_2 = -\sqrt{3} + i$.
b. Find the quotient of $z_1 = -1 + i$ and $z_2 = \sqrt{3} - i$. Write the answers in standard form.

Solution

a. Write z_1 and z_2 in trigonometric form. Then use the Product Property of complex numbers. See Figure 8.47.

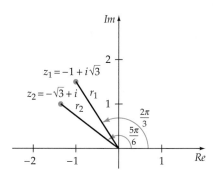

Figure 8.47

$$z_1 z_2 = 2 \operatorname{cis} \frac{2\pi}{3} \cdot 2 \operatorname{cis} \frac{5\pi}{6} = 4 \operatorname{cis}\left(\frac{2\pi}{3} + \frac{5\pi}{6}\right) = 4 \operatorname{cis} \frac{3\pi}{2}$$

$$= 4\left(\cos \frac{3\pi}{2} + i \sin \frac{3\pi}{2}\right) = 4(0 - i) = -4i$$

b. Write z_1 and z_2 in trigonometric form. Then use the Quotient Property of complex numbers. This time we have used degree measure. See Figure 8.48.

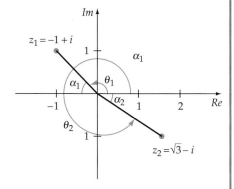

Figure 8.48

$$\frac{z_1}{z_2} = \frac{\sqrt{2} \operatorname{cis} 135°}{2 \operatorname{cis} 330°} = \frac{\sqrt{2}}{2} \operatorname{cis}(135° - 330°)$$

$$= \frac{\sqrt{2}}{2} \operatorname{cis}(-195°) = \frac{\sqrt{2}}{2}[\cos(-195°) + i \sin(-195°)]$$

$$= \frac{\sqrt{2}}{2}(\cos 195° - i \sin 195°)$$

$$\approx -0.6830 + 0.1830i$$

■ *Try Exercise* **52**, *page 447.*

DeMoivre's Theorem

DeMoivre's Theorem is a procedure for finding powers and roots of complex numbers when the complex numbers are expressed in trigonometric form. This theorem can be illustrated by repeated multiplication of a complex number.

Let $z = r \operatorname{cis} \theta$. Then z^2 can be written as

$$z \cdot z = r \operatorname{cis} \theta \cdot r \operatorname{cis} \theta$$

$$z^2 = r^2 \operatorname{cis} 2\theta$$

The product of $z^2 \cdot z$ is

$$z^2 \cdot z = r^2 \operatorname{cis} 2\theta \cdot r \operatorname{cis} \theta$$

$$z^3 = r^3 \operatorname{cis} 3\theta$$

If we continue this process, the results suggest a formula for the nth power of a complex number that is known as DeMoivre's Theorem.

DeMoivre's Theorem

If $z = r \operatorname{cis} \theta$ and n is a positive integer, then

$$z^n = r^n \operatorname{cis} n\theta$$

EXAMPLE 4   **Use DeMoivre's Theorem**

Find $(1 + i)^8$ using DeMoivre's Theorem. Write the answer in standard form.

Solution   Convert $1 + i$ to trigonometric form and then use DeMoivre's Theorem.

$$(1 + i)^8 = (\sqrt{2} \operatorname{cis} 45°)^8 = (\sqrt{2})^8 \operatorname{cis} 8(45) = 16 \operatorname{cis} 360°$$

$$= 16(\cos 360° + i \sin 360°) = 16(1 + 0i) = 16$$

■ *Try Exercise **64**, page 447.*

DeMoivre's Theorem can be extended to include finding the nth roots of any number. Recall that if $w^n = z$, then w is an nth root of z. All the nth roots of a complex number can be found by using DeMoivre's Theorem.

Let $w^n = z$, for $z = r \operatorname{cis} \theta$ and $w = R \operatorname{cis} \alpha$. Then, by DeMoivre's Theorem,

$$w^n = R^n \operatorname{cis} n\alpha$$

Recall that the sine and cosine functions are periodic functions with a period of 2π or 360°. Therefore, for k an integer,

$$z = r \operatorname{cis} \theta = r \operatorname{cis}(\theta + 360°k)$$

Substituting for w^n and z in the equation $w^n = z$ yields

$$R^n \operatorname{cis} n\alpha = r \operatorname{cis}(\theta + 360°k)$$

Two complex numbers written in trigonometric form are equal if and only if their moduli are equal and their arguments are equal. Thus,

$$R^n = r \qquad \text{and} \qquad n\alpha = \theta + 360°k$$

$$R = r^{1/n} \qquad\qquad \alpha = \frac{\theta + 360°k}{n}$$

Since $w = R \operatorname{cis} \alpha$, by substituting $r^{1/n}$ for R and $\left(\dfrac{\theta + 360°k}{n}\right)$ for α, we have

$$w = r^{1/n} \operatorname{cis} \frac{\theta + 360°k}{n}$$

DeMoivre's Theorem for Finding Roots

If $z = r \text{ cis } \theta$ is a complex number, then there are n distinct nth roots of z given by the formula

$$w_k = r^{1/n} \text{ cis } \frac{\theta + 360°k}{n} \quad \text{for } k = 0, 1, 2, \ldots, n - 1$$

EXAMPLE 5 **Find Cube Roots by DeMoivre's Theorem**

Find the three cube roots of 27.

Solution Write 27 in trigonometric form: $27 = 27 \text{ cis } 0°$. Then, from DeMoivre's Theorem, the cube roots of 27 are

$$w_k = 27^{1/3} \text{ cis } \frac{0° + 360°k}{3} \quad \text{for } k = 0, 1, 2$$

Substitute for k to find the cube roots of 27.

$w_0 = 27^{1/3} \text{ cis } 0° = 3(\cos 0° + i \sin 0°)$ • $k = 0;\ \dfrac{0° + 360°(0)}{3} = 0°$

   $= 3$

$w_1 = 27^{1/3} \text{ cis } 120° = 3(\cos 120° + i \sin 120°)$ • $k = 1;\ \dfrac{0° + 360°(1)}{3} = 120°$

   $= -\dfrac{3}{2} + \dfrac{3\sqrt{3}}{2}i$

$w_2 = 27^{1/3} \text{ cis } 240° = 3(\cos 240° + i \sin 240°)$ • $k = 2;\ \dfrac{0° + 360°(2)}{3} = 240°$

   $= -\dfrac{3}{2} - \dfrac{3\sqrt{3}}{2}i$

For $k = 3$, $\dfrac{0° + 1080°}{3} = 360°$. The angles start repeating; thus there are only three cube roots of 27. These roots are graphed in Figure 8.49.

Check:

$$3^3 = 27$$

$$(3 \text{ cis } 120°)^3 = 27 \text{ cis } 360°$$

$$= 27(\cos 360° + i \sin 360°) = 27(1 + 0) = 27$$

$$(3 \text{ cis } 240°)^3 = 27 \text{ cis } 720°$$

$$= 27(\cos 720° + i \sin 720°) = 27(1 + 0) = 27$$

■ *Try Exercise* **78**, *page 447.*

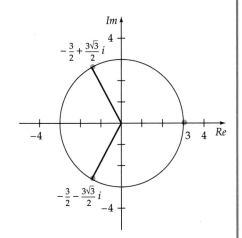

Figure 8.49

Note from Figure 8.49 that the arguments of the cube roots of 27 are 0°, 120°, and 240° and that $|w| = 3$. This means that the complex numbers representing the cube roots of 27 are equally spaced around a circle centered at the origin with a radius 3.

EXAMPLE 6   Find the Fifth Roots of a Complex Number

Find the five fifth roots of $z = 1 + i\sqrt{3}$.

Solution   Write z in trigonometric form: $z = r \text{ cis } \theta$.

$$r = \sqrt{1^2 + (\sqrt{3})^2} = 2$$

$$z = 2 \text{ cis } 60° \quad \bullet \ \theta = \tan^{-1}\frac{\sqrt{3}}{1} = 60°$$

From DeMoivre's Theorem, the modulus of each root is $\sqrt[5]{2}$, and the arguments are determined by $\dfrac{60° + 360°k}{5}$, $k = 0, 1, 2, 3, 4$.

$$w_k = \sqrt[5]{2} \text{ cis } \frac{60 + 360°k}{5} \quad \bullet \ k = 0, 1, 2, 3, 4$$

Substitute for k to find the five fifth roots of z.

$$w_0 = \sqrt[5]{2} \text{ cis } 12° \quad \bullet \ k = 0; \ \frac{60° + 360°(0)}{5} = 12°$$

$$w_1 = \sqrt[5]{2} \text{ cis } 84° \quad \bullet \ k = 1; \ \frac{60° + 360°(1)}{5} = 84°$$

$$w_2 = \sqrt[5]{2} \text{ cis } 156° \quad \bullet \ k = 2; \ \frac{60° + 360°(2)}{5} = 156°$$

$$w_3 = \sqrt[5]{2} \text{ cis } 228° \quad \bullet \ k = 3; \ \frac{60° + 360°(3)}{5} = 228°$$

$$w_4 = \sqrt[5]{2} \text{ cis } 300° \quad \bullet \ k = 4; \ \frac{60° + 360°(4)}{5} = 300°$$

■ *Try Exercise **84**, page 447.*

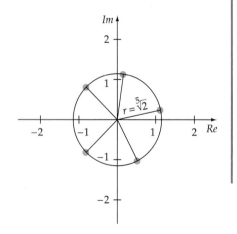

Figure 8.50

Remark   The five roots are graphed in Figure 8.50. The radius of the circle is $\sqrt[5]{2} \approx 1.15$, and the complex numbers are spaced equally around the circle with center $(0, 0)$.

EXERCISE SET 8.4

In Exercises 1 to 8, graph the complex numbers. Find the absolute value of each complex number.

1. $z = -2 - 2i$ **2.** $z = 4 - 4i$ **3.** $z = \sqrt{3} - i$

4. $z = 1 + i\sqrt{3}$ **5.** $z = -2i$ **6.** $z = -5$

7. $z = 3 - 5i$ **8.** $z = -5 - 4i$

In Exercises 9 to 16, write the complex number in trigonometric form.

9. $z = 1 - i$ **10.** $z = -4 - 4i$ **11.** $z = \sqrt{3} - i$

12. $z = 1 + i\sqrt{3}$ **13.** $z = 3i$ **14.** $z = -2i$

15. $z = -5$ **16.** $z = 3$

In Exercises 17 to 34, write the complex number in standard form.

17. $z = 2(\cos 45° + i \sin 45°)$

18. $z = 3(\cos 240° + i \sin 240°)$

19. $z = (\cos 315° + i \sin 315°)$

20. $z = 5(\cos 120° + i \sin 120°)$

21. $z = 6 \text{ cis } 135°$ **22.** $z = \text{cis } 315°$

23. $z = 8 \text{ cis } 0°$ **24.** $z = 5 \text{ cis } 90°$

25. $z = 2\left(\cos \dfrac{5\pi}{6} + i \sin \dfrac{5\pi}{6}\right)$ **26.** $z = 4\left(\cos \dfrac{5\pi}{3} + i \sin \dfrac{5\pi}{3}\right)$

27. $z = 3\left(\cos \dfrac{3\pi}{2} + i \sin \dfrac{3\pi}{2}\right)$ **28.** $z = 5(\cos \pi + i \sin \pi)$

29. $z = 8 \text{ cis } \dfrac{3\pi}{4}$ **30.** $z = 9 \text{ cis } \dfrac{4\pi}{3}$

31. $z = 9 \text{ cis } \dfrac{11\pi}{6}$ **32.** $z = \text{cis } \dfrac{3\pi}{2}$

33. $z = 2 \text{ cis } 2$ **34.** $z = 5 \text{ cis } 4$

In Exercises 35 to 42, multiply the complex numbers. Write the answer in trigonometric form.

35. $2 \text{ cis } 30° \cdot 3 \text{ cis } 225°$ **36.** $4 \text{ cis } 120° \cdot 6 \text{ cis } 315°$

37. $3(\cos 122° + i \sin 122°) \cdot 4(\cos 213° + i \sin 213°)$

38. $8(\cos 88° + i \sin 88°) \cdot 12(\cos 112° + i \sin 112°)$

39. $5\left(\cos \dfrac{2\pi}{3} + i \sin \dfrac{2\pi}{3}\right) \cdot 2\left(\cos \dfrac{2\pi}{5} + i \sin \dfrac{2\pi}{5}\right)$

40. $5 \text{ cis } \dfrac{11\pi}{12} \cdot 3 \text{ cis } \dfrac{4\pi}{3}$ **41.** $4 \text{ cis } 2.4 \cdot 6 \text{ cis } 4.1$

42. $7 \text{ cis } 0.88 \cdot 5 \text{ cis } 1.32$

In Exercises 43 to 50, divide the complex numbers. Write the answers to standard form.

43. $\dfrac{32 \text{ cis } 30°}{4 \text{ cis } 150°}$ **44.** $\dfrac{15 \text{ cis } 240°}{3 \text{ cis } 135°}$

45. $\dfrac{27(\cos 315° + i \sin 315°)}{9(\cos 225° + i \sin 225°)}$ **46.** $\dfrac{9(\cos 25° + i \sin 25°)}{3(\cos 175° + i \sin 175°)}$

47. $\dfrac{12 \text{ cis } \dfrac{2\pi}{3}}{4 \text{ cis } \dfrac{11\pi}{6}}$ **48.** $\dfrac{10 \text{ cis } \dfrac{\pi}{3}}{5 \text{ cis } \dfrac{\pi}{4}}$

49. $\dfrac{25(\cos 3.5 + i \sin 3.5)}{5(\cos 1.5 + i \sin 1.5)}$ **50.** $\dfrac{18 (\cos 0.56 + i \sin 0.56)}{6(\cos 1.22 + i \sin 1.22)}$

In Exercises 51 to 54, perform the indicated operation in trigonometric form. Write the solution to standard form.

51. $\dfrac{1 + i\sqrt{3}}{1 - i\sqrt{3}}$ **52.** $\dfrac{1 + i}{1 - i}$

53. $\dfrac{\sqrt{3} + i\sqrt{3}}{(1 - i\sqrt{3})(2 - 2i)}$ **54.** $\dfrac{(2 - 2i\sqrt{3})(1 - i\sqrt{3})}{\sqrt{3} + i}$

In Exercises 55 to 68, find the indicated power. Write the answers in trigonometric form.

55. $(2 \text{ cis } 30°)^8$ **56.** $(\text{cis } 240°)^{12}$

57. $(2 \text{ cis } 240°)^5$ **58.** $(2 \text{ cis } 45°)^{10}$

59. $[2(\cos 225° + i \sin 225°)]^5$ **60.** $[3(\cos 330° + i \sin 330°)]^4$

61. $[2(\cos 120° + i \sin 120°)]^6$ **62.** $[4(\cos 150° + i \sin 150°)]^3$

63. $(1 - i)^{10}$ **64.** $(1 + i\sqrt{3})^8$

65. $(2 + 2i)^7$ **66.** $(2\sqrt{3} - 2i)^5$

67. $\left(\dfrac{\sqrt{2}}{2} + i\dfrac{\sqrt{2}}{2}\right)^6$ **68.** $\left(-\dfrac{\sqrt{2}}{2} + i\dfrac{\sqrt{2}}{2}\right)^{12}$

In Exercises 69 to 82, find all of the indicated roots. Write all answers in standard form.

69. Square roots of $9 \text{ cis } 135°$

70. Square roots of $16 \text{ cis } 45°$

71. Sixth roots of $64(\cos 120° + i \sin 120°)$

72. Fifth roots of $(\cos 315° + i \sin 315°)$

73. Fifth roots of $\cos 90° + i \sin 90°$

74. Fourth roots of $\cos 0° + i \sin 0°$

75. Cube roots of 1   **76.** Cube roots of i

77. Fourth roots of $1 + i$ **78.** Fifth roots of $-1 + i$

79. Cube roots of $2 - 2i\sqrt{3}$ **80.** Cube roots of $-2 + 2i\sqrt{3}$

81. Square roots of $-16 + 16i\sqrt{3}$

82. Square roots of $-1 + i\sqrt{3}$

In Exercises 83 to 88, find all roots of the equations. Write the answers in trigonometric form.

83. $x^4 + i = 0$ **84.** $x^3 - 2i = 0$

85. $x^3 + 64 = 0$ **86.** $x^5 + 32i = 0$

87. $x^4 - (1 - i\sqrt{3}) = 0$ **88.** $x^3 + (2\sqrt{3} - 2i) = 0$

Supplemental Exercises

89. Show that the conjugate of $z = r(\cos \theta + i \sin \theta)$ is equal to $\bar{z} = r(\cos \theta - i \sin \theta)$.

90. If $z = r(\cos \theta + i \sin \theta)$, show that
$$z^{-1} = r^{-1}(\cos \theta - i \sin \theta).$$

91. If $z = r(\cos \theta + i \sin \theta)$, show that
$$z^{-2} = r^{-2}(\cos 2\theta - i \sin 2\theta).$$

Note that Exercises 90 and 91 suggest that the general expression $z^{-n} = r^{-n}(\cos n\theta - i \sin n\theta)$.

92. Use the results of Exercise 91 to find z^{-4} for $z = 1 - i\sqrt{3}$.

93. Raise $(\cos\theta + i\sin\theta)$ to the second power by using DeMoivre's Theorem. Now square $(\cos\theta + i\sin\theta)$ as a binomial. Equate the real and imaginary parts of the two complex numbers and show that

a. $\cos 2\theta = \cos^2\theta - \sin^2\theta$ **b.** $\sin 2\theta = 2\sin\theta\cos\theta$

94. Raise $(\cos\theta + i\sin\theta)$ to the fourth power by using

DeMoivre's Theorem. Now find the fourth power of the binomial $(\cos\theta + i\sin\theta)$. Equate the real and imaginary parts of the two complex numbers and show that

a. $\cos 4\theta = \cos^2 2\theta - \sin^2 2\theta$

b. $\sin 4\theta = 4\cos^3\theta\sin\theta - 4\cos\theta\sin^3\theta$

Exploring Concepts with Technology

Optimal Branching of Arteries

The physiologist Jean Louis Poiseuille (1799–1869) developed several laws concerning the flow of blood. One of his laws states that the resistance R of a blood vessel of length l and radius r is given by

$$R = k\frac{l}{r^4} \tag{1}$$

The number k is a variation constant that depends on the viscosity of the blood. The following figure shows a large artery with radius r_1 and a smaller artery with radius r_2. The branching angle between the arteries is θ. Make use of Poiseuille's Law, Equation (1), to show that the resistance R of the blood along the path $P_1 P_2 P_3$ is

$$R = k\left(\frac{a - b\cot\theta}{r_1^4} + \frac{b\csc\theta}{r_2^4}\right) \tag{2}$$

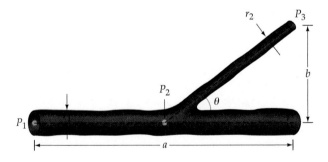

Use a graphing calculator or computer to graph R with $a = 8$ cm, $b = 4$ cm, $r_1 = 0.4$ cm, and $r_2 = (3/4)r_1 = 0.3$ cm. Then estimate (to the nearest degree) the angle θ that minimizes R, given $a = 8$ cm, $b = 4$ cm, and $r_2 = (3/4)r_1$. Explain why it is not necessary to know the specific value of k. By using calculus, it can be demonstrated that R is minimized when

$$\cos\theta = \left(\frac{r_2}{r_1}\right)^4 \tag{3}$$

This equation is remarkable because it is much simpler than Equation (2) and because it does not involve the distance a or b. Solve equation (3) for θ, with $r_2 = (3/4)r_1$. How does this value of θ compare with the value of θ you obtained by graphing?

Chapter Review

8.1 The Law of Sines

- The Law of Sines is used to solve general triangles when two angles and a side are given or when two sides and an angle opposite one of them are given.

$$\frac{a}{\sin A} = \frac{b}{\sin B} = \frac{c}{\sin C}$$

8.2 The Law of Cosines and Area

- The Law of Cosines $a^2 = b^2 + c^2 - 2bc \cos A$ is used to solve general triangles when two sides and the included angle or three sides of the triangle are given.

- Area K of a triangle ABC is $K = \dfrac{1}{2}bc \sin A = \dfrac{b^2 \sin C \sin A}{2 \sin B}$.

- Area for a triangle in which three sides are given (Heron's formula):

$$K = \sqrt{s(s-a)(s-b)(s-c)}, \quad \text{where } s = \frac{1}{2}(a+b+c).$$

8.3 Vectors

- A vector is a quantity with magnitude and direction. Two vectors are equal if they have the same magnitude and the same direction. The resultant of two or more vectors is the single vector that has the equivalent effect of the vectors.

- Vectors can be added by parallelogram addition, triangle addition, or addition of the x- and y-components.

- Multiplication of a vector by a scalar is given by $a(x\mathbf{i} + y\mathbf{j}) = ax\mathbf{i} + ay\mathbf{j}$.

- The dot product of $\mathbf{U} = a\mathbf{i} + b\mathbf{j}$ and $\mathbf{V} = c\mathbf{i} + b\mathbf{j}$ is given by

$$\mathbf{U} \cdot \mathbf{V} = (a\mathbf{i} + b\mathbf{j}) \cdot (c\mathbf{i} + d\mathbf{j}) = ac + bd$$

- The cosine of the angle α between two vectors is given by

$$\cos \alpha = \frac{\mathbf{U} \cdot \mathbf{V}}{\|\mathbf{U}\|\|\mathbf{V}\|}.$$

8.4 Complex Numbers and DeMoivre's Theorem

- The standard form of a complex number is $z = a + bi$.

- The trigonometric form of a complex number is $z = r(\cos \theta + i \sin \theta) = r \operatorname{cis} \theta$.

- The product of two complex numbers in trigonometric form is given by

$$z_1 z_2 = r_1 r_2 [\operatorname{cis}(\theta_1 + \theta_2)]$$

- The quotient of two complex numbers in trigonometric form is given by

$$\frac{z_1}{z_2} = \frac{r_1}{r_2}[\operatorname{cis}(\theta_1 - \theta_2)]$$

- **DeMoivre's Theorem**   If $z = r \operatorname{cis} \theta$ is and n is a positive integer, then $z^n = r^n \operatorname{cis} n\theta$. If $w^n = z$, then $w_k = r^{1/n} \operatorname{cis}\left(\dfrac{\theta + 360°k}{n}\right)$, $k = 0, 1, 2, \dots, n - 1$.

Essays and Projects

1. State Fermat's Principle and Snell's Law. The refractive index of a glass ring is found to be 1.82. The refractive index of a particular diamond ring is 2.38. Use Snell's Law to explain what this means in terms of the reflective properties of the two rings.

2. State Mollweide's Formula and show by an example how it can be used to check the results of solving a triangle.

3. State the Law of Tangents and illustrate with a specific example how to use this law to solve a triangle.

4. Newton's Formula, $\dfrac{a + b}{c} = \dfrac{\cos \frac{1}{2}(A - B)}{\sin \frac{1}{2}C}$, utilizes the measures of the three angles and the measures of the three sides of a triangle. Thus, the formula is useful in checking the results of solving a triangle. Illustrate the use of Newtons Formula by checking the results of solving an oblique triangle.

5. The longest rod that can be carried horizontally around a corner from a hall 3 meters wide into one that is 5 meters wide is the minimum of the length L of the dashed line shown in the figure.

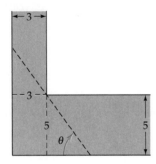

Use similar triangles to show that the length L is a function of the angle θ, given by

$$L(\theta) = \dfrac{5}{\sin \theta} + \dfrac{3}{\cos \theta}$$

Graph L and estimate the minimum value of L.

6. A projectile is fired at an angle of inclination θ from the horizon with an initial velocity v_0. Its trajectory (neglecting air resistance) is a parabolic path given by

$$y = (\tan \theta)x - \dfrac{16 \sec^2\theta}{v_0^2}x^2 \tag{1}$$

a. If $\theta = \dfrac{\pi}{6}$, and $v_0 = 110$ feet per second, find the maximum height attained by the projectile.

b. Set $y = 0$, and solve Equation (1) for x to determine the horizontal distance the trajectory will travel.

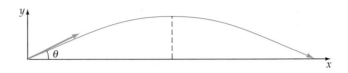

c. Use a graph of your result from part b. to determine the angle θ that produces the maximum horizontal distance for a trajectory.

7. If a projectile is fired at an angle of inclination of θ from the horizon from the base of a hill that is inclined at an angle of ϕ from the horizon, then the distance d that the projectile will travel up the inclined slope is

$$d(\theta) = \dfrac{v_0^2 \cos \theta \sin(\theta - \phi)}{16 \cos^2\phi}$$

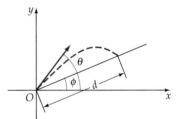

Graph $d(\theta)$ with $v_0 = 110$ feet per second and $\phi = 30°$. Use the graph to estimate the value of θ that will maximize d.

8. Propose a way to use a transit and tape measure for measuring the height of a tall building. How would you measure the height of a penthouse, flagpole, or other structure on the top of the building? How would an error in measurement affect your results?

9. The early Egyptians were faced with the problem of making square corners for their temples and pyramids. One way to do this is by using a rope with 12 equally spaced knots. Write an essay explaining how the rope could be used to make the square corners. Find some other lengths that could have been used to make the corners.

True/False Exercises

For Exercises 1 to 16, answer true or false. If the statement is false, give an example to show that the statement is false.

1. The Law of Cosines can be used to solve any triangle given two sides and an angle.

2. The Law of Sines can be used to solve any triangle given two angles and any side.

3. In any triangle, the largest side is opposite the largest angle.

4. If two vectors have the same magnitude, then they are equal.

5. It is possible for the sum of two nonzero vectors to equal zero.

6. The expression $a^2 = b^2 + c^2 + 2bc \cos D$ is true for triangle ABC, in which angle D is the supplement of angle A.

7. The measure of angle α formed by two vectors is greater than or equal to $0°$ and less than or equal to $180°$.

8. If A, B, and C are the angles of a triangle, then

$$\sin(A + B + C) = 0$$

9. Real numbers are complex numbers.

10. Let $\mathbf{V} = a\mathbf{i} + b\mathbf{j}$. Then $\mathbf{V} \cdot \mathbf{V} = a^2\mathbf{i} + b^2\mathbf{j}$.

11. If $\mathbf{V}$ and $\mathbf{W}$ are vectors with $\mathbf{V} \cdot \mathbf{W} = 0$, then $\mathbf{V} = 0$ or $\mathbf{W} = 0$.

12. The n roots of a complex number can be graphed on a circle and are equally spaced around the circle.

13. Let $z = r(\cos \theta + i \sin \theta)$. Then $z^2 = r^2(\cos^2\theta + i \sin^2\theta)$.

14. $|a + bi| = \sqrt{a^2 + b^2}$

15. $i = \cos \pi + i \sin \pi$

16. $z = \cos 45° + i \sin 45°$ is a square root of i.

Chapter Review Exercises

In Exercises 1 to 10, solve the triangles.

1. $A = 37°, b = 14, C = 90°$

2. $B = 77.4°, c = 11.8, C = 90°$

3. $a = 12, b = 15, c = 20$

4. $a = 24, b = 32, c = 28$

5. $a = 18, b = 22, C = 35°$

6. $b = 102, c = 150, A = 82°$

7. $A = 105°, a = 8, c = 10$

8. $C = 55°, c = 80, b = 110$

9. $A = 55°, B = 80°, c = 25$

10. $B = 25°, C = 40°, c = 40$

In Exercises 11 to 18, find the area of each triangle.

11. $a = 24, b = 30, c = 36$ **12.** $a = 9.0, b = 7.0, c = 12$

13. $a = 60, b = 44, C = 44°$ **14.** $b = 8.0, c = 12, A = 75°$

15. $b = 50, c = 75, C = 15°$ **16.** $b = 18, a = 25, A = 68°$

17. $A = 110°, a = 32, b = 15$

18. $C = 45°, c = 22, b = 18$

In Exercises 19 to 20, find the components of each vector with the given initial and terminal points. Write an equivalent vector in terms of its components.

19. $P_1(-2, 4); P_2(3, 7)$ **20.** $P_1(-4, 0); P_2(-3, 6)$

In Exercises 21 to 24, find the magnitude and direction of each vector.

21. $\mathbf{v} = \langle -4, 2 \rangle$ **22.** $\mathbf{v} = \langle 6, -3 \rangle$

23. $\mathbf{u} = -2\mathbf{i} + 3\mathbf{j}$ **24.** $\mathbf{u} = -4\mathbf{i} - 7\mathbf{j}$

In Exercises 25 to 28, find a unit vector in the direction of the given vector.

25. $\mathbf{w} = \langle -8, 5 \rangle$ **26.** $\mathbf{w} = \langle 7, -12 \rangle$

27. $\mathbf{v} = 5\mathbf{i} + \mathbf{j}$ **28.** $\mathbf{v} = 3\mathbf{i} - 5\mathbf{j}$

In Exercises 29 and 30, perform the indicated operation where $\mathbf{u} = \langle 3, 2 \rangle$, and $\mathbf{v} = \langle -4, -1 \rangle$.

29. $\mathbf{v} - \mathbf{u}$ **30.** $2\mathbf{u} - 3\mathbf{v}$

In Exercises 31 and 32, perform the indicated operation where $\mathbf{u} = 10\mathbf{i} + 6\mathbf{j}$ and $\mathbf{v} = 8\mathbf{i} - 5\mathbf{j}$.

31. $-\mathbf{u} + \frac{1}{2}\mathbf{v}$ **32.** $\frac{2}{3}\mathbf{v} - \frac{3}{4}\mathbf{u}$

33. A plane is flying at an airspeed of 400 mph at a heading of $204°$. A wind of 45 mph is blowing from the east. Find the ground speed of the plane.

34. A 40-pound force keeps a 320-pound object from sliding down an inclined ramp. Find the angle of the ramp.

In Exercises 35 to 38, find the dot product of the vectors.

35. $\mathbf{u} = \langle 3, 7 \rangle; \mathbf{v} = \langle -1, 3 \rangle$

36. $\mathbf{v} = \langle -8, 5 \rangle; \mathbf{u} = \langle 2, -1 \rangle$

37. $\mathbf{v} = -4\mathbf{i} - \mathbf{j}; \mathbf{u} = 2\mathbf{i} + \mathbf{j}$

38. $\mathbf{u} = -3\mathbf{i} + 7\mathbf{j}; \mathbf{v} = -2\mathbf{i} + 2\mathbf{j}$

In Exercises 39 to 42, find the angle between the vectors.

39. $\mathbf{u} = \langle 7, -4 \rangle; \mathbf{v} = \langle 2, 3 \rangle$

40. $\mathbf{v} = \langle -5, 2 \rangle; \mathbf{u} = \langle 2, -4 \rangle$

41. $\mathbf{v} = 6\mathbf{i} - 11\mathbf{j}; \mathbf{u} = 2\mathbf{i} + 4\mathbf{j}$

42. $\mathbf{u} = \mathbf{i} - 5\mathbf{j}; \mathbf{v} = \mathbf{i} + 5\mathbf{j}$

In Exercises 43 and 44, find $\text{proj}_w v$.

43. $v = \langle -2, 5 \rangle$; $w = \langle 5, 4 \rangle$

44. $v = 4i - 7j$; $w = -2i - 5j$

45. A 120-pound box is dragged 14 feet along a level floor. Find the work done if a force of 60 pounds at an angle of 38° is used.

In Exercises 46 and 47, find the modulus and the argument and graph the complex numbers.

46. $z = 2 - 3i$ **47.** $z = -5 + i\sqrt{3}$

In Exercises 48 and 49, write the complex numbers in trigonometric form.

48. $z = 2 - 2i$ **49.** $z = -\sqrt{3} + 3i$

In Exercises 50 and 51, write the complex number in standard form.

50. $z = 5 \text{ cis } 315°$ **51.** $z = 6 \text{ cis } \dfrac{4\pi}{3}$

In Exercises 52 to 55, multiply the complex numbers. Write the answers to standard form.

52. $5 \text{ cis } 162° \cdot 2 \text{ cis } 63°$ **53.** $3 \text{ cis } 12° \cdot 4 \text{ cis } 126°$

54. $7 \text{ cis } \dfrac{2\pi}{3} \cdot 4 \text{ cis } \dfrac{\pi}{4}$ **55.** $3 \text{ cis } 1.8 \cdot 5 \text{ cis } 2.5$

In Exercises 56 to 59, divide the complex numbers. Write the answers in trigonometric form.

56. $\dfrac{6 \text{ cis } 50°}{2 \text{ cis } 150°}$ **57.** $\dfrac{30 \text{ cis } 165°}{10 \text{ cis } 55°}$

58. $\dfrac{40 \text{ cis } 66°}{8 \text{ cis } 125°}$ **59.** $\dfrac{\sqrt{3} - i}{1 + i}$

In Exercises 60 to 63, find the indicated power. Write the answers in standard form.

60. $(3 \text{ cis } 45°)^5$ **61.** $\left(\text{cis } \dfrac{11\pi}{6} \right)^8$

62. $(1 - i\sqrt{3})^7$ **63.** $(-2 - 2i)^{10}$

In Exercises 64 to 67, find all of the indicated roots. Write the answers in trigonometric form.

64. Cube roots of $27i$ **65.** Fourth roots of $8i$

66. Fourth roots of $256 \text{ cis } 120°$

67. Fifth roots of $-1 - i$

Chapter Test

1. Solve triangle ABC if $A = 70°$, $C = 16°$, and $c = 14$.

2. Find B in triangle ABC if $A = 140°$, $b = 13$, and $a = 45$.

3. In triangle ABC, $C = 42°$, $a = 20$, and $b = 12$. Find side c.

4. In triangle ABC, $a = 32$, $b = 24$, and $c = 18$. Find angle B.

5. Given angle $C = 110°$, side $a = 7$, and side $b = 12$, find the area of triangle ABC.

6. Given angle $B = 42°$, angle $C = 75°$, and side $b = 12$, find the area of triangle ABC.

7. Given side $a = 17$, side $b = 55$, and side $c = 42$, find the area of triangle ABC.

8. A vector has a magnitude of 12 and direction 220°. Write an equivalent vector in the form $v = a_1 i + a_2 j$.

9. Find $3u - 5v$ given the vectors $u = 2i - 3j$ and $v = 5i + 4j$.

10. Find the dot product of $u = -2i + 3j$ and $v = 5i + 3j$.

11. Find the angle to the nearest degree between the vectors $u = \langle 3, 5 \rangle$ and $v = \langle -6, 2 \rangle$.

12. Write $z = -3\sqrt{2} + 3i$ in trigonometric form.

13. Write $z = 5 \text{ cis } 315°$ in standard form.

14. Multiply: $3 \text{ cis } 80° \cdot 8 \text{ cis } 210°$. Write the answer in standard form.

15. Divide: $\dfrac{25 \text{ cis } 115°}{10 \text{ cis } 210°}$. Write the answer in trigonometric form.

16. Find $(\sqrt{2} - i)^5$. Write the answer in standard form.

17. Find the three cube roots of $27i$. Write your answer in standard form.

18. One ship leaves a port at 1:00 P.M. traveling at 12 mph at a heading of 65°. At 2:00 P.M. another ship leaves the port traveling at 18 mph at a heading of 142°. Find the distance between the ships at 3:00 P.M.

19. Two fire lookouts are located 12 miles apart. Lookout A is at a bearing of N32°W of lookout B. A fire was sighted at a bearing of S82°E from A and N72°E from B. Find the distance of the fire from lookout B.

20. A triangular commercial piece of real estate is priced at $8.50 per square foot. Find the cost of the lot measuring 112 feet by 165 feet by 140 feet.

Topics in Analytic Geometry

CASE IN POINT *Kidney Stones and Ellipses*

The conic sections, some of the curves we will study in this chapter, were studied by the ancient Greeks. One of the conic sections is an ellipse. Besides applications of the ellipse to astronomy and optics, the ellipse now has an application to medicine. The ellipse is used in a nonsurgical treatment of kidney stones. To understand this application, we will introduce some properties of an ellipse.

An ellipse is an oval-shaped curve. Inside the ellipse there are two points called foci of the ellipse. Sound or light waves emitted from one focus are reflected off the surface of the ellipse to the other focus. In an analogous way, think of a billiard table shaped like an ellipse. A ball struck from one focus would pass through the other focus no matter what the direction of the initial shot.

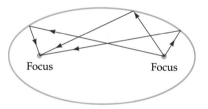

Focus   Focus

The treatment of kidney stones is based on this reflective property of an ellipse. An electrode is placed at one focus of an ellipse, and the patient is placed such that the kidney stone is at the other focus. When the electrode is discharged, ultrasound waves are produced. These waves hit the walls of the ellipse and are reflected to the kidney stone. Because of the reflective property of the ellipse, there is very little energy loss, and it is as though the electrode actually discharged at the kidney stone. The energy of the discharge pulverizes the kidney stone into fragments that can pass through the system.

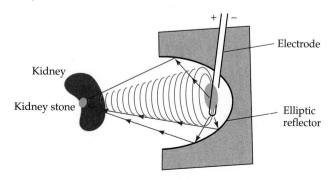

Kidney

Kidney stone

+ | | −

Electrode

Elliptic reflector

9.1   Parabolas

The graph of a parabola, circle, ellipse, or a hyperbola can be formed by the intersection of a plane and a cone. Hence these figures are referred to as conic sections. See Figure 9.1.

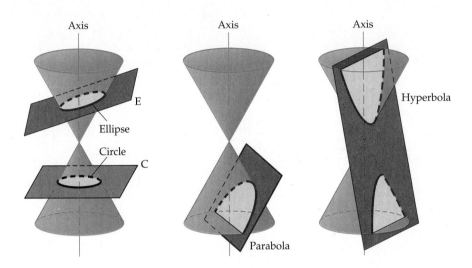

Figure 9.1
Cones intersected by planes.

A plane perpendicular to the axis of the cone intersects the cone in a circle (plane C). The plane E, tilted so that it is not perpendicular to the axis, intersects the cone in an ellipse. When the plane is parallel to a line on the surface of the cone, the plane intersects the cone in a parabola. When the plane intersects both cones, a hyperbola is formed.

Parabolas with Vertex at (0, 0)

Besides the geometric description of a conic section just given, a conic can be defined as a set of points. This method uses some specified conditions about the curve to determine which points in a coordinate system are points of the graph. For example, a parabola can be defined by the following set of points.

Definition of a Parabola

A **parabola** is the set of points in the plane that are equidistant from a fixed line (the **directrix**) and a fixed point (the **focus**) not on the directrix.

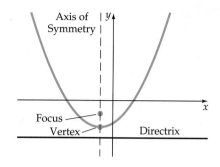

Figure 9.2

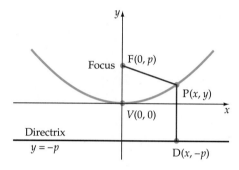

Figure 9.3

The line that passes through the focus and is perpendicular to the directrix is called the **axis of symmetry** of the parabola. The midpoint of the line segment between the focus and directrix on the axis of symmetry is the **vertex** of the parabola as shown in Figure 9.2.

Using this definition of a parabola, we can determine an equation of a parabola. Suppose that the coordinates of the vertex of a parabola are $V(0, 0)$ and the axis of symmetry is the y-axis. The equation of the directrix is $y = -p$, $p > 0$. The focus lies on the axis of symmetry and is the same distance from the vertex as the vertex is from the directrix. Thus the coordinates of the focus are $F(0, p)$ as shown in Figure 9.3.

Let $P(x, y)$ be any point P on the parabola. Then, using the distance formula and the fact that the distance between any point P on the parabola and the focus is equal to the distance from the point P to the directrix (Why?[1]), we can write the equation

$$d(P, F) = d(P, D)$$

By the distance formula,

$$\sqrt{(x - 0)^2 + (y - p)^2} = y + p$$

Now, squaring each side and simplifying, we get

$$(\sqrt{(x - 0)^2 + (y - p)^2})^2 = (y + p)^2$$
$$x^2 + y^2 - 2py + p^2 = y^2 + 2py + p^2$$
$$x^2 = 4py$$

This is an equation of a parabola with vertex at the origin and a vertical axis of symmetry. The equation of a parabola with a horizontal axis of symmetry is derived in a similar manner.

Standard Form of the Equation of a Parabola with Vertex at the Origin

Vertical Axis of Symmetry
The standard form of the equation of a parabola with vertex $(0, 0)$ and a vertical axis of symmetry is $x^2 = 4py$. The focus is $(0, p)$, and the equation of the directrix is $y = -p$.

Horizontal Axis of Symmetry
The standard form of the equation of a parabola with vertex $(0, 0)$ and a horizontal axis of symmetry is $y^2 = 4px$. The focus is $(p, 0)$, and the equation of the directrix is $x = -p$.

Remark In the equation $x^2 = 4py$, $x^2 \geq 0$. Therefore $4py \geq 0$. Thus if $p > 0$, then $y \geq 0$ and the parabola opens up. If $p < 0$, then $y \leq 0$ and the parabola opens down. A similar analysis shows that for $y^2 = 4px$, the parabola opens to the right when $p > 0$ and opens to the left when $p < 0$.

---

[1] From the definition of a parabola, the distance from a point on the parabola to the focus is the same as the distance from the point to the directrix.

EXAMPLE 1   **Find the Focus and Directrix of a Parabola**

Find the focus and directrix of the parabola given by the equation $y = -\frac{1}{2}x^2$.

Solution   Because the x term is squared, the standard form of the equation is $x^2 = 4py$. Write the given equation in standard form.

$$y = -\frac{1}{2}x^2$$

$$x^2 = -2y.$$

Comparing this equation with $x^2 = 4py$ gives

$$4p = -2$$

or

$$p = -\frac{1}{2}$$

Because p is negative, the parabola opens down and the focus is below the vertex $(0, 0)$ as shown in Figure 9.4. The coordinates of the focus are $\left(0, -\frac{1}{2}\right)$. The equation of the directrix is $y = \frac{1}{2}$.

■ *Try Exercise **4**, page 459.*

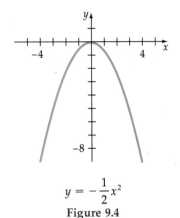

$$y = -\frac{1}{2}x^2$$

Figure 9.4

EXAMPLE 2   **Find the Equation of a Parabola in Standard Form**

Find the equation of the parabola in standard form with vertex at the origin and focus at $(-2, 0)$.

Solution   Because the vertex is at $(0, 0)$ and the focus is at $(-2, 0)$, $p = -2$. The graph of the parabola opens toward the focus, so in this case, the parabola opens to the left. The equation of the parabola in standard form that opens to the left is $y^2 = 4px$. Substitute -2 for p in this equation and simplify.

$$y^2 = 4(-2)x = -8x$$

The equation is $y^2 = -8x$. See Figure 9.5.

■ *Try Exercise **28**, page 459.*

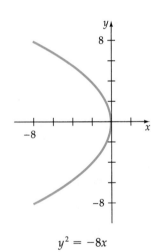

$$y^2 = -8x$$

Figure 9.5

Parabolas with the Vertex at (h, k)

The equation of a parabola with a vertical or horizontal axis and with the vertex at a point (h, k) can be found by using the translations discussed previously. Consider a coordinate system with coordinate axes labeled x' and y' placed so that its origin is at (h, k) of the xy-coordinate system.

The relationship between an ordered pair in the $x'y'$-coordinate system and the xy-coordinate system is given by the transformation equations

$$x' = x - h$$
$$y' = y - k$$

(1)

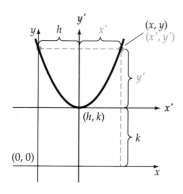

Figure 9.6

Now consider a parabola with vertex at (h, k) as shown in Figure 9.6. Place a new coordinate system labeled x' and y' with its origin at (h, k). The equation of a parabola in the $x'y'$-coordinate system is

$$(x')^2 = 4py' \qquad (2)$$

Using the transformation equations (1), we can substitute the expressions for x' and y' into Equation (2). The standard form of the equation of the parabola with vertex (h, k) and a vertical axis of symmetry is

$$(x - h)^2 = 4p(y - k)$$

Similarly, we can derive the standard form of the equation of the parabola with vertex (h, k) and a horizontal axis of symmetry.

Standard Form of the Equation of a Parabola with Vertex at (h, k)

Vertical Axis of Symmetry
The standard form of the equation of the parabola with vertex $V(h, k)$ and a vertical axis of symmetry is

$$(x - h)^2 = 4p(y - k)$$

The focus is $(h, k + p)$, and the equation of the directrix is $y = k - p$. See Figure 9.7.

Horizontal Axis of Symmetry
The standard form of the equation of the parabola with vertex (h, k) and a horizontal axis of symmetry is

$$(y - k)^2 = 4p(x - h)$$

The focus is $(h + p, k)$, and the equation of the directrix is $x = h - p$.

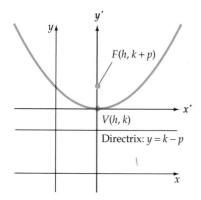

Figure 9.7

EXAMPLE 3 **Find the Focus and Directrix of a Parabola**

Find the equation of the directrix and the coordinates of the vertex and focus of the parabola given by the equation $3x + 2y^2 + 8y - 4 = 0$.

Solution Rewrite the equation so that the y terms are on one side of the equation, and then complete the square on y.

$3x + 2y^2 + 8y - 4 = 0$

$\qquad 2y^2 + 8y = -3x + 4$

$\qquad 2(y^2 + 4y) = -3x + 4$

$2(y^2 + 4y + 4) = -3x + 4 + 8$ • Complete the square. Note that $2 \cdot 4 = 8$ is added to each side.

$\qquad 2(y + 2)^2 = -3(x - 4)$ • Simplify and then factor.

$\qquad (y + 2)^2 = -\dfrac{3}{2}(x - 4)$ • Write the equation in standard form.

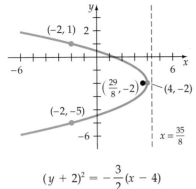

$$(y + 2)^2 = -\frac{3}{2}(x - 4)$$
Figure 9.8

Comparing this equation to $(y - k)^2 = 4p(x - h)$, we have a parabola that opens to the left with vertex $(4, -2)$ and $4p = -\frac{3}{2}$. Thus $p = -\frac{3}{8}$.

The coordinates of the focus are $\left(4 + \left(-\frac{3}{8}\right), -2\right) = \left(\frac{29}{8}, -2\right)$. The equation of the directrix is $x = 4 - \left(-\frac{3}{8}\right) = \frac{35}{8}$.

Choosing some values for y and finding the corresponding values for x, we plot a few points. Because the line $y = -2$ is the axis of symmetry, for each point on one side of the axis of symmetry, there is a corresponding point on the other side. Two points are $(-2, 1)$ and $(-2, -5)$. See Figure 9.8.

■ *Try Exercise* **20**, *page 459.*

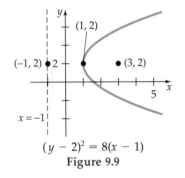

$$(y - 2)^2 = 8(x - 1)$$
Figure 9.9

EXAMPLE 4   **Find the Equation in Standard Form of a Parabola**

Find the equation in standard form of the parabola with directrix $x = -1$ and focus $(3, 2)$.

Solution   The vertex is the midpoint of the line segment joining $(3, 2)$ and the point $(-1, 2)$ on the directrix.

$$(h, k) = \left(\frac{-1 + 3}{2}, \frac{2 + 2}{2}\right) = (1, 2)$$

The standard form of the equation is $(y - k)^2 = 4p(x - h)$. The distance from the vertex to the focus is 2. Thus $4p = 4(2) = 8$, and the equation of the parabola in standard form is $(y - 2)^2 = 8(x - 1)$. See Figure 9.9.

■ *Try Exercise* **30**, *page 459.*

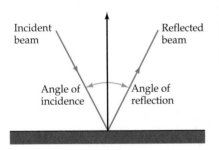

Figure 9.10

A principle of physics states that when light is reflected from a point P on a surface, the angle of incidence (incoming ray) equals the angle of reflection (outgoing ray). See Figure 9.10. This principle applied to parabolas has some useful consequences.

Optical Property of a Parabola

> The line tangent to a parabola at a point P makes equal angles with the line through P and parallel to the axis of symmetry and the line through P and the focus of the parabola (see Figure 9.11).

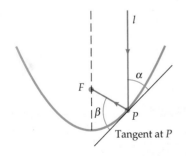

$$\alpha = \beta$$
Figure 9.11

The reflecting mirror of a telescope is designed in the shape of a parabola. The incoming parallel rays of light are reflected from the surface of the mirror and to the focus. See Figure 9.12.

Flashlights and car headlights also make use of this property. The light bulb is positioned at the focus of the parabolic reflector, which causes the reflected light to be reflected outward in parallel rays. See Figure 9.13.

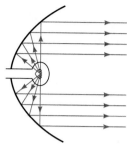

Figure 9.12

Parabolic
mirror

Figure 9.13

EXERCISE SET 9.1

In Exercises 1 to 26, find the vertex, focus, and directrix of the parabola given by each equation. Sketch the graph.

1. $x^2 = -4y$

2. $2y^2 = x$

3. $y^2 = \dfrac{1}{3}x$

4. $x^2 = -\dfrac{1}{4}y$

5. $(x - 2)^2 = 8(y + 3)$

6. $(y + 1)^2 = 6(x - 1)$

7. $(y + 4)^2 = -4(x - 2)$

8. $(x - 3)^2 = -(y + 2)$

9. $(y - 1)^2 = 2x + 8$

10. $(x + 2)^2 = 3y - 6$

11. $(2x - 4)^2 = 8y - 16$

12. $(3x + 6)^2 = 18y - 36$

13. $x^2 + 8x - y + 6 = 0$

14. $x^2 - 6x + y + 10 = 0$

15. $x + y^2 - 3y + 4 = 0$

16. $x - y^2 - 4y + 9 = 0$

17. $2x - y^2 - 6y + 1 = 0$

18. $3x + y^2 + 8y + 4 = 0$

19. $x^2 + 3x + 3y - 1 = 0$

20. $x^2 + 5x - 4y - 1 = 0$

21. $2x^2 - 8x - 4y + 3 = 0$

22. $6x - 3y^2 - 12y + 4 = 0$

23. $2x + 4y^2 + 8y - 5 = 0$

24. $4x^2 - 12x + 12y + 7 = 0$

25. $3x^2 - 6x - 9y + 4 = 0$

26. $2x - 3y^2 + 9y + 5 = 0$

27. Find the equation in standard form of the parabola with vertex at the origin and focus $(0, -4)$.

28. Find the equation in standard form of the parabola with vertex at the origin and focus $(5, 0)$.

29. Find the equation in standard form of the parabola with vertex at $(-1, 2)$ and focus $(-1, 3)$.

30. Find the equation in standard form of the parabola with vertex at $(2, -3)$ and focus $(0, -3)$.

31. Find the equation in standard form of the parabola with focus $(3, -3)$ and directrix $y = -5$.

32. Find the equation in standard form of the parabola with focus $(-2, 4)$ and directrix $x = 4$.

33. Find the equation in standard form of the parabola that

has vertex $(-4, 1)$, has its axis of symmetry parallel to the y-axis, and passes through the point $(-2, 2)$.

34. Find the equation in standard form of the parabola that has vertex $(3, -5)$, has its axis of symmetry parallel to the x-axis, and passes through the point $(4, 3)$.

 Graphing Calculator Exercises*

Sketch a graph of each equation and find the coordinates of the points of intersection of the two graphs to the nearest ten-thousandth.

35. $y = 2x^2 - x - 1$
$y = x$

36. $y = x^2 + 2x - 4$
$y = x - 1$

37. $y = 2x^2 - 1$
$y = x^2 + x + 3$

38. $y = 2x^2 - x - 1$
$y = x^2 - 4$

Using the trace function of a graphing utility, find the maximum value of the function defined by each equation. Express the value to the nearest ten-thousandth.

39. $f(x) = -\sqrt{2}x^2 - 4x + 1$

40. $f(x) = -\pi x^2 + \sqrt{7}x - 2$

Using the trace function of a graphing utility, find the minimum value of the function defined by each equation. Express the value to the nearest ten-thousandth.

41. $f(x) = \dfrac{x^2}{\sqrt{3}} - \dfrac{x}{\sqrt{5}} - \dfrac{1}{\sqrt{7}}$

42. $f(x) = \sqrt[3]{7}x^2 + 5x - \sqrt{5}$

Supplemental Exercises

In Exercises 43 to 45, use the following definition of latus rectum: The line segment that has endpoints on a parabola,

*Additional graphing calculator exercises appear in the Graphing Workbook as described in the front of this textbook.

passes through the focus of the parabola, and is perpendicular to the axis of symmetry is called the **latus rectum** of the parabola.

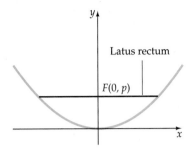

Latus rectum

$F(0, p)$

43. Find the length of the latus rectum for the parabola $x^2 = 4y$.

44. Find the length of the latus rectum for the parabola $y^2 = -8x$.

45. Find the length of the latus rectum for any parabola in terms of $|p|$, the distance from the vertex of the parabola to the focus.

The result of Exercise 45 can be stated as the following theorem: **Theorem:** Two points on a parabola will be $2|p|$ units

on each side of the axis of symmetry on the line through the focus and perpendicular to that axis.

46. Use the theorem to sketch a graph of the parabola given by the equation $(x - 3)^2 = 2(y + 1)$.

47. Use the theorem to sketch a graph of the parabola given by the equation $(y + 4)^2 = -(x - 1)$.

48. Use the theorem to sketch a graph of the parabola given by the equation $4x - y^2 + 8y = 0$.

49. Show that the point on the parabola closest to the focus is the vertex. (*Hint:* Consider the parabola $x^2 = 4py$ and a point on the parabola (a, b). Find the square of the distance between the point (a, b) and the focus. You may want to review the technique of minimizing a quadratic expression.)

50. By using the definition for a parabola, find the equation in standard form of the parabola with $V(0, 0)$, $F(-c, 0)$, and directrix $x = c$.

51. Sketch a graph of $4(y - 2) = x|x| - 1$.

52. Find the equation of the directrix of the parabola with vertex at the origin and focus at the point $(1, 1)$.

53. Find the equation of the parabola with vertex at the origin and focus at the point $(1, 1)$. (*Hint:* You will need the answer to Exercise 52 and the definition of a parabola.)

9.2 Ellipses

An ellipse is another of the conic sections formed when a plane intersects a right circular cone. If β is the angle at which the plane intersects the axis of the cone and α is the angle shown in Figure 9.14, an ellipse is formed when $\alpha < \beta < 90°$. If $\beta = 90°$, then a circle is formed.

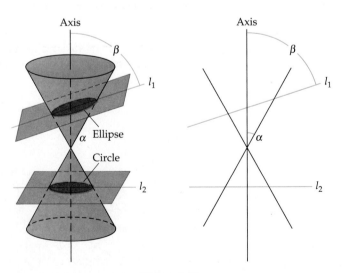

Figure 9.14

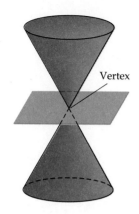

Figure 9.15

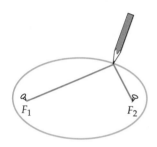

Figure 9.16

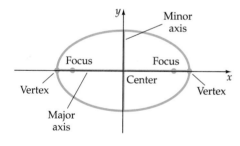

Figure 9.17

Remark If the plane intersects the cone at the vertex of the cone so that the resulting figure is a point, the point is a **degenerate ellipse.** See Figure 9.15.

As was the case for a parabola, there is a definition for an ellipse in terms of a certain set of points in the plane.

Definition of an Ellipse

> An **ellipse** is the set of all points in the plane, the sum of whose distances from two fixed points (foci) is a positive constant.

We can use this definition to draw an ellipse, equipped only with a piece of string and two tacks (see Figure 9.16). Tack the ends of the string to the foci, and trace a curve with a pencil held tight against the string. The resulting curve is an ellipse. The positive constant mentioned in the definition of the ellipse is the length of the string.

Ellipses with Center at (0, 0)

The graph of an ellipse has two axes of symmetry (see Figure 9.17). The longer axis is called the **major axis.** The foci of the ellipse are on the major axis. The shorter axis is called the **minor axis.** It is customary to denote the length of the major axis as $2a$ and the length of the minor axis as $2b$. The length of the **semiaxes** are one-half the axes. Thus the length of the semimajor axis is denoted by a and the length of the semiminor axis by b. The **center** of the ellipse is the midpoint of the major axis. The endpoints of the major axis are the **vertices** (plural of vertex) of the ellipse.

Consider the point $V_1(a, 0)$, which is one vertex on an ellipse, and the point $F_2(c, 0)$ and $F_1(-c, 0)$, which are the foci of the ellipse shown in Figure 9.18. The distance from V_1 to F_1 is $a + c$. Similarly, the distance from V_1 to F_2 is $a - c$. From the definition of an ellipse, the sum of distances from any point on the ellipse to the foci is a positive constant. By adding the expressions $a + c$ and $a - c$, we have

$$(a + c) + (a - c) = 2a$$

Thus the positive constant referred to in the definition of an ellipse is $2a$, the length of the major axis.

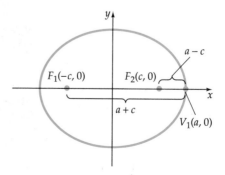

Figure 9.18

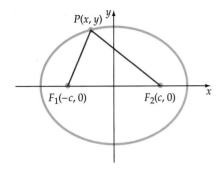

P(x, y)

$F_1(-c, 0)$ $F_2(c, 0)$

Figure 9.19

Now let $P(x, y)$ be any point on the ellipse (see Figure 9.19). By using the definition of an ellipse, we have

$$d(P, F_1) + d(P, F_2) = 2a$$

$$\sqrt{(x + c)^2 + y^2} + \sqrt{(x - c)^2 + y^2} = 2a$$

Subtract the second radical from each side of the equation, and then square each side.

$$[\sqrt{(x + c)^2 + y^2}]^2 = [2a - \sqrt{(x - c)^2 + y^2}]^2$$

$$(x + c)^2 + y^2 = 4a^2 - 4a\sqrt{(x - c)^2 + y^2} + (x - c)^2 + y^2$$

$$x^2 + 2cx + c^2 + y^2 = 4a^2 - 4a\sqrt{(x - c)^2 + y^2} + x^2 - 2cx + c^2 + y^2$$

$$4cx - 4a^2 = -4a\sqrt{(x - c)^2 + y^2}$$

$$[-cx + a^2]^2 = [a\sqrt{(x - c)^2 + y^2}]^2 \quad \bullet \text{ Divide by } -4, \text{ and then square each side.}$$

$$c^2x^2 - 2cxa^2 + a^4 = a^2x^2 - 2cxa^2 + a^2c^2 + a^2y^2$$

$$-a^2x^2 + c^2x^2 - a^2y^2 = -a^4 + a^2c^2 \quad \bullet \text{ Rewrite with } x \text{ and } y \text{ terms on the left side.}$$

$$-(a^2 - c^2)x^2 - a^2y^2 = -a^2(a^2 - c^2) \quad \bullet \text{ Factor and let } b^2 = a^2 - c^2.$$

$$-b^2x^2 - a^2y^2 = -a^2b^2 \quad \bullet \text{ Divide each side by } -a^2b^2.$$

$$\frac{x^2}{a^2} + \frac{y^2}{b^2} = 1 \quad \bullet \text{ An equation of an ellipse with center at } (0, 0).$$

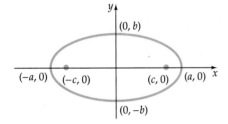

(0, b)

$(-a, 0)$ $(-c, 0)$ $(c, 0)$ $(a, 0)$

(0, −b)

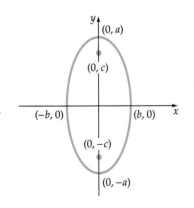

(0, a)

(0, c)

$(-b, 0)$ $(b, 0)$

(0, −c)

(0, −a)

Figure 9.20

Standard Forms of the Equation of an Ellipse with Center at the Origin

Major Axis on the x-axis

The standard form of the equation of an ellipse with the center at the origin and major axis on the x-axis (see Figure 9.20) is given by

$$\frac{x^2}{a^2} + \frac{y^2}{b^2} = 1, \quad a > b$$

The coordinates of the vertices are $(a, 0)$ and $(-a, 0)$, and the coordinates of the foci are $(c, 0)$ and $(-c, 0)$, where $c^2 = a^2 - b^2$.

Major Axis on the y-axis

The standard form of the equation of an ellipse with the center at the origin and major axis on the y-axis (see Figure 9.20) is given by

$$\frac{x^2}{b^2} + \frac{y^2}{a^2} = 1, \quad a > b$$

The coordinates of the vertices are $(0, a)$ and $(0, -a)$, and the coordinates of the foci are $(0, c)$ and $(0, -c)$, where $c^2 = a^2 - b^2$.

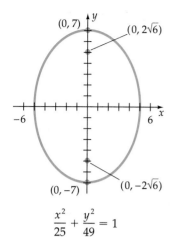

$$\frac{x^2}{25} + \frac{y^2}{49} = 1$$

Figure 9.21

Remark By looking at the standard form of the equations of an ellipse and noting that $a > b$, observe that the orientation of the major axis is determined by the larger denominator. When the x^2 term has the larger denominator, the major axis is on the x-axis. When the y^2 term has the larger denominator, the major axis is on the y-axis.

EXAMPLE 1 **Find the Vertices and Foci of an Ellipse**

Find the vertices and foci of the ellipse given by the equation $\frac{x^2}{25} + \frac{y^2}{49} = 1$. Sketch the graph.

Solution Because the y^2 term has the larger denominator, the major axis is on the y-axis.

$$a^2 = 49 \qquad b^2 = 25 \qquad c^2 = a^2 - b^2$$
$$a = 7 \qquad b = 5 \qquad = 49 - 25 = 24$$
$$c = \sqrt{24} = 2\sqrt{6}$$

The vertices are $(0, 7)$ and $(0, -7)$. The foci are $(0, 2\sqrt{6})$ and $(0, -2\sqrt{6})$. See Figure 9.21.

■ *Try Exercise 20, page 467.*

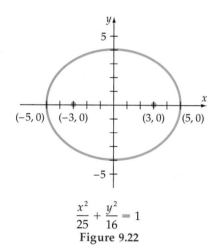

$$\frac{x^2}{25} + \frac{y^2}{16} = 1$$

Figure 9.22

An ellipse with foci $(3, 0)$ and $(-3, 0)$ and major axis of length 10 is shown in Figure 9.22. To find the equation of the ellipse in standard form, we must find a^2 and b^2. Because the foci are on the major axis, the major axis is on the x-axis. The length of the major axis is $2a$. Thus $2a = 10$. Solving for a, we have $a = 5$ and $a^2 = 25$.

Because the foci are $(3, 0)$ and $(-3, 0)$ and the center of the ellipse is the midpoint between the two foci, the distance from the center of the ellipse to a focus is 3. Therefore, $c = 3$. To find b^2, use the equation

$$c^2 = a^2 - b^2$$
$$9 = 25 - b^2$$
$$b^2 = 16$$

The equation of the ellipse is $\frac{x^2}{25} + \frac{y^2}{16} = 1$.

Ellipses with the Center at (h, k)

The equation of an ellipse with center (h, k) and with horizontal or vertical major axes can be found by using a translation of coordinates. On a coordinate system with axes labeled x' and y', the standard form of the equation of an ellipse with center at the origin of the $x'y'$-coordinate system is

$$\frac{(x')^2}{a^2} + \frac{(y')^2}{b^2} = 1$$

Now place the origin of the $x'y'$-coordinate system at (h, k) in an xy-coordinate system. See Figure 9.23.

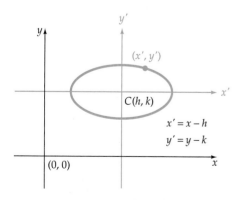

Figure 9.23

The relationship between an ordered pair in the $x'y'$-coordinate system and the xy-coordinate system is given by transformation equations

$$x' = x - h$$

$$y' = y - k$$

Substitute the expressions for x' and y' into the equation of an ellipse. The equation of the ellipse with center at (h, k) is

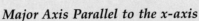

$$\frac{(x - h)^2}{a^2} + \frac{(y - k)^2}{b^2} = 1$$

Standard Form of Ellipses with Center at (h, k)

Major Axis Parallel to the x-axis
The standard form of the equation of an ellipse with the center at (h, k) and major axis parallel to the x-axis (see Figure 9.24) is given by

$$\frac{(x - h)^2}{a^2} + \frac{(y - k)^2}{b^2} = 1 \quad a > b.$$

The coordinates of the vertices are $(h + a, k)$ and $(h - a, k)$, and the coordinates of the foci are $(h + c, k)$ and $(h - c, k)$, where $c^2 = a^2 - b^2$.

Major Axis Parallel to the y-axis
The standard form of the equation of an ellipse with the center at (h, k) and major axis parallel to the y-axis (see Figure 9.24) is given by

$$\frac{(x - h)^2}{b^2} + \frac{(y - k)^2}{a^2} = 1 \quad a > b.$$

The coordinates of the vertices are $(h, k + a)$ and $(h, k - a)$, and the coordinates of the foci are $(h, k + c)$ and $(h, k - c)$, where $c^2 = a^2 - b^2$.

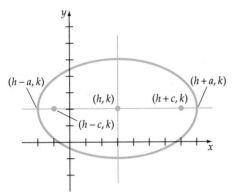

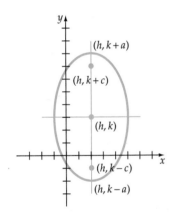

Figure 9.24

EXAMPLE 2 **Find the Vertices and Foci of an Ellipse**

Find the vertices and foci of the ellipse $4x^2 + 9y^2 - 8x + 36y + 4 = 0$. Sketch the graph.

Solution Write the equation of the ellipse in standard form by completing the square.

$$4x^2 + 9y^2 - 8x + 36y + 4 = 0$$

$$4x^2 - 8x + 9y^2 + 36y = -4 \qquad \text{• Rearrange terms.}$$

$$4(x^2 - 2x) + 9(y^2 + 4y) = -4 \qquad \text{• Factor.}$$

$$4(x^2 - 2x + 1) + 9(y^2 + 4y + 4) = -4 + 4 + 36 \qquad \text{• Complete the square}$$

$$4(x - 1)^2 + 9(y + 2)^2 = 36 \qquad \text{• Factor.}$$

$$\frac{(x - 1)^2}{9} + \frac{(y + 2)^2}{4} = 1 \qquad \text{• Divide by 36.}$$

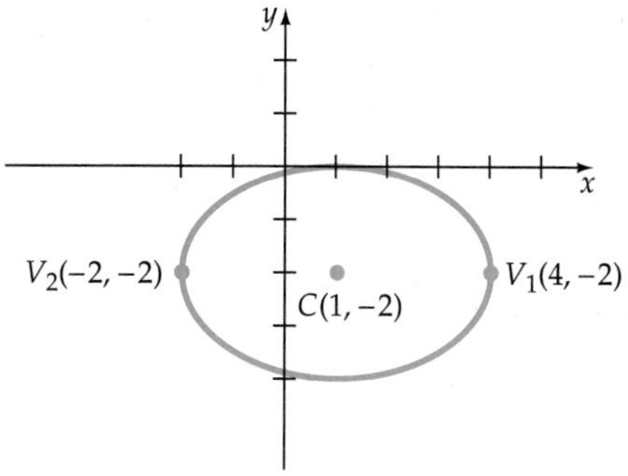

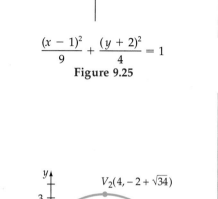

$$\frac{(x-1)^2}{9} + \frac{(y+2)^2}{4} = 1$$

Figure 9.25

From the equation of the ellipse in standard form, the coordinates of the center of the ellipse are $(1, -2)$. Because the larger denominator is 9, the major axis is parallel to the x-axis and $a^2 = 9$. Thus $a = 3$. The vertices are $(4, -2)$ and $(-2, -2)$.

To find the coordinates of the foci, we find c.

$$c^2 = a^2 - b^2 = 9 - 4 = 5$$

$$c = \sqrt{5}$$

The foci are $(1 + \sqrt{5}, -2)$ and $(1 - \sqrt{5}, -2)$. See Figure 9.25.

■ *Try Exercise **26**, page 467.*

EXAMPLE 3 Find the Equation of an Ellipse

Find the standard form of the equation of the ellipse with center at $(4, -2)$, foci $(4, 1)$ and $(4, -5)$, and minor axis of length 10 as shown in Figure 9.26.

Solution Because the foci are on the major axis, the major axis is parallel to the y-axis. The distance from the center of the ellipse to a focus is c. The distance between $(4, -2)$ and $(4, 1)$ is 3. Therefore $c = 3$.

Recall that the length of the minor axis is $2b$. Thus $2b = 10$ and $b = 5$. To find a^2, use the equation $c^2 = a^2 - b^2$.

$$9 = a^2 - 25$$

$$a^2 = 34$$

Thus the equation in standard form is

$$\frac{(x-4)^2}{25} + \frac{(y+2)^2}{34} = 1$$

■ *Try Exercise **42**, page 468.*

Figure 9.26

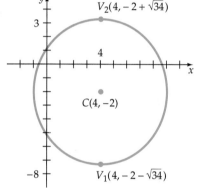

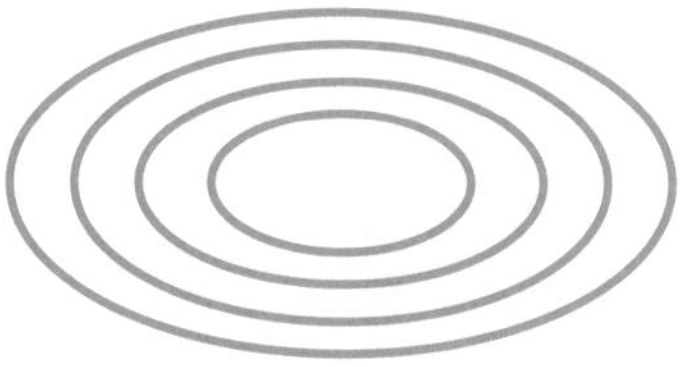

Eccentricity = 0.87

Figure 9.27

Eccentricity of an Ellipse

The graph of an ellipse can be very long and thin, or it can be much like a circle. The **eccentricity** of an ellipse is a measure of its "roundness."

Eccentricity (e) of an Ellipse

The eccentricity e of an ellipse is the ratio of c to a, where c is the distance from the center to the focus and a is the length of the semimajor axis. (See Figure 9.27.) That is,

$$e = \frac{c}{a}$$

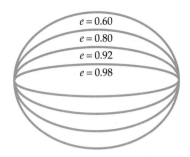

$e = 0.60$

$e = 0.80$

$e = 0.92$

$e = 0.98$

Figure 9.28

Because $c < a$, for an ellipse, $0 < e < 1$. When $e \approx 0$, the graph is almost a circle. When $e \approx 1$, the graph is long and thin. See Figure 9.28.

EXAMPLE 4   **Find the Eccentricity of an Ellipse**

Find the eccentricity of the ellipse $8x^2 + 9y^2 = 18$.

Solution   First, write the equation of the ellipse in standard form. Divide each side of the equation by 18.

$$\frac{8x^2}{18} + \frac{9y^2}{18} = 1$$

$$\frac{4x^2}{9} + \frac{y^2}{2} = 1$$

$$\frac{x^2}{9/4} + \frac{y^2}{2} = 1 \qquad \cdot \frac{4}{9} = \frac{1}{9/4}$$

The last step is necessary because the standard form of the equation has coefficients of 1 in the numerator. Thus

$$a^2 = \frac{9}{4}$$

$$a = \frac{3}{2}$$

Use the equation $c^2 = a^2 - b^2$ to find c.

$$c^2 = \frac{9}{4} - 2 = \frac{1}{4}$$

$$c = \sqrt{\frac{1}{4}} = \frac{1}{2}$$

Now find the eccentricity.

$$e = \frac{c}{a} = \frac{1/2}{3/2} = \frac{1}{3}$$

The eccentricity is $\frac{1}{3}$.

■ *Try Exercise 48, page 468.*

Planet	Eccentricity
Mercury	0.206
Venus	0.007
Earth	0.017
Mars	0.093
Jupiter	0.049
Saturn	0.051
Uranus	0.046
Neptune	0.005
Pluto	0.250

The planets revolve around the sun in elliptical orbits. The eccentricities of planets in our solar system are given at the left.

Which planet has the most nearly circular orbit? Why?[2]

[2] Neptune has the smallest eccentricity, so it is the planet with the most nearly circular orbit.

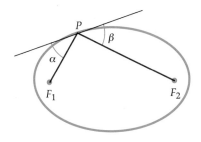

$\alpha = \beta.$
Figure 9.29

Acoustic Property of an Ellipse

Sound waves, although different from light waves, have a similar reflective property. When sound is reflected from a point P on a surface, the angle of incidence equals the angle of reflection. Applying this principle to a room with an elliptical ceiling results in what are called whispering galleries. These galleries are based on the following theorem.

The Reflective Property of an Ellipse

> The lines from the foci to a point on the ellipse make equal angles with the tangent line at that point. See Figure 9.29.

The Rotunda of the Capitol Building in Washington, D.C., is a whispering gallery. Two people standing at the foci of the elliptical ceiling can whisper and yet hear each other even though they are a considerable distance apart. The whisper from one person is reflected to the person standing at the other focus.

EXERCISE SET 9.2

In Exercises 1 to 32, find the vertices and foci of the ellipse given by each equation. Sketch the graph.

1. $\dfrac{x^2}{16} + \dfrac{y^2}{25} = 1$ 　　**2.** $\dfrac{x^2}{49} + \dfrac{y^2}{36} = 1$

3. $\dfrac{x^2}{9} + \dfrac{y^2}{4} = 1$ 　　**4.** $\dfrac{x^2}{64} + \dfrac{y^2}{25} = 1$

5. $\dfrac{x^2}{7} + \dfrac{y^2}{9} = 1$ 　　**6.** $\dfrac{x^2}{5} + \dfrac{y^2}{4} = 1$

7. $\dfrac{4x^2}{9} + \dfrac{y^2}{16} = 1$ 　　**8.** $\dfrac{x^2}{9} + \dfrac{9y^2}{16} = 1$

9. $\dfrac{(x-3)^2}{25} + \dfrac{(y+2)^2}{16} = 1$ 　**10.** $\dfrac{(x+3)^2}{9} + \dfrac{(y+1)^2}{16} = 1$

11. $\dfrac{(x+2)^2}{9} + \dfrac{y^2}{25} = 1$ 　**12.** $\dfrac{x^2}{25} + \dfrac{(y-2)^2}{81} = 1$

13. $\dfrac{(x-1)^2}{21} + \dfrac{(y-3)^2}{4} = 1$ 　**14.** $\dfrac{(x+5)^2}{9} + \dfrac{(y-3)^2}{7} = 1$

15. $\dfrac{9(x-1)^2}{16} + \dfrac{(y+1)^2}{9} = 1$ 　**16.** $\dfrac{(x+6)^2}{25} + \dfrac{25y^2}{144} = 1$

17. $3x^2 + 4y^2 = 12$ 　　**18.** $5x^2 + 4y^2 = 20$

19. $25x^2 + 16y^2 = 400$ 　　**20.** $25x^2 + 12y^2 = 300$

21. $64x^2 + 25y^2 = 400$ 　　**22.** $9x^2 + 64y^2 = 144$

23. $4x^2 + y^2 - 24x - 8y + 48 = 0$

24. $x^2 + 9y^2 + 6x - 36y + 36 = 0$

25. $5x^2 + 9y^2 - 20x + 54y + 56 = 0$

26. $9x^2 + 16y^2 + 36x - 16y - 104 = 0$

27. $16x^2 + 9y^2 - 64x - 80 = 0$

28. $16x^2 + 9y^2 + 36y - 108 = 0$

29. $25x^2 + 16y^2 + 50x - 32y - 359 = 0$

30. $16x^2 + 9y^2 - 64x - 54y + 1 = 0$

31. $8x^2 + 25y^2 - 48x + 50y + 47 = 0$

32. $4x^2 + 9y^2 + 24x + 18y + 44 = 0$

In Exercises 33 to 44, find the equation in standard form of each ellipse, given the information provided.

33. Center $(0, 0)$, major axis of length 10, foci at $(4, 0)$ and $(-4, 0)$

34. Center $(0, 0)$, minor axis of length 6, foci at $(0, 4)$ and $(0, -4)$

35. Vertices $(6, 0)$, $(-6, 0)$; ellipse passes through $(0, -4)$, and $(0, 4)$

36. Vertices $(7, 0)$, $(-7, 0)$; ellipse passes through $(0, 5)$, and $(0, -5)$

37. Major axis of length 12 on the x-axis, center at $(0, 0)$; ellipse passes through $(2, -3)$

38. Minor axis of length 8, center at $(0, 0)$; ellipse passes through $(-2, 2)$

39. Center $(-2, 4)$, vertices $(-6, 4)$ and $(2, 4)$, foci $(-5, 4)$ and $(1, 4)$

40. Center $(0, 3)$, minor axis of length 4, foci $(0, 0)$ and $(0, 6)$

41. Center $(2, 4)$, major axis parallel to the y-axis and of length 10; ellipse passes through the point $(3, 3)$

42. Center $(-4, 1)$, minor axis parallel to the y-axis and of length 8; ellipse passes through $(0, 4)$

43. Vertices $(5, 6)$ and $(5, -4)$, foci $(5, 4)$ and $(5, -2)$

44. Vertices $(-7, -1)$ and $(5, -1)$, foci $(-5, -1)$ and $(3, -1)$

In Exercises 45 to 52, use the eccentricity of the ellipse to find the equation in standard form of each of the following ellipses.

45. Eccentricity $2/5$, major axis on the x-axis and of length 10, center at $(0, 0)$

46. Eccentricity $3/4$, foci at $(9, 0)$ and $(-9, 0)$

47. Foci at $(0, -4)$ and $(0, 4)$, eccentricity $2/3$

48. Foci at $(0, -3)$ and $(0, 3)$, eccentricity $1/4$

49. Eccentricity $2/5$, foci $(-1, 3)$ and $(3, 3)$

50. Eccentricity $1/4$, foci $(-2, 4)$ and $(-2, -2)$

51. Eccentricity $2/3$, major axis of length 24 on the y-axis, centered at $(0, 0)$

52. Eccentricity $3/5$, major axis of length 15 on the x-axis, center at $(0, 0)$

Supplemental Exercises

53. Explain why the graph of the equation
$4x^2 + 9y^2 - 8x + 36y + 76 = 0$ is or is not an ellipse.

54. Explain why the graph of the equation
$4x^2 + 9y - 16x - 2 = 0$ is or is not an ellipse. Sketch the graph of this equation.

In Exercises 55 to 58, find the equation in standard form of an ellipse by using the definition of an ellipse.

55. Find the equation of the ellipse with foci at $(-3, 0)$ and $(3, 0)$ that passes through the point $(3, 9/2)$.

56. Find the equation of the ellipse with foci at $(0, 4)$ and $(0, -4)$ that passes through the point $(9/5, 4)$.

57. Find the equation of the ellipse with foci at $(-1, 2)$ and $(3, 2)$ that passes through the point $(3, 5)$.

58. Find the equation of the ellipse with foci at $(-1, 1)$ and $(-1, 7)$ that passes through the point $(3/4, 1)$.

In Exercises 59 and 60, find the latus rectum of the given ellipse. The line segment with endpoints on the ellipse that is perpendicular to the major axis and passes through a focus is the **latus rectum** of the ellipse.

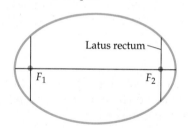

59. Find the length of the latus rectum of the ellipse given by

$$\frac{(x - 1)^2}{9} + \frac{(y + 1)^2}{16} = 1$$

60. Find the length of the latus rectum of the ellipse given by

$$9x^2 + 16y^2 - 36x + 96y + 36 = 0$$

61. Show that for any ellipse, the length of the latus rectum is $2b^2/a$.

62. Use the definition of an ellipse to find the equation of an ellipse with center at $(0, 0)$ and foci $(0, c)$ and $(0, -c)$.

Recall that a parabola has a directrix that is a line perpendicular to the axis of symmetry. An ellipse has two directrices, both of which are perpendicular to the major axis and outside the ellipse. For an ellipse with center at the origin and whose major axis is the x-axis, the equations of the directrices are $x = a^2/c$ and $x = -a^2/c$.

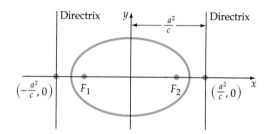

63. Find the directrix of the ellipse in Exercise 3.

64. Find the directrix of the ellipse in Exercise 4.

65. Let $P(x, y)$ be a point on the ellipse $\frac{x^2}{12} + \frac{y^2}{8} = 1$. Show that the distance from the point P to the focus $(2, 0)$ divided by the distance from the point P to the directrix $x = 6$ equals the eccentricity. (*Hint:* Solve the equation of the ellipse for y^2. Substitute this value for y^2 after applying the distance formula.)

66. Let $P(x, y)$ be a point on the ellipse $\frac{x^2}{25} + \frac{y^2}{16} = 1$. Show that the distance from the point P to the focus $(3, 0)$ divided by the distance from the point to the directrix $x = 25/3$ equals the eccentricity. (*Hint:* Solve the equation of the ellipse for y^2. Substitute this value for y^2 after applying the distance formula.)

67. Generalize the results of Exercises 65 and 66. That is, show that if $P(x, y)$ is a point on the ellipse $\frac{x^2}{a^2} + \frac{y^2}{b^2} = 1$, where $F(c, 0)$ is the focus and $x = a^2/c$ is the directrix, then the following equation is true: $e = d(P, F)/d(P, D)$. (*Hint:* Solve the equation of the ellipse for y^2. Substitute this value for y^2 after applying this distance formula.)

9.3   Hyperbolas

The hyperbola is a conic section formed when a plane intersects a right circular cone at a certain angle. If β is the angle at which the plane intersects the axis of the cone and α is the angle shown in Figure 9.30, a hyperbola is formed when $0° < \beta < \alpha$ or when the plane is parallel to the axis of the cone.

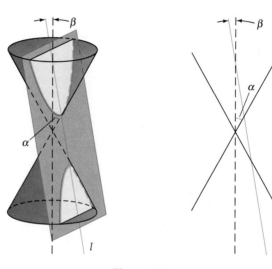

Figure 9.30

Remark   If the plane intersects the cone along the axis of the cone, the resulting curve is two intersecting straight lines. This is the **degenerate** form of a hyperbola. See Figure 9.31.

As with the other conic sections, there is a definition of a hyperbola in terms of a certain set of points in the plane.

Definition of a Hyperbola

A **hyperbola** is the set of all points in the plane, the difference between whose distances from two fixed points (foci) is a positive constant.

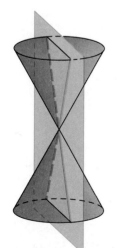

Degenerate hyperbola

Figure 9.31

Remark   This definition differs from that of an ellipse in that the ellipse was defined in terms of the *sum* of two distances, whereas the hyperbola is defined in terms of the *difference* of two distances.

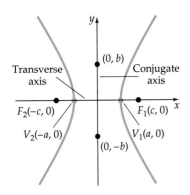

Figure 9.32

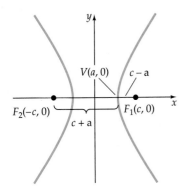

Figure 9.33

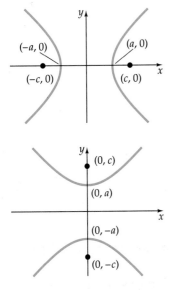

Figure 9.34

Hyperbolas with Center at (0, 0)

The **transverse axis** is the line segment joining the intercepts (see Figure 9.32). The midpoint of the transverse axis is called the **center** of the hyperbola. The **conjugate** axis passes through the center of the hyperbola and is perpendicular to the transverse axis.

The length of the transverse axis is customarily denoted $2a$, and the distance between the two foci is denoted $2c$. The length of the conjugate axis is denoted $2b$.

The **vertices** of a hyperbola are the points where the hyperbola intersects the transverse axis.

To determine the positive constant stated in the definition of a hyperbola, consider the point $V(a, 0)$, which is one vertex of a hyperbola, and the points $F_1(c, 0)$ and $F_2(-c, 0)$, which are the foci of the hyperbola (see Figure 9.33). The difference between the distance from $V(a, 0)$ to F_1, $c - a$, and the distance from $V(a, 0)$ to $F_2(-c, 0)$, $c + a$ must be a constant. By subtracting these distances, we find

$$|(c - a) - (c + a)| = |-2a| = 2a$$

Thus the constant is $2a$ and is the length of the transverse axis. The absolute value is used to ensure that the distance is a positive number.

Standard Forms of the Equation of a Hyperbola with Center at the Origin

Transverse Axis on the x-axis
The standard form of the equation of a hyperbola with the center at the origin and transverse axis on the x-axis (see Figure 9.34) is given by

$$\frac{x^2}{a^2} - \frac{y^2}{b^2} = 1$$

The coordinates of the vertices are $(a, 0)$ and $(-a, 0)$, and the coordinates of the foci are $(c, 0)$ and $(-c, 0)$, where $c^2 = a^2 + b^2$.

Transverse Axis on the y-axis
The standard form of the equation of a hyperbola with the center at the origin and transverse axis on the y-axis (see Figure 9.34) is given by

$$\frac{y^2}{a^2} - \frac{x^2}{b^2} = 1$$

The coordinates of the vertices are $(0, a)$ and $(0, -a)$, and the coordinates of the foci are $(0, c)$ and $(0, -c)$, where $c^2 = a^2 + b^2$.

$$\frac{x^2}{16} - \frac{y^2}{9} = 1$$

Figure 9.35

Remark By looking at the equations, note that it is possible to determine the transverse axis by finding which term in the equation is positive. When the x^2 term is positive, the transverse axis is on the x-axis. When the y^2 term is positive, the transverse axis is on the y-axis.

Consider the hyperbola given by the equation $\frac{x^2}{16} - \frac{y^2}{9} = 1$. Because the x^2 term is positive, the transverse axis is on the x-axis, $a^2 = 16$, and thus $a = 4$. The vertices are $(4, 0)$ and $(-4, 0)$. To find the foci, we determine c.

$$c^2 = a^2 + b^2 = 16 + 9 = 25$$
$$c = \sqrt{25} = 5$$

The foci are $(5, 0)$ and $(-5, 0)$. The graph is shown in Figure 9.35.

Each hyperbola has two asymptotes that pass through the center of the hyperbola. The asymptotes of the hyperbola are a useful guide to sketching the graph of the hyperbola.

Asymptotes of a Hyperbola with Center at the Origin

The **asymptotes** of the hyperbola defined by $\frac{x^2}{a^2} - \frac{y^2}{b^2} = 1$ are given by the equations $y = \frac{b}{a}x$ and $y = -\frac{b}{a}x$. The asymptotes of the hyperbola defined by $\frac{y^2}{a^2} - \frac{x^2}{b^2} = 1$ are given by the equations $y = \frac{a}{b}x$ and $y = -\frac{a}{b}x$. See Figure 9.36.

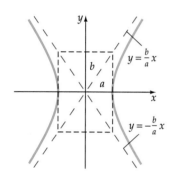

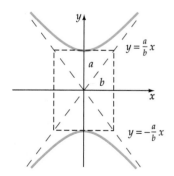

Figure 9.36

We can outline a proof for the equations of the asymptotes by using the equation of a hyperbola in standard form.

$$\frac{x^2}{a^2} - \frac{y^2}{b^2} = 1$$

$$y^2 = b^2\left(\frac{x^2}{a^2} - 1\right) \qquad \bullet \text{ Solve for } y^2.$$

$$= \frac{b^2}{a^2}(x^2 - a^2) \qquad \bullet \text{ Factor out } 1/a^2.$$

$$= \frac{b^2}{a^2}x^2\left(1 - \frac{a^2}{x^2}\right) \qquad \bullet \text{ Factor out } x^2.$$

$$y = \pm\frac{b}{a}x\sqrt{1 - \frac{a^2}{x^2}} \qquad \bullet \text{ Take the square root of each side.}$$

As $|x|$ becomes larger and larger, $1 - \frac{a^2}{x^2}$ approaches 1. (Why?[3]) For large

---

[3] Because a^2 is a constant, a^2/x^2 approaches 0 as $|x|$ becomes very large. Thus $1 - a^2/x^2$ approaches $1 - 0$, or 1.

values of $|x|$, $y \approx \pm \dfrac{b}{a} x$, and thus $y = \pm \dfrac{b}{a} x$ are asymptotes for the hyperbola. A similar outline of a proof can be given for hyperbolas with the transverse axis on the y-axis.

Remark   One method for remembering the equations of the asymptotes is to write the equation of a hyperbola in standard form but to replace 1 by 0 and then solve for y.

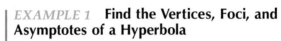

$$\frac{x^2}{a^2} - \frac{y^2}{b^2} = 0 \quad \text{so} \quad y^2 = \frac{b^2}{a^2} x^2, \text{ or } y = \pm \frac{b}{a} x$$

$$\frac{y^2}{a^2} - \frac{x^2}{b^2} = 0 \quad \text{so} \quad y^2 = \frac{a^2}{b^2} x^2, \text{ or } y = \pm \frac{a}{b} x$$

EXAMPLE 1   **Find the Vertices, Foci, and Asymptotes of a Hyperbola**

Find the foci, vertices, and asymptotes of the hyperbola given by the equation $\dfrac{y^2}{9} - \dfrac{x^2}{4} = 1$. Sketch the graph.

Solution   Because the y^2 term is positive, the transverse axis is on the y-axis. We know $a^2 = 9$; thus $a = 3$. The vertices are $V_1(0, 3)$ and $V_2(0, -3)$.

$$c^2 = a^2 + b^2 = 9 + 4$$
$$c = \sqrt{13}$$

The foci are $F_1(0, \sqrt{13})$ and $F_2(0, -\sqrt{13})$.

Because $a = 3$ and $b = 2$ ($b^2 = 4$), the equations of the asymptotes are

$$y = \frac{3}{2} x \quad \text{and} \quad y = -\frac{3}{2} x.$$

To sketch the graph, we draw a rectangle that has its center at the origin and has dimensions equal to the lengths of the transverse and conjugate axes. The asymptotes are extensions of the diagonals of the rectangle. See Figure 9.37.

■ *Try Exercise 4, page 477.*

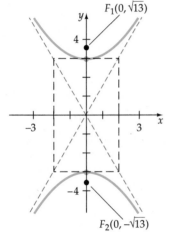

$\dfrac{y^2}{9} - \dfrac{x^2}{4} = 1$

Figure 9.37

Hyperbolas with the Center at the Point (h, k)

Using a translation of coordinates similar to that used for ellipses, we can write the equation of a hyperbola with its center at the point (h, k). Given coordinate axes labeled x' and y', an equation of a hyperbola with center at the origin is

$$\frac{(x')^2}{a^2} - \frac{(y')^2}{b^2} = 1 \tag{1}$$

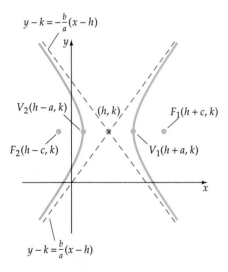

Figure 9.38

Now place the origin of this coordinate system at the point (h, k) of the xy-coordinate system as shown in Figure 9.38. The relationship between an ordered pair in the $x'y'$-coordinate system and the xy-coordinate system is given by the transformation equations

$$x' = x - h$$
$$y' = y - k$$

Substitute the expressions for x' and y' into Equation (1). The equation of the hyperbola with center at (h, k) is

$$\frac{(x - h)^2}{a^2} - \frac{(y - k)^2}{b^2} = 1$$

Standard Form of Hyperbolas with Center at (h, k)

Transverse Axis Parallel to the x-axis
The standard form of the equation of a hyperbola with center (h, k) and transverse axis parallel to the x-axis (see Figure 9.39) is given by

$$\frac{(x - h)^2}{a^2} - \frac{(y - k)^2}{b^2} = 1$$

The coordinates of the vertices are $V_1(h + a, k)$ and $V_2(h - a, k)$. The coordinates of the foci are $F_1(h + c, k)$ and $F_2(h - c, k)$ where $c^2 = a^2 + b^2$. The equations of the asymptotes are $y - k = \frac{b}{a}(x - h)$ and $y - k = -\frac{b}{a}(x - h)$.

Transverse Axis Parallel to the y-axis
The standard form of the equation of a hyperbola with center (h, k) and transverse axis parallel to the y-axis (see Figure 9.39) is given by

$$\frac{(y - k)^2}{a^2} - \frac{(x - h)^2}{b^2} = 1$$

The coordinates of the vertices are $V_1(h, k + a)$ and $V_2(h, k - a)$. The coordinates of the foci are $F_1(h, k + c)$ and $F_2(h, k - c)$ where $c^2 = a^2 + b^2$. The equations of the asymptotes are $y - k = \frac{a}{b}(x - h)$ and $y - k = -\frac{a}{b}(x - h)$.

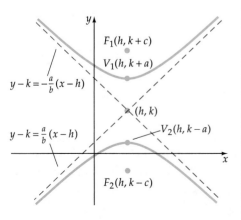

Figure 9.39

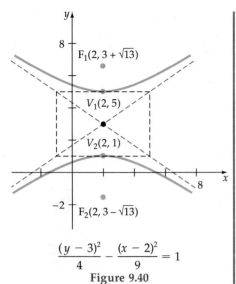

$$\frac{(y-3)^2}{4} - \frac{(x-2)^2}{9} = 1$$

Figure 9.40

EXAMPLE 2   Find the Vertices, Foci, and Asymptotes of a Hyperbola

Find the vertices, foci, and asymptotes of the hyperbola given by the equation $4x^2 - 9y^2 - 16x + 54y - 29 = 0$. Sketch the graph.

Solution   Write the equation of the hyperbola in standard form by completing the square.

$$4x^2 - 9y^2 - 16x + 54y - 29 = 0$$

$$4x^2 - 16x - 9y^2 + 54y = 29 \qquad \bullet \text{ Rearrange terms.}$$

$$4(x^2 - 4x) - 9(y^2 - 6y) = 29 \qquad \bullet \text{ Factor.}$$

$$4(x^2 - 4x + 4) - 9(y^2 - 6y + 9) = 29 + 16 - 81 \qquad \bullet \text{ Complete the square.}$$

$$4(x - 2)^2 - 9(y - 3)^2 = -36 \qquad \bullet \text{ Factor.}$$

$$\frac{(y - 3)^2}{4} - \frac{(x - 2)^2}{9} = 1 \qquad \bullet \text{ Divide by } -36.$$

The coordinates of the center are $(2, 3)$. Because the term containing $(y - 3)^2$ is positive, the transverse axis is parallel to the y-axis. We know $a^2 = 4$; thus $a = 2$. The vertices are $(2, 5)$ and $(2, 1)$. See Figure 9.40.

$$c^2 = a^2 + b^2 = 4 + 9$$

$$c = \sqrt{13}$$

The foci are $(2, 3 + \sqrt{13})$ and $(2, 3 - \sqrt{13})$. We know $b^2 = 9$; thus $b = 3$. The equations of the asymptotes are

$$y = \frac{2}{3}x + \frac{5}{3} \quad \text{and} \quad y = -\frac{2}{3}x + \frac{13}{3}.$$

■ *Try Exercise* **26,** *page 477.*

Eccentricity of a Hyperbola

The graph of a hyperbola can be very wide or very narrow. The **eccentricity** of a hyperbola is a measure of its "wideness."

Eccentricity (e) of a Hyperbola

The eccentricity e of a hyperbola is the ratio of c to a, where c is the distance from the center to a focus and a is the length of the semi-transverse axis.

$$e = \frac{c}{a}$$

For a hyperbola, $c > a$ and therefore $e > 1$. As the eccentricity of the hyperbola increases, the graph becomes wider and wider as shown in Figure 9.41.

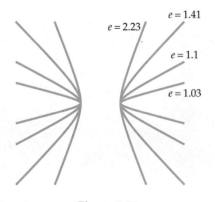

$e = 1.41$
$e = 2.23$
$e = 1.1$
$e = 1.03$

Figure 9.41

EXAMPLE 3   Find the Equation of a Hyperbola Given Its Eccentricity

Find the standard form of the equation of the hyperbola that has eccentricity 3/2, center at the origin, and a focus (6, 0).

Solution   Because the focus is located at (6, 0) and the center is at the origin, $c = 6$. An extension of the transverse axis contains the foci, so the transverse axis is on the x-axis.

$$e = \frac{3}{2} = \frac{c}{a}$$

$$\frac{3}{2} = \frac{6}{a} \quad \bullet \text{ Substitute the value for } c.$$

$$a = 4 \quad \bullet \text{ Solve for } a.$$

To find b^2, use the equation $c^2 = a^2 + b^2$ and the values for c and a.

$$c^2 = a^2 + b^2$$

$$36 = 16 + b^2$$

$$b^2 = 20$$

The equation of the hyperbola is $\dfrac{x^2}{16} - \dfrac{y^2}{20} = 1$. See Figure 9.42.

■ *Try Exercise 48, page 477.*

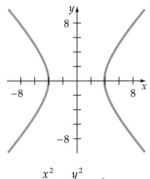

$$\frac{x^2}{16} - \frac{y^2}{20} = 1$$

Figure 9.42

Orbits of Comets

In Section 9.2 we noted that orbits of the planets are elliptical. Some comets have elliptical orbits also, the most notable being Halley's comet, whose eccentricity is 0.97.

Other comets have hyperbolic orbits with the sun at a focus. These comets pass by the sun only once. The velocity of a comet determines whether its orbit is elliptical or hyperbolic. See Figure 9.43.

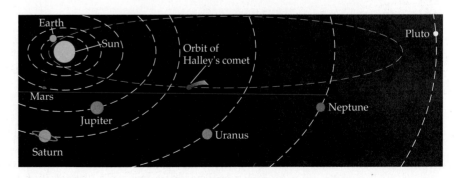

Figure 9.43

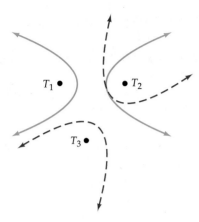

Figure 9.44

Hyperbolas as an Aid to Navigation

Consider two radio transmitters, T_1 and T_2, placed some distance apart. A ship with electronic equipment measures the difference in the time it takes signals from the transmitters to reach the ship. Because the difference in time is proportional to the distance of the ship from the transmitter, the ship must be located on the hyperbola with foci at the two transmitters.

Using a third transmitter, T_3, we can find a second hyperbola with foci T_2 and T_3. The ship lies on the intersection of the two hyperbolas as shown in Figure 9.44.

Reflective Property of a Hyperbola

A ray of light directed toward one focus of a hyperbolic mirror is reflected toward the other focus. See Figures 9.45 and 9.46.

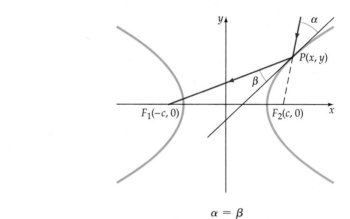

$$\alpha = \beta$$

Figure 9.45

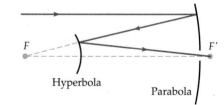

Figure 9.46

EXERCISE SET 9.3

In Exercises 1 to 32, find the center, vertices, foci, and asymptotes for the hyperbola given by each equation. Sketch the graph.

1. $\dfrac{x^2}{16} - \dfrac{y^2}{25} = 1$

2. $\dfrac{x^2}{16} - \dfrac{y^2}{9} = 1$

3. $\dfrac{y^2}{4} - \dfrac{x^2}{25} = 1$

4. $\dfrac{y^2}{25} - \dfrac{x^2}{36} = 1$

5. $\dfrac{x^2}{7} - \dfrac{y^2}{9} = 1$

6. $\dfrac{x^2}{5} - \dfrac{y^2}{4} = 1$

7. $\dfrac{4x^2}{9} - \dfrac{y^2}{16} = 1$

8. $\dfrac{x^2}{9} - \dfrac{9y^2}{16} = 1$

9. $\dfrac{(x - 3)^2}{16} - \dfrac{(y + 4)^2}{9} = 1$

10. $\dfrac{(x + 3)^2}{25} - \dfrac{y^2}{4} = 1$

11. $\dfrac{(y + 2)^2}{4} - \dfrac{(x - 1)^2}{16} = 1$

12. $\dfrac{(y - 2)^2}{36} - \dfrac{(x + 1)^2}{49} = 1$

13. $\dfrac{(x + 2)^2}{9} - \dfrac{y^2}{25} = 1$

14. $\dfrac{x^2}{25} - \dfrac{(y - 2)^2}{81} = 1$

15. $\dfrac{9(x - 1)^2}{16} - \dfrac{(y + 1)^2}{9} = 1$

16. $\dfrac{(x + 6)^2}{25} - \dfrac{25y^2}{144} = 1$

17. $x^2 - y^2 = 9$

18. $4x^2 - y^2 = 16$

19. $16y^2 - 9x^2 = 144$

20. $9y^2 - 25x^2 = 225$

21. $9y^2 - 36x^2 = 4$

22. $16x^2 - 25y^2 = 9$

23. $x^2 - y^2 - 6x + 8y - 3 = 0$

24. $4x^2 - 25y^2 + 16x + 50y - 109 = 0$

25. $9x^2 - 4y^2 + 36x - 8y + 68 = 0$

26. $16x^2 - 9y^2 - 32x - 54y + 79 = 0$

27. $4x^2 - y^2 + 32x + 6y + 39 = 0$

28. $x^2 - 16y^2 + 8x - 64y + 16 = 0$

29. $9x^2 - 16y^2 - 36x - 64y + 116 = 0$

30. $2x^2 - 9y^2 + 12x - 18y + 18 = 0$

31. $4x^2 - 9y^2 + 8x - 18y - 6 = 0$

32. $2x^2 - 9y^2 - 8x + 36y - 46 = 0$

In Exercises 33 to 46, find the equation in standard form of the hyperbola that satisfies the stated conditions.

33. Vertices $(3, 0)$ and $(-3, 0)$, foci $(4, 0)$ and $(-4, 0)$

34. Vertices $(0, 2)$ and $(0, -2)$, foci $(0, 3)$ and $(0, -3)$

35. Foci $(0, 5)$ and $(0, -5)$, asymptotes $y = 2x$ and $y = -2x$

36. Foci $(4, 0)$ and $(-4, 0)$, asymptotes $y = x$ and $y = -x$

37. Vertices $(0, 3)$ and $(0, -3)$ and passing through $(2, 4)$

38. Vertices $(5, 0)$ and $(-5, 0)$ and passing through $(-1, 3)$

39. Asymptotes $y = \dfrac{1}{2}x$ and $y = -\dfrac{1}{2}x$, vertices $(0, 4)$ and $(0, -4)$

40. Asymptotes $y = \dfrac{2}{3}x$ and $y = -\dfrac{2}{3}x$, vertices $(6, 0)$ and $(-6, 0)$

41. Vertices $(6, 3)$ and $(2, 3)$, foci $(7, 3)$ and $(1, 3)$

42. Vertices $(-1, 5)$ and $(-1, -1)$, foci $(-1, 7)$ and $(-1, -3)$

43. Foci $(1, -2)$ and $(7, -2)$, slope of an asymptote $5/4$

44. Foci $(-3, -6)$ and $(-3, -2)$, slope of an asymptote 1

45. Passing through $(9, 4)$, slope of an asymptote $1/2$, center $(7, 2)$, transverse axis parallel to the y-axis

46. Passing through $(6, 1)$, slope of an asymptote 2, center $(3, 3)$, transverse axis parallel to the x-axis

In Exercises 47 to 52, use the eccentricity to find the equation in standard form of a hyperbola.

47. Vertices $(1, 6)$ and $(1, 8)$, eccentricity 2

48. Vertices $(2, 3)$ and $(-2, 3)$, eccentricity $5/2$

49. Eccentricity 2, foci $(4, 0)$ and $(-4, 0)$

50. Eccentricity $4/3$, foci $(0, 6)$ and $(0, -6)$

51. Center $(4, 1)$, conjugate axis of length 4, eccentricity $4/3$
Hint: There are two answers.

52. Center $(-3, -3)$, conjugate axis of length 6, eccentricity 2
Hint: There are two answers.

In Exercises 53 to 60, identify the graph of each equation as a parabola, ellipse, or hyperbola. Sketch the graph.

53. $4x^2 + 9y^2 - 16x - 36y + 16 = 0$

54. $2x^2 + 3y - 8x + 2 = 0$

55. $5x - 4y^2 + 24y - 11 = 0$

56. $9x^2 - 25y^2 - 18x + 50y = 0$

57. $x^2 + 2y - 8x = 0$

58. $9x^2 + 16y^2 + 36x - 64y - 44 = 0$

59. $25x^2 + 9y^2 - 50x - 72y - 56 = 0$

60. $(x - 3)^2 + (y - 4)^2 = (x + 1)^2$

61. Find the equation of the path of Halley's comet in astronomical units by letting the sun (one focus) be at the origin and the other focus on the positive x-axis. The length of the major axis of the orbit of Halley's comet is approximately 36 astronomical units (36 AU) and the length of the minor axis is 9 AU (1 AU = 92,600,000 miles).

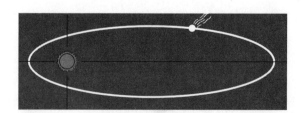

62. A foot suspension bridge is 100 feet long and is supported by cables that hang in the shape of a parabola.

Find the equation of the parabola if the positive x-axis is along the footpath, as shown in the figure.

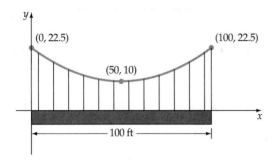

63. Two radio towers are positioned along the coast of California, 200 miles apart. A signal is sent simultaneously from each tower to a ship off the coast. The signal from tower B is received by the ship 500 microseconds after the signal from tower A. The radio signal travels 0.186 miles per microsecond. Find the equation of the hyperbola on which the ship is located.

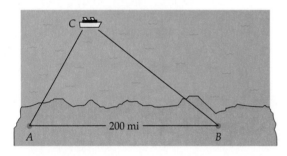

64. A softball player releases a softball at a height of 5 feet above the ground with a speed of 88 feet per second (60 mph). The initial trajectory of the ball is 45°. Neglecting air resistance, the path of the ball is a parabola given by $y = -0.004x^2 + x + 5$. How far will the ball travel in a horizontal direction before hitting the ground? *Hint:* Let $y = 0$ and solve for x.

Supplemental Exercises

In Exercises 65 to 68, use the definition for a hyperbola to find the equation of the hyperbola in standard form.

65. Foci $(2, 0)$ and $(-2, 0)$; and passes through the point $(2, 3)$

66. Foci $(0, 3)$ and $(0, -3)$; and passes through the point $(5/2, 3)$

67. Foci $(0, 4)$ and $(0, -4)$; and passes through the point $(7/3, 4)$

68. Foci $(5, 0)$ and $(-5, 0)$; and passes through the point $(5, 9/4)$

Recall that an ellipse has two directrices that are lines perpendicular to the line containing the foci. A hyperbola also has two directrices that are perpendicular to the transverse axis and outside the hyperbola. For a hyperbola with center at the origin and transverse axis on the x-axis, the equations of the directrices are $x = a^2/c$ and $x = -a^2/c$. In Exercises 69 to 73, use this information to solve each exercise.

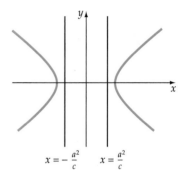

$$x = -\frac{a^2}{c} \qquad x = \frac{a^2}{c}$$

69. Find the directrices for the hyperbola in Exercise 29.

70. Find the directrices for the hyperbola in Exercise 30.

71. Let $P(x, y)$ be a point on the hyperbola $\frac{x^2}{9} - \frac{y^2}{16} = 1$. Show that the distance from the point P to the focus $(5, 0)$ divided by the distance from the point P to the directrix $x = 9/5$ equals the eccentricity.

72. Let $P(x, y)$ be a point on the hyperbola $\frac{x^2}{7} - \frac{y^2}{9} = 1$. Show that the distance from the point P to the focus $(4, 0)$ divided by the distance from the point to the directrix $x = 7/4$ equals the eccentricity.

73. Generalize the results of Exercises 71 and 72. That is, show that if $P(x, y)$ is a point on the hyperbola $\frac{x^2}{a^2} - \frac{y}{b^2} = 1$, $F(c, 0)$ is the focus, and $x = a^2/c$ is the directrix, then the following equation is true: $e = d(P, F)/d(P, D)$.

74. Derive the equation of a hyperbola with center at the origin, foci at $(0, c)$ and $(0, -c)$, and vertices $(0, a)$ and $(0, -a)$.

75. Sketch a graph of $\frac{x|x|}{16} - \frac{y|y|}{9} = 1$.

76. Sketch a graph of $\frac{x|x|}{16} + \frac{y|y|}{9} = 1$.

Introduction to Polar Coordinates

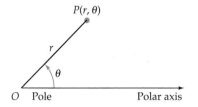

Figure 9.47

Until now, we have used a *rectangular coordinate system* to locate a point in the coordinate plane. An alternative method is to use a *polar coordinate system,* wherein a point is located in a manner similar to giving a distance and an angle from some fixed direction.

The Polar Coordinate System

A **polar coordinate system** is formed by drawing a horizontal ray. The ray is called the **polar axis,** and the beginning point is called the **pole.** A point $P(r, \theta)$ in the plane is located by specifying a distance r from the pole and an angle θ measured from the polar axis to the line segment OP. The angle can be measured in degrees or radians. See Figure 9.47.

The coordinates of the pole are $(0, \theta)$, where θ is an arbitrary angle. Positive angles are measured counterclockwise from the polar axis. Negative angles are measured clockwise from the axis. Positive values of r are measured along the ray that makes an angle θ from the polar axis. Negative values of r are measured along the ray that makes an angle of $\theta + 180°$ from the polar axis. See Figures 9.48 and 9.49.

In a rectangular coordinate system, there is a one-to-one correspondence between the points in the plane and the ordered pairs (x, y). This is not true for a polar coordinate system. For polar coordinates, the relationship is one-to-many. Infinitely many ordered-pair descriptions correspond to each point $P(r, \theta)$ in a polar coordinate system.

For example, consider a point whose coordinates are $P(3, 45°)$. Because there are $360°$ in one complete revolution around a circle, the point P could also be written as $(3, 405°)$, $(3, 765°)$, $(3, 1125°)$, and generally as $(3, 45° + n \cdot 360°)$, where n is an integer. It is also possible to describe the point $P(3, 45°)$ by $(-3, 225°)$, $(-3, -135°)$, and $(3, -315°)$, to name just a few.

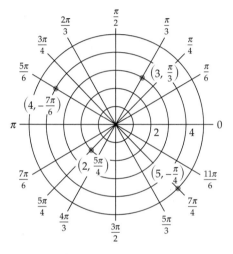

Figure 9.48

Remark The relationship between an ordered pair and a point is not one-to-many. That is, given an ordered pair (r, θ), there is exactly one point in the plane that corresponds to that pair.

Graphs of Equations in a Polar Coordinate System

A **polar equation** is an equation in r and θ. A **solution** to a polar equation is an ordered pair (r, θ) that satisfies the equation. The **graph** of a polar equation is the set of all points whose ordered pairs are solutions of the equation.

The graph of the polar equation $\theta = \pi/6$ is a line. Because θ is independent of r, θ is $\pi/6$ radians from the polar axis for all values of r. The graph

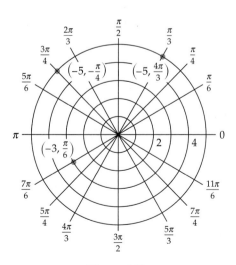

Figure 9.49

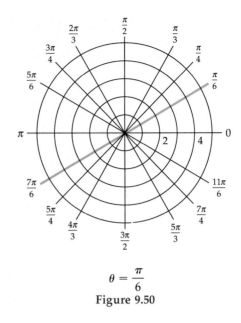

$$\theta = \frac{\pi}{6}$$

Figure 9.50

is a line that makes an angle of $\pi/6$ radians (30°) from the polar axis. See Figure 9.50.

General Equations of a Line

> The graph of $\theta = \alpha$ is a line through the pole at an angle of α from the polar axis. The graph of $r\sin\theta = a$ is a horizontal line passing through the point $(a, \pi/2)$. The graph of $r\cos\theta = a$ is a vertical line passing through the point $(a, 0)$. See Figure 9.51.

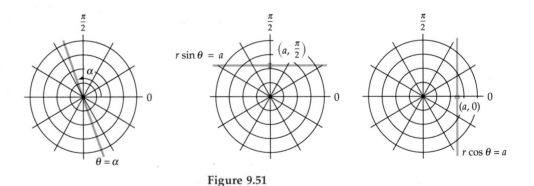

Figure 9.51

Figure 9.52 is the graph of the polar equation $r = 2$. Because r is independent of θ, r is 2 units from the pole for all values of θ. The graph is a circle of radius 2 with center at the pole.

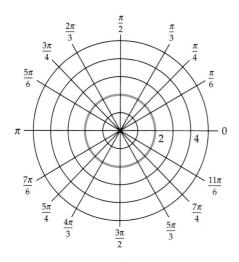

$r = 2$

Figure 9.52

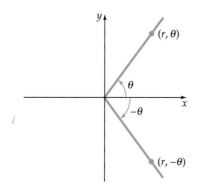

Figure 9.53

The Graph of $r = a$

> The graph of $r = a$ is a circle with center at the pole and radius a.

Suppose that whenever the ordered pair (r, θ) lies on the graph of a polar equation, $(r, -\theta)$ also lies on the graph. From Figure 9.53, the graph will have symmetry with respect to the polar axis. Thus one test for symmetry is to replace θ by $-\theta$ in the polar equation. If the resulting equation is equivalent to the original equation, the graph is symmetric with respect to the polar axis.

The following table shows the types of symmetry and their associated tests. For each type, if the recommended substitution results in an equivalent equation, the graph will have the indicated symmetry.

Tests for Symmetry

Substitution	Symmetry with Respect to
$-\theta$ for θ	The line $\theta = 0$
$\pi - \theta$ for θ, $-r$ for r	The line $\theta = 0$
$\pi - \theta$ for θ	The line $\theta = \pi/2$
$-\theta$ for θ, $-r$ for r	The line $\theta = \pi/2$
$-r$ for r	The pole
$\pi + \theta$ for θ	The pole

Caution The graph of a polar equation may have a symmetry even though a test for that symmetry fails. For example, as we will see later, the graph of $r = \sin 2\theta$ is symmetric with respect to the line $\theta = 0$. However, using the symmetry test of substituting $-\theta$ for θ, we have

$$\sin 2(-\theta) = -\sin 2\theta = -r \neq r$$

Thus this test fails to show symmetry with respect to the line $\theta = 0$. The symmetry test of substituting $\pi - \theta$ for θ and $-r$ for r establishes symmetry with respect to the line $\theta = 0$. (Why?[4]) See Figure 9.54.

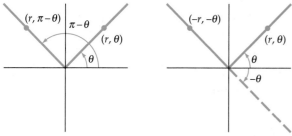

Symmetry with respect to the line $\theta = \pi/2$

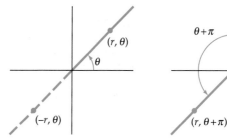

Symmetry with respect to the pole

Figure 9.54

[4] $\sin[2(\pi - \theta)] = \sin(2\pi - 2\theta) = \sin(-2\theta) = -\sin 2\theta = -r$.

EXAMPLE 1   Graph a Polar Equation That is Symmetric with Respect to the Polar Axis

Show that the graph of $r = 4 \cos \theta$ is symmetric to the polar axis. Graph the equation.

Solution   Test for symmetry with respect to the polar axis. Replace θ by $-\theta$.

$$r = 4 \cos(-\theta) = 4 \cos \theta \quad \bullet \; \cos(-\theta) = \cos \theta$$

Because replacing θ by $-\theta$ results in the original equation $r = 4 \cos \theta$, the graph is symmetric with respect to the polar axis.

To graph the equation, begin choosing various values of θ and finding the corresponding values of r. However, before doing so, consider two further observations that will reduce the number of points you must choose.

First, because cosine is a periodic function with period 2π, it is only necessary to choose points between 0 and 2π (0 and 360°). Second, when $\frac{\pi}{2} < \theta < \frac{3\pi}{2}$, $\cos \theta$ is negative, which means that any θ between these values will produce a negative r. Thus the point will be in the first or fourth quadrant. That is, we need consider only angles θ in the first or fourth quadrants. However, since the graph is symmetric with respect to the polar axis, it is only necessary to choose values of θ between 0 and $\pi/2$.

θ	0	$\pi/6$	$\pi/4$	$\pi/3$	$\pi/2$	$-\pi/6$	$-\pi/4$	$-\pi/3$	$-\pi/2$
r	4.0	3.5	2.8	2.0	0.0	3.5	2.8	2.0	0.0

By symmetry spans the last four columns.

The graph of $r = \cos \theta$ is a circle with the center at (2, 0). See Figure 9.55.

■ *Try Exercise* **14,** *page 489.*

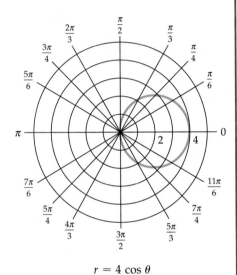

$r = 4 \cos \theta$
Figure 9.55

General Equations of a Circle

The graph of the equation $r = a$ is a circle with center at the pole and radius a. The graph of the equation $r = a \cos \theta$ is a circle that is symmetric with respect to the line $\theta = 0$. The graph of $r = a \sin \theta$ is a circle that is symmetric with respect to the line $\theta = \pi/2$. See Figure 9.56.

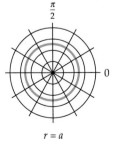

$r = a$

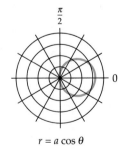

$r = a \cos \theta$

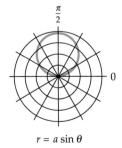

$r = a \sin \theta$

Figure 9.56

Just as there are specifically named curves in an *xy*-coordinate system (for example, parabola and ellipse), there are named curves in a *rθ*-coordinate system. Two of the many types are the *limaçon* and the *rose curve*.

General Equation of a Limaçon

The graph of the equation $r = a + b \cos \theta$ is a **limaçon** that is symmetric with respect to the line $\theta = 0$. The graph of the equation $r = a + b \sin \theta$ is a limaçon that is symmetric with respect to the line $\theta = \frac{\pi}{2}$. In the special case when $|a| = |b|$, the graph is called a **cardioid**.

The graph of $r = a + b \cos \theta$ is shown in Figure 9.57 for various values of *a* and *b*.

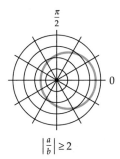

$\left|\frac{a}{b}\right| \geq 2$

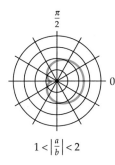

$1 < \left|\frac{a}{b}\right| < 2$

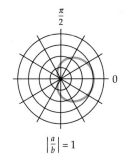

$\left|\frac{a}{b}\right| = 1$

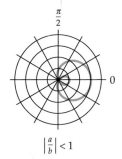

$\left|\frac{a}{b}\right| < 1$

Figure 9.57

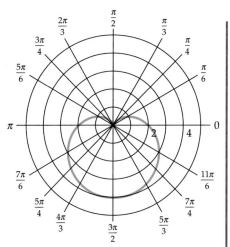

$r = 2 - 2 \sin \theta$
Figure 9.58

EXAMPLE 2   **Sketch the Graph of a Cardioid**

Sketch the graph of $r = 2 - 2 \sin \theta$.

Solution   From the general equation of a limaçon $r = a + b \sin \theta$ with $|a| = |b|$ ($|2| = |-2|$), the graph of $r = 2 - 2 \sin \theta$ is a cardioid that is symmetric to the line $\theta = \frac{\pi}{2}$.

Choose some values of θ and find the corresponding values for r. Plot these points and use symmetry with respect to the line $\theta = \pi/2$ to sketch the graph. See Figure 9.58.

θ	$-\pi/2$	$-\pi/3$	$-\pi/4$	$-\pi/6$	0	$\pi/6$	$\pi/4$	$\pi/3$	$\pi/2$
r	4.0	3.7	3.4	3.0	2.0	1.0	0.6	0.3	0.0

■ *Try Exercise* **18**, *page 489.*

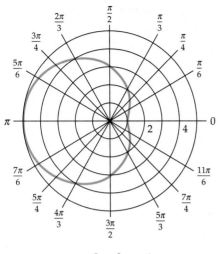

$r = 3 - 2 \cos \theta$
Figure 9.59

EXAMPLE 3   Sketch the Graph of a Limaçon

Sketch the graph of $r = 3 - 2 \cos \theta$.

Solution   From the general equation of a limaçon $r = a + b \cos \theta$, the graph of $r = 3 - 2 \cos \theta$ is a limaçon that is symmetric to the line $\theta = 0$.

Choose some values of θ and find the corresponding values of r. Plot these points and use symmetry to draw the graph. See Figure 9.59.

θ	0	$\pi/6$	$\pi/4$	$\pi/3$	$\pi/2$	$2\pi/3$	$3\pi/4$	$5\pi/6$	π
r	1.0	1.3	1.6	2.0	3.0	4.0	4.4	4.7	5.0

■ *Try Exercise* **20**, *page 489.*

The graph of a rose curve is like the petals of a flower.

General Equations of Rose Curves

The graphs of the equation $r = a \cos n\theta$ and $r = a \sin n\theta$ are rose curves. When n is an even number, the number of petals is $2n$. When n is an odd number, the number of petals is n. See Figure 9.60.

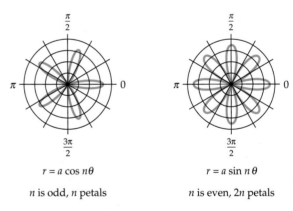

$r = a \cos n\theta$ $r = a \sin n\theta$

n is odd, n petals n is even, $2n$ petals

Figure 9.60

EXAMPLE 4   Sketch the Graph of a Rose Curve

Sketch the graph of $r = 2 \sin 3\theta$.

Solution   From the general equation of a rose curve $r = a \sin n\theta$ with $a = 2$ and $n = 3$, the graph of $r = 2 \sin 3\theta$ is a rose curve that is symmetric to the line $\theta = \dfrac{\pi}{2}$. Because n is an odd number ($n = 3$), there will be 3 leaves in the graph. The graph is a three-leaf rose.

Choose some values for θ and find the corresponding values of r. Use symmetry to sketch the graph. See Figure 9.61.

θ	0	$\pi/18$	$\pi/6$	$5\pi/18$	$\pi/3$	$7\pi/18$	$\pi/2$
r	0.0	1.0	2.0	1.0	0.0	−1.0	−2.0

■ *Try Exercise* **28,** *page 489.*

EXAMPLE 5   Sketch the Graph of a Rose Curve

Graph $r = 4 \cos 2\theta$.

Solution   From the general equation of a rose curve $r = a \cos n\theta$ with $a = 4$ and $n = 2$, the graph of $r = 4 \cos 2\theta$ is a rose curve that is symmetric to the line $\theta = 0$ and $\theta = \dfrac{\pi}{2}$. Because n is an even number ($n = 2$), there will be $2(2) = 4$ leaves in the graph.

Choose some values of θ and find the corresponding values of r. Use symmetry to sketch the graph. The graph is a **four-leaf rose.** See Figure 9.62.

θ	0	$\pi/12$	$\pi/6$	$\pi/4$	$\pi/3$	$5\pi/12$	$\pi/2$
r	4	3.5	2	0	−2	−3.5	−4

■ *Try Exercise* **32,** *page 489.*

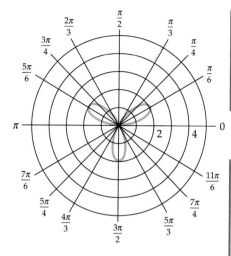

$r = 2 \sin 3\theta$
Figure 9.61

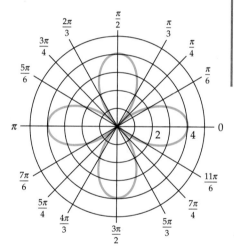

$r = 4 \cos 2\theta$
Figure 9.62

Transformations Between Rectangular and Polar Coordinates

A transformation between coordinate systems is a set of equations that relate the coordinates of a point in one system with the coordinates in a second system. By superimposing a rectangular coordinate system on a polar system, we can derive the set of transformation equations.

Construct a polar coordinate system and a rectangular system so that the pole coincides with the origin and the polar axis coincides with the positive x-axis. Let a point P have coordinates (x, y) in one system and (r, θ) in the other ($r > 0$).

From the definitions of $\sin \theta$ and $\cos \theta$ we have

$$\frac{x}{r} = \cos \theta \qquad \text{or} \qquad x = r \cos \theta$$

and

$$\frac{y}{r} = \sin \theta \qquad \text{or} \qquad y = r \sin \theta$$

It can be shown that these equations are also true when $r < 0$.

Thus given the point (r, θ) in a polar coordinate system, the coordinates of the point in the xy-coordinate system (see Figure 9.63) are given by

$$x = r \cos \theta \qquad y = r \sin \theta$$

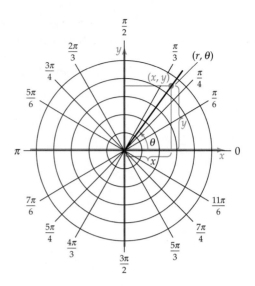

Figure 9.63

For example, to find the point in the xy-coordinate system that corresponds to the point $(4, 2\pi/3)$ in the $r\theta$-coordinate system, substitute into the equations and solve for x and y.

$$x = 4 \cos\left(\frac{2\pi}{3}\right) = 4\left(-\frac{1}{2}\right) = -2$$

$$y = 4 \sin\left(\frac{2\pi}{3}\right) = 4\left(\frac{\sqrt{3}}{2}\right) = 2\sqrt{3}$$

The point $(4, 2\pi/3)$ in the $r\theta$-coordinate system is $(-2, 2\sqrt{3})$ in the xy-coordinate system.

To find the polar coordinates of a given point in the xy-coordinate system, use the Pythagorean Theorem and the definition of the tangent function. Let $P(x, y)$ be a point in the plane, and let r be the distance from the origin to the point P. Then

$$r = \sqrt{x^2 + y^2}$$

From the definition of the tangent function of an angle in a right triangle,

$$\tan \theta = \frac{y}{x}$$

Thus θ is the angle whose tangent is y/x. The quadrant for θ is chosen in accordance with the following:

- $x > 0$ and $y > 0$, θ is a first-quadrant angle.
- $x < 0$ and $y > 0$, θ is a second-quadrant angle.
- $x < 0$ and $y < 0$, θ is a third-quadrant angle.
- $x > 0$ and $y < 0$, θ is a fourth-quadrant angle.

The equations of transformations between a polar and a rectangular coordinate system are summarized as follows.

Transformations between Polar and Rectangular Coordinates

Given the point (r, θ) in the polar coordinate system, the transformation equations to change from polar to rectangular coordinates are

$$x = r \cos \theta \qquad y = r \sin \theta$$

Given the point (x, y) in the rectangular coordinate system, the transformation equations to change from rectangular to polar coordinates are

$$r = \sqrt{x^2 + y^2} \qquad \tan \theta = \frac{y}{x}, x \neq 0$$

where $r \geq 0, 0 \leq \theta < 2\pi$, and θ is chosen so that the point lies in the appropriate quadrant. If $x = 0$, then $\theta = \pi/2$ or $\theta = 3\pi/2$.

EXAMPLE 6   **Transform from Polar to Rectangular Coordinates**

Find the rectangular coordinates of the points whose polar coordinates are   a. $(6, 3\pi/4)$ b. $(-4, 30°)$.

   Solution   Use the two transformation equations $x = r \cos \theta$ and $y = r \sin \theta$.

a. $x = 6 \cos\left(\dfrac{3\pi}{4}\right) = 6 \cdot \left(-\dfrac{\sqrt{2}}{2}\right) = -3\sqrt{2}$

   $y = 6 \sin\left(\dfrac{3\pi}{4}\right) = 6 \cdot \dfrac{\sqrt{2}}{2} = 3\sqrt{2}$

   The rectangular coordinates of $(6, 3\pi/4)$ are $(-3\sqrt{2}, 3\sqrt{2})$.

b. $x = -4 \cos(30°) = -4 \cdot \dfrac{\sqrt{3}}{2} = -2\sqrt{3}$

   $y = -4 \sin(30°) = -4 \cdot \dfrac{1}{2} = -2$

   The rectangular coordinates of $(-4, 30°)$ are $(-2\sqrt{3}, -2)$.

■ *Try Exercise* **44**, *page 489.*

EXAMPLE 7   **Transform from Rectangular to Polar Coordinates**

Find the polar coordinates of the points whose rectangular coordinates are   a. $(-3, 4)$ b. $(-2, -2\sqrt{3})$.

Solution Use the transformation equations $r = \sqrt{x^2 + y^2}$ and $\tan \theta = \dfrac{y}{x}$.

a. $r = \sqrt{(-3)^2 + 4^2} = \sqrt{9 + 16} = \sqrt{25} = 5$

$\tan \theta = \dfrac{4}{-3} = -\dfrac{4}{3}$

From this and the fact that $(-3, 4)$ lies in the second quadrant, $\theta \approx 127°$. The approximate polar coordinates of $(-3, 4)$ are $(5, 127°)$.

b. $r = \sqrt{(-2)^2 + (-2\sqrt{3})^2} = \sqrt{4 + 12} = \sqrt{16} = 4$

$\tan \theta = \dfrac{-2\sqrt{3}}{-2} = \sqrt{3}$

From this and the fact that $(-2, -2\sqrt{3})$ lies in the third quadrant, $\theta = 4\pi/3$. The polar coordinates of $(-2, -2\sqrt{3})$ are $(4, 4\pi/3)$.

■ *Try Exercise 48, page 489.*

Using the transformation equations, it is possible to write a polar coordinate equation in rectangular form or a rectangular coordinate equation in polar form.

EXAMPLE 8   Write a Rectangular Coordinate Equation in Polar Form

Find a polar form of the equation $x^2 + y^2 - 2x = 3$.

Solution

$$x^2 + y^2 - 2x = 3$$

$$(r \cos \theta)^2 + (r \sin \theta)^2 - 2r \cos \theta = 3 \quad \bullet \text{ Use the transformation equations.}$$

$$r^2(\cos^2\theta + \sin^2\theta) - 2r \cos \theta = 3 \quad \bullet \text{ Factor.}$$

$$r^2 - 2r \cos \theta = 3$$

A polar form of $x^2 + y^2 - 2x = 3$ is $r^2 - 2r \cos \theta = 3$.

■ *Try Exercise 56, page 489.*

EXAMPLE 9   Write a Polar Coordinate Equation in Rectangular Form

Find a rectangular form of the equation $r^2 \cos 2\theta = 3$.

Solution

$$r^2 \cos 2\theta = 3$$

$$r^2(1 - 2\sin^2\theta) = 3 \quad \bullet \ \cos 2\theta = 1 - 2\sin^2\theta$$

$$r^2 - 2r^2\sin^2\theta = 3$$

$$x^2 + y^2 - 2r^2\left(\frac{y^2}{r^2}\right) = 3 \quad \bullet \ r^2 = x^2 + y^2; \ \sin\theta = \frac{y}{r}$$

$$x^2 + y^2 - 2y^2 = 3$$

$$x^2 - y^2 = 3$$

A rectangular form of $r^2 \cos 2\theta = 3$ is $x^2 - y^2 = 3$.

■ *Try Exercise* **64**, *page 489.*

EXERCISE SET 9.4

In Exercises 1 to 8, plot the point on a polar coordinate system.

1. $(2, 60°)$ **2.** $(3, -90°)$ **3.** $(1, 315°)$

4. $(2, 400°)$ **5.** $\left(-2, \dfrac{\pi}{4}\right)$ **6.** $\left(4, \dfrac{7\pi}{6}\right)$

7. $\left(-3, \dfrac{5\pi}{3}\right)$ **8.** $(-3, \pi)$

In Exercises 9 to 40, sketch the graphs of the polar equations.

9. $r = 3$ **10.** $r = 5$

11. $\theta = 2$ **12.** $\theta = -\dfrac{\pi}{3}$

13. $r = 6\cos\theta$ **14.** $r = 4\sin\theta$

15. $r = 3 + 3\cos\theta$ **16.** $r = 4 - 4\sin\theta$

17. $r = 2 - 3\sin\theta$ **18.** $r = 2 - 2\cos\theta$

19. $r = 4 + 3\sin\theta$ **20.** $r = 2 + 4\sin\theta$

21. $r = -2 + 2\cos\theta$ **22.** $r = -1 - \cos\theta$

23. $r = 2 - 4\sin\theta$ **24.** $r = 4(1 - \sin\theta)$

25. $r = 3\sin 2\theta$ **26.** $r = 2\cos 2\theta$

27. $r = 4\cos 3\theta$ **28.** $r = 5\sin 3\theta$

29. $r = 2\sin 5\theta$ **30.** $r = 3\cos 5\theta$

31. $r = 3\cos 4\theta$ **32.** $r = 6\sin 4\theta$

33. $r = 3\sec\theta$ **34.** $r = 4\csc\theta$

35. $r = 5\csc\theta$ **36.** $r = 2\sec\theta$

37. $r = \theta, \ \theta \geq 0$ **38.** $r = -\theta, \ \theta \geq 0$

39. $r = 2^\theta, \ \theta \geq 0$ **40.** $r = \dfrac{1}{\theta}, \ \theta > 0$

In Exercises 41 to 48, transform the given coordinates to the indicated ordered pair.

41. $(1, -\sqrt{3})$ to (r, θ) **42.** $(-2\sqrt{3}, 2)$ to (r, θ)

43. $\left(-3, \dfrac{2\pi}{3}\right)$ to (x, y) **44.** $\left(2, -\dfrac{\pi}{3}\right)$ to (x, y)

45. $\left(0, -\dfrac{\pi}{2}\right)$ to (x, y) **46.** $\left(3, \dfrac{5\pi}{6}\right)$ to (x, y)

47. $(3, 4)$ to (r, θ) **48.** $(12, -5)$ to (r, θ)

In Exercises 49 to 60, find the rectangular form of each of the equations.

49. $r = 3\cos\theta$ **50.** $r = 2\sin\theta$

51. $r = 3\sec\theta$ **52.** $r = 4\csc\theta$

53. $r = 4$ **54.** $\theta = \dfrac{\pi}{4}$

55. $r = \tan\theta$ **56.** $r = \cot\theta$

57. $r = \dfrac{2}{1 + \cos\theta}$ **58.** $r = \dfrac{2}{1 - \sin\theta}$

59. $r(\sin\theta - 2\cos\theta) = 6$ **60.** $r(2\cos\theta + \sin\theta) = 3$

In Exercises 61 to 68, find the polar form of each of the equations.

61. $y = 2$ **62.** $x = -4$

63. $x^2 + y^2 = 4$ **64.** $2x - 3y = 6$

65. $x^2 = 8y$ **66.** $y^2 = 4y$

67. $x^2 - y^2 = 25$ **68.** $x^2 + 4y^2 = 16$

Graphing Calculator Exercises*

Sketch a graph of each of the following equations.

69. $r = 3 \cos\left(\theta + \dfrac{\pi}{4}\right)$

70. $r = 2 \sin\left(\theta - \dfrac{\pi}{6}\right)$

71. $r = 2 \sin\left(2\theta - \dfrac{\pi}{3}\right)$

72. $r = 3 \cos\left(2\theta + \dfrac{\pi}{4}\right)$

73. $r = 2 + 2 \sin\left(\theta - \dfrac{\pi}{6}\right)$

74. $r = 3 - 2 \cos\left(\theta + \dfrac{\pi}{3}\right)$

75. $r = 1 + 3 \cos\left(\theta + \dfrac{\pi}{3}\right)$

76. $r = 2 - 4 \sin\left(\theta - \dfrac{\pi}{4}\right)$

Supplemental Exercises

77. $P_1(r_1, \theta_1)$ and $P_2(r_2, \theta_2)$ are two points in the $r\theta$-plane. Use the Law of Cosines to show that the distance between the

*Additional graphing calculator exercises appear in the Graphing Workbook as described in the front of this textbook.

two points $d(P_1, P_2)$ is given by

$$[d(P_1, P_2)]^2 = r_1^2 + r_2^2 - 2r_1r_2 \cos(\theta_1 - \theta_2).$$

78. Prove that the graph of $r = a \sin \theta + b \cos \theta$, $ab \neq 0$, is a circle in polar coordinates.

For Exercises 79 to 86, sketch a graph of the polar equation.

79. $r^2 = 4 \cos 2\theta$ (lemniscate)

80. $r^2 = -2 \sin 2\theta$ (lemniscate)

81. $r = 2(1 + \sec \theta)$ (conchoid)

82. $r = 2 \cos 2\theta \sec \theta$ (strophoid)

83. $r\theta = 2$ (spiral)

84. $r = 2 \sin \theta \cos^2 2\theta$ (bifolium)

85. $r = |\theta|$

86. $r = \ln \theta$

Exploring Concepts with Technology

Graphing an Ellipse or a Hyperbola with a Graphing Calculator

A number of the newer calculators or graphing utilities for computers have a feature that allows the equation of a function to be entered and then the graph of that equation displayed. The equations of the ellipse and hyperbola, however, are not equations that represent functions. Therefore, these equations cannot be directly entered into the calculator or graphing utility.

Consider the graph of the ellipse $\dfrac{x^2}{16} + \dfrac{y^2}{25} = 1$ shown in Figure 9.64. Because a vertical line can intersect the ellipse at more than one point, the graph is not the graph of a function. Solving the equation of the ellipse for y, we have

$$\frac{y^2}{25} = 1 - \frac{x^2}{16}$$

$$y^2 = 25\left(1 - \frac{x^2}{16}\right) \quad \text{or} \quad y = \pm 5\sqrt{1 - \frac{x^2}{16}}$$

Note from the equation for y that for each value of x in $(-4, 4)$, the value of y is positive or negative. This observation is essential for entering the

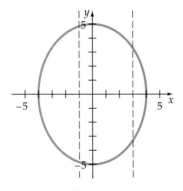

Figure 9.64

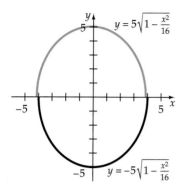

$y = 5\sqrt{1 - \frac{x^2}{16}}$

$y = -5\sqrt{1 - \frac{x^2}{16}}$

Figure 9.65

equation to be graphed. The equation for the ellipse is entered as *two* equations:

$$y = 5\sqrt{1 - \frac{x^2}{16}} \quad \text{and} \quad y = -5\sqrt{1 - \frac{x^2}{16}}$$

One equation represents the top half of the ellipse, the second equation the bottom portion. Figure 9.65 shows the graph of each of these equations. Note that each equation represents the graph of a function. The same technique can be used for a hyperbola. Consider the hyperbola whose equation is

$$x^2 - 4y^2 + 6x + 16y - 43 = 0$$

To solve this equation for y, we begin by completing the square.

$$x^2 - 4y^2 + 6x + 16y - 43 = 0$$

$$x^2 + 6x + 9 - 4(y^2 - 4y + 4) = 9 - 16 + 43$$

$$(x + 3)^2 - 4(y - 2)^2 = 36$$

$$-4(y - 2)^2 = 36 - (x + 3)^2$$

$$(y - 2)^2 = \frac{1}{4}[(x + 3)^2 - 36]$$

$$y - 2 = \pm\sqrt{\frac{1}{4}[(x + 3)^2 - 36]} = \pm\frac{1}{2}\sqrt{[(x + 3)^2 - 36]}$$

$$y = \frac{1}{2}\sqrt{[(x + 3)^2 - 36]} + 2 \quad \text{or} \quad y = -\frac{1}{2}\sqrt{[(x + 3)^2 - 36]} + 2$$

To graph the hyperbola, enter each equation.

Graph some of the exercises in your text by the method shown here.

Chapter Review

9.1 Conic Sections

- A parabola is the set of points in the plane that are equidistant from a fixed line (the directrix) and a fixed point (the focus) not on the directrix.

- The equations of a parabola with vertex at (h, k) and axis of symmetry parallel to a coordinate axis are given by

$(x - h)^2 = 4p(y - k)$ Focus: $(h, k + p)$, directrix: $y = k - p$

$(y - k)^2 = 4p(x - h)$ Focus: $(h + p, k)$, directrix: $x = h - p$

9.2 Ellipses

- An ellipse is the set of all points in the plane, the sum of whose distances from two fixed points (foci) is a positive constant.

- The equations of an ellipse with center at (h, k) and major axis parallel to the coordinate axes are given by

$\dfrac{(x - h)^2}{a^2} + \dfrac{(y - k)^2}{b^2} = 1$ Foci: $(h \pm c, k)$, vertices: $(h \pm a, k)$

$\dfrac{(x - h)^2}{b^2} + \dfrac{(y - k)^2}{a^2} = 1$ Foci: $(h, k \pm c)$, vertices: $(h, k \pm a)$

For each equation, $a > b$ and $c^2 = a^2 - b^2$.

- The eccentricity e of an ellipse is given by $e = c/a$.

9.3 Hyperbolas

- A hyperbola is the set of all points in the plane, the difference of whose distances from two fixed points (foci) is a positive constant.

- The equations of a hyperbola with center at (h, k) and transverse axis parallel to a coordinate axis are given by

$$\frac{(x - h)^2}{a^2} - \frac{(y - k)^2}{b^2} = 1 \quad \text{Foci: } (h \pm c, k), \text{ vertices: } (h \pm a, k)$$

$$\frac{(y - k)^2}{a^2} - \frac{(x - h)^2}{b^2} = 1 \quad \text{Foci: } (h, k \pm c), \text{ vertices: } (h, k \pm a)$$

   For each equation, $c^2 = a^2 + b^2$.
- The eccentricity e of a hyperbola is given by $e = c/a$.

9.4   Introduction to Polar Coordinates

- A polar coordinate system is formed by drawing a ray (polar axis) and concentric circles with center at the beginning of the ray. The pole is the origin of a polar coordinate system.

- A point is specified by coordinates (r, θ), where r is a directed distance from the pole, and θ is an angle measured from the polar axis.

- The transformation equations between a polar coordinate system and a rectangular coordinate system are

   Polar to rectangular: $x = r \cos \theta \qquad y = r \sin \theta$

   Rectangular to polar: $r = \sqrt{x^2 + y^2} \qquad \tan \theta = \frac{y}{x}$

Essays and Projects

1. The German astronomer Johannes Kepler (1571–1630) derived three laws that describe how the planets orbit the sun. Write an essay that includes biographical information about Kepler and a statement of Kepler's Laws. In addition, use Kepler's Laws to answer the following questions.   a. Where is a planet located in its orbit around the sun when it achieves its greatest velocity?   b. What is the period of Mars if it has a mean distance from the sun of 1.52 astronomical units. (*Hint:* Use the earth as a reference with a period of 1 year and a mean distance from the sun of 1 astronomical unit.]

2. The position of the planet Neptune was discovered by using celestial mechanics and mathematics. Write an essay that tells how, when, and by whom Neptune was discovered.

3. The *apogee* is the greatest distance a planet travels from the sun. The *perigee* is the closest a planet comes to the sun. The length of the major axis of the orbit of Mercury is 7.2×10^7 miles. Write a mathematical essay that determines the apogee and the perigee of Mercury (correct to two significant digits). (*Hint:* Use the eccentricity of Mercury as found on page 466.)

4. The Mount Palomar telescope is the world's largest reflective telescope. Write an essay describing the construction of the mirror for this telescope. Include in your essay a graph of the cross section of the mirrored surface of the telescope. Determine the equation of your graph, and use the equation to determine the focal length of the mirror (the distance from the mirror to the focus.)

5. Kepler used a pencil, a string, and the seventeenth-century version of a T-square to draw parabolas. Consult a technical drawing text to discover Kepler's method. Write an essay that explains the method.

6. Explain how to draw a hyperbola by means of the techniques given in a technical drawing text.

7. Carpenters use an instrument called a trammel to draw ellipses. Write an essay that explains what a trammel is and how it is used. What are the advantages and disadvantages of this method of drawing an ellipse? Why does this method approximate an ellipse?

8. What formula is used to find the area of an ellipse with semimajor axis of length a and semiminor axis of length b? Write a historical essay about the development of this formula.

9. The circumference of a circle with radius r is given by $2\pi r$. There is no simple algebraic expression for the circumference of the ellipse given by

$$\frac{x^2}{a^2} + \frac{y^2}{b^2} = 1$$

However, the famous mathematician Srinivasa Ramanujan (1887–1920) discovered a formula that closely approximates the circumference. Read about the mathematical work done by Ramanujan, and then use Ramanujan's formula to estimate the circumference of the ellipse given by $\frac{x^2}{16} + \frac{y^2}{9} = 1$.

10. A *hyperbolic paraboloid* is a three-dimensional figure. Some of its cross sections are parabolas and some hyperbolas. Make a drawing of a hyperbolic paraboloid. Explain the relationship that exists between the equations of the para-

bolic cross sections and the relationship that exists between the equations of the hyperbolic cross sections.

11. Make a sketch of a *hyperboloid of one sheet*. Explain the different cross sections of the hyperboloid of one sheet. Do some research on nuclear power plants and explain why nuclear cooling towers are designed in the shape of hyperboloids of one sheet.

12. Graph

$$y = \frac{[\![x]\!]}{|x|}, \quad -5 \le x \le 5$$

Explain this graph, using the phrase *equilateral hyperbolas*.

True/False Exercises

In Exercises 1 to 10, answer true or false. If the answer is false, give an example to show that the statement is false.

1. The graph of a parabola is the same shape as one branch of a hyperbola.

2. The major axis of an ellipse is always longer than the minor axis.

3. The transverse axis of a hyperbola is always longer than the conjugate axis.

4. If two ellipses have the same foci, they have the same graph.

5. A hyperbola is similar to a parabola in that both curves have asymptotes.

6. If a hyperbola with center at the origin and a parabola with vertex at the origin have the same focus, $(0, c)$, the two graphs always intersect.

7. The graphs of all the conic sections are not the graphs of a function.

8. If F_1 and F_2 are the two foci of an ellipse and P is a point on the ellipse, then $d(P, F_1) + d(P, F_2) = 2a$, where a is the length of the semimajor axis of the ellipse.

9. The eccentricity of a hyperbola is always greater than 1.

10. Each ordered pair (r, θ) in a polar coordinate system specifies exactly one point.

Chapter Review Exercises

In Exercises 1 to 12, find the foci and vertices of each conic. If the conic is a hyperbola, find the asymptotes. Sketch the graph.

1. $x^2 - y^2 = 4$

2. $y^2 = 16x$

3. $x^2 + 4y^2 - 6x + 8y - 3 = 0$

4. $3x^2 - 4y^2 + 12x - 24y - 36 = 0$

5. $3x - 4y^2 + 8y + 2 = 0$

6. $3x + 2y^2 - 4y - 7 = 0$

7. $9x^2 + 4y^2 + 36x - 8y + 4 = 0$

8. $11x^2 - 25y^2 - 44x + 50y - 256 = 0$

9. $4x^2 - 9y^2 - 8x + 12y - 144 = 0$

10. $9x^2 + 16y^2 + 36x - 16y - 104 = 0$

11. $4x^2 + 28x + 32y + 81 = 0$

12. $x^2 - 6x - 9y + 27 = 0$

In Exercises 13 to 20, find the equation in standard form of the conic that satisfies the given conditions.

13. Ellipse with vertices at $(7, 3)$ and $(-3, 3)$; length of minor axis 8

14. Hyperbola with vertices at $(4, 1)$ and $(-2, 1)$; eccentricity 4/3

15. Hyperbola with foci $(-5, 2)$ and $(1, 2)$; length of transverse axis 4

16. Parabola with focus $(2, -3)$ and directrix $x = 6$.

17. Parabola with vertex $(0, -2)$ and passing through the point $(3, 4)$

18. Ellipse with eccentricity 2/3 and foci $(-4, -1)$ and $(0, -1)$

19. Hyperbola with vertices $(\pm 6, 0)$ and asymptotes whose equations are $y = \pm(1/9)x$

20. Parabola passing through the points $(1, 0)$, $(2, 1)$, and $(0, 1)$ with axis of symmetry parallel to the y-axis.

21. Find the equation of the parabola traced by a point $P(x, y)$ that moves in such a way that the distance between $P(x, y)$ and the line $x = 2$ equals the distance between $P(x, y)$ and the point $(-2, 3)$.

22. Find the equation of the parabola traced by a point $P(x, y)$ that moves in such a way that the distance between $P(x, y)$ and the line $y = 1$ equals the distance between $P(x, y)$ and the point $(-1, 2)$.

23. Find the equation of the ellipse traced by a point $P(x, y)$ that moves in such a way that the sum of its distances to $(-3, 1)$ and $(5, 1)$ is 10.

24. Find the equation of the ellipse traced by a point $P(x, y)$ that moves in such a way that the sum of its distances to $(3, 5)$ and $(3, -1)$ is 8.

In Exercises 25 to 34, sketch the graph of the polar equations.

25. $r = 4 \cos 3\theta$

26. $r = 1 + \cos \theta$

27. $r = 2(1 - 2 \sin \theta)$

28. $r = 4 \sin 4\theta$

29. $r = 5 \sin \theta$

30. $r = 3 \sec \theta$

31. $r = 4 \csc \theta$

32. $r = 4 \sin \theta$

33. $r = 3 + 2 \cos \theta$

34. $r = 4 + 2 \sin \theta$

In Exercises 35 to 38, change the equations to polar equations.

35. $y^2 = 16x$

36. $x^2 + y^2 + 4x + 3y = 0$

37. $3x - 2y = 6$

38. $xy = 4$

In Exercises 39 to 42, change the equations to rectangular equations.

39. $r = \dfrac{4}{1 - \cos \theta}$

40. $r = 3 \cos \theta - 4 \sin \theta$

41. $r^2 = \cos 2\theta$

42. $\theta = 1$

Chapter Test

1. Find the vertex, focus, and directrix of the parabola given by the equation $y = \frac{1}{8}x^2$.

2. Find the vertex, focus, and directrix of the parabola given by the equation $x^2 + 4x - 12y + 16 = 0$.

3. Find the equation in standard form of the parabola with directrix $x = 3$ and focus $(-1, -2)$.

4. Graph the parabola with focus $(0, -1)$ and directrix $y = -5$.

5. Find the vertices and foci of the ellipse given by the equation $\dfrac{x^2}{9} + \dfrac{y^2}{64} = 1$.

6. Graph $\dfrac{x^2}{16} + \dfrac{y^2}{1} = 1$.

7. Find the vertices and foci of the ellipse given by the equation $25x^2 - 150x + 9y^2 + 18y + 9 = 0$.

8. Find the equation in standard form of the ellipse with center $(0, -3)$, foci $(-6, -3)$ and $(6, -3)$, and minor axis of length 6.

9. Find the eccentricity of the ellipse given by the equation $9x^2 + 25y^2 = 81$.

10. Graph $\dfrac{y^2}{25} - \dfrac{x^2}{16} = 1$.

11. Find the vertices, foci, and asymptotes of the hyperbola given by the equation $\dfrac{x^2}{36} - \dfrac{y^2}{64} = 1$.

12. Graph $16y^2 + 32y - 4x^2 - 24x = 84$.

13. Find the vertices and foci of the hyperbola given by the equation $\dfrac{(y - 4)^2}{36} - \dfrac{(x + 5)^2}{9} = 1$.

14. Find the equation in standard form of the hyperbola with vertices at $(-2, -3)$ and $(-6, -3)$ and foci $(-4 + \sqrt{34}, -3)$ and $(-4 - \sqrt{34}, -3)$.

15. Graph $r = 4$.

16. Graph $r = 3 \sin \theta$.

17. Graph $r = 2 - \cos \theta$.

18. Find the rectangular coordinates of the point whose polar coordinates are $(5, 7\pi/3)$.

19. Find the rectangular form of $r - r \cos \theta = 4$.

20. Find the polar form of $x^2 = 4y + 4$.

10 Systems of Equations

CASE IN POINT *Boardwalk? Park Place? How about New York?: A Winning Strategy for Monopoly*

During the Depression, an unemployed engineer named Charles Darrow invented Monopoly. In 1935 he sold the game to Parker Brothers. Today, Monopoly is still a very popular board game. In fact, Parker Brothers has sponsored world championship Monopoly games in Atlantic City, home of Baltic and Mediterranean Avenues.

In the early 1980s, Stephen Heppe, at the time a student, became interested in winning Monopoly strategies. He wanted to know which properties on the Monopoly board paid greater rates of return for each dollar invested. Answering Heppe's question required solving a system of linear equations. Heppe's system of equations contained 123 equations with 123 variables. Here are some of the results from solving this system of equations.

A player is less likely to land on Mediterranean Avenue during the course of the game than on any other property.

The chances of a player landing on Illinois Avenue are greater than those of landing on any other property.

Besides knowing which properties have the greatest chance of being occupied, Heppe also wanted to know which properties paid the greatest return for each dollar invested in houses or hotels. Some of his conclusions:

New York with a hotel has the highest rate of return.

The lowest rate of return for a property with a hotel is offered by Mediterranean.

Assuming all the railroads are owned, the B&O railroad has the greatest rate of return of all the railroads. This is because it is more likely that a player will land on this railroad than on the other railroads.

For more information on the mathematics of monopoly, see "Matrix Mathematics: How to Win at Monopoly" by Dr. Crypton in the September 1985 issue of *Science Digest.*

10.1 Systems of Linear Equations in Two Variables

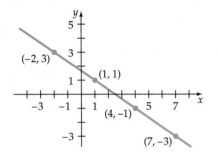

(-2, 3)
(1, 1)
(4, -1)
(7, -3)

$2x + 3y = 5$
Figure 10.1

Recall that an equation of the form $Ax + By = C$ is a linear equation in two variables. A solution of a linear equation in two variables is an ordered pair (x, y) that makes the equation a true statement. For example, $(-2, 3)$ is a solution of the equation

$$2x + 3y = 5 \quad \text{since} \quad 2(-2) + 3(3) = 5.$$

The graph of a linear equation, a straight line, is the set of points whose ordered pairs satisfy the equation. Figure 10.1 is the graph of $2x + 3y = 5$.

A **system of equations** is two or more equations considered together. The following system of equations is a **linear system of equations** in two variables.

$$\begin{cases} 2x + 3y = 4 \\ 3x - 2y = -7 \end{cases}$$

A **solution** of a system of equations is an ordered pair that is a solution of both equations.

In Figure 10.2, the graphs of the two equations in the system of equations above intersect at the point $(-1, 2)$. Since that point lies on both lines, $(-1, 2)$ is a solution of both equations and thus is a solution of the system of equations. The point $(5, -2)$ is a solution of the first equation but not the second equation. Therefore $(5, -2)$ is not a solution of the system of equations.

The graphs of two linear equations in two variables can intersect, be parallel, or be the same line. When the graphs intersect at a single point, the system is called a **consistent** system of equations. The system is called a **dependent** system of equations when the equations represent the same line. In this case, the system has an infinite number of solutions. When the graphs of the two lines are parallel, the system is called **inconsistent** and has no solution. See Figure 10.3.

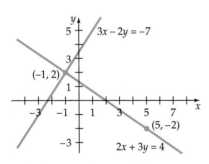

$3x - 2y = -7$
(-1, 2)
(5, -2)
$2x + 3y = 4$

Figure 10.2

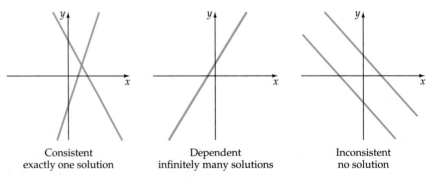

Consistent
exactly one solution

Dependent
infinitely many solutions

Inconsistent
no solution

Figure 10.3

Substitution Method for Solving a System of Linear Equations

The **substitution method** is one procedure for solving a system of equations. This method is illustrated in Example 1.

EXAMPLE 1 **Solve a System of Equations by the Substitution Method**

Solve

$$\begin{cases} 3x - 5y = 7 & (1) \\ y = 2x & (2) \end{cases}$$

Solution The solutions of the equation $y = 2x$ are the ordered pairs $(x, 2x)$. For the system of equations to have a solution, ordered pairs of the form $(x, 2x)$ must also be a solution of the equation $3x - 5y = 7$. To determine if ordered pairs of this form are solutions of Equation (1), substitute $(x, 2x)$ into Equation (1) and solve for x. Think of this as *substituting* $2x$ for y.

$$3x - 5y = 7$$
$$3x - 5(2x) = 7 \qquad \bullet \text{ Substitute } 2x \text{ for } y.$$
$$3x - 10x = 7$$
$$-7x = 7$$
$$x = -1$$
$$y = 2(-1) = -2 \quad \bullet \text{ Substitute } -1 \text{ for } x \text{ in Equation (2).}$$

The only ordered-pair solution of the system of equations is $(-1, -2)$. When a system of equations has a unique solution, the system of equations is consistent. See Figure 10.4.

────────────────────────────── ■ *Try Exercise* **6**, *page 502*.

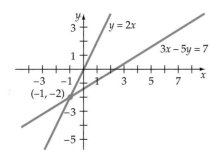

Figure 10.4
An Independent System of Equations

EXAMPLE 2 **Identify an Inconsistent System of Equations**

Solve

$$\begin{cases} 3x - y = 6 & (1) \\ 6x - 2y = 5 & (2) \end{cases}$$

Solution Solve Equation (1) for y.

$$3x - y = 6$$
$$y = 3x - 6$$

The solutions of the equation $y = 3x - 6$ are the ordered pairs $(x, 3x - 6)$. For the system of equations to have a solution, ordered pairs of the form $(x, 3x - 6)$ must also be a solution of the equation $6x - 2y = 5$.

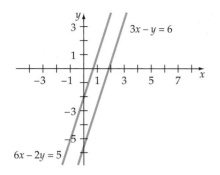

Figure 10.5
An Inconsistent System of Equations

To determine whether ordered pairs of this form are solutions of Equation (2), substitute $(x, 3x - 6)$ into Equation (2) and solve for x.

$$6x - 2(3x - 6) = 5 \quad \bullet \text{ Substitute } 3x - 6 \text{ for } y \text{ in Equation (2).}$$

$$12 = 5 \quad \bullet \text{ A false equation}$$

Arrival at the false equation $12 = 5$ means that no ordered pair that is a solution of Equation (1) is also a solution of Equation (2). The system of equations has no ordered pairs in common and thus has no solution. The system of equations is inconsistent. See Figure 10.5.

■ *Try Exercise* **18,** *page 502.*

EXAMPLE 3   **Solve a Dependent System of Equations**

Solve

$$\begin{cases} 4x - 8y = 16 & (1) \\ x - 2y = 4 & (2) \end{cases}$$

Solution   Solve Equation (2) for y.

$$x - 2y = 4$$

$$y = \frac{1}{2}x - 2$$

The solution of $y = \frac{1}{2}x - 2$ is the set of ordered pairs $(x, \frac{1}{2}x - 2)$. For the system of equations to have a solution, the ordered pairs $(x, \frac{1}{2}x - 2)$ must also be a solution of the equation $4x - 8y = 16$.

To determine if any ordered pair of this form is a solution of Equation (1), substitute $(x, \frac{1}{2}x - 2)$ into Equation (1) and solve for x.

$$4x - 8\left(\frac{1}{2}x - 2\right) = 16 \quad \bullet \text{ Substitute } \frac{1}{2}x - 2 \text{ for } y \text{ in Equation (2).}$$

$$16 = 16 \quad \bullet \text{ A true equation}$$

Arrival at the true equation $16 = 16$ means that the ordered pairs $(x, \frac{1}{2}x - 2)$, which are solutions of Equation (2), are also solutions of Equation (1). Because x can be replaced by any real number c, there are an infinite number of ordered pairs $(c, \frac{1}{2}c - 2)$ that are solutions of the system of equations. The system of equations is dependent. See Figure 10.6.

■ *Try Exercise* **20,** *page 502.*

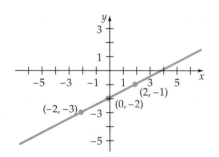

Figure 10.6
A Dependent System of Equations

Some of the specific ordered-pair solutions in Example 3 can be found by choosing various values for c.

The table at the left shows the ordered pairs that result from choosing c as -2, 0, and 2. The ordered pairs $(-2, -3)$, $(0, -2)$, and $(2, -1)$ are specific solutions of the system of equations. These points are the same points that are on the graphs of Equation (1) and Equation (2) of the system of equations, as shown in Figure 10.6.

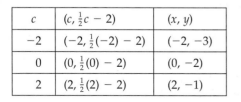

c	$(c, \frac{1}{2}c - 2)$	(x, y)
-2	$(-2, \frac{1}{2}(-2) - 2)$	$(-2, -3)$
0	$(0, \frac{1}{2}(0) - 2)$	$(0, -2)$
2	$(2, \frac{1}{2}(2) - 2)$	$(2, -1)$

Before leaving Example 3, note that there is more than one way to represent the ordered-pair solutions. To illustrate this point, solve Equation (2) for x.

$$x - 2y = 4 \qquad \bullet \text{ Equation (2)}$$

$$x = 2y + 4 \qquad \bullet \text{ Solve for } x.$$

Because y can be replaced by any real number b, there are an infinite number of ordered pairs $(2b + 4, b)$ that are solutions of the system of equations. Choosing b as -3, -2, and -1, gives the same ordered pairs: $(-2, -3)$, $(0, -2)$, and $(2, -1)$. There is always more than one way to describe the ordered pairs when writing the solution of a dependent system of equations. For Example 3, either the ordered pairs $(c, \frac{1}{2}c - 2)$ or the ordered pairs $(2b + 4, b)$ would generate all the solutions of the system of equations.

Elimination Method for Solving a System of Equations

A second technique for solving a system of equations is similar to the strategy for solving first-degree equations in one variable. The system of equations is replaced by a series of equivalent systems until the solution is obvious.

Two systems of equations are **equivalent** if each system has exactly the same solutions. The systems

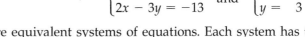

$$\begin{cases} 3x + 5y = 9 \\ 2x - 3y = -13 \end{cases} \quad \text{and} \quad \begin{cases} x = -2 \\ y = 3 \end{cases}$$

are equivalent systems of equations. Each system has the solution $(-2, 3)$, as shown in Figure 10.7.

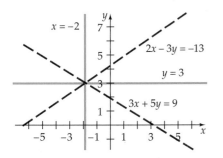

Figure 10.7

Operations That Produce Equivalent Systems of Equations

1. Interchange any two equations.
2. Replace an equation with a nonzero multiple of that equation.
3. Replace an equation with the sum of an equation and a nonzero constant multiple of another equation in the system.

Because the order in which the equations are written does not affect the system of equations, interchanging the equations does not affect its solution. The second operation restates the property that says that multiplying each side of an equation by the same nonzero constant does not change the solutions of the equation.

The third operation can be illustrated as follows. Consider the system of equations

$$\begin{cases} 3x + 2y = 10 & (1) \\ 2x - 3y = -2 & (2) \end{cases}$$

Multiply each side of Equation (2) by 2. (Any nonzero number would work.) Add the resulting equation to Equation (1).

$$
\begin{aligned}
3x + 2y &= 10 &&\bullet \text{ Equation (1)} \\
4x - 6y &= -4 &&\bullet \text{ 2 times Equation (2)} \\
\hline
7x - 4y &= 6 \quad (3) &&\bullet \text{ Add the equations.}
\end{aligned}
$$

Replace Equation (1) with the new equation (3) to produce the following equivalent system of equations.

$$
\begin{cases}
7x - 4y = 6 & (3) \\
2x - 3y = -2 & (2)
\end{cases}
$$

The third property states that the resulting system of equations has the same solutions as the original system and is therefore equivalent to the original system of equations. Figure 10.8 shows the graph of $7x - 4y = 6$. Note that the line passes through the same point at which the lines of the original system of equations intersect, the point $(2, 2)$.

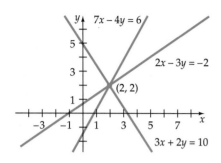

Figure 10.8

EXAMPLE 4   Solve a System of Equations by the Elimination Method

Solve

$$
\begin{cases}
3x - 4y = 10 & (1) \\
2x + 5y = -1 & (2)
\end{cases}
$$

Solution Use the operations that produce equivalent equations to eliminate a variable from one of the equations. We will eliminate x from Equation (2) by multiplying each equation by a different constant so as to have a new system of equations in which the coefficients of x are additive inverses.

$$
\begin{aligned}
6x - 8y &= 20 &&\bullet \text{ 2 times Equation (1)} \\
-6x - 15y &= 3 &&\bullet -3 \text{ times Equation (2)} \\
\hline
-23y &= 23 &&\bullet \text{ Add the equations.} \\
y &= -1 &&\bullet \text{ Solve for } y.
\end{aligned}
$$

Solve Equation (1) for x by substituting -1 for y.

$$
3x - 4(-1) = 10
$$
$$
3x = 6
$$
$$
x = 2
$$

The solution of the system of equations is $(2, -1)$. See Figure 10.9.

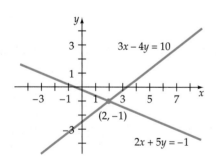

Figure 10.9

■ *Try Exercise* **24**, *page 502.*

The method just described is called the **elimination method** for solving a system of equations, because it involves *eliminating* a variable from one of the equations.

EXAMPLE 5 **Solve a Dependent System of Equations**

Solve

$$\begin{cases} x - 2y = 2 & (1) \\ 3x - 6y = 6 & (2) \end{cases}$$

Solution Eliminate x by multiplying Equation (2) by $-\frac{1}{3}$ and then adding the result to Equation (2).

$$\begin{array}{ll} x - 2y = 2 & \bullet \text{ Equation (1)} \\ \underline{-x + 2y = -2} & \bullet -\frac{1}{3} \text{ times Equation (2)} \\ 0 = 0 & \bullet \text{ Add the two equations.} \end{array}$$

Replace Equation (2) by $0 = 0$.

$$\begin{cases} x - 2y = 2 \\ 0 = 0 \end{cases} \quad \bullet \text{ This is an equivalent system of equations.}$$

Because the equation $0 = 0$ is an identity, an ordered pair that is a solution of Equation (1) is also a solution of $0 = 0$. Thus the solutions are the solutions of $x - 2y = 2$. Solving for y, we find that $y = \frac{1}{2}x - 1$. Because x can be replaced by any real number c, the solutions of the system of equations are the ordered pairs $\left(c, \frac{1}{2}c - 1\right)$. See Figure 10.10.

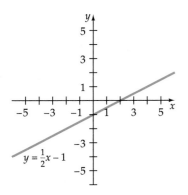

$y = \frac{1}{2}x - 1$

Figure 10.10

■ *Try Exercise* **28**, *page 503.*

Remark Referring again to Example 5 and solving Equation (1) for x, we have $x = 2y + 2$. Because y can be any real number b, the ordered-pair solutions of the system of equations can be written as $(2b + 2, b)$.

If one equation of the system of equations is replaced by a false equation, the system of equations has no solution. For example, the system of equations

$$\begin{cases} x + y = 4 \\ 0 = 5 \end{cases}$$

has no solution because the second equation is false for any choice of x and y.

Applications of Systems of Equations

As application problems become more difficult, it becomes impossible to represent all unknowns in terms of a single variable. In such cases, a system of equations can be used.

EXAMPLE 6 **Solve an Application**

A rowing team rowing with the current traveled 18 miles in 2 hours. Against the current, the team rowed 10 miles in 2 hours. Find the rate of the rowing team in calm water and the rate of the current.

Solution   Let r_1 represent the rate of the boat in calm water, and let r_2 represent the rate of the current.

The rate of the boat *with the current* is $r_1 + r_2$.
The rate of the boat *against the current* is $r_1 - r_2$.

Since the rowing team traveled 18 miles in 2 hours with the current, we use the equation $d = rt$.

$$d = r \cdot t$$
$$18 = (r_1 + r_2) \cdot 2 \quad \bullet \; d = 18, t = 2$$
$$9 = r_1 + r_2 \qquad \bullet \; \text{Divide each side by 2.}$$

Since the team rowed 10 miles in 2 hours against the current, we write

$$10 = (r_1 - r_2) \cdot 2 \quad \bullet \; d = 10, t = 2$$
$$5 = r_1 - r_2 \qquad \bullet \; \text{Divide each side by 2.}$$

Thus we have a system of two linear equations in the variables r_1 and r_2.

$$\begin{cases} 9 = r_1 + r_2 \\ 5 = r_1 - r_2 \end{cases}$$

Solving the system by using the elimination method, we find that r_1 is 7 mph and r_2 is 2 mph. Thus the rate of the boat in calm water is 7 mph and the rate of the current is 2 mph. You should verify these solutions.

■ *Try Exercise* **44**, *page 503.*

EXERCISE SET 10.1

In Exercises 1 to 20, solve the following systems of equations by the substitution method.

1. $\begin{cases} 2x - 3y = 16 \\ x = 2 \end{cases}$

2. $\begin{cases} 3x - 2y = -11 \\ y = 1 \end{cases}$

3. $\begin{cases} 3x + 4y = 18 \\ y = -2x + 3 \end{cases}$

4. $\begin{cases} 5x - 4y = -22 \\ y = 5x - 2 \end{cases}$

5. $\begin{cases} -2x + 3y = 6 \\ x = 2y - 5 \end{cases}$

6. $\begin{cases} 8x + 3y = -7 \\ x = 3y + 15 \end{cases}$

7. $\begin{cases} 6x + 5y = 1 \\ x - 3y = 4 \end{cases}$

8. $\begin{cases} -3x + 7y = 14 \\ 2x - y = -13 \end{cases}$

9. $\begin{cases} 7x + 6y = -3 \\ y = \dfrac{2}{3}x - 6 \end{cases}$

10. $\begin{cases} 9x - 4y = 3 \\ x = \dfrac{4}{3}y + 3 \end{cases}$

11. $\begin{cases} y = 4x - 3 \\ y = 3x - 1 \end{cases}$

12. $\begin{cases} y = 5x + 1 \\ y = 4x - 2 \end{cases}$

13. $\begin{cases} y = 5x + 4 \\ x = -3y - 4 \end{cases}$

14. $\begin{cases} y = -2x - 6 \\ x = -2y - 2 \end{cases}$

15. $\begin{cases} 3x - 4y = 2 \\ 4x + 3y = 14 \end{cases}$

16. $\begin{cases} 6x + 7y = -4 \\ 2x + 5y = 4 \end{cases}$

17. $\begin{cases} 3x - 3y = 5 \\ 4x - 4y = 9 \end{cases}$

18. $\begin{cases} 3x - 4y = 8 \\ 6x - 8y = 9 \end{cases}$

19. $\begin{cases} 4x + 3y = 6 \\ y = -\dfrac{4}{3}x + 2 \end{cases}$

20. $\begin{cases} 5x + 2y = 2 \\ y = -\dfrac{5}{2}x + 1 \end{cases}$

In Exercises 21 to 40, solve the following systems of equations by the elimination method.

21. $\begin{cases} 3x - y = 10 \\ 4x + 3y = -4 \end{cases}$

22. $\begin{cases} 3x + 4y = -5 \\ x - 5y = -8 \end{cases}$

23. $\begin{cases} 4x + 7y = 21 \\ 5x - 4y = -12 \end{cases}$

24. $\begin{cases} 3x - 8y = -6 \\ -5x + 4y = 10 \end{cases}$

25. $\begin{cases} 5x - 3y = 0 \\ 10x - 6y = 0 \end{cases}$

26. $\begin{cases} 3x + 2y = 0 \\ 2x + 3y = 0 \end{cases}$

27. $\begin{cases} 6x + 6y = 1 \\ 4x + 9y = 4 \end{cases}$

28. $\begin{cases} 4x + 5y = 2 \\ 8x - 15y = 9 \end{cases}$

29. $\begin{cases} 3x + 6y = 11 \\ 2x + 4y = 9 \end{cases}$

30. $\begin{cases} 4x - 2y = 9 \\ 2x - y = 3 \end{cases}$

31. $\begin{cases} \dfrac{5}{6}x - \dfrac{1}{3}y = -6 \\ \dfrac{1}{6}x + \dfrac{2}{3}y = 1 \end{cases}$

32. $\begin{cases} \dfrac{3}{4}x + \dfrac{2}{5}y = 1 \\ \dfrac{1}{2}x - \dfrac{3}{5}y = -1 \end{cases}$

33. $\begin{cases} \dfrac{3}{4}x + \dfrac{1}{3}y = 1 \\ \dfrac{1}{2}x + \dfrac{2}{3}y = 0 \end{cases}$

34. $\begin{cases} \dfrac{3}{5}x - \dfrac{2}{3}y = 7 \\ \dfrac{2}{5}x - \dfrac{5}{6}y = 7 \end{cases}$

35. $\begin{cases} 2\sqrt{3}x - 3y = 3 \\ 3\sqrt{3}x + 2y = 24 \end{cases}$

36. $\begin{cases} 4x - 3\sqrt{5}y = -19 \\ 3x + 4\sqrt{5}y = 17 \end{cases}$

37. $\begin{cases} 3\pi x - 4y = 6 \\ 2\pi x + 3y = 5 \end{cases}$

38. $\begin{cases} 2x - 5\pi y = 3 \\ 3x + 4\pi y = 2 \end{cases}$

39. $\begin{cases} 3\sqrt{2}x - 4\sqrt{3}y = -6 \\ 2\sqrt{2}x + 3\sqrt{3}y = 13 \end{cases}$

40. $\begin{cases} 2\sqrt{2}x + 3\sqrt{5}y = 7 \\ 3\sqrt{2}x - \sqrt{5}y = -17 \end{cases}$

In Exercises 41 to 55, solve by using a system of equations.

41. Flying with the wind, a plane traveled 450 miles in 3 hours. Flying against the wind, the plane traveled the same distance in 5 hours. Find the rate of the plane in calm air and the rate of the wind.

42. A plane flew 800 miles in 4 hours while flying with the wind. Against the wind, it took the plane 5 hours to travel the 800 miles. Find the rate of the plane in calm air and the rate of the wind.

43. A motorboat traveled a distance of 120 miles in 4 hours while traveling with the current. Against the current, the same trip took 6 hours. Find the rate of the boat in calm water and the rate of the current.

44. A canoeist can row 12 miles with the current in 2 hours. Rowing against the current, it takes the canoeist 4 hours to travel the same distance. Find the rate of the canoeist in calm water and the rate of the current.

45. A metallurgist made two purchases. The first purchase, which cost $1080, included 30 kilograms of an iron alloy and 45 kilograms of a lead alloy. The second purchase at the same prices cost $372 and included 15 kilograms of the iron alloy and 12 kilograms of the lead alloy. Find the cost per kilogram of the iron and lead alloys.

46. For $14.10, a chemist purchased 10 liters of hydrochloric acid and 15 liters of silver nitrate. A second purchase at

the same prices cost $18.16 and included 12 liters of hydrochloric acid and 20 liters of silver nitrate. Find the cost per liter of each of the two chemicals.

47. A coin bank contains only nickels and dimes. The value of the coins is $1.30. If the nickels were dimes and the dimes were nickels, the value of the coins would be $1.55. Find the original number of nickels and dimes in the bank.

48. The coin drawer of a cash register contains dimes and quarters. The value of the coins is $4.35. If the dimes were quarters and the quarters were dimes, the value of the coins would be $3.00. Find the original number of dimes and quarters in the cash register.

49. The sum of the digits of a two-digit number is 14. If the digits are reversed, the new number is 18 less than the original number. Find the original number.

50. The sum of the digits of a two-digit number is 11. If the digits are reversed, the new number is 63 more than the original number. Find the original number.

51. A broker invests $25,000 of a client's money in two different municipal bonds. The annual rate of return on one bond is 6 percent, and the annual rate of return on the second bond is 6.5 percent. The investor receives a total annual interest payment from the two bonds of $1555. Find the amount invested in each bond.

52. An investment of $3000 is placed in stocks and bonds. The annual rate of return for the stocks is 4.5 percent, and the rate of return on the bonds is 8 percent. The annual interest payment from the stocks and bonds is $177. Find the amount invested in bonds.

53. A goldsmith has two gold alloys. The first alloy is 40% gold; the second alloy is 60% gold. How many grams of each should be mixed to produce 20 grams of an alloy that is 52% gold?

54. One acetic acid solution is 70% water and another is 30% water. How many liters of each solution should be mixed to produce 20 liters of a solution that is 40% water?

55. A chemist wants to make 50 milliliters of a 16% acid solution. How many milliliters each of a 13% acid solution and an 18% acid solution should be mixed to produce the desired solution?

Supplemental Exercises

In Exercises 56 to 65, solve for x and y. Use the fact that if $z_1 = a_1 + b_1 i$ and $z_2 = a_2 + b_2 i$ are two complex numbers, then $z_1 = z_2$ if and only if $a_1 = a_2$ and $b_1 = b_2$.

56. $(2 + i)x + (3 - i)y = 7$

57. $(3 + 2i)x + (4 - 3i)y = 2 - 16i$

58. $(4 - 3i)x + (5 + 2i)y = 11 + 9i$

59. $(2 + 6i)x + (4 - 5i)y = -8 - 7i$

60. $(-3 - i)x - (4 + 2i)y = 1 - i$

61. $(5 - 2i)x + (-3 - 4i)y = 12 - 35i$

62. $\begin{cases} 2x + 5y = 11 + 3i \\ 3x + y = 10 - 2i \end{cases}$

63. $\begin{cases} 4x + 3y = 11 + 6i \\ 3x - 5y = 1 + 19i \end{cases}$

64. $\begin{cases} 2x + 3y = 11 + 5i \\ 3x - 3y = 9 - 15i \end{cases}$

65. $\begin{cases} 5x - 4y = 15 - 41i \\ 3x + 5y = 9 + 5i \end{cases}$

In Exercises 66 to 69, use the system of equations

$$\begin{cases} a_1x + b_1y = c_1 \\ a_2x + b_2y = c_2 \end{cases}$$

66. Assuming $a_1b_2 - a_2b_1 \neq 0$, solve the system for x and y.

67. Prove that the system of equations has a unique solution if and only if $a_1b_2 - a_2b_1 \neq 0$. (*Hint:* See the answer to Exercise 66.)

68. Prove that the system of equations has no solution if and only if

$$\frac{a_1}{a_2} = \frac{b_1}{b_2}, \quad \frac{a_1}{a_2} \neq \frac{c_1}{c_2}, \quad \frac{b_1}{b_2} \neq \frac{c_1}{c_2}$$

69. Prove that the system of equations has infinitely many solutions if and only if

$$\frac{a_1}{a_2} = \frac{b_1}{b_2}, \quad \frac{a_1}{a_2} = \frac{c_1}{c_2}, \quad \frac{b_1}{b_2} = \frac{c_1}{c_2}$$

10.2 Systems of Linear Equations in More Than Two Variables

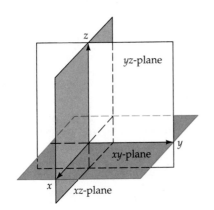

Figure 10.11

An equation of the form $Ax + By + Cz = D$, with A, B, and C not all zero, is a linear equation in three variables. A solution of an equation in three variables is an **ordered triple** (x, y, z).

The ordered triple $(2, -1, -3)$ is one of the solutions of the equation $2x - 3y + z = 4$. The ordered triple $(3, 1, 1)$ is another solution. In fact, an infinite number of ordered triples are solutions of the equation.

Graphing an equation in three variables requires a third coordinate axis perpendicular to the xy-plane. This third axis is commonly called the **z-axis**. The result is a three-dimensional coordinate system called the xyz-coordinate system (Figure 10.11). To help visualize a three-dimensional coordinate system, think of a corner of a room: the floor is the xy-plane, one wall is the yz-plane, and the other wall is the xz-plane.

Graphing an ordered triple requires three moves, the first along the x-axis, the second along the y-axis, and the third along the z-axis. Figure 10.12 is the graph of the points $(-5, -4, 3)$ and $(4, 5, -2)$.

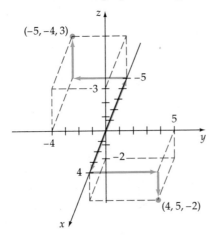

Figure 10.12

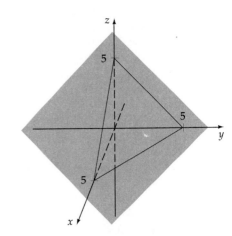

Figure 10.13

The graph of a linear equation in three variables is a plane. That is, if all the solutions of a linear equation in three variables were plotted in an xyz-coordinate system, the graph would look like a large piece of paper with infinite extent. Figure 10.13 is the graph of $x + y + z = 5$.

There are different ways in which three planes can be oriented in an xyz-coordinate system. Figure 10.14 illustrates several ways.

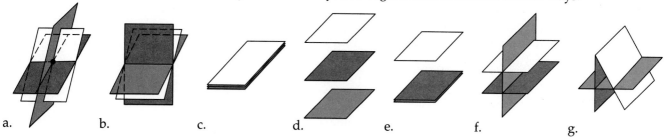

a.   b.   c.   d.   e.   f.   g.

Figure 10.14

For a linear system of equations in three variables to have a solution, the graphs of the planes must intersect at a single point, they must intersect along a common line, or all equations must have a graph that is the same plane. In Figure 10.14, the graphs in (a), (b), and (c) represent systems of equations that have a solution. The system of equations represented in Figure 10.14a is a consistent system of equations. Figures 10.14b and 10.14c are graphs of a dependent system of equations. The remaining graphs are examples of inconsistent systems of equations.

A system of equations in more than two variables can be solved by using the substitution method or the elimination method. To illustrate the substitution method, consider the system of equations

$$\begin{cases} x - 2y + z = 7 & (1) \\ 2x + y - z = 0 & (2) \\ 3x + 2y - 2z = -2 & (3) \end{cases}$$

Solve Equation (1) for x and substitute the result into Equations (2) and (3).

$$x = 2y - z + 7 \qquad \bullet \text{ Equation (4)}$$

$$2(2y - z + 7) + y - z = 0 \qquad \bullet \text{ Substitute } 2y - z + 7 \text{ for } x \text{ in Equation (2).}$$

$$4y - 2z + 14 + y - z = 0 \qquad \bullet \text{ Simplify.}$$

$$5y - 3z = -14 \qquad (5)$$

$$3(2y - z + 7) + 2y - 2z = -2 \qquad \bullet \text{ Substitute } 2y - z + 7 \text{ for } x \text{ in Equation (3).}$$

$$6y - 3z + 21 + 2y - 2z = -2 \qquad \bullet \text{ Simplify.}$$

$$8y - 5z = -23 \qquad (6)$$

Now solve the system of equations formed from Equations (5) and (6).

$$\begin{cases} 5y - 3z = -14 & \text{multiply by 8} \longrightarrow & 40y - 24z = -112 \\ 8y - 5z = -23 & \text{multiply by } -5 \longrightarrow & \underline{-40y + 25z = \quad 115} \\ & & z = \quad 3 \end{cases}$$

Substitute 3 for z into Equation (5) and solve for y.

$$5y - 3z = -14 \quad \bullet \text{ Equation (5)}$$
$$5y - 3(3) = -14$$
$$5y - 9 = -14$$
$$5y = -5$$
$$y = -1$$

Substitute -1 for y and 3 for z into Equation (4) and solve for x.

$$x = 2y - z + 7 = 2(-1) - (3) + 7 = 2$$

The ordered-triple solution is $(2, -1, 3)$. The graphs of the three planes intersect at a single point.

There are many approaches one can take to determine the solution of a system of equations by the elimination method. For consistency, we will always follow a plan that produces an equivalent system of equations in **triangular form**. Three examples of systems of equations in triangular form are

$$\begin{cases} 2x - 3y + z = -4 \\ \quad\quad 2y + 3z = 9 \\ \quad\quad\quad\quad -2z = -2 \end{cases} \quad \begin{cases} w + 3x - 2y + 3z = 0 \\ \quad\quad 2x - y + 4z = 8 \\ \quad\quad\quad\quad -3y - 2z = -1 \\ \quad\quad\quad\quad\quad\quad 3z = 9 \end{cases} \quad \begin{cases} 3x - 4y + z = 1 \\ \quad\quad 3y + 2z = 3 \end{cases}$$

Once a system of equations is written in triangular form, the solution can be found by *back substitution*—that is, by solving the last equation of the system and substituting *back* into the previous equation. This process is continued until the value of each variable has been found.

As an example of solving a system of equations by back substitution, consider the following system of equations in triangular form.

$$\begin{cases} 2x - 4y + z = -3 & (1) \\ \quad\quad 3y - 2z = 9 & (2) \\ \quad\quad\quad\quad 3z = -9 & (3) \end{cases}$$

Solve Equation (3) for z. Substitute the value of z into Equation (2) and solve for y.

$$3z = -9 \quad \bullet \text{ Equation (3)} \qquad\qquad 3y - 2z = 9 \quad \bullet \text{ Equation (2)}$$
$$z = -3 \qquad\qquad\qquad\qquad\qquad 3y - 2(-3) = 9 \quad \bullet \, z = -3$$
$$\qquad\qquad\qquad\qquad\qquad\qquad\qquad 3y = 3$$
$$\qquad\qquad\qquad\qquad\qquad\qquad\qquad y = 1$$

Replace z by -3 and y by 1 in Equation (1) and then solve for x.

$$2x - 4y + z = -3 \quad \bullet \text{ Equation (1)}$$
$$2x - 4(1) + (-3) = -3$$
$$2x - 7 = -3$$
$$x = 2$$

The solution is the ordered triple $(2, 1, -3)$.

EXAMPLE 1   **Solve a System of Equations**

Solve the system of equations by the elimination method.

$$\begin{cases} x + 2y - z = 1 & (1) \\ 2x - y + z = 6 & (2) \\ x + 3y - 2z = -1 & (3) \end{cases}$$

Solution   Eliminate x from Equation (2) by multiplying Equation (1) by -2 and adding to Equation (2). Replace Equation (2) with the new equation.

$$\begin{cases} x + 2y - z = 1 \\ -5y + 3z = 4 \quad (4) \\ x + 3y - 2z = -1 \end{cases}$$

$$\begin{array}{ll} -2x - 4y + 2z = -2 & \bullet \ -2 \text{ times Equation (1)} \\ \underline{2x - y + z = 6} & \bullet \ \text{Equation (2)} \\ -5y + 3z = 4 & \bullet \ \text{Add the equations} \end{array}$$

Eliminate x from Equation (3) by multiplying Equation (1) by -1 and adding it to Equation (3). Replace Equation (3) with the new equation.

$$\begin{cases} x + 2y - z = 1 \\ -5y + 3z = 4 \quad (4) \\ y - z = -2 \quad (5) \end{cases}$$

$$\begin{array}{ll} -x - 2y + z = -1 & \bullet \ -1 \text{ times Equation (1)} \\ \underline{x + 3y - 2z = -1} & \bullet \ \text{Equation (3)} \\ y - z = -2 & \bullet \ \text{Add the equations.} \end{array}$$

Eliminate y from Equation (5) by multiplying Equation (5) by 5 and adding it to Equation (4). Replace Equation (5) by the new equation.

$$\begin{cases} x + 2y - z = 1 \\ -5y + 3z = 4 \\ -2z = -6 \quad (6) \end{cases}$$

$$\begin{array}{ll} 5y - 5z = -10 & \bullet \ 5 \text{ times Equation (5)} \\ \underline{-5y + 3z = 4} & \bullet \ \text{Equation (4)} \\ -2z = -6 & \bullet \ \text{Add the equations.} \end{array}$$

The system of equations is now in standard form. Solve the system of equations by back substitution.

Solve Equation (6) for z. Substitute the value of z into Equation (4) and solve for y.

$$\begin{array}{ll} -2z = -6 & \bullet \ \text{Equation (6)} \\ z = 3 & \end{array} \qquad \begin{array}{ll} -5y + 3z = 4 & \bullet \ \text{Equation (4)} \\ -5y + 3(3) = 4 & \\ -5y = -5 & \\ y = 1 & \end{array}$$

Replace z by 3 and y by 1 in Equation (1) and then solve for x.

$$\begin{array}{ll} x + 2y - z = 1 & \bullet \ \text{Equation (1)} \\ x + 2(1) - 3 = 1 & \bullet \ y = 1, z = 3 \\ x = 2 & \end{array}$$

The solution is the ordered triple (2, 1, 3). The system of equations is consistent.

■ *Try Exercise* **12**, *page 512.*

EXAMPLE 2   **Solve a Dependent System of Equations**

Solve

$$\begin{cases} x + 4y + 2z = 1 & (1) \\ 2x + 7y + 3z = 8 & (2) \\ 4x + 15y + 7z = 10 & (3) \end{cases}$$

Solution   Eliminate x from Equation (2) by multiplying Equation (1) by -2 and then adding it to Equation (2). Replace Equation (2).

$$\begin{cases} x + 4y + 2z = 1 \\ -y - z = 6 \quad (4) \\ 4x + 15y + 7z = 10 \end{cases}$$

$\begin{aligned} -2x - 8y - 4z &= -2 \quad \bullet \; -2 \text{ times Equation (1)} \\ 2x + 7y + 3z &= 8 \quad \bullet \; \text{Equation (2)} \\ \hline -y - z &= 6 \quad \bullet \; \text{Add the equations.} \end{aligned}$

Eliminate x from Equation (3) by multiplying Equation (1) by -4 and then adding it to Equation (3). Replace Equation (3) by the new equation.

$$\begin{cases} x + 4y + 2z = 1 \\ -y - z = 6 \quad (4) \\ -y - z = 6 \quad (5) \end{cases}$$

$\begin{aligned} -4x - 16y - 8z &= -4 \quad \bullet \; -4 \text{ times Equation (1)} \\ 4x + 15y + 7z &= 10 \quad \bullet \; \text{Equation (3)} \\ \hline -y - z &= 6 \quad \bullet \; \text{Add the equations.} \end{aligned}$

Eliminate y from Equation (5) by multiplying Equation (4) by -1 and then adding to Equation (5). Replace Equation (5) by the new equation.

$$\begin{cases} x + 4y + 2z = 1 \quad (1) \\ -y - z = 6 \quad (4) \\ 0 = 0 \quad (6) \end{cases}$$

$\begin{aligned} y + z &= -6 \quad \bullet \; -1 \text{ times Equation (4)} \\ -y - z &= 6 \quad \bullet \; \text{Equation (5)} \\ \hline 0 &= 0 \quad \bullet \; \text{Add the equations.} \end{aligned}$

Because any ordered triple (x, y, z) is a solution of Equation (6), the solutions of the system of equations will be the ordered triples that are solutions of Equations (1) and (4).

Solve Equation (4) for y.

$$y = -z - 6$$

Substitute $-z - 6$ for y in Equation (1) and solve for x.

$$x + 4y + 2z = 1 \quad \bullet \; \text{Equation (1)}$$
$$x + 4(-z - 6) + 2z = 1 \quad \bullet \; \text{Replace } y \text{ by } -z - 6.$$
$$x - 4z - 24 + 2z = 1 \quad \bullet \; \text{Solve for } x.$$
$$x - 2z - 24 = 1$$
$$x = 2z + 25$$

By choosing any real number c for z, we have $y = -c - 6$ and $x = 2c + 25$. The ordered-triple solutions are $(2c + 25, -c - 6, c)$. The system of equations is dependent.

■ *Try Exercise* **16,** *page 512.*

As in the case of a dependent system of equations in two variables, there is more than one way to represent the solutions of the dependent system of equations in Example 2. For instance, let $b = 2c + 25$, the x-component of the ordered triple $(2c + 25, -c - 6, c)$, and then solve for c.

$$b = 2c + 25 \quad \text{or} \quad c = \frac{b - 25}{2}$$

Substitute this value of c into each component of the ordered triple.

$$\left(2\left(\frac{b - 25}{2}\right) + 25, -\frac{b - 25}{2} - 6, \frac{b - 25}{2}\right) = \left(b, -\frac{b - 13}{2}, \frac{b - 25}{2}\right)$$

Thus the solutions of the system of equations can also be written as

$$\left(b, -\frac{b - 13}{2}, \frac{b - 25}{2}\right).$$

EXAMPLE 3 **Identify an Inconsistent System of Equations**

Solve

$$\begin{cases} x + 2y + 3z = 4 & (1) \\ 2x - y - z = 3 & (2) \\ 3x + y + 2z = 5 & (3) \end{cases}$$

Solution Eliminate x from Equation (2) by multiplying Equation (1) by -2 and adding it to Equation (2). Replace Equation 2. Eliminate x from Equation (3) by multiplying Equation (1) by -3 and adding it to Equation (3). Replace Equation (3). The equivalent system is

$$\begin{cases} x + 2y + 3z = 4 & (1) \\ -5y - 7z = -5 & (4) \\ -5y - 7z = -7 & (5) \end{cases}$$

Eliminate y from Equation (5) by multiplying Equation (4) by -1 and adding it to Equation (5). Replace Equation (5). The equivalent system is

$$\begin{cases} x + 2y + 3z = 4 & (1) \\ -5y - 7z = -5 & (4) \\ 0 = -2 & (6) \end{cases}$$

This system of equations contains a false equation. The system is inconsistent and has no solutions.

■ *Try Exercise* **18**, *page 513.*

Nonsquare Systems of Equations

The linear systems of equations that we have solved so far contain the same number of variables as equations. These are *square systems of equations*. If there are fewer equations than variables — a *nonsquare system of equations* — the system has either no solution or an infinite number of solutions.

EXAMPLE 4 **Solve a Nonsquare System of Equations**

Solve

$$\begin{cases} x - 2y + 2z = 3 & (1) \\ 2x - y - 2z = 15 & (2) \end{cases}$$

Solution Eliminate x from Equation (2) by multiplying Equation (1) by -2 and adding it to Equation (2). Replace Equation (2).

$$\begin{cases} x - 2y + 2z = 3 & (1) \\ 3y - 6z = 9 & (3) \end{cases}$$

Solve Equation (3) for y.

$$3y - 6z = 9$$

$$y = 2z + 3$$

Substitute $2z + 3$ for y into Equation (1) and solve for x.

$$x - 2y + 2z = 3$$

$$x - 2(2z + 3) + 2z = 3 \quad \bullet \; y = 2z + 3$$

$$x = 2z + 9$$

For each value of z selected, there correspond values for x and y. If z is any real number c, then the solutions of the system are the ordered triples $(2c + 9, 2c + 3, c)$.

■ *Try Exercise* **20,** *page 513.*

Homogeneous Systems of Equations

A linear system of equations for which the constant term is zero for all equations is called a **homogeneous** system of equations. Two examples of homogeneous systems of equations are

$$\begin{cases} 3x + 4y = 0 \\ 2x + 3y = 0 \end{cases} \qquad \begin{cases} 2x - 3y + 5z = 0 \\ 3x + 2y + z = 0 \\ x - 4y + 5z = 0 \end{cases}$$

The solution $(0, 0)$ is always a solution of a homogeneous system of equations in two variables, and $(0, 0, 0)$ is always a solution of a homogeneous system of equations in three variables. This solution is called the **trivial** solution.

Sometimes a homogeneous system of equations may have solutions other than the trivial solution. For example, $(1, -1, -1)$ is a solution to the homogeneous system of three equations in three variables above.

If a homogeneous system of equations has a unique solution, the graphs intersect only at the origin. If the homogeneous system of equations has infinitely many solutions, the graphs intersect along a line or plane that passes through the origin.

Solutions to a homogeneous system of equations can be found by using the substitution method or the elimination method.

EXAMPLE 5   **Solve a Homogeneous System of Equations**

Solve

$$\begin{cases} x + 2y - 3z = 0 & (1) \\ 2x - y + z = 0 & (2) \\ 3x + y - 2z = 0 & (3) \end{cases}$$

Solution   Eliminate x from Equations (2) and (3).

$$\begin{cases} x + 2y - 3z = 0 & (1) \\ -5y + 7z = 0 & (4) \\ -5y + 7z = 0 & (5) \end{cases}$$

Eliminate y from Equations (4) and (5). Replace Equation (5).

$$\begin{cases} x + 2y - 3z = 0 & (1) \\ -5y + 7z = 0 & (4) \\ 0 = 0 & (6) \end{cases}$$

Since the last equation is an identity, the solutions of the system are the solutions of Equations (1) and (4).

Solve Equation (4) for y.

$$y = \frac{7}{5}z$$

Substitute the expression for y into Equation (1) and solve for x.

$$x + 2y - 3z = 0 \quad \bullet \text{ Equation (1)}$$

$$x + 2\left(\frac{7}{5}z\right) - 3z = 0 \quad \bullet \; y = \frac{7}{5}z$$

$$x = \frac{1}{5}z$$

Letting z be any real number c, we find the solutions of the system are

$$\left(\frac{1}{5}c, \frac{7}{5}c, c\right).$$

■ *Try Exercise* **32**, *page 513.*

One application of a system of equations is "curve fitting." Given a set of points in the plane, try to find an equation whose graph passes through those points, or "fits" those points.

EXAMPLE 6   Solve an Application of a System of Equations to Curve Fitting

Find an equation of the form $y = ax^2 + bx + c$ whose graph passes through the points whose coordinates are $(1, 4)$, $(-1, 6)$, and $(2, 9)$.

Solution   Substitute each of the given ordered pairs into the equation $y = ax^2 + bx + c$. Write the resulting system of equations.

$$
\begin{cases}
4 = a(1)^2 + b(1) + c \\
6 = a(-1)^2 + b(-1) + c \\
9 = a(2)^2 + b(2) + c
\end{cases}
\quad
\begin{array}{l}
\text{or} \\
\text{or} \\
\text{or}
\end{array}
\quad
\begin{cases}
a + b + c = 4 & \text{(1)} \\
a - b + c = 6 & \text{(2)} \\
4a + 2b + c = 9 & \text{(3)}
\end{cases}
$$

Solve the resulting system of equations for a, b, and c.

Eliminate a from Equation (2) by multiplying Equation (1) by -1 and adding it to Equation (2). Now eliminate a from Equation (3) by multiplying Equation (1) by -4 and adding it to Equation (3). The result is

$$
\begin{cases}
a + b + c = 4 \\
 -2b = 2 \\
 -2b - 3c = -7
\end{cases}
$$

Although this system of equations is not in triangular form, we can solve the second equation for b and use this value to find a and c.

Solving by substitution, we obtain $a = 2$, $b = -1$, $c = 3$. The equation of the form $y = ax^2 + bx + c$ whose graph passes through the three points is $y = 2x^2 - x + 3$. See Figure 10.15.

■ *Try Exercise 36, page 513.*

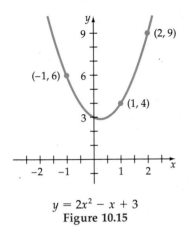

$y = 2x^2 - x + 3$

Figure 10.15

EXERCISE SET 10.2

In Exercises 1 to 24, solve each system of equations.

1.
$$
\begin{cases}
2x - y + z = 8 \\
2y - 3z = -11 \\
3y + 2z = 3
\end{cases}
$$

2.
$$
\begin{cases}
3x + y + 2z = -4 \\
-3y - 2z = -5 \\
2y + 5z = -4
\end{cases}
$$

3.
$$
\begin{cases}
x + 3y - 2z = 8 \\
2x - y + z = 1 \\
3x + 2y - 3z = 15
\end{cases}
$$

4.
$$
\begin{cases}
x - 2y + 3z = 5 \\
3x - 3y + z = 9 \\
5x + y - 3z = 3
\end{cases}
$$

5.
$$
\begin{cases}
3x + 4y - z = -7 \\
x - 5y + 2z = 19 \\
5x + y - 2z = 5
\end{cases}
$$

6.
$$
\begin{cases}
2x - 3y - 2z = 12 \\
x + 4y + z = -9 \\
4x + 2y - 3z = 6
\end{cases}
$$

7.
$$
\begin{cases}
2x - 5y + 3z = -18 \\
3x + 2y - z = -12 \\
x - 3y - 4z = -4
\end{cases}
$$

8.
$$
\begin{cases}
4x - y + 2z = -1 \\
2x + 3y - 3z = -13 \\
x + 5y + z = 7
\end{cases}
$$

9.
$$
\begin{cases}
x + 2y - 3z = -7 \\
2x - y + 4z = 11 \\
4x + 3y - 4z = -3
\end{cases}
$$

10.
$$
\begin{cases}
x - 3y + 2z = -11 \\
3x + y + 4z = 4 \\
5x - 5y + 8z = -18
\end{cases}
$$

11.
$$
\begin{cases}
2x - 5y + 2z = -4 \\
3x + 2y + 3z = 13 \\
5x - 3y - 4z = -18
\end{cases}
$$

12.
$$
\begin{cases}
3x + 2y - 5z = 6 \\
5x - 4y + 3z = -12 \\
4x + 5y - 2z = 15
\end{cases}
$$

13.
$$
\begin{cases}
2x + y - z = -2 \\
3x + 2y + 3z = 21 \\
7x + 4y + z = 17
\end{cases}
$$

14.
$$
\begin{cases}
3x + y + 2z = 2 \\
4x - 2y + z = -4 \\
11x - 3y + 4z = -6
\end{cases}
$$

15.
$$
\begin{cases}
3x - 2y + 3z = 11 \\
2x + 3y + z = 3 \\
5x + 14y - z = 1
\end{cases}
$$

16.
$$
\begin{cases}
2x + 3y + 2z = 14 \\
x - 3y + 4z = 4 \\
-x + 12y - 6z = 2
\end{cases}
$$

17. $\begin{cases} 2x - 3y + 6z = 3 \\ x + 2y - 4z = 5 \\ 3x + 4y - 8z = 7 \end{cases}$
18. $\begin{cases} 2x + 3y - 6z = 4 \\ 3x - 2y - 9z = -7 \\ 2x + 5y - 6z = 8 \end{cases}$

19. $\begin{cases} 2x - 3y + 5z = 14 \\ x + 4y - 3z = -2 \end{cases}$
20. $\begin{cases} x - 3y + 4z = 9 \\ 3x - 8y - 2z = 4 \end{cases}$

21. $\begin{cases} 6x - 9y + 6z = 7 \\ 4x - 6y + 4z = 9 \end{cases}$
22. $\begin{cases} 4x - 2y + 6z = 5 \\ 2x - y + 3z = 2 \end{cases}$

23. $\begin{cases} 5x + 3y + 2z = 10 \\ 3x - 4y - 4z = -5 \end{cases}$
24. $\begin{cases} 3x - 4y - 7z = -5 \\ 2x + 3y - 5z = 2 \end{cases}$

In Exercises 25 to 32, solve each homogeneous system of equations.

25. $\begin{cases} x + 3y - 4z = 0 \\ 2x + 7y + z = 0 \\ 3x - 5y - 2z = 0 \end{cases}$
26. $\begin{cases} x - 2y + 3z = 0 \\ 3x - 7y - 4z = 0 \\ 4x - 4y + z = 0 \end{cases}$

27. $\begin{cases} 2x - 3y + z = 0 \\ 2x + 4y - 3z = 0 \\ 6x - 2y - z = 0 \end{cases}$
28. $\begin{cases} 5x - 4y - 3z = 0 \\ 2x + y + 2z = 0 \\ x - 6y - 7z = 0 \end{cases}$

29. $\begin{cases} 3x - 5y + 3z = 0 \\ 2x - 3y + 4z = 0 \\ 7x - 11y + 11z = 0 \end{cases}$
30. $\begin{cases} 5x - 2y - 3z = 0 \\ 3x - y - 4z = 0 \\ 4x - y - 9z = 0 \end{cases}$

31. $\begin{cases} 4x - 7y - 2z = 0 \\ 2x + 4y + 3z = 0 \\ 3x - 2y - 5z = 0 \end{cases}$
32. $\begin{cases} 5x + 2y + 3z = 0 \\ 3x + y - 2z = 0 \\ 4x - 7y + 5z = 0 \end{cases}$

In Exercises 33 to 42, solve a system of equations.

33. Find an equation of the form $y = ax^2 + bx + c$ whose graph passes through the points $(2, 3)$, $(-2, 7)$, and $(1, -2)$.

34. Find an equation of the form $y = ax^2 + bx + c$ whose graph passes through the points $(1, -2)$, $(3, -4)$, and $(2, -2)$.

35. Find the equation of the circle whose graph passes through the points $(5, 3)$, $(-1, -5)$, and $(-2, 2)$. (*Hint:* Use the equation $x^2 + y^2 + ax + by + c = 0$.)

36. Find the equation of the circle whose graph passes through the points $(0, 6)$, $(1, 5)$, and $(-7, -1)$. (*Hint:* See Exercise 35.)

37. Find the center and radius of the circle whose graph passes through the points $(-2, 10)$, $(-12, -14)$, and $(5, 3)$. (*Hint:* See Exercise 35.)

38. Find the center and radius of the circle whose graph passes through the points $(2, 5)$, $(-4, -3)$, and $(3, 4)$. (*Hint:* See Exercise 35.)

39. A coin bank contains only nickels, dimes, and quarters. The value of the coins is $2. There are twice as many nickels as dimes and one more dime than quarters. Find the number of each coin in the bank.

40. A coin bank contains only nickels, dimes, and quarters. The value of the coins is $5.50. The number of nickels is six more than twice the number of quarters. The number of dimes is one-third the number of nickels. Find the number of each coin in the bank.

41. The sum of the digits of a positive three-digit number is 19. The tens digit is four less than twice the hundreds digit. The number is decreased by 99 when the digits are reversed. Find the number.

42. The sum of the digits of a positive three-digit number is 10. The hundreds digit is one less than twice the ones digit. The number is decreased by 198 when the digits are reversed. Find the number.

Supplemental Exercises

In Exercises 43 to 48, solve each system of equations.

43. $\begin{cases} 2x + y - 3z + 2w = -1 \\ 2y - 5z - 3w = 9 \\ 3y - 8z + w = -4 \\ 2y - 2z + 3w = -3 \end{cases}$

44. $\begin{cases} 3x - y + 2z - 3w = 5 \\ 2y - 5z + 2w = -7 \\ 4y - 9z + w = -19 \\ 3y + z - 2w = -12 \end{cases}$

45. $\begin{cases} x - 3y + 2z - w = 2 \\ 2x - 5y - 3z + 2w = 21 \\ 3x - 8y - 2z - 3w = 12 \\ -2x + 8y + z + 2w = -13 \end{cases}$

46. $\begin{cases} x - 2y + 3z + 2w = 8 \\ 3x - 7y - 2z + 3w = 18 \\ 2x - 5y + 2z - w = 19 \\ 4x - 8y + 3z + 2w = 29 \end{cases}$

47. $\begin{cases} x + 2y - 2z + 3w = 2 \\ 2x + 5y + 2z + 4w = 9 \\ 4x + 9y - 2z + 10w = 13 \\ -x - y + 8z - 5w = 3 \end{cases}$

48. $\begin{cases} x - 2y + 3z - 2w = -1 \\ 3x - 7y - 2z - 3w = -19 \\ 2x - 5y + 2z - w = -11 \\ -x + 3y - 2z - w = 3 \end{cases}$

In Exercises 49 and 50, use the system of equations

$$\begin{cases} x - 3y - 2z = A^2 \\ 2x - 5y + Az = 9 \\ 2x - 8y + z = 18 \end{cases}$$

49. Find all values of A for which the system has no solutions.

50. Find all values of A for which the system has a unique solution.

In Exercises 51 to 53, use the system of equations

$$\begin{cases} x + 2y + z = A^2 \\ -2x - 3y + Az = 1 \\ 7x + 12y + A^2z = 4A^2 - 3 \end{cases}$$

51. Find all values of A for which the system has a unique solution.

52. Find all values of A for which the system has an infinite number of solutions.

53. Find all values of A for which the system has no solution.

54. Find an equation of a plane that contains the points $(2, 1, 1)$, $(-1, 2, 12)$, and $(3, 2, 0)$. (*Hint:* The equation of a plane can be written as $z = ax + by + c$.)

55. Find an equation of a plane that contains the points $(1, -1, 5)$, $(2, -2, 9)$, and $(-3, -1, -1)$. (*Hint:* The equation of a plane can be written as $z = ax + by + c$.)

10.3 Systems of Nonlinear Equations

A **nonlinear system of equations** is one in which one or more equations of the system are not linear equations. Figure 10.16 shows examples of nonlinear systems of equations and the corresponding graphs of the equations. Each point of intersection of the graphs is a solution of the system of equations. In the third example, the graphs do not intersect; therefore the system of equations has no real number solution.

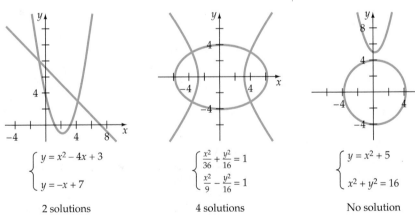

$$\begin{cases} y = x^2 - 4x + 3 \\ y = -x + 7 \end{cases}$$

2 solutions

$$\begin{cases} \dfrac{x^2}{36} + \dfrac{y^2}{16} = 1 \\ \dfrac{x^2}{9} - \dfrac{y^2}{16} = 1 \end{cases}$$

4 solutions

$$\begin{cases} y = x^2 + 5 \\ x^2 + y^2 = 16 \end{cases}$$

No solution

Figure 10.16

To solve a system of nonlinear equations, use the substitution method or the elimination method. When solving a nonlinear system that contains a linear equation, the substitution method is usually easier.

EXAMPLE 1   **Solve a Nonlinear System by the Substitution Method**

Solve

$$\begin{cases} y = x^2 - x - 1 & (1) \\ 3x - y = 4 & (2) \end{cases}$$

Solution   We will use the substitution method. Using the equation $y = x^2 - x - 1$, substitute the expression for y into $3x - y = 4$.

$$3x - y = 4$$

$$3x - (x^2 - x - 1) = 4 \quad \bullet \; y = x^2 - x - 1$$

$$-x^2 + 4x + 1 = 4 \quad \bullet \; \text{Simplify.}$$

$$x^2 - 4x + 3 = 0 \quad \bullet \; \text{Write the quadratic equation in standard form.}$$

$$(x - 3)(x - 1) = 0 \quad \bullet \; \text{Solve for } x.$$

$$x - 3 = 0 \quad \text{or} \quad x - 1 = 0$$

$$x = 3 \quad \text{or} \quad x = 1$$

Substitute these values into the Equation (1) and solve for y.

$$y = 3^2 - 3 - 1 = 5, \quad \text{or} \quad y = 1^2 - 1 - 1 = -1$$

The solutions are $(3, 5)$ and $(1, -1)$. See Figure 10.17.

————————————————————— ■ *Try Exercise* **8,** *page 517.*

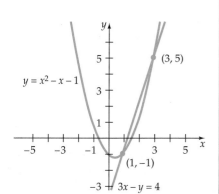

Figure 10.17

EXAMPLE 2   **Solve a Nonlinear System by the Elimination Method**

Solve

$$\begin{cases} 4x^2 + 3y^2 = 48 & (1) \\ 3x^2 + 2y^2 = 35 & (2) \end{cases}$$

Solution   We will eliminate the x^2 term. Multiply Equation (1) by -3 and Equation (2) by 4. Then add the two equations.

$$-12x^2 - 9y^2 = -144$$

$$\underline{12x^2 + 8y^2 = \;\;\; 140}$$

$$-y^2 = \;\;\; -4$$

$$y^2 = \;\;\;\;\; 4$$

$$y = \;\;\; \pm 2$$

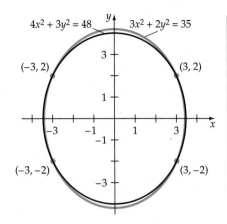

$4x^2 + 3y^2 = 48$ $3x^2 + 2y^2 = 35$

$(-3, 2)$ $(3, 2)$

$(-3, -2)$ $(3, -2)$

Figure 10.18

Substitute 2 for y into Equation (1) and solve for x.

$$4x^2 + 3(2)^2 = 48$$
$$4x^2 = 36$$
$$x^2 = 9$$
$$x = \pm 3$$

Because $(-2)^2 = 2^2$, replacing y by -2 yields the same values of x: $x = 3$ or $x = -3$. The solutions are $(3, 2)$, $(3, -2)$, $(-3, 2)$, and $(-3, -2)$. See Figure 10.18.

■ *Try Exercise* **16,** *page 517.*

EXAMPLE 3   Identify an Inconsistent System of Equations

Solve

$$\begin{cases} 4x^2 + 9y^2 = 36 & (1) \\ x^2 - y^2 = 25 & (2) \end{cases}$$

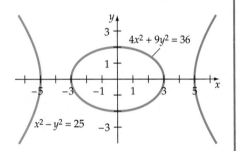

$4x^2 + 9y^2 = 36$

$x^2 - y^2 = 25$

Figure 10.19

Solution   Using the elimination method, we will eliminate the x^2 term from each equation. Multiplying Equation (2) by -4 and then adding, we have

$$\begin{array}{r} 4x^2 + 9y^2 = 36 \\ -4x^2 + 4y^2 = -100 \\ \hline 13y^2 = -64 \end{array}$$

Because the equation $13y^2 = -64$ has no real number solutions, the system of equations has no real solutions. The graphs of the equations do not intersect. See Figure 10.19.

■ *Try Exercise* **20,** *page 517.*

EXAMPLE 4   Solve a Nonlinear System of Equations

Solve

$$\begin{cases} (x + 3)^2 + (y - 4)^2 = 20 \\ (x + 4)^2 + (y - 3)^2 = 26 \end{cases}$$

Solution   Expand the binomials in each equation. Then subtract the two equations and simplify.

$$\begin{array}{rl} x^2 + 6x + 9 + y^2 - 8y + 16 = 20 & (1) \\ x^2 + 8x + 16 + y^2 - 6y + 9 = 26 & (2) \\ \hline -2x - 7 \qquad\quad - 2y + 7 = -6 \\ x + y = 3 \end{array}$$

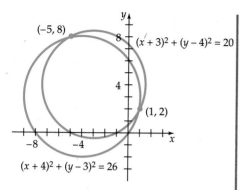

Figure 10.20

Now solve the resulting equation for y.

$$y = -x + 3$$

Substitute $-x + 3$ for y into Equation (1) and solve for x.

$$x^2 + 6x + 9 + (-x + 3)^2 - 8(-x + 3) + 16 = 20$$

$$2(x^2 + 4x - 5) = 0$$

$$2(x + 5)(x - 1) = 0$$

$$x = -5 \quad \text{or} \quad x = 1$$

Substitute -5 and 1 for x into the equation $y = -x + 3$ and solve for y.

$$y = 8 \quad \text{or} \quad y = 2$$

The solutions of the system of equations are $(-5, 8)$ and $(1, 2)$. See Figure 10.20.

■ *Try Exercise 28, page 517.*

EXERCISE SET 10.3

In Exercises 1 to 32, solve the system of equations.

1. $\begin{cases} y = x^2 - x \\ y = 2x - 2 \end{cases}$

2. $\begin{cases} y = x^2 + 2x - 3 \\ y = x + 1 \end{cases}$

3. $\begin{cases} y = 2x^2 - 3x - 3 \\ y = x - 4 \end{cases}$

4. $\begin{cases} y = -x^2 + 2x - 4 \\ y = \dfrac{1}{2}x + 1 \end{cases}$

5. $\begin{cases} y = x^2 - 2x + 3 \\ y = x^2 - x - 2 \end{cases}$

6. $\begin{cases} y = 2x^2 - x + 1 \\ y = x^2 + 2x + 5 \end{cases}$

7. $\begin{cases} x + y = 10 \\ xy = 24 \end{cases}$

8. $\begin{cases} x - 2y = 3 \\ xy = -1 \end{cases}$

9. $\begin{cases} 2x - y = 1 \\ xy = 6 \end{cases}$

10. $\begin{cases} x - 3y = 7 \\ xy = -4 \end{cases}$

11. $\begin{cases} 3x^2 - 2y^2 = 1 \\ y = 4x - 3 \end{cases}$

12. $\begin{cases} x^2 + 3y^2 = 7 \\ x + 4y = 6 \end{cases}$

13. $\begin{cases} y = x^3 + 4x^2 - 3x - 5 \\ y = 2x^2 - 2x - 3 \end{cases}$

14. $\begin{cases} y = x^3 - 2x^2 + 5x + 1 \\ y = x^2 + 7x - 5 \end{cases}$

15. $\begin{cases} 2x^2 + y^2 = 9 \\ x^2 - y^2 = 3 \end{cases}$

16. $\begin{cases} 3x^2 - 2y^2 = 19 \\ x^2 - y^2 = 5 \end{cases}$

17. $\begin{cases} x^2 - 2y^2 = 8 \\ x^2 + 3y^2 = 28 \end{cases}$

18. $\begin{cases} 2x^2 + 3y^2 = 5 \\ x^2 - 3y^2 = 4 \end{cases}$

19. $\begin{cases} 2x^2 + 4y^2 = 5 \\ 3x^2 + 8y^2 = 14 \end{cases}$

20. $\begin{cases} 2x^2 + 3y^2 = 11 \\ 3x^2 + 2y^2 = 19 \end{cases}$

21. $\begin{cases} x^2 - 2x + y^2 = 1 \\ 2x + y = 5 \end{cases}$

22. $\begin{cases} x^2 + y^2 + 3y = 22 \\ 2x + y = -1 \end{cases}$

23. $\begin{cases} (x - 3)^2 + (y + 1)^2 = 5 \\ x - 3y = 7 \end{cases}$

24. $\begin{cases} (x + 2)^2 + (y - 2)^2 = 13 \\ 2x + y = 6 \end{cases}$

25. $\begin{cases} x^2 - 3x + y^2 = 4 \\ 3x + y = 11 \end{cases}$

26. $\begin{cases} x^2 + y^2 - 4y = 4 \\ 5x - 2y = 2 \end{cases}$

27. $\begin{cases} (x - 1)^2 + (y + 2)^2 = 14 \\ (x + 2)^2 + (y - 1)^2 = 2 \end{cases}$

28. $\begin{cases} (x + 2)^2 + (y - 3)^2 = 10 \\ (x - 3)^2 + (y + 1)^2 = 13 \end{cases}$

29. $\begin{cases} (x + 3)^2 + (y - 2)^2 = 20 \\ (x - 2)^2 + (y - 3)^2 = 2 \end{cases}$

30. $\begin{cases} (x - 4)^2 + (y - 5)^2 = 8 \\ (x + 1)^2 + (y + 2)^2 = 34 \end{cases}$

31. $\begin{cases} (x - 1)^2 + (y + 1)^2 = 2 \\ (x + 2)^2 + (y - 3)^2 = 3 \end{cases}$

32. $\begin{cases} (x + 1)^2 + (y - 3)^2 = 4 \\ (x - 3)^2 + (y + 2)^2 = 2 \end{cases}$

Graphing Calculator Exercises*

For Exercises 33 to 40 approximate the real number solutions of each system of equations to the nearest ten-thousandths.

33. $\begin{cases} y = 2^x \\ y = x + 1 \end{cases}$

34. $\begin{cases} y = \log_2 x \\ y = x - 3 \end{cases}$

35. $\begin{cases} y = e^{-x} \\ y = x^2 \end{cases}$

36. $\begin{cases} y = \ln x \\ y = -x + 4 \end{cases}$

37. $\begin{cases} y = \sqrt{x} \\ y = \dfrac{1}{x - 1} \end{cases}$

38. $\begin{cases} y = \dfrac{6}{x + 1} \\ y = \dfrac{x}{x - 1} \end{cases}$

39. $\begin{cases} y = |x| \\ y = 2^{-x^2} \end{cases}$

40. $\begin{cases} y = \dfrac{2^x + 2^{-x}}{2} \\ y = \dfrac{2^x - 2^{-x}}{2} \end{cases}$

Supplemental Exercises

In Exercises 41 to 46, solve the system of equations for *rational number* ordered pairs.

41. $\begin{cases} y = x^2 + 4 \\ x = y^2 - 24 \end{cases}$

42. $\begin{cases} y = x^2 - 5 \\ x = y^2 - 13 \end{cases}$

43. $\begin{cases} x^2 - 3xy + y^2 = 5 \\ x^2 - xy - 2y^2 = 0 \end{cases}$

(*Hint:* Factor the second equation. Now use the principle of zero products and the substitution principle.)

44. $\begin{cases} x^2 + 2xy - y^2 = 1 \\ x^2 + 3xy + 2y^2 = 0 \end{cases}$

(*Hint:* See Exercise 43.)

45. $\begin{cases} 2x^2 - 4xy - y^2 = 6 \\ 4x^2 - 3xy - y^2 = 6 \end{cases}$

(*Hint:* Subtract the two equations.)

46. $\begin{cases} 3x^2 + 2xy - 5y^2 = 11 \\ x^2 + 3xy + y^2 = 11 \end{cases}$

(*Hint:* Subtract the two equations.)

47. Show that the line $y = mx$ intersects the hyperbola $\dfrac{x^2}{a^2} - \dfrac{y^2}{b^2} = 1$ if and only if $|m| < \left| \dfrac{a}{b} \right|$.

48. Show that the line $y = mx$, $m \ge 0$, intersects the circle $(x - a)^2 + y^2 = a^2$, $a > 0$, at $(0,0)$ and the point P whose coordinates are given by

$$P\left(\frac{2a}{1 + m^2}, \frac{2am}{1 + m^2} \right)$$

Now show that the slope of the line through P and $(2a, 0)$ is $-1/m$. What theorem from geometry does this prove?

49. In Example 4 of this section, the equation $x + y = 3$ was found by subtracting the original two equations of the system. Show that the graph of this line passes through the solutions of the system of equations. Do you understand that this would always happen? Think about operations that produce equivalent equations.

10.4 Partial Fractions

An algebraic application of systems of equations is a technique known as *partial fractions*. In Chapter 1, we reviewed the problem of adding two rational expressions. For example,

$$\frac{5}{x - 1} + \frac{1}{x + 2} = \frac{6x + 9}{(x - 1)(x + 2)}$$

Now we will take an opposite approach. That is, given a rational expression, find simpler rational expressions whose sum is the given expression. The method by which a more complicated rational expression is written as

*Additional graphing calculator exercises appear in the Graphing Workbook as described in the front of this textbook.

a sum of rational expressions is called **partial fraction decomposition.**
This technique is based on the following theorem.

Partial Fraction Decomposition Theorem

If $$f(x) = \frac{p(x)}{q(x)}$$

is a rational expression in which the degree of the numerator is less
than the degree of the denominator, and $p(x)$ and $q(x)$ have no common
factors, then $f(x)$ can be written as a partial fraction decomposition in the form

$$f(x) = f_1(x) + f_2(x) + \cdots + f_n(x)$$

where each $f_i(x)$ has one of the following forms:

$$\frac{A}{(px + q)^m} \quad \text{or} \quad \frac{Bx + C}{(ax^2 + bx + c)^m}$$

The procedure for finding a partial fraction decomposition of a rational
expression depends on factorization of the denominator of the rational expression. There are four cases.

Case 1   Nonrepeated Linear Factors

The partial fraction decomposition will contain an expression of the form
$A/(x + a)$ for each nonrepeated linear factor of the denominator. Example:

$$\frac{3x - 1}{x(3x + 4)(x - 2)} \qquad \bullet \text{ Each linear factor of the denominator occurs only once.}$$

Partial fraction decomposition:

$$\frac{3x - 1}{x(3x + 4)(x - 2)} = \frac{A}{x} + \frac{B}{3x + 4} + \frac{C}{x - 2}$$

Case 2   Repeated Linear Factors

The partial fraction decomposition will contain an expression of the form

$$\frac{A_1}{(x + a)} + \frac{A_2}{(x + a)^2} + \cdots + \frac{A_m}{(x + a)^m}$$

for each repeated linear factor. Example:

$$\frac{4x + 5}{(x - 2)^2(2x + 1)} \qquad \bullet (x - 2)^2 = (x - 2)(x - 2), \text{ a repeated linear factor.}$$

Partial fraction decomposition:

$$\frac{4x + 5}{(x - 2)^2(2x + 1)} = \frac{A_1}{x - 2} + \frac{A_2}{(x - 2)^2} + \frac{B}{2x + 1}$$

Case 3   Nonrepeated Quadratic Factors

The partial fraction decomposition will contain an expression of the form

$$\frac{Ax + B}{ax^2 + bx + c}$$

for each quadratic factor irreducible over the real numbers. Example:

$$\frac{x - 4}{(x^2 + x + 1)(x - 4)}$$ • $x^2 + x + 1$ is irreducible over the real numbers.

Partial fraction decomposition:

$$\frac{x - 4}{(x^2 + x + 1)(x - 4)} = \frac{Ax + B}{x^2 + x + 1} + \frac{C}{x - 4}$$

Case 4   Repeated Quadratic Factors

The partial fraction decomposition will contain an expression of the form

$$\frac{A_1x + B_1}{ax^2 + bx + c} + \frac{A_2x + B_2}{(ax^2 + bx + c)^2} + \cdots + \frac{A_mx + B_m}{(ax^2 + bx + c)^m}$$

for each quadratic factor irreducible over the real numbers. Example:

$$\frac{2x}{(x - 2)(x^2 + 4)^2}$$ • $(x^2 + 4)^2$ is a repeated quadratic factor.

Partial fraction decomposition:

$$\frac{2x}{(x - 2)(x^2 + 4)^2} = \frac{A_1x + B_1}{x^2 + 4} + \frac{A_2x + B_2}{(x^2 + 4)^2} + \frac{C}{x - 2}$$

Remark   All the denominators in these examples are shown in factored form. If the denominator is not in factored form, you must factor first before proceeding any further.

There are various methods for finding the constants of a partial fraction decomposition. One such method is based on a property of polynomials.

Equality of Polynomials

If the two polynomials $p(x) = a_nx^n + a_{n-1}x^{n-1} + \cdots + a_1x + a_0$ and $r(x) = b_nx^n + b_{n-1}x^{n-1} + \cdots + b_1x + b_0$ are of degree n, then $p(x) = r(x)$ if and only if $a_0 = b_0, a_1 = b_1, a_2 = b_2, \ldots, a_n = b_n$.

EXAMPLE 1   **Find a Partial Fraction Decomposition for Nonrepeated Factors**

Find a partial fraction decomposition of $\dfrac{x + 11}{x^2 - 2x - 15}$.

Solution   First factor the denominator.

$$x^2 - 2x - 15 = (x + 3)(x - 5)$$

The factors are nonrepeated linear factors. Therefore, the partial fraction decomposition will have the form

$$\frac{x + 11}{(x + 3)(x - 5)} = \frac{A}{x + 3} + \frac{B}{x - 5} \qquad (1)$$

To solve for A and B, multiply each side of the equation by the least common multiple of the denominators, $(x + 3)(x - 5)$.

$$x + 11 = A(x - 5) + B(x + 3)$$

$$x + 11 = (A + B)x + (-5A + 3B) \quad \bullet \text{ Combine like terms.}$$

Using the equality of polynomials theorem, equate coefficients of like powers. The result will be the system of equations

$$\begin{cases} 1 = \quad A + \ B \quad \bullet \text{ Recall that } x = 1 \cdot x \\ 11 = -5A + 3B \end{cases}$$

Solving the system of equations for A and B, we have $A = -1$ and $B = 2$. Substituting -1 for A and 2 for B into the form of the partial fraction decomposition (1), we obtain

$$\frac{x + 11}{(x + 3)(x - 5)} = \frac{-1}{x + 3} + \frac{2}{x - 5}$$

You should add the two expressions to verify the equality.

■ *Try Exercise* **14**, *page 524.*

EXAMPLE 2   **Find a Partial Fraction Decomposition for Repeated Linear Factors**

Find the partial fraction decomposition of $\dfrac{x^2 + 2x + 7}{x(x - 1)^2}$.

Solution   The denominator has one nonrepeated factor and one repeated factor. The partial fraction decomposition will have the form

$$\frac{x^2 + 2x + 7}{x(x - 1)^2} = \frac{A}{x} + \frac{B}{x - 1} + \frac{C}{(x - 1)^2}$$

Multiplying each side by the LCD $x(x - 1)^2$, we have

$$x^2 + 2x + 7 = A(x - 1)^2 + B(x - 1)x + Cx$$

Expanding the right side and combining like terms gives

$$x^2 + 2x + 7 = (A + B)x^2 + (-2A - B + C)x + A$$

Using the equality of polynomials theorem, equate coefficients of like powers. This will result in the system of equations

$$\begin{cases} 1 = A + B \\ 2 = -2A - B + C \\ 7 = A \end{cases}$$

The solution is $A = 7$, $B = -6$, and $C = 10$. Thus the partial fraction decomposition is

$$\frac{x^2 + 2x + 7}{x(x - 1)^2} = \frac{7}{x} + \frac{-6}{x - 1} + \frac{10}{(x - 1)^2}$$

■ *Try Exercise **22**, page 524.*

EXAMPLE 3   Find a Partial Fraction Decomposition with a Quadratic Factor

Find the partial fraction decomposition of $\dfrac{3x + 16}{(x - 2)(x^2 + 7)}$.

Solution   Because $(x - 2)$ is a nonrepeated linear factor and $x^2 + 7$ is an irreducible quadratic over the real numbers, the partial fraction decomposition will have the form

$$\frac{3x + 16}{(x - 2)(x^2 + 7)} = \frac{A}{x - 2} + \frac{Bx + C}{x^2 + 7}$$

Multiplying each side by the LCD $(x - 2)(x^2 + 7)$ yields

$$3x + 16 = A(x^2 + 7) + (Bx + C)(x - 2)$$

Expanding the right side and combining like terms, we have

$$3x + 16 = (A + B)x^2 + (-2B + C)x + (7A - 2C)$$

Using the equality of polynomials theorem, equate coefficients of like powers. This will result in the system of equations

$$\begin{cases} 0 = A + B \\ 3 = -2B + C \\ 16 = 7A - 2C \end{cases} \quad \bullet \text{ Think of } 3x + 16 \text{ as } 0x^2 + 3x + 16$$

The solution is $A = 2$, $B = -2$, and $C = -1$. Thus the partial fraction decomposition is

$$\frac{3x + 16}{(x - 2)(x^2 + 7)} = \frac{2}{x - 2} + \frac{-2x - 1}{x^2 + 7}$$

■ *Try Exercise **24**, page 524.*

EXAMPLE 4   **Find a Partial Fraction Decomposition with Repeated Quadratic Factors**

Find the partial fraction decomposition of $\dfrac{4x^3 + 5x^2 + 7x - 1}{(x^2 + x + 1)^2}$.

Solution   The quadratic factor $(x^2 + x + 1)$ is irreducible over the real numbers and is a repeated factor. The partial fraction decomposition will be of the form

$$\frac{4x^3 + 5x^2 + 7x - 1}{(x^2 + x + 1)^2} = \frac{Ax + B}{x^2 + x + 1} + \frac{Cx + D}{(x^2 + x + 1)^2}$$

Multiplying each side by the LCD $(x^2 + x + 1)^2$ and collecting like terms, we obtain

$$4x^3 + 5x^2 + 7x - 1 = (Ax + B)(x^2 + x + 1) + Cx + D$$
$$= Ax^3 + Ax^2 + Ax + Bx^2 + Bx + B + Cx + D$$
$$= Ax^3 + (A + B)x^2 + (A + B + C)x + (B + D)$$

Equating coefficients of like powers gives the system of equations

$$\begin{cases} 4 = A \\ 5 = A + B \\ 7 = A + B + C \\ -1 = B + D \end{cases}$$

Solving this system, we have $A = 4$, $B = 1$, $C = 2$, and $D = -2$. Thus the partial fraction decomposition is

$$\frac{4x^3 + 5x^2 + 7x - 1}{(x^2 + x + 1)^2} = \frac{4x + 1}{x^2 + x + 1} + \frac{2x - 2}{(x^2 + x + 1)^2}.$$

_____ ■ *Try Exercise* **30,** *page 524.*

The partial fraction decomposition theorem requires that the degree of the numerator be less than the degree of the denominator. If this is *not* the case, use long division to first write the rational expression as a polynomial plus a remainder.

EXAMPLE 5   **Find a Partial Fraction Decomposition When the Degree of the Numerator Exceeds the Degree of the Denominator**

Find the partial fraction decomposition of $F(x) = \dfrac{x^3 - 4x^2 - 19x - 35}{x^2 - 7x}$.

Solution   Because the degree of the denominator is less than the degree of the numerator, use long division first to obtain

$$F(x) = x + 3 + \frac{2x - 35}{x^2 - 7x}$$

The partial fraction decomposition of $\dfrac{2x - 35}{x^2 - 7x}$ will have the form

$$\frac{2x - 35}{x^2 - 7x} = \frac{2x - 35}{x(x - 7)} = \frac{A}{x} + \frac{B}{x - 7}$$

Multiplying each side by $x(x - 7)$ and combining like terms, we have

$$2x - 35 = (A + B)x + (-7A)$$

Equating coefficients of like powers yields

$$\begin{cases} 2 = A + B \\ -35 = -7A \end{cases}$$

The solution of this system is $A = 5$ and $B = -3$. The partial fraction decomposition is

$$\frac{x^3 - 4x^2 - 19x - 35}{x^2 - 7x} = x + 3 + \frac{5}{x} + \frac{-3}{x - 7}$$

■ *Try Exercise **34**, page 525.*

EXERCISE SET 10.4

In Exercises 1 to 10, evaluate the constants A, B, C, and D.

1. $\dfrac{x + 15}{x(x - 5)} = \dfrac{A}{x} + \dfrac{B}{x - 5}$

2. $\dfrac{5x - 6}{x(x + 3)} = \dfrac{A}{x} + \dfrac{B}{x + 3}$

3. $\dfrac{1}{(2x + 3)(x - 1)} = \dfrac{A}{2x + 3} + \dfrac{B}{x - 1}$

4. $\dfrac{6x - 5}{(x + 4)(3x + 2)} = \dfrac{A}{x + 4} + \dfrac{B}{3x + 2}$

5. $\dfrac{x + 9}{x(x - 3)^2} = \dfrac{A}{x} + \dfrac{B}{(x - 3)} + \dfrac{C}{(x - 3)^2}$

6. $\dfrac{2x - 7}{(x + 1)(x - 2)^2} = \dfrac{A}{x + 1} + \dfrac{B}{x - 2} + \dfrac{C}{(x - 2)^2}$

7. $\dfrac{4x^2 + 3}{(x - 1)(x^2 + x + 5)} = \dfrac{A}{x - 1} + \dfrac{Bx + C}{x^2 + x + 5}$

8. $\dfrac{x^2 + x + 3}{(x^2 + 7)(x - 3)} = \dfrac{Ax + B}{x^2 + 7} + \dfrac{C}{x - 3}$

9. $\dfrac{x^3 + 2x}{(x^2 + 1)^2} = \dfrac{Ax + B}{x^2 + 1} + \dfrac{Cx + D}{(x^2 + 1)^2}$

10. $\dfrac{3x^3 + x^2 - x - 5}{(x^3 + 2x + 5)^2} = \dfrac{Ax + B}{x^2 + 2x + 5} + \dfrac{Cx + D}{(x^2 + 2x + 5)^2}$

In Exercises 11 to 36, find the partial fraction decomposition of the given rational expression.

11. $\dfrac{8x + 12}{x(x + 4)}$

12. $\dfrac{x - 14}{x(x - 7)}$

13. $\dfrac{3x + 50}{x^2 - 7x - 18}$

14. $\dfrac{7x + 44}{x^2 + 10x + 24}$

15. $\dfrac{16x + 34}{4x^2 + 16x + 15}$

16. $\dfrac{-15x + 37}{9x^2 - 12x - 5}$

17. $\dfrac{x - 5}{(3x + 5)(x - 2)}$

18. $\dfrac{1}{(x + 7)(2x - 5)}$

19. $\dfrac{x^3 + 3x^2 - 4x - 8}{x^2 - 4}$

20. $\dfrac{x^3 - 13x - 9}{x^2 - x - 12}$

21. $\dfrac{3x^2 + 49}{x(x + 7)^2}$

22. $\dfrac{x - 18}{x(x - 3)^2}$

23. $\dfrac{5x^2 - 7x + 2}{x^3 - 3x^2 + x}$

24. $\dfrac{9x^2 - 3x + 49}{x^3 - x^2 + 10x - 10}$

25. $\dfrac{2x^3 + 9x^2 + 26x + 41}{(x + 3)^2(x^2 + 1)}$

26. $\dfrac{12x^3 - 37x^2 + 48x - 36}{(x - 2)^2(x^2 + 4)}$

27. $\dfrac{3x - 7}{(x - 4)^2}$

28. $\dfrac{5x - 53}{(x - 11)^2}$

29. $\dfrac{3x^3 - x^2 + 34x - 10}{(x^2 + 10)^2}$

30. $\dfrac{2x^3 + 9x + 1}{x^4 + 14x^2 + 49}$

31. $\dfrac{1}{k^2 - x^2}$, where k is a constant

32. $\dfrac{1}{x(k + lx)}$, where k and l are constants

33. $\dfrac{x^3 - x^2 - x - 1}{x^2 - x}$

34. $\dfrac{2x^3 + 5x^2 + 3x - 8}{2x^2 + 3x - 2}$

35. $\dfrac{2x^3 - 4x^2 + 5}{x^2 - x - 1}$

36. $\dfrac{x^4 - 2x^3 - 2x^2 - x + 3}{x^2(x - 3)}$

Supplemental Exercises

In Exercises 37 to 42, find the partial fraction decomposition of the given rational expression.

37. $\dfrac{x^2 - 1}{(x - 1)(x + 2)(x - 3)}$

38. $\dfrac{x^2 + x}{x^2(x - 4)}$

39. $\dfrac{-x^4 - 4x^2 + 3x - 6}{x^4(x - 2)}$

40. $\dfrac{3x^2 - 2x - 1}{(x^2 - 1)^2}$

41. $\dfrac{2x^2 + 3x - 1}{x^3 - 1}$

42. $\dfrac{x^3 - 2x^2 + x - 2}{x^4 - x^3 + x - 1}$

There is short-cut for finding *some* partial fraction decompositions of quadratic polynomials that do not factor over the real numbers. Exercises 43 and 44 show one method and some examples.

43. Show that for real numbers a and b with $a \neq b$,

$$\frac{1}{(b - a)[p(x) + a]} + \frac{1}{(a - b)[p(x) + b]} = \frac{1}{[p(x) + a][p(x) + b]}$$

44. Use the result of Exercise 43 to find the partial fraction decomposition of

a. $\dfrac{1}{(x^2 + 4)(x^2 + 1)}$

b. $\dfrac{3}{(x^2 + 1)(x^2 + 9)}$

c. $\dfrac{1}{(x^2 + x + 1)(x^2 + x + 2)}$

d. $\dfrac{4}{(x^2 + 2x + 4)(x^2 + 2x + 9)}$

Expressing a rational expression with real coefficients as a partial fraction decomposition requires factoring the denominator into the product of linear and quadratic factors. But is this always possible? For example, suppose the denominator of some rational expression was

$$x^4 + \pi x^3 - \sqrt{7}x^2 - 6x + \sqrt[3]{9}$$

Can this be factored into a product of linear and quadratic factors?

The answer is yes to each question we asked, but it may be very difficult to find the factorization. Working Exercises 45 to 48 prove that any polynomial with real coefficients can always be factored into a product of linear and quadratic factors with real coefficients. You may want to review the material in Chapter 4 before trying to prove these exercises.

In Exercises 45 to 48, let $p(x)$ be a polynomial of degree n with real coefficients.

45. Prove that if c is a real number zero of $p(x)$, then $x - c$ is a linear factor of $p(x)$. (*Hint:* See the factor theorem in Chapter 4.)

46. Prove that if $c = a + bi$ is a complex zero of $p(x)$, then $a - bi$, the complex conjugate of c, is also a complex zero of $p(x)$.

47. Use the result of Exercise 46 to show that if $a + bi$ is a complex zero of $p(x)$, then $x^2 - 2ax + (a^2 + b^2)$ is a quadratic factor of $p(x)$.

48. Combine the results of Exercises 45 and 47 to prove that any polynomial with real coefficients can be factored over the real numbers as the product of linear and quadratic factors.

10.5 **Inequalities in Two Variables and Systems of Inequalities**

Two examples of inequalities in two variables are

$$2x + 3y > 6 \quad \text{and} \quad xy \leq 1$$

A solution of an inequality in two variables is an ordered pair (x, y) that satisfies the inequality. For example, $(-2, 4)$ is a solution of the first inequality since $2(-2) + 3(4) > 6$. The ordered pair $(2, 1)$ is not a solution of the second inequality since $(2)(1) \nleq 1$.

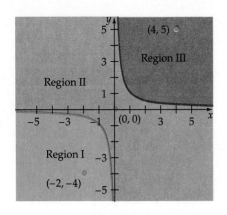

$xy \geq 2$
Figure 10.21

The **solution set of an inequality** in two variables is the set of all ordered pairs that satisfy the inequality. The **graph** of an inequality is the graph of the solution set.

To sketch the graph of an inequality, first replace the inequality symbol by an equality sign and sketch the graph of the equation. Use a dashed graph for $<$ or $>$ to indicate that the curve is not part of the solution set. Use a solid graph for $\leq$ or $\geq$ to show that the graph *is* part of the solution set.

It is important to test an ordered pair in each region of the plane defined by the graph. If the ordered pair satisfies the inequality, shade that entire region. Do this for each region into which the graph divides the plane. For example, consider the inequality $xy \geq 2$. Figure 10.21 shows the three regions of the plane defined by this inequality. Because the inequality is $\geq$, a solid graph is used.

Choose an ordered pair in each of the three regions and determine whether that ordered pair satisfies the inequality. In region I, choose $(-2, -4)$. Because $(-2)(-4) \geq 2$, Region I is part of the solution set. In Region II, choose a point, say $(0, 0)$. Because $0 \cdot 0 \ngeq 2$, Region II is not part of the solution set. In Region III, choose $(4, 5)$. Because $4 \cdot 5 \geq 2$, Region III is part of the solution set.

You may choose the coordinates of any point not on the graph of the equation as a test ordered pair; $(0, 0)$ is usually a good choice.

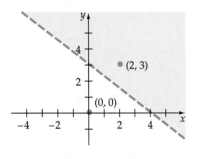

$3x + 4y > 12$
Figure 10.22

EXAMPLE 1   Graph a Linear Inequality

Graph $3x + 4y > 12$.

Solution   Graph the line $3x + 4y = 12$ using a dashed line.

Test the ordered pair $(0, 0)$: $3(0) + 4(0) = 0 \ngtr 12$

Since $(0, 0)$ does not satisfy the inequality, do not shade this region.

Test the ordered pair $(2, 3)$: $3(2) + 4(3) = 18 > 12$

Since $(2, 3)$ satisfies the inequality, the half-plane that includes $(2, 3)$ is the solution set. See Figure 10.22.

■ *Try Exercise 6, page 529.*

In general, the solution set of a *linear inequality in two variables* will be one of the regions of the plane separated by a line. Each region is called a **half-plane.**

EXAMPLE 2   Graph a Nonlinear Inequality

Graph $y \leq x^2 + 2x - 3$.

Solution   Graph the parabola $y = x^2 + 2x - 3$ using a solid curve.

Test the ordered pair $(0, 0)$: $0 \nleq 0^2 + 2(0) - 3$

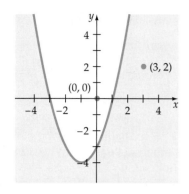

$y \leq x^2 + 2x - 3$
Figure 10.23

Since (0, 0) does not satisfy the inequality, do not shade this region.

Test the ordered pair (3, 2): $2 \leq (3)^2 + 2(3) - 3$

Since (3, 2) satisfies the inequality, shade this region of the plane. See Figure 10.23.

——————————————————————————— ■ *Try Exercise **12**, page 529.*

EXAMPLE 3   Graph an Absolute Value Inequality

Graph $y \geq |x| + 1$.

Solution   Graph the equation $y = |x| + 1$ using a solid graph.

Test the ordered pair (0, 0): $0 \ngeq |0| + 1$

Since $0 \ngeq 1$, (0, 0) does not belong to the solution set, do not shade that portion of the plane that contains (0, 0).

Test the ordered pair (0, 4): $4 \geq |0| + 1$

Since (0, 4) satisfies the inequality, shade this region. See Figure 10.24.

——————————————————————————— ■ *Try Exercise **20**, page 529.*

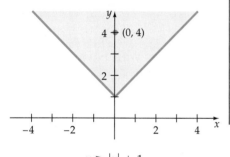

$y \geq |x| + 1$
Figure 10.24

System of Inequalities in Two Variables

The solution set of a system of inequalities is the intersection of the solution sets of each inequality. To graph the solution set of a system of inequalities, first graph the solution set of each inequality. The solution set of the system of inequalities is the region of the plane represented by the intersection of the shaded regions.

EXAMPLE 4   Graph a System of Linear Inequalities

Graph the solution set of the system of inequalities

$$\begin{cases} 3x - 2y > 6 \\ 2x - 5y \leq 10 \end{cases}$$

Solution   Graph the line $3x - 2y = 6$ using a dashed line. Test the ordered pair (0, 0). Because $3(0) - 2(0) \ngtr 6$, (0, 0) does not belong in the solution set. Do not shade the region that contains (0, 0).

Graph the line $2x - 5y = 10$ using a solid line. Test the ordered pair (0, 0). Because $2(0) - 5(0) \leq 10$, shade the region that contains (0, 0).

The solution set is the region of the plane represented by the intersection of the solution sets of each inequality. See Figure 10.25.

——————————————————————————— ■ *Try Exercise **26**, page 529.*

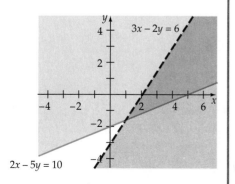

$2x - 5y = 10$

Figure 10.25

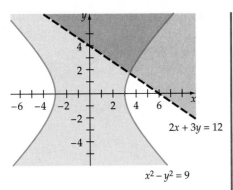

Figure 10.26

EXAMPLE 5 Graph a Nonlinear System of Inequalities

Graph the solution set of the system of inequalities

$$\begin{cases} x^2 - y^2 \leq 9 \\ 2x + 3y > 12 \end{cases}$$

Solution Graph the hyperbola $x^2 - y^2 = 9$ by using a solid graph. Test the ordered pair (0, 0). Because $0^2 - 0^2 \leq 9$, shade the region containing the origin. By choosing points in the other regions, you should show that those regions are not part of the solution set.

Graph the line $2x + 3y = 12$ by using a dashed graph. Test the ordered pair (0, 0). Because $2(0) + 3(0) \not> 12$, shade the half-plane above the line.

The solution set is the region of the plane represented by the intersection of the solution sets of each inequality. See Figure 10.26.

■ *Try Exercise* **36,** *page 529.*

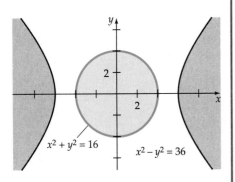

Figure 10.27

EXAMPLE 6 Identify a System of Inequalities with No Solution

Graph the solution set of the system of inequalities

$$\begin{cases} x^2 + y^2 \leq 16 \\ x^2 - y^2 \geq 36 \end{cases}$$

Solution Graph the circle $x^2 + y^2 = 16$ by using a solid graph. Test the ordered pair (0, 0). Because $0^2 + 0^2 \leq 16$, shade the inside of the circle.

Graph the hyperbola $x^2 - y^2 = 36$ by using a solid graph. Test the ordered pair (0, 0). Because $0^2 - 0^2 \not\geq 36$, shade the two branches of the hyperbola.

Because the solution sets of the inequalities do not intersect, the system has no solution. The solution set is the empty set. See Figure 10.27.

■ *Try Exercise* **40,** *page 529.*

EXAMPLE 7 Graph a System of Four Inequalities

Graph the solution set of the system of inequalities

$$\begin{cases} 2x - 3y \leq 2 \\ 3x + 4y \geq 12 \\ x \geq -1, y \geq 2 \end{cases}$$

Solution First graph the inequalities $x \geq -1$ and $y \geq 2$. Because $x \geq -1$ and $y \geq 2$, the solution set for this system will be above the line $y = 2$ and to the right of the line $x = -1$. See Figure 10.28a.

Graph the solution set of $2x - 3y = 2$ by using a solid graph. Because $2(0) - 3(0) \leq 2$, shade the region above the line.

Graph the solution set of $3x + 4y = 12$ by using a solid graph. Because $3(0) + 4(0) \not\geq 12$, shade the region above the line.

The solution set of the system of equations is the region where the graphs of the solution sets of all four inequalities intersect (Figure 10.28b).

a.

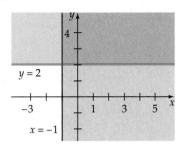

b.

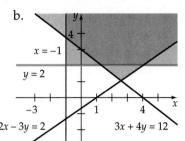

Figure 10.28

■ *Try Exercise* **42**, *page 529.*

EXERCISE SET 10.5

In Exercises 1 to 22, sketch the graph of each inequality.

1. $y \le -2$ **2.** $x + y > -2$

3. $y \ge 2x + 3$ **4.** $y < -2x + 1$

5. $2x - 3y < 6$ **6.** $3x + 4y \le 4$

7. $4x + 3y \le 12$ **8.** $5x - 2y < 8$

9. $y < x^2$ **10.** $x > y^2$

11. $y \ge x^2 - 2x - 3$ **12.** $y < 2x^2 - x - 3$

13. $(x - 2)^2 + (y - 1)^2 < 16$

14. $(x + 2)^2 + (y - 3)^2 > 25$

15. $\dfrac{(x - 3)^2}{9} - \dfrac{(y + 1)^2}{16} > 1$ **16.** $\dfrac{(x + 1)^2}{25} - \dfrac{(y - 3)^2}{16} \le 1$

17. $4x^2 + 9y^2 - 8x + 18y \ge 23$

18. $25x^2 - 16y^2 - 100x - 64y < 64$

19. $y < 2^{x-1}$ **20.** $y > \log_3 x$

21. $y \le \log_2(x - 1)$ **22.** $y > 3^x + 1$

In Exercises 23 to 46, sketch the graph of the solution set of each system of inequalities.

23. $\begin{cases} 1 \le x < 3 \\ -2 < y \le 4 \end{cases}$ **24.** $\begin{cases} -2 < x < 4 \\ \qquad y \ge -1 \end{cases}$

25. $\begin{cases} 3x + 2y \ge 1 \\ x + 2y < -1 \end{cases}$ **26.** $\begin{cases} 2x - 5y < -6 \\ 3x + y < 8 \end{cases}$

27. $\begin{cases} 2x - y \ge -4 \\ 4x - 2y \le -17 \end{cases}$ **28.** $\begin{cases} 4x + 2y > 5 \\ 6x + 3y > 10 \end{cases}$

29. $\begin{cases} 4x - 3y < 14 \\ 2x + 5y \le -6 \end{cases}$ **30.** $\begin{cases} 3x + 5y \ge -8 \\ 2x - 3y \ge 1 \end{cases}$

31. $\begin{cases} y < 2x + 3 \\ y > 2x - 2 \end{cases}$ **32.** $\begin{cases} y > 3x + 1 \\ y < 3x - 2 \end{cases}$

33. $\begin{cases} y < 2x - 1 \\ y \ge x^2 + 3x - 7 \end{cases}$ **34.** $\begin{cases} y \le 2x + 7 \\ y > x^2 + 3x + 1 \end{cases}$

35. $\begin{cases} x^2 + y^2 \le 49 \\ 9x^2 + 4y^2 \ge 36 \end{cases}$ **36.** $\begin{cases} y < 2x - 1 \\ y > x^2 - 2x + 2 \end{cases}$

37. $\begin{cases} (x - 1)^2 + (y + 1)^2 \le 16 \\ (x - 1)^2 + (y + 1)^2 \ge 4 \end{cases}$

38. $\begin{cases} (x + 2)^2 + (y - 3)^2 > 25 \\ (x + 2)^2 + (y - 3)^2 < 16 \end{cases}$

39. $\begin{cases} \dfrac{(x - 4)^2}{16} - \dfrac{(y + 2)^2}{9} > 1 \\ \dfrac{(x - 4)^2}{25} + \dfrac{(y + 2)^2}{9} < 1 \end{cases}$

40. $\begin{cases} \dfrac{(x + 1)^2}{36} + \dfrac{(y - 2)^2}{25} < 1 \\ \dfrac{(x + 1)^2}{25} + \dfrac{(y - 2)^2}{36} < 1 \end{cases}$

41. $\begin{cases} 2x - 3y \ge -5 \\ x + 2y \le 7 \\ x \ge -1, y \ge 0 \end{cases}$ **42.** $\begin{cases} 5x + y \le 9 \\ 2x + 3y \le 14 \\ x \ge -2, y \ge 2 \end{cases}$

43. $\begin{cases} 3x + 2y \geq 14 \\ x + 3y \geq 14 \\ x \leq 10, y \leq 8 \end{cases}$ 44. $\begin{cases} 4x + y \geq 13 \\ 3x + 2y \geq 16 \\ x \leq 15, y \leq 12 \end{cases}$

45. $\begin{cases} 3x + 4y \leq 12 \\ 2x + 5y \leq 10 \\ x \geq 0, y \geq 0 \end{cases}$ 46. $\begin{cases} 5x + 3y \leq 15 \\ x + 4y \leq 8 \\ x \geq 0, y \geq 0 \end{cases}$

Supplemental Exercises

In Exercises 47 to 58, sketch the graph of the inequality.

47. $y < |x|$

48. $y \geq |2x - 4|$

49. $|y| \geq |x|$

50. $|y| \leq |x - 1|$

51. $|x + y| \leq 1$

52. $|x - y| > 1$

53. $|x| + |y| \leq 1$

54. $|x| - |y| > 1$

55. $y > [\![x]\!]$, where $[\![x]\!]$ is the greatest integer function.

56. $y > x - [\![x]\!]$, where $[\![x]\!]$ is the greatest integer function.

57. Sketch the graphs of $xy > 1$ and $y > \dfrac{1}{x}$. Note that the two graphs are not the same, yet the second inequality can be derived from the first by dividing each side by x. Explain.

58. Sketch the graph of $\dfrac{x}{y} < 1$ and the graph of $x < y$. Note that the two graphs are not the same, yet the second inequality can be derived from the first by multiplying each side by y. Explain.

10.6 Linear Programming

Consider a business analyst who is trying to maximize the profit from the production of a product or an engineer who is trying to minimize the amount of energy an electrical circuit needs to operate. Generally, problems that seek to maximize or minimize a situation are called **optimization problems.** One strategy for solving certain of these problems was developed in the 1940s and is called **linear programming.**

A linear programming problem involves a **linear objective function,** which is the function that must be maximized or minimized. This objective function is subject to some **constraints**, which are inequalities, or equations that restrict the values of the variables. To illustrate these concepts, suppose a manufacturer produces two types of computer monitors: monochrome and color. Past sales experience shows that at least twice as many monochrome monitors are sold as color monitors. Suppose further that the manufacturing plant is capable of producing twelve monitors a day. Let x represent the number of monochrome monitors produced, and let y represent the number of color monitors produced. Then

$$\begin{cases} x \geq 2y \\ x + y \leq 12 \end{cases} \quad \bullet \text{ These are the constraints.}$$

These two inequalities place a constraint or restriction on the manufacturer. For example, the manufacturer cannot produce five color monitors, because that would require producing at least ten monochrome monitors, and $5 + 10 \nleq 12$.

Suppose a profit of \$50 is earned on each monochrome monitor sold and \$75 is earned on each color monitor sold. Then the manufacturer's profit, P, is given by the equation

$$P = 50x + 75y \quad \bullet \text{ Objective function}$$

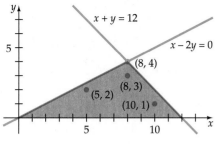

Figure 10.29

The equation $P = 50x + 75y$ defines the objective function. The goal of this linear programming problem is to determine how many of each monitor should be produced to maximize the manufacturer's profit and at the same time satisfy the constraints.

Because the manufacturer cannot produce fewer than zero units of either monitor, there are two other implied constraints, $x \geq 0$ and $y \geq 0$. Our linear programming problem now looks like

Objective function: $P = 50x + 75y$

Constraints:
$$\begin{cases} x - 2y \geq 0 \\ x + y \leq 12 \\ x \geq 0, y \geq 0 \end{cases}$$

To solve this problem, graph the solution set of the constraints. The solution set of the constraints is called the **set of feasible solutions.** Ordered pairs in this set are used to evaluate the objective function to determine which ordered pair maximizes the profit. For example, (5, 2), (8, 3), and (10, 1) are three ordered pairs in the set. See Figure 10.29. For these ordered pairs, the profit would be

$$P = 50(5) + 75(2) \quad = 400 \quad \bullet \; x = 5, y = 2$$
$$P = 50(8) + 75(3) \quad = 625 \quad \bullet \; x = 8, y = 3$$
$$P = 50(10) + 75(1) = 575 \quad \bullet \; x = 10, y = 1$$

It would be quite time-consuming to check every ordered pair in the set of feasible solutions to find which maximizes profit. Fortunately, we can find that ordered pair by solving the objective function $P = 50x + 75y$ for y.

$$y = -\frac{2}{3}x + \frac{P}{75}$$

In this form, the objective function is a linear equation whose graph has slope $-\frac{2}{3}$ and y-intercept $P/75$. If P is as large as possible (P a maximum), then the y-intercept will be as large as possible. Thus the maximum profit will occur on the line that has a slope of $-\frac{2}{3}$ and has the largest possible y-intercept and intersects the set of feasible solutions.

From Figure 10.30, the largest possible y-intercept occurs when the line passes through the point with coordinates (8, 4). At this point, the profit is

$$P = 50(8) + 75(4) = 700$$

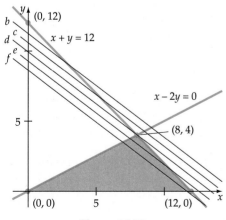

Figure 10.30

The manufacturer will maximize profit by producing 8 monochrome monitors and 4 color monitors each day. The profit will be $700 per day.

In general, the goal of any linear programming problem is to maximize or minimize the objective function, subject to the constraints. Minimization problems occur, for example, when a manufacturer wants to minimize the cost of operations.

Suppose that a cost minimization problem results in the following objective function and constraints.

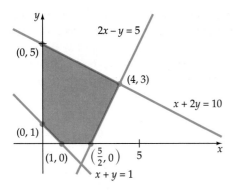

Figure 10.31

Objective function: $C = 3x + 4y$

Constraints: $\begin{cases} x + y \geq 1 \\ 2x - y \leq 5 \\ x + 2y \leq 10 \\ x \geq 0, y \geq 0 \end{cases}$

Figure 10.31 is the graph of the solution set of the constraints. The task is to find the ordered pair that satisfies all the constraints and that will give the smallest value of C. We again could solve the objective function for y and, since we want to minimize C, find the smallest y-intercept. However, a theorem from linear programming simplifies our task even more. The proof of this theorem, omitted here, is based on the techniques we used to solve our examples.

Fundamental Linear Programming Theorem

> If an objective function has an optimal solution, then that solution will be at a vertex of the set of feasible solutions.

Following is a list of the values of C at the vertices. The minimum value of the objective function occurs at the point whose coordinates are (1, 0).

(x, y)	$C = 3x + 4y$	
(1, 0)	$C = 3(1) + 4(0) = 3$	• Minimum
$\left(\dfrac{5}{2}, 0\right)$	$C = 3\left(\dfrac{5}{2}\right) + 4(0) = 7.5$	
(4, 3)	$C = 3(4) + 4(3) = 24$	• Maximum
(0, 5)	$C = 3(0) + 4(5) = 20$	
(0, 1)	$C = 3(0) + 4(1) = 4$	

The maximum value of the objective function can also be determined from the list. It occurs at (4, 3).

It is important to realize that the maximum or minimum value of an objective function depends on the objective function and on the set of feasible solutions. For example, using the same set of feasible solutions as in Figure 10.31 but changing the objective function to $C = 2x + 5y$ changes the maximum value of C to 25 at the ordered pair (0, 5). You should verify this result by making a list similar to the one shown here.

EXAMPLE 1   **Solve a Minimization Problem**

Minimize the objective function $C = 4x + 7y$ with the constraints:

$$\begin{cases} 3x + y \geq 6 \\ x + y \geq 4 \\ x + 3y \geq 6 \\ x \geq 0, y \geq 0 \end{cases}$$

Solution   Determine the set of feasible solutions by graphing the solution set of the inequalities. Note that the set of feasible solutions is an unbounded set.

Find the vertices of the region by solving the following systems of equations. These systems are formed by the equations of the lines that intersect to form a vertex of the set of feasible solutions.

$$\begin{cases} 3x + y = 6 \\ x + y = 4 \end{cases} \qquad \begin{cases} x + 3y = 6 \\ x + y = 4 \end{cases}$$

The solutions of the two systems are $(1, 3)$ and $(3, 1)$, respectively. The points $(0, 6)$ and $(6, 0)$ are the vertices on the y- and x-axes.

Evaluate the objective function at each of the four vertices of the feasible solutions.

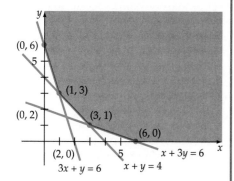

Figure 10.32

$$(x, y) \quad C = 4x + 7y$$

$(0, 6) \quad C = 4(0) + 7(6) = 42$

$(1, 3) \quad C = 4(1) + 7(3) = 25$

$(3, 1) \quad C = 4(3) + 7(1) = 19 \quad \bullet \text{ Minimum}$

$(6, 0) \quad C = 4(6) + 7(0) = 24$

The minimum value of the objective function is 19 at $(3, 1)$. See Figure 10.32.

■ *Try Exercise* **12,** *page 536.*

Linear programming can be used to determine the best allocation of the resources available to a company. In fact, the word *programming* refers to a "program to allocate resources."

EXAMPLE 2   **Solve an Applied Minimization Problem**

A manufacturer of animal food makes two grain mixtures, G_1 and G_2. Each kilogram of G_1 contains 300 grams of vitamins, 400 grams of protein, and 100 grams of carbohydrate. Each kilogram of G_2 contains 100 grams of vitamins, 300 grams of protein, and 200 grams of carbohydrate. Minimum nutritional guidelines require that a feed mixture made from these grains contain at least 900 grams of vitamins, 2200 grams of protein, and 800 grams of carbohydrate. G_1 costs \$2.00 per kilogram to produce and G_2 costs \$1.25 per kilogram to produce. Find the number of kilograms of each grain mixture that should be produced to minimize cost.

Solution   Let

$$x = \text{the number of kilograms of } G_1$$

$$y = \text{the number of kilograms of } G_2$$

The objective function is the cost function $C = 2x + 1.25y$.

Because x kilograms of G_1 contain $300x$ grams of vitamins and y kilograms of G_2 contain $100y$ grams of vitamins, the total amount of vitamins contained in x kilograms of G, and y kilograms of G_2 is $300x + 100y$. Since at least 900 grams of vitamins are necessary, $300x + 100y \geq 900$. Following similar reasoning, we have the constraints

$$\begin{cases} 300x + 100y \geq 900 \\ 400x + 300y \geq 2200 \\ 100x + 200y \geq 800 \\ x \geq 0, y \geq 0 \end{cases}$$

Two of the vertices of the set of feasible solutions can be found by solving two systems of equations. These systems are formed by the equations of the lines that intersect to form a vertex of the set of feasible solutions.

$$\begin{cases} 300x + 100y = 900 \\ 400x + 300y = 2200 \end{cases}$$ • The vertex is $(1, 6)$.

$$\begin{cases} 100x + 200y = 800 \\ 400x + 300y = 2200 \end{cases}$$ • The vertex is $(4, 2)$.

The vertices on the x- and y-axes are the x- and y-intercepts $(8, 0)$ and $(0, 9)$.

Substitute the coordinates of the vertices into the objective function.

(x, y) $C = 2x + 1.25y$

$(0, 9)$ $C = 2(0) + 1.25(9) = 11.25$

$(1, 6)$ $C = 2(1) + 1.25(6) = 9.50$ • Minimum

$(4, 2)$ $C = 2(4) + 1.25(2) = 10.50$

$(8, 0)$ $C = 2(8) + 1.25(0) = 16.00$

The minimum value of the objective function is $9.50. It occurs when the company produces a feed mixture that contains 1 kilogram of G_1 and 6 kilograms of G_2. See Figure 10.33.

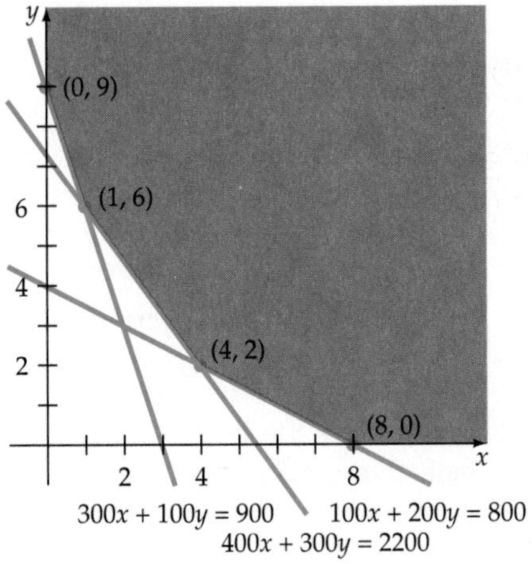

■ *Try Exercise **22**, page 536.*

$300x + 100y = 900$ $100x + 200y = 800$
$400x + 300y = 2200$

Figure 10.33

EXAMPLE 3   Solve an Applied Maximization Problem

A chemical firm produces two types of industrial solvents, S_1 and S_2. Each solvent is a mixture of three chemicals. Each kiloliter of S_1 requires 12 liters of chemical 1, 9 liters of chemical 2, and 30 liters of chemical 3. Each kiloliter of S_2 requires 24 liters of chemical 1, 5 liters of chemical 2, and 30 liters of chemical 3. The profit per kiloliter of S_1 is $100, and the profit per kiloliter of S_2 is $85. The inventory of the company shows 480 liters of chemical 1, 180 liters of chemical 2, and 720 liters of chemical 3. Assuming the company can sell all the solvent it makes, find the number of kiloliters of each solvent that the company should make to maximize profit.

Solution   Let

$x =$ the number of kiloliters of S_1

$y =$ the number of kiloliters of S_2

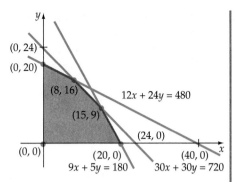

Figure 10.34

The objective function is the profit function $P = 100x + 85y$.

Because x kiloliters of S_1 require $12x$ liters of chemical 1 and y kiloliters of S_2 require $24y$ liters of chemical 1, the total amount of chemical 1 needed is $12x + 24y$. Since there are 480 liters of chemical 1 in inventory, $12x + 24y \leq 480$. Following similar reasoning, we have the constraints

$$\begin{cases} 12x + 24y \leq 480 \\ 9x + 5y \leq 180 \\ 30x + 30y \leq 720 \\ x \geq 0, \quad y \geq 0 \end{cases}$$

Two of the vertices of the set of feasible solutions can be found by solving two systems of equations. These systems are formed by the equations of the lines that intersect to form a vertex of the set of feasible solutions.

$$\begin{cases} 12x + 24y = 480 \\ 30x + 30y = 720 \end{cases}$$ • The vertex is (8, 16).

$$\begin{cases} 9x + 5y = 180 \\ 30x + 30y = 720 \end{cases}$$ • The vertex is (15, 9).

The vertices on the x- and y-axes are the x- and y-intercepts, (20, 0) and (0, 20). See Figure 10.34

Substitute the coordinates of the vertices into the objective function.

(x, y) $P = 100x + 85y$

(0, 20) $P = 100(0) + 85(20) = 1700$

(8, 16) $P = 100(8) + 85(16) = 2160$

(15, 9) $P = 100(15) + 85(9) = 2265$ • Maximum

(20, 0) $P = 100(20) + 85(0) = 2000$

The maximum value of the objective function is \$2265 when the company produces 15 kiloliters of S_1 and 9 kiloliters of S_2.

■ *Try Exercise **24**, page 537.*

EXERCISE SET 10.6

In Exercises 1 to 20, solve the linear programming problems. Assume $x \geq 0$ and $y \geq 0$.

1. Minimize $C = 4x + 2y$ with the constraints

$$\begin{cases} x + y \geq 7 \\ 4x + 3y \geq 24 \\ x \leq 10, y \leq 10 \end{cases}$$

2. Minimize $C = 5x + 4y$ with the constraints

$$\begin{cases} 3x + 4y \geq 32 \\ x + 4y \geq 24 \\ x \leq 12, y \leq 15 \end{cases}$$

3. Maximize $C = 6x + 7y$ with the constraints

$$\begin{cases} x + 2y \leq 16 \\ 5x + 3y \leq 45 \end{cases}$$

4. Maximize $C = 6x + 5y$ with the constraints

$$\begin{cases} 2x + 3y \le 27 \\ 7x + 3y \le 42 \end{cases}$$

5. Minimize $C = 5x + 6y$ with the constraints

$$\begin{cases} 4x - 3y \le 2 \\ 2x + 3y \ge 10 \end{cases}$$

6. Maximize $C = 4x + 5y$ with the constraints

$$\begin{cases} 2x - y \le 0 \\ 0 \le y \le 10 \\ 0 \le x \le 10 \end{cases}$$

7. Maximize $C = x + 6y$ with the constraints

$$\begin{cases} 5x + 8y \le 120 \\ 7x + 16y \le 192 \end{cases}$$

8. Minimize $C = 4x + 5y$ with the constraints

$$\begin{cases} x + 3y \ge 30 \\ 3x + 4y \ge 60 \end{cases}$$

9. Minimize $C = 4x + y$ with the constraints

$$\begin{cases} 3x + 5y \ge 120 \\ x + y \ge 32 \end{cases}$$

10. Maximize $C = 7x + 2y$ with the constraints

$$\begin{cases} x + 3y \le 108 \\ 7x + 4y \le 280 \end{cases}$$

11. Maximize $C = 2x + 7y$ with the constraints

$$\begin{cases} x + y \le 10 \\ x + 2y \le 16 \\ 2x + y \le 16 \end{cases}$$

12. Minimize $C = 4x + 3y$ with the constraints

$$\begin{cases} 2x + y \ge 8 \\ 2x + 3y \ge 16 \\ x + 3y \ge 11 \\ x \le 20, y \le 20 \end{cases}$$

13. Minimize $C = 3x + 2y$ with the constraints

$$\begin{cases} 3x + y \ge 12 \\ 2x + 7y \ge 21 \\ x + y \ge 8 \end{cases}$$

14. Maximize $C = 2x + 6y$ with the constraints

$$\begin{cases} x + y \le 12 \\ 3x + 4y \le 40 \\ x + 2y \le 18 \end{cases}$$

15. Maximize $C = 3x + 4y$ with the constraints

$$\begin{cases} 2x + y \le 10 \\ 2x + 3y \le 18 \\ x - y \le 2 \end{cases}$$

16. Minimize $C = 3x + 7y$ with the constraints

$$\begin{cases} x + y \ge 9 \\ 3x + 4y \ge 32 \\ x + 2y \ge 12 \end{cases}$$

17. Minimize $C = 3x + 2y$ with the constraints

$$\begin{cases} x + 2y \ge 8 \\ 3x + y \ge 9 \\ x + 4y \ge 12 \end{cases}$$

18. Maximize $C = 4x + 5y$ with the constraints

$$\begin{cases} 3x + 4y \le 250 \\ x + y \le 75 \\ 2x + 3y \le 180 \end{cases}$$

19. Maximize $C = 6x + 7y$ with the constraints

$$\begin{cases} x + 2y \le 900 \\ x + y \le 500 \\ 3x + 2y \le 1200 \end{cases}$$

20. Minimize $C = 11x + 16y$ with the constraints

$$\begin{cases} x + 2y \ge 45 \\ x + y \ge 40 \\ 2x + y \ge 45 \end{cases}$$

21. A farmer is planning to raise wheat and barley. Each acre of wheat yields a profit of \$50, and each acre of barley yields a profit of \$70. To sow the crop, two machines, a tractor and tiller, are rented. The tractor is available for 200 hours, and the tiller is available for 100 hours. Sowing an acre of barley requires 3 hours of tractor time and 2 hours of tilling. Sowing an acre of wheat requires 4 hours of tractor time and 1 hour of tilling. How many acres of each crop should be planted to maximize the farmer's profit?

22. An ice cream supplier has two machines that produce vanilla and chocolate ice cream. To meet one of its contractual obligations, the company must produce at least 60 gallons of vanilla ice cream and 100 gallons of chocolate ice cream per hour. One machine makes 4 gallons of vanilla and 5 gallons of chocolate ice cream per hour. The second machine makes 3 gallons of vanilla and 10 gallons of chocolate ice cream per hour. It costs \$28 per hour to run machine 1 and \$25 per hour to run machine 2. How many hours should each machine be operated to fulfill the contract at the least expense?

23. A manufacturer makes two types of golf clubs: a starter model and a professional model. The starter model requires 4 hours in the assembly room and 1 hour in the finishing room. The professional model requires 6 hours in the assembly room and 1 hour in the finishing room. The total number of hours available in the assembly room is 108. There are 24 hours available in the finishing room. The profit for each starter model is $35, and the profit for each professional model is $55. Assuming all the sets produced can be sold, find how many of each set should be manufactured to maximize profit.

24. A company makes two types of telephone answering machines: the standard model and the deluxe model. Each machine passes through three processes: P_1, P_2, and P_3. One standard answering machine requires 1 hour in P_1, 1 hour in P_2, and 2 hours in P_3. One deluxe answering machine requires 3 hours in P_1, 1 hour in P_2, and 1 hour in P_3. Because of employee work schedules, P_1 is available for 24 hours, P_2 is available for 10 hours, and P_3 is available for 16 hours. If the profit is $25 for each standard model and $35 for each deluxe model, how many units of each type should the company produce to maximize profit?

Supplemental Exercises

25. A dietitian formulates a special diet from two food groups: *A* and *B*. Each ounce of food group *A* contains 3 units of vitamin A, 1 unit of vitamin C, and 1 unit of vitamin D. Each unit of food group *B* contains 1 unit of vitamin A, 1 unit of vitamin C, and 3 units of vitamin D. Each ounce of food group *A* cost 40 cents, and each ounce of food group *B* costs 10 cents. The dietary constraints are such that at least 24 units of vitamin A, 16 units of vitamin C, and 30 units of vitamin D are required. Find the amount of each food group that should be used to minimize the cost. What is the minimum cost?

26. Among the many products it produces, an oil refinery makes two specialized petroleum distillates: Pymex *A* and Pymex *B*. Each distillate passes through three stages: S_1, S_2, and S_3. Each liter of Pymex *A* requires 1 hour in S_1, 3 hours in S_2, and 3 hours in S_3. Each liter of Pymex *B* requires 1 hour in S_1, 4 hours in S_2, and 2 hours in S_3. There are 10 hours available for S_1, 36 hours available for S_2, and 27 hours available for S_3. The profit per liter of Pymex *A* is $12, and the profit per liter of Pymex *B* is $9. How many liters of each distillate should be produced to maximize profit? What is the maximum profit?

27. An engine reconditioning company works on 4- and 6-cylinder engines. Each 4-cylinder engine requires 1 hour for cleaning, 5 hours for overhauling, and 3 hours for testing. Each 6-cylinder engine requires 1 hour for cleaning, 10 hours for overhauling, and 2 hours for testing. The cleaning station is available for at most 9 hours, the overhauling equipment is available for at most 80 hours, and the testing equipment is available for at most 24 hours. For each reconditioned 4-cylinder engine, the company makes a profit of $150. A reconditioned 6-cylinder engine yields a profit of $250. The company can sell all the reconditioned engines it produces. How many of each type should be produced to maximize profit? What is the maximum profit?

28. A producer of animal feed makes two food products: F_1 and F_2. The products contain three major ingredients: M_1, M_2, and M_3. Each ton of F_1 requires 200 pounds of M_1, 100 pounds of M_2, and 100 pounds of M_3. Each ton of F_2 requires 100 pounds of M_1, 200 pounds of M_2, and 400 pounds of M_3. There are at least 5000 pounds of M_1 available, at least 7000 pounds of M_2 available, and at least 10,000 pounds of M_3 available. Each ton of F_1 costs $450 to make, and each ton of F_2 costs $300 to make. How many tons of each food product should the feed producer make to minimize cost? What is the minimum cost?

Exploring Concepts with Technology

Ill-Conditioned Systems of Equations

Solving systems of equations algebraically as we did in this chapter is not practical for systems of equations that contain a large number of variables. In those cases, a computer solution is the only hope. Computer solutions are not without some problems, however.

Consider the system of equations

$$\begin{cases} 0.24567x + 0.49133y = 0.73700 \\ 0.84312x + 1.68623y = 2.52935 \end{cases}$$

It is easy to verify that the solution of this system of equations is (1, 1). However, change the constant 0.73700 to 0.73701 (add 0.00001) and the constant 2.52935 to 2.52936 (add 0.00001), and the solution is now (3, 0). Thus a very small change in the constant terms produced a dramatic change in the solution. A system of equations of this sort is said to be *ill-conditioned*.

These types of systems are important because computers generally cannot store numbers beyond a certain number of significant digits. Your calculator, for example, probably allows you to enter no more than 10 significant digits. If an exact number cannot be entered, then an approximation to that number is necessary. When a computer is solving an equation or system of equations, the hope is that approximations of the coefficients it uses will give reasonable approximations to the solutions. For ill-conditioned systems of equations, this is not always true.

In the system of equations above, small changes in the constant terms caused a large change in the solution. It is possible that small changes in the coefficients of the variables will also cause large changes in the solution.

In the two systems of equations that follow, examine the effects of approximating the fractional coefficients on the solutions. Try approximating each fraction to the nearest hundredth, to the nearest thousandth, to the nearest ten-thousandth, and then to the limits of your calculator. The exact solution of the first system of equations is (27, −192, 210). The exact solution of the second system of equations is (−64, 900, −2520, 1820).

$$\begin{cases} x + \dfrac{1}{2}y + \dfrac{1}{3}z = 1 \\ \dfrac{1}{2}x + \dfrac{1}{3}y + \dfrac{1}{4}z = 2 \\ \dfrac{1}{3}x + \dfrac{1}{4}y + \dfrac{1}{5}z = 3 \end{cases} \qquad \begin{cases} x + \dfrac{1}{2}y + \dfrac{1}{3}z + \dfrac{1}{4}w = 1 \\ \dfrac{1}{2}x + \dfrac{1}{3}y + \dfrac{1}{4}z + \dfrac{1}{5}w = 2 \\ \dfrac{1}{3}x + \dfrac{1}{4}y + \dfrac{1}{5}z + \dfrac{1}{6}w = 3 \\ \dfrac{1}{4}x + \dfrac{1}{5}y + \dfrac{1}{6}z + \dfrac{1}{7}w = 4 \end{cases}$$

Note how the solutions change as the approximations change and thus how important it is to know if a system of equations is ill-conditioned. For systems that are not ill-conditioned, approximations of the input yield reasonable approximations of the output. For ill-conditioned systems of equations, that is not always true.

Chapter Review

10.1   Systems of Linear Equations in Two Variables

- A system of equations is two or more equations considered together. A solution of a system of equations in two variables is an ordered pair that satisfies each equation of the system. Equivalent systems of equations have the same solution set.

- Operations that Produce Equivalent Systems of Equations

  1. Interchange any two equations.
  2. Replace an equation with a nonzero multiple of that equation.
  3. Replace an equation with the sum of an equation and a

nonzero constant multiple of another equation in the system.

- A system of linear equations is consistent if it has exactly one solution. A linear system is dependent if it has infinitely many solutions. An inconsistent system of equations has no solution.

10.2   Systems of Linear Equations in More Than Two Variables

- An equation of the form $ax + by + cz = d$, with a, b, and c not all zero, is a linear equation in three variables.
- The graph of a linear equation in three variables is a plane.
- A linear system of equations for which the constant term is zero for all equations of the system is called a homogeneous system of equations.

10.3   Systems of Nonlinear Equations

- A nonlinear system of equations is a system in which one or more of the equations are nonlinear.

10.4   Partial Fractions

- A rational expression can be written as the sum of terms whose denominators are factors of the denominator of the rational expression. This is called a partial fraction decomposition.

10.5   Inequalities in Two Variables and Systems of Inequalities

- The graph of an inequality in two variables frequently separates the plane into two or more regions.
- The solution set of a system of inequalities is the intersection of the solution sets of each inequality.

10.6   Linear Programming

- A linear programming problem consists of a number of constraints that are linear equations or inequalities and a linear objective function.
- The fundamental theorem of linear programming states that the objective function attains its maximum and its minimum at one of the vertices of the set of feasible solutions.

Essays and Projects

1. Linear programming has been used successfully to solve a wide range of problems in fields as diverse as providing health care and hardening nuclear silos. Write an essay on linear programming and some of the applications of this procedure in solving practical problems. Include in your essay the contributions of George Danzig, Narendra Karmarkar, and L. G. Khachian.

2. In this chapter we graphed first-degree equations in three variables. If we were to attempt to graph an equation in four variables, we would need a fourth axis perpendicular to the three axes of an xyz-coordinate system. It seems impossible to imagine a fourth dimension, but incorporating it is really a quite practical matter in mathematics. In fact, there are some systems that require an infinite-dimensional coordinate system. To gain some insight into the concept of dimension, read the book *Flatland* by Edwin A. Abbott and then write an essay explaining what this book has to do with dimension.

3. One concept of a fourth dimension is time. In the early 1930s, Albert Einstein developed what is called the general theory of relativity. He suggested that space and time were not separate but rather were linked in some fundamental way. Physicists began to talk of a *space–time* continuum rather than just of space and of time. The geometry of space–time is not at all like Euclidian geometry. Read some popular accounts of space–time and write an essay describing it and explaining how space–time differs from Euclidian geometry.

4. There have been a number of attempts to describe the abilities of a four-dimensional human. Read some of these accounts and then write an essay on some of the actions a four-dimensional person could perform. Answer the following question in your essay. Can a four-dimensional person remove money from a locked safe without first opening the safe?

5. The xy-coordinate system described in this chapter consisted of two coordinate lines that intersected at right angles. It is not necessary that coordinate lines intersect at right angles for a coordinate system to exist. Draw two coordinate lines that intersect at 0 for each line but for which the angle between the two axes is 45°. You now have a *parallelogram* coordinate system rather than a *rectangular* coordinate system. Explain the last sentence. Now experiment in this system. For example, is the graph of $3x + 4y = 12$ still a straight line in the *parallelogram* coordinate system?

True/False Exercises

In Exercises 1 to 10, answer true or false. If the statement is false, give an example to show that the statement is false.

1. A system of equations will always have a solution as long as the number of equations is equal to the number of variables.

2. A system of two different quadratic equations can have at most four solutions.

3. A homogeneous system of equations is one in which all the variables have the same exponent.

4. The intersection of two planes is a line in an xyz-coordinates system.

5. It is possible to find a partial fraction decomposition for any rational expression with real coefficients.

6. Two systems of equations with the same solution set have the same equations in their respective systems.

7. The systems of equations

$$\begin{cases} x = 0 \\ y = 0 \end{cases} \quad \text{and} \quad \begin{cases} y = x \\ y = -x \end{cases}$$

are equivalent systems of equations.

8. For a linear programming problem, one or more constraints are used to define the set of feasible solutions.

9. A system of three linear equations in three variables for which two of the planes are parallel and the third plane intersects the first two is a dependent system of equations.

10. The inequality $xy < 1$ and the inequality $y < 1/x$ are equivalent inequalities.

Chapter Review Exercises

In Exercises 1 to 30, solve each system of equations.

1. $\begin{cases} 2x - 4y = -3 \\ 3x + 8y = -12 \end{cases}$

2. $\begin{cases} 4x - 3y = 15 \\ 2x + 5y = -12 \end{cases}$

3. $\begin{cases} 3x - 4y = -5 \\ y = \dfrac{2}{3}x + 1 \end{cases}$

4. $\begin{cases} 7x + 2y = -14 \\ y = -\dfrac{5}{2}x - 3 \end{cases}$

5. $\begin{cases} y = 2x - 5 \\ x = 4y - 1 \end{cases}$

6. $\begin{cases} y = 3x + 4 \\ x = 4y - 5 \end{cases}$

7. $\begin{cases} 6x + 9y = 15 \\ 10x + 15y = 25 \end{cases}$

8. $\begin{cases} 4x - 8y = 9 \\ 2x - 4y = 5 \end{cases}$

9. $\begin{cases} 2x - 3y + z = -9 \\ 2x + 5y - 2z = 18 \\ 4x - y + 3z = -4 \end{cases}$

10. $\begin{cases} x - 3y + 5z = 1 \\ 2x + 3y - 5z = 15 \\ 3x + 6y + 5z = 15 \end{cases}$

11. $\begin{cases} x + 3y - 5z = -12 \\ 3x - 2y + z = 7 \\ 5x + 4y - 9z = -17 \end{cases}$

12. $\begin{cases} 2x - y + 2z = 5 \\ x + 3y - 3z = 2 \\ 5x - 9y + 8z = 13 \end{cases}$

13. $\begin{cases} 3x + 4y - 6z = 10 \\ 2x + 2y - 3z = 6 \\ x - 6y + 9z = -4 \end{cases}$

14. $\begin{cases} x - 6y + 4z = 6 \\ 4x + 3y - 4z = 1 \\ 5x - 9y + 8z = 13 \end{cases}$

15. $\begin{cases} 2x + 3y - 2z = 0 \\ 3x - y - 4z = 0 \\ 5x + 13y - 4z = 0 \end{cases}$

16. $\begin{cases} 3x - 5y + z = 0 \\ x + 4y - 3z = 0 \\ 2x + y - 2z = 0 \end{cases}$

17. $\begin{cases} x - 2y + z = 1 \\ 3x + 2y - 3z = 1 \end{cases}$

18. $\begin{cases} 2x - 3y + z = 1 \\ 4x + 2y + 3z = 21 \end{cases}$

19. $\begin{cases} y = x^2 - 2x - 3 \\ y = 2x - 7 \end{cases}$

20. $\begin{cases} y = 2x^2 + x \\ y = 2x + 1 \end{cases}$

21. $\begin{cases} y = 3x^2 - x + 1 \\ y = x^2 + 2x - 1 \end{cases}$

22. $\begin{cases} y = 4x^2 - 2x - 3 \\ y = 2x^2 + 3x - 6 \end{cases}$

23. $\begin{cases} (x + 1)^2 + (y - 2)^2 = 4 \\ 2x + y = 4 \end{cases}$

24. $\begin{cases} (x - 1)^2 + (y + 1)^2 = 5 \\ y = 2x - 3 \end{cases}$

25. $\begin{cases} (x - 2)^2 + (y + 2)^2 = 4 \\ (x + 2)^2 + (y + 1)^2 = 17 \end{cases}$

26. $\begin{cases} (x + 1)^2 + (y - 2)^2 = 1 \\ (x - 2)^2 + (y + 2)^2 = 20 \end{cases}$

27. $\begin{cases} x^2 - 3xy + y^2 = -1 \\ 3x^2 - 5xy - 2y^2 = 0 \end{cases}$

28. $\begin{cases} 2x^2 + 2xy - y^2 = -1 \\ 6x^2 + xy - y^2 = 0 \end{cases}$

29. $\begin{cases} 2x^2 - 5xy + 2y^2 = 56 \\ 14x^2 - 3xy - 2y^2 = 56 \end{cases}$

30. $\begin{cases} 2x^2 + 7xy + 6y^2 = 1 \\ 6x^2 + 7xy + 2y^2 = 1 \end{cases}$

In Exercises 31 to 36, find the partial fraction decomposition.

31. $\dfrac{7x - 5}{x^2 - x - 2}$

32. $\dfrac{x + 1}{(x - 1)^2}$

33. $\dfrac{2x - 2}{(x^2 + 1)(x + 2)}$

34. $\dfrac{5x^2 - 10x + 9}{(x - 2)^2(x + 1)}$

35. $\dfrac{11x^2 - x - 2}{x^3 - x}$

36. $\dfrac{x^4 + x^3 + 4x^2 + x + 3}{(x^2 + 1)^2}$

In Exercises 37 to 48, graph the solution set of each inequality.

37. $4x - 5y < 20$

38. $2x + 7y \geq -14$

39. $y \geq 2x^2 - x - 1$

40. $y < x^2 - 5x - 6$

41. $(x - 2)^2 + (y - 1)^2 > 4$ **42.** $(x + 3)^2 + (y + 1)^2 \leq 9$

43. $\dfrac{(x - 3)^2}{16} - \dfrac{(y + 2)^2}{25} \leq 1$

44. $\dfrac{(x + 1)^2}{9} - \dfrac{(y - 3)^2}{4} < -1$

45. $(2x - y + 1)(x - 2y - 2) > 0$

46. $(2x - 3y - 6)(x + 2y - 4) < 0$

47. $x^2y^2 < 1$

48. $xy \geq 0$

In Exercises 49 to 60, graph the solution set of each system of inequalities.

49. $\begin{cases} 2x - 5y < 9 \\ 3x + 4y \geq 2 \end{cases}$

50. $\begin{cases} 3x + y > 7 \\ 2x + 5y < 9 \end{cases}$

51. $\begin{cases} 2x + 3y > 6 \\ 2x - y > -2 \\ x < 3 \end{cases}$

52. $\begin{cases} 2x + 5y > 10 \\ x - y > -2 \\ x \leq 4 \end{cases}$

53. $\begin{cases} 2x + 3y \leq 18 \\ x + y \leq 7 \\ x \geq 0, y \geq 0 \end{cases}$

54. $\begin{cases} 3x + 5y \geq 25 \\ 2x + 3y \geq 16 \\ x \geq 0, y \geq 0 \end{cases}$

55. $\begin{cases} 3x + y \geq 6 \\ x + 4y \geq 14 \\ 2x + 3y \geq 16 \\ x \geq 0, y \geq 0 \end{cases}$

56. $\begin{cases} 3x + 2y \geq 14 \\ x + y \geq 6 \\ 11x + 4y \leq 48 \\ x \geq 0, y \geq 0 \end{cases}$

57. $\begin{cases} y < x^2 - x - 2 \\ y \geq 2x - 4 \end{cases}$

58. $\begin{cases} y > 2x^2 + x - 1 \\ y > x + 3 \end{cases}$

59. $\begin{cases} x^2 + y^2 - 2x + 4y > 4 \\ y < 2x^2 - 1 \end{cases}$

60. $\begin{cases} x^2 - y^2 - 4x - 2y < -4 \\ x^2 + y^2 - 4x + 4y > 8 \end{cases}$

In Exercises 61 to 66, solve the linear programming problem. In each problem, assume $x \geq 0$ and $y \geq 0$.

61. Objective function: $P = 2x + 2y$
  Constraints: $\begin{cases} x + 2y \leq 14 \\ 5x + 2y \leq 30 \end{cases}$
  Maximize the objective function.

62. Objective function: $P = 4x + 5y$
  Constraints: $\begin{cases} 2x + 3y \leq 24 \\ 4x + 3y \leq 36 \end{cases}$
  Maximize the objective function.

63. Objective function: $P = 4x + y$
  Constraints: $\begin{cases} 5x + 2y \geq 16 \\ x + 2y \geq 8 \\ x \leq 20, y \leq 20 \end{cases}$
  Minimize the objective function.

64. Objective function: $P = 2x + 7y$
  Constraints: $\begin{cases} 4x + 3y \geq 24 \\ 4x + 7y \geq 40 \\ x \leq 10, y \leq 10 \end{cases}$
  Minimize the objective function.

65. Objective function: $P = 6x + 3y$
  Constraints: $\begin{cases} 5x + 2y \geq 20 \\ x + y \geq 7 \\ x + 2y \geq 10 \\ x \leq 15, y \leq 15 \end{cases}$
  Minimize the objective function.

66. Objective function: $P = 5x + 4y$
  Constraints: $\begin{cases} x + y \leq 10 \\ 2x + y \leq 13 \\ 3x + y \leq 18 \end{cases}$
  Maximize the objective function.

67. Find an equation of the form $y = ax^2 + bx + c$ whose graph passes through the points $(1, 0)$, $(-1, 5)$, and $(2, 3)$.

68. Find an equation of the circle that passes through the points $(4, 2)$, $(0, 1)$, and $(3, -1)$.

69. Find an equation of a plane that passes through the points $(2, 1, 2)$, $(3, 1, 0)$, and $(-2, -3, -2)$. Use the equation $z = ax + by + c$.

70. How many liters of a 20% acid solution should be mixed with 10 liters of a 10% acid solution so that the resulting solution is a 16% acid solution?

71. Flying with the wind, a small plane traveled 855 miles in 5 hours. Flying against the same wind, the plane traveled 575 miles in the same time. Find the rate of the wind and the rate of the plane in calm air.

72. A collection of ten coins has a value of $1.25. The collection consists of only nickels, dimes, and quarters. How many of each coin are in the collection? (*Hint:* There is more than one solution.)

73. Consider the ordered triple (a, b, c). Find all real number values for $a, b,$ and c so that the product of any two numbers equals the remaining number.

Chapter Test

In Exercises 1 to 8, solve each system of equations. If a system of equations is inconsistent, so state.

1. $\begin{cases} 3x + 2y = -5 \\ 2x - 5y = -16 \end{cases}$

2. $\begin{cases} x - \dfrac{1}{2}y = 3 \\ 2x - y = 6 \end{cases}$

3. $\begin{cases} x + 3y - z = 8 \\ 2x - 7y + 2z = 1 \\ 4x - y + 3z = 13 \end{cases}$

4. $\begin{cases} 3x - 2y + z = 2 \\ x + 2y - 2z = 1 \\ 4x - z = 3 \end{cases}$

5. $\begin{cases} 2x - 3y + z = -1 \\ x + 5y - 2z = 5 \end{cases}$

6. $\begin{cases} 4x + 2y + z = 0 \\ x - 3y - 2z = 0 \\ 3x + 5y + 3z = 0 \end{cases}$

7. $\begin{cases} y = x + 3 \\ y = x^2 + x - 1 \end{cases}$

8. $\begin{cases} y = x^2 - x - 3 \\ y = 2x^2 + 2x - 1 \end{cases}$

In Exercises 9 to 12, graph each inequality.

9. $3x - 4y > 8$

10. $y \le x^2 - 2x - 3$

11. $x^2 + 4y^2 \ge 16$

12. $x + y^2 < 0$

In Exercises 13 to 16, graph each system of inequalities. If the solution set is empty, so state.

13. $\begin{cases} 2x - 5y \le 16 \\ x + 3y \ge -3 \end{cases}$

14. $\begin{cases} x^2 + y^2 > 9 \\ x^2 + y^2 < 4 \end{cases}$

15. $\begin{cases} x + y \ge 8 \\ 2x + y \ge 11 \\ x \ge 0, y \ge 0 \end{cases}$

16. $\begin{cases} 2x + 3y \le 12 \\ x + y \le 5 \quad x \ge 0, y \ge 0 \\ 3x + 2y \le 11 \end{cases}$

In Exercises 17 and 18, find the partial fraction decomposition.

17. $\dfrac{3x - 5}{x^2 - 3x - 4}$

18. $\dfrac{2x + 1}{x(x^2 + 1)}$

19. A farmer has 160 acres available on which to plant oats and barley. It costs $15 per acre for oat seed and $13 per acre for barley seed. The labor cost is $15 per acre for oats and $20 per acre for barley. The farmer has $2200 available to purchase seed and has set aside $2600 for labor. The profit per acre for oats is $120 and the profit per acre for barley is $150. How many acres of oats should the farmer plant to maximize profit?

20. Find the equation of the circle that passes through the points $(3, 5)$, $(-3, -3)$ and $(4, 4)$.

11 Matrices and Determinants

CASE IN POINT Linear Programming

Wassily Leontief, a Harvard economist, won the Nobel prize in economics for his work in mathematical economics. The theory he developed uses matrices to model the economy of a country. The table below, called an *input–output flow table*, depicts the transfer of goods and services within a particular economy over a period of time.

A large portion of the table contains the *matrix of interindustry flows*. Each row of the table describes how the output of that sector of the economy is distributed to each of the other sectors of the economy. Of particular importance is the output of a sector of the economy that is used by that same sector. From the table, of the $43,339 million of output from the livestock and livestock products sector, $11,316 million was used by the livestock sector itself.

Industry number	Industry name	Livestock and livestock products 1	Agricultural crops 2	Forestry and fishery products 3	Agricultural, forestry and fishery services 4	Iron and ferroalloy ores mining 5	Total interindustry use	Personal consumption expenditures 80	Gross private fixed capital formation 81	Exports 82	Imports 83	Federal Government purchases 84	State and local government purchases 85	Total final demand	Total industry output
1	Livestock and livestock products	11,316	870		188		41,958	1,454		111	−259	4	71	1,381	43,339
2	Agricultural crops	10,088	1,129		52		27,160	4,580		4,763	−630	−982	189	7,920	35,080
3	Forestry and fishery products			9	11		2,696	848		105	−1,152	−533	7	−725	1,971
4	Agricultural, forestry and fishery services	1,222	1,289	49	106		3,361	125		19	−2	14	49	205	3,566
5	Iron and ferroalloy ores mining					24	1,775			102	−619	−25		−542	1,233
6	Nonferrous metal ores mining					3	2,466		199	32	−414	−7		−190	2,276
7	Coal mining		1			4	4,749	125		496	−1	48	22	690	5,439
8	Crude petroleum and natural gas						19,439		53	1	−2,763			−2,709	16,730
9	Stone and clay mining and quarrying	1	88		1	6	3,046	5		90	−179	−2	−49	−135	2,911
10	Chemical and fertilizer mineral mining		46				459	3		79	−96	−2	49	33	492
11	Maintenance and repair construction	224	304		55	33	26,995			6		2,079	7,337	9,422	36,417
12	Ordnance and accessories						399	457	80	325		5,742	16	6,539	6,938
13	Food and kindred products	5,324	1	46	26		46,802	73,276		2,862	−4,857	310	2,229	73,820	120,622
14	Tobacco manufacturers						2,377	6,087		839	−71		−1	6,854	9,231
15	Broad and narrow fabrics, yarn and thread mills		10				16,950	639		432	−895	26	58	260	17,210
16	Miscellaneous textile goods and floor coverings	13	54	52	43		4,391	1,513	606	156	−586	16	3	1,708	6,099
17	Apparel						9,767	22,563		260	−2,638	131	82	20,398	30,165
18	Miscellaneous fabricated textile products		19	23	13		2,824	2,775		103	−134	145	49	2,938	5,762
19	Lumber and wood products except containers	6	3			3	22,424	384	5	850	−2,073	30	27	−777	21,647
20	Wood containers	1	121		17		446			3	−4	13		12	458

From "The World Economy of the Year 2000," by Wassily W. Leontief. Copyright © 1980 *Scientific American, Inc.* All rights reserved.

11.1 Gauss–Jordan Elimination Method

A **matrix** is a rectangular array of numbers. Each number in a matrix is called an **element** of the matrix. The matrix below, with three rows and four columns, is called a 3×4 (read 3 by 4) matrix.

$$\begin{bmatrix} 2 & 5 & -2 & 5 \\ -3 & 6 & 4 & 0 \\ 1 & 3 & 7 & 2 \end{bmatrix}$$

A matrix of m rows and n columns is said to be of **order $m \times n$** or **dimension $m \times n$**. A **square matrix of order n** is a matrix with n roots and n columns. The matrix above has order 3×4. We will use the notation a_{ij} to refer to the element of a matrix in the ith row and jth column. For the matrix given above, $a_{23} = 4$, $a_{31} = 1$, and $a_{13} = -2$.

The elements $a_{11}, a_{22}, a_{33}, \ldots, a_{mm}$ form the **main diagonal** of a matrix. The elements 2, 6, and 7 form the main diagonal of the matrix shown above.

A matrix can be created from a system of linear equations. Consider the system of linear equations

$$\begin{cases} 2x - 3y + z = 2 \\ x \quad\quad - 3z = 4 \\ 4x - y + 4z = 3 \end{cases}$$

Using only the coefficients and constants of this system, we can write the 3×4 matrix

$$\begin{bmatrix} 2 & -3 & 1 & 2 \\ 1 & 0 & -3 & 4 \\ 4 & -1 & 4 & 3 \end{bmatrix}$$

This matrix is called the **augmented matrix** of the system of equations. The matrix formed by the coefficients of the system is the **coefficient matrix**. The matrix formed from the constants is the **constant matrix** for the system. The coefficient matrix and constant matrix for the given system are

$$\text{Coefficient matrix:} \begin{bmatrix} 2 & -3 & 1 \\ 1 & 0 & -3 \\ 4 & -1 & 4 \end{bmatrix} \quad \text{Constant matrix:} \begin{bmatrix} 2 \\ 4 \\ 3 \end{bmatrix}$$

Remark When a term is missing from one of the equations of the system (as in the second equation), the coefficient of that term is 0, and a 0 is entered in the matrix.

We can write a system of equations from an augmented matrix.

$$\text{Augmented matrix:} \begin{bmatrix} 2 & -1 & 4 & 3 \\ 1 & 1 & 0 & 2 \\ 3 & -2 & -1 & 2 \end{bmatrix} \xrightarrow{\text{system}} \begin{cases} 2x - y + 4z = 3 \\ x + y \quad\quad = 2 \\ 3x - 2y - z = 2 \end{cases}$$

In certain cases, an augmented matrix represents a system of equations that we can solve by back substitution. Consider the following augmented matrix and the equivalent system of equations.

$$\begin{bmatrix} 1 & -3 & 4 & 5 \\ 0 & 1 & 2 & -4 \\ 0 & 0 & 1 & -1 \end{bmatrix} \xrightarrow[\text{equivalent system}]{} \begin{cases} x - 3y + 4z = 5 \\ y + 2z = -4 \\ z = -1 \end{cases}$$

Solving this system by using back substitution, we find that the solution is $(3, -2, -1)$. The matrix above is in *echelon form.*

Echelon Form

A matrix is in **echelon form** if all the following conditions are satisfied.

1. The first nonzero number in any row is a 1.

2. Rows are arranged so that the column containing the first nonzero number in any row is to the left of the column containing the first nonzero number of the next row.

3. All rows consisting entirely of zeros appear at the bottom of the matrix.

Following are three examples of matrices in echelon form.

$$\begin{bmatrix} 1 & -3 & 4 & 2 \\ 0 & 1 & -2 & -1 \\ 0 & 0 & 0 & 0 \end{bmatrix}, \quad \begin{bmatrix} 1 & 2 & -1 & 3 \\ 0 & 1 & 2 & -1 \end{bmatrix}, \quad \begin{bmatrix} 1 & -1 & 3 & 2 \\ 0 & 1 & 2 & 5 \\ 0 & 0 & 1 & -2 \end{bmatrix}$$

We can write an augmented matrix in echelon form by using what are called **elementary row operations.** These operations are a rewording, in matrix terminology, of the operations that produce equivalent equations.

Elementary Row Operations

Given the augmented matrix for a system of linear equations, each of the following elementary row operations produces a matrix of an equivalent system of equations.

1. Interchanging any two rows

2. Multiplying all the elements in a row by the same nonzero number

3. Replacing a row by the sum of that row and a nonzero multiple of any other row

It is convenient to specify each operation symbolically as follows:

1. Interchanging the ith and jth rows: $R_i \longleftrightarrow R_j$

2. Multiplying the ith row by k, a nonzero constant: kR_i

3. Replacing the jth row by the sum of that row and a nonzero multiple of the ith row: $kR_i + R_j$

To demonstrate these operations, we will use the 3×3 matrix

$$\begin{bmatrix} 2 & 1 & -2 \\ 3 & -2 & 2 \\ 1 & -2 & 3 \end{bmatrix}$$

$$\begin{bmatrix} 2 & 1 & -2 \\ 3 & -2 & 2 \\ 1 & -2 & 3 \end{bmatrix} \xrightarrow{R_1 \longleftrightarrow R_3} \begin{bmatrix} 1 & -2 & 3 \\ 3 & -2 & 2 \\ 2 & 1 & -2 \end{bmatrix}$$ • Interchange row 1 and row 3.

$$\begin{bmatrix} 2 & 1 & -2 \\ 3 & -2 & 2 \\ 1 & -2 & 3 \end{bmatrix} \xrightarrow{-3R_2} \begin{bmatrix} 2 & 1 & -2 \\ -9 & 6 & -6 \\ 1 & -2 & 3 \end{bmatrix}$$ • Multiply row 2 by -3.

$$\begin{bmatrix} 2 & 1 & -2 \\ 3 & -2 & 2 \\ 1 & -2 & 3 \end{bmatrix} \xrightarrow{-2R_3 + R_1} \begin{bmatrix} 0 & 5 & -8 \\ 3 & -2 & 2 \\ 1 & -2 & 3 \end{bmatrix}$$ • Multiply row 3 by -2 and add to row 1. Replace row 1 by the sum

The **Gauss–Jordan** elimination method is an algorithm[1] that uses elementary row operations to solve a system of linear equations. The goal of this method is to rewrite an augmented matrix in echelon form.

We will now demonstrate how to solve a system of two equations in two variables by the Gauss–Jordan method. Consider the system of equations

$$\begin{cases} 2x + 5y = -1 \\ 3x - 2y = 8 \end{cases}$$

The augmented matrix for this system is

$$\begin{bmatrix} 2 & 5 & -1 \\ 3 & -2 & 8 \end{bmatrix}$$

The goal of the Gauss–Jordan method is to rewrite the augmented matrix in echelon form by using elementary row operations. The row operations are chosen so that first, there is a 1 as a_{11}; second, there is a 0 as a_{21}; and third, there is a 1 as a_{22}.

Begin by multiplying row 1 by $\frac{1}{2}$. The result is a 1 as a_{11}.

$$\begin{bmatrix} 2 & 5 & -1 \\ 3 & -2 & 8 \end{bmatrix} \xrightarrow{\left(\frac{1}{2}\right)R_1} \begin{bmatrix} 1 & \frac{5}{2} & -\frac{1}{2} \\ 3 & -2 & 8 \end{bmatrix}$$

Now multiply row 1 by -3 and add the result to row 2. Replace row 2. The result is a 0 as a_{21}.

$$\begin{bmatrix} 1 & \frac{5}{2} & -\frac{1}{2} \\ 3 & -2 & 8 \end{bmatrix} \xrightarrow{-3R_1 + R_2} \begin{bmatrix} 1 & \frac{5}{2} & -\frac{1}{2} \\ 0 & -\frac{19}{2} & \frac{19}{2} \end{bmatrix}$$

[1] An algorithm is a procedure used in calculations. The word is derived from Al-Khwarismi, the name of the author of an Arabic algebra book written around A.D. 825.

Now multiply row 2 by $-2/19$. The result is a 1 as a_{22}. The matrix is now in row echelon form.

$$\begin{bmatrix} 1 & \frac{5}{2} & -\frac{1}{2} \\ 0 & -\frac{19}{2} & \frac{19}{2} \end{bmatrix} \xrightarrow{\left(-\frac{2}{19}\right)R_2} \begin{bmatrix} 1 & \frac{5}{2} & -\frac{1}{2} \\ 0 & 1 & -1 \end{bmatrix}$$

The system of equations written from the echelon form of the matrix is

$$\begin{cases} x + \dfrac{5}{2}y = -\dfrac{1}{2} \\ \qquad\quad y = -1 \end{cases}$$

To solve by back substitution, replace y in the first equation by -1 and solve for x.

$$x + \left(\frac{5}{2}\right)(-1) = -\frac{1}{2}$$

$$x = 2$$

The solution of the system of equations (1) is $(2, -1)$.

Remark   The order in which a matrix is rewritten in echelon form is important. Using elementary row operations, change a_{11} to 1 and change the remaining elements in the first column to 0. Move to a_{22}. Change a_{22} to 1 and change the remaining elements below a_{22} in column 2 to 0. Move to the next column and repeat the procedure. Continue moving down the main diagonal and repeating the procedure until you reach a_{mm} or until all remaining elements on the main diagonal are zero.

To conserve space, we will occasionally perform more than one elementary row operation in one step. For example, the notation

$$\begin{array}{c} 3R_1 + R_2 \\ -2R_1 + R_3 \\ \xrightarrow{\hspace{2cm}} \end{array}$$

means that two elementary row operations were performed. First multiply row 1 by 3 and add it to row 2. Replace row 2. Now multiply row 1 by -2 and add it to row 3. Replace row 3.

EXAMPLE 1   **Solve a System of Equations by the Gauss–Jordan Method**

Solve by using the Gauss–Jordan method.

$$\begin{cases} 3t - 8u + 8v + 7w = 41 \\ t - 2u + 2v + w = 9 \\ 2t - 2u + 6v - 4w = -1 \\ 2t - 2u + 3v - 3w = 3 \end{cases}$$

Solution Write the augmented matrix and then use elementary row operations to rewrite the matrix in echelon form.

$$\begin{bmatrix} 3 & -8 & 8 & 7 & 41 \\ 1 & -2 & 2 & 1 & 9 \\ 2 & -2 & 6 & -4 & -1 \\ 2 & -2 & 3 & -3 & 3 \end{bmatrix} \xrightarrow{R_1 \longleftrightarrow R_2} \begin{bmatrix} 1 & -2 & 2 & 1 & 9 \\ 3 & -8 & 8 & 7 & 41 \\ 2 & -2 & 6 & -4 & -1 \\ 2 & -2 & 3 & -3 & 3 \end{bmatrix}$$

$$\begin{matrix} -3R_1 + R_2 \\ -2R_1 + R_3 \\ -2R_1 + R_4 \\ \longrightarrow \end{matrix} \begin{bmatrix} 1 & -2 & 2 & 1 & 9 \\ 0 & -2 & 2 & 4 & 14 \\ 0 & 2 & 2 & -6 & -19 \\ 0 & 2 & -1 & -5 & -15 \end{bmatrix} \xrightarrow{\left(-\frac{1}{2}\right)R_2} \begin{bmatrix} 1 & -2 & 2 & 1 & 9 \\ 0 & 1 & -1 & -2 & -7 \\ 0 & 2 & 2 & -6 & -19 \\ 0 & 2 & -1 & -5 & -15 \end{bmatrix}$$

$$\begin{matrix} -2R_2 + R_3 \\ -2R_2 + R_4 \\ \longrightarrow \end{matrix} \begin{bmatrix} 1 & -2 & 2 & 1 & 9 \\ 0 & 1 & -1 & -2 & -7 \\ 0 & 0 & 4 & -2 & -5 \\ 0 & 0 & 1 & -1 & -1 \end{bmatrix} \xrightarrow{R_4 \longleftrightarrow R_3} \begin{bmatrix} 1 & -2 & 2 & 1 & 9 \\ 0 & 1 & -1 & -2 & -7 \\ 0 & 0 & 1 & -1 & -1 \\ 0 & 0 & 4 & -2 & -5 \end{bmatrix}$$

$$\begin{matrix} -4R_3 + R_4 \\ \longrightarrow \end{matrix} \begin{bmatrix} 1 & -2 & 2 & 1 & 9 \\ 0 & 1 & -1 & -2 & -7 \\ 0 & 0 & 1 & -1 & -1 \\ 0 & 0 & 0 & 2 & -1 \end{bmatrix} \xrightarrow{\left(\frac{1}{2}\right)R_4} \begin{bmatrix} 1 & -2 & 2 & 1 & 9 \\ 0 & 1 & -1 & -2 & -7 \\ 0 & 0 & 1 & -1 & -1 \\ 0 & 0 & 0 & 1 & -\frac{1}{2} \end{bmatrix}$$

The last matrix is in echelon form. The system of equations written from the matrix is

$$\begin{cases} t - 2u + 2v + w = 9 \\ u - v - 2w = -7 \\ v - w = -1 \\ w = -\dfrac{1}{2} \end{cases}$$

Solve by back substitution. The solution is $\left(-\dfrac{13}{2}, -\dfrac{19}{2}, -\dfrac{3}{2}, -\dfrac{1}{2} \right)$.

■ *Try Exercise* **14,** *page 551.*

EXAMPLE 2   **Solve a Dependent System of Equations**

Solve using the Gauss–Jordan method.

$$\begin{cases} x - 3y + 4z = 1 \\ 2x - 5y + 3z = 6 \\ x - 2y - z = 5 \end{cases}$$

Solution Write the augmented matrix and then use elementary row operations to rewrite the matrix in echelon form.

$$\begin{bmatrix} 1 & -3 & 4 & 1 \\ 2 & -5 & 3 & 6 \\ 1 & -2 & -1 & 5 \end{bmatrix} \xrightarrow[\substack{-2R_1 + R_2 \\ -R_1 + R_3}]{} \begin{bmatrix} 1 & -3 & 4 & 1 \\ 0 & 1 & -5 & 4 \\ 0 & 1 & -5 & 4 \end{bmatrix}$$

$$\xrightarrow[]{-R_2 + R_3} \begin{bmatrix} 1 & -3 & 4 & 1 \\ 0 & 1 & -5 & 4 \\ 0 & 0 & 0 & 0 \end{bmatrix}$$

$$\begin{cases} x - 3y + 4z = 1 \\ y - 5z = 4 \end{cases} \bullet \text{ Equivalent system}$$

Any solution of the system of equations is a solution of $y - 5z = 4$. Solving this equation for y, we have $y = 5z + 4$.

$$x - 3y + 4z = 1$$
$$x - 3(5z + 4) + 4z = 1 \qquad \bullet\ y = 5z + 4$$
$$x = 11z + 13 \quad \bullet \text{ Solve for } x.$$

Both x and y are expressed in terms of z. Let z be any real number c. The solutions of the system of equations are $(11c + 13, 5c + 4, c)$.

——————————————————————————— ■ *Try Exercise* **18**, *page 551.*

EXAMPLE 3 Identify an Inconsistent System of Equations

Solve using the Gauss–Jordan method.

$$\begin{cases} x - 3y + z = 5 \\ 3x - 7y + 2z = 12 \\ 2x - 4y + z = 3 \end{cases}$$

Solution Write the augmented matrix and then use elementary row operations to write the matrix in echelon form.

$$\begin{bmatrix} 1 & -3 & 1 & 5 \\ 3 & -7 & 2 & 12 \\ 2 & -4 & 1 & 3 \end{bmatrix} \xrightarrow[\substack{-3R_1 + R_2 \\ -2R_1 + R_3}]{} \begin{bmatrix} 1 & -3 & 1 & 5 \\ 0 & 2 & -1 & -3 \\ 0 & 2 & -1 & -7 \end{bmatrix}$$

$$\xrightarrow{\left(\frac{1}{2}\right)R_2} \begin{bmatrix} 1 & -3 & 1 & 5 \\ 0 & 1 & -\frac{1}{2} & -\frac{3}{2} \\ 0 & 2 & -1 & -7 \end{bmatrix} \xrightarrow{-2R_2 + R_3} \begin{bmatrix} 1 & -3 & 1 & 5 \\ 0 & 1 & -\frac{1}{2} & -\frac{3}{2} \\ 0 & 0 & 0 & -4 \end{bmatrix}$$

$$\begin{cases} x - 3y + z = 5 \\ y - \frac{1}{2}z = -\frac{3}{2} \quad \bullet \text{ Equivalent system} \\ 0z = -4 \end{cases}$$

Because the equation $0z = -4$ has no solution, the system of equations has no solution.

——————————————————————————— ■ *Try Exercise* **20**, *page 551.*

EXAMPLE 4   **Solve a Nonsquare System of Equations**

Solve the system of equations using the Gauss–Jordan method.

$$\begin{cases} x_1 - 2x_2 - 3x_3 - 2x_4 = 1 \\ 2x_1 - 3x_2 - 4x_3 - 2x_4 = 3 \\ x_1 + x_2 + x_3 - 7x_4 = -7 \end{cases}$$

Solution   Write the augmented matrix and then use elementary row operations to rewrite the matrix in echelon form.

$$\begin{bmatrix} 1 & -2 & -3 & -2 & 1 \\ 2 & -3 & -4 & -2 & 3 \\ 1 & 1 & 1 & -7 & -7 \end{bmatrix} \begin{array}{c} -2R_1 + R_2 \\ -1R_1 + R_3 \\ \xrightarrow{\hspace{1cm}} \end{array} \begin{bmatrix} 1 & -2 & -3 & -2 & 1 \\ 0 & 1 & 2 & 2 & 1 \\ 0 & 3 & 4 & -5 & -8 \end{bmatrix}$$

$$\begin{array}{c} -3R_2 + R_3 \\ -\dfrac{1}{2}R_3 \\ \xrightarrow{\hspace{1cm}} \end{array} \begin{bmatrix} 1 & -2 & -3 & -2 & 1 \\ 0 & 1 & 2 & 2 & 1 \\ 0 & 0 & 1 & \frac{11}{2} & \frac{11}{2} \end{bmatrix}$$

$$\begin{cases} x_1 - 2x_2 - 3x_3 - 2x_4 = 1 \\ x_2 + 2x_3 + 2x_4 = 1 \quad \bullet \text{ Equivalent system} \\ x_3 + \tfrac{11}{2}x_4 = \tfrac{11}{2} \end{cases}$$

Now express each of the variables in terms of x_4. Solve the third equation for x_3.

$$x_3 = -\frac{11}{2}x_4 + \frac{11}{2}$$

Substitute this value into the second equation and solve for x_2.

$$x_2 + 2\left(-\frac{11}{2}x_4 + \frac{11}{2} \right) + 2x_4 = 1 \quad \bullet \text{ Simplify}$$

$$x_2 = 9x_4 - 10$$

Substitute the values for x_2 and x_3 into the first equation and solve for x_1.

$$x_1 - 2(9x_4 - 10) - 3\left(-\frac{11}{2}x_4 + \frac{11}{2} \right) - 2x_4 = 1$$

$$x_1 = \frac{7}{2}x_4 - \frac{5}{2}. \quad \bullet \text{ Simplify}$$

If x_4 is any real number c, the solution is of the form

$$\left(\frac{7}{2}c - \frac{5}{2}, \ 9c - 10, \ -\frac{11}{2}c + \frac{11}{2}, \ c \right).$$

■ *Try Exercise* **36**, *page 552.*

EXERCISE SET 11.1

In Exercises 1 to 4, write the augmented matrix, the coefficient matrix, and the constant matrix.

1. $\begin{cases} 2x - 3y + z = 1 \\ 3x - 2y + 3z = 0 \\ x \qquad + 5z = 4 \end{cases}$

2. $\begin{cases} -3y + 2z = 3 \\ 2x - y \qquad = -1 \\ 3x - 2y + 3z = 4 \end{cases}$

3. $\begin{cases} 2x - 3y - 4z + w = 2 \\ 2y + z = 2 \\ x - y + 2z = 4 \\ 3x - 3y - 2z = 1 \end{cases}$

4. $\begin{cases} x - y + 2z + 3w = -2 \\ 2x \qquad + z - 2w = 1 \\ 3x \qquad - 2w = 3 \\ -x + 3y - z = 3 \end{cases}$

In Exercises 5 to 12, use elementary row operations to write the matrix in echelon form.

5. $\begin{bmatrix} 2 & -1 & 3 & -2 \\ 1 & -1 & 2 & 2 \\ 3 & 2 & -1 & 3 \end{bmatrix}$

6. $\begin{bmatrix} 1 & 2 & 4 & 1 \\ 2 & 2 & 7 & 3 \\ 3 & 6 & 8 & -1 \end{bmatrix}$

7. $\begin{bmatrix} 4 & -5 & -1 & 2 \\ 3 & -4 & 1 & -2 \\ 1 & -2 & -1 & 3 \end{bmatrix}$

8. $\begin{bmatrix} -2 & 1 & -1 & 3 \\ 2 & 2 & 4 & 6 \\ 3 & 1 & -1 & 2 \end{bmatrix}$

9. $\begin{bmatrix} 1 & -2 & 3 & -4 \\ 3 & -6 & 10 & -14 \\ 5 & -8 & 19 & -21 \\ 2 & -4 & 7 & -10 \end{bmatrix}$

10. $\begin{bmatrix} 2 & -1 & 3 & 2 \\ 1 & 2 & -1 & 3 \\ 3 & 5 & -2 & 2 \\ 4 & 3 & 1 & 8 \end{bmatrix}$

11. $\begin{bmatrix} 1 & -3 & 4 & 2 & 1 \\ 2 & -3 & 5 & -2 & -1 \\ -1 & 2 & -3 & 1 & 3 \end{bmatrix}$

12. $\begin{bmatrix} 2 & -1 & 3 & 2 & 2 \\ 1 & -2 & 2 & 1 & -1 \\ 3 & -5 & -1 & -2 & 3 \end{bmatrix}$

In Exercises 13 to 38, solve the system of equations by the Gauss–Jordan method.

13. $\begin{cases} x + 2y - 2z = -2 \\ 5x + 9y - 4z = -3 \\ 3x + 4y - 5z = -3 \end{cases}$

14. $\begin{cases} x - 3y + z = 8 \\ 2x - 5y - 3z = 2 \\ x + 4y + z = 1 \end{cases}$

15. $\begin{cases} 3x + 7y - 7z = -4 \\ x + 2y - 3z = 0 \\ 5x + 6y + z = -8 \end{cases}$

16. $\begin{cases} 2x - 3y + 2z = 13 \\ 3x - 4y - 3z = 1 \\ 3x + y - z = 2 \end{cases}$

17. $\begin{cases} x + 2y - 2z = 3 \\ 5x + 8y - 6z = 14 \\ 3x + 4y - 2z = 8 \end{cases}$

18. $\begin{cases} 3x - 5y + 2z = 4 \\ x - 3y + 2z = 4 \\ 5x - 11y + 6z = 12 \end{cases}$

19. $\begin{cases} 3x + 2y - z = 1 \\ 2x + 3y - z = 1 \\ x - y + 2z = 3 \end{cases}$

20. $\begin{cases} 2x + 5y + 2z = -1 \\ x + 2y - 3z = 5 \\ 5x + 12y + z = 10 \end{cases}$

21. $\begin{cases} x - 3y + 2z = 0 \\ 2x - 5y - 2z = 0 \\ 4x - 11y + 2z = 0 \end{cases}$

22. $\begin{cases} x + y - 2z = 0 \\ 3x + 4y - z = 0 \\ 5x + 6y - 5z = 0 \end{cases}$

23. $\begin{cases} 2x + y - 3z = 4 \\ 3x + 2y + z = 2 \end{cases}$

24. $\begin{cases} 3x - 6y + 2z = 2 \\ 2x + 5y - 3z = 2 \end{cases}$

25. $\begin{cases} 2x + 2y - 4z = 4 \\ 2x + 3y - 5z = 4 \\ 4x + 5y - 9z = 8 \end{cases}$

26. $\begin{cases} 3x - 10y + 2z = 34 \\ x - 4y + z = 13 \\ 5x - 2y + 7z = 31 \end{cases}$

27. $\begin{cases} x + 3y + 4z = 11 \\ 2x + 3y + 2z = 7 \\ 4x + 9y + 10z = 20 \\ 3x - 2y + z = 1 \end{cases}$

28. $\begin{cases} x - 4y + 3z = 4 \\ 3x - 10y + 3z = 4 \\ 5x - 18y + 9z = 10 \\ 2x + 2y - 3z = -11 \end{cases}$

29. $\begin{cases} t + 2u - 3v + w = -7 \\ 3t + 5u - 8v + 5w = -8 \\ 2t + 3u - 7v + 3w = -11 \\ 4t + 8u - 10v + 7w = -10 \end{cases}$

30. $\begin{cases} t + 4u + 2v - 3w = 11 \\ 2t + 10u + 3v - 5w = 17 \\ 4t + 16u + 7v - 9w = 34 \\ t + 4u + v - w = 4 \end{cases}$

31. $\begin{cases} 2t - u + 3v + 2w = 2 \\ t - u + 2v + w = 2 \\ 3t - 2v - 3w = 13 \\ 2t + 2u - 2w = 6 \end{cases}$

32. $\begin{cases} 4t + 7u - 10v + 3w = -29 \\ 3t + 5u - 7v + 2w = -20 \\ t + 2u - 3v + w = -9 \\ 2t - u + 2v - 4w = 15 \end{cases}$

33. $\begin{cases} 3t + 10u + 7v - 6w = 7 \\ 2t + 8u + 6v - 5w = 5 \\ t + 4u + 2v - 3w = 2 \\ 4t + 14u + 9v - 8w = 8 \end{cases}$

34. $\begin{cases} t - 3u + 2v + 4w = 13 \\ 3t - 8u + 4v + 13w = 35 \\ 2t - 7u + 8v + 5w = 28 \\ 4t - 11u + 6v + 17w = 56 \end{cases}$

35. $\begin{cases} t - u + 2v - 3w = 9 \\ 4t + 11v - 10w = 46 \\ 3t - u + 8v - 6w = 27 \end{cases}$

36. $\begin{cases} t - u + 3v - 5w = 10 \\ 2t - 3u + 4v + w = 7 \\ 3t + u - 2v - 2w = 6 \end{cases}$

37. $\begin{cases} 3t - 4u + v = 2 \\ t + u - 2v + 3w = 1 \end{cases}$

38. $\begin{cases} 2t + 3v - 4w = 2 \\ t + 2u - 4v + w = -3 \end{cases}$

Graphing Calculator Exercises*

Some graphing calculators and computer programs contain a program which will assist you in solving a system of linear equations by rewriting in triangular form. Try one of these programs for Exercises 39 to 44.

39. $\begin{cases} x_1 + 2x_2 - x_3 + 2x_4 + 3x_5 = 11 \\ x_1 - x_2 + 2x_3 - x_4 + 2x_5 = 0 \\ 2x_1 + x_2 - x_3 + 2x_4 - x_5 = 4 \\ 3x_1 + 2x_2 - x_3 + x_4 - 2x_5 = 2 \\ 2x_1 + x_2 - x_3 - 2x_4 + x_5 = 4 \end{cases}$

40. $\begin{cases} x_1 - 2x_2 + 2x_3 - 3x_4 + 2x_5 = 5 \\ x_1 - 3x_2 - x_3 + 2x_4 - x_5 = -4 \\ 3x_1 + x_2 - 2x_3 + x_4 + 3x_5 = 9 \\ 2x_1 - x_2 + 3x_3 - x_4 - 2x_5 = 2 \\ -x_1 + 2x_2 - 2x_3 + 3x_4 - x_5 = -4 \end{cases}$

41. $\begin{cases} x_1 + 2x_2 - 3x_3 - x_4 + 2x_5 = -10 \\ -x_1 - 3x_2 + x_3 + x_4 - x_5 = 4 \\ 2x_1 + 3x_2 - 5x_3 + 2x_4 + 3x_5 = -20 \\ 3x_1 + 4x_2 - 7x_3 + 3x_4 - 2x_5 = -16 \\ 2x_1 + x_2 - 6x_3 + 4x_4 - 3x_5 = -12 \end{cases}$

42. $\begin{cases} x_1 - 2x_2 + 2x_3 - 3x_4 + x_5 = 5 \\ 2x_1 - 3x_2 + 4x_3 - 5x_4 - x_5 = 13 \\ x_1 + x_2 - 2x_3 + 2x_4 + 2x_5 = -11 \\ 3x_1 - 2x_2 + 2x_3 - 2x_4 - 2x_5 = 7 \\ 4x_1 - 4x_2 + 4x_3 - 5x_4 - x_5 = 12 \end{cases}$

43. Find a cubic function whose graph passes through the points $(0, 2)$, $(1, 0)$, $(-2, -12)$, and $(3, 8)$. (*Hint:* Use the equation $y = ax^3 + bx^2 + cx + d$.)

44. Find a cubic function whose graph passes through the points $(0, 0)$, $(1, 1)$, $(2, 6)$, and $(-1, 0)$. (*Hint:* Use the equation $y = ax^3 + bx^2 + cx + d$.)

Supplemental Exercises

In Exercises 45 to 47, use the system of equations

$$\begin{cases} x + 3y - a^2z = a^2 \\ 2x + 3y + az = 2 \\ 3x + 4y + 2z = 3 \end{cases}$$

45. Find all values of a for which the system of equations has a unique solution.

46. Find all values of a for which the system of equations has infinitely many solutions.

47. Find all values of a for which the system of equations has no solutions.

48. Find an equation of the plane that passes through the points $(1, 2, 6)$, $(-1, 1, 7)$, and $(4, 2, 0)$. Use the equation $z = ax + by + c$.

49. Find an equation of the plane that passes through the points $(-1, 0, -4)$, $(2, 1, 5)$, and $(-1, 1, -1)$. Use the equation $z = ax + by + c$.

11.2 **The Algebra of Matrices**

Besides being convenient for solving a system of equations, matrices are appropriate for solving problems in business and science. To apply matrices in these areas effectively, we must develop some of the algebra of matrices.

Throughout this chapter we denote a matrix either by using a capital letter or by surrounding the corresponding lower-case letter with brackets.

*Additional graphing calculator exercises appear in the Graphing Workbook as described in the front of this textbook.

For example, a matrix could be denoted as

$$A \qquad \text{or} \qquad [a_{ij}]$$

Caution Although $[a_{ij}]$ is a *matrix*, a_{ij} is the *element* in the ith row and jth column of the matrix.

An important concept involving matrices is the principle of equality.

Definition of Equality of Two Matrices

> Two matrices $A = [a_{ij}]$ and $B = [b_{ij}]$ are equal if and only if
>
> $$a_{ij} = b_{ij}$$
>
> for every i and j.

For example, $\begin{bmatrix} a & -2 & b \\ 3 & c & 1 \end{bmatrix}$ and $\begin{bmatrix} 3 & x & -4 \\ 3 & -1 & y \end{bmatrix}$ are equal if and only if $a = 3$, $x = -2$, $b = -4$, $c = -1$, and $y = 1$.

Remark The definition of equality implies that the two matrices have the same order. Why?[2]

Definition of Addition of Matrices

> If A and B are matrices of order $m \times n$, then the sum of the matrices is the $m \times n$ matrix given by
>
> $$A + B = [a_{ij} + b_{ij}]$$

This definition states that the sum of two matrices is found by adding their corresponding elements. Note from the definition that both matrices have the same order. The sum of two matrices of different order is not defined.

Given $A = \begin{bmatrix} 2 & -2 & 3 \\ 1 & 3 & -4 \end{bmatrix}$ and $B = \begin{bmatrix} 5 & -2 & 6 \\ -2 & 3 & 5 \end{bmatrix}$, then

$$A + B = \begin{bmatrix} 2 & -2 & 3 \\ 1 & 3 & -4 \end{bmatrix} + \begin{bmatrix} 5 & -2 & 6 \\ -2 & 3 & 5 \end{bmatrix}$$

$$= \begin{bmatrix} 2 + 5 & -2 + (-2) & 3 + 6 \\ 1 + (-2) & 3 + 3 & -4 + 5 \end{bmatrix} = \begin{bmatrix} 7 & -4 & 9 \\ -1 & 6 & 1 \end{bmatrix}$$

Now let $C = \begin{bmatrix} 2 & -3 \\ 4 & 1 \end{bmatrix}$ and $D = \begin{bmatrix} 3 & 2 & 0 \\ 1 & -5 & 0 \end{bmatrix}$.

Then $C + D$ is not defined because the matrices do not have the same order.

---

[2] If the two matrices were of different order, there would be an element of one matrix for which there was no corresponding element of the second matrix.

To define the subtraction of two matrices, we first define the additive inverse of a matrix.

Additive Inverse of a Matrix

Given the matrix $A = [a_{ij}]$, the additive inverse of A is $-A = [-a_{ij}]$.

For example, if $A = \begin{bmatrix} -2 & 3 & -1 \\ 0 & -1 & 4 \end{bmatrix}$, then the additive inverse of A is

$$-A = -\begin{bmatrix} -2 & 3 & -1 \\ 0 & -1 & 4 \end{bmatrix} = \begin{bmatrix} 2 & -3 & 1 \\ 0 & 1 & -4 \end{bmatrix}.$$

Subtraction of two matrices is defined in terms of the additive inverse of a matrix.

Definition of Subtraction of Matrices

Given two matrices A and B of order $m \times n$, then $A - B$ is the sum of A and the additive inverse of B.

$$A - B = A + (-B)$$

As an example, let $A = \begin{bmatrix} 2 & -3 \\ -1 & 2 \\ 2 & 4 \end{bmatrix}$ and $B = \begin{bmatrix} -1 & 2 \\ -4 & 1 \\ 3 & -2 \end{bmatrix}$. Then

$$A - B = \begin{bmatrix} 2 & -3 \\ -1 & 2 \\ 2 & 4 \end{bmatrix} - \begin{bmatrix} -1 & 2 \\ -4 & 1 \\ 3 & -2 \end{bmatrix}$$

$$= \begin{bmatrix} 2 & -3 \\ -1 & 2 \\ 2 & 4 \end{bmatrix} + \begin{bmatrix} 1 & -2 \\ 4 & -1 \\ -3 & 2 \end{bmatrix} = \begin{bmatrix} 3 & -5 \\ 3 & 1 \\ -1 & 6 \end{bmatrix}$$

Of special importance is the *zero matrix*, which is the matrix that consists of all zeros. The zero matrix is the additive identity for matrices.

Definition of the Zero Matrix

The $m \times n$ **zero matrix**, denoted by O, is the matrix whose elements are all zeros.

Three examples of zero matrices are

$$\begin{bmatrix} 0 & 0 & 0 \\ 0 & 0 & 0 \end{bmatrix} \qquad \begin{bmatrix} 0 & 0 & 0 & 0 \\ 0 & 0 & 0 & 0 \\ 0 & 0 & 0 & 0 \end{bmatrix} \qquad \begin{bmatrix} 0 & 0 \\ 0 & 0 \end{bmatrix}$$

Properties of Matrix Addition

Given matrices A, B, C and the zero matrix O, each of order $m \times n$, then the following properties hold.

Commutative	$A + B = B + A$
Associative	$A + (B + C) = (A + B) + C$
Additive inverse	$A + (-A) = O$
Additive identity	$A + O = O + A = A$

Two types of products involve matrices. The first product is the product of a real number and a matrix. The second product is the product of two matrices.

Definition of the Product of a Real Number and a Matrix

Given the $m \times n$ matrix $A = [a_{ij}]$ and a real number c, then $cA = [ca_{ij}]$.

Remark   Finding the product of a real number and a matrix is referred to as **scalar multiplication.**

As an example of this definition, consider the matrix

$$A = \begin{bmatrix} 2 & -3 & 1 \\ 3 & 1 & -2 \\ 1 & -1 & 4 \end{bmatrix}$$

and the constant $c = -2$. Then

$$-2A = -2 \begin{bmatrix} 2 & -3 & 1 \\ 3 & 1 & -2 \\ 1 & -1 & 4 \end{bmatrix} = \begin{bmatrix} -2(2) & -2(-3) & -2(1) \\ -2(3) & -2(1) & -2(-2) \\ -2(1) & -2(-1) & -2(4) \end{bmatrix} = \begin{bmatrix} -4 & 6 & -2 \\ -6 & -2 & 4 \\ -2 & 2 & -8 \end{bmatrix}$$

This definition is also used to factor a constant from a matrix.

$$\begin{bmatrix} \frac{3}{2} & -\frac{5}{4} & \frac{1}{4} \\ \frac{3}{4} & \frac{1}{2} & \frac{5}{2} \end{bmatrix} = \frac{1}{4} \begin{bmatrix} 6 & -5 & 1 \\ 3 & 2 & 10 \end{bmatrix}$$

Properties of Scalar Multiplication

Given real numbers a, b, and c and matrices $A = [a_{ij}]$ and $B = [b_{ij}]$ each of order $m \times n$, then

$$(b + c)A = bA + cA$$

$$c(A + B) = cA + cB$$

$$a(bA) = (ab)A$$

EXAMPLE 1   **Find the Sum of Two Matrices**

Given $A = \begin{bmatrix} -2 & 3 \\ 4 & -2 \\ 0 & 4 \end{bmatrix}$ and $B = \begin{bmatrix} 8 & -2 \\ -3 & 2 \\ -4 & 7 \end{bmatrix}$, find $2A + 5B$.

Solution $2A + 5B = 2\begin{bmatrix} -2 & 3 \\ 4 & -2 \\ 0 & 4 \end{bmatrix} + 5\begin{bmatrix} 8 & -2 \\ -3 & 2 \\ -4 & 7 \end{bmatrix}$

$= \begin{bmatrix} -4 & 6 \\ 8 & -4 \\ 0 & 8 \end{bmatrix} + \begin{bmatrix} 40 & -10 \\ -15 & 10 \\ -20 & 35 \end{bmatrix} = \begin{bmatrix} 36 & -4 \\ -7 & 6 \\ -20 & 43 \end{bmatrix}$

■ *Try Exercise* **6**, *page 562.*

The product of two matrices can be developed by considering long-distance telephone rates. The rates charged by a telephone company depend on the time of day a call is made. For this particular company, the schedule is

Day	**Evening**	**Night**
5:00 A.M.–5:00 P.M.	5:00 P.M.–11:00 P.M.	11:00 P.M.–5:00 A.M.
$.23 per minute	$.17 per minute	$.08 per minute

During one month, the number of long-distance minutes used by a customer of this telephone company was

Day	33 minutes
Evening	48 minutes
Night	15 minutes

The total cost for long-distance telephone service for that month is the sum of the products of the cost per minute and the number of minutes.

$$\text{Total cost} = 0.23(33) + 0.17(48) + 0.08(15) = \$16.95$$

In matrix terms, the cost per minute can be written as the *row* matrix [0.23   0.17   0.08]. The number of minutes of long-distance service used can be written as the *column* matrix $\begin{bmatrix} 33 \\ 48 \\ 15 \end{bmatrix}$. The product of the row matrix and the column matrix is

$$[0.23 \quad 0.17 \quad 0.08]\begin{bmatrix} 33 \\ 48 \\ 15 \end{bmatrix} = 0.23(33) + 0.17(48) + 0.08(15) = 16.95$$

In general, if A is a row matrix of order $1 \times n$,

$$A = [a_1 \quad a_2 \quad \cdots \quad a_n]$$

and B is the column matrix of order $n \times 1$,

$$B = \begin{bmatrix} b_1 \\ b_2 \\ \cdot \\ \cdot \\ \cdot \\ b_n \end{bmatrix}$$

then the product of A and B, written AB, is

$$AB = [a_1 \quad a_2 \quad a_3 \quad \cdots \quad a_n] \begin{bmatrix} b_1 \\ b_2 \\ \cdot \\ \cdot \\ \cdot \\ b_n \end{bmatrix} = a_1b_1 + a_2b_2 + a_3b_3 + \cdots + a_nb_n$$

For example, if $A = [2 \quad -1 \quad 4]$ and $B = \begin{bmatrix} -3 \\ 2 \\ 6 \end{bmatrix}$, then

$$AB = 2(-3) + (-1)(2) + 4(6) = 16$$

Remark Note that the number of elements in the row matrix A equals the number of elements in the column matrix B. If this is not the case, AB is not defined. As an example, if $A = [3 \quad 4 \quad -1]$ and $B = \begin{bmatrix} 2 \\ 5 \end{bmatrix}$, then AB is not defined.

Now consider three phone companies (T_1, T_2, and T_3) with different rate structures and two customers (C_1 and C_2).

	Telephone Company Rates (cost per minute)				Customer Time Chart (minutes)	
	Day	*Night*	*Evening*		C_1	C_2
T_1	$.23	$.17	$.08	Day	45	52
T_2	$.27	$.12	$.10	Evening	73	60
T_3	$.26	$.15	$.09	Night	21	8

In terms of matrices, let the telephone companies' rate structure be denoted by T and the customers' time usage by C. Then

$$T = \begin{bmatrix} 0.23 & 0.17 & 0.08 \\ 0.27 & 0.12 & 0.10 \\ 0.26 & 0.15 & 0.09 \end{bmatrix} \quad \text{and} \quad C = \begin{bmatrix} 45 & 52 \\ 73 & 60 \\ 21 & 8 \end{bmatrix}$$

Let P denote the product TC. This product is determined by extending the concept of the product of a row-and-column matrix. Multiply each row of T and each column of C.

$$P = \begin{bmatrix} 0.23 & 0.17 & 0.08 \\ 0.27 & 0.12 & 0.10 \\ 0.26 & 0.15 & 0.09 \end{bmatrix} \begin{bmatrix} 45 & 52 \\ 73 & 60 \\ 21 & 8 \end{bmatrix}$$

$$P = \begin{bmatrix} [0.23 \quad 0.17 \quad 0.08] \begin{bmatrix} 45 \\ 73 \\ 21 \end{bmatrix} & [0.23 \quad 0.17 \quad 0.08] \begin{bmatrix} 52 \\ 60 \\ 8 \end{bmatrix} \\ [0.27 \quad 0.12 \quad 0.10] \begin{bmatrix} 45 \\ 73 \\ 21 \end{bmatrix} & [0.27 \quad 0.12 \quad 0.10] \begin{bmatrix} 52 \\ 60 \\ 8 \end{bmatrix} \\ [0.26 \quad 0.15 \quad 0.09] \begin{bmatrix} 45 \\ 73 \\ 21 \end{bmatrix} & [0.26 \quad 0.15 \quad 0.09] \begin{bmatrix} 52 \\ 60 \\ 8 \end{bmatrix} \end{bmatrix}$$

$$= \begin{bmatrix} 0.23(45) + 0.17(73) + 0.08(21) & 0.23(52) + 0.17(60) + 0.08(8) \\ 0.27(45) + 0.12(73) + 0.10(21) & 0.27(52) + 0.12(60) + 0.10(8) \\ 0.26(45) + 0.15(73) + 0.09(21) & 0.26(52) + 0.15(60) + 0.09(8) \end{bmatrix}$$

$$= \begin{bmatrix} 24.44 & 22.80 \\ 23.01 & 22.04 \\ 24.54 & 23.24 \end{bmatrix}$$

Each entry in P is the total cost for long-distance service that each customer would incur for each of the three telephone companies. For example, $p_{11} = 24.44$ represents the amount company T_1 would charge customer C_1. The entry in row 3 column 2 ($p_{32} = 23.24$) represents the amount company T_3 would charge customer C_2. In each case, the subscripts on an element of P denote the company and customer, respectively.

Using this application as a model, we now define the product of two matrices. The definition is an extension of the definition of the product of a row matrix and a column matrix.

Definition of the Product of Two Matrices

Let $A = [a_{ij}]$ be a matrix of order $m \times n$ and $B = [b_{ij}]$ be a matrix of order $n \times p$. Then the product AB is the matrix of order $m \times p$ given by $AB = [c_{ij}]$, where each element c_{ij} is

$$c_{ij} = [a_{i1} \quad a_{i2} \quad a_{i3} \cdots a_{in}] \begin{bmatrix} b_{1j} \\ b_{2j} \\ b_{3j} \\ \vdots \\ b_{nj} \end{bmatrix}$$

$$= a_{i1}b_{1j} + a_{i2}b_{2j} + a_{i3}b_{3j} + \cdots + a_{in}b_{nj}$$

Remark This definition may appear complicated, but basically, to multiply two matrices, multiply each row of the first matrix by each column of the second matrix.

For the product of two matrices to be possible, the number of columns of the first matrix must equal the number of rows of the second matrix.

$$\underset{m \times n}{A} \quad \cdot \quad \underset{n \times p}{B} \quad = \quad \underset{m \times p}{C}$$

Must be equal —
Order of product matrix

The product matrix has as many rows as the first matrix and as many columns as the second matrix. For example, let

$$A = \begin{bmatrix} 2 & -3 & 0 \\ 1 & 4 & -1 \end{bmatrix} \quad \text{and} \quad B = \begin{bmatrix} 1 & 0 \\ 4 & -2 \\ 3 & 5 \end{bmatrix}$$

Then A has order 2×3 and B has order 3×2. Thus the order of AB is 2×2.

$$\begin{bmatrix} 2 & -3 & 0 \\ 1 & 4 & -1 \end{bmatrix}_{2\times3} \begin{bmatrix} 1 & 0 \\ 4 & -2 \\ 3 & 5 \end{bmatrix}_{3\times2} = \begin{bmatrix} [2 \ -3 \ 0]\begin{bmatrix}1\\4\\3\end{bmatrix} & [2 \ -3 \ 0]\begin{bmatrix}0\\-2\\5\end{bmatrix} \\ [1 \ 4 \ -1]\begin{bmatrix}1\\4\\3\end{bmatrix} & [1 \ 4 \ -1]\begin{bmatrix}0\\-2\\5\end{bmatrix} \end{bmatrix}_{2\times2}$$

$$= \begin{bmatrix} 2(1) + (-3)(4) + 0(3) & 2(0) + (-3)(-2) + 0(5) \\ 1(1) + 4(4) + (-1)(3) & 1(0) + 4(-2) + (-1)(5) \end{bmatrix}_{2\times2}$$

$$= \begin{bmatrix} -10 & 6 \\ 14 & -13 \end{bmatrix}_{2\times2}$$

EXAMPLE 2 **Find the Product of Two Matrices**

Find the following products.

a. $\begin{bmatrix} 2 & 3 \\ -3 & 1 \\ 1 & -3 \end{bmatrix} \begin{bmatrix} 1 & 2 & -2 & 3 \\ -1 & 0 & 3 & -4 \end{bmatrix}$

b. $\begin{bmatrix} 1 & -1 & 3 \\ 2 & 2 & -1 \\ 0 & -2 & 3 \end{bmatrix} \begin{bmatrix} 4 & -2 & 0 \\ -1 & 3 & 1 \\ 2 & -3 & 1 \end{bmatrix}$

Solution

a.
$$\begin{bmatrix} 2 & 3 \\ -3 & 1 \\ 1 & -3 \end{bmatrix}\begin{bmatrix} 1 & 2 & -2 & 3 \\ -1 & 0 & 3 & -4 \end{bmatrix}$$

$$=\begin{bmatrix} 2(1)+3(-1) & 2(2)+3(0) & 2(-2)+3(3) & 2(3)+3(-4) \\ (-3)(1)+1(-1) & (-3)(2)+1(0) & (-3)(-2)+1(3) & (-3)3+1(-4) \\ 1(1)+(-3)(-1) & 1(2)+(-3)(0) & 1(-2)+(-3)(3) & 1(3)+(-3)(-4) \end{bmatrix}$$

$$=\begin{bmatrix} -1 & 4 & 5 & -6 \\ -4 & -6 & 9 & -13 \\ 4 & 2 & -11 & 15 \end{bmatrix}$$

b.
$$\begin{bmatrix} 1 & -1 & 3 \\ 2 & 2 & -1 \\ 0 & -2 & 3 \end{bmatrix}\begin{bmatrix} 4 & -2 & 0 \\ -1 & 3 & 1 \\ 2 & -3 & 1 \end{bmatrix}$$

$$=\begin{bmatrix} 4+1+6 & -2+(-3)+(-9) & 0+(-1)+3 \\ 8+(-2)+(-2) & -4+6+3 & 0+2+(-1) \\ 0+2+6 & 0+(-6)+(-9) & 0+(-2)+3 \end{bmatrix}$$

$$=\begin{bmatrix} 11 & -14 & 2 \\ 4 & 5 & 1 \\ 8 & -15 & 1 \end{bmatrix}$$

■ *Try Exercise* **16**, *page 562.*

Generally, matrix multiplication is not commutative. That is, given two matrices A and B, $AB \neq BA$. In some cases, as in Example 2a, if the matrices were reversed, the product would not be defined.

$$\begin{bmatrix} 1 & 2 & -2 & 3 \\ -1 & 0 & 3 & -4 \end{bmatrix}_{2\times4}\begin{bmatrix} 2 & 3 \\ -3 & 1 \\ 1 & -3 \end{bmatrix}_{3\times2}$$

Columns ≠ Rows

Even in those cases where multiplication is defined, the products AB and BA may not be equal. Finding the product of part (b) of Example 2 with the matrices reversed illustrates this point.

$$\begin{bmatrix} 4 & -2 & 0 \\ -1 & 3 & 1 \\ 2 & -3 & 1 \end{bmatrix}\begin{bmatrix} 1 & -1 & 3 \\ 2 & 2 & -1 \\ 0 & -2 & 3 \end{bmatrix}=\begin{bmatrix} 0 & -8 & 14 \\ 5 & 5 & -3 \\ -4 & -10 & 12 \end{bmatrix}\neq\begin{bmatrix} 11 & -14 & 2 \\ 4 & 5 & 1 \\ 8 & -15 & 1 \end{bmatrix}$$

Although matrix multiplication is not a commutative operation, the associative property of multiplication and the distributive property do hold for matrices.

Properties of Matrix Multiplication

Associative property Given matrices A, B, and C of orders $m \times n$, $n \times p$, and $p \times q$, respectively, then

$$A(BC) = (AB)C$$

Distributive property Given matrices A_1 and A_2 of order $m \times n$ and matrices B_1 and B_2 of order $n \times p$, then

$$A_1(B_1 + B_2) = A_1B_1 + A_1B_2 \quad \bullet \text{ Left distributive property}$$

and

$$(A_1 + A_2)B_1 = A_1B_1 + A_2B_1 \quad \bullet \text{ Right distributive property}$$

The square matrix that has a 1 for each element on the main diagonal and zeros elsewhere is called the *identity matrix*.

Identity Matrix

The **identity matrix** of order n, denoted I_n, is the $n \times n$ matrix

$$I_n = \begin{bmatrix} 1 & 0 & 0 & \cdots & 0 \\ 0 & 1 & 0 & \cdots & 0 \\ 0 & 0 & 1 & \cdots & 0 \\ \vdots & \vdots & \vdots & \cdots & \vdots \\ 0 & 0 & 0 & \cdots & 1 \end{bmatrix}_{n \times n}$$

The identity matrix has properties similar to the real number 1. For example, the product of a matrix A and I is A.

$$\begin{bmatrix} 2 & -3 & 0 \\ 4 & 7 & -5 \\ 9 & 8 & -6 \end{bmatrix} \begin{bmatrix} 1 & 0 & 0 \\ 0 & 1 & 0 \\ 0 & 0 & 1 \end{bmatrix} = \begin{bmatrix} 2 & -3 & 0 \\ 4 & 7 & -5 \\ 9 & 8 & -6 \end{bmatrix}$$

Multiplicative Identity Property for Matrices

If A is a square matrix of order n, and I_n is the identity matrix of order n, then $AI_n = I_nA = A$.

A system of equations can be expressed as the products of matrices. Consider the matrix equation

$$\begin{bmatrix} 2 & -3 & 4 \\ 3 & 0 & 1 \\ 1 & -2 & -5 \end{bmatrix}_{3 \times 3} \begin{bmatrix} x \\ y \\ z \end{bmatrix}_{3 \times 1} = \begin{bmatrix} 9 \\ 4 \\ -2 \end{bmatrix}_{3 \times 1} \tag{1}$$

Multiplying the two matrices on the left side of the equation, we have

$$\begin{bmatrix} 2x - 3y + 4z \\ 3x \quad + z \\ x - 2y - 5z \end{bmatrix}_{3\times1} = \begin{bmatrix} 9 \\ 4 \\ -2 \end{bmatrix}_{3\times1}$$

Because we multiplied a 3×3 matrix by a 3×1 matrix, the result is a 3×1 matrix. Now, using the definition of matrix equality, we have

$$\begin{cases} 2x - 3y + 4z = \quad 9 \\ 3x \quad\quad + z = \quad 4 \\ x - 2y - 5z = -2 \end{cases}$$

• The matrix Equation (1) is equivalent to this system of equations.

EXERCISE SET 11.2

In Exercises 1 to 8, find **a.** $A + B$,   **b.** $A - B$,   **c.** $2B$, **d.** $2A - 3B$.

1. $A = \begin{bmatrix} 2 & -1 \\ 3 & 3 \end{bmatrix}$, $B = \begin{bmatrix} -1 & 3 \\ 2 & 1 \end{bmatrix}$

2. $A = \begin{bmatrix} 0 & -2 \\ 2 & 3 \end{bmatrix}$, $B = \begin{bmatrix} 5 & -1 \\ 3 & 0 \end{bmatrix}$

3. $A = \begin{bmatrix} 0 & -1 & 3 \\ 1 & 0 & -2 \end{bmatrix}$, $B = \begin{bmatrix} -3 & 1 & 2 \\ 2 & 5 & -3 \end{bmatrix}$

4. $A = \begin{bmatrix} 2 & -2 & 4 \\ 0 & -3 & -4 \end{bmatrix}$, $B = \begin{bmatrix} 1 & -5 & 6 \\ 4 & -2 & -3 \end{bmatrix}$

5. $A = \begin{bmatrix} -3 & 4 \\ 2 & -3 \\ -1 & 0 \end{bmatrix}$, $B = \begin{bmatrix} 4 & 1 \\ 1 & -2 \\ 3 & -4 \end{bmatrix}$

6. $A = \begin{bmatrix} 2 & -2 \\ 3 & 4 \\ 1 & 0 \end{bmatrix}$, $B = \begin{bmatrix} -1 & 8 \\ 2 & -2 \\ -4 & 3 \end{bmatrix}$

7. $A = \begin{bmatrix} -2 & 3 & -1 \\ 0 & -1 & 2 \\ -4 & 3 & 3 \end{bmatrix}$, $B = \begin{bmatrix} 1 & -2 & 0 \\ 2 & 3 & -1 \\ 3 & -1 & 2 \end{bmatrix}$

8. $A = \begin{bmatrix} 0 & 2 & 0 \\ 1 & -3 & 3 \\ 5 & 4 & -2 \end{bmatrix}$, $B = \begin{bmatrix} -1 & 2 & 4 \\ 3 & 3 & -2 \\ -4 & 4 & 3 \end{bmatrix}$

In Exercises 9 to 16, find AB and BA if possible.

9. $A = \begin{bmatrix} 2 & -3 \\ 1 & 4 \end{bmatrix}$, $B = \begin{bmatrix} -2 & 4 \\ 2 & -3 \end{bmatrix}$

10. $A = \begin{bmatrix} 3 & -2 \\ 4 & 1 \end{bmatrix}$, $B = \begin{bmatrix} -1 & -1 \\ 0 & 4 \end{bmatrix}$

11. $A = \begin{bmatrix} 3 & -1 \\ 2 & 3 \end{bmatrix}$, $B = \begin{bmatrix} 4 & 1 \\ 2 & -3 \end{bmatrix}$

12. $A = \begin{bmatrix} -3 & 2 \\ 2 & -2 \end{bmatrix}$, $B = \begin{bmatrix} 0 & 2 \\ -2 & 4 \end{bmatrix}$

13. $A = \begin{bmatrix} 2 & -1 \\ 0 & 3 \\ 1 & -2 \end{bmatrix}$, $B = \begin{bmatrix} 1 & -2 & 3 \\ 2 & 0 & 1 \end{bmatrix}$

14. $A = \begin{bmatrix} -1 & 3 \\ 2 & 1 \\ -3 & -2 \end{bmatrix}$, $B = \begin{bmatrix} 0 & -1 & 2 \\ 1 & 2 & -4 \end{bmatrix}$

15. $A = \begin{bmatrix} 2 & -1 & 3 \\ 0 & 2 & -1 \\ 0 & 0 & 2 \end{bmatrix}$, $B = \begin{bmatrix} 2 & 0 & 0 \\ 1 & -1 & 0 \\ 2 & -1 & -2 \end{bmatrix}$

16. $A = \begin{bmatrix} -1 & 2 & 0 \\ 2 & -1 & 1 \\ -2 & 2 & -1 \end{bmatrix}$, $B = \begin{bmatrix} 2 & -1 & 0 \\ 1 & 5 & -1 \\ 0 & -1 & 3 \end{bmatrix}$

In Exercises 17 to 24, find AB if possible.

17. $A = \begin{bmatrix} 1 & -2 & 3 \end{bmatrix}$, $B = \begin{bmatrix} 1 & 0 \\ 2 & -1 \\ 1 & 2 \end{bmatrix}$

18. $A = \begin{bmatrix} -2 & 3 \\ 1 & -2 \\ 0 & 2 \end{bmatrix}$, $B = \begin{bmatrix} 3 \\ -2 \end{bmatrix}$

19. $A = \begin{bmatrix} 2 & -1 \\ 3 & 3 \end{bmatrix}$, $B = \begin{bmatrix} 1 & -2 \\ 3 & 1 \\ 0 & -2 \end{bmatrix}$

20. $A = \begin{bmatrix} 2 & 0 & -1 \\ 3 & 4 & -3 \end{bmatrix}$, $B = \begin{bmatrix} 3 & -1 & 0 \\ 2 & 4 & 5 \end{bmatrix}$

21. $A = \begin{bmatrix} 2 & 3 \\ -4 & -6 \end{bmatrix}$, $B = \begin{bmatrix} 3 & 6 \\ -2 & -4 \end{bmatrix}$

22. $A = \begin{bmatrix} 2 & -1 & 3 \\ -1 & 2 & 1 \end{bmatrix}$, $B = \begin{bmatrix} 1 & 3 & 2 \\ 2 & -1 & 0 \\ 3 & 1 & 2 \end{bmatrix}$

23. $A = \begin{bmatrix} 1 & 2 & -2 & 3 \\ 0 & -2 & 1 & -3 \end{bmatrix}$, $B = \begin{bmatrix} -2 & 0 \\ 4 & -2 \end{bmatrix}$

24. $A = \begin{bmatrix} 2 & -2 & 4 \\ 1 & 0 & -1 \\ 2 & 1 & 3 \end{bmatrix}$, $B = \begin{bmatrix} 2 & 1 & -3 & 0 \\ 0 & -2 & 1 & -2 \\ 1 & -1 & 0 & 2 \end{bmatrix}$

In Exercises 25 to 28, given the matrices

$$A = \begin{bmatrix} -1 & 3 \\ 2 & -1 \\ 3 & 1 \end{bmatrix} \quad \text{and} \quad B = \begin{bmatrix} 0 & -2 \\ 1 & 3 \\ 4 & -3 \end{bmatrix}$$

find the 3×2 matrix X that is a solution of the equation.

25. $3X + A = B$ **26.** $2A - 3X = 5B$

27. $2X - A = X + B$ **28.** $3X + 2B = X - 2A$

In Exercises 29 to 32, use the matrices

$$A = \begin{bmatrix} 2 & -3 \\ 1 & -1 \end{bmatrix} \quad \text{and} \quad B = \begin{bmatrix} 3 & -1 & 0 \\ 2 & -2 & -1 \\ 1 & 0 & 2 \end{bmatrix}$$

If A is a square matrix, then $A^n = A \cdot A \cdot A \cdots A$, where the matrix A is repeated n times.

29. Find A^2.   **30.** Find A^3.

31. Find B^2.   **32.** Find B^3.

In Exercises 33 to 38, find the system of equations that is equivalent to the given matrix equation.

33. $\begin{bmatrix} 3 & -8 \\ 4 & 3 \end{bmatrix} \begin{bmatrix} x \\ y \end{bmatrix} = \begin{bmatrix} 11 \\ 1 \end{bmatrix}$ **34.** $\begin{bmatrix} 2 & 7 \\ 3 & -4 \end{bmatrix} \begin{bmatrix} x \\ y \end{bmatrix} = \begin{bmatrix} 1 \\ 16 \end{bmatrix}$

35. $\begin{bmatrix} 1 & -3 & -2 \\ 3 & 1 & 0 \\ 2 & -4 & 5 \end{bmatrix} \begin{bmatrix} x \\ y \\ z \end{bmatrix} = \begin{bmatrix} 6 \\ 2 \\ 1 \end{bmatrix}$

36. $\begin{bmatrix} 2 & 0 & 5 \\ 3 & -5 & 1 \\ 4 & -7 & 6 \end{bmatrix} \begin{bmatrix} x \\ y \\ z \end{bmatrix} = \begin{bmatrix} 9 \\ 7 \\ 14 \end{bmatrix}$

37. $\begin{bmatrix} 2 & -1 & 0 & 2 \\ 4 & 1 & 2 & -3 \\ 6 & 0 & 1 & -2 \\ 5 & 2 & -1 & -4 \end{bmatrix} \begin{bmatrix} x_1 \\ x_2 \\ x_3 \\ x_4 \end{bmatrix} = \begin{bmatrix} 5 \\ 6 \\ 10 \\ 8 \end{bmatrix}$

38. $\begin{bmatrix} 5 & -1 & 2 & -3 \\ 4 & 0 & 2 & 0 \\ 2 & -2 & 5 & -4 \\ 3 & 1 & -3 & 4 \end{bmatrix} \begin{bmatrix} x_1 \\ x_2 \\ x_3 \\ x_4 \end{bmatrix} = \begin{bmatrix} -2 \\ 2 \\ -1 \\ 2 \end{bmatrix}$

39. Matrix A gives the stock on hand of four products in a warehouse at the beginning of the week, and matrix B gives the stock on hand for the same four items at the end of the week. Find and interpret $A - B$.

	Blue	Green	Red	
	530	650	815	Pens
$A =$	190	385	715	Pencils
	485	600	610	Ink
	150	210	305	Colored Lead

	Blue	Green	Red	
	480	500	675	Pens
$B =$	175	215	345	Pencils
	400	350	480	Ink
	70	95	280	Colored Lead

40. Matrix A gives the number of employees in the divisions of a company in the west coast branch, and matrix B gives the same information for the east coast branch. Find and interpret $A + B$.

	Engineering	Administration	Data Processing	
	315	200	415	Division I
$A =$	285	175	300	Division II
	275	195	250	Division III

	Engineering	Administration	Data Processing	
	200	175	350	Division I
$B =$	150	90	180	Division II
	105	50	175	Division III

41. The total unit sales matrix for three computer stores is given by

	Monitors	Printers	Computers	Drives	
	25	31	35	12	Store A
$S =$	20	12	30	15	Store B
	16	19	25	18	Store C

The unit pricing matrix in dollars for the three stores is given by

$$P = \begin{array}{c} \\ \\ \\ \\ \end{array} \begin{array}{ccc} \text{Store A} & \text{Store B} & \text{Store C} \\ \left[\begin{array}{ccc} 250 & 225 & 315 \\ 180 & 210 & 225 \\ 400 & 425 & 450 \\ 89 & 95 & 78 \end{array}\right] & & \end{array} \begin{array}{c} \text{Monitor} \\ \text{Printer} \\ \text{Computer} \\ \text{Drive} \end{array}$$

Find the gross income matrix.

42. The total unit sales matrix at three soccer games in a summer league for children is given by

$$S = \begin{array}{c} \\ \\ \\ \end{array} \begin{array}{cccc} \text{Soft Drinks} & \text{Hot Dogs} & \text{Candy} & \text{Popcorn} \\ \left[\begin{array}{cccc} 52 & 50 & 75 & 20 \\ 45 & 48 & 80 & 20 \\ 62 & 70 & 78 & 25 \end{array}\right] \end{array} \begin{array}{c} \text{Game 1} \\ \text{Game 2} \\ \text{Game 3} \end{array}$$

The unit pricing matrix in dollars for wholesale cost of each item and the retail price of each item is given by

$$P = \begin{array}{cc} \text{Wholesale} & \text{Retail} \\ \left[\begin{array}{cc} 0.25 & 0.50 \\ 0.30 & 0.75 \\ 0.15 & 0.45 \\ 0.10 & 0.50 \end{array}\right] \end{array} \begin{array}{c} \text{Soft Drinks} \\ \text{Hot Dogs} \\ \text{Candy} \\ \text{Popcorn} \end{array}$$

Use matrix multiplication to find the total cost and total revenue at each game.

Graphing Calculator Exercises*

Some graphing calculators and some computer programs can perform matrix operations. Try one of these programs for Exercises 43 to 48.

$$\text{Let } A = \begin{bmatrix} 2 & -1 & 3 & 5 & -1 \\ 2 & 0 & 2 & -1 & 1 \\ -1 & -3 & 2 & 3 & 3 \\ 5 & -4 & 1 & 0 & 3 \\ 0 & 2 & -1 & 4 & 3 \end{bmatrix}$$

$$\text{and } B = \begin{bmatrix} 0 & -2 & 1 & 7 & 2 \\ -3 & 0 & 2 & 3 & 1 \\ -2 & 1 & 1 & 4 & 5 \\ 6 & 4 & -4 & 2 & -3 \\ 3 & -2 & -5 & 1 & 3 \end{bmatrix}.$$

Perform the indicated operations.

43. AB 44. BA

45. A^3 46. B^3

47. $A^2 + B^2$ 48. $AB - BA$

Supplemental Exercises

The elements of a matrix can be complex numbers. In Exercises 49 to 58, let

$$A = \begin{bmatrix} 2 + 3i & 1 - 2i \\ 1 + i & 2 - i \end{bmatrix} \text{ and } B = \begin{bmatrix} 1 - i & 2 + 3i \\ 3 + 2i & 4 - i \end{bmatrix}$$

Perform the indicated operations.

49. $3A$ 50. $-2B$ 51. $2iB$ 52. $3iA$

53. $A + B$ 54. $A - B$ 55. AB 56. BA

57. A^2 58. B^2

Matrices with complex number elements play a role in the theory of the atom. The following three matrices, called Pauli spin matrices, were used by Linus Pauli in his early study of the electron. Use these matrices in Exercises 59 to 61.

$$\sigma_1 = \begin{bmatrix} 0 & 1 \\ 1 & 0 \end{bmatrix}, \quad \sigma_2 = \begin{bmatrix} 0 & -i \\ i & 0 \end{bmatrix}, \quad \sigma_3 = \begin{bmatrix} 1 & 0 \\ 0 & -1 \end{bmatrix}$$

59. Show that $(\sigma_i)^2 = I_2$ for $i = 1, 2,$ and 3.

60. Show that $\sigma_1 \cdot \sigma_2 = i\sigma_3$.

61. Show that $\sigma_1 \cdot \sigma_2 + \sigma_2 \cdot \sigma_1 = O$.

62. Given two real numbers a and b and a matrix A of order 2×2, prove that $(a + b)A = aA + bA$.

63. Given two real numbers a and b and a matrix A of order 2×2, prove that $a(bA) = (ab)A$.

11.3 The Inverse of a Matrix

Recall that the multiplicative inverse of a nonzero real number c is $1/c$, the number whose product with c is 1. For example, the multiplicative inverse of $\frac{2}{3}$ is $\frac{3}{2}$ because $\frac{2}{3} \cdot \frac{3}{2} = 1$.

For some square matrices we can define a multiplicative inverse.

*Additional graphing calculator exercises appear in the Graphing Workbook as described in the front of this textbook.

Multiplicative Inverse of a Matrix

If A is a square matrix of order n, then the inverse of matrix A, denoted by A^{-1}, has the property that

$$A \cdot A^{-1} = A^{-1} \cdot A = I_n$$

where I_n is the identity matrix of order n.

Remark As we will see shortly, not all square matrices have a multiplicative inverse. Are there any real numbers that do not have a multiplicative inverse?[3]

A procedure for finding the inverse (we will simply say *inverse* for *multiplicative inverse*) uses elementary row operations. The procedure will be illustrated by finding the inverse of a 2×2 matrix.

Let $A = \begin{bmatrix} 2 & 7 \\ 1 & 4 \end{bmatrix}$. To the matrix A we will merge the identity matrix I_2 to the right of A and denote this new matrix by $[A:I_2]$.

$$[A:I_2] = \begin{bmatrix} 2 & 7 & 1 & 0 \\ 1 & 4 & 0 & 1 \end{bmatrix}$$

$$A \underline{\hspace{2cm}}\uparrow \qquad \uparrow\underline{\hspace{2cm}} I_2$$

Now we use elementary row operations in a manner similar to that of the Gauss–Jordan method. The goal is to produce

$$[I_2:A^{-1}] = \begin{bmatrix} 1 & 0 & b_{11} & b_{12} \\ 0 & 1 & b_{21} & b_{22} \end{bmatrix}$$

$$I_2 \underline{\hspace{2cm}}\uparrow \qquad \uparrow\underline{\hspace{2cm}} A^{-1}$$

In this form, the inverse matrix is the matrix that is to the right of the identity matrix. That is,

$$A^{-1} = \begin{bmatrix} b_{11} & b_{12} \\ b_{21} & b_{22} \end{bmatrix}$$

To find A^{-1}, we first use a series of elementary row operations that will result in a 1 in the first row and the first column.

$$\begin{bmatrix} 2 & 7 & 1 & 0 \\ 1 & 4 & 0 & 1 \end{bmatrix} \xrightarrow{\frac{1}{2}R_1} \begin{bmatrix} 1 & \frac{7}{2} & \frac{1}{2} & 0 \\ 1 & 4 & 0 & 1 \end{bmatrix}$$

$$\xrightarrow{-1R_1 + R_2} \begin{bmatrix} 1 & \frac{7}{2} & \frac{1}{2} & 0 \\ 0 & \frac{1}{2} & -\frac{1}{2} & 1 \end{bmatrix} \xrightarrow{2R_2} \begin{bmatrix} 1 & \frac{7}{2} & \frac{1}{2} & 0 \\ 0 & 1 & -1 & 2 \end{bmatrix}$$

$$\xrightarrow{-\frac{7}{2}R_2 + R_1} \begin{bmatrix} 1 & 0 & 4 & -7 \\ 0 & 1 & -1 & 2 \end{bmatrix}$$

---

[3] The real number zero does not have a multiplicative inverse.

The inverse matrix is the matrix to the right of the identity matrix. Therefore,

$$A^{-1} = \begin{bmatrix} 4 & -7 \\ -1 & 2 \end{bmatrix}$$

Each elementary row operation is chosen to advance the process of transforming the original matrix into the identity matrix.

EXAMPLE 1   **Find the Inverse of a 3 × 3 Matrix**

Find the inverse of the matrix $A = \begin{bmatrix} 1 & -1 & 2 \\ 2 & 0 & 6 \\ 3 & -5 & 7 \end{bmatrix}$.

Solution

$$\begin{bmatrix} 1 & -1 & 2 & 1 & 0 & 0 \\ 2 & 0 & 6 & 0 & 1 & 0 \\ 3 & -5 & 7 & 0 & 0 & 1 \end{bmatrix}$$

• Merge the given matrix with the identity matrix I_3.

$\begin{matrix} -2R_1 + R_2 \\ -3R_1 + R_3 \\ \xrightarrow{\hspace{1cm}} \end{matrix}$ $\begin{bmatrix} 1 & -1 & 2 & 1 & 0 & 0 \\ 0 & 2 & 2 & -2 & 1 & 0 \\ 0 & -2 & 1 & -3 & 0 & 1 \end{bmatrix}$

• Since a_{11} is already 1, we next produce zeros in a_{21} and a_{31}.

$\begin{matrix} (1/2)R_2 \\ \xrightarrow{\hspace{1cm}} \end{matrix}$ $\begin{bmatrix} 1 & -1 & 2 & 1 & 0 & 0 \\ 0 & 1 & 1 & -1 & \frac{1}{2} & 0 \\ 0 & -2 & 1 & -3 & 0 & 1 \end{bmatrix}$

• Produce a 1 in a_{22}.

$\begin{matrix} 2R_2 + R_3 \\ \xrightarrow{\hspace{1cm}} \end{matrix}$ $\begin{bmatrix} 1 & -1 & 2 & 1 & 0 & 0 \\ 0 & 1 & 1 & -1 & \frac{1}{2} & 0 \\ 0 & 0 & 3 & -5 & 1 & 1 \end{bmatrix}$

• Produce a 0 in a_{32}.

$\begin{matrix} (1/3)R_3 \\ \xrightarrow{\hspace{1cm}} \end{matrix}$ $\begin{bmatrix} 1 & -1 & 2 & 1 & 0 & 0 \\ 0 & 1 & 1 & -1 & \frac{1}{2} & 0 \\ 0 & 0 & 1 & -\frac{5}{3} & \frac{1}{3} & \frac{1}{3} \end{bmatrix}$

• Produce a 1 in a_{33}.

$\begin{matrix} -1R_3 + R_2 \\ -2R_3 + R_1 \\ \xrightarrow{\hspace{1cm}} \end{matrix}$ $\begin{bmatrix} 1 & -1 & 0 & \frac{13}{3} & -\frac{2}{3} & -\frac{2}{3} \\ 0 & 1 & 0 & \frac{2}{3} & \frac{1}{6} & -\frac{1}{3} \\ 0 & 0 & 1 & -\frac{5}{3} & \frac{1}{3} & \frac{1}{3} \end{bmatrix}$

• Now work upward. Produce a 0 in a_{23} and a_{13}.

$\begin{matrix} R_2 + R_1 \\ \xrightarrow{\hspace{1cm}} \end{matrix}$ $\begin{bmatrix} 1 & 0 & 0 & 5 & -\frac{1}{2} & -1 \\ 0 & 1 & 0 & \frac{2}{3} & \frac{1}{6} & -\frac{1}{3} \\ 0 & 0 & 1 & -\frac{5}{3} & \frac{1}{3} & \frac{1}{3} \end{bmatrix}$

• Produce a 0 in a_{12}.

The inverse matrix is $A^{-1} = \begin{bmatrix} 5 & -\frac{1}{2} & -1 \\ \frac{2}{3} & \frac{1}{6} & -\frac{1}{3} \\ -\frac{5}{3} & \frac{1}{3} & \frac{1}{3} \end{bmatrix}$.

You should verify that this matrix satisfies the condition of an inverse matrix. That is, show that $A^{-1} \cdot A = A \cdot A^{-1} = I_3$.

■ *Try Exercise* **6,** *page 571.*

A **singular matrix** is a matrix that does not have a multiplicative inverse. A matrix that has a multiplicative inverse is a **nonsingular matrix.** As you apply the procedure above to a singular matrix, there will come a point where there are zeros in a row of the *original* matrix. When that condition exists, the matrix does not have an inverse.

EXAMPLE 2   **Identify a Singular Matrix**

Show that the matrix $\begin{bmatrix} 1 & -1 & -1 \\ 2 & -3 & 0 \\ 1 & -2 & 1 \end{bmatrix}$ is a singular matrix.

Solution

$$\begin{bmatrix} 1 & -1 & -1 & 1 & 0 & 0 \\ 2 & -3 & 0 & 0 & 1 & 0 \\ 1 & -2 & 1 & 0 & 0 & 1 \end{bmatrix} \xrightarrow[\substack{-2R_1 + R_2 \\ -1R_1 + R_3}]{} \begin{bmatrix} 1 & -1 & -1 & 1 & 0 & 0 \\ 0 & -1 & 2 & -2 & 1 & 0 \\ 0 & -1 & 2 & -1 & 0 & 1 \end{bmatrix}$$

$$\xrightarrow{-1 \cdot R_2} \begin{bmatrix} 1 & -1 & -1 & 1 & 0 & 0 \\ 0 & 1 & -2 & 2 & -1 & 0 \\ 0 & -1 & 2 & -1 & 0 & 1 \end{bmatrix} \xrightarrow{R_2 + R_3} \begin{bmatrix} 1 & -1 & -1 & -1 & 0 & 0 \\ 0 & 1 & -2 & 2 & -1 & 0 \\ 0 & 0 & 0 & 1 & -1 & 1 \end{bmatrix}$$

There are zeros in a row of the original matrix. The original matrix does not have an inverse.

■ *Try Exercise 10, page 571.*

Systems of linear equations can be solved by finding the inverse of the coefficient matrix. Consider the system of equations

$$\begin{cases} 3x_1 + 4x_2 = -1 \\ 3x_1 + 5x_2 = 1 \end{cases} \qquad (1)$$

Using matrix multiplication and the concept of equality of matrices, we can write this system as a matrix equation.

$$\begin{bmatrix} 3 & 4 \\ 3 & 5 \end{bmatrix} \begin{bmatrix} x_1 \\ x_2 \end{bmatrix} = \begin{bmatrix} -1 \\ 1 \end{bmatrix} \qquad (2)$$

If we let

$$A = \begin{bmatrix} 3 & 4 \\ 3 & 5 \end{bmatrix}, \quad X = \begin{bmatrix} x_1 \\ x_2 \end{bmatrix}, \quad B = \begin{bmatrix} -1 \\ 1 \end{bmatrix}$$

then Equation (2) can be written as $AX = B$.

The inverse of the coefficient matrix A is $A^{-1} = \begin{bmatrix} \frac{5}{3} & -\frac{4}{3} \\ -1 & 1 \end{bmatrix}$.

To solve the system of equations, multiply each side of the equation $AX = B$ by the inverse A^{-1}.

$$\begin{bmatrix} \frac{5}{3} & -\frac{4}{3} \\ -1 & 1 \end{bmatrix} \begin{bmatrix} 3 & 4 \\ 3 & 5 \end{bmatrix} \begin{bmatrix} x_1 \\ x_2 \end{bmatrix} = \begin{bmatrix} \frac{5}{3} & -\frac{4}{3} \\ -1 & 1 \end{bmatrix} \begin{bmatrix} -1 \\ 1 \end{bmatrix}$$

$$\begin{bmatrix} x_1 \\ x_2 \end{bmatrix} = \begin{bmatrix} -3 \\ 2 \end{bmatrix}$$

Thus $x_1 = -3$ and $x_2 = 2$. The solution to the system (1) is $(-3, 2)$.

EXAMPLE 3   **Solve a System of Equations by Using the Inverse of the Coefficient Matrix**

Find the solution of the system of equations by using the inverse of the coefficient matrix.

$$\begin{cases} x_1 + + 7x_3 = 20 \\ 2x_1 + x_2 - x_3 = -3 \\ 7x_1 + 3x_2 + x_3 = 2 \end{cases} \quad (1)$$

Solution   Write the system as a matrix equation.

$$\begin{bmatrix} 1 & 0 & 7 \\ 2 & 1 & -1 \\ 7 & 3 & 1 \end{bmatrix} \begin{bmatrix} x_1 \\ x_2 \\ x_3 \end{bmatrix} = \begin{bmatrix} 20 \\ -3 \\ 2 \end{bmatrix} \quad (2)$$

The inverse of the coefficient matrix is $\begin{bmatrix} -\frac{4}{3} & -7 & \frac{7}{3} \\ 3 & 16 & -5 \\ \frac{1}{3} & 1 & -\frac{1}{3} \end{bmatrix}$.

Multiply each side of the matrix equation (2) by the inverse.

$$\begin{bmatrix} -\frac{4}{3} & -7 & \frac{7}{3} \\ 3 & 16 & -5 \\ \frac{1}{3} & 1 & -\frac{1}{3} \end{bmatrix} \begin{bmatrix} 1 & 0 & 7 \\ 2 & 1 & -1 \\ 7 & 3 & 1 \end{bmatrix} \begin{bmatrix} x_1 \\ x_2 \\ x_3 \end{bmatrix} = \begin{bmatrix} -\frac{4}{3} & -7 & \frac{7}{3} \\ 3 & 16 & -5 \\ \frac{1}{3} & 1 & -\frac{1}{3} \end{bmatrix} \begin{bmatrix} 20 \\ -3 \\ 2 \end{bmatrix}$$

$$\begin{bmatrix} x_1 \\ x_2 \\ x_3 \end{bmatrix} = \begin{bmatrix} -1 \\ 2 \\ 3 \end{bmatrix}$$

Thus $x_1 = -1$, $x_2 = 2$, and $x_3 = 3$. The solution to (1) is $(-1, 2, 3)$.

■ *Try Exercise* **20**, *page 571.*

The advantage of using the inverse matrix to solve a system of equations is not apparent unless it is necessary to solve repeatedly a system of equations with the same coefficient matrix but different constant matrices. *Input–output analysis* is one such application of this method.

Input–Output Analysis

In an economy, some of the output of an industry is used by the industry to produce its product. For example, an electric company uses water and electricity to produce electricity, and a water company uses water and electricity to produce drinking water. **Input–output analysis** attempts to determine the necessary output of industries to satisfy each other's demands plus the demands of consumers. Wassily Leontieff, a Harvard economist, was awarded the Nobel prize for his work in this field.

An **input–output matrix** is used to express the interdependence among industries in an economy. Each column of this matrix gives the dollar values of the inputs an industry needs to produce $1 worth of output.

To illustrate the concepts, we will assume an economy with only three industries: agriculture, transportation and oil. Suppose that to produce $1 of agricultural products requires $.05 of agriculture, $.02 of transportation, and $.05 of oil. To produce $1 of transportation requires $.10 of agriculture, $.08 of transportation, and $.10 of oil. To produce $1 of oil requires $.10 of agriculture, $.15 of transportation, and $.13 of oil. The input–output matrix A is

Input Requirements of

		Agriculture	Transportation	Oil
	Agriculture	0.05	0.10	0.10
from	Transportation	0.02	0.08	0.15
	Oil	0.05	0.10	0.13

Consumers (other than the industries themselves) want to purchase some of the output from these industries. The amount of output that the consumer will want is called the **final demand** on the economy. This is represented by a column matrix.

Suppose in our example that the final demand is $3 billion of agriculture, $1 billion of transportation, and $2 billion of oil. The final demand matrix is

$$\begin{bmatrix} 3 \\ 1 \\ 2 \end{bmatrix} = D$$

We represent the total output of each industry (in billions of dollars) as follows:

$x =$ total output of agriculture

$y =$ total output of transportation

$z =$ total output of oil

The object of input–output analysis is to determine the values of x, y, and z that will satisfy the amount the consumer demands. To find these values, consider agriculture. The amount of agriculture left for the consumer (demand d) is

$d = x -$ (amount of agriculture used by industries)   (1)

To find the amount of agriculture used by the three industries in our economy, refer to the input–output matrix. Production of x billion dollars of agriculture takes $.05x$ of agriculture, production of y billion dollars of transportation takes $.10y$ of agriculture, and production of z billion dollars of oil takes $.10z$ of agriculture. Thus,

$$\begin{matrix} \text{Amount of agriculture} \\ \text{used by industries} \end{matrix} = .05x + .10y + .10z \qquad (2)$$

Combining Equations (1) and (2), we have

$d = x - (.05x + .10y + .10z)$

$3 = .95x - .10y - .10z$ • d is $3 billion for agriculture.

We could continue this way for each of the other industries. The result would be a system of equations. Instead, however, we will use a matrix approach.

If X = total output of the three industries of the economy, then

$$X = \begin{bmatrix} x \\ y \\ z \end{bmatrix}$$

The product of A, the input–output matrix, and X is

$$AX = \begin{bmatrix} .05 & .10 & .10 \\ .02 & .08 & .15 \\ .05 & .10 & .13 \end{bmatrix} \begin{bmatrix} x \\ y \\ z \end{bmatrix}$$

This matrix represents the dollar amount of products used in the production for all three industries. Thus the amount available for consumer demand is $X - AX$. As a matrix equation, we can write

$$X - AX = D$$

Solving this equation for X, we determine the output necessary to meet the needs of our industries and the consumer.

$IX - AX = D$ • I is the identity matrix. Thus $IX = X$.

$(I - A)X = D$ • Right distributive property

$X = (I - A)^{-1}D$ • Assuming the inverse of $(I - A)$ exists

The last equation states that the solution to an input–output problem can be found by multiplying the demand matrix D by the inverse of $(I - A)$. In our example, we have

$$I - A = \begin{bmatrix} 1 & 0 & 0 \\ 0 & 1 & 0 \\ 0 & 0 & 1 \end{bmatrix} - \begin{bmatrix} .05 & .10 & .10 \\ .02 & .08 & .15 \\ .05 & .10 & .13 \end{bmatrix} = \begin{bmatrix} .95 & -.10 & -.10 \\ -.02 & .92 & -.15 \\ -.05 & -.10 & .87 \end{bmatrix}$$

$$(I - A)^{-1} \approx \begin{bmatrix} 1.063 & .131 & .145 \\ .034 & 1.112 & .200 \\ .065 & .135 & 1.180 \end{bmatrix}$$

The consumer demand is

$$X = (I - A)^{-1}D$$

$$X \approx \begin{bmatrix} 1.063 & .131 & .145 \\ .034 & 1.112 & .200 \\ .065 & .135 & 1.180 \end{bmatrix} \begin{bmatrix} 3 \\ 1 \\ 2 \end{bmatrix} = \begin{bmatrix} 3.61 \\ 1.61 \\ 2.69 \end{bmatrix}$$

This matrix indicates that $3.61 billion of agriculture, $1.61 billion of transportation, and $2.69 billion of oil must be produced by the industries to satisfy consumers' demands and the industries' internal requirements.

If we change the final demand matrix,

$$D = \begin{bmatrix} 2 \\ 2 \\ 3 \end{bmatrix}$$

then the total output of the economy can be found as

$$X \approx \begin{bmatrix} 1.063 & .131 & .155 \\ .034 & 1.112 & .200 \\ .065 & .135 & 1.180 \end{bmatrix} \begin{bmatrix} 2 \\ 2 \\ 3 \end{bmatrix} = \begin{bmatrix} 2.82 \\ 2.89 \\ 3.94 \end{bmatrix}$$

Thus agriculture must produce \$2.82 billion, transportation must produce \$2.89 billion, and oil must produce \$3.94 billion to satisfy the given consumer demand and the industries' internal requirements.

EXERCISE SET 11.3

In Exercises 1 to 14, find the inverse of the given matrix.

1. $\begin{bmatrix} 1 & -3 \\ -2 & 5 \end{bmatrix}$

2. $\begin{bmatrix} 1 & 2 \\ -2 & -3 \end{bmatrix}$

3. $\begin{bmatrix} 1 & 4 \\ 2 & 10 \end{bmatrix}$

4. $\begin{bmatrix} -2 & 3 \\ -6 & -8 \end{bmatrix}$

5. $\begin{bmatrix} 1 & 2 & -1 \\ 2 & 5 & 1 \\ 3 & 6 & -2 \end{bmatrix}$

6. $\begin{bmatrix} 1 & 3 & -2 \\ -1 & -5 & 6 \\ 2 & 6 & -3 \end{bmatrix}$

7. $\begin{bmatrix} 1 & 2 & -1 \\ 2 & 6 & 1 \\ 3 & 6 & -4 \end{bmatrix}$

8. $\begin{bmatrix} 2 & 1 & -1 \\ 6 & 4 & -1 \\ 4 & 2 & -3 \end{bmatrix}$

9. $\begin{bmatrix} 2 & 4 & -4 \\ 1 & 3 & -4 \\ 2 & 4 & -3 \end{bmatrix}$

10. $\begin{bmatrix} 1 & -2 & 2 \\ 2 & -3 & 1 \\ 3 & -6 & 6 \end{bmatrix}$

11. $\begin{bmatrix} 1 & -1 & 2 & 1 \\ 2 & -1 & 5 & 1 \\ 3 & -3 & 7 & 5 \\ -2 & 3 & -4 & -1 \end{bmatrix}$

12. $\begin{bmatrix} 1 & 1 & -1 & 2 \\ 3 & 2 & -1 & 5 \\ 2 & 2 & -1 & 5 \\ 4 & 4 & -4 & 7 \end{bmatrix}$

13. $\begin{bmatrix} 1 & -1 & 1 & 3 \\ 2 & -1 & 4 & 8 \\ 1 & 1 & 6 & 10 \\ -1 & 5 & 5 & 4 \end{bmatrix}$

14. $\begin{bmatrix} 1 & -1 & 1 & 2 \\ 2 & -1 & 6 & 6 \\ 3 & -1 & 12 & 12 \\ -2 & -1 & -14 & -10 \end{bmatrix}$

In Exercises 15 to 24, solve the system of equations by using inverse matrix methods.

15. $\begin{cases} x + 4y = 6 \\ 2x + 7y = 11 \end{cases}$

16. $\begin{cases} 2x + 3y = 5 \\ x + 2y = 4 \end{cases}$

17. $\begin{cases} x - 2y = 8 \\ 3x + 2y = -1 \end{cases}$

18. $\begin{cases} 3x - 5y = -18 \\ 2x - 3y = -11 \end{cases}$

19. $\begin{cases} x + y + 2z = 4 \\ 2x + 3y + 3z = 5 \\ 3x + 3y + 7z = 14 \end{cases}$

20. $\begin{cases} x + 2y - z = 5 \\ 2x + 3y - z = 8 \\ 3x + 6y - 2z = 14 \end{cases}$

21. $\begin{cases} x + 2y + 2z = 5 \\ -2x - 5y - 2z = 8 \\ 2x + 4y + 7z = 19 \end{cases}$

22. $\begin{cases} x - y + 3z = 5 \\ 3x - y + 10z = 16 \\ 2x - 2y + 5z = 9 \end{cases}$

23. $\begin{cases} w + 2x + z = 6 \\ 2w + 5x + y + 2z = 10 \\ 2w + 4x + y + z = 8 \\ 3w + 6x + 4z = 16 \end{cases}$

24. $\begin{cases} w - x + 2y = 5 \\ 2w - x + 6y + 2z = 16 \\ 3w - 2x + 9y + 4z = 28 \\ w - 2x - z = 2 \end{cases}$

25. A vacation resort offers a helicopter tour of an island. The price for an adult ticket is \$20; the price for a children's ticket is \$15. The records of the tour operator show that 100 people took the tour on Saturday and 120 people took the tour on Sunday. The total receipts for Saturday were \$1900, and on Sunday the receipts were \$2275. Find the number of adults and children who took the tour on Saturday and on Sunday.

26. A company sells a standard and a deluxe model tape recorder. Each standard tape recorder costs \$45 to manufacture, and each deluxe model costs \$60 to manufacture. The January manufacturing budget for 90 of these recorders was $4650; the February budget for 100 recorders was $5250. Find the number of each type of recorder manufactured in January and in February.

27. The following table shows the active chemical content of three different soil additives.

Grams per 100 Grams

Additive	Ammonium Nitrate	Phosphorous	Iron
1	30	10	10
2	40	15	10
3	50	5	5

A soil chemist wants to prepare two chemical samples. The first sample contains 380 grams of ammonium nitrate, 95 grams of phosphorous, and 85 grams of iron. The second sample requires 380 grams of ammonium nitrate, 110 grams of phosphorous, and 90 grams of iron. How many grams of each additive are required for sample 1, and how many grams of each additive are required for sample 2?

28. The following table shows the carbohydrate, fat, and protein content of three food types.

Grams per 100 Grams

Food Type	Carbohydrate	Fat	Protein
I	13	10	13
II	4	4	3
III	1	0	10

A nutritionist must prepare two diets from these three food groups. The first diet must contain 23 grams of carbohydrate, 18 grams of fat, and 39 grams of protein. The second diet must contain 35 grams of carbohydrate, 28 grams of fat, and 42 grams of protein. How many grams of each food type are required for the first diet, and how many grams of each food type are required for the second diet?

Graphing Calculator Exercises*

In exercises 29 to 32, use a calculator or computer to find the inverse of each matrix.

29. $\begin{bmatrix} 2 & -2 & 3 & 1 \\ 5 & 2 & -2 & 3 \\ 6 & -1 & 2 & 3 \\ 2 & 3 & -1 & 5 \end{bmatrix}$

30. $\begin{bmatrix} 3 & -1 & 0 & 1 \\ 2 & -2 & 3 & 0 \\ -1 & -3 & 5 & 3 \\ 5 & 3 & -2 & 1 \end{bmatrix}$

*Additional graphing calculator exercises appear in the Graphing Workbook as described in the front of this textbook.

31. $\begin{bmatrix} -\frac{2}{7} & 4 & -\frac{1}{6} \\ -2 & \sqrt{2} & -3 \\ \sqrt{3} & 3 & -\sqrt{5} \end{bmatrix}$

32. $\begin{bmatrix} 6 & \pi & -\frac{4}{7} \\ -5 & \sqrt{7} & 2 \\ \frac{5}{6} & -\sqrt{3} & \sqrt{10} \end{bmatrix}$

33. A simplified economy has three industries: manufacturing, transportation, and service. The input–output matrix for this economy is

$$\begin{bmatrix} 0.20 & 0.15 & 0.10 \\ 0.10 & 0.30 & 0.25 \\ 0.20 & 0.10 & 0.10 \end{bmatrix}$$

Find the gross output needed to satisfy the consumer demand of $120 million worth of manufacturing, $60 million of transportation, and $55 million worth of service.

34. A four-sector economy consists of manufacturing, agriculture, service, and transportation. The input–output matrix for this economy is

$$\begin{bmatrix} 0.10 & 0.05 & 0.20 & 0.15 \\ 0.20 & 0.10 & 0.30 & 0.10 \\ 0.05 & 0.30 & 0.20 & 0.40 \\ 0.10 & 0.20 & 0.15 & 0.20 \end{bmatrix}$$

Find the gross output needed to satisfy a consumer demand of $80 million worth of manufacturing, $100 million worth of agriculture, $50 million worth of service, and $80 million worth of transportation.

35. A conglomerate is composed of three industries: coal, iron, and steel. To produce $1 of coal requires $.05 of coal, $.02 of iron, and $.10 of steel. To produce $1 of iron requires $.20 of coal, $.03 of iron, and $.12 of steel. To produce $1 of steel requires $.15 of coal, $.25 of iron, and $.05 of steel. How much should each industry produce to allow for a consumer demand of $30 million worth of coal, $5 million worth of iron, and $25 million worth of steel?

36. A conglomerate has three divisions: plastics, semiconductors, and computers. For each $1 of output, the plastics division needs $.01 worth of plastics, $.03 worth of semiconductors, and $.10 worth of computers. Each $1 of output from the semiconductor division requires $.08 worth from plastics, $.05 worth from semiconductors, and $.15 worth from computers. For each $1 of output, the computer division needs $.20 worth from plastics, $.20 worth from semiconductors, and $.10 worth from computers. The conglomerate estimates consumer demand of $100 million from the plastics division, $75 million from the semiconductor division, and $150 million from the computer division. At what level should each division produce to satisfy this demand?

Supplemental Exercises

37. Let

$$A = \begin{bmatrix} 2 & -3 \\ -6 & 9 \end{bmatrix} \quad \text{and} \quad B = \begin{bmatrix} -3 & 15 \\ -2 & 10 \end{bmatrix}$$

Show that $AB = O$, the 2×2 zero matrix. This illustrates that for matrices, if $AB = O$, it is not necessarily so that $A = O$ or $B = O$.

38. Show that if a matrix A has an inverse and $AB = O$, then $B = O$.

39. Let

$$A = \begin{bmatrix} 2 & -1 \\ -4 & 2 \end{bmatrix}, B = \begin{bmatrix} 3 & 4 \\ 1 & 5 \end{bmatrix}, \text{ and } C = \begin{bmatrix} 4 & 7 \\ 3 & 11 \end{bmatrix}$$

Show that $AB = AC$ but that $B \ne C$. This illustrates that the cancellation rule of real numbers may not apply to matrices.

40. (Continuation of Exercise 39.) Show that if A is a matrix that has an inverse and $AB = AC$, then $B = C$.

41. Show that if

$$A = \begin{bmatrix} a & b \\ c & d \end{bmatrix} \quad \text{and} \quad ad - bc \ne 0$$

then $\quad A^{-1} = \dfrac{1}{ad - bc} \begin{bmatrix} d & -b \\ -c & a \end{bmatrix}$

42. Use the result of Exercise 41 to show that a square matrix of order 2 has an inverse if and only if $ad - bc \ne 0$.

43. Use the result of Exercise 41 to find the inverse of each matrix.

a. $\begin{bmatrix} 2 & -3 \\ 4 & -5 \end{bmatrix}$, **b.** $\begin{bmatrix} 5 & 6 \\ 3 & 4 \end{bmatrix}$, **c.** $\begin{bmatrix} 0 & -1 \\ 4 & 4 \end{bmatrix}$

44. Let

$$A = \begin{bmatrix} 3 & -2 \\ 1 & 1 \end{bmatrix} \quad \text{and} \quad B = \begin{bmatrix} 2 & -1 \\ 2 & 3 \end{bmatrix}$$

Use Exercise 41 to show that

$$A^{-1} = \frac{1}{5} \begin{bmatrix} 1 & 2 \\ -1 & 3 \end{bmatrix} \quad \text{and} \quad B^{-1} = \frac{1}{8} \begin{bmatrix} 3 & 1 \\ -2 & 2 \end{bmatrix}$$

Now show that $(AB)^{-1} = B^{-1} \cdot A^{-1}$.

45. Generalize the last result in Exercise 44. That is, show that if A and B are square matrices of order n and each have an inverse matrix, then $(AB)^{-1} = B^{-1} \cdot A^{-1}$. (*Hint:* Begin with the equation $(AB)(AB)^{-1} = I$, where I is the identity matrix. Now multiply each side of the equation by A^{-1} and then by B^{-1}.)

11.4 Determinants

Associated with each square matrix A is a number called the *determinant* of A. We will denote the determinant of the matrix A by $\det(A)$ or by $|A|$. For the remainder of this chapter, we assume that all matrices are square matrices.

The Determinant of a 2 × 2 Matrix

The **determinant** of the matrix $A = [a_{ij}]$ of order 2 is

$$|A| = \begin{vmatrix} a_{11} & a_{12} \\ a_{21} & a_{22} \end{vmatrix} = a_{11}a_{22} - a_{21}a_{12}$$

Caution   Be careful not to confuse the notation for a matrix and that for a determinant. The symbol [ ] (brackets) is used for a matrix; the symbol | | (vertical bars) is used for the determinant of a matrix.

An easy way to remember the formula for the determinant of a 2 × 2 matrix is to recognize that the determinant is the difference between the products of the diagonal elements. That is,

$$\begin{vmatrix} a_{11} & a_{12} \\ a_{21} & a_{22} \end{vmatrix} = a_{11}a_{22} - a_{21}a_{12}$$

EXAMPLE 1   **Find the Value of a Determinant**

Find the value of the determinant of the matrix $A = \begin{bmatrix} 5 & 3 \\ 2 & -3 \end{bmatrix}$.

Solution

$$|A| = \begin{vmatrix} 5 & 3 \\ 2 & -3 \end{vmatrix} = 5(-3) - 2(3) = -15 - 6 = -21$$

■ *Try Exercise **2**, page 580.*

To define the determinant of a matrix of order greater than 2, we first need two other definitions.

The Minor of a Matrix

The **minor** M_{ij} of the element a_{ij} of a square matrix A of order $n \geq 3$ is the determinant of the matrix of order $n - 1$ obtained by deleting the ith row and the jth column of A.

Consider the matrix $A = \begin{bmatrix} 2 & -1 & 5 \\ 4 & 3 & -7 \\ 8 & -7 & 6 \end{bmatrix}$. The minor M_{23} is the determinant of the matrix A formed by deleting row 2 and column 3 from A.

$$M_{23} = \begin{vmatrix} 2 & -1 \\ 8 & -7 \end{vmatrix} \quad \bullet \quad \begin{vmatrix} 2 & -1 & 5 \\ 4 & 3 & -7 \\ 8 & -7 & 6 \end{vmatrix}$$

$$= 2(-7) - 8(-1) = -14 + 8 = -6$$

The minor M_{31} is the determinant of the matrix A formed by deleting row 3 and column 1 from A.

$$M_{31} = \begin{vmatrix} -1 & 5 \\ 3 & -7 \end{vmatrix} \quad \bullet \quad \begin{vmatrix} 2 & -1 & 5 \\ 4 & 3 & -7 \\ 8 & -7 & 6 \end{vmatrix}$$

$$= (-1)(-7) - 3(5) = 7 - 15 = -8$$

The second definition we need is that of the *cofactor* of a matrix.

Cofactor of a Matrix

> The **cofactor** C_{ij} of the element a_{ij} of a square matrix A is given by $C_{ij} = (-1)^{i+j}M_{ij}$, where M_{ij} is the minor of a_{ij}.

Remark When $i + j$ is an even integer, $(-1)^{i+j} = 1$. When $i + j$ is an odd integer, $(-1)^{i+j} = -1$. Thus

$$C_{ij} = \begin{cases} M_{ij}, & i + j \text{ is an even integer} \\ -M_{ij}, & i + j \text{ is an odd integer} \end{cases}$$

EXAMPLE 2 **Find the Minor and Cofactor of a Matrix**

Given $A = \begin{bmatrix} 4 & 3 & -2 \\ 5 & -2 & 4 \\ 3 & -2 & -6 \end{bmatrix}$, find M_{32} and C_{12}.

Solution

$$M_{32} = \begin{vmatrix} 4 & -2 \\ 5 & 4 \end{vmatrix} = 4(4) - 5(-2) = 16 + 10 = 26$$

$$C_{12} = (-1)^{1+2}M_{12} = -M_{12} = -\begin{vmatrix} 5 & 4 \\ 3 & -6 \end{vmatrix} = -(-30 - 12) = 42$$

■ *Try Exercise* **14,** *page 580.*

Cofactors are used to evaluate the determinant of a matrix of order 3 or greater. The technique used to evaluate a determinant by using cofactors is called *expanding by cofactors.*

Determinants by Expanding by Cofactors

> Given the square matrix A of order 3 or greater, the value of the determinant of A is the sum of the products of the elements of any row or column and their cofactors. For the rth row of A, the value of the determinant of A is
>
> $$|A| = a_{r1}C_{r1} + a_{r2}C_{r2} + a_{r3}C_{r3} + \cdots + a_{rn}C_{rn}$$
>
> For the cth column of A, the determinant of A is
>
> $$|A| = a_{1c}C_{1c} + a_{2c}C_{2c} + a_{3c}C_{3c} + \cdots + a_{nc}C_{nc}$$

This theorem states that the value of a determinant can be found by expanding by cofactors of *any* row or column. The value of the determinant is the same in each case. To illustrate the method, consider the matrix $A = \begin{bmatrix} 2 & 3 & -1 \\ 4 & -2 & 3 \\ 1 & -3 & 4 \end{bmatrix}$. Expanding the determinant of A by some row, say row 2, gives

$$|A| = \begin{vmatrix} 2 & 3 & -1 \\ 4 & -2 & 3 \\ 1 & -3 & 4 \end{vmatrix} = 4C_{21} + (-2)C_{22} + 3C_{23}$$

$$= 4(-1)^{2+1}M_{21} + (-2)(-1)^{2+2}M_{22} + 3(-1)^{2+3}M_{23}$$

$$= (-4)\begin{vmatrix} 3 & -1 \\ -3 & 4 \end{vmatrix} + (-2)\begin{vmatrix} 2 & -1 \\ 1 & 4 \end{vmatrix} + (-3)\begin{vmatrix} 2 & 3 \\ 1 & -3 \end{vmatrix}$$

$$= (-4)9 + (-2)9 + (-3)(-9) = -27$$

Expanding the determinant of A by some column, say column 3, gives

$$|A| = \begin{vmatrix} 2 & 3 & -1 \\ 4 & -2 & 3 \\ 1 & -3 & 4 \end{vmatrix} = (-1)C_{13} + 3C_{23} + 4C_{33}$$

$$= (-1)(-1)^{1+3}M_{13} + 3(-1)^{2+3}M_{23} + 4(-1)^{3+3}M_{33}$$

$$= (-1)\begin{vmatrix} 4 & -2 \\ 1 & -3 \end{vmatrix} + (-3)\begin{vmatrix} 2 & 3 \\ 1 & -3 \end{vmatrix} + 4\begin{vmatrix} 2 & 3 \\ 4 & -2 \end{vmatrix}$$

$$= (-1)(-10) + (-3)(-9) + 4(-16) = -27$$

The value of the determinant of A is the same whether we expanded by cofactors of the elements of a row or by cofactors of the elements of a column. When evaluating a determinant, choose the most convenient row or column, which usually is the row or column containing the most zeros.

EXAMPLE 3   Evaluate a Determinant by Cofactors

Evaluate the determinant of $A = \begin{bmatrix} 5 & -3 & -1 \\ -2 & 1 & -1 \\ 1 & 0 & 2 \end{bmatrix}$ by expanding by cofactors.

Solution   Because $a_{32} = 0$, expand using row 3 or column 2. Row 3 will be used here.

$$|A| = 1C_{31} + 0C_{32} + 2C_{33} = 1(-1)^{3+1}M_{31} + 0(-1)^{3+2}M_{32} + 2(-1)^{3+3}M_{33}$$

$$= 1\begin{vmatrix} -3 & -1 \\ 1 & -1 \end{vmatrix} + 0 + 2\begin{vmatrix} 5 & -3 \\ -2 & 1 \end{vmatrix} = 1[3 - (-1)] + 0 + 2[5 - 6] = 4 - 2 = 2$$

■ *Try Exercise* **20**, *page 580.*

Remark   Example 3 illustrates the reason a row or column with zeros is chosen when expanding by cofactors. The product of zero and the cofactor will be zero, which is easy to evaluate.

Evaluating by cofactors is very time-consuming for determinants of large orders unless a row or column contains many zeros. A determinant of order 10 has more than 3 million addends, and each addend is the product of 10 numbers. To avoid the extensive computational problems for large-order determinants, a modification of elementary row operations is used to rewrite the determinant as one with many zeros.

Effects of Elementary Row Operations on the Value of a Determinant of a Matrix

If A is a square matrix of order n, then the following elementary row operations produce the indicated change in the determinant of A.

1. Interchanging any two rows of A changes the sign of $|A|$.
2. Multiplying a row of A by a constant k multiplies the determinant of A by k.
3. Adding a multiple of a row of A to another row does not change the value of the determinant of A.

Remark The properties of determinants just stated remain true when the word "row" is replaced by "column." In that case, we would have elementary *column* operations.

To illustrate these properties, consider the matrix $A = \begin{bmatrix} 2 & 3 \\ 1 & -2 \end{bmatrix}$. The determinant of A is $|A| = 2(-2) - 1(3) = -7$. Now consider each of the elementary row operations.

Interchange the rows of A and evaluate the determinant.

$$\begin{vmatrix} 1 & -2 \\ 2 & 3 \end{vmatrix} = 1(3) - 2(-2) = 3 + 4 = 7 = -|A|$$

Multiply row 2 of A by -3 and evaluate the determinant.

$$\begin{vmatrix} 2 & 3 \\ -3 & 6 \end{vmatrix} = 2(6) - (-3)3 = 12 + 9 = 21 = -3|A|$$

Multiply row 1 of A by -2 and add it to row 2. Then evaluate the determinant.

$$\begin{vmatrix} 2 & 3 \\ -3 & -8 \end{vmatrix} = 2(-8) - (-3)(3) = -16 + 9 = -7 = |A|.$$

These elementary row operations are often used to rewrite a matrix in *diagonal form*. A matrix is in **diagonal form** if all elements below or above the main diagonal are zero. The matrices

$$A = \begin{bmatrix} 2 & -2 & 3 & 1 \\ 0 & -2 & 4 & 2 \\ 0 & 0 & 6 & 9 \\ 0 & 0 & 0 & -5 \end{bmatrix} \quad \text{and} \quad B = \begin{bmatrix} 3 & 0 & 0 & 0 \\ 2 & -3 & 0 & 0 \\ 6 & 4 & -2 & 0 \\ 8 & 3 & 4 & 2 \end{bmatrix}$$

are in diagonal form.

Determinant of a Matrix in Diagonal Form

Let A be a square matrix of order n in diagonal form. The determinant of A is the product of the elements on the main diagonal.

$$|A| = a_{11}a_{22}a_{33} \cdots a_{nn}$$

For matrices A and B given on page 577,

$$|A| = 2(-2)(6)(-5) = 120$$

$$|B| = 3(-3)(-2)(2) = 36$$

EXAMPLE 4   Evaluate a Determinant by Elementary Row Operations

Evaluate the determinant by rewriting in diagonal form.

$$\begin{vmatrix} 2 & 1 & -1 & 3 \\ 2 & 2 & 0 & 1 \\ 4 & 5 & 4 & -3 \\ 2 & 2 & 7 & -3 \end{vmatrix}$$

Solution   Rewrite the matrix in diagonal form by using elementary row operations.

$$\begin{vmatrix} 2 & 1 & -1 & 3 \\ 2 & 2 & 0 & 1 \\ 4 & 5 & 4 & -3 \\ 2 & 2 & 7 & -3 \end{vmatrix} \begin{matrix} -1R_1 + R_2 \\ -2R_1 + R_3 \\ -1R_1 + R_4 \\ = \end{matrix} \begin{vmatrix} 2 & 1 & -1 & 3 \\ 0 & 1 & 1 & -2 \\ 0 & 3 & 6 & -9 \\ 0 & 1 & 8 & -6 \end{vmatrix}$$

Factor 3, from row 3. $$3 \begin{vmatrix} 2 & 1 & -1 & 3 \\ 0 & 1 & 1 & -2 \\ 0 & 1 & 2 & -3 \\ 0 & 1 & 8 & -6 \end{vmatrix} \begin{matrix} -1R_2 + R_3 \\ -1R_2 + R_4 \\ = \end{matrix} 3 \begin{vmatrix} 2 & 1 & -1 & 3 \\ 0 & 1 & 1 & -2 \\ 0 & 0 & 1 & -1 \\ 0 & 0 & 7 & -4 \end{vmatrix}$$

$$\begin{matrix} -7R_3 + R_4 \\ = \end{matrix} 3 \begin{vmatrix} 2 & 1 & -1 & 3 \\ 0 & 1 & 1 & -2 \\ 0 & 0 & 1 & -1 \\ 0 & 0 & 0 & 3 \end{vmatrix} = 3(2)(1)(1)(3) = 18$$

■ *Try Exercise 42, page 581.*

Remark   The last example used only elementary row operations to reduce the matrix to diagonal form. Elementary column operations could also have been used, as could a combination of row and column operations.

In some cases it is possible to recognize when the determinant of a matrix is zero.

Conditions for a Zero Determinant

If A is a square matrix, then $|A| = 0$ when any one of the following is true.

1. A row (column) consists entirely of zeros.
2. Two rows (columns) are identical.
3. One row (column) is a constant multiple of a second row (column).

Proof: To prove part 2 of this theorem, let A be the given matrix and $D = |A|$. Now interchange the two identical rows. Then $|A| = -D$. (Why?[4]) Thus

$$D = -D$$

Zero is the only real number that is its own additive inverse, and hence $D = |A| = 0$.

The proofs of the other two properties are left as exercises.

The last property of determinants that we will discuss is a product property.

Product Property of Determinants

If A and B are square matrices of order n, then

$$|AB| = |A||B|$$

Recall that a singular matrix is one that does not have a multiplicative inverse. The Product Property of Determinants can be used to determine whether a matrix has an inverse.

Consider a matrix A with an inverse A^{-1}. Then, by the last theorem,

$$|A \cdot A^{-1}| = |A||A^{-1}|$$

But $A \cdot A^{-1} = I$, the identity matrix, and $|I| = 1$. (Why?[5]) Therefore,

$$1 = |A||A^{-1}|$$

From the last equation, $|A| \neq 0$. And, in particular,

$$|A^{-1}| = \frac{1}{|A|}$$

These results are summarized in the following theorem.

---

[4] Interchanging two rows of a matrix changes the sign of the determinant of the matrix.

[5] The identity matrix is in diagonal form with 1s on the main diagonal. Thus $|I|$ is a product of 1s, or $|I| = 1$.

Existence of the Inverse of a Square Matrix

If A is a square matrix of order n, then A has a multiplicative inverse if and only if $|A| \neq 0$. Furthermore,

$$|A^{-1}| = \frac{1}{|A|}$$

Remark We proved only part of this theorem. It remains to show that given $|A| \neq 0$, then A has an inverse. This proof can be found in most texts on linear algebra.

EXERCISE SET 11.4

In Exercises 1 to 8, evaluate the determinants.

1. $\begin{vmatrix} 2 & -1 \\ 3 & 5 \end{vmatrix}$ 2. $\begin{vmatrix} 2 & 9 \\ -6 & 2 \end{vmatrix}$ 3. $\begin{vmatrix} 5 & 0 \\ 2 & -3 \end{vmatrix}$

4. $\begin{vmatrix} 0 & -8 \\ 3 & 4 \end{vmatrix}$ 5. $\begin{vmatrix} 4 & 6 \\ 2 & 3 \end{vmatrix}$ 6. $\begin{vmatrix} -3 & 6 \\ 4 & -8 \end{vmatrix}$

7. $\begin{vmatrix} 0 & 9 \\ 0 & -2 \end{vmatrix}$ 8. $\begin{vmatrix} -3 & 9 \\ 0 & 0 \end{vmatrix}$

In Exercises 9 to 12, evaluate the indicated minor and cofactor for the determinant

$$\begin{vmatrix} 5 & -2 & -3 \\ 2 & 4 & -1 \\ 4 & -5 & 6 \end{vmatrix}$$

9. M_{11}, C_{11} 10. M_{21}, C_{21} 11. M_{32}, C_{32} 12. M_{33}, C_{33}

In Exercises 13 to 16, evaluate the indicated minor and cofactor for the determinant

$$\begin{vmatrix} 3 & -2 & 3 \\ 1 & 3 & 0 \\ 6 & -2 & 3 \end{vmatrix}$$

13. M_{22}, C_{22} 14. M_{13}, C_{13} 15. M_{31}, C_{31} 16. M_{23}, C_{23}

In Exercises 17 to 26, evaluate the determinant by expanding by cofactors.

17. $\begin{vmatrix} 2 & -3 & 1 \\ 2 & 0 & 2 \\ 3 & -2 & 4 \end{vmatrix}$ 18. $\begin{vmatrix} 3 & 1 & -2 \\ 2 & -5 & 4 \\ 3 & 2 & 1 \end{vmatrix}$

19. $\begin{vmatrix} -2 & 3 & 2 \\ 1 & 2 & -3 \\ -4 & -2 & 1 \end{vmatrix}$ 20. $\begin{vmatrix} 3 & -2 & 0 \\ 2 & -3 & 2 \\ 8 & -2 & 5 \end{vmatrix}$

21. $\begin{vmatrix} 2 & -3 & 10 \\ 0 & 2 & -3 \\ 0 & 0 & 5 \end{vmatrix}$ 22. $\begin{vmatrix} 6 & 0 & 0 \\ 2 & -3 & 0 \\ 7 & -8 & 2 \end{vmatrix}$

23. $\begin{vmatrix} 0 & -2 & 4 \\ 1 & 0 & -7 \\ 5 & -6 & 0 \end{vmatrix}$ 24. $\begin{vmatrix} 5 & -8 & 0 \\ 2 & 0 & -7 \\ 0 & -2 & -1 \end{vmatrix}$

25. $\begin{vmatrix} 4 & -3 & 3 \\ 2 & 1 & -4 \\ 6 & -2 & -1 \end{vmatrix}$ 26. $\begin{vmatrix} -2 & 3 & 9 \\ 4 & -2 & -6 \\ 0 & -8 & -24 \end{vmatrix}$

In Exercises 27 to 40, without expanding, give a reason for each equality.

27. $\begin{vmatrix} 2 & -1 & 3 \\ 0 & 0 & 0 \\ 3 & 4 & 1 \end{vmatrix} = 0$ 28. $\begin{vmatrix} 2 & 3 & 0 \\ 1 & -2 & 0 \\ 4 & 1 & 0 \end{vmatrix} = 0$

29. $\begin{vmatrix} 1 & 4 & -1 \\ 2 & 4 & 12 \\ 3 & 1 & 4 \end{vmatrix} = 2 \begin{vmatrix} 1 & 4 & -1 \\ 1 & 2 & 6 \\ 3 & 1 & 4 \end{vmatrix}$

30. $\begin{vmatrix} 1 & -3 & 4 \\ 4 & 6 & 1 \\ 0 & -9 & 3 \end{vmatrix} = -3 \begin{vmatrix} 1 & 1 & 4 \\ 4 & -2 & 1 \\ 0 & 3 & 3 \end{vmatrix}$

31. $\begin{vmatrix} 1 & 5 & -2 \\ 2 & -1 & 4 \\ 3 & 0 & -2 \end{vmatrix} = \begin{vmatrix} 1 & 5 & -2 \\ 0 & -11 & 8 \\ 3 & 0 & -2 \end{vmatrix}$

32. $\begin{vmatrix} 1 & 1 & -3 \\ 2 & 2 & 5 \\ 1 & -2 & 4 \end{vmatrix} = \begin{vmatrix} 1 & 1 & -3 \\ 2 & 2 & 5 \\ 0 & -3 & 7 \end{vmatrix}$

33. $\begin{vmatrix} 4 & -3 & 2 \\ 6 & 2 & 1 \\ -2 & 2 & 4 \end{vmatrix} = 2 \begin{vmatrix} 2 & -3 & 2 \\ 3 & 2 & 1 \\ -1 & 2 & 4 \end{vmatrix}$

34. $\begin{vmatrix} 2 & -1 & 3 \\ 3 & 0 & 1 \\ -4 & 2 & -6 \end{vmatrix} = 0$

35. $\begin{vmatrix} 2 & -4 & 5 \\ 0 & 3 & 4 \\ 0 & 0 & -2 \end{vmatrix} = -12$

36. $\begin{vmatrix} 3 & 0 & 0 \\ 2 & -1 & 0 \\ 3 & 4 & 5 \end{vmatrix} = -15$

37. $\begin{vmatrix} 3 & 5 & -2 \\ 2 & 1 & 0 \\ 9 & -2 & -3 \end{vmatrix} = - \begin{vmatrix} 9 & -2 & -3 \\ 2 & 1 & 0 \\ 3 & 5 & -2 \end{vmatrix}$

38. $\begin{vmatrix} 6 & 0 & -2 \\ 2 & -1 & -3 \\ 1 & 5 & -7 \end{vmatrix} = - \begin{vmatrix} 0 & 6 & -2 \\ -1 & 2 & -3 \\ 5 & 1 & -7 \end{vmatrix}$

39. $a^3 \begin{vmatrix} 1 & 1 & 1 \\ a & a & a \\ a^2 & a^2 & a^2 \end{vmatrix} = \begin{vmatrix} a & a & a \\ a^2 & a^2 & a^2 \\ a^3 & a^3 & a^3 \end{vmatrix}$

40. $\begin{vmatrix} 1 & 1 & 1 \\ 2 & 2 & 2 \\ 3 & 3 & 3 \end{vmatrix} = 0$

In Exercises 41 to 50, evaluate the determinant by first rewriting the determinant in diagonal form.

41. $\begin{vmatrix} 2 & 4 & 1 \\ 1 & 2 & -1 \\ 1 & 2 & 2 \end{vmatrix}$

42. $\begin{vmatrix} 3 & -2 & -1 \\ 1 & 2 & 4 \\ 2 & -2 & 3 \end{vmatrix}$

43. $\begin{vmatrix} 1 & 2 & -1 \\ 2 & 3 & 1 \\ 3 & 4 & 3 \end{vmatrix}$

44. $\begin{vmatrix} 1 & 2 & 5 \\ -1 & 1 & -2 \\ 3 & 1 & 10 \end{vmatrix}$

45. $\begin{vmatrix} 0 & -1 & 1 \\ 1 & 0 & -2 \\ 2 & 2 & 0 \end{vmatrix}$

46. $\begin{vmatrix} 2 & -1 & 3 \\ 1 & 1 & 1 \\ 3 & -4 & 5 \end{vmatrix}$

47. $\begin{vmatrix} 1 & 2 & -1 & 2 \\ 1 & -2 & 0 & 3 \\ 3 & 0 & 1 & 5 \\ -2 & -4 & 1 & 6 \end{vmatrix}$

48. $\begin{vmatrix} 1 & -1 & -1 & 2 \\ 0 & 2 & 4 & 6 \\ 1 & 1 & 4 & 12 \\ 1 & -1 & 0 & 8 \end{vmatrix}$

49. $\begin{vmatrix} 1 & 2 & 3 & -1 \\ 6 & 5 & 9 & 8 \\ 2 & 4 & 12 & -1 \\ 1 & 2 & 6 & -1 \end{vmatrix}$

50. $\begin{vmatrix} 1 & 2 & 0 & -2 \\ -1 & 1 & 3 & 5 \\ 2 & 1 & 4 & 0 \\ -2 & 5 & 2 & 6 \end{vmatrix}$

Graphing Calculator Exercises*

Some graphing calculators contain a program that calculates the value of the determinant of a matrix. Try one of these programs for Exercises 51 to 54.

51. $\begin{bmatrix} 2 & -2 & 3 & 1 \\ 5 & 2 & -2 & 3 \\ 6 & -1 & 2 & 3 \\ 2 & 3 & -1 & 5 \end{bmatrix}$

52. $\begin{bmatrix} 3 & -1 & 0 & 1 \\ 2 & -2 & 3 & 0 \\ -1 & -3 & 5 & 3 \\ 5 & 3 & -2 & 1 \end{bmatrix}$

53. $\begin{bmatrix} -\frac{2}{7} & 4 & -\frac{1}{6} \\ -2 & \sqrt{2} & -3 \\ \sqrt{3} & 3 & -\sqrt{5} \end{bmatrix}$

54. $\begin{bmatrix} 6 & \pi & -\frac{4}{7} \\ -5 & \sqrt{7} & 2 \\ \frac{5}{6} & -\sqrt{3} & \sqrt{10} \end{bmatrix}$

Supplemental Exercises

The area of a triangle with vertices (x_1, y_1), (x_2, y_2), and (x_3, y_3) can be given as the absolute value of the determinant

$$\frac{1}{2} \begin{vmatrix} x_1 & y_1 & 1 \\ x_2 & y_2 & 1 \\ x_3 & y_3 & 1 \end{vmatrix}$$

Use this formula to find the area of the triangles whose coordinates are given in Exercises 55 to 58.

55. $(2, 3)$, $(-1, 0)$, $(4, 8)$

56. $(-3, 4)$, $(1, 5)$, $(5, -2)$

57. $(4, 9)$, $(8, 2)$, $(-3, -2)$

58. $(0, 4)$, $(-5, 7)$, $(2, 9)$

59. Given a square matrix of order 3 where one row is a constant multiple of a second row, show that the determinant of the matrix is zero. (*Hint:* Use an elementary row operation and part 2 of the theorem for conditions for a zero determinant.)

60. Given a square matrix of order 3 with a zero as every element in a column, show that the determinant of the matrix is zero. (*Hint:* Expand the determinant by cofactors using the column of zeros.)

61. Show that the determinant $\begin{vmatrix} x & y & 1 \\ x_1 & y_1 & 1 \\ x_2 & y_2 & 1 \end{vmatrix} = 0$ is the equation of a line through the points (x_1, y_1) and (x_2, y_2).

62. Use Exercise 61 to find the equation of the line passing through the points $(2, 3)$ and $(-1, 4)$.

63. Use Exercise 61 to find the equation of the line passing through the points $(-3, 4)$, and $(2, -3)$.

64. Show that $\begin{vmatrix} a_1 & b_1 \\ a_2 & b_2 \end{vmatrix} = \begin{vmatrix} a_1 & b_1 \\ ka_1 + a_2 & kb_1 + b_2 \end{vmatrix}$.
What property of determinants does this illustrate?

*Additional graphing calculator exercises appear in the Graphing Workbook as described in the front of this textbook.

582   CHAPTER 11   MATRICES AND DETERMINANTS

65. Surveyors use a formula to find the area of a plot of land. *Surveyor's Area Formula:* If the vertices (x_1, y_1), (x_2, y_2), (x_3, y_3), ... , (x_n, y_n) of a simple polygon are listed counterclockwise around the perimeter, the area of the polygon is

$$A = \frac{1}{2}\left\{ \begin{vmatrix} x_1 & x_2 \\ y_1 & y_2 \end{vmatrix} + \begin{vmatrix} x_2 & x_3 \\ y_2 & y_3 \end{vmatrix} + \begin{vmatrix} x_3 & x_4 \\ y_3 & y_4 \end{vmatrix} + \cdots + \begin{vmatrix} x_n & x_1 \\ y_n & y_1 \end{vmatrix} \right\}$$

Use the Surveyor's Area Formula to find the area of the polygon with vertices $(8, -4)$, $(25, 5)$, $(15, 9)$, $(17, 20)$, $(0, 10)$.

11.5 Cramer's Rule

An application of determinants is to solve a system of linear equations. Consider the system

$$\begin{cases} a_{11}x_1 + a_{12}x_2 = b_1 \\ a_{21}x_1 + a_{22}x_2 = b_2 \end{cases}$$

To eliminate x_2 from this system, we first multiply the top equation by a_{22} and the bottom equation by a_{12}. Then subtract.

$$a_{22}a_{11}x_1 + a_{22}a_{12}x_2 = a_{22}b_1$$

$$a_{12}a_{21}x_1 + a_{12}a_{22}x_2 = a_{12}b_2$$

$$(a_{22}a_{11} - a_{12}a_{21})x_1 \qquad = a_{22}b_1 - a_{12}b_2$$

$$x_1 = \frac{a_{22}b_1 - a_{12}b_2}{a_{22}a_{11} - a_{12}a_{21}}$$

or
$$x_1 = \frac{\begin{vmatrix} b_1 & a_{12} \\ b_2 & a_{22} \end{vmatrix}}{\begin{vmatrix} a_{11} & a_{12} \\ a_{21} & a_{22} \end{vmatrix}}, \qquad \begin{vmatrix} a_{11} & a_{12} \\ a_{21} & a_{22} \end{vmatrix} \neq 0$$

We can find x_2 in a similar manner. The results are given in Cramer's Rule for a system of two linear equations.

Cramer's Rule for a System of Two Linear Equations

Let

$$\begin{cases} a_{11}x_1 + a_{12}x_2 = b_1 \\ a_{21}x_1 + a_{22}x_2 = b_2 \end{cases}$$

be the system of equations for which the determinant of the coefficient matrix is not zero. The solution of the system of equations is the ordered pair whose coordinates are

$$x_1 = \frac{\begin{vmatrix} b_1 & a_{12} \\ b_2 & a_{22} \end{vmatrix}}{\begin{vmatrix} a_{11} & a_{12} \\ a_{21} & a_{22} \end{vmatrix}} \qquad \text{and} \qquad x_2 = \frac{\begin{vmatrix} a_{11} & b_1 \\ a_{21} & b_2 \end{vmatrix}}{\begin{vmatrix} a_{11} & a_{12} \\ a_{21} & a_{22} \end{vmatrix}}.$$

Note that the denominator is the determinant of the coefficient matrix of the variables. The numerator of x_1 is formed by replacing column 1 of the coefficient determinant with the constants b_1 and b_2. The numerator of x_2 is formed by replacing column 2 of the coefficient determinant with the constants b_1 and b_2.

EXAMPLE 1 **Solve a System of Equations by Using Cramer's Rule**

Solve the following system of equations using Cramer's Rule.

$$\begin{cases} 5x_1 - 3x_2 = 6 \\ 2x_1 + 4x_2 = -7 \end{cases}$$

Solution

$$x_1 = \frac{\begin{vmatrix} 6 & -3 \\ -7 & 4 \end{vmatrix}}{\begin{vmatrix} 5 & -3 \\ 2 & 4 \end{vmatrix}} = \frac{3}{26}, \qquad x_2 = \frac{\begin{vmatrix} 5 & 6 \\ 2 & -7 \end{vmatrix}}{\begin{vmatrix} 5 & -3 \\ 2 & 4 \end{vmatrix}} = -\frac{47}{26}$$

The solution is $\left(\dfrac{3}{26}, -\dfrac{47}{26} \right)$.

■ *Try Exercise* **4**, *page 585.*

Cramer's Rule can be used for a system of three linear equations in three variables. For example, consider the system of equations

$$\begin{cases} 2x - 3y + z = 2 \\ 4x + 2z = -3 \\ 3x + y - 2z = 1 \end{cases} \qquad (1)$$

To solve this system of equations, we extend the concepts behind the solution for a system of two linear equations. The solution of the system has the form (x, y, z), where

$$x = \frac{D_x}{D}, \qquad y = \frac{D_y}{D}, \qquad z = \frac{D_z}{D}$$

The determinant D is the determinant of the coefficient matrix. The determinants D_x, D_y, and D_z are the determinants of the matrices formed by replacing the first, second, and third columns, respectively, by the constants. For the system of equations (1),

$$x = \frac{D_x}{D} \qquad y = \frac{D_y}{D} \qquad z = \frac{D_z}{D}$$

where $\quad D = \begin{vmatrix} 2 & -3 & 1 \\ 4 & 0 & 2 \\ 3 & 1 & -2 \end{vmatrix} = -42 \qquad D_x = \begin{vmatrix} 2 & -3 & 1 \\ -3 & 0 & 2 \\ 1 & 1 & -2 \end{vmatrix} = 5$

$$D_y = \begin{vmatrix} 2 & 2 & 1 \\ 4 & -3 & 2 \\ 3 & 1 & -2 \end{vmatrix} = 49, \qquad D_z = \begin{vmatrix} 2 & -3 & 2 \\ 4 & 0 & -3 \\ 3 & 1 & 1 \end{vmatrix} = 53$$

Thus

$$x = -\frac{5}{42} \qquad y = -\frac{49}{42} \qquad z = -\frac{53}{42}$$

The solution of the system (1) is

$$\left(-\frac{5}{42}, -\frac{49}{42}, -\frac{53}{42}\right)$$

Cramer's Rule can be extended to a system of n linear equations in n variables.

Cramer's Rule

Let

$$\begin{cases} a_{11}x_1 + a_{12}x_2 + a_{13}x_3 + \cdots + a_{1n}x_n = b_1 \\ a_{21}x_1 + a_{22}x_2 + a_{23}x_3 + \cdots + a_{2n}x_n = b_2 \\ a_{31}x_1 + a_{32}x_2 + a_{33}x_3 + \cdots + a_{3n}x_n = b_3 \\ \quad \vdots \qquad\quad \vdots \qquad\quad \vdots \qquad\qquad\quad \vdots \\ a_{n1}x_1 + a_{n2}x_2 + a_{n3}x_3 + \cdots + a_{nn}x_n = b_n \end{cases}$$

be a system of n equations in n variables. The solution of the system is given by $(x_1, x_2, x_3, \ldots, x_n)$, where

$$x_1 = \frac{D_1}{D}, \quad x_2 = \frac{D_2}{D}, \quad \ldots, \quad x_i = \frac{D_i}{D}, \quad \ldots, \quad x_n = \frac{D_n}{D}$$

and D is the determinant of the coefficient matrix, $D \neq 0$. D_i is the determinant formed by replacing the ith column of the coefficient matrix with the column of constants $b_1, b_2, b_3, \ldots, b_n$.

Because the determinant of the coefficient matrix must be nonzero for us to use Cramer's Rule, this method is not appropriate for systems of linear equations with no solution or infinitely many solutions. In fact, the only time a system of linear equations has a unique solution is when the coefficient determinant is not zero, a fact summarized in the following theorem.

Systems of Linear Equations with Unique Solutions

A system of n linear equations in n variables has a unique solution if and only if the determinant of the coefficient matrix is not zero.

Cramer's Rule is also useful when we want to determine the value of only a single variable in a system of equations.

EXAMPLE 2 **Determine the Value of a Single Variable in a System of Linear Equations**

Find x_3 for the system of equations

$$\begin{cases} 4x_1 \qquad\quad + 3x_3 - 2x_4 = 2 \\ 3x_1 + x_2 + 2x_3 - x_4 = 4 \\ x_1 - 6x_2 - 2x_3 + 2x_4 = 0 \\ 2x_1 + 2x_2 \qquad\quad - x_4 = -1 \end{cases}$$

Solution Find D and D_3.

$$D = \begin{vmatrix} 4 & 0 & 3 & -2 \\ 3 & 1 & 2 & -1 \\ 1 & -6 & -2 & 2 \\ 2 & 2 & 0 & -1 \end{vmatrix} = 39 \qquad D_3 = \begin{vmatrix} 4 & 0 & 2 & -2 \\ 3 & 1 & 4 & -1 \\ 1 & -6 & 0 & 2 \\ 2 & 2 & -1 & -1 \end{vmatrix} = 96$$

Thus $x_3 = \dfrac{96}{39} = \dfrac{32}{13}$.

■ *Try Exercise **24**, page 586.*

EXERCISE SET 11.5

In Exercises 1 to 20, solve each system of equations by using Cramer's Rule.

1. $\begin{cases} 3x_1 + 4x_2 = 8 \\ 4x_1 - 5x_2 = 1 \end{cases}$

2. $\begin{cases} x_1 - 3x_2 = 9 \\ 2x_1 - 4x_2 = -3 \end{cases}$

3. $\begin{cases} 5x_1 + 4x_2 = -1 \\ 3x_1 - 6x_2 = 5 \end{cases}$

4. $\begin{cases} 2x_1 + 5x_2 = 9 \\ 5x_1 + 7x_2 = 8 \end{cases}$

5. $\begin{cases} 7x_1 + 2x_2 = 0 \\ 2x_1 + x_2 = -3 \end{cases}$

6. $\begin{cases} 3x_1 - 8x_2 = 1 \\ 4x_1 + 5x_2 = -2 \end{cases}$

7. $\begin{cases} 3x_1 - 7x_2 = 0 \\ 2x_1 + 4x_2 = 0 \end{cases}$

8. $\begin{cases} 5x_1 + 4x_2 = -3 \\ 2x_1 - x_2 = 0 \end{cases}$

9. $\begin{cases} 1.2x_1 + 0.3x_2 = 2.1 \\ 0.8x_1 - 1.4x_2 = -1.6 \end{cases}$

10. $\begin{cases} 3.2x_1 - 4.2x_2 = 1.1 \\ 0.7x_1 + 3.2x_2 = -3.4 \end{cases}$

11. $\begin{cases} 3x_1 - 4x_2 + 2x_3 = 1 \\ x_1 - x_2 + 2x_3 = -2 \\ 2x_1 + 2x_2 + 3x_3 = -3 \end{cases}$

12. $\begin{cases} 5x_1 - 2x_2 + 3x_3 = -2 \\ 3x_1 + x_2 - 2x_3 = 3 \\ x_1 - 2x_2 + 3x_3 = -1 \end{cases}$

13. $\begin{cases} x_1 + 4x_2 - 2x_3 = 0 \\ 3x_1 - 2x_2 + 3x_3 = 4 \\ 2x_1 + x_2 - 3x_3 = -1 \end{cases}$

14. $\begin{cases} 4x_1 - x_2 + 2x_3 = 6 \\ x_1 + 3x_2 - x_3 = -1 \\ 2x_1 + 3x_2 - 2x_3 = 5 \end{cases}$

15. $\begin{cases} 2x_2 - 3x_3 = 1 \\ 3x_1 - 5x_2 + x_3 = 0 \\ 4x_1 + 2x_3 = -3 \end{cases}$

16. $\begin{cases} 2x_1 + 5x_2 = 1 \\ x_1 - 3x_3 = -2 \\ 2x_1 - x_2 + 2x_3 = 4 \end{cases}$

17. $\begin{cases} 4x_1 - 5x_2 + x_3 = -2 \\ 3x_1 + x_2 = 4 \\ x_1 - x_2 + 3x_3 = 0 \end{cases}$

18. $\begin{cases} 3x_1 - x_2 + x_3 = 5 \\ x_1 + 3x_3 = -2 \\ 2x_1 + 2x_2 - 5x_3 = 0 \end{cases}$

19. $\begin{cases} 2x_1 + 2x_2 - 3x_3 = 0 \\ x_1 - 3x_2 + 2x_3 = 0 \\ 4x_1 - x_2 + 3x_3 = 0 \end{cases}$

20. $\begin{cases} x_1 + 3x_2 = -2 \\ 2x_1 - 3x_2 + x_3 = 1 \\ 4x_1 + 5x_2 - 2x_3 = 0 \end{cases}$

In Exercises 21 to 26, solve for the indicated variable.

21. Solve for x_2. $\begin{cases} 2x_1 - 3x_2 + 4x_3 - x_4 = 1 \\ x_1 + 2x_2 + 2x_4 = -1 \\ 3x_1 + x_2 - 2x_4 = 2 \\ x_1 - 3x_2 + 2x_3 - x_4 = 3 \end{cases}$

22. Solve for x_4. $\begin{cases} 3x_1 + x_2 - 2x_3 + 3x_4 = 4 \\ 2x_1 - 3x_2 + 2x_3 = -2 \\ x_1 + x_2 - 2x_3 + 2x_4 = 3 \\ 2x_1 + 3x_3 - 2x_4 = 4 \end{cases}$

23. Solve for x_1.

$$\begin{cases} x_1 - 3x_2 + 2x_3 + 4x_4 = 0 \\ 3x_1 + 5x_2 - 6x_3 + 2x_4 = -2 \\ 2x_1 - x_2 + 9x_3 + 8x_4 = 0 \\ x_1 + x_2 + x_3 - 8x_4 = -3 \end{cases}$$

24. Solve for x_3.

$$\begin{cases} 2x_1 + 5x_2 - 5x_3 - 3x_4 = -3 \\ x_1 + 7x_2 + 8x_3 - x_4 = 4 \\ 4x_1 + x_3 + x_4 = 3 \\ 3x_1 + 2x_2 - x_3 = 0 \end{cases}$$

25. Solve for x_4.

$$\begin{cases} 3x_2 - x_3 + 2x_4 = 1 \\ 5x_1 + x_2 + 3x_3 - x_4 = -4 \\ x_1 - 2x_2 + 9x_4 = 5 \\ 2x_1 + 2x_3 = 3 \end{cases}$$

26. Solve for x_1.

$$\begin{cases} 4x_1 + x_2 - 3x_4 = 4 \\ 5x_1 + 2x_2 - 2x_3 + x_4 = 7 \\ x_1 - 3x_2 + 2x_3 - 2x_4 = -6 \\ 3x_3 + 4x_4 = -7 \end{cases}$$

Supplemental Exercises

27. A solution of the system of equations

$$\begin{cases} 2x_1 - 3x_2 + x_3 = 9 \\ x_1 + x_2 - 2x_3 = -3 \\ 4x_1 - x_2 - 3x_3 = 3 \end{cases}$$

is $(1, -2, 1)$. However, this solution cannot be found by using Cramer's Rule. Explain.

28. Verify the solution for x_2 given in Cramer's Rule for a System of Two Equations by solving the system of equations

$$\begin{cases} a_{11}x_1 + a_{12}x_2 = b_1 \\ a_{21}x_1 + a_{22}x_2 = b_2 \end{cases}$$

for x_2 by using the elimination method.

29. For what value of k does the system of equations

$$\begin{cases} kx + 3y = 7 \\ kx - 2y = 5 \end{cases}$$

have a unique solution?

30. For what values of k does the system of equations

$$\begin{cases} kx + 4y = 5 \\ 9x - ky = 2 \end{cases}$$

have a unique solution?

31. For what values of k does the system of equations

$$\begin{cases} x + 2y - 3z = 4 \\ 2x + ky - 4z = 5 \\ x - 2y + z = 6 \end{cases}$$

have a unique solution?

32. For what values of k does the system of equations

$$\begin{cases} kx_1 + x_2 = 1 \\ x_2 - 4x_3 = 1 \\ x_1 + kx_3 = 1 \end{cases}$$

have a unique solution?

33. Find real values for r and s so that $ru + sv = w$, where u, v, and w are complex numbers and $u = 2 + 3i$, $v = 4 - 2i$, and $w = -6 + 15i$.

34. Find real values for r and s such that $ru + sv = w$, where $u = 3 - 4i$, $v = 1 + 2i$, and $w = 4 - 22i$.

Exploring Concepts with Technology

Stochastic Matrices

Matrices can be used to predict how percents of populations will change over time. Consider two neighborhood supermarkets, Super A and Super B. Each week Super A loses 5% of its customers to Super B, and each week Super B loses 8% of its customers to Super A. If this trend continues and if Super A currently has 40% of the neighborhood customers and Super B the remaining 60% of the neighborhood customers, what percent of the neighborhood will each have after n weeks?

We will approach this problem by examining the changes on a week-by-week basis. Because Super A loses 5% of its customers each week, it retains 95% of its customers. Since it has 40% of the neighborhood customers now, after one week it will have 95% of its 40% share, or 38% $(0.95 \cdot 0.40)$ of the customers. In that same week, it gains 8% of the customers of Super B. Since Super B has 60% of the neighborhood customers, Super A's gain is 4.8% $(0.08 \cdot 0.60)$. After one week, Super A now has $38\% + 4.8\% = 42.8\%$ of the neighborhood customers. Super B has the remaining 57.2% of the customers.

The changes for the second week are calculated similarly. Super A retains 95% of its 42.8% and gains 8% of Super B's 57.2%. After week 2, Super A has

$$0.95 \cdot 0.428 + 0.08 \cdot 0.572 \approx 0.452$$

or approximately 45.2% of the neighborhood customers. Super B has the remaining 54.8%.

We could continue in this way, but using matrices is a more convenient way to proceed. Let $T = \begin{bmatrix} 0.95 & 0.05 \\ 0.08 & 0.92 \end{bmatrix}$, where column 1 represents the percent retained by Super A and column 2 represents the percent retained by Super B. Let $X = [0.40 \quad 0.60]$ be the current market shares of Super A and Super B, respectively. Now form the product XT.

$$[0.40 \quad 0.60] \begin{bmatrix} 0.95 & 0.05 \\ 0.08 & 0.92 \end{bmatrix} = [0.428 \quad 0.572]$$

For the second week, multiply the market share after week 1 by T.

$$[0.428 \quad 0.572] \begin{bmatrix} 0.95 & 0.05 \\ 0.08 & 0.92 \end{bmatrix} \approx [0.452 \quad 0.548]$$

The last product can also be expressed as

$$[0.452 \quad 0.548] = \overbrace{[0.428 \quad 0.572] \begin{bmatrix} 0.95 & 0.05 \\ 0.08 & 0.92 \end{bmatrix}}^{[0.428 \quad 0.572]} = [0.40 \quad 0.60] \begin{bmatrix} 0.95 & 0.05 \\ 0.08 & 0.92 \end{bmatrix} \begin{bmatrix} 0.95 & 0.05 \\ 0.08 & 0.92 \end{bmatrix}$$

$$= [0.40 \quad 0.60] \begin{bmatrix} 0.95 & 0.05 \\ 0.08 & 0.92 \end{bmatrix}^2 = XT^2$$

Note that the exponent of T corresponds to the fact that 2 weeks have passed. In general, the market share after n weeks is XT^n. The matrix T is called a **stochastic matrix.** A stochastic matrix is characterized by the fact that each element of the matrix is nonnegative and the sum of the elements in each row is 1.

Calculate the market share of Super A and Super B after 20 weeks, 40 weeks, 60 weeks, and 100 weeks. What observations do you draw from your calculations? We started this problem with the assumption that Super A had 40% of the market and Super B had 60% of the market. Suppose, however, that originally Super A had 99% of the market and Super B 1%. Does this affect the market share each will have after 100 weeks? If

Super A had 1% of the market and Super B had 99% of the market, what is the market share of each after 100 weeks?

To take another example, suppose each of three department stores is vying for the business of the other two stores. In one month, Store A loses 15% of its customers to Store B and 8% of its customers to Store C. Store B loses 10% of its customers to Store A and 12% to Store C. Store C loses 5% to Store A and 9% to Store B. Assuming these three stores have 100% of the market and the trend continues, what market share will each have after 100 months?

Chapter Review

11.1   Gauss–Jordan Elimination Method

- A matrix is a rectangular array of numbers. A matrix with m rows and n columns is of order $m \times n$ or dimension $m \times n$.

- For a system of equations, it is possible to form a coefficient matrix, an augmented matrix, and a constant matrix.

- An **Echelon Form** matrix is in echelon form if all the following conditions are satisfied:

  1. The first nonzero number in any row is a 1.

  2. Rows are arranged so that the column containing the first nonzero number is to the left of the column containing the first nonzero number of the next row.

  3. All rows consisting entirely of zeros appear at the bottom of the matrix.

- The Gauss–Jordan elimination method is used to solve a system of equations.

- **Elementary Row Operations**
  The elementary row operations for a matrix are

  1. Interchanging two rows

  2. Multiplying all the elements in a row by the same nonzero number

  3. Replacing a row by the sum of that row and a nonzero multiple of any other row

11.2   The Algebra of Matrices

- Two matrices A and B are equal if and only if $a_{ij} = b_{ij}$ for every i and j.

- The sum of two matrices of the same order is the matrix whose elements are the sum of the corresponding elements of the two matrices.

- The $m \times n$ zero matrix is the matrix whose elements are all zeros.

- Taking the product of a real number and a matrix is called scalar multiplication.

- In order for us to multiply two matrices, the number of columns of the first matrix must equal the number of rows of the second matrix.

- In general, matrix multiplication is not commutative.

- The multiplicative identity matrix is the matrix with 1s on the main diagonal and zeros everywhere else.

11.3   The Inverse of a Matrix

- The multiplicative inverse of a square matrix A, denoted by A^{-1}, has the property that

$$A \cdot A^{-1} = A^{-1} \cdot A = I_n$$

where I_n is the multiplicative identity matrix.

- A singular matrix is one that does not have a multiplicative inverse.

- Input–output analysis attempts to determine the necessary output of industries to satisfy each other's demands plus the demands of consumers.

11.4   Determinants

- Associated with each square matrix is a number called the determinant of the matrix.

- The minor of the element a_{ij} of a square matrix is the determinant of the matrix obtained by deleting the ith row and the jth column of A.

- The cofactor of the element a_{ij} of a square matrix is $(-1)^{i+j} M_{ij}$, where M_{ij} is the minor of the element.

- The value of a determinant can be found by multiplying the elements of any row or column by their respective cofactors and then adding the results. This is called expanding by cofactors.

11.5   Cramer's Rule

- Cramer's Rule is a method of solving a system of n equations in n variables.

Essays and Projects

1. Matrices are very prevalent in applied mathematics. Write an essay in which you outline some of these applications. You should be able to find applications to physics, economics, business management, games, and computer science—just to name a few disciplines.

2. Write an essay on the early history of matrices and determinants. Include in your essay the contributions of James Sylvester, Arthur Cayley, Gabriel Cramer, and William Rowan Hamilton. Also explain why the word "matrix" was chosen and give its etymology.

3. If you have some computer programming experience, write a program to find the inverse of a matrix by using elementary row operations. After you complete your program, try to find the inverse of one of the so called Hilbert matrices,

$$\begin{bmatrix} 1 & \frac{1}{2} & \frac{1}{3} & \frac{1}{4} \\ \frac{1}{2} & \frac{1}{3} & \frac{1}{4} & \frac{1}{5} \\ \frac{1}{3} & \frac{1}{4} & \frac{1}{5} & \frac{1}{6} \\ \frac{1}{4} & \frac{1}{5} & \frac{1}{6} & \frac{1}{7} \end{bmatrix}. \text{ The inverse is } \begin{bmatrix} 16 & -120 & 240 & -140 \\ -120 & 1200 & -2700 & 1680 \\ 240 & -2700 & 6480 & -4200 \\ -140 & 1680 & -4200 & 2800 \end{bmatrix}.$$

Now find the inverse of this matrix. Discuss the problems you had, if any, or explain what potential problems exist. If you used elementary row operations to find the inverse matrix, explain how you could modify your program to find the determinant of the matrix.

4. Each elementary row operation can be expressed as a matrix. For 2 × 2 matrices, the elementary row operation matrices are

$$R_1 \leftrightarrow R_2: \begin{bmatrix} 0 & 1 \\ 1 & 0 \end{bmatrix} \qquad cR_1: \begin{bmatrix} c & 0 \\ 0 & 1 \end{bmatrix} \qquad cR_2: \begin{bmatrix} 1 & 0 \\ 0 & c \end{bmatrix}$$

$$cR_1 + R_2: \begin{bmatrix} 1 & 0 \\ c & 1 \end{bmatrix} \qquad cR_2 + R_1: \begin{bmatrix} 1 & c \\ 0 & 1 \end{bmatrix}$$

Here are some examples starting with $A = \begin{bmatrix} 2 & 7 \\ 1 & 4 \end{bmatrix}$. At each step, the resulting matrix is used.

$$\frac{1}{2}R_1: \begin{bmatrix} \frac{1}{2} & 0 \\ 0 & 1 \end{bmatrix}\begin{bmatrix} 2 & 7 \\ 1 & 4 \end{bmatrix} = \begin{bmatrix} 1 & \frac{7}{2} \\ 1 & 4 \end{bmatrix}$$

$$-1R_1 + R_2: \begin{bmatrix} 1 & 0 \\ -1 & 1 \end{bmatrix}\begin{bmatrix} 1 & \frac{7}{2} \\ 1 & 4 \end{bmatrix} = \begin{bmatrix} 1 & \frac{7}{2} \\ 0 & \frac{1}{2} \end{bmatrix}$$

$$2R_2: \begin{bmatrix} 1 & 0 \\ 0 & 2 \end{bmatrix}\begin{bmatrix} 1 & \frac{7}{2} \\ 0 & \frac{1}{2} \end{bmatrix} = \begin{bmatrix} 1 & \frac{7}{2} \\ 0 & 1 \end{bmatrix}$$

$$-\frac{7}{2}R_2 + R_1: \begin{bmatrix} 1 & -\frac{7}{2} \\ 0 & 1 \end{bmatrix}\begin{bmatrix} 1 & \frac{7}{2} \\ 0 & 1 \end{bmatrix} = \begin{bmatrix} 1 & 0 \\ 0 & 1 \end{bmatrix}$$

The elementary row operation matrices used for this illustration were chosen so as to produce the identity matrix. Multiply all the elementary row operation matrices (the shaded ones) in the order in which they were used. Explain why this product is A^{-1}. In this example, each elementary row operation matrix was placed on the left of the previous matrix. Was this necessary? Write all the elementary row operation matrices for a 3 × 3 matrix. Use the elementary row operation of multiplying a row by a constant and the product property of determinants, prove that multiplying a row of a matrix by a constant multiplies the value of the determinant of the matrix by that same constant.

True/False Exercises

In Exercises 1 to 15, answer true or false. If the answer is false, give an example.

1. If $A = \begin{bmatrix} 2 & 3 \\ 1 & 4 \end{bmatrix}$, then $A^2 = \begin{bmatrix} 4 & 9 \\ 1 & 16 \end{bmatrix}$.

2. Every matrix has an additive inverse.

3. Every square matrix has a multiplicative inverse.

4. Let the matrices A, B, and C be square matrices of order n. If $AB = AC$, then $B = C$.

5. It is possible to find the determinant of every square matrix.

6. If A and B are square matrices of order n, then

$$\det(A + B) = \det(A) + \det(B).$$

7. Cramer's Rule can be used to solve any system of three equations in three variables.

8. If A and B are matrices of order n, then $AB - BA = O$.

9. A nonsingular matrix has a multiplicative inverse.

10. If A, B, and C are square matrices of order n, then the product ABC depends on which two matrices are multiplied first. That is, $(AB)C$ produces a different result from $A(BC)$.

11. The Gauss–Jordan method for solving a system of linear equations can be applied only to systems of equations that have the same number of variables as equations.

12. If A is a square matrix of order n, then $\det(2A) = 2\det(A)$.

13. If A and B are matrices, then the product AB is defined when the number of columns of A equals the number of rows of B.

14. If A and B are square matrices of order n and $AB = O$ (the zero matrix), then $A = O$ or $B = O$.

15. If $A = \begin{bmatrix} 3 & 6 \\ -1 & -2 \end{bmatrix}$, then $A^5 = A$.

Chapter Review Exercises

In Exercises 1 to 18, perform the indicated operations. Let

$$A = \begin{bmatrix} 2 & -1 & 3 \\ 3 & 2 & -1 \end{bmatrix}, \quad B = \begin{bmatrix} 0 & -2 \\ 4 & 2 \\ 1 & -3 \end{bmatrix}, \quad C = \begin{bmatrix} 2 & 6 & 1 \\ 1 & 2 & -1 \\ 2 & 4 & -1 \end{bmatrix},$$

and $D = \begin{bmatrix} -3 & 4 & 2 \\ 4 & -2 & 5 \end{bmatrix}$.

1. $3A$
2. $-2B$
3. $-A + D$
4. $2A - 3D$

5. AB
6. DB
7. BA
8. BD

9. C^2
10. C^3
11. BAC
12. ADB

13. $AB - BA$
14. $DB - BD$

15. $(A - D)C$
16. $AC - DC$

17. C^{-1}
18. $|C|$

In Exercises 19 to 34, solve the system of equations by using the Gauss–Jordan method.

19. $\begin{cases} 2x - 3y = 7 \\ 3x - 4y = 10 \end{cases}$

20. $\begin{cases} 3x + 4y = -9 \\ 2x + 3y = -7 \end{cases}$

21. $\begin{cases} 4x - 5y = 12 \\ 3x + y = 9 \end{cases}$

22. $\begin{cases} 2x - 5y = 10 \\ 5x + 2y = 4 \end{cases}$

23. $\begin{cases} x + 2y + 3z = 5 \\ 3x + 8y + 11z = 17 \\ 2x + 6y + 7z = 12 \end{cases}$

24. $\begin{cases} x - y + 3z = 10 \\ 2x - y + 7z = 24 \\ 3x - 6y + 7z = 21 \end{cases}$

25. $\begin{cases} 2x - y - z = 4 \\ x - 2y - 2z = 5 \\ 3x - 3y - 8z = 19 \end{cases}$

26. $\begin{cases} 3x - 7y + 8z = 10 \\ x - 3y + 2z = 0 \\ 2x - 8y + 7z = 5 \end{cases}$

27. $\begin{cases} 4x - 9y + 6z = 54 \\ 3x - 8y + 8z = 49 \\ x - 3y + 2z = 17 \end{cases}$

28. $\begin{cases} 3x + 8y - 5z = 6 \\ 2x + 9y - z = -8 \\ x - 4y - 2z = 16 \end{cases}$

29. $\begin{cases} x + y + 2z = -5 \\ 2x + 3y + 5z = -13 \\ 2x + 5y + 7z = -19 \end{cases}$

30. $\begin{cases} x - 2y + 3z = 9 \\ 3x - 5y + 8z = 25 \\ x - z = 5 \end{cases}$

31. $\begin{cases} w + 2x - y + 2z = 1 \\ 3w + 8x + y + 4z = 1 \\ 2w + 7x + 3y + 2z = 0 \\ w + 3x - 2y + 5z = 6 \end{cases}$

32. $\begin{cases} w - 3x - 2y + z = -1 \\ 2w - 5x + 3z = 1 \\ 3w - 7x + 3y = -18 \\ 2w - 3x - 5y - 2z = -8 \end{cases}$

33. $\begin{cases} w + 3x + y - 4z = 3 \\ w + 4x + 3y - 6z = 5 \\ 2w + 8x + 7y - 5z = 11 \\ 2w + 5x - 6z = 4 \end{cases}$

34. $\begin{cases} w + 4x - 2y + 3z = 6 \\ 2w + 9x - y + 5z = 13 \\ w + 7x + 6y + 5z = 9 \\ 3w + 14x + 7z = 20 \end{cases}$

In Exercises 35 to 46, find the inverse, if it exists, of the given matrix.

35. $\begin{bmatrix} 2 & -2 \\ 3 & -2 \end{bmatrix}$

36. $\begin{bmatrix} 3 & 4 \\ 2 & 3 \end{bmatrix}$

37. $\begin{bmatrix} -2 & 3 \\ 2 & 4 \end{bmatrix}$

38. $\begin{bmatrix} 5 & -4 \\ 3 & 2 \end{bmatrix}$

39. $\begin{bmatrix} 1 & 2 & 1 \\ 2 & 6 & 4 \\ 3 & 8 & 6 \end{bmatrix}$

40. $\begin{bmatrix} 1 & -3 & 2 \\ 3 & -8 & 7 \\ 2 & -3 & 6 \end{bmatrix}$

41. $\begin{bmatrix} 3 & -2 & 7 \\ 2 & -1 & 5 \\ 3 & 0 & 10 \end{bmatrix}$

42. $\begin{bmatrix} 4 & 9 & -11 \\ 3 & 7 & -8 \\ 2 & 6 & -3 \end{bmatrix}$

43. $\begin{bmatrix} 1 & -1 & 2 & 3 \\ 2 & -1 & 6 & 5 \\ 3 & -1 & 9 & 6 \\ 2 & -2 & 4 & 7 \end{bmatrix}$

44. $\begin{bmatrix} 1 & 2 & -2 & 1 \\ 3 & 7 & -3 & 1 \\ 2 & 7 & 4 & 3 \\ 1 & 4 & 2 & 4 \end{bmatrix}$

45. $\begin{bmatrix} 3 & 7 & -1 & 8 \\ 2 & 5 & 0 & 5 \\ 3 & 6 & -4 & 8 \\ 2 & 4 & -4 & 4 \end{bmatrix}$

46. $\begin{bmatrix} 3 & 1 & 5 & -5 \\ 2 & 1 & 4 & -3 \\ 3 & 0 & 4 & -3 \\ 4 & 1 & 8 & 1 \end{bmatrix}$

In Exercises 47 to 50, solve the given system of equations for each set of constants. Use the inverse matrix method.

47. $\begin{cases} 3x + 4y = b_1 \\ 2x + 3y = b_2 \end{cases}$

   a. $b_1 = 2, b_2 = -3$
   b. $b_1 = -2, b_2 = 4$

48. $\begin{cases} 2x - 5y = b_1 \\ 3x - 7y = b_2 \end{cases}$

   a. $b_1 = -3, b_2 = 4$
   b. $b_1 = 2, b_2 = -5$

49. $\begin{cases} 2x + y - z = b_1 \\ 4x + 4y + z = b_2 \\ 2x + 2y - 3z = b_3 \end{cases}$

   a. $b_1 = -1, b_2 = 2, b_3 = 4$
   b. $b_1 = -2, b_2 = 3, b_3 = 0$

50. $\begin{cases} 3x - 2y + z = b_1 \\ 3x - y + 3z = b_2 \\ 6x - 4y + z = b_3 \end{cases}$

   a. $b_1 = 0, b_2 = 3, b_3 = -2$
   b. $b_1 = 1, b_2 = 2, b_3 = -4$

In Exercises 51 to 58, evaluate the determinants by using elementary row or column operations.

51. $\begin{vmatrix} 2 & 6 & 4 \\ 1 & 2 & 1 \\ 3 & 8 & 6 \end{vmatrix}$

52. $\begin{vmatrix} 3 & 0 & 10 \\ 3 & -2 & 7 \\ 2 & -1 & 5 \end{vmatrix}$

53. $\begin{vmatrix} 3 & -8 & 7 \\ 2 & -3 & 6 \\ 1 & -3 & 2 \end{vmatrix}$

54. $\begin{vmatrix} 4 & 9 & -11 \\ 2 & 6 & -3 \\ 3 & 7 & -8 \end{vmatrix}$

55. $\begin{vmatrix} 1 & -1 & 2 & 1 \\ 2 & -1 & 6 & 3 \\ 3 & -1 & 8 & 7 \\ 3 & 0 & 9 & 9 \end{vmatrix}$

56. $\begin{vmatrix} 1 & 2 & -2 & 3 \\ 3 & 7 & -3 & 11 \\ 2 & 3 & -5 & 11 \\ 2 & 6 & 1 & 8 \end{vmatrix}$

57. $\begin{vmatrix} 1 & 2 & -2 & 1 \\ 2 & 5 & -3 & 1 \\ 2 & 0 & -10 & 1 \\ 3 & 8 & -4 & 1 \end{vmatrix}$

58. $\begin{vmatrix} 1 & 3 & -2 & 0 \\ 3 & 11 & -4 & 4 \\ 2 & 9 & -8 & 2 \\ 3 & 12 & -10 & 2 \end{vmatrix}$

In Exercises 59 to 64, solve the system of equations by using Cramer's Rule.

59. $\begin{cases} 2x_1 - 3x_2 = 2 \\ 3x_1 + 5x_2 = 2 \end{cases}$

60. $\begin{cases} 3x_1 + 4x_2 = -3 \\ 5x_1 - 2x_2 = 2 \end{cases}$

61. $\begin{cases} 2x_1 + x_2 - 3x_3 = 2 \\ 3x_1 + 2x_2 + x_3 = 1 \\ x_1 - 3x_2 + 4x_3 = -2 \end{cases}$

62. $\begin{cases} 3x_1 + 2x_2 - x_3 = 0 \\ x_1 + 3x_2 - 2x_3 = 3 \\ 4x_1 - x_2 - 5x_3 = -1 \end{cases}$

63. $\begin{cases} 2x_2 + 5x_3 = 2 \\ 2x_1 - 5x_2 + x_3 = 4 \\ 4x_1 + 3x_2 = 2 \end{cases}$

64. $\begin{cases} 2x_1 - 3x_2 - 4x_3 = 2 \\ x_1 - 2x_2 + 2x_3 = -1 \\ 2x_1 + 7x_2 - x_3 = 2 \end{cases}$

In Exercises 65 and 66, use Cramer's Rule to solve for the indicated variable.

65. Solve for x_3.

$\begin{cases} x_1 - 3x_2 + x_3 + 2x_4 = 3 \\ 2x_1 + 7x_2 - 3x_3 + x_4 = 2 \\ -x_1 + 4x_2 + 2x_3 - 3x_4 = -1 \\ 3x_1 + x_2 - x_3 - 2x_4 = 0 \end{cases}$

66. Solve for x_2.

$\begin{cases} 2x_1 + 3x_2 - 2x_3 + x_4 = -2 \\ x_1 - x_2 - 3x_3 + 2x_4 = 2 \\ 3x_1 + 3x_2 - 4x_3 - x_4 = 4 \\ 5x_1 - 5x_2 - x_3 + 2x_4 = 7 \end{cases}$

In Exercises 67 and 68, solve the input–output problems.

67. An electronics conglomerate has three divisions, which produce computers, monitors, and disk drives. For each $1 of output, the computer division needs $.05 worth of computers, $.02 worth of monitors, and $.03 of disk drives. For each $1 of output, the monitor division needs $.06 worth of computers, $.04 worth of monitors, and $.03 of disk drives. For each $1 of output, the disk drive division requires $.08 worth of computers, $.04 worth of monitors, and $.05 worth of disk drives. Sales estimates are $30 million for the computer division, $12 million for the monitor division, and $21 million for the disk drive division. At what level should each division produce to satisfy this demand?

68. A manufacturing conglomerate has three divisions, which produce paper, lumber, and prefabricated walls. For each $1 of output, the lumber division needs $.07 worth of lumber, $.03 worth of paper, and $.03 of prefabricated walls. For each $1 of output, the paper division needs $.04 worth of lumber, $.07 worth of paper, and $.03 worth of prefabricated wall. For each $1 of output, the prefabricated walls division requires $.07 worth of lumber, $.04 worth of paper, and $.02 worth of prefabricated walls. Sales estimates are

$27 million for the lumber division, $18 million for the paper division, and $10 million for the prefabricated walls division. At what level should each division produce to satisfy this demand?

Chapter Test

1. Write the augmented matrix, coefficient matrix, and constant matrix for the system of equations

$$\begin{cases} 2x + 3y - 3z = 4 \\ 3x \quad\quad + 2z = -1 \\ 4x - 4y + 2z = 3 \end{cases}$$

2. Write a system of equations that is equivalent to the augmented matrix $\begin{bmatrix} 3 & -2 & 5 & -1 & 9 \\ 2 & 3 & -1 & 4 & 8 \\ 1 & 0 & 3 & 2 & -1 \end{bmatrix}$.

In Exercises 3 to 5, solve the system of equations by using the Gauss–Jordan method.

3. $\begin{cases} x - 2y + 3z = 10 \\ 2x - 3y + 8z = 23 \\ -x + 3y - 2z = -9 \end{cases}$

4. $\begin{cases} 2x + 6y - z = 1 \\ x + 3y - z = 1 \\ 3x + 10y - 2z = 1 \end{cases}$

5. $\begin{cases} w + 2x - 3y + 2z = 11 \\ 2w + 5x - 8y + 5z = 28 \\ -2w - 4x + 7y - z = -18 \end{cases}$

In Exercises 6 to 15, let $A = \begin{bmatrix} -1 & 3 & 2 \\ 1 & 4 & -1 \end{bmatrix}$, $B = \begin{bmatrix} 2 & -1 & 3 \\ 4 & -2 & -1 \\ 3 & 2 & 2 \end{bmatrix}$,

and $C = \begin{bmatrix} 1 & -2 & 3 \\ 2 & -3 & 8 \\ -1 & 3 & -2 \end{bmatrix}$. Perform each possible operation. If an operation is not possible, so state.

6. $-3A$

7. $A + B$

8. $3B - 2C$

9. AB

10. $AB - A$

11. CA

12. $BC - CB$

13. A^2

14. B^2

15. C^{-1}

16. Find the minor and cofactor of b_{21} for matrix B.

17. Find the determinant of B by expanding by cofactors of row 3.

18. Find the determinant of C by using elementary row operations.

19. Find the value of z for the following system of equations by using Cramer's Rule.

$$\begin{cases} 3x + 2y - z = 12 \\ 2x - 3y + 2z = -1 \\ 5x + 6y + 3z = 4 \end{cases}$$

20. A simplified economy has three major industries: mining, manufacturing, and transportation. The input–output matrix for this economy is

$$\begin{bmatrix} .15 & .23 & .11 \\ .08 & .10 & .05 \\ .16 & .11 & .07 \end{bmatrix}$$

Set up, but do not solve, a matrix equation that when solved will determine the gross output needed to satisfy consumer demand of $50 million of mining, $32 million of manufacturing, and $8 million of transportation.

12

Sequences, Series, and Probability

CASE IN POINT *A Snowflake with Infinite Perimeter*

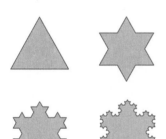

Figure 12.1

We started this book with a discussion of infinities. Now that we have reached the last chapter, it seems appropriate to end with a discussion of infinity.

We begin with an equilateral triangle each side of which is 1 unit long. The perimeter, then, is 3 units. Now construct an identical but smaller triangle onto the middle third of each side. The snowflake now has 12 line segments each of length $\frac{1}{3}$ unit. The perimeter of the snowflake is $12 \cdot \frac{1}{3}$ or 4 units. Repeat the procedure and construct identical but smaller triangles on the middle third of each of the 12 sides of the snowflake. The result is 48 line segments each of length $\frac{1}{9}$ unit. The perimeter is $48 \cdot \frac{1}{9} = \frac{16}{3}$ units. Continuing this procedure, we can show that the perimeter of each succeeding snowflake is $\frac{4}{3}$ the perimeter of the preceding one. Thus the perimeter continues to grow without bound and becomes infinite. See Figure 12.1.

Now examine the area of each snowflake. Let A be the area of the original triangle. In the second stage, each new triangle has an area that is $\frac{1}{9}$ of A. Because there are 3 new triangles, the area of the snowflake is the original area plus the area of these 3 triangles: $A + 3(\frac{1}{9})A = \frac{4}{3}A$.

For the next snowflake, each new triangle has an area equal to $\frac{1}{81}A$. The total area of the 12 new triangles is $12(\frac{1}{81}A) = \frac{4}{27}A$. The area of the snowflake is now $\frac{4}{3}A + \frac{4}{27}A = \frac{40}{27}A$.

At each stage after the first, the area added is $\frac{4}{9}$ the preceding area. Continuing in this way, it is possible to show that the area of each succeeding snowflake approaches $\frac{8}{5}A$. That is, the area is finite.

Thus we have a snowflake with infinite perimeter and finite area. Creating figures such as this snowflake is part of the realm of fractals.

593

12.1   Infinite Sequences and Summation Notation

The *ordered* list of numbers 2, 4, 8, 16, 32, ... is called an infinite sequence. The list is ordered simply because order makes a difference. The sequence 2, 8, 4, 16, 32, ... contains the same numbers, but in a different order. Therefore, it is a different infinite sequence.

An infinite sequence can be thought of as a pairing between positive numbers and real numbers. For example, 1, 4, 9, 16, 25, 36, ..., n^2... pairs a natural number with its square.

$$
\begin{array}{ccccccccc}
1 & 2 & 3 & 4 & 5 & 6 & \cdots & n & \cdots \\
\downarrow & \downarrow & \downarrow & \downarrow & \downarrow & \downarrow & & \downarrow & \\
1 & 4 & 9 & 16 & 25 & 36 & \cdots & n^2 & \cdots
\end{array}
$$

This pairing of numbers enables us to define an infinite sequence as a function with domain the positive integers.

Infinite Sequence

> An **infinite sequence** is a function whose domain is the positive integers and whose range is a set of real numbers.

Remark   Although the positive integers do not include zero, it is occasionally convenient to include zero in the domain of an infinite sequence. Also, we will frequently use the word *sequence* instead of the phrase *infinite sequence.*

As an example of a sequence, let $f(n) = 2n - 1$. The range of this function is

$$
\begin{array}{cccccc}
f(1), & f(2), & f(3), & f(4), \ldots, & f(n), & \ldots \\
1, & 3, & 5, & 7, \ldots, & 2n - 1, & \ldots
\end{array}
$$

The elements in the range of a sequence are called the **terms** of the sequence. For our example, the terms are 1, 3, 5, 7, ..., $2n - 1$, ... The **first term** of the sequence is 1, the **second term** is 3, and so on. The **nth term,** or the **general term,** is $2n - 1$.

Rather than use functional notation for sequences, it is customary to use a subscript notation. Thus a_n represents the nth term of a sequence. Using this notation, we would write

$$a_n = 2n - 1$$

Thus $a_1 = 1, a_2 = 3, a_3 = 5, a_4 = 7$.

EXAMPLE 1   **Find the Terms of a Sequence**

a. Find the first three terms of the sequence $a_n = \dfrac{1}{n(n+1)}$.

b. Find the eighth term of the sequence $a_n = \dfrac{2^n}{n^2}$.

Solution

a. $a_1 = \dfrac{1}{1(1+1)} = \dfrac{1}{2}, a_2 = \dfrac{1}{2(2+1)} = \dfrac{1}{6}, a_3 = \dfrac{1}{3(3+1)} = \dfrac{1}{12}$

b. $a_8 = \dfrac{2^8}{8^2} = \dfrac{256}{64} = 4$

■ *Try Exercise 6, page 599.*

An **alternating sequence** is one in which the signs of the terms *alternate* between positive and negative values. The sequence defined by $a_n = (-1)^{n+1} \cdot 1/n$ is an alternating sequence.

$$a_1 = (-1)^{1+1} \cdot \frac{1}{1} = 1 \qquad a_2 = (-1)^{2+1} \cdot \frac{1}{2} = -\frac{1}{2} \qquad a_3 = (-1)^{3+1} \cdot \frac{1}{3} = \frac{1}{3}$$

The first six terms of the sequence are

$$1, -\frac{1}{2}, \frac{1}{3}, -\frac{1}{4}, \frac{1}{5}, -\frac{1}{6}$$

A **recursively** defined sequence is one in which each succeeding term of the sequence is defined by using some of the preceding terms. For example, let $a_1 = 1, a_2 = 1, a_{n+1} = a_{n-1} + a_n$.

$$a_3 = a_1 + a_2 = 1 + 1 = 2 \quad \bullet \, n = 2$$
$$a_4 = a_2 + a_3 = 1 + 2 = 3 \quad \bullet \, n = 3$$
$$a_5 = a_3 + a_4 = 2 + 3 = 5 \quad \bullet \, n = 4$$
$$a_6 = a_4 + a_5 = 3 + 5 = 8 \quad \bullet \, n = 5$$

This recursive sequence 1, 1, 2, 3, 5, 8, ... is called the Fibonacci sequence, named after Leonardo Fibonacci (1180?–?1250), an Italian mathematician.

EXAMPLE 2   **Find Terms of a Sequence Defined Recursively**

Let $a_1 = 1, a_n = na_{n-1}$. Find a_2, a_3, and a_4.

Solution

$$a_2 = 2a_1 = 2 \cdot 1 = 2; \qquad a_3 = 3a_2 = 3 \cdot 2 = 6; \qquad a_4 = 4a_3 = 4 \cdot 6 = 24$$

■ *Try Exercise 28, page 599.*

It is possible to find a nonrecursive formula for the general term a_n of the sequence defined recursively in Example 2 by

$$a_1 = 1 \qquad a_n = na_{n-1}$$

Consider the term a_5 of that sequence.

$$
\begin{aligned}
a_5 &= 5a_4 \\
&= 5 \cdot 4a_3 \qquad &\bullet \ a_4 = 4a_3 \\
&= 5 \cdot 4 \cdot 3a_2 \qquad &\bullet \ a_3 = 3a_2 \\
&= 5 \cdot 4 \cdot 3 \cdot 2a_1 \qquad &\bullet \ a_2 = 2a_1 \\
&= 5 \cdot 4 \cdot 3 \cdot 2 \cdot 1 \qquad &\bullet \ a_1 = 1
\end{aligned}
$$

Continuing in this manner for a_n, we have

$$
\begin{aligned}
a_n &= na_{n-1} \\
&= n(n-1)a_{n-2} \\
&= n(n-1)(n-2)a_{n-3} \\
&\ \ \vdots \\
&= n(n-1)(n-2)(n-3) \cdots 2 \cdot 1
\end{aligned}
$$

The number $n \cdot (n-1) \cdots 3 \cdot 2 \cdot 1$ is called **n factorial** and is written $n!$.

The Factorial of a Number

If n is a positive integer, then $n!$, which is read "n factorial," is

$$n! = n \cdot (n-1) \cdots 3 \cdot 2 \cdot 1$$

We also define

$$0! = 1$$

Remark It may seem strange to define $0! = 1$, but we shall see later that it is a reasonable definition.

Examples of factorials include

$$5! = 5 \cdot 4 \cdot 3 \cdot 2 \cdot 1 = 120$$

$$10! = 10 \cdot 9 \cdot 8 \cdot 7 \cdot 6 \cdot 5 \cdot 4 \cdot 3 \cdot 2 \cdot 1 = 3,628,800$$

Note that we can write $12!$ as

$$12! = 12 \cdot 11! = 12 \cdot 11 \cdot 10! = 12 \cdot 11 \cdot 10 \cdot 9!$$

In general,

$$n! = n \cdot (n-1)!$$

EXAMPLE 3   **Evaluate Factorial Expressions**

Evaluate each factorial expression.   a. $\dfrac{8!}{5!}$ b. $6! - 4!$

Solution

a. $\dfrac{8!}{5!} = \dfrac{8 \cdot 7 \cdot 6 \cdot 5!}{5!} = 8 \cdot 7 \cdot 6 = 336$

b. $6! - 4! = (6 \cdot 5 \cdot 4 \cdot 3 \cdot 2 \cdot 1) - (4 \cdot 3 \cdot 2 \cdot 1) = 720 - 24 = 696$

■ *Try Exercise* **42,** *page 599.*

Another important way of obtaining a sequence is by adding the terms of a given sequence. For example, consider the sequence whose general term is given by $a_n = 1/2^n$. The terms of this sequence are

$$\frac{1}{2}, \frac{1}{4}, \frac{1}{8}, \frac{1}{16}, \frac{1}{32}, \cdots, \frac{1}{2^n}, \cdots$$

From this sequence we can generate a new sequence that is the sum of the terms of $1/2^n$.

$$S_1 = \frac{1}{2}$$

$$S_2 = \frac{1}{2} + \frac{1}{4} = \frac{3}{4}$$

$$S_3 = \frac{1}{2} + \frac{1}{4} + \frac{1}{8} = \frac{7}{8}$$

$$S_4 = \frac{1}{2} + \frac{1}{4} + \frac{1}{8} + \frac{1}{16} = \frac{15}{16}$$

and, in general,

$$S_n = \frac{1}{2} + \frac{1}{4} + \frac{1}{8} + \frac{1}{16} + \cdots + \frac{1}{2^n}$$

The term S_n is called the ***n*th partial sum** of the infinite sequence, and the sequence $S_1, S_2, S_3, \ldots, S_n$ is called the **sequence of partial sums.**

A convenient notation used for partial sums is called **summation notation.** The sum of the first n terms of a sequence a_n is represented by using the Greek letter Σ (sigma).

$$\sum_{i=1}^{n} a_i = a_1 + a_2 + a_3 + \cdots + a_n$$

This sum is called a **series.** The letter i is called the **index of the summation;** n is the **upper limit** of the summation; 1 is the **lower limit** of the summation.

EXAMPLE 4   **Evaluating Series**

Evaluate each series.   a. $\displaystyle\sum_{i=1}^{4} \frac{i}{i+1}$ b. $\displaystyle\sum_{j=2}^{5} (-1)^j j^2$

Solution

a. $\displaystyle\sum_{i=1}^{4} \frac{i}{i+1} = \frac{1}{2} + \frac{2}{3} + \frac{3}{4} + \frac{4}{5} = \frac{163}{60}$

b. $\displaystyle\sum_{j=2}^{5} (-1)^j j^2 = (-1)^2 2^2 + (-1)^3 3^2 + (-1)^4 4^2 + (-1)^5 5^2$

$$= 4 - 9 + 16 - 25 = -14$$

■ *Try Exercise **52**, page 599.*

Remark   Example 4 illustrates that the index of the sum does not have to be i; any letter can be used. Furthermore, the lower limit of the summation does not have to be a 1.

Properties of Summation Notation

If a_n and b_n are sequences and c a real number, then

1. $\displaystyle\sum_{i=1}^{n} (a_i \pm b_i) = \sum_{i=1}^{n} a_i \pm \sum_{i=1}^{n} b_i$

2. $\displaystyle\sum_{i=1}^{n} c a_i = c \sum_{i=1}^{n} a_i$

3. $\displaystyle\sum_{i=1}^{n} c = nc$

The proof of property (1) depends on the commutative and associative properties of real numbers.

$$\sum_{i=1}^{n} (a_i \pm b_i) = (a_1 \pm b_1) + (a_2 \pm b_2) + \cdots + (a_n \pm b_n)$$

$$= (a_1 + a_2 + \cdots + a_n) \pm (b_1 + b_2 + \cdots + b_n)$$

$$= \sum_{i=1}^{n} a_i \pm \sum_{i=1}^{n} b_i$$

Property (2) is proved by using the distributive property; this is left as an exercise.

To prove property (3), let $a_n = c$. That is, each a_n is equal to the same constant c. (This is called a **constant sequence**.) Then

$$\sum_{i=1}^{n} a_n = a_1 + a_2 + \cdots + a_n = \underbrace{c + c + \cdots + c}_{n \text{ terms}} = nc$$

EXERCISE SET 12.1 _____

In Exercises 1 to 24, find the first three terms and the eighth term of the sequence that has the given nth term.

1. $a_n = n(n - 1)$

2. $a_n = 2n$

3. $a_n = 1 - \dfrac{1}{n}$

4. $a_n = \dfrac{n + 1}{n}$

5. $a_n = \dfrac{(-1)^{n+1}}{n^2}$

6. $a_n = \dfrac{(-1)^{n+1}}{n(n + 1)}$

7. $a_n = \dfrac{(-1)^{2n-1}}{3n}$

8. $a_n = \dfrac{(-1)^n}{2n - 1}$

9. $a_n = \left(\dfrac{2}{3}\right)^n$

10. $a_n = \left(\dfrac{-1}{2}\right)^n$

11. $a_n = 1 + (-1)^n$

12. $a_n = 1 + (-0.1)^n$

13. $a_n = (1.1)^n$

14. $a_n = \dfrac{n}{n^2 + 1}$

15. $a_n = \dfrac{(-1)^{n+1}}{\sqrt{n}}$

16. $a_n = \dfrac{3^{n-1}}{2^n}$

17. $a_n = n!$

18. $a_n = \dfrac{n!}{(n - 1)!}$

19. $a_n = \log n$

20. $a_n = \ln n$ (natural logarithm)

21. a_n is the digit in the nth place in the decimal expansion of $\frac{1}{7}$.

22. a_n is the digit in the nth place in the decimal expansion of $\frac{1}{13}$.

23. $a_n = 3$

24. $a_n = -2$

In Exercises 25 to 34, find the first three terms of the recursively defined sequences.

25. $a_1 = 5, a_n = 2a_{n-1}$

26. $a_1 = 2, a_n = 3a_{n-1}$

27. $a_1 = 2, a_n = na_{n-1}$

28. $a_1 = 1, a_n = n^2 a_{n-1}$

29. $a_1 = 2, a_n = (a_{n-1})^2$

30. $a_1 = 4, a_n = \dfrac{1}{a_{n-1}}$

31. $a_1 = 2, a_n = 2na_{n-1}$

32. $a_1 = 2, a_n = (-3)na_{n-1}$

33. $a_1 = 3, a_n = (a_{n-1})^{1/n}$

34. $a_1 = 2, a_n = (a_{n-1})^n$

35. $a_1 = 1, a_2 = 3, a_n = \dfrac{1}{2}(a_{n-1} + a_{n-2})$. Find a_3, a_4, and a_5.

36. $a_1 = 1, a_2 = 4, a_n = (a_{n-1})(a_{n-2})$. Find a_3, a_4, and a_5.

In Exercises 37 to 44, evaluate the factorial expression.

37. $7! - 6!$

38. $(4!)^2$

39. $\dfrac{9!}{7!}$

40. $\dfrac{10!}{5!}$

41. $\dfrac{8!}{3!5!}$

42. $\dfrac{12!}{4!8!}$

43. $\dfrac{100!}{99!}$

44. $\dfrac{100!}{98!2!}$

In Exercises 45 to 58, evaluate the series.

45. $\displaystyle\sum_{i=1}^{5} i$

46. $\displaystyle\sum_{i=1}^{4} i^2$

47. $\displaystyle\sum_{i=1}^{5} i(i - 1)$

48. $\displaystyle\sum_{i=1}^{7} (2i + 1)$

49. $\displaystyle\sum_{k=1}^{4} \dfrac{1}{k}$

50. $\displaystyle\sum_{k=1}^{6} \dfrac{1}{k(k + 1)}$

51. $\displaystyle\sum_{j=1}^{8} 2j$

52. $\displaystyle\sum_{i=1}^{6} (2i + 1)(2i - 1)$

53. $\displaystyle\sum_{i=3}^{5} (-1)^i 2^i$

54. $\displaystyle\sum_{i=3}^{5} \dfrac{(-1)^i}{2^i}$

55. $\displaystyle\sum_{n=1}^{7} \log\left(\dfrac{n + 1}{n}\right)$

56. $\displaystyle\sum_{n=2}^{8} \ln\left(\dfrac{n}{n + 1}\right)$

57. $\displaystyle\sum_{k=0}^{8} \dfrac{8!}{k!(8 - k)!}$

58. $\displaystyle\sum_{k=0}^{7} \dfrac{1}{k!}$

In Exercises 59 to 66, write the given series in summation notation.

59. $\dfrac{1}{1} + \dfrac{1}{4} + \dfrac{1}{9} + \dfrac{1}{16} + \dfrac{1}{25} + \dfrac{1}{36}$

60. $2 + 4 + 6 + 8 + 10 + 12 + 14$

61. $2 - 4 + 8 - 16 + 32 - 64 + 128$

62. $1 - 8 + 27 - 64 + 125$

63. $7 + 10 + 13 + 16 + 19$

64. $30 + 26 + 22 + 18 + 14 + 10$

65. $\dfrac{1}{2} + \dfrac{1}{4} + \dfrac{1}{8} + \dfrac{1}{16}$

66. $1 - \dfrac{2}{3} + \dfrac{4}{9} - \dfrac{8}{27} + \dfrac{16}{81} - \dfrac{32}{243}$

Supplemental Exercises

67. Newton's approximation to the square root of a number is given by the recursive sequence

$$a_1 = \dfrac{N}{2} \qquad a_n = \dfrac{1}{2}\left(a_{n-1} + \dfrac{N}{a_{n-1}}\right)$$

Approximate $\sqrt{7}$ by computing a_4. Compare this result with the calculator value of $\sqrt{7} \approx 2.6457513$.

68. Use the formula in Exercise 67 to approximate $\sqrt{10}$ by finding a_5.

69. Let $a_1 = N$ and $a_n = \sqrt{a_{n-1}}$. Find a_{20} when $N = 7$. (*Hint:*

Enter seven into your calculator and then press the $\boxed{\sqrt{}}$ key nineteen times. Make a conjecture as to the value of a_{100}.)

70. Evaluate $\sum_{k=1}^{n} \dfrac{(-1)^{k+1}}{2k-1}$ for $n = 20$. Compare your answer to $\pi/4$. As n increases, the series better approximates $\pi/4$.

71. Let $a_1 = 1$ and $a_n = na_{n-1}$. Evaluate $1 + \sum_{k=1}^{4} \dfrac{1}{a_k}$. Compare your answer to the value of e, the base of the natural logarithms. As n increases, the series better approximates e.

72. It is not possible to define a sequence by giving a finite number of terms of the sequence. For example, the question "What is the next term in the sequence 2, 4, 6, 8, ...?" does not have a unique answer. Verify this statement by finding a formua for a_n such that the first four terms of the sequence are 2, 4, 6, 8 and the next term is 43. *Hint:*
$$a_n = \dfrac{(n-1)(n-2)(n-3)(n-4)}{4!} + 2n \text{ generates the sequence}$$
2, 4, 6, 8, 15 for $n = 1, 2, 3, 4,$ and 5.

73. Extend the result of Exercise 72 by finding a formula for a_n that will give the first four terms as 2, 4, 6, 8 and the fifth term x, where x is any real number.

74. Let $a_n = i^n$, where i is the imaginary unit. Find the first eight terms of the sequence defined by a_n. Find a_{237}.

75. Let $a_n = \left[\dfrac{1}{2}(-1 + i\sqrt{3}) \right]^n$. Find the first six terms of the sequence defined by a_n. Find a_{99}.

76. By using a calculator, evaluate $\sqrt{2\pi n} \left(\dfrac{n}{e} \right)^n$, where e is the base of the natural logarithms for $n = 10, 20,$ and 30. This formula is called Stirling's formula and is used as an approximation for $n!$. For $n > 20$, the error in the approximation is less than 0.1 percent.

77. Prove that $\sum_{i=1}^{n} ca_i = c \sum_{i=1}^{n} a_i$, where c is a constant.

12.2   Arithmetic Sequences and Series

Note that in the sequence
$$2, 5, 8, 11, 14, \ldots, 3n - 1, \ldots$$

the difference between successive terms is always 3. Such a sequence is an *arithmetic sequence* or an *arithmetic progression*. These sequences have the following property: The difference between successive terms is the same constant. This constant is called the *common difference*. For the sequence above, the common difference is 3.

In general, an arithmetic sequence can be defined as follows.

Arithmetic Sequence

Let d be a real number. A sequence a_n is an **arithmetic sequence** if
$$a_{i+1} - a_i = d \quad \text{for all } i$$
The number d is the **common difference** for the sequence.

Further examples of arithmetic sequences include
$$3, 8, 13, 18, \ldots, 5n - 2, \ldots$$
$$11, 7, 3, -1, \ldots, -4n + 15, \ldots$$
$$1, 2, 3, 4, \ldots, n, \ldots$$

Consider an arithmetic sequence in which the first term is a_1 and the common difference is d. By adding the common difference to each successive term of the arithmetic sequence, we can find a formula for the nth term.

$$a_1 = a_1$$

$$a_2 = a_1 + d$$

$$a_3 = a_2 + d = a_1 + d + d = a_1 + 2d$$

$$a_4 = a_3 + d = a_1 + 2d + d = a_1 + 3d$$

Note the relationship between the term number and the multiplier of d. The multiplier is one less than the term number.

Formula for the nth Term of an Arithmetic Sequence

> The **nth term of an arithmetic sequence** with common difference of d is given by
>
> $$a_n = a_1 + (n - 1)d$$

EXAMPLE 1 **Find the nth Term of an Arithmetic Sequence**

a. Find the twenty-fifth term of the arithmetic sequence whose first three terms are $-12, -6, 0$.

b. The fifteenth term of an arithmetic sequence is -3 and the first term is 25. Find the tenth term.

Solution

a. Find the common difference: $d = a_2 - a_1 = -6 - (-12) = 6$. Use the formula $a_n = a_1 + (n - 1)d$ with $n = 25$.

$$a_{25} = -12 + (25 - 1)(6)$$

$$= -12 + 24(6) = -12 + 144 = 132$$

b. Solve the equation $a_n = a_1 + (n - 1)d$ for d, given that $n = 15$, $a_1 = 25$, and $a_{15} = -3$.

$$-3 = 25 + (14)d$$

$$d = -2$$

Now find the tenth term.

$$a_{10} = 25 + (9)(-2) = 7$$

■ *Try Exercise* **16**, *page 605.*

Consider the arithmetic sequence given by

$$1, 3, 5, \ldots, 2n - 1, \ldots$$

Adding successive terms of this sequence, we generate a sequence of partial sums. The sum of the terms of an arithmetic sequence is called an **arithmetic series.**

$$S_1 = 1$$

$$S_2 = 1 + 3 = 4$$

$$S_3 = 1 + 3 + 5 = 9$$

$$S_4 = 1 + 3 + 5 + 7 = 16$$

$$S_5 = 1 + 3 + 5 + 7 + 9 = 25$$

$$\vdots \qquad \vdots$$

$$S_n = 1 + 3 + \cdots + (2n - 1)$$

The first five terms of this sequence are 1, 4, 9, 16, 25. It appears from this example that the sum of the first n odd integers is n^2. Shortly, we will be able to prove this result by using the following formula.

Formula for nth Partial Sum of an Arithmetic Sequence

The **nth partial sum S_n of an arithmetic sequence** a_n with common difference d is

$$S_n = \frac{n}{2}(a_1 + a_n)$$

Proof: We write S_n in both forward and reverse order.

$$S_n = a_1 + a_2 + a_3 + \cdots + a_{n-2} + a_{n-1} + a_n$$

$$S_n = a_n + a_{n-1} + a_{n-2} + \cdots + a_3 + a_2 + a_1$$

Add the two partial sums

$$2S_n = (a_1 + a_n) + (a_2 + a_{n-1}) + (a_3 + a_{n-2}) + \cdots$$
$$+ (a_{n-2} + a_3) + (a_{n-1} + a_2) + (a_n + a_1)$$

Consider the term $(a_3 + a_{n-2})$. Using the formula for the nth term of an arithmetic sequence, we have

$$a_3 = a_1 + (3 - 1)d = a_1 + 2d$$

$$a_{n-2} = a_1 + [(n - 2) - 1]d = a_1 + nd - 3d$$

Thus

$$a_3 + a_{n-2} = (a_1 + 2d) + (a_1 + nd - 3d)$$

$$= a_1 + (a_1 + nd - d) = a_1 + [a_1 + (n - 1)d]$$

$$= a_1 + a_n$$

In a similar manner, we can show that each term in parentheses in the equation for $2S_n$ equals $(a_1 + a_n)$. Since there are n such terms, we have

$$2S_n = n(a_1 + a_n)$$

$$S_n = \frac{n}{2}(a_1 + a_n).$$

There is an alternative form for the formula for the sum of n terms of an arithmetric sequence.

Alternative Formula for the Sum of an Arithmetic Series

The nth partial sum S_n of an arithmetic sequence with common difference d is

$$S_n = \frac{n[2a_1 + (n - 1)d]}{2}$$

The proof of this theorem is left as an exercise.

EXAMPLE 2   **Find a Partial Sum of an Arithmetic Sequence**

a. Find the sum of the first 100 positive odd integers
b. Find the sum of the first 50 terms of the arithmetic sequence whose first three terms are $2, \frac{13}{4}$, and $\frac{9}{2}$.

Solution   Use the formula $S_n = \frac{n}{2}[2a_1 + (n - 1)d]$.

a. We have $a_1 = 1$, $d = 2$, and $n = 100$. Thus

$$S_{100} = \frac{100}{2}[2(1) + (100 - 1)2] = 10,000$$

b. We have $a_1 = 2$, $d = \frac{5}{4}$, and $n = 50$. Thus

$$S_{50} = \frac{50}{2}\left[2(2) + (50 - 1)\frac{5}{4}\right] = \frac{6525}{4}$$

■ *Try Exercise* **22**, *page 605.*

The first n positive integers $1, 2, 3, 4, \ldots, n$ are part of an arithmetic sequence with a common difference of 1, $a_1 = 1$, and $a_n = n$. A formula for the sum of the first n positive integers can be found by using the formula for the nth partial sum of an arithmetic sequence.

$$S_n = \frac{n}{2}(a_1 + a_n)$$

Replacing a_1 by 1 and a_n by n,

$$S_n = \frac{n}{2}(1 + n) = \frac{n(n + 1)}{2}$$

This proves the following theorem.

Sum of the First n Positive Integers

The sum of the first n positive integers is given by

$$S_n = \frac{n(n + 1)}{2}$$

To find the sum of the first 85 positive integers, use $n = 85$.

$$S_{85} = \frac{85(85 + 1)}{2} = 3655$$

The **arithmetic mean** of two numbers a and b is $(a + b)/2$. The three numbers a, $(a + b)/2$, b form an arithmetic sequence. In general, given two numbers a and b, it is possible to insert k numbers $c_1, c_2, \ldots, c_k$ in such a way that the sequence

$$a, c_1, c_2, \ldots, c_k, b$$

is an arithmetic sequence. This is called *inserting k arithmetic means between a and b.*

EXAMPLE 3   Insert Arithmetic Means

Insert three arithmetic means between 3 and 13.

Solution   After we insert the three terms, the sequence will be

$$a = 3, c_2, c_3, c_4, b = 13$$

The first term of the sequence is 3, the fifth term is 13, and n is 5. Thus

$$a_n = a_1 + (n - 1)d$$
$$13 = 3 + 4d$$
$$d = \frac{5}{2}$$

The three arithmetic means are

$$c_1 = a + d = 3 + \frac{5}{2} = \frac{11}{2}$$

$$c_2 = a + 2d = 3 + 2\left(\frac{5}{2}\right) = 8$$

$$c_3 = a + 3d = 3 + 3\left(\frac{5}{2}\right) = \frac{21}{2}$$

■ *Try Exercise **34**, page 605.*

EXERCISE SET 12.2

In Exercises 1 to 14, find the ninth, twenty-fourth, and nth terms of the arithmetic sequence.

1. 6, 10, 14, ...

2. 7, 12, 17, ...

3. 6, 4, 2, ...

4. 11, 4, −3, ...

5. −8, −5, −2, ...

6. −15, −9, −3, ...

7. 1, 4, 7, ...

8. −4, 1, 6, ...

9. $a, a + 2, a + 4, \ldots$

10. $a - 3, a + 1, a + 5, \ldots$

11. log 7, log 14, log 28, ...

12. ln 4, ln 16, ln 64, ...

13. $\log a, \log a^2, \log a^3, \ldots$

14. $\log_2 5, \log_2 5a, \log_2 5a^2, \ldots$

15. The fourth and fifth terms of an arithmetic sequence are 13 and 15. Find the twentieth term.

16. The sixth and eighth terms of an arithmetic sequence are −14 and −20. Find the fifteenth term.

17. The fifth and seventh terms of an arithmetic sequence are −19 and −29. Find the seventeenth term.

18. The fourth and seventh terms of an arithmetic sequence are 22 and 34. Find the twenty-third term.

In Exercises 19 to 32, find the nth partial sum of the arithmetic sequence.

19. $a_n = 3n + 2; n = 10$

20. $a_n = 4n - 3; n = 12$

21. $a_n = 3 - 5n; n = 15$

22. $a_n = 1 - 2n; n = 20$

23. $a_n = 6n; n = 12$

24. $a_n = 7n; n = 14$

25. $a_n = n + 8; n = 25$

26. $a_n = n - 4; n = 25$

27. $a_n = -n; n = 30$

28. $a_n = 4 - n; n = 40$

29. $a_n = n + x; n = 12$

30. $a_n = 2n - x; n = 15$

31. $a_n = nx; n = 20$

32. $a_n = -nx; n = 14$

In Exercises 33 to 36, insert k arithmetic means between the given numbers.

33. −1 and 23; $k = 5$

34. 7 and 19; $k = 5$

35. 3 and $\dfrac{1}{2}$; $k = 4$

36. $\dfrac{11}{3}$ and 6; $k = 4$

37. Show that the sum of the first n positive odd integers is n^2.

38. Show that the sum of the first n positive even integers is $n^2 + n$.

39. Logs are stacked so that there are 25 logs in the bottom row, 24 logs in the second row, and so on, decreasing by 1 log each row. How many logs are stacked in the sixth row? How many logs are in the six rows?

40. The seating section in a theater has 27 seats in the first row, 29 seats in the second row, and so on, increasing by 2 seats each row for a total of 10 rows. How many seats are in the tenth row, and how many seats are there in the section?

41. A contest offers 15 prizes. The first prize is $5000, and each successive prize is $250 less than the preceding prize. What is the value of the fifteenth prize? What is the total amount of money distributed in prizes?

42. An exercise program calls for walking 15 minutes each day for a week. Each week thereafter, the amount of time spent walking increases by 5 minutes per day. In how many weeks will a person be walking 60 minutes each day?

43. An object dropped from a cliff will fall 16 feet the first second, 48 feet the second second, 80 feet the third second, and so on, increasing by 32 feet each second. What is the total distance the object will fall in 7 seconds?

44. The distance a ball rolls down a ramp each second is given by the arithmetic sequence whose nth term is $2n - 1$ feet. Find the distance the ball rolls during the tenth second and the total distance the ball travels in 10 seconds.

Supplemental Exercises

45. The sum of the interior angles of a triangle is 180°; the sum is 360° for a quadrilateral, 540° for a pentagon, and so on. Assuming this pattern continues, find the sum of the interior angles for a dodecagon (twelve sides).

46. If $f(x)$ is a linear function, show that $f(n)$, where n is a positive integer, is an arithmetic sequence.

47. Find the formula for a_n in terms of a_1 and n for the sequence that is defined recursively by $a_1 = 3, a_n = a_{n-1} + 5$.

48. Find a formula for a_n in terms of a_1 and n for the sequence that is defined recursively by $a_1 = 4, a_n = a_{n-1} - 3$.

49. Suppose a_n and b_n are two sequences such that $a_1 = 4$, $a_n = b_{n-1} + 5$ and $b_1 = 2, b_n = a_{n-1} + 1$. Show that a_n and b_n are arithmetic sequences. Find a_{100}.

50. Suppose a_n and b_n are two sequences such that $a_1 = 1$, $a_n = b_{n-1} + 7$ and $b_1 = -2, b_n = a_{n-1} + 1$. Show that a_n and b_n are arithmetic sequences. Find a_{50}.

51. Prove the Alternative Formula for the Sum of an Arithmetic Series.

12.3   Geometric Sequences and Series

Arithmetic sequences are characterized by a common *difference* between successive terms. A *geometric sequence* is characterized by a common *ratio* between successive terms.

The sequence

$$3, 6, 12, 24, \ldots, 3(2^{n-1}), \ldots$$

is a geometric sequence. Note that the ratio of any two successive terms is 2.

$$\frac{6}{3} = 2 \qquad \frac{12}{6} = 2 \qquad \frac{24}{12} = 2$$

Geometric Sequence

> Let r be a constant real number. A sequence is a **geometric sequence** if
>
> $$\frac{a_{i+1}}{a_i} = r \quad \text{for all positive integers } i \text{ and } r \neq 0$$

EXAMPLE 1   **Determine Whether a Sequence Is a Geometric Sequence**

Which of the following are geometric sequences?

a. $4, -2, 1, \ldots, 4\left(-\dfrac{1}{2}\right)^{n-1}, \ldots$

b. $1, 4, 9, \ldots, n^2, \ldots$

Solution   To determine whether the sequence is a geometric sequence, calculate the ratio of successive terms.

a. $\dfrac{a_{i+1}}{a_i} = \dfrac{4\left(-\dfrac{1}{2}\right)^i}{4\left(-\dfrac{1}{2}\right)^{i-1}} = -\dfrac{1}{2}.$ • Because the ratio of successive terms is a constant, the sequence is a geometric sequence.

b. $\dfrac{a_{i+1}}{a_i} = \dfrac{(i+1)^2}{i^2} = \left(1 + \dfrac{1}{i}\right)^2$ • Because the ratio of successive terms is not a constant, the sequence is not a geometric sequence.

■ *Try Exercise 6, page 612.*

Consider a geometric sequence in which the first term is a_1 and the common ratio is r. By multiplying each successive term of the geometric sequence by the common ratio, we can derive a formula for the nth term.

$$a_1 = a_1$$

$$a_2 = a_1 r$$

$$a_3 = a_2 r = (a_1 r)r = a_1 r^2$$

$$a_4 = a_3 r = (a_1 r^2)r = a_1 r^3$$

Note the relationship between the number of the term and the number that is the exponent on r. The exponent on r is one less than the number of the term. With this observation, we can write a formula for the nth term of a geometric sequence.

nth Term of a Geometric Sequence

The **nth term of a geometric sequence** with first term a_1 and common ratio r is

$$a_n = a_1 r^{n-1}$$

EXAMPLE 2 Find the nth Term of a Geometric Sequence

Find the nth term of the geometric sequence whose first three terms are

a. $4, 8/3, 16/9, \ldots$ b. $5, -10, 20, \ldots$

Solution

a. $r = \dfrac{8/3}{4} = \dfrac{2}{3}$ and $a_1 = 4$. Thus $a_n = 4\left(\dfrac{2}{3}\right)^{n-1}$.

b. $r = \dfrac{-10}{5} = -2$ and $a_1 = 5$. Thus $a_n = 5(-2)^{n-1}$.

■ *Try Exercise 18, page 612.*

Adding the terms of a geometric sequence, we can define the nth partial sum of a geometric sequence in a manner similar to that of an arithmetic sequence. Consider the geometric sequence $1, 2, 4, 8, \ldots, 2^{n-1}, \ldots$.

$$S_1 = 1$$

$$S_2 = 1 + 2 = 3$$

$$S_3 = 1 + 2 + 4 = 7$$

$$S_4 = 1 + 2 + 4 + 8 = 15$$

$$\vdots \qquad \vdots$$

$$S_n = 1 + 2 + 4 + 8 + \cdots + 2^{n-1}$$

The first four terms of the sequence of partial sums are $1, 3, 7, 15$.

To find a general formula for S_n, the nth term of the sequence of partial sums of a geometric sequence, let

$$S_n = a_1 + a_1 r + a_1 r^2 + \cdots + a_1 r^{n-1}$$

Multiply each side of this equation by r.

$$S_n = a_1 + a_1 r + a_1 r^2 + \cdots + a_1 r^{n-2} + a_1 r^{n-1}$$

$$r S_n = \qquad a_1 r + a_1 r^2 + \cdots + a_1 r^{n-2} + a_1 r^{n-1} + a_1 r^n$$

Subtract the two equations.

$$S_n - r S_n = a_1 - a_1 r^n$$

$$S_n (1 - r) = a_1 (1 - r^n)$$

If $r \neq 1$, then

$$S_n = \frac{a_1 (1 - r^n)}{1 - r}$$

This proves the following theorem.

The nth Partial Sum of a Geometric Sequence

The nth partial sum of a geometric sequence with first term a_1 and common ratio r is

$$S_n = \frac{a_1 (1 - r^n)}{1 - r} \qquad r \neq 1$$

Remark   If $r = 1$, then the sequence is given by $a_1, a_1, a_1, a_1, \ldots, a_1, \ldots,$ and the nth partial sum equals na_1. Why?[1]

EXAMPLE 3   Find the nth Partial Sum of a Geometric Sequence

Find the partial sum of the given geometric sequences.

a. $5, 15, 45, \ldots, 5(3)^{n-1}, \ldots; n = 4$ b. $\displaystyle\sum_{n=1}^{17} 3\left(\frac{3}{4}\right)^{n-1}$

Solution

a.   We have $a_1 = 5$, $r = 3$, and $n = 4$. Thus

$$S_4 = \frac{5[1 - 3^4]}{1 - 3} = \frac{5(-80)}{-2} = 200$$

b.   When $n = 1$, $a_1 = 3$. The first term is 3. The second term is $\frac{9}{4}$. Therefore, the common ratio is $r = \frac{3}{4}$. Thus $S_{17} = \frac{3[1 - (3/4)^{17}]}{1 - (3/4)} \approx 11.909797$.

■ *Try Exercise 40, page 612.*

---

[1] When $r = 1$, the sequence is the constant sequence a_1. The nth partial sum of a constant sequence is na_1.

Infinite Geometric Series

Following are two examples of geometric sequences for which $|r| < 1$.

$$3, \frac{3}{4}, \frac{3}{16}, \frac{3}{64}, \frac{3}{256}, \frac{3}{1024}, \cdots \qquad \bullet \; r = \frac{1}{4}$$

$$2, -1, \frac{1}{2}, -\frac{1}{4}, \frac{1}{8}, -\frac{1}{16}, \frac{1}{32}, \cdots \qquad \bullet \; r = -\frac{1}{2}$$

Note that when the absolute value of the common ratio of a geometric sequence is less than 1, the terms of the geometric sequence approach zero as n increases. We write, for $|r| < 1$, $|r|^n \rightarrow 0$ as $n \rightarrow \infty$.

Consider again the geometric sequence

$$3, \frac{3}{4}, \frac{3}{16}, \frac{3}{64}, \frac{3}{256}, \frac{3}{1024}, \cdots$$

The nth partial sums for $n = 3, 6, 9$, and 12 are given in the following table, along with the value of r^n.

n	S_n	r^n
3	3.93750000	0.01562500
6	3.99902344	0.00024414
9	3.99998474	0.00000381
12	3.99999976	0.00000006

As n increases, S_n is closer to 4 and r^n is closer to zero. By finding more values of S_n for larger values of n, we would find that $S_n \rightarrow 4$ as $n \rightarrow \infty$. As n becomes larger and larger, S_n is the nth partial sum of more and more terms of the sequence. The sum of *all* the terms of a sequence is called an **infinite series.** If the sequence is a geometric sequence, we have an **infinite geometric series.**

Sum of an Infinite Geometric Sequence

If a_n is a geometric sequence with $|r| < 1$ and first term a_1, then the sum of the infinite geometric series is

$$S = \frac{a_1}{1 - r}$$

A formal proof of this formula requires topics that typically are studied in calculus. We can, however, give an intuitive argument. Start with the formula for the nth partial sum of a geometric sequence.

$$S_n = \frac{a_1(1 - r^n)}{1 - r}$$

When $|r| < 1$, $|r|^n \approx 0$ when n is large. Thus

$$S_n = \frac{a_1(1 - r^n)}{1 - r} \approx \frac{a_1(1 - 0)}{1 - r} = \frac{a_1}{1 - r}$$

An infinite series is represented by $\sum\limits_{n=1}^{\infty} a_n$. One application of infinite geometric series concerns repeating decimals. Consider the repeating decimal

$$0.\overline{6} = \frac{6}{10} + \frac{6}{100} + \frac{6}{1000} + \frac{6}{10,000} + \cdots$$

The right-hand side is a geometric series with $a_1 = \frac{6}{10}$ and common ratio $r = \frac{1}{10}$. Thus

$$S = \frac{6/10}{1 - (1/10)} = \frac{6/10}{9/10} = \frac{2}{3}$$

The repeating decimal $0.\overline{6} = \frac{2}{3}$. We can write any repeating decimal as a ratio of two integers by using the formula for the sum of an infinite geometric series.

EXAMPLE 4   **Find the Value of an Infinite Geometric Series**

a. Evaluate the infinite geometric series $\sum\limits_{n=0}^{\infty} \left(-\frac{2}{3}\right)^n$.

b. Write $0.3\overline{45}$ as the ratio of two integers in lowest terms.

Solution

a. To find the first term, we let $n = 0$. Then $a_1 = \left(-\frac{2}{3}\right)^0 = 1$. The common ratio $r = -\frac{2}{3}$. Thus

$$S = \frac{1}{1 - (-2/3)} = \frac{1}{(5/3)} = \frac{3}{5}$$

b. $0.3\overline{45} = \frac{3}{10} + \left[\frac{45}{1000} + \frac{45}{100,000} + \frac{45}{10,000,000} + \cdots\right]$

The terms in the brackets form an infinite geometric series. Evaluate that series with $a_1 = \frac{45}{1000}$ and $r = \frac{1}{100}$, and then add the term $\frac{3}{10}$.

$$\frac{45}{1000} + \frac{45}{100,000} + \frac{45}{10,000,000} + \cdots = \frac{45/1000}{1 - (1/100)} = \frac{1}{22}$$

$$0.3\overline{45} = \frac{3}{10} + \frac{1}{22} = \frac{19}{55}$$

■ *Try Exercise **62**, page 613.*

Caution   The sum of an infinite geometric series is not defined for $|r| \geq 1$. For example, the geometric series

$$2 + 4 + 8 + 16 + 32 + \cdots$$

increases without bound. However, applying the formula $S = a_1/(1 - r)$ with $r = 2$ produces $S = -2$, which is not correct.

In an earlier chapter we discussed compound interest by using exponential functions. As an extension of this idea, suppose that A dollars are deposited on December 31 for each of the next 5 years into an account earning i percent annual interest compounded annually. Using the compound interest formula, we can find the total value of all the deposits. The table shows the growth of the investment.

Deposit Number	Value of Each Deposit	
1	$A(1 + i)^4$	Value of first deposit after 4 years
2	$A(1 + i)^3$	Value of second deposit after 3 years
3	$A(1 + i)^2$	Value of third deposit after 2 years
4	$A(1 + i)$	Value of fourth deposit after 1 year
5	A	Value of fifth deposit

The total value of the investment after the last deposit, called the **future value** of the investment, is the sum of the values of each deposit.

$$P_5 = A + A(1 + i) + A(1 + i)^2 + A(1 + i)^3 + A(1 + i)^4$$

This is a geometric series with first term A and common ratio $1 + i$. Thus, using the formula for the nth partial sum of a geometric sequence

$$S = \frac{a_1(1 - r^n)}{1 - r}$$

we have

$$P_5 = \frac{A[1 - (1 + i)^5]}{1 - (1 + i)} = \frac{A[(1 + i)^5 - 1]}{i}$$

Deposits of equal amounts at equal intervals of time are called **annuities.** When the amounts are deposited at the end of a compounding period (as in our example), we have an **ordinary annuity.**

Future Value of an Ordinary Annuity

Let $r = i/n$ and $m = nt$, where i is the annual interest rate, n is the number of compounding periods per year, and t is the number of years. Then the future value of an ordinary annuity is given by

$$P = \frac{A[(1 + r)^m - 1]}{r}$$

EXAMPLE 5   **Find the Future Value of an Ordinary Annuity**

An employee savings plan allows any employee to deposit $25 at the end of each month into a savings account earning 6 percent annual interest compounded monthly. Find the future value of this savings plan if an employee makes the deposits for 10 years.

Solution   We are given $A = 25$, $i = 0.06$, $n = 12$, and $t = 10$. Thus,

$$r = \frac{i}{n} = \frac{0.06}{12} = 0.005 \quad \text{and} \quad m = nt = 12(10) = 120$$

$$P = \frac{25[(1 + 0.005)^{120} - 1]}{0.005} \approx 4096.9837$$

The future value is $4096.98.

■ *Try Exercise* **70,** *page 613.*

EXERCISE SET 12.3

In Exercises 1 to 12, determine which sequences are geometric. For geometric sequences, find the common ratio.

1. $4, 16, 64, \ldots, 4^n, \ldots$

2. $1, 6, 36, \ldots, 6^{n-1}, \ldots$

3. $1, \dfrac{1}{2}, \dfrac{1}{3}, \ldots, \dfrac{1}{n}, \ldots$

4. $\dfrac{1}{2}, \dfrac{1}{4}, \dfrac{1}{8}, \ldots, \dfrac{1}{2^n}, \ldots$

5. $2^x, 2^{2x}, 2^{3x}, \ldots, 2^{nx}, \ldots$

6. $e^x, -e^{2x}, e^{3x}, \ldots, (-1)^{n-1}e^{nx}, \ldots$

7. $3, 6, 12, \ldots, 3(2^{n-1}), \ldots$

8. $5, -10, 20, \ldots, 5(-2)^{n-1}, \ldots$

9. $x^2, x^4, x^6, \ldots, x^{2n}, \ldots$

10. $3x, 6x^2, 9x^3, \ldots, 3nx^n, \ldots$

11. $\ln 5, \ln 10, \ln 15, \ldots, \ln 5n, \ldots$

12. $\log x, \log x^2, \log x^4, \ldots, \log x^{2^{n-1}}, \ldots$

In Exercises 13 to 32, find the nth term of the geometric sequence.

13. $2, 8, 32, \ldots$

14. $1, 5, 25, \ldots$

15. $-4, 12, -36, \ldots$

16. $-3, 6, -12, \ldots$

17. $6, 4, \dfrac{8}{3}, \ldots$

18. $8, 6, \dfrac{9}{2}, \ldots$

19. $-6, 5, -\dfrac{25}{6}, \ldots$

20. $-2, \dfrac{4}{3}, -\dfrac{8}{9}, \ldots$

21. $9, -3, 1, \ldots$

22. $8, -\dfrac{4}{3}, \dfrac{2}{9}, \ldots$

23. $1, -x, x^2, \ldots$

24. $2, 2a, 2a^2, \ldots$

25. $c^2, c^5, c^8, \ldots$

26. $-x^2, x^4, -x^6, \ldots$

27. $\dfrac{3}{100}, \dfrac{3}{10,000}, \dfrac{3}{1,000,000}, \ldots$

28. $\dfrac{7}{10}, \dfrac{7}{10,000}, \dfrac{7}{10,000,000}, \ldots$

29. $0.5, 0.05, 0.005, \ldots$

30. $0.4, 0.004, 0.00004, \ldots$

31. $0.45, 0.0045, 0.000045, \ldots$

32. $0.234, 0.000234, 0.000000234, \ldots$

33. Find the third term of a geometric sequence whose first term is 2 and whose fifth term is 162.

34. Find the fourth term of a geometric sequence whose third term is 1 and whose eighth term is $\dfrac{1}{32}$.

35. Find the second term of a geometric sequence whose third term is 4/3 and whose sixth term is $-32/81$.

36. Find the fifth term of a geometric sequence whose fourth term is 8/9 and whose seventh term is 64/243.

In Exercises 37 to 46, find the sum of the geometric series.

37. $\displaystyle\sum_{n=1}^{5} 3^n$

38. $\displaystyle\sum_{n=1}^{7} 2^n$

39. $\displaystyle\sum_{n=1}^{6} \left(\dfrac{2}{3}\right)^n$

40. $\displaystyle\sum_{n=1}^{14} \left(\dfrac{4}{3}\right)^n$

41. $\displaystyle\sum_{n=0}^{8} \left(-\dfrac{2}{5}\right)^n$

42. $\displaystyle\sum_{n=0}^{7} \left(-\dfrac{1}{3}\right)^n$

43. $\displaystyle\sum_{n=1}^{10} (-2)^{n-1}$

44. $\displaystyle\sum_{n=0}^{7} 2(5)^n$

45. $\displaystyle\sum_{n=0}^{9} 5(3)^n$

46. $\displaystyle\sum_{n=0}^{10} 2(-4)^n$

In Exercises 47 to 56, find the sum of the infinite geometric series.

47. $\displaystyle\sum_{n=1}^{\infty} \left(\frac{1}{3}\right)^n$

48. $\displaystyle\sum_{n=1}^{\infty} \left(\frac{3}{4}\right)^n$

49. $\displaystyle\sum_{n=1}^{\infty} \left(-\frac{2}{3}\right)^n$

50. $\displaystyle\sum_{n=1}^{\infty} \left(-\frac{3}{5}\right)^n$

51. $\displaystyle\sum_{n=1}^{\infty} \left(\frac{9}{100}\right)^n$

52. $\displaystyle\sum_{n=1}^{\infty} \left(\frac{7}{10}\right)^n$

53. $\displaystyle\sum_{n=1}^{\infty} (0.1)^n$

54. $\displaystyle\sum_{n=1}^{\infty} (0.5)^n$

55. $\displaystyle\sum_{n=0}^{\infty} (-0.4)^n$

56. $\displaystyle\sum_{n=0}^{\infty} (-0.8)^n$

In Exercises 57 to 68, write each rational number as the quotient of two integers in simplest form.

57. $0.\overline{3}$

58. $0.\overline{5}$

59. $0.\overline{45}$

60. $0.\overline{63}$

61. $0.\overline{123}$

62. $0.3\overline{95}$

63. $0.4\overline{22}$

64. $0.\overline{355}$

65. $0.25\overline{4}$

66. $0.37\overline{2}$

67. $1.20\overline{84}$

68. $2.25\overline{90}$

69. Find the future value of an ordinary annuity that calls for depositing $100 at the end of every 6 months for 8 years into an account that earns 9 percent interest compounded semiannually.

70. To save for the replacement of a computer, a business deposits $250 at the end of each month into an account that earns 8 percent annual interest compounded monthly. Find the future value of the ordinary annuity in 4 years.

Supplemental Exercises

71. If the sequence a_n is a geometric sequence, make a conjecture about the sequence $\log a_n$ and give a proof.

72. If the sequence a_n is an arithmetic sequence, make a conjecture about the sequence 2^{a_n} and give a proof.

73. A remarkable set of numbers called the Cantor set, after Georg Cantor, is formed in the following way. Begin with the interval of real numbers $[0, 1]$. Remove the middle one-third. Now remove the middle one-third of the remaining interval. Continue indefinitely by removing the middle one-third of each remaining interval. Find the

sum of the lengths that have been removed. The remaining points form the Cantor set.

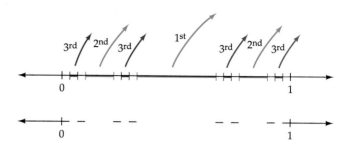

74. Consider a square with a side of length 1. Construct another square inside the first one by connecting the midpoints of the first square. What is the area of the inscribed square? Continue constructing squares in the same way. Find the area of the nth inscribed square.

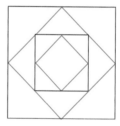

75. The product $P_n = a_1 \cdot a_2 \cdot a_3 \cdots a_n$ is called the nth partial product of a sequence. Find a formula for the nth partial product of the geometric sequence whose nth term is ar^{n-1}.

76. Let $f(x) = ab^x$, $a, b > 0$. Show that if x is restricted to positive integers n, then $f(n)$ is a geometric sequence.

77. A ball is dropped from a height of 5 feet. The ball rebounds 80 percent of the distance after each fall. Use an infinite geometric series to find the total distance the ball traveled.

78. The bob of a pendulum swings through an arc of 30 inches on its first swing. Each successive swing is 90 percent of the length of the previous swing. Find the total distance the bob travels before coming to rest.

79. Some people can trace their ancestry back ten generations, which means two parents, four grandparents, eight great-grandparents, and so on. How many grandparents does such a family tree include?

80. Find the perimeter of the snowflake curve given on page 593.

81. Find the area of the snowflake curve given on page 593.

12.4 Mathematical Induction

Consider the sequence

$$\frac{1}{1\cdot 2},\ \frac{1}{2\cdot 3},\ \frac{1}{3\cdot 4},\ \ldots,\ \frac{1}{n(n+1)},\ \ldots$$

and the sequence of partial sums for this sequence:

$$S_1 = \frac{1}{1\cdot 2} = \frac{1}{2}$$

$$S_2 = \frac{1}{1\cdot 2} + \frac{1}{2\cdot 3} = \frac{2}{3}$$

$$S_3 = \frac{1}{1\cdot 2} + \frac{1}{2\cdot 3} + \frac{1}{3\cdot 4} = \frac{3}{4}$$

$$S_4 = \frac{1}{1\cdot 2} + \frac{1}{2\cdot 3} + \frac{1}{3\cdot 4} + \frac{1}{4\cdot 5} = \frac{4}{5}$$

This pattern suggests the conjecture that

$$S_n = \frac{1}{1\cdot 2} + \frac{1}{2\cdot 3} + \frac{1}{3\cdot 4} + \cdots + \frac{1}{n(n+1)} = \frac{n}{n+1}$$

How can we be sure that the pattern does not break down when $n = 50$ or maybe $n = 2000$ or some other large number? As we will show, this conjecture is true for all values of n.

As a second example, consider the conjecture that the expression $n^2 - n + 41$ is a prime number for all positive integers. To test this conjecture, we will try various values of n.

n	$n^2 - n + 41$	
1	41	Prime
2	43	Prime
3	47	Prime
4	53	Prime
5	61	Prime

The results suggest that the conjecture is true. But again, how can we be sure? In fact, this conjecture is false when $n = 41$. In that case we have

$$n^2 - n + 41 = (41)^2 - 41 + 41 = (41)^2$$

and $(41)^2$ is not a prime.

The last example illustrates that just verifying a conjecture for a few values of n does not constitute a proof of the conjecture. To prove theorems about statements involving positive integers, a process called *mathematical induction* is used. This process is based on an axiom called the *induction axiom*.

Induction Axiom

> Suppose S is a set of positive integers with the following two properties:
>
> 1. 1 is an element of S.
> 2. If the positive integer k is in S, then $k + 1$ is in S.
>
> Then S contains all the positive integers.

Part 2 of this axiom states that if some positive integer, say 8, is in S, then $8 + 1$ or 9 is in S. But since 9 is in S, part 2 says that $9 + 1$ or 10 is in S, and so on. Part 1 states that 1 is in S. Thus 2 is in S; thus 3 is in S; thus 4 is in S; .... Therefore all the positive integers are in S.

The induction axiom is used to prove the *Principle of Mathematical Induction*.

Principle of Mathematical Induction

> Let P_n be a statement about a positive integer n. If
>
> 1. P_1 is true, and
> 2. The truth of P_k implies the truth of P_{k+1}
>
> then P_n is true for all positive integers.

Remark   Part 2 of the Principle of Mathematical Induction is referred to as the **induction hypothesis.** When applying this step, we assume the statement P_k is true and then try to prove that P_{k+1} is also true.

As an example, we will prove that the first conjecture we made in this section is true for all positive integers. Every induction proof has the two distinct parts stated in the theorem. First we must show the result is true for $n = 1$. Second, we assume the statement is true for some positive integer k and, using that assumption, prove the statement is true for $n = k + 1$.

Prove that

$$S_n = \frac{1}{1 \cdot 2} + \frac{1}{2 \cdot 3} + \frac{1}{3 \cdot 4} + \cdots + \frac{1}{n(n + 1)} = \frac{n}{n + 1}$$

for all positive integers n.

Proof:

1. For $n = 1$,

$$S_1 = \frac{1}{1(1 + 1)} = \frac{1}{1 \cdot 2} = \frac{1}{2} = \frac{1}{1 + 1}$$

The statement is true for $n = 1$.

2. Assume the statement is true for some positive integer k.

$$S_k = \frac{1}{1 \cdot 2} + \frac{1}{2 \cdot 3} + \frac{1}{3 \cdot 4} + \cdots + \frac{1}{k(k+1)} = \frac{k}{k+1}$$ • Induction hypothesis

Now verify that the formula is true when $n = k + 1$. That is, verify that

$$S_{k+1} = \frac{k+1}{(k+1)+1} = \frac{k+1}{k+2}$$ • This is the goal of the induction proof.

It is helpful, when proving a theorem about sums, to note that

$$S_{k+1} = S_k + a_{k+1}$$

Begin by noting that $a_k = \dfrac{1}{k(k+1)}$; thus, $a_{k+1} = \dfrac{1}{(k+1)(k+2)}$.

$$S_{k+1} = S_k \qquad + a_{k+1}$$

$$= \frac{k}{k+1} + \frac{1}{(k+1)(k+2)}$$ • By the induction hypothesis and substituting for a_{k+1}.

$$= \frac{k(k+2)}{(k+1)(k+2)} + \frac{1}{(k+1)(k+2)}$$

$$= \frac{k(k+2) + 1}{(k+1)(k+2)}$$

$$= \frac{k^2 + 2k + 1}{(k+1)(k+2)} = \frac{(k+1)^2}{(k+1)(k+2)}$$

$$S_{k+1} = \frac{k+1}{k+2}$$

Because we have verified the two parts of the Principle of Mathematical Induction, we can conclude that the statement is true for all positive integers.

EXAMPLE 1   **Prove by Mathematical Induction**

Prove that $1^2 + 2^2 + 3^2 + \cdots + n^2 = \dfrac{n(n+1)(2n+1)}{6}$.

Solution Verify the two parts of the Principle of Mathematical Induction.

1. Let $n = 1$.

$$S_1 = 1^2 = 1 = \frac{1(1+1)(2 \cdot 1 + 1)}{6}$$

2. Assume the statement is true for some positive integer k.

$$S_k = 1^2 + 2^2 + 3^2 + \cdots + k^2 = \frac{k(k+1)(2k+1)}{6}$$ • Induction hypothesis

Verify that the statement is true when $n = k + 1$. Show that

$$S_{k+1} = \frac{(k + 1)(k + 2)(2k + 3)}{6}$$

Because $a_k = k^2$, $a_{k+1} = (k + 1)^2$.

$$S_{k+1} = \qquad S_k \qquad + a_{k+1}$$

$$= \frac{k(k + 1)(2k + 1)}{6} + (k + 1)^2$$

$$= \frac{k(k + 1)(2k + 1)}{6} + \frac{6(k + 1)^2}{6} = \frac{k(k + 1)(2k + 1) + 6(k + 1)^2}{6}$$

$$= \frac{(k + 1)[k(2k + 1) + 6(k + 1)]}{6} = \frac{(k + 1)(2k^2 + 7k + 6)}{6}$$

$$= \frac{(k + 1)(k + 2)(2k + 3)}{6}$$

By the Principle of Mathematical Induction, the statement is true for all positive integers.

——————————————————————— ■ *Try Exercise 8, page 619.*

Mathematical induction can also be used to prove statements about sequences, products, and inequalities.

EXAMPLE 2 **Prove a Product Formula by Mathematical Induction**

Prove that

$$\left(1 + \frac{1}{1}\right)\left(1 + \frac{1}{2}\right)\left(1 + \frac{1}{3}\right)\cdots\left(1 + \frac{1}{n}\right) = n + 1$$

Solution

1. Verify for $n = 1$.

$$\left(1 + \frac{1}{1}\right) = 2 = 1 + 1$$

2. Assume the statement is true for some positive integer k.

$$P_k = \left(1 + \frac{1}{1}\right)\left(1 + \frac{1}{2}\right)\left(1 + \frac{1}{3}\right)\cdots\left(1 + \frac{1}{k}\right) = k + 1 \quad \bullet \text{ Induction hypothesis}$$

Verify that the statement is true when $n = k + 1$. That is, prove $P_{k+1} = k + 2$.

$$P_{k+1} = \left(1 + \frac{1}{1}\right)\left(1 + \frac{1}{2}\right)\left(1 + \frac{1}{3}\right)\cdots\left(1 + \frac{1}{k}\right)\left(1 + \frac{1}{k + 1}\right)$$

$$= P_k\left(1 + \frac{1}{k + 1}\right) = (k + 1)\left(1 + \frac{1}{k + 1}\right) = k + 1 + 1$$

$$P_{k+1} = k + 2$$

By the Principle of Mathematical Induction the statement is true for all positive integers.

■ *Try Exercise* **12,** *page 619.*

EXAMPLE 3   **Prove an Inequality by Mathematical Induction**

Prove that $1 + 2n \leq 3^n$ for all positive integers.

Solution

1. Let $n = 1$. Then $1 + 2(1) = 3 \leq 3^1$. The statement is true when n is 1.
2. Assume the statement is true for some positive integer k.

$$1 + 2k \leq 3^k \quad \bullet \text{ Induction hypothesis}$$

Now prove the statement is true for $n = k + 1$. That is, prove that $1 + 2(k + 1) \leq 3^{k+1}$.

$$3^{k+1} = 3^k(3)$$
$$\geq (1 + 2k)(3) \quad \bullet \text{ Why?}^2$$
$$= 6k + 3$$
$$> 2k + 2 + 1 \quad \bullet \; 6k > 2k, \text{ and } 3 = 2 + 1.$$
$$= 2(k + 1) + 1$$

Thus $1 + 2(k + 1) \leq 3^{k+1}$. By the Principle of Mathematical Induction, $1 + 2k \leq 3^k$ for all positive integers.

■ *Try Exercise* **16,** *page 619.*

The Principle of Mathematical Induction can be extended to cases where the beginning index is greater than 1.

Extended Principle of Mathematical Induction

Let P_n be a statement about a positive integer n. If

1. P_j is true for some positive integer j, and
2. For $k \geq j$ the truth of P_k implies the truth of P_{k+1} then P_n is true for all positive integers $n \geq j$.

EXAMPLE 4   **Prove an Inequality by Mathematical Induction**

For $n \geq 3$, prove that $n^2 > 2n + 1$.

---

^{2} By the induction hypothesis, $1 + 2k \leq 3^k$.

Solution

1. Let $n = 3$. Then $3^2 = 9$; $2(3) + 1 = 7$. Thus $n^2 > 2n + 1$ for $n = 3$.
2. Assume the statement is true for some positive integer $k \geq 3$.

$$k^2 > 2k + 1 \quad \bullet \text{ Induction hypothesis}$$

Verify that the statement is true when $n = k + 1$. That is, show that

$$(k + 1)^2 > 2(k + 1) + 1 = 2k + 3$$

$$(k + 1)^2 = k^2 + 2k + 1$$

$$> (2k + 1) + 2k + 1 \quad \bullet \text{ Induction hypothesis}$$

$$> 2k + 1 + 1 + 1 \quad \bullet \; 2k > 1$$

$$= 2k + 3$$

Thus $(k + 1)^2 > 2k + 3$. By the Extended Principle of Mathematical Induction, $n^2 > 2n + 1$ for all $n \geq 3$.

■ *Try Exercise* **20**, *page 620.*

EXERCISE SET 12.4

In Exercises 1 to 12, use mathematical induction to prove each statement.

1. $\displaystyle\sum_{k=1}^{n} 3k - 2 = 1 + 4 + 7 + \cdots + 3n - 2 = \dfrac{n(3n - 1)}{2}$

2. $\displaystyle\sum_{k=1}^{n} 2k = 2 + 4 + 6 + \cdots + 2n = n(n + 1)$

3. $\displaystyle\sum_{k=1}^{n} k^3 = 1 + 8 + 27 + \cdots + n^3 = \dfrac{n^2(n + 1)^2}{4}$

4. $\displaystyle\sum_{k=1}^{n} 2^k = 2 + 4 + 8 + \cdots + 2^n = 2(2^n - 1)$

5. $\displaystyle\sum_{k=1}^{n} 4k - 1 = 3 + 7 + 11 + \cdots + 4n - 1 = n(2n + 1)$

6. $\displaystyle\sum_{k=1}^{n} 3^k = 3 + 9 + 27 + \cdots + 3^n = \dfrac{3(3^n - 1)}{2}$

7. $\displaystyle\sum_{k=1}^{n} (2k - 1)^3 = 1 + 27 + 125 + \cdots + (2n - 1)^3$

$= n^2(2n^2 - 1)$

8. $\displaystyle\sum_{k=1}^{n} k(k + 1) = 2 + 6 + 12 + \cdots + n(n + 1)$

$= \dfrac{n(n + 1)(n + 2)}{3}$

9. $\displaystyle\sum_{k=1}^{n} \dfrac{1}{(2k - 1)(2k + 1)} = \dfrac{1}{1 \cdot 3} + \dfrac{1}{3 \cdot 5} + \dfrac{1}{5 \cdot 7} + \cdots$

$+ \dfrac{1}{(2n - 1)(2n + 1)} = \dfrac{n}{2n + 1}$

10. $\displaystyle\sum_{k=1}^{n} \dfrac{1}{2k(2k + 2)} = \dfrac{1}{2 \cdot 4} + \dfrac{1}{4 \cdot 6} + \dfrac{1}{6 \cdot 8} + \cdots$

$+ \dfrac{1}{2n(2n + 2)} = \dfrac{n}{4(n + 1)}$

11. $\displaystyle\sum_{k=1}^{n} k^4 = 1 + 16 + 81 + \cdots + n^4$

$= \dfrac{n(n + 1)(2n + 1)(3n^2 + 3n - 1)}{30}$

12. $P_n = \left(1 - \dfrac{1}{2}\right)\left(1 - \dfrac{1}{3}\right)\left(1 - \dfrac{1}{4}\right) \cdots \left(1 - \dfrac{1}{n + 1}\right) = \dfrac{1}{n + 1}$

In Exercises 13 to 20, use mathematical induction to prove the inequalities.

13. $\left(\dfrac{3}{2}\right)^n > n + 1, \; n \geq 4$ **14.** $\left(\dfrac{4}{3}\right)^n > n, \; n \geq 7$

15. If $0 < a < 1$, show that $a^{n+1} < a^n$ for all positive integers.

16. If $a > 1$, show that $a^{n+1} > a^n$ for all positive integers.

17. $1 \cdot 2 \cdot 3 \cdot \cdots \cdot n > 2^n, \; n \geq 4$

18. $\dfrac{1}{\sqrt{1}} + \dfrac{1}{\sqrt{2}} + \dfrac{1}{\sqrt{3}} + \cdots + \dfrac{1}{\sqrt{n}} \geq \sqrt{n}$

19. For $a > 0$, show that $(1 + a)^n \geq 1 + na$.

20. $\log_{10} n < n$ for all positive integers. [*Hint:* Because $\log_{10} x$ is an increasing function, $\log_{10}(n + 1) \leq \log_{10}(n + n)$.]

In Exercises 21 to 30, use mathematical induction to prove the statements.

21. 2 is a factor of $n^2 + n$ for all positive integers.

22. 3 is a factor of $n^3 - n$ for all positive integers.

23. 4 is a factor of $5^n - 1$ for all positive integers. (*Hint:* $5^{k+1} - 1 = 5 \cdot 5^k - 5 + 4$.)

24. 5 is a factor of $6^n - 1$ for all positive integers.

25. $(xy)^n = x^n y^n$ for all positive integers.

26. $\left(\dfrac{x}{y}\right)^n = \dfrac{x^n}{y^n}$ for all positive integers.

27. For $a \neq b$, show that $(a - b)$ is a factor of $a^n - b^n$, where n is a positive integer. [*Hint:*
$$a^{k+1} - b^{k+1} = (a \cdot a^k - ab^k) + (ab^k - b \cdot b^k).]$$

28. For $a \neq -b$, show that $(a + b)$ is a factor of $a^{2n+1} + b^{2n+1}$, where n is a positive integer. [*Hint:*
$$a^{2k+3} + b^{2k+3} = (a^{2k+2} + b^{2k+2})(a + b) - ab(a^{2k+1} + b^{2k+1}).]$$

29. $\displaystyle\sum_{k=1}^{n} ar^{k-1} = \dfrac{a(1 - r^n)}{1 - r}$ for $r \neq 1$

30. $\displaystyle\sum_{k=1}^{n} (ak + b) = \dfrac{n[(n + 1)a + 2b]}{2}$

Supplemental Exercises

In Exercises 31 to 35, use mathematical induction to prove the statements.

31. Using a calculator, find the smallest integer N for which $\log N! > N$. Now prove $\log n! > n$ for all $n > N$.

32. Let a_n be a sequence for which there is a number r and an integer N for which $a_{n+1}/a_n < r$ for $n \geq N$. Show that $a_{N+k} < a_N r^k$ for each positive integer k.

33. For constant positive integers m and n, show that $(x^m)^n = x^{mn}$.

34. Prove that $\displaystyle\sum_{i=0}^{n} \dfrac{1}{i!} \leq 3 - \dfrac{1}{n}$ for all positive integers.

35. Prove that $\left(\dfrac{n + 1}{n}\right)^n < n$ for all integers $n \geq 3$.

36. When we are giving a proof by mathematical induction, it is important that both parts of the principle of mathematical induction be verified. For example, consider the formula
$$2 + 4 + 8 + \cdots + 2^n \overset{?}{=} 2^{n+1} + 1$$

a. Show that if we assume the formula is true for some positive integer k, then the formula is true for $k + 1$.

b. Show that the formula is not true for $n = 1$. Now show that the formula is not valid for any n by showing that the left side is always an even number and the right side is always an odd number.

Thus, although the second part of the principle of induction is valid, the first part is not and the formula is not correct.

37. The Tower of Hanoi is a game that consists of three pegs and n disks of distinct diameter arranged on one of the pegs so that the largest disk is on the bottom, then the next largest, and so on. The object of the game is to move all the disks from one peg to a second peg. The rules require that only one disk be moved at a time and that a larger disk may not be placed on a smaller disk. All pegs may be used. Show that it is possible to complete the game in $2^n - 1$ moves.

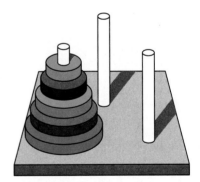

38. A legend says that in the center of the universe, high priests have the task of moving 64 golden disks from one of three diamond needles by using the rules of the game outlined in Exercise 37. When they have completed the transfer, the universe will cease to exist. If one move is made each second, and the priest started 5 billion years ago (the approximate age of the earth), how many more years, assuming the legend to be true, will the universe exist?

12.5 The Binomial Theorem

In certain situations in mathematics, it is necessary to write $(a + b)^n$ as the sum of its terms. Because $(a + b)$ is a binomial, this process is called **expanding the binomial**. For small values of n, it is relatively easy to write the expansion by using multiplication.

Earlier in the text, we found

$$(a + b)^1 = a + b$$

$$(a + b)^2 = a^2 + 2ab + b^2$$

$$(a + b)^3 = a^3 + 3a^2b + 3ab^2 + b^3$$

Building on these expansions, we can write a few more.

$$(a + b)^4 = a^4 + 4a^3b + 6a^2b^2 + 4ab^3 + b^4$$

$$(a + b)^5 = a^5 + 5a^4b + 10a^3b^2 + 10a^2b^3 + 5ab^4 + b^5$$

We could continue to build on previous expansions and eventually have quite a comprehensive list of binomial expansions. Instead, however, we will look for a theorem that will enable us to expand $(a + b)^n$ directly without multiplying.

Look at the variable parts of each expansion above. Note that for each $n = 1, 2, 3, 4, 5$

- The first term is a^n. The exponent on a decreases by 1 for each successive term.

- The exponent on b increases by 1 for each successive term. The last term is b^n.

- The degree of each term is n.

To find a pattern for the coefficients in each expansion, first note that there are $n + 1$ terms and that the coefficient of the first and last term is 1. To find the remaining coefficients, consider the expansion of $(a + b)^5$.

$$(a + b)^5 = a^5 + 5a^4b + 10a^3b^2 + 10a^2b^3 + 5ab^4 + b^5$$

$$\frac{5}{1} = 5; \quad \frac{5 \cdot 4}{2 \cdot 1} = 10; \quad \frac{5 \cdot 4 \cdot 3}{3 \cdot 2 \cdot 1} = 10; \quad \frac{5 \cdot 4 \cdot 3 \cdot 2}{4 \cdot 3 \cdot 2 \cdot 1} = 5$$

Observe from these patterns that there is a strong relationship to factorials. In fact, we can express each coefficient by using factorial notation.

$$\frac{5!}{1! \, 4!} = 5; \quad \frac{5!}{2! \, 3!} = 10; \quad \frac{5!}{3! \, 2!} = 10; \quad \frac{5!}{4! \, 1!} = 5$$

In each denominator, the first factorial is the exponent of b and the second factorial is the exponent of a.

In general, we will conjecture that the coefficient of the term $a^{n-k}b^k$ in the expansion of $(a + b)^n$ is $\dfrac{n!}{k!\,(n-k)!}$. Each coefficient of a term of a binomial expansion is called a **binomial coefficient** and is denoted by $\dbinom{n}{k}$.

Formula for a Binomial Coefficient

The coefficient of the term whose variable part is $a^{n-k}b^k$ in the expansion of $(a + b)^n$ is

$$\binom{n}{k} = \frac{n!}{k!\,(n-k)!}$$

Remark   The first term of the expansion of $(a + b)^n$ can be thought of as $a^n b^0$. In that case, we can calculate the coefficient of that term as

$$\binom{n}{0} = \frac{n!}{0!\,(n-0)!} = \frac{n!}{1 \cdot n!} = 1$$

EXAMPLE 1   **Evaluate a Binomial Coefficient**

Evaluate each binomial coefficient.

a. $\dbinom{9}{6}$ b. $\dbinom{10}{10}$

   Solution

a. $\dbinom{9}{6} = \dfrac{9!}{6!\,3!} = \dfrac{9 \cdot 8 \cdot 7 \cdot 6!}{6! \cdot 3 \cdot 2 \cdot 1} = 84$

b. $\dbinom{10}{10} = \dfrac{10!}{10!\,0!} = 1.$ • Remember that $0! = 1$.

── ■ *Try Exercise* **4**, *page 625.*

We are now ready to state the Binomial Theorem for positive integers.

Binomial Theorem for Positive Integers

If n is a positive integer, then

$$(a + b)^n = \sum_{i=0}^{n} \binom{n}{i} a^{n-i} b^i$$

$$= \binom{n}{0} a^n + \binom{n}{1} a^{n-1} b + \binom{n}{2} a^{n-2} b^2 + \cdots + \binom{n}{n} b^n$$

Proof: The proof of this theorem is by mathematical induction.

1. For $n = 1$, the statement reduces to $a + b = a + b$.

2. Assume the statement is true when $n = k$ and verify the statement for $n = k + 1$.

The induction hypothesis states that

$$(a + b)^k = \binom{k}{0}a^k + \binom{k}{1}a^{k-1}b + \binom{k}{2}a^{k-2}b^2 + \cdots + \binom{k}{i-1}a^{k-i+1}b^{i-1}$$

$$+ \binom{k}{i}a^{k-i}b^i + \cdots + \binom{k}{k-1}ab^{k-1} + \binom{k}{k}b^k$$

Now multiply each side of this equation by $(a + b)$ and collect like terms. We suggest that you try this for $(a + b)^4(a + b)$ to see how terms will be arranged.

$$(a + b)^{k+1} = \binom{k}{0}a^{k+1} + \left[\binom{k}{1} + \binom{k}{0}\right]a^k b + \left[\binom{k}{2} + \binom{k}{1}\right]a^{k-1}b^2 + \cdots$$

$$+ \left[\binom{k}{i} + \binom{k}{i-1}\right]a^{k-i+1}b^i + \cdots + \binom{k}{k}b^{k+1}$$

The coefficient of the general term $a^{k-i+1}b^i$ can be simplified.

$$\binom{k}{i} + \binom{k}{i-1} = \frac{k!}{i!\,(k - i)!} + \frac{k!}{(i - 1)!\,(k - i + 1)!}$$

$$= \frac{k!\,(k - i + 1)}{i!\,(k - i + 1)!} + \frac{k!\,i}{i!\,(k - i + 1)!} = \frac{k!\,[(k - i + 1) + i]}{i!\,(k - i + 1)!}$$

$$= \frac{k!\,(k + 1)}{i!\,(k - i + 1)!} = \frac{(k + 1)!}{i!\,(k - i + 1)!} = \binom{k + 1}{i}$$

Thus we have shown that the general term of $(a + b)^{k+1}$ is the same term as given in the theorem, with n replaced by $k + 1$. Accordingly, the theorem is true for all positive integers.

EXAMPLE 2   **Expand the Sum of Two Terms**

Expand $(2x^2 + 3)^4$.

Solution

$$(2x^2 + 3)^4 = \binom{4}{0}(2x^2)^4 + \binom{4}{1}(2x^2)^3(3) + \binom{4}{2}(2x^2)^2(3)^2$$

$$+ \binom{4}{3}(2x^2)(3)^3 + \binom{4}{4}(3)^4$$

$$= 16x^8 + 96x^6 + 216x^4 + 216x^2 + 81$$

■ *Try Exercise **18**, page 625.*

EXAMPLE 3   **Expand a Difference of Two Terms**

Expand $(\sqrt{x} - 2y)^5$.

Solution

$$(\sqrt{x} - 2y)^5 = \binom{5}{0}(\sqrt{x})^5 + \binom{5}{1}(\sqrt{x})^4(-2y) + \binom{5}{2}(\sqrt{x})^3(-2y)^2$$

$$+ \binom{5}{3}(\sqrt{x})^2(-2y)^3 + \binom{5}{4}(\sqrt{x})(-2y)^4 + \binom{5}{5}(-2y)^5$$

$$= x^{5/2} - 10x^2y + 40x^{3/2}y^2 - 80xy^3 + 80x^{1/2}y^4 - 32y^5$$

■ *Try Exercise* **20,** *page 625.*

Remark   If one of the terms a or b in $(a + b)^n$ is negative, the terms of the expansion alternate signs.

The Binomial Theorem can also be used to find a specific term in the expansion of $(a + b)^n$.

Formula for the ith Term of a Binomial Expansion

The ith term of the expansion of $(a + b)^n$ is given by

$$\binom{n}{i-1}a^{n-i+1}b^{i-1}$$

Remark   Note that the exponent on b is *one less* than the term number.

EXAMPLE 4   **Find the ith Term of a Binomial Expansion**

Find the fourth term in the expansion of $(2x^3 - 3y^2)^5$.

Solution   With $a = 2x^3$ and $b = -3y^2$, and using the last theorem with $i = 4$ and $n = 5$, we have

$$\binom{5}{3}(2x^3)^2(-3y^2)^3 = -1080x^6y^6$$

The fourth term is $-1080x^6y^6$.

■ *Try Exercise* **34,** *page 625.*

A pattern for the coefficients of the terms of an expanded binomial can be found by writing the coefficients in a triangular array known as **Pascal's Triangle.** See Figure 12.2.

Each row begins and ends with the number 1. Any other number in a row is the sum of the two closest numbers above it. For example, $4 + 6 = 10$. Thus each succeeding row can be found from the preceding row. This triangle can be used to expand a binomial for small values of n.

$$(a + b)^1:$$ 1   1

$$(a + b)^2:$$ 1   2   1

$$(a + b)^3:$$ 1   3   3   1

$$(a + b)^4:$$ 1   4   6   4   1

$$(a + b)^5:$$ 1   5   10   10   5   1

$$(a + b)^6:$$ 1   6   15   20   15   6   1

Figure 12.2

For instance, the seventh row of Pascal's Triangle is

1   7   21   35   35   21   7   1

Therefore,

$$(a + b)^7 = a^7 + 7a^6b + 21a^5b^2 + 35a^4b^3 + 35a^3b^4 + 21a^2b^5 + 7ab^6 + b^7$$

EXERCISE SET 12.5

In Exercises 1 to 8, evaluate the binomial coefficients.

1. $\binom{7}{4}$ **2.** $\binom{8}{6}$ **3.** $\binom{9}{2}$ **4.** $\binom{10}{5}$

5. $\binom{12}{9}$ **6.** $\binom{6}{5}$ **7.** $\binom{11}{0}$ **8.** $\binom{14}{14}$

In Exercises 9 to 28, expand the binomial.

9. $(x - y)^6$ **10.** $(a - b)^5$ **11.** $(x + 3)^5$

12. $(x - 5)^4$ **13.** $(2x - 1)^7$ **14.** $(2x + y)^6$

15. $(x + 3y)^6$ **16.** $(x - 4y)^5$ **17.** $(2x - 5y)^4$

18. $(3x + 2y)^4$ **19.** $\left(x + \dfrac{1}{x}\right)^6$ **20.** $(2x - \sqrt{y})^7$

21. $(x^2 - 4)^7$ **22.** $(x - y^3)^6$ **23.** $(2x^2 + y^3)^5$

24. $(2x - y^3)^6$ **25.** $\left(\dfrac{2}{x} - \dfrac{x}{2}\right)^4$ **26.** $\left(\dfrac{a}{b} + \dfrac{b}{a}\right)^3$

27. $(s^{-2} + s^2)^6$ **28.** $(2r^{-1} + s^{-1})^5$

In Exercises 29 to 36, find the indicated term without expanding.

29. $(3x - y)^{10}$; eighth term

30. $(x + 2y)^{12}$; fourth term

31. $(x + 4y)^{12}$; third term

32. $(2x - 1)^{14}$; thirteenth term

33. $(\sqrt{x} - \sqrt{y})^9$; fifth term

34. $(x^{-1/2} + x^{1/2})^{10}$; sixth term

35. $\left(\dfrac{a}{b} + \dfrac{b}{a}\right)^{11}$; ninth term

36. $\left(\dfrac{3}{x} - \dfrac{x}{3}\right)^{13}$; seventh term

37. Find the term that contains b^8 in the expansion of $(2a - b)^{10}$.

38. Find the term that contains s^7 in the expansion of $(3r + 2s)^9$.

39. Find the term that contains y^8 in the expansion of $(2x + y^2)^6$.

40. Find the term that contains b^9 in the expansion of $(a - b^3)^8$.

41. Find the middle term of $(3a - b)^{10}$.

42. Find the middle term of $(a + b^2)^8$.

43. Find the two middle terms of $(s^{-1} + s)^9$.

44. Find the two middle terms of $(x^{1/2} - y^{1/2})^7$.

In Exercises 45 to 50, use the Binomial Theorem to simplify the powers of the complex numbers.

45. $(2 - i)^4$ **46.** $(3 + 2i)^3$

47. $(1 + 2i)^5$ **48.** $(1 - 3i)^5$

49. $\left(\dfrac{\sqrt{2}}{2} + i\dfrac{\sqrt{2}}{2}\right)^8$ **50.** $\left(\dfrac{1}{2} + i\dfrac{\sqrt{3}}{2}\right)^6$

Supplemental Exercises

51. Let n be a positive integer. Expand and simplify $\dfrac{(x + h)^n - x^n}{h}$, where x is any real number and $h \neq 0$.

52. Show that $\dbinom{n}{k} = \dbinom{n}{n - k}$ for all positive integers n and k with $0 \leq k \leq n$.

53. Show that $\displaystyle\sum_{k=0}^{n} \dbinom{n}{k} = 2^n$. (*Hint:* Use the Binomial Theorem with $x = 1$, $y = 1$.)

54. Prove that $\dbinom{n}{k} + \dbinom{n}{k + 1} = \dbinom{n + 1}{k + 1}$, n and k integers, $0 \leq k \leq n$.

55. Prove that $\displaystyle\sum_{i=0}^{n} (-1)^i \dbinom{n}{i} = 0$.

56. Approximate $(0.98)^8$ by evaluating the first three terms of $(1 - 0.02)^8$.

57. Approximate $(1.02)^8$ by evaluating the first three terms of $(1 + 0.02)^8$.

There is an extension of the Binomial Theorem called the *Multinomial Theorem*. This theorem is used in determining probabilities. *Multinomial Theorem:*   If n, r, and k are positive integers, then the coefficient of $a^r b^k c^{n-r-k}$ in the expansion of $(a + b + c)^n$ is

$$\frac{n!}{r!\, k!\, (n - r - k)!}$$

In Exercises 58 to 61, use the Multinomial Theorem to find the indicated coefficient.

58. Find the coefficient of $a^2 b^3 c^5$ in the expansion of $(a + b + c)^{10}$.

59. Find the coefficient of $a^5 b^2 c^2$ in the expansion of $(a + b + c)^9$.

60. Find the coefficient of $a^4 b^5$ in the expansion of $(a + b + c)^9$.

61. Find the coefficient of $a^3 c^5$ in the expansion of $(a + b + c)^8$.

12.6 Permutations and Combinations

An electronics store offers a three-component stereo system for $250. A buyer must choose one amplifier, one tuner, and one pair of speakers. If the store has two models of amplifiers, three models of tuners, and two speaker models, how many different stereo systems could a consumer purchase?

This problem belongs to a class of problems called *counting problems.* The problem is to *count* the number of ways the conditions of the problem can be satisfied. One way to do this is to make a list and then count the items on the list. We will organize the list in a table using A_1 and A_2 for the amplifiers; T_1, T_2, and T_3 for the tuners; and S_1 and S_2 for the speakers. See Figure 12.3.

By counting the possible systems that can be purchased, we find there are 12 different systems. Another way to arrive at this result is to find the product of the number of options available.

$$
\begin{array}{ccccccc}
\text{Number of} & \times & \text{number of} & \times & \text{number of} & = & \text{number of} \\
\text{amplifiers} & & \text{tuners} & & \text{speakers} & & \text{systems} \\
2 & \times & 3 & \times & 2 & = & 12
\end{array}
$$

In some states, a standard car license plate begins with a nonzero digit, followed by three letters, followed by three more digits. What is the maximum number of car license plates of this type that could be issued? If we begin a

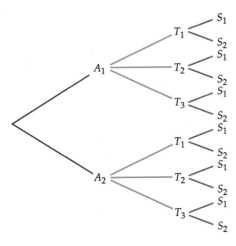

Figure 12.3

list of the possible license plates, it soon becomes apparent that listing them all would be very time-consuming and impractical.

1AAA000,   1AAA001,   1AAA002,   1AAA003, ...

Instead, the following counting principle is used. This principle forms the basis for all counting problems.

Fundamental Counting Principle

Let $T_1, T_2, T_3, \ldots, T_n$ be a sequence of n conditions. Suppose that T_1 can occur in m_1 ways, T_2 can occur in m_2 ways, T_3 can occur in m_3 ways, and so on until finally T_n can occur in m_n ways. Then the number of ways of satisfying the conditions $T_1, T_2, T_3, \ldots, T_n$ in succession is given by the product

$$m_1 m_2 m_3 \cdots m_n$$

Applying the counting principle to the license plate problem, we have

Condition	Number of Ways
T_1: a nonzero digit	$m_1 = 9$
T_2: a letter of the alphabet	$m_2 = 26$
T_3: a letter of the alphabet	$m_3 = 26$
T_4: a letter of the alphabet	$m_4 = 26$
T_5: a digit	$m_5 = 10$
T_6: a digit	$m_6 = 10$
T_7: a digit	$m_7 = 10$

Number of car
license plates $= 9 \cdot 26 \cdot 26 \cdot 26 \cdot 10 \cdot 10 \cdot 10 = 158{,}184{,}000$

EXAMPLE 1   **Apply the Fundamental Counting Principle**

An automobile dealer offers three midsized cars. A customer selecting one of these cars must choose one of three different engines, one of five different colors, and one of four different interior packages. How many different selections can the customer make?

Solution

T_1: midsized car $m_1 = 3$

T_2: engine $m_2 = 3$

T_3: color $m_3 = 5$

T_4: interior $m_4 = 4$

Number of different selections $= 3 \cdot 3 \cdot 5 \cdot 4 = 180$.

■ *Try Exercise* **12,** *page 631.*

An application of the fundamental counting principle is to determine the number of arrangements of distinct elements in a definite order.

Permutation

A **permutation** is an arrangement of distinct objects in a definite order.

For example, *abc* and *bca* are two of the possible permutations of the three elements *a*, *b*, *c*.

Consider a race with 10 runners. In how many different orders can the runners finish first, second, and third (assuming no ties)?

Any one of the 10 runners could finish first:	$m_1 = 10$
Any one of the remaining 9 runners could be second:	$m_2 = 9$
Any one of the remaining 8 runners could be third:	$m_3 = 8$

By the fundamental counting principle, there are $10 \cdot 9 \cdot 8 = 720$ possible first-, second-, and third-place finishes for the 10 runners. Using the language of permutations, we would say "There are 720 permutations of 10 objects (the runners) taken 3 (the possible finishes) at a time."

Permutations occur so frequently in counting problems that a formula rather than the counting principle is often used.

Formula for a Permutation of *n* Distinct Objects Taken *r* at a Time

The number of permutations of *n* distinct objects taken *r* at a time is

$$P(n, r) = \frac{n!}{(n - r)!}$$

EXAMPLE 2   **Find the Number of Permutations**

In how many ways can a president, vice president, secretary, and treasurer be selected from a committee of fifteen people?

Solution   There are fifteen distinct people to place in four positions. Thus $n = 15$ and $r = 4$.

$$P(15, 4) = \frac{15!}{(15 - 4)!} = \frac{15!}{11!} = \frac{15 \cdot 14 \cdot 13 \cdot 12 \cdot 11!}{11!} = 32,760$$

■ *Try Exercise* **16**, *page 632.*

EXAMPLE 3   **Find the Number of Seating Permutations**

Six people attend a movie and all sit in the same row with six seats.

a. Find the number of ways the group can sit together.

b. Find the number of ways the group can sit together if two people in the group must sit together.

c. Find the number of ways the group can sit together if two people in the group refuse to sit together.

Solution

a. There are six distinct people to place in six distinct positions. Thus $n = 6$ and $r = 6$.

$$P(6, 6) = \frac{6!}{(6 - 6)!} = \frac{6!}{0!} = \frac{6!}{1} = 720$$

b. Think of the two people who must sit together as a single object and count the number of arrangements of the *five* objects (AB), C, D, E, F. Thus $n = 5$ and $r = 5$.

$$P(5, 5) = \frac{5!}{(5 - 5)!} = \frac{5!}{0!} = \frac{5!}{1} = 120$$

There are also 120 arrangements with A and B reversed (BA), C, D, E, F. Thus the total number of arrangements is $120 + 120 = 240$.

c. From part (a), there are 720 possible seating arrangements. From part (b), there are 240 arrangements with two specific people next to each other. Thus there are $720 - 240 = 480$ arrangements where two specific people are not seated together.

■ *Try Exercise 22, page 632.*

Up to this point, we have been counting the number of distinct arrangements of objects. In some cases we may be interested in determining the number of ways of selecting objects without regard to the order of the selection. For example, suppose we want to select a committee of three people from five candidates denoted by A, B, C, D, and E. One possible committee is A, C, D. If we select D, C, A, we still have the same committee because the order of the selection is not important. An arrangement of objects for which the order of the selection is not important is a **combination.**

Formula for the Combination of *n* Objects Taken *r* at a Time

The number of combinations of n objects taken r at a time is

$$C(n, r) = \frac{n!}{r!(n - r)!}$$

EXAMPLE 4   **Find the Number of Combinations**

A standard deck of playing cards consists of fifty-two cards. How many five-card hands can be chosen from this deck?

Solution   We have $n = 52$ and $r = 5$. Thus

$$C(52, 5) = \frac{52!}{5!(52 - 5)!} = \frac{52!}{5!\, 47!} = \frac{52 \cdot 51 \cdot 50 \cdot 49 \cdot 48 \cdot 47!}{5 \cdot 4 \cdot 3 \cdot 2 \cdot 1 \cdot 47!} = 2{,}598{,}960$$

■ *Try Exercise* **20**, *page 632.*

EXAMPLE 5   **Find the Number of Combinations**

A chemist has nine samples of a solution, of which four are type A and five are type B. If the chemist chooses three of the solutions at random, in how many ways will the chemist have **a.** exactly one type A solution, **b.** more than one type A solution?

Solution

a. The chemist has chosen three solutions, one of which is type A. If one is type A, then two are type B. The number of ways of choosing one type A solution from four type A solutions is $C(4, 1)$.

$$C(4, 1) = \frac{4!}{1!(4 - 1)!} = \frac{4!}{1!\, 3!} = 4$$

The number of ways of choosing two type B solutions from five type B solutions is $C(5, 2)$.

$$C(5, 2) = \frac{5!}{2!\,(5 - 2)!} = \frac{5!}{2!\, 3!} = 10$$

By the counting principle, there are

$$C(4, 1) \cdot C(5, 2) = 4 \cdot 10 = 40$$

ways to have one type A and two type B solutions.

b. More than one type A solution means two type A and one type B or three type A and zero type B solutions. First we calculate the number of ways of choosing two type A and one type B solutions.

$$C(4, 2) = \frac{4!}{2!\, 2!} = 6 \quad \bullet \text{ Number of ways of choosing two type } A$$
$$\text{solutions from four type } A \text{ solutions}$$

$$C(5, 1) = \frac{5!}{1!\, 4!} = 5 \quad \bullet \text{ Number of ways of choosing one type } B$$
$$\text{solution from five type } B \text{ solutions}$$

The number of ways to have two type A and one type B is

$$C(4, 2) \cdot C(5, 1) = 6 \cdot 5 = 30$$

Now we calculate the number of ways of choosing three type A and zero type B solutions.

$$C(4, 3) = \frac{4!}{3!\,1!} = 4 \quad \bullet \text{ Number of ways of choosing three type } A \text{ solutions from four type } A \text{ solutions.}$$

$$C(5, 0) = \frac{5!}{0!\,5!} = 1 \quad \bullet \text{ Number of ways of choosing zero type } B \text{ solutions from five type } B \text{ solutions}$$

The number of ways to have three type A and zero type B is

$$C(4, 3) \cdot C(5, 0) = 4 \cdot 1 = 4$$

The number of ways to have more than one type A solution is

$$C(4, 2) \cdot C(5, 1) + C(4, 3) \cdot C(5, 0) = 30 + 4 = 34$$

■ *Try Exercise* **30**, *page 632.*

The difficult part of counting is determining whether to use the counting principle, the permutation formula, or the combination formula. Following is a summary of guidelines.

Guidelines for Solving Counting Problems

1. The counting principle will always work but is not always the easiest method to apply.
2. When reading a problem, ask yourself, "Is the order of the selection process important?" If the answer is yes, the arrangements are permutations. If the answer is no, the arrangements are combinations.

EXERCISE SET 12.6

In Exercises 1 to 10, evaluate each of the quantities.

1. $P(6, 2)$ **2.** $P(8, 7)$ **3.** $C(8, 4)$ **4.** $C(9, 2)$

5. $P(8, 0)$ **6.** $P(9, 9)$ **7.** $C(7, 7)$ **8.** $C(6, 0)$

9. $C(10, 4)$ **10.** $P(10, 4)$

11. A computer manufacturer offers a computer system with three different disk drives, two different monitors, and two different keyboards. How many different computer systems could a consumer purchase from this manufacturer?

12. A computer monitor produces color by blending colors on *palettes*. If a computer monitor has four palettes and each palette has four colors, how many blended colors can be formed?

13. A large conference room has four doors. At the entrance to each door there is a single light switch. How many different configurations of "on" and "off" are possible for the light switches?

14. An integer is stored in a computer's memory as a series of zeros and ones. Each memory unit contains 8 spaces for a zero or a one. The first space is used for the sign of the number, and the remaining 7 spaces are used for the integer. How many positive integers can be stored in one memory unit of this computer?

15. In how many different ways can six employees be assigned to six different jobs?

16. First-, second-, and third-place prizes are to be awarded in a dance contest in which twelve contestants are entered. In how many ways can the prizes be awarded?

17. There are five mailboxes outside a post office. In how many ways can three letters be deposited into the five boxes?

18. How many different committees of three people can be selected from nine people?

19. A professor provides a class 25 possible essay questions for an upcoming test. Of the 25 questions, the professor will ask 5 of the questions on the exam. How many different tests can the professor prepare?

20. Twenty-six people enter a tennis tournament. How many different first-round matches are possible if each player can be matched with any other player?

21. A company has more than 676 employees. Explain why there must be at least 2 employees who have the same first and last initials.

22. A car holds six passengers, three in the front seat and three in the back seat. How many different seating arrangements of six people are possible if one person refuses to sit in front and one person refuses to sit in back?

23. A committee of six people is chosen from six senators and eight representatives. How many committees are possible if there are to be three senators and three representatives on the committee?

24. The numbers 1, 2, 3, 4, 5, 6 are to be arranged. How many different arrangements are possible under each of the following conditions?

   a. All the even numbers come first?

   b. The arrangements are such that the numbers alternate between even and odd.

25. A true-false examination contains ten questions. In how many ways can a person answer the questions on this test by just guessing? Assume that all questions are answered.

26. A twenty-question, four-option multiple-choice examination is given as a pre-employment test. In how many ways could a prospective employee answer the questions on this test by just guessing? Assume that all questions are answered.

27. A state lottery game requires a person to select six differ-ent numbers from forty numbers. The order of the selection is not important. In how many ways can this be done?

28. A student must answer eight of ten questions on an exam. How many different choices can the student make?

29. A warehouse receives a shipment of ten computers, of which three are defective. Five computers are then randomly selected from the ten and delivered to a store.

   a. In how many ways can the store receive no defective computers?

   b. In how many ways can the store receive one defective computer?

   c. In how many ways can the store receive all three defective computers?

30. Fifteen students of whom seven are seniors are selected as semifinalists for a literary award. Of the fifteen students, ten finalists will be selected.

   a. In how many ways can ten finalists be selected from the fifteen students?

   b. In how many ways can the ten finalists contain three seniors?

   c. In how many ways can the ten finalists contain at least five seniors?

31. A television manufacturer uses a code for the serial number of a television. The first symbol is the letter A, B, or C and represents the location of the manufacturing plant. The next two symbols are 01, 02, ..., 12 and represents the month in which the set was manufactured. The next symbol is a 5, 6, 7, 8, or 9 and represents the year the set was manufactured. The last seven symbols are digits. How many serial numbers are possible?

32. Five cards are chosen at random from an ordinary deck of playing cards. In how many ways can the cards be chosen under each of the following conditions?

   a. All are hearts.   b. All are the same suit.
   c. Exactly three are kings.
   d. Two or more are aces.

33. A quality control inspector receives a shipment of ten computer disk drives and randomly selects three of the drives for testing. If two of the disk drives in the shipment are defective, find the number of ways in which the inspector could select at most one defective drive.

34. A basketball team has twelve members. In how many ways can five players be chosen if   a. the selection is random;   b. the two tallest players are always among the five selected?

35. The numbers 1, 2, 3, 4, 5, 6 are arranged in random order. In how many ways can the numbers 1 and 2 appear next to one another and in the order 1, 2?

36. Seven identical balls are randomly placed in seven available containers in such a way that two balls are in one container. Of the remaining six containers, each receives at most one ball. Find the number of ways in which this can be accomplished.

37. Seven points lie in a plane in such a way that no three points lie on the same line. How many lines are determined by seven points?

38. A chess tournament has twelve participants. How many games must be scheduled if every player must play every other player exactly once?

39. Eight couples attend a benefit at which two prizes are given. In how many ways can two names be randomly drawn so that the prizes are not awarded to the same couple?

40. Suppose there are twelve distinct points on a circle. How many different triangles can be formed with vertices at the given points?

41. In how many ways can a student answer a twenty-question true-false test if the student marks ten of the questions true and ten of the questions false?

42. From a group of fifteen people a committee of eight is formed. From the committee a president, secretary, and treasurer are selected. Find the number of ways in which the two consecutive operations can be carried out.

43. From a group of twenty people a committee of twelve is formed. From the committee of twelve, a subcommittee of four people is chosen. Find the number of ways in which the two consecutive operations can be carried out.

Supplemental Exercises

44. Generalize Exercise 37. That is, given n points in a plane, no three of which lie on the same line, how many lines are determined by n points?

45. Seven people are asked the month of their birth. In how many ways will **a.** no two people have a birthday in the same month? **b.** at least two people have a birthday in the same month?

46. From a penny, nickel, dime, and quarter, how many different sums of money can be formed using one or more of the coins?

47. Five sticks of equal length are broken into a short piece and a long piece. The ten pieces are randomly arranged in five pairs. In how many ways will each pair consist of a long stick and a short stick? (This exercise actually has a practical side. When cells are exposed to harmful radiation, some chromosomes break. If two long sides unite or two short sides unite, the cell dies.)

48. Four random digits are drawn (repetitions are allowed). Among the four digits, in how many ways will two or more repetitions occur?

49. An aimless tourist, standing on a street corner, tosses a coin. If the result is heads, the tourist walks one block north. If the result is tails, the tourist walks one block south. At the new corner, the coin is tossed again and the same rule applied. If the coin is tossed ten times, in how many ways will the tourist be back at the original corner? This problem is an elementary example of what is called a random walk. Random walk problems have many applications in physics, chemistry, and economics.

12.7 Introduction to Probability

Many events in the world around us have random character, such as the chances of an accident occurring on a certain freeway, the chances of winning a state lottery, and the chances that the nucleus of an atom will undergo fission. By repeatedly observing such events, it is often possible to recognize certain patterns. **Probability** is the mathematical study of random patterns.

When a weather reporter predicts 30 percent rain, the forecaster is saying that similar weather conditions have led to rain 30 times out of 100. When a fair coin is tossed, we expect heads to occur $\frac{1}{2}$, or 50 percent, of the time. The numbers 30 percent (or 0.3) and $\frac{1}{2}$ are the probabilities of the events.

An activity with an observable outcome is called an **experiment**. Examples of experiments include

1. Flipping a coin and observing the side facing upward
2. Observing the incidence of a disease in a certain population
3. Observing the length of time a person waits in a checkout line in a grocery store

The **sample space** of an experiment is the set of *all possible* outcomes of that experiment.

Consider the experiment of tossing one coin three times and recording the upward side of the coin. The sample space is

$$S = \{HHH, HHT, HTH, THH, HTT, THT, TTH, TTT\}$$

EXAMPLE 1   **List the Elements of a Sample Space**

Suppose that among five batteries, two are defective. Two batteries are randomly drawn from the five and tested for defects. List the elements in the sample space.

Solution   Label the nondefective batteries N_1, N_2, N_3 and the defective batteries D_1, D_2. The sample space is

$$S = \{N_1 D_1, N_2 D_1, N_3 D_1, N_1 D_2, N_2 D_2, N_3 D_2, N_1 N_2, N_1 N_3, N_2 N_3, D_1 D_2\}$$

■ *Try Exercise* **6**, *page 639.*

An **event** E is any subset of a sample space. For the sample space defined in Example 1, several of the events we could define are

E_1: There are no defective batteries.

E_2: At least one battery is defective.

E_3: Both batteries are defective.

Because an event is a subset of the sample space, each of these events can be expressed as a set.

$$E_1 = \{N_1 N_2, N_1 N_3, N_2 N_3\}$$
$$E_2 = \{N_1 D_1, N_2 D_1, N_3 D_1, N_1 D_2, N_2 D_2, N_3 D_2, D_1 D_2\}$$
$$E_3 = \{D_1 D_2\}$$

There are two methods by which elements are drawn from a sample space, with replacement and without replacement. *With replacement* means that after the element is drawn, it is returned to the sample space. The same element could be selected on the next drawing. When elements are drawn *without replacement,* the element is not returned to the sample space and therefore is not available for any subsequent drawing.

EXAMPLE 2   **List the Elements of an Event**

Two numbers are drawn from the digits 1, 2, 3, 4 both with replacement and without replacement. Express each event as a set.

a. E_1: The second number is greater than or equal to the first number.

b. E_2: Both numbers are less than zero.

Solution

a.   With replacement: $E_1 = \{11, 12, 13, 14, 22, 23, 24, 33, 34, 44\}$

Without replacement: $E_1 = \{12, 13, 14, 23, 24, 34\}$

b. $E_3 = \varnothing$ Choosing from the digits 1, 2, 3, 4, this event is impossible. The impossible event is denoted by the empty set or null set.

■ *Try Exercise* **14,** *page 639.*

The probability of an event is defined in terms of the concepts of sample space and event.

Probability of an Event

Let $n(S)$ and $n(E)$ represent the number of elements in the sample space S and the event E, respectively. The probability of event E, $P(E)$, is

$$P(E) = \frac{n(E)}{n(S)}$$

Remark   Because E is a subset of S, $n(E) \leq n(S)$. Thus $P(E) \leq 1$. If E is an impossible event, then $E = \varnothing$ and $n(E) = 0$. Thus $P(E) = 0$. If E is the event that *always* occurs, then $E = S$ and $n(E) = n(S)$. Thus $P(E) = 1$. Combining these elements, we have, for any event E,

$$0 \leq P(E) \leq 1.$$

EXAMPLE 3   **Calculate the Probability of an Event**

A coin is tossed three times. What is the probability of each of the following outcomes?

a. E_1: Two or more heads will appear.

b. E_2: At least one tail will appear.

Solution   First determine the number of elements in the sample space. The sample space for this experiment is

$$S = \{HHH, HHT, HTH, THH, HTT, THT, TTH, TTT\}$$

Therefore $n(S) = 8$. Now determine the number of elements in each event. Then calculate the probability of the event by using $P(E) = \dfrac{n(E)}{n(S)}$.

a. $E_1 = \{HHH, HHT, HTH, THH\}$

$$P(E_1) = \frac{n(E_1)}{n(S)} = \frac{4}{8} = \frac{1}{2}$$

b. $E_2 = \{HHT, HTH, THH, HTT, THT, TTH, TTT\}$

$$P(E_2) = \frac{n(E_2)}{n(S)} = \frac{7}{8}$$

■ *Try Exercise* **22,** *page 640.*

Calculating probabilities by listing and then counting the elements of a sample space is not always practical. Instead, we will use the counting principles developed in the last section to determine the number of elements in the sample space and in an event.

EXAMPLE 4   **Use the Counting Principles to Calculate a Probability**

A state lottery game allows a person to choose five numbers from the integers 1 to 40. Repetitions of numbers are not allowed. If three or more numbers match the numbers chosen by the lottery, the player wins a prize. Find the probability that a player will match   **a.** exactly three numbers **b.** exactly four numbers.

Solution   The sample space S is the number of ways in which five numbers can be chosen from forty numbers. This is a combination because the order of the drawing is not important.

$$n(S) = C(40, 5) = \frac{40!}{5!\,35!} = 658{,}008$$

We will call the five numbers chosen by the state lottery "lucky" and the remaining thirty-five numbers "unlucky."

a.   Let E_1 be the event a player has three lucky and therefore two unlucky numbers. The three lucky numbers are chosen from the five lucky numbers. There are $C(5, 3)$ ways to do this. The two unlucky numbers are chosen from the thirty-five unlucky numbers. There are $C(35, 2)$

ways to do this. By the counting principle, the number of ways the event E_1 can occur is

$$n(E_1) = C(5, 3) \cdot C(35, 2) = 10 \cdot 595 = 5950$$

$$P(E_1) = \frac{n(E_1)}{n(S)} = \frac{C(5, 3) \cdot C(35, 2)}{C(40, 5)} = \frac{5950}{658,008} \approx 0.009042$$

b. Let E_2 be the event a player has four lucky numbers and one unlucky number. The number of ways a person can select four lucky numbers and one unlucky number is $C(5, 4) \cdot C(35, 1)$.

$$P(E_2) = \frac{n(E_2)}{n(S)} = \frac{C(5, 4) \cdot C(35, 1)}{C(40, 5)} = \frac{175}{658,008} \approx 0.000266$$

■ *Try Exercise 32, page 640.*

The expression "one or the other of two events occurs" is written as the union of the two sets. For example, if an experiment leads to the sample space $S = \{1, 2, 3, 4, 5, 6\}$ and the events are

Draw a number less than four, $E_1 = \{1, 2, 3\}$

Draw an even number, $E_2 = \{2, 4, 6\}$

then the event $E_1 \cup E_2$ is described by drawing a number less than four *or* an even number. Thus

$$E_1 \cup E_2 = \{1, 2, 3\} \cup \{2, 4, 6\} = \{1, 2, 3, 4, 6\}$$

Two events E_1 and E_2 that cannot occur at the same time are **mutually exclusive** events. Using set notation, if $E_1 \cap E_2 = \varnothing$, then E_1 and E_2 are mutually exclusive.

For example, using the same sample space $\{1, 2, 3, 4, 5, 6\}$, if

$$E_3 = \{1, 3, 5\} \quad \text{Draw an odd number.}$$

Then $E_2 \cap E_3 = \varnothing$ and the events E_2 and E_3 are mutually exclusive. On the other hand,

$$E_1 \cap E_2 = \{2\}$$

so the events E_1 and E_2 are not mutually exclusive.

One of the axioms of probability involves the union of mutually exclusive events.

A Probability Axiom

If E_1 and E_2 are mutually exclusive events, then

$$P(E_1 \cup E_2) = P(E_1) + P(E_2)$$

If the events are not mutually exclusive, the addition rule for probabilities can be used.

Addition Rule for Probabilities

If E_1 and E_2 are two events, then

$$P(E_1 \cup E_2) = P(E_1) + P(E_2) - P(E_1 \cap E_2).$$

The probability axiom and addition rule are useful when calculating probabilities of events connected by the word *or*.

Using the calculations of Example 4, we can find the probability that a player will have three or four lucky numbers in the lottery. Because the events E_1 and E_2 as defined in Example 4 are mutually exclusive,

$$P(E_1 \cup E_2) = P(E_1) + P(E_2) = 0.009042 + 0.000266 = 0.009308$$

As an example of nonmutually exclusive events, draw a card at random from a deck of ordinary playing cards. Find the probability of drawing an ace or a heart.

$$S = \{52 \text{ ordinary playing cards}\}$$

Let $E_1 = \{\text{an ace}\}$ and $E_2 = \{\text{a heart}\}$. Then

$$P(E_1) = \frac{n(E_1)}{n(S)} = \frac{4}{52} = \frac{1}{13}, \qquad P(E_2) = \frac{n(E_2)}{n(S)} = \frac{13}{52} = \frac{1}{4}$$

We have $E_1 \cup E_2 = \{\text{an ace } or \text{ a heart}\}$ and $E_1 \cap E_2 = \{\text{ace of hearts}\}$. First, we find $P(E_1 \cap E_2)$.

$$P(E_1 \cap E_2) = \frac{n(E_1 \cap E_2)}{n(S)} = \frac{1}{52}$$

Now we can find $P(E_1 \cup E_2)$.

$$P(E_1 \cup E_2) = P(E_1) + P(E_2) - P(E_1 \cap E_2) = \frac{1}{13} + \frac{1}{4} - \frac{1}{52} = \frac{16}{52} = \frac{4}{13}$$

Two events are **independent** if the outcome of the first event does not influence the outcome of the second event. As an example, consider tossing a fair coin twice. The outcome of the first toss has no bearing on the outcome of the second toss. The two events are independent.

Now consider drawing two cards in succession from a regular deck of playing cards. The probability that the second card drawn will be an ace depends on the card drawn first.

Probability Rule for Independent Events

If E_1 and E_2 are two independent events, then the probability that both E_1 *and* E_2 will occur is

$$P(E_1) \cdot P(E_2)$$

EXAMPLE 5   **Calculate the Probability of Independent Events**

A television survey suggests that 40 percent of a city population watch a nightly news program. Three people are randomly selected from the city. Assuming the events are independent, find the probability that all three watch a nightly news program?

Solution   If E is the event that one person watches a nightly news program, then we are given that $P(E) = 0.4$. The probability that all three people watch a news program is

$$P(E) \cdot P(E) \cdot P(E) = (0.4)(0.4)(0.4) = 0.064$$

■ *Try Exercise 28, page 640.*

Following are five guidelines for calculating probabilities.

Guidelines for Calculating a Probability

1. The word "or" usually means to add the probabilities of each event.

2. The word "and" usually means to multiply the probabilities of each event.

3. The phrase "at least n" means n *or* more. At least 5 is 5 or more.

4. The phrase "at most n" means n *or* less. At most 5 is 5 or less.

5. "Exactly n" means just that. Exactly 5 heads in 7 tosses of a coin means 5 heads *and therefore* 2 tails.

EXERCISE SET 12.7

In Exercises 1 to 10, list the elements in the sample space defined by the given experiment.

1. Two people are selected from two senators and three representatives.

2. A letter is chosen at random from the word "Tennessee."

3. A fair coin is tossed and then a random integer between 1 and 4, inclusive, is selected.

4. A fair coin is tossed four times.

5. Two identical tennis balls are randomly placed in three tennis ball cans.

6. Two people are selected from among one Republican, one Democrat, and one Independent.

7. Three cards are randomly chosen from the ace of hearts, ace of spades, ace of clubs, and ace of diamonds.

8. Three letters addressed to A, B, and C are randomly put into the three envelopes addressed to A, B, and C.

9. Two vowels are randomly chosen from a, e, i, o, and u.

10. Three computer disks are randomly chosen from one defective disk and three nondefective disks.

In Exercises 11 to 15, use the sample space defined by the experiment of tossing a fair coin four times. Express each event as a subset of the sample space.

11. There are no tails.

12. There are exactly two heads.

13. There are at most two heads.

14. There are more than two heads.

15. There are twelve tails.

In Exercises 16 to 20, use the sample space defined by the experiment of choosing two random numbers, in succession, from the integers 1, 2, 3, 4, 5, and 6. The numbers are chosen with replacement. Express each event as a subset of the sample space.

16. The sum of the numbers is 7.

17. The two numbers are the same.

18. The first number is greater than the second number.

19. The second number is a 4.

20. The sum of the two numbers is greater than 1.

In Exercises 21 to 44, calculate the probabilities of the events.

21. From a deck of regular playing cards, one card is chosen at random. Find the probability that **a.** the card is a king, **b.** the card is a spade.

22. A single number is chosen from the digits 1, 2, 3, 4, 5, and 6. Find the probability that the number is an even number or is divisible by 3.

23. An economist predicts that the probability of an increase in gross domestic product (GDP) is 0.64 and that the probability of an increase in inflation is 0.55. The economist also predicts that the probability of an increase in GDP and inflation is 0.22. Find the probability of an increase in GDP or an increase in inflation.

24. Four digits are selected from the digits 1, 2, 3, and 4, and a number is formed. Find the probability that the number is greater than 3000, assuming digits can be repeated.

25. Four digits are selected from the digits 1, 2, 3, and 4, and a number is formed. Find the probability that the number is greater than 3000, assuming digits cannot be repeated.

26. An owner of a construction firm has bid for the contracts on two buildings. If the contractor estimates that the probability of getting the first contract is $\frac{1}{2}$, of getting the second contract is $\frac{1}{5}$, and of getting both contracts is $\frac{1}{10}$, find the probability that the contractor will get at least one of the two building contracts.

27. A box contains 500 envelopes, of which 50 have $100 in cash, 75 have $50 in cash, and 125 have $25. If an envelope is selected at random from this box, what is the probability that it will contain at least $50?

28. A missile radar detection system consists of two radar screens. The probability that any one of the radar screens will detect an incoming missile is 0.95. Radar detections are independent events. If a missile enters the detection space of this radar, what is the probability that at least one of the radar screens will detect it?

29. An oil drilling venture involves drilling four wells in different parts of the country. For each well, the probability that it will be profitable is 0.10, and the probability that it will be unprofitable is 0.90. If these events are independent, what is the probability of drilling at least one unprofitable well?

30. A software firm is considering marketing two newly designed spreadsheet programs, *A* and *B*. To test the appeal of the programs, the firm installs the programs in four corporations. After 2 weeks, the firm asks the corporations to evaluate each program. If the corporations have no preference, what is the probability that all four will choose product *A*?

31. A manufacturer of computer disks claims that 1 disk in 1000 will have a defect. Of the next three disks manufactured, what is the probability that all three are defective?

32. A shipment of ten calculators contains two defective calculators. Two calculators are chosen from the shipment. What is the probability that **a.** both are defective, **b.** at least one is defective?

33. A manufacturing company uses an acceptance scheme on articles it produces before shipping the articles. Crates of twenty items are readied for shipment, and three articles are randomly selected and tested for defectives. If no defectives are found, the shipment is sent. If any defectives are found, the entire box is tested. What is the probability of sending a shipment that contains three defective articles?

34. A magician claims to be able to read minds. To test this claim, five cards numbered 1 to 5 are used. A subject selects two cards from the five and concentrates on the numbers. What is the probability that the magician can correctly identify the two numbers by just guessing?

35. Consider a lottery that sells 100 tickets and awards two prizes. If you purchase 5 tickets, what is the probability of winning a prize?

36. A quality control inspector receives a shipment of twenty computer consoles. From the lot of twenty, the inspector randomly chooses three consoles. If two of the consoles in the shipment are defective, what is the probability that the inspector will choose at least one defective console?

37. A jury of twelve people is selected from thirty people: fifteen women and fifteen men. What is the probability that the jury will have six men and six women?

38. Suppose a small town has two plumbers, *A* and *B*. On a certain day, three residents select a plumber at random. What is the probability that all three residents choose plumber *A*?

39. Six persons are arranged in a line. What is the probability that two specific people, say *A* and *B*, are standing next to each other?

40. Five random digits are chosen with replacement. What is the probability that the digit 0 does not occur?

41. A fruit grower claims that three-fourths of the current orange crop have suffered frost damage. Find the probability that among four oranges, exactly three have frost damage.

42. Suppose that airplane engines operate independently and that the probability that any one engine will fail is 0.03. A plane can make a safe flight if at least one-half of its engines operate. Is a safe flight more likely to occur in a two-engine or a four-engine plane? Why?

43. Three girls and three boys are randomly placed in six seats. What is the probability that boys and girls will be in alternating seats?

44. A committee of four is chosen from three accountants and five actuaries. Find the probability that the committee consists of two accountants and two actuaries.

Supplemental Exercises

45. A club has nine members. One member starts a rumor by telling it to a second club member, who repeats the rumor to a third person, and so on. At each stage, the recipient of the rumor is chosen at random from the nine club members. What is the probability that the rumor will be told three times without its returning to the originator?

46. As a test for extrasensory perception (ESP), ten cards, five black and five white, are shuffled, and then a person looks at each card. In another room, the ESP subject attempts to guess whether the card is black or white. The ESP subject must guess black five times and white five times. If the ESP subject has no extrasensory perception, what is the probability that the subject will correctly name eight of the ten cards?

47. The telephone extensions at a university are four-digit numbers. If two telephone numbers are randomly chosen from the telephone book, what is the probability that the last two digits (and no others) match?

48. Each arrangement of the letters of the word "Tennessee" is written on a piece of paper, and all the pieces of paper are placed in a bowl. One piece of paper is selected at random. What is the probability that the first letter is a T?

Exploring Concepts with Technology

Mathematical Expectation

Expectation E is a number used to determine the fairness of a gambling game. It is defined as the probability *P* of winning a bet times the amount *A* available to win.

$$E = P \cdot A$$

A game is called fair if the expectation of the game equals the amount bet. For example, if you and a friend each bet $1 on who can guess the side facing up on the flip of a coin, then the expectation is $E = \frac{1}{2} \cdot \$2 = \$1$. Because the amount of your bet equals the expectation, the game is fair.

When a game is unfair, it benefits one of the players. If you bet $1 and your friend bets $2 on who can guess the flip of a coin, your expectation is $E = \frac{1}{2} \cdot \$3 = \$1.50$. Because your expectation is greater than the amount you bet, the game is advantageous to you. Your friend's expectation is also $1.50, which is less than the amount your friend bet. This is a disadvantage to your friend.

Keno is a game of chance played in many casinos. In this game, a large basket contains 80 balls numbered from 1 to 80. From these balls, the casino randomly chooses 20 balls. The number of ways in which the dealer can choose 20 balls from 80 is the number of combinations of 80 things chosen 20 at a time, or $C(80, 20)$.

In one particular game, a gambler can bet $1 and mark five numbers. The gambler will win a prize if three of the five numbers marked are included in the 20 numbered balls chosen by the dealer. By the counting principle, there are $C(20, 3) \cdot C(60, 2) = 2,017,800$ ways the gambler can do this. The probability of this event is $\dfrac{C(20,\,3)\cdot C(60,\,2)}{C(80,\,5)}\approx 0.0839$. The amount the gambler wins for this event is $2 (The $1 bet plus $1 from the casino), so the expectation of the gambler is approximately $.17 $(0.0839\cdot \$2)$.

Each casino has different rules and different methods of awarding prizes. The table below gives the prizes for a $2 bet for some of the possible choices a gambler can make at four casinos. Complete the Expectation column. In each case, the Mark column indicates how many numbers the gambler marked, and the Catch column shows how many of the numbers marked by the gambler were also chosen by the dealer.

Casino 1

Mark	Catch	Win	Expectation
6	4	$8	
6	5	$176	
6	6	$2960	

Casino 2

Mark	Catch	Win	Expectation
6	4	$6	
6	5	$160	
6	6	$3900	

Casino 3

Mark	Catch	Win	Expectation
6	4	$8	
6	5	$180	
6	6	$3000	

Casino 4

Mark	Catch	Win	Expectation
6	4	$6	
6	5	$176	
6	6	$3000	

Adding the expectations in each column gives you the total expectation for marking six numbers. Find the total expectation for each casino. Which casino offers the gambler the greatest expectation?

Chapter Review

12.1   Infinite Sequences and Summation Notation

- An infinite sequence is a function whose domain is the positive integers and whose range is a set of real numbers.

- An alternating sequence is one in which the signs of the terms *alternate* between plus and minus values.

- A recursive sequence is defined by using some of the preceding terms of the sequence.

- If n is a positive integer, then n factorial, $n!$, is the product of the first n positive integers.

$$n! = n(n-1)(n-2)\cdots 3\cdot 2\cdot 1$$

- If a_n is a sequence, then $S_n = \displaystyle\sum_{i=1}^{n} a_i$ is the nth partial sum of the sequence.

12.2   Arithmetic Sequences and Series

- Given that d is a constant, the sequence a_n is an arithmetic sequence if $a_{i+1} - a_i = d$ for all i. The constant d is called the common difference for the arithmetic sequence.

- The nth term of an arithmetic sequence is $a_n = a_1 + (n-1)d$.

- If a_n is an arithmetic sequence, then the nth partial sum of the sequence is given by

$$S_n = \sum_{i=1}^{n} a_i = \frac{n}{2}\,(a_1 + a_n)$$

- The nth partial sum of an arithmetic sequence is called a finite arithmetic series.

12.3   Geometric Sequences and Series

- Given that $r \neq 0$ is a constant, the sequence a_n is a geometric sequence if $a_{i+1}/a_i = r$ for all i. The ratio r is called the common ratio for the geometric sequence.

- The nth term of the geometric sequence is $a_n = a_1 r^{n-1}$, where a_1 is the first term of the sequence and r is the common ratio.

- If a_n is a geometric sequence, then the nth partial sum of the sequence is given by

$$S_n = \sum_{i=1}^{n} a_i = \frac{a_1(1 - r^n)}{1 - r} \quad r \neq 1$$

- If $|r| < 1$, then the sum of an infinite geometric series is given by

$$S = \sum_{i=1}^{\infty} a_i = \frac{a_1}{1 - r}$$

12.4   Mathematical Induction

- *Principle of Mathematical Induction*
  Let P_n be a statement that involves positive integers. If

  1. P_1 is true, and
  2. The truth of P_k implies the truth of P_{k+1}

  then P_n is true for all positive integers.

12.5   The Binomial Theorem

- *Binomial Theorem for Positive Integers*
  If n is a positive integer, then

$$(a + b)^n = \sum_{i=0}^{n} \binom{n}{i} a^{n-i} b^i$$

- The ith term of $(a + b)^n$ is

$$\binom{n}{i - 1} a^{n-i+1} b^{i-1}$$

12.6   Permutations and Combinations

- The fundamental counting principle is used to count the number of ways in which a sequence of n conditions can occur.

- A permutation is an arrangement of distinct objects in a definite order. The formula for the permutations of n distinct objects taken r at a time is

$$P(n,r) = \frac{n!}{(n - r)!}$$

- A combination is an arrangement of objects for which the order of the selection is not important. The formula for the number of combinations of n objects taken r at a time is

$$C(n,r) = \frac{n!}{r!(n - r)!}$$

12.7   Introduction to Probability

- Probability is the mathematical study of random events. The sample space of an experiment is the set of all possible outcomes of that experiment. An event is any subset of a sample space.

- If S is the sample space of an experiment and E is an event in the sample space, then the probability of an event is given by

$$P(E) = \frac{n(E)}{n(S)}$$

where $n(E)$ and $n(S)$ are the number of elements in E and S, respectively.

Essays and Projects

1. Exercise 73 in Section 12.3 explains the Cantor set, which was named after Georg Cantor. The set has quite unusual properties and is frequently employed as a counterexample for some theorems. This set exemplifies some of the problems that occur when we deal with infinity. Write an essay on the mathematical concept of infinity. Include in your essay the contributions made by Cantor and Dedekind. Also include the concepts of *cardinality, countably infinite,* and the *continuum.* What is the cardinality of the interval $[0, 1]$? of the set of real numbers $(-\infty, \infty)$? of the integers?

2. Recursive sequences were defined in Section 12.1. Building on this idea, some computer programming languages provide for *recursive subroutines.* In fact, the recursive definition of *factorial* that we gave in Section 12.1 is used in these languages as an example of a recursive subroutine. Write

an essay on computer languages that provide for recursive subroutines. What are the benefits of recursive subroutines? What are the disadvantages?

3. The introduction to probability given in this chapter provided some insight into calculating probabilities. However, some probability problems are so complex that a straightforward calculation is impossible or highly impractical. In these cases, the *Monte Carlo* method might be appropriate. Write an essay on the Monte Carlo method and some of its applications.

4. Geometric series have important applications to the mathe-

matical treatment of business finance. Write an essay that provides examples of the uses of series in finance. Some of the areas you might investigate are compound interest, annuities, and the pricing of bonds.

5. Now that you have completed this course, it is time to assess the value of taking this course and mathematics in general. Write an essay that explains the benefits of understanding mathematics. In your essay, defend or attack (if you dare) the position that the study of mathematics, at least through the level you have completed, should be required of all students graduating from college.

True/False Exercises

In Exercises 1 to 15, answer true or false. If the statement is false, give an example to show that the statement is false.

1. $0! \cdot 4! = 0$

2. $\left(\sum_{i=1}^{3} a_i \right) \left(\sum_{i=1}^{3} b_i \right) = \sum_{i=1}^{3} a_i b_i$

3. $\dfrac{n(n - 1)(n - 2) \cdots (n - k + 1)}{k!} = C(n, k)$

4. No two terms of a sequence can be equal.

5. $1, 8, 27, 64, \ldots, k^3, \ldots$ is a geometric sequence.

6. $a_1 = 2, a_{n+1} = a_n - 3$ defines an arithmetic sequence.

7. $0.\overline{9} = 1$

8. Adding all the terms of an infinite sequence produces an infinite sum.

9. Because the first step of an induction proof is normally easy, this step can be omitted.

10. In the expansion of $(a + b)^8$, the exponent on a for the fifth term is 5.

11. The counting principle states that if there are n ways to satisfy one condition and m ways to satisfy a second condition, then there are $n + m$ ways to satisfy both conditions.

12. The number of permutations of n things taken r at a time is given by $n!/r!$.

13. If E is an event in a sample space, then $0 \le P(E) \le 1$, where $P(E)$ is the probability of E.

14. If A and B are mutually exclusive events, then $P(A \cap B) = 1$.

15. If a coin is tossed five times, then the probability of observing HHHHH is the same as the probability of observing HTHHT.

Chapter Review Exercises

In Exercises 1 to 20, find the third and seventh terms of the sequence defined by a_n.

1. $a_n = n^2$

2. $a_n = n!$

3. $a_n = 3n + 2$

4. $a_n = 1 - 2n$

5. $a_n = 2^{-n}$

6. $a_n = 3^n$

7. $a_n = \dfrac{1}{n!}$

8. $a_n = \dfrac{1}{n}$

9. $a_n = \left(\dfrac{2}{3} \right)^n$

10. $a_n = \left(-\dfrac{4}{3} \right)^n$

11. $a_1 = 2, a_n = 3a_{n-1}$

12. $a_1 = -1, a_n = 2a_{n-1}$

13. $a_1 = 1, a_n = -na_{n-1}$

14. $a_1 = 2, a_n = n^2 a_{n-1}$

15. $a_1 = 4, a_n = a_{n-1} + 2$

16. $a_1 = 3, a_n = a_{n-1} - 3$

17. $a_1 = 1, a_2 = 2, a_n = a_{n-1} a_{n-2}$

18. $a_1 = 1$, $a_2 = 2$, $a_n = a_{n-1}/a_{n-2}$

19. $a_1 = -1$, $a_n = 3na_{n-1}$ **20.** $a_1 = 2$, $a_n = -2na_{n-1}$

21–40. Classify the sequences defined in Exercises 1 to 20 as arithmetic, geometric, or neither.

In Exercises 41 to 56, find the indicated sum of the series.

41. $\sum\limits_{n=1}^{9} (2n - 3)$

42. $\sum\limits_{i=1}^{11} (1 - 3i)$

43. $\sum\limits_{k=1}^{8} (4k + 1)$

44. $\sum\limits_{i=1}^{10} (i^2 + 3)$

45. $\sum\limits_{n=1}^{6} 3 \cdot 2^n$

46. $\sum\limits_{i=1}^{5} 2 \cdot 4^{i-1}$

47. $\sum\limits_{k=1}^{9} (-1)^k 3^k$

48. $\sum\limits_{i=1}^{8} (-1)^{i+1} 2^i$

49. $\sum\limits_{i=1}^{10} \left(\dfrac{2}{3}\right)^i$

50. $\sum\limits_{i=1}^{11} \left(\dfrac{3}{2}\right)^i$

51. $\sum\limits_{n=1}^{9} \dfrac{(-1)^{n+1}}{n^2}$

52. $\sum\limits_{k=1}^{5} \dfrac{(-1)^{k+1}}{k!}$

53. $\sum\limits_{n=1}^{\infty} \left(\dfrac{1}{4}\right)^n$

54. $\sum\limits_{i=1}^{\infty} \left(-\dfrac{5}{6}\right)^i$

55. $\sum\limits_{k=1}^{\infty} \left(-\dfrac{4}{5}\right)^k$

56. $\sum\limits_{j=0}^{\infty} \left(\dfrac{1}{5}\right)^j$

In Exercises 57 to 64, prove each statement by mathematical induction.

57. $\sum\limits_{i=1}^{n} (5i + 1) = \dfrac{n(5n + 7)}{2}$ **58.** $\sum\limits_{i=1}^{n} (3 - 4i) = n(1 - 2n)$

59. $\sum\limits_{i=0}^{n} \left(-\dfrac{1}{2}\right)^i = \dfrac{2(1 - (-1/2)^{n+1})}{3}$

60. $\sum\limits_{i=0}^{n} (-1)^i = \dfrac{1 - (-1)^{n+1}}{2}$

61. $n^n \geq n!$ **62.** $n! > 4^n$, $n \geq 9$

63. 3 is a factor of $n^3 + 2n$ for all positive integers.

64. Let $a_1 = \sqrt{2}$ and $a_n = (\sqrt{2})^{a_{n-1}}$. Prove that $a_n < 2$ for all positive integers n.

In Exercises 65 to 68, use the Binomial Theorem to expand each binomial.

65. $(4a - b)^5$ **66.** $(x + 3y)^6$

67. $(\sqrt{a} + 2\sqrt{b})^8$ **68.** $\left(2x - \dfrac{1}{2x}\right)^7$

69. Find the fifth term in the expansion of $(3x - 4y)^7$.

70. Find the eighth term in the expansion of $(1 - 3x)^9$.

71. A computer password consists of eight letters. How many passwords are possible? Assume there is no difference between lower-case and upper-case letters.

72. The serial number on an airplane begins with the letter N, followed by six numerals, followed by one letter. How many serial numbers are possible?

73. From a committee of fifteen members, a president, vice president, and treasurer are elected. In how many ways can this be accomplished?

74. The emergency staff for a hospital consists of four supervisors and twelve regular employees. How many shifts of four people can be formed if each shift must contain exactly one supervisor?

75. From twelve people, a committee of five people is formed. In how many ways can this be accomplished if there are two people among the twelve who refuse to serve together on the committee?

76. A shipment of ten calculators contains two defective ones. A quality control inspector randomly chooses four of the calculators for testing. What is the probability that the inspector will choose one defective calculator?

77. A nickel, dime, and quarter are tossed. What is the probability that the nickel and dime will show heads and the quarter will show tails? What is the probability that only one of the coins will show tails?

78. A deck of ten cards contains five red and five black cards. If four cards are drawn from the deck, what is the probability that two are red and two are black?

79. For the 1000 numbers 000 to 999, what is the probability that the middle digit is greater than the other two digits?

80. Two numbers are chosen, with replacement, from the digits 1, 2, 3, 4, 5, and 6, and their sum is recorded. Now two more digits are selected and their sum noted. This process continues until the sum is 7 or the original sum is obtained. If the original sum was 9, what is the probability of having another sum of 9 before having a sum of 7? (*Hint:* Assume the events are independent. The probability can be found by summing an infinite geometric series.)

81. Which of the following has the greater probability: drawing an ace and a ten-card (ten, jack, queen, or king) from a regular deck of fifty-two playing cards or drawing an ace and a ten-card from two decks of regular playing cards?

82. From the digits 1, 2, 3, 4, and 5, two numbers are chosen without replacement. What is the probability that the second number is greater than the first number?

83. A room contains twelve people who are wearing badges numbered 1 to 12. If three people are randomly selected, what is the probability that the person wearing badge 6 will be included?

Chapter Test

In Exercises 1 to 3, find the third and fifth terms of the sequence defined by a_n.

1. $a_n = \dfrac{2^n}{n!}$ **2.** $a_n = \dfrac{(-1)^{n+1}}{2n}$ **3.** $a_1 = 3, a_n = 2a_{n-1}$

In Exercises 4 to 6, classify the sequence as an arithmetic sequence, a geometric sequence, or neither.

4. $a_n = -2n + 3$ **5.** $a_n = 2n^2$ **6.** $a_n = \dfrac{(-1)^{n-1}}{3^n}$

In Exercises 7 to 9, find the indicated sum of the series.

7. $\displaystyle\sum_{i=1}^{6} \dfrac{1}{i}$ **8.** $\displaystyle\sum_{j=1}^{10} \dfrac{1}{2^j}$ **9.** $\displaystyle\sum_{k=1}^{20} (3k - 2)$

10. The third term of an arithmetic sequence is 7 and the eighth term is 22. Find the twentieth term.

11. Find the sum of the infinite geometric series given by $\displaystyle\sum_{k=1}^{\infty} \left(\dfrac{3}{8}\right)^k$.

12. Write $0.\overline{15}$ as the quotient of integers in simplest form.

In Exercises 13 and 14, prove the statement by mathematical induction.

13. $\displaystyle\sum_{i=1}^{n} (2 - 3i) = \dfrac{n(1 - 3n)}{2}$ **14.** $n! > 3^n, \quad n \geq 7$

15. Write the binomial expansion of $(x - 2y)^5$.

16. Write the binomial expansion of $\left(x + \dfrac{1}{x}\right)^6$.

17. Find the sixth term in the expansion of $(3x + 2y)^8$.

18. Three cards are randomly chosen from a regular deck of playing cards. In how many ways can the cards be chosen so that the three cards are different?

19. A serial number consists of seven characters. The first three characters are upper-case letters of the alphabet. The next two characters are selected from the digits 1 through 9. The last two characters are upper-case letters of the alphabet. How many serial numbers are possible if no letter or number can be used twice in the same serial number?

20. Five cards are randomly selected from a deck of cards containing eight black cards and ten red cards. What is the probability that three black cards and two red cards are selected?

Appendix I | Using Tables

Logarithmic Tables

If a calculator is not available, then the values of logarithmic functions can be found by using the tables in this book.

Let x be a number written in scientific notation. That is, let $x = a \cdot 10^k$ where $1 \le a < 10$ and k is an integer. Applying some properties of logarithms, we have

$$\log x = \log(a \cdot 10^k)$$
$$= \log a + \log 10^k \quad \bullet \text{ The Product Property}$$
$$= \log a + k \quad \bullet \log_b b^p = p.$$

This last equation states that given any real number x, the real number $\log x$ is the sum of the logarithm of a number between one and ten, and an integer k. The real number $\log a$ is called the **mantissa** of $\log x$ and k is called the **characteristic** of $\log x$.

The Table of Common Logarithms in Appendix II gives the common logarithms accurate to 4 decimal places of numbers between 1.00 and 9.99 in increments of 0.01. For example, from the section of the Table of Common Logarithms at the left, $\log 2.43 \approx 0.3856$.

TABLE A1 Portion of a Common Logarithm Table

x	0	1	2	3
2.0	.3010	.3032	.3054	.3075
2.1	.3222	.3243	.3263	.3284
2.2	.3424	.3444	.3464	.3483
2.3	.3617	.3636	.3655	.3674
2.4	.3802	.3820	.3838	.3856
2.5	.3979	.3997	.4014	.4031
2.6	.4150	.4166	.4183	.4200

EXAMPLE 1 **Evaluate Logarithms**

Find   a. log 54,600   b. log 54.6   c. log 0.00546

Solution

a. $\log 54{,}600 = \log(5.46 \times 10^4)$ $\quad \bullet$ Write 54,600 in scientific notation.

$= \log 5.46 + \log 10^4 \quad \bullet$ The Product Property

$\approx 0.7372 + 4 = 4.7372 \quad \bullet \log 10^4 = 4$

b. $\log 54.6 = \log(5.46 \times 10^1)$

$= \log 5.46 + \log 10^1$

$\approx 0.7372 + 1 = 1.7372$

c. $\log 0.00546 = \log(5.46 \times 10^{-3})$

$= \log 5.46 + \log 10^{-3}$

$\approx 0.7372 + (-3)$

A1

In Example 1c, adding the characteristic to the mantissa would result in a negative logarithm. This form of a logarithm is inconvenient to use with logarithmic tables because these tables contain only positive mantissas. The value of log 0.00546 can be written in many forms. For example,

$$\log 0.00546 \approx 0.7372 + (-3)$$

$$\log 0.00546 \approx 0.7372 + (7 - 10) = 7.7372 - 10$$

$$\log 0.00546 \approx 0.7372 + (1 - 4) = 1.7372 - 4$$

Of these forms, log 0.00546 = 7.7372 − 10 is the most common.

Caution When a logarithm has a negative characteristic like $0.7372 + (-3)$, sometimes $0.7372 + (-3)$ is incorrectly written as -3.7372. The correct result is -2.2628.

As a final note, when using a calculator, log 0.00546 will be displayed as -2.26280736. By adding and subtracting 10 to this number, we can write the number with a positive mantissa.

$$\log 0.00546 \approx -2.26280736$$

$$= (-2.26280736 + 10) - 10$$

$$= 7.73719264 - 10$$

Adding and subtracting 10 is a somewhat arbitrary choice. Any integer that will produce a positive mantissa could be used. For example,

$$\log 0.00546 \approx -2.26280736$$

$$= (-2.26280736 + 3) - 3$$

$$= 0.73719264 - 3$$

We now examine the process of finding x given log x. For example, given log $x = 0.3522$, locate 0.3522 in the *body* of the Table of Common Logarithms. Find the number that corresponds to this mantissa. The number is 2.25. Thus log 2.25 $\approx$ 0.3522 and 2.25 is called the antilogarithm of 0.3522. We write antilog 0.3522 $\approx$ 2.25.

TABLE A2   Portion of a Common Logarithm Table

x	3	4	5	6
2.0	.3075	.3096	.3118	.3139
2.1	.3284	.3304	.3324	.3345
2.2	.3483	.3502	.3522	.3541
2.3	.3674	.3692	.3711	.3729
2.4	.3856	.3874	.3892	.3909
2.5	.4031	.4048	.4065	.4082
2.6	.4200	.4216	.4232	.4249

EXAMPLE 2 **Evaluate Antilogarithms**

Find   a. antilog 3.4639   b. antilog −1.4881

Solution

a. antilog 3.4639 = antilog(0.4639 + 3)   • Write the number as the sum of the mantissa and the characteristic.

$$\approx 2.91 \times 10^3$$ • Use the table to find the number that corresponds to the mantissa. The characteristic is the exponent on 10.

$$= 2910$$

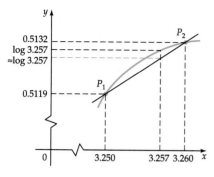

Figure A.1

b. The first step is to write the antilogarithm with a positive mantissa.

$$\text{antilog}(-1.4881) = \text{antilog}(-1.4881 + 2 - 2)$$

$$= \text{antilog}(0.5119 - 2)$$

$$\approx 3.25 \times 10^{-2} = 0.0325$$

Linear Interpolation

Many functions can be approximated, over small intervals, by a straight line. Using this technique, we can approximate logarithms and antilogarithms of numbers that are not given in the Table of Common Logarithms.

A portion of the graph of $y = \log x$ is shown in Figure A.1. The points $P_1(3.250, 0.5119)$ and $P_2(3.260, 0.5132)$ are two points on the graph of $y = \log x$. The value of log 3.257 is on the curve. An approximation for this value is on the straight line connecting P_1 and P_2. Because the slopes are the same between any two points on a straight line, we can write

$$\frac{y_2 - y_1}{x_2 - x_1} = \frac{y - y_1}{x - x_1}.$$

Substituting the coordinates of P_1, P_2, and using $x = 3.257$, we have

$$\frac{0.5132 - 0.5119}{3.260 - 3.250} = \frac{y - 0.5119}{3.257 - 3.250}.$$

Solving for y, we have $\qquad \dfrac{0.0013}{0.01} = \dfrac{y - 0.5199}{0.007}$

$$0.00091 = y - 0.5119$$

$$0.51281 = y$$

log $3.257 \approx 0.5128$ rounded to the nearest ten thousandth.

A more convenient method of using linear interpolation is illustrated in the following example.

EXAMPLE 3 **Evaluate Logarithms Using Linear Interpolation**

Find log 0.02903.

Solution Arrange the numbers and their corresponding mantissas in a table. Indicate the differences between the numbers and the mantissas as shown using d to represent the unknown difference between the mantissa of 2.900 and 2.910.

$$\log 0.02903 = \log(2.903 \times 10^{-2}) = \log 2.903 + \log 10^{-2}$$

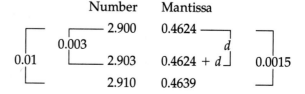

	Number	Mantissa
	2.900	0.4624
	2.903	0.4624 + d
	2.910	0.4639

Write a proportion and solve for d.

$$\frac{0.003}{0.01} = \frac{d}{0.0015}$$

$$0.0005 = d$$

Add $d = 0.0005$ to the mantissa 0.4624.

$$\log 0.02903 = \log 2.903 + \log 10^{-2}$$
$$\approx (0.4624 + 0.0005) + (-2)$$
$$= 0.4629 + (-2) = 8.4629 - 10$$

Therefore $\log 0.02903 \approx 8.4629 - 10$.

Linear interpolation can also be used to find an antilogarithm.

EXAMPLE 4   Evaluate Antilogarithms Using Linear Interpolation

Find antilog$(9.8465 - 10)$.

Solution

	Mantissa	Antilog
	0.8463	7.02
	0.8465	7.02 + d
	0.8470	7.03

(0.0007 bracket; 0.0002 bracket; d; 0.01 bracket)

Write a proportion and solve for d.

$$\frac{0.0002}{0.0007} = \frac{d}{0.01}$$

$$0.003 \approx d$$

Thus,

$$\text{antilog}(9.8465 - 10) \approx (7.02 + d) \times 10^{-1} = (7.02 + 0.003) \times 10^{-1}$$
$$= 7.023 \times 10^{-1} = 0.7023$$

Using Trigonometric Tables

The values of trigonometric functions can be found by using the Table of Values of Trigonometric Functions found in Appendix III. These tables give the values of the trigonometric functions, accurate to four significant digits, in increments of 10′.

To use the tables of trigonometric functions, first note that the table has degrees listed in both the right- and left-hand columns and the trigonometric functions are listed in both the top and bottom rows of the tables. For angles listed at the left, use the top row of functions. For angles listed at the right, use the bottom row of functions. For example, to determine the value of sin 27°40', locate 27°40' in the left-hand column of the table and read across until you locate the column at the top headed by "sin." Thus, sin 27°40' ≈ 0.4643.

To determine the value of cos 72°20', locate 72°20' in the right-hand column of the table and read across until you locate the column at the bottom headed by "cos." Thus, cos 72°20' ≈ 0.3035.

The Table of Trigonometric Functions only gives the values of the trigonometric functions from 0° to 90°. To find the value of a trigonometric function for other angles, first find the reference angle. Then attach the appropriate sign depending on the quadrant where the terminal side of the angle lies.

EXAMPLE 1 Evaluate Trigonometric Functions Using the Reference Angle

Find   a.   tan 153°40'   b.   sec(−87°10')   c.   sin 3.9910

Solution

a.   tan 153°40' = −tan 26°20'   • 180° − 153°40' = 26°20'

         ≈ −0.4950

b.   sec(−87°10') = sec 87°10' ≈ 20.23

c.   In this case, we are finding the sine of a real number. Find the reference number in the column headed by θ.

         sin 3.9910 ≈ −sin 0.8494 ≈ −0.7509   • 3.9910 − π ≈ 0.8494

We now examine the process of finding θ given a trigonometric function of θ. For example, given cos θ = 0.5736 with θ an angle in the first quadrant, locate 0.5736 in the body of the table in the cosine column. Thus θ ≈ 55° and cos 55° ≈ 0.5736. Because the trigonometric functions are periodic, it is necessary to specify some bounds on θ.

EXAMPLE 2 Solve a Trigonometric Equation

Solve the equation for 0° ≤ θ < 360°.

a.   sin θ = 0.5225   b.   tan θ = −1.611

Solution

a.   Let θ' be the reference angle so that sin θ' = 0.5225. From the table, θ' ≈ 31°30'. Because sin θ is positive in quadrants I and II,

         θ ≈ 31°30'   or θ ≈ 180° − 31°30' = 148°30'.

b.   Let θ' be the reference angle so that tan θ' = 1.611. From the table, θ' ≈ 58°10'. Because tan θ is negative in quadrants II and IV,

         θ ≈ 180° − 58°10' = 121°50'   or θ ≈ 360° − 58°10' = 301°50'.

Using the method of linear interpolation, we can approximate trigonometric function values that are not given in the tables.

EXAMPLE 3   **Evaluate Trigonometric Functions**

Find   a. cos 58°28'   b. tan(−0.5828)   c. csc 231°12'

Solution

a.

$$10' \left[8' \left[\begin{array}{l} \cos 58°20' \approx 0.5250 \\ \cos 58°28' \approx 0.5250 + d \end{array} \right] d \right. \left. \begin{array}{r} \\ -0.0025 \end{array} \right] \\ \cos 58°30' \approx 0.5225$$

Write a proportion and solve for d.

$$\frac{8}{10} = \frac{d}{-0.0025}$$

$$d = -0.0020$$

cos 58°28' ≈ 0.5250 + (−0.0020) = 0.5230.

b. Let $\theta' = 0.5828$ be the reference number. Look for 0.5828 in the θ column.

$$0.0029 \left[0.0010 \left[\begin{array}{l} \tan 0.5818 \approx 0.6577 \\ \tan 0.5828 \approx 0.6577 + d \end{array} \right] d \right. \left. \begin{array}{r} \\ 0.0042 \end{array} \right] \\ \tan 0.5847 \approx 0.6619$$

Write a proportion and solve for d.

$$\frac{0.0010}{0.0029} = \frac{d}{0.0042}$$

$$d \approx 0.0014$$

tan(−0.5828) = −tan 0.5828 ≈ −(0.6577 + 0.0014) = −0.6591.

c. Let $\theta' = 51°12'$ be the reference angle for 231°12'.

$$10' \left[2' \left[\begin{array}{l} \csc 51°10' \approx 1.284 \\ \csc 51°12' \approx 1.284 + d \end{array} \right] d \right. \left. \begin{array}{r} \\ -0.0030 \end{array} \right] \\ \csc 51°20' \approx 1.281$$

Write and solve a proportion for d.

$$\frac{2}{10} = \frac{d}{-0.0030}$$

$$d \approx -0.0006$$

csc 231°12' = −csc 51°12' = −(1.284 − 0.0006) = −1.283.

We can also use interpolation to find θ given the value of the trigonometric function of θ.

EXAMPLE 4   **Solve a Trigonometric Equation**

Solve the equation $\cot \theta = 1.044$ where $0° \le \theta < 360°$.

Solution   Search the body of the trigonometric table to find the two values of θ', where θ' is the reference angle, for which $\cot \theta$ is closest to 1.044.

$$
10' \left[d \left[\begin{array}{l} \cot 43°40' \approx 1.048 \\ \cot(43°40' + d) \approx 1.044 \end{array} \right. \left. \begin{array}{l} -0.004 \end{array} \right] -0.006 \\ \cot 43°50' \approx 1.042 \right.
$$

Write a proportion and solve for d.

$$\frac{d}{10} = \frac{-0.004}{-0.0060}$$

$$d \approx 7'$$

Thus $\theta' \approx 43°40' + 7' = 43°47'$. Because $\cot \theta$ is positive in the first and third quadrants,

$$\theta \approx 43°47' \quad \text{or} \quad \theta \approx 180° + 43°47' = 223°47'.$$

EXAMPLE 5   **Solve a Trigonometric Equation**

Solve the equation $\sin \theta = -0.5672$ where $0 \le \theta < 2\pi$.

Solution   Find two values of θ, where θ' is the reference number for θ, for which $\sin \theta$ is closest to 0.5672.

$$
0.0029 \left[d \left[\begin{array}{l} \sin 0.6021 \approx 0.5664 \\ \sin(0.6021 + d) \approx 0.5672 \end{array} \right. \left. \begin{array}{l} 0.0008 \end{array} \right] 0.0024 \\ \sin 0.6050 \approx 0.5688 \right.
$$

Write a proportion and solve for d.

$$\frac{d}{0.0029} = \frac{0.0008}{0.0024}$$

$$d \approx 0.0010$$

Thus $\theta' \approx 0.6021 + 0.0010 = 0.6031$. Because $\sin \theta$ is negative in the third and fourth quadrants,

$$\theta \approx \pi + 0.6031 \approx 3.7447 \quad \text{or} \quad \theta \approx 2\pi - 0.6031 \approx 5.6801.$$

Appendix II | Logarithmic and Exponential Tables

Common Logarithms

x	0	1	2	3	4	5	6	7	8	9
1.0	.0000	.0043	.0086	.0128	.0170	.0212	.0253	.0294	.0334	.0374
1.1	.0141	.0453	.0492	.0531	.0569	.0607	.0645	.0682	.0719	.0755
1.2	.0792	.0828	.0864	.0899	.0934	.0969	.1004	.1038	.1072	.1106
1.3	.1139	.1173	.1206	.1239	.1271	.1303	.1335	.1367	.1399	.1430
1.4	.1461	.1492	.1523	.1553	.1584	.1614	.1644	.1673	.1703	.1732
1.5	.1761	.1790	.1818	.1847	.1875	.1903	.1931	.1959	.1987	.2014
1.6	.2041	.2068	.2095	.2122	.2148	.2175	.2201	.2227	.2253	.2279
1.7	.2304	.2330	.2355	.2380	.2405	.2430	.2455	.2480	.2504	.2529
1.8	.2553	.2577	.2601	.2625	.2648	.2672	.2695	.2718	.2742	.2765
1.9	.2788	.2810	.2833	.2856	.2878	.2900	.2923	.2945	.2967	.2989
2.0	.3010	.3032	.3054	.3075	.3096	.3118	.3139	.3160	.3181	.3201
2.1	.3222	.3243	.3263	.3284	.3304	.3324	.3345	.3365	.3385	.3404
2.2	.3424	.3444	.3464	.3483	.3502	.3522	.3541	.3560	.3579	.3598
2.3	.3617	.3636	.3655	.3674	.3692	.3711	.3729	.3747	.3766	.3784
2.4	.3802	.3820	.3838	.3856	.3874	.3892	.3909	.3927	.3945	.3962
2.5	.3979	.3997	.4014	.4031	.4048	.4065	.4082	.4099	.4116	.4133
2.6	.4150	.4166	.4183	.4200	.4216	.4232	.4249	.4265	.4281	.4298
2.7	.4314	.4330	.4346	.4362	.4378	.4393	.4409	.4425	.4440	.4456
2.8	.4472	.4487	.4502	.4518	.4533	.4548	.4564	.4579	.4594	.4609
2.9	.4624	.4639	.4654	.4669	.4683	.4698	.4713	.4728	.4742	.4757
3.0	.4771	.4786	.4800	.4814	.4829	.4843	.4857	.4871	.4886	.4900
3.1	.4914	.4928	.4942	.4955	.4969	.4983	.4997	.5011	.5024	.5038
3.2	.5051	.5065	.5079	.5092	.5105	.5119	.5132	.5145	.5159	.5172
3.3	.5185	.5198	.5211	.5224	.5237	.5250	.5263	.5276	.5289	.5302
3.4	.5315	.5328	.5340	.5353	.5366	.5378	.5391	.5403	.5416	.5428
3.5	.5441	.5453	.5465	.5478	.5490	.5502	.5514	.5527	.5539	.5551
3.6	.5563	.5575	.5587	.5599	.5611	.5623	.5635	.5647	.5658	.5670
3.7	.5682	.5694	.5705	.5717	.5729	.5740	.5752	.5763	.5775	.5786
3.8	.5798	.5809	.5821	.5832	.5843	.5855	.5866	.5877	.5888	.5899
3.9	.5911	.5922	.5933	.5944	.5955	.5966	.5977	.5988	.5999	.6010
4.0	.6021	.6031	.6042	.6053	.6064	.6075	.6085	.6096	.6107	.6117
4.1	.6128	.6138	.6149	.6160	.6170	.6180	.6191	.6201	.6212	.6222
4.2	.6232	.6243	.6253	.6263	.6274	.6284	.6294	.6304	.6314	.6325
4.3	.6335	.6345	.6355	.6365	.6375	.6385	.6395	.6405	.6415	.6425
4.4	.6435	.6444	.6454	.6464	.6474	.6484	.6493	.6503	.6513	.6522
4.5	.6532	.6542	.6551	.6561	.6571	.6580	.6590	.6599	.6609	.6618
4.6	.6628	.6637	.6646	.6656	.6665	.6675	.6684	.6693	.6702	.6712
4.7	.6721	.6730	.6739	.6749	.6758	.6767	.6776	.6785	.6794	.6803
4.8	.6812	.6821	.6830	.6839	.6848	.6857	.6866	.6875	.6884	.6893
4.9	.6902	.6911	.6920	.6928	.6937	.6946	.6955	.6964	.6972	.6981

Common Logarithms (continued)

x	0	1	2	3	4	5	6	7	8	9
5.0	.6990	.6998	.7007	.7016	.7024	.7033	.7042	.7050	.7059	.7067
5.1	.7076	.7084	.7093	.7101	.7110	.7118	.7126	.7135	.7143	.7152
5.2	.7160	.7168	.7177	.7185	.7193	.7202	.7210	.7218	.7226	.7235
5.3	.7243	.7251	.7259	.7267	.7275	.7284	.7292	.7300	.7308	.7316
5.4	.7324	.7332	.7340	.7348	.7356	.7364	.7372	.7380	.7388	.7396
5.5	.7404	.7412	.7419	.7427	.7435	.7443	.7451	.7459	.7466	.7474
5.6	.7482	.7490	.7497	.7505	.7513	.7520	.7528	.7536	.7543	.7551
5.7	.7559	.7566	.7574	.7582	.7589	.7597	.7604	.7612	.7619	.7627
5.8	.7634	.7642	.7649	.7657	.7664	.7672	.7679	.7686	.7694	.7701
5.9	.7709	.7716	.7723	.7731	.7738	.7745	.7752	.7760	.7767	.7774
6.0	.7782	.7789	.7796	.7803	.7810	.7818	.7825	.7832	.7839	.7846
6.1	.7853	.7860	.7868	.7875	.7882	.7889	.7896	.7903	.7910	.7917
6.2	.7924	.7931	.7938	.7945	.7952	.7959	.7966	.7973	.7980	.7987
6.3	.7993	.8000	.8007	.8014	.8021	.8028	.8035	.8041	.8048	.8055
6.4	.8062	.8069	.8075	.8082	.8089	.8096	.8102	.8109	.8116	.8122
6.5	.8129	.8136	.8142	.8149	.8156	.8162	.8169	.8176	.8182	.8189
6.6	.8195	.8202	.8209	.8215	.8222	.8228	.8235	.8241	.8248	.8254
6.7	.8261	.8267	.8274	.8280	.8287	.8293	.8299	.8306	.8312	.8319
6.8	.8325	.8331	.8338	.8344	.8351	.8357	.8363	.8370	.8376	.8382
6.9	.8388	.8395	.8401	.8407	.8414	.8420	.8426	.8432	.8439	.8445
7.0	.8451	.8457	.8463	.8470	.8476	.8482	.8488	.8494	.8500	.8506
7.1	.8513	.8519	.8525	.8531	.8537	.8543	.8549	.8555	.8561	.8567
7.2	.8573	.8579	.8585	.8591	.8597	.8603	.8609	.8615	.8621	.8627
7.3	.8633	.8639	.8645	.8651	.8657	.8663	.8669	.8675	.8681	.8686
7.4	.8692	.8698	.8704	.8710	.8716	.8722	.8727	.8733	.8739	.8745
7.5	.8751	.8756	.8762	.8768	.8774	.8779	.8785	.8791	.8797	.8802
7.6	.8808	.8814	.8820	.8825	.8831	.8837	.8842	.8848	.8854	.8859
7.7	.8865	.8871	.8876	.8882	.8887	.8893	.8899	.8904	.8910	.8915
7.8	.8921	.8927	.8932	.8938	.8943	.8949	.8954	.8960	.8965	.8971
7.9	.8976	.8982	.8987	.8993	.8998	.9004	.9009	.9015	.9020	.9025
8.0	.9031	.9036	.9042	.9047	.9053	.9058	.9063	.9069	.9074	.9079
8.1	.9085	.9090	.9096	.9101	.9106	.9112	.9117	.9122	.9128	.9133
8.2	.9138	.9143	.9149	.9154	.9159	.9165	.9170	.9175	.9180	.9186
8.3	.9191	.9196	.9201	.9206	.9212	.9217	.9222	.9227	.9232	.9238
8.4	.9243	.9248	.9253	.9258	.9263	.9269	.9274	.9279	.9284	.9289
8.5	.9294	.9299	.9304	.9309	.9315	.9320	.9325	.9330	.9335	.9340
8.6	.9345	.9350	.9355	.9360	.9365	.9370	.9375	.9380	.9385	.9390
8.7	.9395	.9400	.9405	.9410	.9415	.9420	.9425	.9430	.9435	.9440
8.8	.9445	.9450	.9455	.9460	.9465	.9469	.9474	.9479	.9484	.9489
8.9	.9494	.9499	.9504	.9509	.9513	.9518	.9523	.9528	.9533	.9538
9.0	.9542	.9547	.9552	.9557	.9562	.9566	.9571	.9576	.9581	.9586
9.1	.9590	.9595	.9600	.9605	.9609	.9614	.9619	.9624	.9628	.9633
9.2	.9638	.9643	.9647	.9652	.9657	.9661	.9666	.9671	.9675	.9680
9.3	.9685	.9689	.9694	.9699	.9703	.9708	.9713	.9717	.9722	.9727
9.4	.9731	.9736	.9741	.9745	.9750	.9754	.9759	.9763	.9768	.9773
9.5	.9777	.9782	.9786	.9791	.9795	.9800	.9805	.9809	.9814	.9818
9.6	.9823	.9827	.9832	.9836	.9841	.9845	.9850	.9854	.9859	.9863
9.7	.9868	.9872	.9877	.9881	.9886	.9890	.9894	.9899	.9903	.9908
9.8	.9912	.9917	.9921	.9926	.9930	.9934	.9939	.9943	.9948	.9952
9.9	.9956	.9961	.9965	.9969	.9974	.9978	.9983	.9987	.9991	.9996

Natural Exponential Function

x	e^x	e^{-x}	x	e^x	e^{-x}	x	e^x	e^{-x}
.00	1.00000	1.00000	.40	1.49182	.67032	.80	2.22554	.44032
.01	1.01005	.99005	.41	1.50682	.66365	.85	2.33965	.42741
.02	1.02020	.98020	.42	1.52196	.65705	.90	2.45960	.40657
.03	1.03045	.97045	.43	1.53726	.65051	.95	2.58571	.38674
.04	1.04081	.96079	.44	1.55271	.64404	1.00	2.71828	.36788
.05	1.05127	.95123	.45	1.56831	.63763	1.10	3.00416	.33287
.06	1.06184	.94176	.46	1.58407	.63128	1.20	3.32011	.30119
.07	1.07251	.93239	.47	1.59999	.62500	1.30	3.66929	.27253
.08	1.08329	.92312	.48	1.61607	.61878	1.40	4.05519	.24659
.09	1.09417	.91393	.49	1.63232	.61263	1.50	4.48168	.22313
.10	1.10517	.90484	.50	1.64872	.60653	1.60	4.95302	.20189
.11	1.11628	.89583	.51	1.66529	.60050	1.70	5.47394	.18268
.12	1.12750	.88692	.52	1.68203	.59452	1.80	6.04964	.16529
.13	1.13883	.87810	.53	1.69893	.58860	1.90	6.68589	.14956
.14	1.15027	.86936	.54	1.71601	.58275	2.00	7.38905	.13533
.15	1.16183	.86071	.55	1.73325	.57695	2.10	8.16616	.12245
.16	1.17351	.85214	.56	1.75067	.57121	2.20	9.02500	.11080
.17	1.18530	.84366	.57	1.76827	.56553	2.30	9.97417	.10025
.18	1.19722	.83527	.58	1.78604	.55990	2.40	11.02316	.09071
.19	1.20925	.82696	.59	1.80399	.55433	2.50	12.18248	.08208
.20	1.22140	.81873	.60	1.82212	.54881	3.00	20.08551	.04978
.21	1.23368	.81058	.61	1.84043	.54335	3.50	33.11545	.03020
.22	1.24608	.80252	.62	1.85893	.53794	4.00	54.59815	.01832
.23	1.25860	.79453	.63	1.87761	.53259	4.50	90.01713	.01111
.24	1.27125	.78663	.64	1.89648	.52729	5.00	148.41316	.00674
.25	1.28403	.77880	.65	1.91554	.52205	5.50	224.69193	.00409
.26	1.29693	.77105	.66	1.93479	.51685	6.00	403.42879	.00248
.27	1.30996	.76338	.67	1.95424	.51171	6.50	665.14163	.00150
.28	1.32313	.75578	.68	1.97388	.50662	7.00	1096.63316	.00091
.29	1.33643	.74826	.69	1.99372	.50158	7.50	1808.04241	.00055
.30	1.34986	.74082	.70	2.01375	.49659	8.00	2980.95799	.00034
.31	1.36343	.73345	.71	2.03399	.49164	8.50	4914.76884	.00020
.32	1.37713	.72615	.72	2.05443	.48675	9.00	8130.08392	.00012
.33	1.39097	.71892	.73	2.07508	.48191	9.50	13359.72683	.00007
.34	1.40495	.71177	.74	2.09594	.47711	10.00	22026.46579	.00005
.35	1.41907	.70469	.75	2.11700	.47237			
.36	1.43333	.69768	.76	2.13828	.46767			
.37	1.44773	.69073	.77	2.15977	.46301			
.38	1.46228	.68386	.78	2.18147	.45841			
.39	1.47698	.67706	.79	2.20340	.45384			

Natural Logarithms

x	$\ln x$	x	$\ln x$	x	$\ln x$
		4.5	1.5041	9.0	2.1972
0.1	−2.3026	4.6	1.5261	9.1	2.2083
0.2	−1.6094	4.7	1.5476	9.2	2.2192
0.3	−1.2040	4.8	1.5686	9.3	2.2300
0.4	−0.9163	4.9	1.5892	9.4	2.2407
0.5	−0.6931	5.0	1.6094	9.5	2.2513
0.6	−0.5108	5.1	1.6292	9.6	2.2618
0.7	−0.3567	5.2	1.6487	9.7	2.2721
0.8	−0.2231	5.3	1.6677	9.8	2.2824
0.9	−0.1054	5.4	1.6864	9.9	2.2925
1.0	0.0000	5.5	1.7047	10	2.3026
1.1	0.0953	5.6	1.7228	11	2.3979
1.2	0.1823	5.7	1.7405	12	2.4849
1.3	0.2624	5.8	1.7579	13	2.5649
1.4	0.3365	5.9	1.7750	14	2.6391
1.5	0.4055	6.0	1.7918	15	2.7081
1.6	0.4700	6.1	1.8083	16	2.7726
1.7	0.5306	6.2	1.8245	17	2.8332
1.8	0.5878	6.3	1.8405	18	2.8904
1.9	0.6419	6.4	1.8563	19	2.9444
2.0	0.6931	6.5	1.8718	20	2.9957
2.1	0.7419	6.6	1.8871	25	3.2189
2.2	0.7885	6.7	1.9021	30	3.4012
2.3	0.8329	6.8	1.9169	35	3.5553
2.4	0.8755	6.9	1.9315	40	3.6889
2.5	0.9163	7.0	1.9459	45	3.8067
2.6	0.9555	7.1	1.9601	50	3.9120
2.7	0.9933	7.2	1.9741	55	4.0073
2.8	1.0296	7.3	1.9879	60	4.0943
2.9	1.0647	7.4	2.0015	65	4.1744
3.0	1.0986	7.5	2.0149	70	4.2485
3.1	1.1314	7.6	2.0281	75	4.3175
3.2	1.1632	7.7	2.0412	80	4.3820
3.3	1.1939	7.8	2.0541	85	4.4427
3.4	1.2238	7.9	2.0669	90	4.4998
3.5	1.2528	8.0	2.0794	100	4.6052
3.6	1.2809	8.1	2.0919	110	4.7005
3.7	1.3083	8.2	2.1041	120	4.7875
3.8	1.3350	8.3	2.1163	130	4.8676
3.9	1.3610	8.4	2.1282	140	4.9416
4.0	1.3863	8.5	2.1401	150	5.0106
4.1	1.4110	8.6	2.1518	160	5.0752
4.2	1.4351	8.7	2.1633	170	5.1358
4.3	1.4586	8.8	2.1748	180	5.1930
4.4	1.4816	8.9	2.1861	190	5.2470

Appendix III | Trigonometric Tables

Trigonometric Functions in Degrees and Radians

θ (degrees)	θ (radians)	sin θ	cos θ	tan θ	cot θ	csc θ	sec θ	θ (radians)	θ (degrees)
0°00'	.0000	.0000	1.0000	.0000	—	—	1.000	1.5708	90°00'
10	.0029	.0029	1.0000	.0029	343.8	343.8	1.000	1.5679	50
20	.0058	.0058	1.0000	.0058	171.9	171.9	1.000	1.5650	40
30	.0087	.0087	1.0000	.0087	114.6	114.6	1.000	1.5621	30
40	.0116	.0116	.9999	.0116	85.94	85.95	1.000	1.5592	20
50	.0145	.0145	.9999	.0145	68.75	68.76	1.000	1.5563	10
1°00'	.0175	.0175	.9998	.0175	57.29	57.30	1.000	1.5533	89°00'
10	.0204	.0204	.9998	.0204	49.10	49.11	1.000	1.5504	50
20	.0233	.0233	.9997	.0233	42.96	42.98	1.000	1.5475	40
30	.0262	.0262	.9997	.0262	38.19	38.20	1.000	1.5446	30
40	.0291	.0291	.9996	.0291	34.37	34.38	1.000	1.5417	20
50	.0320	.0320	.9995	.0320	31.24	31.26	1.000	1.5388	10
2°00'	.0349	.0349	.9994	.0349	28.64	28.65	1.001	1.5359	88°00'
10	.0378	.0378	.9993	.0378	26.43	26.45	1.001	1.5330	50
20	.0407	.0407	.9992	.0407	24.54	24.56	1.001	1.5301	40
30	.0436	.0436	.9990	.0437	22.90	22.93	1.001	1.5272	30
40	.0465	.0465	.9989	.0466	21.47	21.49	1.001	1.5243	20
50	.0495	.0494	.9988	.0495	20.21	20.23	1.001	1.5213	10
3°00'	.0524	.0523	.9986	.0524	19.08	19.11	1.001	1.5184	87°00'
10	.0553	.0552	.9985	.0553	18.07	18.10	1.002	1.5155	50
20	.0582	.0581	.9983	.0582	17.17	17.20	1.002	1.5126	40
30	.0611	.0610	.9981	.0612	16.35	16.38	1.002	1.5097	30
40	.0640	.0640	.9980	.0641	15.60	15.64	1.002	1.5068	20
50	.0669	.0669	.9978	.0670	14.92	14.96	1.002	1.5039	10
4°00'	.0698	.0698	.9976	.0699	14.30	14.34	1.002	1.5010	86°00'
10	.0727	.0727	.9974	.0729	13.73	13.76	1.003	1.4981	50
20	.0756	.0756	.9971	.0758	13.20	13.23	1.003	1.4952	40
30	.0785	.0785	.9969	.0787	12.71	12.75	1.003	1.4923	30
40	.0814	.0814	.9967	.0816	12.25	12.29	1.003	1.4893	20
50	.0844	.0843	.9964	.0846	11.83	11.87	1.004	1.4864	10
5°00'	.0873	.0872	.9962	.0875	11.43	11.47	1.004	1.4835	85°00'
10	.0902	.0901	.9959	.0904	11.06	11.10	1.004	1.4806	50
20	.0931	.0929	.9957	.0934	10.71	10.76	1.004	1.4777	40
30	.0960	.0958	.9954	.0963	10.39	10.43	1.005	1.4748	30
40	.0989	.0987	.9951	.0992	10.08	10.13	1.005	1.4719	20
50	.1018	.1016	.9948	.1022	9.788	9.839	1.005	1.4690	10
6°00'	.1047	.1045	.9945	.1051	9.514	9.567	1.006	1.4661	84°00'
10	.1076	.1074	.9942	.1080	9.255	9.309	1.006	1.4632	50
20	.1105	.1103	.9939	.1110	9.010	9.065	1.006	1.4603	40
30	.1134	.1132	.9936	.1139	8.777	8.834	1.006	1.4573	30
40	.1164	.1161	.9932	.1169	8.556	8.614	1.007	1.4544	20
50	.1193	.1190	.9929	.1198	8.345	8.405	1.007	1.4515	10
		cos θ	sin θ	cot θ	tan θ	sec θ	csc θ	(θ radians)	θ (degrees)

θ (degrees)	θ (radians)	sin θ	cos θ	tan θ	cot θ	csc θ	sec θ	θ (radians)	θ (degrees)
7°00'	.1222	.1219	.9925	.1228	8.144	8.206	1.008	1.4486	83°00'
10	.1251	.1248	.9922	.1257	7.953	8.016	1.008	1.4457	50
20	.1280	.1276	.9918	.1287	7.770	7.834	1.008	1.4428	40
30	.1309	.1305	.9914	.1317	7.596	7.661	1.009	1.4399	30
40	.1338	.1334	.9911	.1346	7.429	7.496	1.009	1.4370	20
50	.1367	.1363	.9907	.1376	7.269	7.337	1.009	1.4341	10
8°00'	.1396	.1392	.9903	.1405	7.115	7.185	1.010	1.4312	82°00'
10	.1425	.1421	.9899	.1435	6.968	7.040	1.010	1.4283	50
20	.1454	.1449	.9894	.1465	6.827	6.900	1.011	1.4254	40
30	.1484	.1478	.9890	.1495	6.691	6.765	1.011	1.4224	30
40	.1513	.1507	.9886	.1524	6.561	6.636	1.012	1.4195	20
50	.1542	.1536	.9881	.1554	6.435	6.512	1.012	1.4166	10
9°00'	.1571	.1564	.9877	.1584	6.314	6.392	1.012	1.4137	81°00'
10	.1600	.1593	.9872	.1614	6.197	6.277	1.013	1.4108	50
20	.1629	.1622	.9868	.1644	6.084	6.166	1.013	1.4079	40
30	.1658	.1650	.9863	.1673	5.976	6.059	1.014	1.4050	30
40	.1687	.1679	.9858	.1703	5.871	5.955	1.014	1.4021	20
50	.1716	.1708	.9853	.1733	5.769	5.855	1.015	1.3992	10
10°00'	.1745	.1736	.9848	.1763	5.671	5.759	1.015	1.3963	80°00'
10	.1774	.1765	.9843	.1793	5.576	5.665	1.016	1.3934	50
20	.1804	.1794	.9838	.1823	5.485	5.575	1.016	1.3904	40
30	.1833	.1822	.9833	.1853	5.396	5.487	1.017	1.3875	30
40	.1862	.1851	.9827	.1883	5.309	5.403	1.018	1.3846	20
50	.1891	.1880	.9822	.1914	5.226	5.320	1.018	1.3817	10
11°00'	.1920	.1908	.9816	.1944	5.145	5.241	1.019	1.3788	79°00'
10	.1949	.1937	.9811	.1974	5.066	5.164	1.019	1.3759	50
20	.1978	.1965	.9805	.2004	4.989	5.089	1.020	1.3730	40
30	.2007	.1994	.9799	.2035	4.915	5.016	1.020	1.3701	30
40	.2036	.2022	.9793	.2065	4.843	4.945	1.021	1.3672	20
50	.2065	.2051	.9787	.2095	4.773	4.876	1.022	1.3643	10
12°00'	.2094	.2079	.9781	.2126	4.705	4.810	1.022	1.3614	78°00'
10	.2123	.2108	.9775	.2156	4.638	4.745	1.023	1.3584	50
20	.2153	.2136	.9769	.2186	4.574	4.682	1.024	1.3555	40
30	.2182	.2164	.9763	.2217	4.511	4.620	1.024	1.3526	30
40	.2211	.2193	.9757	.2247	4.449	4.560	1.025	1.3497	20
50	.2240	.2221	.9750	.2278	4.390	4.502	1.026	1.3468	10
13°00'	.2269	.2250	.9744	.2309	4.331	4.445	1.026	1.3439	77°00'
10	.2298	.2278	.9737	.2339	4.275	4.390	1.027	1.3410	50
20	.2327	.2306	.9730	.2370	4.219	4.336	1.028	1.3381	40
30	.2356	.2334	.9724	.2401	4.165	4.284	1.028	1.3352	30
40	.2385	.2363	.9717	.2432	4.113	4.232	1.029	1.3323	20
50	.2414	.2391	.9710	.2462	4.061	4.182	1.030	1.3294	10
		cos θ	sin θ	cot θ	tan θ	sec θ	csc θ	(θ radians)	θ (degrees)

Trigonometric Functions in Degrees and Radians (continued)

θ (degrees)	θ (radians)	sin θ	cos θ	tan θ	cot θ	sec θ	csc θ	θ (radians)	θ (degrees)
14°00'	.2443	.2419	.9703	.2493	4.011	1.031	4.134	1.3265	76°00'
10	.2473	.2447	.9696	.2524	3.962	1.031	4.086	1.3235	50
20	.2502	.2476	.9689	.2555	3.914	1.032	4.039	1.3206	40
30	.2531	.2504	.9681	.2586	3.867	1.033	3.994	1.3177	30
40	.2560	.2532	.9674	.2617	3.821	1.034	3.950	1.3148	20
50	.2589	.2560	.9667	.2648	3.776	1.034	3.906	1.3119	10
15°00'	.2618	.2588	.9659	.2679	3.732	1.035	3.864	1.3090	75°00'
10	.2647	.2616	.9652	.2711	3.689	1.036	3.822	1.3061	50
20	.2676	.2644	.9644	.2742	3.647	1.037	3.782	1.3032	40
30	.2705	.2672	.9636	.2773	3.606	1.038	3.742	1.3003	30
40	.2734	.2700	.9628	.2805	3.566	1.039	3.703	1.2974	20
50	.2763	.2728	.9621	.2836	3.526	1.039	3.665	1.2945	10
16°00'	.2793	.2756	.9613	.2867	3.487	1.040	3.628	1.2915	74°00'
10	.2822	.2784	.9605	.2899	3.450	1.041	3.592	1.2886	50
20	.2851	.2812	.9596	.2931	3.412	1.042	3.556	1.2857	40
30	.2880	.2840	.9588	.2962	3.376	1.043	3.521	1.2828	30
40	.2909	.2868	.9580	.2994	3.340	1.044	3.487	1.2799	20
50	.2938	.2896	.9572	.3026	3.305	1.045	3.453	1.2770	10
17°00'	.2967	.2924	.9563	.3057	3.271	1.046	3.420	1.2741	73°00'
10	.2996	.2952	.9555	.3089	3.237	1.047	3.388	1.2712	50
20	.3025	.2979	.9546	.3121	3.204	1.048	3.356	1.2683	40
30	.3054	.3007	.9537	.3153	3.172	1.049	3.326	1.2654	30
40	.3083	.3035	.9528	.3185	3.140	1.049	3.295	1.2625	20
50	.3113	.3062	.9520	.3217	3.108	1.050	3.265	1.2595	10
18°00'	.3142	.3090	.9511	.3249	3.078	1.051	3.236	1.2566	72°00'
10	.3171	.3118	.9502	.3281	3.047	1.052	3.207	1.2537	50
20	.3200	.3145	.9492	.3314	3.018	1.053	3.179	1.2508	40
30	.3229	.3173	.9483	.3346	2.989	1.054	3.152	1.2479	30
40	.3258	.3201	.9474	.3378	2.960	1.056	3.124	1.2450	20
50	.3287	.3228	.9465	.3411	2.932	1.057	3.098	1.2421	10
19°00'	.3316	.3256	.9455	.3443	2.904	1.058	3.072	1.2392	71°00'
10	.3345	.3283	.9446	.3476	2.877	1.059	3.046	1.2363	50
20	.3374	.3311	.9436	.3508	2.850	1.060	3.021	1.2334	40
30	.3403	.3338	.9426	.3541	2.824	1.061	2.996	1.2305	30
40	.3432	.3365	.9417	.3574	2.798	1.062	2.971	1.2275	20
50	.3462	.3393	.9407	.3607	2.773	1.063	2.947	1.2246	10
20°00'	.3491	.3420	.9397	.3640	2.747	1.064	2.924	1.2217	70°00'
10	.3520	.3448	.9387	.3673	2.723	1.065	2.901	1.2188	50
20	.3549	.3475	.9377	.3706	2.699	1.066	2.878	1.2159	40
30	.3578	.3502	.9367	.3739	2.675	1.068	2.855	1.2130	30
40	.3607	.3529	.9356	.3772	2.651	1.069	2.833	1.2101	20
50	.3636	.3557	.9346	.3805	2.628	1.070	2.812	1.2072	10
		cos θ	sin θ	cot θ	tan θ	csc θ	sec θ	θ (radians)	θ (degrees)

θ (degrees)	θ (radians)	sin θ	cos θ	tan θ	cot θ	sec θ	csc θ	θ (radians)	θ (degrees)
21°00'	.3665	.3584	.9336	.3839	2.605	1.071	2.790	1.2043	69°00'
10	.3694	.3611	.9325	.3872	2.583	1.072	2.769	1.2014	50
20	.3723	.3638	.9315	.3906	2.560	1.074	2.749	1.1985	40
30	.3752	.3665	.9304	.3939	2.539	1.075	2.729	1.1956	30
40	.3782	.3692	.9293	.3973	2.517	1.076	2.709	1.1926	20
50	.3811	.3719	.9283	.4006	2.496	1.077	2.689	1.1897	10
22°00'	.3840	.3746	.9272	.4040	2.475	1.079	2.669	1.1868	68°00'
10	.3869	.3773	.9261	.4074	2.455	1.080	2.650	1.1839	50
20	.3898	.3800	.9250	.4108	2.434	1.081	2.632	1.1810	40
30	.3927	.3827	.9239	.4142	2.414	1.082	2.613	1.1781	30
40	.3956	.3854	.9228	.4176	2.394	1.084	2.595	1.1752	20
50	.3985	.3881	.9216	.4210	2.375	1.085	2.577	1.1723	10
23°00'	.4014	.3907	.9205	.4245	2.356	1.086	2.559	1.1694	67°00'
10	.4043	.3934	.9194	.4279	2.337	1.088	2.542	1.1665	50
20	.4072	.3961	.9182	.4314	2.318	1.089	2.525	1.1636	40
30	.4102	.3987	.9171	.4348	2.300	1.090	2.508	1.1606	30
40	.4131	.4014	.9159	.4383	2.282	1.092	2.491	1.1577	20
50	.4160	.4041	.9147	.4417	2.264	1.093	2.475	1.1548	10
24°00'	.4189	.4067	.9135	.4452	2.246	1.095	2.459	1.1519	66°00'
10	.4218	.4094	.9124	.4487	2.229	1.096	2.443	1.1490	50
20	.4247	.4120	.9112	.4522	2.211	1.097	2.427	1.1461	40
30	.4276	.4147	.9100	.4557	2.194	1.099	2.411	1.1432	30
40	.4305	.4173	.9088	.4592	2.177	1.100	2.396	1.1403	20
50	.4334	.4200	.9075	.4628	2.161	1.102	2.381	1.1374	10
25°00'	.4363	.4226	.9063	.4663	2.145	1.103	2.366	1.1345	65°00'
10	.4392	.4253	.9051	.4699	2.128	1.105	2.352	1.1316	50
20	.4422	.4279	.9038	.4734	2.112	1.106	2.337	1.1286	40
30	.4451	.4305	.9026	.4770	2.097	1.108	2.323	1.1257	30
40	.4480	.4331	.9013	.4806	2.081	1.109	2.309	1.1228	20
50	.4509	.4358	.9001	.4841	2.066	1.111	2.295	1.1199	10
26°00'	.4538	.4384	.8988	.4877	2.050	1.113	2.281	1.1170	64°00'
10	.4567	.4410	.8975	.4913	2.035	1.114	2.268	1.1141	50
20	.4596	.4436	.8962	.4950	2.020	1.116	2.254	1.1112	40
30	.4625	.4462	.8949	.4986	2.006	1.117	2.241	1.1083	30
40	.4654	.4488	.8936	.5022	1.991	1.119	2.228	1.1054	20
50	.4683	.4514	.8923	.5059	1.977	1.121	2.215	1.1025	10
27°00'	.4712	.4540	.8910	.5095	1.963	1.122	2.203	1.0996	63°00'
10	.4741	.4566	.8897	.5132	1.949	1.124	2.190	1.0966	50
20	.4771	.4592	.8884	.5169	1.935	1.126	2.178	1.0937	40
30	.4800	.4617	.8870	.5206	1.921	1.127	2.166	1.0908	30
40	.4829	.4643	.8857	.5243	1.907	1.129	2.154	1.0879	20
50	.4858	.4669	.8843	.5280	1.894	1.131	2.142	1.0850	10
		cos θ	sin θ	cot θ	tan θ	csc θ	sec θ	θ (radians)	θ (degrees)

Table (35°–41°)

θ (degrees)	θ (radians)	sin θ	cos θ	tan θ	cot θ	sec θ	csc θ	θ (radians)	θ (degrees)
35°00'	.6109	.5736	.8192	.7002	1.428	1.221	1.743	.9599	55°00'
10	.6138	.5760	.8175	.7046	1.419	1.223	1.736	.9570	50
20	.6167	.5783	.8158	.7089	1.411	1.226	1.729	.9541	40
30	.6196	.5807	.8141	.7133	1.402	1.228	1.722	.9512	30
40	.6225	.5831	.8124	.7177	1.393	1.231	1.715	.9483	20
50	.6254	.5854	.8107	.7221	1.385	1.233	1.708	.9454	10
36°00'	.6283	.5878	.8090	.7265	1.376	1.236	1.701	.9425	54°00'
10	.6312	.5901	.8073	.7310	1.368	1.239	1.695	.9396	50
20	.6341	.5925	.8056	.7355	1.360	1.241	1.688	.9367	40
30	.6370	.5948	.8039	.7400	1.351	1.244	1.681	.9338	30
40	.6400	.5972	.8021	.7445	1.343	1.247	1.675	.9308	20
50	.6429	.5995	.8004	.7490	1.335	1.249	1.668	.9279	10
37°00'	.6458	.6018	.7986	.7536	1.327	1.252	1.662	.9250	53°00'
10	.6487	.6041	.7969	.7581	1.319	1.255	1.655	.9221	50
20	.6516	.6065	.7951	.7627	1.311	1.258	1.649	.9192	40
30	.6545	.6088	.7934	.7673	1.303	1.260	1.643	.9163	30
40	.6574	.6111	.7916	.7720	1.295	1.263	1.636	.9134	20
50	.6603	.6134	.7898	.7766	1.288	1.266	1.630	.9105	10
38°00'	.6632	.6157	.7880	.7813	1.280	1.269	1.624	.9076	52°00'
10	.6661	.6180	.7862	.7860	1.272	1.272	1.618	.9047	50
20	.6690	.6202	.7844	.7907	1.265	1.275	1.612	.9018	40
30	.6720	.6225	.7826	.7954	1.257	1.278	1.606	.8988	30
40	.6749	.6248	.7808	.8002	1.250	1.281	1.601	.8959	20
50	.6778	.6271	.7790	.8050	1.242	1.284	1.595	.8930	10
39°00'	.6807	.6293	.7771	.8098	1.235	1.287	1.589	.8901	51°00'
10	.6836	.6316	.7753	.8146	1.228	1.290	1.583	.8872	50
20	.6865	.6338	.7735	.8195	1.220	1.293	1.578	.8843	40
30	.6894	.6361	.7716	.8243	1.213	1.296	1.572	.8814	30
40	.6923	.6383	.7698	.8292	1.206	1.299	1.567	.8785	20
50	.6952	.6406	.7679	.8342	1.199	1.302	1.561	.8756	10
40°00'	.6981	.6428	.7660	.8391	1.192	1.305	1.556	.8727	50°00'
10	.7010	.6450	.7642	.8441	1.185	1.309	1.550	.8698	50
20	.7039	.6472	.7623	.8491	1.178	1.312	1.545	.8668	40
30	.7069	.6494	.7604	.8541	1.171	1.315	1.540	.8639	30
40	.7098	.6517	.7585	.8591	1.164	1.318	1.535	.8610	20
50	.7127	.6539	.7566	.8642	1.157	1.322	1.529	.8581	10
41°00'	.7156	.6561	.7547	.8693	1.150	1.325	1.524	.8552	49°00'
10	.7185	.6583	.7528	.8744	1.144	1.328	1.519	.8523	50
20	.7214	.6604	.7509	.8796	1.137	1.332	1.514	.8494	40
30	.7243	.6626	.7490	.8847	1.130	1.335	1.509	.8465	30
40	.7272	.6648	.7470	.8899	1.124	1.339	1.504	.8436	20
50	.7301	.6670	.7451	.8952	1.117	1.342	1.499	.8407	10
(degrees)	(radians)	cos θ	sin θ	cot θ	tan θ	csc θ	sec θ	(radians)	(degrees)

Table (28°–34°)

θ (degrees)	θ (radians)	sin θ	cos θ	tan θ	cot θ	sec θ	csc θ	θ (radians)	θ (degrees)
28°00'	.4887	.4695	.8829	.5317	1.881	1.133	2.130	1.0821	62°00'
10	.4916	.4720	.8816	.5354	1.868	1.134	2.118	1.0792	50
20	.4945	.4746	.8802	.5392	1.855	1.136	2.107	1.0763	40
30	.4974	.4772	.8788	.5430	1.842	1.138	2.096	1.0734	30
40	.5003	.4797	.8774	.5467	1.829	1.140	2.085	1.0705	20
50	.5032	.4823	.8760	.5505	1.816	1.142	2.074	1.0676	10
29°00'	.5061	.4848	.8746	.5543	1.804	1.143	2.063	1.0647	61°00'
10	.5091	.4874	.8732	.5581	1.792	1.145	2.052	1.0617	50
20	.5120	.4899	.8718	.5619	1.780	1.147	2.041	1.0588	40
30	.5149	.4924	.8704	.5658	1.767	1.149	2.031	1.0559	30
40	.5178	.4950	.8689	.5696	1.756	1.151	2.020	1.0530	20
50	.5207	.4975	.8675	.5735	1.744	1.153	2.010	1.0501	10
30°00'	.5236	.5000	.8660	.5774	1.732	1.155	2.000	1.0472	60°00'
10	.5265	.5025	.8646	.5812	1.720	1.157	1.990	1.0443	50
20	.5294	.5050	.8631	.5851	1.709	1.159	1.980	1.0414	40
30	.5323	.5075	.8616	.5890	1.698	1.161	1.970	1.0385	30
40	.5352	.5100	.8601	.5930	1.686	1.163	1.961	1.0356	20
50	.5381	.5125	.8587	.5969	1.675	1.165	1.951	1.0327	10
31°00'	.5411	.5150	.8572	.6009	1.664	1.167	1.942	1.0297	59°00'
10	.5440	.5175	.8557	.6048	1.653	1.169	1.932	1.0268	50
20	.5469	.5200	.8542	.6088	1.643	1.171	1.923	1.0239	40
30	.5498	.5225	.8526	.6128	1.632	1.173	1.914	1.0210	30
40	.5527	.5250	.8511	.6168	1.621	1.175	1.905	1.0181	20
50	.5556	.5275	.8496	.6208	1.611	1.177	1.896	1.0152	10
32°00'	.5585	.5299	.8480	.6249	1.600	1.179	1.887	1.0123	58°00'
10	.5614	.5324	.8465	.6289	1.590	1.181	1.878	1.0094	50
20	.5643	.5348	.8450	.6330	1.580	1.184	1.870	1.0065	40
30	.5672	.5373	.8434	.6371	1.570	1.186	1.861	1.0036	30
40	.5701	.5398	.8418	.6412	1.560	1.188	1.853	1.0007	20
50	.5730	.5422	.8403	.6453	1.550	1.190	1.844	.9977	10
33°00'	.5760	.5446	.8387	.6494	1.540	1.192	1.836	.9948	57°00'
10	.5789	.5471	.8371	.6536	1.530	1.195	1.828	.9919	50
20	.5818	.5495	.8355	.6577	1.520	1.197	1.820	.9890	40
30	.5847	.5519	.8339	.6619	1.511	1.199	1.812	.9861	30
40	.5876	.5544	.8323	.6661	1.501	1.202	1.804	.9832	20
50	.5905	.5568	.8307	.6703	1.492	1.204	1.796	.9803	10
34°00'	.5934	.5592	.8290	.6745	1.483	1.206	1.788	.9774	56°00'
10	.5963	.5616	.8274	.6787	1.473	1.209	1.781	.9745	50
20	.5992	.5640	.8258	.6830	1.464	1.211	1.773	.9716	40
30	.6021	.5664	.8241	.6873	1.455	1.213	1.766	.9687	30
40	.6050	.5688	.8225	.6916	1.446	1.216	1.758	.9657	20
50	.6080	.5712	.8208	.6959	1.437	1.218	1.751	.9628	10
(degrees)	(radians)	cos θ	sin θ	cot θ	tan θ	csc θ	sec θ	(radians)	(degrees)

θ (degrees)	θ (radians)	sin θ	cos θ	tan θ	cot θ	sec θ	csc θ		
42°00′	.7330	.6691	.7431	.9004	1.111	1.346	1.494	.8378	**48°00′**
10	.7359	.6713	.7412	.9057	1.104	1.349	1.490	.8348	50
20	.7389	.6734	.7392	.9110	1.098	1.353	1.485	.8319	40
30	.7418	.6756	.7373	.9163	1.091	1.356	1.480	.8290	30
40	.7447	.6777	.7353	.9217	1.085	1.360	1.476	.8261	20
50	.7476	.6799	.7333	.9271	1.079	1.364	1.471	.8232	10
43°00′	.7505	.6820	.7314	.9325	1.072	1.367	1.466	.8203	**47°00′**
10	.7534	.6841	.7294	.9380	1.066	1.371	1.462	.8174	50
20	.7563	.6862	.7274	.9435	1.060	1.375	1.457	.8145	40
30	.7592	.6884	.7254	.9490	1.054	1.379	1.453	.8116	30
40	.7621	.6905	.7234	.9545	1.048	1.382	1.448	.8087	20
50	.7650	.6926	.7214	.9601	1.042	1.386	1.444	.8058	10
44°00′	.7679	.6947	.7193	.9657	1.036	1.390	1.440	.8029	**46°00′**
10	.7709	.6967	.7173	.9713	1.030	1.394	1.435	.7999	50
20	.7738	.6988	.7153	.9770	1.024	1.398	1.431	.7970	40
30	.7767	.7009	.7133	.9827	1.018	1.402	1.427	.7941	30
40	.7796	.7030	.7112	.9884	1.012	1.406	1.423	.7912	20
50	.7825	.7050	.7092	.9942	1.006	1.410	1.418	.7883	10
45°00′	.7854	.7071	.7071	1.000	1.000	1.414	1.414	.7854	**45°00′**
		cos θ	sin θ	cot θ	tan θ	csc θ	sec θ	θ (radians)	θ (degrees)

Solutions to Selected Exercises

EXERCISE SET 1.1, page 8

2. a. Integers: 21, 53

b. Rational numbers: $5.\overline{17}$, -4.25, $\frac{1}{4}$, 21, 53, 0.45454545. . .

c. Irrational numbers: π

d. Real numbers: All of the given numbers are real numbers.

e. Prime numbers: 53

f. Composite numbers: 21

4. $\{0, 1, 2, 3, 4\} \cap \{1, 3, 6, 10\} = \{1, 3\}$

16. Commutative property of addition

18. Symmetric property of equality

30. $\dfrac{2a}{5} + \dfrac{3a}{7} = \dfrac{2a \cdot 7}{5 \cdot 7} + \dfrac{3a \cdot 5}{7 \cdot 5} = \dfrac{29a}{35}$

EXERCISE SET 1.2, page 14

16. $[-2, 1)$

56. $|x + 6| + |x - 2| = x + 6 - (x - 2)$
$$= x + 6 - x + 2$$
$$= 8$$

64. $|-5 - 8| = 13$ **74.** $|x - 20| < 4$

EXERCISE SET 1.3, page 21

4. $-2^4 = -16$

18. $\left(\dfrac{4 \cdot 5^{-1}}{2^{-3}}\right)^{-2} = \dfrac{5^2}{4^2 \cdot 2^6} = \dfrac{25}{16 \cdot 64} = \dfrac{25}{1024}$

30. $\left(\dfrac{3pq^2}{-2pq^3r^2}\right)^4 = \left(\dfrac{3}{-2qr^2}\right)^4 = \dfrac{81}{16q^4r^8}$

48. $25{,}600 = 2.56 \times 10^4$

EXERCISE SET 1.4, page 27

24. $(5y^2 - 7y + 3) + (2y^2 + 8y + 1) = 7y^2 + y + 4$

32. $(5x - 7)(3x^2 - 8x - 5)$
$$= 15x^3 - 40x^2 - 25x - 21x^2 + 56x + 35$$
$$= 15x^3 - 61x^2 + 31x + 35$$

60. $(4x^2 - 3y)(4x^2 + 3y) = (4x^2)^2 - (3y)^2 = 16x^4 - 9y^2$

72. $-x^2 - 5x + 4 = -(-5)^2 - 5(-5) + 4$
$$= -25 + 25 + 4 = 4$$

82. $\dfrac{1}{6}n^3 - \dfrac{1}{2}n^2 + \dfrac{1}{3}n = \dfrac{1}{6}(21)^3 - \dfrac{1}{2}(21)^2 + \dfrac{1}{3}(21)$
$$= 1330 \text{ committees}$$

84. a. $4.3 \times 10^{-6}(1000)^2 - 2.1 \times 10^{-4}(1000)$
$$= 4.09 \text{ s}$$

b. $4.3 \times 10^{-6}(5000)^2 - 2.1 \times 10^{-4}(5000)$
$$= 106.45 \text{ s}$$

c. $4.3 \times 10^{-6}(10{,}000)^2 - 2.1 \times 10^{-4}(10{,}000)$
$$= 427.9 \text{ s}$$

EXERCISE SET 1.5, page 37

6. $6a^3b^2 - 12a^2b + 72ab^3 = 6ab(a^2b - 2a + 12b^2)$

18. $b^2 + 12b - 28 = (b + 14)(b - 2)$

22. $57y^2 + y - 6 = (19y - 6)(3y + 1)$

30. $b^2 - 4ac = 8^2 - 4(16)(-35)$
$$= 2304 = 48^2$$

The trinomial is factorable.

38. $81b^2 - 16c^2 = (9b - 4c)(9b + 4c)$

48. $b^2 - 24b + 144 = (b - 12)^2$

54. $b^3 + 64 = (b + 4)(b^2 - 4b + 16)$

64. $81y^4 - 16 = (9y^2 - 4)(9y^2 + 4)$

$$= (3y - 2)(3y + 2)(9y^2 + 4)$$

74. $4y^2 - 4yz + z^2 - 9 = (2y - z)^2 - 9 = (2y - z)^2 - 3^2$

$$= (2y - z - 3)(2y - z + 3)$$

EXERCISE SET 1.6, page 45

2. $\dfrac{2x^2 - 5x - 12}{2x^2 + 5x + 3} = \dfrac{(2x + 3)(x - 4)}{(2x + 3)(x + 1)} = \dfrac{x - 4}{x + 1}$

16. $\dfrac{x^2 - 16}{x^2 + 7x + 12} \cdot \dfrac{x^2 - 4x - 21}{x^2 - 4x}$

$$= \dfrac{(x - 4)(x + 4)(x + 3)(x - 7)}{(x + 3)(x + 4)x(x - 4)} = \dfrac{x - 7}{x}$$

30. $\dfrac{3y - 1}{3y + 1} - \dfrac{2y - 5}{y - 3} = \dfrac{(3y - 1)(y - 3)}{(3y + 1)(y - 3)} - \dfrac{(2y - 5)(3y + 1)}{(y - 3)(3y + 1)}$

$$= \dfrac{(3y^2 - 10y + 3) - (6y^2 - 13y - 5)}{(3y + 1)(y - 3)}$$

$$= \dfrac{-3y^2 + 3y + 8}{(3y + 1)(y - 3)}$$

42. $\dfrac{3 - \dfrac{2}{a}}{5 + \dfrac{3}{a}} = \dfrac{\left(3 - \dfrac{2}{a}\right)a}{\left(5 + \dfrac{3}{a}\right)a} = \dfrac{3a - 2}{5a + 3}$

60. $\dfrac{e^{-2} - f^{-1}}{ef} = \dfrac{\dfrac{1}{e^2} - \dfrac{1}{f}}{ef} = \dfrac{\dfrac{f - e^2}{e^2 f}}{ef} = \dfrac{f - e^2}{e^2 f} \div \dfrac{ef}{1}$

$$= \dfrac{f - e^2}{e^2 f} \cdot \dfrac{1}{ef} = \dfrac{f - e^2}{e^3 f^2}$$

64. a. $\dfrac{v_1 + v_2}{1 + \dfrac{v_1 v_2}{c^2}} = \dfrac{1.2 \times 10^8 + 2.4 \times 10^8}{1 + \dfrac{(1.2 \times 10^8)(2.4 \times 10^8)}{(6.7 \times 10^8)^2}} \approx 3.4 \times 10^8$

b. $\dfrac{v_1 + v_2}{1 + \dfrac{v_1 \cdot v_2}{c^2}} = \dfrac{c^2(v_1 + v_2)}{c^2\left(1 + \dfrac{v_1 \cdot v_2}{c^2}\right)} = \dfrac{c^2(v_1 + v_2)}{c^2 + v_1 \cdot v_2}$

EXERCISE SET 1.7, page 54

2. $49^{1/2} = 7$

6. $16^{3/2} = (2^4)^{3/2} = 2^6 = 64$

38. $\left(\dfrac{r^3 s^{-2}}{rs^4}\right)^{1/2} = (r^{3-1} s^{-2-4})^{1/2} = (r^2 s^{-6})^{1/2} = rs^{-3} = \dfrac{r}{s^3}$

46. $\sqrt[3]{-64} = -4$

52. $(\sqrt[4]{25})^2 = 25^{2/4} = 25^{1/2} = 5$

70. $\sqrt{75} = \sqrt{3 \cdot 5^2} = \sqrt{3} \cdot \sqrt{5^2} = 5\sqrt{3}$

98. $5\sqrt[3]{3} + 2\sqrt[3]{81} = 5\sqrt[3]{3} + 2\sqrt[3]{3^4} = 5\sqrt[3]{3} + 6\sqrt[3]{3} = 11\sqrt[3]{3}$

102. $(\sqrt{7} + 4)(\sqrt{7} - 1) = 7 - \sqrt{7} + 4\sqrt{7} - 4 = 3 + 3\sqrt{7}$

116. $\dfrac{2}{\sqrt[3]{4}} = \dfrac{2}{\sqrt[3]{4}} \cdot \dfrac{\sqrt[3]{2}}{\sqrt[3]{2}} = \dfrac{2\sqrt[3]{2}}{\sqrt[3]{8}} = \dfrac{2\sqrt[3]{2}}{2} = \sqrt[3]{2}$

124. $\dfrac{5}{\sqrt{y} - \sqrt{3}} = \dfrac{5}{\sqrt{y} - \sqrt{3}} \cdot \dfrac{\sqrt{y} + \sqrt{3}}{\sqrt{y} + \sqrt{3}} = \dfrac{5(\sqrt{y} + \sqrt{3})}{y - 3}$

$$= \dfrac{5\sqrt{y} + 5\sqrt{3}}{y - 3}$$

EXERCISE SET 1.8, page 61

2. $3 + \sqrt{-25} = 3 + i\sqrt{25} = 3 + 5i$

16. $(-3 + i) - (-8 + 2i) = -3 + i + 8 - 2i = 5 - i$

24. $(5 - 3i)(-2 - 4i) = -10 - 20i + 6i + 12i^2$

$$= -10 - 14i - 12 = -22 - 14i$$

34. $\dfrac{5 - 7i}{5 + 7i} = \dfrac{5 - 7i}{5 + 7i} \cdot \dfrac{5 - 7i}{5 - 7i} = \dfrac{25 - 35i - 35i + 49i^2}{25 - 49i^2}$

$$= \dfrac{25 - 70i - 49}{25 + 49} = \dfrac{-24 - 70i}{74}$$

$$= -\dfrac{12}{37} - \dfrac{35}{37}i$$

54. $i^{28} = (i^4)^7 = 1$

68. $\sqrt{-3}\sqrt{-121} = i\sqrt{3} \cdot i\sqrt{121} = i\sqrt{3} \cdot 11i$

$$= 11i^2\sqrt{3} = -11\sqrt{3}$$

EXERCISE SET 2.1, page 74

2. $-3y + 20 = 2$

$$-3y = -18$$

$$y = 6$$

12. $\dfrac{1}{2}x + 7 - \dfrac{1}{4}x = \dfrac{19}{2}$

$$4\left(\dfrac{1}{2}x + 7 - \dfrac{1}{4}x\right) = 4\left(\dfrac{19}{2}\right)$$

$$2x + 28 - x = 38$$

$$x = 38 - 28$$

$$x = 10$$

20. $5(x + 4)(x - 4) = (x - 3)(5x + 4)$

$$5(x^2 - 16) = 5x^2 - 11x - 12$$

$$5x^2 - 80 = 5x^2 - 11x - 12$$

$$-80 + 12 = -11x$$

$$-68 = -11x$$

$$\dfrac{68}{11} = x$$

30. $2x + \dfrac{1}{3} = \dfrac{6x + 1}{3}$ • Rewrite the left side.

$$\dfrac{6x + 1}{3} = \dfrac{6x + 1}{3}$$

This equation is an identity.

38. $\dfrac{4}{y + 2} = \dfrac{7}{y - 4}$ $y \neq 2, y \neq 4$

$4(y - 4) = 7(y + 2)$

$4y - 16 = 7y + 14$

$4y - 7y = 14 + 16$

$-3y = 30$

$y = -10$

66. $\dfrac{2}{x - 2} = \dfrac{y}{x - 3}$ $x \neq 2, x \neq 3$

$2x - 6 = y(x - 2)$

$2x - 6 = xy - 2y$

$2x - xy = 6 - 2y$

$x(2 - y) = 6 - 2y$

$x = \dfrac{6 - 2y}{2 - y}, y \neq 2$

EXERCISE SET 2.2, page 82

2. $P = S - Sdt$

$Sdt = S - P$

$t = \dfrac{S - P}{Sd}$

22. $P = 2l + 2w,$ $w = \dfrac{1}{2}l + 1$

$110 = 2l + 2\left(\dfrac{1}{2}l + 1\right)$

$110 = 2l + l + 2$

$36 = l$

$l = 36$ meters

$w = \dfrac{1}{2}l + 1 = 19$ meters

30. Let $t_1 =$ the time it takes to travel to the island.
Let $t_2 =$ the time it takes to make the return trip.

$t_1 + t_2 = 7.5$

$t_2 = 7.5 - t_1$

$15t_1 = 10t_2$

$15t_1 = 10(7.5 - t_1)$

$15t_1 = 75 - 10t_1$

$25t_1 = 75$

$t_1 = 3$ hours

$D = 15t_1 = 3(15) = 45$ nautical miles

36. Let $x =$ the score on the final examination.

$$\dfrac{90 + 74 + 82 + 90 + 2x}{6} = 85$$

$336 + 2x = 510$

$x = 87$

A final examination score of 87 will produce an average of 85.

38. Let $x =$ the number of glasses of orange juice that were sold.

Profit = Income − Cost

$2337 = 0.75x - 0.18x$

$2337 = 0.57x$

$4100 = x$

4100 glasses of orange juice were sold.

42.

5%	x
7%	$7500 - x$

$0.05x + 0.07(7500 - x) = 405$

$0.05x + 525 - 0.07x = 405$

$- 0.02x = -120$

$x = 6000$

$7500 - x = 1500$

$6000 was invested at 5%. $1500 was invested at 7%.

46. Let $x =$ the number of liters of the 40% solution to be mixed with the 24% solution.

0.40	x
0.24	4
0.30	$4 + x$

$0.40x + 0.24(4) = 0.30(4 + x)$

$0.40x + 0.96 = 1.2 + 0.30x$

$0.10x = 0.24$

$x = 2.4$

2.4 liters of 40% sulfuric acid should be mixed with 4 liters of a 24% sulfuric acid solution, to produce the 30° solution.

56. Let $x =$ the number of hours needed to print the report if both the printers are used.
Printer A prints ⅓ of the report every hour.
Printer B prints ¼ of the report every hour.
Thus

$$\dfrac{1}{3}x + \dfrac{1}{4}x = 1$$

$4x + 3x = 12 \cdot 1$

$7x = 12$

$x \approx 1.71$

It would take 1.71 hours to print the report.

EXERCISE SET 2.3, page 93

4. $12w^2 - 41w + 24 = 0$

$(4w - 3)(3w - 8) = 0$

$4w - 3 = 0$ or $3w - 8 = 0$

$w = \dfrac{3}{4}$ $w = \dfrac{8}{3}$

18. $y^2 = 225$

$y = \pm\sqrt{225}$

$y = \pm 15$

34. $x^2 - 6x = 0$

$x^2 - 6x + 9 = 9$

$(x - 3)^2 = 9$

$x - 3 = \pm\sqrt{9}$

$x = 3 \pm 3$

$x = 3 + 3$ or $x = 3 - 3$

$x = 6$ $x = 0$

38. $2x^2 + 10x - 3 = 0$

$2x^2 + 10x = 3$

$2(x^2 + 5x) = 3$

$x^2 + 5x = \dfrac{3}{2}$

$x^2 + 5x + \dfrac{25}{4} = \dfrac{3}{2} + \dfrac{25}{4}$

$\left(x + \dfrac{5}{2}\right)^2 = \dfrac{31}{4}$

$x + \dfrac{5}{2} = \pm\sqrt{\dfrac{31}{4}}$

$x = -\dfrac{5}{2} \pm \dfrac{\sqrt{31}}{2}$

$x = \dfrac{-5 + \sqrt{31}}{2}$ or $x = \dfrac{-5 - \sqrt{31}}{2}$

48. $2x^2 + 4x - 1 = 0$

$x = \dfrac{-4 \pm \sqrt{4^2 - 4(2)(-1)}}{4}$

$x = \dfrac{-4 \pm \sqrt{16 + 8}}{4} = \dfrac{-4 \pm \sqrt{24}}{4}$

$x = \dfrac{-4 \pm 2\sqrt{6}}{4} = \dfrac{-2 \pm \sqrt{6}}{2}$

$x = \dfrac{-2 + \sqrt{6}}{2}$ $x = \dfrac{-2 - \sqrt{6}}{2}$

58. $x^2 + 3x - 11 = 0$

$b^2 - 4ac = 3^2 - 4(1)(-11)$

$= 9 + 44 = 53 > 0$

Thus the equation has two distinct real roots.

68. $(10 \text{ feet} + \tfrac{1}{4} \text{ in})^2 = (10 \text{ feet})^2 + x^2$

$\sqrt{(120.25)^2 - (120)^2} = x$ • (change feet to inches)

$7.75 = x$

To the nearest inch, the concrete will rise 8 inches.

74. Let P = perimeter and A = area.

$P = 4w + 2l = 400$ $A = 4800 = lw$

$2w + l = 200$ $l = \dfrac{4800}{w}$

$2w + \dfrac{4800}{w} = 200$

$2w^2 + 4800 = 200w$

$w^2 - 100w + 2400 = 0$

$(w - 60)(w - 40) = 0$

$w = 60$ or $w = 40$

$l = \dfrac{4800}{60} = 80$ or $l = \dfrac{4800}{40} = 120$

There are two solutions: 60 yd $\times$ 80 yd or 40 yd $\times$ 120 yd.

EXERCISE SET 2.4, page 101

6. $x^4 - 36x^2 = 0$

$x^2(x^2 - 36) = 0$

$x^2(x - 6)(x + 6) = 0$

$x = 0, \, x = 6, \, x = -6$

14. $\sqrt{10 - x} = 4$ Check: $\sqrt{10 - (-6)} = 4$

$10 - x = 16$ $\sqrt{16} = 4$

$-x = 6$ $4 = 4$

$x = -6$

The solution is -6.

16. $x = \sqrt{5 - x} + 5$

$(x - 5)^2 = (\sqrt{5 - x})^2$

$x^2 - 10x + 25 = 5 - x$

$x^2 - 9x + 20 = 0$

$(x - 5)(x - 4) = 0$

$x = 5$ or $x = 4$

Check: $5 = \sqrt{5 - 5} + 5$ $4 = \sqrt{5 - 4} + 5$

$5 = 0 + 5$ $4 = 1 + 5$

$5 = 5$ $4 = 6$ No

The solution is 5.

20.
$$(\sqrt{x+7} - 2)^2 = (\sqrt{x-9})^2$$
$$x + 7 - 4\sqrt{x+7} + 4 = x - 9$$
$$-4\sqrt{x+7} = -20$$
$$(\sqrt{x+7})^2 = (5)^2$$
$$x + 7 = 25$$
$$x = 18$$

Check: $\sqrt{18+7} - 2 = \sqrt{18-9}$
$$\sqrt{25} - 2 = \sqrt{9}$$
$$5 - 2 = 3$$
$$3 = 3$$

The solution is 18.

32.
$$(4z + 7)^{1/3} = 2$$
$$[(4z+7)^{1/3}]^3 = 2^3$$
$$4z + 7 = 8$$
$$4z = 1$$
$$z = \frac{1}{4}$$

Check: $\left[4\left(\dfrac{1}{4}\right) + 7\right]^{1/3} = 2$
$$8^{1/3} = 2$$
$$2 = 2$$

The solution is $\dfrac{1}{4}$.

42. $x^4 - 10x^2 + 9 = 0$. Let $u = x^2$.
$$u^2 - 10u + 9 = 0$$
$$(u - 9)(u - 1) = 0$$
$$u = 9 \quad u = 1$$
$$x^2 = 9 \quad x^2 = 1$$
$$x = \pm3 \quad x = \pm1$$

The solutions are 3, −3, 1, and −1.

52. $6x^{2/3} - 7x^{1/3} - 20 = 0$. Let $u = x^{1/3}$.
$$6u^2 - 7u - 20 = 0$$
$$(3u + 4)(2u - 5) = 0$$
$$u = -\frac{4}{3} \quad \text{or} \quad u = \frac{5}{2}$$
$$x^{1/3} = -\frac{4}{3} \qquad x^{1/3} = \frac{5}{2}$$
$$(x^{1/3})^3 = \left(-\frac{4}{3}\right)^3 \qquad (x^{1/3})^3 = \left(\frac{5}{2}\right)^3$$
$$x = -\frac{64}{27} \qquad x = \frac{125}{8}$$

EXERCISE SET 2.5, page 110

8. $-4(x - 5) \geq 2x + 15$
$$-4x + 20 \geq 2x + 15$$
$$-6x \geq -5$$
$$x \leq \frac{5}{6}$$

12. Company A: $19 + 0.12x$ x is the number of miles driven

Company B: $12 + 0.21x$
$$19 + 0.12x < 12 + 0.21x$$
$$77.7 < x$$

Company A is less expensive if you drive at least 78 miles.

22. $41 \leq \quad F \quad \leq 68$
$$41 \leq \frac{9}{5}C + 32 \leq 68$$
$$9 \leq \frac{9}{5}C \leq 36$$
$$\frac{5}{9}(9) \leq \left(\frac{5}{9}\right)\left(\frac{9}{5}\right)C \leq \frac{5}{9}(36)$$
$$5 \leq \quad C \quad \leq 20$$

30. $x^2 + 5x + 6 < 0$
$$(x + 2)(x + 3) = 0$$
$$x = -2, \quad \text{or } x = -3 \quad \bullet \text{ critical values}$$

Use a test number from each of the following intervals, $(-\infty, -3)$, $(-3, -2)$, $(-2, \infty)$ to determine where $x^2 + 5x + 6 < 0$.

```
 +++++++|-|+++++
<++++++++++++++>
      -3-2   0
```

The solution is $(-3, -2)$.

38. Find the critical values of $-3x^2 + 312x \geq 5925$
$$3x^2 - 312x + 5925 = 0$$
$$3(x - 25)(x - 79) = 0$$
$$x = 25 \quad x = 79 \quad \bullet \text{ critical values}$$

The solution set is $[25, 79]$. Thus the monthly revenue is greater than or equal to $5925 when $\$25 \leq x \leq \$79$.

40. $\dfrac{x - 2}{x + 3} > 0$
$$x = 2 \quad x = -3 \quad \bullet \text{ critical values}$$

The solution set is $(-\infty, -3) \cup (2, \infty)$.

48.
$$\frac{3x + 1}{x - 2} \geq 4$$

$$\frac{3x + 1}{x - 2} - 4 \geq 0$$

$$\frac{3x + 1 - 4(x - 2)}{x - 2} \geq 0$$

$$\frac{-x + 9}{x - 2} \geq 0$$

$x = 2$ and $x = 9$ • critical values

The solution set is $(2, 9]$.

EXERCISE SET 2.6, page 115

10. $|2x - 3| = 21$

$2x - 3 = 21$ or $2x - 3 = -21$

$\quad 2x = 24 \qquad\qquad 2x = -18$

$\quad\quad x = 12 \qquad\qquad x = -9$

The solution set is $\{12, -9\}$.

30. $|2x - 9| < 7$

$-7 < 2x - 9 < 7$

$\quad 2 < \quad 2x \quad < 16$

$\quad 1 < \quad x \quad < 8$

The solution set is $(1, 8)$.

34. $|2x - 5| \geq 1$

$2x - 5 \leq -1$ or $2x - 5 \geq 1$

$\quad 2x \leq 4 \qquad\qquad 2x \geq 6$

$\quad\quad x \leq 2 \qquad\qquad x \geq 3$

The solution set is $(-\infty, 2] \cup [3, \infty)$.

46. $|x - c| < d$

$-d < x - c < d$

$-d + c < \quad x \quad < d + c$

The solution set is $(-d + c, d + c)$.

48. $|x^2 - 2| > 1$

$x^2 - 2 = 1 \qquad x^2 - 2 = -1$

$\quad x^2 = 3 \qquad\qquad x^2 = 1$

$\quad x = \pm\sqrt{3} \qquad\quad x = \pm 1$ • critical values

The solution set is $(-\infty, -\sqrt{3}) \cup (-1, 1) \cup (\sqrt{3}, \infty)$.

SECTION 3.1, page 129

26.

30.

32.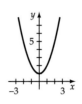

40. $\left(0, -\dfrac{15}{4}\right)$, $(5, 0)$

60. $(x - 5)^2 + (y + 3)^2 = 4^2$

68.
$$x^2 - 6x + y^2 - 4y = -12$$

$$x^2 - 6x + 9 + y^2 - 4y + 4 = -12 + 9 + 4$$

$$(x - 3)^2 + (y - 2)^2 = 1^2$$

Center $(3, 2)$, radius 1

EXERCISE SET 3.2, page 138

2. Given $g(x) = 2x^2 + 3$

a. $g(3) = 2(3)^2 + 3 = 18 + 3 = 21$

b. $g(-1) = 2(-1)^2 + 3 = 2 + 3 = 5$

c. $g(0) = 2(0)^2 + 3 = 0 + 3 = 3$

d. $g\left(\dfrac{1}{2}\right) = 2\left(\dfrac{1}{2}\right)^2 + 3 = \dfrac{1}{2} + 3 = \dfrac{7}{2}$

e. $g(c) = 2(c)^2 + 3 = 2c^2 + 3$

f. $g(c + 5) = 2(c + 5)^2 + 3 = 2c^2 + 20c + 50 + 3$

$$= 2c^2 + 20c + 53$$

14.
$$x^2 - 2y = 2$$

$$-2y = -x^2 + 2$$

$$y = \frac{1}{2}x^2 - 1$$

y is a function of x because each x value will yield one and only one y value.

22. The domain is the set of all real numbers.

40. This is the graph of a function, because every vertical line intersects the graph in one point.

52. Domain: $\{x \mid -\infty < x < \infty\}$
Range: $\{y \mid y \leq 3\}$

80.

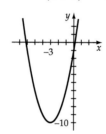

92. $v(t) = 44{,}000 - 4200t$, $0 \le t \le 8$.

100. $t = \dfrac{\sqrt{1 + x^2}}{2} + \dfrac{3 - x}{8}$ hours

EXERCISE SET 3.3, page 152

2. $m = \dfrac{1 - 4}{5 - (-2)} = -\dfrac{3}{7}$

16.

$m = -1$

$b = 1$

28. $y - 5 = -2(x - 0)$

$y = -2x + 5$

40. $m = \dfrac{-8 - (-6)}{2 - 5} = \dfrac{-2}{-3} = \dfrac{2}{3}$

$y - (-6) = \dfrac{2}{3}(x - 5)$

$y = \dfrac{2}{3}x - \dfrac{10}{3} - 6$

$y = \dfrac{2}{3}x - \dfrac{28}{3}$

48. The graph of $x + y = 10$ has a slope of -1.

$y + 1 = (-1)(x - 2)$

$y = -x + 2 - 1$

$y = -x + 1$

52. The graph of $2x - y = 7$ has a slope of 2. Thus we use a slope of $-\frac{1}{2}$.

$y - 4 = -\dfrac{1}{2}(x + 3)$

$y = -\dfrac{1}{2}x - \dfrac{3}{2} + \dfrac{8}{2}$

$y = -\dfrac{1}{2}x + \dfrac{5}{2}$

60. a. Annual depreciation $= \dfrac{40{,}090 - 8500}{15} = 2106$

$V(n) = 40{,}090 - 2106n$

b. $V(4) = 40{,}090 - 2106(4) = 40{,}090 - 8424 = \$31{,}666$

62. $P(x) = 124x - (78.5x + 5005)$

$P(x) = 45.5x - 5005$

$45.5x - 5005 = 0$

$45.5x = 5005$

$x = 110.$ • *The break-even point*

EXERCISE SET 3.4, page 160

10. $f(x) = x^2 + 6x - 1$

$\quad = x^2 + 6x + 9 + (-1 - 9)$

$\quad = (x + 3)^2 - 10,$ vertex $(-3, -10)$

20. $-\dfrac{b}{2a} = -\dfrac{-6}{2(1)} = 3$

$y = f(3) = 3^2 - 6(3) = -9$

Vertex $= (3, -9)$

$f(x) = (x - 3)^2 - 9$

30. $f(x) = -x^2 - 6x$

$\quad = -(x^2 + 6x)$

$\quad = -(x^2 + 6x + 9) + 9$

$\quad = -(x + 3)^2 + 9$

Maximum value of f is 9 when $x = -3$.

40. Let $x =$ the number of rackets to be purchased.

The price per racket is given by

$60 - [0.10(x - 40)]$ $41 \le x \le 400$

The total revenue R for the company is

$R = (60 - [0.10(x - 40)])x$

$R = (60 - 0.10x + 4)x$

$R = (64 - 0.10x)x$

$R = 64x - 0.10x^2$

The maximum value of R is obtained when

$x = -\dfrac{b}{2a} = -\dfrac{64}{2(-0.10)} = 320$

56. Let x = the number of parcels

  a. $R(x) = xp = x(22 - 0.01x) = -0.01x^2 + 22x$

  b. $P(x) = R(x) - C(x)$

$$= (-0.01x^2 + 22x) - (2025 + 7x)$$

$$= -0.01x^2 + 15x - 2025$$

  c. $-\dfrac{b}{2a} = -\dfrac{15}{2(-0.01)} = 750$

$$P(750) = -0.01(750)^2 + 15(750) - 2025$$

$$= \$3600$$

  d. $p(750) = 22 - 0.01(750) = \$14.50$

  e. The break-even point(s) occur when $R(x) = C(x)$.

$$-0.01x^2 + 22x = 2025 + 7x$$

$$0 = 0.01x^2 - 15x + 2025$$

$$x = \frac{-(-15) \pm \sqrt{(-15)^2 - 4(0.01)(2025)}}{2(0.01)}$$

$$x = 150 \text{ or } x = 1350 \text{ are the break-even points.}$$

Thus the minimum number of parcels the air freight company must ship to break even is 150.

58. $h(t) = -16t^2 + 64t + 80$

$$t = \frac{-b}{2a} = \frac{-64}{2(-16)} = 2$$

$$h(2) = -16(2)^2 + 64(2) + 80$$

$$= -64 + 128 + 80 = 144$$

  a. The vertex (2, 144) gives us the maximum height of 144 feet.

  b. The vertex of the graph of h is (2, 144) thus the time when it achieves this maximum height is at time $t = 2$ seconds.

  c. $-16t^2 + 64t + 80 = 0$

$$-16(t^2 - 4t - 5) = 0$$

$$-16(t + 1)(t - 5) = 0$$

$$t = -1 \qquad t - 5 = 0$$

$$\text{no} \qquad t = 5$$

The projectile will have a height of 0 feet at time $t = 5$ seconds.

EXERCISE SET 3.5, page 171

14. Symmetrical with respect to the x-axis, since replacing y with $-y$ leaves the equation unaltered.

24. Yes. Since $(-y) = (-x)^3 - (-x)$ simplifies to $-y = -x^3 + x$ which is equivalent to the original equation $y = x^3 - x$.

32.

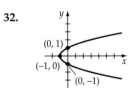

44. Even, since $h(-x) = (-x)^2 + 1 = x^2 + 1 = h(x)$.

58.

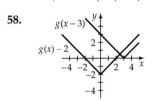

60.

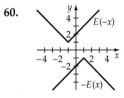

62.

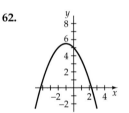

64. a.

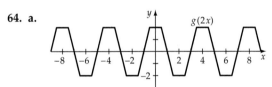

  b.

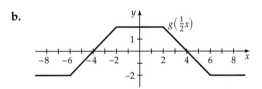

EXERCISE SET 3.6, page 179

10. $f(x) + g(x) = \sqrt{x - 4} - x$ domain $\{x \mid x \geq 4\}$

$f(x) - g(x) = \sqrt{x - 4} + x$ domain $\{x \mid x \geq 4\}$

$f(x)g(x) = -x\sqrt{x - 4}$ domain $\{x \mid x \geq 4\}$

$f(x)/g(x) = -\dfrac{\sqrt{x - 4}}{x}$ domain $\{x \mid x \geq 4\}$

14. $(f + g)(x) = (x^2 - 3x + 2) + (2x - 4) = x^2 - x - 2$

$(f + g)(-7) = (-7)^2 - (-7) - 2 = 49 + 7 - 2 = 54$

30. $\dfrac{f(x+h)-f(x)}{h} = \dfrac{[4(x+h)-5]-(4x-5)}{h}$

$$= \dfrac{4x+4(h)-5-4x+5}{h}$$

$$= \dfrac{4(h)}{h} = 4$$

38. $(g \circ f)(x) = g[f(x)] = g[2x-7]$

$$= 3[2x-7]+2 = 6x-19$$

$(f \circ g)(x) = f[g(x)] = f[3x+2]$

$$= 2[3x+2]-7 = 6x-3$$

54. $(h \circ g)(x) = -3x^4 + 30x^3 - 75x^2 + 4$

$(h \circ g)(0) = -3(0)^4 + 30(0)^3 - 75(0)^2 + 4 = 4$

EXERCISE SET 3.7, page 186

2. $(f \circ g)(x) = f[g(x)] = f[2x+6]$

$$= \dfrac{1}{2}[2x+6]-3 = x+3-3 = x$$

$(g \circ f)(x) = g[f(x)] = g\left[\dfrac{1}{2}x-3\right]$

$$= 2\left[\dfrac{1}{2}x-3\right]+6 = x-6+6 = x$$

10. $g(x) = \dfrac{2}{3}x + 4$

$$y = \dfrac{2}{3}x + 4$$

$$x = \dfrac{2}{3}y + 4$$

$$x - 4 = \dfrac{2}{3}y$$

$$\dfrac{3}{2}x - 6 = y$$

Thus $g^{-1}(x) = \dfrac{3}{2}x - 6.$

18. $G(x) = \dfrac{3x}{x-5},\quad x \neq 5$

$$y = \dfrac{3x}{x-5}$$

$$x = \dfrac{3y}{y-5}$$

$$xy - 5x = 3y$$

$$xy - 3y = 5x$$

$$y = \dfrac{5x}{x-3}$$

Thus $G^{-1}(x) = \dfrac{5x}{x-3},\ x \neq 3.$

32. $f(x) = x^2 + 6x - 6,\ x \geq -3$

domain f is $\{x \,|\, x \geq -3\}$, range f is $\{y \,|\, y \geq -15\}$,

$$y = x^2 + 6x - 6$$

$$x = y^2 + 6y - 6$$

$$x + 6 = y^2 + 6y$$

$$x + 15 = y^2 + 6y + 9 \quad \bullet \text{ Complete the square.}$$

$$x + 15 = (y + 3)^2$$

Choose the positive root, since range f^{-1} is $\{y \,|\, y \geq -3\}$.

$$\sqrt{x + 15} = y + 3$$

$$-3 + \sqrt{x + 15} = y$$

Thus $f^{-1}(x) = -3 + \sqrt{x + 15}$, domain f^{-1} is $\{x \,|\, x \geq -15\}$, range f^{-1} is $\{y \,|\, y \geq -3\}$

36.

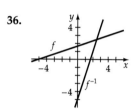

EXERCISE SET 3.8, page 191

22. $d = kw$

$$6 = k \cdot 80$$

$$\dfrac{6}{80} = k$$

$$k = \dfrac{3}{40}$$

Therefore $d = \dfrac{3}{40} \cdot 100$

$$= 7.5 \text{ inches}$$

24. $r = kv^2$

$$140 = k \cdot 60^2$$

$$\dfrac{140}{60^2} = k$$

$$\dfrac{7}{180} = k$$

Thus $r = \dfrac{7}{180} \cdot 65^2 \approx 164 \text{ feet}$

28. $I = \dfrac{k}{d^2}$

$50 = \dfrac{k}{10^2}$

$5000 = k$

Thus $I = \dfrac{5000}{d^2}$

$= \dfrac{5000}{15^2}$

$= \dfrac{5000}{225} \approx 22.2$ footcandles

30. $i = kpr$

$78 = k2600 \cdot (0.04)$

$k = \dfrac{78}{2600(0.04)} = 0.75$

Thus $i = 0.75 \cdot 4400 \cdot (0.06) = \$198$.

32. $L = kwd^2$

$200 = k \cdot 2 \cdot 6^2$

$\dfrac{25}{9} = \dfrac{200}{2 \cdot 6^2} = k$

Thus $L = \dfrac{25}{9} \cdot 4 \cdot 4^2 = \dfrac{1600}{9} \approx 177.8$ pounds.

34. $L = \dfrac{kd^4}{h^2}$

$6 = \dfrac{k \cdot 2^4}{10^2}$

$k = \dfrac{6 \cdot 10^2}{2^4} = \dfrac{600}{16} = 37.5$

Thus $L = \dfrac{37.5(3^4)}{14^2} \approx 15.5$ tons.

EXERCISE SET 4.1, page 208

6.
$$2x^2 - x - 5\overline{\smash{\big)}\,2x^4 - x^3 - 23x^2 + 9x + 45}$$
$$\underline{2x^4 - x^3 - 5x^2}$$
$$-18x^2 + 9x + 45$$
$$\underline{-18x^2 + 9x + 45}$$
$$0$$

quotient: $x^2 - 9$

12.
$$\begin{array}{r|rrrr}
5 & 5 & 6 & -8 & 1 \\
  &   & 25 & 155 & 735 \\
\hline
  & 5 & 31 & 147 & 736
\end{array}$$

$\dfrac{5x^3 + 6x^2 - 8x + 1}{x - 5} = 5x^2 + 31x + 147 + \dfrac{736}{x - 5}$

32.
$$\begin{array}{r|rrrr}
3 & 2 & -1 & 3 & -1 \\
  &   & 6 & 15 & 54 \\
\hline
  & 2 & 5 & 18 & 53
\end{array}$$

$P(c) = P(3) = 53$

42.
$$\begin{array}{r|rrrr}
-6 & 1 & 4 & -27 & -90 \\
   &   & -6 & 12 & 90 \\
\hline
   & 1 & -2 & -15 & 0
\end{array}$$

A remainder of 0 implies that $x + 6$ is a factor of $P(x)$.

EXERCISE SET 4.2, page 214

2. Since $a_n = -2$ is negative and $n = 3$ is odd, the graph of P goes up to the far left and down to the far right.

12. a.
$$\begin{array}{r|rrrr}
1 & 2 & -5 & 4 & 10 \\
  &   & 2 & -3 & 1 \\
\hline
  & 2 & -3 & 1 & 11
\end{array}$$
• $g(1) = 11$

b.
$$\begin{array}{r|rrrr}
-1 & 2 & -5 & 4 & 10 \\
   &   & -2 & 7 & -11 \\
\hline
   & 2 & -7 & 11 & -1
\end{array}$$
• $g(-1) = -1$

c.
$$\begin{array}{r|rrrr}
2 & 2 & -5 & 4 & 10 \\
  &   & 4 & -2 & 4 \\
\hline
  & 2 & -1 & 2 & 14
\end{array}$$
• $g(2) = 14$

d.
$$\begin{array}{r|rrrr}
3 & 2 & -5 & 4 & 10 \\
  &   & 6 & 3 & 21 \\
\hline
  & 2 & 1 & 7 & 31
\end{array}$$
• $g(3) = 31$

e.
$$\begin{array}{r|rrrr}
0 & 2 & -5 & 4 & 10 \\
  &   & 0 & 0 & 0 \\
\hline
  & 2 & -5 & 4 & 10
\end{array}$$
• $g(0) = 10$

f.
$$\begin{array}{r|rrrr}
-5 & 2 & -5 & 4 & 10 \\
   &   & -10 & 75 & -395 \\
\hline
   & 2 & -15 & 79 & -385
\end{array}$$
• $g(-5) = -385$

32. Since P has 2, -3, and -1 as zeros, the x-intercepts of the graph of P are $(2, 0)$, $(-3, 0)$, $(-1, 0)$.

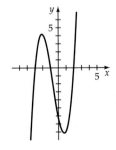

48.

$$\begin{array}{r|rrrr} 0 & 4 & -1 & -6 & 1 \\ & & 0 & 0 & 0 \\ \hline & 4 & -1 & -6 & 1 \end{array} \quad \bullet \ P(0) = 1$$

$$\begin{array}{r|rrrr} 1 & 4 & -1 & -6 & 1 \\ & & 4 & 3 & -3 \\ \hline & 4 & 3 & -3 & -2 \end{array} \quad \bullet \ P(1) = -2$$

Because $P(0)$ and $P(1)$ have opposite signs, P must have a real zero between 0 and 1.

EXERCISE SET 4.3, page 222

12. $p = \pm 1, \pm 2, \pm 3, \pm 5, \pm 6, \pm 10, \pm 15, \pm 30$

$q = \pm 1$

$\dfrac{p}{q} = \pm 1, \pm 2, \pm 3, \pm 5, \pm 6, \pm 10, \pm 15, \pm 30$

24.

$$\begin{array}{r|rrrr} 3 & 1 & 0 & -19 & -28 \\ & & 3 & 9 & -30 \\ \hline & 1 & 3 & -10 & -58 \end{array} \qquad \begin{array}{r|rrrr} -2 & 1 & 0 & -19 & -28 \\ & & -2 & 4 & 30 \\ \hline & 1 & -2 & -15 & 2 \end{array}$$

$$\begin{array}{r|rrrr} 4 & 1 & 0 & -19 & -28 \\ & & 4 & 16 & -12 \\ \hline & 1 & 4 & -3 & -40 \end{array} \qquad \begin{array}{r|rrrr} -3 & 1 & 0 & -19 & -28 \\ & & -3 & 9 & 30 \\ \hline & 1 & -3 & -10 & 2 \end{array}$$

$$\begin{array}{r|rrrr} 5 & 1 & 0 & -19 & -28 \\ & & 5 & 25 & 30 \\ \hline & 1 & 5 & 6 & 2 \end{array} \qquad \begin{array}{r|rrrr} -4 & 1 & 0 & -19 & -28 \\ & & -4 & 16 & 12 \\ \hline & 1 & -4 & -3 & -16 \end{array}$$

$$\begin{array}{r|rrrr} -5 & 1 & 0 & -19 & -28 \\ & & -5 & 25 & -30 \\ \hline & 1 & -5 & 6 & -58 \end{array}$$

5 is an upper bound and -5 is a lower bound.

36. One positive real zero since the polynomial has one variation in sign.

$$(-x)^3 - 19(-x) - 30 = -x^3 + 19x - 30$$

2 or no negative real zeros because $-x^3 + 19x - 30$ has two variations in sign.

48. One positive and two or no negative real zeros (see Exercise 36).

$$\begin{array}{r|rrrr} 5 & 1 & 0 & -19 & -30 \\ & & 5 & 25 & 30 \\ \hline & 1 & 5 & 6 & 0 \end{array}$$

The reduced polynomial is

$$x^2 + 5x + 6 = (x + 3)(x + 2),$$

which has -3 and -2 as zeros. The zeros of $x^3 - 19x - 30$ are 5, -2, and -3.

EXERCISE SET 4.4, page 227

2.

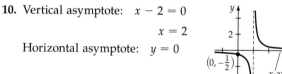

$$\begin{array}{r|rrrr} 5 + 3i & 3 & -29 & 92 & 34 \\ & & 15 + 9i & -97 + 3i & -34 \\ \hline & 3 & -14 + 9i & -5 + 3i & 0 \end{array}$$

$$\begin{array}{r|rrr} 5 - 3i & 3 & -14 + 9i & -5 + 3i \\ & & 15 - 9i & 5 - 3i \\ \hline & 3 & 1 & 0 \end{array}$$

The reduced polynomial $3x + 1$ has $-\dfrac{1}{3}$ as a zero.

The zeros of $3x^3 - 29x^2 + 92x + 34$ are $5 + 3i$, $5 - 3i$ and $-\dfrac{1}{3}$.

20. $6x^3 - 23x^2 - 4x = x(6x^2 - 23x - 4)$

$$= x(6x + 1)(x - 4)$$

40. Since P has real coefficients, use the Conjugate Pair Theorem.

$$P = (x - [3 + 2i])(x - [3 - 2i])(x - 7)$$

$$= (x - 3 - 2i)(x - 3 + 2i)(x - 7)$$

$$= (x^2 - 6x + 13)(x - 7)$$

$$= x^3 - 13x^2 + 55x - 91$$

EXERCISE SET 4.5, page 239

2.

$$x^2 - 4 = 0$$

$$(x - 2)(x + 2) = 0$$

$$x = 2 \quad \text{or} \quad x = -2$$

The vertical asymptotes are $x = 2$ and $x = -2$.

6. The horizontal asymptote is $y = 0$ (x-axis) because the degree of the denominator is larger than the degree of the numerator.

10. Vertical asymptote: $x - 2 = 0$

$$x = 2$$

Horizontal asymptote: $y = 0$

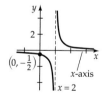

26. Vertical asymptote: $x^2 - 6x + 9 = 0$

$$(x - 3)(x - 3) = 0$$

$$x = 3$$

Horizontal asymptote is $y = \frac{1}{1} = 1$ (the Theorem on Horizontal Asymptotes) because numerator and denominator both have degree 2.

34.

$$x^2 - 3x + 5 \overline{\smash{\big)}\, x^3 - 2x^2 + 3x + 4}$$

$$\underline{x^3 - 3x^2 + 5x}$$

$$x^2 - 2x + 4$$

$$\underline{x^2 - 3x + 5}$$

$$x - 1$$

$$F(x) = x + 1 + \frac{x - 1}{x^2 - 3x + 5}$$

Slant asymptote: $y = x + 1$

40. $2x + 5 = 0$

$$2x = -5$$

$$x = -\frac{5}{2}$$

$$\frac{1}{2}x - \frac{13}{4}$$

$$2x + 5 \overline{\smash{\big)}\, x^2 - 4x - 5}$$

$$\underline{x^2 + \frac{5}{2}x}$$

$$-\frac{13}{2}x - 5$$

$$\underline{-\frac{13}{2}x - \frac{65}{4}}$$

$$\frac{45}{4}$$

$$F(x) = \frac{1}{2}x - \frac{13}{4} + \frac{45/4}{2x + 5}$$

Vertical asymptote: $x = -\frac{5}{2}$

Slant asymptote: $y = \frac{1}{2}x - \frac{13}{4}$

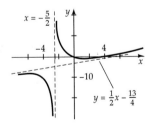

50.

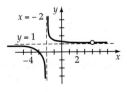

EXERCISE SET 4.6, page 245

2. $P(1) = (1)^3 + 18(1) - 30 = -11$

$P(2) = (2)^3 + 18(2) - 30 = 14$ opposite signs

P has a zero between 1 and 2 (zero location theorem).

8. $P(0) = -10$ $P(1) = 9$

$$P\left(\frac{1}{2}\right) = \left(\frac{1}{2}\right)^3 + 18\left(\frac{1}{2}\right) - 10 = -0.875$$

$$P\left(\frac{3}{4}\right) = (0.75)^3 + 18(0.75) - 10 = 3.921875$$

$$P\left(\frac{5}{8}\right) = \left(\frac{5}{8}\right)^3 + 18\left(\frac{5}{8}\right) - 10 = 1.494140625$$

P has a zero at approximately $\dfrac{\frac{1}{2} + \frac{5}{8}}{2} = 0.5625$.

14. $P(3) = 1$

$P(2.9) = -1.411$

$P(2.99) \approx 0.751$

$P(2.96) \approx 0.014$

$P(2.95) \approx -0.228$

P has a zero at approximately $\dfrac{2.96 + 2.95}{2} = 2.955$.

EXERCISE SET 5.1, page 257

26.

36.

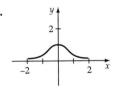

38.

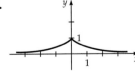

44.

46.

EXERCISE SET 5.2, page 266

2. $\log_{10} 1000 = 3$

$$1000 = 10^3$$

12. $3^5 = 243$

$\log_3 243 = 5$

22. $\log_{10} \dfrac{1}{1000} = n$

$$10^n = \dfrac{1}{1000}$$

$$10^n = 10^{-3}$$

$$n = -3$$

32. $\log_b x^2 y^3 = \log_b x^2 + \log_b y^3 = 2 \log_b x + 3 \log_b y$

42. $\log_7 20 = \log_7 2^2 \cdot 5 = 2 \log_7 2 + \log_7 5$
$$= 2(0.3562) + (0.8271) = 1.5395$$

52. $5 \log_3 x - 4 \log_3 y + 2 \log_3 z$
$$= \log_3 x^5 - \log_3 y^4 + \log_3 z^2$$
$$= \log_3 \dfrac{x^5 z^2}{y^4}$$

62. $\log_5 37 = \dfrac{\log 37}{\log 5} \approx 2.2436$

EXERCISE SET 5.3, page 272

2.

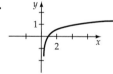

10.

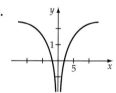

12.

14.

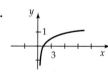

20.

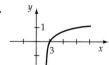

42. $I = 100{,}000 I_0$

$$R = \log \dfrac{100{,}000 I_0}{I_0}$$

$$R = \log 100{,}000$$

$$R = 5$$

44. Let I = the intensity of the smaller earthquake.
Then $1000I$ = the intensity of the larger earthquake.

$$R_1 = \log \dfrac{I}{I_0}$$

$$R_2 = \log 1000 \dfrac{I}{I_0} = \log 1000 + \log \dfrac{I}{I_0} = 3 + R_1$$

The larger earthquake has a Richter scale measure that is 3 more than the smaller earthquake.

EXERCISE SET 5.4, page 278

2. $3^x = 243$

$$3^x = 3^5$$

$$x = 5$$

12. $6^x = 50$

$$\log(6^x) = \log 50$$

$$x \log 6 = \log 50$$

$$x = \dfrac{\log 50}{\log 6} \approx 2.18$$

24. $\log_3 x + \log_3(x + 6) = 3$

$$\log_3 x(x + 6) = 3$$

$$3^3 = x(x + 6)$$

$$27 = x^2 + 6x$$

$$x^2 + 6x - 27 = 0$$

$$(x + 9)(x - 3) = 0$$

$$x = 3 \quad \text{or} \quad -9$$

Since $\log_3 x$ is only defined for $x > 0$, the only solution is $x = 3$.

32. $\ln x = \dfrac{1}{2} \ln\left(2x + \dfrac{5}{2}\right) + \dfrac{1}{2} \ln 2$

$$\ln x = \dfrac{1}{2} \ln 2\left(2x + \dfrac{5}{2}\right)$$

$$\ln x = \dfrac{1}{2} \ln (4x + 5)$$

$$\ln x = \ln(4x + 5)^{1/2}$$

$$x = \sqrt{4x + 5}$$

$$x^2 = 4x + 5$$

$$0 = x^2 - 4x - 5$$

$$0 = (x - 5)(x + 1)$$

$$x = 5 \quad \text{or} \quad -1$$

Check: $\ln 5 = \dfrac{1}{2} \ln\left(10 + \dfrac{5}{2}\right) + \dfrac{1}{2} \ln 2$

$$1.6094 \approx 1.2629 + 0.3466$$

Since $\ln(-1)$ is not defined, -1 is not a solution. Thus the only solution is $x = 5$.

42.
$$\frac{10^x + 10^{-x}}{2} = 8$$

$$10^x(10^x + 10^{-x}) = (16)10^x$$

$$10^{2x} + 1 = 16(10^x)$$

$$10^{2x} - 16(10^x) + 1 = 0$$

Let $u = 10^x$.

$$u^2 - 16u + 1 = 0$$

$$u = \frac{16 \pm \sqrt{16^2 - 4(1)(1)}}{2} = 8 \pm 3\sqrt{7}$$

$$10^x = 8 \pm 3\sqrt{7}$$

$$x = \log(8 \pm 3\sqrt{7})$$

$$x = \log(8 \pm 3\sqrt{7}) \approx \pm 1.20241$$

50. $pH = -\log(1.26 \times 10^{-12}) \approx 11.9$

Thus the ammonia is a basic solution.

52. $pH = 3.1$

$$-\log[H_3O^+] = 3.1$$

$$H_3O^+ = 10^{-3.1} \approx 7.9 \times 10^{-4}$$

EXERCISE SET 5.5, page 287

4. a. $P = 12,500, r = 0.08, t = 10, n = 1.$

$$B = 12,500\left(1 + \frac{0.08}{1}\right)^{10} \approx \$26,986.56$$

b. $n = 365$

$$B = 12,500\left(1 + \frac{0.08}{365}\right)^{3650} \approx \$27,816.82$$

c. $n = 8760$

$$B = 12,500\left(1 + \frac{0.08}{8760}\right)^{87600} \approx \$27,819.16$$

6. $P = 32,000, r = 0.08, t = 3.$

$$B = Pe^{rt} = 32,000e^{3(0.08)} \approx \$40,679.97$$

10. $t = \dfrac{\ln 3}{r}$ $r = 0.055$

$$t = \frac{\ln 3}{0.055}$$

$$t \approx 20 \text{ years}$$

16. $N(t) = N_0 e^{kt}$

$N(0) = 53,700 = N_0$ (by definition)

$N(6) = 58,100$

Use $t = 6$ and $N_0 = 53,700$.

$$58,100 = 53,700e^{k(6)}$$

$$\frac{58,100}{53,700} = e^{6k}$$

$$\ln\left(\frac{58,100}{53,700}\right) = 6k$$

$$\frac{\ln\left(\frac{58,100}{53,700}\right)}{6} = k$$

$$k \approx 0.0131$$

$$N(t) = 53,700e^{0.0131t}$$

18. $P(t) = 15,600(e^{0.09t})$

a. $P(10) = 15,600(e^{0.9}) \approx 38,400$

b. We must determine the value of t when the population $N(t) = 15,600 \cdot 2 = 31,200$.

$$31,200 = 15,600e^{0.09t}$$

$$\frac{31,200}{15,600} = e^{0.09t}$$

$$t = \frac{\ln\left(\frac{31,200}{15,600}\right)}{0.09} \approx 8$$

The population will double by 1992.

20. $N(138) = N_0 e^{138k}$

$$0.5N_0 = N_0 e^{138k}$$

$$0.5 = e^{138k}$$

$$\ln 0.5 = 138k$$

$$k = \frac{\ln 0.5}{138} \approx -0.005023$$

$$N(t) = N_0 e^{-0.005023t}$$

22. $N(t) = N_0(0.5)^{t/138}$

$$N(730) = N_0(0.5)^{730/138} \approx N_0(0.0256)$$

After 2 years (730 days), only 2.56% of the polonium remains.

24. $N(t) = N_0(0.5)^{t/5730}$

$0.65N_0 = N_0(0.5)^{t/5730}$

$t = 5730 \dfrac{\ln 0.65}{\ln 0.5} \approx 3560$

The bone is approximately 3560 years old.

28. $N(3.4 \times 10^{-5}) = 10 \log\left(\dfrac{3.4 \times 10^{-5}}{10^{-16}}\right)$

$= 10(\log 3.4 + \log 10^{11})$

$= 10 \log 3.4 + 110 \approx 115.3$ decibels

EXERCISE SET 6.1, page 303

16. $35°42'' = 35° + 42'' = 35° + 42''\left(\dfrac{1°}{3600''}\right)$

$= 35° + 0.012° = 35.012°$

22. $55.44° = 55° + 0.44°\left(\dfrac{60'}{1°}\right) = 55° + 26.4'$

$= 55° + 26' + 0.4'\left(\dfrac{60''}{1'}\right)$

$= 55° + 26' + 24'' = 55°26'24''$

30. $-330° = -330°\left(\dfrac{\pi}{180°}\right) = -\dfrac{11\pi}{6}$

34. $\dfrac{11\pi}{18} = \dfrac{11\pi}{18}\left(\dfrac{180°}{\pi}\right) = 110°$

50. $s = r\theta = 5\,(144°)\left(\dfrac{\pi}{180°}\right) \approx 12.57$ meters

54. Let θ_2 be the angle through which the pulley with a diameter of 0.8 m turns. Let θ_1 be the angle through which the pulley with a diameter 1.2 m turns. Let $r_2 = 0.4$ m be the radius of the smaller pulley and let $r_1 = 0.6$ m be the radius of the larger pulley. Then

$r_2\theta_2 = r_1\theta_1$

$0.4\theta_2 = 0.6(240°)$

$\theta_2 = \dfrac{0.6}{0.4}(240°)\left(\dfrac{\pi}{180°}\right) = 2\pi$

56. Earth makes 1 revolution ($\theta = 2\pi$) in 1 day.

$t = 24 \cdot 3600 = 86{,}400$ seconds

$\omega = \dfrac{\theta}{t} = \dfrac{2\pi}{86{,}400} \approx 7.27 \times 10^{-5}$ radian/s

60. Change 55 mph to ft/sec.

$\dfrac{55\text{ mi}}{1\text{ hr}} \cdot \dfrac{5280\text{ ft}}{1\text{ mi}} \cdot \dfrac{1\text{ hr}}{3600\text{ s}} = \dfrac{242}{3}$ ft/sec

Change 14 inches to feet. 14 inches $= \dfrac{7}{6}$ ft

$\omega = \dfrac{v}{r}$

$= \dfrac{242/3}{7/6}$

≈ 69.14 radians/s

EXERCISE SET 6.2, page 312

6.

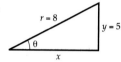

$x = \sqrt{8^2 - 5^2}$

$x = \sqrt{64 - 25} = \sqrt{39}$

$\sin\theta = \dfrac{y}{r} = \dfrac{5}{8}$ $\csc\theta = \dfrac{r}{y} = \dfrac{8}{5}$

$\cos\theta = \dfrac{x}{r} = \dfrac{\sqrt{39}}{8}$ $\sec\theta = \dfrac{r}{x} = \dfrac{8}{\sqrt{39}} = \dfrac{8\sqrt{39}}{39}$

$\tan\theta = \dfrac{y}{x} = \dfrac{5}{\sqrt{39}} = \dfrac{5\sqrt{39}}{39}$ $\cot\theta = \dfrac{x}{y} = \dfrac{\sqrt{39}}{5}$

16. Since $\tan\theta = \dfrac{y}{x} = \dfrac{4}{3}$ • Let $y = 4$ and $x = 3$.

$r = \sqrt{3^2 + 4^2} = 5$

$\sec\theta = \dfrac{r}{x} = \dfrac{5}{3}$

34. $\sin\dfrac{\pi}{3}\cos\dfrac{\pi}{4} - \tan\dfrac{\pi}{4} = \dfrac{\sqrt{3}}{2} \cdot \dfrac{\sqrt{2}}{2} - 1$

$= \dfrac{\sqrt{6}}{4} - 1 = \dfrac{\sqrt{6} - 4}{4}$

62.

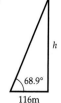

$\tan 68.9 = \dfrac{h}{116}$

$h = 116 \tan 68.9°$

$h \approx 301$ m

64.

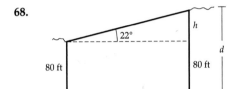

$$\sin 5° = \frac{80}{d}$$

$$d = \frac{80}{\sin 5°}$$

$$d \approx 917.9 \text{ ft}$$

Change 9 mph to ft/min

$$r = 9 \frac{\text{mi}}{\text{hr}} = \frac{9 \text{ mi}}{1 \text{ hr}} \cdot \frac{5280 \text{ ft}}{1 \text{ mi}} \cdot \frac{1 \text{ hr}}{60 \text{ min}}$$

$$= \frac{9(5280)}{60} \frac{\text{ft}}{\text{min}} = 792 \frac{\text{ft}}{\text{min}}.$$

$$t = \frac{d}{r}$$

$$t = \frac{917.9 \text{ ft}}{792 \text{ ft/min}}$$

$$t \approx 1.16 \text{ min}$$

68.

$$\tan 22° = \frac{h}{240}$$

$$h = 240 \tan 22°$$

$$d = 80 + h$$

$$d = 80 + 240 \tan 22°$$

$$d \approx 180 \text{ ft (to 2 significant digits)}$$

EXERCISE SET 6.3, page 320

6. $x = -6, y = -9, r = \sqrt{(-6)^2 + (-9)^2} = \sqrt{117} = 3\sqrt{13}.$

$$\sin \theta = \frac{y}{r} = \frac{-9}{3\sqrt{13}} = -\frac{3}{\sqrt{13}} = -\frac{3\sqrt{13}}{13} \qquad \csc \theta = -\frac{\sqrt{13}}{3}$$

$$\cos \theta = \frac{x}{r} = \frac{-6}{3\sqrt{13}} = -\frac{2}{\sqrt{13}} = -\frac{2\sqrt{13}}{13} \qquad \sec \theta = -\frac{\sqrt{13}}{2}$$

$$\tan \theta = \frac{y}{x} = \frac{-9}{-6} = \frac{3}{2} \qquad \cot \theta = \frac{2}{3}$$

18. $\sec \theta = \frac{2\sqrt{3}}{3} = \frac{r}{x}$ Let $r = 2\sqrt{3}, x = 3,$

$$y = \pm\sqrt{(2\sqrt{3})^2 - 3^2} = \pm\sqrt{3},$$

$$y = -\sqrt{3} \quad y < 0 \text{ in quadrant IV.}$$

$$\sin \theta = \frac{-\sqrt{3}}{2\sqrt{3}} = -\frac{1}{2}$$

54. $\cos \frac{7\pi}{4} \tan \frac{4\pi}{3} + \cos \frac{7\pi}{6} = \frac{\sqrt{2}}{2} \cdot (\sqrt{3}) + \left(-\frac{\sqrt{3}}{2}\right)$

$$= \frac{\sqrt{6}}{2} - \frac{\sqrt{3}}{2} = \frac{\sqrt{6} - \sqrt{3}}{2}$$

EXERCISE SET 6.4, page 328

10. $t = -\frac{7\pi}{4}; w(t) = P(x, y)$ where

$$y = \sin t \qquad\qquad x = \cos t$$

$$= \sin\left(-\frac{7\pi}{4}\right) \qquad = \cos\left(-\frac{7\pi}{4}\right)$$

$$= -\sin \frac{7\pi}{4} \qquad = \cos \frac{7\pi}{4}$$

$$= \frac{\sqrt{2}}{2} \qquad\qquad = \frac{\sqrt{2}}{2}$$

$$w\left(-\frac{7\pi}{4}\right) = \left(\frac{\sqrt{2}}{2}, \frac{\sqrt{2}}{2}\right).$$

16. The reference angle for $-\frac{5\pi}{6}$ is $\frac{\pi}{6}$.

$$\sec\left(-\frac{5\pi}{6}\right) = -\sec \frac{\pi}{6} \quad \bullet \sec t < 0 \text{ for } t \text{ in Quadrant III}$$

$$= -\frac{2\sqrt{3}}{3}$$

36. $F(-x) = \tan(-x) + \sin(-x)$

$$= -\tan x - \sin x \quad \bullet \tan x \text{ and } \sin x \text{ are}$$
$$\qquad\qquad\qquad\qquad \text{odd functions.}$$

$$= -(\tan x + \sin x)$$

$$= -F(x)$$

The function defined by $F(x) = \tan x + \sin x$ is an odd function.

42. $\qquad \tan t = \frac{y}{x}$

$$\tan(t - \pi) = \frac{-y}{-x} = \frac{y}{x} \quad \bullet \text{ From the unit circle.}$$

Therefore, $\tan t = \tan(t - \pi)$

60. $\frac{1}{1 - \sin t} + \frac{1}{1 + \sin t} = \frac{1 + \sin t + 1 - \sin t}{(1 - \sin t)(1 + \sin t)}$

$$= \frac{2}{1 - \sin^2 t}$$

$$= \frac{2}{\cos^2 t}$$

86. $1 + \tan^2 t = \sec^2 t$

$$\tan^2 t = \sec^2 t - 1$$

$$\tan t = \pm\sqrt{\sec^2 t - 1}$$

Because $\frac{3\pi}{2} < t < 2\pi$, $\tan t$ is negative. Thus

$$\tan t = -\sqrt{\sec^2 t - 1}$$

EXERCISE SET 6.5, page 336

20. $y = -\dfrac{3}{2} \sin x$, $a = \left| -\dfrac{3}{2} \right| = \dfrac{3}{2}$, period $= 2\pi$.

30. $y = \sin \dfrac{3\pi}{4} x$, $a = 1$, period $= \dfrac{2\pi}{b} = \dfrac{2\pi}{\dfrac{3\pi}{4}} = \dfrac{8}{3}$

32. $y = \cos 3\pi x$, $a = 1$, period $= \dfrac{2\pi}{b} = \dfrac{2\pi}{3\pi} = \dfrac{2}{3}$

38. $y = \dfrac{1}{2} \sin \dfrac{\pi x}{3}$, $a = \dfrac{1}{2}$, period $= \dfrac{2\pi}{b} = \dfrac{2\pi}{\pi/3} = 6$

46. $y = -\dfrac{3}{4} \cos 5x$, $a = \left| -\dfrac{3}{4} \right| = \dfrac{3}{4}$, period $= \dfrac{2\pi}{b} = \dfrac{2\pi}{5}$.

52. $y = -\left| 3 \sin \dfrac{2}{3} x \right|$

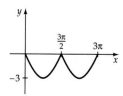

EXERCISE SET 6.6, page 344

22. $y = \dfrac{1}{3} \tan x$, period $= \dfrac{\pi}{b} = \pi$

30. $y = -3 \tan 3x$, period $= \dfrac{\pi}{b} = \dfrac{\pi}{3}$.

32. $y = \dfrac{1}{2} \cot 2x$, period $= \dfrac{\pi}{b} = \dfrac{\pi}{2}$.

38. $y = 3 \csc \dfrac{\pi x}{2}$, period $= \dfrac{2\pi}{b} = \dfrac{2\pi}{\pi/2} = 4$.

42. $y = \sec \dfrac{x}{2}$, period $= \dfrac{2\pi}{b} = \dfrac{2\pi}{1/2} = 4\pi$.

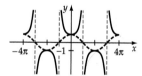

EXERCISE SET 6.7, page 348

20. $y = \cos \left(2x - \dfrac{\pi}{3} \right)$

$a = 1$, period $= \pi$, phase shift $= -\dfrac{c}{b} = -\dfrac{-\pi/3}{2} = \dfrac{\pi}{6}$

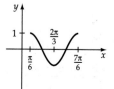

22. $y = \tan(x - \pi)$

Period $= \pi$, phase shift $= -\dfrac{c}{b} = -\dfrac{-\pi}{1} = \pi$.

40. $y = 2\sin\left(\dfrac{\pi x}{2} + 1\right) - 2$, $a = 2$, $p = 4$,

phase shift $= -\dfrac{c}{b} = -\dfrac{1}{\pi/2} = -\dfrac{2}{\pi}$

42. $y = -3\cos(2\pi x - 3) + 1$, $p = 1$

phase shift $= -\dfrac{c}{b} = \dfrac{3}{2\pi}$.

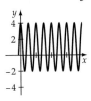

48. $y = \csc\dfrac{x}{3} + 4$, $p = 6\pi$.

52. $y = \dfrac{x}{2} + \cos x$

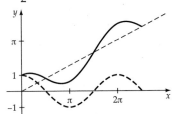

56. $y = -\sin x + \cos x$.

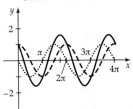

EXERCISE SET 6.8, page 355

20. Amplitude $= 3$, frequency $= 1/\pi$, period $= \pi$.
Since $2\pi/b = \pi$, we have $b = 2$. Thus

$$y = 3\cos 2t$$

28. Amplitude $= |-1.5|$.

$$f = \dfrac{1}{2\pi}\sqrt{\dfrac{k}{m}} = \dfrac{1}{2\pi}\sqrt{\dfrac{3}{27}} = \dfrac{1}{2\pi}\cdot\dfrac{1}{3} = \dfrac{1}{6\pi}, \qquad \text{period} = 6\pi$$

$$y = a\cos 2\pi ft = -1.5\cos\left[2\pi\left(\dfrac{1}{6\pi}\right)t\right] = -1.5\cos\dfrac{1}{3}t$$

30. Amplitude $= 4$.

$$f = \dfrac{1}{2\pi}\sqrt{\dfrac{g}{l}} = \dfrac{1}{2\pi}\sqrt{\dfrac{32}{20}} = \dfrac{1}{2\pi}\cdot\dfrac{4}{\sqrt{10}} = \dfrac{\sqrt{10}}{5\pi}$$

$$\text{period} = \dfrac{5\pi}{\sqrt{10}} = \dfrac{\pi\sqrt{10}}{2}$$

$$y = a\cos 2\pi ft = 4\cos\left[2\pi\left(\dfrac{\sqrt{10}}{5\pi}\right)t\right] = 4\cos\dfrac{2\sqrt{10}}{5}t$$

EXERCISE SET 7.1, page 366

14. $\sin x \cot x \sec x = \sin x \cdot \dfrac{\cos x}{\sin x} \cdot \dfrac{1}{\cos x} = 1$

24. $\quad \sin^4 x - \cos^4 x = (\sin^2 x + \cos^2 x)(\sin^2 x - \cos^2 x)$

$$= 1(\sin^2 x - \cos^2 x)$$

$$= \sin^2 x - \cos^2 x$$

36. $\dfrac{2\sin x \cot x + \sin x - 4\cot x - 2}{2\cot x + 1}$

$$= \dfrac{(\sin x)(2\cot x + 1) - 2(2\cot x + 1)}{2\cot x + 1}$$

$$= \dfrac{(2\cot x + 1)(\sin x - 2)}{2\cot x + 1}$$

$$= \sin x - 2$$

46. $\dfrac{\dfrac{1}{\sin x} + \dfrac{1}{\cos x}}{\dfrac{1}{\sin x} - \dfrac{1}{\cos x}} = \dfrac{\dfrac{1}{\sin x} + \dfrac{1}{\cos x}}{\dfrac{1}{\sin x} - \dfrac{1}{\cos x}} \cdot \dfrac{\sin x \cos x}{\sin x \cos x}$

$$= \dfrac{\cos x + \sin x}{\cos x - \sin x}$$

$$= \dfrac{\cos x + \sin x}{\cos x - \sin x} \cdot \dfrac{\cos x - \sin x}{\cos x - \sin x}$$

$$= \dfrac{\cos^2 x - \sin^2 x}{\cos^2 x - 2\sin x \cos x + \sin^2 x}$$

$$= \dfrac{\cos^2 x - \sin^2 x}{1 - 2\sin x \cos x}$$

56.
$$\frac{\dfrac{1}{\tan x} + \cot x}{\dfrac{1}{\tan x} + \tan x} = \frac{\dfrac{1}{\tan x} + \cot x}{\dfrac{1}{\tan x} + \tan x} \cdot \frac{\tan x}{\tan x}$$

$$= \frac{1 + 1}{1 + \tan^2 x}$$

$$= \frac{2}{\sec^2 x}$$

EXERCISE SET 7.2, page 373

32. $\tan \alpha = \dfrac{24}{7}$, with $\theta° < \alpha < 90°$, $\sin \alpha = \dfrac{24}{25}$, $\cos \alpha = \dfrac{7}{25}$,

$\sin \beta = -\dfrac{8}{17}$, with $180° < \beta < 270°$

$\cos \beta = -\dfrac{15}{17}$, $\tan \beta = \dfrac{8}{15}$

a. $\sin(\alpha + \beta) = \sin \alpha \cos \beta + \cos \alpha \sin \beta$

$$= \left(\frac{24}{25}\right)\left(-\frac{15}{17}\right) + \left(\frac{7}{25}\right)\left(-\frac{8}{17}\right)$$

$$= -\frac{360}{425} - \frac{56}{425} = -\frac{416}{425}$$

b. $\cos(\alpha + \beta) = \cos \alpha \cos \beta - \sin \alpha \sin \beta$.

$$= \left(\frac{7}{25}\right)\left(-\frac{15}{17}\right) - \left(\frac{24}{25}\right)\left(-\frac{8}{17}\right)$$

$$= -\frac{105}{425} + \frac{192}{425} = \frac{87}{425}$$

c. $\tan(\alpha - \beta) = \dfrac{\tan \alpha - \tan \beta}{1 + \tan \alpha \tan \beta}$.

$$= \frac{\dfrac{24}{7} - \dfrac{8}{15}}{1 + \left(\dfrac{24}{7}\right)\left(\dfrac{8}{15}\right)} = \frac{\dfrac{24}{7} - \dfrac{8}{15}}{1 + \dfrac{192}{105}} \cdot \frac{105}{105}$$

$$= \frac{360 - 56}{105 + 192}$$

$$= \frac{304}{297}$$

44. $\cos(\theta + \pi) = \cos \theta \cos \pi - \sin \theta \sin \pi$

$$= (\cos \theta)(-1) - (\sin \theta)(0)$$

$$= -\cos \theta$$

56. $\cos 5x \cos 3x + \sin 5x \sin 3x = \cos(5x - 3x)$

$$= \cos 2x$$

$$= \cos(x + x)$$

$$= \cos x \cos x - \sin x \sin x$$

$$= \cos^2 x - \sin^2 x$$

60. $\sin(\alpha - \beta) - \sin(\alpha + \beta)$

$$= \sin \alpha \cos \beta - \cos \alpha \sin \beta - (\sin \alpha \cos \beta + \cos \alpha \sin \beta)$$

$$= -2 \cos \alpha \sin \beta$$

EXERCISE SET 7.3, page 380

26. $\cos \theta = \dfrac{24}{25}$ with $270° < \theta < 360°$

$\sin \theta = -\sqrt{1 - \left(\dfrac{24}{25}\right)^2}$ $\tan \theta = \dfrac{-7/25}{24/25}$

$\quad = -\dfrac{7}{25}$ $\quad = -\dfrac{7}{24}$

$\sin 2\theta = 2 \sin \theta \cos \theta$ $\cos 2\theta = \cos^2 \theta - \sin^2 \theta$

$$= 2\left(\frac{-7}{25}\right)\left(\frac{24}{25}\right)$$ $$= \left(\frac{24}{25}\right)^2 - \left(-\frac{7}{25}\right)^2$$

$$= -\frac{336}{625}$$ $$= \frac{527}{625}$$

$\tan 2\theta = \dfrac{2 \tan \theta}{1 - \tan^2 \theta}$

$$= \frac{2(-7/24)}{1 - (-7/24)^2}$$

$$= \frac{-7/12}{1 - 49/576} \cdot \frac{576}{576}$$

$$= -\frac{336}{527}$$

54. $\dfrac{1}{1 - \cos 2x} = \dfrac{1}{1 - 1 + 2 \sin^2 x}$

$$= \frac{1}{2 \sin^2 x} = \frac{1}{2} \csc^2 x$$

58. $\dfrac{2 \sin x \cos x}{\cos^2 x - \sin^2 x} = \dfrac{\sin 2x}{\cos 2x} = \tan 2x$

72. $\cos^2 \dfrac{x}{2} = \left[\pm\sqrt{\dfrac{1 + \cos x}{2}}\right]^2 = \dfrac{1 + \cos x}{2}$

$$= \frac{1 + \cos x}{2} \cdot \frac{\sec x}{\sec x} = \frac{\sec x + 1}{2 \sec x}$$

78. $\tan^2 \dfrac{x}{2} = \left(\dfrac{1 - \cos x}{\sin x}\right)^2$

$= \dfrac{(1 - \cos x)^2}{\sin^2 x}$

$= \dfrac{(1 - \cos x)^2}{1 - \cos^2 x}$

$= \dfrac{(1 - \cos x)^2}{(1 - \cos x)(1 + \cos x)}$

$= \dfrac{1 - \cos x}{1 + \cos x}$

$= \dfrac{\dfrac{1}{\cos x} - \dfrac{\cos x}{\cos x}}{\dfrac{1}{\cos x} + \dfrac{\cos x}{\cos x}}$

$= \dfrac{\sec x - 1}{\sec x + 1}$

EXERCISE SET 7.4, page 386

22. $\cos 3\theta + \cos 5\theta = 2 \cos \dfrac{3\theta + 5\theta}{2} \cos \dfrac{3\theta - 5\theta}{2}$

$= 2 \cos 4\theta \cos(-\theta)$

$= 2 \cos 4\theta \cos \theta$

36. $\sin 5x \cos 3x = \dfrac{1}{2}[\sin(5x + 3x) + \sin(5x - 3x)]$

$= \dfrac{1}{2}(\sin 8x + \sin 2x)$

$= \dfrac{1}{2}(2 \sin 4x \cos 4x + 2 \sin x \cos x)$

$= \sin 4x \cos 4x + \sin x \cos x$

44. $\dfrac{\cos 5x - \cos 3x}{\sin 5x + \sin 3x} = \dfrac{-2 \sin \dfrac{5x + 3x}{2} \sin \dfrac{5x - 3x}{2}}{2 \sin \dfrac{5x + 3x}{2} \cos \dfrac{5x - 3x}{2}}$

$= -\dfrac{\sin 4x \sin x}{\sin 4x \cos x} = -\tan x$

62. $a = 1$, $b = \sqrt{3}$, $k = \sqrt{(\sqrt{3})^2 + (1)^2} = 2$. Thus α is a first quadrant angle.

$$\sin \alpha = \dfrac{\sqrt{3}}{2}$$

$$\text{and } \cos \alpha = \dfrac{1}{2}, \quad \text{Thus } \alpha = \dfrac{\pi}{3}$$

$$y = k \sin(x + \alpha)$$

$$y = 2 \sin\left(x + \dfrac{\pi}{3}\right)$$

70. From Exercise 62, we know

$$y = \sin x + \sqrt{3} \cos x = 2 \sin\left(x + \dfrac{\pi}{3}\right)$$

EXERCISE SET 7.5, page 397

2. $y = \sin^{-1} \dfrac{\sqrt{2}}{2}$

$\sin y = \dfrac{\sqrt{2}}{2}$ for $-\dfrac{\pi}{2} \le y \le \dfrac{\pi}{2}$

$y = \dfrac{\pi}{4}$

44. $y = \tan\left(\tan^{-1} \dfrac{1}{2}\right) = \dfrac{1}{2}$

62. $y = \tan^{-1} \dfrac{\sqrt{3}}{3} \approx 0.5236$

70. Let $x = \cos^{-1} 3/5$. Thus $\cos x = 3/5$ and $\sin x = \sqrt{1 - (3/5)^2} = 4/5$.

$y = \tan\left(\cos^{-1} \dfrac{3}{5}\right) = \tan x = \dfrac{\sin x}{\cos x} = \dfrac{4/5}{3/5} = \dfrac{4}{3}$

84. $\tan^{-1} x = \sin^{-1} \dfrac{24}{25}$

$\tan(\tan^{-1} x) = \tan\left(\sin^{-1} \dfrac{24}{25}\right)$

$x = \tan\left(\sin^{-1} \dfrac{24}{25}\right)$

$x = \dfrac{24}{7}$

94. $\sin^{-1}x + \cos^{-1}\dfrac{4}{5} = \dfrac{\pi}{6}$

$$\sin^{-1}x = \dfrac{\pi}{6} - \cos^{-1}\dfrac{4}{5}$$

$$\sin(\sin^{-1}x) = \sin\left(\dfrac{\pi}{6} - \cos^{-1}\dfrac{4}{5}\right)$$

$$x = \sin\dfrac{\pi}{6}\cos\left(\cos^{-1}\dfrac{4}{5}\right) - \cos\dfrac{\pi}{6}\sin\left(\cos^{-1}\dfrac{4}{5}\right)$$

$$= \dfrac{1}{2}\cdot\dfrac{4}{5} - \dfrac{\sqrt{3}}{2}\cdot\dfrac{3}{5} = \dfrac{4 - 3\sqrt{3}}{10}$$

102. Let $\alpha = \cos^{-1}x$ and $\beta = \cos^{-1}(-x)$. Thus $\cos\alpha = x$ and $\cos\beta = -x$. We have $\sin\alpha = \sqrt{1 - x^2}$ and we have $\sin\beta = \sqrt{1 - x^2}$ because α is in quadrant I and β is in quadrant II.

$\cos^{-1}x + \cos^{-1}(-x) = \alpha + \beta$

$$= \cos^{-1}[\cos(\alpha + \beta)]$$

$$= \cos^{-1}(\cos\alpha\cos\beta - \sin\alpha\sin\beta)$$

$$= \cos^{-1}[x(-x) - \sqrt{1 - x^2}\cdot\sqrt{1 - x^2}]$$

$$= \cos^{-1}(-x^2 - 1 + x^2)$$

$$= \cos^{-1}(-1) = \pi$$

106. $y = \dfrac{1}{2}\sin^{-1}x$

$\sin 2y = x$

$$-\dfrac{\pi}{2} < 2y < \dfrac{\pi}{2}$$

Range: $\left\{y\ \Big|\ -\dfrac{\pi}{4} < y < \dfrac{\pi}{4}\right\}$

Domain: $\{x\,|-1 < x < 1\}$

EXERCISE SET 7.6, page 403

14. $2\cos^2x + 1 = -3\cos x$ for $0 \le x < 2\pi$
$2\cos^2x + 3\cos x + 1 = 0$
$(2\cos x + 1)(\cos x + 1) = 0$

$2\cos x + 1 = 0$ $\cos x + 1 = 0$

$\cos x = -\dfrac{1}{2}$ $\cos x = -1$

$x = \pi$

$x = \dfrac{2\pi}{3}, \dfrac{4\pi}{3}$

The solutions are $\dfrac{2\pi}{3}, \pi, \dfrac{4\pi}{3}$.

52. $\sin x + 2\cos x = 1$ for $0° \le x < 360°$

$$\sin x = 1 - 2\cos x$$

$$(\sin x)^2 = (1 - 2\cos x)^2$$

$$\sin^2x = 1 - 4\cos x + 4\cos^2x$$

$$1 - \cos^2x = 1 - 4\cos x + 4\cos^2x$$

$$0 = 5\cos^2x - 4\cos x$$

$$0 = \cos x(5\cos x - 4)$$

$\cos x = 0$ or $5\cos x - 4 = 0$

$x = 90°, 270°$ $\cos x = \dfrac{4}{5}$

$$x \approx 36.9°, 323.1°$$

The solutions are $90°$ and $323.1°$.

56. $2\cos^2x - 5\cos x - 5 = 0$

$$\cos x = \dfrac{5 \pm \sqrt{(-5)^2 - 4(2)(-5)}}{2(2)} = \dfrac{5 \pm \sqrt{65}}{4}$$

$\cos x \approx 3.26$ $\cos x \approx -0.7656$

no solution $x \approx 140.0°, 220.0°$

The solutions are $140°$ and $220°$.

66. $\cos 2x = -\dfrac{\sqrt{3}}{2}$

$2x = \dfrac{5\pi}{6} + 2k\pi$ or $2x = \dfrac{7\pi}{6} + 2k\pi$, k an integer

$x = \dfrac{5\pi}{12} + k\pi$ or $x = \dfrac{7\pi}{12} + k\pi$

84. $2\sin x\cos x - 2\sqrt{2}\sin x - \sqrt{3}\cos x + \sqrt{6} = 0$

$2\sin x(\cos x - \sqrt{2}) - \sqrt{3}(\cos x - \sqrt{2}) = 0$

$(\cos x - \sqrt{2})(2\sin x - \sqrt{3}) = 0$

$\cos x = \sqrt{2}$ or $\sin x = \dfrac{\sqrt{3}}{2}$

no solution $x = \dfrac{\pi}{3}, \dfrac{2\pi}{3}$

The solutions are $\dfrac{\pi}{3}$ and $\dfrac{2\pi}{3}$.

EXERCISE SET 8.1, page 417

4. $A = 180° - 78° - 28° = 74°$

$$\frac{b}{\sin B} = \frac{c}{\sin C} \qquad\qquad \frac{a}{\sin A} = \frac{c}{\sin C}$$

$$\frac{b}{\sin 28°} = \frac{44}{\sin 78°} \qquad\qquad \frac{a}{\sin 74°} = \frac{44}{\sin 78°}$$

$$b = \frac{44 \sin 28°}{\sin 78°} \approx 21 \qquad a = \frac{44 \sin 74°}{\sin 78°} \approx 43$$

18. $\sin 22.6° = \dfrac{h}{13.8}$

$$h = 13.8 \sin 22.6 \approx 5.30$$

Since $h < 5.55$, two solutions exist.

$$\frac{a}{\sin A} = \frac{b}{\sin B}$$

$$\frac{13.8}{\sin A} = \frac{5.55}{\sin 22.6}$$

$$\sin A = 0.9555$$

$$A \approx 72.9° \quad \text{or} \quad 107.1°$$

$$A = 72.9°$$

$$C = 180° - 72.9° - 22.6°$$

$$C = 84.5°$$

$$\frac{c}{\sin 84.5°} = \frac{5.55}{\sin 22.6°}$$

$$c = \frac{5.55 \sin 84.5°}{\sin 22.6°} \approx 14.4$$

$$A = 107.1°$$
$$C = 180° - 107.1° - 22.6°$$
$$C = 50.3°$$

$$\frac{c}{\sin 50.3°} = \frac{5.55}{\sin 22.6°}$$

$$c = \frac{5.55 \sin 50.3°}{\sin 22.6°} \approx 11.1$$

Case 1: $A = 72.9°$, $C = 84.5°$, and $c = 14.4$
Case 2: $A = 107.1°$, $C = 50.3°$, and $c = 11.1$

26. $A = 56°$, $\alpha = 35°$

$$D = 180° - 56° = 124°$$

$$\frac{AC}{\sin D} = \frac{DC}{\sin \alpha}$$

$$\frac{AC}{\sin 124°} = \frac{6.0}{\sin 35°}$$

$$AC \approx 8.7 \text{ m}$$

32. $\alpha = 65°$

$$B = 65° + 8° = 73°$$

$$A = 180° - 50° - 65° = 65°$$

$$C = 180° - 65° - 73° = 42°$$

$$\frac{b}{\sin B} = \frac{c}{\sin C}$$

$$\frac{b}{\sin 73°} = \frac{20}{\sin 42°}$$

$$b = \frac{20 \sin 73°}{\sin 42°}$$

$$b \approx 29 \text{ mi}$$

EXERCISE SET 8.2, page 423

12. $c^2 = a^2 + b^2 - 2ab \cos C$

$$c^2 = 14.2^2 + 9.30^2 - 2(14.2)(9.30)\cos 9.20°$$

$$c = \sqrt{14.2^2 + 9.30^2 - 2(14.2)(9.30)\cos 9.20°}$$

$$c \approx 5.24$$

18. $\cos A = \dfrac{b^2 + c^2 - a^2}{2bc}$

$$\cos A = \frac{132^2 + 160^2 - 108^2}{2(132)(160)} \approx 0.7424$$

$$A \approx \cos^{-1}(0.7424) \approx 42.1°$$

26. $K = \dfrac{1}{2} ac \sin B$

$$K = \frac{1}{2}(32)(25)\sin 127°$$

$$K \approx 320 \text{ square units}$$

28. $A = 180° - 102° - 27° = 51°$

$$K = \frac{a^2 \sin B \sin C}{2 \sin A}$$

$$K = \frac{8.5^2 \sin 102° \sin 27°}{2 \sin 51°}$$

$$K \approx 21 \text{ square units}$$

36. $s = \dfrac{1}{2}(a + b + c)$

$$= \frac{1}{2}(10.2 + 13.3 + 15.4) = 19.45$$

$$K = \sqrt{s(s - a)(s - b)(s - c)}$$

$$= \sqrt{19.45(19.45 - 10.2)(19.45 - 13.3)(19.45 - 15.4)}$$

$$\approx 66.9 \text{ square units}$$

48. $\alpha = 270° - 254° = 16°$

$A = 16° + 90° + 32° = 138°$

$b = 4 \cdot 16 = 64$

$c = 3 \cdot 22 = 66$

$a^2 = b^2 + c^2 - 2bc \cos A$

$a^2 = 64^2 + 66^2 - 2(64)(66) \cos 138°$

$a = \sqrt{64^2 + 66^2 - 2(64)(66) \cos 138°}$

$a \approx 120$ mi

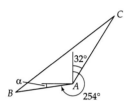

56. $S = \dfrac{1}{2}(324 + 412 + 516) = 626$

$K = \sqrt{626(626 - 324)(626 - 412)(626 - 516)}$

$K = \sqrt{4{,}450{,}284{,}080}$

$\text{cost} = 4.15(\sqrt{4{,}450{,}284{,}080}) \approx \$276{,}848$

EXERCISE SET 8.3, page 437

6. $a = 3 - 3 = 0$

$b = 0 - (-2) = 2$

A vector equivalent to $P_1 P_2$ is $\boldsymbol{\sigma} = \langle 0, 2 \rangle$

8. $|\mathbf{v}| = \sqrt{6^2 + 10^2}$

$= \sqrt{36 + 100}$

$= \sqrt{136} = 2\sqrt{34}$

≈ 11.7

A unit vector in the direction of $\mathbf{v}$ is

$\mathbf{u} = \left\langle \dfrac{6}{2\sqrt{34}}, \dfrac{10}{2\sqrt{34}} \right\rangle = \left\langle \dfrac{3\sqrt{34}}{34}, \dfrac{5\sqrt{34}}{34} \right\rangle$

20. $\dfrac{3}{4}\mathbf{u} - 2\mathbf{v} = \dfrac{3}{4}\langle -2, 4 \rangle - 2\langle -3, -2 \rangle$

$= \left\langle -\dfrac{3}{2}, 3 \right\rangle - \langle -6, -4 \rangle$

$= \left\langle \dfrac{9}{2}, 7 \right\rangle$

26. $3\mathbf{u} + 2\mathbf{v} = 3(3\mathbf{i} - 2\mathbf{j}) + 2(-2\mathbf{i} + 3\mathbf{j})$

$= (9\mathbf{i} - 6\mathbf{j}) + (-4\mathbf{i} + 6\mathbf{j})$

$= (9 - 4)\mathbf{i} + (-6 + 6)\mathbf{j}$

$= 5\mathbf{i} + 0\mathbf{j}$

36. $a_1 = 2 \cos 8\pi/7 \approx -1.8$

$a_2 = 2 \sin 8\pi/7 \approx -0.9$

$\mathbf{v} = a_1\mathbf{i} + a_2\mathbf{j} = -1.8\mathbf{i} - 0.9\mathbf{j}$

40.

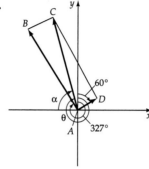

$\mathbf{AB} = 18 \sin 327°\mathbf{i} + 18 \cos 327°\mathbf{j}$

$\approx -9.8\mathbf{i} + 15.1\mathbf{j}$

$\mathbf{AO} = 4 \sin 60°\mathbf{i} + 4 \cos 60°\mathbf{j}$

$\approx 3.5\mathbf{i} - 2\mathbf{j}$

$\mathbf{AC} = \mathbf{AB} + \mathbf{AD}$

$= -9.8\mathbf{i} + 15.1\mathbf{j} + 3.5\mathbf{i} - 2\mathbf{j}$

$= -6.3\mathbf{i} + 13.1\mathbf{j}$

$|\mathbf{AC}| = \sqrt{(-6.3)^2 + (13.1)^2} \approx 14.5$

$\alpha = \tan^{-1}\left|\dfrac{13.1}{-6.3}\right| = \tan^{-1}\dfrac{13.1}{6.3} \approx 64.3°$

$\theta = 270° + 64.3° \approx 334.3°$

The course of the boat is 14.5 mph at a heading of 334.3°.

42. $\alpha = \theta$

$\sin \alpha = \dfrac{120}{800}$

$\alpha = 8.6°$

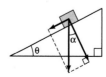

50. $\mathbf{v} \cdot \mathbf{w} = (5\mathbf{i} + 3\mathbf{j}) \cdot (4\mathbf{i} - 2\mathbf{j})$

$= 5(4) + 3(-2) = 20 - 6 = 14$

60. $\cos\theta = \dfrac{\mathbf{v}\cdot\mathbf{w}}{\|\mathbf{v}\|\,\|\mathbf{w}\|}$

$\cos\theta = \dfrac{(3\mathbf{i}-4\mathbf{j})\cdot(6\mathbf{i}-12\mathbf{j})}{\sqrt{3^2+(-4)^2}\,\sqrt{6^2+(-12)^2}}$

$\cos\theta = \dfrac{3(6)+(-4)(-12)}{\sqrt{25}\,\sqrt{180}}$

$\cos\theta = \dfrac{66}{5\sqrt{180}} \approx 0.9839$

$\theta \approx 10.3°$

62. $\operatorname{proj}_{\mathbf{w}}\mathbf{v} = \dfrac{\mathbf{v}\cdot\mathbf{w}}{\|\mathbf{w}\|}$

$\operatorname{proj}_{\mathbf{w}}\mathbf{v} = \dfrac{\langle -7,5\rangle\cdot\langle -4,1\rangle}{\sqrt{(-4)^2+1^2}} = \dfrac{28+5}{\sqrt{17}} = \dfrac{33}{\sqrt{17}} = \dfrac{33\sqrt{17}}{17}$

70. $w = \|\mathbf{F}\|\,\|\mathbf{S}\|\cos\alpha$

$w = 100\cdot 25\cdot\cos 42°$

$w \approx 1857.9 \text{ ft-lbs}$

EXERCISE SET 8.4, page 446

12. $\qquad r = \sqrt{1^2+(\sqrt{3})^2} = 2$

$\alpha = \tan^{-1}\left|\dfrac{\sqrt{3}}{1}\right| = \tan^{-1}\sqrt{3} = 60°$

$\theta = 60°$

$z = 2\operatorname{cis}60°$

26. $z = 4\left(\cos\dfrac{5\pi}{3}+i\sin\dfrac{5\pi}{3}\right) = 4\left(\dfrac12-\dfrac{\sqrt{3}}{2}i\right) = 2-2i\sqrt{3}$

52. $z_1 = 1+i \qquad\qquad \alpha_1 = \tan^{-1}\left|\dfrac{1}{1}\right| = 45°$

$r_1 = \sqrt{1^2+1^2}\qquad\qquad \theta_1 = 45°$

$r_1 = \sqrt{2}$

$z_1 = \sqrt{2}(\cos 45°+i\sin 45°)$

$z_2 = 1-i \qquad\qquad \alpha_2 = \tan^{-1}\left|\dfrac{-1}{1}\right| = 45°$

$r_2 = \sqrt{1^2+(-1)^2}$

$r_2 = \sqrt{2}\qquad\qquad \theta_2 = 315°$

$z_2 = \sqrt{2}(\cos 315°+i\sin 315°)$

$\dfrac{z_1}{z_2} = \dfrac{\sqrt{2}(\cos 45°+i\sin 45°)}{\sqrt{2}(\cos 315°+i\sin 315°)} = \dfrac{\sqrt{2}\operatorname{cis}45°}{\sqrt{2}\operatorname{cis}315°}$

$\dfrac{z_1}{z_2} = \cos(45°-315°)+i\sin(45°-315°) = \operatorname{cis}(-270°)$

$\dfrac{z_1}{z_2} = \cos 270°-i\sin 270° = 0-(-i) = i$

64. $z = 1+i\sqrt{3}$

$r = \sqrt{1^2+(\sqrt{3})^2}\qquad \alpha = \tan^{-1}\left|\dfrac{\sqrt{3}}{1}\right|$

$r = 2 \qquad\qquad\qquad \alpha = 60°$

$z = 2\operatorname{cis}60° \qquad\qquad \theta = 60°$

$(1+i\sqrt{3})^8 = (2\operatorname{cis}60°)^8 = 2^8\operatorname{cis}(8\cdot 60) = 256\operatorname{cis}480°$

$\qquad\qquad = 256\operatorname{cis}120°$

78. $-1+i = \sqrt{2}(\cos 135°+i\sin 135°)$

$w_k = (\sqrt{2})^{1/5}\left(\cos\dfrac{135°+360°k}{5}+i\sin\dfrac{135°+360°k}{5}\right)$

$\qquad\qquad\qquad\qquad\qquad\qquad k = 0, 1, 2, 3, 4$

$k = 0$

$w_0 = 2^{1/10}\left(\cos\dfrac{135°}{5}+i\sin\dfrac{135°}{5}\right)$

$w_0 \approx 0.955+0.487i$

$k = 1$

$w_1 = 2^{1/10}\left(\cos\dfrac{135°+360°}{5}+i\sin\dfrac{135°+360°}{5}\right)$

$w_1 \approx -0.168+1.059i$

$k = 2$

$w_2 = 2^{1/10}\left(\cos\dfrac{135°+360°\cdot 2}{5}+i\sin\dfrac{135°+360°\cdot 2}{5}\right)$

$w_2 \approx -1.059+0.168i$

$k = 3$

$w_3 = 2^{1/10}\left(\cos\dfrac{135°+360°\cdot 3}{5}+i\sin\dfrac{135°+360°\cdot 3}{5}\right)$

$w_3 \approx -0.487-0.955i$

$k = 4$

$w_4 = 2^{1/10}\left(\cos\dfrac{135°+360°\cdot 4}{5}+i\sin\dfrac{135°+360°\cdot 4}{5}\right)$

$w_4 \approx 0.758-0.758i$

84. $x^3-2i = 0$

$x^3 = 2i \qquad\qquad 2i = 2\operatorname{cis}90°$

$x = (2i)^{1/3}$

$x_k = 2^{1/3}\operatorname{cis}\dfrac{90°+360°k}{3} \qquad k = 0, 1, 2$

$x_0 \approx 1.26\operatorname{cis}90°/3 = 1.26\operatorname{cis}30°$

$x_1 \approx 1.26\operatorname{cis}\dfrac{90°+360°}{3} = 1.26\operatorname{cis}150°$

$x_2 \approx 1.26\operatorname{cis}\dfrac{90°+360°\cdot 2}{3} = 1.26\operatorname{cis}270°$

EXERCISE SET 9.1, page 459

4. Comparing $x^2 = 4py$ with $x^2 = -\dfrac{1}{4}y$, we have

$$4p = -\frac{1}{4} \quad \text{or} \quad p = -\frac{1}{16}$$

vertex $(0, 0)$, focus $\left(0, -\dfrac{1}{16}\right)$, directrix $y = \dfrac{1}{16}$

20. $x^2 + 5x - 4y - 1 = 0$

$$x^2 + 5x = 4y + 1$$

$$x^2 + 5x + \frac{25}{4} = 4y + 1 + \frac{25}{4} \quad \bullet \text{ Complete the square}$$

$$\left(x + \frac{5}{2}\right)^2 = 4\left(y + \frac{29}{16}\right)$$

$$4p = 4 \qquad \bullet \text{ Compare to } (x - h)^2 = 4p(y - k)^2$$

$$p = 1$$

vertex $\left(-\dfrac{5}{2}, -\dfrac{29}{16}\right)$, focus $\left(-\dfrac{5}{2}, -\dfrac{13}{16}\right)$,

directrix $y = -\dfrac{45}{16}$

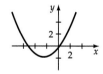

28. vertex $(0, 0)$, focus $(5, 0)$, $p = 5$ since focus is $(p, 0)$

$$y^2 = 4px$$

$$y^2 = 4(5)x$$

$$y^2 = 20x$$

30. vertex $(2, -3)$, focus $(0, -3)$. $(h, k) = (2, -3)$ therefore $h = 2$ and $k = -3$. Focus is $(h + p, k) = (2 + p, -3) = (0, -3)$. Therefore $2 + p = 0$ and $p = -2$.

$$(y - k)^2 = 4p(x - h)$$

$$(y + 3)^2 = 4(-2)(x - 2)$$

$$(y + 3)^2 = -8(x - 2)$$

EXERCISE SET 9.2, page 467

20. $25x^2 + 12y^2 = 300$

$$\frac{x^2}{12} + \frac{y^2}{25} = 1 \quad \bullet \ a^2 = 25, \ b^2 = 12, \ c^2 = \sqrt{25 - 12} = \sqrt{13}$$

center $(0, 0)$,
vertices $(0, 5)$ and $(0, -5)$
foci $(0, \sqrt{13})$ and $(0, -\sqrt{13})$.

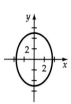

26. $9x^2 + 16y^2 + 36x - 16y - 104 = 0$

$$9x^2 + 36x + 16y^2 - 16y - 104 = 0$$

$$9(x^2 + 4x) + 16(y^2 - y) = 104$$

$$9(x^2 + 4x + 4) + 16\left(y^2 - y + \frac{1}{4}\right) = 104 + 36 + 4$$

$$9(x + 2)^2 + 16\left(y - \frac{1}{2}\right)^2 = 144$$

$$\frac{(x + 2)^2}{16} + \frac{\left(y - \dfrac{1}{2}\right)^2}{9} = 1$$

center $\left(-2, \tfrac{1}{2}\right)$, $a = 4$, $b = 3$,
$c = \sqrt{4^2 - 3^2} = \sqrt{7}$
vertices $\left(2, \tfrac{1}{2}\right)$ and $\left(-6, \tfrac{1}{2}\right)$,
foci $\left(-2 + \sqrt{7}, \tfrac{1}{2}\right)$ and
$\left(-2 - \sqrt{7}, \tfrac{1}{2}\right)$

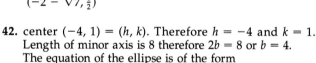

42. center $(-4, 1) = (h, k)$. Therefore $h = -4$ and $k = 1$. Length of minor axis is 8 therefore $2b = 8$ or $b = 4$. The equation of the ellipse is of the form

$$\frac{(x - h)^2}{a^2} + \frac{(y - k)^2}{b^2} = 1$$

$$\frac{(x + 4)^2}{a^2} + \frac{(y - 1)^2}{16} = 1 \quad \bullet \ h = -4, \ k = 1, \ b = 4$$

$$\frac{(0 + 4)^2}{a^2} + \frac{(4 - 1)^2}{16} = 1 \quad \bullet \ (x, y) = (0, 4). \text{ Thus } x = 0 \text{ and } y = 4$$

$$\frac{16}{a^2} + \frac{9}{16} = 1 \quad \bullet \text{ Solve for } a^2.$$

$$\frac{16}{a^2} = \frac{7}{16}$$

$$a^2 = \frac{256}{7}$$

$$\frac{(x + 4)^2}{256/7} + \frac{(y - 1)^2}{16} = 1$$

48. Because the foci are $(0, -3)$ and $(0, 3)$, $c = 3$ and center is $(0, 0)$, the midpoint of the line segment between $(0, -3)$ and $(0, 3)$.

$$e = \frac{c}{a}$$

$$\frac{1}{4} = \frac{3}{a} \qquad \bullet \; e = \frac{1}{4}$$

$$a = 12$$

$$3^2 = 12^2 - b^2 \qquad \bullet \; c^2 = a^2 - b^2$$

$$b^2 = 144 - 9 = 135 \quad \bullet \text{ Solve for } b^2$$

The equation of the ellipse is $\dfrac{x^2}{135} + \dfrac{y^2}{144} = 1$.

EXERCISE SET 9.3, page 476

4. $\dfrac{y^2}{25} - \dfrac{x^2}{36} = 1$

$a^2 = 25 \qquad b^2 = 36$

$a = 5 \qquad b = 6$

$c^2 = a^2 + b^2 = 25 + 36 = 61$

$c = \sqrt{61}$

Transverse axis is on y-axis because y^2 term is positive. Center $(0, 0)$, foci $(0, \sqrt{61})$, $(0, -\sqrt{61})$, and asymptotes $y = \dfrac{5}{6}x$ and $y = -\dfrac{5}{6}x$.

26. $16x^2 - 9y^2 - 32x - 54y + 79 = 0$

$16(x^2 - 2x + 1) - 9(y^2 + 6y + 9) = -79 + 16 - 81 = -144$

$$\frac{(y + 3)^2}{16} - \frac{(x - 1)^2}{9} = 1$$

Transverse axis is parallel to y-axis because y^2 term is positive. Center is at $(1, -3)$; because $a^2 = 16$, $a = 4$
vertices: $(h, k + a) = (1, 1)$
$\qquad\qquad (h, k - a) = (1, -7)$
$$c^2 = a^2 + b^2 = 16 + 9 = 25$$
$$c = \sqrt{25} = 5$$

Foci: $(h, k + c) = (1, 2)$
$\qquad (h, k - c) = (1, -8)$

Since $b^2 = 9$ or $b = 3$, Asymptotes: $y + 3 = \dfrac{4}{3}(x - 1)$

and, $y + 3 = -\dfrac{4}{3}(x - 1)$.

48. Because the vertices are $(2, 3)$ and $(-2, 3)$, $a = 2$ and center is $(0, 3)$.

$$e = \frac{c}{a}$$

$$\frac{5}{2} = \frac{c}{2} \qquad \bullet \; e = \frac{5}{2}$$

$$c = 5$$

$$5^2 = 2^2 + b^2$$

$$b^2 = 25 - 4 = 21$$

The equation of the hyperbola is $\dfrac{x^2}{4} - \dfrac{(y - 3)^2}{21} = 1$.

EXERCISE SET 9.4, page 489

14. 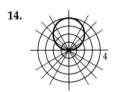   **18.**

20.    **28.**

32.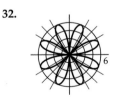

44. $x = r \cos \theta \qquad\qquad y = r \sin \theta$

$\qquad = (2)\left[\cos\left(-\dfrac{\pi}{3}\right)\right] \qquad = (2)\left[\sin\left(-\dfrac{\pi}{3}\right)\right]$

$\qquad = (2)\left(\dfrac{1}{2}\right) = 1 \qquad = (2)\left(-\dfrac{\sqrt{3}}{2}\right) = -\sqrt{3}$

The rectangular coordinates of the point are $(1, -\sqrt{3})$.

48. $r = \sqrt{x^2 + y^2}$ $\alpha = \tan^{-1}\left|\dfrac{y}{x}\right|$ • α is the reference angle.

$= \sqrt{(12)^2 + (-5)^2}$ $= \tan^{-1}\left|-\dfrac{5}{12}\right| \approx 22.6°$

$= \sqrt{144 + 25}$

$= \sqrt{169}$

$= 13$ $\theta = 360° - 22.6° = 337.4°$

The approximate polar coordinates of the point are $(13, 337.4°)$.

56. $r = \cot \theta$

$r = \dfrac{\cos \theta}{\sin \theta}$ • $\cot \theta = \dfrac{\cos \theta}{\sin \theta}$

$r \sin \theta = \cos \theta$

$r(r \sin \theta) = r \cos \theta$ • multiply both sides by r.

$(\sqrt{x^2 + y^2})y = x$ • $y = r \sin \theta;\ x = r \cos \theta$

$(x^2 + y^2)y^2 = x^2$ • Square each side

$y^4 + x^2y^2 - x^2 = 0$

64. $2x - 3y = 6$

$2r \cos \theta - 3r \sin \theta = 6$ • $x = r \cos \theta;\ y = r \sin \theta$

$r(2 \cos \theta - 3 \sin \theta) = 6$

$r = \dfrac{6}{2 \cos \theta - 3 \sin \theta}$

EXERCISE SET 10.1, page 502

6. $\begin{cases} 8x + 3y = -7 \\ \quad\ \ x = 3y + 15 \end{cases}$

$8(3y + 15) + 3y = -7$

$24y + 120 + 3y = -7$

$27y = -127$

$y = -\dfrac{127}{27}$

$x = 3\left(-\dfrac{127}{27}\right) + 15 = \dfrac{8}{9}$

The solution is $\left(\dfrac{8}{9}, -\dfrac{127}{27}\right)$.

18. $\begin{cases} 3x - 4y = 8 \\ 6x - 8y = 9 \end{cases}$

$8y = 6x - 9$

$y = \dfrac{3}{4}x - \dfrac{9}{8}$

$3x - 4\left(\dfrac{3}{4}x - \dfrac{9}{8}\right) = 8$

$3x - 3x + \dfrac{9}{2} = 8$

$\dfrac{9}{2} = 8$

This is a false equation. Therefore, the system of equations is inconsistent and has no solution.

20. $\begin{cases} 5x + 2y = 2 \\ \qquad y = -\dfrac{5}{2}x + 1 \end{cases}$

$5x + 2\left(-\dfrac{5}{2}x + 1\right) = 2$

$5x - 5x + 2 = 2$

$2 = 2$

The system of equations is dependent.

Let $x = c$. Then $y = -\dfrac{5}{2}c + 1$. Thus the solutions are $\left(c, -\dfrac{5}{2}c + 1\right)$.

24. $\begin{cases} 3x - 8y = -6 & (1) \\ -5x + 4y = 10 & (2) \end{cases}$

$\begin{array}{r} 3x - 8y = -6 \\ -10x + 8y = 20 \\ \hline -7x\qquad = 14 \end{array}$ • 2 times Eq. (2)

$x = -2$

$3(-2) - 8y = -6$ • Substitute -2 for x in Eq. (1). Solve for y.

$-8y = 0$

$y = 0$

The solution is $(-2, 0)$.

28. $\begin{cases} 4x + 5y = 2 & (1) \\ 8x - 15y = 9 & (2) \end{cases}$

$12x + 15y = 6$ • 3 times Eq. (1)

$\dfrac{8x - 15y = 9}{20x \qquad = 15}$

$x = \dfrac{3}{4}$

$4\left(\dfrac{3}{4}\right) + 5y = 2$ • Substitute $\dfrac{3}{4}$ for x in Eq. (1)

$3 + 5y = 2$ • Solve for y.

$y = -\dfrac{1}{5}$

The solution is $\left(\dfrac{3}{4}, -\dfrac{1}{5}\right)$.

44. Let $r =$ the rate of the canoeist.
Let $w =$ the rate of the current.
Rate of canoeist with the current: $r + w$
Rate of canoeist against the current: $r - w$

$r \cdot t = d$

$(r + w) \cdot 2 = 12$ (1)

$(r - w) \cdot 4 = 12$ (2)

$r + w = 6$ • Divide Eq. 1 by 2

$\dfrac{r - w = 3}{2r \quad = 9}$ • Divide Eq. 2 by 4

$r = 4.5$

$4.5 + w = 6$

$w = 1.5$

Rate of canoeist = 4.5 mph
Rate of current = 1.5 mph

EXERCISE SET 10.2, page 512

12. $\begin{cases} 3x + 2y - 5z = 6 & (1) \\ 5x - 4y + 3z = -12 & (2) \\ 4x + 5y - 2z = 15 & (3) \end{cases}$

$15x + 10y - 25z = 30$ 5 times (1)

$\dfrac{-15x + 12y - 9z = 36}{22y - 34z = 66}$ -3 times (2)

$11y - 17z = 33$ (4)

$12x + 8y - 20z = 24$ 4 times (1)

$\dfrac{-12x - 15y + 6z = -45}{-7y - 14z = -21}$ -3 times (3)

$y + 2z = 3$ (5)

$11y - 17z = 33$ (4)

$\dfrac{-11y - 22z = -33}{-39z = 0}$ -11 times (5)

$z = 0$ (6)

$11y - (17 \cdot 0) = 33$

$y = 3$

$3x + 2(3) - 5(0) = 6$

$x = 0$

The solution is $(0, 3, 0)$.

16. $\begin{cases} 2x + 3y + 2z = 14 & (1) \\ x - 3y + 4z = 4 & (2) \\ -x + 12y - 6z = 2 & (3) \end{cases}$

$2x + 3y + 2z = 14$ (1)

$\dfrac{-2x + 6y - 8z = -8}{9y - 6z = 6}$ -2 times (2)

$3y - 2z = 2$ (4)

$2x + 3y + 2z = 14$ (1)

$\dfrac{-2x + 24y - 12z = 4}{27y - 10z = 18}$ 2 times (3)

   (5)

$-27y + 18z = -18$ -9 times (4)

$\dfrac{27y - 10z = 18}{8z = 0}$ (5)

$z = 0$ (6)

$3y - 2(0) = 2$ • Substitute $z = 0$ in Eq. (4).

$y = \dfrac{2}{3}$

$2x + 3\left(\dfrac{2}{3}\right) + 2(0) = 14$ • Substitute $y = \dfrac{2}{3}$ and $z = 0$ in Eq. (1).

$x = 6$

The solution is $\left(6, \dfrac{2}{3}, 0\right)$.

18. $\begin{cases} 2x + 3y - 6z = 4 & (1) \\ 3x - 2y - 9z = -7 & (2) \\ 2x + 5y - 6z = 8 & (3) \end{cases}$

$6x + 9y - 18z = 12$ 3 times   (1)

$\dfrac{-6x + 4y + 18z = 14}{13y = 26}$ -2 times (2)

$y = 2$ (4)

$$2x + 3y - 6z = 4 \quad (1)$$
$$\underline{-2x - 5y + 6z = -8} \quad -1 \text{ times } (3)$$
$$-2y = -4$$
$$y = 2 \quad (5)$$

$$y = 2 \quad (4)$$
$$\underline{-y = -2} \quad \bullet \; -1 \text{ times } (5)$$
$$0 = 0 \quad (6)$$

The equations are dependent. Let $z = c$.

$$2x + 3(2) - 6c = 4 \quad \bullet \text{ Substitute } y = 2,\, z = c \text{ in Eq. (1).}$$

$$x = 3c - 1$$

The solutions are $(3c - 1, 2, c)$.

20. $\begin{cases} x - 3y + 4z = 9 \quad (1) \\ 3x - 8y - 2z = 4 \quad (2) \end{cases}$

$$-3x + 9y - 12z = -27 \quad -3 \text{ times } \quad (1)$$
$$\underline{3x - 8y - 2z = 4} \quad (2)$$
$$y - 14z = -23 \quad (3)$$

$$y = 14z - 23 \quad \bullet \text{ Solve Eq. (3) for } y.$$

$$x - 3(14z - 23) + 4z = 9 \qquad \bullet \text{ Substitute } 14z - 23$$
$$\text{for } y \text{ in Eq. (1).}$$

$$x = 38z - 60 \quad \bullet \text{ Solve for } x.$$

Let $z = c$. The solutions are $(38c - 60, 14c - 23, c)$.

32. $\begin{cases} 5x + 2y + 3z = 0 \quad (1) \\ 3x + y - 2z = 0 \quad (2) \\ 4x - 7y + 5z = 0 \quad (3) \end{cases}$

$$15x + 6y + 9z = 0 \quad 3 \text{ times } (1)$$
$$\underline{-15x - 5y + 10z = 0} \quad -5 \text{ times } (2)$$
$$y + 19z = 0 \quad (4)$$

$$20x + 8y + 12z = 0 \quad 4 \text{ times } (1)$$
$$\underline{-20x + 35y - 25z = 0} \quad -5 \text{ times } (3)$$
$$43y - 13z = 0 \quad (5)$$

$$-43y - 817z = 0 \quad -43 \text{ times } (4)$$
$$\underline{43y - 13z = 0} \quad (5)$$
$$-830z = 0$$

$$z = 0 \quad (6)$$

Solving by back substitution, the only solution is $(0, 0, 0)$.

36. $\begin{cases} x^2 + y^2 + ax + by + c = 0 \\ 0 + 36 + a(0) + b(6) + c = 0 \quad \bullet \text{ let } x = 0,\, y = 6 \\ 1 + 25 + a(1) + b(5) + c = 0 \quad \bullet \text{ let } x = 1,\, y = 5 \\ 49 + 1 + a(-7) + b(-1) + c = 0 \quad \bullet \text{ let } x = -7,\, y = -1 \end{cases}$

$$6b + c = -36 \quad (1)$$
$$a + 5b + c = -26 \quad (2)$$
$$-7a - b + c = -50 \quad (3)$$

$$7a + 35b + 7c = -182 \quad 7 \text{ times } (2)$$
$$\underline{-7a - b + c = -50} \quad (3)$$
$$34b + 8c = -232$$
$$17b + 4c = -116 \quad (4)$$

$$-24b - 4c = 144 \quad -4 \text{ times } (1)$$
$$\underline{17b + 4c = -116} \quad (4)$$
$$-7b = 28$$
$$b = -4$$

$$17(-4) + 4c = -116 \quad \bullet \text{ Substitute } -4 \text{ for } b \text{ in Eq. (4).}$$

$$c = -12$$

$$-7a - (-4) - 12 = -50 \qquad \bullet \text{ Substitute } -4 \text{ for } b \text{ and } -12$$
$$\text{for } c \text{ in Eq. (3).}$$

$$a = 6$$

The equation whose graph passes through the three points is $x^2 + y^2 + 6x - 4y - 12 = 0$.

EXERCISE SET 10.3, page 517

8. $\begin{cases} x - 2y = 3 \quad (1) \\ xy = -1 \quad (2) \end{cases}$

$$x = 2y + 3 \qquad \bullet \text{ Solve Eq. (1) for } x.$$

$$(2y + 3)y = -1 \quad \bullet \text{ Replace } x \text{ by } 2y + 3 \text{ in Eq. (2).}$$

$$2y^2 + 3y + 1 = 0 \quad \bullet \text{ Solve for } y.$$

$$(2y + 1)(y + 1) = 0$$

$$y = -\frac{1}{2} \quad \text{or} \quad y = -1$$

Substitute for y in Eq. (1).

$$x - 2\left(-\frac{1}{2}\right) = 3 \qquad x - 2(-1) = 3$$

$$x = 2 \qquad\qquad\qquad x = 1$$

The solutions are $(2, -1/2)$ and $(1, -1)$.

16. $\begin{cases} 3x^2 - 2y^2 = 19 & (1) \\ x^2 - y^2 = 5 & (2) \end{cases}$

$\begin{array}{l} 3x^2 - 2y^2 = 19 \quad (1) \\ \underline{-3x^2 + 3y^2 = -15} \quad \bullet \text{ Multiply Eq. (2) by } -3. \\ \qquad\quad y^2 = 4 \quad \bullet \text{ Add the equations.} \\ \qquad\quad y = \pm 2 \quad \bullet \text{ Solve for } y. \end{array}$

when $y = -2$, $x^2 - (-2)^2 = 5$ $\bullet$ From Eq. (2)

$$x^2 - 4 = 5$$
$$x^2 = 9$$
$$x = \pm 3$$

when $y = 2$, $x^2 - 2^2 = 5$

$$x^2 - 4 = 5$$
$$x^2 = 9$$
$$x = \pm 3$$

The solutions are $(-2, -3)$, $(-2, 3)$, $(2, -3)$ and $(2, 3)$.

20. $\begin{cases} 2x^2 + 3y^2 = 11 & (1) \\ 3x^2 + 2y^2 = 19 & (2) \end{cases}$

Use the elimination method to eliminate y^2

$\begin{array}{l} 4x^2 + 6y^2 = \quad 22 \quad \bullet \text{ 2 times Eq. (1)} \\ \underline{-9x^2 - 6y^2 = -57} \quad \bullet -3 \text{ times Eq. (2)} \\ -5x^2 \qquad\quad = -35 \end{array}$

$$x = \pm 7$$

Substitute for x in Eq. (1).

$$2(\pm 7)^2 + 3y^2 = 11$$
$$3y^2 = 11 - 14$$
$$y^2 = -3$$

$y^2 = -13$ has no real number solutions. The graphs of the equations do not intersect. The system is inconsistent and has no solution.

28. $\begin{cases} (x + 2)^2 + (y - 3)^2 = 10 \\ (x - 3)^2 + (y + 1)^2 = 13 \end{cases}$

$\begin{array}{l} x^2 + 4x + 4 + y^2 - 6y + 9 = 10 \quad (1) \\ \underline{x^2 - 6x + 9 + y^2 + 2y + 1 = 13} \quad (2) \\ \quad 10x - 5 \qquad -8y + 8 = -3 \qquad \bullet \text{ Subtract.} \\ \qquad\quad 10x - 8y = -6 \end{array}$

$$y = \frac{5x + 3}{4} \quad (3) \quad \bullet \text{ Solve for } y.$$

$$(x + 2)^2 + \left(\frac{5x - 9}{4}\right)^2 = 10 \quad \bullet \text{ Substitute for } y.$$

$$x^2 + 4x + 4 + \frac{25x^2 - 90x + 81}{16} = 10 \quad \bullet \text{ Solve for } x.$$

$$16x^2 + 64x + 64 + 25x^2 - 90x + 81 = 160$$

$$41x^2 - 26x - 15 = 0$$

$$(41x + 15)(x - 1) = 0$$

$$x = -\frac{15}{41} \quad \text{or} \quad x = 1$$

$$y = \frac{5}{4}\left(-\frac{15}{41}\right) + \frac{3}{4} \quad \text{or} \quad y = \frac{5(1) + 3}{4} \quad \begin{array}{l} \bullet \text{ Substitute} \\ \text{ for } x \\ \text{ into Eq. (3).} \end{array}$$

$$y = \frac{12}{41} \qquad\qquad \text{or} \quad y = 2$$

The solutions are $\left(-\dfrac{15}{41}, \dfrac{12}{41}\right)$ and $(1, 2)$.

EXERCISE SET 10.4, page 524

14. $\dfrac{7x + 44}{x^2 + 10x + 24} = \dfrac{7x + 44}{(x + 4)(x + 6)} = \dfrac{A}{x + 4} + \dfrac{B}{x + 6}$

$$7x + 44 = A(x + 6) + B(x + 4)$$

$$7x + 44 = (A + B)x + (6A + 4B)$$

$$\begin{cases} 7 = A + B \\ 44 = 6A + 4B \end{cases}$$

The solution is $A = 8$, $B = -1$.

$$\frac{7x + 44}{x^2 + 10x + 24} = \frac{8}{x + 4} + \frac{-1}{x + 6}$$

22. $\dfrac{x - 18}{x(x - 3)^2} = \dfrac{A}{x} + \dfrac{B}{x - 3} + \dfrac{C}{(x - 3)^2}$

$$x - 18 = A(x - 3)^2 + Bx(x - 3) + Cx$$

$$x - 18 = Ax^2 - 6Ax + 9A + Bx^2 - 3Bx + Cx$$

$$x - 18 = (A + B)x^2 + (-6A - 3B + C)x + 9A$$

$$\begin{cases} 0 = A + B \\ 1 = -6A - 3B + C \\ -18 = 9A \end{cases}$$

The solution is $A = -2$, $B = 2$, $C = -5$.

$$\frac{x - 18}{x(x - 3)^2} = \frac{-2}{x} + \frac{2}{x - 3} + \frac{-5}{(x - 3)^2}$$

24. $x^3 - x^2 + 10x - 10 = (x - 1)(x^2 + 10)$

$$\frac{9x^2 - 3x + 49}{(x - 1)(x^2 + 10)} = \frac{A}{x - 1} + \frac{Bx + C}{x^2 + 10}$$

$$9x^2 - 3x + 49 = A(x^2 + 10) + (Bx + C)(x - 1)$$

$$9x^2 - 3x + 49 = (A + B)x^2 + (-B + C)x + (10A - C)$$

$$\begin{cases} 9 = A + B \\ -3 = -B + C \\ 49 = 10A - C \end{cases}$$

The solution is $A = 5$, $B = 4$, $C = 1$.

$$\frac{9x^2 - 3x + 49}{x^3 - x^2 + 10x - 10} = \frac{5}{x - 1} + \frac{4x + 1}{x^2 + 10}$$

30. $\dfrac{2x^3 + 9x + 1}{(x^2 + 7)^2} = \dfrac{Ax + B}{x^2 + 7} + \dfrac{Cx + D}{(x^2 + 7)^2}$

$$2x^3 + 9x + 1 = (Ax + B)(x^2 + 7) + Cx + D$$

$$2x^3 + 9x + 1 = Ax^3 + Bx^2 + (7A + C)x + (7B + D)$$

$$\begin{cases} 2 = A \\ 0 = B \\ 9 = 7A + C \\ 1 = 7B + D \end{cases}$$

The solutions are $A = 2$, $B = 0$, $C = -5$, $D = 1$.

$$\frac{2x^3 + 9x + 1}{x^4 + 14x^2 + 49} = \frac{2x}{x^2 + 7} + \frac{-5x + 1}{(x^2 + 7)^2}$$

34. $\quad \dfrac{x + 1}{2x^2 + 3x - 2\,\big)\,2x^3 + 5x^2 + 3x - 8}$

$$\begin{array}{r} 2x^3 + 3x^2 - 2x \\ \hline 2x^2 + 5x - 8 \\ 2x^2 + 3x - 2 \\ \hline 2x - 6 \end{array}$$

$$\frac{2x^3 + 5x^2 + 3x - 8}{2x^2 + 3x - 2} = x + 1 + \frac{2x - 6}{2x^2 + 3x - 2}$$

$$\frac{2x - 6}{(2x - 1)(x + 2)} = \frac{A}{2x - 1} + \frac{B}{x + 2}$$

$$2x - 6 = A(x + 2) + B(2x - 1)$$

$$2x - 6 = Ax + 2A + 2Bx - B$$

$$2x - 6 = (A + 2B)x + (2A - B)$$

$$\begin{cases} 2 = A + 2B \\ -6 = 2A - B \end{cases}$$

The solutions are $A = -2$, $B = 2$.

$$\frac{2x^3 + 5x^2 + 3x - 8}{2x^2 + 3x - 2} = x + 1 + \frac{-2}{2x - 1} + \frac{2}{x + 2}$$

EXERCISE SET 10.5, page 529

6.

12.

20.

26.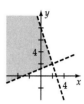

36.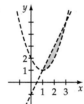

40.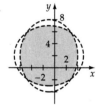

42.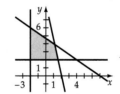

EXERCISE SET 10.6, page 535

12. $C = 4x + 3y$

	C	
(0, 8)	24	
(2, 4)	20	minimum
(5, 2)	26	
(11, 0)	44	
(20, 0)	80	
(20, 20)	140	
(0, 20)	60	

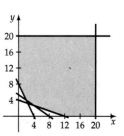

22. x = hours of Machine 1 use
y = hours of Machine 2 use
Cost = $28x + 25y$

Constraints: $4x + 3y \geq 60$ $x \geq 0, y \geq 0$

$5x + 10y \geq 100$

	Cost	
(0, 20)	500	
(12, 4)	436	minimum
(20, 0)	560	

To achieve the minimum cost, use Machine 1 for 12 hours and Machine 2 for 4 hours.

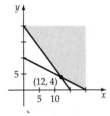

24. Let x = number of standard models.
Let y = number of deluxe models.
Profit = $25x + 35y$

Constraints: $x + 3y \leq 24$ $x \geq 0, y \geq 0$

$x + y \leq 10$

$2x + y \leq 16$

	Profit	
(0, 0)	0	
(0, 8)	280	
(6, 4)	290	
(3, 7)	320	maximum
(8, 0)	200	

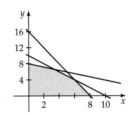

To maximize profits, produce 3 standard models and 7 deluxe models.

EXERCISE SET 11.1, page 551

14. $\begin{bmatrix} 1 & -3 & 1 & 8 \\ 2 & -5 & -3 & 2 \\ 1 & 4 & 1 & 1 \end{bmatrix} \xrightarrow[\substack{-2R_1 + R_2 \\ -1R_1 + R_3}]{} \begin{bmatrix} 1 & -3 & 1 & 8 \\ 0 & 1 & -5 & -14 \\ 0 & 7 & 0 & -7 \end{bmatrix}$

$\xrightarrow[-7R_2 + R_3]{} \begin{bmatrix} 1 & -3 & 1 & 8 \\ 0 & 1 & -5 & -14 \\ 0 & 0 & 35 & 91 \end{bmatrix}$

$\xrightarrow[(1/35)R_3]{} \begin{bmatrix} 1 & -3 & 1 & 8 \\ 0 & 1 & -5 & -14 \\ 0 & 0 & 1 & \frac{13}{5} \end{bmatrix}$

$\begin{cases} x - 3y + z = 8 \\ \qquad y - 5z = -14 \\ \qquad\qquad z = \dfrac{13}{5} \end{cases}$

By back substitution, the solution is $\left(\dfrac{12}{5}, -1, \dfrac{13}{5} \right)$.

18. $\begin{bmatrix} 3 & -5 & 2 & 4 \\ 1 & -3 & 2 & 4 \\ 5 & -11 & 6 & 12 \end{bmatrix} \xrightarrow{R_2 \longleftrightarrow R_1} \begin{bmatrix} 1 & -3 & 2 & 4 \\ 3 & -5 & 2 & 4 \\ 5 & -11 & 6 & 12 \end{bmatrix}$

$\xrightarrow[\substack{-3R_1 + R_2 \\ -5R_1 + R_3}]{} \begin{bmatrix} 1 & -3 & 2 & 4 \\ 0 & 4 & -4 & -8 \\ 0 & 4 & -4 & -8 \end{bmatrix} \xrightarrow{(1/4)R_2} \begin{bmatrix} 1 & -3 & 2 & 4 \\ 0 & 1 & -1 & -2 \\ 0 & 4 & -4 & -8 \end{bmatrix}$

$\xrightarrow[-4R_2 + R_3]{} \begin{bmatrix} 1 & -3 & 2 & 4 \\ 0 & 1 & -1 & -2 \\ 0 & 0 & 0 & 0 \end{bmatrix} \begin{cases} x - 3y + 2z = 4 & (1) \\ \quad y - z = -2 & (2) \\ \qquad\qquad 0 = 0 & (3) \end{cases}$

$y - z = -2 \quad \text{or} \quad y = z - 2$

$x - 3(z - 2) + 2z = 4$ • Substitute
$z - 2$ for y in (1).

$x - 3z + 6 + 2z = 4$

$x = z - 2$

Let $z = c$. The solutions are $(c - 2, c - 2, c)$.

20. $\begin{bmatrix} 2 & 5 & 2 & -1 \\ 1 & 2 & -3 & 5 \\ 5 & 12 & 1 & 10 \end{bmatrix} \xrightarrow{R_2 \longleftrightarrow R_1} \begin{bmatrix} 1 & 2 & -3 & 5 \\ 2 & 5 & 2 & -1 \\ 5 & 12 & 1 & 10 \end{bmatrix}$

$\xrightarrow[\substack{-2R_1 + R_2 \\ -5R_1 + R_3}]{} \begin{bmatrix} 1 & 2 & -3 & 5 \\ 0 & 1 & 8 & -11 \\ 0 & 2 & 16 & -15 \end{bmatrix}$

$\xrightarrow[-2R_2 + R_3]{} \begin{bmatrix} 1 & 2 & -3 & 5 \\ 0 & 1 & 8 & -11 \\ 0 & 0 & 0 & 7 \end{bmatrix}$

$\begin{cases} x + 2y - 3z = 5 \\ \quad y + 8z = -11 \\ \qquad\qquad 0 = 7 \end{cases}$

Because $0 = 7$ is a false equation, the system of equations has no solution.

36. $\begin{bmatrix} 1 & -1 & 3 & -5 & 10 \\ 2 & -3 & 4 & 1 & 7 \\ 3 & 1 & -2 & -2 & 6 \end{bmatrix}$

$\xrightarrow[-3R_1 + R_3]{-2R_1 + R_2} \begin{bmatrix} 1 & -1 & 3 & -5 & 10 \\ 0 & -1 & -2 & 11 & -13 \\ 0 & 4 & -11 & 13 & -24 \end{bmatrix}$

$\xrightarrow{-1R_2} \begin{bmatrix} 1 & -1 & 3 & -5 & 10 \\ 0 & 1 & 2 & -11 & 13 \\ 0 & 4 & -11 & 13 & -24 \end{bmatrix}$

$\xrightarrow{-4R_2 + R_3} \begin{bmatrix} 1 & -1 & 3 & -5 & 10 \\ 0 & 1 & 2 & -11 & 13 \\ 0 & 0 & -19 & 57 & -76 \end{bmatrix}$

$\xrightarrow{(-1/19)R_3} \begin{bmatrix} 1 & -1 & 3 & -5 & 10 \\ 0 & 1 & 2 & -11 & 13 \\ 0 & 0 & 1 & -3 & 4 \end{bmatrix}$

$\begin{cases} t - u + 3v - 5w = 10 & (1) \\ u + 2v - 11w = 13 & (2) \\ v - 3w = 4 & (3) \end{cases}$

$v = 3w + 4$

$u + 2(3w + 4) - 11w = 13$ • Substitute $3w + 4$ for v in (2).

$u = 5w + 5$

$t - (5w + 5) + 3(3w + 4) - 5w = 10$ • Substitute $5w + 5$ for u and $3w + 4$ for v in (1).

$t = w + 3$

Let w be any real number c. The solution of the system of equations is $(c + 3, 5c + 5, 3c + 4, c)$.

EXERCISE SET 11.2, page 562

6. a. $A + B = \begin{bmatrix} 2 & -2 \\ 3 & 4 \\ 1 & 0 \end{bmatrix} + \begin{bmatrix} -1 & 8 \\ 2 & -2 \\ -4 & 3 \end{bmatrix} = \begin{bmatrix} 1 & 6 \\ 5 & 2 \\ -3 & 3 \end{bmatrix}$

b. $A - B = \begin{bmatrix} 2 & -2 \\ 3 & 4 \\ 1 & 0 \end{bmatrix} - \begin{bmatrix} -1 & 8 \\ 2 & -2 \\ -4 & 3 \end{bmatrix} = \begin{bmatrix} 3 & -10 \\ 1 & 6 \\ 5 & -3 \end{bmatrix}$

c. $2B = 2 \begin{bmatrix} -1 & 8 \\ 2 & -2 \\ -4 & 3 \end{bmatrix} = \begin{bmatrix} -2 & 16 \\ 4 & -4 \\ -8 & 6 \end{bmatrix}$

d. $2A - 3B = 2 \begin{bmatrix} 2 & -2 \\ 3 & 4 \\ 1 & 0 \end{bmatrix} - 3 \begin{bmatrix} -1 & 8 \\ 2 & -2 \\ -4 & 3 \end{bmatrix} = \begin{bmatrix} 7 & -28 \\ 0 & 14 \\ 14 & -9 \end{bmatrix}$

16. $AB = \begin{bmatrix} -1 & 2 & 0 \\ 2 & -1 & 1 \\ -2 & 2 & -1 \end{bmatrix} \begin{bmatrix} 2 & -1 & 0 \\ 1 & 5 & -1 \\ 0 & -1 & 3 \end{bmatrix}$

$= \begin{bmatrix} (-1)(2) + & (2)(1) + & (0)(0) \\ (2)(2) + (-1)(1) + & (1)(0) \\ (-2)(2) + & (2)(1) + (-1)(0) \end{bmatrix}$

$\begin{matrix} (-1)(-1) + & (2)(5) + & (0)(-1) \\ (2)(-1) + (-1)(5) + & (1)(-1) \\ (-2)(-1) + & (2)(5) + (-1)(-1) \end{matrix}$

$\left. \begin{matrix} (-1)(0) + & (2)(-1) + & (0)(3) \\ (2)(0) + (-1)(-1) + & (1)(3) \\ (-2)(0) + & (2)(-1) + (-1)(3) \end{matrix} \right]$

$= \begin{bmatrix} 0 & 11 & -2 \\ 3 & -8 & 4 \\ -2 & 13 & -5 \end{bmatrix}$

$BA = \begin{bmatrix} 2 & -1 & 0 \\ 1 & 5 & -1 \\ 0 & -1 & 3 \end{bmatrix} \begin{bmatrix} -1 & 2 & 0 \\ 2 & -1 & 1 \\ -2 & 2 & -1 \end{bmatrix}$

$= \begin{bmatrix} (2)(-1) + (-1)(2) + & (0)(-2) \\ (1)(-1) + & (5)(2) + (-1)(-2) \\ (0)(-1) + (-1)(2) + & (3)(-2) \end{bmatrix}$

$\begin{matrix} (2)(2) + (-1)(-1) + & (0)(2) \\ (1)(2) + & (5)(-1) + (-1)(2) \\ (0)(2) + (-1)(-1) + & (3)(2) \end{matrix}$

$\left. \begin{matrix} (2)(0) + (-1)(1) + & (0)(-1) \\ (1)(0) + & (5)(1) + (-1)(-1) \\ (0)(0) + (-1)(1) + & (3)(-1) \end{matrix} \right]$

$= \begin{bmatrix} -4 & 5 & -1 \\ 11 & -5 & 6 \\ -8 & 7 & -4 \end{bmatrix}$

EXERCISE SET 11.3, page 571

6.
$$\begin{bmatrix} 1 & 3 & -2 & 1 & 0 & 0 \\ -1 & -5 & 6 & 0 & 1 & 0 \\ 2 & 6 & -3 & 0 & 0 & 1 \end{bmatrix}$$

$$\begin{matrix} R_1 + R_2 \\ -2R_1 + R_3 \\ \sim\sim\sim 1 \end{matrix} \begin{bmatrix} 1 & 3 & -2 & 1 & 0 & 0 \\ 0 & -2 & 4 & 1 & 1 & 0 \\ 0 & 0 & 1 & -2 & 0 & 1 \end{bmatrix}$$

$$\begin{matrix} (-1/2)R_2 \\ \sim\sim\sim 1 \end{matrix} \begin{bmatrix} 1 & 3 & -2 & 1 & 0 & 0 \\ 0 & 1 & -2 & -\frac{1}{2} & -\frac{1}{2} & 0 \\ 0 & 0 & 1 & -2 & 0 & 1 \end{bmatrix}$$

$$\begin{matrix} 2R_3 + R_2 \\ 2R_3 + R_1 \\ \sim\sim\sim 1 \end{matrix} \begin{bmatrix} 1 & 3 & 0 & -3 & 0 & 2 \\ 0 & 1 & 0 & -\frac{9}{2} & -\frac{1}{2} & 2 \\ 0 & 0 & 1 & -2 & 0 & 1 \end{bmatrix}$$

$$\begin{matrix} -3R_2 + R_1 \\ \sim\sim\sim 1 \end{matrix} \begin{bmatrix} 1 & 0 & 0 & \frac{21}{2} & \frac{3}{2} & -4 \\ 0 & 1 & 0 & -\frac{9}{2} & -\frac{1}{2} & 2 \\ 0 & 0 & 1 & -2 & 0 & 1 \end{bmatrix}$$

The inverse matrix is $\begin{bmatrix} \frac{21}{2} & \frac{3}{2} & -4 \\ -\frac{9}{2} & -\frac{1}{2} & 2 \\ -2 & 0 & 1 \end{bmatrix}$.

10.
$$\begin{bmatrix} 1 & -2 & 2 & 1 & 0 & 0 \\ 2 & -3 & 1 & 0 & 1 & 0 \\ 3 & -6 & 6 & 0 & 0 & 1 \end{bmatrix}$$

$$\begin{matrix} -2R_1 + R_2 \\ -3R_1 + R_3 \\ \sim\sim\sim 1 \end{matrix} \begin{bmatrix} 1 & -2 & 2 & 1 & 0 & 0 \\ 0 & 1 & -3 & -2 & 1 & 0 \\ 0 & 0 & 0 & -3 & 0 & 1 \end{bmatrix}$$

Because if there are zeros below the rightmost 1 along the main diagonal, the matrix does not have an inverse.

20.
$$\begin{bmatrix} 1 & 2 & -1 \\ 2 & 3 & -1 \\ 3 & 6 & -2 \end{bmatrix}\begin{bmatrix} x \\ y \\ z \end{bmatrix} = \begin{bmatrix} 5 \\ 8 \\ 14 \end{bmatrix}$$

The inverse of the coefficient matrix is $\begin{bmatrix} 0 & 2 & -1 \\ -1 & -1 & 1 \\ -3 & 0 & 1 \end{bmatrix}$.

Multiplying each side of the equation by the inverse, we have

$$\begin{bmatrix} x \\ y \\ z \end{bmatrix} = \begin{bmatrix} 0 & 2 & -1 \\ -1 & -1 & 1 \\ -3 & 0 & 1 \end{bmatrix}\begin{bmatrix} 5 \\ 8 \\ 14 \end{bmatrix} = \begin{bmatrix} 2 \\ 1 \\ -1 \end{bmatrix}$$

The solution is $(2, 1, -1)$.

EXERCISE SET 11.4, page 580

2. $\begin{vmatrix} 2 & 9 \\ -6 & 2 \end{vmatrix} = 2 \cdot 2 - (-6)(9) = 4 + 54 = 58$

14. $M_{13} = \begin{vmatrix} 1 & 3 \\ 6 & -2 \end{vmatrix} = 1(-2) - 6(3) = -2 - 18 = -20$

$C_{13} = (-1)^{1+3} \cdot M_{13} = 1 \cdot M_{13} = 1(-20) = -20$

20. Expanding with cofactors of row one yields.

$$\begin{vmatrix} 3 & -2 & 0 \\ 2 & -3 & 2 \\ 8 & -2 & 5 \end{vmatrix} = 3C_{11} + (-2)C_{12} + 0 \cdot C_{13}$$

$$= 3\begin{vmatrix} -3 & 2 \\ -2 & 5 \end{vmatrix} + 2\begin{vmatrix} 2 & 2 \\ 8 & 5 \end{vmatrix} + 0\begin{vmatrix} 2 & -3 \\ 8 & -2 \end{vmatrix}$$

$$= 3(-15 + 4) + 2(10 - 16) + 0$$

$$= 3(-11) + 2(-6) = -33 + (-12)$$

$$= -45$$

42. Let $D = \begin{vmatrix} 3 & -2 & -1 \\ 1 & 2 & 4 \\ 2 & -2 & 3 \end{vmatrix}$. Then

$$\begin{matrix} R_1 \text{ i } R_2 \\ D \end{matrix} = -\begin{vmatrix} 1 & 2 & 4 \\ 3 & -2 & -1 \\ 2 & -2 & 3 \end{vmatrix} \begin{matrix} -3R_1 + R_2 \\ -2R_1 + R_3 \\ = \end{matrix} -\begin{vmatrix} 1 & 2 & 4 \\ 0 & -8 & -13 \\ 0 & -6 & -5 \end{vmatrix}$$

$$\begin{matrix} -\frac{1}{8}R_2 \\ = \frac{8}{1} \end{matrix}\begin{vmatrix} 1 & 2 & 4 \\ 0 & 1 & \frac{13}{8} \\ 0 & -6 & -5 \end{vmatrix} \begin{matrix} 6R_2 + R_3 \\ = \frac{8}{1} \end{matrix}\begin{vmatrix} 1 & 2 & 4 \\ 0 & 1 & \frac{13}{8} \\ 0 & 0 & \frac{19}{4} \end{vmatrix}$$

$$= 8(1)(1)\left(\frac{19}{4}\right) = 38$$

EXERCISE SET 11.5, page 585

4. $x_1 = \dfrac{\begin{vmatrix} 9 & 5 \\ 8 & 7 \end{vmatrix}}{\begin{vmatrix} 2 & 5 \\ 5 & 7 \end{vmatrix}} = \dfrac{63 - 40}{14 - 25} = \dfrac{23}{-11} = -\dfrac{23}{11}$

$x_2 = \dfrac{\begin{vmatrix} 2 & 9 \\ 5 & 8 \end{vmatrix}}{\begin{vmatrix} 2 & 5 \\ 5 & 7 \end{vmatrix}} = \dfrac{16 - 45}{14 - 25} = \dfrac{-29}{-11} = \dfrac{29}{11}$

The solution is $\left(-\dfrac{23}{11}, \dfrac{29}{11}\right)$.

24. $x_3 = \dfrac{\begin{vmatrix} 2 & 5 & -3 & -3 \\ 1 & 7 & 4 & -1 \\ 4 & 0 & 3 & 1 \\ 3 & 2 & 0 & 0 \end{vmatrix}}{\begin{vmatrix} 2 & 5 & -5 & -3 \\ 1 & 7 & 8 & -1 \\ 4 & 0 & 1 & 1 \\ 3 & 2 & -1 & 0 \end{vmatrix}} = \dfrac{157}{168}$

EXERCISE SET 12.1, page 599

6. $a_n = \dfrac{(-1)^{n+1}}{n(n+1)}, \ a_1 = \dfrac{(-1)^{1+1}}{1(1+1)} = \dfrac{1}{2},$

$a_2 = \dfrac{(-1)^{2+1}}{2(2+1)} = -\dfrac{1}{6}, \ a_3 = \dfrac{(-1)^{3+1}}{3(3+1)} = \dfrac{1}{12},$

$a_8 = \dfrac{(-1)^{8+1}}{8(8+1)} = -\dfrac{1}{72}$

28. $a_1 = 1, \ a_2 = 2^2 \cdot a_1 = 4 \cdot 1 = 4, \ a_3 = 3^2 \cdot a_2 = 9 \cdot 4 = 36$

42. $\dfrac{12!}{4!\,8!} = \dfrac{12 \cdot 11 \cdot 10 \cdot 9 \cdot 8!}{4!\,8!} = \dfrac{12 \cdot 11 \cdot 10 \cdot 9}{4 \cdot 3 \cdot 2 \cdot 1} = 495$

52. $\displaystyle\sum_{i=1}^{6} (2i+1)(2i-1) = \sum_{i=1}^{6} (4i^2 - 1)$

$\qquad = (4 \cdot 1^2 - 1) + (4 \cdot 2^2 - 1)$

$\qquad \quad + (4 \cdot 3^2 - 1) + (4 \cdot 4^2 - 1)$

$\qquad \quad + (4 \cdot 5^2 - 1) + (4 \cdot 6^2 - 1)$

$\qquad = 3 + 15 + 35 + 63 + 99 + 143$

$\qquad = 358$

EXERCISE SET 12.2, page 605

16. $a_6 = -14, \ a_8 = -20$

$\qquad a_8 = a_6 + 2d$

$\qquad \dfrac{a_8 - a_6}{2} = d \quad \bullet \text{ Solve for } d.$

$\qquad \dfrac{-20 - (-14)}{2} = d$

$\qquad -3 = \dfrac{-6}{2} = d$

$\qquad\qquad a_n = a_1 + (n-1)d$

$\qquad\qquad -14 = a_1 + (6-1)(-3)$

$\qquad\qquad -14 = a_1 + (-15)$

$\qquad\qquad a_1 = 1$

$\qquad a_{15} = 1 + (15-1)(-3) = 1 + (14)(-3) = -41$

22. $s_{20} = \dfrac{20}{2}(a_1 + a_{20})$

$\qquad a_1 = 1 - 2(1) = -1$

$\qquad a_{20} = 1 - 2(20) = -39$

$\qquad s_{20} = 10(-1 + (-39))$

$\qquad\quad = 10(-40)$

$\qquad\quad = -400$

34. $a = 7, \ c_1, c_2, c_3, c_4, c_5, \ b = 19$

$\qquad\qquad a_n = a_1 + (n-1)d$

$\qquad\qquad 19 = 7 + 6d \quad \bullet \text{ There are 7 terms,}$
$\qquad\qquad\qquad\qquad\qquad \text{thus } n = 7.$

$\qquad\qquad d = 2$

$\qquad\qquad c_1 = a_1 + d = 7 + 2 = 9$

$\qquad\qquad c_2 = a_1 + 2d = 7 + 4 = 11$

$\qquad\qquad c_3 = a_1 + 3d = 7 + 6 = 13$

$\qquad\qquad c_4 = a_1 + 4d = 7 + 8 = 15$

$\qquad\qquad c_5 = a_1 + 5d = 7 + 10 = 17$

EXERCISE SET 12.3, page 612

6. $\dfrac{a_{n+1}}{a_n} = \dfrac{(-1)^n e^{(n+1)x}}{(-1)^{n-1} e^{nx}} = -e^{(n+1)x - nx} = -e^x$

Because x is a constant, $-e^x$ is a constant and the sequence is a geometric sequence.

18. $\dfrac{a_2}{a_1} = \dfrac{6}{8} = \dfrac{3}{4} = r$

$\qquad a_n = a_1 r^{n-1}$

$\qquad a_n = 8\left(\dfrac{3}{4}\right)^{n-1}$

40. $r = \dfrac{4}{3}, \ a_1 = \dfrac{4}{3}, \ n = 14$

$\qquad S_n = \dfrac{a_1(1 - r^n)}{1 - r}$

$\qquad S_{14} = \dfrac{\dfrac{4}{3}\left(1 - \left(\dfrac{4}{3}\right)^{14}\right)}{1 - \dfrac{4}{3}} \approx 220.49$

62. $0.3\overline{95} = \dfrac{3}{10} + \dfrac{95}{1000} + \dfrac{95}{100{,}000} + \cdots = \dfrac{3}{10} + \dfrac{95/1000}{1 - 1/100}$

$\qquad = \dfrac{3}{10} + \dfrac{95}{990} = \dfrac{392}{990} = \dfrac{196}{495}$

70. $P = \dfrac{A[(1 + r)^m - 1]}{r}; \quad A = 250, r = 0.08/12, m = 12(4)$

$P = \dfrac{250[(1 + 0.08/12)^{48} - 1]}{0.08/12} \approx 14087.48$

EXERCISE SET 12.4, page 619

8. $S_n = 2 + 6 + 12 + \cdots + n(n + 1) = \dfrac{n(n + 1)(n + 2)}{3}$

when $n = 1$, $S_1 = 1(1 + 1) = 2$; $\dfrac{1(1 + 1)(1 + 2)}{3} = 2$

Therefore the statement is true for $n = 1$.
Assume true for $n = k$.

$S_k = 2 + 6 + 12 + \cdots + k(k + 1)$

$= \dfrac{k(k + 1)(k + 2)}{3}$ Induction Hypothesis

Prove the statement is true for $n = k + 1$. That is, prove

$S_{k+1} = \dfrac{(k + 1)(k + 2)(k + 3)}{3}.$

Since $a_k = k(k + 1)$, $a_{k+1} = (k + 1)(k + 2)$

$S_{k+1} = S_k + a_{k+1} = \dfrac{k(k + 1)(k + 2)}{3} + (k + 1)(k + 2)$

$= \dfrac{k(k + 1)(k + 2) + 3(k + 1)(k + 2)}{3}$

$= \dfrac{(k + 1)[k(k + 2) + 3(k + 2)]}{3}$

$= \dfrac{(k + 1)[k^2 + 2k + 3k + 6]}{3}$

$= \dfrac{(k + 1)(k^2 + 5k + 6)}{3} = \dfrac{(k + 1)(k + 2)(k + 3)}{3}$

12. $P_n = \left(1 - \dfrac{1}{2}\right)\left(1 - \dfrac{1}{3}\right)\cdots\left(1 - \dfrac{1}{n + 1}\right) = \dfrac{1}{n + 1}$

Let $n = 1$, then $P_1 = \left(1 - \dfrac{1}{2}\right) = \dfrac{1}{2}$; $\dfrac{1}{1 + 1} = \dfrac{1}{2}$

The statement is true for $n = 1$.
Assume the statement is true for $n = k$.

$P_k = \left(1 - \dfrac{1}{2}\right)\left(1 - \dfrac{1}{3}\right)\cdots\left(1 - \dfrac{1}{k + 1}\right) = \dfrac{1}{k + 1}$

Prove the statement is true for $n = k + 1$. That is, prove

$P_{k+1} = \left(1 - \dfrac{1}{2}\right)\left(1 - \dfrac{1}{3}\right)\cdots\left(1 - \dfrac{1}{k + 1}\right)\left(1 - \dfrac{1}{k + 2}\right)$

$= \dfrac{1}{k + 2}$

Since $a_k = \left(1 - \dfrac{1}{k + 1}\right)$, $a_{k+1} = \left(1 - \dfrac{1}{k + 2}\right)$,

$P_{k+1} = P_k \cdot a_{k+1} = \dfrac{1}{k + 1} \cdot \left(1 - \dfrac{1}{k + 2}\right)$

$= \dfrac{1}{k + 1} \cdot \dfrac{k + 1}{k + 2} = \dfrac{1}{k + 2}$

16. If $a > 1$, show that $a^{n+1} > a^n$ for all positive integers.
Because $a > 1$, $a \cdot a > a \cdot 1$ or $a^2 > a$. Thus the statement is true when $n = 1$.
Assume the statement is true for $n = k$:

$a^{k+1} > a^k$ Induction Hypothesis

Prove the statement is true for $n = k + 1$. That is, prove

$a^{k+2} > a^{k+1}.$

Because $a^{k + 1} > a^k$ and $a > 0$,

$a(a^{k+1}) > a(a^k)$

$a^{k+2} > a^{k+1}$

20. 1. Let $n = 1$. Since $\log_{10} 1 = 0$

$\log_{10} 1 < 1$

The inequality is true for $n = 1$.

2. Assume $\log_{10} k < k$ is true for some positive integer k (Induction Hypothesis).
Prove the inequality is true for $n = k + 1$. That is, prove $\log_{10}(k + 1) < k + 1$ is true when $n = k + 1$.

$\log_{10}(k + 1) \leq \log_{10}(k + k) = \log_{10} 2k = \log_{10} 2 + \log_{10} k < 1 + k$

Thus $\log_{10}(k + 1) < k + 1$. By the Principle of Mathematical Induction, $\log_{10} n < n$ for all positive integers n.

EXERCISE SET 12.5, page 625

4. $\dbinom{10}{5} = \dfrac{10!}{5!\,5!} = \dfrac{10 \cdot 9 \cdot 8 \cdot 7 \cdot 6 \cdot 5!}{5!\,5!} = \dfrac{10 \cdot 9 \cdot 8 \cdot 7 \cdot 6}{5 \cdot 4 \cdot 3 \cdot 2 \cdot 1}$

$= 252$

18. $(3x + 2y)^4 = (3x)^4 + 4(3x)^3(2y) + 6(3x)^2(2y)^2$

$+ 4(3x)(2y)^3 + (2y)^4$

$= 81x^4 + 216x^3y + 216x^2y^2 + 96xy^3 + 16y^4$

20. $(2x - \sqrt{y})^7 = \binom{7}{0}(2x)^7 + \binom{7}{1}(2x)^6(-\sqrt{y})$

$\quad + \binom{7}{2}(2x)^5(-\sqrt{y})^2 + \binom{7}{3}(2x)^4(-\sqrt{y})^3$

$\quad + \binom{7}{4}(2x)^3(-\sqrt{y})^4 + \binom{7}{5}(2x)^2(-\sqrt{y})^5$

$\quad + \binom{7}{6}(2x)(-\sqrt{y})^6 + \binom{7}{7}(-\sqrt{y})^7$

$\quad = 128x^7 - 448x^6\sqrt{y} + 672x^5y - 560x^4y\sqrt{y}$

$\quad + 280x^3y^2 - 84x^2y^2\sqrt{y}$

$\quad + 14xy^3 - y^3\sqrt{y}$

34. $\binom{10}{6-1}(x^{-1/2})^{10-6+1}(x^{1/2})^{6-1} = \binom{10}{5}(x^{-1/2})^5(x^{1/2})^5 = 252$

EXERCISE SET 12.6, page 631

12. Since there are 4 palettes and each palette contains 4 colors, by the counting principle there are $4 \cdot 4 \cdot 4 \cdot 4 = 256$ possible colors.

16. There are three possible finishes, first, second, and third, for the 12 contestants. Because the order of finish is important, these are the permutations of the 12 contestants selected 3 at a time.

$$P(12, 3) = \frac{12!}{(12-3)!} = \frac{12!}{9!} = 12 \cdot 11 \cdot 10 = 1320$$

There are 1320 possible finishes.

20. Player A matched against Player B is the same tennis match as Player B matched against Player A. Therefore, this is a combination of 26 players selected 2 at a time.

$$C(26, 2) = \frac{26!}{2!(26-2)!} = \frac{26!}{2! \, 24!} = \frac{26 \cdot 25 \cdot 24!}{2 \cdot 1 \cdot 24!} = 325$$

There are 325 possible first-round matches.

22. Let one seat in the front and one seat in the back be occupied by the passengers who wanted a front or back seat. This leaves 4 remaining seats which can be filled by the 4 remaining passengers. Seating arrangements are permutations. The number of seating arrangements is

$$P(4, 4) = \frac{4!}{(4-4)!} = \frac{4!}{0!} = \frac{4!}{1} = 24$$

There are 24 seating arrangements.

30. a. The number of ways 10 finalists can be selected from 15 semifinalists is the combinations of 15 students selected 10 at a time.

$$\binom{15}{10} = 3003$$

There are 3003 ways the finalists can be chosen.

b. The number of ways the 10 finalists can contain three seniors is the product of the combination of 7 seniors selected 3 at a time and the combination of 8 remaining students selected 7 at a time.

$$\binom{7}{3}\binom{8}{7} = 35 \cdot 8 = 280$$

There are 280 ways the finalists can contain 3 seniors.

c. At least five seniors means 5 or 6 or 7 seniors are finalists (there are only 7 seniors). Since the events are related by 'or', sum the number of ways each event can occur.

$$\binom{7}{5}\binom{8}{5} + \binom{7}{6}\binom{8}{4} + \binom{7}{7}\binom{8}{3}$$

$$= 21 \cdot 56 + 7 \cdot 70 + 1 \cdot 56 = 1176 + 490 + 56 = 1722$$

There are 1722 ways the finalists can contain at least 5 seniors.

EXERCISE SET 12.7, page 639

6. {(R, D), (R, I), (D, I)}

14. {HHHT, HHTH, HTHH, THHH, HHHH}

22. Let $E = \{2, 4, 6\}$, $T = \{3, 6\}$, $S = \{1, 2, 3, 4, 5, 6\}$.

$$E \cup T = \{2, 3, 4, 6\}$$

$$P(E \cup T) = \frac{N(E \cup T)}{N(S)} = \frac{4}{6} = \frac{2}{3}$$

28. Let A and B be the radar stations and let D indicate detection and N not detected. Here are the possibilities.

A	B	probability
D	D	(0.95) (0.95)
D	N	(0.95) (0.05)
N	D	(0.05) (0.95)

The probability of detection is the sum of these probabilities.

Probability of detection $= 0.9025 + 0.0475 + 0.0475$

$$= 0.9975$$

32. a. $\dfrac{C(2, 2)}{C(10, 2)} = \dfrac{1}{45}$

b. $\dfrac{C(2, 1)C(8, 1)}{C(10, 2)} + \dfrac{C(2, 2)C(8, 0)}{C(10, 2)} = \dfrac{16}{45} + \dfrac{1}{45} = \dfrac{17}{45}$

Answers to Odd-Numbered Exercises

EXERCISE SET 1.1, page 8

1. a. $-3, 4, 11, 57$ are integers   **b.** $-3, 4, \frac{1}{5}, 11, 3.14, 57$ are rational numbers   **c.** $0.25225222522225\ldots$ is an irrational number   **d.** All are real numbers.   **e.** 11 is a prime number   **f.** 4, 57 are composite numbers   **3.** $\{1, 3\}$ **5.** $\{1, 3\}$ **7.** $\{0, 2, 4\}$
9. $\{0, 1, 2, 3, 4, 5, 11\}$ **11.** $\{1, 3, 5, 6, 10, 11\}$ **13.** $\{0, 1, 2, 3, 4, 6, 8, 10\}$ **15.** Associative property of addition
17. Identity property of multiplication   **19.** Reflexive property of equality   **21.** Transitive property of equality
23. Inverse property of addition   **25.** Commutative property of addition   **27.** Transitive property of equality   **29.** $-3a/7$
31. $-7a/20$ **33.** $-10/21$ **35.** $-18/5$ **37.** $-2a/15$ **39. a.** 26/55 of the pool   **b.** 26x/165 of the pool   **41.** 8/59
43. $2 - 1 \neq 1 - 2$ **45.** $(3 - 1) - 5 = -3; 3 - (1 - 5) = 7$ **47.** false   **49.** true   **51.** false   **53.** true   **55.** true
57. a. $0.\overline{72}$ **b.** 0.825 **c.** $0.\overline{285714}$ **d.** $0.1\overline{35}$ **59.** $0.16620626, 0.16662040, 0.16666204, 0.16667, \dfrac{\sqrt{x+9}-3}{x}$ is approaching $0.1\overline{6} = 1/6$.
61. $A = \{4, 6, 8, 9, 10\}$ **63.** $C = \{53, 59\}$ **65.** All properties except for the identity property of addition and inverse properties
67. all of the properties
69. a. $T > L$ and L is the largest prime, so T can't be prime.   **b.** Since the prime numbers $2, 3, 5, 7, \ldots L$ all are factors of $T - 1$, none can be factors of T, so T can't be a composite number.
71. The Theorem states that a number of the form $1111\ldots1$ *can* be prime only if the number of 1's is prime. It does not say that it will be a prime number.

EXERCISE SET 1.2, page 14

1. $-4 \quad -2 \quad \frac{7}{4} \quad 2.5$ **3.** $5/2 < 4$ **5.** $2/3 > 0.6666$ **7.** $1.75 < 2.23$ **9.** $0.\overline{36} = \dfrac{4}{11}$ **11.** $\dfrac{10}{5} = 2$ **13.** $\pi > 3.14159$

15. $(3, 5)$ **17.** $(-\infty, 3)$ **19.** $[0, 3)$ **21.** $(-\infty, -3) \cup [2, \infty)$

23. $(3, 4)$ **25.** $(-1, 3]$ **27.** $-4 \leq x \leq 1$ **29.** $1 < x < 5$

31. $x \geq 2.5$ **33.** $x < 2$ **35.** $x \leq 2$ or $x > 3$ **37.** $x < 3$ or $x > 3$

39. $(-\infty, 3]$ **41.** $-1 < x \leq 2, (-1, 2]$ **43.** $1 \leq x \leq 4$ **45.** $-2 \leq x < \pi$

47. 4   **49.** 27.4   **51.** -11 **53.** $y^2 + 10$ **55.** $1 + \pi$ **57.** 9   **59.** $x + 7$ **61.** 7   **63.** 8   **65.** 50   **67.** 33   **69.** $\pi + 3$
71. $\dfrac{5}{12}$ **73.** $|x - 5| < 4$ **75.** $|a - 2|$ **77.** $|m - n|$ or $|n - m|$ **79.** $|z - 5| > 4$ **81.** $(-\infty, 3) \cup (3, \infty)$ **83.** $(-3, 3)$ **85.** False
87. True   **91.** $I \leq 120$ **93.** $2 \leq A < 3$ **95.** $|x - 2| < |x - 6|$ **97.** $|x - 3| > |x + 7|$ **99.** $2 < |x - 4| < 7$ **101.** $|x - a| < \delta$

EXERCISE SET 1.3, page 21

1. -64 **3.** -64 **5.** 1   **7.** 1   **9.** 243   **11.** $1/27$ **13.** 32   **15.** 64   **17.** $125/63$ **19.** $81/16$ **21.** $4/2025$ **23.** $32/5$
25. 1   **27.** $6x^7y^4$ **29.** $8c^9/125$ **31.** $36x^4/y^4$ **33.** $2y^2/9x$ **35.** $y/3x$ **37.** y^6/x^4 **39.** $(b^2 + a)/(ab^2)$ **41.** $b/(ac^2)$ **43.** 16
45. 1   **47.** 7.34×10^1 **49.** 1.9×10^6 **51.** 1.63×10^{11} **53.** 3.2×10^{-5} **55.** 7×10^{-3} **57.** 8.21×10^{-8} **59.** 6500
61. 0.0000731   **63.** 80,000,000,000   **65.** 0.000217   **67.** 100,000,000,000   **69.** 3.75   **71.** three   **73.** six   **75.** one   **77.** three
79. two   **81.** four   **83.** 2.158924997   **85. a.** $2^8 = 256, 2^{16} = 65{,}536$ **87.** 1.97×10^4 seconds   **89.** \$4873.50   **91.** x^3y^{n+1}
93. $x^{4n-2}y^4$ **95.** y^{n+1}/x^{5n} **97.** $3^{(3^3)} > (3^3)^3$ **99.** $\dfrac{V_c}{V_s} = \dfrac{3}{2}$

EXERCISE SET 1.4, page 27

1. D   **3.** H   **5.** G   **7.** B   **9.** J   **11. a.** $x^2 + 2x - 7$ **b.** 2 **c.** 1, 2, -7 **d.** 1 **e.** $x^2, 2x, -7$
13. a. $x^3 - 1$ **b.** 3 **c.** 1, -1 **d.** 1 **e.** $x^3, -1$ **15. a.** $2x^4 + 3x^3 + 4x^2 + 5$ **b.** 4 **c.** 2, 3, 4, 5 **d.** 2 **e.** $2x^4, 3x^3, 4x^2, 5$ **17.** 3
19. 5   **21.** 2   **23.** $5x^2 + 11x + 3$ **25.** $9w^3 + 8w^2 - 2w + 6$ **27.** $-2r^2 + 3r - 12$ **29.** $-3u^2 - 2u + 4$ **31.** $8x^3 + 18x^2 - 67x + 40$
33. $6x^4 - 19x^3 + 26x^2 - 29x + 10$ **35.** $10x^2 + 22x + 4$ **37.** $y^2 + 3y + 2$ **39.** $4z^2 - 19z + 12$ **41.** $a^2 + 3a - 18$ **43.** $b^2 + 2b - 24$
45. $10x^2 - 57xy + 77y^2$ **47.** $18x^2 + 55xy + 25y^2$ **49.** $12w^2 - 40wx + 33x^2$ **51.** $6p^2 - 11pq - 35q^2$ **53.** $12d^2 + 4d - 8$ **55.** $r^3 + s^3$
57. $60c^3 - 49c^2 + 4$ **59.** $9x^2 - 25$ **61.** $9x^4 - 6x^2y + y^2$ **63.** $16w^2 + 8wz + z^2$ **65.** $x^2 - 4x + 4 + 2xy - 4y + y^2$
67. $x^2 + 10x + 25 - y^2$ **69.** 29   **71.** -17 **73.** -1 **75.** 33   **77. a.** 48.46 **b.** 51.44   **79 a.** 10.994998 **b.** 10.99949998
81. 11,175 matches   **83.** 14.8, 90.4   **85.** $a^3 + 3a^2b + 3ab^2 + b^3$ **87.** $x^3 - 3x^2 + 3x - 1$ **89.** $8x^3 - 36x^2y + 54xy^2 - 27y^3$
91. a. n **b.** n **c.** n **d.** no degree   **93. a.** 41, 43, 47, 53 **b.** $n = 41$ **95. a.** $x^5 - 1$, **b.** $x^6 - 1$

EXERCISE SET 1.5, page 37

1. $5(x + 4)$ **3.** $-3x(5x + 4)$ **5.** $2xy(5x + 3 - 7y)$ **7.** $(x - 3)(2a + 4b)$ **9.** $(3x + 1)(x^2 + 2)$ **11.** $(x - 1)(ax + b)$
13. $(3w + 2)(2w^2 - 5)$ **15.** $(x + 3)(x + 4)$ **17.** $(a - 12)(a + 2)$ **19.** $(6x + 1)(x + 4)$ **21.** $(17x + 4)(3x - 1)$
23. $(3x + 8y)(2x - 5y)$ **25.** $(x^2 + 5)(x^2 + 1)$ **27.** $(6x^2 + 5)(x^2 + 3)$ **29.** factorable   **31.** not factorable   **33.** not factorable
35. $(x - 3)(x + 3)$ **37.** $(2a - 7)(2a + 7)$ **39.** $(1 - 10x)(1 + 10x)$ **41.** $(x^2 - 3)(x^2 + 3)$ **43.** $(x + 3)(x + 7)$ **45.** $(x + 5)^2$
47. $(a - 7)^2$ **49.** $(2x + 3)^2$ **51.** $(z^2 + 2w^2)^2$ **53.** $(x - 2)(x^2 + 2x + 4)$ **55.** $(2x - 3y)(4x^2 + 6xy + 9y^2)$
57. $(2 - x^2)(4 + 2x^2 + x^4)$ **59.** $(x - 3)(x^2 - 3x + 3)$ **61.** $2(3x - 1)(3x + 1)$ **63.** $(2x - 1)(2x + 1)(4x^2 + 1)$
65. $a(3x - 2y)(4x - 5y)$ **67.** $b(3x + 4)(x - 1)(x + 1)$ **69.** $2b(6x + y)^2$ **71.** $(w - 3)(w^2 - 12w + 39)$ **73.** $(x + 3y - 1)(x + 3y + 1)$
75. not factorable   **77.** $(2x - 5)^2(3x + 5)$ **79.** $(2x - y)(2x + y + 1)$ **81.** 8   **83.** 64   **85.** $(x^n - 1)(x^n + 1)(x^{2n} + 1)$
87. $\pi(R - r)(R + r)$ **89.** $r^2(4 - \pi)$ **91. a.** I + II + III **b.** II + III + V **c.** I + II + III = II + III + V

EXERCISE SET 1.6, page 45

1. $\dfrac{x + 4}{3}$ **3.** $\dfrac{x - 3}{x - 2}$ **5.** $\dfrac{a^2 - 2a + 4}{a - 2}$ **7.** $-\dfrac{x + 8}{x + 2}$ **9.** $-\dfrac{4y^2 + 7}{y + 7}$ **11.** $-\dfrac{8}{a^3b}$ **13.** $\dfrac{10}{27q^2}$ **15.** $\dfrac{x(3x + 7)}{2x + 3}$ **17.** $\dfrac{x + 3}{2x + 3}$

19. $\dfrac{(2y + 3)(3y - 4)}{(2y - 3)(y + 1)}$ **21.** $\dfrac{1}{a - 8}$ **23.** $\dfrac{3p - 2}{r}$ **25.** $\dfrac{8x(x - 4)}{(x - 5)(x + 3)}$ **27.** $\dfrac{3y - 4}{y + 4}$ **29.** $\dfrac{7z(2z - 5)}{(2z - 3)(z - 5)}$ **31.** $\dfrac{-2x^2 + 14x - 3}{(x - 3)(x + 3)(x + 4)}$

33. $\dfrac{(2x - 1)(x + 5)}{x(x - 5)}$ **35.** $\dfrac{-q^2 + 12q + 5}{(q - 3)(q + 5)}$ **37.** $\dfrac{3x^2 - 7x - 13}{(x + 3)(x + 4)(x - 3)(x - 4)}$ **39.** $\dfrac{(x + 2)(3x - 1)}{x^2}$ **41.** $\dfrac{4x + 1}{x - 1}$ **43.** $\dfrac{x - 2y}{y(y - x)}$

45. $\dfrac{(5x + 9)(x + 3)}{(x + 2)(4x + 3)}$ **47.** $\dfrac{(b + 3)(b - 1)}{(b - 2)(b + 2)}$ **49.** $\dfrac{x - 1}{x}$ **51.** $2 - m^2$ **53.** $\dfrac{-x^2 + 5x + 1}{x^2}$ **55.** $\dfrac{-x - 7}{x^2 + 6x - 3}$ **57.** $\dfrac{2x - 3}{x + 3}$

59. $\dfrac{a + b}{ab(a - b)}$ **61.** $\dfrac{(b - a)(b + a)}{ab(a^2 + b^2)}$ **63. a.** 136.55 mph **b.** $\dfrac{2v_1v_2}{v_1 + v_2}$ **65.** $\dfrac{2x + 1}{x(x + 1)}$ **67.** $\dfrac{3x^2 - 4}{x(x - 2)(x + 2)}$ **69.** $\dfrac{x^2 + 9x + 25}{(x + 5)^2}$

71. $\dfrac{x(1 - 4xy)}{(1 - 2xy)(1 + 2xy)}$ **73.** $R\left[\dfrac{(1 + i)^n - 1}{i(1 + i)^n}\right]$ **75.** 0.5   **77.** 0.6   **79.** 0.618

EXERCISE SET 1.7, page 54

1. 3   **3.** -3 **5.** 8   **7.** -16 **9.** 16   **11.** 1/4   **13.** 1/9   **15.** 3/4   **17.** 8/125   **19.** 100   **21.** 7   **23.** 5   **25.** $x^{11/10}$
27. $4a^2$ **29.** $3xy^3$ **31.** $a^{1/2}b^{3/10}$ **33.** $a^2 + 7a$ **35.** $p - q$ **37.** $m^2n^{3/2}$ **39.** 1   **41.** $r^{(m-n)/(nm)}$ **43.** 2   **45.** -6 **47.** 3/4   **49.** 2   **51.** 2   **53.** 6   **55.** 49   **57.** $\sqrt{3x}$
59. $5\sqrt[4]{xy}$ **61.** $\sqrt[3]{25w^2}$ **63.** $(17k)^{1/3}$ **65.** $a^{2/5}$ **67.** $\left(\dfrac{7a}{3}\right)^{1/2}$ **69.** $3\sqrt{5}$ **71.** $2\sqrt[3]{3}$ **73.** $-3\sqrt[3]{3}$ **75.** $-2\sqrt[3]{4}$ **77.** $15\sqrt{5}$
79. $10\sqrt[3]{35}$ **81.** $2xy\sqrt{6x}$ **83.** $-2ay^2\sqrt[3]{2y}$ **85.** $\sqrt{3x}$ **87.** $\sqrt[3]{4m^2n}$ **89.** $\sqrt[4]{9x^3y}$ **91.** $7\sqrt{2}$ **93.** $6\sqrt{3}$ **95.** $-3\sqrt{2}$ **97.** 0
99. $3x\sqrt{2xy}$ **101.** $29 + 11\sqrt{5}$ **103.** $2x - 9$ **105.** $50y + 10\sqrt{6yz} + 3z$ **107.** $x + 10\sqrt{x - 3} + 22$ **109.** $2x + 14\sqrt{2x + 5} + 54$
111. $\sqrt{2}$ **113.** $\sqrt{10}/6$ **115.** $3\sqrt[3]{4}/2$ **117.** $2\sqrt[3]{x}/x$ **119.** $\sqrt{5}/3$ **121.** $\dfrac{\sqrt{6xy}}{9y}$ **123.** $\dfrac{3\sqrt{5} - 3\sqrt{x}}{5 - x}$ **125.** $-\dfrac{2\sqrt{7} + 7}{3}$
127. $\dfrac{5}{3\sqrt{5}}$ **129.** $\dfrac{4}{21 - 7\sqrt{5}}$ **131.** $\dfrac{a^2\sqrt{30}}{3}$ **133.** $\dfrac{3\sqrt{2xy}}{2y}$ **135.** $\dfrac{\sqrt{30y}}{2y}$ **137.** $\dfrac{\sqrt[3]{12x^2y}}{x}$ **139.** 1.52   **141.** -4.44
143. 20.9   **145.** 279   **147.** -6.31 **149. a.** 81% **b.** 66%   **151.** 8/5   **153.** $-19/12$ **155.** $(x - \sqrt{7})(x + \sqrt{7})$
157. $(x + 3\sqrt{2})^2$ **159.** $\sqrt{3^2 + 4^2} \neq 3 + 4$ **161.** Step VI

EXERCISE SET 1.8, page 61

1. $2 + 3i$ **3.** $4 - 11i$ **5.** $8 + i\sqrt{3}$ **7.** $7 + 4i$ **9.** $9i$ **11.** $5 + 12i$ **13.** $4 - 3i$ **15.** $-2 - 5i$ **17.** $12 - 2i$

19. $-2 + 11i$ **21.** $16 + 16i$ **23.** $23 + 2i$ **25.** 74   **27.** $-117 - i$ **29.** $\dfrac{1}{2} - \dfrac{1}{2}i$ **31.** $\dfrac{7}{58} + \dfrac{3}{58}i$ **33.** $\dfrac{5}{13} + \dfrac{12}{13}i$

35. $\dfrac{1}{61} + \dfrac{11}{61}i$ **37.** $1 - 6i$ **39.** $-16 - 30i$ **41.** $-29 - 17i$ **43.** $-2 - 2i$ **45.** -16 **47.** $75i$ **49.** $-i$ **51.** i

53. -1 **55.** -1 **57.** $-i$ **59.** i **61.** -1 **63.** $-i$ **65.** -2 **67.** $-8\sqrt{5}$ **69.** 11 **71.** $9 + 40i$ **73.** $\dfrac{1}{2} \pm \dfrac{\sqrt{3}}{2}i$

75. $-1 \pm i$ **77.** $-\dfrac{3}{2} \pm \dfrac{\sqrt{3}}{2}i$ **79.** $-\dfrac{1}{4} \pm \dfrac{\sqrt{23}}{4}i$ **81.** 5 **83.** $\sqrt{29}$ **85.** $\sqrt{65}$ **87.** 3 **95.** no   **97.** $66 + 6\sqrt{5} - 14\sqrt{3}$

99. 1 **101.** 0 **103.** $(x + 3i)(x - 3i)$ **105.** $(2x + 9i)(2x - 9i)$ **107.** 0 **109.** 0

CHAPTER 1 TRUE/FALSE EXERCISES, page 66

1. true   **2.** False, If $a = \frac{1}{2}$, then $(\frac{1}{2})^2 = \frac{1}{4} < \frac{1}{2}$.   **3.** True   **4.** False, $\sqrt{2} + (-\sqrt{2}) = 0$ which is a rational number
5. False, $(2 \oplus 4) \oplus 6 \neq 2 \oplus (4 \oplus 6)$ **6.** False, $x > a$ is written as (a, ∞) **7.** False, $\sqrt{(-2)^2} \neq -2$.   **8.** True   **9.** True   **10.** True

CHAPTER 1 REVIEW EXERCISES, page 66

1. integer, rational number, real number, prime number   **3.** rational number, real number   **5.** $\{1, 2, 3, 5, 7, 11\}$
7. distributive property   **9.** Associative property of multiplication   **11.** Identity property of addition
13. symmetric property of equality   **15.** $(-4, 2]$ **17.** $-3 \le x < 2$ **19.** 7 **21.** $4 - \pi$ **23.** 17

25. -36 **27.** $12x^8y^3$ **29.** 6.2×10^5 **31.** $35{,}000$ **33.** $-a^2 - 2a - 1$ **35.** $6x^4 + 5x^3 - 13x^2 + 22x - 20$ **37.** $3(x + 5)^2$
39. $4(5a^2 - b^2)$ **41.** $\dfrac{3x - 2}{x + 4}$ **43.** $\dfrac{2x + 3}{2x - 5}$ **45.** $\dfrac{x(3x + 10)}{(x + 3)(x - 3)(x + 4)}$ **47.** $\dfrac{2x - 9}{3x - 17}$ **49.** 5 **51.** $x^{17/12}$ **53.** $x^{3/4}y^2$
55. $4ab^3\sqrt{3b}$ **57.** $6x\sqrt{2y}$ **59.** $\dfrac{3y\sqrt{15y}}{5}$ **61.** $\dfrac{7\sqrt[3]{4x}}{2}$ **63.** $-3y^2\sqrt[3]{5x^2y}$ **65.** $3 - 8i$, conjugate $3 + 8i$ **67.** $5 + 2i$
69. $25 - 19i$ **71.** 1 **73.** 1

CHAPTER 1 TEST, page 67

1. The distributive property   **2.** $\{0, 1, 2, 3, 4, 5, 6, 7, 8, 9\}$ **3.** -3 **4.** 7 **5.** $\dfrac{4}{9x^4y^2}$ **6.** $\dfrac{96bc^2}{a^5}$ **7.** 1.37×10^{-3}
8. $x^3 - 2x^2 + 5xy - 2x^2y - 2y^2$ **9.** -94 **10.** $(7x - 1)(x + 5)$ **11.** $(a - 4b)(3x - 2)$ **12.** $2x(2x - y)(4x^2 + 2xy + y^2)$
13. $(x + y)(x - 1)(x^2 + x + 1)$ **14.** $\dfrac{(x - 2)(x^2 + x + 1)}{(x + 1)(x - 1)}$ **15.** $\dfrac{(x - 6)(x + 1)}{(x + 3)(x - 2)(x - 3)}$ **16.** $\dfrac{x(x + 2)}{x - 3}$ **17.** $\dfrac{3a^2 - 3ab - 10a + 5b}{a(2a - b)}$
18. $\dfrac{x(2x - 1)}{2x + 1}$ **19.** $\dfrac{x^{5/6}}{y^{9/4}}$ **20.** $7xy\sqrt[3]{3xy}$ **21.** $\dfrac{\sqrt[4]{8x}}{2}$ **22.** $\dfrac{3\sqrt{x} - 6}{x - 4}$ **23.** $2 + 11i$ **24.** $11 - 10i$ **25.** $\dfrac{6}{13} - \dfrac{4}{13}i$

EXERCISE SET 2.1, page 74

1. 15 **3.** -4 **5.** $9/2$ **7.** $108/23$ **9.** $2/9$ **11.** 12 **13.** 16 **15.** 9 **17.** 75 **19.** $1/2$ **21.** $22/13$ **23.** $95/18$ **25.** 1200
27. 2500 **29.** identity   **31.** conditional equation   **33.** contradiction   **35.** contradiction   **37.** 31 **39.** 2 **41.** no solution

43. no solution   **45.** $7/2$ **47.** 6 **49.** -4 **51.** -12 **53.** 1 **55.** no solution   **57.** $x = -\dfrac{3}{2}y + 3$ **59.** $x = \dfrac{d - 5c}{2c - 5}, c \neq 5/2$

61. $x = \dfrac{ab - ay}{b}, b \neq 0$ **63.** $x = \dfrac{y - y_1 + mx_1}{m}, m \neq 0$ **65.** $x = \dfrac{3 - 4l - 4w}{l + w}, l \neq -w, x \neq -4$ **67.** $x = m + n, m \neq n$ **69.** 5.47

71. no solution   **73.** 0.000879 **75.** no   **77.** yes   **81.** $\dfrac{10\sqrt{7}}{7}$ **83.** $-\dfrac{7\sqrt{3}}{6}$ **85.** $\dfrac{1}{a - b}$

EXERCISE SET 2.2, page 82

1. $h = \dfrac{3V}{\pi r^2}$ **3.** $t = \dfrac{I}{Pr}$ **5.** $m_1 = \dfrac{Fd^2}{Gm_2}$ **7.** $v_0 = \dfrac{s + 16t^2}{t}$ **9.** $T_w = \dfrac{-Q_w + m_w c_w T_f}{m_w c_w}$ **11.** $d = \dfrac{a_n - a_1}{n - 1}$ **13.** $r = \dfrac{S - a_1}{S}$

15. $f_1 = \dfrac{w_1 f - w_2 f_2 + w_2 f}{w_1}$ **17.** $v_{LC} = \dfrac{f_{LC}v - f_v v}{f_v}$ **19.** 100 **21.** 30 feet by 57 feet   **23.** 12 cm, 36 cm, 36 cm   **25.** $872, 873$
27. $18, 20$ **29.** 240 meters   **31.** 2 hours   **33.** $\sqrt{52}$ **35.** 98 **37.** 850 **39.** $\$937.50$
41. $\$7600$ invested at 8%, $\$6400$ invested at 6.5%   **43.** $\$3750$ **45.** $18\frac{2}{11}$ grams   **47.** 64 liters   **49.** 1200 at $\$14$ and 1800 at $\$25$
51. $6\frac{2}{3}$ lb of the $\$12$ coffee and $13\frac{1}{3}$ lb of the $\$9$ coffee   **53.** 10 grams   **55.** 7.875 hours   **57.** $13\frac{1}{3}$ hours
59. $\$10.05$ for book, $\$0.05$ for bookmark   **61.** 6.25 feet   **63.** 40 lb   **65.** $9\frac{3}{7}$ hours   **67.** 1384 feet   **69.** $\frac{2}{9}$ hour   **71.** 84 years old

EXERCISE SET 2.3, page 93

1. 5, −3   **3.** −24, 3/8   **5.** 0, 7/3   **7.** 4/3, −2/5   **9.** 1/2, −4   **11.** 8, 2   **13.** 3, 8/3   **15.** 7   **17.** ±9   **19.** ±2√6

21. ±2i   **23.** 11, −1   **25.** 7/2   **27.** −1/2   **29.** −3 ± 2√2   **31.** 5, −3   **33.** 0, −10   **35.** $\dfrac{-3 \pm \sqrt{13}}{2}$ **37.** $\dfrac{-2 \pm \sqrt{6}}{2}$

39. $\dfrac{4 \pm \sqrt{13}}{3}$ **41.** $\dfrac{-3 \pm 2\sqrt{6}}{3}$ **43.** −3, 5   **45.** $\dfrac{-1 \pm \sqrt{5}}{2}$ **47.** $\dfrac{-2 \pm \sqrt{2}}{2}$ **49.** $\dfrac{5 \pm i\sqrt{11}}{6}$ **51.** $\dfrac{-3 \pm \sqrt{41}}{4}$

53. $-\dfrac{\sqrt{2}}{2}, -\sqrt{2}$ **55.** $\dfrac{3 \pm i\sqrt{11}}{2}$ **57.** 81, 2 distinct real numbers   **59.** −116, 2 distinct nonreal numbers

61. 0, 1 real number   **63.** ±24   **65.** 49/4   **67.** 76.4 in.   **69.** 26.8 cm.   **71.** 10 cm by 3.5 cm   **73.** 100 feet by 150 feet

75. 10, 12   **77.** $\dfrac{5 + \sqrt{29}}{2}$ **79.** 35 mph for the first part and 45 mph for the last part   **81.** 12 hours   **83.** 2.02 s   **85.** yes

87. yes   **89.** yes   **91.** $\dfrac{v_0 \pm \sqrt{v_0^2 - 2g(s - s_0)}}{g}$ **93.** $\dfrac{2 \pm \sqrt{4 + 3x}}{x}$ **95.** $-\dfrac{y}{6} \pm \dfrac{yi}{6}\sqrt{47}$ **97.** $\dfrac{-1 \pm \sqrt{33 + 4x}}{2}$

101. a. $l\left(\dfrac{-1 + \sqrt{5}}{2}\right)$ **b.** 62.4 ft   **103.** 22

EXERCISE SET 2.4, page 101

1. 0, ±5   **3.** 2, ±1   **5.** 0, ±3   **7.** 0, −5, 8   **9.** 0, ±4   **11.** 2, −1 ± i√3   **13.** 40   **15.** 3   **17.** 7   **19.** 7   **21.** 9
23. 5/2   **25.** 1, −6   **27.** 4   **29.** 1, 5   **31.** 2   **33.** 23, −31   **35.** 2, −1/8   **37.** 0, 1/256   **39.** −1, −59/3   **41.** ±√7, ±√2
43. ±2, ±√6/2   **45.** $\sqrt[3]{2}, -\sqrt[3]{3}$ **47.** $-\sqrt[3]{36}/3, \sqrt[3]{98}/7$ **49.** 1, 16   **51.** −1/27, 64   **53.** ±√15/3   **55.** ±1   **57.** 1/2, −1/5

59. 256/81, 16   **61.** ±0.62, ±1.62   **63.** ±0.34, ±2.98   **65.** $x = \pm\sqrt{9 - y^2}$ **67.** $x = y + 2\sqrt{yz} + z$ **69.** $x = \dfrac{7 - 2y^2}{2y}$ **71.** 9, 36

73. 3 inches   **75.** 10.5 mm   **77.** 87 feet   **79. a.** 8.93 in   **b.** 5√3 in   **81.** $s = \left(\dfrac{-275 + 5\sqrt{3025 + 176T}}{2}\right)^2$

EXERCISE SET 2.5, page 110

1. $x < 4$ **3.** $x < -6$ **5.** $x \le -3$ **7.** $x \ge -13/8$ **9.** $x < 2$ **11.** If you write more than 57 checks a month.
13. at least 34 sales   **15.** $-3/4 < x \le 4$ **17.** $11/3 \ge x \ge 1/3$ **19.** $11/4 > x \ge -3/8$ **21.** $20 \le C \le 40$
23. {12, 14, 16}, {14, 16, 18}   **25.** $(-\infty, -7) \cup (0, \infty)$ **27.** $[-4, 4]$ **29.** $(-5, -2)$ **31.** $(-\infty, -4] \cup [7, \infty)$ **33.** $[-1/2, 4/3]$
35. $(-\infty, -5/4] \cup [3/2, \infty)$ **37.** $(0, 210)$ **39.** $(-4, 1)$ **41.** $(-\infty, -8) \cup [5, \infty)$ **43.** $(-\infty, -7/2) \cup [0, \infty)$ **45.** $(-\infty, -1) \cup (2, 4)$
47. $(-\infty, 5) \cup [12, \infty)$ **49.** $(-2/3, 0) \cup (5/2, \infty)$ **51.** $(-\infty, 5)$ **53.** $[4, \infty)$ **55.** $[-9, \infty)$ **57.** $[-3, 3]$
59. $(-\infty, -4] \cup [4, \infty)$ **61.** $(-\infty, -3] \cup [5, \infty)$ **63.** $(-\infty, \infty)$ **65.** $(-\infty, 3) \cup (3, 6) \cup (6, \infty)$ **67.** $(-3, \infty)$ **69.** $[-\sqrt{3}, 0] \cup [\sqrt{3}, \infty)$
71. $(-\infty, \infty)$ **73.** $[-2, 0] \cup [5, \infty)$ **75.** $(-\infty, -2\sqrt{6}] \cup [2\sqrt{6}, \infty)$ **77.** $(-\infty, -2\sqrt{14}] \cup [2\sqrt{14}, \infty)$ **79.** $1 < t < 3$
81. a. $x > 4$ **b.** $0 < x < 1$ **c.** $x > \sqrt{2}$

EXERCISE SET 2.6, page 115

1. {−4, 4}   **3.** {7, 3}   **5.** {−5, −7}   **7.** {−34, 6}   **9.** {8, −3}   **11.** {2, −8}   **13.** {20, −12}   **15.** no solution   **17.** {12, −18}
19. $\left\{\dfrac{a + b}{2}, \dfrac{a - b}{2}\right\}$ **21.** {a + δ, a − δ}   **23.** (−4, 4)   **25.** (−8, 10)   **27.** $(-\infty, -33) \cup (27, \infty)$ **29.** $(-\infty, -3/2) \cup (5/2, \infty)$
31. $(-\infty, -8] \cup [2, \infty)$ **33.** $[-4/3, 8]$ **35.** $(-\infty, -4] \cup [28/5, \infty)$ **37.** $(-\infty, \infty)$ **39.** {4}   **41.** no solution   **43.** $(-\infty, \infty)$
45. (3 − b, 3 + b)   **47.** $(-\sqrt{2}, 0) \cup (0, \sqrt{2})$ **49.** $(-4, -2) \cup (2, 4)$ **51.** $(-\infty, -5] \cup [-4, -3] \cup [-2, \infty)$
53. $(-\infty, -\sqrt{26}] \cup [-\sqrt{17}, \sqrt{17}] \cup [\sqrt{26}, \infty)$ **55.** false, not true for $x < 0$.   **57.** true   **59.** true   **61.** $\{x \mid x \ge -4\}$
63. $\{x \mid x \le -7\}$ **65.** $\{x \mid x \ge -7/2\}$ **67.** $\{x \mid x = 3, -5\}$ **69.** $(-5, -1) \cup (1, 5)$ **71.** $(-7, -3] \cup [3, 7)$ **73.** $(a - δ, a) \cup (a, a + δ)$
75. $(2, 4) \cup (8, 10)$ **77.** $(-\infty, -20/3] \cup [28/3, \infty)$ **79.** $(1/2, \infty)$ **81. a.** $|x - 3| < 8$ **b.** $|x - j| < k$
83. a. $|s - 4.25| < 0.01$ **b.** $4.24 < s < 4.26$

CHAPTER 2 TRUE/FALSE EXERCISES, page 118

1. false, $(-3)^2 = 9$ **2.** False, one has a solution set {3}, the other has solution set {3, −4}   **3.** True   **4.** True
5. false, $100 > 1$ but $\frac{1}{100} \not> \frac{1}{1}$ **6.** false, the discriminant is $b^2 - 4ac$ **7.** false, $\sqrt{1} + \sqrt{1} = 1 + 1 = 2$ but $1 + 1 = 2 \ne 4$ **8.** True
9. False, $3x^2 - 48 = 0$ has roots of 4 and −4   **10.** True

CHAPTER 2 REVIEW EXERCISES, page 118

1. 3/2   **3.** 1/2   **5.** −38/15   **7.** 3, 2   **9.** $(1 \pm \sqrt{13})/6$ **11.** 0, 5/3   **13.** $\pm 2\sqrt{3}/3$, $\pm\sqrt{10}/2$ **15.** 3, −5   **17.** 4   **19.** −4
21. −2, −4   **23.** 5, 1   **25.** 2, −3   **27.** −2, −1   **29.** 14, $-31\frac{1}{2}$ **31.** $(-\infty, 2]$ **33.** $[-5, 2]$ **35.** $145/9 \le C \le 35$
37. $(-\infty, 0] \cup [3, 4]$ **39.** $(-\infty, -3) \cup (4, \infty)$ **41.** $(-\infty, 5/2] \cup (3, \infty)$ **43.** $(2/3, 2)$ **45.** $(-2, 0) \cup (0, 2)$ **47.** $(1, 2) \cup (2, 3)$
49. $h = \dfrac{V}{\pi r^2}$ **51.** $b_1 = \dfrac{2A - hb_2}{h}$ **53.** $m = \dfrac{e}{c^2}$ **55.** 80   **57.** 24 nautical miles   **59.** $1750 in the 4% account, $3750 in the 6% account
61. $864   **63.** 18 h   **65.** 13 feet

CHAPTER 2 TEST, page 120

1. 9.6 or $\dfrac{48}{5}$ **2.** $-\dfrac{14}{3}$ **3.** $x = \dfrac{c - cd}{a - c}$, $a \ne c$ **4.** $-2 \pm \sqrt{5}$ **5.** $\dfrac{-1 \pm 2\sqrt{7}}{3}$ **6.** $-4, 1, -\dfrac{1}{2} - \dfrac{\sqrt{3}}{2}i, -\dfrac{1}{2} + \dfrac{\sqrt{3}}{2}i$
7. $\dfrac{5}{2} + \dfrac{5\sqrt{23}}{2}i, \dfrac{5}{2} - \dfrac{5\sqrt{23}}{2}i$ **8.** $\dfrac{8}{27}, -64$ **9.** 67, −61   **10.** $x \le 5/2$ **11.** $[-4, -1) \cup [3, \infty)$ **12.** $2 \le x \le \dfrac{13}{3}$ **13.** −1, −6
14. $(-7, -1)$ **15.** $(-\infty, -5/3] \cup [3, \infty)$ **16.** 2 mph   **17.** $6500 at 8.2%; $2500 at 6.5%   **18.** 2.25 L   **19.** 15 h
20. More than 100 miles

EXERCISE SET 3.1, page 129

1.
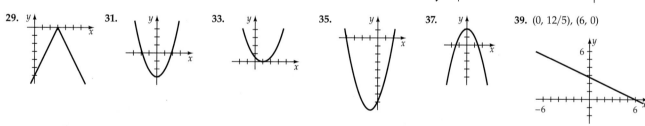
    3.   **5.** $7\sqrt{5}$ **7.** $\sqrt{1261}$ **9.** $\sqrt{89}$ **11.** $\sqrt{38 - 12\sqrt{6}}$ **13.** $2\sqrt{a^2 + b^2}$

15. $-x\sqrt{10}$ **17.** $(12, 0), (-4, 0)$ **19.** $(3, 2)$ **21.** $(6, 4)$ **23.** $(-0.875, 3.91)$ **25.**   **27.**

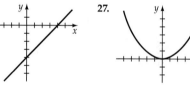

29.   **31.**   **33.**   **35.**   **37.**   **39.** $(0, 12/5), (6, 0)$

41. $(0, \sqrt{5}), (0, -\sqrt{5}), (5, 0)$ **43.** $(0, 4), (0, -4), (-4, 0)$ **45.** $(0, \pm 2), (\pm 2, 0)$ **47.** $(0, \pm 4), (\pm 4, 0)$

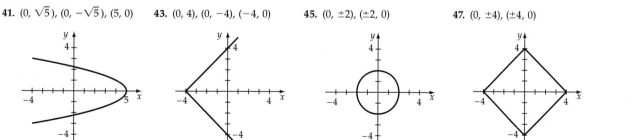

49. center $(0, 0)$, radius 6   **51.** center $(0, 0)$, radius 10   **53.** center $(1, 3)$, radius 7   **55.** center $(-2, -5)$, radius 5
57. center $(8, 0)$, radius 1/2   **59.** $(x - 4)^2 + (y - 1)^2 = 2^2$ **61.** $(x - 1/2)^2 + (y - 1/4)^2 = (\sqrt{5})^2$ **63.** $(x - 0)^2 + (y - 0)^2 = 5^2$
65. $(x - 1)^2 + (y - 3)^2 = 5^2$ **67.** center $(3, 0)$, radius 2   **69.** center $(2, 5)$, radius 3   **71.** center $(7, -4)$, radius 3
73. center $(-1/2, 0)$, radius 4   **75.** center $(1/2, -1/3)$, radius 1/6

77. **79.** **81.** **83.** **85.** **87.**

89. $(13, 5)$ **91.** $(7, -6)$ **93.** yes   **95.** $x^2 - 6x + y^2 - 8y = 0$ **97.** $9x^2 + 25y^2 = 225$ **99. a.** yes **b.** yes **c.** yes **d.** yes
101. $(x + 1)^2 + (y - 7)^2 = 5^2$ **103.** $(x - 7)^2 + (y - 11)^2 = 11^2$ **105.** $(x + 3)^2 + (y - 3)^2 = 3^2$

EXERCISE SET 3.2, page 138

1. a. 5 **b.** -4 **c.** -1 **d.** 1 **e.** $3k - 1$ **f.** $3k + 5$ **3. a.** $\sqrt{5}$ **b.** 3 **c.** 3 **d.** $\sqrt{21}$ **e.** $\sqrt{r^2 + 2r + 6}$ **f.** $\sqrt{c^2 + 5}$
5. a. $1/2$ **b.** $1/2$ **c.** $5/3$ **d.** $1/\pi$ **e.** $1/(c^2 + 4)$ **f.** $1/|2 + h|$ **7. a.** -4 **b.** $-2\sqrt{3}$ **c.** $-\sqrt{14}$ **d.** $-\sqrt{39}/2$ **e.** $-\sqrt{16 - a^2}$ **f.** $-\sqrt{16 - a^2}$
9. *a.* 1 **b.** 1 **c.** -1 **d.** -1 **e.** 1 **f.** -1 **11.** yes   **13.** no   **15.** no   **17.** yes   **19.** no   **21.** all real numbers
23. all real numbers   **25.** $\{x \mid x \neq -2\}$ **27.** $\{x \mid x \geq -7\}$ **29.** $\{x \mid -2 \leq x \leq 2\}$ **31.** $\{x \mid x > -4\}$ **33.** yes   **35.** yes   **37.** yes
39. yes   **41.** no   **43.** yes   **45.** no   **47.** yes   **49.** domain: $(-\infty, \infty)$, range: $[-1, \infty)$ **51.** domain: $(-\infty, \infty)$, range: $(-\infty, \infty)$
53. domain: $(-\infty, \infty)$, range: $[0, \infty)$ **55.** domain: $(-\infty, \infty)$, range: $[0, \infty)$ **57.** domain: $(-\infty, \infty)$, range: $[-2, \infty)$
59. decreasing on $(-\infty, 0]$; increasing on $[0, \infty)$ **61.** increasing on $(-\infty, \infty)$
63. decreasing on $(-\infty, -3]$; increasing on $[-3, 0]$; decreasing on $[0, 3]$; increasing on $[3, \infty)$ **65.** constant on $(-\infty, 0]$; increasing on $[0, \infty)$
67. decreasing on $(-\infty, 0]$; constant on $[0, 1]$; increasing on $[1, \infty)$ **69.** no   **71.** yes   **73.** no   **75.** no   **77.** no

79. 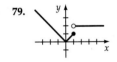 **81.** 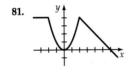 **83.** 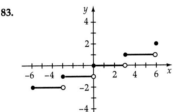 **85.**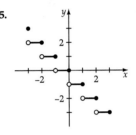

87. 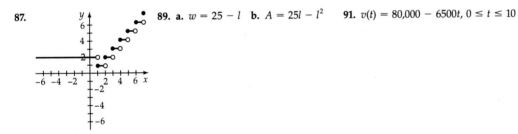 **89. a.** $w = 25 - l$ **b.** $A = 25l - l^2$ **91.** $v(t) = 80{,}000 - 6500t,\ 0 \leq t \leq 10$

93. $c(x) = [19.95 - 0.05(x - 50)]x,\ 50 < x < 200$ **95. a.** $C(x) = 2000.00 + 22.80x$ **b.** $R(x) = 37.00x$ **c.** $P(x) = 14.20x - 2000.00$
97. $h(r) = 15 - 5r$ **99.** $d = \sqrt{9t^2 + 2500}$ **101.** 0.56, 0.66, 0.77, 0.88, 0.99   **103.** 275, 375, 385, 390, 394   **105.** 3, -2

107. 1 is not in the range of f **109.** 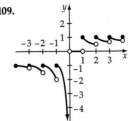 **111.** 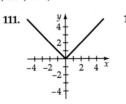 **113.** 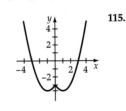 **115.**

117. 4   **119.** 2   **121.** 0   **123. a.** 36 **b.** 13 **c.** 12 **d.** 30 **e.** $13k - 2$ **f.** $8k - 11$ **125.** $4\sqrt{21}$ **127.** 9   **129.** 17   **131.** 1, -3

133.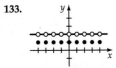

EXERCISE SET 3.3, page 152

1. $-3/2$ **3.** $-1/2$ **5.** the line does not have slope   **7.** 6   **9.** 9/19   **11.** $\dfrac{f(3 + h) - f(3)}{h}$ **13.** $\dfrac{f(h) - f(0)}{h}$

15.   **17.** 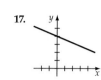   **19.**   **21.** 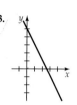   **23.** 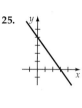   **25.**

27. $y = x + 3$ **29.** $y = 3x - 1$ **31.** $y = \dfrac{3}{4}x + \dfrac{1}{2}$ **33.** $y = (0)x + 4 = 4$ **35.** $y = 2x + 1$ **37.** $y = -4x - 10$

39. $y = -\dfrac{3}{4}x + \dfrac{13}{4}$ **41.** $y = \dfrac{12}{5}x - \dfrac{29}{5}$ **43.** $y = \dfrac{4}{5}x - \dfrac{2}{5}$ **45.** $y = -\dfrac{25}{7}x + \dfrac{543}{70}$ **47.** $y = -\dfrac{3}{4}x + \dfrac{15}{4}$ **49.** $y = 2x + 10$

51. $y = x + 1$ **53.** $y = -\dfrac{5}{3}x - \dfrac{7}{3}$ **55.** $\dfrac{f(x_1 + h) - f(x_1)}{h}$ **57.** $F = \dfrac{9}{5}C + 32$ or $C = \dfrac{5}{9}(F - 32)$

59. a. $V(n) = 8280 - 690n, 0 \le n \le 12$ **b.** \$6210   **61.** $P(x) = 40.50x - 1782, x = 44$, the break-even point

63. $P(x) = 79x - 10{,}270, x = 130$, the break-even point   **65.a.** \$275   **b.** \$283   **c.** \$355   **d.** \$8

67. a. $C(t) = 19{,}500.00 + 6.75t$ **b.** $R(t) = 55.00t$ **c.** $P(t) = 48.25t - 19{,}500.00$ **d.** approximately 405 days   **71.** $y = -2x + 11$

73. $y = -\dfrac{13}{18}x + \dfrac{1}{18}$ **75.** $5x + 3y = 15$ **77.** $2x + 4y = 1$ **79.** $3x + y = 17$

EXERCISE SET 3.4, page 160

1. d   **3.** b   **5.** g   **7.** c   **9.** $f(x) = (x + 2)^2 - 3$ **11.** $f(x) = (x - 4)^2 - 11$ **13.** $f(x) = \left(x - \left(-\dfrac{3}{2}\right)\right)^2 - \dfrac{5}{4}$

15. $f(x) = -(x - 2)^2 + 6$ **17.** $f(x) = -3\left(x - \dfrac{1}{2}\right)^2 + \dfrac{31}{4}$ **19.** vertex: $(5, -25)$, $f(x) = (x - 5)^2 - 25$ **21.** vertex: $(0, -10)$, $f(x) = x^2 - 10$

23. vertex: $(3, 10)$, $f(x) = -(x - 3)^2 + 10$ **25.** vertex: $\left(\dfrac{3}{4}, \dfrac{47}{8}\right)$, $f(x) = 2\left(x - \dfrac{3}{4}\right)^2 + \dfrac{47}{8}$ **27.** vertex: $\left(\dfrac{1}{8}, \dfrac{17}{16}\right)$, $f(x) = -4\left(x - \dfrac{1}{8}\right)^2 + \dfrac{17}{16}$

29. -16, minimum   **31.** 11, maximum   **33.** $-1/8$, minimum   **35.** -11, minimum   **37.** 35, maximum

39. a. 27 feet   **b.** $22\frac{5}{16}$ feet   **c.** 20.1 feet from the center   **41. a.** 16   **b.** \$400   **43. a.** $w = \dfrac{600 - 2l}{3}$ **b.** $A = 200l - \dfrac{2}{3}l^2$

c. $w = 100$ ft, $l = 150$ ft   **45.** y-intercept $(0, 0)$; x-intercepts $(0, 0)$ and $(-6, 0)$ **47.** y-intercept $(0, -6)$; no x-intercepts

49. 740 units yields a maximum revenue of \$109,520.   **51.** 85 units yields a maximum profit of \$24.25.

53. $P(x) = -0.1x^2 + 50x - 1840$, Break-even points: $x = 40$ or $x = 460$ **55. a.** $R(x) = -0.25x^2 + 30.00x$ **b.** $P(x) = -0.25x^2 + 27.50x - 180$

c. \$576.25   **d.** 55   **57. a.** $t = 4$ seconds   **b.** 256 ft   **c.** $t = 8$ seconds   **59.** $f(x) = \dfrac{3}{4}x^2 - 3x + 4$ **61. a.** $w = 16 - x$ **b.** $A = 16x - x^2$

63. Equation has two real zeros for all values of b.   **65.** Increases the height of each point on the graph by c units.   **67.** 4, 4

EXERCISE SET 3.5, page 171

1.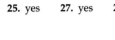
$\bullet$ C(-5, 3)   B(5, 3) $\bullet$

A(-5, -3)   P(5, -3) $\bullet$

3.
R(-2, 3)   A(2, 3)

B(-2, -3)   C(2, -3)

5.
B(-4, 5)   C(4, 5)

T(-4, -5)   A(4, -5)

7.

9.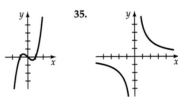

11.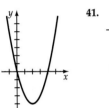

13. a. no **b.** yes   **15. a.** no **b.** no   **17. a.** yes **b.** yes   **19. a.** yes **b.** yes   **21. a.** yes **b.** yes   **23.** no

25. yes   **27.** yes   **29.** yes   **31.** 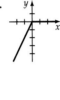   **33.**   **35.**   **37.**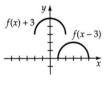

39. 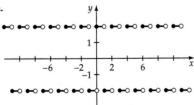   **41.**   **43.** even   **45.** odd   **47.** even   **49.** even   **51.** even   **53.** even   **55.** neither

57.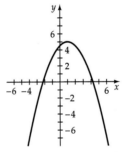
$f(x) + 3$ $f(x - 3)$

59.
$F(-x)$
$-F(x)$

61.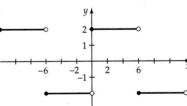

63. a. 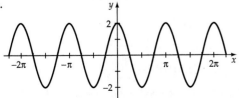   **b.**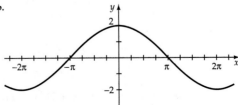

65. a.   **b.**

67.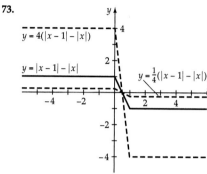

69.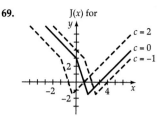

71.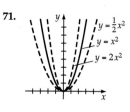

73.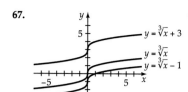

75. a. 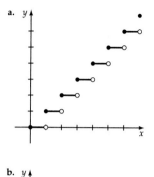 **b.** **c.**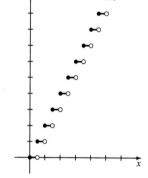

77. a. $f(x) = \dfrac{2}{(x+1)^2 + 1} + 1$ **b.** $f(x) = -\dfrac{2}{(x-2)^2 + 1}$ **79.** $h(x) = 3x - 3$ **81.** $h(x) = -x^3 - x^2 + 6x$

EXERCISE SET 3.6, page 179

1. $f(x) + g(x) = x^2 - x - 12$,   domain all real numbers
  $f(x) - g(x) = x^2 - 3x - 18$,   domain all real numbers
    $f(x) \cdot g(x) = x^3 + x^2 - 21x - 45$,   domain all real numbers
      $\dfrac{f(x)}{g(x)} = x - 5$,   domain $\{x \mid x \neq -3\}$

3. $f(x) + g(x) = 3x + 12$,   domain all real numbers
  $f(x) - g(x) = x + 4$,   domain all real numbers
    $f(x) \cdot g(x) = 2x^2 + 16x + 32$,   domain all real numbers
      $\dfrac{f(x)}{g(x)} = 2$,   domain $\{x \mid x \neq -4\}$

5. $f(x) + g(x) = x^3 + 2x^2 + 8x$,   domain all real numbers
  $f(x) - g(x) = x^3 + 2x^2 + 6x$,   domain all real numbers
    $f(x) \cdot g(x) = x^4 + 2x^3 + 7x^2$,   domain all real numbers
      $\dfrac{f(x)}{g(x)} = x^2 + 2x + 7$,   domain $\{x \mid x \neq 0\}$

7. $f(x) + g(x) = 4x^2 + 7x - 12$,   domain all real numbers
  $f(x) - g(x) = x - 2$,   domain all real numbers
    $f(x) \cdot g(x) = 4x^4 + 14x^3 - 12x^2 - 41x + 35$,   domain all real numbers
      $\dfrac{f(x)}{g(x)} = 1 + \dfrac{x-2}{2x^2 + 3x - 5}$,   domain $\{x \mid x \neq 1, x \neq -5/2\}$

9. $f(x) + g(x) = \sqrt{x-3} + x$,   domain $\{x \mid x \geq 3\}$
  $f(x) - g(x) = \sqrt{x-3} - x$,   domain $\{x \mid x \geq 3\}$
    $f(x) \cdot g(x) = x\sqrt{x-3}$,   domain $\{x \mid x \geq 3\}$
      $\dfrac{f(x)}{g(x)} = \dfrac{\sqrt{x-3}}{x}$,   domain $\{x \mid x \geq 3\}$

11. $f(x) + g(x) = \sqrt{4-x^2} + 2 + x$,   domain $\{x \mid -2 \leq x \leq 2\}$
  $f(x) - g(x) = \sqrt{4-x^2} - 2 - x$,   domain $\{x \mid -2 \leq x \leq 2\}$
    $f(x) \cdot g(x) = (\sqrt{4-x^2})(2 + x)$,   domain $\{x \mid -2 \leq x \leq 2\}$
      $\dfrac{f(x)}{g(x)} = \dfrac{\sqrt{4-x^2}}{2 + x}$,   domain $\{x \mid -2 < x \leq 2\}$

13. 18   **15.** $-9/4$ **17.** 30   **19.** 12   **21.** 300   **23.** $-384/125$ **25.** $-5/2$ **27.** $-1/4$ **29.** 2   **31.** $2x + h$
33. $4x + 2h + 4$ **35.** $-8x - 4h$ **37.** $(g \circ f)(x) = 6x + 3$, $(f \circ g)(x) = 6x - 16$ **39.** $(g \circ f)(x) = x^2 + 4x + 1$
  $(f \circ g)(x) = x^2 + 8x + 11$

41. $(g \circ f)(x) = -5x^3 - 10x$ **43.** $(g \circ f)(x) = \dfrac{1-5x}{x+1}$ **45.** $(g \circ f)(x) = \dfrac{\sqrt{1-x^2}}{\dfrac{1}{|x|}}$ **47.** $(g \circ f)(x) = -\dfrac{2|5-x|}{3}$
  $(f \circ g)(x) = -125x^3 - 10x$
    $(f \circ g)(x) = \dfrac{2}{3x-4}$ $(f \circ g)(x) = \dfrac{\frac{1}{|x|}}{x-1}$ $(f \circ g)(x) = \dfrac{3|x|^3}{|5x+2|}$

49. 66   **51.** 51   **53.** -4 **55.** 41   **57.** $-3848/625$ **59.** $6 + 2\sqrt{3}$ **61.** $16c^2 + 4c - 6$ **63.** $9k^4 + 36k^3 + 45k^2 + 18k - 4$
65. 25.39   **67.** 5.474×10^{-6} **75. a.** $(s \circ m)(x) = 87 + 49{,}300/x$ **b.** \$89   **77. a.** 2   **b.** 3/2   **c.** 5/3   **d.** 8/5

EXERCISE SET 3.7, page 186

9. $f^{-1}(x) = \dfrac{x-1}{4}$ **11.** $F^{-1}(x) = \dfrac{1-x}{6}$ **13.** $j^{-1}(t) = \dfrac{t-1}{2}$ **15.** $f^{-1}(v) = \sqrt[3]{1-v}$ **17.** $f^{-1}(x) = -\dfrac{4x}{x+3}$, $x \neq -3$

19. $M^{-1}(t) = \dfrac{5}{1-t}$, $t \neq 1$ **21.** $r^{-1}(t) = -\sqrt{\dfrac{1}{t}}$, $t > 0$ **23.** $J^{-1}(x) = \sqrt{x-4}$, $x \geq 4$

25. $f^{-1}(x) = \sqrt{x-3}$, domain $\{x \mid x \geq 3\}$, range $\{y \mid y \geq 0\}$ **27.** $f^{-1}(x) = x^2$, domain $\{x \mid x \geq 0\}$, range $\{y \mid y \geq 0\}$
f has domain $\{x \mid x \geq 0\}$, range $\{y \mid y \geq 3\}$ f has domain $\{x \mid x \geq 0\}$, range $\{y \mid y \geq 0\}$

29. $f^{-1}(x) = \sqrt{9-x^2}$, domain $\{x \mid 0 \leq x \leq 3\}$, range $\{y \mid 0 \leq y \leq 3\}$ **31.** $f^{-1}(x) = 2 + \sqrt{x+3}$, domain $\{x \mid x \geq -3\}$, range $\{y \mid y \geq 2\}$
f has domain $\{x \mid 0 \leq x \leq 3\}$, range $\{y \mid 0 \leq y \leq 3\}$ f has domain $\{x \mid x \geq 2\}$, range $\{y \mid y \geq -3\}$

33. $f^{-1}(x) = -4 - \sqrt{x+25}$, domain $\{x \mid x \geq -25\}$, range $\{y \mid y \leq -4\}$ **35.** 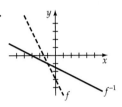 **37.**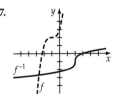
f has domain $\{x \mid x \leq -4\}$, range $\{y \mid y \geq -25\}$

39. 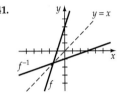 **41.** 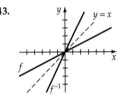 **43.** 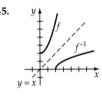 **45.**   **47.**

49. $f^{-1}(x) = \dfrac{x-b}{a}$, $a \neq 0$ **51.** $f^{-1}(x) = \dfrac{x+1}{1-x}$, $x \neq 1$ **53.** no **55.** yes **57.** yes **59.** no **61.** 5 **63.** 4

67. $\left(-\dfrac{7}{5}, \dfrac{26}{5}\right)$, $\left(\dfrac{a + 2bm - am^2}{m^2 + 1}, ma + m\left(\dfrac{a + 2bm - am^2}{m^2 + 1}\right) - b\right)$

EXERCISE SET 3.8, page 191

1. $d = kt$ **3.** $y = k/x$ **5.** $m = knp$ **7.** $V = klwh$ **9.** $A = ks^2$ **11.** $F = km_1m_2/d^2$ **13.** $y = kx$, $k = 4/3$
15. $r = kt^2$, $k = 1/81$ **17.** $T = krs^2$, $k = 7/25$ **19.** $V = klwh$, $k = 1$ **21.** 1.02 liters **23.** 437.5 lbs/ft^2
25. a. approximately 3.3 seconds **b.** approximately 3.7 ft **27.** 112 db **29.** $330 **31. a.** 9 times larger **b.** 3 times larger
c. 27 times larger **33.** 6 times larger **35.** approximately 3.2 mi/sec **37.** approximately 410 vibrations per second
39. approximately 3950 lbs **41.** approximately 1657 mi **43.** 11 centimeters **45.** 2.0 liters

CHAPTER 3 TRUE/FALSE EXERCISES, page 197

1. False. Let $f(x) = x^2$. Then $f(3) = f(-3) = 9$, however $3 \neq -3$. **2.** False. $f(x) = x^2$ does not have an inverse function.
3. False. Let $f(x) = 2x$, $g(x) = 3x$. Then $f(g(0)) = 0$ and $g(f(0)) = 0$, however f and g are not inverse functions.
4. True **5.** False. Let $f(x) = 3x$. $[f(x)]^2 = 9x^2$ whereas $f[f(x)] = f(3x) = 3(3x) = 9x$.
6. False. Let $f(x) = x^2$, then $f(1) = 1$, $f(2) = 4$ thus $\dfrac{f(2)}{f(1)} = 4 \neq \dfrac{2}{1}$. **7.** True **8.** False. $f(-1 + 3) = f(2) = 2$. $f(-1) + f(3) = 1 + 3 = 4$.
9. True **10.** True

CHAPTER 3 REVIEW EXERCISES, page 198

1. $\sqrt{181}$ **3.** $(-1/2, 10)$ **5.** center $(3, -4)$, radius 9 **7.** $(x-2)^2 + (y+3)^2 = 5^2$ **9. a.** 2 **b.** 10 **c.** $3t^2 + 4t - 5$
d. $3x^2 + 6xh + 3h^2 + 4x + 4h - 5$ **e.** $9t^2 + 12t - 15$ **f.** $27t^2 + 12t - 5$ **11. a.** 5 **b.** -11 **c.** $x^2 - 12x + 32$ **d.** $x^2 + 4x - 8$

13. $8x + 4h - 3$ **15.**

17.

19.

increasing on $[3, \infty)$
decreasing on $(-\infty, 3]$

increasing on $[-2, 2]$
constant on $(-\infty, -2] \cup [2, \infty)$

increasing $(-\infty, \infty)$

21. Domain $\{x \mid x \text{ is a real number}\}$ **23.** Domain $\{x \mid -5 \le x \le 5\}$ **25.** $y = -2x + 1$ **27.** $y = \dfrac{3}{4}x + \dfrac{19}{2}$ **29.** $f(x) = (x + 3)^2 + 1$

31. $f(x) = -(x + 4)^2 + 19$ **33.** $f(x) = -3\left(x - \dfrac{2}{3}\right)^2 - \dfrac{11}{3}$ **35.** $(1, 8)$ **37.** $(5, 161)$ **39.** $(0, 1050)$ **41.**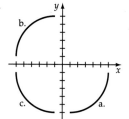

43. Symmetric to the y-axis   **45.** Symmetric to the origin   **47.** Symmetric to the x-axis, y-axis, and the origin.
49. Symmetric to the x-axis, y-axis, and the origin.

51.

53.

55.

a. domain is set of all real numbers
  range $\{y \mid y \le 4\}$
b. even

a. domain is set of all real numbers
  range $\{y \mid y \ge 4\}$
b. even

a. domain is set of all real numbers
  range is set of all real numbers
b. odd

57. $F(x) = (x + 2)^2 - 11$

59. $P(x) = 3(x - 0)^2 - 4$

61. $W(x) = -4(x + 3/4)^2 + 33/4$

63.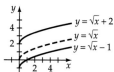

$y = \sqrt{x} + 2$
$y = \sqrt{x}$
$y = \sqrt{x} - 1$

65.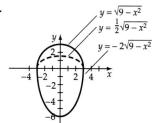

$y = \sqrt{9 - x^2}$
$y = \frac{1}{2}\sqrt{9 - x^2}$
$y = -2\sqrt{9 - x^2}$

67.

69. $f(x) + g(x) = x^2 + x - 6$, the domain is the set of all real numbers
$f(x) - g(x) = x^2 - x - 12$, the domain is the set of all real numbers
$f(x) \cdot g(x) = x^3 + 3x^2 - 9x - 27$, the domain is the set of all real numbers
$\dfrac{f(x)}{g(x)} = x - 3$, domain $\{x \mid x \neq -3\}$

71. yes   **73.** yes   **75.** $f^{-1}(x) = \dfrac{x + 4}{3}$

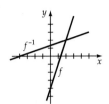

77. $h^{-1}(x) = -2x - 4$ **79.** 25, 25   **81.** $t \approx 3.7$ seconds

CHAPTER 3 TEST, page 200

1. (1, 1)   **2.** $(0, \sqrt{2}), (0, -\sqrt{2}), (-4.0)$ 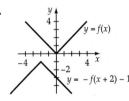   **3.** 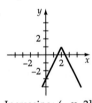   **4.** center $(2, -1)$; radius: 3

5. Domain: $\{x \mid x \geq 4 \text{ or } x \leq -4\}$ **6.** $v = 60{,}000 - 8000t;\ 0 \leq t \leq 6$ **7.**   **8.**

Increasing: $(-\infty, 2]$
Decreasing: $[2, \infty)$

Domain: all real, Range: $\{y \mid y \geq 2\}$

9. 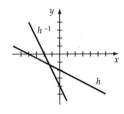   **10.** 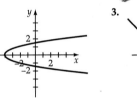   **11.** b   **12.** $y = -\dfrac{2}{3}x + \dfrac{2}{3}$ **13.** -12 minimum   **14.** $x^2 + x - 3;\ \dfrac{x^2 - 1}{x - 2},\ x \neq 2$

15. $2x + h$ **16.** $x - 2\sqrt{x - 2} - 1$ **17.** $f^{-1}(x) = \sqrt{x + 9}$ **18.** $f^{-1}(x) = \dfrac{1}{2}x + \dfrac{3}{2}$ **19.** 8.4375 in.   **20.** 5.69 Lumens.

Domain of f: $\{x \mid x \geq 0\}$
Range of f: $\{y \mid y \geq -9\}$
Domain of f^{-1}: $\{x \mid x \geq -9\}$
Range of f^{-1}: $\{y \mid y \geq 0\}$

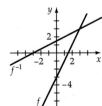

EXERCISE SET 4.1, page 208

1. $5x^2 - 9x + 10 + \dfrac{-10}{x + 3}$ **3.** $x^3 + 5x^2 - 9x - 45$ **5.** $x^2 - \dfrac{100}{3} + \dfrac{100x + 10}{3(3x^2 + x + 1)}$ **7.** $4x^2 + 1 + \dfrac{11}{5x^2 - 2}$

9. $x + 4 + \dfrac{6x - 3}{x^2 + x - 4}$ **11.** $4x^2 + 3x + 12 + \dfrac{17}{x - 2}$ **13.** $4x^2 - 4x + 2 + \dfrac{1}{x + 1}$ **15.** $x^4 + 4x^3 + 6x^2 + 24x + 101 + \dfrac{403}{x - 4}$

17. $x^4 + x^3 + x^2 + x + 1$ **19.** $8x^2 + 6$ **21.** $x^7 + 2x^6 + 5x^5 + 10x^4 + 21x^3 + 42x^2 + 85x + 170 + \dfrac{344}{x - 2}$

23. $x^5 - 3x^4 + 9x^3 - 27x^2 + 81x - 242 + \dfrac{716}{x + 3}$ **25.** $3x - 3.1 + \dfrac{4.07}{x - 0.3}$ **27.** $2x^2 - 11x - 17 + \dfrac{3}{x}$ **29.** $1 + \dfrac{6}{x + 2}$ **31.** 25

33. 45   **35.** -2230 **37.** -80 **39.** -187 **41.** yes   **43.** yes   **45.** yes   **47.** yes   **49.** yes   **51.** yes

63. $(x + 3)(x - 1)(x - 2)$ **65.** $(x + 3)(x + 2)(x + 1)(x - 4)$ **69.** 13   **71.** By the Factor Theorem, $P(x)$ has a factor of $x - c$ if and only if $P(c) = 0$. However, $P(c) = 4c^2 + 7c^2 + 12$ which is greater than 0 for any real number c.

EXERCISE SET 4.2, page 214

1. up to far left, up to far right   **3.** down to far left, up to far right   **5.** down to far left, down to far right
7. down to far left, up to far right   **9.** up to far left, down to far right   **11. a.** -7 **b.** 33 **c.** -2 **d.** 58 **e.** 9 **f.** 169
13. a. -77 **b.** 97 **c.** 15 **d.** 5 **e.** -18 **f.** 38   **15. a.** 22 **b.** 74 **c.** -286 **d.** 176 **e.** -748 **f.** -1592
17. a. 126 **b.** 86 **c.** 14 **d.** 126 **e.** 134 **f.** 110   **19. a.** -8 **b.** 82 **c.** 168 **d.** 782 **e.** 6482/81 **f.** 23,192   **21.** $-7/2, -2, 3$
23. $0, 2/5, 1$ **25.** $-5, -7/3, 11/2$ **27.** $-3, 0, 2$ **29.** 1

31.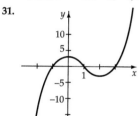

33.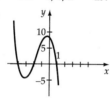

35.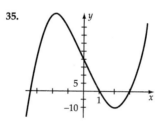

37.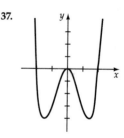

39.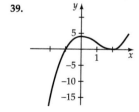

41.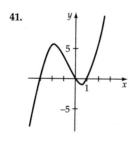

43.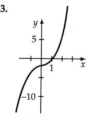

45.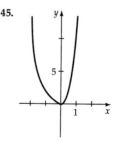

53.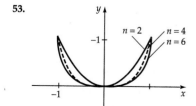

55.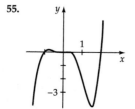

57.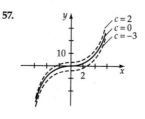

59.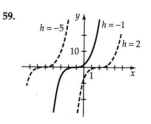

EXERCISE SET 4.3, page 222

1. 3 (multiplicity two), -5 (multiplicity one)   **3.** 0 (multiplicity two), $-5/3$ (multiplicity two)
5. 2 (multiplicity one), -2 (multiplicity one), -3 (multiplicity two)   **7.** 5 (multiplicity two), -2 (multiplicity two)
9. -3 (multiplicity one), 3 (multiplicity one), -1 (multiplicity one), 1 (multiplicity one)   **11.** $\pm1, \pm2, \pm4, \pm8$
13. $\pm1, \pm2, \pm3, \pm4, \pm6, \pm12, \pm1/2, \pm3/2$ **15.** $\pm1, \pm2, \pm4, \pm1/2, \pm1/3, \pm2/3, \pm4/3, \pm1/6$ **17.** $\pm1, \pm3, \pm9, \pm1/2, \pm3/2, \pm9/2$
19. $\pm1, \pm7, \pm1/2, \pm7/2, \pm1/4, \pm7/4$ **21.** $\pm1, \pm2, \pm4, \pm8, \pm16, \pm32$ **23.** upper bound 2, lower bound -5 **25.** upper bound 4, lower bound -4
27. upper bound 1, lower bound -4 **29.** upper bound 1, lower bound -5 **31.** upper bound 4, lower bound -2
33. upper bound 2, lower bound -1 **35.** one positive, two or zero negative   **37.** two or zero positive, one negative
39. one positive, three or one negative   **41.** one positive, two or zero negative   **43.** three or one positive, one negative
45. one positive, no negative   **47.** $2, -1, -4$ **49.** $3, -4, 1/2$ **51.** $1/2, -1/3, -2$ **53.** $1, -1, -9/2$
55. $1/2, 4, \sqrt{3}, -\sqrt{3}$ **57.** $2, -1$ **59.** $0, -2, 1 + \sqrt{2}, 1 - \sqrt{2}$ **61.** $-1, 2$ **63.** $-3/2, 1, 8$ **71.** yes   **73.** yes

EXERCISE SET 4.4, page 227

1. $1 - i, 1/2$ **3.** $i, -3$ **5.** $-i\sqrt{2}, 1, \sqrt{5}, -\sqrt{5}$ **7.** $1 + \dfrac{1}{2}i, 1 - \dfrac{1}{2}i$ **9.** $1 - 3i, 1 + 2i, 1 - 2i$ **11.** $-i, 3, -1$

13. $2, -3, 2i, -2i$ **15.** $1/2, -3, 1 + 5i, 1 - 5i$ **17.** $1, 3 + 2i, 3 - 2i$ **19.** $x(x - 2)(x + 1)$ **21.** $x(x^2 + 9)$
23. $(x^2 + 6)(x + 2)(x - 2)$ **25.** $(x^2 + 2)(x^2 + 1)$ **27.** $(x + 1)(x - 3)(x^2 + 4)$ **29.** $x^3 - 3x^2 - 10x + 24$
31. $x^3 - 3x^2 + 4x - 12$ **33.** $x^4 - 10x^3 + 63x^2 - 214x + 290$ **35.** $x^5 - 22x^4 + 212x^3 - 1012x^2 + 2251x - 1830$
37. $4x^3 - 19x^2 + 224x - 159$ **39.** $x^3 + 13x + 116$ **41.** $x^4 - 18x^3 + 131x^2 - 458x + 650$ **43.** $3x^3 - 12x^2 + 3x + 18$
45. $-2x^4 + 4x^3 + 36x^2 - 140x + 150$ **47.** one real zero   **49.** two real zeros   **51.** Since $x^3 - x^2 - ix^2 - 9x + 9 + 9i$ does

not have real coefficients, the theorem does not apply.   **53.** $P(x) = (x - 2)^3(x^2 + 9)$ **55.** $P(x) = \dfrac{1}{2}x^5 - 4x^4 + \dfrac{25}{2}x^3 - 19x^2 + 14x - 4$

EXERCISE SET 4.5, page 239

1. vertical asymptotes $x = 0, x = -3$ **3.** vertical asymptote $x = 4/3, x = -1/2$ **5.** horizontal asymptote $y = 4$
7. horizontal asymptote $y = 30$ **9.** vertical asymptote $x = -4$ **11.** vertical asymptote $x = 3$
                                            horizontal asymptote $y = 0$ horizontal asymptote $y = 0$

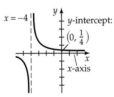

13. vertical asymptote $x = 0$ **15.** vertical asymptote $x = -4$ **17.** vertical asymptote $x = 2$
horizontal asymptote $y = 0$ horizontal asymptote $y = 1$ horizontal asymptote $y = -1$

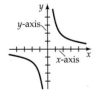

19. vertical asymptotes $x = 3, x = -3$ **21.** vertical asymptote $x = -3, x = 1$ **23.** vertical asymptotes $x = 3, x = -3$
horizontal asymptote $y = 0$ horizontal asymptote $y = 0$ horizontal asymptote $y = 0$

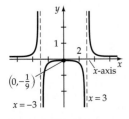

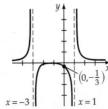

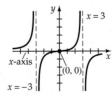

25. vertical asymptote $x = -2$
horizontal asymptote $y = 1$

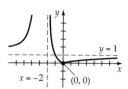

27. vertical asymptote   none
horizontal asymptote $y = 0$

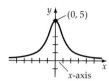

29. vertical asymptotes $x = 3, x = -3$
horizontal asymptote $y = 2$

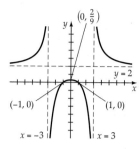

31. vertical asymptotes $x = -1 + \sqrt{2}, x = -1 - \sqrt{2}$
horizontal asymptote $y = 1$

33. $y = 3x - 7$ **35.** $y = x$ **37.** vertical asymptote $x = 0$
slant asymptote $y = x$

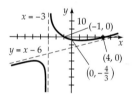

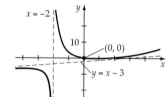

39. vertical asymptote $x = -3$
slant asymptote $y = x - 6$

41. vertical asymptote $x = 4$
slant asymptote $y = 2x + 13$

43. vertical asymptote $x = -2$
slant asymptote $y = x - 3$

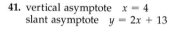

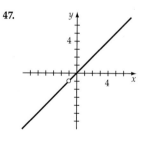

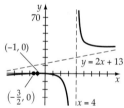

45. vertical asymptotes $x = 2, x = -2$
slant asymptote $y = x$

47.   **49.**   **51.**

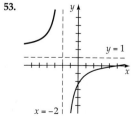

53.   **55.**   **57. a.** $1333.33   **b.** \$8000   **c.**

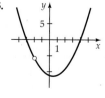

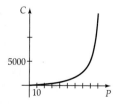

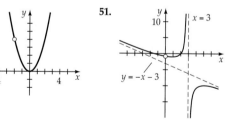

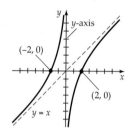

59. a. 611   **b.** 1777   **c.** $y = 2000$ **d.**

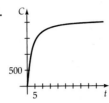

18 weeks

61. a. As the radius of the blood vessel gets smaller, the resistance gets larger.   **63.** $(-2, 2)$ **65.** $(0, 1), (-4, 1)$
b. As the radius of the blood vessel gets larger, the resistance approaches zero.

c.

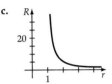

EXERCISE SET 4.6, page 245

7. 2.0625   **9.** 1.1875   **11.** -1.5625 **13.** 0.855   **15.** 1.275   **17.** 2.125   **19.** 1.4   **21.** 1.2

CHAPTER 4 TRUE/FALSE EXERCISES, page 248

1. false, $x - i$ has a zero of i, but it does not have a zero of $-i$.   **2.** false, Descartes' Rule of Signs indicates
that $x^3 - x^2 + x - 1$ has 3 or 1 positive zeros.   **3.** true   **4.** true   **5.** false, $f(x) = \dfrac{x}{x^2 + 1}$ does not have a vertical asymptote.
6. false, because $f(x) = \dfrac{(x - 2)^2}{(x - 3)(x - 2)} = \dfrac{x - 2}{x - 3}, x \neq 2$.   **7.** true   **8.** true   **9.** true   **10.** true   **11.** true   **12.** true   **13.** true
14. false, $x^2 + 1$ does not have a real zero.

CHAPTER 4 REVIEW EXERCISES, page 249

1. $x + 4 + \dfrac{-5x - 29}{x^2 + x + 3}$ **3.** $-x + \dfrac{3x^2 - 12x - 3}{x^3 + x}$ **5.** $3x^2 - 5x - 1$ **7.** $4x^2 + x + 8 + \dfrac{22}{x - 3}$ **9.** $3x^2 - 6x + 7 + \dfrac{-13}{x + 2}$

11. $3x^2 + 5x - 11$ **13.** 77   **15.** 33   **21.**   **23.**   **25.**   **27.** $\pm 1, \pm 2, \pm 3, \pm 6$

29. $\pm 1, \pm 2, \pm 3, \pm 4, \pm 6, \pm 12, \pm 1/3, \pm 2/3, \pm 4/3, \pm 1/5,$ **31.** ± 1
$\pm 2/5, \pm 3/5, \pm 4/5, \pm 6/5, \pm 12/5, \pm 1/15, \pm 2/15, \pm 4/15$
33. no positive and three or one negative   **35.** one positive and one negative   **37.** $1, -2, -5$ **39.** $-2, -1/2, -4/3$ **41.** 1
43. $2x^3 - 3x^2 - 23x + 12$ **45.** $x^4 - 3x^3 + 27x^2 - 75x + 50$ **47.** vertical asymptote $x = -2$ **49.** vertical asymptote $x = -1$
horizontal asymptote $y = 3$ slant asymptote $y = 2x + 3$
59. 1.375   **61.** 0.875

51.   **53.**   **55.**   **57.**

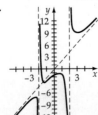

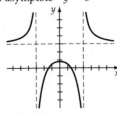

CHAPTER 4 TEST, page 250

1. $3x^2 - x + 6 - \dfrac{13}{x+2}$ **2.** 43   **4.** up to the far left and down to the far right   **5.** $0, \dfrac{2}{3}, -3$ **6.** between 1 and 2

7.   2 (multiplicity two)   **8.** $\pm 1, \pm 3, \pm \dfrac{1}{2}, \pm \dfrac{3}{2}, \pm \dfrac{1}{3}, \pm \dfrac{1}{6}$ **9.** upper bound 4, lower bound -5

   -2 (multiplicity two)

   $\dfrac{3}{2}$ (multiplicity one)

   -1 (multiplicity three)

10. 4.2, or 0 positive zeros, no negative zero.   **11.** $\dfrac{1}{2}, 3, -2$ **12.** $i, 2, -\dfrac{1}{2}$ **13.** $0, 1, 2+i, 2-i$ **14.** $x^4 - 5x^3 + 8x^2 - 6x$

15. Vertical asymptotes at $x = 3$ and $x = 2$ **16.** Horizontal asymptote $y = \dfrac{3}{2}$ **17.**

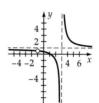

18.

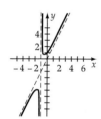

19.

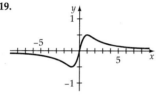

20. 1.8

EXERCISE SET 5.1, page 257

1. 4.72880   **3.** 442.335   **5.** 2.17458   **7.** 164.022   **9.** 5.65223   **11.** 0.969476   **13.** 70.4503   **15.** 19.8130   **17.** 14.0940
19. 15.1543   **21.** 3353.33   **23.** 8103.08

25.

27.

29.

31.
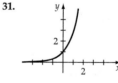

33.

35.
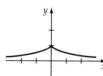

37.

39.

41.

43.

45.

47.

49.
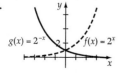

51.
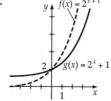

53.

Value of n	Value of $\left(1 + \dfrac{1}{n}\right)^n$
5	2.48832
50	2.691588029
500	2.715568521
5000	2.718010050
50,000	2.718254646
500,000	2.718279110
5,000,000	2.718281557

55. a. 0.6925   **b.** *f* approaches *y* = 1   **57.** *y* = 1

domain: $(-\infty, \infty)$
range: $(-1, 1)$
f is an odd function

59.

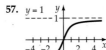

domain: $(-\infty, \infty)$
range: $[0, 2.2)$
f is an even function

61.

domain: $(-\infty, \infty)$
range: $(0.5, \infty)$
f is neither even or odd

63.

domain: $(-\infty, 0]$
range: $[0, 1)$
f is neither even or odd

65. a.

domain: $(-\infty, \infty)$
range: $[2, \infty)$

b.

domain: $(-\infty, \infty)$
range: $(-\infty, \infty)$

67.

domain: $(0, \infty)$
range: $(-\infty, \infty)$

69. a. -2
b. 4
c. does not exist
h is not an exponential function
because *b* is not a positive constant.

71.

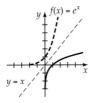
$f(x) = e^x$
$y = x$

73. $F(x) = \dfrac{e^x - e^{-x}}{2}$ is an odd function, that is, prove $F(-x) = -F(x)$.

Proof: $F(x) = \dfrac{e^x - e^{-x}}{2}$

$F(-x) = \dfrac{e^{-x} - e^x}{2}$

$F(-x) = -\dfrac{(-e^{-x} + e^x)}{2}$

$F(-x) = -\dfrac{(e^x - e^{-x})}{2}$

$F(-x) = -F(x)$

75. a. 20,000   **b.** 40,000   **c.** 320,000

77. e^{π} **79.**

EXERCISE SET 5.2, page 266

1. $10^2 = 100$ **3.** $5^3 = 125$ **5.** $3^4 = 81$ **7.** $b^t = r$ **9.** $3^{-3} = \frac{1}{27}$ **11.** $\log_2 16 = 4$ **13.** $\log_7 343 = 3$ **15.** $\log_{10} 10{,}000 = 4$
17. $\log_b j = k$ **19.** $\log_b b = 1$ **21.** 6   **23.** 5   **25.** 3   **27.** -2 **29.** 0   **31.** $\log_b x + \log_b y + \log_b z$ **33.** $\log_3 x - 4 \log_3 z$
35. $\frac{1}{2} \log_b x - 3 \log_b y$ **37.** $\log_b x + \frac{2}{3} \log_b y - \frac{1}{3} \log_b z$ **39.** $\frac{1}{2} \log_7(x + z^2) - \log_7(x^2 - y)$ **41.** 0.9208   **43.** 1.1292   **45.** -0.4709
47. 1.7479   **49.** 1.3562   **51.** $\log_{10} x^2(x + 5)$ **53.** $\log_b \sqrt{\dfrac{(x - y)^3(x + y)}{z}}$ **55.** $\log_8(x + y)$ **57.** $\ln x^2(x - 3)^4$ **59.** $\ln \dfrac{xz}{y}$
61. 1.5395   **63.** 0.86719   **65.** -1.7308 **67.** -2.3219 **69.** 0.87357   **71.** 3.06   **73.** 3190   **75.** 0.00334   **77.** 7.40
79. 2.00   **81.** 0.300   **83.** $[1, 10^{1000}]$ **85.** $[e^e, e^{e^3}]$ **87.** $(0,1)$ **91.** reflexive property of equality, definition of $\log_b x = n$

EXERCISE SET 5.3, page 272

1. 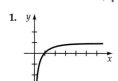 **3.** 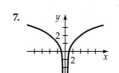 **5.** 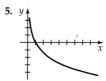 **7.**   **9.**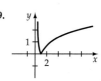

11. 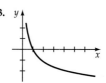 **13.** 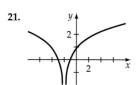 **15.**   **17.** 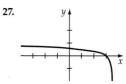 **19.**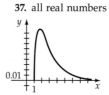

21. 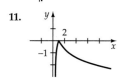 **23.**  **25.** 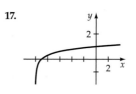 **27.**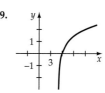

29. $\{x \mid x > 0, x \neq 1\}$ **31.** $\{x \mid x < -3 \text{ or } x > 3\}$ **33.** $\{x \mid x \neq 0\}$ **35.** $\{y \mid y \neq 0\}$ **37.** all real numbers   **39.** all real numbers
41. a. 75   **b.** 56.5   **c.** 13   **43.** 8.6   **45.** $f(x)$ and $g(x)$ are inverse functions   **47.**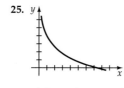

49. domain $\{x \mid x \geq 1\}$, range $\{y \mid y \geq 0\}$ **51.** domain $\{x \mid -1 < x < 1\}$, range $\{y \mid y \geq 100\}$ **53.** domain $\{x \mid x > 1\}$, range all real numbers
55. a. 3   **b.** 1.386   **c.** 3.296   **57.** The domain of F is $\{x \mid x \neq 0\}$, but the domain of G is $\{x \mid x > 0\}$ **59.** 8.8

EXERCISE SET 5.4, page 278

1. 6   **3.** 3   **5.** $-3/2$ **7.** $-6/5$ **9.** 3   **11.** 2.64   **13.** -4.36 **15.** 0.163   **17.** -0.251 **19.** 2.30   **21.** 5.09
23. $2 + 2\sqrt{2}$ **25.** 199/95   **27.** 3   **29.** 10^{10} **31.** 25   **33.** 2   **35.** $3, -3$ **37.** $2\sqrt{2}$ **39.** 5   **41.** $\log(20 + \sqrt{401})$
43. $\frac{1}{2} \log \frac{3}{2}$ **45.** $\ln(15 \pm 4\sqrt{14})$ **47.** $\ln(1 + \sqrt{65}) - \ln 8$ **49.** 2.2   **51.** 3.2×10^{-5}
53. a. 8500, 10,285   **b.** in 6 years   **55. a.** 60°F   **b.** in 27 minutes

57. no solutions   **59.** 2 solutions   **61.** 2 solutions   **63.** 4 solutions

  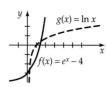 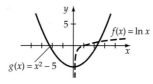 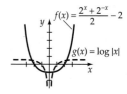

65. The second step, because $\log 0.5 < 0$, thus the inequality sign must be reversed.   **67.** $x = \dfrac{y}{y-1}$ **69.** $e^{0.336} \approx 1.4$ **71.** $\left(0, \dfrac{1}{\ln 2}\right)$

EXERCISE SET 5.5, page 287

1. a. \$9724.05   **b.** \$11,256.80   **3. a.** \$48,885.72   **b.** \$49,282.20   **c.** \$49,283.29   **5.** \$24,730.82   **7.** 8.8 years   **9.** $t = \dfrac{\ln 3}{r}$
11. 14 years   **13. a.** 2200 bacteria   **b.** 17,600 bacteria   **15.** $N(t) = 22,600e^{0.01368t}$ **17. a.** 18,400   **b.** 1994   **19.** $N(t) = 100e^{-0.000418t}$
21. 39.1%   **23.** 2161 years old   **25. a.** 0.056   **b.** 42°   **c.** 54 minutes   **d.** will never reach 34°   **27.** 10 times   **29.** 3.01
31. a. 211 hours   **b.** 1386 hours   **33.** 3.1 years   **35. a.** 21.7, 0.87   **b.** 1086, 0.88   **c.** 72, 382, 0.92   **37.** 5 years   **39.** 1000 times
41. a. 96   **b.** 3385   **c.** 13,395   **d.** 39,751   **43.** 7%

CHAPTER 5 TRUE/FALSE EXERCISES, page 292

1. true   **2.** true   **3.** true
4. false, because f is not defined for negative values of x, and thus $g(f(x))$ is undefined for negative values of x.
5. false, $h(x)$ is not an increasing function for $0 < b < 1$.   **6.** false, $h(x)$ is not an increasing function for $0 < b < 1$.
7. true   **8.** true   **9.** true   **10.** true   **11.** false, $\log x + \log y = \log xy$ **12.** true   **13.** true   **14.** true

CHAPTER 5 REVIEW EXERCISES, page 292

1. 2   **3.** 3   **5.** -2 **7.** -3 **9.** ±1000 **11.** 7   **13.** 15.6729   **15.** 5.47395   **17.** 13.6458   **19.**

21.   **23.**   **25.**   **27.** 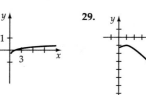   **29.**   **31.**

33. $4^3 = 64$ **35.** $(\sqrt{2}^4) = 4$ **37.** $\log_5 125 = 3$ **39.** $\log_{10} 1 = 0$
41. $2 \log_b x + 3 \log_b y - \log_b z$ **43.** $\ln x + 3 \ln y$ **45.** $\log(x^2 \sqrt[3]{x+1})$ **47.** $\ln\left(\dfrac{\sqrt{2xy}}{z^3}\right)$ **49.** 2.86754   **51.** -0.117233
53. 295   **55.** 1.41×10^{22} **57.** $\dfrac{\ln 30}{\ln 4}$ **59.** 4   **61.** 4   **63.** $\dfrac{\ln 3}{2 \ln 4}$ **65.** 10^{1000} **67.** 1,000,005   **69.** 81   **71.** 4
73. 4.2   **75. a.** \$20,323.79   **b.** \$20,339.99   **77.** \$4438.10   **79.** $N(t) = e^{0.8047t}$ **81.** $N(t) = 3.783e^{0.0558t}$

CHAPTER 5 TEST, page 294

1. 4.88936   **2.** 3.85743   **3.** 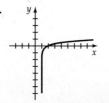   **4.** 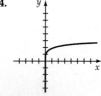   **5.**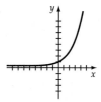

6. $b^c = 5x - 3$ **7.** $\log_3 y = \dfrac{x}{2}$ **8.** $2 \log_b x + 4 \log_b y - 3 \log_b z$ **9.** $2 \log_b z - 3 \log_b y - \dfrac{1}{2} \log_b x$ **10.** $\log_{10} \dfrac{2x + 3}{(x - 2)^3}$

11. 1.0219   **12.** 1.7925   **13.** 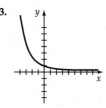   **14.** 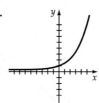   **15.**

16. 1   **17.** 1.9206   **18.** 1   **19.** \$29,539.62   **20.** \$3109.26

EXERCISE SET 6.1, page 303

1. 75°, 165°   **3.** 19°45′, 109°45′   **5.** 33°26′45″, 123°26′45″   **7.** $\frac{\pi}{2} - 1$, $\pi - 1$ **9.** $\frac{\pi}{4}, \frac{3\pi}{4}$ **11.** $0, \frac{\pi}{2}$ **13.** 78.133°   **15.** 16.012°

17. 47.338°   **19.** 110°24′   **21.** −66°43′12″   **23.** −7°3′   **25.** $\frac{\pi}{12}$ **27.** $\frac{7\pi}{4}$ **29.** $-\frac{5\pi}{4}$ **31.** 30°   **33.** 67.5°   **35.** 660°

37. 85.94°   **39.** 2.32   **41.** 300.80°   **43.** 4, 229.2°   **45.** 2.38, 136.6°   **47.** 6.28 in.   **49.** 18.33 cm   **51.** 3π   **53.** $\frac{5\pi}{12}$ radians or 75°

55. $\frac{\pi}{30}$ radians/s   **57.** $\frac{5\pi}{3}$ radians/s   **59.** $\frac{3\pi}{2}$ radians/s   **61.** 40 mi/h   **63.** 13 in²   **65.** 4680 cm²   **67.** 436 m²

69. a. $\frac{6}{5}\pi$ radians **b.** 23.6 ft   **71.** 1039 m, 3.5 m/s   **73.** 18.8 ft/s   **75.** 23.1 in²

EXERCISE SET 6.2, page 312

1. $\sin\theta = \frac{12}{13}$ $\csc\theta = \frac{13}{12}$ **3.** $\sin\theta = \frac{4}{7}$ $\csc\theta = \frac{7}{4}$ **5.** $\sin\theta = \frac{5\sqrt{29}}{29}$ $\csc\theta = \frac{\sqrt{29}}{5}$

$\cos\theta = \frac{5}{13}$ $\sec\theta = \frac{13}{5}$ $\cos\theta = \frac{\sqrt{33}}{7}$ $\sec\theta = \frac{7\sqrt{33}}{33}$ $\cos\theta = \frac{2\sqrt{29}}{29}$ $\sec\theta = \frac{\sqrt{29}}{2}$

$\tan\theta = \frac{12}{5}$ $\cot\theta = \frac{5}{12}$ $\tan\theta = \frac{4\sqrt{33}}{33}$ $\cot\theta = \frac{\sqrt{33}}{4}$ $\tan\theta = \frac{5}{2}$ $\cot\theta = \frac{2}{5}$

7. $\sin\theta = \frac{\sqrt{21}}{7}$ $\csc\theta = \frac{\sqrt{21}}{3}$ **9.** $\sin\theta = \frac{2\sqrt{30}}{15}$ $\csc\theta = \frac{\sqrt{30}}{4}$ **11.** 3/4   **13.** 4/5   **15.** 3/4   **17.** 12/13   **19.** 13/5

$\cos\theta = \frac{2\sqrt{7}}{7}$ $\sec\theta = \frac{\sqrt{7}}{2}$ $\cos\theta = \frac{\sqrt{105}}{15}$ $\sec\theta = \frac{\sqrt{105}}{7}$

$\tan\theta = \frac{\sqrt{3}}{2}$ $\cot\theta = \frac{2\sqrt{3}}{3}$ $\tan\theta = \frac{2\sqrt{14}}{7}$ $\cot\theta = \frac{\sqrt{14}}{4}$

21. 3/2   **23.** $\sqrt{2}$ **25.** −3/4   **27.** 5/4   **29.** $\sqrt{3} - \sqrt{6}$ **31.** $\sqrt{3}$ **33.** $\frac{3\sqrt{2} + 2\sqrt{3}}{6}$ **35.** $\frac{3 - \sqrt{3}}{3}$ **37.** $2\sqrt{2} - \sqrt{3}$

39. 0.6249   **41.** 0.4488   **43.** 0.8221   **45.** 1.0053   **47.** 0.4816   **49.** 1.0729   **51.** 0.3153   **53.** 1.2331   **55.** 9.5 ft   **59.** 4:28 PM
61. 5.1 ft   **63.** 1.7 mi   **65.** 1400 ft   **67.** 612 ft   **69.** 5.60 × 10² ft   **71.** 5.2 m   **73.** 49.9 m ≤ h ≤ 52.1 m   **75.** 8.5 ft   **77.** 63.7 ft
83. $A_{10} \approx 2.938926261$, $A_{50} \approx 3.133330839$, $A_{100} \approx 3.139525976$, $A_{1000} \approx 3.141571983$, $A_{10,000} \approx 3.141592447$, $A_n \ll \pi$ as n increases because the area of the polygon approaches the area of the circle, which is π.

EXERCISE SET 6.3, page 320

1. $\sin\theta = 3\sqrt{13}/13$ $\csc\theta = \sqrt{13}/3$ **3.** $\sin\theta = 3\sqrt{13}/13$ $\csc\theta = \sqrt{13}/3$ **5.** $\sin\theta = -5\sqrt{89}/89$ $\csc\theta = -\sqrt{89}/5$
$\cos\theta = 2\sqrt{13}/13$ $\sec\theta = \sqrt{13}/2$ $\cos\theta = -2\sqrt{13}/13$ $\sec\theta = -\sqrt{13}/2$ $\cos\theta = -8\sqrt{89}/89$ $\sec\theta = -\sqrt{89}/8$
$\tan\theta = 3/2$ $\cot\theta = 2/3$ $\tan\theta = -3/2$ $\cot\theta = -2/3$ $\tan\theta = 5/8$ $\cot\theta = 8/5$
7. $\sin\theta = 0$ $\csc\theta$ is undefined   **9.** quadrant I   **11.** quadrant IV   **13.** quadrant III   **15.** $\tan\theta = \sqrt{3}/3$ **17.** $\cot\theta = -1$
$\cos\theta = -1$ $\sec\theta = -1$
$\tan\theta = 0$ $\cot\theta$ is undefined
19. $\tan\theta = -\sqrt{3}/3$ **21.** $\csc\theta = 2\sqrt{3}/3$ **23.** $\cot\theta = -\sqrt{3}/3$ **25.** 0.7986   **27.** −0.4384   **29.** −1.2690   **31.** −0.5878
33. −1.7013   **35.** −3.8552   **37.** −6.3138   **39.** −3.2361   **41.** 8.1443   **43.** −0.8511   **45.** 1.6304   **47.** 0.9957   **49.** 0   **51.** 1

53. −3/2   **55.** 1   **57.** 30°, 150°   **59.** 150°, 210°   **61.** 225°, 315°   **63.** $\frac{3\pi}{4}, \frac{7\pi}{4}$ **65.** $\frac{5\pi}{6}, \frac{11\pi}{6}$ **67.** $\frac{\pi}{3}, \frac{2\pi}{3}$ **79.** (0.2079, 0.9781)

81. (−0.9900, 0.1411)   **83.** (0.3746, −0.9272)   **85. a.** 1   **b.** varies between 0 and 1 inclusive   **c.** varies between 0 and 1 inclusive.

EXERCISE SET 6.4, page 328

1. $\left(\frac{\sqrt{3}}{2}, \frac{1}{2}\right)$ **3.** $\left(-\frac{\sqrt{3}}{2}, -\frac{1}{2}\right)$ **5.** $\left(\frac{1}{2}, -\frac{\sqrt{3}}{2}\right)$ **7.** $\left(\frac{\sqrt{3}}{2}, -\frac{1}{2}\right)$ **9.** (−1, 0)   **11.** $\left(-\frac{1}{2}, -\frac{\sqrt{3}}{2}\right)$ **13.** $-\frac{\sqrt{3}}{3}$ **15.** $-\frac{1}{2}$

17. $-\frac{2\sqrt{3}}{3}$ **19.** −1   **21.** $-\frac{2\sqrt{3}}{3}$ **23.** 0.9391   **25.** −1.1528   **27.** −0.2679   **29.** 0.8090   **31.** 48.0889   **33.** odd   **35.** neither

37. even   **39.** odd   **49.** $\sin t$ **51.** $\sec t$ **53.** $-\tan^2 t$ **55.** $-\cot t$ **57.** $\cos^2 t$ **59.** $2\csc^2 t$ **61.** $\csc^2 t$ **63.** 1   **65.** $-\frac{\sin^2 t}{\cos t}$

67. $\csc t \sec t$ **69.** $1 - 2 \sin t + \sin^2 t$ **71.** $1 - 2 \sin t \cos t$ **73.** $\cos^2 t$ **75.** $2 \csc t$ **77.** $(\cos t - \sin t)(\cos t + \sin t)$
79. $(\tan t + 2)(\tan t - 3)$ **81.** $(2 \sin t + 1)(\sin t - 1)$ **83.** $(\cos t - \sin t)(\cos t + \sin t)$ **85.** $\sin t = \sqrt{1 - \cos^2 t}$
87. $\csc t = \sqrt{1 + \cot^2 t}$ **89.** $\dfrac{\sqrt{2}}{2}$ **91.** $-\dfrac{\sqrt{3}}{3}$ **93. a.** CD   **b.** OD   **c.** AB   **99.** -1 **101.** 6

EXERCISE SET 6.5, page 336

1. $2, 2\pi$ **3.** $1, \pi$ **5.** $1/2, 1$ **7.** $2, 4\pi$ **9.** $1/2, 2\pi$ **11.** $1, 8\pi$ **13.** $2, 6$ **15.** $3, 3\pi$ **17.**

19.   **21.**   **23.**   **25.**   **27.**

29.   **31.**   **33.**   **35.**   **37.**

39.   **41.**   **43.**   **45.**   **47.**

49.   **51.**   **53.** 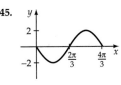   **55.** $y = \cos 2x$ **57.** $y = 2 \sin \dfrac{2}{3} x$

59. $y = -2 \cos \pi x$ **61.**   **63.** 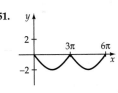 $g(x) = 2 \cos x$ $f(x) = 2 \cos \dfrac{x}{2}$ **65.**

67.   **69.**   **71.**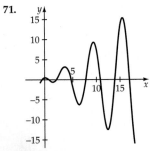

73.

x	-0.1	-0.05	-0.01	-0.001	0.001	0.01	0.05	0.1
$\dfrac{\sin x}{x}$	0.99833	0.99958	0.99998	0.99999	0.99999	0.99998	0.99958	0.99833

The graph of f approaches 1 as x approaches 0.

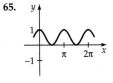

75.

max = e

min = $\dfrac{1}{e} \approx 0.3679$

period = 2π

77. $y = 2 \sin \dfrac{2}{3}x$ **79.** $y = 4 \sin \pi x$ **81.** $y = 3 \cos 4x$ **83.** $y = 3 \cos \dfrac{4\pi}{5}x$

85. $a = 60,\ p = 20,\ B = \dfrac{\pi}{10}$

$f(t) = 60 \cos \dfrac{\pi}{10}t$

87. $f(t) = 4 \sin 3t,\ 3.1$ ft

EXERCISE SET 6.6, page 344

1. $\pi/2 + k\pi$, k an integer   **3.** $\pi/2 + k\pi$, k an integer   **5.** 2π **7.** π **9.** 2π **11.** $2\pi/3$ **13.** $\pi/3$ **15.** 8π **17.** 1

19. 4   **21.** 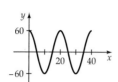   **23.**   **25.** 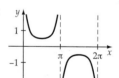   **27.**

29. 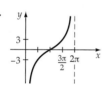   **31.**   **33.** 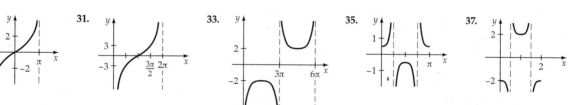   **35.** 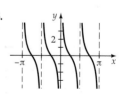   **37.**

39.   **41.** 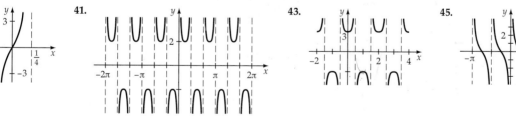   **43.** 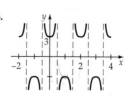   **45.**

47. 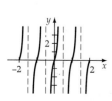   **49.** $y = \cot \dfrac{3}{2}x$ **51.** $y = \csc \dfrac{2}{3}x$ **53.** $y = \sec \dfrac{3}{4}x$ **55.**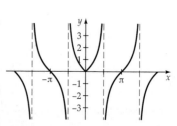

57.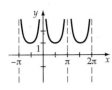

59.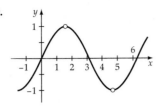

61.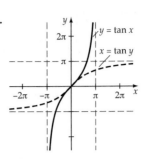

63. $y = \tan 3x$ **65.** $y = \sec \dfrac{8}{3}x$

67. $y = \cot \dfrac{\pi}{2}x$ **69.** $y = \csc \dfrac{4\pi}{3}x$

EXERCISE SET 6.7, page 348

1. $2, \pi/2, 2\pi$ **3.** $1, \pi/8, \pi$ **5.** $4, -\pi/4, 3\pi$ **7.** $5/4, 2\pi/3, 2\pi/3$ **9.** $\pi/8, \pi/2$ **11.** $-3\pi, 6\pi$ **13.** $\pi/16, \pi$ **15.** $-12\pi, 4\pi$

17.   **19.** 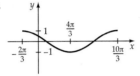 **21.**   **23.**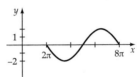

25.   **27.**   **29.**   **31.**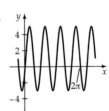

33.   **35.**   **37.** 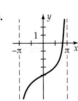 **39.** 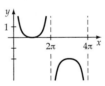 **41.**

43.   **45.** 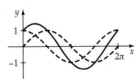 **47.** 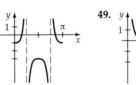 **49.** 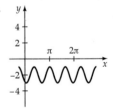 **51.**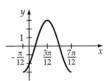

53.   **55.** 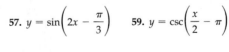 **57.** $y = \sin\left(2x - \dfrac{\pi}{3}\right)$ **59.** $y = \csc\left(\dfrac{x}{2} - \pi\right)$

61. $y = \sec\left(x - \dfrac{\pi}{2}\right)$ **63.** 25 ppm **65.** $y = 7\cos 10\pi t + 5$ **67.** $s = 400\tan\left(\dfrac{\pi}{5}t\right)$, t in sec **69.** $y = 3\cos\dfrac{\pi}{6}t + 9$, $y = 12$ ft

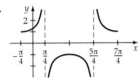 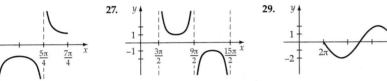 

71.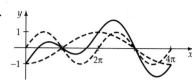

73.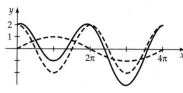

75.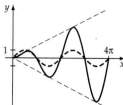

77.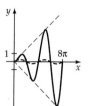

79.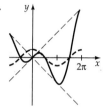

81.

83.

85. $y = 2 \sin\left(2x - \dfrac{2\pi}{3}\right)$ **87.** $y = \tan\left(\dfrac{x}{2} - \dfrac{\pi}{4}\right)$ **89.** $y = \sec\left(\dfrac{x}{2} - \dfrac{3\pi}{8}\right)$

91. 1   **93.** $\cos^2 x + 2$ **95.** 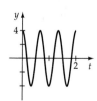   **97.**

EXERCISE SET 6.8, page 355

1. $2, \pi, 1/\pi$ **3.** $3, 3\pi, 1/(3\pi)$ **5.** $4, 2, 1/2$ **7.** $3/4, 4, 1/4$ **9.** $y = 4 \cos 3\pi t$ **11.** $y = \dfrac{3}{2} \cos \dfrac{4\pi}{3} t$ **13.** $y = 2 \sin 2t$

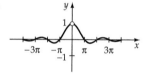

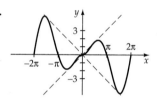

15. $y = \sin \pi t$ **17.** $y = 2 \sin 2\pi t$ **19.** $y = \dfrac{1}{2} \cos 4t$ **21.** $y = 2.5 \cos \pi t$ **23.** $y = \dfrac{1}{2} \cos \dfrac{2\pi}{3} t$ **25.** $y = 4 \cos 4t$

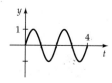

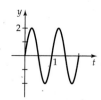

27. 4π, $1/(4\pi)$, 2

$$y = -2\cos\frac{1}{2}t$$

29. $\dfrac{\pi\sqrt{3}}{2}, \dfrac{2\sqrt{3}}{3\pi}$

$$y = \cos\frac{4\sqrt{3}}{3}t$$

31. $\sqrt{10}\,\pi$, $\sqrt{10}/(10\pi)$, 6

$$y = 6\cos\frac{\sqrt{10}}{5}t$$

33.

35.

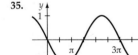

37. $y = \dfrac{1}{2}\cos 2\sqrt{2}\,t$,

$f = \sqrt{2}/\pi$

39. period = 2/3

$$y = \frac{1}{2}\cos 3\pi t$$

for displacement
in feet.

41. Period increases by a factor of $\sqrt{2}$

CHAPTER 6 TRUE/FALSE EXERCISES, page 360

1. False, the initial side must be along the positive x-axis   **2.** True   **3.** True   **4.** False, in the third quadrant $\cos\theta < 0$ and $\tan\theta > 0$
5. False, $\sec^2\theta - \tan^2\theta = 1$ is an identity.   **6.** False, the tangent function has no amplitude.   **7.** False, the period is 2π.   **8.** True

9. False, $\sin 45° + \cos(90° - 45°) = \sin 45° + \cos 45° = \dfrac{\sqrt{2}}{2} + \dfrac{\sqrt{2}}{2} = \sqrt{2}$.

10. False, $\sin\left(\dfrac{\pi}{2} + \dfrac{\pi}{2}\right) = \sin\pi = 0$, $\sin\dfrac{\pi}{2} + \sin\dfrac{\pi}{2} = 1 + 1 = 2$ **11.** False, $\sin^2\dfrac{\pi}{6} = \left(\dfrac{1}{2}\right)^2 = \dfrac{1}{4}$, $\sin\left(\dfrac{\pi}{6}\right)^2 = \sin\dfrac{\pi^2}{36} \approx 0.2707$

12. False, the phase shift is $\dfrac{\pi/3}{2} = \dfrac{\pi}{6}$.   **13.** True   **14.** False, 1 rad $\approx 57.3°$

CHAPTER 6 REVIEW EXERCISES, page 360

1. $37.5\overline{6}°$ **3.** $114.59°$ **5.** 3.93 m   **7.** 55 radians/s   **9.** $\dfrac{\sqrt{5}}{2}$ **11.** $\dfrac{3\sqrt{5}}{5}$ **13. a.** $-\dfrac{2\sqrt{3}}{3}$ **b.** 1   **c.** -1 **d.** $-\dfrac{1}{2}$

15. a. $-\dfrac{1}{2}$ **b.** $\dfrac{\sqrt{3}}{3}$ **17. a.** $\dfrac{\sqrt{2}}{2}$ **b.** -1 **19.** even   **23.** $\tan\phi$ **25.** $\tan^2\phi$ **27.** 0   **29.** no amplitude, period $\dfrac{\pi}{3}$, phase shift 0

31. amplitude 1, period π, phase shift $\dfrac{\pi}{3}$ **33.** no amplitude, period 2π, phase shift $\dfrac{\pi}{4}$

35.

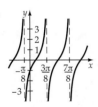

37.

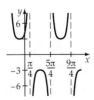

39.

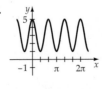

41.

43.

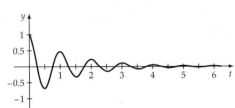

45.

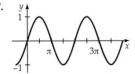

47.

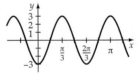

49.

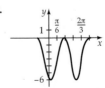

51.

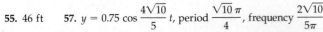

53. 12.3 ft   **55.** 46 ft   **57.** $y = 0.75\cos\dfrac{4\sqrt{10}}{5}t$, period $\dfrac{\sqrt{10}\,\pi}{4}$, frequency $\dfrac{2\sqrt{10}}{5\pi}$

CHAPTER 6 TEST, page 362

1. $\dfrac{5\pi}{6}$ **2.** $\dfrac{\pi}{12}$ **3.** 13.1 cm   **4.** 12π rad/s   **5.** 80 cm/s   **6.** $\dfrac{\sqrt{58}}{7}$ **7.** 1.0864   **8.** $\dfrac{\sqrt{3} - 6}{6}$ **9.** $\left(\dfrac{\sqrt{3}}{2}, -\dfrac{1}{2}\right)$ **10.** $\sin^2 t$

11. $\dfrac{\pi}{3}$ **12.** amplitude 3, period π, phase shift $-\dfrac{\pi}{4}$

13. 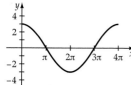   **14.** 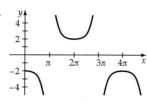   **15.**   **16.**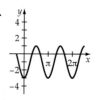

17. 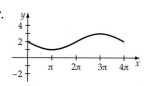   **18.**    **19.** 25.4 m   **20.** $y = 13 \sin \dfrac{2\pi}{5} t$

EXERCISE SET 7.1, page 366

1. No   **3.** No   **5.** No   **7.** No   **9.** No   **11.** No   **67.** 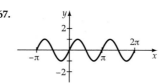   **69.**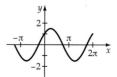

71. $\cos x = \pm\sqrt{1 - \sin^2 x}$ **73.** $\sec x = \pm\dfrac{\sqrt{1 - \sin^2 x}}{1 - \sin^2 x}$

EXERCISE SET 7.2, page 373

1. $\dfrac{\sqrt{6} + \sqrt{2}}{4}$ **3.** $\dfrac{\sqrt{6} + \sqrt{2}}{4}$ **5.** 0   **7.** $\dfrac{-\sqrt{6} + \sqrt{2}}{4}$ **9.** $-\dfrac{\sqrt{6} + \sqrt{2}}{4}$ **11.** $2 + \sqrt{3}$ **13.** 0   **15.** 1/2   **17.** $\sqrt{3}$

19. $\sin 5x$ **21.** $\cos x$ **23.** $\sin 4x$ **25.** $\cos 2x$ **27.** $\sin x$ **29.** $\tan 7x$ **31. a.** $-77/85$ **b.** 84/85   **c.** 13/84

33. a. $-63/65$ **b.** $-56/65$ **c.** $-63/16$ **35. a.** 63/65   **b.** 56/65   **c.** 33/56   **37. a.** $-77/85$ **b.** $-84/85$ **c.** $-13/84$

39. a. $-33/65$ **b.** $-16/65$ **c.** 63/16   **41. a.** $-56/65$ **b.** $-63/65$ **c.** 16/63

67. 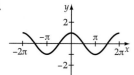   **69.**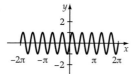

EXERCISE SET 7.3, page 380

1. $\sin 4\alpha$ **3.** $\cos 10\beta$ **5.** $\cos 6\alpha$ **7.** $\tan 6\alpha$ **9.** $\dfrac{\sqrt{2 + \sqrt{3}}}{2}$ **11.** $\sqrt{2} + 1$ **13.** $-\dfrac{\sqrt{2 + \sqrt{2}}}{2}$ **15.** $\dfrac{\sqrt{2 - \sqrt{2}}}{2}$

17. $\dfrac{\sqrt{2 - \sqrt{2}}}{2}$ **19.** $\dfrac{\sqrt{2 - \sqrt{3}}}{2}$ **21.** $-2 - \sqrt{3}$ **23.** $\dfrac{\sqrt{2 + \sqrt{3}}}{2}$ **25.** $\sin 2\theta = -24/25$, $\cos 2\theta = 7/25$, $\tan 2\theta = -24/7$

27. $\sin 2\theta = -240/289$, $\cos 2\theta = 161/289$, $\tan 2\theta = -240/161$ **29.** $\sin 2\theta = -336/625$, $\cos 2\theta = -527/625$, $\tan 2\theta = 336/527$

31. $\sin 2\theta = 240/289$, $\cos 2\theta = -161/289$, $\tan 2\theta = -240/161$ **33.** $\sin 2\theta = -720/1681$, $\cos 2\theta = 1519/1681$, $\tan 2\theta = -720/1519$

35. $\sin 2\theta = 240/289$, $\cos 2\theta = -161/289$, $\tan 2\theta = -240/161$ **37.** $\sin \alpha/2 = 5\sqrt{26}/26$, $\cos \alpha/2 = \sqrt{26}/26$, $\tan \alpha/2 = 5$

39. $\sin \alpha/2 = 5\sqrt{34}/34$, $\cos \alpha/2 = -3\sqrt{34}/34$, $\tan \alpha/2 = -5/3$ **41.** $\sin \alpha/2 = \sqrt{5}/5$, $\cos \alpha/2 = 2\sqrt{5}/5$, $\tan \alpha/2 = 1/2$
43. $\sin \alpha/2 = \sqrt{2}/10$, $\cos \alpha/2 = -7\sqrt{2}/10$, $\tan \alpha/2 = -1/7$ **45.** $\sin \alpha/2 = \sqrt{17}/17$, $\cos \alpha/2 = 4\sqrt{17}/17$, $\tan \alpha/2 = 1/4$
47. $\sin \alpha/2 = 5\sqrt{34}/34$, $\cos \alpha/2 = -3\sqrt{34}/34$, $\tan \alpha/2 = -5/3$ **95.**

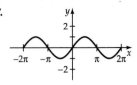

97.

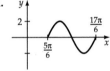

EXERCISE SET 7.4, page 386

1. $\sin 3x - \sin x$ **3.** $\dfrac{1}{2}[\sin 8x - \sin 4x]$ **5.** $\sin 8x + \sin 2x$ **7.** $\dfrac{1}{2}[\cos 4x - \cos 6x]$ **9.** 1/4   **11.** $-\sqrt{2}/4$ **13.** $-1/4$

15. $\dfrac{\sqrt{3} - 2}{4}$ **17.** $2 \sin 3\theta \cos \theta$ **19.** $2 \cos 2\theta \cos \theta$ **21.** $-2 \sin 4\theta \sin 2\theta$ **23.** $2 \cos 4\theta \cos 3\theta$ **25.** $2 \sin 7\theta \cos 2\theta$

27. $-2 \sin \dfrac{3}{2}\theta \sin \dfrac{1}{2}\theta$ **29.** $2 \sin \dfrac{3}{4}\theta \sin \dfrac{\theta}{4}$ **31.** $2 \cos \dfrac{5}{12}\theta \sin \dfrac{1}{12}\theta$ **49.** $y = \sqrt{2} \sin(x - 135°)$ **51.** $y = \sin(x - 60°)$

53. $y = \dfrac{\sqrt{2}}{2} \sin(x - 45°)$ **55.** $y = 17 \sin(x + 61.9°)$ **57.** $y = 8.5 \sin(x - 20.6°)$ **59.** $y = \sqrt{2} \sin(x + 3\pi/4)$ **61.** $y = \sin(x + \pi/6)$

63. $y = 9.8 \sin(x + 1.99)$ **65.** $y = 5\sqrt{2} \sin(x + 3\pi/4)$ **67.**

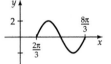

69.

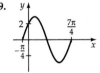

71.

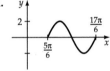

73.

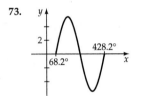

75.

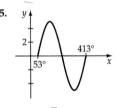

77.

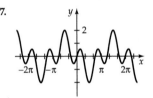

79.

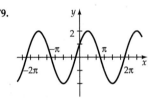

81.

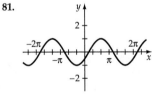

97. $\sqrt{2}$, 90°, 4π

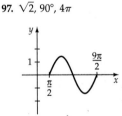

99. 2, $\pi/12$, π

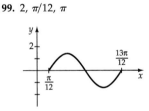

101. 2, $-1/3$, 2

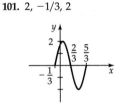

105. 90°

EXERCISE SET 7.5, page 397

1. $\pi/2$ **3.** $5\pi/6$ **5.** $-\pi/4$ **7.** $\pi/3$ **9.** $\pi/3$ **11.** $-\pi/4$ **13.** $-60°$ **15.** 120°   **17.** 30°   **19.** 30°   **21.** 0.4729
23. 1.8087   **25.** -0.9620 **27.** 0.7943   **29.** $-16.1°$ **31.** 56.3°   **33.** $-63.9°$ **35.** 134.3°   **37.** 106.7°   **39.** 84.0°   **41.** 1/2
43. 2   **45.** 3/5   **47.** 1   **49.** 1/2   **51.** $\pi/6$ **53.** $\pi/4$ **55.** not defined   **57.** 0.4636   **59.** $-\pi/6$ **61.** $-\pi/3$ **63.** 2.1735
65. 2.3986   **67.** 0.6675   **69.** 24/25   **71.** 13/5   **73.** $\pi/3$ **75.** $\pi/6$ **77.** $-\pi/4$ **79.** 0   **81.** 24/25   **83.** 12/13   **85.** 2
87. $\dfrac{2 - \sqrt{2}}{2}$ **89.** $\dfrac{2 + \sqrt{15}}{6}$ **91.** $\dfrac{1}{5}(3\sqrt{7} - 4\sqrt{3})$ **93.** $7\sqrt{2}/10$ **95.** $\dfrac{\sqrt{6} - \sqrt{2}}{4}$ **97.** $\sqrt{1 - x^2}$ **99.** $\dfrac{\sqrt{x^2 - 1}}{x}$

105.   **107.**   **109.**   **111.**   **113.**

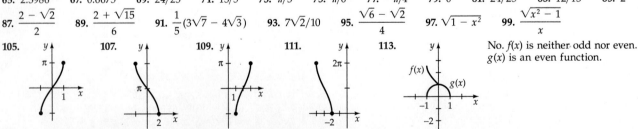

No. $f(x)$ is neither odd nor even.
$g(x)$ is an even function.

115.   **117.** 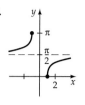   **119.**   **121.** 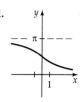   **127.** $y = \dfrac{1}{3}\tan 5x$

129. $y = 3 + \cos\left(x - \dfrac{\pi}{3}\right)$ **131.** 0.1014

EXERCISE SET 7.6, page 403

1. $\pi/4, 7\pi/4$ **3.** $\pi/3, 4\pi/3$ **5.** $\pi/4, \pi/2, 3\pi/4, 3\pi/2$ **7.** $\pi/2, 3\pi/2$ **9.** $\pi/6, \pi/4, 3\pi/4, 11\pi/6$ **11.** $\pi/4, 3\pi/4$
13. $\pi/6, \pi/2, 5\pi/6$ **15.** $\pi/6, 5\pi/6, 7\pi/6, 11\pi/6$ **17.** $0, \pi/4, 3\pi/4, \pi, 5\pi/4, 7\pi/4$ **19.** $\pi/6, 5\pi/6, 4\pi/3, 5\pi/3$ **21.** $0, \pi/2, \pi, 3\pi/2$
23. 41.4°, 318.6°   **25.** no solution   **27.** 68.0°, 292°   **29.** no solution   **31.** 12.8°, 167.2°   **33.** 15.5°, 164.5°
35. 0°, 33.7°, 180°, 213.7°   **37.** no solution   **39.** no solution   **41.** 0°, 120°, 240°   **43.** 70.5°, 289.5°   **45.** 68.2°, 116.6°, 248.2°, 296.6°
47. 19.5°, 90°, 160.5°, 270°   **49.** 60°, 90°, 300°   **51.** 53.1°, 180°   **53.** 72.4°, 220.2°   **55.** 50.1°, 129.9°, 205.7°, 334.3°   **57.** no solution
59. 22.5°, 157.5°   **61.** $\pi/8 + k\pi/2$ where k is an integer   **63.** $\pi/10 + 2k\pi/5$ where k is an integer
65. $0 + 2k\pi, \pi/3 + 2k\pi, \pi + 2k\pi, 5\pi/3 + 2k\pi$ where k is an integer   **67.** $\pi/2 + k\pi, 5\pi/6 + k\pi$ where k is an integer
69. $0 + 2k\pi$ where k is an integer   **71.** $0, \pi$ **73.** $0, \pi/6, \pi/2, 5\pi/6, \pi, 7\pi/6, 3\pi/2, 11\pi/6$ **75.** $0, \pi/2, 3\pi/2$
77. $0, \pi/3, 2\pi/3, \pi, 4\pi/3, 5\pi/3$ **79.** $4\pi/3, 5\pi/3$ **81.** $0, \pi/4, 3\pi/4, \pi, 5\pi/4, 7\pi/4$ **83.** $\pi/6, 5\pi/6, \pi$ **85.** 0.7393   **87.** $-3.2956, 3.2956$
89. 1.16   **91.** $\pi/6, \pi/2$ **93.** $5\pi/3, 0$ **95.** $0, \pi/4, \pi/2, 3\pi/4, \pi, 5\pi/4, 3\pi/2, 7\pi/4$ **97.** $0, \pi/2, \pi, 3\pi/2$
99. $\pi/6, \pi/2, 5\pi/6, 7\pi/6, 3\pi/2, 11\pi/6$ **101.** $A = \dfrac{1}{2}r^2\theta$ **103.** $A = \dfrac{1}{2}r^2(\theta - \sin\theta)$ **105.** 0.93 ft   1.39 ft

CHAPTER 7 TRUE/FALSE EXERCISES, page 408

1. false, $\dfrac{\tan 45°}{\tan 60°} \neq \dfrac{45°}{60°}$ **2.** false, if $y = 0$ $\tan\dfrac{x}{0}$ is not defined   **3.** false, $\sin^{-1}x \neq \dfrac{1}{\sin x}$

4. false, if $\alpha = \dfrac{\pi}{2}$, $\sin 2\alpha = \sin\pi = 0$ but $2\sin\dfrac{\pi}{2} = 2$ **5.** false, $\sin(30° + 60°) \neq \sin 30° + \sin 60°$

6. False, $\sin x = 0$ has an infinite number of solutions $x = k\pi$, but $\sin x = 0$ is not an identity.

7. false, $\tan 45° = \tan 225°$ but $45° \neq 225°$ **8.** false, $\cos^{-1}\left(\cos\dfrac{3\pi}{2}\right) = \cos^{-1}(0) = \dfrac{\pi}{2} \neq \dfrac{3\pi}{2}$ **9.** false, $\cos(\cos^{-1}2) \neq 2$

10. false, if $\alpha = 1$ we get $\dfrac{\pi}{2} = \dfrac{1}{\csc 1} \approx 0.8415$ **11.** false, $\sin(180° - \theta) = \sin\theta$

12. false, since $\sin^2\theta \geq 0$ for all θ, but $\sin\theta^2$ can easily be < 0.

CHAPTER 7 REVIEW EXERCISES, page 409

1. $\dfrac{\sqrt{6} - \sqrt{2}}{4}$ **3.** $\dfrac{\sqrt{6} - \sqrt{2}}{4}$ **5.** $-\dfrac{\sqrt{6} + \sqrt{2}}{4}$ **7.** $\dfrac{\sqrt{2} - \sqrt{2}}{2}$ **9.** $\sqrt{2} + 1$ **11. a.** 0   **b.** $\sqrt{3}$ **c.** 1/2

13. a. $\sqrt{3}/2$ **b.** $-\sqrt{3}$ **c.** $-\dfrac{\sqrt{2} - \sqrt{3}}{2}$ **15.** $\sin 6x$ **17.** $\sin 3x$ **19.** $\cos\beta$ **21.** 0.2740   **23.** $-\sqrt{3}/2$ **25.** $2\sin 3\theta\sin\theta$
27. $2\sin 4\theta\cos 2\theta$ **47.** 13/5   **49.** 3/2   **51.** 4/5   **53.** 30°, 150°, 240°, 300°
55. $\pi/2 + 2k\pi, 3.8713 + 2k\pi, 5.553 + 2k\pi$ where k is an integer   **57.** $\pi/12, 5\pi/12, 13\pi/12, 17\pi/12$

59. $2, -\dfrac{\pi}{6}$ **61.** $2, -\dfrac{4\pi}{3}$ **63.**   **65.**

CHAPTER 7 TEST, page 410

5. $\dfrac{-\sqrt{6} + \sqrt{2}}{4}$ **6.** $-\dfrac{\sqrt{2}}{10}$ **8.** $\sin 9x$ **9.** $-\dfrac{7}{25}$ **12.** $\dfrac{2 - \sqrt{3}}{4}$ **13.** $y = \sin\left(x + \dfrac{5\pi}{6}\right)$ **14.** $150°$ **15.** 0.701 **16.** $\dfrac{5}{13}$

17.

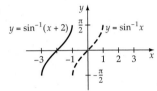

18. $41.8°, 138.2°$ **19.** $0°, 180°$ **20.** $90°, 120°, 240°$

EXERCISE SET 8.1, page 417

1. $C = 77°, b = 16, c = 17$ **3.** $B = 38°, a = 18, c = 10$ **5.** $C = 15°, B = 33°, C = 7.7$ **7.** $C = 32.6°, c = 21.6, a = 39.8$
9. $B = 47.7°, a = 57.4, b = 76.3$ **11.** $A = 58.5°, B = 7.3°, a = 81.5$ **13.** $C = 59°, B = 84°, b = 46$ or $C = 121°, B = 22°, b = 17$
15. no solution   **17.** no solution   **19.** $C = 19.8°, B = 145.4°, b = 10.7$ or $C = 160.2°, B = 5.0°, b = 1.64$ **21.** no solution
23. $C = 51.21°, A = 11.47°, a = 59.00$ **25.** 114 ft   **27.** 126 yd   **29.** 96 ft   **31.** 33 ft   **33.** 8.1 mi   **35.** 1170 mi   **37.** 257 m

EXERCISE SET 8.2, page 423

1. 12.5   **3.** 150.5   **5.** 29   **7.** 9.5   **9.** 10.5   **11.** 40.1   **13.** 90.7   **15.** 38.7°   **17.** 89.6°   **19.** 47.9°   **21.** 116.7°
23. 80.3°   **25.** 139   **27.** 52.9   **29.** 58.8   **31.** 299   **33.** 36.3   **35.** 7.3   **37.** 709 mi   **39.** 347 mi   **41.** 40 cm
43. 9.7 in, 24.9 in   **45.** 55.4 cm   **47.** 2764 ft   **49.** $47{,}520 \text{ m}^2$ **51.** 203 m^2 **53.** 162 in^2 **55.** \$41,479   **57.** 6.23 acres
59. 12.5°   **61.** 72.7°   **63.** 52.0 cm   **69.** 137 in^3

EXERCISE SET 8.3, page 437

1. $\langle 7, -1 \rangle$ **3.** $\langle -7, -5 \rangle$ **5.** $\langle 0, 8 \rangle$ **7.** $5, 126.9°, \left\langle -\dfrac{3}{5}, \dfrac{4}{5} \right\rangle$ **9.** $44.7, 296.6°, \left\langle \dfrac{\sqrt{5}}{5}, \dfrac{-2\sqrt{5}}{5} \right\rangle$ **11.** $4.5, 296.6°, \left\langle \dfrac{\sqrt{5}}{5}, \dfrac{-2\sqrt{5}}{5} \right\rangle$

13. $45.7, 336.8°, \left\langle \dfrac{7\sqrt{58}}{58}, \dfrac{-3\sqrt{58}}{58} \right\rangle$ **15.** $\langle -6, 12 \rangle$ **17.** $\langle -1, 10 \rangle$ **19.** $\left\langle -\dfrac{11}{6}, \dfrac{7}{3} \right\rangle$ **21.** $2\sqrt{5}$ **23.** $2\sqrt{109}$ **25.** $-8i + 12j$

27. $14i - 6j$ **29.** $\dfrac{11}{12}i + \dfrac{1}{2}j$ **31.** $\sqrt{113}$ **33.** $4.5i + 2.3j$ **35.** $2.8i + 2.8j$ **37.** 378 mph, 120°   **39.** 254.4 mph, 86.3°

41. 292.7 lb   **43.** 24.7 lb   **45.** -3 **47.** 0   **49.** 1   **51.** 0   **53.** 79.7°   **55.** 45°   **57.** 90°   **59.** 180°   **61.** $\dfrac{46}{5}$

63. $\dfrac{14\sqrt{29}}{29}$ **65.** $\sqrt{5}$ **67.** $-\dfrac{11\sqrt{5}}{5}$ **69.** 954 ft-lbs   **71.** 779 ft-lbs   **73.**

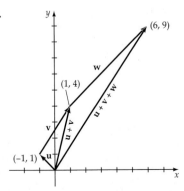

75. The vector from $P_1(3, -1)$ to $P_2(5, -4)$ is equivalent to $2i - 3j$.   **77.** $\vec{v} \cdot \vec{w} = 0$, the lines are perpendicular
79. A perpendicular vector is $\langle 7, 2 \rangle$.   **81.** 6.4 lb   **83.** No   **87.** The same amount of work is done.

EXERCISE SET 8.4, page 446

1.–7.

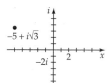

$-2-2i$ • $\bullet$ $\sqrt{3}-i$
    $-2i$

$\bullet$ $3-5i$

9. $\sqrt{2}$ cis 315°   **11.** 2 cis 330°   **13.** 3 cis 90°   **15.** 5 cis 180°   **17.** $\sqrt{2}+i\sqrt{2}$ **19.** $\dfrac{\sqrt{2}}{2}-\dfrac{i\sqrt{2}}{2}$

21. $-3\sqrt{2}+3i\sqrt{2}$ **23.** 8   **25.** $-\sqrt{3}+i$ **27.** $-3i$ **29.** $-4\sqrt{2}+4i\sqrt{2}$ **31.** $\dfrac{9\sqrt{3}}{2}-\dfrac{9}{2}i$ **33.** $-0.832+1.819i$

35. 6 cis 255°   **37.** 12(cos 335° + i sin 335°)   **39.** $10\left(\cos\dfrac{16\pi}{15}+i\sin\dfrac{16\pi}{15}\right)$ **41.** 24 cis 6.5   **43.** $-4-4i\sqrt{3}$ **45.** 3i

47. $-\dfrac{3\sqrt{3}}{2}+\dfrac{3i}{2}$ **49.** $-2.081+4.546i$ **51.** $-\dfrac{1}{2}+\dfrac{\sqrt{3}}{2}i$ **53.** $-\dfrac{3}{8}+\dfrac{\sqrt{3}}{8}i$ **55.** 256 cis 240°   **57.** 32 cis 120°

59. 32(cos 45° + i sin 45°)   **61.** 64(cos 0° + i sin 0°)   **63.** 32 cis 270°   **65.** $1024\sqrt{2}$ cis 315°   **67.** cis 270°
69. $1.148+2.772i$ **71.** $1.879+0.684i$ **73.** $0.951+0.309i$ **75.** 1   **77.** $1.070+0.213i$ **79.** $-0.276+1.563i$
    $-1.148-2.772i$ $0.347+1.970i$ i $-0.5+0.866i$ $-0.213+1.070i$ $-1.216-1.020i$
            $-1.532+1.286i$ $-0.951+0.309i$ $-0.5-0.866i$ $-1.070-0.213i$ $1.492-0.543i$
            $-1.879-0.684i$ $-0.588-0.809i$ $0.213-1.070i$
            $-0.347-1.970i$ $0.588-0.809i$
            $1.532-1.286i$

81. $2\sqrt{2}+2i\sqrt{6}$ **83.** cis 67.5°   **85.** 4 cis 60°   **87.** 1.19 cis 75°
    $-2\sqrt{2}-2i\sqrt{6}$ cis 157.5°   4 cis 180°   1.19 cis 165°
            cis 247.5°   4 cis 300°   1.19 cis 255°
            cis 337.5°   1.19 cis 345°

CHAPTER 8 TRUE/FALSE EXERCISES, page 450

1. false, cannot solve a triangle given the angle opposite one of the given sides.   **2.** true   **3.** true   **4.** false, $2i \neq 2j$ **5.** true
6. true   **7.** true   **8.** true   **9.** true   **10.** false, $\mathbf{v}\cdot\mathbf{v}=a^2+b^2$ **11.** false, let $\mathbf{v}=\mathbf{i}+\mathbf{j}$, $\mathbf{w}=\mathbf{i}-\mathbf{j}$ then $\mathbf{v}\cdot\mathbf{w}=1-1=0$
12. true   **13.** false, $z^2=r^2(\cos 2\theta + i \sin 2\theta)$ **14.** true   **15.** false; $i=\cos\dfrac{\pi}{2}+i\sin\dfrac{\pi}{2}$ **16.** true

CHAPTER 8 REVIEW EXERCISES, page 451

1. $B=53°$, $a=11$, $c=18$ **3.** $B=48°$, $C=95°$, $A=37°$ **5.** $c=13$, $A=55°$, $B=90°$ **7.** No triangle is formed
9. $C=45°$, $a=29$, $b=35$ **11.** 357   **13.** 917   **15.** 792   **17.** 167   **19.** $\langle 5, 3\rangle$ **21.** 4.5, 153.4°   **23.** 3.6, 123.7°

25. $\left\langle -\dfrac{8\sqrt{89}}{89}, \dfrac{5\sqrt{89}}{89}\right\rangle$ **27.** $\dfrac{5\sqrt{26}}{26}\mathbf{i}+\dfrac{\sqrt{26}}{26}\mathbf{j}$ **29.** $\langle -7, -3\rangle$ **31.** $-6\mathbf{i}-17\mathbf{j}/2$ **33.** 420 mph, 210°   **35.** 18   **37.** -9 **39.** 86°

41. 125°   **43.** $\dfrac{10\sqrt{41}}{41}$ **45.** 662 ft-lbs   **47.** 5.29, 161°

$\bullet$
$-5+i\sqrt{3}$

$-2i$ | 2 | x

49. $2\sqrt{3}$ cis 120°

51. $-3-3i\sqrt{3}$ **53.** $-8.918+8.030i$ **55.** $-6.012-13.742i$ **57.** 3 cis 110°   **59.** $\sqrt{2}$ cis 285°   **61.** $-0.5+0.866i$ **63.** $32,768i$
65. 1.68 cis 22.5°   **67.** 1.07 cis 45°
    1.68 cis 112.5°   1.07 cis 117°
    1.68 cis 202.5°   1.07 cis 189°
    1.68 cis 292.5°   1.07 cis 261°
            1.07 cis 333°

CHAPTER 8 TEST, page 452

1. $B = 94°$, $a = 47.7$, $b = 50.7$ **2.** 10.7°   **3.** 14.9   **4.** 47.9°   **5.** 39.5   **6.** 92.6   **7.** 233.9   **8.** $-9.2\mathbf{i} - 7.7\mathbf{j}$ **9.** $-19\mathbf{i} - 29\mathbf{j}$
10. -1 **11.** 103°   **12.** $3\sqrt{3}$ cis 145°   **13.** $3.536 - 3.536i$ **14.** $8.208 - 22.553i$ **15.** 2.5 cis(−95°)   **16.** $-15.556 - 0.998i$
17. $2.598 + 1.5i$, $-2.598 + 1.5i$, or $-3i$ **18.** 26.6 mi   **19.** 21.0 mi   **20.** \$65,815

EXERCISE SET 9.1, page 459

1. vertex: (0, 0)
focus: (0, −1)
directrix: $y = 1$

3. vertex: (0, 0)
focus: (1/12, 0)
directrix: $x = -1/12$

5. vertex: (2, −3)
focus: (2, −1)
directrix: $y = -5$

7. vertex: (2, −4)
focus: (1, −4)
directrix: $x = 3$

9. vertex: (−4, 1)
focus: (−7/2, 1)
directrix: $x = -9/2$

11. vertex: (2, 2)
focus: (2, 5/2)
directrix: $y = 3/2$

13. vertex: (−4, −10)
focus: (−4, −39/4)
directrix: $y = -41/4$

15. vertex: (−7/4, 3/2)
focus: (−2, 3/2)
directrix: $x = -3/2$

17. vertex: (−5, −3)
focus: (−9/2, −3)
directrix: $x = -11/2$

19. vertex: (−3/2, 13/12)
focus: (−3/2, 1/3)
directrix: $y = 11/6$

21. vertex: (2, −5/4)
focus: (2, −3/4)
directrix: $y = -7/4$

23. vertex: (9/2, −1)
focus: (35/8, −1)
directrix: $x = 37/8$

25. vertex: (1, 1/9)
focus: (1, 31/36)
directrix: $y = -23/36$

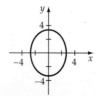

27. $x^2 = -16y$ **29.** $(x + 1)^2 = 4(y - 2)$

31. $(x - 3)^2 = 4(y + 4)$ **33.** $(x + 4)^2 = 4(y - 1)$ **35.** (−0.367, −0.367) and (1.367, 1.367)   **37.** (−1.562, 3.877) and (2.562, 12.123)
39. 3.8284   **41.** −0.4646   **43.** 4   **45.** $4|p|$ **47.**

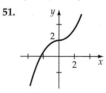

51.

53. $x^2 + y^2 - 8x - 8y - 2xy = 0$

EXERCISE SET 9.2, page 467

1. vertices: (0, 5), (0, −5)
foci: (0, 3), (0, −3)

3. vertices: (3, 0), (−3, 0)
foci: $(\sqrt{5}, 0)$, $(-\sqrt{5}, 0)$

5. vertices: (0, 3), (0, −3)
foci: $(0, \sqrt{2})$, $(0, -\sqrt{2})$

7. vertices: (0, 4), (0, −4)
foci: $\left(0, \dfrac{\sqrt{55}}{2}\right)$, $\left(0, -\dfrac{\sqrt{55}}{2}\right)$

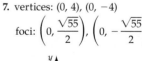

9. vertices: $(8, -2), (-2, -2)$
foci: $(6, -2), (0, -2)$

11. vertices: $(-2, 5), (-2, -5)$
foci: $(-2, 4), (-2, -4)$

13. vertices: $(1 + \sqrt{21}, 3), (1 - \sqrt{21}, 3)$
foci: $(1 + \sqrt{17}, 3), (1 - \sqrt{17}, 3)$

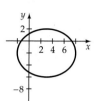 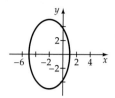 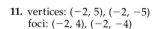

15. vertices: $(1, 2), (1, -4)$
foci: $\left(1, -1 + \dfrac{\sqrt{65}}{3}\right), \left(1, -1 - \dfrac{\sqrt{65}}{3}\right)$

17. vertices $(2, 0), (-2, 0)$
foci: $(1, 0), (-1, 0)$

19. vertices: $(0, 5), (0, -5)$
foci: $(0, 3), (0, -3)$

21. vertices $(0, 4), (0, -4)$
foci: $(0, \sqrt{39}/2), (0, -\sqrt{39}/2)$

23. vertices: $(3, 6), (3, 2)$
foci: $(3, 4 + \sqrt{3}), (3, 4 - \sqrt{3})$

25. vertices: $(-1, -3), (5, -3)$
foci: $(0, -3), (4, -3)$

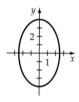 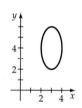

27. vertices: $(2, 4), (2, -4)$
foci $(2, \sqrt{7}), (2, -\sqrt{7})$

29. vertices: $(-1, 6), (-1, -4)$
foci: $(-1, 4), (-1, -2)$

31. vertices: $(11/2, -1) (1/2, -1)$
foci: $(3 + \sqrt{17}/2, -1), (3 - \sqrt{17}/2, -1)$

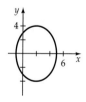 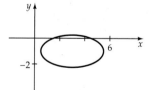

33. $\dfrac{x^2}{25} + \dfrac{y^2}{9} = 1$

35. $\dfrac{x^2}{36} + \dfrac{y^2}{16} = 1$

37. $\dfrac{x^2}{36} + \dfrac{y^2}{81/8} = 1$

39. $\dfrac{(x + 2)^2}{16} + \dfrac{(y - 4)^2}{7} = 1$

41. $\dfrac{(x - 2)^2}{25/24} + \dfrac{(y - 4)^2}{25} = 1$

43. $\dfrac{(x - 5)^2}{16} + \dfrac{(y - 1)^2}{25} = 1$

45. $\dfrac{x^2}{25} + \dfrac{y^2}{21} = 1$

47. $\dfrac{x^2}{20} + \dfrac{y^2}{36} = 1$

49. $\dfrac{(x - 1)^2}{25} + \dfrac{(y - 3)^2}{21} = 1$

51. $\dfrac{x^2}{80} + \dfrac{y^2}{144} = 1$

53. The information does not describe an ellipse.

55. $\dfrac{x^2}{36} + \dfrac{y^2}{27} = 1$

57. $\dfrac{(x - 1)^2}{16} + \dfrac{(y - 2)^2}{12} = 1$

59. $9/2$

63. $x = \pm\dfrac{9\sqrt{5}}{5}$

EXERCISE SET 9.3, page 476

1. center: (0, 0)
vertices: (±4, 0)
foci: (±√41, 0)
asymptotes: $y = ±5x/4$

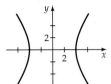

3. center: (0,0)
vertices: (0, ±2)
foci: (0, ±√29)
asymptotes: $y = ±2x/5$

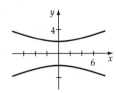

5. center: (0, 0)
vertices: (±√7, 0)
foci: (±4, 0)
asymptotes: $y = ±\dfrac{3\sqrt{7}}{7}x$

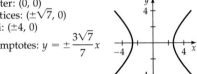

7. center: (0, 0)
vertices: $\left(±\dfrac{3}{2}, 0\right)$
foci: $\left(±\dfrac{\sqrt{73}}{2}, 0\right)$
asymptotes: $y = ±\dfrac{8}{3}x$

9. center: (3, −4)
vertices: (7, −4), (−1, −4)
foci: (8, −4), (−2, −4)
asymptotes: $y + 4 = ±3(x − 3)/4$

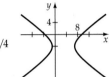

11. center: (1, −2)
vertices: (1, 0), (1, −4)
foci: (1, −2 ± 2√5)
asymptotes: $y + 2 = ±(x − 1)/2$

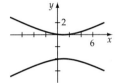

13. center: (−2, 0)
vertices: (1, 0), (−5, 0)
foci: (−2 ± √34, 0)
asymptotes: $y = ±\dfrac{5}{3}(x + 2)$

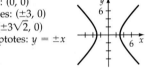

15. center: (1, −1)
vertices: $\left(\dfrac{7}{3}, −1\right), \left(−\dfrac{1}{3}, −1\right)$
foci: $\left(1 ± \dfrac{\sqrt{97}}{3}, −1\right)$
asymptotes: $y + 1 = ±\dfrac{9}{4}(x − 1)$

17. center: (0, 0)
vertices: (±3, 0)
foci: (±3√2, 0)
asymptotes: $y = ±x$

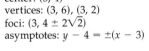

19. center: (0, 0)
vertices: (0, ±3)
foci: (0, ±5)
asymptotes: $y = ±3x/4$

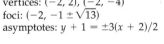

21. center: (0, 0)
vertices: (0, ±2/3)
foci: (0, ±√5/3)
asymptotes: $y = ±2x$

23. center: (3, 4)
vertices: (3, 6), (3, 2)
foci: (3, 4 ± 2√2)
asymptotes: $y − 4 = ±(x − 3)$

25. center: (−2, −1)
vertices: (−2, 2), (−2, −4)
foci: (−2, −1 ± √13)
asymptotes: $y + 1 = ±3(x + 2)/2$

27. center: (−4, 3)
vertices: (−2, 3), (−6, 3)
foci: (−4 ± 2√5, 3)
asymptotes: $y − 3 = ±2(x + 4)$

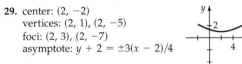

29. center: (2, −2)
vertices: (2, 1), (2, −5)
foci: (2, 3), (2, −7)
asymptote: $y + 2 = ±3(x − 2)/4$

31. center: (−1, −1)
vertices: (−1/2, −1), (−3/2, −1)
foci: ((−6 ± √13)/6, −1)
asymptotes: $y + 1 = ±2(x + 1)/3$

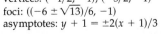

33. $\dfrac{x^2}{9} − \dfrac{y^2}{7} = 1$

35. $\dfrac{y^2}{20} − \dfrac{x^2}{5} = 1$

37. $\dfrac{y^2}{9} − \dfrac{x^2}{36/7} = 1$

39. $\dfrac{y^2}{16} − \dfrac{x^2}{64} = 1$

41. $\dfrac{(x-4)^2}{4} - \dfrac{(y-3)^2}{5} = 1$ **43.** $\dfrac{41(x-4)^2}{144} - \dfrac{41(y+2)^2}{225} = 1$ **45.** $\dfrac{(y-2)^2}{3} - \dfrac{(x-7)^2}{12} = 1$ **47.** $\dfrac{(y-7)^2}{1} - \dfrac{(x-1)^2}{3} = 1$

49. $\dfrac{x^2}{4} - \dfrac{y^2}{12} = 1$ **51.** $\dfrac{7(x-4)^2}{36} - \dfrac{(y-1)^2}{4} = 1$ and $\dfrac{7(y-1)^2}{36} - \dfrac{(x-4)^2}{4} = 1$ **53.** ellipse   **55.** parabola

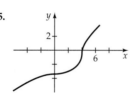

57. parabola   **59.** ellipse   **61.** $\dfrac{(x - \sqrt{1215}/2)^2}{324} + \dfrac{y^2}{81/4} = 1$ **63.** $\dfrac{x^2}{2500} - \dfrac{y^2}{7500} = 1$ **65.** $\dfrac{x^2}{1} - \dfrac{y^2}{3} = 1$

67. $\dfrac{y^2}{9} - \dfrac{x^2}{7} = 1$ **69.** $y + 2 = \pm 9/5$ **75.**

EXERCISE SET 9.4, page 489

1.–7.   **9.** 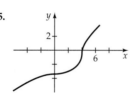   **11.**   **13.**   **15.**

17.    **19.** 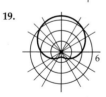   **21.** 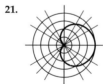   **23.**    **25.**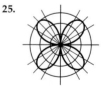

27.    **29.**   **31.** 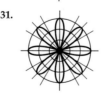   **33.** 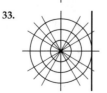   **35.**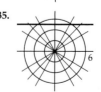

37.    **39.**   **41.** $(2, -60°)$ **43.** $\left(\dfrac{3}{2}, -\dfrac{3\sqrt{3}}{2}\right)$ **45.** $(0, 0)$ **47.** $(5, 53.1°)$ **49.** $x^2 + y^2 - 3x = 0$

51. $x = 3$ **53.** $x^2 + y^2 = 16$ **55.** $x^4 - y^2 + x^2 y^2 = 0$ **57.** $y^2 + 4x - 4 = 0$ **59.** $y = 2x + 6$ **61.** $r = 2\csc\theta$ **63.** $r = 2$

65. $r \cos^2\theta = 8 \sin\theta$ **67.** $r^2(\cos 2\theta) = 25$ **69.**   **71.** 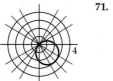   **73.**

75.   **79.**   **81.**   **83.**   **85.**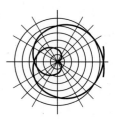

CHAPTER 9 TRUE/FALSE EXERCISES, page 493

1. false, a parabola has no asymptotes   **2.** true   **3.** false, by keeping foci fixed and varying asymptotes, we can make conjugate axis any size needed.   **4.** false, it depends on the total distance from a point to the foci.   **5.** false, parabolas have no asymptotes
6. true   **7.** false, a parabola can be a function   **8.** true   **9.** true   **10.** true

CHAPTER 9 REVIEW EXERCISES, page 493

1. vertices: $(\pm2, 0)$
foci: $(\pm2\sqrt{2}, 0)$
asymptotes: $y = \pm x$

3. vertices: $(-1, -1)$, $(7, -1)$
foci: $(3 \pm 2\sqrt{3}, -1)$
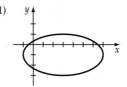

5. vertex: $(-2, 1)$
foci: $(-29/16, 1)$

7. vertices: $(-2, -2)$, $(-2, 4)$
foci: $(-2, 1 \pm \sqrt{5})$

9. vertices: $(-5, 2/3)$, $(7, 2/3)$
foci: $(1 \pm 2\sqrt{13}, 2/3)$
asymptotes: $y - 2/3 = \pm2(x - 1)/3$
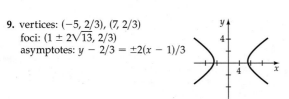

11. vertex: $(-7/2, -1)$
focus: $(-7/2, -3)$
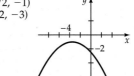

13. $\dfrac{(x - 2)^2}{25} + \dfrac{(y - 3)^2}{16} = 1$ **15.** $\dfrac{(x + 2)^2}{4} - \dfrac{(y - 2)^2}{5} = 1$ **17.** $x^2 = 3(y + 2)/2$ or $(y + 2)^2 = 12x$ **19.** $\dfrac{x^2}{36} - \dfrac{y^2}{4/9} = 1$

21. $(y - 3)^2 = -8x$ **23.** $\dfrac{(x - 1)^2}{25} + \dfrac{(y - 1)^2}{9} = 1$ **25.**   **27.** 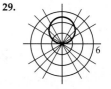   **29.**

31. 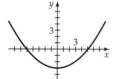   **33.**   **35.** $r \sin^2 \theta = 16 \cos \theta$ **37.** $3r \cos \theta - 2r \sin \theta = 6$ **39.** $y^2 = 8x + 16$

41. $x^4 + y^4 + 2x^2 y^2 - x^2 + y^2 = 0$

CHAPTER 9 TEST, page 494

1. focus: $(0, 2)$, vertex: $(0, 0)$, directrix: $y = -2$ **2.** focus: $(-2, 4)$, vertex: $(-2, 1)$, directrix: $y = -2$ **3.** $(y + 2)^2 = -8(x - 1)$

4.   **5.** vertices: $(0, 8)$, $(0, -8)$ **6.** 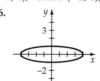   **7.** vertices: $(3, 4)$, $(3, -6)$
foci: $(0, \sqrt{55})$, $(0, -\sqrt{55})$ foci: $(3, 3)$, $(3, -5)$

8. $\dfrac{x^2}{45} + \dfrac{(y + 3)^2}{9} = 1$ **9.** $\dfrac{4}{5}$ **10.** 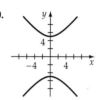   **11.** vertices: $(6, 0)$, $(-6, 0)$ **12.**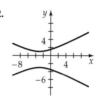
foci: $(-10, 0)$, $(10, 0)$
asymptotes: $y = \pm \dfrac{4}{3}x$

13. vertices: $(-5, 10)$, $(-5, -2)$ **14.** $\dfrac{(x + 4)^2}{4} - \dfrac{(y + 3)^2}{30} = 1$ **15.**   **16.**
foci: $(-5, 4 + 3\sqrt{5})$, $(-5, 4 - 3\sqrt{5})$

17. 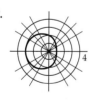   **18.** $\left(\dfrac{5}{2}, \dfrac{5\sqrt{3}}{2} \right)$ **19.** $y^2 = 8x + 16$ **20.** $r^2 \cos^2 \theta - 4r \sin \theta = 4$

EXERCISE SET 10.1, page 502

1. $(2, -4)$ **3.** $(-6/5, 27/5)$ **5.** $(3, 4)$ **7.** $(1, -1)$ **9.** $(3, -4)$ **11.** $(2, 5)$ **13.** $(-1, -1)$ **15.** $(62/25, 34/25)$ **17.** no solution
19. $(c, -4c/3 + 2)$ **21.** $(2, -4)$ **23.** $(0, 3)$ **25.** $(3c/5, c)$ **27.** $(-1/2, 2/3)$ **29.** no solution   **31.** $(-6, 3)$ **33.** $(2, -3/2)$
35. $(2\sqrt{3}, 3)$ **37.** $(38/17\pi, 3/17)$ **39.** $(\sqrt{2}, \sqrt{3})$ **41.** plane: 120 mph, wind: 30 mph   **43.** boat: 25 mph, current: 5 mph
45. \$12 for iron, \$16 for lead   **47.** 12 nickels, 7 dimes   **49.** 86   **51.** \$14,000 at 6%, \$11,000 at 6.5%
53. 8 gm of 40% gold, 12 gm of 60% gold   **55.** 20 ml of 13% solution, 30 ml of 18% solution   **57.** $x = -58/17$, $y = 52/17$
59. $x = -2$, $y = -1$ **61.** $x = 153/26$, $y = 151/26$ **63.** $x = 2 + 3i$, $y = 1 - 2i$ **65.** $x = 3 - 5i$, $y = 4i$

EXERCISE SET 10.2, page 512

1. $(2, -1, 3)$ **3.** $(2, 0, -3)$ **5.** $(2, -3, 1)$ **7.** $(-5, 1, -1)$ **9.** $(3, -5, 0)$ **11.** $(0, 2, 3)$ **13.** $(5c - 25, 48 - 9c, c)$

15. $(3, -1, 0)$ **17.** no solution   **19.** $((50 - 22c)/11, (11c - 18)/11, c)$ **21.** no solution   **23.** $\left(\dfrac{25 + 4c}{29}, \dfrac{55 - 26c}{29}, c \right)$

25. $(0, 0, 0)$ **27.** $(5c/14, 4c/7, c)$ **29.** $(-11c, -6c, c)$ **31.** $(0, 0, 0)$ **33.** $y = 2x^2 - x - 3$ **35.** $x^2 + y^2 - 4x + 2y - 20 = 0$

37. center $(-7, -2)$, radius 13   **39.** 5 dimes, 10 nickels, 4 quarters   **41.** 685   **43.** $(3, 5, 2, -3)$ **45.** $(1, -2, -1, 3)$

47. $(14a - 7b - 8, -6a + 2b + 5, a, b)$ **49.** $A = -13/2$ **51.** $A \neq -3$, $A \neq 1$ **53.** $A = -3$ **55.** $3x - 5y - 2z = -2$

EXERCISE SET 10.3, page 517

1. (1, 0), (2, 2)   **3.** $\left(\dfrac{2 + \sqrt{2}}{2}, \dfrac{-6 + \sqrt{2}}{2}\right), \left(\dfrac{2 - \sqrt{2}}{2}, \dfrac{-6 - \sqrt{2}}{2}\right)$ **5.** (5, 18)   **7.** (4, 6), (6, 4)   **9.** (−3/2, −4), (2, 3)

11. (19/29, −11/29), (1, 1)   **13.** (−2, 9), (1, −3), (−1, 1)   **15.** (−2, 1), (−2, −1), (2, 1), (2, −1)   **17.** (4, 2), (−4, 2), (4, −2), (−4, −2)

19. no solution   **21.** (12/5, 1/5), (2, 1)   **23.** (26/5, −3/5), (1, −2)   **25.** (39/10, −7/10), (3, 2)

27. $\left(\dfrac{-3 + \sqrt{3}}{2}, \dfrac{1 + \sqrt{3}}{2}\right), \left(\dfrac{-3 - \sqrt{3}}{2}, \dfrac{1 - \sqrt{3}}{2}\right)$ **29.** (19/13, 22/13), (1, 4)   **31.** no solution   **33.** (0, 1), (1, 2)   **35.** 0.7035, 0.4949

37. (1.7549, 1.3247)   **39.** (−0.7071, 0.7071), (0.7071, 0.7071)   **41.** (1, 5)   **43.** (−1, 1), (1, −1)   **45.** (1, −2), (−1, 2)

EXERCISE SET 10.4, page 524

1. $A = -3, B = 4$ **3.** $A = -2/5, B = 1/5$ **5.** $A = 1, B = -1, C = 4$ **7.** $A = 1, B = 3, C = 2$ **9.** $A = 1, B = 0, C = 1, D = 0$

11. $\dfrac{3}{x} + \dfrac{5}{x + 4}$ **13.** $\dfrac{7}{x - 9} + \dfrac{-4}{x + 2}$ **15.** $\dfrac{5}{2x + 3} + \dfrac{3}{2x + 5}$ **17.** $\dfrac{20}{11(3x + 5)} + \dfrac{-3}{11(x - 2)}$ **19.** $x + 3 + \dfrac{1}{x - 2} + \dfrac{-1}{x + 2}$

21. $\dfrac{1}{x} + \dfrac{2}{x + 7} + \dfrac{-28}{(x + 7)^2}$ **23.** $\dfrac{2}{x} + \dfrac{3x - 1}{x^2 - 3x + 1}$ **25.** $\dfrac{2}{x + 3} + \dfrac{-1}{(x + 3)^2} + \dfrac{4}{x^2 + 1}$ **27.** $\dfrac{3}{x - 4} + \dfrac{5}{(x - 4)^2}$ **29.** $\dfrac{3x - 1}{x^2 + 10} + \dfrac{4x}{(x^2 + 10)^2}$

31. $\dfrac{1}{2k(k - x)} + \dfrac{1}{2k(k + x)}$ **33.** $x + \dfrac{1}{x} + \dfrac{-2}{x - 1}$ **35.** $2x - 2 + \dfrac{3}{x^2 - x - 1}$ **37.** $\dfrac{1}{5(x + 2)} + \dfrac{4}{5(x - 3)}$

39. $\dfrac{1}{x} + \dfrac{2}{x^2} + \dfrac{3}{x^4} + \dfrac{-2}{x - 2}$ **41.** $\dfrac{4}{3(x - 1)} + \dfrac{2x + 7}{3(x^2 + x + 1)}$

EXERCISE SET 10.5, page 529

1. 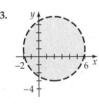   **3.** 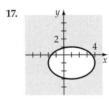   **5.**   **7.**   **9.**

11.   **13.**   **15.**   **17.**   **19.**

21.   **23.**   **25.** 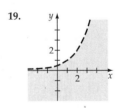   **27.** no solution   **29.**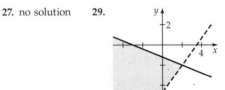

31. 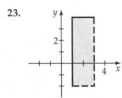   **33.** 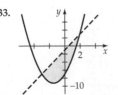   **35.** 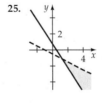 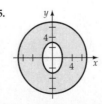   **37.** 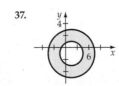   **39.**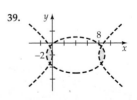

41. 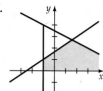 **43.** 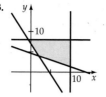 **45.** 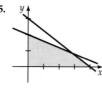 **47.**   **49.**

51. 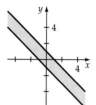 **53.** 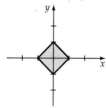 **55.**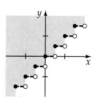

57. a. 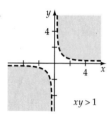 **b.** 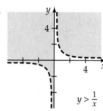   If x is a negative number, then the inequality is reversed when multiplying both sides of the inequality by a negative number.

$xy > 1$ $y > \dfrac{1}{x}$

EXERCISE SET 10.6, page 535

1. minimum at (0, 8): 16   **3.** minimum at (6, 5): 71   **5.** minimum at (0, 10/3): 20   **7.** maximum at (0, 12): 72
9. minimum at (0, 32): 32   **11.** maximum at (0, 8): 56   **13.** minimum at (2, 6): 18   **15.** maximum at (3, 4): 25
17. minimum at (2, 3): 12   **19.** maximum at (100, 400): 3400   **21.** 20 acres of wheat and 40 acres of barley
23. 0 Starter sets and 18 Pro sets   **25.** 24 ounces of group B and 0 ounces of group A, yields a minimum cost of $2.40.
27. two 4-cylinder engines and seven 6-cylinder engines, yields a maximum profit of $2050.

CHAPTER 10 TRUE/FALSE EXERCISES, page 540

1. false, $\begin{cases} x + y = 1 \\ x + y = 2 \end{cases}$ has no solution.   **2.** true   **3.** false, a homogeneous system is one where one constant term in each equation is zero.

4. true   **5.** true   **6.** false, $\begin{cases} x + y = 2 \\ x + 2y = 3 \end{cases}$ and $\begin{cases} 2x + 3y = 5 \\ 2x - 2y = 0 \end{cases}$ are two systems with the same solution but no common equations.

7. true   **8.** true   **9.** false, it is inconsistent.   **10.** false, since $(-1, 1)$ satisfies the first but not the second and $(-2, -1)$ satisfies the first but not the second.

CHAPTER 10 REVIEW EXERCISES, page 540

1. $(-18/7, -15/28)$ **3.** $(-3, -1)$ **5.** $(3, 1)$ **7.** $((5 - 3c)/2, c)$ **9.** $(1/2, 3, -1)$ **11.** $((7c - 3)/11, (16c - 43)/11, c)$
13. $(2, (3c + 2)/2, c)$ **15.** $(14c/11, -2c/11, c)$ **17.** $((c + 1)/2, (3c - 1)/4, c)$ **19.** $(2, -3)$ **21.** no solution   **23.** $(1/5, 18/5), (1, 2)$
25. $(2, 0), (18/17, -64/17)$ **27.** $(2, 1), (-2, -1)$ **29.** $(2, -3), (-2, 3)$ **31.** $\dfrac{3}{x - 2} + \dfrac{4}{x + 1}$ **33.** $\dfrac{6x - 2}{5(x^2 + 1)} + \dfrac{-6}{5(x + 2)}$

35. $\dfrac{2}{x} + \dfrac{4}{x - 1} + \dfrac{5}{x + 1}$ **37.** 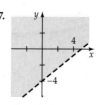   **39.**   **41.** 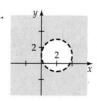   **43.**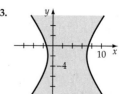

45.   **47.** 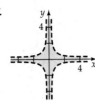   **49.**   **51.** 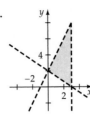   **53.**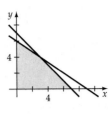

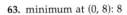

55.   **57.** 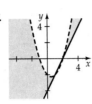   **59.** 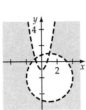   **61.** maximum at (4, 5): 18   **63.** minimum at (0, 8): 8

65. minimum at (2, 5): 27   **67.** $y = \dfrac{11}{6}x^2 - \dfrac{5}{2}x + \dfrac{2}{3}$ **69.** $z = -2x + 3y + 3$ **71.** wind: 28 mph, plane: 143 mph

73. (0, 0, 0), (1, 1, 1), (1, −1, −1), (−1, −1, 1), (−1, 1, −1)

CHAPTER 10 TEST, page 542

1. (−3, 2)   **2.** $\left(\dfrac{6 + c}{2}, c\right)$ **3.** $\left(\dfrac{173}{39}, \dfrac{29}{39}, -\dfrac{4}{3}\right)$ **4.** $\left(\dfrac{c + 3}{4}, \dfrac{7c + 1}{8}, c\right)$ **5.** $\left(\dfrac{c + 10}{13}, \dfrac{5c + 11}{13}, c\right)$ **6.** $\left(\dfrac{c}{14}, \dfrac{-9c}{14}, c\right)$

7. (2, 5), (−2, 1)   **8.** (−2, 3), (−1, −1)   **9.**   **10.**   **11.**

12. 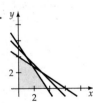   **13.**   **14.** No graph, the solution set is the empty set.   **15.**

16. 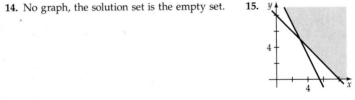   **17.** $\dfrac{7}{5(x - 4)} + \dfrac{8}{5(x + 1)}$ **18.** $\dfrac{1}{x} + \dfrac{-x + 2}{x^2 + 1}$ **19.** $\dfrac{680}{7}$ acres of oats   **20.** $x^2 + y^2 - 2y - 24 = 0$

EXERCISE SET 11.1, page 551

1. $\begin{bmatrix} 2 & -3 & 1 & 1 \\ 3 & -2 & 3 & 0 \\ 1 & 0 & 5 & 4 \end{bmatrix}$ $\begin{bmatrix} 2 & -3 & 1 \\ 3 & -2 & 3 \\ 1 & 0 & 5 \end{bmatrix}$ $\begin{bmatrix} 1 \\ 0 \\ 4 \end{bmatrix}$

3. $\begin{bmatrix} 2 & -3 & -4 & 1 & 2 \\ 0 & 2 & 1 & 0 & 2 \\ 1 & -1 & 2 & 0 & 4 \\ 3 & -3 & -2 & 0 & 1 \end{bmatrix}$ $\begin{bmatrix} 2 & -3 & -4 & 1 \\ 0 & 2 & 1 & 0 \\ 1 & -1 & 2 & 0 \\ 3 & -3 & -2 & 0 \end{bmatrix}$ $\begin{bmatrix} 2 \\ 2 \\ 4 \\ 1 \end{bmatrix}$

5. $\begin{bmatrix} 1 & -1 & 2 & 2 \\ 0 & 1 & -1 & -6 \\ 0 & 0 & 1 & -27/2 \end{bmatrix}$

7. $\begin{bmatrix} 1 & -2 & -1 & 3 \\ 0 & 1 & 2 & -\frac{11}{2} \\ 0 & 0 & 1 & -\frac{13}{6} \end{bmatrix}$ **9.** $\begin{bmatrix} 1 & -2 & 3 & -4 \\ 0 & 1 & 2 & -1/2 \\ 0 & 0 & 1 & -2 \\ 0 & 0 & 0 & 0 \end{bmatrix}$ **11.** $\begin{bmatrix} 1 & -3 & 4 & 2 & 1 \\ 0 & 1 & -1 & -2 & -1 \\ 0 & 0 & 0 & 1 & 3 \end{bmatrix}$ **13.** (2, −1, 1)   **15.** (1, −2, −1)

17. $(2 - 2c, 2c + 1/2, c)$ **19.** $(1/2, 1/2, 3/2)$ **21.** $(16c, 6c, c)$ **23.** $(7c + 6, -11c - 8, c)$ **25.** $(c + 2, c, c)$ **27.** no solution
29. $(2, -2, 3, 4)$ **31.** $(21/10, -8/5, 2/5, -5/2)$ **33.** $(3, -3/2, 1, -1)$ **35.** $(27c/2 + 39, 5c/2 + 10, -4c - 10, c)$
37. $(c_1 - 12c_2/7 + 6/7, c_1 - 9c_2/7 + 1/7, c_1, c_2)$ **39.** $(1, 0, -2, 1, 2)$ **41.** $(77c + 151)/3, (-25c - 50)/3, (14c + 34)/3, -3c - 7, c)$
43. $f(x) = x^3 - 2x^2 - x + 2$ **45.** all values of a except $a \neq 1$ and $a \neq -6$ **47.** $a = -6$ **49.** $z = 2x + 3y - 2$

EXERCISE SET 11.2, page 562

1. a. $\begin{bmatrix} 1 & 2 \\ 5 & 4 \end{bmatrix}$ **b.** $\begin{bmatrix} 3 & -4 \\ 1 & 2 \end{bmatrix}$ **c.** $\begin{bmatrix} -2 & 6 \\ 4 & 2 \end{bmatrix}$ **d.** $\begin{bmatrix} 7 & -11 \\ 0 & 3 \end{bmatrix}$

3. a. $\begin{bmatrix} -3 & 0 & 5 \\ 3 & 5 & -5 \end{bmatrix}$ **b.** $\begin{bmatrix} 3 & -2 & 1 \\ -1 & -5 & 1 \end{bmatrix}$ **c.** $\begin{bmatrix} -6 & 2 & 4 \\ 4 & 10 & -6 \end{bmatrix}$ **d.** $\begin{bmatrix} 9 & -5 & 0 \\ -4 & -15 & 5 \end{bmatrix}$

5. a. $\begin{bmatrix} 1 & 5 \\ 3 & -5 \\ 2 & -4 \end{bmatrix}$ **b.** $\begin{bmatrix} -7 & 3 \\ 1 & -1 \\ -4 & 4 \end{bmatrix}$ **c.** $\begin{bmatrix} 8 & 2 \\ 2 & -4 \\ 6 & -8 \end{bmatrix}$ **d.** $\begin{bmatrix} -18 & 5 \\ 1 & 0 \\ -11 & 12 \end{bmatrix}$

7. a. $\begin{bmatrix} -1 & 1 & -1 \\ 2 & 2 & 1 \\ -1 & 2 & 5 \end{bmatrix}$ **b.** $\begin{bmatrix} -3 & 5 & -1 \\ -2 & -4 & 3 \\ -7 & 4 & 1 \end{bmatrix}$ **c.** $\begin{bmatrix} 2 & -4 & 0 \\ 4 & 6 & -2 \\ 6 & -2 & 4 \end{bmatrix}$ **d.** $\begin{bmatrix} -7 & 12 & -2 \\ -6 & -11 & 7 \\ -17 & 9 & 0 \end{bmatrix}$ **9.** $\begin{bmatrix} -10 & 17 \\ 6 & -8 \end{bmatrix}, \begin{bmatrix} 0 & 22 \\ 1 & -18 \end{bmatrix}$

11. $\begin{bmatrix} 10 & 6 \\ 14 & -7 \end{bmatrix}, \begin{bmatrix} 14 & -1 \\ 0 & -11 \end{bmatrix}$ **13.** $\begin{bmatrix} 0 & -4 & 5 \\ 6 & 0 & 3 \\ -3 & -2 & 1 \end{bmatrix}, \begin{bmatrix} 5 & -13 \\ 5 & -4 \end{bmatrix}$ **15.** $\begin{bmatrix} 9 & -2 & -6 \\ 0 & -1 & 2 \\ 4 & -2 & -4 \end{bmatrix}, \begin{bmatrix} 4 & -2 & 6 \\ 2 & -3 & 4 \\ 4 & -4 & 3 \end{bmatrix}$ **17.** $[0, 8]$

19. The product is not possible   **21.** $\begin{bmatrix} 0 & 0 \\ 0 & 0 \end{bmatrix}$ **23.** The product is not possible   **25.** $\begin{bmatrix} 1/3 & -5/3 \\ -1/3 & 4/3 \\ 1/3 & -4/3 \end{bmatrix}$ **27.** $\begin{bmatrix} -1 & 1 \\ 3 & 2 \\ 7 & -2 \end{bmatrix}$

29. $\begin{bmatrix} 1 & -3 \\ 1 & -2 \end{bmatrix}$ **31.** $\begin{bmatrix} 7 & -1 & 1 \\ 1 & 2 & 0 \\ 5 & -1 & 4 \end{bmatrix}$ **33.** $\begin{cases} 3x - 8y = 11 \\ 4x + 3y = 1 \end{cases}$ **35.** $\begin{cases} x - 3y - 2z = 6 \\ 3x + y = 2 \\ 2x - 4y + 5z = 1 \end{cases}$ **37.** $\begin{cases} 2x_1 - x_2 + 2x_4 = 5 \\ 4x_1 + x_2 + 2x_3 - 3x_4 = 6 \\ 6x_1 + x_3 - 2x_4 = 10 \\ 5x_1 + 2x_2 - x_3 - 4x_4 = 8 \end{cases}$

39. $A - B = \begin{bmatrix} 50 & 150 & 140 \\ 15 & 170 & 370 \\ 85 & 250 & 130 \\ 80 & 115 & 25 \end{bmatrix}$

$A - B$ is the number sold of each item during the week.

41. $\begin{bmatrix} 26,898 & 28,150 & 31,536 \\ 20,495 & 21,195 & 23,670 \\ 19,022 & 19,925 & 21,969 \end{bmatrix}$ **43.** $\begin{bmatrix} 24 & 21 & -12 & 32 & 0 \\ -7 & -8 & 3 & 21 & 20 \\ 32 & 10 & -32 & 1 & 5 \\ 19 & -15 & -17 & 30 & 20 \\ 29 & 9 & -28 & 13 & -6 \end{bmatrix}$ **45.** $\begin{bmatrix} 46 & -100 & 36 & 273 & 93 \\ 82 & -93 & 19 & 27 & 97 \\ 73 & -10 & -23 & 109 & 83 \\ 212 & -189 & 52 & 37 & 156 \\ 68 & -22 & 54 & 221 & 58 \end{bmatrix}$

47. $\begin{bmatrix} 51 & 22 & -35 & 50 & 6 \\ 19 & 10 & -8 & 10 & 2 \\ 24 & 18 & -51 & 34 & 27 \\ 4 & 15 & 12 & 62 & 1 \\ 36 & 4 & -30 & 23 & 5 \end{bmatrix}$ **49.** $\begin{bmatrix} 6 + 9i & 3 - 6i \\ 3 + 3i & 6 - 3i \end{bmatrix}$ **51.** $\begin{bmatrix} 2 + 2i & -6 + 4i \\ -4 + 6i & 2 + 8i \end{bmatrix}$ **53.** $\begin{bmatrix} 3 + 2i & 3 + i \\ 4 + 3i & 6 - 2i \end{bmatrix}$

55. $\begin{bmatrix} 12 - 3i & -3 + 3i \\ 10 + i & 6 - i \end{bmatrix}$ **57.** $\begin{bmatrix} -2 + 11i & 8 - 6i \\ 2 + 6i & 6 - 5i \end{bmatrix}$

EXERCISE SET 11.3, page 571

1. $\begin{bmatrix} -5 & -3 \\ -2 & -1 \end{bmatrix}$ **3.** $\begin{bmatrix} 5 & -2 \\ -1 & 1/2 \end{bmatrix}$ **5.** $\begin{bmatrix} -16 & -2 & 7 \\ 7 & 1 & -3 \\ -3 & 0 & 1 \end{bmatrix}$ **7.** $\begin{bmatrix} 15 & -1 & -4 \\ -11/2 & 1/2 & 3/2 \\ 3 & 0 & -1 \end{bmatrix}$ **9.** $\begin{bmatrix} 7/2 & -2 & -2 \\ -5/2 & 1 & 2 \\ -1 & 0 & 1 \end{bmatrix}$

11. $\begin{bmatrix} 19/2 & -1/2 & -3/2 & 3/2 \\ 7/4 & 1/4 & -1/4 & 3/4 \\ -7/2 & 1/2 & 1/2 & -1/2 \\ 1/4 & -1/4 & 1/4 & 1/4 \end{bmatrix}$ **13.** $\begin{bmatrix} 2 & 3/5 & -7/5 & 4/5 \\ 4 & -7/5 & -2/5 & 4/5 \\ -6 & 14/5 & -1/5 & -3/5 \\ 3 & -8/5 & 2/5 & 1/5 \end{bmatrix}$ **15.** $(2, 1)$ **17.** $(7/4, -25/8)$ **19.** $(1, -1, 2)$ **21.** $(23, -12, 3)$

23. $(0, 4, -6, -2)$ **25.** on Saturday 80 adults, 20 children   **27.** Sample 1: 500 g of additive 1, 200 g of additive 2, 300 g of additive 3
on Sunday 95 adults, 25 children   Sample 2: 400 g of additive 1, 400 g of additive 2, 200 g of additive 3

29.
$$\begin{bmatrix} 0.020 & 0.143 & 0.102 & -0.143 \\ 0.565 & -0.381 & 0.490 & 0.048 \\ 0.136 & -0.381 & 0.347 & 0.048 \\ 0.320 & 0.095 & -0.265 & 0.238 \end{bmatrix}$$
31.
$$\begin{bmatrix} -0.150 & -0.217 & 0.302 \\ 0.248 & -0.024 & 0.013 \\ 0.217 & -0.200 & -0.195 \end{bmatrix}$$

33. $194.40 million manufacturing, $156.65 million transportation, $121.70 million services
35. $39.65 million coal, $14.15 million iron, $32.15 million steel

47. a. $\begin{bmatrix} -5/2 & 3/2 \\ -2 & 1 \end{bmatrix}$ **b.** $\begin{bmatrix} 2 & -3 \\ -3/2 & 5/2 \end{bmatrix}$ **c.** $\begin{bmatrix} 1 & \frac{1}{4} \\ -1 & 0 \end{bmatrix}$

EXERCISE SET 11.4, page 580

1. 13   **3.** -15 **5.** 0   **7.** 0   **9.** 19, 19   **11.** 1, -1 **13.** $-9, -9$ **15.** $-9, -9$ **17.** 10   **19.** 53   **21.** 20   **23.** 46
25. 0   **27.** row 2 consists of zeros. Therefore the determinant is zero.   **29.** 2 was factored from row 2
31. Row 1 was multiplied by -2 and added to row 2.   **33.** 2 was factored from column 1.
35. The matrix is in diagonal form. The value of the determinant is the product of the terms on the main diagonal.
37. Row 1 and Row 3 were interchanged. Therefore, the sign of the determinant was changed.
39. Each row of the determinant was multiplied by a.   **41.** 0   **43.** 0   **45.** 6   **47.** -90 **49.** 21   **51.** 3   **53.** -38.932 **55.** 9/2
57. $46\frac{1}{2}$ **63.** $7x + 5y = -1$ **65.** 263.5

EXERCISE SET 11.5, page 585

1. $x_1 = 44/31, x_2 = 29/31$ **3.** $x_1 = 1/3, x_2 = -2/3$ **5.** $x_1 = 2, x_2 = -7$ **7.** $x_1 = 0, x_2 = 0$ **9.** $x_1 = 41/32, x_2 = 15/8$
11. $x_1 = 21/17, x_2 = -3/17, x_3 = -29/17$ **13.** $x_1 = 32/49, x_2 = 13/49, x_3 = 6/7$ **15.** $x_1 = -29/64, x_2 = -25/64, x_3 = -19/32$
17. $x_1 = 50/53, x_2 = 62/53, x_3 = 4/53$ **19.** $x_1 = 0, x_2 = 0, x_3 = 0$ **21.** $x_2 = -35/19$ **23.** $x_1 = -121/131$ **25.** $x_4 = 4/3$
27. The determinant of the coefficient matrix is zero. Thus, Cramer's Rule cannot be used. The system of equations has infinitely many
solutions.   **29.** all values of k except $k = 0$ **31.** all values of k except $k = 2$ **33.** $r = 3, s = -3$

CHAPTER 11 TRUE/FALSE EXERCISES, page 589

1. false, $A^2 = A \cdot A = \begin{bmatrix} 7 & 18 \\ 6 & 19 \end{bmatrix}$ **2.** true   **3.** false, a singular matrix does not have a multiplicative inverse

4. False, as an example, $A = \begin{bmatrix} 2 & -1 \\ -4 & 2 \end{bmatrix}$, $B = \begin{bmatrix} 3 & 4 \\ 1 & 5 \end{bmatrix}$ and $C = \begin{bmatrix} 4 & 7 \\ 3 & 11 \end{bmatrix}$. $AB = AC$ but $B \neq C$.

5. True   **6.** False, for example, if $A = \begin{bmatrix} 1 & 4 \\ -2 & 3 \end{bmatrix}$ and $B = \begin{bmatrix} 2 & 0 \\ -1 & 5 \end{bmatrix}$, then $\det(A) + \det(B) \neq \det(A + B)$.

7. False, if the determinant of the coefficient matrix is zero, Cramer's Rule cannot be used to solve the system of equations.
8. False, since matrix multiplication is not commutative, that is, $AB \neq BA$, $AB - BA \neq 0$.
9. False, the matrix must not only be nonsingular but also be square to have a multiplicative inverse.
10. False, by the Associative Property of Matrix Multiplication, given A, B, and C square matrices of order n, $(AB)C = A(BC)$.
11. False, if the number of equations is less than the number of variables, the Gauss-Jordan method can be used to solve the system of linear equations. If the system of equations has a solution, the solutions will be given in terms of one or more of the variables.
12. False, for example for a 2×2 matrix, $\det(2A) = 2 \cdot 2 \det(A)$ and for a 3×3 matrix, $\det(2A) = 4 \cdot 2 \det(A)$.

13. True   **14.** False, for example, given $A = \begin{bmatrix} -3 & 2 \\ -6 & 4 \end{bmatrix}$ and $B = \begin{bmatrix} 2 & 4 \\ 3 & 6 \end{bmatrix}$, then $AB = \begin{bmatrix} 0 & 0 \\ 0 & 0 \end{bmatrix} = 0$, but $A \neq 0$ and $B \neq 0$.   **15.** True

CHAPTER 11 REVIEW EXERCISES, page 590

1. $\begin{bmatrix} 6 & -3 & 9 \\ 9 & 6 & -3 \end{bmatrix}$ **3.** $\begin{bmatrix} -5 & 5 & -1 \\ 1 & -4 & 6 \end{bmatrix}$ **5.** $\begin{bmatrix} -1 & -15 \\ 7 & 1 \end{bmatrix}$ **7.** $\begin{bmatrix} -6 & -4 & 2 \\ 14 & 0 & 10 \\ -7 & -7 & 6 \end{bmatrix}$ **9.** $\begin{bmatrix} 12 & 28 & -5 \\ 2 & 6 & 0 \\ 6 & 16 & -1 \end{bmatrix}$ **11.** $\begin{bmatrix} -12 & -36 & -4 \\ 48 & 124 & 4 \\ -9 & -32 & -6 \end{bmatrix}$

13. not possible   **15.** $\begin{bmatrix} 7 & 24 & 9 \\ -10 & -22 & 1 \end{bmatrix}$ **17.** $\begin{bmatrix} -1 & -5 & 4 \\ 1/2 & 2 & -3/2 \\ 0 & -2 & 1 \end{bmatrix}$ **19.** $(2, -1)$ **21.** $(3, 0)$ **23.** $(3, 1, 0)$ **25.** $(1, 0, -2)$

27. $(3, -4, 1)$ **29.** $(-c - 2, -c - 3, c)$ **31.** $(1, -2, 2, 3)$ **33.** $(-37c + 2, 16c, -7c + 1, c)$ **35.** $\begin{bmatrix} -1 & 1 \\ -3/2 & 1 \end{bmatrix}$ **37.** $\begin{bmatrix} -2/7 & 3/14 \\ 1/7 & 1/7 \end{bmatrix}$

39. $\begin{bmatrix} 2 & -2 & 1 \\ 0 & 3/2 & -1 \\ -1 & -1 & 1 \end{bmatrix}$ **41.** $\begin{bmatrix} -10 & 20 & -3 \\ -5 & 9 & -1 \\ 3 & -6 & 1 \end{bmatrix}$ **43.** $\begin{bmatrix} -1 & -7 & 4 & 2 \\ -6 & -3 & 2 & 3 \\ 1 & 2 & -1 & -1 \\ -2 & 0 & 0 & 1 \end{bmatrix}$ **45.** The matrix does not have an inverse.

47. a. $(18, -13)$ **b.** $(-22, 16)$ **49. a.** $(-18/7, 23/7, -6/7)$ **b.** $(-31/14, 20/7, 3/7)$ **51.** -2 **53.** -1 **55.** 0 **57.** 0
59. $x_1 = 16/19, x_2 = -2/19$ **61.** $x_1 = 13/44, x_2 = 1/4, x_3 = -17/44$ **63.** $x_1 = 54/69, x_2 = -26/69, x_3 = 38/69$ **65.** $x_3 = 115/126$
67. \$51.75 million computer division, \$14.13 million disk drive division, \$23.64 million moniter division.

CHAPTER 11 TEST, page 592

1. $\begin{bmatrix} 2 & 3 & -3 & 4 \\ 3 & 0 & 2 & -1 \\ 4 & -4 & 2 & 3 \end{bmatrix}, \begin{bmatrix} 2 & 3 & -3 \\ 3 & 0 & 2 \\ 4 & -4 & 2 \end{bmatrix}, \begin{bmatrix} 4 \\ -1 \\ 3 \end{bmatrix}$ **2.** $\begin{cases} 3x - 2y + 5z - w = 9 \\ 2x + 3y - z + 4w = 8 \\ x + 3z + 2w = -1 \end{cases}$ **3.** $(2, -1, 2)$ **4.** $(3, -1, -1)$

5. $(3c - 5, -7c + 14, 4 - 3c, c)$ **6.** $\begin{bmatrix} 3 & -9 & -6 \\ -3 & -12 & 3 \end{bmatrix}$ **7.** $A + B$ is not defined.   **8.** $\begin{bmatrix} 4 & 1 & 3 \\ 8 & 0 & -19 \\ 11 & 0 & 10 \end{bmatrix}$ **9.** $\begin{bmatrix} 16 & -1 & -2 \\ 15 & -11 & -3 \end{bmatrix}$

10. $\begin{bmatrix} 17 & -4 & -4 \\ 14 & -15 & -2 \end{bmatrix}$ **11.** CA is not defined.   **12.** $\begin{bmatrix} -6 & -1 & -19 \\ -15 & -25 & -27 \\ 1 & 3 & 31 \end{bmatrix}$ **13.** A^2 is not defined.   **14.** $\begin{bmatrix} 9 & 6 & 13 \\ -3 & -2 & 12 \\ 20 & -3 & 11 \end{bmatrix}$

15. $\begin{bmatrix} 18 & -5 & 7 \\ 4 & -1 & 2 \\ -3 & 1 & -1 \end{bmatrix}$ **16.** $M_{21} = -8, C_{21} = 8$ **17.** 49 **18.** -1 **19.** $-\dfrac{140}{41}$

20. $\left(\begin{bmatrix} 1 & 0 & 0 \\ 0 & 1 & 0 \\ 0 & 0 & 1 \end{bmatrix} - \begin{bmatrix} 0.15 & 0.23 & 0.11 \\ 0.08 & 0.10 & 0.05 \\ 0.16 & 0.11 & 0.07 \end{bmatrix} \right)^{-1} \begin{bmatrix} 50 \\ 32 \\ 8 \end{bmatrix}$

EXERCISE SET 12.1, page 599

1. $0, 2, 6, a_8 = 56$ **3.** $0, 1/2, 2/3, a_8 = 7/8$ **5.** $1, -1/4, 1/9, a_8 = -1/64$ **7.** $-1/3, -1/6, -1/9, a_8 = -1/24$
9. $2/3, 4/9, 8/27, a_8 = 256/6561$ **11.** $0, 2, 0, a_8 = 2$ **13.** $1.1, 1.21, 1.331, a_8 = 2.14358881$ **15.** $1, -\sqrt{2}/2, \sqrt{3}/3, a_8 = -\sqrt{2}/4$
17. $1, 2, 6, a_8 = 40320$ **19.** $0, 0.3010, 0.4771, a_8 = 0.9031$ **21.** $1, 4, 2, a_8 = 4$ **23.** $3, 3, 3, a_8 = 3$ **25.** $5, 10, 20$ **27.** $2, 4, 12$
29. $2, 4, 16$ **31.** $2, 8, 48$ **33.** $3, \sqrt{3}, \sqrt[6]{3}$ **35.** $2, 5/2, 9/4$ **37.** 4320 **39.** 72 **41.** 56 **43.** 100 **45.** 15 **47.** 40 **49.** $25/12$

51. 72 **53.** -24 **55.** 0.9031 **57.** 256 **59.** $\sum_{i=1}^{6} \dfrac{1}{i^2}$ **61.** $\sum_{i=1}^{7} 2^i(-1)^{i+1}$ **63.** $\sum_{i=0}^{4} (7 + 3i)$ **65.** $\sum_{i=1}^{4} \dfrac{1}{2^i}$ **67.** 2.6457520

69. $a_{20} \approx 1.0000037, a_{100} \approx 1$ **71.** $2.7083\overline{3}$ **73.** $a_n = (x - 2n)\left[\dfrac{n(n - 1)(n - 2)(n - 3)(n - 4)}{n!} \right] + 2n$

75. $\dfrac{1}{2}(-1 + i\sqrt{3}), \dfrac{1}{2}(-1 - i\sqrt{3}), 1, \dfrac{1}{2}(-1 + i\sqrt{3}), \dfrac{1}{2}(-1 - i\sqrt{3}), 1, a_{99} = 1$

EXERCISE SET 12.2, page 605

1. $a_9 = 38, a_{24} = 98, a_n = 4n + 2$ **3.** $a_9 = -10, a_{24} = -40, a_n = 8 - 2n$ **5.** $a_9 = 16, a_{24} = 61, a_n = 3n - 11$
7. $a_9 = 25, a_{24} = 70, a_n = 3n - 2$ **9.** $a_9 = a + 16, a_{24} = a + 46, a_n = a + 2n - 2$
11. $a_9 = \log 7 + 8 \log 2, a_{24} = \log 7 + 23 \log 2, a_n = \log 7 + (n - 1) \log 2$ **13.** $a_9 = 9 \log a, a_{24} = 24 \log a, a_n = n \log a$
15. 45 **17.** -79 **19.** 185 **21.** -555 **23.** 468 **25.** 525 **27.** -465 **29.** $78 + 12x$ **31.** $210x$ **33.** $3, 7, 11, 15, 19$
35. $5/2, 2, 3/2, 1$ **39.** 20 on 6th row, 135 in the 6 rows   **41.** \$1500, \$48,750   **43.** 784 ft   **45.** $1800°$ **47.** $a_n = 5n - 2$ **49.** $a_{100} = 301$

EXERCISE SET 12.3, page 612

1. yes, $r = 4$ **3.** no   **5.** yes, $r = 2^x$ **7.** yes, $r = 2$ **9.** yes, $r = x^2$ **11.** no   **13.** 2^{2n-1} **15.** $4(-3)^{n-1}$ **17.** $6\left(\dfrac{2}{3} \right)^{n-1}$

19. $-6\left(-\dfrac{5}{6} \right)^{n-1}$ **21.** $\left(-\dfrac{1}{3} \right)^{n-3}$ **23.** $(-x)^{n-1}$ **25.** c^{3n-1} **27.** $3\left(\dfrac{1}{100} \right)^n$ **29.** $5(0.1)^n$ **31.** $45(0.01)^n$ **33.** 18 **35.** -2

37. 363 **39.** $\dfrac{1330}{729}$ **41.** $\dfrac{279,091}{390,625}$ **43.** -341 **45.** $147,620$ **47.** $1/2$ **49.** $-2/5$ **51.** $9/91$ **53.** $1/9$ **55.** $5/7$ **57.** $1/3$

59. 5/11   **61.** 41/333   **63.** 422/999   **65.** 229/900   **67.** 1994/1650   **69.** $2271.93
71. Since log r is a constant, the sequence log a_n is an arithmetic sequence.   **73.** The total amount removed is 1.   **75.** $a_1^n r^{[(n-1)n]/2}$
77. 45 ft   **79.** 2044   **81.** 8A/5

EXERCISE SET 12.5, page 625

1. 35   **3.** 36   **5.** 220   **7.** 1   **9.** $x^6 - 6x^5y + 15x^4y^2 - 20x^3y^3 + 15x^2y^4 - 6xy^5 + y^6$ **11.** $x^5 + 15x^4 + 90x^3 + 270x^2 + 405x + 243$
13. $128x^7 - 448x^6 + 672x^5 - 560x^4 + 280x^3 - 84x^2 + 14x - 1$ **15.** $x^6 + 18x^5y + 135x^4y^2 + 540x^3y^3 + 1215x^2y^4 + 1458xy^5 + 729y^6$
17. $16x^4 - 160x^3y + 600x^2y^2 - 1000xy^3 + 625y^4$ **19.** $x^6 + 6x^4 + 15x^2 + 20 + \dfrac{15}{x^2} + \dfrac{6}{x^4} + \dfrac{1}{x^6}$
21. $x^{14} - 28x^{12} + 336x^{10} - 2240x^8 + 8960x^6 - 21{,}504x^4 + 28{,}672x^2 - 16{,}384$ **23.** $32x^{10} + 80x^8y^3 + 80x^6y^6 + 40x^4y^9 + 10x^2y^{12} + y^{15}$
25. $\dfrac{16}{x^4} - \dfrac{16}{x^2} + 6 - x^2 + \dfrac{x^4}{16}$ **27.** $s^{-12} + 6s^{-8} + 15s^{-4} + 20 + 15s^4 + 6s^8 + s^{12}$ **29.** $-3240x^3y^7$ **31.** $1056x^{10}y^2$ **33.** $126x^2y^2\sqrt{x}$
35. $165b^5/a^5$ **37.** $180a^2b^8$ **39.** $60x^2y^8$ **41.** $-61{,}236a^5b^5$ **43.** $126s^{-1}, 126s$ **45.** $-7 - 24i$ **47.** $41 - 38i$ **49.** 1
51. $nx^{n-1} + \dfrac{n(n-1)x^{n-2}h}{2} + \dfrac{n(n-1)(n-2)x^{n-3}h^2}{6} + \cdots + h^{n-1}$ **57.** 1.1712   **59.** 756   **61.** 56

EXERCISE SET 12.6, page 631

1. 30   **3.** 70   **5.** 1   **7.** 1   **9.** 210   **11.** 12   **13.** 16   **15.** 720   **17.** 125   **19.** 53,130
21. There are 676 ways to arrange 26 letters taken two at a time. Now if there are more than 676 employees then at least two employees will
have the same first and last initials.   **23.** 76   **25.** 1024   **27.** 3,838,380   **29. a.** 21   **b.** 105   **c.** 21   **31.** 1.8×10^9
33. 112   **35.** 120   **37.** 21   **39.** 112   **41.** 184,756   **43.** 62,355,150   **45. a.** 3,991,680   **b.** 31,840,128   **47.** 120   **49.** 252

EXERCISE SET 12.7, page 639

1. {S_1R_1, S_1R_2, S_1R_3, S_2R_1, S_2R_2, S_2R_3, R_1R_2, R_1R_3, R_2R_3, S_1S_2}   **3.** {H1, H2, H3, H4, T1, T2, T3, T4}
5. Let A, B, C represent cans and 1, 2 represent the balls. 00 means no balls in the can. 01 means ball 1 in the can. 02 means ball 2 in the can.
{$A00$, $A01$, $A02$, $A12$, $B00$, $B01$, $B02$, $B12$, $C00$, $C01$, $C02$, $C12$}
7. {HSC, HSD, HCD, SCD}   **9.** {ae, ai, ao, au, ei, eo, eu, io, iu, ou}   **11.** {HHHH}
13. {TTTT, HTTT, THTT, TTHT, TTTH, TTHH, THTH, HTHT, THHT, HTTH, HHTT}   **15.** $\varnothing$ **17.** {(1, 1), (2, 2), (3, 3), (4, 4), (5, 5), (6, 6)}
19. {(1, 4), (2, 4), (3, 4), (4, 4), (5, 4), (6, 4)}   **21. a.** 1/13   **b.** 1/4   **23.** 0.97   **25.** 1/2   **27.** 1/4   **29.** 0.9999   **31.** $\dfrac{1}{1{,}000{,}000{,}000}$
33. 34/57   **35.** 97/990   **37.** $\dfrac{11{,}011}{38{,}019}$ **39.** 1/3   **41.** 27/64   **43.** 1/10   **45.** $\left(\dfrac{7}{8}\right)^2$ **47.** $\dfrac{81}{10{,}000}$

CHAPTER 12 TRUE/FALSE EXERCISES, page 644

1. false. $0! \cdot 4! = 1 \cdot 4 \cdot 3 \cdot 2 \cdot 1 = 24$ **2.** false, $\left(\sum\limits_{i=1}^{3} i\right)\left(\sum\limits_{i=1}^{3} i\right) \neq \sum\limits_{i=1}^{3} i^2$ **3.** true   **4.** false, the constant sequence has all terms equal
5. false, $\dfrac{(k+1)^3}{k^3} = (1 + 1/k)^3$ is not a constant   **6.** true   **7.** true   **8.** false, $\sum\limits_{i=1}^{\infty} \dfrac{1}{2^i} = 1$
9. false, see Exercise 36, Section 4   **10.** false, the exponent is 4   **11.** false, there are $m \cdot n$ ways   **12.** False, $P(n, r) = \dfrac{n!}{(n-r)!}$
13. true   **14.** false, $P(A \cap B) = P(\varnothing) = 0$ **15.** true

CHAPTER 12 REVIEW EXERCISES, page 644

1. $a_3 = 9, a_7 = 49$ **3.** $a_3 = 11, a_7 = 23$ **5.** $a_3 = 1/8, a_7 = 1/128$ **7.** $a_3 = 1/6, a_7 = 1/5040$ **9.** $a_3 = 8/27, a_7 = 128/2187$
11. $a_3 = 18, a_7 = 1458$ **13.** $a_3 = 6, a_7 = 5040$ **15.** $a_3 = 8, a_7 = 16$ **17.** $a_3 = 2, a_7 = 256$ **19.** $a_3 = -54, a_7 = -3{,}674{,}160$
21. neither   **23.** arithmetic   **25.** geometric   **27.** neither   **29.** geometric   **31.** geometric   **33.** neither   **35.** arithmetic
37. neither   **39.** neither   **41.** 63   **43.** 152   **45.** 378   **47.** $-14{,}763$ **49.** 1.9653   **51.** 0.8280   **53.** 1/3   **55.** $-4/9$
65. $1024a^5 - 1280a^4b + 640a^3b^2 - 160a^2b^3 + 20ab^4 - b^5$
67. $a^4 + 16a^{7/2}b^{1/2} + 112a^3b + 448a^{5/2}b^{3/2} + 1120a^2b^2 + 1792a^{3/2}b^{5/2} + 1792ab^3 + 1024a^{1/2}b^{7/2} + 256b^4$ **69.** $241{,}920x^3y^4$ **71.** 26^8 **73.** 2730
75. 672   **77.** 1/8, 3/8   **79.** 0.285   **81.** Drawing an ace and 10 card from one deck   **83.** 1/4

CHAPTER 12 TEST, page 646

1. $a_3 = 4/3, a_5 = 4/15$ **2.** $a_3 = 1/6, a_5 = 1/10$ **3.** $a_3 = 12, a_5 = 48$ **4.** arithmetic   **5.** neither   **6.** geometric   **7.** 49/20
8. 1023/1024   **9.** 590   **10.** 58   **11.** 3/5   **12.** 5/33   **15.** $x^5 - 10x^4y + 40x^3y^2 - 80x^2y^3 + 80xy^4 - 32y^5$
16. $x^6 + 6x^4 + 15x^2 + 20 + \dfrac{15}{x^2} + \dfrac{6}{x^4} + \dfrac{1}{x^6}$ **17.** $48{,}384x^3y^5$ **18.** 132,600   **19.** 568,339,200   **20.** 0.294118

Index

Definitions of Trigonometric Functions

Angle θ

$$\sin \theta = \frac{b}{r} \qquad \csc \theta = \frac{r}{b}$$

$$\cos \theta = \frac{a}{r} \qquad \sec \theta = \frac{r}{a}$$

$$\tan \theta = \frac{b}{a} \qquad \cot \theta = \frac{a}{b}$$

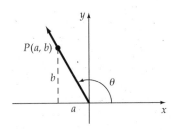

Real Number *t* (*Circular Functions*)

$$\sin t = y \qquad \csc t = \frac{1}{y}$$

$$\cos t = x \qquad \sec t = \frac{1}{x}$$

$$\tan t = \frac{y}{x} \qquad \cot t = \frac{x}{y}$$

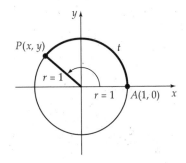

Oblique Triangles

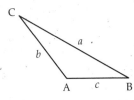 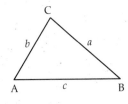

Law of Sines

$$\frac{a}{\sin A} = \frac{b}{\sin B} = \frac{c}{\sin C}$$

Law of Cosines

$$c^2 = a^2 + b^2 - 2ab \cos C$$

$$a^2 = b^2 + c^2 - 2bc \cos A$$

$$b^2 = a^2 + c^2 - 2ac \cos B$$

Area *K* of the Triangle

$$K = \frac{1}{2}ab \sin C = \frac{1}{2}ac \sin B = \frac{1}{2}bc \sin A$$

$$K = \frac{a^2 \sin B \sin C}{2 \sin A} = \frac{b^2 \sin A \sin C}{2 \sin B} = \frac{c^2 \sin A \sin B}{2 \sin C}$$

Heron's Formula

$$K = \sqrt{s(s-a)(s-b)(s-c)} \quad \text{where } s = \frac{a+b+c}{2}$$

Reciprocal Identities

$$\csc \theta = \frac{1}{\sin \theta}$$

$$\sec \theta = \frac{1}{\cos \theta}$$

$$\cot \theta = \frac{1}{\tan \theta}$$